ASTRONOMY AND ASTROPHYSICS LIBRARY

Series Editors: I. Appenzeller, Heidelberg, Germany
G. Börner, Garching, Germany
M. Harwit, Washington, DC, USA
R. Kippenhahn, Göttingen, Germany
J. Lequeux, Paris, France
P. A. Strittmatter, Tucson, AZ, USA
V. Trimble, College Park, MD, and Irvine, CA, USA

Springer
Berlin
Heidelberg
New York
Barcelona
Hong Kong
London
Milan
Paris
Singapore
Tokyo

Physics and Astronomy   ONLINE LIBRARY

http://www.springer.de/phys/

Eberhard Grün   Bo Å. S. Gustafson
Stan Dermott   Hugo Fechtig (Eds.)

Interplanetary Dust

With 242 Figures Including 4 Color Plates,
and 46 Tables

Springer

Professor Eberhard Grün
Max-Planck-Institut für Kernphysik
Saupfercheckweg 1
69117 Heidelberg, Germany

Professor Bo Å. S. Gustafson
Department of Astronomy
University of Florida
Gainesville, FL 32611-2055, USA

Professor Stan Dermott
Department of Astronomy
University of Florida
Gainesville, FL 32611-2055, USA

Professor Hugo Fechtig
Max-Planck-Institut für Kernphysik
Saupfercheckweg 1
69117 Heidelberg, Germany

Cover picture: A wedge of interplanetary dust. The dusk twilight sky (pink) towards the northwest shows zodiacal light (blue), framed by the Pleiades (upper left), Comet Hale-Bopp (upper right), and Mercury in Aries (left of center above the horizon). This photograph was taken by Marco Fulle, 5 April 1997, 18:50 UT, from the slopes of Stromboli (750 Meters above sea level), Sicily.

ISSN 0941-7834
ISBN 3-540-42067-3 Springer-Verlag Berlin Heidelberg New York

Library of Congress Cataloging-in-Publication Data applied for.

Die Deutsche Bibliothek - CIP-Einheitsaufnahme
Interplanetary dust: with 46 tables/Eberhard Grün ... (ed.). –
Berlin; Heidelberg; New York; Barcelona; Hong Kong; London;
Milan; Paris; Singapore; Tokyo: Springer, 2001
(Astronomy and astrophysics library)
(Physics and astronomy online library)
ISBN 3-540-42067-3

This work is subject to copyright. All rights are reserved, whether the whole or part of the material is concerned, specifically the rights of translation, reprinting, reuse of illustrations, recitation, broadcasting, reproduction on microfilm or in any other way, and storage in data banks. Duplication of this publication or parts thereof is permitted only under the provisions of the German Copyright Law of September 9, 1965, in its current version, and permission for use must always be obtained from Springer-Verlag. Violations are liable for prosecution under the German Copyright Law.

Springer-Verlag Berlin Heidelberg New York
a member of BertelsmannSpringer Science+Business Media GmbH

http://www.springer.de

© Springer-Verlag Berlin Heidelberg 2001
Printed in Germany

The use of general descriptive names, registered names, trademarks, etc. in this publication does not imply, even in the absence of a specific statement, that such names are exempt from the relevant protective laws and regulations and therefore free for general use.

Typesetting: Camera-ready version from the authors/editors
Cover design: *design & production* GmbH, Heidelberg

Printed on acid-free paper   SPIN: 10836746   55/3141/tr - 5 4 3 2 1 0

To the memory of

Herbert A. Zook
(1932–2001)

A scientist who inspired many chapters of the interplanetary dust story.

Preface

Conceived more than six years ago, this book took much effort to develop, and several updates in parts of the book became necessary because significant advances in the field of Interplanetary Dust had occurred. Now, at the beginning of the new millennium, this book provides up-to-date coverage of all major aspects of dust in the Solar System. The volume is conceived as a source book for researchers in the field as well as a graduate-level textbook. In order to achieve the highest standard the individual chapters are written by experts in the field, preserving the somewhat different style and language of the sub-topic.

The book follows the comprehensive review of the "Cosmic Dust" field assembled by Tony McDonnell more than 20 years ago. That book covered dust in its various physical appearances as the common theme, but it described phenomena that appeared rather unrelated. The topics ranged from zodiacal light over lunar craters to dust particles collected in the atmosphere, from interstellar dust to comets, and from dust dynamics to laboratory simulation of dusty phenomena.

Twenty years later, the field has matured far enough to warrant a new presentation in the form of a comprehensive review. Again, four major sub-fields are at the center of the book: astronomical observations of dust in the Solar System and beyond, in situ measurements of dust in various locations of the planetary system, laboratory studies of interplanetary dust and of dusty phenomena, and theoretical investigations. However, this time, the field has become more integrated because of the many interrelations that have been developed between the sub-fields.

Astronomical observations have extended the wavelength range from optical to thermal infrared wavelengths for which dust is the major source. As a consequence, new information has been gathered about the structure of the interplanetary dust cloud, specifically in the outer Solar System. Dust around the Earth and other planets provides analogs to the interplanetary dust cloud where interrelationships between sources, sinks and dynamical effects can be studied more closely. New sophisticated instruments have been developed that provide in situ compositional analyses of dust. Micro-analytic laboratory tools are available to analyze collected dust grains structurally and compositionally on nano- to micrometer scales. Multi-disciplinary campaigns have been organized to study time-critical cometary and meteor phenomena. Sophisticated laboratory studies together with advanced theoretical developments provide the basis for the interpretation of astrophysical observations of dust. The description of the interplanetary dust cloud has evolved to synthetic models representing a variety of observations, and even evolutionary models of components of the zodiacal cloud are being developed. The discovery of interstellar grains traversing the planetary system and the detection of dust clouds around other stars opens the door to a whole new look at the Solar System dust cloud and beyond.

However, the field is far from being complete. Despite the great advances made in recent years in the understanding of the interplanetary dust environment, there remain many important questions to be answered. Starting close to the Sun, nature separates meteoroid material according to its volatility. Analysis of the spatial distribution of matter close to the Sun will immediately give us information on the volatility of their constituents.

Closer to home, the dust environment of the Earth is of interest, because mankind is affecting this environment due to our space activities. However, also the natural dust environment is of technological and scientific interest. Hazards from meteor streams (like the Leonids) will require continuous attention. Meteoroids that pass the Earth are of scientific interest because they are messengers from distant worlds, like asteroids, comets and even interstellar space. Once we know where dust grains originate from, compositional analysis of grains can tell us many things about these worlds. Therefore, the goal of such dust studies is to identify the sources of dust particles together with their in-depth analysis. The future of dust measurements in Earth orbit lies in three areas: (1) environmental monitoring; (2) use of dust telescopes to separate and analyze dust populations of different origin; and (3) collection and sample return of dust from various sources for in-depth analysis in laboratories.

Enhanced dust densities in the Martian environment have long been suspected. The comparison of future dust observations at Mars with those in the dusty rings of Saturn and the other giant planets will tell us what effects solar radiation pressure (strongest at Mars) and planetary magnetospheres (negligible at Mars) have. In addition, during a future human colonization of Mars, dust from the Martian satellites may play a similar hazardous role as space debris in the Earth environment.

Analysis of particulates from comet Halley brought us new and important information that has relevance to the understanding of the formation of our planetary system. Currently the Stardust mission is on its way to analyze, collect and return dust from comet Tempel 2. This will give us a second example from the large variety of comets. It is the analysis of this variety that will tell us about the spatial and compositional variations in the protoplanetary nebula through the comets that may have sampled different regions of this nebula.

The satellite and ring systems of the outer planets are models for the early Solar System with satellites and ring particles in intimate interactions. While Saturn's ring system is of high complexity, the rings of Jupiter, Uranus and Neptune show other features that have not yet been found elsewhere. The Pluto–Charon system may have rings with yet other features. To understand the common characteristics of all these rings and the reasons why they are so different requires detailed measurements of the rings and their environments.

Detection of Edgeworth–Kuiper belt objects (EKOs) of up to a few hundred kilometers diameter confirmed the existence of objects outside the planetary system. Mutual collisions among EKOs as well as impacts of interstellar grains generate dust locally. The action of the Poynting–Robertson effect together with resonances with the outer giant planets, interaction with the solar wind and neutral interstellar gas, may have lead to radial and azimuthal structure of the distribution of dust at the edge of the planetary system. The detection of infrared excess at main sequence stars started renewed interest in the outer extensions of our own Solar System dust cloud. Especially, the observation of dust disks around β Pictoris stimulated this interest. Observations of the Edgeworth–Kuiper dust belt in our Solar System can, therefore, be used as a model for extra-solar dust clouds and can help to reveal information about other planetary systems.

The effects of the solar cycle-dependent heliosphere reach into interstellar space, out to about 300 AU from the Sun where it interferes with the small ($\leq 0.1\,\mu$m) particles entering the heliosphere and modulates their flow. Their origin, however, may be different from that of bigger grains that are accessible in Earth orbit. We know that evolved stars continuously lose mass. This "stardust" provides the seeds for interstellar dust grains that grow in cool interstellar clouds by accretion of atoms and molecules and by agglomeration. An unbiased look into this interstellar dust factory will provide us with information on processes that are difficult to quantify by astronomical observations alone. Therefore, in situ dust analysis will be an important method when automated probes leave our Solar System.

This collection of problems demonstrates that dust analysis will remain an important topic of planetary and astrophysical research and it is hoped that this book will stimulate interest in the study of these challenging issues.

Further developments, errata, and related websites can be found at http://www.springer.de/books/errata/interplanetary-dust/.

The book has benefited from the support of a large number of people, many of whom are acknowledged in the individual chapters. We thank the chapter authors for their efforts, persistence, and dedication to this project. Assistance in the preparation of the book from Amara Graps, Dagmar Koch, Harald Krüger, Markus Landgraf, and Richard Moissl is acknowledged. The unrewarded help of chapter reviewers is greatly appreciated: Josh Colwell, Bruce Draine, Gerhard Drolshagen, Dan Durda, Priscilla Frisch, Mayo Greenberg, Bob Hawkes, Doug Hamilton, Martha Hanner, Mihaly Horányi, Don Humes, Eduard Igenbergs, Jochen Kissel, Wolfgang Klöck, Mark Matney, Michel Maurette, Tony McDonnell, Derral Mulholland, Klaus Paul, Jiri Svestka, Tony Tuzzolino and Iwan Williams.

Heidelberg, Gainesville, April 2001

Eberhard Grün
Bo Gustafson
Stanley Dermott
Hugo Fechtig

Contents

Color Plates xxi
Contributors xxvii

Historical Perspectives 1
Hugo Fechtig, Christoph Leinert, Otto E. Berg

I. Introductory Overview 1
II. Early Reports on the Zodiacal Light 2
III. Zodiacal Light Observations Until the Beginning
of the Space Age 10
IV. After the Beginning of the Space Age 17
    IV.A. Rise and Fall of the Earth's Dust Belt 17
    IV.B. Zodiacal Light Studied from Near-Earth Space 26
V. Microcraters on Lunar Surface Samples
and the Lunar Ejecta and Micrometeorite Experiment 30
VI. Experiments on Satellites and Space Probes 36
VII. Important Results of the Dust Experiments PIA/PUMA and
DIDSY on the Missions GIOTTO and VeGa to Comet Halley . 43
VIII. Outlook 45
References 47

Optical and Thermal Properties of Interplanetary Dust 57
*Anny-Chantal Levasseur-Regourd, Ingrid Mann,
René Dumont, Martha S. Hanner*

I. Zodiacal Scattered Light 57
    I.A. Historical Survey 58
    I.B. Zodiacal Light Measurements 59
    I.C. Main Trends in the Data 60
    I.D. Zodiacal Brightness from 1 AU 63
    I.E. Zodiacal Polarisation from 1 AU 66
II. F-Corona Scattered Light 67
    II.A. Solar Corona Observations 67
    II.B. Brightness and Polarisation of the F-Corona 69
III. Zodiacal and F-Coronal Thermal Emission 72
    III.A. Thermal Emission Measurements 72
    III.B. Zodiacal Thermal Emission from 1 AU 73
    III.C. Thermal Emission from the F-Corona 74
IV. Local Scattering and Thermal Properties 76

IV.A. Need for Inversion	76
IV.B. Volume Scattering and Emitting Functions	77
IV.C. Inversion with Homogeneity Assumption	79
IV.D. Local Rigorous Inversion	79
IV.E. Local Inversion Through Mathematical Methods	80
IV.F Models of the Near Infrared F-Corona	85
V. Conclusions and Perspectives	86
References	88

Cometary Dust ... 95
Zdenek Sekanina, Martha S. Hanner,
Elmar K. Jessberger, Marina N. Fomenkova

I. Introduction	95
II. Dust Dynamical Properties	96
II.A. Ejection and Motion of Dust Grains	96
II.B. Dust Features in Cometary Heads	101
II.C. Dust Tails and Their Structure	108
II.D. Dust in Periodic Comet Shoemaker-Levy 9	115
III. Dust Optical and Physical Properties	120
III.A. Thermal Emission	120
III.B. Silicates	127
III.C. Infrared Spectral Features of Hydrocarbons	130
III.D. Scattering by Dust	132
III.E. Icy Grains	136
IV. Dust Chemical and Isotopic Composition	137
IV.A. Facts from Ion Spectra	138
IV.B. Bulk Composition	143
IV.C. Mineralogical Composition	144
V. The Future	145
References	147

Near Earth Environment ... 163
Tony McDonnell, Neil McBride, Simon F. Green,
Paul R. Ratcliff, David J. Gardner, Andrew D. Griffiths

I. Introduction	163
II. The Earth as a Target	165
II.A. Natural Meteoroids	165
II.B. Meteoroid Properties and Dynamics	166
II.C. The Sporadic Background	174
II.D. The Annual Meteor Showers	174
II.E. Atmospheric Effects	181
III. Space Debris	189
IV. Modelling Tools	193
IV.A. The Grün Interplanetary Dust Model at 1 AU	193

IV.B. Spacecraft Geometry 193
IV.C. The NASA Orbital Debris Environment Model 200
IV.D. ESABASE 200
IV.E. MASTER 201
IV.F. The Divine Interplanetary Dust Model 202
V. Measurements 202
V.A. Measurement Techniques 202
V.B. Results 210
VI. Summary 223
References 224

Discoveries from Observations and Modeling of the 1998/99 Leonids 233
Peter Jenniskens

I. Introduction 233
II. Meteoroid Streams and Meteor Storms 233
III. Observing Campaigns 238
IV. Meteoroid Morphology and Composition 239
V. The Impact Hazard 241
VI. Interaction of Meteoroids with the Atmosphere 242
VII. Atmospheric Phenomena 246
References 248

Properties of Interplanetary Dust: Information from Collected Samples 253
Elmar K. Jessberger, Thomas Stephan, Detlef Rost,
Peter Arndt, Mischa Maetz, Frank J. Stadermann,
Don E. Brownlee, John P. Bradley, Gero Kurat

I. Introduction 253
II. Antarctic and Greenland Micrometeorites 255
II.A. Mineralogy and Petrography of MMs 255
II.B. Major, Minor and Trace Element Chemistry of MMs 257
II.C. Isotope Abundances in MMs 259
II.D. Rare Gas Abundances in MMs 259
II.E. Conclusions from MM Studies 259
III. Stratospheric Interplanetary Dust 260
III.A. Shape and External Morphology 260
III.B. Density 261
III.C. Optical and Infrared Properties 265
III.D. Classification and Mineralogy 267
III.E. Elemental Composition 270
III.F. Isotopic Composition 282
IV. Origins 285
References 288

In situ Measurements of Cosmic Dust 295
Eberhard Grün, Michael Baguhl,
Håkan Svedhem, Herbert A. Zook

    I. Introduction 295
    II. Characteristics of In-Situ Dust Measurements in Space 299
        II.A. Dust Missions and Detectors 299
        II.B. Reliability of Impact Detection
            and Impact Rate Measurements 301
        II.C. Small Number Statistics 303
        II.D. Detection Geometry and Orbit Determination 305
    III. Measurements at 1 AU 309
        III.A. Early Meteoroid Flux Measurements
            in the Earth-Moon System 309
        III.B. HEOS-2 313
        III.C. Hiten 314
    IV. Measurements Within the Zodiacal Cloud 317
        IV.A. Helios 317
        IV.B. Pioneers 8 and 9 320
        IV.C. Galileo 322
        IV.D. Ulysses 323
    V. Measurements in the Outer Solar System 324
        V.A. Pioneers 10 and 11 324
        V.B. Jupiter Dust Streams 327
        V.C. Interstellar Dust 331
    VI. Characteristics of the Interplanetary Dust Complex
        as Measured by Spacecraft 333
        VI.A. Gravity and Radiation Pressure Effects 333
        VI.B. Electromagnetic Effects 336
    VII. Future Developments 339
    References 342

Synthesis of Observations 347
Peter Staubach, Eberhard Grün, Mark J. Matney

    Preamble 347
    I. Introduction 347
        I.A. Physical Processes 349
        I.B. Properties of Interplanetary Dust 350
        I.C. Model Assumptions 351
    II. Early Modeling 352
        II.A. Cour-Palais (1969) 353
        II.B. Kessler (1970) 354
        II.C. Grün et al. (1985) 355
        II.D. Zook (1991) 356

II.E. Comparison	356
III. Basic Formulation	357
III.A. Phase Space Density	357
III.B. Orbital Parameter Distributions	358
III.C. Concentrations	358
III.D. Particle Fluxes	359
III.E. Directional Flux and Impact Speed	361
III.F. Radiation Pressure Effects and Hyperbolic Orbits	362
IV. Meteoroid Data Sets	363
IV.A. Meteors	363
IV.B. Lunar Microcraters	364
IV.C. Zodiacal Light and Thermal Emission	364
IV.D. Early Spacecraft Detectors	364
IV.E. Ulysses	365
IV.F. Galileo	366
V. Divine's Original Model Populations	367
VI. Comparison of Divine's Model with Observations	369
VI.A. Interplanetary Flux Model (Size Distribution)	369
VI.B. Meteors (Radial Distribution)	369
VI.C. Zodiacal Light and Thermal Emission	369
VI.D. Spaceprobe Data	371
VII. New Results	371
VII.A. Interstellar Dust Population	372
VII.B. Meteoroid Populations Affected by Radiation Pressure	372
VII.C. Predicted Fluxes onto the Cassini Detector	376
VIII. Future Developments	377
VIII.A. New Meteor Data and Analysis	377
VIII.B. Small Meteoroid Populations	378
VIII.C. Formulation of the Dust Environment of Earth-Orbiting Satellites	378
VIII.D. Directional Flux onto a Satellite Surface	380
VIII.E. Meteoroid Fluxes on LDEF	380
References	382

Instrumentation 385
Siegfried Auer

I. Introduction	385
II. Detection and Characterization of Dust Particles	386
II.A. Detection of Scattered and Emitted Light	387
II.B. Charge	389
II.C. Impact Light Flash	391
II.D. Impact Ionization	393
II.E. Thin-Foil Penetration	400
II.F. Momentum	406

II.G.	Velocity, Trajectory, and Orbit	407
II.H.	Deceleration for Intact Capture	410
II.I.	Mass, Density, and Diameter	411
II.J.	Chemical and Isotopic Composition	411
III.	Flight Instrumentation	412
III.A.	Explorer 16, Pegasus, and Pioneer 10: Large-Area Penetration Detectors	412
III.B.	Pioneer 8: Reliable Coincidence Detector	413
III.C.	Heos 2: The First Speed-and-Mass Sensor for Small Dust Particles	416
III.D.	Helios: The First Dust Composition Analyzer	417
III.E.	VeGa 1/2 and Giotto to Comet Halley	420
III.F.	Galileo/Ulysses: Large-Area Multi-Coincidence Dust Detector System (DDS)	421
III.G.	Hiten: Dust Counter (MDC) with a Transient Recorder	423
III.H.	Cassini: Multi-Parameter Cosmic Dust Analyzer (CDA)	425
III.I.	Very-High-Resolution Cometary Dust Composition Analyzer (COSIMA)	427
IV.	Laboratory Simulation	428
IV.A.	Acceleration of Dust Particles	430
IV.B.	Dust Charging in an Electrodynamic Quadrupole	435
References		439

Physical Processes on Interplanetary Dust 445
Tadashi Mukai, Jürgen Blum, Akiko M. Nakamura,
Robert E. Johnson, Ove Havnes

I.	Introduction	445
II.	Collisional Growth of Solid Particles	447
II.A.	Two-Particle Collisions	448
II.B.	Aggregation Phenomena	452
II.C.	Coagulation and Aggregation Studies in the Laboratory	456
III.	Collisional Fragmentation	458
III.A.	Impact Process	458
III.B.	Fragmentation and Strength	460
III.C.	Size Distribution of Fragments	462
III.D.	Shape Distribution of Fragments	464
III.E.	Velocity and Spin Distribution of Fragments	464
IV.	Sublimation	467
IV.A.	Equilibrium	468
IV.B.	Vapor Pressure Versus Temperature	468
IV.C.	Sublimation Rate	472
IV.D.	Interplanetary Dust Grain Temperatures	474
IV.E.	Comets	477
IV.F.	Reaction Force	477

V. Sputtering	478
V.A. Plasma Parameters	478
V.B. Materials	482
V.C. UV Irradiation	483
V.D. Plasma-Induced Sputtering and Alteration	484
VI. Charging	488
VI.A. Charging of Single Isolated Dust Particles	488
VI.B. Collective Effects on Dust Charging	494
VII. Lifetimes	497
References	500

Interactions with Electromagnetic Radiation: Theory and Laboratory Simulations 509
Bo Å. S. Gustafson, J. Mayo Greenberg,
Ludmilla Kolokolova, Yu-lin Xu, Ralf Stognienko

I. Introduction	509
II. A Physical Dust Model	511
III. Optical Constants	514
III.A. Bulk Materials	515
III.B. Aggregates and Other Inhomogeneous Materials	518
IV. Scattering Solutions	521
IV.A. Mie Theory and Related Boundary Solutions	528
IV.B. Extension of Boundary Conditions to N-Spheres	529
IV.C. T-Matrix Solutions	531
IV.D. Internal Field Solutions	532
IV.E. Experiments	535
V. Results	538
V.A. Theory-based studies	538
V.B. Experiment-Based Studies	545
V.C. Radiation Pressure	552
VI. Closing Remarks	555
References	559

Orbital Evolution of Interplanetary Dust 569
Stanley F. Dermott, Keith Grogan, Daniel D. Durda,
Sumita Jayaraman, Thomas J.J. Kehoe,
Stephen J. Kortenkamp, Mark C. Wyatt

I. Introduction	569
II. Forces and Collisions	571
II.A. Radiation Forces	573
II.B. Poynting-Robertson (P-R) Light Drag	574
II.C. Collisions	575
III. Orbital Evolution	578
III.A. P-R Drag Affected Orbits	582

III.B. Numerical Simulations	583
III.C. SIMUL - Visualizing the Orbital Distribution	588
III.D. Cometary Particles	589
IV. Dust Bands	592
IV.A. IRAS Observations	596
IV.B. Modeling the Dust Bands	597
IV.C. The Importance of Secular Perturbations	601
IV.D. Equilibrium vs. Non-Equilibrium	604
V. Background Cloud	606
V.A. Tilt, Warp and Offset	606
V.B. Physical Understanding of the Asymmetries	611
V.C. Application to Circumstellar Disks	613
VI. Resonant Ring	614
VII. Accretion of IDPs	624
VII.A. Long-Term Variations	628
VIII. Conclusions	631
References	635

Dusty Rings and Circumplanetary Dust: Observations and Simple Physics 641

Joseph A. Burns, Douglas P. Hamilton, Mark R. Showalter

I. Introduction	642
II. Description	644
II.A. Physical Models	644
II.B. Observational Methods	647
II.C. Physical Properties of the Dusty Rings	652
III. Physical and Dynamical Processes Acting on Circumplanetary Dust	673
III.A. Electrical Charging	673
III.B. Forces	677
III.C. Size Distributions	686
III.D. Destruction and Generation of Grains	687
III.E. Interactions with Nearby Satellites	690
IV. Celestial Mechanics and Orbital Evolution	693
IV.A. Introduction	693
IV.B. Resonances	694
IV.C. Orbit-Averaged Equations of Motion	699
IV.D. Approximate Analytic Solutions	705
V. Putting It Together	709
V.A. Jovian Rings	710
V.B. Saturn's E Ring	711
V.C. The Dust Bands of Uranus and Neptune	713
VI. Expected Advances	714
References	715

Interstellar Dust and Circumstellar Dust Disks 727
Johann Dorschner

  I. Landmarks in Interstellar Dust Research 727
    I.A. From Early Conjectures to a Physical Theory 727
    I.B. The Classical Dust Model 728
    I.C. Interstellar Polarization 728
    I.D. Refractory Dust Grains 729
    I.E. Diagnostic Dust Bands and Laboratory Astrophysics 729
  II. Dust and Galactic Evolution 731
    II.A. The Multi-Phase Interstellar Medium 731
    II.B. Molecular Clouds and Star-Forming Regions 732
    II.C. Dust Populations and the Lifecycle of Dust 733
  III. Dust in Diffuse Interstellar Clouds 734
    III.A. Basic Observational Phenomena 734
    III.B. Dust Models 744
  IV. Dust in Molecular Clouds and Star-Forming Regions 752
    IV.A. Basic Observational Phenomena 752
    IV.B. Processes in Molecular Clouds and Star-Forming Regions . 759
  V. Dust in Stellar Outflows 761
    V.A. Oxidic Stardust 761
    V.B. Carbonaceous Stardust 764
    V.C. Other Stardust Components 765
  VI. Dust in Young Circumstellar Disks and Planetary Systems 767
    VI.A. Observational Evidence for Young Circumstellar Disks 767
    VI.B. Vega-Phenomenon Dust 770
  References 774

Glossary 787

Index 793

Color Plates

Dust jets in Comet Hale-Bopp xxii

Secondary ion images of IDP L2006G1 xxiii

Secondary ion images of IDP U2071H9 xxiv

Secondary ion images of IDP L2006E10 xxv

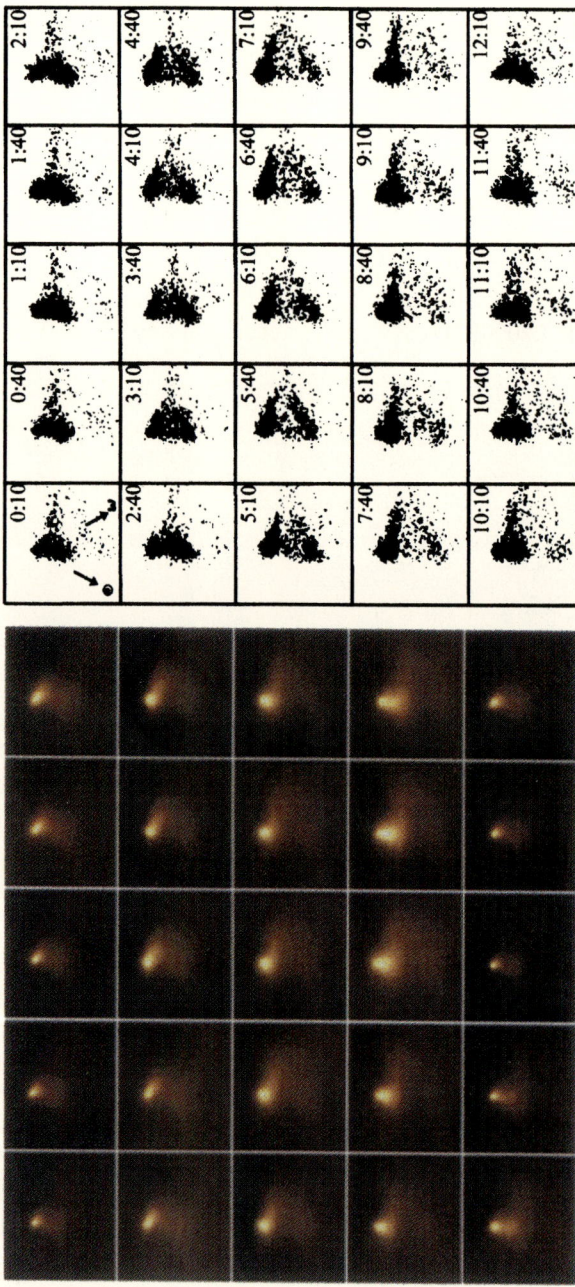

Plate 1. (Chapter on Cometary Dust by Sekanina et al., p. 107). Modeling the diurnal evolution of a dust jet in comet Hale-Bopp (C/1995 O1). *Left panel*: selected frames from the jet's animation sequence taken by Jorda et al. (1997) with a 105 cm telescope at Pic du Midi between 3:50 and 15:35 UTC on 28 February 1997. A complete rotation cycle (~11.35 hours) is covered. The image in the upper left corner corresponds to ~12:15 UTC. The temporal separation between neighboring frames, each 32,400 km on a side at the comet, is ~30 minutes. (Original images courtesy of L. Jorda, J. Lecacheux, and F. Colas.) *Right panel*: computer-generated images simulating the observations in the left panel on the assumption that the jet emanated from a discrete source on the nucleus. There is a one-to-one correspondence between the frames in the two panels. The times are reckoned from the source's activation time. The dust ejecta were expanding with velocities of up to 600 m/s. The first frame shows the projected directions of the spin vector (ω) and the Sun (circled dot). (From Sekanina 1998.)

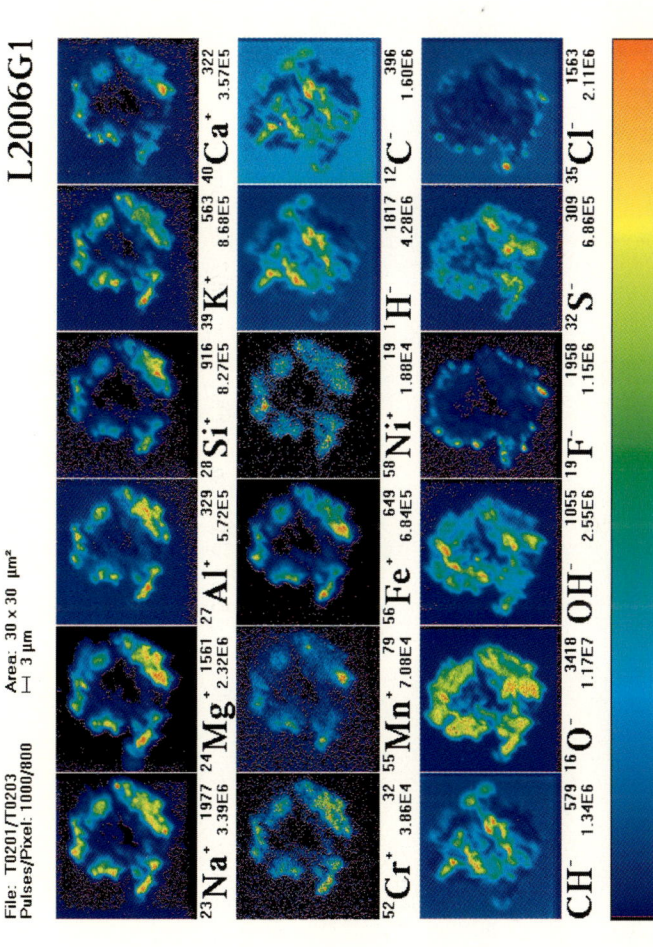

Plate 2. (Chapter on Properties of Interplanetary Dust by Jessberger et al., p. 278). TOF-SIMS secondary ion images for the IDP L2006G1. Positive and negative ions were measured consecutively. The field of view ($30 \times 30~\mu m^2$) and the number of primary ion pulses per pixel (1000 for positive, 800 for negative secondary ions) are given on top. The sample was scanned (128/128 pixels) with a Ga$^+$ primary ion beam. The subscripts of the individual ion images give the secondary ion species (e.g., ^{23}Na$^+$), the maximum number of counts/pixel (1977) and the integrated number of counts for the total image ($3.39 \cdot 10^6$). For all secondary ion images the same linear color scale was used, where black corresponds to zero counts and the maximum count rate is red.

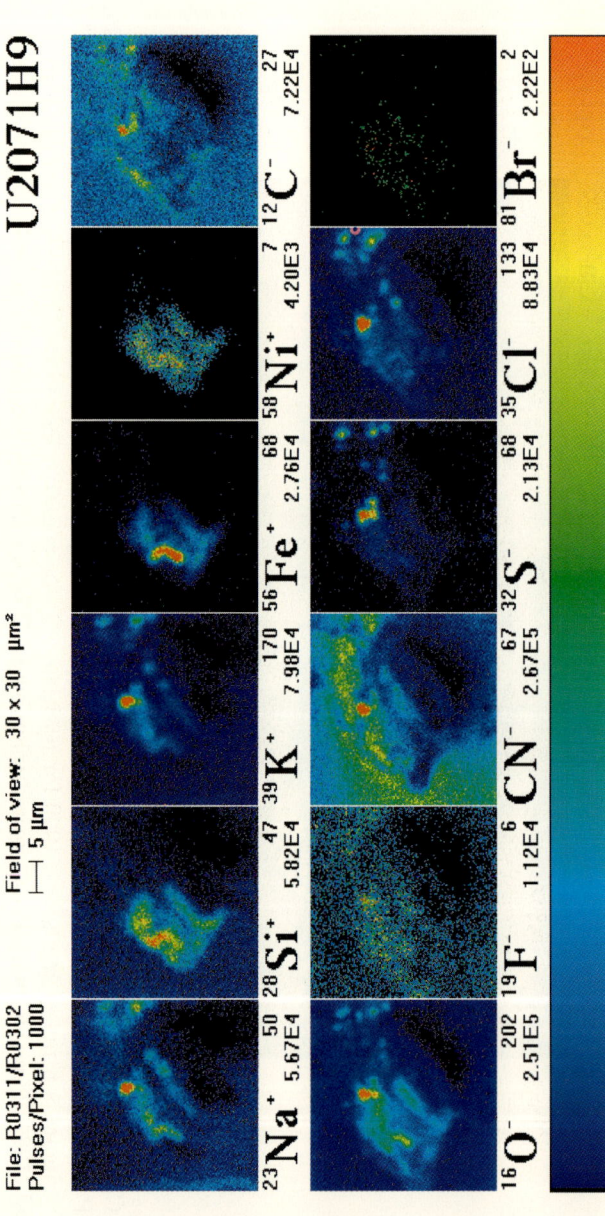

Plate 3. (Chapter on Properties of Interplanetary Dust by Jessberger et al., p. 279). TOF-SIMS secondary ion images of the original surface of U2071H9, an (Fe,Ni)S rich IDP. The halogens F, Cl, and Br are detectable on the very surface (from Rost et al., 1999).

COLOR PLATES   XXV

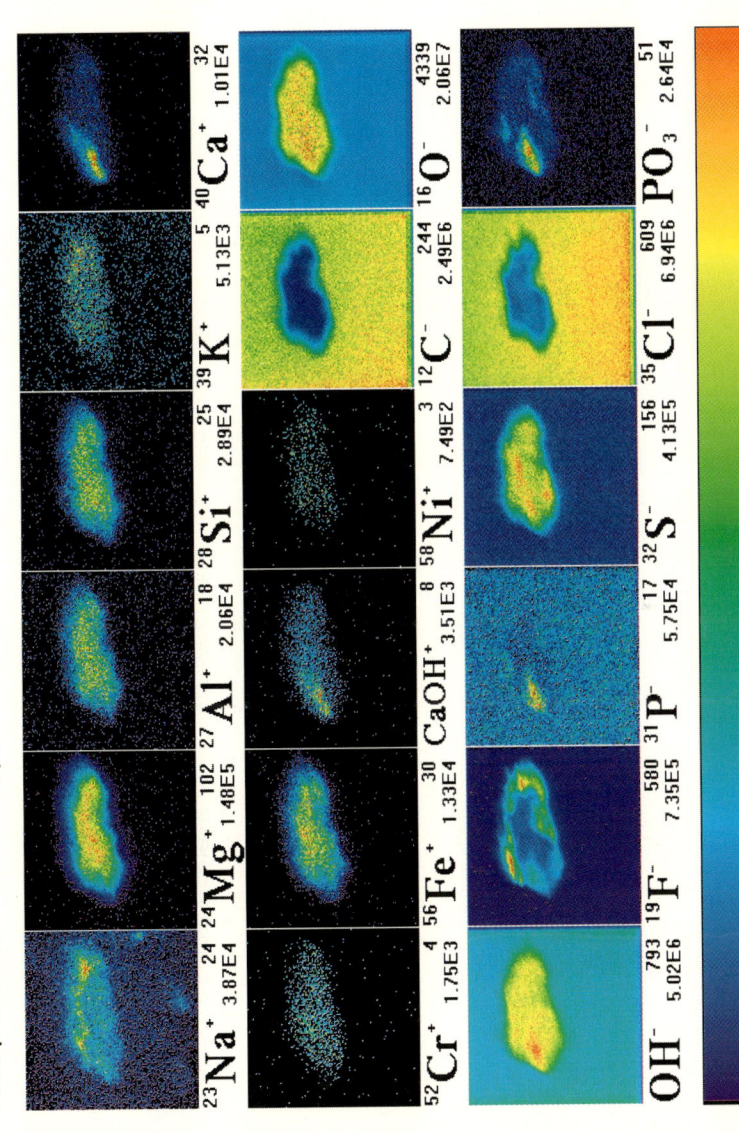

Plate 4. (Chapter on Properties of Interplanetary Dust by Jessberger et al., p. 280). TOF-SIMS secondary ion images (for explanation, see Plate 2) of the IDP L2006E10 reveal a Ca and P rich area, most probably an apatite grain.

Contributors

Peter Arndt
Obere Kirchgasse 4
74918 Angelbachtal, Germany
peter.arndt@sap.com

Siegfried Auer
46 Palmer Road
Bayse, VA 22810, USA
siegfried_auer@yahoo.com

Michael Baguhl
Astrium GmbH
An der Bundesstraße 31
88039 Friedrichshafen, Germany
michael.baguhl@astrium-space.com

Otto E. Berg
216 Lombardy Court
Middletown, MD 21769, USA
lilott@juno.com

Jürgen Blum
Astrophysical Institute
and University Observatory
Friedrich Schiller University Jena
Schillergäßchen 2
07745 Jena, Germany
blum@astro.uni-jena.de

John P. Bradley
MVA Inc.
5500 Oakbrook Pkwy. #200
Norcross Georgia 30093, USA
john.bradley@mse.gatech.edu

Donald Brownlee
Astronomy Department
University of Washington
Seattle, WA 98195, USA
brownlee@astro.washington.edu

Joseph A. Burns
328 Space Sciences Bldg.
Cornell University
Ithaca, NY 14853, USA
jab16@cornell.edu

Stanley F. Dermott
Department of Astronomy
University of Florida
211 Bryant Space Science Center
P.O. Box 112055
Gainesville, FL 32611-2055, USA
dermott@astro.ufl.edu

Johann Dorschner
Astrophysikalisches Institut
Schillergäßchen 3
07745 Jena, Germany
dorsch@astro.uni-jena.de

René Dumont
Observatoire de Bordeaux
33270 Floirac, France

Daniel D. Durda
Southwest Research Institute
1050 Walnut Street, Suite 426
Boulder, CO 80302, USA
durda@boulder.swri.edu

Hugo Fechtig
Max-Planck-Institut für Kernphysik
Saupfercheckweg 1
69117 Heidelberg, Germany
h.fechtig@t-online.de

Marina N. Fomenkova
Center for Astrophysics
and Space Science
University of California San Diego
La Jolla, CA 92093, USA
marifo@cassir.ucsd.edu

David J. Gardner
University of Kent
Canterbury, Kent CT2 7NT, UK

Simon F. Green
Space Science Research Group
Planetary and Space Science
Research Institute
The Open University
Milton Keynes MK7 6AA, UK
s.f.green@open.ac.uk

J. Mayo Greenberg
Huygens Laboratorium
Postbus 9504,
2300 RA Leiden, The Netherlands
greenber@strw.LeidenUniv.nl

Andrew D. Griffiths
University of Kent
Canterbury, Kent CT2 7NT, UK

Keith Grogan
Laboratory for Astronomy
and Solar Physics
NASA Goddard Space Flight Center
Code 681
Greenbelt, MD 20771-0003, USA
grogan@stis.gsfc.nasa.gov

Eberhard Grün
Max-Planck-Institut für Kernphysik
Saupfercheckweg 1
69117 Heidelberg, Germany
eberhard.gruen@mpi-hd.mpg.de

Bo Å. S. Gustafson
Department of Astronomy
211 Bryant Space Science Center
P.O. Box 112055
University of Florida
Gainesville, FL 32611-2055, USA
gustaf@astro.ufl.edu

Douglas P. Hamilton
Astronomy Department
University of Maryland
College Park MD 20742, USA
hamilton@astro.umd.edu

Martha S. Hanner
Jet Propulsion Laboratory
California Institute of Technology
Pasadena, CA 91109, USA
msh@scn1.jpl.nasa.gov

Ove Havnes
Department of Physics
Faculty of Science
University of Tromsø
9037 Tromsø, Norway
ove.havnes@phys.uit.no

Sumita Jayaraman
Vanguard Research, Inc.
Astronomy and Geophysics Divison
5321 Scotts Valley Drive, Suite 204
Scotts Valley, CA 95066, USA
sumita@vrisv.com

Peter Jenniskens
The SETI Institute
NASA Ames Research Center
Mail Stop 239-4
Moffett Field, CA 94035-1000, USA
pjenniskens@mail.arc.nasa.gov

Elmar K. Jessberger
Institut fuer Planetologie
Westfälische Wilhelms-Universität
48149 Münster, Germany
ekj@nwz.uni-muenster.de

Robert E. Johnson
J. L. Newcomb Professor
Engineering Physics
and Department of Astronomy
Thornton Hall B103
University of Virginia
Charlottesville, VA 22903, USA
rej@virginia.edu

Thomas J. J. Kehoe
Department of Astronomy
University of Florida
211 Bryant Space Science Center
P.O. Box 112055
Gainesville, FL 32611-2055, USA
kehoe@astro.ufl.edu

Ludmilla Kolokolova
University of Florida,
Gainesville, FL 32611-2055, USA
ludmilla@astro.ufl.edu

Stephen J. Kortenkamp
Department of Astronomy
University of Maryland
1204 CSS Bldg, Stadium Dr
College Park, MD 20742-2421, USA
kortenka@astro.umd.edu

Gero Kurat
Naturhistorisches Museum
Postfach 417
1014 Wien, Austria
gero.kurat@univie.ac.at

A. Chantal Levasseur-Regourd
Université Paris VI / Aéronomie
Route de Gatines
91371 Verrières, France
aclr@aerov.jussieu.fr

Christoph Leinert
Max-Planck-Institut für Astronomie
Königstuhl 17
69117 Heidelberg, Germany
leinert@mpia-hd.mpg.de

Mischa Maetz
Max-Planck-Institut für Kernphysik
Postfach 10 39 80
69029 Heidelberg, Germany
mischa.maetz@mpi-hd.mpg.de

Ingrid Mann
ESA Space Science Dept
ESTEC
2200 AG Noordwijk, The Netherlands
imann@esa.int

Mark J. Matney
Lockheed Martin Space Operations
2400 NASA Road One
Houston, Texas 77058, USA
mark.matney1@jsc.nasa.gov

Neil McBride
Space Science Research Group
Planetary and Space Science
Research Institute
The Open University
Milton Keynes, MK7 6AA, UK
n.m.mcbride@open.ac.uk

J.A.M. McDonnell
Planetary and Space Science
Research Institute
The Open University
Walton Hall
Milton Keynes, MK7 6AA, UK
j.a.m.mcdonnell@open.ac.uk

Tadashi Mukai
Graduate School of Sci. & Tech. and
Dept. of Earth and Planetary Sci.,
Faculty of Science
Kobe University, Nada
Kobe 657-8501, Japan
mukai@kobe-u.ac.jp

Akiko M. Nakamura
Graduate School of Sci. & Tech.
Kobe University
1-1 Rokkodai-cho, Nada-ku
Kobe, 657-8501, Japan
amnakamu@kobe-u.ac.jp

Paul R. Ratcliff
University of Kent
Canterbury, Kent CT2 7NT, UK

Detlef Rost
Institut für Planetologie
Wilhelm-Klemm-Str. 10
48149 Münster, Germany
Rostd@uni-muenster.de

Zdenek Sekanina
Jet Propulsion Laboratory
California Institute of Technology
Pasadena, CA 91109, USA
zs@sek.jpl.nasa.gov

Mark R. Showalter
245-3, NASA-Ames Research Center
Moffett Field, CA 94035, USA
showalter@ringside.arc.nasa.gov

Frank J. Stadermann
Laboratory for Space Sciences
Physics Department, CB 1105
Washington University
One Brookings Drive
St. Louis, Missouri 63130, USA
fjs@howdy.wustl.edu

Peter Staubach
DREGIS Dresdner Global
IT-Services GmbH
C/S Zentrale & Unix Services
Stresemannallee 36, 508
60596 Frankfurt, Germany
peter.staubach@dregis.com

Thomas Stephan
Institut für Planetologie
Wilhelm-Klemm-Str. 10
48149 Münster, Germany
stephan@uni-muenster.de

Ralf Stognienko
University of Jena
07745 Jena, Germany
ralf.stognienko@jena-optronik.de

Håkan Svedhem
ESA ESTEC
2200 AG Noordwijk zh
The Netherlands
hakan.svedhem@esa.int

Mark C. Wyatt
UK Astronomy Technology Centre
Royal Observatory, Blackford Hill
Edinburgh EH9 3HJ, UK
wyatt@roe.ac.uk

Yu-lin Xu
University of Florida
Gainesville, FL 32611-2055, USA
shu@astro.ufl.edu

Herbert A. Zook
NASA Johnson Space Center
Houston, Texas 77058, USA

Historical Perspectives

Hugo Fechtig[1], Christoph Leinert[2], Otto E. Berg[3]

[1] Max-Planck Institut für Kernphysik, Heidelberg, Germany
[2] Max-Planck Institut für Astronomie, Heidelberg, Germany
[3] NASA/Goddard Space Flight Center, Greenbelt, USA

I. INTRODUCTORY OVERVIEW

This chapter outlines the historical development of the field. There are two intentions in this: first, to follow a path through the scientific developments from the early zodiacal light observations to the late spaceprobe experiments, which directly explore the properties of the solar system dust grains. Secondly and not to forget the human side, we look at the personal aspects of the dust researchers, their organization on the scientific international scene. We also reflect on the science politics of national funding agencies, which partly has led towards and partly against a fruitful scientific work.

This overview is weighted towards the observational side and the exploration of the space environment. Theoretical developments and concepts will be presented in the science chapters of this book. Similarly those results are emphasized which will not be treated extensively in the following review chapters. The selection of references, also, was not intended to be complete and to do justice to everybody but to mark the path we try to delineate through the multitude of seemingly uncorrelated activities.

The scientific development starts with historical observations of the zodiacal light even as early as in the 17th century. With the beginning of the space age the direct study of interplanetary dust particles became a reality; the period during the early 1960s was dominated by dust collection programs using high altitude sounding rockets, and by microphone and other detectors flown on Earth satellites. Zodiacal light research also profited from the newly available space platforms. However, the resulting situation at the end of the sixties was unsatisfying: the measured extremely high flux values in the near Earth neighborhood had given rise to the proposition of a huge dust concentration around the Earth, the so-called "dust belt". (Not to be confounded with the artificial dust belt around the Earth arising later from rocket debris). Later shown to be erroneous, these spectacular findings nevertheless helped to rise interest and to raise funds for further dust research programs.

The arrival of lunar samples on the Earth by NASA's Apollo program has given our field a solid ground: low fluxes comparable to those derived from zodiacal light observations. So did reliable detection techniques, like the impact ionization detector first developed at the Goddard Space Flight Center and flown on Pioneer 8 and 9, which has confirmed earlier measurements by Marshall Space Center using pressurized-cell detectors and capacitor detectors. Further experiments on HEOS 2, Helios, Pioneer 10 and 11 and–currently– on Galileo and Ulysses have reliably explored the dust environment between 0.3 AU and 18 AU in the ecliptic plane. The zodiacal light experiments on the Pioneer 10 and 11 and the Helios A and B space probes meanwhile determined the overall spatial distribution of the interplanetary dust cloud.

A special success was the Halley flyby in 1986. Dust detectors, some combined with time-of-flight mass spectrometers have discovered the atomic mass composition of cometary dust grains. New momentum was also given to interplanetary dust research by the great infrared measurements of the IRAS and COBE satellites.

The "international cosmic dust family" has mainly gathered under the umbrella of COSPAR: it was Curtis L. Hemenway who has founded the Cosmic Dust Panel III C in 1966/67 as a subgroup of Planetary Research. From that time on (but even before on a more individual basis), Panel III C of COSPAR has met every year until 1978 and, after the reorganization of COSPAR, every second year until today. Many personal friendships have developed, and we must say that this international small group has cooperated in many facets since then.

In other international bodies dust researchers were active as well, e.g. in the commissions 15 (physical study of comets, minor bodies, and meteorites), 21 (light of the night sky) and 22 (meteors and interplanetary dust) of the IAU. And after the arrival of the Apollo lunar surface samples in 1969 the yearly Lunar and Planetary Science conferences and, since about 10 years, the European Geophysical Society activities have also attracted our field. Zodiacal light continues to be an important reference for interplanetary dust research, in particular since its infrared emission is being measured by dedicated infrared satellites; but the topics of interplanetary dust research are now broadening, a promising way for its further development.

II. EARLY REPORTS ON THE ZODIACAL LIGHT

"New discoveries are not as remarkable in the beginning as they are going to be afterwards: it is the continuation of the observations which makes them complete and leads us to recognize their grandeur and their consequences." (Translated from the french original)

These are the introductory words Giovanni Domenico Cassini, astronomer at the Royal Observatory in Paris under Louis XIV, used in 1693 as an excuse for presenting to the academy of sciences his full report "Découverte de la lumiere celeste qui paroist dans le zodiaque"; a shorter note *"Nouveau phe-*

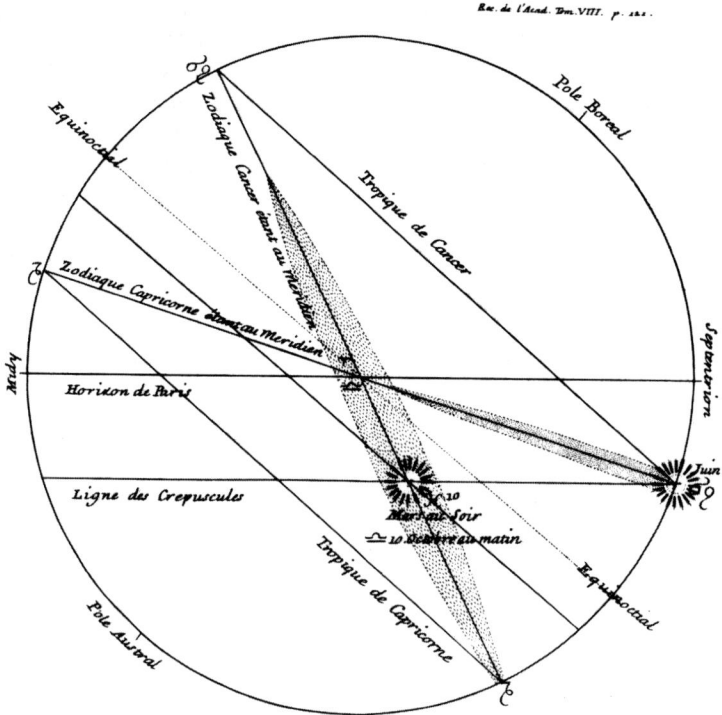

Figure 1. Sketch given by Cassini to explain why the zodiacal light in the evening at middle northern latitudes can best be seen in spring, when the ecliptic rises steeply through the Sun and above the horizon.

nomene rare et singulier d'une Lumiere Celeste, qui a paru au commencement du Printemps de cette année 1683" had already been published in the Journal des Sçavans on June 10, 1683. Written more than 300 years ago, these words may, and hopefully will, still appeal to many of those contributing new findings today to the study of interplanetary dust.

Cassini's paper (see Fig. 1) is the first which brings zodiacal light, and hence interplanetary dust to scientific discussion. It is remarkable in many respects: for the careful documentation of the observations and the awareness of their limitations and their errors; for the formulation of hypotheses where conclusions would not be compelling; in the broad range of new astronomical knowledge used in the geometrical as well as in the physical discussion; in the perseverance in following the subject. The latter led to 11 years of his own study and 'service' observations by people traveling to the far east, only unfortunately—as Cassini noted—southern hemisphere observations were missing. But there also is a tendency to accept part of the apparent changes in spatial extent as real, even after having discussed the different atmospheric effects which could be (and we would today say "are") responsible for them; and there was a degree eagerness to show that his earlier work agreed with, or already had essentially contained new findings by others.

It is really remarkable how convincing an observational evidence he presents to prove the temporal disappearance of the zodiacal light in the years 1665–1681, a fact having an interesting analogy in the temporal absence of sunspots but considered erroneous from our present point of view. It is noteworthy also that the renowned astronomer Cassini preferred an explanation of his discovery in terms of phenomena recently found in other fields of astronomy. He suggested that the material was distributed in the shape of Saturn's rings (recognized as such by Christian Huygens in 1659) and ejected from the Sun like the matter then thought to be responsible for sunpots and faculae on the solar disk. His disciple Niccolo Fatio de Duilliers, on the other hand, continuing the zodiacal light observations after his move to his home near Geneva in 1684, at once jumped to the hypothesis now proven to be correct: the zodiacal light is caused by the reflection of sunlight by particles circling the Sun as mini-planets. It was the failure—quite understandable today—to detect individual particles by telescopic view and the huge number of them required to produce the zodiacal light—as opposed to the small number of known planets or moons—which hindered Cassini from making this also his favorite hypothesis. Not always do the best arguments lead to the most correct interpretations.

Cassini's paper explicitly deals with an astonishingly large and varied number of aspects of interplanetary dust research. In the first place these are the three-dimensional distribution of interplanetary matter (as we would now call the topic); its appearance close to the Sun ("perhaps coma-like"); its plane of symmetry which, as he notes, is deviating from the ecliptic, with Venus' orbit and the plane of the solar equator given as best fitting solutions (possibly with the latter one being representative for the inner region, the first one for the remaining space out to 1 AU; and in general with the plane of symmetry seeming to follow the planes of the planets). But it also includes discussions of parallax measurements (not feasible with Cassini's instrumentation), the lifetime ("long") and variability of zodiacal light (thought of as a behavior in analogy to sunpots), and its origin (conjecture: ejected from the Sun) and dynamics (within the framework of then modern astrophysics). It deals with the question of scattering of sunlight versus emission and finally mentions the possibility that the matter in interplanetary space is of the same kind as that seen in the tails of comets. Some of these topics are still of current interest, some regained it in recent years. With this background, the impressive progress in our knowledge of interplanetary dust since Cassini's time—summarized in this book—appears as it could primarily be due just to the impressive developments in instrumentation and techniques we can put to our service today.

As a phenomenon surpassing in its brightest parts the Milky Way (see Fig. 2), the zodiacal light could hardly have escaped the notice of ancient observers and, indeed, there do appear to be early reports on it. Ancient Egypt representations are said to show it in triangular form, either perpendicular or

Figure 2. The zodiacal light as seen on March 5, 1856, over the city of Freiburg in Breisgau, stretching up to the Pleiades, as during the discovery by Cassini (from Müller 1865).

inclined with respect to the horizon (Asaad 1979). Levasseur (1976) mentions that it certainly was known to the Chaldeans (600 B.C.). Several references to the graecoroman period where expressions like "glowing beam" are found to describe the light seen, are given by Cassini himself. And to the observing astronomer, used to watch for the coming dawn, the biblical wording "wings of the morning"(in German "Flügel der Morgenröte", Psalm 139,9) appears as a poetic description of the phenomenon of zodiacal light.

Among the later reports, the apparent references to the zodiacal light in the anthology 'Rubaiyyat' of the Persian twelfth century scientist and poet Omar Khayyam appear to be due to incorrect translation (see Burkhardt 1986), while the old ($\approx$ 1509) reference to sightings of a light in the morning, (presumably the zodiacal light) by the Aztecs is well documented (von Humboldt 1816). The existence of zodiacal light also was announced in printed form 22 years before Cassini by Childrey as an annex to his "Britannica Baconica or the natural rarities of England, Scotland & Whales" (1661). Cassini also had found out about this reference. He concludes that Childrey's words *"In February... you shall see a plainly discernible way of the twilight striking up towards the Pleiades... And I believe it ... will be constantly visible at that time of the year. But what the cause of it in nature should be I cannot yet imagine but leave it to further inquiry"* do describe the same phenomenon. But Cassini does not miss noting that they neither give the important geometrical information that the zodiacal light stretches from the Sun along the ecliptic nor do they point to the annual motion of the zodiacal light with the Sun. The latter fact, basic for understanding the nature of the zodiacal light, was contained implicitly already in Cassini's letter of 1683 announcing the discovery of zodiacal light but it seems to have been explicitly noted first by Fatio in a letter to Cassini in the following year. Cassini—as it appears—purposely is vague on this point. Historically the question who discovered the zodiacal light may be ill-posed (who discovered the moon?) but scientifically, it seems fair to leave the credit for the discovery of the zodiacal light with Cassini; however, we should not forget the important contributions by his disciple Fatio.

With time slowly it became clear that the zodiacal light covered a large fraction of the sky. Originally, Cassini saw it to extend to $\approx 60°$ from the Sun, in February 1687 in one case to 100°; Mairan (1733) definitely confirmed the extension of the zodiacal light into the anti-solar hemisphere; Jones (1856) under favorable conditions in the tropical zones saw the zodiacal light extend to even 160° from the Sun.

The discovery of the *Gegenschein*, a faint enhancement of the zodiacal light around the anti-solar point (see Fig. 3), followed in the middle of the nineteenth century by Brorsen (1854). There have been repeated reports on the discovery of this phenomenon; an earlier announcement (Pézénas 1731) most likely referred to an aurora, while Humboldt and Jones appear to have seen the "false zodiacal light" (=atmospheric backscattering of the bright cones of zodiacal light, see Roosen 1971). Later discoveries were reported by Backhouse (1857) and by Barnard (1883) who coined the English term "counter- glow" and gave the phenomenon enough publicity to prevent further rediscoveries. The history of the interpretations of the Gegenschein is even more varied than its discovery, maybe because the position at the anti-solar point suggested some relation to the Earth. Brorsen (1859) saw the Gegenschein and the Earth in analogy to comets which have their gaseous tail pointing away from the Sun; Searle (1882) suggested a cloud in the libration point L3 of the Sun-Earth system as source of the enhanced brightness but later (1893) was convinced that enhanced backscattering by the particles of the interplanetary dust cloud would be a simpler and preferable explanation.

At the end of the 19th century, with the existence of the zodiacal light and the underlying cloud of interplanetary dust particles well established, and with a theory of the Gegenschein based on just this dust cloud available, the topic could have come to rest. But new theories abounded up until at least the year 1961.

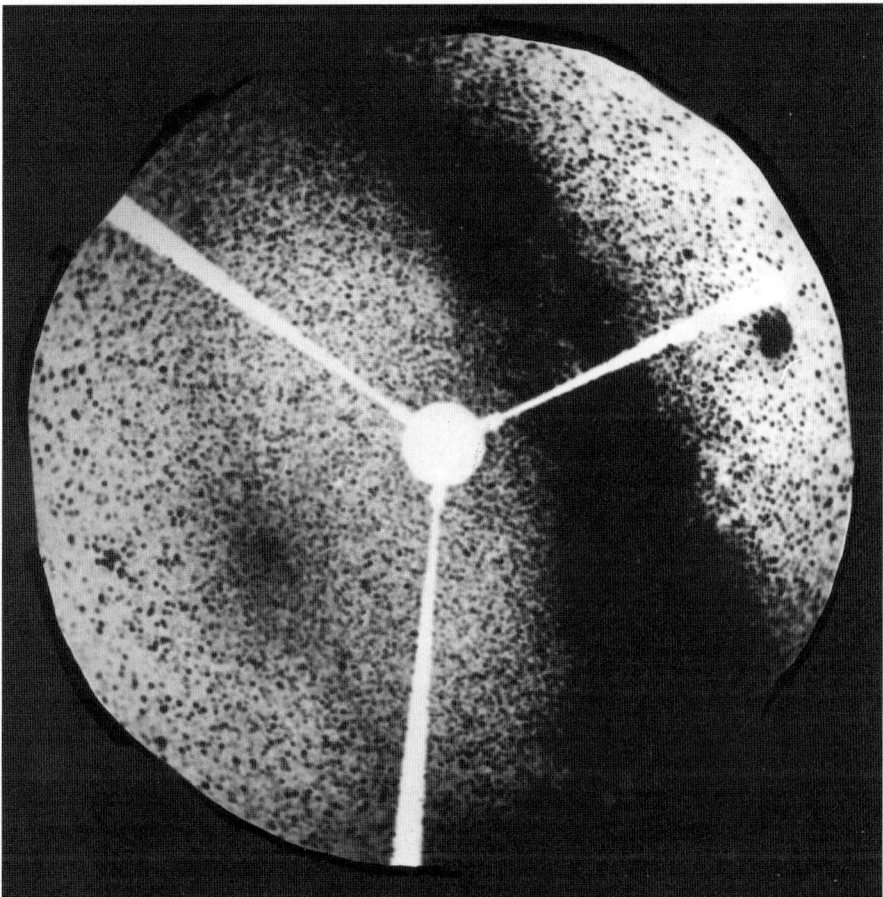

Figure 3. The Gegenschein, photographed at ESO, La Silla, with the wide-field camera of the University of Bochum (Germany). The strong band of light in the right part is the Milky Way, with the Large Magellanic Cloud seen to the right of it. In the left part the Gegenschein appears as a distinctly enhanced patch in the faintly visible band of the zodiacal light (by courtesy of W. Schlosser, Bochum).

From Roosen's (1970) bibliography we take without claim for completeness the following examples: a contrast effect against the Earth's shadow, lensing of sunlight through the Earth's atmosphere, reflected moonlight, a dust blanket around the Earth, material in the asteroid belt or in a heliocentric ring, the geomagnetic tail, a dust tail of the Earth, an ionic tail, interaction of a tail of the Earth with meteoroidal bodies, a geocentric ring of debris from previous space flights. Discussing and rejecting all of these additional hypotheses certainly made the case stronger for the interpretation of the Gegenschein by interplanetary dust, but at quite some effort. And all this theoretical activity hardly conformed to Newton's guideline "hypotheses non fingo"—"I do not invent hypotheses"! For all practical purposes the discussions were settled

 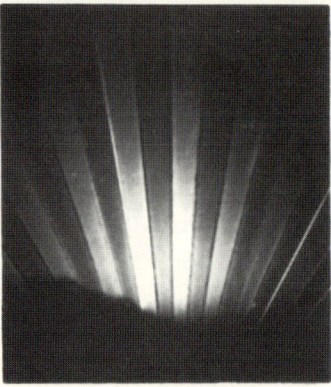

Figure 4. Demonstration of the strong polarization of zodiacal light. Two photographs of the zodiacal light as photographed on August 2, 1958, from the very high altitude station Chacaltaya at 5130 m in the Andes are shown. The left one is a direct image, the right one is taken through an assembly of polaroid strips, with the polarization vector oriented alternatively horizontally and vertically (from Blackwell and Ingham 1961a).

in favor of the interplanetary dust hypothesis, when the Imaging Photopolarimeter on Pioneer 10 saw the Gegenschein on March 14, 1972 from a position in interplanetary space, more than 10 million km from the Earth (Hanner and Weinberg 1973).

The nineteenth century also sees the first study of the *polarization* of zodiacal light. A.W.Wright from Yale College specifically constructed a wide-field polariscope allowing him to investigate such faint brightnesses as encountered in the zodiacal light. The presence of linear polarization in this apparatus showed by the appearance of narrow dark or bright stripes, somewhat similar in this to the Polaroid stripes used in Blackwell and Ingham's photograph of the zodiacal light shown in Fig. 4.

Wright's (1874) conclusions from his successful attempts were the following:

1. *The zodiacal light is polarized in a plane through the Sun* [i.e. with the electric vector perpendicular to this plane]
2. *The amount of polarization is, with a high degree of probability, as much as 15 %, but can hardly be as much as 20 per cent* [at 30°–40° from the Sun].
3. *The spectrum of the light is not perceptibly different from that of sunlight except in intensity* [based on his spectroscopic measurements with other instruments].
4. *The light is derived from the Sun and is reflected from solid matter.*
5. *This solid matter consists of small bodies (meteoroids) revolving about the Sun in orbits crowded together toward the ecliptic.*

It is surprising how well these observations *"made in the upper floor of one of the college buildings, the windows of which look towards the southwest..."* agree with results obtained from modern space probe observations, with e.g. Helios giving a degree of polarization in the visual at 36° from the Sun of 17 %–19 %. This coincidence must, in part, be fortuitous because Wright neglected the other sources of sky brightness and the polarizing effects of atmospheric scattering. Certainly though his conclusions, giving a quite direct demonstration of the fact that the zodiacal light is scattered sunlight, have stood the test of time. His last two points are not just inferred or taken from the literature but build on his laboratory measurements on rough solid surfaces, including a fragment of the Pultusk meteorite, which on the average showed a polarization similar to that of the zodiacal light. The broad scope of investigation used in this pioneering visual experiment still presents a good example for a successful polarimetric study of the zodiacal light.

To proceed further, more accurate quantitative measurement techniques were necessary; these were available much later, namely in the photographic observations beginning with Dufay (1925) and the following photo-electric polarization measurements of Huruhata (1948). We will discuss a few of the later polarization measurements in section III, because it is in the time of maturing ground-based photometric surveys that polarization contributed most to the developing picture of the interplanetary dust cloud (a thorough overview on then existing polarization measurements of zodiacal light has been given by Weinberg (1974)). Since then the importance of zodiacal light polarimetry has declined: the size distribution and nature of interplanetary dust particles, earlier the domain of polarimetry, now can better be assessed by in-situ observations, and mid-infrared emission, which otherwise gave rise to a renaissance of astronomical esteem for zodiacal light is, essentially, unpolarized.

Up to the end of the nineteenth century, *spectral information* on the zodiacal light was scarce, the observations of Piazzi-Smyth (1872) and others showing *"that the spectrum is continuous, and not perceptibly different from that of faint sunlight"*. And viewed in whole, remarkably little effort has gone into the spectroscopic study of zodiacal light, except for some difficult high-resolution studies aiming at the kinematics of interplanetary dust (see section III). The chapter *Spectrum* is missing for this reason in recent reviews on zodiacal light. According to Roosen (1971) the first really good spectrograms of zodiacal light were made at Lowell Observatory by Slipher (1933).

It is telling that Fig. 5 still appears to be one of the best available spectrograms of zodiacal light in the easily accessible literature. This branch of zodiacal light research is underrepresented, probably because the zodiacal light is faint and the spectra are not easily interpreted in terms of physical properties of interplanetary dust. Infrared satellites like IRTS and ISO, which have the instrumentation to perform infrared spectroscopy at a range of spectral resolutions, may reverse this trend.

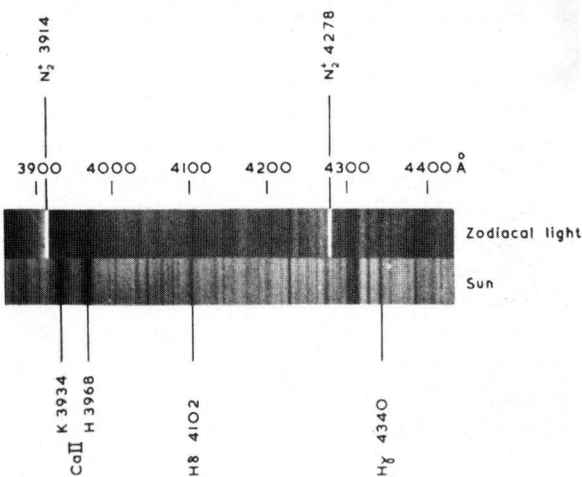

Figure 5. Spectrum of the zodiacal light compared to the solar spectrum. Four main lines of the solar spectrum and two airglow lines are indicated (from Blackwell and Ingham 1961a).

III. ZODIACAL LIGHT OBSERVATIONS UNTIL THE BEGINNING OF THE SPACE AGE

Around 1964 two monumental theses (Weinberg 1964; Dumont 1965) appeared in the literature and ended the long-during struggle for correct ground-based photometry of zodiacal light in presence of the many other disturbing light sources like tropospheric scattering and extinction, airglow and contribution of unresolved stars. Soon afterwards zodiacal light measurements had almost exclusively become a topic of space experiments. This section mostly deals with the activity up to the time of those two theses. We will essentially skip over visual observations and concentrate on the photoelectric ones but include photographic work where important for the development. Again no attempt is made to reach completeness and the choice of work presented reflects the preferences and perhaps the limited overview of the authors.

The presence of interplanetary dust close to the Sun was recognized by Grotrian (1934) in the analysis of spectra of the solar corona taken during an eclipse 11 years earlier. The spectra best were described as superposition of a pure continuum (called K corona, due to Thomson scattering of sunlight by coronal electrons) and of the solar Fraunhofer spectrum reflected by dust particles at least 3 μm in size. This second part, the so-called F corona, was suggested to be due to just the inward extension of the cloud of particles producing the zodiacal light. Calculations showed that for a spatial distribution $n(r) \sim r^{-1.5}$, where $r =$ heliocentric distance, the brightnesses of F corona and zodiacal light would join smoothly, if with Seeliger (1901) he assumed that

Figure 6. The transition region between the F corona and the zodiacal light seen over the lunar horizon. The ecliptic is approximately vertical and passes near Regulus and Mercury, the two bright objects in the upper part of the picture. The inner zodiacal light (or outer corona) is seen from $5°$–$20°$ elongation, with the symmetry distorted by a coronal streamer. The photograph is not sensitive enough to also show the continuation of the zodiacal light towards larger angles from the Sun. (NASA photograph AG 15-98-13 311 by Worden during the Apollo 15 flight).

most of the brightness increase would be due to the spatial distribution of the particles close to the Sun. His assumption and conclusion both remain accepted today. This hypothesis also explained the radial dependence of polarization in the corona.

While Grotrian still thought that the gap between F corona ($\approx 1°$) and the zodiacal light ($\approx 30°$) could not be bridged by observations, the astronauts on the Apollo flights, well outside the Earth's atmosphere, could do just this (Fig. 6). Thus with Grotrian's study it had finally been made clear that the zodiacal light stretches all along the ecliptic and hence interplanetary dust must fill *all of the ecliptic plane* within some range of distances from the Sun.

Less well known but equally interesting from today's point of view are his speculations that dust particles might get smaller towards the Sun by sublimation, and probably will be blown out from the Sun by radiation pressure. Grotrian wanted to use this concept to explain red-shifts of 20–30 km s^{-1} found in some coronal observations. He could not imagine that such "β-meteoroids" would become the topic of a controversial discussion much later (section VI). His aversion, however, to accept that mass density in the interplanetary dust cloud could vary with heliocentric distance forced him to assume mm- to cm-sizes for interplanetary dust near 1 AU. Here his intuition misled him.

To show convincingly that the zodiacal light covered *all of the celestial sphere* required photoelectric observations, although the photographic polarimetric work of Dufay (1925) already had given strong indications for the omnipresence of zodiacal light. Such a study was performed from 1934 on by Elvey

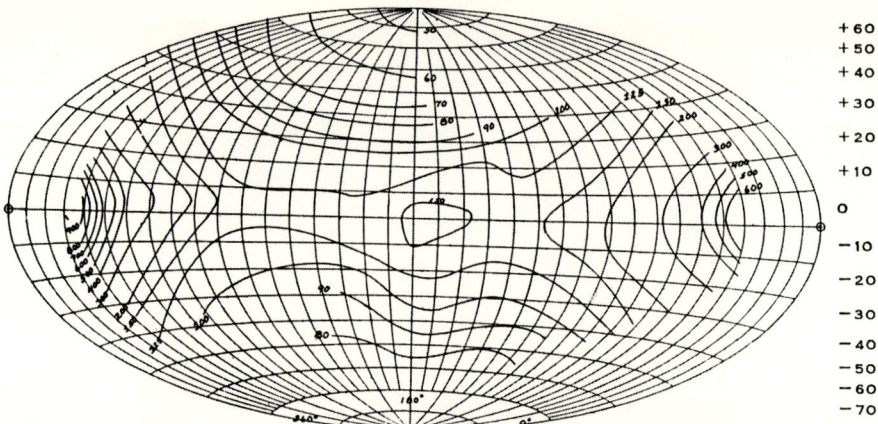

Figure 7. First isophote map of the zodiacal light, obtained by Elvey and Roach (1937) from measurements in November and December 1934. The full sky is projected into the figure, with the north ecliptic pole at top, the ecliptic running horizontally, with the Gegenschein at the center and the Sun to be pictured both at the extreme right and extreme left.

and Roach (1937) from Mount Locke (1860 m) and led to the first isophote map of zodiacal light (Fig. 7). Apart from being high by about a factor of ≈ 1.5, this map looks quite reasonable and also finds a ratio of in- ecliptic to polar brightness at an elongation of 90° from the Sun of about a factor of four, not very different from the factor of ≈ 3 used today.

At that time then, with an approximate knowledge of the two-dimensional distribution of zodiacal light, with the concept of dust particles much larger than the wavelength of light and of a strong increase in the spatial density of scatterers towards the Sun, the starting point seemed set for a systematic approach towards our present picture of interplanetary dust. Steady, small improvements and corrections concerning spatial distribution, size distribution and albedo of interplanetary dust were all which would have been needed. The remainder of this section, however, shows a somewhat different story.

First, Allen (1947) and van de Hulst (1947) questioned Grotrian's (1934) interpretation. Approximating the *scattering function* of dust particles by the sum of the diffraction pattern for a sphere and of a more or less isotropic scattering component, they suggested instead that it is the diffraction peak on micron- to sub-millimeter-sized particles which explains by far most of the coronal brightness increase towards the Sun. Van de Hulst goes on, on the assumption of constant space density of interplanetary dust, to derive a size distribution of $n(a) \sim a^{-2.6}$ for radii $a < 350$ μm. This leads to a surprisingly accurate estimate of the mass density at 1 AU of $5 \cdot 10^{-21}$ g cm^{-3}. Comparison to the brightness of zodiacal light yields a very low albedo of < 0.005, but a value of albedo of 0.1 is said to be more realistic and also acceptable, given the uncertainties in spatial distribution of interplanetary dust. Elsässer (1955) took in addition solar limb darkening into account and revised the size distribution to

$n(a) \sim a^{-2.0}$. With equivalent conclusions concerning albedo as van de Hulst, he inferred a metallic nature of the dust particles. He used, though, a spatial distribution of dust both constant and also limited to heliocentric distances > 0.4 AU. The Poynting-Robertson effect in its application to the solar system (Wyatt and Whipple 1950) is nowadays mostly discussed as physical reason for an increase of spatial density towards the Sun at least as steep as $1/r$, but was then taken only as mechanism for clearing interplanetary space preferentially of the small particles. The fact that it can do so only via piling up the particles close to the Sun, was not yet seen. In hindsight one wonders why the very low resulting values of albedo were not taken as a hint that the particular treatment chosen by these three authors overestimated the effect of diffraction. Admittedly the derived size distribution and the values of albedo assumed as realistic were essentially correct, but as it appears, for the wrong reason.

Polarization measurements also led to a modification of the assumed physical sources of zodiacal light. Behr and Siedentopf (1953) undertook a new careful measurement from a very high site in the Swiss Alps (3576 m) to improve on the large uncertainties in Huruhata's (1951) original measurements. Following a suggestion by Whipple and Gossner (1949) they interpreted the polarization as due to scattering by interplanetary electrons, where a density of 600 cm^{-3} at 1 AU was needed to explain the observations. [In explaining the out-of-ecliptic decrease in these results, Elsässer (1954) used ellipsoidal isodensites and Gaussian decreases depending only on height above the ecliptic, typical of the two main types of out-of-ecliptic spatial distributions discussed today.] The need for electrons as polarizing agents became even stronger when subsequent measurements (Elsässer 1958a, 1958b) resulted in degrees of polarization of up to 37 %. These high values of polarization were supported by the photographic measurements of Blackwell and Ingham (1961a) who went up even to the high altitude station Chacaltaya at 5130 m in the Andes, in order to get as free as possible from disturbing atmospheric effects. In addition, Beckers (1959) had apparently detected the reduction in depth of the Fraunhofer lines in the zodiacal light due to the extra continuum due to scattering on the fast-moving electrons. But Blackwell and Ingham (1961b) disagreed: they had found no extra continuum in their spectra, with a resulting upper limit for the electron density at 1 AU of 120 cm^{-3} (later from improved spectra $\approx$ 20 cm^{-3}) and pointed out the possibility of high degrees of polarization by reflection from dust grains. The controversy went back and forth, being resolved finally by the first interplanetary measurements of solar wind by Mariner 2, which determined the plasma density at 1 AU to about 5 cm^{-3}. At about the same time, measurements in narrower wavelength bands with improved techniques of separation of the components (Weinberg 1964; Dumont 1965, the theses mentioned above) also had led to the more moderate polarizations in the zodiacal light of 15 %–20 % now considered correct.

Now the explanations reverted back to scattering by dust only. The availability of computers made exact scattering calculations for spheres possible (*Mie theory*) which, however, were not then extended to particle sizes beyond

20 μm because of limits in computing time. From the many *models* reviewed by Giese (1962) it was now very small (tenth-micron) dielectric particles with constant spatial distribution which represented particularly well both brightness and polarization of zodiacal light (Weinberg 1964). This was quite a swing of the pendulum, decreasing the estimated effective size of the interplanetary particles by about a factor of 100. However, such sub-micron particles would make the zodiacal light bluer than the Sun, contrary to the reddening found in subsequent observations (see section IV.B). Therefore, Giese (1973) also tested models with larger particles, and he found that for a slow increase of number-density towards the Sun (less than $\propto r^{0.5}$) half-micron sized absorbing or ten-micron sized very slightly absorbing particles would also fit the observations. Only after space experiments had shown that the effective size of the scattering interplanetary dust particles is at least 30 μm and that the increase in spatial density towards the Sun is steeper than $1/r$ (see section V), was it recognized that these new assumptions could explain most of the relevant observations (Giese and Grün 1976); Giese et al. (1978) demonstrated from laboratory scattering experiments that the absorbing fluffy particles could explain the observed polarization; and Röser and Staude (1978), using Mie theory, showed in a remarkable paper that the successful predictions of large-particle models included even the brightness of the F corona and the broad-band spectrum of the zodiacal light from 1500 Å to 60 μm. The pendulum had swung back. Lamy and Perrin (1980), based on a different derivation of a flux curve from lunar micro-crater counts, argued for a last time that small particles could give a substantial contribution to the brightness of the zodiacal light, but found little support. Thus the starting point of the further scientific discussions contained in this book was not reached in a particular straightforward way; we may ask ourselves whether our contributions to interplanetary dust research will look any more straightforward when summarized 30 or 50 years from now.

A few special topics of zodiacal light research before the era of space observations should still be mentioned. The *polarization of the F corona* was found to be quite small (0.05 % to 2.8 % from 5 $R_\odot$ to 40 $R_\odot$) by Blackwell et al. (1967), while extrapolations from zodiacal light observations and simple modeling would suggest values around 10 %. Their models of interplanetary dust thus had to force the albedo close to the Sun to very small values (in the extreme case to 0.0027 at 4 $R_\odot$) in order to reproduce the low measured polarizations. With respect to this coronal polarization measurement the author of this section has completely ambivalent reactions: high respects on one side, and at the same time he wishes it could be proven wrong, because it is so hard to understand. Since it is a difficult measurement, requiring both polarimetric and simultaneous spectroscopic observations, it has not been repeated so far, but the instruments on satellite SOHO could and should revisit this question which is of some importance for the understanding of circumsolar and maybe interplanetary dust.

In principle, optical observations can give information on the *kinematics* of interplanetary dust, but this requires high spectral resolution observations difficult for this low surface brightness source and typically requiring the use of staggered Fabry-Pérot instruments. Nevertheless, the observed systematic Doppler shifts of a few tenth of an Å (e.g. Hicks et al. 1974; East and Reay 1984) show that most of the interplanetary dust particles are in prograde orbits ($> 95\%$ according to James and Smeethe 1970) and there is even an indication of recessional movement in the Gegenschein (East and Reay 1984), possibly due to the outward motion of β-meteoroids. As with other zodiacal light measurements, these are large-scale averages over individual interplanetary dust particles, weak in detailed information but strong in giving a definite answer what the average has to be. The reason for the discrepant results by Fried (1978), which could have been explained only by particles in hyperbolic orbits, never became really clear; these results could be erroneous.

Temporal variations of zodiacal light, or at least reports on such variations, are as old as zodiacal light research itself; this became evident above in the review of Cassini's (1683) fundamental paper. In the decade around 1970, the community of zodiacal light observers was split about equally into optimists emphasizing the fascinating physical interactions to be revealed only through the study of such variations and the sceptics like Weinberg (1970) stressing the various possible pitfalls, from differences and uncertainties in calibration through variations in airglow brightness and atmospheric transparency and scattering to the problems associated with a unique separation of the components of the light of the night sky. The sometimes sharp mutual criticism between members of different attitude found its way into the literature at various occasions.

Quite a number of variations or peculiarities were reported (see, e.g. Leinert 1975), namely long term variations of zodiacal light brightness and polarization with the solar cycle by up to a factor of two; a correlation of zodiacal light intensity with the total number and brightness of yearly appearing comets; variability with lunar phase (Divari 1964); a short term brightness increase of zodiacal light after a class 3^+ solar flare on July 8/9, 1958, possibly by fluorescence of interplanetary dust particles exposed to energetic solar particles; variations in Gegenschein brightness, possibly due to a passing dust cloud; variations in degree and orientation of polarization—indicative of elongated interplanetary dust particles with varying degrees of alignment according to the varying streams in the solar wind; an enhancement by increased dust concentrations in the libration point L4 of the Earth-moon system; brightness increases due to local concentrations in meteor streams (Levasseur and Blamont 1973, 1976).

Given the fact that naturally each sequence of measurements must contain fluctuations on the level of a few σ, doubts are allowed concerning many of these reports. An 11 year space experiment on the space probe Helios A, with a sensitivity to see the additional light from plasma clouds ejected by

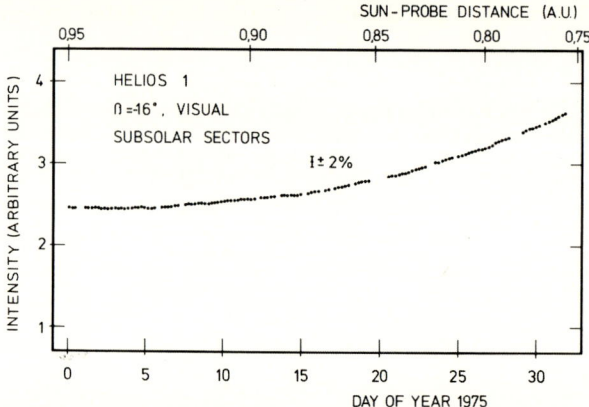

Figure 8. Example for smoothness and short term stability of the zodiacal light as seen from Helios A (= Helios 1) in a 5.6° × 1° wide field-of-view. The gradual brightness increase is due to the approach of Helios towards the sun (from Leinert et al. 1977, copyright American Geophysical Union).

the Sun (Richter et al. 1982; Jackson and Leinert 1985), did not notice any measurable variations (Leinert and Pitz 1989; see also Fig. 8), the view that the zodiacal light is quite stable gained wide support. What remains are the variations of zodiacal light caused by a change of position of the—earthbound or interplanetary—observing instrument, closer to or farther from the Sun; in, below or above the symmetry plane of highest concentration of interplanetary dust (see Fig. 21). But also, there is no doubt about variations related to the new structures (cometary trails, asteroidal bands, particles trapped along the Earth's orbit) which appeared in the mid- and far-infrared sky surveys of the satellites IRAS and COBE. And the apparently changing concentrations of circumsolar dust detected by infrared coronal observations from ground and from balloon (see Maihara et al. 1985) could be real, too. Once again the development appears to follow the old pattern of thesis, antithesis and synthesis.

The fact that interplanetary dust needs replenishing had already been stressed by Poynting (1903) in his pioneering paper. But Whipple (1955) first found an intelligent way to quantify this process. He determined the *mass loss* of the interplanetary dust cloud due to the Poynting-Robertson effect from the total brightness of zodiacal light. He argued that since both quantities depend on total cross section, they should be proportional, and this led him to the first estimate of the mass lost to the interplanetary dust cloud of $\approx$ 1 ton/day. This is lower than current best estimates but these include also effects other than Poynting-Robertson losses. This is just one example for the sometimes unexpected relations between the interplanetary dust cloud and its optical manifestation, the zodiacal light.

IV. AFTER THE BEGINNING OF THE SPACE AGE: FIRST SPACE EXPERIMENTS TO MEASURE THE ZODIACAL LIGHT AND TO DIRECTLY INVESTIGATE INDIVIDUAL DUST GRAINS

IV.A. Rise and Fall of the Earth's Dust Belt

The earliest reported in-situ measurement of Interplanetary Dust Particle (IDP) flux occurred in 1950 using a V-2 rocket as the transport medium (Bohn and Nadig 1950). The single (and simple) objective of the experiment was to determine the number of extraterrestrial dust particles a surface would encounter in space exploration beyond Earth's protective atmosphere. The flux was determined by the number of amplified signals received from a ceramic microphone attached directly to the rocket skin. Although the extremely high flux reported was met with severe skepticism (bordering on scorn) by astronomers and investigators, engaged in zodiacal light and related studies alike, the reported results could hardly be ignored as a technique with potential because they were, indeed, the first direct measurements of interplanetary dust particle flux as distinguished from results derived by ground observations. As a result of the pioneering experiment and, not too surprisingly, interplanetary dust particles suddenly enjoyed a certain amount of ecumenical respect among astronomers and astrophysicists! Previous to the in-situ measurements by Bohn and Nadig (1950), dust particles had only been represented by dark patches in the night sky or by zodiacal light studies of scattered light at Earth's horizon. Suddenly, it represented itself as a solid, single particle having speeds calculated to be extremely high and now profusely invading Earth's upper atmosphere.

With the advent of the Space Age in 1950, several National Laboratories recognized the opportunities and importance of in-situ detections and measurements of interplanetary dust parameters and, accordingly, organized experimental groups specifically assigned to that interest. Most of those experimental groups were funded by the newly established National Aeronautics and Space Administration (NASA) which had dual interests in that field: (1) to study the particulates as they contributed to the overall interplanetary space ensemble; and (2) to consider the particulates as a possible hazard to man and instruments in space. NASA organized three individual experimental dust groups within its own administration: W.M. Alexander, O.E. Berg and coworkers at the Goddard Space Flight Center (GSFC); R. Nauman and coworkers at Marshall Space Center (MSC); and D.E. Gault, N. Farlow and coworkers at the Ames Research Center (ARC). Outside of NASA, but funded thereby, was one group at the Air Force Cambridge Research Laboratories (AFCRL) under R.A. Soberman and R.A. Skrivanek, and another group at the Dudley Observatory under C.L. Hemenway. In addition to the NASA-funded experimental groups were several groups outside of the USA: in England, France, and Russia. In 1963 the intrigue of evaluating the nature of interplanetary dust particle caught the interest of the Max-Planck-Institute für Kernphysik (MPIK) in Heidelberg, Germany. They were already well established and well-renowned in the experimental study of larger extraterrestrial particulates, the meteorites. Their first intention at that time was to collect particles in the upper atmosphere and analyze their compositions by using the neutron-activation-analysis method.

Skepticism aside, the reported results initiated an immediate international interest in the extraterrestrial particles and set the stage for numerous subsequent experiments (primarily acoustical devices) aboard rockets, probes, and satellites. Essentially all of the early acoustical experiments reported comparably high fluxes seemingly validating the V-2 results. Concurrently, other experimenters conducting a series of particle collection experiments from balloon altitudes (appr. 30 km) and rocket probes above 75 km reported results which seemed to corroborate the early acoustical data (Hemenway et al. 1967; Hemenway and Soberman 1962). Although the reported results of other contemporaneously performed dynamic sensor measurements and collections indicated a dust flux several orders of magnitude lower than the results of the microphone sensor experiments, their results were generally rejected or ignored if they tended to disagree significantly with the "established flux curve"! Representative of the dust experiments yielding dissenting results were:

1. Pressurized-cell detectors on the Earth satellite Explorer 16 (d'Aiutolo 1964);
2. Capacitor-type sensors on the Earth satellites Pegasus (Clifton and Naumann 1966) and Ariel II (Jennison et al. 1967);
3. OGO satellite data (Alexander et al. 1971);
4. Photomultipliers, adapted to measure the kinetic energy of the visible light flash produced by the dust particles' impact. They registered a dust flux comparable to the zodiacal light values (Berg and Meredith 1966; Berg and Secretan 1967). The inspiration and feasibility of using the photomultiplier for interplanetary dust particle experiments was justified on the basis of theoretical studies by Öpik (1958) suggesting that approximately one percent of the particle's kinetic energy was dissipated as visible light at the moment of impact—sufficient energy to register impacts of dust particles having a mass of 10^{-14} g or greater and a speed of 5 or more km s^{-1}.

The old adage—"*hindsight is always* 20/20"—seems appropriately applicable to the early in-situ studies of interplanetary dust. Now we see clearly that, for a number of years, the acoustical sensor results were heavily favored for reasons that were not based on sound scientific logic:

1. The microphone results had achieved inordinate respect as "*firsts*" and were, accordingly and generally, accepted throughout the technical world as "gospel revelation";
2. The microphone sensor was rugged, simple, and its response was assumed to be dependably reliably related to the dust particle's momentum; and
3. The acoustical results were mutually corroborative as performed by several unassociated experimenters.

Thus, over several early years of dust experimentation, the sheer volumes of data and concomitant results from the omnipresent microphone seemed to present overwhelming statistics and confirmation, evoking fear in the hearts of the general experimenter! If the acoustical results did indeed represent a reasonable portrait of the dust environment, the extremely high flux of particles also implied and represented a major hazard to any and all ventures beyond Earth's atmosphere.

So convincing were this high flux concept and the "established flux curve", that even a "Giant of astronomers", Fred Whipple, was quoted as saying, *"Means should be taken to protect astronauts from being subjected to the noise of impinging meteoritic dust. The psychological effects may be serious, even though the physical hazard is small"* (Whipple 1961).

A good summary of the situation at the end of the 1960s is given by the flux diagram of Fig. 9. (The authors are of the opinion that it would not be expedient to reference the numerous papers and publications describing the experiments and results of the early acoustical experiments. For those who may have a special interest in those results, we refer to an appropriate review by W.M. Alexander et al. 1963.)

Plotted is the logarithm of the cumulative flux per m^2 and sec as a function of log(particle mass). The references for the various symbols are given in the legend of Fig. 9. The high fluxes in the mass range between 10^{-16} to 10^{-6} g appeared to be mutually confirmed by the Venus Flytrap (VFT) dust collections and by detectors (McCracken et al. 1961). Note, on the other hand, fluxes derived from zodiacal light measurements are lower by 5 to 8 orders of magnitude (Elsässer 1963; Ingham 1963). Remember, that it is almost impossible to be wrong in a zodiacal light measurement by more than a factor of $1.5 - 2$. To derive as high as possible a particle flux from a measured brightness, one could assume the particles to be almost black (albedo $= 0.05$), very dense (like iron), and on hyperbolic unbound orbits. Taken together, this would increase the derived particle flux by about one order of magnitude only with respect to standard assumptions. Zodiacal light researchers therefore never accepted the high fluxes for all of interplanetary space but at most for a local near-Earth environment. But these general indirect conclusions did not very much impress those getting direct responses, signal by signal, back from their dust detectors.

A certain measure of the intrigue for evaluating the interplanetary dust particle was the swiftly evolving rumor of a *"dust belt"* around Planet Earth. It is difficult to identify the original source for the promotion of that rumor, but it was certainly encouraged by the quasi-acceptance of the very high fluxes and the obvious discrepancy to the low level derived from zodiacal light measurements, which could not be completely neglected. A paper entitled, "The Earth's dust belt: fact or fiction?" (Shapiro et al. 1966) served well to set the stage for a multi-year controversy that spawned international conferences with tempers energetic enough to form a dust belt of its own! But before continuing with this remarkable story, let us have a look at some parallel developments in the field.

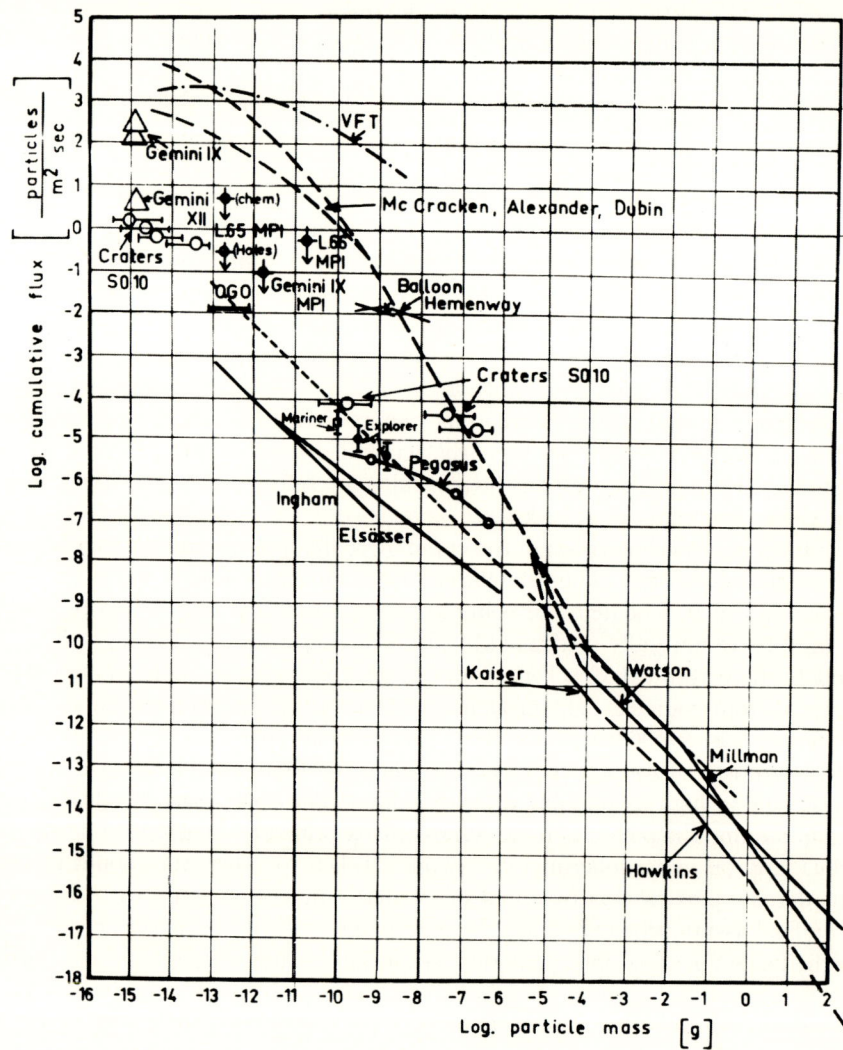

Figure 9. Log. cumulative flux vs. log. particle mass. References in Fig. 9 corresponding to general reference list: Balloon Hemenway: Hemenway et al. 1967; Craters S 010: Hemenway et al. 1968; Elsässer: Elsässer 1963; Explorer: d'Aiutolo 1964; Gemini IX: Weihrauch et al. 1968, Hemenway et al. 1968; Gemini XII: Hemenway et al. 1968; Hawkins: Hawkins and Upton 1958; Ingham: Ingham 1963; Kaiser: Kaiser 1961; L 65 MPI: Gerloff et al. 1967; L 66 MPI: Weihrauch et al. 1968; Mariner: Alexander 1962; McCracken, Alexander, Dubin: McCracken et al. 1961; Millman: Millman 1957; OGO: Alexander et al. 1971; Pegasus: Naumann 1966, Johnson et al. 1966, Clifton and Naumann 1966; VFT: Hemenway and Soberman 1962, Soberman and Hemenway 1965; Watson: Watson 1956.

Concurrently within this first period of in-situ cosmic dust studies, but with specific interests essentially unrelated, were experiments into the nature and effects of hypervelocity impacts upon materials, i.e. impacts by discrete particles having speeds in excess of the speed of sound in the material of the target. They were hypervelocity particle impact studies principally associated with military interest. The surprisingly devastating effect of these particles upon targets was directly applicable to interplanetary dust impacts also, because their speeds were also primarily within the hypervelocity range. Accordingly, the "meteor hazard" was exacerbated and rose to prominence as the major hazard to space travel and associated instrumentation. Unfortunately, due to military interests in hypervelocity experiments to date, the projectiles used in the experiments were orders of magnitudes larger in mass, so it became necessary to scale down results to accommodate the impact effects of micron-sized dust impacts. However, an ingenious modification to a 2 MeV Van de Graaf accelerator by Shelton et al. 1960 and Friichtenicht 1962 provided single, micron-sized iron spheres with measured speeds up to more than 50 km s^{-1} and of derived mass. The advent of the dust particle accelerator provided accurate testing and calibration of hypervelocity dust experiments and encouraged the design and optimum development of a variety of new and different sensor systems. Further, the craters or impact signatures observed in the laboratory by these "man-made micrometeoroids" opened a new horizon of research into the registration and collection of interplanetary dust-induced signatures.

Several experimenters argued and acted upon the proposition that, "if" (and what an enormous "*if*" it eventually turned out to be!) the extremely high reported fluxes were authentic, then it would be also reasonable to collect impact signatures of the hypervelocity interplanetary dust upon prepared surfaces. Furthermore, the high flux coupled with the readily calculated deceleration of these particles, as they traversed the Earth's atmosphere, would constitute a concentration of particles at altitudes below (roughly) 100 km, providing a reasonable set of conditions for the intact capture and collection of the space particulates. Again, several experimenters responded to that proposition (Hemenway and Soberman 1962; Soberman and Hemenway 1965; Berg et al. 1965; Farlow et al. 1966).

A different type of particle was found in a rocket collection experiment flown through the noctilucent clouds (NLC's). Figure 10 shows an electron microscope image of a "collected NLC grain" (Hemenway et al. 1964). The authors have assumed that NLC-grains consist of a silicate core and an icy mantle which, when collected on a warmer substrate, will melt and produce a pattern on a nitrocellulose substrate like that seen in Fig. 10. The Heidelberg group has flown detectors through NLC's and, from those results, the authors (Rauser and Fechtig 1972) have concluded that interplanetary dust grains from outer space become temporarily ice-coated as they traverse the temperature-minimum at 85 km altitude of the NLC's. Thus, sunlight upon the ice crystals reveals their existence, and only within this altitude range. These particles continuously enter Earth's atmosphere. Below NLC-altitudes, the ice sublimates.

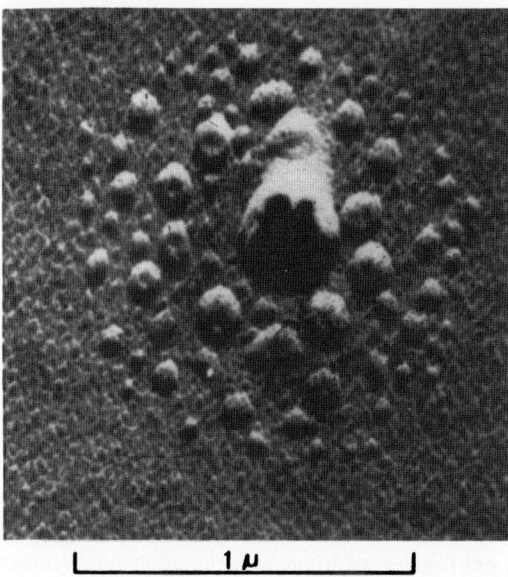

Figure 10. Electron microscope image of a "Noctilucent Cloud particle" according to Hemenway et al. 1964.

Unfortunately, the attempts to repeat the early results achieved by Hemenway et al. (1964) were unsuccessful (Hallgren et al. 1973). Only one positive NLC collection was achieved during the summer of 1968. Two participating groups have reported a modest collection of sub-micron-sized, irregularly formed particles (Farlow et al. 1970; Fechtig and Feuerstein 1970). Later experiments by other groups have shown that the dominating role for the appearance of NLC's is played by ion-clusters as nuclei for icy grains rather than by interplanetary dust (Arnold and Joos 1979).

The first dust collections in the atmosphere were performed by Hodge and Rinehard 1958 and by Junge and Mason 1961. In both cases the intention was to collect interplanetary dust for which it was, at least at that time, difficult to prove an extraterrestrial origin. Junge and Mason 1961 have collected between 20 and 30 km altitudes and there they found a large number of sulfide particles. These "Junge"-particles proved to be condensation products from terrestrial sources and the layer in which these sulfide grains were found is since then known as the "Junge"-layer.

A number of collections were performed on the Earth described in detail by Brownlee (1978). But it was always difficult to identify the extraterrestrial component of the collected particles. A breakthrough in dust collections was achieved by Brownlee and coworkers using a U-2 aircraft (Brownlee 1978). On these flights in altitudes of about 20 km a large amount of air could be screened: many dust grains were collected using suitable impact collectors. The subsequent laboratory analyses have shown that part of the collected dust is of extraterrestrial origin. These so-called interplanetary dust particles were analyzed for their chemical composition (Brownlee 1978; Recent results were published by Arndt et al. 1996). A strong proof for the extraterrestrial origin was found by incorporated solar wind (Rajan et al. 1977; Nier and Schlutter 1993) and by the presence of nuclear tracks (Bradley et al. 1984).

We now come back to the Earth's dust belt, an idea with such an important number of proponents that it could not easily be shaken. As far back as 1966, Nilsson had published a paper reporting results from a critical experiment using microphone detectors on a satellite flight. He found that the microphone crystals emit noise when exposed to varying temperatures. The rates of noise pulses were consistent with flight data which were earlier interpreted as real dust grain impacts. Despite this, considered by some a devastating blow, a spirited controversy still persisted for years between the high flux group adhering to acoustical results and a low flux group consisting of experimenters using other sensor systems. The demise of the high flux concept finally came rather suddenly and unequivocally with the first results of a cosmic dust experiment ensemble in the heliocentric space probes Pioneer 8 and 9. The ensemble included a coincidence ionization sensor system (with controls) and an acoustical sensor system (with controls) exposed to the same environment. Even though the ionization sensors were orders of magnitude more sensitive to the impact characteristics of interplanetary dust, they registered a flux 7 orders of magnitude lower than that of the neighboring acoustical sensor system (Berg and Gerloff 1971). That fact alone made the microphone sensor suspect, but totally devastating to the reliability of the microphone as a dust impact sensor was the unambiguous evidence that it was responding to high energy ions! Among the experiments aboard the same Pioneer 8 and 9 spacecraft was an independent experiment designed to measure proton energies of 10 MeV or greater (Rice University). Over a period of several years of data from each of the 2 Pioneer spacecraft experiments, there was perfect correlation between events registered on the Rice University proton sensors and the Pioneer microphone sensors, although the ionization sensors were not affected! It was shown later in laboratory experiments that the microphones were responsive to the heated path of the high energy protons' traversal through the ceramic crystal, generating a ringing, damped signal to the amplifiers. The correlation was irrefutable. The acoustical sensors were recording events not related to dust impacts at all! The infamous high flux curve would soon be abandoned! (For further important information about the dust belt discussion see the publications by McDonnell (1971) as well as by Nazarova (1968) and Nazarova and Rybakov (1974).)

An additional design factor in the Pioneer experiment served to provide credibility to the low flux registered by the ionization sensor ensemble. In each of the Pioneer missions, the ionization ensemble was comprised of 30 identical and quasi-independent coincident sensors. The distribution of recorded impacts upon the 60 sensors was normal for an exposure to a common space atmosphere. Meanwhile, the ionization sensor controls which were exposed to the same environment as the main ionization sensors, but protected by a plastic cover, registered zero impacts. All these data were perfectly reassuring. But the first reaction of the Principal Investigator of the Pioneer dust experiments to the drastically lower impact rate strikingly showed the still widespread acceptance of the high-flux concept at that time. He really thought, "—that the ionization sensors had crapped out, and that only the microphones were operating!".

The demise of the high flux was not only of academic interest, unfortunately. With some logic, NASA reduced its support for dust experimentation as it became evident that the probability of significant damage and/or danger from interplanetary dust impacts was extremely low and thus statistically acceptable. Extremely small particles ($< 100~\mu$m diameter), although frequent, generally present no significant hazard to astronauts in their life-support suits.

As the Space Age evolved, participants interested in interplanetary dust recognized a need for an international commission or panel dedicated to their specific interests. The initial steps toward formation of such a panel were led by Curtis L. Hemenway of the Dudley Observatory and occurred in 1966 in Vienna during the COSPAR Plenary Meeting: The Cosmic Dust Panel III C of COSPAR was established. The first official meeting of the Panel III C took place during the COSPAR Plenary Meeting in London in 1967. From its inception, the "international cosmic dust family" met annually at COSPAR Meetings until 1980. (To the reader who may be interested in the names of experimenters who have regularly participated in COSPAR since the inception of Panel III C, the following names come to mind: Hemenway and Hallgren from Dudley Observatory; Soberman and Skrivanek from AFCRL; Alexander and Berg from GSFC; Nauman from MSC; Gault, Farlow and Vedder from ARC; Hodge and Brownlee from Seattle; Weinberg from Hawaii (later Dudley Observatory); Hanner from Dudley Observatory; Dubin from NASA Headquarters; Nazarova from Moscow; Vilman from Turku; Jennison and McDonnell from Canterbury; Levasseur-Regourd from Paris; Witt and Wilhelm from Stockholm; Dumont from Bordeaux; Lamy from Marseille; Hughes from Sheffield; Giese, Schwehm, Zerull from Bochum; Leinert, Fechtig, Grün and Kissel from Heidelberg. We apologize to all those we have not been able to list as regular participants; their contributions to the field are fully recognized.). After that year, they met every two years. The same group was also concurrently active in the three IAU Commissions 15, 21, and 22. At this writing, there are several additional conferences and meetings which are "inviting" to the dust family (for example Lunar and Planetary Meetings in Houston; European Geophysical Society EGS; Division of Planetary Science DPS).

Nevertheless, the "International Dust Family" has arranged a series of 11 meetings (conferences, later IAU Symposia or Colloquia): see Table 1.

At the beginning there was serious doubt as to the acceptance of this new venture. (I, (H.F.), personally recall a discussion and a decision at the Vienna COSPAR Meeting in 1966, that the family would concentrate on COSPAR, and we would not participate in the forthcoming dust meeting in Honolulu/Hawaii early 1967. However, as it turned out, I was the only one who did not come to

Table 1

Dust Meetings between 1961 and 2000

1	"Astronomy and Physics of Meteors" Symposium of the Smithsonian Astrophysical Observatory Cambridge/Mass. USA Aug. 28–Sept. 1 1961
2	"Cosmic Dust" Conference of The New York Academy of Sciences New York/NY USA Nov. 21–22 1963
3	"Meteor Orbits and Dust" Symposium of the Smithsonian Astrophysical Observatory Cambridge/Mass. USA Aug. 9–13 1965
4	"The Zodiacal Light and the Interplanetary Medium" Symposium of the University of Hawaii Honolulu/ Hawaii USA Jan. 30–Feb. 2 1967
5	"Evolutionary and Physical Properties of Meteoroids" IAU Colloquium No. 13 at the New York State University Albany/NY USA June 14–17 1971
6	"Interplanetary Dust and Zodiacal Light" IAU Colloquium No. 31 at the Max-Planck-Institute f. Kernphysik Heidelberg Germany June 10–13 1975
7	"Solid Particles in the Solar System" IAU Symposium No. 90 at the Herzberg Institute of Astrophysics Ottawa Canada August 27–30 1979
8	"Properties and Interactions of Interplanetary Dust" IAU Colloquium No. 85 at the Laboratoire d'Astronomie Spatiale Marseille France June 9–12 1984
9	"Origin and Evolution of Interplanetary Dust" IAU Colloquium No. 126 at the Kyoto University Kyoto Japan August 27–30 1990
10	"Physics, Chemistry, and Dynamics of Interplanetary Dust" IAU Colloquium No. 150 at the University of Florida Gainesville/FL USA August 14–18 1995
11	"Dust in the Solar System and other Planetary Systems" IAU Colloquium No. 181 at the University of Kent at Canterbury, U.K. April 10–14 2000

Honolulu). Today these series of meetings are fully accepted by the dust family and should definitely be continued.

The dust family also has witnessed the excellent overview of our field in 1979 by Tony McDonnell's book "Cosmic Dust". After 20 years, this volume represents the first equivalent effort.

IV.B. Zodiacal Light Studied from Near-Earth Space

In parallel with collection and impact experiments, although with some delay, there have also been a few dozens of experiments on sounding rockets, balloons or high-flying aircraft devoted to the study of zodiacal light. This was a natural and modern answer to the difficulties in the separation of components in ground-based experiments, but at the price of reducing the observing time to a few minutes or, with balloons and aircraft, to a few hours. When going through the list of these measurements, one is struck by the impression that many of these well prepared efforts are not really remembered today. This may result from the difficulty to build a consistent—and convincing—picture from such snapshot-type observations. Consequently the best complete map of zodiacal light over the sky is still due to ground-based observations (Levasseur-Regourd and Dumont 1980, superseding earlier, respectable attempts by Smith et al. 1965 and Roach 1972). But here we want to recognize the various noteworthy results obtained by these space experiments.

They concentrated, as to be expected, on areas of zodiacal light research difficult or impossible to do from ground: antisolar hemisphere, regions close to the Sun including corona, the ultraviolet and infrared wavelength ranges. The first one may have been an Aerobee rocket flight on September 15, 1964 (Wolstencroft and Rose 1967), which is best remembered for its reported peculiarities in zodiacal light polarization: "negative" polarization (i.e. oriented parallel to the scattering plane) within 15° of the antisolar point and a comparatively large circular polarization ($-2.9° \pm 1.0$ %) around elongations of 50°. While this circular polarization, which would require oriented elongated particles for explanation, is very much in doubt, the existence and extent of negative polarization, which also has been observed from ground (Weinberg and Mann 1968), contains important information on structure and composition of interplanetary dust. And the less spectacular general measurements obtained of linear polarization at 703 nm and of the B-R color of zodiacal light, which gets bluer at large elongations and high latitudes, also deserve being noted.

Space observations of the *solar corona* are affected by the ever-present variations in coronal plasma in the same way as ground-based studies, which makes consistent interpretations difficult. But qualitative information giving insight into dynamics and composition of circumsolar dust has been obtained with the balloon flight of McQueen (1968) which confirmed undoubtedly the existence of enhanced thermal emission at about 4 solar radii (see Fig. 11), and with a supersonic aircraft flight by Léna et al. (1974) which reconfirmed the excess emission near 4 $R_\odot$ at 10 μm and found the emission spectrum of the corona to show the deep minimum at 10.8 μm expected for silicate par-

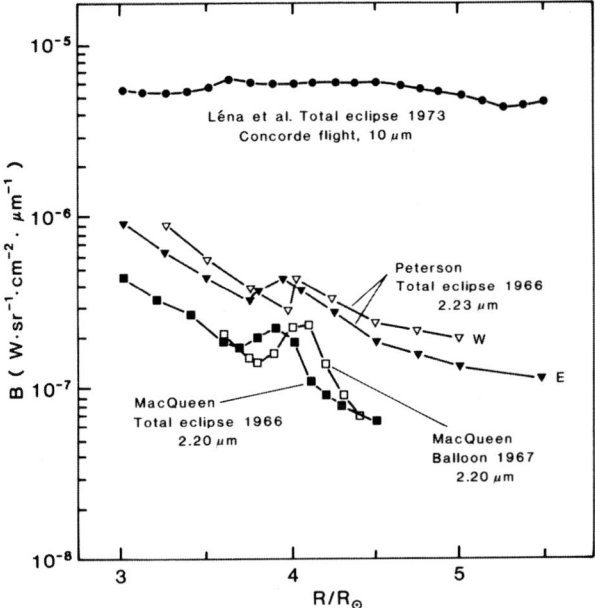

Figure 11. Early infrared measurements of the coronal brightness distribution, showing the enhanced emission at 4 $R_\odot$ (due to a dust ring?) first reported by Peterson (1967), (from Leinert and Grün 1990).

ticles. Mukai and Yamamoto (1979) have given a dynamical explanation for the existence of a circumsolar dust cloud as a result of the interplay between Poynting-Robertson effect, evaporation and radiation pressure. However, the corresponding spatial emission structures in the corona have not been seen by several of the later observers. Certainly the now available infrared detectors should be used to determine again a reliable map of coronal dust emission from 1 μm to 12 μm. The first observations of this kind, although affected by varying thin cirrus clouds, also failed to see the enhancement. Maybe the circumsolar dust cloud is a transient feature, and its appearance is coupled to the injection of dust into circumsolar space by a Sun-grazing comet (Hodapp et al. 1992; the further history of the modeling of circumsolar dust has been reviewed e.g. by Mann (1996)).

An accurate measurement of zodiacal light *color*, which is related to size and material of the scattering particles, best is done from above the atmosphere, and one of the initial experiments, using the Copernicus satellite (Lillie 1972), set the stage with a spectacular announcement: at wavelengths shorter than 200 nm the zodiacal light showed an upturn with respect to the solar spectrum, up to a factor of 200 at 168 nm (see Fig. 12). The outcome after

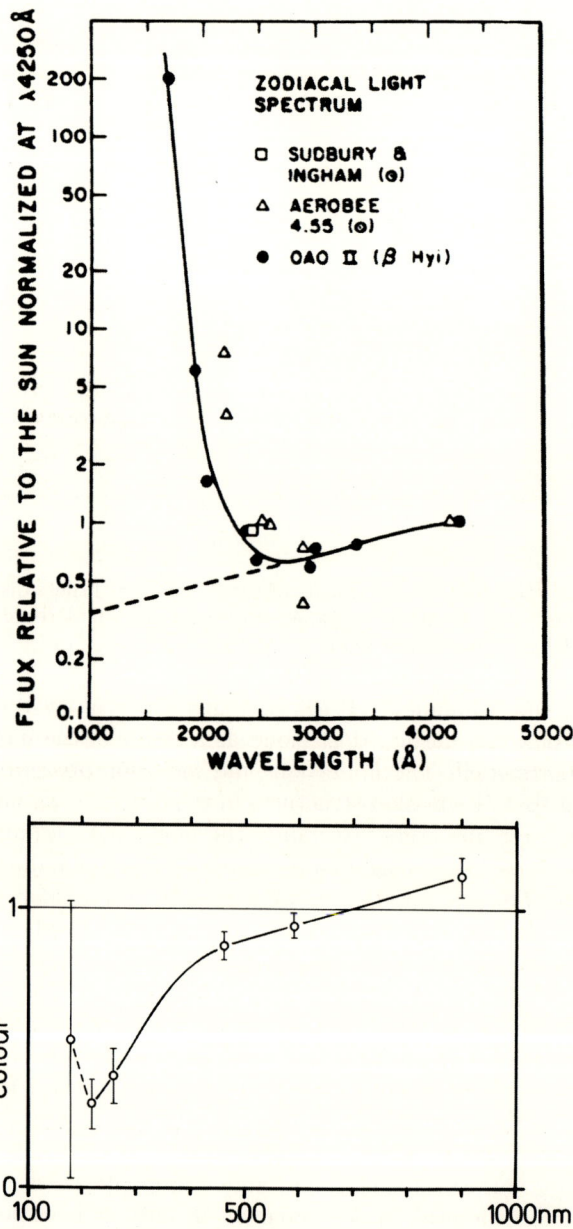

Figure 12. Flux of the zodiacal light relative to the solar spectrum. Upper: Lillie's (1972) amazing result. Lower: results of the rocket experiment by Pitz et al. (1979). Apart from the disagreement in the ultraviolet the data sets both show a reddening of the zodiacal light longward of 300 nm.

quite a number of experiments and discussions was back more or less to normal: according to measurements on the satellites TD-1 (Morgan 1978) and D2B "Aura" (Maucherat-Joubert et al. 1979) and to measurements on two sounding rockets (Pitz et al. 1978; Cebula and Feldman 1982) an ultraviolet upturn of the zodiacal light spectrum shortward of 200 nm cannot be fully excluded, but has to be ten times smaller than claimed by Lillie, if it exists at all. These measurements are not simple because the small UV flux may easily be contaminated by any red leak or by contributions from a different source. It may be that Lillie's measurements were correct, but whatever the origin of the signal was—interplanetary, galactic, or instrumental—his high values of zodiacal light brightness given for wavelengths < 200 nm no longer are accepted today. Beyond 200 nm wavelength the zodiacal light may match the color of the Sun (Frey et al. 1977) or show a reddening of about 10 % per 100 nm (Lillie 1972; Cebula and Feldman 1982; Pitz et al. 1979; Leinert et al. 1974). The remaining discrepancies are no longer very important and the confirmation by the measurements on the Helios space probes (Leinert et al. 1981), namely that the reddening of the zodiacal light decreases with increasing solar elongation, may have reduced them below the level which stimulates further discussion.

Contrary to this, the *near-infrared* colors of zodiacal light remain to be firmly established and are in current discussion (Matsuura et al. 1996). Increased accuracy is needed here by those who want to find out about the amount of cosmic diffuse background radiation in the "cosmological window" at 3 μm (Noda et al. 1992). Fortunately, the DIRBE experiment on the satellite COBE covers also the wavelength range of 1 μm–5 μm as does the near- infrared spectrometer on the infrared satellite IRTS which nicely measured the background spectrum through the minimum around 3 μm (Matsumoto et al. 1996). But the obvious measure of parallel observations in the visual, which would considerably improve the reliability of determinations of the shape of the zodiacal light spectrum from the visual to the near-infrared, have so far only been done in the balloon experiment of Hofmann et al. (1973).

Mid- to far-infrared observations of the zodiacal light have been pioneered by Martin Harwit (Soifer et al. 1971; Briotta et al. 1976), leading e.g. to the first empirical determination of the albedo of interplanetary dust. These observations were quite advanced at their time, but their full objective is only now being fulfilled with the new technology of the infrared satellites IRAS, COBE and ISO. It is remarkable that after the availability of these powerful facilities a mid-infrared rocket experiment still can give a useful contribution. But because those three satellites basically are built to observe at right angles to the Sun ($\epsilon = 90° \pm 30°$), the extension to small ($\epsilon = 22°$) and large ($\epsilon = 180°$) elongation angles obtained by Murdock and Price (1985) is very important.

Thus "small" space research by airplanes, balloons and sounding rockets has made important special contributions to the knowledge of zodiacal light, and could continue to do so in the future.

V. MICROCRATERS ON LUNAR SURFACE SAMPLES AND THE LUNAR EJECTA AND MICROMETEORITE EXPERIMENT

The advent of lunar samples on Earth 1969 and during the 1970s opened a new era in the research of interplanetary dust. For 4.5 billion years, the atmosphere free moon has been a target for impacting projectiles. It was shown that the extremely low lunar surface erosion has conserved km-sized impact craters since the existence of the moon. And even mm-sized dust particles produce impact craters which have been preserved on the lunar surface for probably 10^5 years (Neukum et al. 1972). Typically, there are a few craters in the mm-size range per 10 cm^2 sample surface area, but considerably more in smaller sizes down to μm-diameters. In total, thousands of craters were found and thus, the flux curve was established between 10^{-16} and 10^{-6} g on a sound statistical basis.

Craters as footprints of impacted interplanetary dust grains were studied extensively during the 1970s by a number of groups: Gault and coworkers at Ames; Hörz, Hartung, and Zook, at Houston (Hörz et al. 1975); Brownlee and coworkers at Seattle; McDonnell and coworkers at Canterbury. These groups and our Heidelberg group were direct competitors in this field of research. The period of major interest and activity of the investigations of lunar samples continued throughout the 1970s. The studies not only provided information concerning interplanetary dust flux onto the lunar surface, but correlated other interests related to dust, as well:

1. It was possible to simulate the microcraters on lunar materials by using an electrostatic dust accelerator and light gas guns. Figure 13 shows a comparison between natural lunar microcraters and laboratory produced craters using a 2 MeV dust accelerator (Fechtig et al. 1978; Bloch et al. 1971). Systematic laboratory crater experiments have provided a quantitative connection between projectile speed and mass with the crater diameters produced on lunar analogue material. Norite and glass were the prime laboratory materials used because essentially all lunar craters observed were on either silicates or glass splashes. In order to derive the projectile masses individually it was necessary to know the individual impact speeds. This information was made available on a statistical basis by the Heidelberg's HEOS 2 experimenters (Hoffmann et al. 1975a,b). Thus, it was possible to calculate fluxes for interplanetary dust grains at 1 AU Sun distance between 10^{-16} and 10^{-6} g (see Fig. 14).

2. The crater shape (diameter/depth ratio) of microcraters is mainly determined by the relative densities of the projectile and the target. Hence, an experimental determination of the densities of the projectiles was made possible. Only 25 to 35 % of the dust particles impacting upon the lunar surface indicate material densities of < 1 g cm^{-3}. This implies that at 1 AU Sun distance in the ecliptic, the interplanetary dust particles are presumably more than 70 % of asteroidal origin (Brownlee et al. 1973; Vedder and Mandeville 1974; Nagel and Fechtig 1980).

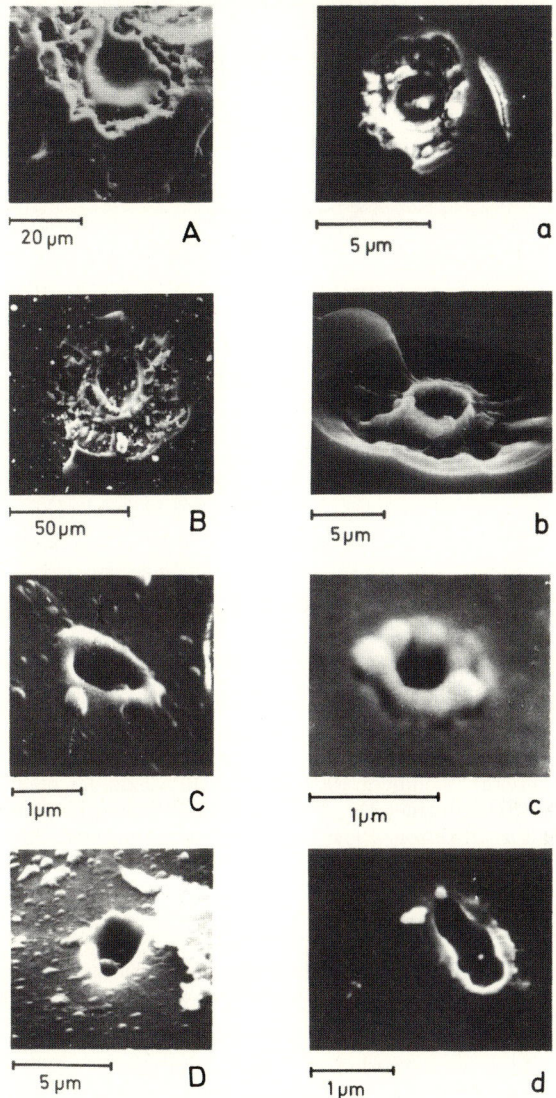

Figure 13. Types of micron- and submicron-size craters found on lunar samples in comparison to craters produced in the laboratory:

Lunar sample No.	Impact condition
(A) 12063,106a Top (cryst. rock)	(a) Fe—Norite, $v = 7.6$ km/sec
(B) 14257 (coarse fines)	(b) Al – Quartz glass, $v = 5$ km/sec
(C) 14257 (coarse fines)	(c) Fe – Duran glass, $v = 50$ km/sec
(D) 12063,106a Top (cryst. rock)	(d) Fe – Duran glass, $v = 30$ km/sec impact angle $45°$

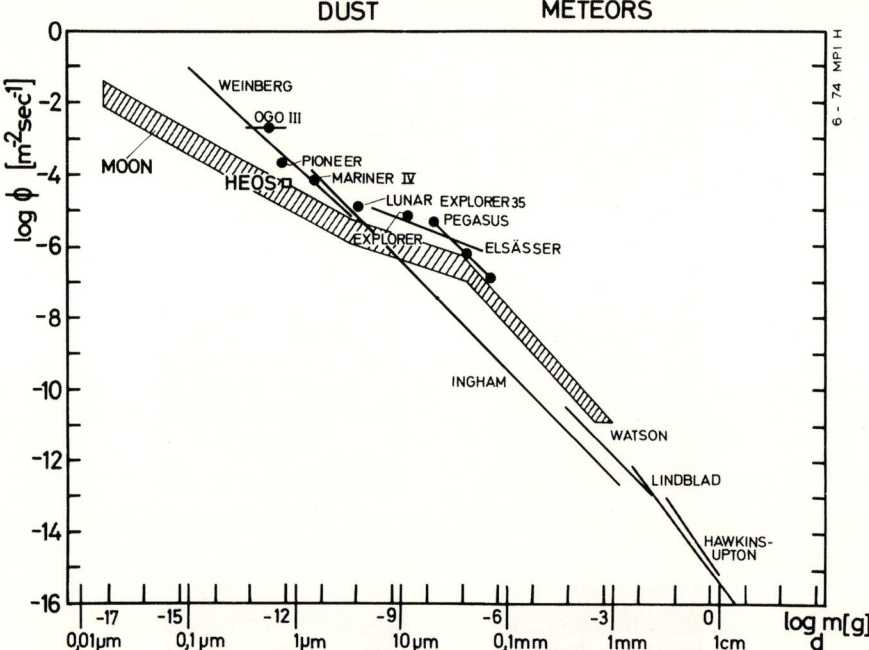

Figure 14. Comparison of micrometeoroid fluxes derived from lunar microcrater measurements with those obtained in various satellite in situ experiments. (Cumulative number per m^2 per sec for particles with masses $> m$) References in Fig. 14 correspond to general reference list: Moon: Fechtig et al. 1977; HEOS: Hoffmann et al. 1975a,b; Weinberg: Weinberg 1964; OGO III: Alexander et al. 1971; Pioneer: Berg and Gerloff 1971; Mariner IV: Alexander 1962; Lunar Explorer 35: Alexander et al. 1971; Explorer: d'Aiutolo 1964; Pegasus: Naumann 1966, Johnson et al. 1966, Clifton and Naumann 1966; Elsässer: Elsässer 1963; Ingham: Ingham 1963; Watson: Watson 1956; Lindblad: Lindblad 1976, Lindblad 1978; Hawkins-Upton: Hawkins and Upton 1958.

3. The erosion of microcraters under lunar conditions could be studied and simulated. A 1 μm diameter crater on a lunar silicate sample will probably completely disappear by sputter erosion after 10^6 years (Ashworth 1978).
4. Several groups have extended their interests and investigations further to include large lunar craters. Pioneering work has been done by Gault 1970. From the number densities of km-sized lunar craters related to the ages of lunar samples, a so-called "crater age" of specific lunar areas has been derived. Between 4 billion and 3 billion years ago, the rates of projectile impacts on the moon decreased exponentially with a half-life of probably 130 million years (Neukum et al. 1975).

The first two of these results have been particularly influential. In a collaborative effort between the Heidelberg group and the Bochum group the brightness of zodiacal light was predicted on the basis of the measured fluxes. And it was found (Giese and Grün 1976), that predicted and measured zodiacal

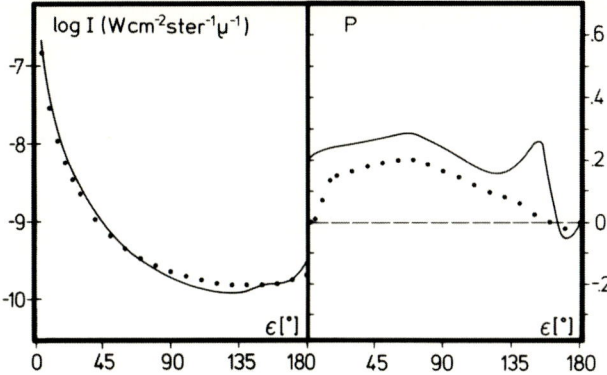

Figure 15. Compatibility of zodiacal light brightness and polarization distribution (·) with predictions (-) based on the measured flux curve of Fechtig et al.(1977) (from Röser and Staude 1978).

light brightness agreed to better than a factor of 2. Certainly the discrepancy between optical and direct detection of interplanetary dust had disappeared, once and (hopefully) for ever (see also section III). Also at this point in time, the new flux measurements had also advanced to the status of an established flux curve, "second generation". There have been further refinements (Grün et al. 1985), but essentially the flux curve has remained the same until today. And later studies showed that the compatibility with optical zodiacal light observations went much further than just to correct brightness estimates (see Fig. 15) and also resulted in reasonable values of particle albedo of 0.1–0.2 (Hanner 1980) resp. 0.09 ± 0.01 (Lumme and Bowell 1985) and in conclusions concerning the structure of the materials based on scattering laboratory experiments (see Fig. 16) (Zerull 1976; Giese et al. 1978; Weiss-Wrana 1983).

The finding that only a fraction of the particles was of low (cometary) density brought new life to the debate of the origin of interplanetary dust, which had essentially come to rest with Whipple's hypothesis of a cometary origin; and who would want to argue against Whipple in this matter? The newcomer F. C. Gillet in his thesis (1966) had at least considered the asteroid belt as a possible source for interplanetary dust and concluded *"... that collisions in the asteroid belt can easily produce enough small particles to sustain the zodiacal light"*; however, this did not lead to a change of opinion in the community in particular because Dohnanyi (1976) performed a more detailed study, and, conversely, found that asteroidal contribution is insufficient; this mostly was taken as support of Whipple's hypothesis. On the other side Röser (1976) and Delsemme (1976) had shown that short period comets supply less than 1 ton/sec of dust, which is not by far enough to maintain the zodiacal dust cloud and which put the majority opinion of cometary origin in question again. But the audience of the Heidelberg conference in 1975 was

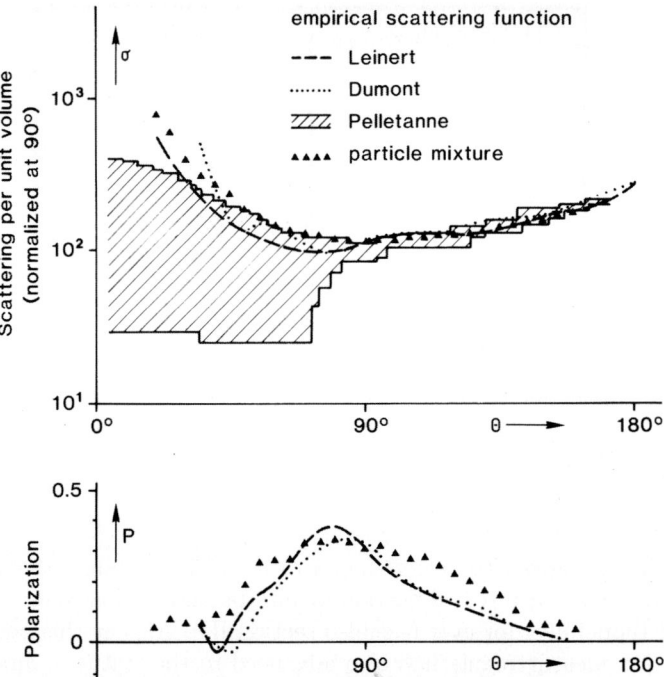

Figure 16. Scattering functions derived for the interplanetary dust at 1 AU compared with the scattering of a mixture of particles, measured in the laboratory from Weiss-Wrana (1983).

happy with Delsemme's assertion (Delsemme 1976) that some twenty thousand years ago Encke may have been a mighty comet feeding the interplanetary dust cloud we see today, and *"... that this proposition of Whipple (1967) is an idea we should not lightly put aside"*. Since then several independent experiments have shown that less than 30 % of the interplanetary dust particles in the ecliptic consist of low density (cometary) material: these experiments are the lunar crater studies just mentioned (Nagel and Fechtig 1980) and the analysis of dust particles by the Helios dust experiment (Grün et al. 1979). This is also in agreement with the observations of fireballs by Ceplecha and McCrosky (1976) and Ceplecha et al. (1993). Taken together with the usual assumption that cometary grains should be loosely structured, this would have forced the conclusion that only a minority of interplanetary dust grains in the ecliptic is of cometary origin. But the turning of opinion in the community happened only—but swiftly—when IRAS had detected the asteroidal band (see Fig. 17). Now a purely asteroidal origin of zodiacal light becomes more and more fashionable, but this extreme position it is not tenable either because of discrepancies in inclination distribution between asteroids and interplanetary dust (Banderman 1968).

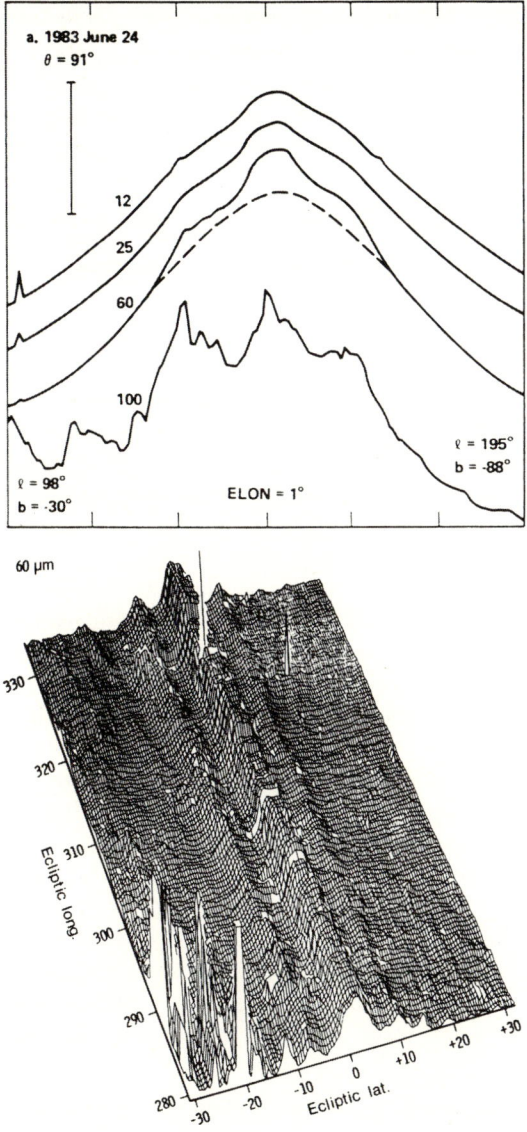

Figure 17. Upper: Detection of the asteroidal band by the IRAS satellite (Low et al. 1984). Shown are scans perpendicular to the ecliptic at 12 μm, 25 μm, 60 μm and 100 μm. The tick marks indicate steps of 10° in ecliptic latitude, the ecliptic being in the middle of the abscissa. The bands show as shoulders in the brightness profiles about ± 10° ecliptic latitude and as extra brightness close to the ecliptic. Irregular structures are due to galactic emission. Lower: The main asteroidal bands in the zodiacal light at ecliptic latitudes of ± 10° and ≈ 0° , as observed by IRAS at 60 μm. For easier recognition—the underlying general zodiacal light emission has been subtracted (from Dermott et al. 1984).

Let us return to the lunar surface again after these more general considerations. The Pioneer 8 and 9 experiments concept served again in 1972 as an ensemble selected for the experimental array of the Apollo 17 mission to the moon. Its design was essentially the same as that of the earlier Pioneer experiments except with 3 separate axes—one axis pointing along the lunar east, one axis along lunar west, and one toward Lunar Zenith. Also, the lunar mission had two objectives. One objective was similar to that of the Pioneer missions—to provide data concerning the nature and extent of interplanetary dust particles impacting on the moon. The other, and major objective, was to provide data concerning the nature and extent of lunar ejecta, or secondary particle activity on the moon. Although the high flux hazard was at this time considered simply wrong, there was critical concern for the erosion and degradation of optical surfaces and long term experiment exposures on the lunar surface as those might be affected by ejecta. Laboratory experiments on hypervelocity projectiles impact studies indicated the generation of secondary particles (or ejecta) propelled quasi-perpendicularly to the axis of the primary particles' trajectory and with speeds as high as 10 times that of the primary particle (Gault and Heitowit 1963; Eichhorn 1975 and 1976)—constituting a major hazard to man and instruments on planetary surfaces. The lunar project was given the acronym "LEAM", from Lunar Ejecta And Micrometeorite experiment. Surprisingly, but propitiously, neither one of the two experimental objectives were met! The preponderance of electrostatically transported lunar soil, (or "fines"), totally shrouded any evidence of interplanetary dust impacts or associated ejecta (Berg et al. 1974; Berg et al. 1976; Berg 1978). The experiment had provided, for the first time, irrefutable data concerning a mechanism for the movement of lunar soil—electrostatic transport—a phenomenon heretofore theoretically proposed (but not generally accepted) as a possible answer to the obvious movement of lunar fines (Gold 1972). As has often been the case as scientists probe the mysteries of space, the planned objectives of an experiment yielded to the revelation of the unexpected! A recognition and cursory evaluation of the unexpected results of the experiment was given by Noel Hinners, Director of the Goddard Space Flight Center— *"If you have a handle on the transport of lunar fines, forget about the objectives!"* (Hinners 1977).

VI. EXPERIMENTS ON SATELLITES AND SPACE PROBES

A significant step forward toward picturing the interplanetary dust portrait was derived from the HEOS 2 satellite data. The HEOS 2 satellite had a polar orbit with an apogee of 244 000 km and a perigee of < 5 000 km. Its period of operation was from February 1970 through August 1972. The sensitive area of the dust experiment was 100 cm^2. The nature of its orbit permitted a continuous scan between 5 000 km and 244 000 km. Further, the design of the HEOS spacecraft made it possible to orient its axis to scan selected regions of space. Hence, it was possible to search for and reliably determine near-Earth dust enrichments.

The HEOS 2 results (Hoffmann et al. 1975a,b) can be summarized as follows:

1. During its period of operation, the experiment recorded 15 *swarms* of particles within 10 Earth radii. *Swarms* are defined as a number of events recorded within 1 hour or less. For comparison, individual particles are normally recorded at a rate of approximately 1 event per 5 days. Swarms are believed to be fragments of larger parent bodies which become charged as they traverse the Earth's magnetosphere through the auroral zones and fragment electrostatically. The existence of such loosely conglomerated larger bodies in space has also been shown by Ceplecha and McCrosky (1976) and Ceplecha et al. (1993).

2. There is clear evidence of gravitational enhancement of μm-sized dust particles within 10 Earth radii by a factor of 3. This is in agreement with calculations by Öpik (1958).

3. The individual particles show a considerable anisotropy of their flux as a function of direction. Particles intercepted when the sensor looks in the direction of the Earth's orbit ("apex"—or "α"-particles) are 10 times as abundant as those in the antiapex direction.

Interestingly, some of the results of the HEOS 2 dust experiment have been confirmed by a dust experiment flown on the British Earth satellite PROSPERO (Bedford et al. 1975).

Soon after, a new class of interplanetary dust particles appeared. Revealed for the first time within the Pioneer 8 and 9 data was a hitherto unknown dust particle now referred to as the "β-meteoroids" (Berg and Grün 1973; Zook and Berg 1975). Once they were known to exist, they were swiftly explained: these are submicron-sized dust grains produced by collisions of dust particles. If dust grains collide, small fragments are produced. Larger ones follow a trajectory led by the Poynting-Robertson effect (the "α-particles"), but submicron-sized fragments are accelerated by solar radiation and follow a trajectory leaving the solar system. For them the ratio β of radiation pressure to solar gravitation is $\beta > 1$, therefore the name "β-meteoroids". The story of their detection presents a noteworthy case study. Given the hyperbolic orbits of these particles, they were preferentially detected when the experiment was pointing towards the Sun. With the misinterpretation of the Earth's dust belt in mind, still very recent history, in a first attempt the β-meteoroids were conservatively classified as solar-induced disturbances. But this time it was the right choice to be optimistic and to believe in the reality of even the small signals!

In this context a very controversially discussed contribution by Hemenway et al. 1972 has been published. Submicron-sized dust grains from a space collection have been interpreted as condensates from the Sun, so-called "stardust". This interpretation was exclusively based on their chemical composition: high vapor pressure oxides of heavy metal elements (for example Pt). The authors claimed that those oxides with extremely high evaporation temperatures (appr. 1800° C) could have survived the extremely high temperatures of the solar corona or being first recondensates. However, there is perhaps another possible explanation: dust grains of the solar system spiralling into the near-solar neighborhood (within a few solar radii), may partly evaporate, particularly when they are composed of several chemical compounds which is the normal case. The compounds with the highest evaporation temperatures survive for a very long time and form submicron- sized particle remnants (β-meteoroids) which are accelerated out of the solar system. Are these the particles named stardust by Hemenway et al. (1972)? Interestingly enough, El Goresy et al. (1978) have found small inclusions in carbonaceous chondrites (so-called "Fremdlinge") which have the same chemical composition as the stardust grains.

The exposed and recovered surfaces of the Long Duration Experiment Facility LDEF, after several years in Earth orbit, have received important and reliable new results for the near Earth interplanetary dust environment. In another chapter of this volume, McDonnell et al. give detailed results from these investigations.

In the meantime, zodiacal light observers have not been lazy either. Maybe they even proceeded too fast. Satellites spend most of their orbit in full sunlight, and in the sixties the experiments on OGO I, OGO III and OSO 6 did not provide sufficient stray light protection to prevent scattered sunlight from spoiling the measurements (see Wolff 1967). Others fared better, like the UV experiments on OAO 2 (= Copernicus), TD-1 and D2B "Aura" mentioned above in the discussion of the results obtained by sounding rockets. Optical satellite experiments also successfully contributed to the study of zodiacal light, e.g. to its general stability and to variations due to spatial structure, where Sparrow and Ney (1973) from OSO B2 and OSO 5 and Levasseur and Blamont (1973) from satellite D2A "Tournesol" emphasized the two opposite sides of the same coin, and did so controversially and with temperament. Multicolor observations were added by a photometer flown 1973 on Skylab (Sparrow et al. 1977), mainly for selected points like the ecliptic, galactic and celestial poles; Atmospheric Explorer C (Torr et al. 1979) had wider sky coverage, but used a presentation difficult to compare with the results of previous experiments, leaving doubts about the obtained accuracy. The next logical step, a dedicated satellite producing well calibrated maps of the optical diffuse sky brightness, and thus also of the zodiacal light and the galactic contributions, at many wavelengths—this project never happened, lacking widespread astronomical interest and support and also lacking obvious immediate gain in physical interpretation. And that solar polar probe, which was to carry a photometer and which could have performed part of these measurements and in addition searched for extragalactic background light, was cancelled by the Reagan administration in October 1981.

Instead the pioneering and monumental sky surveys with the infrared satellites IRAS (launched 1983) and COBE (launched 1989) boosted the field of infrared astronomy and revived the fading interest in zodiacal light, certainly by fascinating discoveries like cometary trails (see Fig. 18), asteroidal bands (Dermott et al. 1984), and a ring of trapped particles around the Earth's orbit

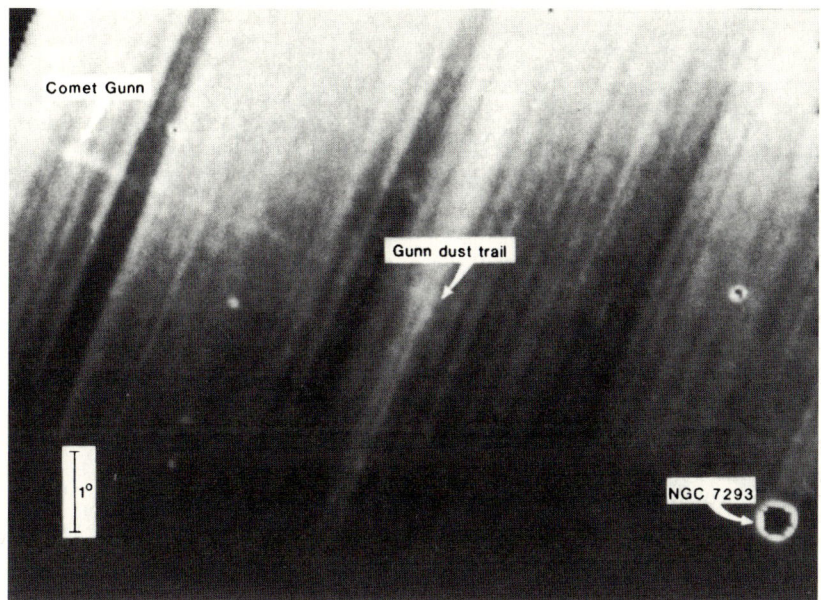

Figure 18. Detection of the dust trail of comet Gunn from the IRAS satellite (Reprinted from Sykes et al. 1986. Copyright 1986 American Association for the Advancement of Science).

(Dermott et al. 1994; Reach et al. 1995), but also by providing infrared maps of zodiacal light almost free of superimposed competing light contributions which clearly show how dominating this interplanetary "background" radiation is at mid-infrared wavelengths.

But the spatial distribution of interplanetary dust and the optical proper ties of the dust particles, both of which determine the zodiacal light distribution on the sky, can not be separated uniquely from a map observed at fixed heliocentric distance. As far as the out-of-ecliptic distribution of interplanetary dust is concerned, this uncomfortable situation remains. Although it is the same dust producing the visible and infrared zodiacal light, different distributions so far resulted from the analysis (see Fig. 19): a decrease of dust density, even stronger than in Fig. 19a, over the solar poles from infrared measurements, and an increase otherwise (a unified description of visible and infrared measurements by the same out-of-ecliptic distribution should be seriously considered, though). However, the separation of the radial dependence is possible in principle for observations done from space probes at varying heliocentric distances. Such an opportunity occurred from 1972 on, when Pioneer 10 and 11, carrying a two-color photopolarimeter (Weinberg et al. 1974; Hanner et al. 1974) in addition to a dust detection experiment (Humes 1980), went out to Jupiter and beyond, and when the space probes Helios A and B moved in to within 0.3 AU from the Sun.

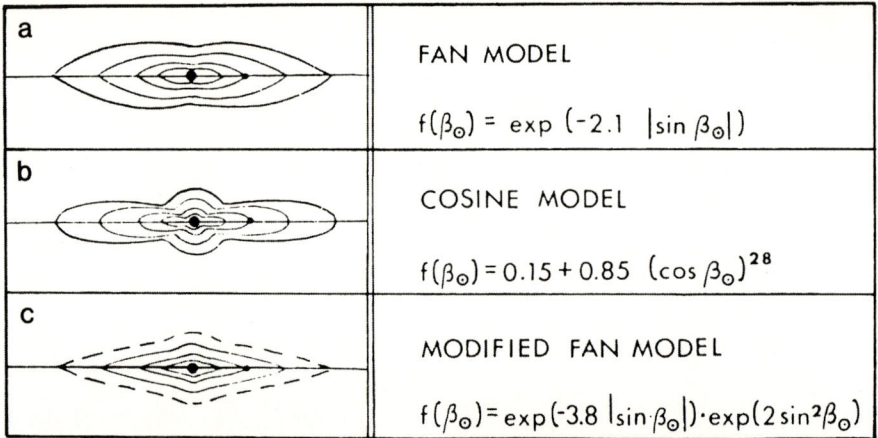

Figure 19. Typical models discussed for the out-of-ecliptic spatial distribution of interplanetary dust (adapted from Giese et al. 1986).

In the outer solar system the photopolarimeter on Pioneer 10 recorded a radial decrease of zodiacal light brightness out to 3.3 AU proportional to r^{-2} to $r^{-2.5}$, where r is the heliocentric distance of the spacecraft. This corresponds to a radial decrease in the dust density or, more precisely, the volume scattering function of $1/r$ to $1/r^{1.5}$. This fits in well with the inner solar system where Helios still provides the only set of direct observations of interplanetary dust.

The reader may excuse the break in style from a more quiet overall report to a more engaged detailed presentation when now going from the Pioneer measurements to the Helios results. Two of the authors (H.F. and Ch.L.) of this chapter have been heavily involved in the Helios mission, which now faces the fate to slowly become part of the past history, and they can not deny their special relation to these experiments.

The german-american Helios space probes carried two dust experiments conducted by the Max-Planck-Institut für Kernphysik (MPIK) and für Astronomie (MPIA) in Heidelberg, which in the years 1974–1984 explored the dust population between 0.3 and 1 AU solar distance. The MPIK experiment consisted of two 100 cm^2 large impact ionization detectors, and the MPIA experiment were three small (4–5 cm diameter) optical telescopes to measure the zodiacal light (A comprehensive summary of the Helios results has been published by Leinert and Grün 1990).

The main results of the MPIK experiment (Grün et al. 1979) can be summarized as follows:

1. The dust population between 0.3 and 1 AU was determined quantitatively. The increase of the dust fluxes with decreasing heliocentric distances was recorded between 1 and 0.5 AU. Closer to the Sun, the increase flattens considerably as shown in Fig. 20. This is due to the extent of the source

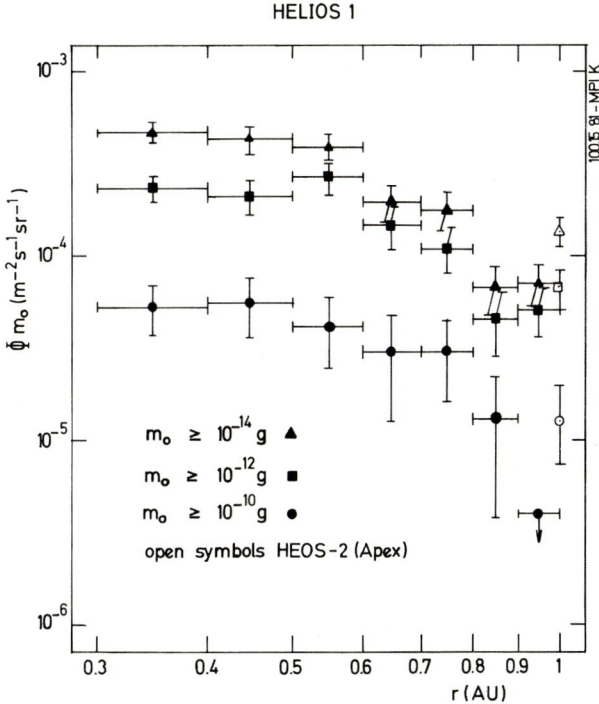

Figure 20. Cumulative flux Φ vs. distance r from the Sun for different mass ranges

region for β-meteoroids in which the number of collisions between dust grains continuously increases with decreasing distance from the Sun.

2. Two different dust populations were found: dust of "normal" densities (2 to 8 g cm^{-3}) orbiting the Sun in elliptical orbits with low eccentricities ($e < 0.6$), and dust of "low" densities (< 2 g cm^{-3}) with high eccentricities ($e > 0.6$). The consensus is that the first (larger) group represents particles from the asteroidal belt, while the second population is of cometary origin.

The zodiacal light experiment on Helios A + B (Leinert et al. 1981) demonstrated that it is possible, by careful straylight suppression, to observe as close as 15° to the Sun with negligible (< 1 %) stray light contribution, corresponding to a straylight suppression by a factor of 10^{-14}. Its main results were:

1. The scattering cross section per unit volume of interplanetary dust increases towards the Sun with a power law somewhat steeper than the dependence $1/r$ (corresponding to a brightness increase $1/r^2$) predicted for circular orbits in equilibrium under action of the Poynting-Robertson effect (see Fig. 21). The degree of polarization of zodiacal light, on the other hand, decreases when approaching the Sun. A possible explanation

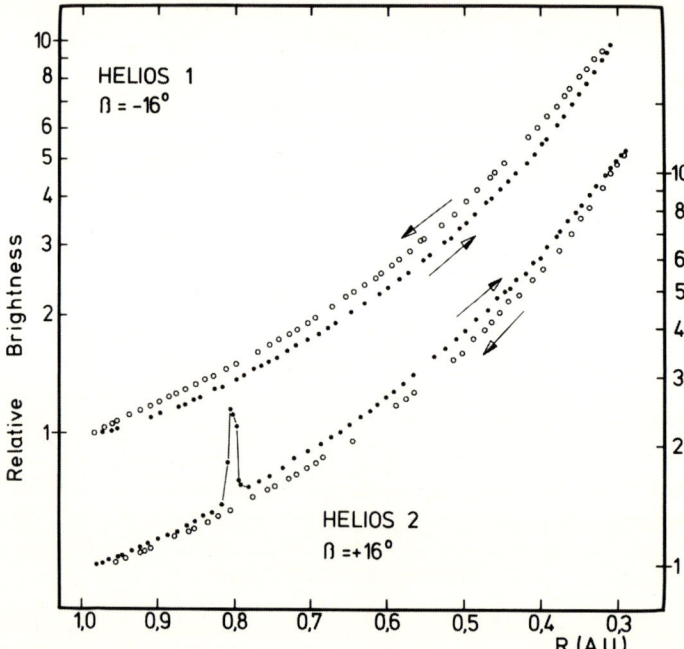

Figure 21. Observed brightness increase of zodiacal light when the space probes Helios 1 and Helios 2 approached the sun to 0.3 AU. The difference between inbound and outbound measurements is due to the inclination of the symmetry plane of interplanetary dust with respect to the ecliptic where the Helios spaceprobes had their orbits, the large brightening observed by Helios 2 at 0.8 AU is due to the passage of comet West through the field-of-view (from Leinert et al. 1984).

is that the actual spatial density of interplanetary dust follows the $1/r$ law predicted under certain conditions by the Poynting-Robertson effect, while an increase in particle albedo with decreasing heliocentric distance should account for both, the steeper increase of scattered zodiacal light brightness and the decrease in polarization. But, as discussed later in this book, the situation may be more complicated: Levasseur-Regourd et al. (1991) conclude that a superposition of two populations is more likely: particles of lower albedo concentrated towards the ecliptic plane, and particles of higher albedo in a more isotropic distribution.

2. The zodiacal light, when observed from the same position in interplanetary space in the same direction, is stable on the level of 1 % over 10 years (Leinert and Pitz 1989). This means that the interplanetary dust is stable on the same level, and that coronal streamers and other strong changes in interplanetary plasma density can be seen as changing additions to the zodiacal light—a zodiacal light photometer as far-reaching plasma probe, a surprising idea which recently materialized in the "Solar mass ejection imager" satellite experiment (Jackson et al. 1995).

Further progress in the field was the result of a dust experiment on the

Pioneer 10 and 11 space probes which were operating between 1 and 20 AU with close flybys of Jupiter and Saturn. Humes (1980) has given an impressive evaluation of the scientific data of his pressurized cells experiment (so-called "beer can" experiment). Each of the 108 pressurized cells could record only one "hit", thus the effective area decreased with time. (Martha Hanner recalls a meeting with the Pioneer Project Manager before launch to discuss the data processing requirements for each experiment. When it was Bill Kinard's turn he said, "Just send us a postcard when we get a hit!") The total number of counts never quite reached 108, but by 20 AU Sun distance, it was getting close. Between 1 AU and the asteroidal belt, a normal and expected decrease of dust events has been recorded. Further out to 20 AU, a constant flux has been measured. During the close flybys with Jupiter and Saturn an increase of the dust fluxes covering several orders of magnitudes has been recorded. Humes shows that most of these particles in the outer solar system have high orbital eccentricities. This observation may mean that the dust population is of cometary origin. In the new light of the detection of interstellar dust particles by the dust experiment on Ulysses (Grün et al. 1993) and of other so far unknown dust sources (Jupiter) in the outer solar system, other interpretations may be necessary (see chapter on In Situ Measurements by Grün et al.). But here we have reached a point which certainly goes beyond a historical introduction.

VII. IMPORTANT RESULTS OF THE EXPERIMENTS PIA/PUMA AND DIDSY ON THE MISSIONS GIOTTO AND VEGA TO COMET HALLEY

Toward the end of the 1970s, NASA made plans to rendezvous with Comet Halley in 1986, using a propulsion technology: either ion drive or solar sailing (because of Halley's retrograde orbit it is not possible to rendezvous using conventional technology). After long discussions within two subsequent study committees, NASA and ESA have agreed on a joint mission. NASA planned to rendezvous with comet Temple 2. ESA planned to provide a flyby payload which could be released en route to comet Temple 2 for a fast flyby through the coma of comet Halley. Unfortunately, NASA's plan was cancelled, but ESA's plan went on ahead.

The european scientific community was split at that time between a Halley flyby mission and a Polar Lunar Orbiter mission (POLO). As a matter of fact the Solar System Working Group of ESA finally had voted in favor of POLO. However the Space Science Advisory Committee of ESA finally overruled this opinion and decided to initiate the Giotto Mission to comet Halley. At about the same time the Soviet Union had started a mission to Venus and comet Halley (VeGa = Venus Galley). Both the Giotto- and VeGa-flyby through the coma of comet Halley were performed in March 1986 with great success. Parallel to the space missions Giotto and VeGa, the International Halley Watch was organized to coordinate world-wide Earth-based observations.

As a personal reminiscence (H.F.) the following might be interesting. On recommendation of Prof. J. Blamont, the director of IKI, Academician Sagdeev, invited us in 1980 to fly a copy of our Giotto dust composition experiment on VeGa. The french space agency CNES was cooperating closely with the Soviets in this project VeGa. A german cooperation at that time was difficult due to the cold war situation and our Research Ministry refused to finance any cooperation. However, it was possible to establish a cooperation between MPIK, CNES and IKI. Due to the prohibition to export high technology tools (so-called CoCom-list), this cooperation was still quite sensitive. The necessary electronics was used in sealed containers ("black boxes"). In these years I used to say: *"either our PI Kissel will win a medal or I will have to go to jail". In 1986, at the time of the flight, the situation had changed to a better one: Kissel received a medal and I was so glad not to have to go to jail!*

The composition experiment PIA/PUMA of the Heidelberg group has been flown on Giotto and VeGa (Kissel and Krueger 1987; detailed results are given in another chapter of this volume). Several thousands of cometary dust grains have been analyzed. With a mass resolution of better than 100 a cometary mass spectrum of a single grain consists of a spectrum of atomic ions between mass 1 and 110; because of the extreme high relative velocity between the comet and the spacecraft, the molecules of the projectile material has been widely decomposed into atoms. As the main result for the composition of cometary dust grains, the calculated abundances of cometary atoms compared with the abundances of atoms of carbonaceous chondrites are comparable within a factor of 2 for heavy elements. The light elements H, C, N and O are considerably enhanced in cometary dust. Carbon, for example, in the average is present in cometary grains as high as about 20 %; in many grains even higher than 30 % (Jessberger and Kissel 1991).

What does that mean? It is unlikely that the light elements stem from cometary ices because grains are quickly heated up in sunlight after release from the cometary nucleus. Several experiments have identified organic molecules in cometary gases. Accepting the suggestions by Greenberg (1983) on the basis of his results of simulation experiments, the increase in light elements could mean the isotopes are produced upon impact of organic-material-containing dust grains on the target of the composition experiment. In other words: if a particle composed of a silicate nucleus coated by an organic mantle material hits this dust experiment, one would expect an ion mass spectrum of the type which was observed.

The DIDSY-experiment (Dust Impact Detector System) on the Giotto spacecraft has directly measured impacting cometary grains on a sensitive detector surface of about 1 m^2 in the mass range between $5 \cdot 10^{-16}$ and $2 \cdot 10^{-7}$ g. (In October 1986 the first scientific meeting on the results of the space missions to comet Halley took place in Heidelberg/Germany. About 500 scientists, including a strong participation from the former Soviet Union, reported on all aspects of cometary science.) We will not discuss these results in detail but selectively emphasize only one result of high importance: The direct determination of the dust to gas ratio of cometary material in the coma of comet Halley by DIDSY. As extensively discussed by McDonnell et al. (1991), this ratio was determined to be between 1.3 and 3: a compromise of 2 was suggested. This value is surprisingly high! So far, a value of dust/gas = 0.3 had always been assumed.

VIII. OUTLOOK

The latest dust experiments are Grün's experiments on the space probes Galileo and Ulysses. Both experiments are identical detectors based on impact plasma detection. Compared to its preceding experiment on HEOS 2, their sensitive detection areas are enlarged by a factor of 10 from 100 cm^2 to 1 000 cm^2 on Galileo and Ulysses. The most surprising results are (1) the detection of dust streams from the Jupiter system; (2) the direct identification of interstellar dust grains. These results are described in this book elsewhere.

The future in the dust field is basically given by the NASA's missions Cassini to Saturn and Stardust to comet P/Wild 2 and by the third ESA cornerstone mission Rosetta as a cometary rendezvous mission to comet P/Wirtanen.

The Cassini dust experiment has mainly the same goals as the Galileo experiment for Jupiter: the exploration of the dust environment and its dynamics and sources in the Saturn system, but with a large area chemical analyzer.

The Rosetta mission of ESA is planned to be a comet rendezvous mission with comet Wirtanen in 2011–2013. A lander is expected to land directly on the surface of the comet nucleus. The dust experiment will explore primarily the composition of the dust in the coma analyzing dust grains. The lander experiments will explore the physical and chemical nature of the surface of the comet nucleus. The first time it should be possible to directly compare both composition results and also to compare the composition of the surface of the comet nucleus with the composition of the carbonaceous meteorites. But the comet-rendezvous-sample-return mission, which one of the authors (H.F.) dreamed of almost for half a lifetime, is not likely to happen in the foreseeable future. However, the planned NASA mission Stardust has the attractive goal to catch cometary dust grains in the coma of comet Wild 2 in the year 2004 at a comparatively low flyby velocity using aerogel as collector material (Tsou 1996). This would be, in fact, the first Earth sample return mission for cometary dust (Brownlee et al. 1996).

In summary, working for a more detailed knowledge of dust populations in all their variety within interplanetary space appears to be one main direction of development of the field.

The other one is the attempt to tie together what we know about circumstellar matter, commonly found around young stars (Beckwith et al. 1990), with the properties of solar system dust, and to understand the solar system as the outcome of the evolution of circumstellar material. The IAU recently (Hanner 1995) has devoted a "joint discussion" at the General Assembly in Den Haag to this topic. And when searching for planets around other stars (the proposed ESA cornerstone mission Darwin and the US project of a Terrestrial Planet Finder), one must know which kind of signatures to expect from circumstellar matter similar to the interplanetary dust cloud. These are the types of relationships which add to continued interest in the historical topic of interplanetary dust.

Acknowledgements

We very much thank our reviewers Martha Hanner and Tony McDonnell for their very constructive comments.

REFERENCES

d'Aiutolo, C. T. 1964. Satellite measurements of the meteoroid environment. In *Annals of the New York Academy of Sciences*, **119**, Cosmic Dust, ed. W. A. Cassidy (New York: published by the Academy), pp. 82–97.

Alexander, W. M. 1962. Cosmic dust. *Science*, **138**, pp. 1098–1099.

Alexander, W. M., Arthur, C. W., and Bohn, J. L. 1971. Lunar Explorer 35 and OGO 3: dust particle measurements in selenocentric and cislunar space from 1967 to 1969. In *Space Research*, **XI**, eds. K. YA. Kondratyev, M. J. Rycroft, and C. Sagan (Berlin: Akademie-Verlag), pp. 279–285.

Alexander, W. M., McCracken, C. W., Secretan, L., and Berg, O. E. 1963. Review of direct measurements of interplanetary dust from satellites and probes. In *Space Research*, **III**, ed. W. Priester (Amsterdam: North Holland Publ. Comp.), pp. 891–917.

Allen, C. W. 1947. The spectrum of the corona at the eclipse of 1940 October 1. *M.N.R.A.S.*, **106**, pp. 137–150.

Arndt, P., Bohsung, J., Maetz, M., and Jessberger, E. K. 1996. The elemental abundances in interplanetary dust particles. *Meteoritics & Planetary Science*, **31**, pp. 817–833.

Arnold, F., and Joos, W. 1979. Rapid growth of atmospheric cluster ions at the cold mesopause. *Geophys. Res. Lett.*, **6**, pp. 763–766.

Asaad, A. S. 1979. Possible explanations for the variation of zodiacal light brightness as observed from ground base and outer space. *J. Astron. Soc. Egypt.*, **1**, pp. 84–95.

Ashworth, D. G. 1978. Lunar and planetary impact erosion. In *Cosmic Dust*, ed. J. A. M. McDonnell (New York: Wiley and Sons), pp. 427–526.

Backhouse, T. W. 1857. On the aspect of the zodiacal light opposite the Sun. *M.N.R.A.S.*, **36**, pp. 46–48.

Bandermann, L. W. 1968. Physical properties of interplanetary dust. Thesis. (University of Maryland).

Barnard, E. E. 1883. Gegenschein. *Sidereal Messenger*, **2**, p. 254.

Beckers, J. M. 1959. The spectrum of the zodiacal light. *Proc. Kon. Ned. Akad.*, Ser. B, **62**, No. 4, pp. 248–262.

Beckwith, St. V. W., Sargent, A. I., Chini, R. S., and Güsten, R. 1990. A survey for circumstellar disks around young stellar objects. *Astron. J.*, **99**, pp. 924–945.

Bedford, D. K., Adams, N. G., and Smith, D. 1975. The flux and spatial distribution of micrometeoroids in the near-Earth environment. *Planet. Space Sci.*, **23**, pp. 1451–1456.

Behr, A., and Siedentopf, H. 1953. Untersuchungen über Zodiakallicht und Gegenschein nach lichtelektrischen Messungen auf dem Jungfraujoch. *Z. Astrophys.*, **32**, pp. 19–50.

Berg, O. E. 1978. A lunar terminator configuration. *Earth Planet. Sci. Lett.*, **39**, pp. 377–381.

Berg, O. E., Alexander, W. M., and Secretan, L. 1965. Physical parameters of cosmic dust obtained from rocket collections. *Proc. Symp. Meteors and Cosmic Dust*, (SAO: Cambridge/Mass).

Berg, O. E., and Gerloff, U. 1971. More than two years of micrometeorite data from two Pioneer satellites. In *Space Research*, **XI**, eds. K. YA. Kondratyev, M. J. Rycroft and C. Sagan (Berlin: Akademie-Verlag), pp. 225–235.

Berg, O. E., and Grün, E. 1973. Evidence of hyperbolic cosmic dust particles. In *Space Research*, **XIII**, eds. M. J. Rycroft and S. K. Runcorn (Berlin: Akademie-Verlag), pp. 1047–1055.

Berg, O. E., and Meredith, L. H. 1966. Micrometeorite impacts to an altitude of 103 km. *J. Geophys. Res.*, **61**, pp. 7511–7514.

Berg, O. E., Richardson, F. F., Rhee, J. W., and Auer, S. 1974. Preliminary results of a cosmic dust experiment on the Moon. *Geophys. Res. Lett.*, **1**, pp. 289–290.

Berg, O. E., and Secretan, L. 1967. Evidence of dust concentration in the mesosphere. The Zodiacal Light and the Interplanetary Medium, NASA SP-150.

Berg, O. E., Wolf, H., and Rhee, J. W. 1976. Lunar soil movement. In *Interplanetary Dust and Zodiacal Light, Lecture Notes*, **48**, eds. H. Elsässer and H. Fechtig (Berlin-Heidelberg-New York: Springer-Verlag), pp. 233–237.

Blackwell, D. E., and Ingham, M. F. 1961a. Observations of the zodiacal light from a very high altitude station, I. The average zodiacal light. *M.N.R.A.S.*, **122**, pp. 113–127.

Blackwell, D. E., and Ingham, M. F. 1961b. Observations of the zodiacal light from a very high altitude station, II. Electron densities in interplanetary space. *M.N.R.A.S.*, **122**, pp. 129–141.

Blackwell, D. E., Ingham, M. F., and Petford, A. D. 1967. The distribution of dust in interplanetary space. *M.N.R.A.S.*, **136**, pp. 313–328.

Bloch, M. R., Fechtig, H., Gentner, W., Neukum, G., and Schneider, E. 1971. Meteorite impact craters, crater simulations, and the meteoroid flux in the early solar system. In *Proc. 2nd Lunar Sci. Conf.*, **3**, ed. A. A. Levinson (Cambridge: MIT-Press), pp. 2639–2652.

Bohn, J. L., and Nadig, F. H. 1950. Acoustical studies with V-2 rockets. Report 8, Research Institute of Temple University.

Bradley, J. P., Brownlee, D. E., and Fraundorf, P. 1984. Discovery of nuclear tracks in interplanetary dust. *Science*, **226**, pp. 1432–1434.

Briotta, D. A., Jr., Pipher, J. L., and Houck, J. R. 1976. Rocket infrared spectroscopy of the zodiacal dust cloud. Report No. AFGL-TR-76-0236.

Brorsen, Th. 1854. Über eine neue Erscheinung am Zodiakkallicht. Unterhaltungen für Dilettanten und Freunde der Astronomie. *Geographie und Witterungskunde*, **8**, pp. 156–160.

Brorsen, Th. 1859. Über die ringförmige Gestalt des Zodiacallichtes. *Astron. Nachrichten*, **49**, pp. 219–220.

Brownlee, D. E. 1978. Microparticle studies by sampling techniques. In *Cosmic Dust*, ed. J. A. M. McDonnell (Chichester–New York–Brisbane–Toronto: John Wiley & Sons), pp. 295–336.

Brownlee, D. E., Burnett, D., Clark, B., Hanner, M. S., Hörz, F., Kissel, J., Newburn, R., Sandford, S., Sekanina, Z., Tsou, P., and Zolensky, M. 1996. Stardust: Comet and interstellar dust sample return mission. In *Physics, Chemistry, and Dynamics of interplanetary dust*, ASP Conference Series, eds. B. Å. S. Gustafson and M. S. Hanner (Astronomical Society, San Francisco), **104**, pp. 223–226.

Brownlee, D. E., Hörz, F., Vedder, J. F., Gault, D. E., and Hartung, J. B. 1973. Some physical parameters of micrometeoroids. In *Proc. 4th Lunar Sci. Conf III.*, ed. W. A. Gose (New York–Oxford–Toronto–Sydney–Braunschweig: Pergamon Press), pp. 3197–3212.

Burkhardt, G. 1986. Kollisionsfreie Dynamik von interplanetaren Staubteilchen. Thesis, Univ. Heidelberg, pp. 5–6.

Cassini, G. D. 1683. Découverte de la lumière celeste qui paroist dans le zodiaque. *Mémoires de l'Académie Royale des Sciences depuis 1666 jusqu'a 1699*, **Tome VIII**, Paris 1699, (Compagnie des Libraires, 1730), pp. 119–209.

Cebula, R. P., and Feldman, P. D. 1982. Ultraviolet spectroscopy of zodiacal light. *Astrophys. J.*, **263**, pp. 987–992.

Ceplecha, Z., and McCrosky, R. E. 1976. Fireball end heights: a diagnostic for the structure of meteoric material. *J. Geophys. Res.*, **81**, pp. 6257–6275.

Ceplecha, Z., Spurny, P., Borovicka, J., and Keclikova, J. 1993. Atmospheric fragmentation of meteoroids. *Astron. Astrophys.*, **279**, pp. 615–626.

Childrey, J. 1661. Britannia Baconia or the Natural Rarities of England, Scotland, & Wales. Printed for the author, (London).

Clifton, S., and Naumann, R. J. 1966. Pegasus satellite measurements of meteoroid penetrations (Feb. 16–Dec. 31,1965). NASA TM X-1316.

Delsemme, A. H. 1976. The production rate of dust by comets. In Lecture Notes in *Physics*, **48**, Interplanetary Dust and Zodiacal Light, eds. H. Elsässer and H. Fechtig (Berlin-Heidelberg-New York: Springer Verlag), pp. 314–318.

Dermott, S. F., Nicholson, P. A., Burns, J. A., and Houck, J. R. 1984. Origin of the solar system dust band discovered by IRAS. *Nature*, **312**, pp. 505–509.

Dermott, S. F., Jayaraman, S., Xu, Y. L., Gustafson, B. Å. S., and Liou, J. C. 1994. A circumsolar ring of asteroidal dust in resonant lock with the Earth. *Nature*, **369**, pp. 719–723.

Divari, N. B. 1964. Lunar effects on zodiacal brightness. *Soviet Astronomy- AJ*, **7**, pp. 547–548; Original: *Astron. Zh.*, **40**, pp. 717–718, 1963.

Dohnanyi, J. S. 1976. Sources of interplanetary dust: asteroids. In Lecture Notes in *Physics*,

48, Interplanetary Dust and Zodiacal Light, eds. H. Elsässer and H. Fechtig (Berlin–Heidelberg–New York: Springer Verlag), pp. 187–205.

Dufay, J. 1925. La polarization de la lumière zodiacale. *Compt. Rend.*, **181**, pp. 399–401.

Dumont, R. 1965. Séparation des composantes atmosphérique, interplanetaire et stellaire du ciel nocturne á 5 000 Å. Application a la photométrie de la lumière zodiacale et du Gegenschein. *Ann. d'Astrophys.*, **28**, pp. 265–320.

East, I. R., and Reay, N. K. 1984. The motion of interplanetary dust particles. I. Radial velocity measurements on Fraunhofer line profiles in the zodiacal light spectrum. *Astron. Astrophys.*, **139**, pp. 512–516.

Eichhorn, G. 1975. Measurements of the light flash produced by high velocity particle impact. *Planet. Space Sci.*, **23**, pp. 1519–1525.

Eichhorn, G. 1976. Analysis of the hypervelocity impact process from impact flash measurements. *Planet. Space Sci.*, **24**, pp. 771–781.

El Goresy, A., Nagel, K., and Ramdohr, P. 1978. Fremdlinge and their noble relatives. In *Proc. Lunar Planet. Sci. Conf. 9th*, ed. R. B. Merrill (New York–Oxford–Toronto–Sydney–Frankfurt: Pergamon Press), pp. 1279–1303.

Elsässer, H. 1954. Die räumliche Verteilung der Zodiakallichtmaterie. *Z. Astrophys.*, **33**, pp. 274–285.

Elsässer, H. 1955. Fraunhoferkorona und Zodiakallicht. *Z. f. Astrophys.*, **37**, pp. 114–124.

Elsässer, H. 1958a. Neue Helligkeits- und Polarisationsmessungen am Zodiakallicht und ihre Interpretation. *Die Sterne*, **10**, pp. 166–169.

Elsässer, H. 1958b. Interplanetare Materie. *Mitt. d. Astr. Inst. d. Univ. Tübingen*, **Nr. 35**, pp. 61–88.

Elsässer, H. 1963. The zodiacal light. *Planetary Space Sci.*, **11**, pp. 1015–1033.

Elvey, C. T., and Roach, F. R. 1937. A photoelectric study of the light from the night sky. *Astrophys. J.*, **85**, pp. 213–241.

Farlow, N. H., Blanchard, M. B., and Ferry, V. G. 1966. Sampling with a LUSTER Sounding Rocket. *J. Geophys. Res.*, **71**, pp. 5689–5693.

Farlow, N. H., Ferry, G. V., and Blanchard, M. B. 1970. Examination of surfaces exposed to a noctilucent cloud on August 1, 1968. *J. Geophys. Res.*, **75**, pp. 6736–6750.

Fechtig, H., and Feuerstein, M. 1970. Particle collection results from a rocket flight on August 1, 1968. *J. Geophys. Res.*, **75**, pp. 6751–6757.

Fechtig, H., Gentner, W., Hartung, J. B., Nagel, K., Neukum, G., Schneider, E., and Storzer, D. 1977. Microcraters on lunar samples. In *The Soviet-American Conference on Cosmochemistry of the Moon and Planets*, eds. J. H. Pomeroy and N. J. Hubbard (Washington: NASA), pp. 585–603.

Fechtig, H., Grün, E., and Kissel J. 1978. Laboratory simulation. In *Cosmic Dust*, ed. J. A. M. McDonnell (New York: Wiley and Sons), pp. 607–669.

Frey, A., Hofmann, W., and Lemke, D. 1977. Spectrum of the zodiacal light in the middle UV. *Astron. Astrophys. J.*, **54**, pp. 853–855.

Fried, J. W. 1978. Doppler shifts in the zodiacal light spectrum. *Astron. Astrophys.*, **68**, pp. 259–264.

Friichtenicht, J. F. 1962. Two-million-volt electrostatic accelerator for hypervelocity research. *Rev. Sci. Instr.*, **33**, pp. 209–212.

Gault, D. E. 1970. Saturation and equilibrium conditions for impact cratering on the lunar surface: criteria and implications. *Radio Sci.*, **5**, pp. 273–291.

Gault, D. E., and Heitowit, E. D. 1963. The partition of energy for hypervelocity impact craters formed in rocks. In *Proc. 6th Hypervelocity Impact Symposium*, (Cleveland/Ohio: Firestone Rubber Company), **2**, pp. 419–427.

Gerloff, U., Weihrauch, J. H., and Fechtig, H. 1967. Electron microscope and microprobe measurements on LUSTER-flight samples. In *Space Research*, **VII**, eds. R. L. Smith-Rose and J. W. King (Amsterdam: North Holland Publ. Comp.), pp. 1412–1420.

Giese, R. H. 1962. Light scattering by small particles and models of interplanetary matter derived from the zodiacal light. *Space Science Review*, **1**, pp. 589–611.

Giese, R. H. 1973. Optical properties of single-component zodiacal light models. *Planet. Space Sci.*, **21**, pp. 513–521.

Giese, R. H., and Grün, E. 1976. The compatibility of recent micrometeoroid flux curves with observations and models of the zodiacal light. In Lecture Notes in *Physics*, **48**, Interplanetary Dust and Zodiacal Light, eds. H. Elsässer and H. Fechtig (Berlin–Heidelberg–New

York: Springer-Verlag), pp. 135–139.
Giese, R. H., Kneisel, B., and Rittich, U. 1986. Three-dimensional zodiacal dust cloud, a comparative study. *Icarus*, **68**, pp. 395–411.
Giese, R. H., Weiss K., Zerull, R. H., and Ono, T. 1978. Large fluffy particles: a possible explanation of the optical properties of interplanetary dust. *Astron. Astrophys.*, **65**, pp. 265–272.
Gillet, F. C. 1966. Zodiacal light and interplanetary dust. Thesis, University Minnesota.
Gold, T. 1972. Erosion, transportation, and the nature of the maria on the moon. *IAU Coll.*, **47**, pp. 55–67.
Greenberg, J. M. 1983. Laboratory dust experiments-tracing the composition of cometary dust. In *Cometary Exploration*, **II**, ed. T. I. Gombosi (Budapest: Hungarian Academy of Sciences), pp. 23–54.
Grotrian, W. 1934. Über das Fraunhofersche Spektrum der Sonnenkorona. *Z. Astrophys.*, **8**, pp. 124–146.
Grün, E., Pailer, N., Fechtig, H., and Kissel, J. 1979. Orbital and physical characteristics of micrometeoroids in the inner solar system observed by Helios 1. *Planet. Space Sci.*, **28**, pp. 333–349.
Grün, E., Zook, H. A., Baguhl, M., Balogh, A., Bame, S. J., Fechtig, H., Forsyth, R., Hanner, M. S., Horanyi, M., Kissel, J., Lindblad B.-A., Linkert, D., Linkert, G., Mann, I., McDonnell, J. A. M., Morfill, G. E., Phillips, J. L., Polanskey, C., Schwehm, G., Siddique, N., Staubach, P., Svestka, J., and Taylor, A. 1993. Discovery of jovian dust streams and interstellar grains by the Ulysses spacecraft. *Nature*, **362**, pp. 428–430.
Grün, E., Zook, H. A., Fechtig, H., and Giese, R. H. 1985. Collisional balance of the meteoritic complex. *Icarus*, **62**, pp. 244–272.
Hallgren, D. S., Hemenway, C. L., Mohnen, V. A., and Tackett, D. C. 1973. Preliminary results from the Noctilucent cloud sampling by a multi-experiment payload. In *Space Research*, **XIII**, eds. M. J. Rycroft and S. K. Runcorn (Berlin: Akademie-Verlag), pp. 1099–1104.
Hanner, M. S. 1980. On the albedo of the interplanetary dust. *Icarus*, **43**, pp. 373–380.
Hanner, M. S. 1995. Dust around young stars: how related to solar system dust? In *Highlights of Astronomy*, ed. I. Appenzeller, (IAU: printed in the Netherlands), **10**, pp. 351–392.
Hanner, M. S., and Weinberg, J. L. 1973. Gegenschein observations from Pioneer 10. *Sky and Telescope*, **45**, pp. 217–218.
Hanner, M. S., Weinberg, J. L., DeShields II, L. M., Green, B. A., and Toller, G. N. 1974. Zodiacal light and the asteroidal belt: the view from Pioneer 10. *J. Geophys. Res.*, **79**, pp. 3671–3675.
Hawkins, G. S., and Upton, E. K. L. 1958. The influx rate of meteors in the Earth's atmosphere. *Astrophys. J.*, **128**, pp. 727–735.
Hemenway, C. L., Fullam, E. F., Skrivanek, R. A., Soberman, R. K., and Witt, G. 1964. Electron microscope studies of noctilucent clouds particles. *Tellus*, **16**, pp. 96–102.
Hemenway, C. L., Hallgren, D. S., and Coon, R. E. 1967. High altitude balloon-top collections of cosmic dust. In *Space Research*, **VII**, eds. R. L. Smith-Rose and J. W. King (Amsterdam: North Holland Publ. Comp.), pp. 1423–1431.
Hemenway, C. L., Hallgren, D. S., and Kerridge, J. F. 1968. Results from the GEMINI S-10 and S-12 micrometeorite experiments. In *Space Research*, **VIII**, eds. A. P. Mitra, L. G. Jacchia and W. S. Newman (Amsterdam: North Holland Publ. Comp.), pp. 521–535.
Hemenway, C. L., Hallgren, D. S., and Schmalberger, D. C. 1972. Stardust. *Nature*, **238**, pp. 256–260.
Hemenway, C. L., and Soberman, R. K. 1962. Studies of micrometeorites obtained from a recoverable sounding rocket. *Astron. J.*, **67**, pp. 256–266.
Hicks, T. R., May, B., and Reay, N. K. 1974. An investigation of the motion of zodiacal dust particles, I. Radial velocity measurements on Fraunhofer line profiles. *M.N.R.A.S.*, **166**, pp. 439–448.
Hinners, N. 1977. Personal communication.
Hodapp, K.-W., MacQueen, R. M., and Hall, D., W. B. 1992. A search during the 1991 solar eclipse for the infrared signature of circumsolar dust. *Nature*, **355**, pp. 707–710.
Hodge, P. W., and Rinehart, T. S. 1958. High altitude collection of extraterrestrial particulate material. *Astron. J.*, **63**, p. 306.
Hoffmann, H.-J., Fechtig, H., Grün, E., and Kissel, J. 1975a. First results of the micromete-

oroid experiment S 215 on the HEOS 2 satellite. *Planet. Space Sci.*, **23**, pp. 215–224.
Hoffmann, H.-J., Fechtig, H., Grün, E., and Kissel, J. 1975b. Temporal Fluctuations and anisotropy of the micrometeoroid flux in the Earth-Moon system measured by HEOS 2. *Planet. Space Sci.*, **23**, pp. 985–991.
Hofmann, W., Lemke, D., Thum, C., and Fahrbach, U. 1973. Observations of the zodiacal light at 2.4 μ m. *Nature*, **243**, pp. 140–141.
Hörz, F., Brownlee, D. E., Fechtig, H., Hartung, J. B., Morrison, D. A., Neukum, G., Schneider, E., Vedder, J. F., and Gault, D. E. 1975. Lunar microcraters: implications for the micrometeoroid complex. *Planet. Space Sci.*, **23**, pp. 151–172.
van de Hulst, H. C. 1947. Zodiacal light in the solar corona. *Astrophys. J.*, **105**, pp. 471–488.
von Humboldt, A. 1816. Voyage de Humboldt et Bonpland. 1. Voyage aux regions equinoctiales du nouveau continent, fait en 1799, 1800, 1801, 1802, 1803 et 1804. 4. Atlas pittorescque (vue des Cordilleres et monuments des peuples de l'Amerique 1810) l'ed. Paris 1814–34, p. 282.
Humes, D. H. 1980. Results of Pioneer 10 and 11 meteoroid experiments: interplanetary and near Saturn. *J. Geophys. Res.*, **85**, pp. 5841–5852.
Huruhata, M. 1948. Polarization of the night sky light and the zodiacal light. *Tokyo Astr. Bull.*, (Tokyo Astron. Obs.), Second Series, **No. 10**, pp.77–79.
Huruhata, M. 1951. Photoelectric study of the zodiacal light. *Publ. Astr. Soc. Japan*, **2**, pp. 156–171.
Ingham, M. F. 1963. Interplanetary matter. *Space Sci. Rev.*, **1**, pp. 576–588.
Jackson, B. C., Buffington, A., Hick, P. L., Kahler, S. W., Altrock, R. C., Gold, R. E., and Webb, D. F. 1995. The solar mass ejection imager. In *Solar Wind Eight*, eds. D. Winterhalter, J. T. Gosling, S. R. Habbal, W. S. Kurth, and M. Neugebauer, *AIP Conference Proceedings* **382**, (Woodbury), pp. 536–539.
Jackson, B. V., and Leinert, Ch. 1985. Helios images of solar mass ejections. *J. Geophys. Res.*, **90**, pp. 10759–10768.
James, J. F., and Smeethe, M. J. 1970. Motion of the interplanetary dust cloud. *Nature*, **227**, pp. 588–589.
Jennison, R. C., McDonnell, J. A. M., and Rodger, I. 1967. The Ariel II micrometeorite penetration measurements. *Proc. Roy. Soc.*, **300**, pp. 251–... .
Jessberger, E. K., and Kissel, J. 1991. Chemical properties of cometary dust and a note on carbon isotopes. In *Comets in the Post-Halley Era*, **2**, eds. R. L. Newburn Jr., M. Neugebauer and J. Rahe (Dordrecht–Boston–London: Kluwer Academic Publ.), pp. 1075–1092.
Johnson, W. G., Heller, G. B., Smith, M. J., Dozier, J. B., and Shelton, P. D. 1966. The meteoroid satellite project PEGASUS—First summary report. NASA TN D-3505.
Jones, G. 1856. Observations on the zodiacal light, (Washington: B. Tucker, Senate Printer).
Junge, C. E., and Mason, J. E. 1961. Stratospheric aerosol studies. *J. Geophys. Res.*, **66**, pp. 2163–2182.
Kaiser, T. R. 1961. The determination of the incident flux of radio meteors vs. sporadic meteors. *M. N. R. A. S.*, **123**, pp. 265–271.
Kissel, J., and Krueger, F. R. 1987. The organic component in dust from Comet Halley as measured by the PUMA mass-spectrometer on board VeGa 1. *Nature*, **326**, pp. 755–760.
Lamy, P. L., and Perrin, J. M. 1980. Zodiacal light models with a bimodal population. In *Solid Particles in the Solar System*, IAU-Symposium No. 90, eds. I. Halliday and B. A. McIntosh (Dordrecht: Reidel Publ. Comp.), pp. 75–80.
Leinert, Ch. 1975. Zodiacal light—a measure of the interplanetary environment. *Space Sci. Rev.*, **18**, pp. 281–339.
Leinert, Ch. and Grün, E. 1990. Interplanetary dust. In *Physics of the inner Heliosphere*, eds. R. Schwenn and E. Marsch (Berlin–Heidelberg–New York: Springer Verlag), pp. 207–275.
Leinert, Ch., Link, H., and Pitz, E. 1974. Rocket photometry in the inner zodiacal light. *Astron. Astrophys.*, **30**, pp. 411–422.
Leinert, Ch., and Pitz, E. 1989. Zodiacal light observed by Helios throughout solar cycle no. 21: stable dust and varying plasma. *Astron. Astrophys.*, **210**, pp. 399–402.
Leinert, Ch., Pitz, E., Hanner, M. S., and Link, H. 1977. Observations of zodiacal light from Helios 1 and 2. *J. Geophys. Res.*, **42**, pp. 699–704.
Leinert, Ch., Pitz, E., and Link, H. 1984. Zodiakallicht – ein Abbild der interplanetaren

Staubwolke. In *Helios*, ed. H. Porsche (Oberpfaffenhofen: DFVLR), pp. 50–57.

Leinert, Ch., Richter, I., Pitz, E., and Planck, B. 1981. The zodiacal light from 1.0 to 0.3 AU as observed by the Helios space probes. *Astron. Astrophys.*, **103**, pp. 177–188.

Léna, P., Viala, Y., Hall, D., and Soufflot, A. 1974. The thermal emission of the dust corona during the eclipse of June 30, 1973, II. Photometric and spectral observations. *Astron. Astrophys.*, **37**, pp. 81–86.

Levasseur, A. C. 1976. Observations atmosphérique et astronomique au voisinage de 6563 Åa bord du satellite D2A: contribution à l'étude de la lumière zodiacale, de la géocouronne, des nébuleuses émissives et des aurores équatoriales. Thesis, Univ. Pierre et Marie Curie, Paris 1976, p. 51.

Levasseur, A. C., and Blamont, J. E. 1973. Satellite observations of intensity variations of the zodiacal light. *Nature*, **246**, pp. 26–28.

Levasseur, A. C., and Blamont, J. E. 1976. Evidence for scattering particles in meteor streams. In Lecture Notes in *Physics*, **48**, Interplanetary Dust and Zodiacal Light, eds. H. Elsässer and H. Fechtig (Berlin–Heidelberg–New York: Springer Verlag), pp. 58–62.

Levasseur-Regourd, A. C., and Dumont, R. 1980. Absolute photometry of zodiacal light. *Astron. Astrophys.*, **84**, pp. 277–279.

Levasseur-Regourd, A. C., Renard, J. B., and Dumont, R. 1991. The zodiacal cloud complex. In *Origin and Evolution of Interplanetary Dust*, eds. A.-C. Levasseur-Regourd and H. Hasegawa (Dordrecht–Boston–London: Kluwer Academic Publ.) pp. 131–138.

Lillie, F. C. 1972. OAO-2 observations of the zodiacal light. In *The Scientific Results from OAO-2*, ed. A. O. Code, NASA SP-310, pp. 95–108.

Lindblad, B. A. 1976. Meteor radar rates and the solar cycle. *Nature*, **259**, pp. 99–101.

Lindblad, B. A. 1978. Meteor radar rates, geomagnetic activity and solar wind sector structure. *Nature*, **273**, pp. 732–734.

Low, F. J., Beintema, D. A., Gautier, T. N., Gillett, F. C., Beichman, C. A., Neugebauer, G., Young, E., Aumann, H. H., Boggess, N., Emerson, J. P., Habing, H. J., Hauser, M. G., Houck, J. R., Rowan-Robinson, M., Soifer, B. T., Walker, R. G., and Wesselius, P. R. 1984. Infrared cirrus: new components of the extended infrared emission. *Astrophys. J. Letters*, **278**, pp. L19–L22.

Lumme, K., and Bowell, E. 1985. Photometric properties of zodiacal light particles. *Icarus*, **62**, pp. 54–71.

Maihara, T., Mizutani, K., Hiromoto, N., Takami, H., and Hasegawa, H. 1985. A balloon observation of the thermal radiation from the circumsolar dust cloud in the 1983 total eclipse. In *Properties and Interactions of Interplanetary Dust*, eds. R. H. Giese and P. Lamy (Dordrecht: Reidel Publ. Comp.), pp. 55–58.

Mairan, J. J. Dortous de 1733. Traité physique et historique de l'aurore boreale. *Suite de Memoires de l'Academie royale des sciences*, année M.DCCXXXI (Imprimerie royale Paris).

Mann, I. 1996. Dust near the Sun. In *Physics, Chemistry and Dynamics of Interplanetary Dust*, eds B. Å. S. Gustafson and M. S. Hanner (Astronomical Society of the Pacific, San Francisco), pp. 315–320.

Matsumoto, T., Kawada, M., Murakami, H., Noda, M., Matsuura, S., Tanaka, M., and Narita, K. 1996. IRTS observations of the near-infrared spectrum of the zodiacal light. *Publ. Astr. Soc. Japan*, **48**, pp. L47–L51.

Matsuura, S., Matsumoto, M., Matsuhara, H., and Noda, M. 1996. Rocket-borne observations of the zodiacal light in the near-infrared. *Icarus*, **115**, pp. 199–208.

Maucherat-Joubert, M., Cruvellier, P., and Deharveng, J. M. 1979. Ultraviolet observations of the zodiacal light from D2B-Aura satellite. *Astron. Astrophys.*, **74**, pp. 218-224.

McCracken, C. W., Alexander, W. M., and Dubin, M. 1961. Direct measurements of interplanetary dust particles in the vicinity of Earth. *Nature*, **192**, pp. 441–442.

McDonnell, J. A. M. 1971. Review of in-situ measurements of cosmic dust particles in space. In *Space Research*, eds. K. Ya. Kontratyev, M. J. Rycroft and C. Sagan (Berlin: Akademie Verlag), pp. 415–433.

McDonnell, J. A. M. ed. 1979. Cosmic Dust. (New York: Wiley and Sons)

McDonnell, J. A. M., Lamy, P. L., and Pankiewicz, G. S. 1991. Physical properties of cometary dust. In *Comets in the Post-Halley Era*, eds. R. L. Newburn, M. Neugebauer, J. Rahe (Dordrecht, Boston, London: Kluwer Academic Publishers), **Vol. 2**, pp. 1043–1073.

McQueen, R. M. 1968. Infrared observations of the outer solar corona. *Astrophys. J.*, **154**, pp. 1059–1076.

Millman, P. M. 1957. The relative numbers of bright and faint meteors. *J. Roy. Astron. Soc. Canada*, **51**, pp. 113–115.

Morgan, D. H. 1978. The zodiacal light at 1550 Å. *Astron. Astrophys.*, **70**, pp. 543–545.

Mukai, T., and Yamamoto, T. 1979. A model of the circumsolar dust cloud. *Publ. Astron. Soc. Japan*, **31**, pp. 585–595.

Müller, J. 1865. Atlas zum Lehrbuch der kosmischen Physik. (Braunschweig: Vieweg-Verlag), zweite Auflage.

Murdock, Th. L., and Price, S. D. 1985. Infrared measurements of zodiacal light. *Astron. J.*, **90**, pp. 375–386.

Nagel, K., and Fechtig, H. 1980. Diameter to depth dependence of impact craters. *Planet. Space Sci.*, **28**, pp. 567–573.

Naumann, R. J. 1966. The near Earth meteoroid environment. NASA TN D-3717.

Nazarova, T. N. 1968. Solid component of interplanetary matter from vehicle observations. *Space Sci. Rev.*, **8**, pp. 455–466.

Nazarova, T. N., and Rybakov, A. K. 1974. The meteoritic particle space density near the Earth and Moon, according to data obtained by simultaneous observations of space vehicles. In *Space Research*, **XIV**, eds. M. J. Rycroft and R. D. Reasenberg (Berlin: Akademie Verlag), p. 773.

Neukum, G., König, B., Fechtig, H., and Storzer, D. 1975. Cratering in the Earth-Moon system: consequences for age determination by crater counting. *Proc. Lunar Sci. Conf. 6th*, compiled by The Lunar Science Institute, Houston/Texas (New York: Pergamon Press), pp. 2597–2620.

Neukum, G., Schneider, E., Mehl, A., Storzer, D., Wagner, G. A., Fechtig, H., and Bloch, M. R. 1972. Lunar craters and exposure ages derived from crater statistics and solar flare tracks. *Proc. 3rd Lunar Sci. Conf.*, **3**, ed. D. R. Criswell (Cambridge: MIT-Press), pp. 2793–2810.

Nier, A. O., and Schlutter, D. J. 1993. The thermal history of interplanetary dust particles collected in the Earth's stratosphere. *Meteoritics*, **28**, pp. 675–681.

Nilsson, C. 1966. Some doubts about the Earth's dust cloud. *Science*, **153**, pp. 1242–1246.

Noda, M., Christov, V. V., Matsuhara, H., Matsumoto, S., Noguchi, K., and Sato, S. 1992. Rocket observations of the near-infrared spectrum of the sky. *Astrophys. J.*, **391**, pp. 456–465.

Öpik, E. J. 1958. Physics of meteor flight in the atmosphere. (New York: Wiley Interscience Publishers, Inc).

Peterson, A. W. 1967. Experimental detection of thermal radiation from interplanetary dust. *Astrophys. J. Letters*, **148**, pp. L37–L39.

Pézénas, E. 1731. Observations Astronomique & Meteorologiques faites à Marseille par le E. Pézénas, Professeur d'Hydrographie, pendant l'année 1730. *Histoire de l'Academie Royale des Sciences avec les Memoires de Mathématique et de Physique pour la même année, tirés des registres de cette Académie*, (Académie des Sciences: Paris), pp. 7–9.

Piazzi-Smyth, C. 1872. Spectroscopic observations of the zodiacal light. *M. N. R. A. S.*, **32**, pp. 277–288.

Pitz, E., Leinert, Ch., Schulz, A., and Link, H. 1978. Ultraviolet zodiacal light observed by the Astro 7 rocket experiment. *Astron. Astrophys.*, **69**, pp. 297–304.

Pitz, E., Leinert, Ch., Schulz, A., and Link, H. 1979. Color and polarization of the zodiacal light from the ultraviolet to the near infrared. *Astron. Astrophys.*, **74**, pp. 15–20.

Poynting, I. H. 1903....... *Phil. Trans. Roy. Soc.*, **A 202**, p. 525, (reprinted in his "Collected Scientific Papers", Cambridge 1920).

Rajan, R. S., Brownlee, D. E., Tomandl, D., Hodge, P. W., Farrar IV, H., and Britten, R. A. 1977. Detection of He in stratospheric particles gives evidence of extraterrestrial origin. *Nature*, **267**, pp. 133–134.

Rauser, P., and Fechtig, H. 1972. Combined dust collection and detection experiment during a noctilucent cloud display above Kiruna, Sweden. In *Space Research*, **XII**, eds. S. A. Bowhill, L. D. Jaffe and M. J. Rycrodt (Berlin: Akademie-Verlag), pp. 391–402.

Reach, T., Franz, B. A., Weiland, J. L., Hauser, M. G., Kelsall, T. N., Wright, E. L., Rawley, G., Stemwedel, S. W., and Spiesman, W. J. 1995. Observational confirmation of the circumstellar dust ring by the COBE satellite. *Nature*, **374**, pp. 521–523.

Richter, I., Leinert, Ch., and Planck, B. 1982. Search for short term variations of zodiacal light and optical detection of interplanetary plasma clouds. *Astron. Astrophys.*, **110**, pp. 115–120.

Roach, F. E. 1972. A photometric model of the zodiacal light. *Astron. J.*, **77**, pp. 887–891.

Roosen, R. G. 1970. An annotated bibliography on the Gegenschein. *Icarus*, **13**, pp. 523–539.

Roosen, R. G. 1971. The Gegenschein. *Rev. Geophys. and Space Phys.*, **9**, pp. 275–304.

Röser, S. 1976. Can short period comets maintain the zodiacal cloud?. In Lecture Notes in *Physics*, **48**, Interplanetary Dust and Zodiacal Light, eds. H. Elsässer and H. Fechtig (Berlin–Heidelberg–New York), pp. 319–322.

Röser, S., and Staude, H. J. 1978. The zodiacal light from 1 500 Å to 60 micron—Mie scattering and thermal emission. *Astron. Astrophys.*, **67**, pp. 381–394.

Searle, A. 1882. On certain zodiacal phenomena. *Astron. Nachr.*, **102**, pp. 263–266.

Searle, A. 1893. The zodiacal light. *Ann. Astr. Obs. Harvard Coll. Obs.*, **19**, pp. 165–245.

Seeliger, H. 1901. Über kosmische Staubmassen und das Zodiacallicht. *Sitzungsbericht d. bayer. Akad. d. Wiss.*, **31**, pp. 265–292.

Shapiro, I. C., Lautman, D. A., and Colombo, G. 1966. The Earth's dust belt: fact or fiction? 1. Forces perturbing dust particle motion. *J. Geophys. Res.*, **71**, pp. 5695–5704.

Shelton, H., Hendricks Jr., C. D., and Wuerker, R. F. 1960. Electrostatic acceleration of microparticles to hypervelocities. *J. Appl. Phys.*, **31**, pp. 1243–1246.

Slipher, V. M. 1933. Spectra of the night sky, the zodiacal light, the Aurora, and the cosmic radiations of the sky. *J. Roy. Astron. Soc. Can.*, **27**, pp. 365–369.

Smith, L. L., Roach, F. E., and Owen, R. W. 1965. The absolute photometry of the zodiacal light. *Plan. Space Sci.*, **13**, pp. 207–217.

Soberman, R. K., and Hemenway, C. L. 1965. Meteoric dust in the upper atmosphere. *J. Geophys. Res.*, **70**, pp. 4943–4949.

Soifer, B. T., Houck, J. R., and Harwit, M. 1971. Rocket infrared observations of the interplanetary medium. *Astrophys. J. Letters*, **168**, pp. L73–L78.

Sparrow, J. G., and Ney, E. P. 1973. Temporal constancy of zodiacal light. *Science*, **181**, pp. 438–440.

Sparrow, J. G., Weinberg, J. L., and Hahn, R. C. 1977. Ten-color Gegenschein zodiacal light photometer. *Appl. Opt.*, **16**, pp. 978–982.

Sykes, M. V., Lebovsky, L. A., Hunten, D. M., and Low, F. 1986. The discovery of dust trails in the orbits of periodic comets. *Science*, **232**, pp. 1115–1117.

Torr, M. R., Torr, D. G., and Stencel, R. 1979. Zodiacal light surface brightness measurements by Atmosphere Explorer-C. *Icarus*, **40**, pp. 40–59.

Tsou, P. 1996. Hypervelocity capture of meteoroids in aerogel. In *Physics, Chemistry, and Dynamics of interplanetary dust*, ASP Conference Series, eds. B. Å. S. Gustafson and M. S. Hanner (Astronomical Society San Francisco), **104**, pp. 237–242.

Vedder, J. F., and Mandeville, J.-C. 1974. Microcraters formed in glass by projectiles of various densities. *J. Geophys. Res.*, **79**, pp. 3247–3256.

Watson, F. G. 1956. Between the planets. Harvard University Press Cambridge.

Weihrauch, J. H., Gerloff, U., and Fechtig, H. 1968. Stereo scan investigations of metal plates exposed on LUSTER 1966, GEMINI 9 and 12. In *Space Research*, **VIII**, eds. A. P. Mitra, L. G. Jacchia and W. S. Newman (Amsterdam: North Holland Publ. Comp.), pp. 566–578.

Weinberg, J. L. 1964. The zodiacal light at 5 300 Å. *Ann. d'Astrophys.*, **27**, pp. 718–738.

Weinberg, J. L. 1970. Current problems in the zodiacal light. In *Space Research*, **X**, eds. T. M. Donahue, P. A. Smith and L. Thomas (Amsterdam: North Holland Publ. Comp.), pp. 233–243.

Weinberg, J. L. 1974. Polarization of the zodiacal light. In *Planets, Stars and Nebulae studied with polarimetry*, ed. T. Gehrels (Tucson: University of Arizona Press), pp. 781–793.

Weinberg, J. L., Hanner, M. S., Beeson, D. E., DeShields II, L. M., and Green, B. A. 1974. Background starlight observed from Pioneer 10. *J. Geophys. Res.*, **121**, pp. 750–770.

Weinberg, J. L., and Mann, H. M. 1968. Negative polarization in the zodiacal light. *Astrophys. J.*, **152**, pp. 665–666.

Weiss-Wrana, K. 1983. Optical properties of interplanetary dust: comparison with light scattering by larger meteoritic and terrestrial grains. *Astron. Astrophys.*, **126**, pp. 240–250.

Whipple, F. L. 1955. A comet model, III. The zodiacal light. *Astrophys. J.*, **121**, pp. 750–

770.

Whipple F. L. 1961. Medical and biological aspects of the energies of space. Ed. P. Campbell (New York: Columbia University Press).

Whipple, F. L. 1967. On maintaining the meteoritic complex. In *The Zodiacal Light and the Interplanetary Medium*, ed. J. L. Weinberg (Washington: US Government Printing Office), NASA SP-150, pp. 409–426.

Whipple, F. L., and Gossner, J. L. 1949. An upper limit to the electron density near the Earth's orbit. *Astrophys. J.*, **109**, pp. 380–390.

Wolff, Ch. 1967. Optical environment about the OGO III satellite. *Science*, **158**, pp. 1045–1046.

Wolstencroft, R. D., and Rose, L. J. 1967. Observations of the zodiacal light from a sounding rocket. *Astrophys. J.*, **147**, pp. 271–292.

Wright, A. W. 1874. On the polarization of the zodiacal light. *American J. of Science and Arts*, (Third Series) **VII**, pp. 451–459.

Wyatt Jr., St. P., and Whipple, F. L. 1950. The Poynting-Robertson effect in meteor orbits. *Astrophy. J.*, **111**, pp. 134–141.

Zerull, R. H. 1976. Scattering measurements of dielectric and absorbing nonspherical particles. *Phys. Atmosph.*, **49**, pp. 168–188.

Zook, A. E., and Berg, O. E. 1975. A source for hyperbolic cosmic dust particles. *Planet. Space Sci.*, **23**, pp. 183–203.

Optical and Thermal Properties of Interplanetary Dust

A. Chantal Levasseur-Regourd[1], Ingrid Mann[2,3], René Dumont[4], Martha S. Hanner[5]

[1] Université Paris VI / Aéronomie CNRS-IPSL, Verrières, France
[2] ESA Space Science Dept., ESTEC, Noordwijk, Netherlands
[3] Inst. of Planetology, Münster Univ., Münster, Germany
[4] Observatoire de Bordeaux, Floirac, France
[5] Jet Propulsion Laboratory, Pasadena, California, USA

Abstract. This chapter reviews our understanding of the interplanetary dust, from its optical and thermal properties. The interplanetary dust cloud, although of extremely low optical thickness, is indeed visible through the faint and evanescent glow of the zodiacal light. Two types of physical processes play a role in the emission of light by the interplanetary dust cloud: the scattering of solar light by dust particles, prevailing in the visible domain, and the thermal radiation, prevailing in the infrared domain.

The first part of the chapter presents our understanding of the zodiacal scattered light, with emphasis on the observational constraints, and on the parameters describing its characteristics. The second part is devoted to the light scattered by dust in the F-corona. The third part summarizes the properties of the thermal emission of the zodiacal cloud and F-corona. The fourth part points out variations in the local properties of the interplanetary dust: the cloud appears to be a complex mixture of particles of different origin, whose mixing ratio changes with location in the cloud, and whose properties change with time, as a result of various physical processes.

I. ZODIACAL SCATTERED LIGHT

Zodiacal light appears, to the naked eye, as a faint solar colored cone of light above the western horizon in the evening about one hour after sunset, or above the eastern horizon in the morning before sunrise. Just as the Milky Way reveals the shape of our Galaxy, the zodiacal light reveals the existence of a huge number of small dust particles in interplanetary space. The brightness, and thus the space density, increase sharply towards the Sun, and, as suggested by the adjective "zodiacal", towards the ecliptic. In the absence of any light contamination, the zodiacal cone is visible when the ecliptic is high above the horizon. It is relatively easy to detect in tropical latitudes, while, for northern hemisphere mid-latitudes, it is mainly detectable in March for the evening sky, and in September for the morning sky.

I.A. Historical Survey

An extensive discussion about early reports on zodiacal light can be found in the chapter by Fechtig et al. in this volume. The first known printed description of the zodiacal light has been given by Joshua Childrey (1661), and the first interpretation has been provided by Jean Dominique Cassini (1683). Although some hints are suspected in ancient texts (e.g., by Aristotle or Seneca), it does not seem to have been mentioned previously in the European literature; the question of the stability of zodiacal light on a scale of more than a few centuries or millennia is thus still completely open.

J. D. Cassini explained that the zodiacal light originates in a cloud that is flattened along the solar equator, and scatters the light better when the line of sight gets closer to the Sun or its equator. This interpretation remains valid, except that the symmetry plane is closer to the ecliptic than to the solar equator.

J. J. Dortous de Mairan (1733) suggested that the cloud extends further than the terrestrial orbit, since the zodiacal cone may be detected above 90° solar elongation. Figure 1 reproduces his drawing of the zodiacal light, as seen from mid-latitudes at the end of winter.

About one century later, Brorsen (1854) noticed a slight brightness enhancement around the antisolar point, barely detectable with the naked eye, the so-called Gegenschein. The zodiacal light indeed extends over the whole celestial sphere: it represents a foreground veil for the observation of faint and extended astronomical sources, from the Earth or from the Earth's orbit.

After it had been established by Arago (1858) that solar light is partially linearly polarized by scattering, and that polarization is most sensitive to fluctuations in the physical properties, a first estimation of the polarization of the zodiacal light was made by Wright (1874).

The first photographic observations have taken place during the first half of the 20th century (Dufay 1925). By the same time, coronagraphy has allowed the inner part of the zodiacal light, i.e. the F-corona, to be observed even in the absence of a solar eclipse (Lyot 1930, 1939).

In the fifties, the development of spectrography has demonstrated that the zodiacal light spectrum is comparable to that of the Sun (Blackwell and Ingham 1961). From polarization and spectral studies, it was finally recognized that the zodiacal light comes from the scattering of solar light by the optically thin interplanetary dust cloud.

It is really during the second half of the 20th century that significant progress has been achieved, due to the development of quantitative measurements and of space borne observations. Quite ironically, the zodiacal light has been simultaneously vanishing in mid-latitude populated regions, as a result of the increasing light pollution of the night sky and atmospheric pollution near the horizon.

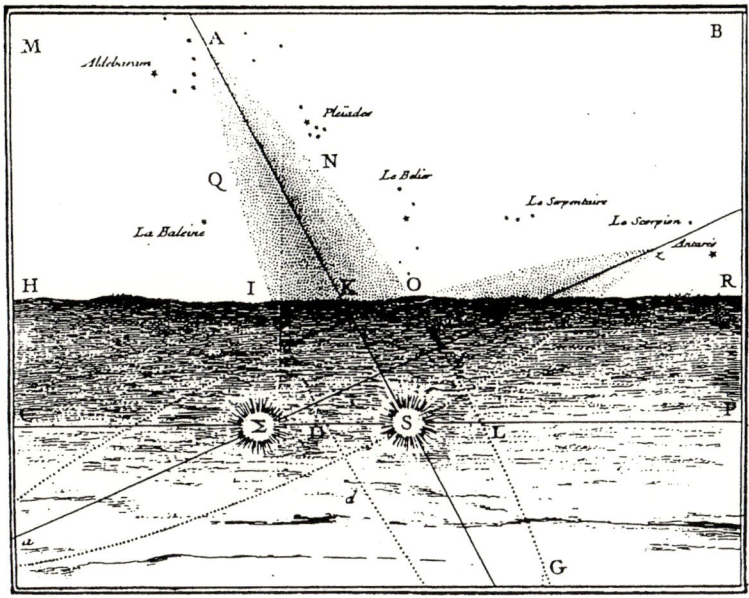

Figure 1. Visibility of zodiacal cones from northern mid-latitudes at the end of the winter. Drawing, from J.J. Dortous de Mairan (1733).

Since the late sixties, periodic meetings, under the auspices of IAU and/or COSPAR, have allowed to keep track of progress in our understanding of the zodiacal light, and more generally of optical/thermal properties of interplanetary dust (Weinberg 1967; Elsässer and Fechtig 1976; Halliday and McIntosh 1980; Giese and Lamy 1985; Levasseur-Regourd and Hasegawa 1991; Gustafson and Hanner 1996). In addition, review papers have been prepared by Leinert (1975), Weinberg and Sparrow (1978), Leinert and Grün (1990), Dumont (1992), Levasseur-Regourd (1992, 1999), and by Leinert et al. (1998).

I.B. Zodiacal Light Measurements

1. Zodiacal light and light from the night sky

The zodiacal light is always entangled with other components of the light of the night sky. What the poet described as "This obscure brightness that falls from the stars" (Corneille 1637), is rather falling from faint diffuse sources that veil the night sky.

These various veils, so called atmospheric nightglow, zodiacal light, galactic light and extra galactic light, are at various distances along our line of sight and radiate light through various processes. However, the mirror symmetry of some of the components with respect to their planes of reference, and the motion of the corresponding coordinate systems (e.g., helio-ecliptic, galactic), may be used to disentangle them.

2. Near Earth's observations

Extensive ground-based measurements of the zodiacal light have been performed by various groups, for instance from Tenerife (R. Dumont) and from Hawaii (J.L. Weinberg). Results have also been obtained from satellites and space stations (e.g., OSO 2, OSO 5, D_2A, D_2B, TD-1, Skylab, Salyut 6, Salyut 7, IRTS, COBE), covering from the near ultraviolet to the near infrared.

In the case of ground-based measurements, the contamination by the nightglow continuum is a most serious problem. It needs to be taken into account through cumbersome methods, such as the intensity correlation between the atmospheric continuum and the oxygen green line, both observed with the multiple elevations method (Dumont 1965). In the case of space measurements, complete sky coverage is impossible to obtain because of viewing constraints. The results obtained in the seventies from Tenerife and D_2A have been found to be in good agreement, the survey by Dumont and Sanchez (1975a, 1975b) remaining the most reliable and complete source (Levasseur-Regourd and Dumont 1980; Fechtig et al. 1981).

3. Deep space observations

Inside 1 AU, observations have been obtained from the Helios 1 and 2 probes, for three fixed ecliptic latitudes (Leinert et al. 1982). Between 1 and 0.3 AU, the brightness increases with decreasing solar distance R of the observing probe in the ecliptic plane, approximately as $R^{-2.3}$.

In the outer ecliptic plane, observations have been performed from the Pioneer 10 and 11 probes (Hanner et al. 1976). They indicate that the Gegenschein mainly corresponds to a backscattering effect, and show that the zodiacal light is detectable at least up to the outer fringe of the asteroid belt. It decreases as $R^{-2.6}$ (Toller and Weinberg 1985), or, with a bimodal population, as $R^{-2.5}$ up to 2.3 AU and $R^{-2.37}$ in the asteroid belt (Hanner et al. 1976). At greater solar distances, the brightness has decreased so much that a radial gradient cannot be measured anymore against the background of integrated starlight and diffuse galactic light.

I.C. Main Trends in the Data

1. Observing geometry and parameters

Due to the symmetries already mentioned, the coordinates used to describe the zodiacal light observations are the helio-ecliptic coordinates of the viewing direction, i.e. the ecliptic latitude β and the elongation angle from the Sun ε or helio-ecliptic longitude $\lambda - \lambda_o$, where λ is the ecliptic longitude of the line of sight, and λ_o is the ecliptic longitude of the Sun. The observatory (which until the time of writing always is in the ecliptic plane) is at the distance R from the Sun. Figure 2 presents the geometry of the observations and of the scattering.

2. Wavelength dependence

Zodiacal light brightness depends upon the wavelength of the observations. The color of the zodiacal light is nearly that of the solar spectrum in the 200 to 3000 nm domain, except for a possible reddening at small solar elongations

(Dumont and Sanchez-Martinez 1973; Leinert et al. 1974; Cebula and Feldman 1982; Leinert et al. 1998).

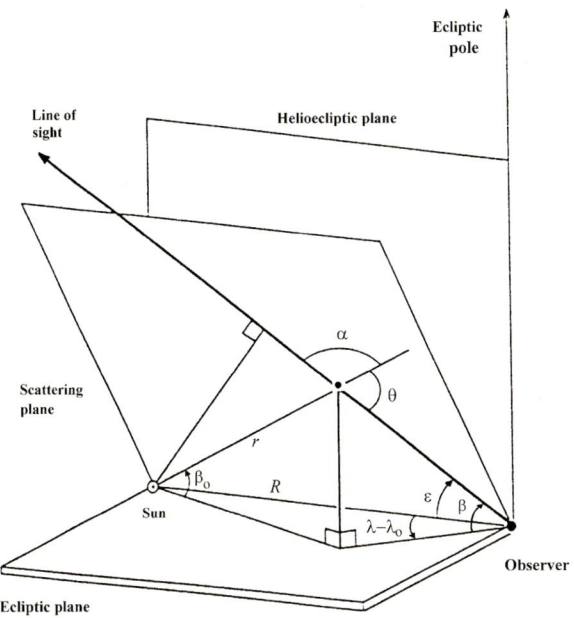

Figure 2. Observing geometry of scattering in the zodiacal cloud.

Since the zodiacal light follows the solar spectrum, it also reproduces the Fraunhofer spectral lines. Their wavelength is shifted by the motion of the dust particles with respect to the Sun and their motion relative to the observer (Mukai and Mann 1993). Observations of the Doppler shifts have been performed by various teams (e.g., Ring et al. 1964; Fried 1978; East and Reay 1985; Robley et al. 1985). The observations and their interpretation remain difficult, since radial velocities are low and averaging effects take place along the line of sight. The measurements suggest however that most of the particles are on elliptical prograde orbits.

3. Spatial or temporal variations of small amplitude

The zodiacal light, as observed from a given observatory, is mainly smooth with some perturbations. No significant correlation is found with the solar cycle (Dumont and Levasseur-Regourd 1978; Leinert and Pitz 1989). For an observer moving with the Earth, second order variations, such as small amplitude variations and seasonal oscillations are noticed.

Faint enhancements of a few percent in the brightness have been pointed out in certain directions, at certain epochs of the year. They could be due to optical detection of meteor streams, to dust accretion near libration points of the Earth-Moon system, and to ejection of solar plasma (Levasseur and Blamont 1973; Baggaley 1977; Mercer et al. 1979; Leinert and Pitz 1989).

However, deconvolving an increase in the spatial density from a change in the scattering properties of the dust is not straightforward.

Although small-scale structures are easier to detect through their thermal emission in the infrared domain (see Subsec. 3 of III.A), cooled CCD cameras have now the potential of detecting faint and diffuse structures in the zodiacal light (Ishiguro et al. 1999). The contribution from meteor streams, previously suspected based on satellite photometric data (Levasseur and Blamont 1976), has been confirmed through the detection of an enhancement in the dust trail of the Leonids parent comet, from ground based high altitude observations (Nakamura et al. 2000).

4. Seasonal oscillations and the plane of symmetry of the dust cloud

Annual variations, up to 10%, take place at high and medium latitudes (Fig. 3). They originate both in the slight inclination of the plane of symmetry of the cloud with respect to the ecliptic plane, and in the eccentricity of the Earth's orbit (Dumont and Levasseur-Regourd 1978; Leinert et al. 1980).

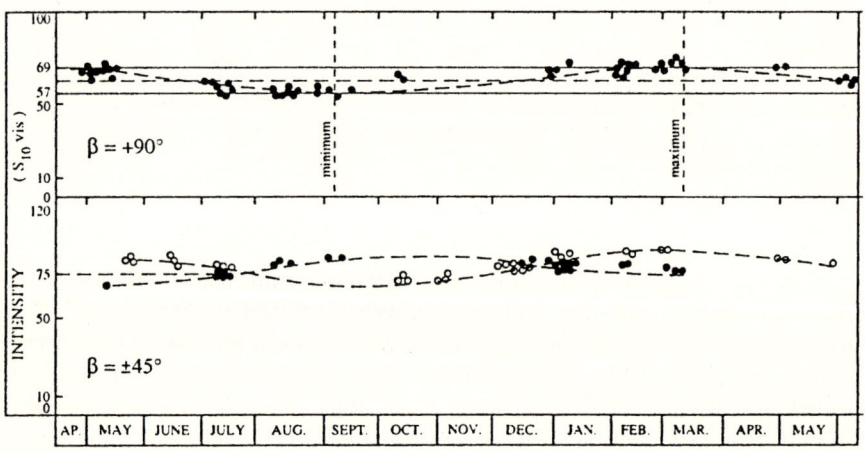

Figure 3. Oscillations in high latitude observations (from Levasseur and Blamont 1973).

The question of the precise location of the tilted symmetry plane is still open. At 1 AU, it is quite close to the invariable plane of the solar system, with an ascending node at $(95 \pm 20)°$, and an inclination of $(1.5 \pm 0.4)°$ (Dumont and Levasseur-Regourd 1978). The location of the Gegenschein, as derived from maps obtained in March and September (Ishiguro et al. 1998), indeed agrees with such ranges, both for the ascending node and for the inclination.

In the inner solar system, the inclination is of the order of $3°$ (Leinert et al. 1980). It is therefore suspected that this "plane" is rather a warped surface of symmetry. Calibrated CCD images (e.g., James et al. 1997) now offer the opportunity, once the airglow contamination is carefully subtracted, of determining the precise location of the symmetry surface.

I.D. Zodiacal Brightness from 1 AU

After corrections are made for the seasonal oscillations, due to the inclination of the zodiacal cloud symmetry surface to the ecliptic plane, and to the eccentricity of the Earth orbit, the brightness measured from the Earth or from its environment is obtained as a function of the helio-ecliptic coordinates $(\beta, \lambda - \lambda_o)$. The plane of reference should indeed be the symmetry plane of the interplanetary dust cloud at 1 AU, rather than the ecliptic plane.

1. Magnitude related units

The zodiacal light brightness was initially estimated through star counts. The values are thus traditionally given in $S_{10}(V)$, i.e. the equivalent number of 10th visual magnitude solar type stars per square degree. Since the zodiacal light has approximately a solar spectrum, this peculiar unit has the advantage of being valid in the whole visual domain.

Levasseur-Regourd and Dumont (1980) tabulated the average brightness distribution over the sky in 5° steps. Table 1 presents a recent update of this work which is also illustrated in Fig. 4a. The table is completed up to 15° solar elongation ε with the values previously proposed remaining the same, except for a slight increase in the region relatively close to the Sun or for high ecliptic latitudes. Not taking into account the above mentioned short-time enhancements; intermediate values are obtained by smooth interpolation. The relative uncertainty remains below 0.05.

TABLE 1

Zodiacal light brightness Z, measured from the Earth, in $S_{10}(V)$. The directions are defined by their ecliptic latitude (β) and helio-ecliptic longitude ($\lambda - \lambda_0$), once the corrections for the slight inclination of the symmetry plane and for the Earth's orbit eccentricity have been made.

$\lambda - \lambda_0$ \ β	0°	5°	10°	15°	20°	25°	30°	45°	60°	75°	90°
0°				24502	1260	770	500	215	117	78	60
5°				300	1200	740	490	212	117	78	60
10°			3700	1930	1070	675	460	206	116	78	60
15°	9000	5300	2690	1450	870	590	410	196	114	78	60
20°	5000	3500	1880	1100	710	495	355	185	110	77	60
25°	3000	2210	1350	860	585	425	320	174	106	76	60
30°	1940	1460	955	660	480	365	285	162	102	74	60
35°	1290	990	710	530	400	310	250	151	98	73	60
40°	925	735	545	415	325	264	220	140	94	72	60
45°	710	570	435	345	278	228	195	130	91	70	60
60°	395	345	275	228	190	163	143	105	81	67	60
75°	264	248	210	177	153	134	118	91	73	64	60
90°	202	296	176	151	130	115	103	81	67	62	60
105°	166	164	154	133	117	104	93	75	64	60	60
120°	147	145	138	120	108	98	88	70	60	58	60
135°	140	139	130	115	105	95	86	70	60	57	60
150°	140	139	129	116	107	99	91	75	62	56	60
165°	153	150	140	129	118	110	102	81	64	56	60
180°	180	166	152	139	127	116	105	82	65	56	60

The zodiacal brightness is close to 2000 $S_{10}(V)$ at 30° solar elongation in the ecliptic plane and outshines the brightest part of the Milky Way. The brightness reaches a minimum equal to 60 $S_{10}(V)$ toward the ecliptic pole. In the ecliptic plane at $\varepsilon = 90°$, it is equal to 202 $S_{10}(V)$ which is 3.3 times the ecliptic pole value. An excess of the order of 40 $S_{10}(V)$ with respect to the general trend is noticed toward the antisolar direction where the brightness reaches 180 $S_{10}(V)$.

2. SI units

For those who are not familiar with $S_{10}(V)$ or other magnitude related units, the data are converted in SI units in Table 2 (Levasseur-Regourd 1996), using:

$$1\ S_{10}(V) = 1.261 \cdot 10^{-8}\ \text{W m}^{-2}\ \text{sr}^{-1} \mu\text{m}^{-1} \text{ at } 0.55\ \mu\text{m}.$$

For other wavelengths, the values need to be multiplied by a corrective factor to account for the solar spectrum profile. Detailed discussions on the units can be found in Sparrow and Weinberg (1976) and in Leinert et al. (1998).

TABLE 2

Zodiacal light brightness Z, measured from the Earth near 0.55μm, in 10^{-8} W m^{-2} sr^{-1} μm^{-1}. The directions are defined by their ecliptic latitude (β) and helio-ecliptic longitude ($\lambda - \lambda_0$), once the corrections for the slight inclination of the symmetry plane and for the Earth's orbit eccentricity have been made.

β \\ $\lambda - \lambda_0$	0°	5°	10°	15°	20°	25°	30°	45°	60°	75°	90°
0°				3090	1590	970	630	271	147	98	76
5°				2900	1510	930	615	267	147	98	76
10°			4660	2430	1450	850	580	260	146	98	76
15°	11350	6680	3390	1830	1100	745	515	247	144	98	76
20°	6300	4410	2370	1390	895	625	447	233	139	97	76
25°	3780	2780	1700	1080	735	535	403	219	134	96	76
30°	2440	1840	1200	830	605	460	359	204	129	93	76
35°	1630	1250	895	670	505	391	315	190	123	92	76
40°	1170	930	685	525	409	333	277	176	118	91	76
45°	895	720	550	435	350	287	246	164	115	88	76
60°	500	435	346	287	239	205	180	132	102	84	76
75°	333	312	265	223	193	169	149	115	92	81	76
90°	255	241	222	190	164	145	130	102	84	78	76
105°	209	207	194	168	147	131	117	94	81	76	76
120°	185	183	174	151	136	123	111	88	76	73	76
135°	176	175	164	145	132	120	108	88	76	72	76
150°	176	175	163	146	135	125	115	94	78	71	76
165°	193	189	176	163	149	139	129	102	81	71	76
180°	227	209	192	175	160	146	132	103	82	71	76

No matter what units are used, it has to be pointed out that zodiacal light is a "foreground noise" that limits space photometry of faint objects. For observations at solar elongations greater than 90°, the celestial hemisphere where the zodiacal foreground is low, the brightness at 0.55 μm remains between 8 and $26 \cdot 10^{-7}$ W m^{-2} sr$^{-1} \mu$m^{-1}. A typical brightness of

$20 \cdot 10^{-7}$ W m^{-2} sr^{-1}μm^{-1}, equivalent to 160 S_{10}(V), corresponds to approximately 200 stars of magnitude 28 per square second of arc. Except at high ecliptic latitudes, it is thus an obvious advantage to make an appropriate choice of the epoch of observation of faint objects, the best one being when the longitude of target is about ± 135° from the longitude of the Sun (Dumont and Levasseur-Regourd 1981).

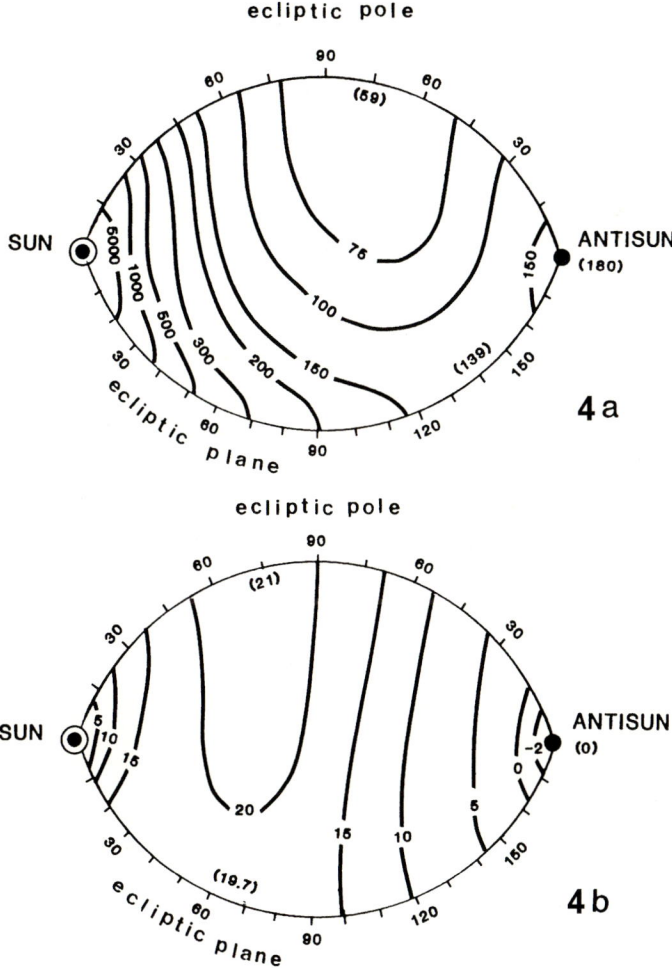

Figure 4. Maps of the zodiacal light as seen from the Earth. The outside circle corresponds to the ecliptic plane, and the bottom line to the plane perpendicular to the Sun-Earth line with the ecliptic pole at its center. Corrections have been made due to the slight inclination of the symmetry plane and to account for the eccentricity of the Earth's orbit. 4a: brightness; 4b: polarization (from Dumont 1992).

I.E. Zodiacal Polarization from 1 AU

A complete description of the zodiacal light polarization would require the knowledge of its Stokes parameters. Fortunately, the incident solar light is not polarized and the scattering medium is optically thin. The resulting scattered light is partially linearly polarized, with the electric field vector either parallel or perpendicular to the scattering plane, defined by the directions of the incident light from the Sun and scattered light from the dust particle (e.g., Hapke 1993). It is then sufficient to determine, together with the brightness, the degree of linear polarization and the orientation of the electric field vector. The linear polarization values do not require any normalization to be compared with values obtained for other scattering media.

Table 3, illustrated in Fig. 4b, provides smoothed values of the average degree of polarization P (Levasseur-Regourd 1996). The error bars are approximately 2% or less. The values are obtained from measurements of the linearly polarized components of the brightness, $Z_\perp$, perpendicular to the scattering plane, and $Z_{//}$, parallel to the scattering plane, using:

$$P = (Z_\perp - Z_{//})/Z = (Z_\perp - Z_{//})/(Z_\perp + Z_{//}).$$

TABLE 3

Zodiacal light polarization P, measured from the Earth in the visible domain, typically near 0.55μm (in percent). The directions are defined by their ecliptic latitude (β) and helio-ecliptic longitude ($\lambda - \lambda_0$), once the corrections for the slight inclination of the symmetry plane and for the Earth's orbit eccentricity have been made.

$\lambda - \lambda_0$ \ β	0°	5°	10°	15°	20°	25°	30°	45°	60°	75°	90°
0°				8	10	11	12	16	19	20	20
5°				9	10	11	12	16	19	20	20
10°			11	11	12	13	14	17	19	20	20
15°	13	13	13	13	13	14	15	17	19	20	20
20°	14	14	14	15	15	15	15	17	19	20	20
25°	15	15	16	16	16	16	16	18	19	20	20
30°	16	16	16	16	16	17	17	18	19	20	20
35°	17	17	17	17	17	17	17	18	20	20	20
40°	17	17	17	17	18	18	18	19	20	20	20
45°	18	18	18	18	18	18	18	19	20	20	20
60°	19	19	19	19	19	20	20	20	20	20	20
75°	18	18	18	18	18	19	19	19	19	19	20
90°	16	16	16	16	16	16	17	18	18	19	20
105°	12	12	12	12	13	13	14	15	17	19	20
120°	8	8	9	9	9	10	11	13	15	18	20
135°	5	5	5	6	6	7	8	11	14	17	20
150°	2	2	2	3	3	4	5	8	12	16	20
165°	-2	-2	-1	-1	0	2	3	7	11	16	20
180°	0	-2	-3	-2	-1	0	2	6	11	16	20

Although P almost does not depend upon the wavelength λ in the visible domain, it decreases slightly with increasing λ, as was shown by COBE in the near infrared (Berriman et al. 1994; Leinert et al. 1998). Such a trend is similar to the trend noticed for light scattered by regoliths, but opposite to the trend noticed for light scattered by cometary dust. At 90° elongation, this dependence can be represented by the relation: $P(\lambda) \approx 0.17 + 0.10 \log(\lambda/0.5 \mu m)$.

The polarization depends mainly upon the helio-ecliptic latitude and reaches its maximum between 40° and 90°. The degree of polarization is of the order of 20% towards the ecliptic pole and close to 0% towards the antisolar direction. In the ecliptic plane at 90° solar elongation, it is approximately 16%. The electric field vector is mostly perpendicular to the scattering plane (Weinberg 1985); from the definition of P, the polarization is thus mostly positive. For elongations greater than 160° (the Gegenschein region), it is parallel to the scattering plane so that negative polarization values are obtained.

II. F-CORONA SCATTERED LIGHT

II.A. Solar Corona Observations

As pointed out by Blackwell et al. in 1967: "It is now accepted that the zodiacal light is an outer extension of the solar corona. The two phenomena are ordinarily observed in different circumstances on different occasions, but if observations could be made outside the Earth's atmosphere, and scattering of sunlight in the instrument eliminated, they would be seen as one." Images of the inner zodiacal light from the lunar orbit (Cooper et al. 1996) indeed illustrate the extension of the zodiacal light towards the solar corona, as seen in Fig. 5.

The coronal brightness is visible to the naked eye during solar eclipses when the Moon occults the Sun. Also, special coronagraphs occult the brightness of the solar disk, to image the surrounding corona. But, the high level of atmospheric and instrumental straylight that the incident solar radiation produces hampers observations with coronagraphs. The sky background that influences coronal observations may vary with daily conditions as well as with the eclipse site. In addition to the straylight caused by atmospheric brightness components from outside the shadow cone of the eclipse, a second component, the solar aureole, varies with elongation.

Mainly originating from solar brightness components scattered by atmospheric aerosols, the solar aureole gives a signal, which increases with decreasing distance R to the Sun. Duerst (1982) derived values of 10^{-11} to 10^{-9} mean solar disk brightness and a radial slope according to $R^{-1.4}$ for the eclipse in India at 600 nm wavelength. MacQueen and Greeley describe the slope of the aureole as $R^{-1.5}$ at 2.12 μm for the Hawaii 1991 solar eclipse.

Coronagraph observations from satellites are not influenced by atmospheric straylight and allow the observation of the transition from the solar corona into the zodiacal light. Externally occulted coronagraphs, e.g., onboard the ESA/NASA Solar Heliospheric Observatory SOHO, presently achieve lower straylight levels and enable coronal observations out to at least 30 solar radii (Brückner et al. 1995).

Figure 5. Clementine space probe photograph of the inner zodiacal light over the lunar horizon (from Cooper et al. 1996). The bright object to the left is Venus, and the Moon is illuminated by the Earthshine.

An additional problem in F-corona studies lies in the separation of its different brightness components (see Fig. 6). While the zodiacal light clearly stems from scattering by interplanetary dust particles, the brightness of the solar corona results from several distinct phenomena: Thomson scattering of the solar radiation by free electrons (K-corona), line emission from ions (L-corona) and light scattering by dust particles (F-corona). The K-corona is predominant at distances within 4 solar radii from the Sun (which corresponds to about 1° elongation) and its brightness is highly variable in space and time, reflecting the plasma distribution near the Sun. The F-corona yields the main brightness at larger distances from the Sun, and can be separated by means of spectral analysis.

The light scattered by dust particles shows the Fraunhofer lines of the solar spectrum with a Doppler shift caused by the dust orbital velocities (Mukai 1993). The K-coronal signal shows no Fraunhofer lines, which disappear from the Doppler shift of the signal due to the broad distribution of thermal electron velocities. As a result, the depth of the Fraunhofer lines in the F-corona should be similar to those in the solar spectrum, whereas the K-coronal brightness is almost continuous.

Another method to subtract the K-corona is based on the polarization produced by the Thomson scattering process. This approach assumes that the F-coronal brightness is produced by diffraction of dust near the observer and hence is unpolarized. Blackwell et al. (1967) give a detailed description of the separation methods.

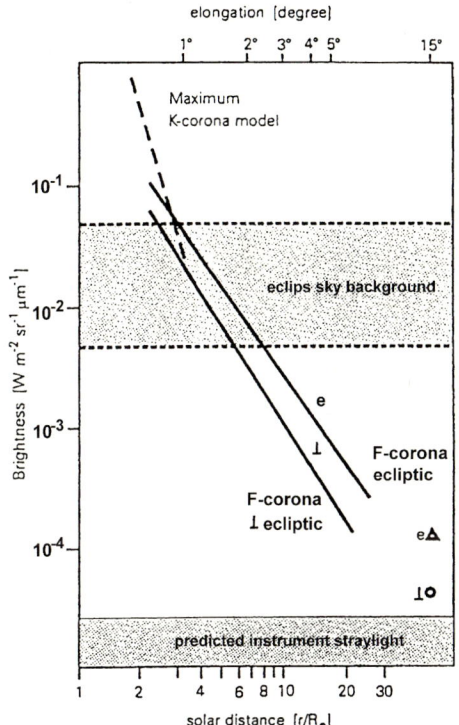

Figure 6. F-corona, compared to K-corona, background eclipse sky and instrument straylight (Leinert et al. 1998).

It should be added that four years of continuous photopolarimetric observations of the solar corona with the SOHO/LASCO coronagraphs have recently allowed Lamy et al. (2000) to construct an empirical model, that predicts the radiance of the F-corona at any time of the year as seen from the SOHO spacecraft.

II.B. Brightness and Polarization of the F-Corona

1. Brightness

While the zodiacal light is concentrated near the ecliptic plane, the more spherical shape of the F-corona is presumably a result of two effects: the contribution from diffraction dominates forward scattering from dust near the observer and the scattering from particles at high latitudes (Mann 1996). Based on both observational data and extrapolation from the zodiacal light, Koutchmy and Lamy (1985) describe the visible F-corona brightness in the 400 to 600 nm range as proportional to $R^{-2.25}$ at the solar equator and to $R^{-2.47}$ at the solar poles.

The change in single particle scattering properties at small angles, however, is expected to change the radial slope of the F-coronal brightness compared to the zodiacal light (Mann 1993). Measurements during the 1980 eclipse (Duerst 1982) yield different exponents of the solar distance power law from 2 to 10 solar radii of -2.44 at the equator and -2.76 at the poles. Observations from the Apollo 16 spacecraft describe the equatorial brightness beyond 20 solar radii as about $R^{-1.93}$ (MacQueen et al. 1973). Taking into account recent measurements, we suggest an exponent of -2.5 at the solar equator and -2.8 at the poles, as can be seen in Fig. 7.

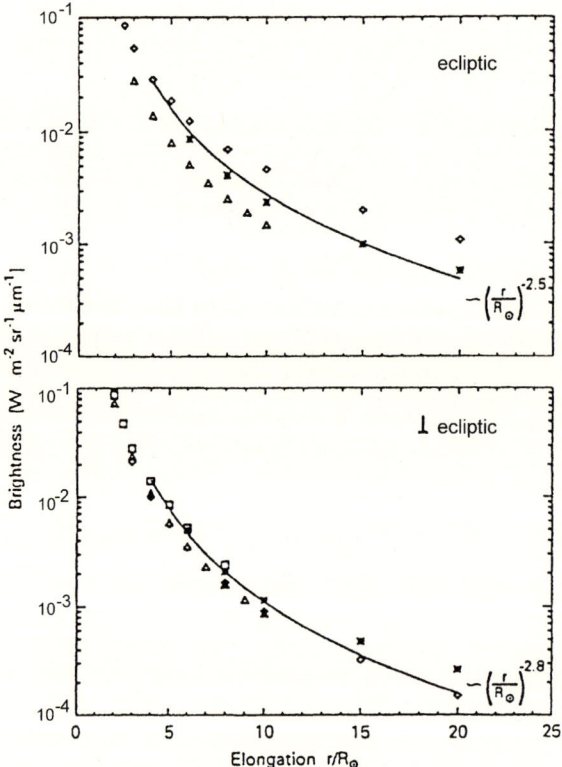

Figure 7. F-corona brightness at 0.55 μm along the ecliptic and along the polar meridian (Leinert et al. 1998). Asterix: Blackwell (1955, 1954 eclipse); diamonds, Michard (1954, 1954 eclipse); triangles, Duerst (1982, 1980 eclipse). The solid line corresponds to the model.

2. Color

Measured color differences in the F-corona are often below the measuring accuracy (Duerst 1982) and especially the comparison of different data sets is hampered by problems of absolute calibration. A detailed comparison of color measurements is given in Kimura et al. (1998), showing that a clear trend in the color for wavelength smaller than 1 μm cannot be derived, while most data at larger wavelength indicate the reddening of the F-coronal brightness. The color is expected to vary within the corona as a result of the superposition of light scattering and thermal emission effects (Mann 1993).

3. Polarization

Even if its brightness contribution decreases with distance from the Sun, the high degree of polarization in the K-corona causes a large uncertainty in the derived polarization of the F-corona brightness. When extrapolated from the zodiacal light, the polarization of the F-corona close to the equator amounts to (4.3 ± 1) % at 5 solar radii, (6.4 ± 1) % at 10 solar radii and (7.2 ± 1) % at 15 solar radii. The classical coronal model suggested in Blackwell et al. (1967) gives almost no F-corona polarization within 10 solar radii of the Sun.

It has been suggested that an irregular slope of the F-corona polarization could either result from the beginning of the dust free zone around the Sun or reflect the existence of a dust ring. Observations of the 1991 eclipse showed no hump in the polarization between 3 and 6.4 solar radii and gave an upper limit of 10 % for the polarization (Tanabe et al. 1992). Figure 8 shows the measured polarization of the total coronal brightness. It is obvious that the solar K-corona has a strong influence on the polarization data. At this point, a clear analysis of the F-corona polarization has to be based on future, more accurate measurements as well as a reliable K-corona separation.

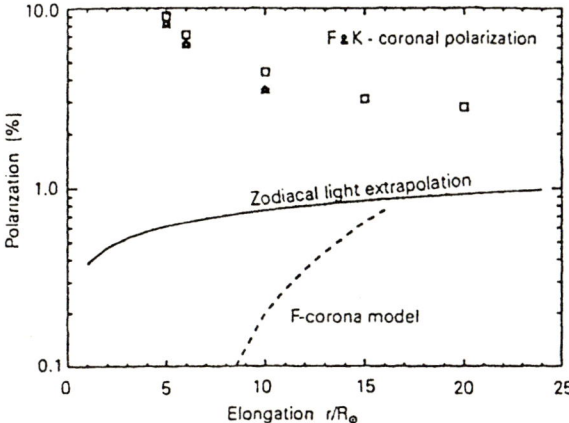

Figure 8. F-corona polarization, observations and models. The solid line gives the extrapolation from the zodiacal light polarization and the dashed line corresponds to the F-corona model (Leinert et al. 1998).

III. ZODIACAL AND F-CORONAL THERMAL EMISSION

III.A. Thermal Emission Measurements

1. Characteristics of the zodiacal thermal emission

Thermal emission from interplanetary dust is the most prominent component of the light of the night sky in the 5 to 100 μm region, at least away from the galactic plane. Also, the F-corona has an increasing contribution to the total coronal brightness in the infrared domain.

The thermal emission, which is isotropic, is less sensitive to the dust optical properties than brightness or polarization; it thus easily reveals fluctuations in dust concentration. Some of the most fascinating discoveries made by the IRAS sky survey have indeed been those of bright local enhancements in the emission, which have provided links between the cometary and asteroidal sources, and the interplanetary dust cloud. Compared to scattered light data, thermal emission data are easier to interpret since they do not depend upon the changing geometry of the phase angle along the line of sight. However, obtaining measurements and performing absolute calibrations is more difficult in the infrared than in the visible domain.

Measurements of the thermal emission in the infrared domain have been performed from balloons (e.g., Salama et al. 1987), rockets (e.g., Murdock and Price 1985) and from IRAS, COBE, IRTS and ISO satellites with helium-cooled sensors (e.g., Hauser et al. 1984; Hauser 1996; Okuda et al. 1997; Reach et al. 1996, Kelsall et al. 1998). The satellite observations have up to the present provided a smaller coverage than the rocket observations in solar elongations, being performed between 64° and 124°, and mostly at 90° from the Sun. Hanner (1991) gave a review of thermal emission measurements.

Observations of the zodiacal thermal emission at a wide range of elongations should soon be available from the Midcourse Space Experiment (MSX). MSX was flown in 1996 and made infrared measurements of a wide range of natural and made-made phenomenon in five infrared spectral bands spanning the wavelength range from 4 to 25 μm during the 10 month cryogenic phase of the mission. The data will be made generally accessible through IPAC under an agreement between NASA and the Ballistic Missile Defense Organization, the MSX sponsor (S.D. Price, personal communication).

2. Time dependence and symmetry plane

The location of the symmetry plane at 1 AU has been derived from thermal emission data based on different approaches (Dumont and Levasseur-Regourd 1987; Hauser 1988; Kwon and Hong 1998). A weighted mean value of $(77.1 \pm 0.4)°$ for the ascending node and $(1.71 \pm 0.02)°$ for the inclination was derived from fits at the four IRAS wavelengths (Reach 1988). A similar value of $(77.7 \pm 0.6)°$ for the node and $(2.03 \pm 0.02)°$ for the inclination was obtained from a fit to the multiwavelength COBE data (Kelsall et al. 1998). The results are consistent with the location derived from zodiacal light data (see Subsec. 3 of I.C.).

Once corrections are made for the eccentricity of the Earth's orbit and for the variation of the distance from the Earth to the plane (or surface) of symmetry, the thermal emission is found not to vary significantly with time, at least over a timescale of about nine months (Reach 1991). This is in agreement with the results previously obtained for the zodiacal light.

3. Spatial or temporal small amplitude enhancements

Long and narrow features in the thermal emission have been detected along arcs covering the perihelion part of the orbits of several short-period comets (e.g., Sykes et al. 1986; Sykes and Walker 1992). These trails, which provide the expected link between the comets and the zodiacal cloud (Whipple 1955), are likely to be made of dark and porous millimeter sized particles.

Infrared surveys have also revealed the thermal emission from dust bands at low ecliptic latitude; this dust has been attributed to major collisions between asteroids belonging to the Themis, Koronis and possibly Eos families (e.g., Dermott et al. 1984; Sykes 1990; Reach et al. 1997). Such a result has been the first clue to the fact that asteroids are also a significant source of dust to the zodiacal cloud. As already mentioned, the dust bands also induce faint local enhancements in the zodiacal brightness, which have been observed both from space (Spiesman et al. 1995) and from the Earth (Ishiguro et al. 1999).

The smoothness of the zodiacal light thermal emission has been studied through a mapping of five $0.5° \times 0.5°$ fields at low, medium and high latitude with the ISO photometer (Abraham et al. 1997). Although the results do not rule out the existence of spatial or temporal enhancements, it is of interest to notice that the fluctuations within the five chosen fields remain below ± 2 %.

A remarkable feature of the infrared zodiacal observations at 1 AU is that the sky seems slightly fainter when viewed in the direction of the Earth's orbital motion, the so-called the leading direction, than in the trailing direction (Reach 1991).

Dynamical calculations of the behavior of asteroidal particles, trapped in orbital resonance with the Earth as they spiral towards the Sun, show that one would expect an asymmetrical ring of such particles, with a relatively large density enhancement trailing the Earth and a smaller one leading the Earth (Dermott et al. 1994). IRAS and COBE observations indeed confirm the existence of an inhomogeneous ring of enhanced dust density just outside the Earth orbit (Reach et al. 1995; Dermott et al. 1996).

III.B. Zodiacal Thermal Emission from 1 AU

1. Infrared spectrum

The zodiacal spectrum in the infrared domain provides clues to the composition of the dust particles. A thermal emission spectrum has been obtained from ISO near the ecliptic plane at 90° elongation. In the 5 to 16.5 μm domain, it corresponds to a blackbody temperature of about 260 K along the line of sight, with a feature typical of the silicate emission band near 10 μm (Reach et al. 1996).

2. Thermal emission intensity

Although the infrared data are less complete than the visual data and their calibration is more difficult, the overall distribution appears to be the same as in the visual domain.

A useful empirical function from IRAS data, which provides an analytical value of the thermal emission intensity I at $\lambda - \lambda_o = (90 \pm 5)°$ for 12, 25, and 60 μm, has been derived by Vrtilek and Hauser (1995). The intensity is a function of the ecliptic latitude and of the day in the year.

As a first approximation, the emission near 90° solar elongation is given as a function of the ecliptic latitude β by

$$I(\beta) = I_o - \delta I \{1 - \delta\beta \,|\cosec(\beta - \beta_o)|\, [1 - \exp(-x - x^2/3)]\},$$

where $x = |(\beta - \beta_o)/\delta\beta|$.

The ecliptic latitude of the maximum emission, close to 0°, is equal to β_o. The intensity at the peak is equal to I_o. The intensity at minimum is equal to $[I_o - \delta I(1 - \delta\beta)]$. The values of ($I_o$ in MJy sr^{-1}, δI in MJy sr^{-1}, $\delta\beta$ in degrees) are of the order of (34, 31, 15), (73, 67, 14), and (29, 27, 12), respectively near 12, 25, and 60 μm.

In a second approximation, this expression is modified to take into account the location of the symmetry plane and of the Earth resonant circumsolar ring, that is to say the day of the year and the leading or trailing direction of the observations (Vrtilek and Hauser 1995).

III.C. Thermal Emission from the F-Corona

1. Near infrared intensity

Observations of the solar F-corona were made in the J, H and K bands, during the 1991 total solar eclipse (Hodapp et al. 1992; Lamy et al. 1992; Tollestrup et al. 1994). Scattered light from volcanic dust, cirrus clouds, and backscattering from clouds below the observing sites limited the observing conditions.

Detailed analysis of the data could nevertheless yield the slope of the F-corona light for regions inside 8 solar radii (Kuhn et al. 1994; MacQueen et al. 1994; Tollestrup et al. 1994). Observations by Hodapp et al. have been approximated with a $R^{-1.9}$ power law near the equator, and a $R^{-2.3}$ law near the pole, for regions inside 8 solar radii (MacQueen and Greeley 1995).

Figure 9 shows different values of the F-corona intensity at 4 solar radii in the near infrared than the solar spectral slope normalized to the F-corona at 0.55 μm (Peterson 1967; MacQueen 1968; Maihara et al. 1985; Mann 1990; MacQueen and Greeley 1995). The differences in the radial slope between the visible and infrared data can be explained by the increasing contribution of the thermal emission components.

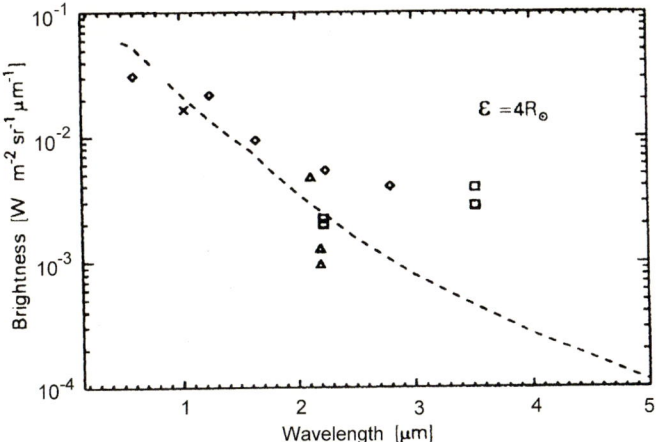

Figure 9. Wavelength dependence of the equatorial F-corona brightness at 4 solar radii (Leinert et al. 1998). The dashed line gives the solar spectrum normalized to 0.55 μm.

2. Observations at longer wavelengths

It was expected that the existence of small silicate particles near the Sun could be detected at the mid infrared spectral range of silicate features (Kaiser 1970). Model calculations, however, predict the possible enhancement of the brightness to be small and the brightness to be less sensitive to the spatial variation of dust than the near infrared corona (Mann 1992).

Observations near 10 μm have barely been made. An observation of the 1970 eclipse from a high altitude site gave a mean coronal brightness at 4 solar radii of $(9 \pm 5) \cdot 10^{-3}$ W m^{-2} sr^{-1} μm^{-1} in the 7.5 to 13 μm spectral interval (Mankin et al. 1974). However, the authors claim that they cannot achieve a clear separation from the atmospheric brightness components, since no measurements of the actual daytime mean sky radiance was made at the eclipse site.

Léna et al. (1974) found an almost constant brightness of approximately $5 \cdot 10^{-2}$ W m^2 sr^{-1} μm^{-1} between 3 and 6 solar radii. These data, which were from a Concorde flight during the June 1973 eclipse, might have been influenced by contamination from the aircraft window.

3. Coronal peaks and dust rings

Belton (1966) anticipated that interplay of the inward forcing Poynting-Robertson drag and the outward directed radiation pressure may lead to the accumulation of dust particles near their sublimation zone in the corona. Subsequent measurements of the near infrared coronal brightness revealed a deviation from a continuous increase within the corona, where the brightness was enhanced by a factor of about 3 to 3.5 (Peterson 1967; MacQueen 1968; Mizutani et al. 1984).

Model calculations by Mukai and Yamamoto (1979) have shown that these "humps" can be explained by the influence of dust rings around the Sun. It is also possible that a hump in the infrared brightness is produced when the line of sight crosses the beginning of the dust free zone (Mann 1992). Kimura et al. (1998) have shown that this geometric effect depends on the contribution of thermal emission to the brightness in comparison to scattered light, and hence may depend on the material composition of dust near the Sun. Most of the observers of the 1991 eclipse have claimed that their data show no clear evidence for the existence of a dust ring around the Sun (i.e., Hodapp et al. 1992; Lamy et al. 1992; Kuhn et al. 1994; Tollestrup et al. 1994; MacQueen and Greeley 1995).

The beginning of a dust free zone around the Sun also could not be observed in the data beyond 3 solar radii (Mann and MacQueen 1993). The observations are better described with a gradual change in the dust number density due to the sublimation of dust between about 9 and less than 3 solar radii from the Sun (Mann and MacQueen 1993). In this context, the presently available data do not allow a study of temporal effects in the F-coronal brightness, such as the appearance of dust clouds from Sun-grazing comets or temporal dust rings.

IV. LOCAL SCATTERING AND THERMAL PROPERTIES

IV.A. Need for Inversion

The results discussed above, although of major interest to estimate the zodiacal foreground contamination, do not provide immediate information about the local properties of the dust: the values of $Z_\perp$, $Z_{//}$, Z, P, I, are integrated along a line of sight which extends from the observer to the outer fringe of the zodiacal cloud. The situation is especially critical for the scattering properties, since the phase angle α (or the supplementary scattering angle $\theta = \pi - \alpha$) varies along the line of sight, except in the antisolar direction where α remains equal to $0°$.

Fitting an observation with a model in order to infer local physical properties of the dust does not give a unique solution. Various inversion techniques have therefore been developed for the available observations. They are aimed at providing the local dust number density $n(M)$, which depends upon the location M of the scattering or emitting volume, the local dust scattering function $\sigma_\lambda(M, \theta)$ at wavelength λ, which depends upon M and upon the scattering angle θ and the local dust emissivity $E_\lambda(M, T)$ at wavelength λ, which depends upon M and upon the local dust temperature T. Optical and thermal properties are then derived from the values of the scattering function, and of the spectral emissivity.

As a first approximation, the small-scale spatial or temporal inhomogeneities mentioned above are disregarded and it is assumed that the dust cloud possesses a cylindrical symmetry. Under these assumptions, the brightness and emissivity integrals for observations restricted to an observer in the symmetry plane at solar distance R, and a line of sight at elongation ε in the symmetry plane can be written as:

$$Z_\lambda(\varepsilon, R) = F(\lambda) \int_0^\infty [n(r)\sigma_\lambda(r,\theta)/r^2]\, dl$$

$$I_\lambda(\varepsilon, R) = \int_0^\infty n(r) E_\lambda(r, T)\, dl,$$

where $F(\lambda)$ is the solar flux at 1 AU, and r is the solar distance of the elementary volume.

IV.B. Volume Scattering and Emitting Functions

1. Volume scattering function

The concept of the volume scattering function, VSF, may be used to describe the average differential cross section of particles per unit volume, including the spatial variation of the particle number density (see for instance Dumont 1973; Hong 1985; Giese et al. 1986; Dumont and Levasseur-Regourd 1985; Perrin and Lamy 1989). The VSF is given as a function of the solar distance r of the elementary volume, the scattering angle θ, and the wavelength λ of the scattered light.

The brightness integral can then be written by changing the integration over l to an integration over θ, from ε to π, with:

$$R \sin \varepsilon = r \sin \theta$$

$$Z_\lambda(\varepsilon, R) = F(\lambda)/(R \sin \varepsilon) \int_\varepsilon^\pi \text{VSF}(r, \theta, \lambda)\, d\theta$$

The volume scattering function gives the average differential scattering cross section of particles per volume element. It includes the spatial variation of the local number density n, as well as the spatial and spectral variations of the scattering properties. The VSF is a result of integration over the size distribution of dust particles. However, this integral is implicitly replaced by the product $(\langle n' \rangle \langle G \rangle A)$ when average properties are derived, where n' is the number density of the optical efficient particles, G is the geometric cross section of one particle, and A is the single-scattering albedo.

2. Volume emitting function

In the case of the thermal emission,

$$I_\lambda(\varepsilon, R) = \int_0^\infty [\text{VEF}(r, \lambda) B(\langle T(r) \rangle, \lambda)/r^2]\, dl,$$

where $B(\langle T(r) \rangle, \lambda)$ is the Planck function for the dust particles at temperature T and wavelength λ. Both the VEF and T, which are averaged for the ensemble of particles within the elementary volume, can be written in terms of global properties. To make it consistent with the brightness of the scattered light and the concept of a graybody:

$$\text{VEF}(r, \lambda) = \langle n' \rangle (r) \langle G \rangle (1 - A(r)),$$

where G is the geometric cross section, and A is the geometric albedo, as defined below. It may be noted that the variation of the emissivity with the solar distance has a much smaller influence on the final thermal brightness than the albedo has on the scattered brightness (Mann 1992).

3. Albedo

The geometric albedo is defined at backscattering conditions, i.e. for $\theta = 180°$ (or $\alpha = 0°$). Such a definition may be extended to other scattering (or phase) angles as the ratio of the energy scattered by the particles at a given angle for solid angle $d\Omega$ to that scattered by a white Lambert disk of the same geometric cross section (Hanner et al. 1981), with:

$$A(\theta) = \pi d\sigma(\theta)/Gd\Omega$$

A constant size distribution with solar distance r keeps the average geometric cross section constant. The change in the average geometric albedo of the particles can then be derived, when assuming a certain radial slope of the number densities.

4. Power law variation of the volume scattering and emitting functions

Assuming a power law variation of the VSF with solar distance, such as:

$$\text{VSF}(r, \theta, \lambda) = \text{VSF}(R_o, \theta, \lambda)(r/R_o)^{-\nu},$$

where R_o equals 1 AU, yields to the brightness integral:

$$Z_\lambda(\varepsilon, R_o = 1\text{AU}) = F_o(\lambda) R_o (\sin \varepsilon)^{\nu+1} \int_\varepsilon^\pi (\sin \theta)^\nu \text{VSF}(r, \theta, \lambda) d\theta.$$

Similarly, for the thermal emission:

$$I_\lambda(\varepsilon, R_o) = R_o n_o (\sin\varepsilon)^{1-\nu} \int_\varepsilon^\pi \langle C \rangle B(\lambda, \langle T \rangle)(\sin \theta)^{\nu-2} d\theta,$$

where $\langle C \rangle$ is the average emission cross section, i.e., the product of the emissivity by the average geometric cross section. Finally, for the number density and for the average geometric albedo:

$$n'(r) = n'(R_o)(r/R_o)^{-\nu'}$$

$$A(r) = A(R_o)(r/R_o)^{-\nu''}$$

$$\text{with } (\nu' + \nu'') = \nu.$$

Such a concept has the benefit that it allows a relatively easy comparison between the scattering functions derived from observations and scattering functions obtained either through model calculations or from laboratory experiments.

IV.C. Inversion with Homogeneity Assumption

If the dust cloud is homogeneous and the dust has the same optical properties everywhere, and if the local dust density is assumed to obey a $r^{-\nu}$ power law in the symmetry plane, then the integral does not depend on the solar distance anymore:

$$Z_\lambda(\varepsilon, R) = C_\lambda(\varepsilon) r^{-\nu-1}.$$

From such an approach, again assuming that the properties of the dust particles do not depend upon their location in the zodiacal cloud, the dust density can be deduced. It is then found to vary approximately as $r^{-1.25\pm0.05}$ (e.g., Dumont 1973).

With the same homogeneity assumption, all-sky inversions have been attempted and successfully carried out by Lamy and Perrin (1986, 1991). From a comparison of Helios observations at 1 AU and 0.3 AU, they conclude that: i) The density distribution rather follows a law such as r^{-1} in the symmetry plane. ii) The shape of the brightness phase function presents a shallow minimum in the 60° to 100° phase angle range, while the shape of the polarization phase function is smooth, with a broad maximum in the 80° to 110° region and has two negative branches, in the backscattering domain and in the forward direction. iii) The maximum in polarization is greater at 1 AU than at 0.3 AU.

Assuming that the dust has the same optical properties everywhere is indeed questionable. When measurements are performed at constant helioecliptic coordinates, the line-of sight degree of linear polarization (which does not require any normalization to a constant distance to the Sun or to the observer to allow comparisons) increases approximately as $r^{+0.3}$ when the distance of the observing probe to the Sun increases (Leinert et al. 1982). The homogeneity assumption should thus be avoided.

IV.D. Local Rigorous Inversion

A rigorous inversion is only feasible, as was demonstrated by Dumont (1973), for a line of sight tangent to the direction of motion of the Earth (or of a moving probe), and for the section of the line of sight where the observer is located. This inversion provides the bulk values of the local properties in the vicinity of the Earth orbit. Typical local values of brightness, polarization, temperature, albedo and space density at 1 AU are presented in Table 4. The very low space density corresponds to 5 to 20 particles of 10 μm size per cubic kilometer, depending on their density (compact grains or fluffy aggregates).

TABLE 4
Variation of the local properties of the dust in the symmetry plane. The properties are described as a function of a function of solar distance with a power law assumption. The optical properties (polarization, albedo and brightness in W m^{-2} sr^{-1}μm^{-1} rad^{-1} at 0.55 μm) are retrieved at $\alpha = 90°$

	Value at 1 AU	Gradient	Domain (AU)
Brightness	$\simeq 23\,10^{-7}$	-1.25 ± 0.02	0.5 to 1.4
Polarization	0.30 ± 0.03	$+0.5 \pm 0.1$	0.5 to 1.4
Temperature	$250\,K \pm 10\,K$	-0.36 ± 0.03	1.1 to 1.4
Albedo	0.07 ± 0.03	-0.34 ± 0.05	1.1 to 1.4
Space density	10^{-19} kg m^{-3}	-0.93 ± 0.07	1.1 to 1.4

The nodes of lesser uncertainty method has been developed to retrieve local information in regions that are not located near the orbit of the Earth without assuming that the dust particles have identical optical properties everywhere in the interplanetary dust cloud (Dumont and Levasseur-Regourd 1985). This method only assumes an axial symmetry towards the ecliptic pole (or more precisely towards the pole of the symmetry surface), a steady state of the cloud (most likely over the period covered by the observations), and a rather monotonous variation, described by mathematical functions, of the local brightness along the line of sight (fair enough once second order variations are dismissed).

IV.E. Local Inversion Through Mathematical Methods

In the symmetry plane, local contributions are retrieved with less uncertainty than elsewhere at the so-called radial and Martian nodes by using two successive locations of the observer on the same line of sight, with solar elongations ε and $\pi - \varepsilon$ respectively (Dumont and Levasseur-Regourd 1985, 1988). The first one provides, at a constant phase angle of 90°, the brightness and polarization at:

$$r = \sin \varepsilon, \text{where } 0.1 < r < 1.5\,\text{AU}.$$

The second node provides, at a constant solar distance of 1.5 AU, these quantities at:

$$\alpha = \sin^{-1}[(\sin \varepsilon)/r], \text{where } 0° < \alpha < 90°.$$

Towards the ecliptic pole, local contributions are retrieved with less uncertainty than elsewhere at the so-called polar node (Levasseur-Regourd et al. 1991; Renard et al. 1995). By using the integrated signal towards the pole and the local value at 1 AU, 90°, a local value is retrieved at:

$$r = 1/\sin \alpha, \text{where } 68° < \alpha < 78°.$$

Local values are also obtained towards the tangential plane, tangent to the terrestrial orbit and perpendicular to the symmetry plane. Altogether, local values are retrieved for elevations in the 0 to 0.4 AU range. By writing the brightness integral in the form of a Volterra integral equation of the first kind, a new method, which corresponds to the search for an analytical kernel of the integral, has recently been developed (Lumme 2000). Although very promising, it requires that observations are performed at different solar distances, as has presently only been done at 0.3 AU and 1 AU from Helios.

1. Solar distance dependence:
a key to a temporal evolution

The results for the solar distance dependence of local brightness, polarization, temperature, albedo and space density with solar distance, obtained through the nodes of lesser uncertainty method at the radial node, are summarized in Table 4 and illustrated in Fig. 10. The local values that are derived agree fairly well, to the first order, with solar distance power laws. Although the uncertainties in the power law exponents (due to the observational errors and to the inversion methods) are far from negligible, it is established that the optical properties of the dust are not the same everywhere and the following conclusions can be reached.

The brightness at 90° phase angle decreases, as expected, with increasing solar distance (Levasseur-Regourd et al. 1991), with a (-1.25 ± 0.02) gradient. A more detailed analysis shows that a slight deviation from a power law is detected for the local brightness in the vicinity of the Earth's orbit, corresponding to an increase in brightness of about 10 $S_{10}(V)$ (Renard et al. 1996). This effect, compatible with the detection of enhancements in the Gegenschein brightness (Maucherat et al. 1986), is likely to be due to the resonant dust ring along the Earth orbit.

The polarization at 90° phase angle increases with increasing solar distance (Levasseur-Regourd et al. 1990), with a (+0.5 ± 0.1) gradient. From the analytical kernel method (Lumme 2000), the same evolution is noticed between 0.3 AU and 1 AU, for phase angles greater than 20°. From the nodes of lesser uncertainty method, it is suggested that drastic changes take place below 0.3 AU, in agreement with the trend obtained by Mann (1996) in the F-corona.

The local temperature with solar distance dependence is found to be less steep, about $r^{-0.35}$ (Dumont and Levasseur-Regourd 1988), than expected from a gray-body case, which would lead to T proportional to r^{-1}. A more precise value of the gradient, of (-0.36 ± 0.03), has been derived by Renard et al. (1995), and found to be in agreement with the results obtained by Reach (1991).

The local albedo, as defined at 90°, decreases with increasing solar distance. A gradient of (-0.34 ± 0.05) is derived from IRAS data (Renard et al. 1995). The trend is compatible with the results obtained in the F-corona (Mann 1996). This could be a clue to the evaporation of dark organic compounds and/or the fragmentation of fluffy aggregates, as the dust particles spiral towards the Sun under Poynting-Robertson drag (Levasseur-Regourd et al. 1991).

The gradient in the local density can be deduced from the previous results. It is found to equal (-0.93 ± 0.07), and thus agrees with a r^{-1} law (Levasseur-

Regourd et al. 1991). Such a gradient is expected for dust particles in circular orbits under Poynting-Robertson drag inside the production region.

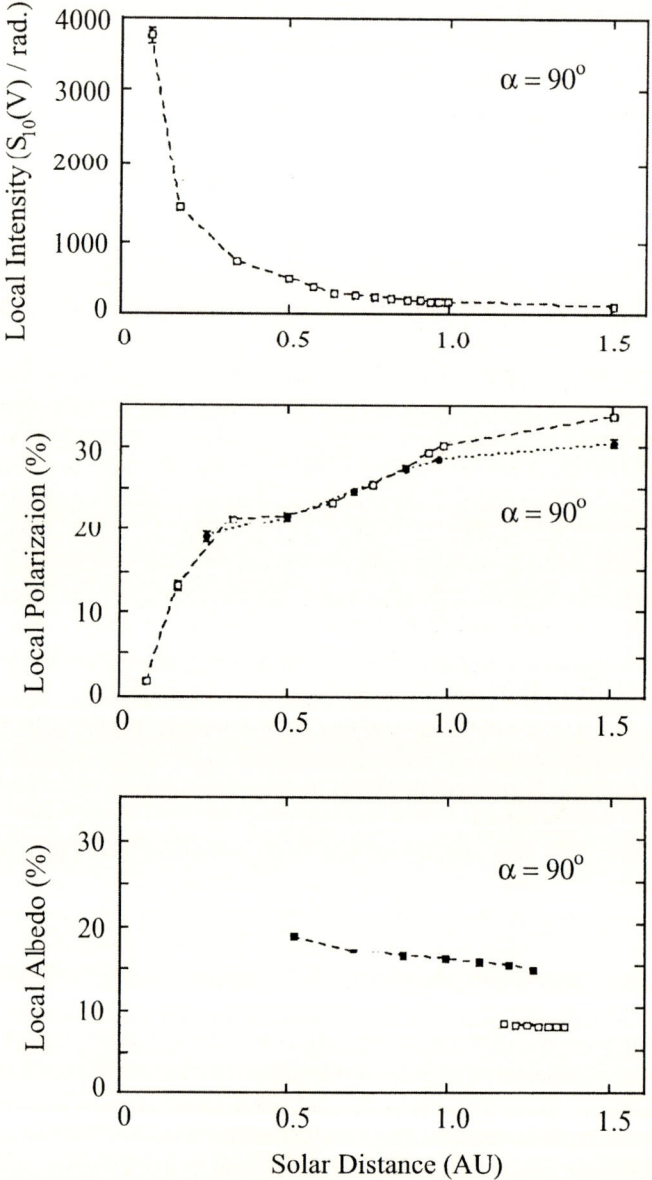

Figure 10. Evolution of the optical properties of the dust in the symmetry plane with solar distance. Upper panel: local brightness. Middle panel: local polarization (open symbols: Fechtig et al.; solid symbols, Dumont and Levasseur-Regourd). Lower panel: local albedo (open symbols: IRAS, solid symbols: Murdock and Price).

2. *Phase angle dependence:*
a key to the structure of the dust particles

Figure 11 presents the results obtained at the Martian node for the phase dependence of the local brightness, local polarization, and local albedo at 1.5 AU from the Sun in the symmetry plane. The overall shape of the phase curves is smooth; in the backscattering region, the brightness presents a moderate enhancement and the polarization is slightly negative (Levasseur-Regourd et al. 1990; Levasseur-Regourd 1996).

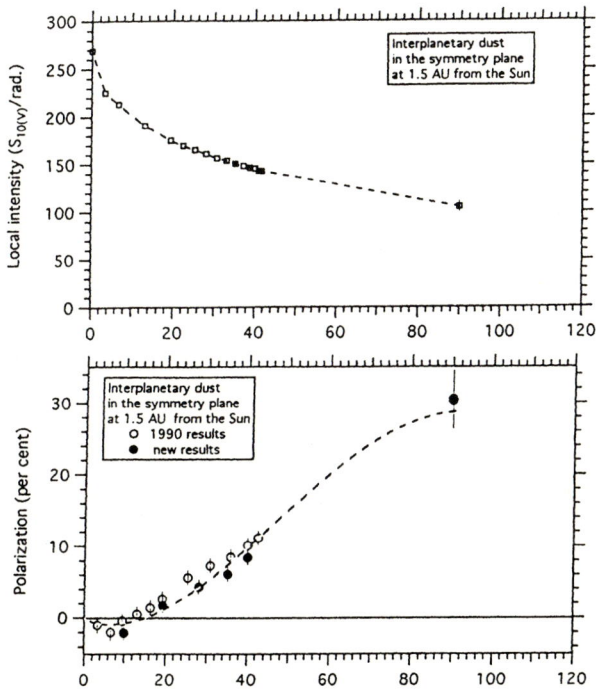

Figure 11. Phase functions of zodiacal light (from Levasseur-Regourd 1996). Upper curve: local brightness. Lower curve: local polarization.

Such phase curves strongly differ from those of Mie spherical particles. Although they can be fitted by arbitrary mixtures of two or three families of spherical particles, such shapes are more likely to be due to scattering by irregular dust particles and/or fluffy aggregates of numerous submicron absorbing particles (Levasseur-Regourd et al. 1997; Lumme et al. 1997). By comparing the polarization phase curve, which is more sensitive than the brightness to changes in the physical properties, to that for other bodies, it may be noticed that the inversion angle at $(15 \pm 5)°$ phase angle, is slightly smaller than for other dust clouds and atmosphereless bodies in the solar system. The slope at inversion, equal to (0.2 ± 0.1) % per degree, is comparable to that of cometary dust (Levasseur-Regourd et al. 1996) or of C type asteroids (Goidet-Devel et al. 1995). Similar polarization phase curves at 0.3 AU and 1 AU are retrieved from the analytical kernel method (Lumme et al. 2000).

The characteristics of such phase curves, which can be approximated using trigonometric functions proportional to:

$$(\sin \alpha)^a (\cos \alpha/2)^b \sin(\alpha - \alpha_o),$$

seem to provide clues to the physical properties of the scattering dust (e.g., albedo, shape, porosity, size and size distribution). The difference between the curves obtained for the interplanetary dust cloud and for the cometary dust (e.g., wavelength dependence, inversion angle value) are tentatively explained through laboratory experiments and computational approaches (Levasseur-Regourd et al. 1999). As an example, it has been found that non-spherical silicate particles of a size comparable to the wavelength, and aggregates of such particles produce negative polarization in the backscattering region (Xing and Hanner 1996). Also large aggregates of small absorbing particles of fractal dimension ≈ 2 present a slightly negative polarization at small phase angles (Levasseur-Regourd et al. 1997; Haudebourg et al. 1999).

The slope at inversion of the polarization phase curve is inversely related to the geometric albedo (e.g., KenKnight et al. 1967). This value thus leads to a geometric albedo in the 0.05 to 0.09 range. With the above mentioned radial dependence, it suggests a geometric albedo at 1 AU of the order of 0.08. On the other hand, the extrapolation of the albedo obtained by rigorous inversion at 90° to 0° phase angle indicates a greater value. A value of (0.15 ± 0.08) has been proposed for the geometric albedo at 1 AU taking into account the large uncertainties in the observational data in the power laws and in the polarization-albedo relationship (Levasseur-Regourd 1998). Such a value agrees with the models of dark absorbing dust particles and/or fluffy aggregates. It also agrees with the classical assumption that both asteroidal dust of higher albedo, and cometary dust of lower albedo, are the most important sources of interplanetary dust between the orbits of Mars and the Earth.

3. Inclination dependence:
a key to the existence of different populations

Maps of the local properties are obtained in the tangential plane, once the values have been tentatively normalized for $\alpha = 90°$ and r = 1 AU. The maps reveal a variation in the particles' properties with the inclination of their orbits, and can be interpreted in terms of two populations of dust particles (Levasseur-Regourd et al. 1991; Renard et al. 1995).

The comparison of different models of the zodiacal cloud had indeed led to the suggestion of a two components dust cloud (Kneissel and Mann 1991), consisting of asteroidal particles in regions near the ecliptic and particles from long period comets in an isotropic distribution, with different optical properties assumed for the two components, and alteration assumed for the cometary component. In a more recent attempt, the difference between the two components is explained by a different type of radial distribution rather than by different particles properties (Mann 1995).

Among the various three-dimensional models of the dust distribution discussed by Giese et al. (1986), we favor those of the "sombrero" type, with the existence of (at least) two dust populations. The first one is the flattened, strictly speaking zodiacal cloud, with a solar distance dependence of about r^{-1} and a latitudinal dependence of about $(\cos \beta_o)$; the second is a spherically symmetrical interplanetary cloud centered on the Sun, whose brightness seems to exceed that of the flattened cloud at elongations below 60° in the helio-ecliptic meridian, and whose solar distance gradient could be steeper, of about r^{-2} (Dumont et al. 1998). The mixing ratio of the particles belonging to the two dust clouds thus changes with the location in the cloud. Periodic comets and dust produced by asteroidal collisions could replenish the flattened cloud; the spherical cloud might be replenished by new non-periodic comets, including comets vaporized in the solar vicinity.

It may be added that a geometrical model has been proposed for the spatial density and for the temperature distribution in the dust cloud in order to reproduce the temperature distribution measured by ISO (Abraham et al. 1999). In the ecliptic plane, the trends for the density distribution, $r^{-0.93}$, and for the temperature distribution, $r^{-0.36}$, are identical to those which have been derived from the nodes of lesser uncertainty method (Levasseur-Regourd et al. 1991; Renard et al. 1995). Out of the ecliptic plane, it is proposed that the density follows an $\exp(-4.8|\sin \beta_o|^{1.3})$ law (Abraham et al. 1999).

IV.F. Models of the Near Infrared F-Corona

Recent observations of the F-corona in the near infrared have stimulated new model calculations. Models that include only scattered light components (MacQueen and Greeley 1995) have to increase the absolute number density of dust near the Sun (compared to the values extrapolated from the zodiacal cloud models) in order to match the observed brightness. Other models allow for a component of thermal emission from the dust (Mann and MacQueen 1993). The total brightness can be readily matched by the latter calculations.

The values, although consistent with the flux model given by Grün et al. (1985) and with common assumptions of the interplanetary dust albedo (Hanner 1980; Levasseur-Regourd et al. 1991), cannot give a detailed proof for albedos and flux models since the description of the VSF bears ambiguities. Further model calculations of the scattered light component (Davidson et al. 1995; MacQueen et al. 1996) have shown that a clear distinction of different size distribution is not possible on basis of the F-coronal data that are presently available. A near solar dust free zone needs to be considered in order to reproduce the observed slope of the brightness (cf. Mann and MacQueen 1993).

The heating and subsequent sublimation builds up an effective dust-free zone around the Sun. The nature and extent of this zone is determined by the dust material composition. As presented in Table 5, a model calculation of the heating and sublimation shows that porous silicate particles, which contain varying amounts of absorbing material could explain the gradual decrease of the dust number density in the solar corona (Mann et al. 1994), and that dynamical effects may also play a role. The steep decrease in the number density at the beginning of the dust free zone produces a signature in the thermal emission brightness (Mann 1992). Such a signature is not observed beyond 3 solar radii. Recent observations indicate a rather gradual beginning of the dust free zone, described in the slope of the VSF (Mann and MacQueen 1993). It is assumed to run with $r^{-1.25}$ from the outer regions to 9 solar radii, and with $r^{-0.25}$ more inward up to 3 solar radii. The increase may not only be due to an increasing number density, but may also result from an increasing albedo since the VSF depends upon the number density and the albedo.

At the time of writing, without clear identification of a signature of a dust free zone, we must rely upon modeling of coronal scattering and emission. Observations of the color of the F-corona over a wide range of wavelengths, encompassing the infrared, are required to help distinguish between the various possibilities.

TABLE 5
Zone of sublimation for 10 μm sized compounds of particles, with different amounts of absorbing graphite in silicate particles

Volume ratio of graphite in the compound	Solar distance of sublimation for graphite	
	Porous	Compact
0 % (pure silicate)	2 - 3 R_o	3 R_o
0.1 % graphite	4 R_o	5 R_o
1 % graphite	9 R_o	6 R_o

V. CONCLUSIONS AND PERSPECTIVES

At present, we have reliable measurements of the zodiacal light brightness and polarization in the ecliptic plane close to 1 AU. Surveys of the brightness and polarization from space probes, inward to 0.3 AU and outward to 3 AU, also exist. In the infrared domain, the thermal emission from 60° to 120° elongation has been mapped from Earth orbit, and some data at smaller elongation have been obtained from rocket flights. The limits of our understanding are due to the lack of data in the inner and outer solar system dust cloud, and out of the ecliptic plane.

As far as the dust near the Sun is concerned, observations from space using coronagraphs show the transition from the solar F-corona to the zodiacal light. However, interpretation is difficult, since the data are integrated along the line of sight, and should be inverted to provide local information.

The observed zodiacal light and zodiacal light emission are also integrated over the line of sight; the scattering or emission by a localized volume element can fortunately be retrieved under certain geometric conditions. These retrievals indicate that the average properties are not the same in all regions of the zodiacal cloud. The average albedo increases with decreasing solar distance, while the polarization decreases with decreasing solar distance. Such changes could indicate an alteration in the physical properties of the grains over time as they spiral towards the Sun and/or a change in the size distribution of the particles, due to collisions. The relative contribution from different populations with different size distributions and optical properties may also vary with solar distance. The deconvolution of particles properties and particles number density distribution would certainly be easier when based on a larger data set, namely covering a wider spectral and spatial range of observations, as well as further knowledge of dust dynamics and size distributions from impact experiments. The open question about the different cometary and asteroidal sources indicate that the dynamics and the collisional evolution may be more complex than simple models would predict.

Another goal of future observations is the study of the outer solar system dust cloud, which may consist of fragments from long period comets, particles from interstellar space, or even fragments from Kuiper belt objects. This leads to a variety of questions which should be addressed in the future study of the interplanetary dust cloud, and which are related to the evolution of circumstellar dust debris shells such as the cloud around Beta Pictoris. In this context, the circumsolar dust cloud can be used as an example to study questions like the link between dust clouds and their parent bodies, their collisional evolution, and the influence of planetary perturbations.

REFERENCES

Abraham, P., Leinert, C., and Lemke, D. 1997. Search for brightness fluctuations in the zodiacal light at 25 µm with ISO. *Astron. Astrophys.* **328**, pp. 702–05.

Abraham, P., Leinert, C., Acosta-Pulido, J., Schmidtobreick, L., and Lemke, D. 1999, Zodiacal light observations with ISOPHOT. In *The Universe as seen by ISO*, ESA SP-427, pp. 145–48.

Arago, F. 1858. Les comètes. (Paris: Gide).

Baggaley, W. J. 1977. The meteoric nightglow. *Mon. Not. R. Astron. Soc.* **181**, pp. 203–210.

Berriman, G. B., Boggess, N. W., Hauser, M. G., Kelsall, T., Lisse, C. M., Moseley, S. H., Reach, W. T., and Silverberg, R. F. 1994. COBE DIRBE near-infrared polarimetry of the zodiacal light: initial results. *Ap. J.* **431**, pp. L63–L66.

Blackwell, D. E., and Ingham, M. F. 1961. Observations of the zodiacal light from a very high altitude station. *Mon. Not. R. Astron. Soc.* **122**, pp. 129–141.

Blackwell, D. E., Ingham, M. F., and Petford, A. D. 1967. The distribution of dust in interplanetary space. *Mon. Not. Royal Astron. Soc.* **136**, pp. 313–328.

Brorsen, T. 1854. Über eine neue Erscheinung am Zodiacallichtes, Geographie und Witterungskunde. **8**, pp. 156–160.

Brückner, G., Howard, R. A., Koomen, M. J., Corendyke, C. M., Michels, D. J., Moses, J. D., Sockert, D. G., Dere, K. P., Lamy, P. L., Llebaria, A., Bout, M. V., Schwenn, R., Simnet, G. M., Bedford, D. K., and Eyles, C. J. 1995. The large angle spectroscopic coronagraph (LASCO). *Solar. Phys.* **162**, pp. 357–402.

Cassini, J. D. 1683. Découverte de la lumière qui paroist dans le zodiaque. (Paris: Académie royale des sciences).

Cebula, R. P., and Feldman, P. D. 1982. Ultraviolet spectroscopy of the zodiacal light. *Ap. J.* **263**, pp. 987–992.

Childrey, J. 1661. Britannia Baconica or the natural rarities of England, Scotland and Whales. (London).

Cooper, B. L., Zook, H. A., and Potter, A. E. 1996. Clementine photographs of the inner zodiacal light. In *Physics, Chemistry and Dynamics of Interplanetary Dust*, eds. B. Å. S. Gustafson and M. S. Hanner (San Francisco: Astron. Soc. of the Pacific Press), pp. 333–336.

Corneille, P. 1637. Le Cid. (Paris).

Davidson, W. C., MacQueen, R. M., and Mann, I. 1995. Scattering models for the solar infrared F-corona brightness. *Planet. Space Sci.* **43**, pp. 1395–1400.

Dermott, S., Nicholson, P., Burns, J., and Houck, J. 1984. Origin of the solar system dust bands discovered by IRAS. *Nature* **312**, pp. 505—509.

Dermott, S. F., Jayaraman, S., Xu, Y. L., Gustafson, B. Å. S., and Liou, J. C. 1994. Circumsolar ring of asteroidal dust in resonant lock with the Earth. *Nature* **369**, pp. 719–723.

Dermott, S. F., Jayaraman, S., Xu, Y. L., Grogan, K., and Gustafson, B. Å. S. 1996. The origin and dynamics of interplanetary dust cloud. In *Unveiling the cosmic infrared background*, ed. E. Dwek (AIP 348), pp. 25–36.

Dortous de Mairan, J. J. 1733. Traité physique et historique de l'aurore boréale. (Paris: Académie royale des sciences).

Duerst, J. 1982. Two colour photopolarimetry of the solar corona of 16 February 1980. *Astron. Astrophys.* **112**, pp. 241–250.

Dufay, J. 1925. La polarisation de la lumière zodiacale. *C. R. Acad. Sc. Paris* **181**, pp. 399–401.

Dumont, R. 1965. Séparation des composantes atmosphérique, interplanétaire et stellaire du ciel nocturne à 5000 Å. Application à la photométrie de la lumière zodiacale et du gegenschein. *Ann. Astrophys.* **28**, pp. 265–320.

Dumont, R. 1973. Phase function and polarization curve of interplanetary scatterers from zodiacal light photopolarimetry. *Planet. Space Sci.* **21**, pp. 2149–2155.

Dumont, R. 1992. Zodiacal light and gegenschein. In The astronomy and astrophysics encyclopedia, ed. S. P. Maran (New York: Van Nostrand Reinhold), pp. 969–971.

Dumont, R., and Sanchez-Martinez, F. 1973. Photométrie de la lumière zodiacale hors de l'écliptique en quadrature et en opposition avec le Soleil. *Astron. Astrophys.* **22**, pp. 321–328.

Dumont, R., and Sanchez, F. 1975a. Zodiacal light photopolarimetry. I. Observations, reductions, disturbing phenomena, accuracy. *Astron. Astrophys.* **38**, pp. 397–403.

Dumont, R., and Sanchez, F. 1975b. Zodiacal light photopolarimetry. II. Gradients along the ecliptic and the phase functions of interplanetary Matter. *Astron. Astrophys.* **38**, pp. 405–412.

Dumont, R., and Levasseur-Regourd, A. C. 1978. Zodiacal light photopolarimetry IV, annual variations of brightness and the symmetry plane of the zodiacal cloud. *Astron. Astrophys.* **64**, pp. 9–16.

Dumont, R., and Levasseur-Regourd, A. C. 1981. Zodiacal light and space observation of faint objects. *Adv. Space. Res.* **1**, 8, pp. 127–130.

Dumont, R., and Levasseur-Regourd, A. C. 1985. Zodiacal light gathered along the line of sight, retrieval of the local scattering coefficient. *Planet. Space Sci.* **33**, pp. 1–9.

Dumont, R., and Levasseur-Regourd, A. C. 1987. The symmetry plane of the zodiacal cloud retrieved from IRAS data. In ERAM 10 (Astro. Inst. Czech. Acad. Sci. 67) **2**, pp. 281–284.

Dumont, R., and Levasseur-Regourd, A. C. 1988. Properties of interplanetary dust from infrared and optical observations. I. Temperature, global volume intensity, albedo and their heliocentric gradients. *Astron. Astrophys.* **191**, pp. 154–160.

Dumont, R, Renard, J. B., Levasseur-Regourd, A. C., and Weinberg, J. L. 1998. Disentangling the main populations of the zodiacal cloud from zodiacal light observations. *Earth Planets Space* **50**, pp. 473–476.

East, I. R., and Reay, N. K. 1985. The motion of interplanetary dust particles. In *Properties and interaction of interplanetary dust*, eds. R. H. Giese and P. Lamy (Dordrecht: D. Reidel), pp. 81–84.

Elsässer, H., and Fechtig, H. (eds.) 1976. Interplanetary dust and zodiacal light. Lectures notes in physics 48 (Berlin: Springer-Verlag).

Fechtig, H., Leinert, C., and Grün, E. 1981. Interplanetary dust and zodiacal light. In *Landolt-Börnstein New Series*, VI/2A, pp. 228–243.

Fried, J. W. 1978. Doppler shifts in the zodiacal light spectrum. *Astron. Astrophys.* **68**, pp. 259–264.

Giese, R. H., and Lamy, P. (eds.) 1985. Properties and interactions of interplanetary dust. (Dordrecht: D. Reidel).

Giese, R. H., Kneissel, B., and Rittich, U. 1986. Three dimensional models of the zodiacal dust cloud: A comparative study. *Icarus* **68**, pp. 395–411.

Goidet-Devel, B, Renard, J. B., and Levasseur-Regourd, A. C. 1995. Polarization of asteroids: synthetic curves and characteristic Parameters. *Planet. Space Sci.* **43**, 6, pp. 779–786.

Grün E., Zook, H. A., Fechtig, H., and Giese, R. H. 1985. Collisional balance of the meteoritic complex. *Icarus* **62**, pp. 244–272.

Gustafson, B. Å. S., and Hanner, M. S. (eds.) 1996. Physics, chemistry, and dynamics of interplanetary dust. (San Francisco: Astron. Soc. of the Pacific Press).

Halliday, I., and McIntosh, B. A. (eds.) 1980. Solid particles in the solar system. (Dordrecht: D. Reidel).

Hanner, M. S. 1980. On the albedo of the interplanetary dust. *Icarus* **43**, pp. 373–380.

Hanner, M. S. 1991. The infrared zodiacal light. In *Origin and evolution of interplanetary dust*, eds. A. C. Levasseur-Regourd and H. Hasegawa (Dordrecht: Kluwer), pp. 171–178.

Hanner, M. S., Sparrow, J. G., Weinberg, J. L., and Beeson, D. E. 1976. Pioneer 10 observations of zodiacal light brightness near the ecliptic: changes with heliocentric distance. In *Interplanetary dust and zodiacal Light. Lecture Notes in Physics 48*, eds. H. Elsässer and H. Fechtig (Berlin: Springer-Verlag), pp. 29–35.

Hanner, M. S., Giese, R. H., Weiss, K., and Zerull, R. H. 1981. On the definition of albedo and application to irregular particles. *Astron. Astrophys.* **104**, pp. 42–46.

Hapke, B. 1993. Theory of reflectance and emittance spectroscopy. (Cambridge: Cambridge University Press).

Haudebourg, V., Cabane, M., and Levasseur-Regourd, A. C. 1999. Theoretical polarimetric responses of fractal aggregates, in relation with experimental studies of dust in the solar system. *Phys. Chem. Earth* **24-5**, pp. 603–608.

Hauser, M. G. 1988. Models for infrared emission from zodiacal dust. In *Comets to cosmology, Lectures notes in physics 297*, ed. A. Lawrence (Berlin: Springer-Verlag), pp. 27–39.

Hauser, M. G. 1996. COBE observations of zodiacal emission. In *Physics, Chemistry and Dynamics of Interplanetary Dust*, eds. B. Å. S. Gustafson and M. S. Hanner (San Francisco: Astron. Soc. of the Pacific Press), pp. 309–314.

Hauser M. G., Gillet, F. C., Low, F. J., Gautier, T. N., Beichman, C. A., Neugebauer, G., Auman, H. H., Baud, B., Bogess, N., Emerson, J. P., Houck, J. R., Soifer, B. T., and Walker, R. G. 1984. IRAS observations of the diffuse infrared background. *Ap. J.* **278**, pp. L15–L18.

Hodapp, K. W., MacQueen, R. M., and Hall, D. N. B. 1992. A search during the 1991 solar eclipse for the infrared signature of circumsolar dust. *Nature* **355**, pp. 707–710.

Hong, S. S. 1985. Henyey-Greenstein representation of the mean volume phase function for zodiacal dust. *Astron. Astrophys.* **146**, pp. 67–75.

Ishiguro, M., Fukushima, H., Kinoshita, D., Mukai, T., Nakamura, R., Watanabe, J. I., Watanabe, T., and James, J. F. 1998. The isophote maps of the gegenschein obtained by CCD observations. *Earth Planets Space* **50**, pp. 477–480.

Ishiguro, M., Nakamura, R., Fujii, Y., Morishige, K., Yano, H., Yasuda, H., Yokogawa, S., and Mukai, T. 1999. First detection of visible zodiacal dust bands from ground-based observations. *Ap. J.* **511**, pp. 432–435.

James, J. F., Mukai, T., Watanabe, T., Ishiguro, M., and Nakamura, R. 1997. The morphology and brightness of the zodiacal and Gegenschein. *Mon. Not. R. Astron. Soc.* **288**, pp. 1022–1026.

Kaiser, C. B. 1970. The thermal emission of the corona. *Ap. J.* **159**, pp. 77–92.

Kelsall, T., Weiland, J. L., Franz, B. A., Reach, W. T., Arendt, R. G., Dwek, E., Freudenreich, H. T., Hauser, M. G., Moseley, S. H., Odegard, N. P., Silverberg, R. F., and Wright, E. L. 1998. The COBE diffuse infrared background experiment search for the cosmic infrared background. II. Model of the interplanetary dust cloud. *Ap. J.* **508**, pp. 44–73.

KenKnight, C. E., Rosenberg, D. L., Wehner, G. K. 1967. Parameters of the optical properties of the lunar surface powder in relation to solar-wind bombardment. *J. Geophys. Res.* **72**, pp. 3105–3129.

Kimura, H., Mann, I., and Mukai, T. 1998. Influence of dust shape and material composition on the solar F-corona. *Planet. Space Sci.* **46**, 8, pp. 911–919.

Kneissel, B., and Mann, I. 1991. Spatial distribution and orbital properties of zodiacal dust. In *Origin and evolution of interplanetary dust*, eds. A. C. Levasseur-Regourd and H. Hasegawa (Dordrecht: Kluwer), pp. 139–146.

Koutchmy, S., Lamy, P. L. 1985. The F-corona and the circumsolar dust. evidences and properties. In *Properties and Interactions of Interplanetary Dust*, eds. R. H. Giese and P. L. Lamy (Dordrecht: D. Reidel), pp. 63–74.

Kuhn, J. R., Lin, H., Lamy, P., Koutchmy, S., and Smartt, N. R. 1994. IR observations of the K and F corona during the 1991 eclipse. In *Infrared solar Physics*, eds. D. M. Rabin, J. T. Jefferies and C. Lindsey (Dordrecht: Kluwer), pp. 185–197.

Kwon, S. M., and Hong, S. S. 1998. Three-dimensional infrared models of the interplanetary dust distribution. *Earth Planets Space* **50**, pp. 501–505.

Lamy, P., and Perrin, J. M. 1986. Volume scattering function and space distribution of the interplanetary dust cloud. *Astron. Astrophys.* **163**, pp. 269–286.

Lamy, P., and Perrin, J. M. 1991. The optical properties of the interplanetary dust. In *Origin and evolution of interplanetary dust*, eds. A. C. Levasseur-Regourd and H. Hasegawa (Dordrecht: Kluwer), pp. 163–170.

Lamy, P., Kuhn, J. R., Lin, H., Koutchmy, S., and Smartt, R. N. 1992. No evidence of a circumsolar dust ring from infrared observations of the 1991 eclipse. *Science* **257**, pp. 1377–1380.

Lamy, P., Llebaria, A., Bout, M., Howard, R., Simnett, G, and Schwenn, R. 2000. A global characterisation of the inner zodiacal cloud (F-corona) from four years of SOHO/LASCO coronagraphic observations. In *Abstracts of IAU-COSPAR Colloquium on Dust in the Solar system and other planetary systems* (University of Kent, UK), p. 16.

Leinert, C. 1975. Zodiacal light, a measure of the interplanetary environment. *Space Science Rev.* **18**, pp. 281–339.

Leinert, C., Link, H., and Pitz, E. 1974. Rocket photometry of the inner zodiacal light. *Astron. Astrophys.* **30**, pp. 411–422.

Leinert, C., Hanner, M., Richter, I, and Pitz, E. 1980. The plane of symmetry of interplanetary dust in the inner solar system. *Astron. Astrophys.* **82**, pp. 328–336.

Leinert, C., Richter, I., Pitz, E., and Hanner, M. 1982. Helios zodiacal light measurements, a tabulated summary. *Astron. Astrophys.* **110**, pp. 355–357.

Leinert, C., and Pitz, E. 1989. Zodiacal light observed by Helios through solar cycle No. 21. *Astron. Astrophys.* **210**, pp. 399–402.

Leinert, C., and Grün, E. 1990. Interplanetary dust. In *Physics of the inner heliosphere I*, eds. R. Schwenn and E. Marsch (Berlin: Springer-Verlag) pp. 207–275.

Leinert, C., Bowyer, S., Haikala, L., Hanner, M., Hauser, M. G., Levasseur- Regourd, A. C., Mann, I., Mattila, K., Reach, W. T., Schlosser, W., Staude, J., Toller, G. N., Weiland, J. L., Weinberg, J. L., and Witt, A. 1998. The 1997 reference of diffuse night sky brightness. *Astron. Astrophys. Suppl.* **127**, pp. 1–99.

Léna, P., Viala, Y., Hall, D., and Soufflot, A. 1974, The thermal emission of the dust corona during the eclipse of June 30, 1973. *Astron. Astrophys.* **37**, pp. 81–86.

Levasseur-Regourd, A. C. 1992. Interplanetary dust, remote sensing. In *The astronomy and astrophysics encyclopedia*, ed. S. P. Maran (New York: Van Nostrand Reinhold), pp. 326–328.

Levasseur-Regourd, A. C. 1996. Optical and thermal properties of zodiacal dust. In *Physics, Chemistry and Dynamics of Interplanetary Dust*, eds. B. Å. S. Gustafson and M. S. Hanner (San Francisco: Astron. Soc. of the Pacific Press), pp. 301–308.

Levasseur-Regourd, A. C. 1998. Zodiacal light, certitudes and questions. *Earth Planets Space* **50**, pp. 607–610.

Levasseur-Regourd, A. C. 1999. Lumière zodiacale et poussières interplanétaires. In *Astérodes, météorites et poussières interplanétaires*, eds. D. Benest and C. Froeschlé (Paris: Eska), pp. 177–202.

Levasseur, A. C., and Blamont, J. 1973. Satellite observations of intensity variations of the zodiacal light. *Nature* **246**, pp. 26–28.

Levasseur, A. C., and Blamont, J. 1976. Evidence for scattering particles in meteor streams. In *Interplanetary dust and zodiacal light. Lectures notes in physics 48*, eds. H. Elsässer and H. Fechtig (Berlin: Springer-Verlag), pp. 58–62.

Levasseur-Regourd, A. C., and Dumont, R. 1980. Absolute photometry of zodiacal light. *Astron. Astrophys.* **84**, pp. 277–279.

Levasseur-Regourd, A. C., Dumont, R., and Renard, J. B. 1990. A comparison between polarimetric properties of cometary dust and interplanetary dust particles. *Icarus* **86**, pp. 264–272.

Levasseur-Regourd, A. C., and Hasegawa, H. (eds.) 1991. Origin and evolution of interplanetary dust. (Dordrecht: Kluwer).

Levasseur-Regourd, A. C., Renard, J. B., and Dumont, R. 1991. The zodiacal cloud complex. In *Origin and evolution of interplanetary dust*, eds. A. C. Levasseur-Regourd and H. Hasegawa (Dordrecht: Kluwer), pp. 131–138.

Levasseur-Regourd, A. C., Hadamcik, H., Renard, J. B. 1996. Evidence for two classes of comets from their polarimetric properties at large phase angles, *Astron. Astrophys.* **313**, pp. 327–333.

Levasseur-Regourd, A. C., Cabane, M., Worms, J. C., Haudebourg, V. 1997. Physical properties of dust in the solar system: relevance of a computational approach and of measurements under microgravity conditions. *Adv. Space. Res.* **20**, 8, pp. 1585–1594.

Levasseur-Regourd, A. C., Cabane M., and Haudebourg, V. 1999. Observational evidence for the scattering properties of interplanetary and cometary dust clouds: an update. *J. Quant. Spect. Rad. Trans.* **63**, pp. 631–641.

Lumme, K. 2000. Scattering properties of interplanetary dust particles. In *Light scattering by non spherical particles*, eds. M. I. Mishchenko, J. W. Hovenier and L. D. Travis (San Diego: Academic Press), pp. 555–583.

Lumme, K., Rahola, J., and Hovenier, J. W. 1997. Light scattering by dense clusters of spheres. *Icarus* **126**, pp. 455–469.

Lyot, B. 1930. La couronne solaire étudiée en dehors des éclipses. *C. R. Acad. Sc. Paris*, **191**, pp. 834–837.

Lyot, B. 1939, A study of the solar corona and prominences without eclipses. *Mon. Not. R. Astron. Soc.* **11**, pp. 580–594.

MacQueen, R. M. 1968. Infrared observations of the outer solar corona. *Ap. J.* **154**, pp. 1059–1076.

MacQueen, R. M., Ross, C. L., and Mattingly, T. 1973. Observations from space of the solar corona. *Planet. Space Sci.* **21**, pp. 2173–2179.

MacQueen R. M., Hodapp, K. W., and Hall, D. N. B. 1994. Infrared coronal observations at the 1991 solar eclipse. In *Infrared solar Physics*, eds. D. M. Rabin, J. T. Jefferies and C. Lindsey (Dordrecht: Kluwer), pp. 199–203.

MacQueen, R. M., and Greeley, B. W. 1995. Solar Corona dust scattering in the infrared. *Ap. J.* **440**, pp. 361–368.

MacQueen, R. M., Davidson, W. C., and Mann, I. 1996. The Role of particles size in producing the F-coronal scattered brightness. In *Physics, Chemistry and Dynamics of Interplanetary Dust*, eds. B. Å. S. Gustafson and M. S. Hanner (San Francisco: Astron. Soc. of the Pacific Press), pp. 349–352.

Maihara, T., Mizutani, K., Hirimoto, N., Takami, H., Hasegawa, H. 1985. A balloon observation of the thermal radiation from the circumsolar dust cloud in the 1983 total eclipse. In *Properties and interactions of interplanetary dust*, eds. R. H. Giese and P. Lamy (Dordrecht: D. Reidel), pp. 55–58.

Mankin, W. G., MacQueen, R. M., and Lee, R. H. 1974. The coronal radiance in the intermediate infrared. *Astron. Astrophys.* **31**, pp. 17–21

Mann, I. 1990. Die Strahlung der Fraunhoferkorona im Hinblick auf raumfahrzeuggetragene Fernerkundungsexperimente für den optischen und infraroten Spektralbereich. Dissertation, Universität Bochum.

Mann, I. 1992. The solar F-corona: calculations of the optical and infrared brightness of circumsolar dust. *Astron. Astrophys.* **261**, pp. 329–335.

Mann, I. 1993. The influence of circumsolar dust on the white light corona. *Planet. Space Sci.* **41**, pp. 301–305.

Mann, I. 1995. Spatial distribution and orbital properties of interplanetary dust at high latitudes. *Space Sci. Rev.* **72**, pp. 477–482.

Mann, I., and MacQueen, R. M. 1993. The solar F-corona at 2.12 μm: Calculations of near solar dust in comparison to 1991 eclipse Observations. *Astron. Astrophys.* **275**, pp. 293–297.

Mann, I., Okamoto, H., Mukai, T., Kimura, H., and Kitada, Y. 1994. Fractal aggregates analogues for near solar dust properties. *Astron. Astrophys.* **291**, pp. 1011–1018.

Maucherat, A., Llebaria, A., and Gonin, J. C. 1986. A general survey of the gegenschein in blue light. *Astron. Astrophys.* **167**, pp. 173–178.

Mercer, R. D., Dunkelman, L., Kinglesmith, D. A., and Alvord, G. G. 1979. Lunar libration region L4 photometry. *Space Res.* XIX, pp. 467–470.

Mizutani, K., Maihara, T., Hiromoto, N., and Takami, H. 1984. Near-infrared observation of the circumsolar dust emission. *Nature* **312**, pp. 134–136.

Mukai, M., and Mann, I. 1993. Analysis of Doppler shifts in the zodiacal light. *Astron. Astrophys.* **271**, pp. 530–534.

Mukai, T., and Yamamoto, T. 1979. A model of the circumsolar dust cloud. *Publ. Astron. Soc. Japan* **31**, pp. 585–595.

Murdock, T. L., and Price, S. D. 1985. Infrared measurements of zodiacal light. *Astron. J.* **90**, pp. 375–386.

Nakamura, R., Fujii, Y., Ishiguro, M., Morishige, K., Yokagawa, S., Jenniskens, P., and Mukai, T. 2000. The discovery of a faint glow of scattered sunlight from the dust trail of the Leonids parent comet 55P/Tempel-Tuttle. *Astron. Astrophys.* **540**, pp. 1172–1176.

Okuda, H., Matsumoto, T., and Roellig, T. L (eds.) 1997. Diffuse infrared radiation and the IRTS. (San Francisco: Astronomical Society of the Pacific Press).

Perrin, J. M., and Lamy, P. 1989. The color of the zodiacal light and the size distribution and composition of interplanetary dust. *Astron. Astrophys.* **226**, pp. 288–296.

Peterson, A. W. 1967. Experimental detection of thermal radiation from interplanetary dust. *Ap. J.* **148**, pp. L37–L39.

Reach, W. T. 1988. Zodiacal emission. I. Dust near the Earth's orbit. *Ap. J.* **335**, pp. 468–485.

Reach, W. T. 1991. Zodiacal emission. II. Dust near ecliptic. *Ap. J.* **369**, pp. 529–543.

Reach, W. T., Franz, B. A., Weiland, J. L., Hauser, M. G., Kelsall, T. N., Wright, E. L, Rawley, G., Stemwedel, S. W., and Spiesman, W. J. 1995. Observational confirmation of a circumsolar dust ring by the COBE Satellite. *Nature* **374**, pp. 521–523.

Reach, W. T., Abergel, A., Boulanger, F., Désert, F. X., Pérault, M., Bernard, J. P., Blommaert, J., Césarsky, C., Césarsky, D., Metcalfe, L., Puget, J. L., Sibille, F., and Vigroux, L. 1996. Mid-infrared spectrum of the zodiacal light. *Astron. Astrophys.* **315**, pp. L381–L384.

Reach, W. T., Franz, B. A., and Weiland, J. L. 1997. The three-dimensional structure of the zodiacal dust bands. *Icarus* **127**, pp. 461–484.

Renard, J. B., Levasseur-Regourd, A. C., and Dumont, R. 1995. Properties of interplanetary dust from infrared observations. II. Brightness, polarization, temperature, albedo and their dependence on the elevation above the ecliptic. *Astron. Astrophys.* **304**, pp. 602–608.

Renard, J. B., Dumont, R., Levasseur-Regourd, A. C., and Hadamcik, E. 1996. Clues in zodiacal light observations for a dust ring along the Earth's orbit. In *Physics, Chemistry and Dynamics of Interplanetary Dust*, eds. B. Å. S. Gustafson and M. S. Hanner (San Francisco: Astron. Soc. of the Pacific Press), pp. 329–332.

Ring, J., Clarke, D., James, J. F., Daehler, M., and Mack, J. E. 1964. Profile of the H_β line in the spectrum of zodiacal light. *Nature* **202**, pp. 167–168.

Robley, R., Bücher, A., Koutchmy, S., and Lamy, P. 1985. Doppler shifts measurements of the zodiacal light at the Pic du Midi observatory. In *Properties and interaction of interplanetary dust*, eds. R. H. Giese and P. Lamy (Dordrecht: D. Reidel), pp. 85–88.

Salama A., Andreani, P., Dall'Oglio, G., DeBernardis, P., Masi, S., Melchiorri, F., Moreno, G., Nisimi, B., and Shivanandan K. 1987. Measurements of near and far infrared zodiacal dust emission. *Astron. J.* **92**, pp. 467–473.

Sparrow, J. G., and Weinberg, J. L. 1976. The $S_{10}(V)$ unit of surface brightness. In *Interplanetary dust and zodiacal light. Lectures notes in physics 48*, eds. H. Elsässer and H. Fechtig (Berlin: Springer-Verlag), pp. 41–44.

Spiesman, W. J., Hauser, M. G., Kelsall, T., Lisse, C. M., Moseley, S. H., Reach, W. T., Silverberg, R. F., Stemwedel, S. W., and Weiland, J. L. 1995. Near and far infrared observations of interplanetary dust bands from the COBE diffuse infrared background. *Ap. J.* **442**, pp. 662–667.

Sykes, M. 1990. Zodiacal dust bands: their relation to asteroid families. *Icarus* **85**, pp. 267–289.

Sykes, M., Lebofsky, L. A., Hunten D. M., and Low F. 1986. The discovery of dust trails in the orbits of periodic comets. *Science* **232**, pp. 1115–1117.

Sykes, M., and Walker, R. 1992. Cometary dust trails. *Icarus* **95**, pp. 180–210.

Tanabe, T., Tsumuraya, F., Baba, N., Alvarez, M., Noguchi, M., Isobe, S. 1992. Optical polarization observations of the solar corona during the total solar eclipse of 1991 July 11. *Publ. Astron. Soc. Japan* **44**, pp. L221–L226.

Toller G. N., and Weinberg, J. L. 1985. The change in near-ecliptic zodiacal light brightness with heliocentric distance. In *Properties and interactions of interplanetary dust*, eds. R. H. Giese and P. Lamy (Dordrecht: D. Reidel), pp. 21–25.

Tollestrup, E. V., Fazio, G. G., Woolaway, J., Blackwell, J., and Brecher, K. 1994. In *Infrared solar Physics*, eds. D. M. Rabin, J. T. Jefferies and C. Lindsey (Dordrecht: Kluwer), pp. 179–183.

Vrtilek, J. M., and Hauser, M. G. 1995. IRAS measurements of diffuse solar system radiation: annual sky brightness variation and geometry of the interplanetary dust cloud. *Ap. J.* **455**, pp. 677–692.

Weinberg, J. L. (ed.) 1967. The zodiacal light and the interplanetary medium. SP-150 (Washington: NASA).

Weinberg, J. L. 1985. Zodiacal light and interplanetary dust. In *Properties and interactions of interplanetary dust*, eds. R. H. Giese and P. Lamy (Dordrecht: D. Reidel), pp. 1–6.

Weinberg, J. L., and Sparrow, J. G. 1978. Zodiacal light as an indicator of interplanetary dust. In *Cosmic dust*, ed. J. A. M. McDonnell (Chichester: John Wiley and sons), pp. 75–122.

Whipple, F. 1955. A comet model, the zodiacal light. *Ap. J.* **121**, pp. 750–770.

Wright, A. W. 1874. On the polarization of the zodiacal light. *American J. of science and arts* VII, pp. 451–459.

Xing Z., and Hanner, M. S. 1996. Modelling the temperature of cometary particles. In *Physics, Chemistry and Dynamics of Interplanetary Dust*, eds. B. Å. S. Gustafson and M. S. Hanner (San Francisco: Astron. Soc. of the Pacific Press), pp. 437–441.

Cometary Dust

Zdenek Sekanina[1], Martha S. Hanner[1],
Elmar K. Jessberger[2], Marina N. Fomenkova[3]

[1] JPL, California Institute of Technology, Pasadena, California, USA
[2] Westfälische Wilhelms-Universität Münster, Germany
[3] Center for Astrophysics and Space Sciences,
University of California at San Diego, La Jolla, California, USA

Abstract. This chapter reviews the history of cometary dust investigations. The individual sections describe the progress achieved in the understanding of the dynamical properties of cometary dust, its optical, thermal, and other physical properties, and its chemistry. The review emphasizes information that was obtained during the three recent major observing campaigns, focused on comets 1P/Halley, Shoemaker-Levy 9 (1994 X = D/1993 F2), and Hale-Bopp (C/1995 O1). Where appropriate, the review relates the discussed topics to relevant issues of cometary nuclei and their physical and chemical properties.

I. INTRODUCTION

Since the most recent encyclopedic review of the problems of cosmic dust, in a book edited by McDonnell (1978), three major events have occurred that have greatly contributed, and still are contributing, to our much increased knowledge of comets in general and of cometary dust in particular. One of the events was a long awaited one, while the other two have been entirely unexpected. We refer, of course, to the return of comet 1P/Halley in 1986; to the most peculiar cometary object ever observed, Shoemaker-Levy 9 (1994 X = D/1993 F2), in 1993-94; and to comet Hale-Bopp (C/1995 O1), one of the most spectacular comets of all time, which was discovered in 1995 and is still under observation at the time of this writing.

Among the highlights from observations of Halley's comet, at least four that involve dust should be listed at the outset: (i) first direct information was obtained on the chemistry of cometary particles, (ii) first evidence was gathered on attogram grains, which bridge the gap between molecules and particulates, (iii) a pre-1986 inference, from morphological studies of a number of comets, that dust ejection is confined to relatively small, isolated active areas on the sunlit side of the nucleus was spectacularly confirmed by closeup imaging from onboard the intercepting spacecraft, and (iv) a major outburst of dust was detected at a record heliocentric distance of 14 AU after perihelion.

The greatest contributions from Shoemaker-Levy 9's observations to our understanding of cometary dust are in the areas of (i) mechanical strength of cometary material (or, rather, the lack thereof), (ii) fragmentation processes both in interplanetary space and upon the object's entry into the Jovian atmosphere, and (iii) physical and chemical interactive atmospheric processes triggered during and after each fragment's penetration into the atmosphere.

The ongoing investigations of comet Hale-Bopp provide a wealth of new information in a number of research areas on cometary dust, including huge amounts of invaluable data on coma morphology, on thermal properties and composition of grains, and on processes of dust emission at large heliocentric distances. Many of these studies have been published in volumes 77–79 of *Earth, Moon, and Planets*, others are still in progress.

In the following, the discussion of cometary dust is divided into three broad categories: (i) its dynamical properties, (ii) its physical properties, including the optical and thermal ones, and (iii) its chemistry. Because of constraints of space, not discussed are the problems of cometary contributions to maintaining the interplanetary dust cloud and relevant implications for studies of cometary dust from laboratory experiments.

II. DUST DYNAMICAL PROPERTIES

Historically, the early efforts that led to information on cometary dust began with Bessel's (1836) pioneering work on the coma morphology of Halley's comet at its 1835 apparition, in which he introduced the concept of a repulsive force from the Sun. The progress continued with Bredikhin's independent studies of tail formation in the late 19th century (for a review, see Jaegermann 1903). Bredikhin is responsible for two additions to cometary terminology that are still extensively used today to describe the dust tails: a *syndyname* (or *syndyne*) as a locus of particulates that are subjected to a constant acceleration by the repulsive force and a *synchrone* (or *isochrone*) as a locus of particulates that are ejected from the comet at the same time. The repulsive force was identified as *solar radiation pressure* about 100 years ago (Arrhenius 1900; Schwarzschild 1901), although Norton (1844) pointed out that L. Euler had already considered the possibility that the repulsive force "consists in an impulsive action of the Sun's rays." For dust tails, this interpretation has universally been accepted.

II.A. Ejection and Motion of Dust Grains

Based on Whipple's (1950) icy-conglomerate model, a consensus has developed that dust particles are released from the cometary nucleus and accelerated to their "terminal" velocities by drag forces exerted by the sublimating ice, with which the refractive material is mixed in cometary nuclei and on their surface. The terminal velocity is reached when the grain becomes dynamically decoupled from the expanding gas, that is, when particle-molecule collisions are no longer dynamically significant. As the gas drag drops, solar radiation pressure becomes, next to solar attraction, the dominant force on most dust particles that are observed. And since the acceleration due to solar attraction affects the motions of dust and the nucleus equally, the motion of a particle relative to the nucleus is determined by solar radiation pressure alone, unless additional (usually minor) nongravitational forces are involved.

1. Gas-dust interaction and particle ejection velocities

A simple formula for the dust ejection velocity derived from the equation of motion involving the drag force was published by Whipple (1951), who assumed that the drag coefficient and the gas velocity were constant in the gas-dust interaction zone. This formula was more recently generalized to nonspherical shapes by Gustafson (1989).

The first elaborate treatment of the problem of dusty gasdynamics in comets was presented by Probstein (1969), who used a free molecular approximation to describe the gas flow and solved the equation of motion for a dust particle together with the relevant conservation equations for the dusty gas flow's mass, momentum, and energy. Probstein demonstrated that the drag coefficient is primarily a function of the Mach number and the specific heat ratio of the gas. For assumed perfect thermal accommodation between gas and dust and for a *single* characteristic grain size, he obtained transonic solutions for the accelerating gas flow, with the terminal velocity of dust particles reached within ~20 radii of the nucleus (where gas and dust essentially decouple), depending on the dust loading of the gas and on the accommodation coefficient. This coefficient varies as the product of a dust particle's size and bulk density, the nucleus radius, and the thermal velocity of the gas, and inversely as the mass flow rate of the gas. At an assumed gas temperature of 200 K, Probstein's results yield dust terminal velocities for submicron-sized grains that converge to 0.74 km/s when the dust loading of the gas flow is negligibly low, to 0.57 km/s when the dust and gas mass flows are equal, and to 0.36 km/s when the dust mass flow exceeds the gas mass flow by a factor of 10. Probstein's two-component solution to the gas-dust interaction was incorporated as an essential component of the Finson-Probstein analysis of the dust tails of comets (Subsec. 1 of II.C).

A stringent test of Probstein's results was provided by subsequent studies of the gradual expansion of distinct dust features in the heads of comets 109P/Swift-Tuttle (Sekanina 1981a) and 1P/Halley (Sekanina and Larson 1984, 1986a) at heliocentric distances near 1 AU. The ejection velocities (equal to Probstein's terminal velocities) of submicron-sized particles were found to vary between 0.4 and 0.7 km/s, in excellent correspondence with the theory. Fitting Probstein's theoretical curves led to a suggestion (Sekanina 1981a) that the relationship between the ejection velocity $v_{\rm eject}$ of a dust particle and the solar radiation pressure acceleration β to which the particle's motion is subjected (Subsec. 2 of II.A) could be approximated by a simple empirical formula:

$$v_{\rm eject} = \frac{a}{1 + b/\sqrt{\beta}}, \qquad (1)$$

where a and b are coefficients expressible in terms of the physical parameters of Probstein's theory.

In spite of the apparent agreement between theory and observation, some aspects of Probstein's treatment (assumption of a single characteristic grain size, oversimplified energy conservation equation, neglect of molecule-molecule collisions near the nucleus, etc.) were found by others to be unrealistic enough to warrant numerous refinements in, and innovations to, Probstein's original approach (e.g., Hellmich 1981; Hellmich and Keller 1981; Marconi and Mendis 1983, 1984; Gombosi et al. 1983, 1985; Gombosi 1986; cf. also a review by Crifo 1991), but these efforts led to no dramatic changes in the determination of the terminal velocities of dust particles. The most recent review paper in which these issues are addressed in detail, especially in reference to comet Hale-Bopp, is that by Combi et al. (1999). Applications of hydrodynamic models to emission scenarios involving localized jets are discussed in Subsec. 2 of II.B.

2. Dust acceleration by solar radiation pressure

The acceleration that solar radiation pressure exerts on dust particles in comets is commonly expressed in units of the acceleration by solar attraction at the same heliocentric distance. This dimensionless ratio, in the current literature usually called β, varies as the particle's projected cross sectional area A and inversely as its mass m:

$$\beta = \frac{Q_{\mathrm{pr}} L_\odot A}{4\pi c G M_\odot m}, \tag{2}$$

where c is the speed of light, G is the gravitation constant, $M_\odot$ and $L_\odot$ are the Sun's mass and total energy emitted per second, and Q_{pr} is the radiation pressure efficiency for the particle, which depends on its size, shape, and optical properties. A large number of studies exist that are dedicated to calculations of Q_{pr} and β and their variations with particle size. Most of these investigations employ the Mie theory and refer to compact spherical grains of particular compositions (e.g., Schwehm and Rohde 1977; Burns et al. 1979), but some also consider models of core-mantle particles (Schwehm 1976) and particles that are composed of mixtures of materials (Lien 1991). Generally, for a spherical particle of radius a (in μm) and density ρ (in g/cm^3) we have

$$\beta = \frac{0.574 \, Q_{\mathrm{pr}}}{\rho a}. \tag{3}$$

The principal results of these calculations are major differences between the values of β for absorbing (such as carbon-rich or metallic) grains on the one hand and for dielectric (such as glassy, basaltic, or icy) particles on the other hand. Absorbing grains have a peak β value greater than unity (for example, $\sim$1.8 for iron, $>$5 for graphite) at a particle radius near 0.1 μm; for extremely tiny grains ($<$0.01 μm) the β value converges to a nonzero constant. By contrast, dielectric grains have a peak β value smaller than unity (mostly near 0.5–0.6) at a particle radius near 0.2–0.3 μm and very tiny particles of this kind are virtually transparent to light, having $\beta \sim 0$.

The intermediate critical value, $\beta = 1$, has a very simple and fundamental meaning. Such particles are subjected to solar radiation pressure that balances solar gravity, so that they move through the solar system with constant velocities along straight lines. For $\beta < 1$ the orbit may be either an ellipse or a concave hyperbola, depending on the motion of the parent comet. If the comet's orbit is a parabola, then the orbit of a particle released with no impulse is a hyperbola. For $\beta > 1$, a particle's orbit is always a convex hyperbola, regardless of the parent comet's motion.

Observations indicate that, except for antitails (Subsec. 3 of II.C), a peak radiation pressure acceleration on dust particles in comets is typically $\beta_{\text{peak}} \simeq 2.5$, including grains in streamers (Subsec. 2 of II.C) and striae (Subsec. 4 of II.C) (e.g., Orlov 1960; Sekanina and Farrell 1980, 1982; Sekanina 1981b, 1986; Akabane 1983; Lamy 1986a; Beisser and Boehnhardt 1987a, b; Notni and Thänert 1988; Nishioka and Watanabe 1990).

3. Motions of charged grains

It is generally recognized that because of their interaction with the radiative and plasma environment, dust particles in cometary heads and tails must be electrostatically charged. The grain charging depends on the physical and electrical properties of the grains, on the nature of their interaction with the surrounding radiation and plasma fields, and on the relative velocity. The most important contributions come from a flux of electrons and ions, from UV-radiation-induced photoemission, and from secondary emission of electrons.

The process of electrostatic charging on cometary dust particles and the equilibrium potentials to which they are expected to be charged have extensively been discussed in the literature (e.g., Notni 1964, 1966; Boehnhardt 1986; Horányi and Mendis 1986; Boehnhardt and Fechtig 1987; Notni and Tiersch 1987; Tiersch and Notni 1989; and in the chapter by Mukai et al.). The results indicate that the potential is usually only a few volts; it is positive in cases involving high photoelectron currents, negative in a high plasma-density environment. Boehnhardt and Fechtig (1987) find that, during the spacecraft encounters, silicate grains in 21P/Giacobini-Zinner and 1P/Halley carried a positive charge of up to 10 volts outside the cometopause, whereas the electrostatic potential of carbon-rich grains varied strongly with the plasma environment conditions, reaching both positive and negative values. Near the nucleus, the potential was negative but very low (~ 0.1 volt).

There are two important effects that charged particles are subjected to: electrostatic fragmentation (cf. Subsec. 4 of II.C) and interaction with the interplanetary magnetic field. Boehnhardt (1986) and Boehnhardt and Fechtig (1987) conclude that only small and fluffy dust particles (0.1 to 1 μm in size and of a tensile strength of 0.001 to 0.01 bar) can be broken up electrostatically in comets. Even though there is an indication, primarily from examined events of cometary splitting, that on large scales the tensile strength of cometary nuclei is in the required range (e.g., Sekanina 1982a, 1996a; Greenberg et al. 1995), the situation on microscopic scales is far less clear and evidence for electrostatic fragmentation remains inconclusive.

The interaction of a charged dust particle with the interplanetary magnetic field leads to the generation of a Lorentz force. The acceleration of a spherical particle of radius a and density ρ by the Lorentz force is

$$\mathbf{L} = \frac{3\epsilon_0 \Phi}{\rho a^2}[(\mathbf{v}-\mathbf{w}) \times \mathbf{B}], \tag{4}$$

where ϵ_0 is the permittivity constant, Φ is the equilibrium electrostatic potential, $\mathbf{v}$ is the vector of the particle's velocity relative to the Sun, $\mathbf{w}$ is the vector of the solar-wind velocity, and $\mathbf{B}$ is the interplanetary magnetic field's strength. In analogy to the dimensionless quantity β for the radiation pressure acceleration (Subsec. 2 of II.A), one can express the acceleration by the Lorentz force with a dimensionless quantity γ, in units of the acceleration due to solar attraction. Applying Parker's (1958, 1963) model of the axially symmetric, quiet-day interplanetary magnetic field, the ratio γ is equal to

$$\gamma = \frac{0.447\,\Phi\kappa B_{\rm rad}}{\rho a^2}, \tag{5}$$

where a is in μm, ρ in g/cm^3, and Φ in volts, $B_{\rm rad}$ (in gauss) is the radial component of the magnetic field at a heliocentric distance of 1 AU, and κ is a function of the vector product in Eq. (4). When $||\mathbf{w}|| \gg ||\mathbf{v}||$, as is almost universally the case, then $\kappa \approx \Omega r \cos b$, with Ω being the Sun's angular rotation velocity (in s^{-1}), r the comet's heliocentric distance (in km), and b its heliographic latitude.

There are two major differences between effects of the Lorentz force and radiation pressure on a dust particle's motion: (i) the Lorentz acceleration $\gamma \propto a^{-2}$, whereas, in the first approximation, the radiation pressure acceleration $\beta \propto a^{-1}$, so that effects of electrostatic charging should become the more important the smaller the particle; and (ii) unlike radiation pressure, the Lorentz force affects the motion in the direction normal to the comet's orbital plane.

Effects of particle charging on the motion of cometary dust have not been subjected to systematic tests. From the fact that solar radiation pressure adequately accounts for the motions of striae (Subsec. 4 of II.C) in the dust tail of comet West (1976 VI = C/1975 V1), Sekanina and Farrell (1980) estimated that submicron particles that made up the features could not be charged to potentials higher than a few volts at the most. An apparent signature, in the tail of comet SOLWIND 1 (1979 XI = C/1979 Q1), of the interaction of charged dust with the coronal magnetic field of the Sun (Sekanina 1982b) is likely to be invalidated by a more recent revision of the object's positional observations and orbital elements (Marsden 1989). At present, the strongest evidence against detectable effects of electrostatic charging on the motion of cometary dust comes from incidental observations during the Earth's transits across the orbital planes of comets. Such episodes are common, taking place twice a year for every object, and they are virtually continuous occurrences for comets of a very low orbit inclination (a significant fraction of the short-period comets). In spite of the omnipresence of such events, no instance is known of a dust-tail orientation out of the orbital plane, which at the time of Earth's transit projects as the great circle in a predictable position angle. Particularly significant is the absence of the Lorentz force inferred from the tail orientations of the SOHO sungrazing comets, because of the strong magnetic field and the comets' high velocity relative to the solar wind along the inbound orbit (Sekanina 2000).

II.B. Dust Features in Cometary Heads

Information on structural detail in the heads of comets has slowly been accumulating over centuries. It appears that the first observation of a distinct coma feature, clearly documented in the literature, is that of Halley's comet by Hevelius (1682) on September 8, 1682. The drawing, displaying a bright curved jet near the nucleus, predates E. Halley's discovery of the object's orbital periodicity by 13 years. Numerous graphical renditions of dust features in dozens of comets — some of them accompanied by micrometric measurements — were published during the 19th century. Examples of these drawings were reproduced by Rahe et al. (1969). As photography was gradually replacing visual telescopic observations of comets in the late 1800s and the early 1900s, the amount of new information on coma morphology began to decrease. The renewed interest in the subject since the 1970s has been motivated primarily by the development of sophisticated computer-processing techniques that allow digital enhancement of structural detail on available images and by major advances in detector performance, especially the availability of the charge coupled device (CCD) arrays. Also instrumental in these revitalization efforts was the pressing need for a better understanding of the nucleus environment in the era of space exploration of comets.

1. Nucleus rotation and discrete emission sources

A growing body of evidence for nucleus rotation and for the presence of discrete emission sources on the nucleus surface was one of the factors that contributed to the broad acceptance of the icy conglomerate model in the 1970s. The pioneering efforts included Whipple's (1978) examination of old visual observations of nearly concentric dust halos in the head of comet Donati (1858 VI = C/1858 L1), as reported in great detail by Bond (1862). Earlier, Larson and Minton (1972) (cf. also Larson 1978) had derived the spin rate of comet Bennett (1970 II = C/1969 Y1) from separations in a system of photographically observed spiral jets in its dust coma. Independent efforts also in progress were aimed at determining the position of the nuclear rotation axis from projected orientations of dust features (such as jets, fans, or spirals) and their motions in the coma (Sekanina 1979, 1981a). An attempt was even made to establish precession of comet 2P/Encke (Whipple and Sekanina 1979). A review of the early morphological studies of cometary dust (Sekanina 1981c) confirmed that outgassing from many, especially short-period, comets is indeed largely confined to discrete areas on the sunlit side of their rotating nuclei and that the appearance of the observed features is determined by the surface distribution of the sources and by the emission mode (continuous vs. erratic) and insolation regime (circumpolar Sun vs. day-and-night). Even though it has subsequently been recognized that Halley's state of rotation is more complex than originally thought, the closeup images of this comet's nucleus — particularly those taken with the Giotto's Halley Multicolor Camera (Keller et al. 1987) — fully confirm the earlier conclusion that dust emanations from the nucleus are largely restricted to isolated sources situated on its sunlit side.

2. Hydrodynamic models for dust emission from discrete sources

Following Halley's apparition of 1986, significant progress has been achieved in the understanding of the outflow mechanism and the thermophysical properties of gas and dust emissions that expand from the nucleus in a highly anisotropic fashion. Observations of several comets, including 1P/Halley and 2P/Encke, show that both the radio and the UV spectra of OH exhibit significant asymmetries in the line-of-sight velocity profiles (e.g., Snyder et al. 1976; Bockelée-Morvan and Gérard 1984; de Pater et al. 1986; A'Hearn and Schleicher 1988).

Stimulated by the closeup appearance of Halley's nucleus and its environment, efforts aimed at formulating axisymmetric hydrodynamic models that offer dynamical solutions for an anisotropic flow of gas and/or dusty gas from a localized active area got underway in the late 1980s (Kitamura 1986, 1987; Kömle and Ip 1987a, b; Körösmezey and Gombosi 1990). The Kitamura and Kömle-Ip models were reviewed by Kömle (1990), while the Kitamura and the Körösmezey-Gombosi models were discussed by Gombosi (1991). These models begin with the mass, momentum, and energy conservation equations for perfect gas and with the continuity and energy balance equations for the dust particles entrained in the gas flow. Some of the more interesting results from these calculations include, according to Körösmezey and Gombosi, the formation of a dust spike and a jet cone, where much of the ejecta is contained. The spike is a combined effect of gas heating by hot dust grains above the source and of the absence of a lateral expansion velocity component along the emission axis itself (in the idealized case of an axisymmetric flow). The cone is a product of the gas pressure vectorial distribution near the nuclear surface outside the jet's axis. Keller and Thomas (1989) conclude that a near-surface "breeze," a lateral transport of material from the source toward the nightside along the pressure gradients strongly increases the dust density in the anti-sunward direction and thereby decreases the sunlit/dark brightness asymmetry. Keller et al. (1990) point out that an important role is apparently played by dust particle fragmentation almost immediately after release from the nucleus. It causes an increase in the sublimation rate from the just generated distributed source as well as a rapid mass loading of the gas flow, thereby slowing down the flow's expansion acceleration, enlarging the subsonic region, and facilitating the near-surface breeze in side directions. A breeze is predicted by the Kitamura and Körösmezey-Gombosi models. However, it is emphasized by Gombosi (1991) that the transport of dust toward the dark side is more pronounced in the models for strong jets that create a lateral shock and lead to the formation of dust cones on the dark side due to emission from a source on the sunlit side. The conclusion is that activity from isolated sources located on the sunward side of the nucleus is compatible with all observed near-nucleus phenomena, as also pointed out recently by Combi et al. (1999) in reference to comet Hale-Bopp.

3. Expansion of dust ejecta through the coma of a rotating comet

The hydrodynamic models describe the evolution of a dust jet in its very early stages and they disregard rotation of the nucleus. Once the interaction between the dust and the expanding gas ceases, the particulate ejecta expand through the coma under the effects of solar attraction and solar radiation pressure (described in Subsec. 2 of II.A). On time scales of hours and days, the nuclear rotation has dramatic effects on the distribution of dust in the coma and thereby on coma morphology.

As observed from Earth, large amounts of freshly ejected dust are first detected as a sharp central condensation or a "false" nucleus, caused by limited seeing. Next usually comes the development of a spiral jet that "unwinds" on the sunward side of the coma from the gradually fading condensation. Sometimes, however, one observes a halo instead, which is entirely separated from the condensation. The feature, whether a spiral or a halo, subsequently evolves into a slowly expanding envelope, whose surface brightness decreases with time until it vanishes completely. The feature's period of visibility is often long enough to notice its being swept into the tail on one or both sides of the nucleus.

In a special case of a point-like emission region located in the nucleus equatorial plane that coincides with the comet's orbit plane, the cometocentric motion of a dust particle ejected with a constant velocity v_{eject} (identified with its "terminal" velocity due to momentum exchange with expanding gas in the early post-ejection period of time) and subjected to a constant acceleration g due to solar radiation pressure is described by the rectangular equatorial coordinates x, y, oriented, respectively, toward the Sun and 90° ahead of it in the direction of rotation (Sekanina and Larson 1984):

$$x(t, \Theta) = f(t, \Theta) \left[v_{\text{eject}} \cos \Theta - \frac{1}{2} g\, f(t, \Theta) \right]$$

$$y(t, \Theta) = f(t, \Theta)\, v_{\text{eject}} \sin \Theta, \tag{6}$$

where Θ is the angle of ejection, reckoned from the sunward direction in the sense of rotation and measured in radians, P is the rotation period, and

$$f(t, \Theta) = (t - t_{\text{b}}) - \frac{P}{2\pi}(\Theta - \Theta_{\text{b}}), \tag{7}$$

with t_b and Θ_b being the time and ejection angle at the onset of emission (at sunrise). Equations (6) hold for any time $t \geq t_b$, but the function $f(t, \Theta)$ is defined only for emission during one rotation, that is, for $\Theta_b \leq \Theta \leq \Theta_e$, where Θ_e is the ejection angle at the end of emission (at sunset).

The striking effect of the spin rate on the observed morphology is exemplified in Fig. 1, which shows a pole-on view of the evolution of a feature made up of dust particles ejected continuously with a constant velocity from an equatorial source between sunrise and sunset. When the nucleus spins fast, the feature soon separates from the central condensation, so that the spiral-jet phase of development has a very short duration. Several rotations after the onset of emission, a system of fairly symmetrical, almost semicircular and approximately concentric envelopes sets off. Their outlines are eventually distorted by directionally nonuniform expansion. Particulates of identical dynamical properties ejected from a slowly rotating nucleus first form a spiral jet, which gradually evolves into a highly asymmetrical envelope. Thus, whether one observes a spiral or a halo depends strongly on the nuclear spin rate.

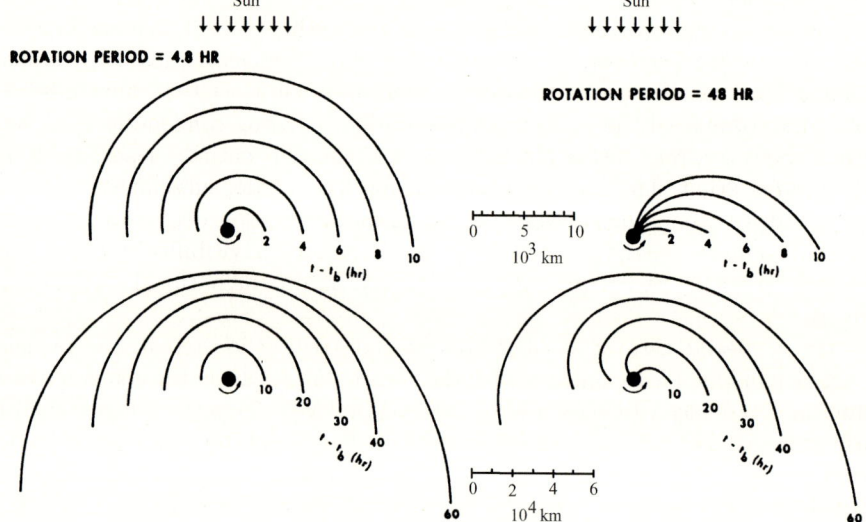

Figure 1. Effects of the spin rate on the evolution of dust ejecta from a point-like source on the equator of a rotating nucleus. The spin axis is normal to the comet's orbit plane (the plane of the figure). The source is active from sunrise ($\Theta_b = -90°$) to sunset ($\Theta_e = +90°$) and all particles are ejected with the same velocity of 500 m/s and subjected to the same solar radiation-pressure acceleration of 0.25 cm/s². The circles show the position of the nucleus and the curves describe the loci of particles ejected at various times after the onset of emission, $t - t_b$. The sense of rotation and the direction to the Sun are also indicated. The left-hand side of the figure depicts the case of a rapidly rotating nucleus; the right-hand side, of a slowly rotating nucleus. Note the scale differences for the early (top) and late (bottom) phases of evolution. (From Sekanina and Larson 1984.)

It is similarly possible to demonstrate effects of the inertial position of the nuclear spin axis on both the activity regime and the resulting dust coma morphology. In particular, comprehensive analysis of a variety of scenarios makes it possible to explain the distinction between fans and jets, as shown in the next subsection.

4. Conceptual models and computer simulation of dust features

Serious attempts to analyze, model, and interpret large-scale morphology of cometary dust atmospheres began only in the late 1970s and the early 1980s. Restricted initially to fitting contours of observed features (e.g., Sekanina and Larson 1984, 1986a, b), this work subsequently developed into an increasingly sophisticated Monte Carlo image simulation procedure (Sekanina 1987a, b, 1991a, 1996b).

Closeup images of Halley's nucleus, particularly those taken with the Giotto camera, convincingly document dust jets streaming away from discrete sources on the sunlit side of the nucleus (Keller et al. 1987). This scenario was originally proposed for 1P/Halley by Sekanina and Larson (1984), who applied their computer simulation model to the comet's ground-based images of 1910. More recently, the model's parameters — which include the nucleus spin vector at the time of dust emission, the surface distribution of dust sources, and the range of particle ejection velocities and solar radiation pressure accelerations — were expanded to introduce random noise into synthetic images (Sekanina 1991b) to account for effects of both physical nature (such as dispersion in the vector field of particle expansion velocities) and incidental nature (such as imperfect seeing). The inclusion of noise substantially enhances the model's capability to simulate faithfully the observed coma appearance of dust comets. In addition, by increasing noise it is possible gradually to "erase" any morphological feature in the computer-generated images, as is illustrated in Fig. 2 for a system of sunward, nearly concentric halos (displayed e.g. by comet Donati C/1858 L1 and very recently by comet Hale-Bopp C/1995 O1). This possibility to erase a feature implies that large-scale morphology of a comet's head indeed is the product of collimation of a dust particle flow from discrete sources, but that the lack of morphology is not necessarily an indicator of the absence of such isolated sources. Jet collimation is also discussed by Keller et al. (1994).

During the 1990s, the Monte Carlo image simulation model was further upgraded. It can now accommodate a great variety of particle size distribution laws and account for short-term (diurnal) variations in the production rate of dust from a rotating source (Sekanina 1993). This new capability of the software package is particularly helpful when one models rapidly changing morphological features associated with sudden dust bursts.

The experience with the image simulation experiments achieved so far should serve as the basis for assessing the future of Monte Carlo modeling investigations into dust coma morphology. On the one hand, the technique, even though essentially trial-and-error in nature, is powerful enough to provide, with fairly restricted sets of reference parameters, an impressive match

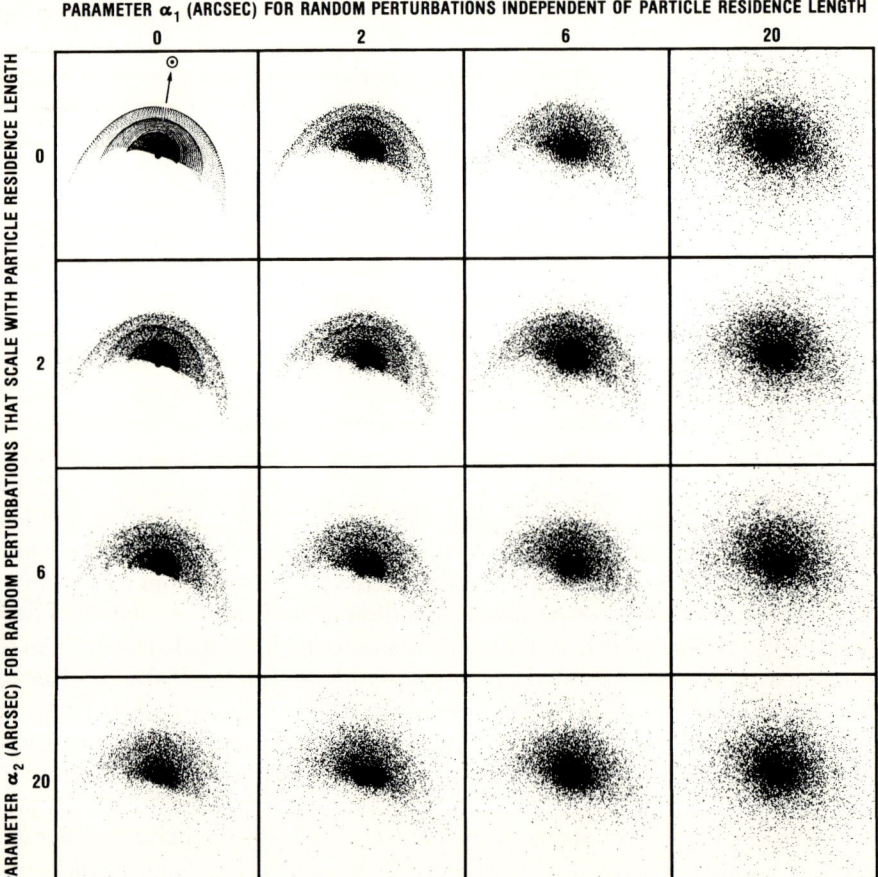

Figure 2. Computer-generated images of a dust coma consisting of a set of sunward, nearly concentric halos, simulating the appearance of comets such as Donati (C/1858 L1) or Hale-Bopp (C/1995 O1). All the images were generated with identical reference parameters, except for the two randomization constants, α_1 and α_2. Random noise that is independent of the particle residence length in the coma increases from the left to the right, while the noise that scales with residence length increases from the top to the bottom, α_2 characterizing this noise at 1 arcmin from the nucleus. Each frame is about 100 arcsec on a side. The image in the upper left corner ($\alpha_1 = \alpha_2 = 0$), which shows the direction to the Sun, consists of noiseless (deterministic) particle loci. The meaning of the parameters α_1 and α_2 is explained in the original study. (From Sekanina 1991b.)

to extremely diverse jet patterns. On the other hand, the possible existence of multiple solutions, which may lessen the merits of the image simulation technique, is clearly of concern. Luckily, it turns out that short-term variations in dust coma morphology represent a major discriminating factor of immense diagnostic value. The significance of this time-lapse approach is particularly well demonstrated in two very recent investigations of morphological features displayed by the dust coma of comet Hale-Bopp (C/1995 O1). One of these studies (Sekanina 1998) offers a quantitative, dynamically attractive interpretation for the diurnal evolution of a dust jet observed in late February 1997 throughout one rotation period of the comet (Plate 1) and describes the jet's transformation into a system of halos on a time scale of several rotations. The other study (Sekanina and Boehnhardt 1999) presents a conceptually innovative model for the comet's porcupine-like appearance during much of 1996.

In recent years, progress in modeling dust coma morphology was reported by several groups, thanks to the variety of spectacular features exhibited by comet Hale-Bopp (e.g., Vasundhara and Chakraborty 1999, Vasundhara et al. 1999, Jorda et al. 1999, Mueller et al. 1999, Samarasinha et al. 1999, Warell et al. 1999). It was also shown that the dynamical properties of dust ejecta are correlated with their thermal properties. For example, only grains with $\beta < 1$ — i.e., silicate particles — can explain the uniform spacing of the halos of comet Hale-Bopp in February-March 1997 (Hayward et al. 2000; cf. also Sec. III.D).

5. Dust emission at large heliocentric distances

The statement, often copied in astronomical textbooks, that cometary activity is confined only to small distances from the Sun (< 2–3 AU) is demonstrably incorrect. The first comet that has repeatedly been found active at heliocentric distances exceeding 5 AU is 29P/Schwassmann-Wachmann 1. While this object can most appropriately be characterized as having continuous activity on which sporadic outbursts are superimposed (Jewitt 1990), isolated flare-ups appear to be the most conspicuous mode of activity for comets far from the Sun.

Since the problem of activity at large distances from the Sun is too broad to discuss here in all of its aspects, we focus in the following on two events: an isolated outburst of 1P/Halley at 14 AU from the Sun after perihelion and the quasi-recurring flare-ups of comet Hale-Bopp (C/1995 O1) at 6–7 AU before perihelion. Both comets have perihelion distances of less than 1 AU. A morphological and photometric study of P/Halley's slowly expanding, crescent-shaped halo at 14 AU suggested that the feature represented a segment of a conical surface populated by particulate ejecta that had been released from a suddenly activated source on the sunlit side of the nucleus (Sekanina et al. 1992). The ejecta's total mass was estimated at $\sim 10^{12}$ g or more and carbon monoxide was suggested as the most likely driver, accelerating the smallest detected grains to a terminal velocity of ~ 45 m/s.

The general scenario is not dramatically different for the recurring bursts of comet Hale-Bopp, even though they took place closer to the Sun and before perihelion. The comet's characteristic appearance during three major events in August–October 1995 consisted of a radial, rectilinear jet that emerged from the nucleus condensation to the northwest and turned abruptly to the east at a distance of several seconds of arc from the center, terminating as a gradually fading spiral arm that vanished in the first quadrant. These features were successfully modeled (Fig. 3) as products of sharply peaking dust injection episodes originating from an isolated source on the nucleus and lasting only a quarter of the rotation cycle (Sekanina 1996c). The peak particle expansion velocities were found to amount to 50 m/s and the total mass of the dust ejected during each event was estimated at a few times 10^{11} g, significantly exceeding the measured production rate of carbon monoxide, the apparent driver for the dust (e.g., Jewitt et al. 1996).

An interesting characteristic of the emission episodes experienced by P/Halley and Hale-Bopp at large heliocentric distances seems to be a very high mass loading of the CO gas flow by dust, estimated at $\gg 10$. It remains to be seen whether this is a common property of comets.

II.C. Dust Tails and Their Structure

Dust tails are in most textbooks described as being structureless. This is not universally true, although their structure cannot compete with prominent features that are so characteristic of plasma tails. Dust tails are also said to be moderately curved, with a sharp leading boundary and a more diffuse trailing boundary. Again, this is often, but by no means always, the case. For extracting information on a comet's activity from its dust tail, the two-dimensional distribution of light and any evidence of morphology are the tail's most important attributes, although its orientation and approximate length and width are in some cases also highly diagnostic, as will be shown in Subsec. 4 of II.D.

1. Dust tail formation and its analysis

The fundamental point about dust tails of comets is that it takes some, often a long, time for them to form. As a result, the history of a comet's dust emission is imprinted and preserved in its dust tail for a limited period of time. One can take advantage of this fact and recover much of the information if one knows how to "read" the tail. Additional benefits of having this kind of opportunity stem from the circumstance that the recovered information may, and often does, refer to times when the comet was not under observation, either because it was not yet discovered or because it was too close to the Sun in the sky to be observable.

Because of the dominance of the intervening effects of solar radiation pressure, the initial conditions — such as the emission anisotropy — become of secondary importance in studies of dust tails and, with rare exceptions, have been neglected: even some of the most refined models assume that dust emission is spherically symmetrical. The three fundamental parametric functions determining the distribution of light in a dust tail are the primary objectives of the modeling efforts: (i) the mass production rate of dust particles and its temporal variations, (ii) the distribution function of the acceleration ratio β, which is closely related to the particle size and mass distribution functions, and (iii) the particle ejection velocity and its temporal variations.

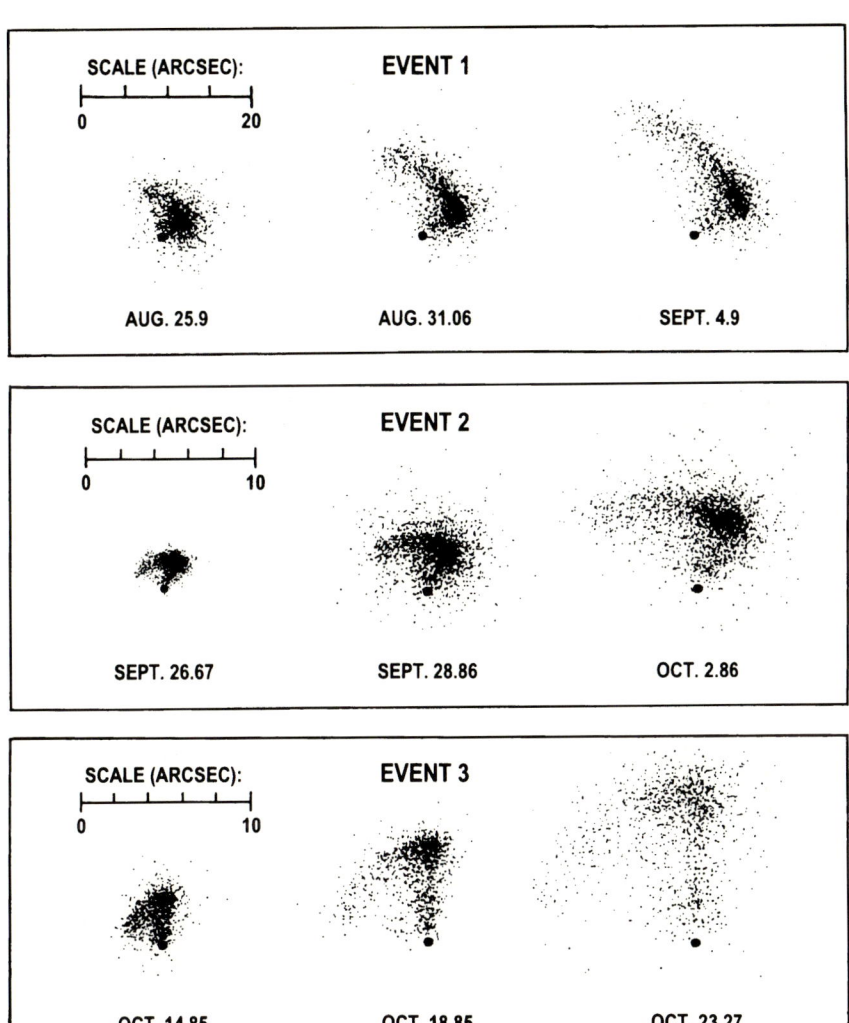

Figure 3. Computer-generated images of comet Hale-Bopp (C/1995 O1), simulating morphology of the dust coma consisting of a radial jet and a spiral arm, for selected times following each of the three emission events that took place between late August and late October 1995. The synthetic image for September 26 can be compared with an image taken then with the Planetary mode Wide Field Planetary Camera-2 of the Hubble Space Telescope (cf. Weaver et al. 1997). Note that the scale for the Event 1 images differs from the scales for the other two events. North is always up and east to the left. (From Sekanina 1996c.)

The history of sophisticated dust tail modeling is brief. It began with a pioneering work by Finson and Probstein (1968a), which treats dust tails as continuous particle-flow phenomena, is based on the concept of synchrones and syndynames, and adopts Probstein's (1969) fluid-dynamic model for the interaction between gas and dust near the nucleus. The Finson-Probstein model essentially integrates contributions from superimposed uniformly expanding shells (because of finite ejection velocities) of particulates that were ejected at various times and subjected to a variety of radiation pressure accelerations, calculates the projected spatial distribution of scattered sunlight in the tail, and varies the three parametric functions until a satisfactory match has been found between the observed and modeled isophotes. In its original version this model was applied to comets Arend-Roland (1957 III = C/1956 R1) by Finson and Probstein (1968b), to comet Bennett (1970 II = C/1969 Y1) by Sekanina and Miller (1973), and to comet Seki-Lines (1962 III = C/1962 C1) by Jambor (1973).

Kimura and Liu (1977) modified the Finson-Probstein approach by introducing the concept of a "neckline" and abandoning the approximation by *uniformly* expanding shells. They argued that from the basic laws of celestial mechanics it follows that any particle ejected before perihelion with a finite velocity component normal to the comet's orbital plane passes through the same plane again after perihelion. If Earth should transit the comet's orbital plane at the same time as these particles, a neckline structure would be observed. This suggestion is identical with the idea suggested more than one decade earlier by Southworth (1963, 1964). Further improvements of the Finson-Probstein method were proposed by Richter and Keller (1987) and by Fulle (1987a), who also developed an independent Monte Carlo approach and examined the roles of emission asymmetry and a Maxwellian distribution of ejection velocities (Fulle 1989, 1992). Fulle and his collaborators proceeded with applications of these innovative techniques to images of dust tails of comets Halley (1986 III = 1P/1982 U1) (Fulle 1987b; Fulle et al. 1987, 1988; Cremonese and Fulle 1989), Bennett (1970 II = C/1969 Y1) (Pansecchi et al. 1987; Fulle 1987c; Fulle and Sedmak 1988), Arend-Roland (1957 III = C/1956 R1) and Seki-Lines (1962 III = 1962 C1) (Fulle 1988a), Kohoutek (1973 XII = C/1973 E1) (Fulle 1988b), 1910 I = C/1910 A1 (Pansecchi and Fulle 1990), Wilson (1987 VII = C/1986 P1) (Cremonese and Fulle 1990), Liller (1988 V = C/1988 A1) (Fulle et al. 1992), Austin (1990 V = C/1989 X1) (Fulle et al. 1993a), Grigg-Skjellerup (1992 XVIII = 26P) (Fulle et al. 1993b), and Swift-Tuttle (1992 XXVIII = 109P/1992 S2) (Fulle et al. 1994). These techniques provide very satisfactory results for dust tails with a smooth light distribution or when they exhibit a neckline structure or an antitail (Subsec. 3 of II.C). They are less suitable for analysis of some of the discrete dust features that are discussed next.

2. Streamers as indicators of outbursts

Major but brief enhancements of dust production are brought about by cometary outbursts. The problem of outbursts and their mechanisms is outside the scope of this review and the reader is referred to numerous papers in which these topics are addressed and/or summarized at some length and from different standpoints (e.g., Hughes 1991; Sekanina 1991a; Rettig et al. 1992; Sekanina et al. 1992). Ejecta released during an outburst may display a variety of features, depending on the ejection circumstances, the heliocentric distance, and the time elapsed between the outburst and observation. If solar radiation pressure has long enough been the dominant force so that most ejecta from the outburst have reached the tail, they become observed as a discrete band or ray called here a streamer.

There is always one streamer per outburst. The streamers have the following properties: (i) they are issued from the nucleus in directions that deviate perceptibly from the extended radius vector toward the negative orbital-velocity vector; (ii) they are usually rather narrow, sometimes slightly cone-shaped, and either rectilinear or moderately curved; and (iii) as a rule, they number no more than several at a given time, all converging to the nucleus, subtending distinct angles with one another. The position angle of each streamer is diagnostic of the time of outburst, while its length provides information on the peak radiation pressure acceleration to which the ejecta were subjected. Unfortunately, streamers have a tendency to fade rapidly with time and since their brightness decreases with increasing distance from the nucleus, their observed length usually provides only a lower bound to the peak radiation pressure effect.

The dynamical behavior of particles that left the comet simultaneously was first considered by Norton (1861), who already pointed out that such particles should be distributed in the tail along a nearly straight line that points approximately at the nucleus. On the other hand, Bredikhin classified streamers as type III tails, which he regarded to be syndynames (Jaegermann 1903). Comet 1901 I (= C/1901 G1), the last studied by him, became later instrumental in bringing about the first major modification to his classification. Moiseyev (1925) found that the streamers were synchrones and, subsequently, Orlov (1928, 1929) regarded all type III tails (or type II_0 tails, as they were referred to for some time; cf. Bobrovnikoff 1951) to be "complete" synchrones (Orlov 1960). Since outbursts occur commonly in comets, streamers are fairly frequent phenomena in their dust tails. Two outstanding examples in the past three decades are comets West (1976 VI = C/1975 V1) with at least five bright and up to seven additional streamers (Sekanina and Farrell 1978; Sekanina 1980; Akabane 1983) and comet Halley with at least six to eight streamers (Lamy 1986a; Sekanina 1986; Beisser and Boehnhardt 1987a,b). In both comets the multiple streamers were observed one to a few weeks post-perihelion and the inferred outbursts occurred within 2 weeks of the perihelion passage.

3. Anomalous tails and antitails

It appears that Harding (1824) and Olbers (1825) were the first to use the term *anomalous tail* in reference to a sunward-pointing extension displayed by the comet of 1823 (C/1823 Y1). The term was also employed by Olbers (1831) in what can be regarded as the first review paper on the subject. While it was suggested by Needham et al. (1957) that several records in ancient annals of the Chinese dynasties might be interpreted as referring to comets with tails on both sides of the nucleus, the first positive account of a tail in an apparent direction of the Sun was given by Kirch (1681) in his description of the comet of 1680 (C/1680 V1), calling it a pseudo tail (*After-Schwanz*).

Bredikhin's (Jaegermann 1903) hypothesis was the most ambitious one among the early efforts aimed at explaining the nature of the sunward-oriented tails, which did not fit any of the three types of his classification. Like Harding and Olbers, Bredikhin called these tails anomalous, but he distinguished two kinds: genuine and pseudo anomalous tails. Unfortunately, his results suffered from the lack of a sound physical model, from unacceptable approximations in his dynamical treatment of the problem, and from inaccurate observations available. His most significant contribution to the understanding of the nature of these tails was the conclusion on their close relationship with meteor streams.

At the present time, the development of anomalous tails is fully understood. As a rule, they are made up of relatively large, often submillimeter- to millimeter-sized, particles ejected from the nucleus long (at least weeks, sometimes months or even years) before observation (e.g., Finson and Probstein 1968a, b). The basic conditions that must be satisfied for an anomalous tail to appear include two geometric constraints (e.g., Sekanina 1976): (i) the Earth must be near the comet's orbital plane, and (ii) at the same time, the angle subtended, at the comet, by the comet-Earth vector and the comet's radius vector must be smaller than the lag angle Λ of the earliest detectable dust emission (which is always smaller than 180° at the comet), in which case the anomalous tail points toward Earth, or between 180° and 180° + Λ, in which case it points away from Earth. About the time of the Earth's transit across the comet's orbital plane, the anomalous tail becomes very sharp and is sometimes called an antitail or a sunward spike to imply that it is a thin sheet of debris confined to the orbital plane. The above geometric constraints were employed in the first successful prediction of an antitail (Sekanina 1973), which involved comet Kohoutek (1973 XII = C/1973 E1). More recent studies (e.g., Richter and Keller 1988, Fulle 1988a, b) have shown that the effect of particle ejection velocity on anomalous tails, neglected in the original Finson-Probstein (1968a) approach, ought to be accounted for in rigorous investigations and that the presence of a neckline structure (Subsec. 1 of II.C) further enhances the sharpness of some antitails.

Every anomalous tail, observed around the time of the Earth's transit across the orbital plane, exhibits a characteristic rotational motion about the Sun's projected direction in the sky. Invariably noticed by observers, it merely reflects the Earth's motion from one side of the comet's orbital plane to the other. The antitail's apparent rotation is clockwise at the ascending node, counterclockwise at the descending node. The antitail's sharp edge, caused by the "crowding" of pertinent synchrones, is always on the side of the radius vector: it is a leading boundary before the transit and a trailing boundary afterwards. Finally, since the lag angle of synchrones increases with their age and depends sensitively on the comet's true anomaly at the observation time, the probability of appearance of an anomalous tail increases dramatically after perihelion, especially for comets with small perihelion distances.

4. Striated tails and particle fragmentation

Unlike streamers, striae are bands in the dust tail that (i) appear less commonly; (ii) are always separated from the nucleus by huge gaps; (iii) are narrow, almost perfectly rectilinear, and nearly parallel to each other; (iv) their orientations are inconsistent with those of synchronic or syndynamic formations and, when extended beyond their visible length, they intersect the radius vector almost always on the sunward side of the nucleus; and (v) tend to cluster into groups, sometimes numbering more than a dozen at a time.

The first comet in whose tail striae were positively identified (e.g., Bond 1862) was Donati (1858 VI = C/1858 L1). Early dynamical studies of striae, especially in the tail of comet 1910 I (= C/1910 A1), led to a conclusion — now known to be incorrect — that they are synchronic formations just like streamers. The most significant observed characteristic of striae was thought to be not their peculiar orientation, but the gap between their sunward end and the nucleus, which explains why Orlov (1960) called them "terminal" synchrones to discriminate between striae and streamers.

After the appearance of another comet with striae in its tail, Mrkos (1957 V = C/1957 P1), the orientation discrepancies between striae and true synchrones could no longer be ignored. Work on this subject intensified more recently as observations of additional comets with striated tails accumulated, especially after the arrival of comet West (1976 VI = C/1975 V1). Two competing models emerged out of these efforts: Notni's (1964) high-speed particle ejection theory and Sekanina and Farrell's (1980) particle fragmentation theory. Notni found that the motions of striae in the tail of comet Mrkos could be fitted on the assumption that, upon their ejection from the nucleus, dust particles interact with comet plasma so strongly that they get accelerated, in a tailward direction, to velocities of 10 km/s or more while still near the nucleus. On the other hand, Sekanina and Farrell explained the formation of striae as a two-step process: parent particles ejected in an outburst are subjected to the same, relatively high, radiation pressure acceleration during their motion through the tail, and subsequently they all fragment at the same time, at distances of up to a few million kilometers from the nucleus. For comet Mrkos, Sekanina and Farrell (1982) found two kinds of striae that consisted, respectively, of absorbing and dielectric grains (Fig. 4). Akabane (1983) employed an essentially identical approach (but a different terminology) in his investigation of comet West. Comparing the two competing models, Notni and Thänert (1988) confirmed that the fragmentation theory is consistent with the motions of striae in both Mrkos and West, but found that the high-speed ejection theory fails for West. The fragmentation model was also successfully applied to comet Seki-Lines (1962 III = C/1962 C1) by Nishioka and Watanabe (1990), to comet 1910 I (= C/1910 A1) by Sekanina and Farrell (1986), and, very recently, to comet Hale-Bopp (C/1995 O1) by Pittichová et al. (1999).

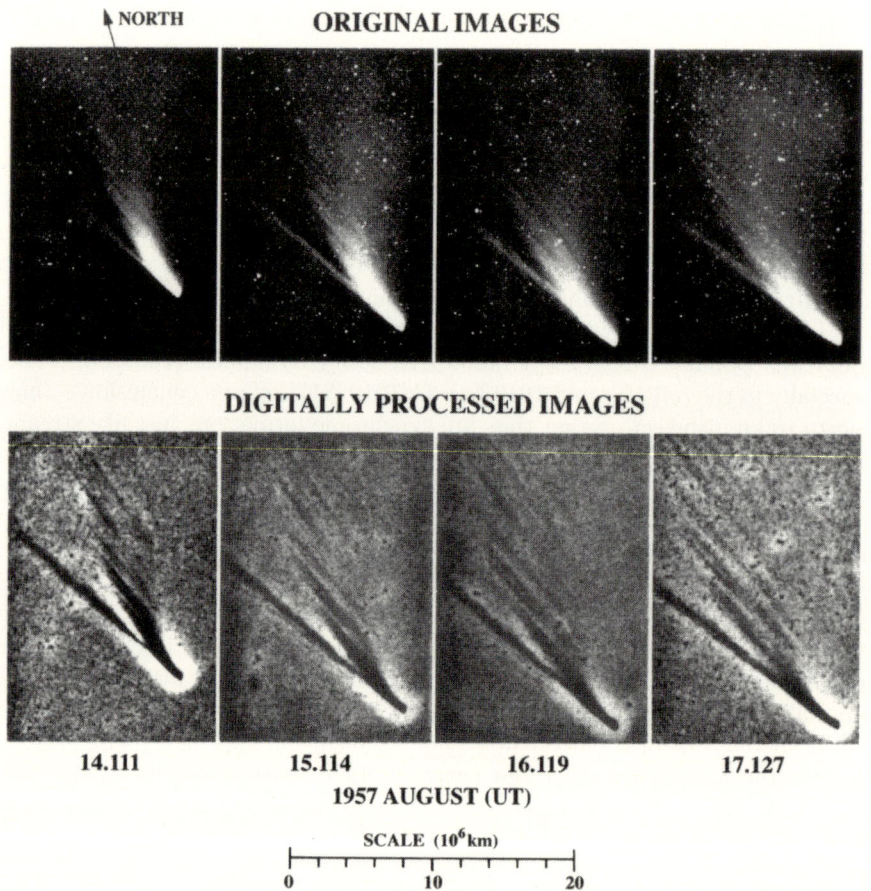

Figure 4. The photographs of comet Mrkos (C/1957 P1) taken by J. A. Farrell with a 19-cm Schmidt camera near Fort Worth, Texas, on four consecutive days in August 1957. The original exposures are in the upper row, the digitally processed images in the lower row. The arrow on the first image points to the north. The scale refers to the distances projected onto the plane of the sky at the comet's nucleus. The long narrow tail directed to the upper left is the plasma tail. The striae of the first kind are the long, narrow streaks situated in the left part of the broad, dust tail. The striae of the second kind are the short, stubby bands located further to the right end of the dust tail. (Adapted from Sekanina and Farrell 1982.)

Nishioka and Watanabe (1990) argued that the constraint on the fragmentation time of parent particles can be fully relaxed, if the fragments have finite lifespans (cf. also Watanabe and Nishioka 1991; Nishioka et al. 1992). However, Sekanina and Pittichová (1999) have shown that in the case of comet Hale-Bopp the condition of a constant fragmentation time can be relaxed at most to a few days. The constraint on the radiation pressure acceleration of parent particles remains firm. Their source might in fact be a single massive piece so extraordinarily porous as to be optically thin, a property that could be dictated observationally by the high acceleration values. Alternatively, as shown by Fröhlich and Notni (1988), optically thick clouds of grains of a limited size spectrum could temporarily be stabilized against dissipation due to solar radiation pressure and they too could satisfy the condition of a high, constant acceleration value. Very recently, however, an innovative model has been proposed for the formation of striations in the dust tail, which does not require the parent particles to be subjected to any repulsive accelerations (Sekanina et al. 2000). The parents are in this scenario released near the comet's aphelion with velocities of a few m/s and in the absence of activity the involved process is essentially spontaneous, perhaps escape of boulder-sized temporary satellites from the comet's gravitational field.

Striae are by no means the only phenomena that imply fragmentation of cometary dust. Large amounts of attogram grains, discovered by Utterback and Kissel (1990) in 1P/Halley, were interpreted by them as products of vigorous fragmentation of dust at distances of up to at least one million kilometers from the comet. Fragmentation was also invoked by Simpson et al. (1986, 1987, 1989) to explain "clusters" and "packets" of dust grains impacting the detectors onboard the Halley flyby spacecraft; by Thomas and Keller (1987) and by Keller et al. (1990) to interpret the radial brightness profiles of jets on closeup, high-resolution images returned by the Giotto's Halley Multicolour Camera and to study the hydrodynamic implications of near-nucleus dust-grain disintegration; by Combi (1994) to understand isophote dust-coma profiles on ground-based images of comets; and by Boehnhardt (1986) and Boehnhardt and Fechtig (1987), among others, to demonstrate potential effects of electrostatic dust charging in comets (Subsec. 3 of II.A).

II.D. Dust in Periodic Comet Shoemaker-Levy 9

This object, under observation from late March 1993, which is nearly nine months after its extremely close approach to Jupiter when it had split into as many as a dozen major fragments, until it collided with Jupiter in the second half of July 1994, was one of the most observed comets ever. For many weeks and months following the initial breakup, the individual fragments continued to split in discrete events which are usually referred to as episodes of *secondary fragmentation* and which gave birth to the so-called off-train condensations. It is argued below that continuing fragmentation was almost certainly responsible for the highly atypical size (and mass) distribution of this comet's dust as well as for its dynamical evolution, two issues of primary interest here.

1. Fragmentation model for the comet's progenitor nucleus

At discovery the brightest part of the comet appeared as a string, or a train, of 21 discrete condensations less than 1 arcmin in length and arranged in an almost perfectly rectilinear configuration (Fig. 5). With time the separations between the condensations were gradually increasing and about one half of them began to show small but detectable deviations from the train (the so-called off-train condensations). Even though the condensations were the comet's most prominent features, significant amounts of material were also situated in between them, along the train's entire length.

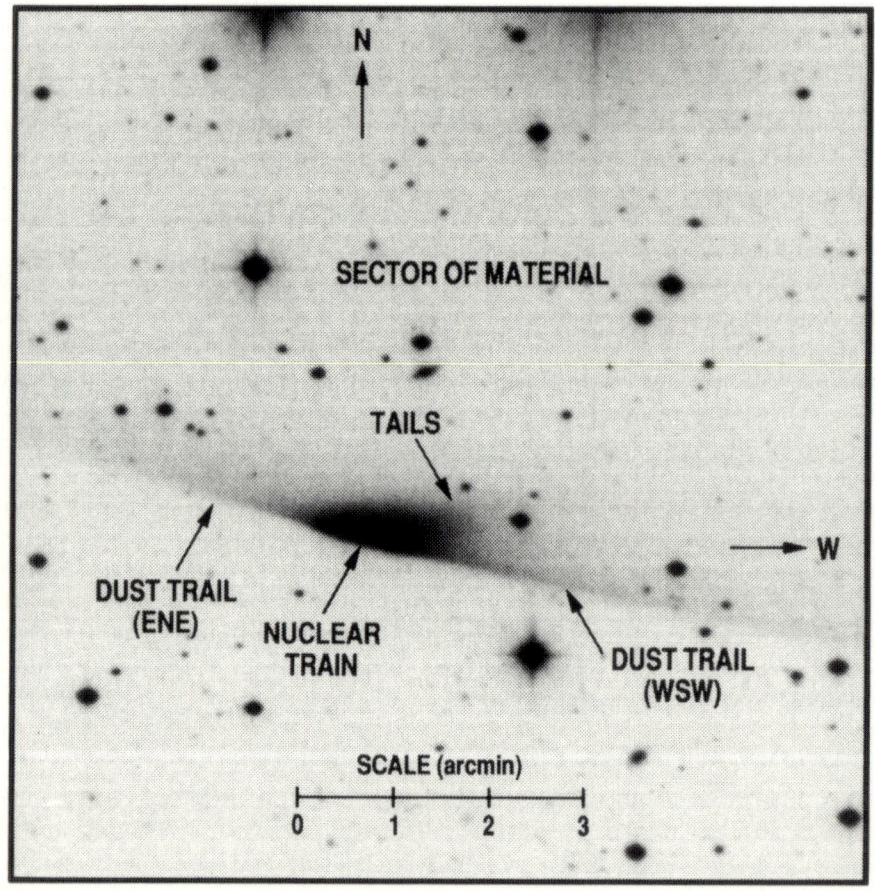

Figure 5. A 440 second exposure of comet Shoemaker-Levy 9 (D/1993 F2) taken by J. V. Scotti with the University of Arizona's 91-cm Spacewatch telescope on March 30, 1993. North is up and east to the left. The nuclear train, the two dust trails, and the wide sector of material with embedded tails are identified. (From Sekanina et al. 1994; original image courtesy of J. V. Scotti, Lunar and Planetary Laboratory.)

The comet displayed three other kinds of morphological feature. Extending on either side of the train were the trails or wings. The west-southwestern trail was perfectly aligned with the train, whereas the east-northeastern branch appeared to be slightly inclined to it. Subtending a relatively small angle with the train and pointing almost exactly to the west was a set of parallel, rectilinear, and narrow tails, whose roots coincided with the individual condensations in the train. These tails were immersed in, and on low-resolution images (such as the one in Fig. 5) blended with, an enormous structureless sector of material, which was stretching to the north of its sharp boundary, delineated by the nuclear train and the two trails.

The formulation of a viable model for secondary-fragmentation events is of key importance for understanding the process of this comet's disintegration. In their comprehensive analysis, Sekanina et al. (1998) conclude that the Jovian tidal forces inflicted extensive cracks throughout the interior of the progenitor nucleus but did not split it apart. The initial disruption apparently consisted of a rapid sequence of individual breakups that gave birth to 10–12 major fragments and were brought about by stresses exerted on the cracked object by its fast rotation in the early post-perijove period of time. The secondary-fragmentation events are then understood as stochastic manifestations of the continuing process of progressive disintegration. The vectorial distribution of separation velocities of the secondary-fragmentation products shows a strong concentration toward a great circle, unquestionably an effect of the approximately conserved angular momentum of the progenitor comet. The velocity vectors are actually distributed within a segment of the great circle, thus implying that the subfragments were released from one side of their parents. The preferential appearance of the off-train condensations on one side of the nuclear train is thereby explained for the first time. The model requires that the points of separation be on the antisolar side of the parent fragments, where thermal stresses should enhance the effects due to the rotation stresses. The spin axis of the progenitor nucleus was situated nearly in its jovicentric orbit plane, which rules out the strengthless aggregate models (Solem 1995; Asphaug and Benz 1996) as plausible breakup scenarios. From the derived separation velocities (of up to 1.7 m/s) of the products of secondary fragmentation, the progenitor nucleus is found to have been approximately 10 km in diameter and spinning rapidly. Constraints set by this scenario on the tensile strength of the nucleus are reasonably consistent with the value derived independently by Greenberg et al. (1995).

2. *Dust content in the nuclear train*

The sizes and dynamics of particulate material in the individual condensations have been subject to much controversy. An internally consistent solution (cf. also Subsec. 4 of II.D) can be offered on the assumption that the mass distribution was dominated by the processes that occurred following its closest approach to Jupiter in July 1992.

From quantitative considerations on radiation pressure accelerations, the minimum diameter of particulates in the innermost regions of the condensations, within 0.5 arcsec (or about 1800–2000 km) of their center, is estimated at about 1 meter in late January 1994 and 2 meters some $5\frac{1}{2}$ months later, assuming a bulk density of 0.2 g/cm^3. The observed brightness of these innermost regions, combined with this lower limit, makes it possible to estimate the mass involved as a function of the particles' albedo, the upper size limit, and the slope of the size distribution function. For a geometric albedo of 4%, for example, the dust cloud of the condensation Q_1, one of the more prominent ones, would at both times be about 1×10^{14} g in mass if the size distribution varied as an inverse fourth power of the size and some 3×10^{15} g, if it varied as an inverse size squared. A conservative value of 0.2 km was used to estimate the upper limit of particle diameter; the actual upper limit must in fact be greater. This mass does not include the contributions from the six fragments >1 km in diameter, detected individually in the condensation Q_1 on both the late January and the early July images (Sekanina 1995), which appear to have contributed collectively 8 to 9×10^{15} g, assuming the same albedo and bulk density as above. Most of this mass, nearly 7×10^{15} g, is found to have been concentrated in the largest fragment, which in this scenario dominated. If, on the other hand, the upper particle diameter limit in the dust cloud were near 1 km and/or the size distribution function were flatter than assumed above, the mass of the cloud would have been comparable with, or greater than, that of the largest fragment.

Relative particle velocities in the proximity of major fragments must have been extremely low, about 0.1 m/s or less, significantly lower than the separation velocities involving the events of secondary fragmentation. Indeed, the fragment Q_2, which split off from Q_1 in April 1993 with a velocity of $\sim$0.4 m/s (Sekanina et al. 1998), appeared as a distinctly separate condensation, located far outside the Q_1's dust cloud, by January 1994. Another line of evidence for very low velocities is presented in Subsec. 4 of II.D.

3. The dust trails (or wings)

Although a number of nuclear models have been presented to explain the comet's peculiar appearance, relatively little attention has been paid to the dust trails. Sekanina et al.'s (1994) model regarded the dust trails as one of manifestations of the initial disruption and the physical and dynamical conditions in the resulting cloud of debris. In particular, these authors showed that the extent and orientation of the trails can be interpreted as centimeter-sized and larger products of ubiquitous particle-particle collisions. The rotation-driven, rapidly "thermalized" particle velocity distribution of the cloud of debris was found to display a long "tail", with a small fraction of the particulates having been accelerated to velocities of up to $\sim$7 m/s in the direction of the orbital motion, necessary to explain the maximum observed lengths of the trails. The slight inclination of the east-northeastern branch to the nuclear train can in this scenario be explained by a deficit of pebble-sized and larger particles, a possible effect due to highly irregular shape of the parent nucleus (Sekanina et al. 1994).

4. The tails and the sector of material

The most diagnostic information provided by the tails is their orientation as a function of time and their characteristic width, length, and appearance. The tails are perfectly parallel, which strongly suggests the same mode of their origin. The temporal variations in the orientation indicate that the release of particulates occurred most probably during the second half of 1992 (Sekanina 1996a). It is not known, however, whether the release was continuous or proceeded in a sequence of events and whether or not it was outgassing-driven. It is tempting to associate the tail formation with both the initial disruption and the events of secondary fragmentation, most of which occurred in the implied time span. It is significant, however, that no tail extension was ever observed to point to the southeast of the train (cf. Fig. 5). Dust emissions from active fragments during long periods of time in 1993 and 1994 should have resulted in fairly persistent features at position angles of 100–110°. Their lack, especially on images taken with the HST, offers rather tight constraints on the dust production rate from the nuclei during some periods of time in 1993 and 1994. These upper limits are as low as 0.2 kg/s, or more than two orders of magnitude lower than the limits derived spectroscopically for the water production rate (30 to 60 kg/s; Weaver et al. 1995, corrected values) from an unsuccessful search for the hydroxyl radical.

The fact that the tails appeared as natural extensions of the condensations suggests that the particulates in both regions were of common origin and that the only major dynamical difference between them is the magnitude of solar radiation pressure acceleration to which they were subjected. This acceleration increases and therefore the characteristic particle size decreases along the tail. Particles located at a distance of about 15,000 km from the "parent" condensation on images taken shortly before the crash, were typically several centimeters in diameter. The tail width, on the other hand, is a measure of a maximum particle velocity in the plane unaffected by solar radiation pressure. The tails broadened with increasing distance from the condensation (e.g., Weaver et al. 1995), implying that smaller particles had a wider velocity distribution. From a sample of the projected linear tail widths near the nuclear train one finds upper limits on a particle velocity in a general range of 0.1 to 0.4 m/s (Sekanina 1996a), comparable with, or lower than, the separation velocities of the products of secondary fragmentation. These estimates confirm that particles in the condensations must have been larger than several centimeters in diameter and must have had velocities not exceeding about 0.1 m/s (Subsec. 2 of II.D).

It would be erroneous to conclude that the split comet Shoemaker-Levy 9 showed no evidence for microscopic grains, which are so abundant in other comets. Although a severe bias against such very small particles in all observations of this comet is inherent because of the discovery some $8\frac{1}{2}$ months after the parent object's disruption, the area in the upper right corner of Fig. 5, the most remote one from the train and the trails, is populated by grains whose diameters were a few tens of microns at the most.

5. The problem of continuing activity

One of the most controversial issues concerning comet Shoemaker-Levy 9 is whether or not the individual fragments continued to be active. To some degree an answer to this question depends on how one defines activity. We identify activity with sublimation of ices (and parallel emission of dust) at rates that are not trivial.

The strongest arguments usually presented in favor of this comet's continuing post-breakup activity are the apparent sphericity of the condensations and a very gradual decrease in their intrinsic brightness with time (Hahn et al. 1996; Rettig and Hahn 1997; Tanigawa et al. 1997). Unquestionably, a cloud of dust ejected from the comet in the immediate proximity of Jupiter in July 1992 would soon become, and then remain, extremely elongated.

The strongest arguments against continuing activity are the failure to detect it and the already mentioned extremely low dust particle velocities as well as the implied considerable particle dimensions.

Of the two mutually exclusive scenarios, the reader can choose the one he finds less vulnerable. We note, however, that continuing spontaneous (i.e., activity-independent) fragmentation of dust in each condensation can account both for the condensation's spherically symmetric shape and for a slow rate of its brightness decrease, as argued by Tanigawa et al. (1997). Considering, on the one hand, overwhelming evidence for ubiquitous disintegration of this comet by chain fragmentation and imagining, on the other hand, the absurdity of a subdecimeter-per-second rate of expansion of a cloud of dust particles entrained in a transonic gas flow, we find the hypothesis of inactive nuclear fragments of comet Shoemaker-Levy 9 far more attractive.

Comet Shoemaker-Levy 9 strikingly illustrates the kind of extreme physical phenomena and extraordinary dynamical scenarios that one is confronted with when investigating truly exceptional cometary objects.

III. DUST OPTICAL AND PHYSICAL PROPERTIES

The thermal emission and scattered radiation from the dust coma allow us to make some general statements about the optical properties, size, and composition of cometary dust grains.

III.A. Thermal Emission

The 3–20 μm thermal emission from the dust coma has been observed with infrared photometers for many comets since the first detection of Ikeya-Seki (1965 VIII = C/1965 S1) (Becklin and Westphal 1966). 1P/Halley was monitored regularly in 1985–1986 (Gehrz and Ney 1992; Tokunaga et al. 1986, 1988; IHW Electronic Archive). Hale-Bopp (C/1995 O1) was observed over a wide range of heliocentric distances, beginning at 4.9 AU preperihelion.

The observed spectral energy distribution (SED) corresponds to color temperatures that are typically 5–30% hotter than the temperature of a theoretical blackbody at the same heliocentric distance. For Halley, for example, Tokunaga et al. (1988) find a relationship $T_c = 315\, r^{-0.502}$, where r is the heliocentric distance in AU and T_c is the 8–20 μm color temperature. In contrast, the strong thermal emission from comet Hale-Bopp exhibited even higher color temperatures, with $T_c/T_{BB} \sim 1.5$ at 7–13 μm (Grün et al. 2000) and $T_c/T_{BB} \sim 1.8$ at 3–5 μm and 5–8 μm near perihelion (Hayward et al. 2000; Williams et al. 1997).

The apparent color temperature is determined by the physical temperatures of the grains and their wavelength-dependent emissivities. The physical temperature of a particle in the solar radiation field depends on the balance between the solar energy absorbed at visual wavelengths and the energy radiated in the infrared:

$$\frac{\pi a^2}{r^2} \int Q_{\rm abs}(a,\lambda)\, S(\lambda)\, d\lambda = 4\pi a^2 \int \pi B(\lambda, T)\, Q_{\rm abs}(a,\lambda)\, d\lambda, \qquad (8)$$

where $S(\lambda)$ is the solar flux at 1 AU, $\pi B(\lambda, T)$ is the Planck function for grain temperature T, $Q_{\rm abs}(a,\lambda)$ is the absorption efficiency factor, which depends on grain size and optical constants, and r is the heliocentric distance in AU.

Small carbonaceous grains absorb strongly at visual wavelengths, but cannot radiate efficiently in the infrared at wavelengths greater than about 10 times their size. Thus, they heat up until the energy radiated at 3–8 μm balances the absorbed energy. In fact, for small absorbing grains, their size controls the temperature, regardless of the specific composition (Hanner 1983). Fig. 6 plots $T(r)$ for spherical glassy carbon grains. Grains smaller than about 2 μm radius are warmer than a theoretical blackbody and a grain of 0.1 μm radius can be several hundred degrees warmer than a blackbody. For these grains, $T(r) \propto r^{-0.35}$ instead of $\propto r^{-0.5}$ expected for a blackbody. Thus, $T_{\rm gr}/T_{BB}$ increases with r. We may have seen evidence of this trend in the high mid-infrared color temperatures recorded for comet Hale-Bopp at large r. Grains larger than a few microns will be warmer than a blackbody only if they are very fluffy and the unit structure is micron-sized or smaller (Hage and Greenberg 1990; Xing and Hanner 1997).

In contrast to small carbon grains, silicate grains radiate efficiently in the mid-infrared; it is the amount of absorption at visual wavelengths that controls their temperature. The absorption at visual wavelengths depends strongly on the iron content of the silicate (Dorschner et al. 1995). Mg-rich silicates have very low absorption (the imaginary part of the refractive index $k \sim 0.0003$ at 0.5 μm for a glass with Fe/Mg = 0.05); whereas $k \sim 0.05$ at 0.5 μm for a pyroxene glass with Fe/Mg = 1 and $k \sim 0.1$ for an olivine glass with Fe/Mg = 1. Figure 7 illustrates the dependence of temperature on k for grains of 0.5 μm radius. One sees that, for $k = 0.001$ the grain temperature is much cooler than T_{BB}. Setting $k = 0.01$ (Fe/Mg ~ 0.5) raises the grain temperature above that of a blackbody. For Fe-rich olivine (Fe/Mg = 1; Dorschner et al. 1995), the temperature is similar to that of a fully absorbing grain; indeed, at $r < 1$ AU,

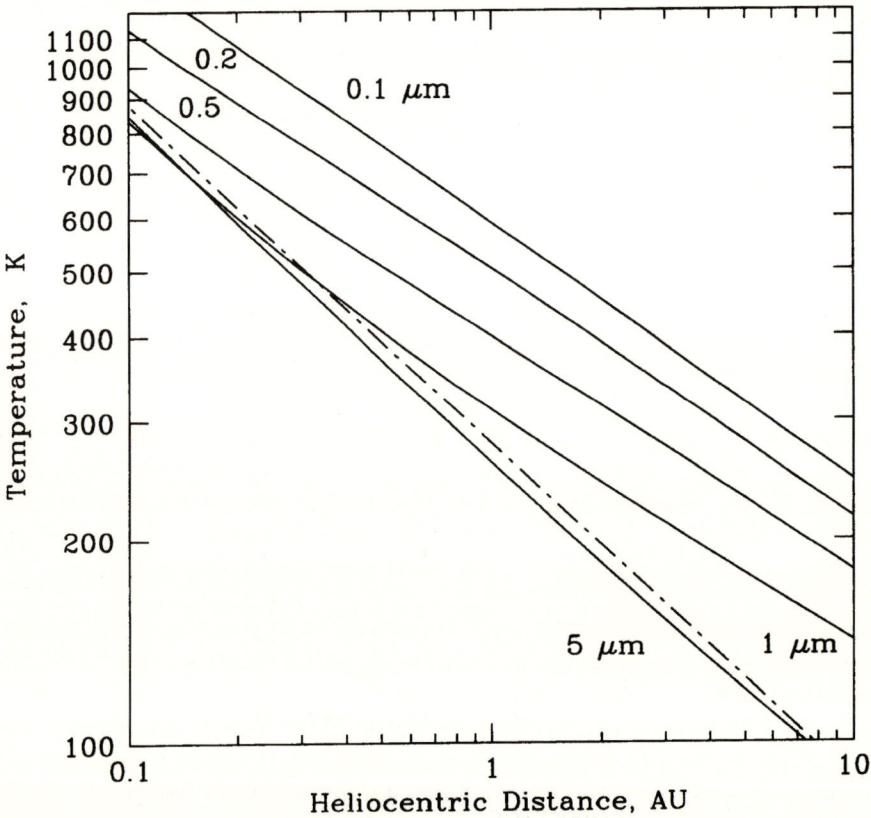

Figure 6. Equilibrium temperature vs. heliocentric distance for glassy carbon spheres, with radii 0.1–5 μm. Dash-dot curve is for perfect blackbody.

a 0.5 μm Fe-rich olivine grain would be hotter than a corresponding carbon grain. The temperature depends only weakly on grain size for the silicate particles. The dust mass spectra from the Halley probes indicated that the cometary silicates were Mg-rich (Jessberger et al. 1988; Lawler et al. 1989; Schulze et al. 1997).

In general, the cometary silicate grains must be sufficiently absorbing to be warm. Otherwise, the large amount of silicates necessary to produce a visible feature above the thermal continuum from the hot absorbing grains would result in a far stronger scattered light continuum than is observed. Either the silicates have a relatively high iron content (in contrast to the Halley results), or the grains are physically attached to absorbing material (the "mixed" particles detected during the Halley flybys). The absorbing material may be in the form of organic refractory mantles, as originally described by Greenberg (1982).

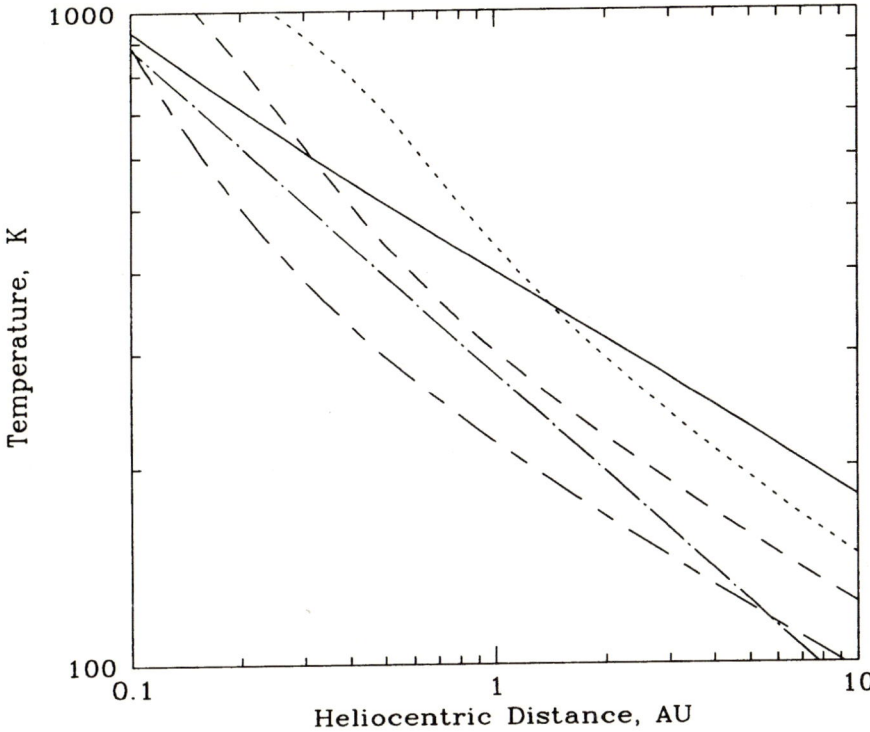

Figure 7. Equilibrium temperature versus heliocentric distance for glassy silicate spheres, radius 0.5 μm, with differing absorption. — — — — — $k = 0.001$; — — — $k = 0.01$; - - - - - olivine glass, Mg/Fe = 1 (Dorschner et al. 1995); ——— glassy carbon; — · — · — blackbody.

Bradley (1994) reported that a major component of chondritic interplanetary dust particles (IDPs) of likely cometary origin consists of Glass with Embedded FeNi Metal and Sulfides (GEMS). The FeNi beads are ∼0.01–0.04 μm in diameter and comprise 5 to less than 10% of the volume (Bradley, private communication). For such inclusions small compared to the wavelength, effective medium theory can be used to compute effective optical constants (Bohren and Huffman 1983). A volume fraction of 5% Fe beads in a Mg-rich silicate leads to an imaginary part of the refractive index $k = 0.01$, sufficient to heat the grains.

The absorptivity of organic residue grains depends on the ratio of H/C and can vary from $k \sim 0.01$ to 0.1 or more (e.g. Khare et al. 1990; Greenberg and Li 1996). Consequently, small organic grains could have temperatures ranging from $\sim T_{BB}$ up to the temperatures of small carbon grains.

We can conclude from the observed cometary SED, independent of a specific model, that the 3–13 μm thermal emission arises predominantly from submicron- to micron-sized grains with strong absorption or from very fluffy grains having structural units $\leq 1\mu$m radius (e.g., Hage and Greenberg 1990).

The observed color temperature corresponds to the envelope of the wavelength-dependent emission from the ensemble of grains and does not represent the physical temperature of the grains. Moreover, the color temperature can vary with wavelength, since the smallest, hottest grains will radiate more strongly at the shorter wavelengths. The comets with the strongest dust emission, such as Halley or Hale-Bopp, have higher color temperature at 3.5–8 μm than at 8–20 μm, indicating an enhanced abundance of hot submicron grains (Tokunaga et al. 1988; Williams et al. 1997).

Although the 3–20 μm thermal emission allows us to determine the amount of small dust in the coma, these data do not help us to assess the amount of mass in large particles with lower ratio of cross section to mass. Thermal emission at far-infrared and submillimeter wavelengths can, in principle, constrain the abundance of large particles in the coma, since the emissivity of the smaller grains will decrease as $\lambda^{-\alpha}$, $1 \leq \alpha \leq 2$ while the emissivity of the large grains remains essentially constant. Several comets were detected at 60 and 100 μm during the IRAS mission (Walker and Aumann 1984). The DIRBE instrument on the COBE satellite measured the spectral energy distribution from 3.5 to 100 μm for comets Okazaki-Levy-Rudenko (1989 XIX = C/1989 Q1), Austin (1990 V = C/1989 XI), and Levy (1990 XX = C/1990 K1) (Lisse et al. 1994, 1998). Only Levy had a color temperature $> 1.10 T_{BB}$ and a drop in the average grain emissivity at 60 and 100 μm, indicating that small grains dominated the emitted radiation. This is consistent with the 10 μm spectra; among these three comets only Levy had a strong silicate feature produced by small grains.

Comet Hale-Bopp was observed at long wavelengths from the ISO satellite. The on-board photometer (PHOT) measured the thermal flux through filters at 7–160 μm, while the two spectrometers (SWS, LWS) recorded the spectrum from 5–160 μm. To fit the slope of the SED at the long wavelengths with a size distribution of the form $n(a) \propto a^{-\alpha}$, requires $\alpha \leq 3.5$ (Grün et al. 2000). For an outflow velocity $v(a) \propto a^{-0.5}$, this result implies that the dust production size distribution from the nucleus has $\alpha \leq 4$ and that the mass is concentrated in large particles.

Jewitt and Matthews (1999) acquired submillimeter continuum images of Hale-Bopp in 1997. The observed submillimeter spectral index of 0.6 suggests that the emitting particles were millimeter sized. A dust production rate of 1–2 x 10^6 kg/s near 1 AU was derived, giving a dust/gas mass ratio of at least 5. Grün et al. (2000) also find this ratio to exceed 5 at $2.8 < r < 4.6$ AU.

Continuum emission at submillimeter wavelengths has been measured in several other comets and can be compared with the 10 μm flux to assess the relative contributions of large and small grains. Jewitt and Luu (1990) detected emission from 23P/Brorsen-Metcalf at 800 and 1100 μm. If one compares the flux with the 10 μm flux recorded 4 days earlier, one finds that the data are consistent with a radiating blackbody having constant emissivity (Lynch et al. 1992a). That is, the large grains emitting at 800 and 1100 μm can account for the 10 μm flux as well; P/Brorsen-Metcalf apparently lacked the population of small grains present in most comets, consistent also with the lack of a silicate feature and the low scattered light continuum. In contrast, four other comets measured by Jewitt and Luu (1992) and Hyakutake (Jewitt and Matthews 1997) have submillimeter fluxes or upper limits that are an order of magnitude lower than the flux extrapolated from 10 μm on the assumption of constant emissivity, as summarized in Table 1. In each case, the 10 μm and submillimeter measurements were less than 14 days apart. We infer that, in these comets, the 10 μm flux arises from smaller grains which do not radiate efficiently at submillimeter wavelengths.

The space probe missions to comet P/Halley provided an opportunity to measure directly the mass distribution in the dust coma of an active comet. The results are summarized by McDonnell et al. (1991). In particular, the Giotto DIDSY experiment covered a wide mass range, extending to millimeter grain sizes. The cumulative mass fluence is presented in Fig. 8, supplemented at masses $< 10^{-14}$ kg by results from the PIA experiment. The slope of 0.9 over much of the mass range is consistent with $n(a) \propto a^{-3.7}$. However, the mass distribution at particle masses $> 10^{-9}$ kg is much flatter, indicating that the total dust mass along the track sampled by Giotto was concentrated in these larger particles. If this distribution is typical of all comets, then much of the mass lost by comets is "hidden" in these larger particles and most estimates of cometary mass loss are too low. The dust production in comet Halley was very variable spatially and temporally. Giotto flew past at a time when the dust production was apparently decreasing. It is possible that the large particles were left from the earlier outburst and that the corresponding smaller, high-velocity particles had already left the inner coma.

Hage and Greenberg (1990) proposed that the DIDSY mass distribution, requiring large grains, can be reconciled with the high color temperature and silicate feature, requiring small grains, if the particles are large, extremely fluffy aggregates of micron- or submicron-sized grains, with porosity $> 95\%$.

Certainly, large particles are present in comets. Meteor streams associated with comets contain particles in this size range. Infrared-bright dust trails associated with comets were discovered by IRAS (Sykes et al. 1986, 1990); these trails consist of submillimeter size particles. Radar observations of comets IRAS–Araki–Alcock (1983 VII = C/1983 H1) (Goldstein et al. 1984; Harmon et al. 1989), Hyakutake (Harmon et al. 1997) and P/Halley (Campbell et al. 1989) detected the presence of large grains surrounding the nucleus. However, the submillimeter data in Table 1 would suggest that, except for P/Brorsen–Metcalf, the cross section is not concentrated in the largest particles.

TABLE 1.
10 micron and submillimeter fluxes from comets.

Comet	R (AU)	T_{BB} (K)	10 μm flux[a] (10^{-12} W/m²/μm)	Submillimeter flux density		λ (mm)
				predicted[b] (mJy)	observed[c] (mJy)	
23P/Brorsen-Metcalf 1989 N1	0.48	401	1.6	80	90	0.8
				42	45	1.1
Okazaki-Levy-Rudenko 1989 Q1	0.67	340	1.93	157	21[e]	0.8
Austin 1989 X1	0.57	368	2.5	160	<37[d]	0.8
4P/Faye 1991	1.59	220	0.3	85	<7.5[d]	1.1
Levy 1990 K1	1.67	215	0.8	258	<13.5[d]	1.1
Hyakutake 1996 B2	1.08	268	3.0[f]	6000	550[g]	0.8

[a] Data from Hanner et al. 1994a, Hanner et al. 1996; scaled to 18″ FOV.
[b] Blackbody extrapolation from 10 μm flux.
[c] Submillimeter data from Jewitt and Luu 1992.
[d] 3σ upper limit.
[e] Mean of 6 days.
[f] Mason et al. 1998, March 23.3, scaled to 16″ FOV.
[g] Jewitt and Mathews 1997, March 23.5.

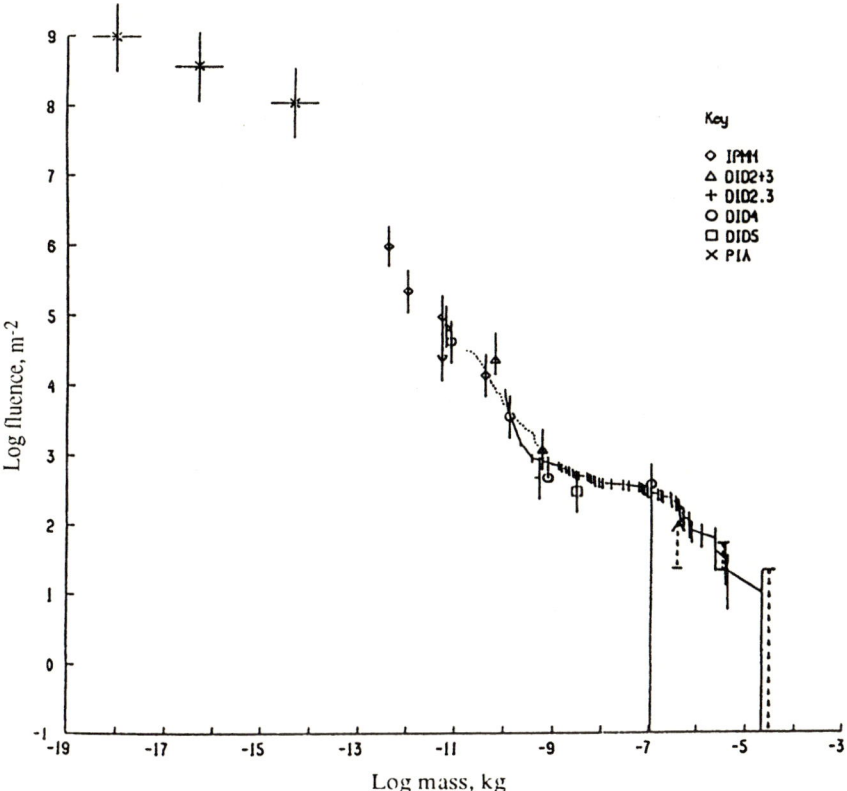

Figure 8. Dust particle fluence vs. mass as measured by the DIDSY and PIA experiments onboard the Giotto Halley probe (McDonnell et al. 1991).

III.B. Silicates

Small silicate grains in the coma produce an emission feature near 10 μm due to stretching vibrations in Si–O bonds. Additional bending mode vibrations occur between 16 and 35 μm. The wavelengths and shapes of these features are diagnostic of the mineral composition. The 10 μm feature lies within the 8–13 μm atmospheric "window" allowing ground-based observations. Although some 20 μm observations can also be made from the ground, the full 16–35 μm region is best studied from above the atmosphere.

Excess 10 μm emission is seen in filter photometry of many new and long period comets (Ney 1974, 1982; Rieke and Lee 1974; Ney and Merrill 1976; Gehrz and Ney 1992). The emission feature is strongest in active comets with strong scattered light continuum at visible wavelengths. The feature was observed in comet Halley throughout the 1985–1986 apparition, from 1.3 AU preperihelion to 1.48 AU post-perihelion; the strength of the feature varied from day to day and with position in the coma, as well as with heliocentric distance (Bregman et al. 1987; Tokunaga et al. 1986, 1988; Hanner et al. 1987; Ryan and Campins 1991; Gehrz and Ney 1992).

Low resolution 8–13 µm spectra with good signal-to-noise have been acquired for about a dozen comets. Six of the comets display a strong structured silicate emission feature. These are long-period comets Bradfield (1987 XXIX = C/1987 P1) (Hanner et al. 1990, Hanner et al. 1994a), Levy (1990 XX = C/1990 K1) (Lynch et al. 1992b), Hyakutake (C/1996 B2), Hale-Bopp (C/1995 O1), (Hanner et al. 1999, Wooden et al. 1999; Hayward et al. 2000), new comet Mueller (1994 I = C/1993 A1) (Hanner et al. 1994b) and 1P/Halley (Bregman et al. 1987; Campins and Ryan 1989).

By far the strongest silicate emission is seen in Hale-Bopp. This comet was also unusual in displaying a strong silicate feature even at 4.6 AU preperihelion (Crovisier et al. 1996). Near perihelion, spectra were obtained by several groups; all spectra show similar structure (see Hanner et al. 1999 for a review). A representative spectrum is presented in Fig. 9; the observed fluxes have been divided by a blackbody fitted at 8 and 12.5–13 µm. One sees that there are three peaks, at 9.2, 10.0, and 11.2 µm and minor structure at 11.9 and 10.5 µm. The spectral shape is very similar to that in Halley (Fig. 9; scaled by a factor of 3) and in the other comets cited above.

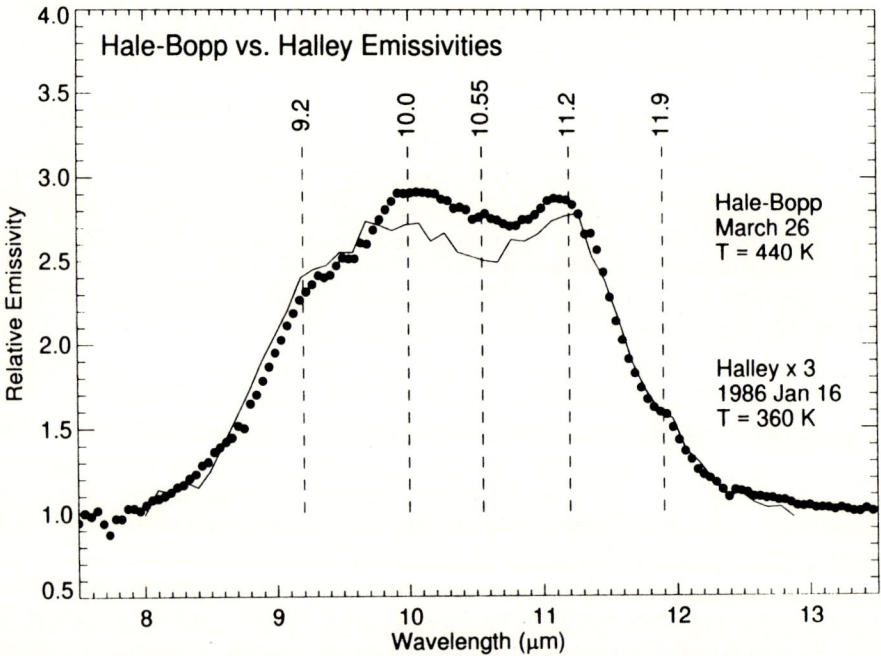

Figure 9. Silicate emission feature in two dusty comets: flux divided by blackbody continuum fit near 8 and 13 µm (Hanner et al. 1999). Filled circles: Hale-Bopp (C/1995 O1) at 0.92 AU; solid line: 1P/Halley at 0.79 AU (Campins and Ryan 1989).

The 11.2 μm peak is attributed to crystalline olivine, based on the good spectral match with the measured spectral emissivity of Mg-rich olivine (Stephens and Russell 1979; Koike et al. 1993). Only a small fraction (15–20%) of the silicate material needs to be in the form of crystalline olivine to produce the observed peak (Hanner et al. 1994a); the mass absorption coefficient of olivine near 11.2 μm is a factor of 3 to 10 times that of glassy silicates (Day 1974, 1976). The 11.9 μm shoulder is also due to crystalline olivine.

The 9.2 μm feature, first recognized in Hale-Bopp, is a signature of pyroxene. A peak wavelength of 9.2 μm corresponds to amorphous, Mg-rich pyroxene (Stephens and Russell 1979; Dorschner et al. 1995). Crystalline pyroxenes generate more variety in their spectra. The peaks usually appear at 10–11 μm, but a peak near 9.3 μm can also occur (Sandford and Walker 1985). The broader 10 μm maximum is characteristic of amorphous olivine (Stephens and Russell 1979), although crystalline olivine has a secondary peak at 10 μm as well.

Spectral models to match the Hale-Bopp spectra with a mixture of silicate minerals have been presented by Brucato et al. (1999), Hanner at al. (1999), Hayward et al. (2000), and Wooden et al. (1999, 2000). Wooden et al. have proposed that the observed changes in spectral shape with heliocentric distance can be explained by temperature differences between more transparent (cooler) Mg-rich pyroxene grains and less transparent (warmer) olivine grains. In their model, the cooler crystalline pyroxenes comprise the major fraction of small silicate grains.

A remarkable 16–45 μm spectrum of comet Hale–Bopp at $r = 2.9$ AU, shown in Fig. 10, was acquired with the ISO SWS spectrometer (Crovisier et al. 1997a,b). Five peaks are clearly visible, corresponding in every case to laboratory spectra of Mg–rich crystalline olivine (Koike et al. 1993). Minor structure is attributed to crystalline pyroxene (Wooden et al. 1999). Yet, airborne spectra of comet Halley at $r = 1.3$ AU (the only other 16–30 μm spectra of a comet) show only weak olivine features at 28.4 and 23.8 μm (Herter et al. 1987; Glaccum et al. 1987).

In summary, the observed spectral features suggest a complex mineralogy for the cometary silicates, including both amorphous and crystalline grains of pyroxene and olivine composition. This mineralogy is consistent with the chondritic aggregate IDPs thought to originate from comets because of their porous structure, fine-grained texture, high carbon content, and relatively high atmospheric entry velocities. Thus, we can apply what has been learned in the laboratory about the chondritic aggregate IDPs to enhance our understanding of cometary dust.

The spectra of four other new comets discussed in Hanner et al. (1994a) are puzzling; each has a unique, and not understood, spectrum. For example, an extremely broad emission feature is present in Wilson (1987 VII = C/1986 P1), suggesting a very amorphous silicate material. We may be witnessing the effect of cosmic ray damage to the outermost layer of the nucleus over the lifetime of the Oort Cloud. Silicate emission was detected for the first time in two short-period comets, 4P/Faye and 19P/Borrelly (Hanner et al. 1996).

The emission is broad in both comets, with no clear evidence for a peak at 11.2 μm. An ISO spectrum of P/Hartley 2 also shows broad weak 10 μm emission with a possible 11.2 μm peak (Crovisier et al. 2000). Other comets, such as 23P/Brorsen-Metcalf (Lynch et al. 1992a) and many other short-period comets lack a silicate feature; the most likely explanation is a lack of small grains rather than a deficiency of silicate material.

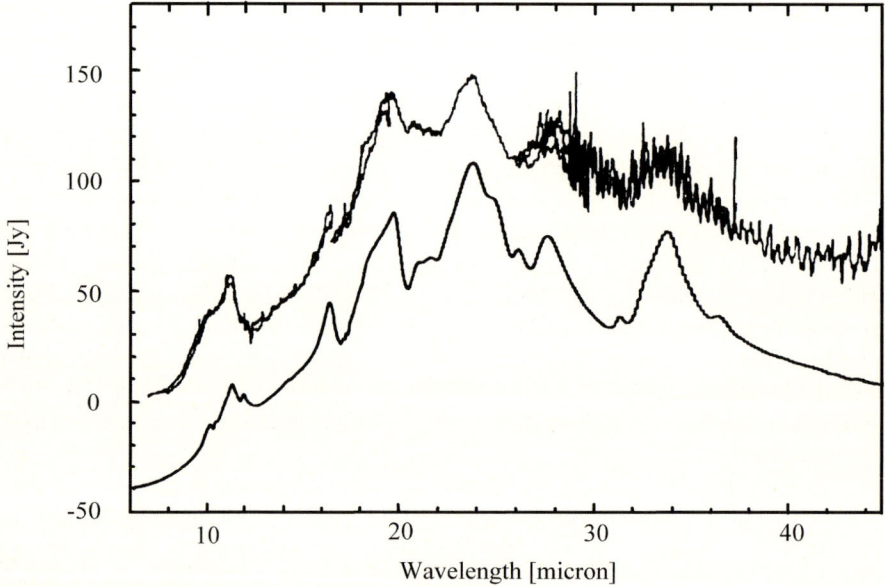

Figure 10. Silicate emission features in comet Hale-Bopp (C/1995 O1). Upper curve: ISO SWS spectrum at $r = 2.82$ AU; lower curve: modelled spectrum of forsterite from laboratory data (Crovisier et al. 1997b).

III.C. Infrared Spectral Features of Hydrocarbons

A broad emission feature centered near 3.36 μm was discovered in spectra of 1P/Halley (Combes et al. 1988; Baas et al. 1986). The emission is attributed to the stretching vibration of C–H bonds in organic molecules and was originally attributed to very small organic-rich dust grains. A similar emission feature has been detected in a number of bright comets since Halley, including new, long-period, and periodic comets (Brooke et al. 1991).

Reuter (1992) showed that methanol vibrational bands at 3.33 μm (ν_2) and 3.37 μm (ν_9) will contribute to the broad 3.36 μm emission. Detailed modeling of the methanol bands leads to a residual feature centered at 3.424–3.43 μm (Bockelée–Morvan et al. 1995; DiSanti et al. 1995). The strength of the residual feature correlates with water production rates and especially with the methanol abundance; thus, it is likely to arise from a gaseous species. Assuming a g-factor comparable to that of methanol, the abundance of the carrier is comparable to that of methanol, about 4% relative to water, making it a significant reservoir of carbon. Overtone and combination bands of methanol may also contribute. Further progress will require high spectral resolution to resolve line structure and to determine the methanol contribution.

A small, but distinct, feature at 3.29 μm is present in the spectra of the dusty comets 1P/Halley (Baas et al. 1986), Levy (C/1990 K1) (Davies et al. 1991) and 109P/Swift-Tuttle (DiSanti et al. 1995). In P/Swift-Tuttle, the feature appeared to be stronger when the dust continuum was stronger (Fig. 11). Typical of aromatic bonds, this feature could arise from polycyclic aromatic hydrocarbons either in molecules or in small solid grains. An interstellar feature at 3.29 μm is associated with other bands at 6.2, 7.7, 8.6, and 11.3 μm. No evidence of the 6.2 and 7.7 μm bands is seen in 5–8 μm spectra of comets Halley (Bregman et al. 1987), Wilson (C/1986 P1) (Lynch et al. 1989), and Hale-Bopp (Crovisier et al. 1997b).

Thus, to date, there is no positive identification of a spectral feature from CHON grains, but further study of the 3.29 μm feature is warranted.

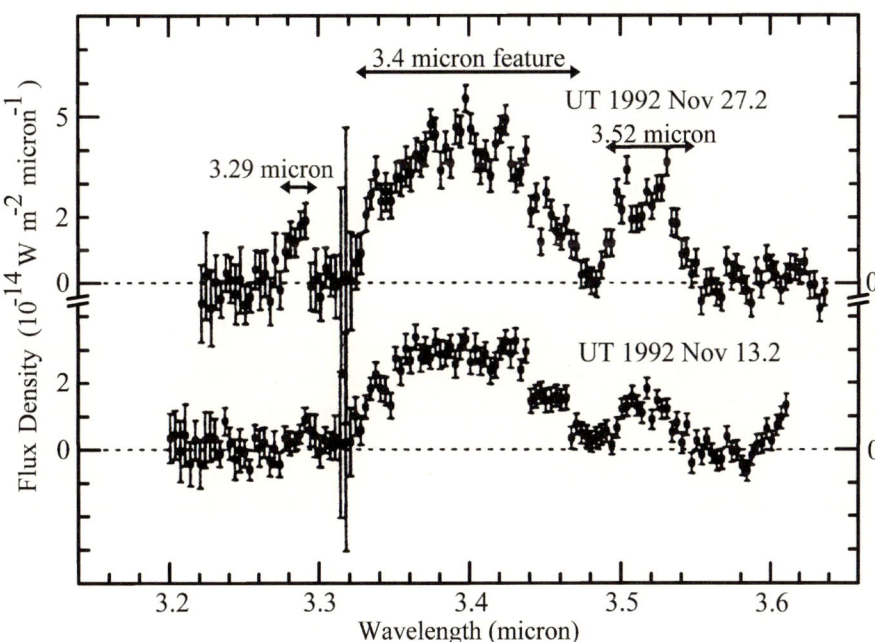

Figure 11. 109P/Swift-Tuttle spectra at $r = 1.0$ AU, as observed through the atmosphere, showing emissions in excess of the continuum (DiSanti et al. 1995). Note the 3.29 μm emission feature.

III.D. Scattering by Dust

The polarization, phase function, color and albedo are related to the size, composition (refractive index) and structure of the scattering particles. A recent review is given by Jockers (1999).

To determine the scattering phase function requires observations over an extended time and heliocentric distance range as the sun-earth-comet geometry changes. Consequently, one has the problem of normalization to account for the changing dust production rate; typically, the visual continuum is normalized to the gas production rate, assuming constant dust/gas ratio. Millis et al. (1982) derived the phase function for the dusty comet 38P/Stephan-Oterma by normalizing to the C_2 production rate. The phase function was a factor of two higher at phase angle θ of 3°–4° than at 30°, corresponding to a slope of $\sim$0.02 mag/deg. Meech and Jewitt (1987) determined a linear slope of 0.02–0.035 mag/deg for four comets observed at $0° < \theta < 25°$. They found no evidence for an opposition surge larger than 20% in P/Halley at $1.37° < \theta < 8.6°$. The phase function for the comets from 0°–30° is steeper than the volume scattering function of zodiacal dust derived by Lamy and Perrin (1986), but less steep than that of dark asteroids (Meech and Jewitt 1987). It is consistent with the measured phase function of fluffy absorbing particles (Hanner et al. 1981). Only two comets, West (C/1975 V1) and Bradfield (1980 XV = C/1980 Y1), have been observed at phase angle 150°–120° (scattering angles 30°–60°) (Ney and Merrill 1976; Ney 1982). The ratio of scattered to thermal energy shows strong forward scattering. The diffraction lobe appears to be wider for the comet dust than for the zodiacal light (Lamy 1986b).

The geometric albedo of a particle is defined as the ratio of the energy scattered at 0° phase to that scattered by a white Lambert disk of the same geometric cross section (Hanner et al. 1981). Since comets are rarely observed at 0° phase, it is convenient to define $A_p(\theta)$ at phase angle θ as the geometric albedo times the normalized phase function. Hanner and Newburn (1989) summarized the $A_p(\theta)$ at J (1.25 μm) and K (2.2 μm) for several comets determined from simultaneous measurements of the scattered and thermal radiation (Fig. 12). The albedos are very low, ranging from 0.025 at large phase angles to 0.05–0.10 near 0° phase in the J bandpass. It appears that comets beyond 3 AU have somewhat higher albedo, a trend also inferred by Hartmann and Cruikshank (1984) from the near-infrared colors. Mason et al. (2000) find significantly higher albedo for the dust in Hale-Bopp than in comet Halley.

A higher albedo for the dust during times of strong jet activity was seen in comet Halley (Tokunaga et al. 1986). This could be due to a shift in the mean grain size or to a component of less-absorbing grains ejected during outburst. Laboratory measurements of the geometric albedo for irregular absorbing particles showed a size dependence from $A_p \sim 0.08$ for particle radius 2 μm to $\sim$0.025 for particle radius 60 μm (Giese et al. 1986). Models of porous aggregates also predict a decrease in the albedo with increasing porosity for particles with constant mass (Hage and Greenberg 1990).

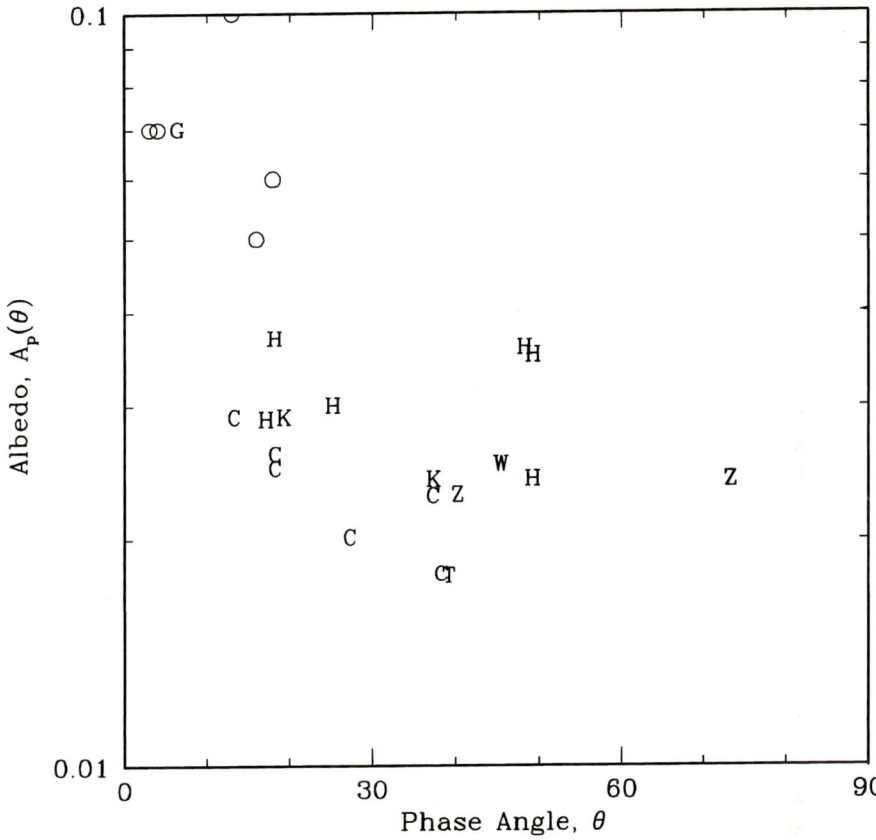

Figure 12. Mean albedo of comet dust vs. phase angle in the J (1.25 μm) bandpass (Hanner and Newburn 1989). The symbols are: H = 1P/Halley, C = 67P/Churyumov-Gerasimenko, K = 22P/Kopff, Z = 21P/Giacobini-Zinner, T = 9P/Tempel 1, G = 65P/Gunn, W = Wilson (C/1986 P1). Open circles are for comets observed at $r > 3$ AU.

Albedo maps of comets 21P/Giacobini-Zinner (Telesco et al. 1986), 1P/Halley (Hammel et al. 1987), and 109P/Swift-Tuttle (Fomenkova et al. 1994) were created by combining CCD images with thermal infrared images. The albedo was not constant across the coma, indicating variation in grain properties. In these 3 cases, the albedo increased radially outward from the nucleus. The lowest albedo occurred on the anti-sunward side of the nucleus in Halley and Giacobini-Zinner.

The color of the scattered light is generally redder than the sun; the reflectivity gradient decreases with wavelength from 5–18% per 0.1 μm at wavelengths 0.35–0.65 μm to 0–2% per 0.1 μm at 1.6–2.2 μm (Jewitt and Meech 1986). Measured near-infrared $J-H$ and $H-K$ colors are plotted in Fig. 13. The near-infrared colors of the dust may depend on heliocentric distance (Hartmann et al. 1982; Hartmann and Cruikshank 1984); however, Hanner and Newburn (1989) noted that only the $H-K$ color was less red in comets observed at $r > 3$ AU, while the $J-H$ color showed no trend with r. The dust is not necessarily the same color as the nucleus. In the case of 10P/Tempel 2, the $J-H$ color of the nucleus was up to 0.3 mag redder than that of the dust (Tokunaga et al. 1992).

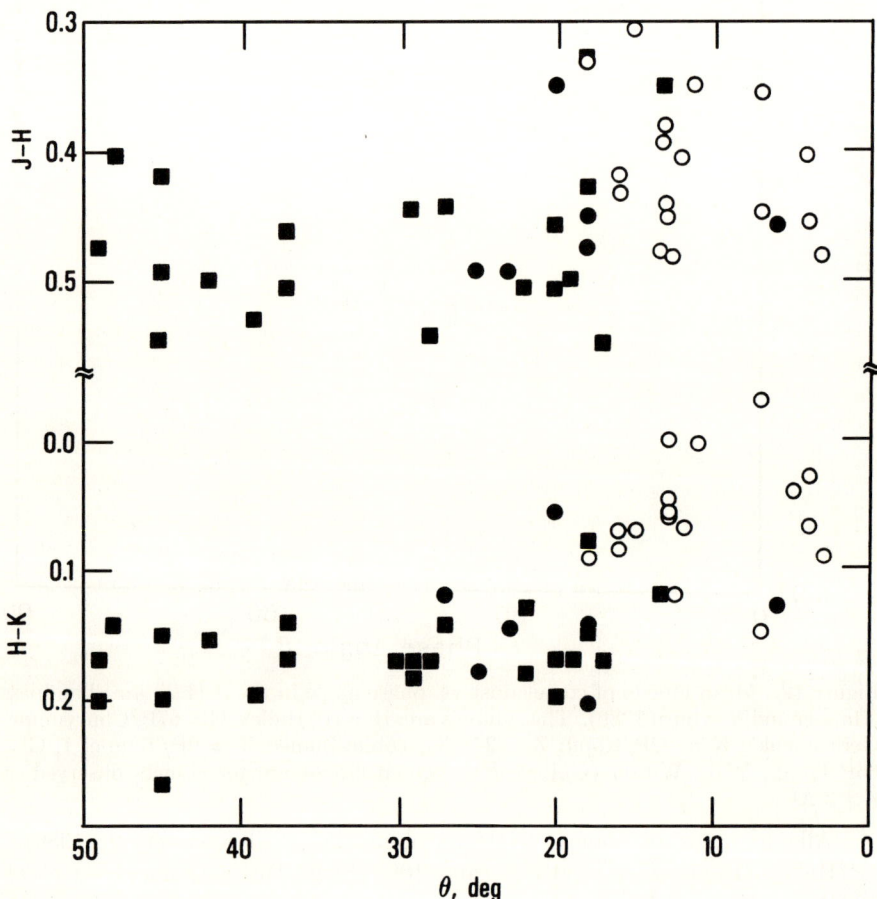

Figure 13. Observed $J-H$ and $H-K$ colors of comets vs. phase angle (Hanner and Newburn 1989). Symbols: squares: $r < 2$ AU; filled circles: $2 < r < 3$ AU; open circles: $r > 3$ AU.

The polarization $P(\theta)$ as a function of phase angle θ has been established by combining measurements from a number of comets (Kikuchi et al. 1987; Dollfus et al. 1988; Chernova et al. 1993; Levasseur-Regourd et al. 1996). Polarization has to be observed through filters that isolate the continuum; some earlier measurements were contaminated by gas emission. Negative polarization of order -2% is seen at small phase angles, with a neutral point at $\theta_c = 21° \pm 2°$, slope $h \sim 0.2$–0.3% per deg at θ_c, and maximum polarization near $90°$.

While the polarization at small phase angles is similar in all the comets, they tend to divide into two groups at larger θ, having $P_{\max} \sim 10-15\%$ and $\sim 25\%$ respectively (Chernova et al. 1993; Levasseur-Regourd et al. 1996). Hale-Bopp was observed at $7° < \theta < 47°$ in 1996–1997 (Furosho et al. 1999; Ganesh et al. 1998; Hadamcik et al. 1999; Hasegawa et al. 1999; Jockers et al. 1999; Kiselev and Velichko 1999; Manset and Bastien 2000). The polarization at all phase angles was distinctly higher than that measured for any previous comet at similar θ, even those in the high $P_{\max}$ class, such as Halley. Even the negative polarization branch appeared weaker in Hale-Bopp.

The higher $P_{\max}$ correlates with a stronger scattered light continuum and stronger silicate emission feature. For particle sizes of a few tenths of a micron, silicate particles tend to have relatively low positive polarization and more negative polarization at visual wavelengths, while carbonaceous particles show stronger positive polarization (Yanamandra-Fisher and Hanner 1999); this is not consistent with the association of a stronger silicate feature with higher $P_{\max}$. Only for grain radii ~ 0.1 μm and smaller does maximum polarization near 90° increase for both silicate and absorbing grains. Thus, the comets in the higher $P_{\max}$ group likely have a higher abundance of small grains ~ 0.1 μm radius or fluffy aggregates of such grains. The polarization at large phase angles $> \theta_c$ increases with wavelength (Dollfus et al. 1988; Hadamcik 1999), as one would expect if the ratio of particle size to wavelength is controlling $P_{\max}$.

Spatially resolved observations of the inner coma often show increased polarization in jets. Widespread use of CCD arrays to construct polarization images of comet Hale-Bopp has allowed a comparison of the polarization with coma morphology (Furosho et al. 1999; Hadamcik 1999; Hadamcik et al. 1999; Jockers et al. 1999; Jones and Gehrz 2000). Visible jets and arcs generally correspond to regions of higher polarization. The jets and arcs in Hale-Bopp also produced a stronger silicate feature and a higher color temperature, both signatures of smaller grains (Hayward et al. 2000).

As discussed in Subsection 2 of II.A., the parameter β, which governs the motion of particles after separation from the gas drag, links the coma morphology to dust optical properties via Q_{pr}, the efficiency factor for radiation pressure. In particular, there is a clear distinction between $\beta_{\max} > 2$ for submicron absorbing grains and $\beta_{\max} = 0.5$–0.6 for submicron silicate grains. For the spiral arcs in the coma of Hale-Bopp in 1997, Hayward et al. (2000) showed that $\beta_{\max} \sim 0.6$ was more consistent with the morphology than $\beta_{\max} \sim 2.5$, implying that the visible spirals traced the locus of small silicate grains.

In their study of polarization in P/Halley, Dollfus et al. (1988) distinguished 3 regions: a bright inner halo (radius $r \sim 100$ km), an inner coma or "fresh" dust ($r < 5000$ km) and an outer coma ($r \sim 10,000$ km). The inner coma consistently produced higher polarization at large phase angles, while the inner halo displayed lower polarization than the other regions at $\theta = 30°$–$40°$. Lower albedo near the nucleus was also apparent in Halley. A near-nucleus region of lower polarization a few thousand km in diameter was seen in Hale-Bopp (Hadamcik 1999). This circumnuclear zone was evident in both 1996 and 1997, although the morphology of the inner coma changed considerably during that time. At small phase angles, this zone exhibited strong negative polarization, as much as −5% at 0.67 μm (Hadamcik 1999). Other comets have also shown lower near-nucleus polarization, including Levy (C/1990 K1) (Renard et al. 1992), 47P/Ashbrook-Jackson (Renard et al. 1996; and 81P/Wild 2 (Hadamcik and Levasseur-Regourd 1999). Various explanations have been proposed, including a circumnuclear region of larger particles or sublimating grain mantles (e.g., Kolokolova et al. 2000).

In recent years, there has been encouraging progress in understanding the polarization by irregular particles, particularly aggregates. Numerical methods have allowed exploration of particle shape and aggregate structure, while direct scattering measurements have contributed data on particle structures too large to be handled easily by the numerical codes. West (1991) and Zerull et al. (1993) demonstrated from computation and laboratory measurements, respectively, that the polarization by fluffy aggregate particles is determined primarily by the polarization properties of the constituent grains. Thus, the above discussion of the relation between $P_{\max}$ and grain size applies also to the constituent grains in fluffy aggregates.

Two groups have recently reported results of experiments to measure the polarization by fluffy aggregates. The PROGRA^2 experiments make use of microgravity during parabolic airplane flights to levitate particles in the path of a laser (Worms et al. 1999; Hadamcik 1999). The microwave scattering laboratory at the University of Florida allows measurements of cometary analogue particles by scaling both particle size and wavelength to the microwave domain (Gustafson and Kolokolova 1999). These on-going studies should lead to better understanding of the cometary polarization, including the negative polarization at small phase angles.

III.E. Icy Grains

A number of new and long-period comets display a substantial coma of solid grains at $r > 4$ AU. The coma may include a significant component of water ice grains, ejected during sublimation of a more volatile species such as CO. With their large surface area, these grains can become a significant source of water production. A'Hearn et al. (1984) proposed that the strong OH production they observed in comet Bowell (1982 I = C/1980 E1) at $r > 4$ AU was due to icy grains. In fact, the OH production rate actually decreased by an order of magnitude between $r > 4$ AU and perihelion at 3.36 AU, implying that the icy component of the grains was not being replenished. Thermal emission from the grains at $r > 4$ AU helped to constrain the temperature and total cross section of the icy grains in comet Bowell, since the sublimation rate is a steep function of temperature while the 20 μm thermal emission varies only slowly with temperature. Hanner and Campins (1986) showed that the OH production and thermal emission were consistent only if the grains were at a temperature of ∼150 K, slightly higher than the blackbody temperature, and indicating that ice and absorbing material were well mixed. They also suggested that the phase change from amorphous to crystalline ice at ~140 K could have been an energy source for sublimation.

Water ice has spectral features near 1.5, 2.2, and 3 µm; of these, the 3 µm band is the strongest. The wavelength depends upon whether one sees the feature in scattering or absorption (Hanner 1981). Broad, shallow absorption bands at 1.5 and 2.05 µm attributed to water ice were detected in the spectrum of comet Hale-Bopp at 7 AU (Davies et al. 1997). By the time the comet reached 4.6 AU, however, the continuum from the large coma of small warm grains was too strong to permit further detections of ice absorption bands. Water ice also has spectral bands near 44 and 65 µm. Excess emission at these wavelengths was detected in the ISO LWS spectrum of Hale-Bopp at $r = 2.8$ AU by Lellouch et al. (1998).

At smaller heliocentric distances, water ice particles are unlikely to be observable in the coma. Although pure water ice grains of radius 10 µm or larger are stable against sublimation at 1 AU, the slightest admixture of absorbing material will raise their temperature into the sublimation regime (Hanner 1981). Thus, the lifetimes of even slightly dirty ice grains in a comet coma at $r < 3$ AU will be a few hours or less, limiting their range to a few hundred kilometers. A steep brightness gradient in the scattered light continuum with radial distance from the nucleus in some comets has been attributed to sublimation of icy grains. However, very strict limits on the size and purity would be necessary in order to invoke sublimation of water ice grains to explain a steeper than $1/r$ brightness gradient on a scale of 10^3–10^4 km (e.g., Hanner 1981). Moreover, a steep brightness gradient has also been found in thermal infrared images (Fomenkova et al. 1993) and the thermal emission certainly does not arise from icy grains. Nevertheless, loss of volatiles and grain fragmentaion do apparently take place in the coma, most likely due to organic components that volatilize at temperatures of order 300 K or higher (e.g., Greenberg and Li 1998). A'Hearn et al. (1986) reported narrow CN jets in comet Halley; evaporation from dust grains is the most plausible explanation for the narrow collimation. An extended source of CO in comet Halley was detected by the Giotto neutral mass spectrometer (Eberhardt et al. 1987) and an extended source of CO in Hale-Bopp was determined from long slit spectra (DiSanti et al. 1999).

IV. DUST CHEMICAL AND ISOTOPIC COMPOSITION

The knowledge of cometary dust composition before the Halley missions was very limited and merely qualitative. The presence of silicates was inferred from photometric observations at 10 µm (Ney 1982) and a few metals were observed in the sun-grazing comets (Arpigny 1979). Other sources of information like meteor spectra (Yavnel 1977) or the laboratory analysis of IDPs (cf. the subject chapter by Jessberger et al. in this book) were too indirect, since the true cometary origin of the material analyzed could not be established. Thus one of the most challenging goals of the three missions to comet Halley was to reveal the elemental, isotopic, and molecular composition of its dust. To that end on all missions there were almost identical impact-ionization time-of-flight mass spectrometers: PUMA-1 and PUMA-2 on the VEGA spacecraft and PIA on Giotto. The principle of their operation is described by Kissel (1986a,b). The present chapter is based on the results of these experiments, especially the results from the instrument PUMA-1 on the probe VEGA-1 that functioned best and supplied the largest volume of top-quality data. Altogether, about 5000 individual grains were measured by all three instruments. We will start with the facts that are well established from the mass spectra and proceed to the less secure inferences. The following aspects always have to be taken into account:

(1) The measurements were carried out at ~ 2 AU from Earth during the brief (about two-hour long) encounters with the coma and, obviously, cannot be repeated.

(2) The impact-ionization mass spectrometer cannot be calibrated in the laboratory since there is no technical means to accelerate analogues of micron-sized cometary particles up to 70–80 km/s, equivalent to the encounter velocity.

(3) Comprehensive theoretical description from "first principles" is unavailable (Inogamov 1987; Jessberger et al. 1988; Lawler et al. 1989; Mukhin et al. 1991; Hornung and Kissel 1994).

(4) The total amount of analyzed cometary material was very small (about a few nanograms), probably comparable to the mass of a single IDP collected in the stratosphere (Fomenkova et al. 1992).

(5) The data provide only very indirect information on the grain structure (Kissel and Krueger 1987a, Fomenkova et al. 1994b).

(6) The composition and structure of the analyzed dust grains may have been altered to some extent relative to their original state in the nucleus because of heating and UV radiation during the few hours they spent in the coma before the encounter.

IV.A. Facts from Ion Spectra

Very early in the analysis of the mass spectra it was noted (Kissel et al. 1986a, b) that in some spectra ions of the light elements H, C, N, O show the highest intensity peaks, while other spectra are dominated by ions of the rock-forming elements like Mg, Si, Ca, and Fe. The close relationship between ions within each of these groups is demonstrated by a common factor analysis (Jessberger et al. 1988). The cometary dust particles represented by those spectra were dubbed CHON and SILICATES, respectively, where CHON denotes refractory organic material (Jessberger et al. 1986; Clark et al. 1987) and SILICATES stands for the major rock forming elements (Jessberger et al. 1988). As will be clear from our later discussion on the mineralogic composition of Halley's dust, the term SILICATES appears to be too narrow. The presence of other compounds — such as Fe-sulfides, metal particles, etc. — suggests the term ROCK component as a more appropriate one, and it will be used throughout the paper. Lawler and Brownlee (1992) showed convincingly, however, that there is no grain consisting purely of CHON or purely of ROCK material and that even the smallest cometary particles appear to be fine scale mixtures of carbonaceous and inorganic phases, the proportion of the end member components varying from grain to grain. Fomenkova et al. (1992) proposed quantitative criteria for the classification of ion spectra. If the abundance ratio of carbon to any rock-forming element is more than 10, the particle is categorized as CHON. If this ratio is less than 0.1, the particle is included in the group ROCK; the rest are MIXED particles. According to these criteria, about a half of PUMA-1 and PUMA-2 spectra combined are MIXED, while CHON and ROCK particles each comprise about 25% of the total data set.

One of the features of the PUMA-1 and PUMA-2 instruments onboard the VEGA spacecraft (but not the PIA onboard Giotto) was the switching of the reflectron voltages every 30 seconds in order to accept ions with initial energies up to 50 eV or up to 150 eV to generate the so-called long and short spectra, respectively.

Figure 14 shows the average Mg-normalized ion abundances for a number of frequently occurring elements in both types of spectra analyzed by Langevin et al. (1987), by Jessberger et al. (1988), by Lawler et al. (1989), and by Mukhin et al. (1991). The elements H, C, and O are roughly ten times more abundant in the short spectra than in the long ones, while the relative abundances of Si, S, Ca, and Fe are about the same. Since the dust composition did not change every 30 seconds (certainly there is no hint of such a change!), the effect implies that the initial energies of CHON ions upon formation are on the average higher than those of the rock-forming ions. This finding led Kissel and Krueger (1987a) to the interpretation that cometary grains are composed of a silicate core covered by an organic refractory mantle, thus substantiating pre-Halley hypotheses (Greenberg 1982).

Analysis of the spatial distributions of the various types of particles in the coma based on the PUMA-1 data (Fomenkova et al. 1994b) established that ROCK particles (depleted in organic material) dominate the outer regions of the coma, while CHON and mixed particles are relatively more abundant closer to the nucleus. This observation indicates that the less refractory fraction of the organic component is lost from the grains as they spend some time in the coma. Partial volatilization of organics releases smaller ROCK grains further out from the nucleus, while providing the extended source for CO, H_2CO, CN, and some other species in the gas coma (Eberhardt et al. 1987; A'Hearn et al. 1986). This interpretation suggests that cometary grains have an "IDP-type" structure: a carbonaceous matrix with tiny silicate granules embedded rather than the core-mantle structure mentioned above. Both ideas can be reconciled by considering the possible size dependence of the grain structure: the core-mantle type for small grains, the IDP type for larger ones.

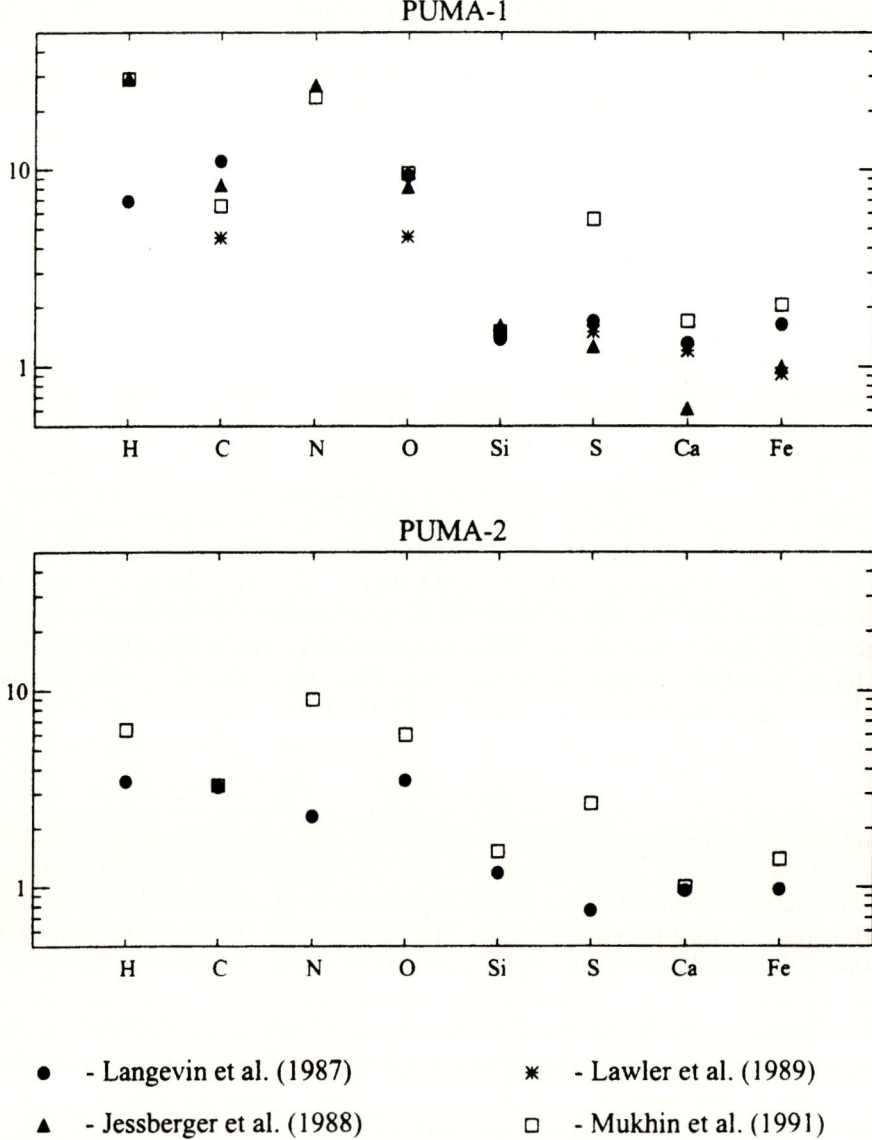

Figure 14. The ratios of average Mg-normalized abundances in the short and long spectra, as determined from the data collected by the instruments PUMA-1 and PUMA-2. All studies essentially agree that ions from the CHON-elements in the short spectra are approximately a factor of ten more abundant than in the long spectra, while ions from the rock-forming elements in both types of spectra are of equal abundance. We do not have an explanation why sulfur in Mukhin et al.'s (1991) study deviates from this general observation.

Mass spectrometric analysis itself yields information on the isotopic composition of the studied elements. However, because the signal-to-noise ratio was not high enough, isotopic information was obtained only for a few among the most abundant elements in the mass spectra of the grains: C, Mg, Si, S, Cl (terrestrial contaminator), and Fe. The isotopic ratios are generally normal, i.e., within large uncertainties they are indistinguishable from the terrestrial values (Šolc et al. 1986, 1987; Sagdeev et al. 1987; Grün and Jessberger 1990; Fomenkova et al. 1992). Specifically, a ^{26}Mg excess — from the now extinct ^{26}Al — larger than a factor of two is excluded (Jessberger et al. 1989). The ^{12}C/^{13}C ratios below the normal value — as encountered in meteoritic stardust — are inaccessible for the PIA and PUMA instruments because of unavoidable molecular interferences. A clear indication of an isotopic anomaly has been found only for light carbon (Jessberger and Kissel 1991): a few ^{12}C/^{13}C ratios are much higher than the normal value of 89 and range up to 5000. Graphite grains exhibiting similarly high ratios were extracted from the carbonaceous chondrite Murchison meteorite (Zinner et al. 1990). Ultralight carbon can be produced in He-burning or explosive H-burning processes and, thus, is indicative of a circumstellar origin of those grains (Anders and Zinner 1993). Therefore, Fomenkova and Chang (1996) hypothesized that cometary dust may contain a certain fraction of circumstellar solids, in addition to refractory interstellar materials. The particle with the isotopically lightest carbon, i.e., with the highest ^{12}C/^{13}C ratio in the PUMA-1 data, is composed of almost pure ^{12}C (99 wt%) and, in addition, contains 0.5 wt% by weight of rocky material which most probably consists of FeS and Fe-poor pyroxene with a trace of FeNi. This is a typical C-grain (Fomenkova et al. 1994b; see below).

There are a few other features of the dust particles that were further deduced from the mass spectra. From the number of projectile (cometary) ions and target (silver) ions the densities and masses of the projectiles have been estimated (Maas et al. 1989). CHON-dominated particles on the average are less dense (~ 1 g/cm^3) than the ROCK-dominated particles (~ 2–3 g/cm^3), the masses of individual dust grains being from 10^{-16} to 10^{-11} g. The dust particle masses were also estimated from the so-called front end channel signals (Fomenkova et al. 1991). These are electrical signals from the target, the accelerating grid, the photomultiplier, and the catcher grid that were measured during the collision of a dust particle with the target of the instrument and stored with the resulting mass spectrum. The mass of cometary particles was found to be comparable to that deduced by Maas et al. (1989): 5×10^{-17} to 5×10^{-12} g. For the density of the order of 0.3–3 g/cm^3 the radii of the cometary particles in question would be 0.2–2 μm.

With all the uncertainties involved in the dust particle mass determination, all authors agree that the analyzed grains were very small: the total mass of the cometary solid matter investigated by the PUMA and PIA instruments was only a few nanograms. Fortunately, the measured particles were not collected from one spot, but originated from different parts of the nucleus, thereby providing a somewhat representative sample of the comet. Note, however, that the proper comparison of the composition and properties of the measured Halley particles with those of IDPs and meteorites can be made only if these extraterrestrial materials are also sampled on the same submicron scale.

Three attempts have been made to gain more insight into the nature of the refractory organic component, CHON. Clark et al. (1987) qualitatively identified different types of particles based on the presence of the various combinations of H, C, O, N ions (i.e., [H,C], [H,C,N], etc.) in the spectra measured by the PIA onboard the Giotto spacecraft. The quantitative analysis, however, was hampered by the malfunction of this instrument's amplifier during the encounter.

Kissel and Krueger (1987a) considered 43 spectra of PUMA-1 and analyzed the residuals that remained after the peaks due to the obvious atomic ions had been removed from the spectra. Some of the residual peaks were identified with complex molecular ions and a number of organic-substance classes was therefore inferred to be present in cometary dust. However, an alternative explanation of these peaks in the mass spectra was later suggested by a number of authors (Sagdeev et al. 1989; Utterback and Kissel 1990). They showed that the peaks were (i) statistically random, (ii) caused by impacts of very small, attogram grains present in the P/Halley's coma very far, almost one million kilometers, from the nucleus.

Fomenkova et al. (1994b) applied cluster analysis to 515 PUMA-1 and PUMA-2 spectra dominated by CHON ions. They detected more than 30 distinguishable types of grains. The majority of Halley's organic particles appears to be multicomponent mixtures of carbon phases and organic compounds. In most cases, the available data do not warrant an unambiguous identification of corresponding compound types, but some general conclusions about the makeup of cometary nonvolatile organics still can be made. For example, elemental carbon grains composed essentially of carbon with other elements at the level of $\sim$1% (C-grains) have been found in cometary dust and they contain $\sim$8–10% of the total solid carbon in Halley. Tielens and Allamandola (1987) estimate that up to 50% of the solid carbon in the interstellar medium is locked up in graphite or amorphous carbon, while in carbonaceous chondrites well characterized carbon phases occur in rather low abundances of $\sim$2% (Zinner et al. 1990). In their more recent model of interstellar dust grains, Li and Greenberg (1997) find that the small carbonaceous particles responsible for the 220 nm hump use about 10–20% of the available carbon.

Other compositionally simple candidates for components of cometary solid organics are aliphatic and aromatic hydrocarbons (PAHs), and polymers of carbon suboxide (C_nCO) and of cyanopolyynes (HC_nN or NC_nN). Cometary dust also appears to contain heteropolymers and/or variable mixtures of various alcohols, aldehydes, ketones, acids and amino acids, and their salts. The simplest members of those homologous series occur in the interstellar medium (Irvine and Knacke 1989). However, the molecular weight of the cometary compounds should be high enough to prevent them from evaporation since the grains in the coma may reach a temperature of a few hundred Kelvin.

A variety of similarly nonvolatile compounds were found in the Murchison meteorite (Cronin and Chang 1993). These meteoritic organics often carry deuterium enrichments indicative of their interstellar origin. Finally, a similar set of products — polyalcohols, ethers, esters, carboxylic acids, and hydrocarbons — was observed in laboratory experiments where organic residues ("cometary ice tholins") were produced by plasma discharge irradiation of water/methanol/carbon dioxide/ethane cocondensed ices (McDonald et al. 1996). These experiments were intended to model the production of complex organics from simple interstellar ices which is considered as one of the possible mechanisms for formation of cometary organics.

We conclude that the compositional variety of types of cometary organic compounds is consistent with the interstellar dust model of comets (Greenberg 1982) and probably reflects the original diversity in the population of interstellar grains and/or the differences in the evolutionary history of precursor dust. It is hoped that the Rosetta mission will test some of those findings and shed more light on the nature of cometary organics.

IV.B. Bulk Composition

Unknown yields of elements ionized by a hypervelocity impact present a problem when calculating bulk elemental abundances from ion abundances registered in the mass spectra. As we stated in the beginning of this section, straightforward quantitative pre-flight calibrations of the instruments in the laboratory were not possible. Therefore, several empirical and theoretical approaches to the problem of conversion of the spectra from ions into atoms have been explored by various authors. Lawler et al. (1989) suggested empirical correction factors that place the peaks of the histograms of corrected Mg/Si, Fe/Si, and Fe/Mg ratios at the proper solar ratios. Jessberger et al. (1988) used yields calculated by Kissel and Krueger (1987b) from the results of experiments with dust particles accelerated up to 40 km/s, ion yields obtained by SIMS methods, and theoretical considerations. The fact that the derived total bulk atom abundances were solar within a factor of less than 2, has been used as a justification for the proposed coefficients. Mukhin et al. (1991), however, showed that if the bulk ion abundances of rock-forming elements in comet Halley dust are calculated without correction factors, but with the mass of dust particles taken into account (i.e., the contribution of each particle to the bulk abundances is proportional to its mass), then the resulting abundances are similarly close to the solar values.

Obvious shortcomings of all these approaches are that (i) none of them considers a possible dependence of ion yields on the mass of a colliding dust particle (that is, on the total number of ions in the post-impact plasma cloud); (ii) possible matrix effects (that is, the collective effects produced by ions of various elements upon each other) are neglected. Recently, a new attempt has been made to model "from first principles" the processes within an expanding plasma cloud under the conditions corresponding to those of the PUMA/PIA measurements (Hornung and Kissel 1994). When completed, these results may lead to a better understanding of the physics of the hypervelocity impact and facilitate further interpretations of the results from the PUMA/PIA experiments.

For now, the firm conclusion of all studies is that the bulk abundances of the major rock-forming elements integrated over all spectra are solar (that is, indistiguishable CI-chondritic — Anders and Grevesse 1989) within a factor of <2, which characterizes the overall accuracy of measurements. The average solar/chondritic elemental abundances substantiate the view that comet Halley accreted from the same nebular reservoir as the rest of the solar system, while the wide diversity in individual grain composition underscores the unequilibrated nature of comet Halley material.

All studies of Halley's dust have found that, in contrast to the rock forming elements, the CHON elements H, C, and N are enriched relative to their CI-chondritic abundances; some studies have found that O is enriched as well. The depletion of these "non-condensible" elements in CI-chondrites relative to the Sun is interpreted as early loss of these elements. Thus, the lesser depletion of H, C, and N in cometary dust testifies that cometary solids are more primitive than CI-chondrites.

Grün and Jessberger (1990) suggested that the abundance of volatile elements in the whole comet, i.e., dust + ice (the latter measured as the gas; Krankowsky and Eberhardt 1990), is solar. From fitting both the C and O abundances to solar, they derived a cosmochemical dust/gas mass ratio of two in excellent agreement with an independent physical dust/gas ratio determination of 2.2 (Hughes 1988). Hydrogen and nitrogen were shown to be depleted in the dust + ice relative to solar by factors of 600 and 2, respectively (Grün and Jessberger 1990; Fomenkova and Chang 1996). An earlier estimate (Geiss 1987) showing the depletion of N by a factor of 3 was based on too low a dust/gas ratio of 0.5. For the improved estimate of the dust-to-gas ratio of $2.2:1$ (Hughes 1988), the partition of the volatile elements between gas and dust is $1:2$ for C and N, and $2:1$ for H and O. Apparently, C and N occur predominantly in cometary dust, while H and O prevail in the gas phase, as expected for a water-rich object (Fomenkova and Chang 1996).

IV.C. Mineralogical Composition

To infer information on the mineralogic composition of the dust grains, one needs to know exact yields — which convert ions to atoms — with uncertainty much smaller than a factor of two. Since they are not available, a number of approaches, as described in Jessberger et al. (1988), Jessberger and Kissel (1991), Fomenkova et al. (1992) and references therein, were undertaken to circumvent the problem. Although these approaches cannot do without circular reasoning, they all basically show that Halley's grains are not in chemical equilibrium. This is demonstrated by the broad distribution of $Fe/(Fe+Mg)$ ratios (Brownlee et al. 1987; Jessberger et al. 1988, 1989; Lawler et al. 1989) and also by comparison of the Mg-Si-Fe distributions of Halley's grains (Jessberger and Kissel 1991; Jessberger 1993) with that of anhydrous and hydrated interplanetary dust particles and meteorites (Bradley et al. 1989; Germani et al. 1990).

Rietmeijer et al. (1989) inferred the presence of hydrated silicates from the PUMA-2 data and Fomenkova et al. (1992) identified Mg-carbonate particles in cometary dust. These minerals are abundant in CI and CM meteorites where they formed as a result of aqueous activity on the parent bodies. Therefore, their presence in comet Halley indicates the possibility of hydration in the history of cometary dust. The hydration of cometary material could not have taken place in the nebula (Prinn and Fegley 1989), but probably occurred *in situ* in the hydrocryogenic mode. Carbonates and layer silicates are minor but persistent phases in anhydrous chondritic porous IDPs as well.

Fomenkova et al. (1992) also analyzed Fe-rich particles and found metal, iron oxide, iron sulfides, Fe-rich silicates, and mixtures of these minerals in Halley's grains. Some of these grains are Ni-poor and some are Ni-rich, the mean ratio of Ni/Fe being 0.14 ± 0.03.

Finally, the absence of Ca-Al-rich grains (the earliest objects formed in the solar system) at least in the analyzed portion of the spectra has been demonstrated by Jessberger et al. (1989). Fomenkova and Chang (1994) have discussed the absence of SiC grains (an interstellar compound found in meteorites) in the PUMA data and estimated the upper limit of SiC in cometary dust as <150 ppm (to compare with 6–8 ppm in meteorites; Anders and Zinner 1993).

In the most recent study of Halley's dust, Schulze et al. (1997) found the mineralogical composition shown in Table 2.

TABLE 2.
Estimated mineralogical composition of Halley's dust (Schulze et al. 1996).

Mineral group	Estimated proportion	Mineral chemistry	Possible minerals
Mg silicates	>20%	Fe-poor, Ca-poor	Mg-rich pyroxene and/or olivine
Fe sulfides	~10%	some Ni-rich	pyrrhotite, pentlandite
Fe metal	1–2%	Ni-poor	kamacite
Fe oxide	<1%		magnetite

Reproduced from Schulze et al. (1997) courtesy of the Astronomical Society of the Pacific Conference Series.

V. THE FUTURE

Progress in all research fields is rapid these days and the science of cometary dust is no exception. Although we have made an effort to update this chapter, new data, their analysis, and their interpretations are being continuously published. Many issues addressed in this chapter are still evolving, and unexpected changes in the consensus, which is here by and large presented, are possible.

Investigation of the dynamical properties of cometary dust has of course the longest tradition and this is reflected in a relatively large number of references to classical investigations, some of which were undertaken fairly long ago. Yet, even in this research area there have recently been major new developments, e.g., the role of localized sources of outgassing on the mechanism of dust emission from the nucleus. An omnipresent process that affects cometary solids is fragmentation, which appears to be occurring on all possible scales. In more specific terms, it is expected that further progress in our understanding the dynamical properties of cometary dust will come from further studies of comet Hale-Bopp, from analysis of the very recently disintegrated comet LINEAR (C/1999 S4), and from investigations of the behavior of the enormous population of the sungrazing comets, to name a few. It is also expected that the trend toward more synergistic studies, involving the dynamical, optical, thermal, and other properties of the dust, will become common.

The origin of the crystalline cometary silicates remains puzzling (Hanner 1999). If they formed in the inner solar nebula, then their presence in comets requires extensive mixing in the solar nebula during the epoch of comet formation. If they are presolar, then one has to explain why the spectral signature of crystalline olivine is not seen in the interstellar medium or in young stellar objects except at their latest evolutionary stage. Further progress may be achieved by future analyses of the silicate grains in IDPs of likely cometary origin.

The PIA/PUMA experiments onboard the Giotto and VEGA missions to comet Halley for the first time provided information on the chemical composition and nature as well as on the isotopic composition of cometary dust. Despite limitations and uncertainties, they brought about a quantum jump compared to what had been known before the missions. They also prepared the baseline for the two next approved cometary missions, ROSETTA and STARDUST, that are expected to yield data with a higher degree of security and much more details. The laboratory and *in-situ* analysis of dust from two more comets will address the issue to what extent the composition of Halley's comet is typical.

Acknowledgements

This research was carried out in part at the Jet Propulsion Laboratory, California Institute of Technology, under contract with the National Aeronautics and Space Administration.

REFERENCES

A'Hearn, M. F., Hoban, S., Birch, P. V., Bowers, C., Martin, R., and Klinglesmith, D. A. 1986. CN jets in comet P/Halley. *Nature*, **324**, pp. 649–651.

A'Hearn, M. F., Millis, R. L., Schleicher, D. G., Osip, D. J., and Birch, P. V. 1995. The ensemble properties of comets: Results from narrowband photometry of 85 comets, 1976-1992. *Icarus*, **118**, pp. 223–270.

A'Hearn, M. F., and Schleicher, D. G. 1988. Comet P/Encke's nongravitational force. *Ap. J.*, **331**, pp. L47–L51.

A'Hearn M. F., Schleicher, D. G., Feldman, P. D., Millis, R. L., and Thompson, D. T. 1984. Comet Bowell 1980b. *Astron. J.*, **89**, pp. 579–591.

Akabane, T. 1983. The secondary tail of comet 1976 VI West. *Publ. Astron. Soc. Japan*, **35**, pp. 565–578.

Anders, E., and Grevesse, N. 1989. Abundances of the elements: Meteoritic and solar. *Geochim. Cosmochim. Acta*, **53**, pp. 197–214.

Anders, E., and Zinner, E. 1993. Interstellar grains in primitive meteorites: diamond, silicon carbide, and graphite. *Meteoritics*, **28**, pp. 490–514.

Arpigny, C. 1979. Relative abundances of the heavy elements in comet Ikeya-Seki (1965 VIII). In *Les éléments et leurs isotopes dans l'univers*, XXIIe Coll. Int. d'Astrophys. (Liège: Université de Liège), pp. 189–197.

Arrhenius, S. A. 1900. Ueber die Ursache der Nordlichter. *Phys. Zeitschr.*, **2**, pp. 81–110.

Asphaug, E., and Benz, W. 1996. Size, density, and structure of comet Shoemaker-Levy 9 inferred from the physics of tidal breakup. *Icarus*, **121**, pp. 225–248.

Baas, F., Geballe, T. R., and Walther, D. M. 1986. Spectroscopy of the 3.4 micron emission feature in comet Halley. *Ap. J.*, **311**, pp. L97–L101.

Becklin, E. E., and Westphal, J. A. 1966. Infrared observations of comet 1965f. *Ap. J.*, **145**, pp. 445–453.

Beisser, K., and Boehnhardt, H. 1987a. Evidence for the nucleus rotation in streamer patterns of comet Halley's dust tail. *Astrophys. Space Sci.*, **139**, pp. 5–12.

Beisser, K., and Boehnhardt, H. 1987b. Dust tail streamers and Halley's nucleus rotation. In *Diversity and Similarity of Comets*, ESA SP-278, eds. E. J. Rolfe and B. Battrick (Noordwijk: ESTEC), pp. 665–670.

Bessel, F. W. 1836. Beobachtungen ueber die physische Beschaffenheit des Halley'schen Kometen und dadurch veranlasste Bemerkungen. *Astron. Nachr.*, **13**, pp. 185–232.

Bobrovnikoff, N. T. 1951. Comets. In *Astrophysics*, ed. J. A. Hynek (New York: McGraw-Hill), pp. 302–356.

Bockelée-Morvan, D., Brooke, T. Y., and Crovisier, J. 1995. On the origin of the 3.2- to 3.6-μm emission features in comets. *Icarus*, **116**, pp. 18–39.

Bockelée-Morvan, D., and Gérard, E. 1984. Radio observations of the hydroxyl radical in comets with high spectral resolution. Kinematics and asymmetries of the OH coma in C/Meier (1978 XXI), C/Bradfield (1979 X), and C/Austin (1982g). *Astron. Astrophys.*, **131**, pp. 111–122.

Boehnhardt, H. 1986. The charge of fluffy dust grains of silicate and carbon near P/Halley and P/Giacobini-Zinner. In *Exploration of Halley's Comet*, ESA SP-250, eds. B. Battrick, E. J. Rolfe, and R. Reinhard (Noordwijk: ESTEC), Vol.2, pp. 207–213.

Boehnhardt, H., and Fechtig, H. 1987. Electrostatic charging and fragmentation of dust near P/Giacobini-Zinner and P/Halley. *Astron. Astrophys.*, **187**, pp. 824–828.

Bohren, C. F., and Huffman, D. R. 1983. *Absorption and Scattering of Light by Small Particles*. (New York: John Wiley & Sons.)

Bond, G. P. 1862. Account of the great comet of 1858. *Ann. Harvard Coll. Obs.*, **3**, pp. 1–372.

Bradley, J. P. 1994. Chemically anomalous, preaccretionally irradiated grains in interplanetary dust from comets. *Science*, **265**, pp. 925–929.

Bradley, J. P., Brownlee, D. E., and Veblen, D. R. 1983. Pyroxene whiskers and platelets in interplanetary dust: evidence of vapour phase growth. *Nature*, **301**, pp. 473–477.

Bradley, J. P., Germani, M., and Brownlee, D. 1989. Automated thin-film analyses of anhydrous interplanetary dust particles in the analytical electron microscope. *Earth Planet. Sci. Lett.*, **93**, pp. 1–13.

Bregman, J. D., Campins, H., Witteborn, F. C., Wooden, D. H., Rank, D. M., Allamandola, L. J., Cohen, M., and Tielens, A. G. G. M. 1987. Airborne and groundbased spectrophotometry of comet P/Halley from 5–13 micrometers. *Astron. Astrophys.*, **187**, pp. 616–620.

Brooke, T. Y., Tokunaga, A. T., and Knacke, R. F. 1991. Detection of the 3.4 μm emission feature in comets P/Brorsen-Metcalf and Okazaki-Levy-Rudenko (1989r) and an observational summary. *Astron. J.*, **101**, pp. 268–278.

Brownlee, D., Wheelock, M., Temple, S., Bradley, J., and Kissel, J. 1987. A quantitative comparison of comet Halley and carbonaceous chondrites at the submicron level. *Lunar Planet. Sci. Conf.*, **18**, pp. 133–134.

Brucato, J. R., Colangeli, L., Mennella, V., Palumbo, P., and Bussoletti, E. 1999. Silicates in Hale-Bopp: hints from laboratory studies. *Plan. Space Sci.*, **47**, pp. 773–779.

Burns, J. A., Lamy, P. L., and Soter, S. 1979. Radiation forces on small particles in the solar system. *Icarus*, **40**, pp. 1–48.

Campbell, D. B., Harmon, J. K., and Shapiro, I. I. 1989. Radar observations of comet Halley. *Astrophys. J.*, **338**, pp. 1094–1105.

Campins, H., Rieke, G. H., and Lebofsky, M. J. 1983. Ice in comet Bowell. *Nature*, **302**, pp. 405–406.

Campins, H., and Ryan, E. V. 1989. The identification of crystalline olivine in cometary silicates. *Astrophys. J.*, **341**, pp. 1059–1066.

Chernova, G. P., Kiselev, N. N., and Jockers, K. 1993. Polarimetric characteristics of dust particles as observed in 13 comets: Comparisons with asteroids. *Icarus*, **103**, pp. 144–158.

Clark, B., Mason, L., and Kissel, J. 1987. Systematics of the "CHON" and other light-element particle populations in comet P/Halley. *Astron. Astrophys.*, **187**, pp. 779–784.

Combes, M., Moroz, V. I., Crovisier, J., Encrenaz, T., Bibring, J.-P., Grigoriev, A. V., Sanko, N. F., Coron, N., Crifo, J. F., Gispert, R., Bockelée-Morvan, D., Nikolsky, Yu. V., Krasnopolsky, V. A., Owen, T., Emerich, C., Lamarre, J. M., and Rocard, F. 1988. The 2.5–12-μm spectrum of comet Halley from the IKS–VEGA experiment. *Icarus*, **76**, pp. 404–436.

Combi, M. R. 1994. The fragmentation of dust in the innermost comae of comets: possible evidence from ground-based images. *Astron. J.*, **108**, pp. 304–312.

Combi, M. R., Kabin, K., DeZeeuw, D. L., Gombosi, T. I., and Powell, K. G. 1999. Dust-gas interrelations in comets: Observations and theory. *Earth Moon Plan.*, **79**, pp. 275–306.

Cremonese, G., and Fulle, M. 1989. Photometrical analysis of the neck-line structure of comet Halley. *Icarus*, **80**, pp. 267–279.

Cremonese, G., and Fulle, M. 1990. The dust tail of comet Wilson 1987 VII. *Astron. J.*, **100**, pp. 1285–1292.

Crifo, J. F. 1991. Hydrodynamic models of the collisional coma. In *Comets in the Post-Halley Era*, eds. R. L. Newburn, Jr., M. Neugebauer, and J. Rahe (Dordrecht: Kluwer), pp. 937–989.

Cronin, J., and Chang, S. 1993. Organic matter in meteorites: molecular and isotopic analysis of the Murchison meteorite. In *The Chemistry of Life's Origin*, eds. J. M. Greenberg, C. X. Mendoza-Gomez, and V. Pirronello (Dordrecht: Kluwer), pp. 209–258.

Crovisier, J., Brooke, T. Y., Hanner, M. S., Keller, H. U., Lamy, P. L., Altieri, B., Bockelée-Morvan, D., Jorda, L., Leech, K., and Lellouch, E. 1996. The infrared spectrum of comet C/1995 O1 (Hale-Bopp) at 4.6 AU from the Sun. *Astron. Astrophys.*, **315**, pp. L385–L388.

Crovisier, J., Brooke, T. Y., Leech, K., Bockelée-Morvan, D., Lellouch, E., Hanner, M. S., Altieri, B., Keller, H. U., Lim, T., Encrenaz, T., Salama, A., Griffin, M., de Graauw, T., van Dishoeck, E., and Knacke, R. F. 2000. The thermal infrared spectra of comets Hale-Bopp and 103P/Hartley 2 observed with the Infrared Space Observatory. In *Thermal Emission Spectroscopy and Analysis of Dust Disks and Regoliths*, Conf. Series, eds. M. L. Sitko, A. L. Sprague, and D. K. Lynch (San Francisco: Astron. Soc. Pacific Press), pp. 109–117.

Crovisier, J., Leech, K., Bockelée-Morvan, D., Brooke, T. Y., Hanner, M. S., Altieri, B., Keller, H. U., and Lellouch, E. 1997a. The spectrum of comet Hale-Bopp (C/1995 O1) observed with the Infrared Space Observatory at 2.9 astronomical units from the sun. *Science*, **275**, pp. 1904–1907.

Crovisier, J., Leech, K., Bockelée-Morvan, D., Brooke, T. Y., Hanner, M. S., Altieri, B., Keller, H. U., and Lellouch, E. 1997b. The infrared spectrum of comet Hale-Bopp. In *First ISO Workshop on Analytical Spectroscopy*, ESA SP-419, eds. A. M. Heras, K. Leech, N. R. Trams, and M. Perry (Noordwijk: ESTEC), pp. 137–140.

Davies, J. K., Green, S. F., and Geballe, T. R. 1991. The detection of a strong 3.28 μm emission feature in comet Levy. *Mon. Not. R. Astron. Soc.*, **251**, pp. 148–151.

Davies, J. K., Roush, T. L., Cruikshank, D. P., Bartholomew, M. J., Geballe, T. R., Owen, T., and de Bergh, C. 1997. The detection of water ice in comet Hale-Bopp. *Icarus*, **127**, pp. 238–245.

Day, K. L. 1974. A possible identification of the 10-micron "silicate" feature. *Ap. J.*, **192**, pp. L15–L17.

Day, K. L. 1976. Further measurements of amorphous silicates. *Ap. J.*, **210**, pp. 614–617.

de Pater, I., Palmer, P., Snyder, L. E., and Ip, W.-H. 1986. VLA observations of comet Halley: The brightness distribution of OH around the comet. In *Exploration of Halley's Comet*, ESA SP-250, eds. B. Battrick, E. J. Rolfe, and R. Reinhard (Noordwijk: ESTEC), Vol.1, pp. 409–412.

DiSanti, M. A., Mumma, M. J., Geballe, T. R., and Davies, J. K. 1995. Systematic observations of methanol and other organics in comet P/Swift-Tuttle: discovery of new spectral structure at 3.42 μm. *Icarus*, **116**, pp. 1–17.

DiSanti, M. A., Mumma, M. J., Dello Russo, N., Magee-Sauer, K., Novak, R., and Rettig, T. W. 1999. Identification of two sources of carbon monoxide in comet Hale-Bopp. *Nature*, **399**, pp. 662–665.

Dollfus, A., Bastien, P., Le Borgne, J.-F., Levasseur-Regourd, A. C., and Mukai, T. 1988. Optical polarimetry of P/Halley: synthesis of the measurements in the continuum. *Astron. Astrophys.*, **206**, pp. 348–356.

Dorschner, J., Begemann, B., Henning, Th., Jager, C., and Mutschke, H. 1995. Steps toward interstellar silicate mineralogy II. Study of Mg-Fe silicate glasses of variable composition. *Astron. Astrophys.*, **300**, pp. 503–520.

Dorschner, J., Friedemann, C., Gürtler, J., and Henning, T. 1988. Optical properties of glassy bronzite and the interstellar silicate bands. *Astron. Astrophys.*, **198**, pp. 223–232.

Eberhardt, P., Krankowsky, D., Schulte, W., Dolder, U., Lämmerzahl, P., Berthelier, J. J., Woweries, J., Stubbemann, U., Hodges, R. R., Hoffman, J. H., and Illiano, J. M. 1987. The CO and N_2 abundance in comet P/Halley. *Astron. Astrophys.*, **187**, pp. 481–484.

Finson, M. L., and Probstein, R. F. 1968a. A theory of dust comets. I. Model and equations. *Ap. J.*, **154**, pp. 327–352.

Finson, M. L., and Probstein, R. F. 1968b. A theory of dust comets. II. Results for Comet Arend-Roland. *Ap. J.*, **154**, pp. 353–380.

Fomenkova, M. N., and Chang, S. 1994. Carbon in comet Halley dust particles. In *Analysis of Interplanetary Dust*, eds. M. E. Zolensky, T. L. Wilson, F. J. M. Rietmeijer, and G. J. Flynn (New York: Am. Inst. Phys.), pp. 193–202.

Fomenkova, M. N., and Chang, S. 1996. The link between cometary and interstellar dust. In *The Cosmic Dust Connection*, ed. J. M. Greenberg (Dordrecht: Kluwer), pp. 459–465.

Fomenkova, M., Larson, S., Jones, B., and Pina, R. 1994a. Albedo maps of comet Swift-Tuttle. *Bull. Am. Astron. Soc.*, **26**, p. 1119. (Abstract.)

Fomenkova, M. N., Chang, S., and Mukhin, L. M. 1994b. Carbonaceous components in the comet Halley dust. *Geochim. Cosmochim. Acta*, **58**, pp. 4503–4512.

Fomenkova, M. N., Evlanov, E. N., Mukhin, L. M., and Prilutskii, O. F. 1991. Determination of mass of comet Halley dust particles. *Lunar Planet. Sci. Conf.*, **22**, pp. 397–398.

Fomenkova, M. N., Jones, B., Pina, R. K., Puetter, R. C., McFadden, L. A., Abney, F., and Gehrz, R. D. 1993. Thermal-infrared high-resolution imaging of comet Austin. *Icarus*, **106**, pp. 489–498.

Fomenkova, M. N., Kerridge, J., Marti, K., and McFadden, L. 1992. Compositional trends in rock-forming elements of comet Halley dust. *Science*, **258**, pp. 266–269.

Fröhlich, H.-E., and Notni, P. 1988. Radiation pressure — a stabilizing agent of dust clouds in comets? *Astron. Nachr.*, **309**, pp. 147–155.

Fulle, M. 1987a. A new approach to the Finson-Probstein method of interpreting cometary dust tails. *Astron. Astrophys.*, **171**, pp. 327–335.

Fulle, M. 1987b. A possible neck-line structure in the dust tail of comet Halley. *Astron. Astrophys.*, **181**, pp. L13–L14.

Fulle, M. 1987c. Meteoroids from comet Bennett 1970 II. *Astron. Astrophys.*, **183**, pp. 392–396.

Fulle, M. 1988a. Meteoroids from comets Arend-Roland 1957 III and Seki-Lines 1962 III. *Astron. Astrophys.*, **189**, pp. 281–291.

Fulle, M. 1988b. Meteoroids from comet Kohoutek 1973 XII. *Astron. Astrophys.*, **201**, pp. 161–168.

Fulle, M. 1989. Evaluation of cometary dust parameters from numerical simulations: comparison with an analytical approach and the role of anisotropic emissions. *Astron. Astrophys.*, **217**, pp. 283–297.

Fulle, M. 1992. A dust-tail model based on Maxwellian velocity distribution. *Astron. Astrophys.*, **265**, pp. 817–824.

Fulle, M., Barbieri, C., and Cremonese, G. 1987. The dust tail of comet Halley. In *Diversity and Similarity of Comets*, ESA SP-278, eds. E. J. Rolfe and B. Battrick (Noordwijk: ESTEC), pp. 639–644.

Fulle, M., Barbieri, C., and Cremonese, G. 1988. The dust tail of comet P/Halley from ground-based CCD images. *Astron. Astrophys.*, **201**, pp. 362–372.

Fulle, M., Böhm, C., Mengoli, G., Muzzi, F., Orlandi, S., and Sette, G. 1994. Current meteor population of comet P/Swift-Tuttle 1992t. *Astron. Astrophys.*, **292**, pp. 304–310.

Fulle, M., Bosio, S., Cremonese, G., Cristaldi, S., Liller, W., and Pansecchi, L. 1993a. The dust environment of comet Austin 1990 V. *Astron. Astrophys.*, **272**, pp. 634–650.

Fulle, M., Cremonese, G., Jockers, K., and Rauer, H. 1992. The dust tail of comet Liller 1988 V. *Astron. Astrophys.*, **253**, pp. 615–624.

Fulle, M., Mennella, V., Rotundi, A., Colangeli, L., Bussoletti, E., and Pasian, F. 1993b. The dust environment of comet P/Grigg-Skjellerup as evidenced from ground-based observations. *Astron. Astrophys.*, **276**, pp. 582–588.

Fulle, M., and Sedmak, G. 1988. Photometrical analysis of the neck-line structure of comet Bennett 1970 II. *Icarus*, **74**, pp. 383–398.

Furusho, R., Suzuki, B., Yamamoto, N., Kawakita, H., Sasaki, T., Shimizu, Y. and Kurakami, T. 1999. Imaging polarimetry and color of the inner coma of comet Hale-Bopp (C/1995 O1). *Pub. Astron. Soc. Japan*, **51**, pp. 367–373.

Ganesh, S., Joshi, U. C., Baliyan, K. S., and Deshpande, M. R. 1998. Polarimetric observations of the comet Hale-Bopp. *Astron. Astrophys. Suppl. Ser.*, **129**, pp. 489–493.

Gehrz, R. D., and Ney, E. P. 1992. 0.7–23 μm photometric observations of P/Halley 1986 III and six recent bright comets. *Icarus*, **100**, pp. 162–186.

Geiss, J. 1987. Composition measurements and the history of cometary matter. *Astron. Astrophys.*, **187**, pp. 859–866.

Germani, M., Bradley, J. P., and Brownlee, D. 1990. Automated thin-film analyses of hydrated interplanetary dust particles in the analytical electron microscope. *Earth Planet. Sci. Lett.*, **101**, pp. 162–179.

Giese, R. H., Killinger, R. T., Kneissel, B., and Zerull, R. H. 1986. Albedo and color of dust grains: Laboratory versus cometary results. In *Exploration of Halley's Comet*, ESA SP-250, eds. B. Battrick, E. J. Rolfe, and R. Reinhard (Noordwijk: ESTEC), Vol.2, pp. 53–57.

Glaccum, W., Moseley, S. H., Campins, H., and Lowenstein, R. F. 1987. Airborne spectrophotometry of P/Halley from 20 to 65 microns. *Astron. Astrophys.*, **187**, pp. 635–638.

Goldstein, R. M., Jurgens, R. F., and Sekanina, Z. 1984. A radar study of comet IRAS-Araki-Alcock 1983d. *Astron. J.*, **89**, pp. 1745–1754.

Gombosi, T. I. 1986. A heuristic model of the comet Halley dust size distribution. In *Exploration of Halley's Comet*, ESA SP-250, eds. B. Battrick, E. J. Rolfe, and R. Reinhard (Noordwijk: ESTEC), Vol.2, pp. 167–171.

Gombosi, T. I. 1991. Multidimensional dusty gasdynamical models of inner cometary atmospheres. In *Comets in the Post-Halley Era*, eds. R. L. Newburn, Jr., M. Neugebauer, and J. Rahe (Dordrecht: Kluwer Acad. Publ.), pp. 991–1001.

Gombosi, T. I., Cravens, T. E., and Nagy, A. F. 1985. Time-dependent dusty gas dynamical flow near cometary nuclei. *Ap. J.*, **293**, pp. 328–341.

Gombosi, T. I., Szegö, K., Gribov, B. E., Sagdeev, R. Z., Shapiro, V. D., Shevchenko, V. I., and Cravens, T. E. 1983. Gas dynamic calculations of dust terminal velocities with realistic dust size distributions. In *Cometary Exploration*, ed. T. I. Gombosi (Budapest: Hung. Acad. Sci.), Vol.2, pp. 99–111.

Greenberg, J. M. 1982. What are comets made of? A model based on interstellar dust. In *Comets*, ed. L. L. Wilkening (Tucson: Univ. of Arizona Press), pp. 131–163.

Greenberg, J. M., and Gustafson, B. Å. S. 1981. A comet fragment model for zodiacal light particles. *Astron. Astrophys.*, **93**, pp. 35–42.

Greenberg, J. M., and Hage, J. I. 1990. From interstellar dust to comets: a unification of observational constraints. *Ap. J.*, **361**, pp. 260–274.

Greenberg, J. M., and Li, A. 1996. What are the true astronomical silicates? *Astron. Astrophys.*, **309**, pp. 258–266.

Greenberg, J. M., and Li, A. 1998. From interstellar dust to comets: the extended CO source in comet Halley. *Astron. Astrophys.*, **332**, pp. 374–384.

Greenberg, J. M., Mizutani, H., and Yamamoto, T. 1995. A new derivation of the tensile strength of cometary nuclei: application to comet Shoemaker-Levy 9. *Astron. Astrophys.*, **295**, pp. L35–L38.

Grün, E., and Jessberger, E. K. 1990. Dust. In *Physics and Chemistry of Comets*, ed. W. F. Huebner (Heidelberg: Springer), pp. 113–176.

Grün, E., Hanner, M. S., Peschke, S. B., Müller, T., Boehnhardt, H., Brooke, T. Y., Campins, H., Crovisier, J., Delahodde, C., Heinrichsen, I., Keller, H. U., Knacke, R. F., Krüger, H., Lamy, P., Leinert, Ch., Lemke, D., Lisse, C. M., Müller, M., Osip, D. J., Solc, M., Stickel, M., Sykes, M., Vanysek, V., and Zarnecki, J. 2000. Broadband infrared photometry of comet Hale-Bopp with ISOPHOT. *Astron. Astrophys.*, in press.

Gustafson, B. Å. S. 1989. Comet ejection and dynamics of nonspherical dust particles and meteoroids. *Ap. J.*, **337**, pp. 945–949.

Gustafson, B. Å. S. and Kolokolova, L. 1999. A systematic study of the light scattering by aggregate particles using the microwave analog technique: Angular and wavelength dependence of intensity and polarization. *J. Geophys. Res.*, **104**, pp. 31,711–31,720.

Hadamcik, E. 1999. Contribution a une classification des comètes a partir d'observations et de simulations en laboratoire. Ph.D. dissertation, University of Paris 6.

Hadamcik, E., and Levasseur-Regourd, A. C. 1999. Lumière diffusée par les particules solides éjectées du noyau de la comète 81P/Wild 2. *C. R. Acad. Sci. Paris*, 2B,C2–316.

Hadamcik, E., Levasseur-Regourd, A. C., and Renard, J. B. 1999. CCD polarimetric imaging of comet Hale-Bopp (C/1995 O1). *Earth Moon Plan.*, **78**, pp. 365–371.

Hage, J. I., and Greenberg, J. M. 1990. A model for the optical properties of porous grains. *Ap. J.*, **361**, pp. 251–259.

Hahn, J. M., Rettig, T. W., and Mumma, M. J. 1996. Comet Shoemaker-Levy 9 dust. *Icarus*, **121**, pp. 291–304.

Hammel, H. B., Telesco, C. M., Campins, H., Decher, R., Storrs, A. D., and Cruikshank, D. P. 1987. Albedo maps of comets P/Halley and P/Giacobini-Zinner. *Astron. Astrophys.*, **187**, pp. 665–668.

Hanner, M. S. 1981. On the detectability of icy grains in the comae of comets. *Icarus*, **47**, pp. 342–350.

Hanner, M. S. 1983. The nature of cometary dust from remote sensing. In *Cometary Exploration*, ed. T. I. Gombosi (Budapest: Hung. Acad. of Sci.), Vol.2, pp. 1–22.

Hanner, M. S. 1999. The silicate material in comets. *Space Sci. Rev.*, **90**, pp. 99–108.

Hanner, M. S., and Campins, H. 1986. Thermal emission from the dust coma of comet Bowell and a model for the grains. *Icarus*, **67**, pp. 51–62.

Hanner, M. S., Gehrz, R. D., Harker, D. E., Hayward, T. L., Lynch, D. K., Mason, C. C., Russell, R. W., Williams, D. M., Wooden, D. H., and Woodward, C. E. 1999. Thermal emission from the dust coma of comet Hale-Bopp and the composition of the silicate grains. *Earth Moon Plan.*, **79**, pp. 247–264.

Hanner, M. S., Giese, R. H., Weiss, K., and Zerull, R. 1981. On the definition of albedo and application to irregular particles. *Astron. Astrophys.*, **104**, pp. 42–46.

Hanner, M. S., Hackwell, J. A., Russell, R. W., and Lynch, D. K. 1994b. Silicate emission feature in the spectrum of comet Mueller 1993a. *Icarus*, **112**, pp. 490–495.

Hanner, M. S., Lynch, D. K., and Russell, R. W. 1994a. The 8–13 micron spectra of comets and the composition of the silicate grains. *Ap. J.*, **425**, pp. 274–285.

Hanner, M. S., Lynch, D. K., Russell, R. W., Hackwell, J. A., and Kellogg, R. 1996. Mid-infrared spectra of comets P/Borrelly, P/Faye, and P/Schaumasse. *Icarus*, **124**, pp. 344–351.

Hanner, M. S., and Newburn, R. L. 1989. Infrared photometry of comet Wilson (1986 l) at two epochs. *Astron. J.*, **97**, pp. 254–261.

Hanner, M. S., Newburn, R. L., Gehrz, R. D., Harrison, T., Ney, E. P., and Hayward, T. L. 1990. The infrared spectrum of comet Bradfield (1987s) and the silicate emission feature. *Ap. J.*, **348**, pp. 312–321.

Hanner, M. S., Tedesco, E., Tokunaga, A. T., Veeder, G. J., Lester, D. F., Witteborn, F. C., Bregman, J. D., Gradie, J., and Lebofsky, L. 1985. The dust coma of periodic comet Churyumov-Gerasimenko (1982 VIII). *Icarus*, **64**, pp. 11–19.

Hanner, M. S., Tokunaga, A. T., Golisch, W. F., Griep, D. M., and Kaminski, C. D. 1987. Infrared emission from P/Halley's dust coma during March 1986. *Astron. Astrophys.*, **187**, pp. 653–660.

Harding, C. L. 1824. Astronomische Nachrichten, Beobachtungen des diesjährigen Kometen, etc. *Berlin. Astron. Jahrbuch für 1827*, pp. 131–135.

Harmon, J. K., Campbell, D. B., Hine, A. A., Shapiro, I. I., and Marsden, B. G. 1989. Radar observations of comet IRAS-Araki-Alcock 1983d. *Ap. J.*, **338,** pp. 1071–1093.

Harmon, J. K., Ostro, S. J., Benner, L. A. M., Rosema, K. D., Jurgens, R. F., Winkler, R., Yeomans, D. K., Choate, D., Cormier, R., Giorgini, J. D., Mitchell, D. L., Chodas, P. W., Rose, R., Kelley, D., Slade, M. A., and Thomas, M. L. 1997. Comet Hyakutake (C/1996 B2): Radar detection of nucleus and coma. *Science*, **278**, pp. 1921–1924.

Hartmann, W. K., and Cruikshank, D. P. 1984. Comet color changes with solar distance. *Icarus*, **57**, pp. 55–62.

Hartmann, W. K., Cruikshank, D. P., and Degewij, J. 1982. Remote comets and related bodies: VJHK colorimetry and surface materials. *Icarus*, **52**, pp. 377–408.

Hasegawa, H., Ichikawa, T., Abe, S., Hamamura, S., Ohnishi, K., and Watanabe, J. 1999. Near-infrared photometric and polarimetric observations of comet Hale-Bopp. *Earth Moon Plan.*, bf 78, pp. 353–358.

Hayward, T. L., Hanner, M. S., and Sekanina, Z. 2000. Thermal infrared imaging and spectroscopy of comet Hale-Bopp. *Ap. J.*, **538**, pp. 428–455.

Hellmich, R. 1981. The influence of the radiation transfer in cometary dust halos on the production rates of gas and dust. *Astron. Astrophys.*, **93**, pp. 341–346.

Hellmich, R., and Keller, H. U. 1981. On the visibility of nuclei of dusty comets. *Icarus*, **47**, pp. 325–332.

Herter, T., Campins, H., and Gull, G. E. 1987. Airborne spectrophotometry of P/Halley from 16 to 30 microns. *Astron. Astrophys.*, **187**, pp. 629–631.

Hevelius, J. 1682. Excerpta ex epistola. II. De cometa anno 1682 mense Augusto & Septimbri viso. *Acta Erudit.*, Dec. 1682, pp. 389–391.

Hewins, R. H. 1988. Experimental studies of chondrules. In *Meteorites and the Early Solar System*, eds. J. F. Kerridge and M. S. Matthews (Tucson: Univ. of Arizona Press), pp. 660–679.

Horányi, M., and Mendis, D. A. 1986. The effects of electrostatic charging on the dust distribution at Halley's comet. *Ap. J.*, **307**, pp. 800–807.

Hornung, K., and Kissel, J. 1994. On shock wave impact ionization of dust particles. *Astron. Astrophys.*, **291**, pp. 324–336.

Hughes, D. W. 1988. Origin of the Solar System. In *Origins*, ed. A. C. Fabian (Cambridge: Cambridge University Press), pp. 26–68.

Hughes, D. W. 1991. Possible mechanisms for cometary outbursts. In *Comets in the Post-Halley Era*, eds. R. L. Newburn, Jr., M. Neugebauer, and J. Rahe (Dordrecht: Kluwer Acad. Publ.), pp. 825–851.

Inogamov, N. A. 1987. Electrostatic screening by self-consistent space charge and the ion dynamics in a time-of-flight mass spectrometer. *J. Eng. Phys.*, **52**, pp. 396–403.

Irvine, W. M., and Knacke, R. F. 1989. The chemistry of interstellar gas and grains. In *Origin and Evolution of Planetary and Satellite Atmospheres*, eds. S. K. Atreya et al. (Tucson: Univ. of Arizona Press), pp. 3–34.

Jaegermann, R. 1903. *Prof. Dr. Th. Bredichin's Mechanische Untersuchungen über Cometenformen*. (St. Petersburg: Voss).

Jambor, B. J. 1973. The split tail of comet Seki-Lines. *Ap. J.*, **185,** pp. 727–734.

Jessberger, E. K. 1993. Über die Zusammensetzung des kometären Staubes. *Mitt. Österreich. Mineralog. Ges.*, **138,** pp. 19–32.

Jessberger, E. K., Christoforidis, A., and Kissel, J. 1988. Aspects of the major element composition of Halley's dust. *Nature*, **332,** pp. 691–695.

Jessberger, E. K., and Kissel, J. 1991. Chemical properties of cometary dust and a note on carbon isotopes. In *Comets in the Post-Halley Era*, eds. R. L. Newburn, Jr., M. Neugebauer, and J. Rahe (Dordrecht: Kluwer Acad. Publ.), pp. 1075–1092.

Jessberger, E. K., Kissel, J., Fechtig, H., and Krueger, F. R. 1986. On the average chemical composition of cometary dust. In *Comet Nucleus Sample Return*, ESA SP-249, ed. O. Melita (Noordwijk: ESTEC), pp. 27–30.

Jessberger, E. K., Kissel, J., and Rahe, J. 1989. The composition of comets. In *Origin and Evolution of Planetary and Satellite Atmospheres*, eds. S. K. Atreya, J. B. Pollack, and M. S. Matthews (Tucson: Univ. of Arizona Press), pp. 167–191.

Jewitt, D. 1990. The persistent coma of comet P/Schwassmann-Wachmann 1. *Ap. J.*, **351,** pp. 277–286.

Jewitt, D., and Luu, J. 1990. The submillimeter radio continuum of comet P/Brorsen-Metcalf. *Ap. J.*, **365,** pp. 738–747.

Jewitt, D., and Luu, J. 1992. Submillimeter continuum emission from comets. *Icarus*, **100,** pp. 187–196.

Jewitt, D. C., and Matthews, H. E. 1997. Submillimeter continuum observations of comet Hyakutake (1996 B2). *Astron. J.*, **113,** pp. 1145–1151.

Jewitt, D., and Matthews, H. 1999. Particulate mass loss from comet Hale-Bopp. *Astron. J.*, **117,** pp. 1056–1062.

Jewitt, D., and Meech, K. J. 1986. Cometary grain scattering versus wavelength, or, "what color is cometary dust?". *Ap. J.*, **310,** pp. 937–952.

Jewitt, D., Senay, M., and Matthews, H. 1996. Observations of carbon monoxide in comet Hale-Bopp. *Science*, **271,** pp. 1110–1113.

Jockers, K. 1999. Observations of scattered light from cometary dust and their interpretation. *Earth Moon Plan.*, **79,** pp. 221–245.

Jockers, K., Rosenbush, V. K., Bonev, T., and Credner, T. 1999. Images of polarization and colour in the inner coma of comet Hale-Bopp. *Earth Moon Plan.*, **78,** pp. 373–379.

Jones, T. J., and Gehrz, R. D. 2000. Infrared imaging polarimetry of comet C/1995 O1 (Hale-Bopp). *Icarus*, **143,** pp. 338–346.

Jorda, L., Lecacheux, J., and Colas, F. 1997. Comet C/1995 O1 (Hale-Bopp). *IAU Circ.* No. 6583.

Jorda, L., Rembor, K., Lecacheux, J., Colom, P., Colas, F., Frappa, E., and Lara, L. M. 1999. The rotational parameters of Hale-Bopp (C/1995 O1) from observations of the dust jets at Pic du Midi Observatory. *Earth Moon Plan.*, **77,** pp. 167–180.

Keller, H. U., Delamere, W. A., Huebner, W. F., Reitsema, H. J., Schmidt, H. U., Whipple, F. L., Wilhelm, K., Curdt, W., Kramm, R., Thomas, N., Arpigny, C., Barbieri, C., Bonnet, R. M., Cazes, S., Coradini, M., Cosmovici, C. B., Hughes, D. W., Jamar, C., Malaise, D., Schmidt, K., Schmidt, W. K. H., and Seige, P. 1987. Comet P/Halley's nucleus and its activity. *Astron. Astrophys.*, **187,** pp. 807–823.

Keller, H. U., Knollenburg, J., and Markiewicz, J. 1994. Collimation of cometary dust jets and filaments. *Planet. Space Sci.*, **42,** pp. 367–382.

Keller, H. U., Marconi, M. L., and Thomas, N. 1990. Hydrodynamic implications of particle fragmentation near cometary nuclei. *Astron. Astrophys.*, **227,** pp. L1–L4.

Keller, H. U., and Thomas, N. 1989. Evidence for near-surface breezes on comet P/Halley. *Astron. Astrophys.*, **226,** pp. L9–L12.

Khare, B. N., Arakawa, E. T., Meisse, C., Thompson, W. R., Sagan, C., Gilmour, I., and Anders, E. 1990. Optical constants of kerogen from 0.15 to 40 μm. In *First International Conference on Laboratory Research for Planetary Atmospheres*, NASA CP-3077 (Washington D.C.: NASA), pp. 340–356.

Kikuchi, S., Mikami, Y., Mukai, T., Mukai, S., and Hough, J. H. 1987. Polarimetry of comet P/Halley. *Astron. Astrophys.*, **187,** pp. 689–692.

Kimura, H., and Liu, C.-p. 1977. On the structure of cometary dust tails. *Chinese Astron.*, **1,** pp. 235–264.

Kirch, G. 1681. *Neuen Himmelszeitung, darinn sonderlich und ausführlich von den zwey neuen grossen im Jahr 1680 erschienenen Cometen, etc.* (Nürnberg), p. 62.

Kiselev, N. N., and Velichko, F. P. 1999. Aperture polarimetry and photometry of comet Hale-Bopp. *Earth Moon Plan.*, **78**, pp. 347–352.

Kissel, J. 1986a. The Giotto particulate impact analyser. In *The Giotto Mission — Its Scientific Investigations*, ESA SP-1077, eds. R. Reinhard and B. Battrick (Noordwijk: ESTEC), pp. 67–83.

Kissel, J. 1986b. Mass spectrometric studies of Halley comet. *Adv. Mass Spectr. 1985*, pp. 175–184.

Kissel, J., Brownlee, D. E., Büchler, K., Clark, B. C., Fechtig, H., Grün, E., Hornung, K., Igenbergs, E. B., Jessberger, E. K., Krueger, F. R., Kuczera, H., McDonnell, J. A. M., Morfill, G. M., Rahe, J., Schwehm, G. H., Sekanina, Z., Utterback, N. G., Völk, H. J., and Zook, H. A. 1986a. Composition of comet Halley dust particles from Giotto observations. *Nature*, **321**, pp. 336–338.

Kissel, J., and Krueger, F. R. 1987a. The organic component in dust from comet Halley as measured by the PUMA mass spectrometer on board Vega 1. *Nature*, **326**, pp. 755–760.

Kissel, J., and Krueger, F. R. 1987b. Ion formation by impact of fast dust particles and comparison with related techniques. *Appl. Phys. A*, **42**, pp. 69–85.

Kissel, J., Sagdeev, R. Z., Bertaux, J. L., Angarov, V. N., Audouze, J., Blamont, J. E., Büchler, K., Evlanov, E. N., Fechtig, H., Fomenkova, M. N., von Hoerner, H., Inogamov, N. A., Khromov, V. N., Knabe, W., Krueger, F. R., Langevin, Y., Leonas, V. B., Levasseur-Regourd, A. C., Managadze, G. G., Podkolzin, S. N., Shapiro, V. D., Tabaldyev, S. R., and Zubkov, B. V. 1986b. Composition of comet Halley dust particles from Vega observations. *Nature*, **321**, pp. 280–282.

Kitamura, Y. 1986. Axisymmetric dusty gas jet in the inner coma of a comet. *Icarus*, **66**, pp. 241–257.

Kitamura, Y. 1987. Axisymmetric dusty gas jet in the inner coma of a comet. II. The case of isolated jets. *Icarus*, **72**, pp. 555–567.

Koike, C., Shibai, H., and Tuchiyama, A. 1993. Extinction of olivine and pyroxene in mid- and far-infrared regions. *Mon. Not. R. Astron. Soc.*, **264**, pp. 654–658.

Kolokolova, L., Gustafson, B. Å. S., and Jockers, K. 2000. Evolution of cometary grains from studies of cometary images. Presented at *Dust in the Solar System and Other Planetary Systems*, IAU Colloq. 181 and COSPAR Colloq. 11, Canterbury UK, April 2000.

Kömle, N. I. 1990. Jet and shell structures in the cometary coma: modelling and observations. In *Comet Halley: Investigations, Results, Interpretations*, ed. J. W. Mason (Chichester: Ellis Horwood), Vol. 1, pp. 231–243.

Kömle, N. I., and Ip, W.-H. 1987a. Anisotropic non-stationary gas flow dynamics in the coma of comet P/Halley. *Astron. Astrophys.*, **187**, pp. 405–410.

Kömle, N. I., and Ip, W.-H. 1987b. A model for the anisotropic structure of the neutral gas coma of a comet. In *Diversity and Similarity of Comets*, ESA SP-278, eds. E. J. Rolfe and B. Battrick (Noordwijk: ESTEC), pp. 247–254.

Körösmezey, A., and Gombosi, T. I. 1990. A time-dependent dusty gas dynamic model of axisymmetric cometary jets. *Icarus*, **84**, pp. 118–153.

Krankowsky, D., and Eberhardt, P. 1990. Evidence for the composition of ices in the nucleus of comet Halley. In *Comet Halley: Worldwide Investigations, Results, Interpretations*, ed. J. Mason (Chichester: Ellis Horwood), pp. 273–289.

Lamy, P. 1986a. Ground-based observations of the dust emission from comet Halley. *Adv. Space Res.*, **5**, pp. (12)317–(12)323.

Lamy, P. L. 1986b. Cometary dust: Observational evidences and properties. In *Asteroids, Comets, Meteors II*, eds. C.-I. Lagerkvist, B. A. Lindblad, H. Lundstedt, and H. Rickman (Uppsala: University of Uppsala Press), pp. 373–388.

Lamy, P. L., and Perrin, J.-M. 1986. Volume scattering function and space distribution of the interplanetary dust cloud. *Astron. Astrophys.*, **163**, pp. 269–286.

Langevin, Y., Kissel, J., Bertaux, J.-L., and Chassefière, E. 1987. First statistical analysis of 5000 mass spectra of cometary grains obtained by PUMA 1 (Vega 1) and PIA (Giotto) impact ionization mass spectrometers in the compressed modes. *Astron. Astrophys.*, **187**, pp. 779–784.

Larson, S. M. 1978. A rotation model for the spiral structure in the coma of comet Bennett (1970 II). *Bull. Am. Astron. Soc.*, **10**, p. 589. (Abstract.)

Larson, S. M., and Minton, R. B. 1972. Photographic observations of comet Bennett, 1970 II. In *Comets: Scientific Data and Missions*, eds. G. P. Kuiper and E. Roemer (Tucson: Univ. of Arizona Press), pp. 183–208.

Lawler, M., and Brownlee, D. 1992. CHON as a component of dust from comet Halley. *Nature*, **359**, pp. 810–812.

Lawler, M. E., Brownlee, D. E., Temple, S., and Wheelock, M. M. 1989. Iron, magnesium, and silicon in dust from Comet Halley. *Icarus*, **80**, pp. 225–242.

Lellouch, E., Crovisier, J., Lim, T., Bockelee-Morvan, D., Leech, K., Hanner, M. S., Altieri, B., Schmitt, B., Trotta, F., and Keller, H. U. 1998. Evidence for water ice and estimate of dust production rate in comet Hale-Bopp at 2.9 AU from the Sun. *Astron. Astrophys.*, **339**, pp. L9–L12.

Levasseur-Regourd, A. C., Hadamcik, E., and Renard, J. B. 1996. Evidence for two classes of comets from their polarimetric properties at large phase angles. *Astron. Astrophys.*, **313**, pp. 327–333.

Li, A., and Greenberg, J. M. 1997. A unified model of interstellar dust. *Astron. Astrophys.*, **323**, pp. 566–584.

Li, A., and Greenberg, J. M. 1998. From interstellar dust to comets: infrared emission from Comet Hale-Bopp (C/1995 O1). *Astrophys. J.*, **498**, pp. L83–L87.

Lien, D. J. 1991. Optical properties of cometary dust. In *Comets in the Post-Halley Era*, eds. R. L. Newburn, Jr., M. Neugebauer, and J. Rahe (Dordrecht: Kluwer Acad. Publ.), pp. 1005–1041.

Lisse, C. M., A'Hearn, M. F., Hauser, M. G., Kelsall, T., Lien, D. J., Moseley, S. H., Reach, W. T., and Silverberg, R. F. 1998. Infrared observations of comets by COBE. *Astrophys. J.*, **496**, pp. 971–991.

Lisse, C. M., Freudenreich, H. T., Hauser, M. G., Kelsall, T., Moseley, S. H., Reach, W. T., and Silverberg, R. F. 1994. Infrared observations of comet Austin (1990 V) by the COBE: Diffuse Infrared Background Experiment. *Astrophys. J.*, **432**, pp. L71–L74.

Lynch, D. K., Hanner, M. S., and Russell, R. W. 1992a. 8–13 μm spectroscopy and IR photometry of comet P/Brorsen-Metcalf (1989o) near perihelion. *Icarus*, **97**, pp. 269-275.

Lynch, D. K., Russell, R. W., Campins, H., Witteborn, F. C., Bregman, J. D., Rank, D. W., and Cohen, M. C. 1989. 5–13 μm airborne observations of comet Wilson 1986 l. *Icarus*, **82**, pp. 379–388.

Lynch, D. K., Russell, R. W., Hackwell, J. A., Hanner, M. S., and Hammel, H. B. 1992b. 8- to 13-μm spectroscopy of comet Levy 1990 XX. *Icarus*, **100**, pp. 197–202.

Maas, D., Krueger, F. R., and Kissel, J. 1989. Mass and density of silicate- and CHON-type dust particles released by comet P/Halley. In *Asteroids, Comets, Meteors III*, eds. C.-I. Lagerkvist, H. Rickman, B. A. Lindblad, and M. Lindgren (Uppsala: Uppsala Univ. Press), pp. 389–392.

Manset, N., and Bastien, P. 2000. Polarimetric observations of comets C/1995 O1 Hale-Bopp and C/1996 B2 Hyakutake. *Icarus*, **145**, pp. 203–219.

Marconi, M. L., and Mendis, D. A. 1983. The atmosphere of a dirty-clathrate cometary nucleus: a two-phase, multifluid model. *Ap. J.*, **273**, pp. 381–396.

Marconi, M. L., and Mendis, D. A. 1984. The effects of the diffuse radiation fields due to multiple scattering and thermal reradiation by dust on the dynamics and thermodynamics of a dusty cometary atmosphere. *Ap. J.*, **287**, pp. 445–454.

Marsden, B. G. 1989. The sungrazing comet group. II. *Astron. J.*, **98**, pp. 2306–2321.

Mason, C. G., Gehrz, R. D., Jones, T. J., Hanner, M. S., Williams, D. M., and Woodward, C. E. 2000. Observations of unusually small dust grains in the coma of comet Hale-Bopp C/1995 O1. *Ap. J.*, in press.

Mason, C. G., Gehrz, R. D., Ney, E. P., Williams, D. M., and Woodward, C. E. 1998. The temporal development of the pre-perihelion infrared spectral energy distribution of comet Hyakutake (C/1996 B2). *Ap. J.*, **507**, pp. 398–403.

McDonald, G. D., Whited, L. J., DeRuiter, C., Khare, B. N., Patnaik, A., and Sagan, C. 1996. Production and chemical analysis of cometary ice tholins. *Icarus*, **122**, pp. 107–117.

McDonnell, J. A. M., ed. 1978. *Cosmic Dust*. (New York: Wiley.)

McDonnell, J. A. M., Lamy, P. L., and Pankiewicz, G. S. 1991. Physical properties of cometary dust. In *Comets in the Post-Halley Era*, eds. R. L. Newburn, Jr., M. Neugebauer, and J. Rahe (Dordrecht: Kluwer Acad. Publ.), pp. 1043–1073.

Meech, K. J., and Jewitt, D. C. 1987. Observations of comet P/Halley at minimum phase angle. *Astron. Astrophys.*, **187**, pp. 585–593.

Merrill, K. M. 1974. 8-13 μm spectrophotometry of comet Kohoutek. *Icarus*, **23**, pp. 566–567.

Millis, R. L., A'Hearn, M. F., and Thompson, D. T. 1982. Narrowband photometry of comet P/Stephan-Oterma and the backscattering properties of cometary grains. *Astron. J.*, **87**, pp. 1310–1317.

Moiseyev, N. D. 1925. Über den Schweif des Kometen 1901 I. *Russ. Astron. J.*, **2**, pp. 73–84.

Mueller, B. E. A., Samarasinha, N. H., and Belton, M. J. S. 1999. Imaging of the structure and evolution of the coma morphology of comet Hale-Bopp (C/1995 O1). *Earth Moon Plan.*, **77**, pp. 181–188.

Mukhin, L. M., Dolnikov, G. A., Evlanov, E. N., Fomenkova, M. N., Prilutskii, O. F., and Sagdeev, R. Z. 1991. Re-evaluation of the chemistry of dust grains in the coma of comet Halley. *Nature*, **350**, pp. 480–481.

Needham, J., Beer, A., and Ho, P.-Y. 1957. "Spiked" comets in ancient China. *Observatory*, **77**, pp. 137–138.

Ney, E. P. 1974. Multiband photometry of comets Kohoutek, Bennett, Bradfield, and Encke. *Icarus*, **23**, pp. 551–560.

Ney, E. P. 1982. Optical and infrared observations of bright comets in the range 0.5 μm to 20 μm. In *Comets*, ed. L. L. Wilkening (Tucson: Univ. of Arizona Press), pp. 323–340.

Ney, E. P., and Merrill, K. M. 1976. Comet West and the scattering function of cometary dust. *Science*, **194**, pp. 1051–1053.

Nishioka, K., Saito, K., Watanabe, J.-i., and Ozeki, T. 1992. Photographic observations of the synchronic band in the tail of comet West 1976 VI. *Publ. Nat. Astron. Obs. Japan*, **2**, pp. 601–621.

Nishioka, K., and Watanabe, J.-i. 1990. Finite lifetime fragment model for synchronic band formation in dust tails of comets. *Icarus*, **87**, pp. 403–411.

Norton, W. A. 1844. On the mode of formation of the tails of comets. *Am. J. Sci. Arts*, **46**, pp. 104–129.

Norton, W. A. 1861. Theoretical determination of the dimensions of Donati's comet. *Am. J. Sci. Arts*, (Ser. 2) **32**, pp. 54–71.

Notni, P. 1964. Eigenschaften und Bewegung der Staubteilchen in Koma und Schweif von Kometen. *Veröff. Sternw. Babelsberg*, 15, No. **1**, pp. 1–51.

Notni, P. 1966. On the forces acting on charged dust particles in cometary atmospheres. *Mém. Soc. R. Sci. Liège*, (Ser. 5) **12**, pp. 379–383.

Notni, P., and Thänert, W. 1988. The striae in the dust tails of great comets — a comparison to various theories. *Astron. Nachr.*, **309**, pp. 133–146.

Notni, P., and Tiersch, H. 1987. Charging of dust particles in comets and in interplanetary space. *Astron. Astrophys.*, **187**, pp. 796–800.

Olbers, H. W. M. 1825. (Extract from a letter.) *Astron. Nachr.* **3**, pp. 5–10.

Olbers, H. W. M. 1831. Ueber anomale Cometenschweife. *Astron. Nachr.*, **8**, pp. 469–472.

Orlov, S. V. 1928. The mechanical theory of cometary forms. *Publ. Inst. Astrophys. Russ.* 3, No. 4, pp. 1–77. (In Russian.)

Orlov, S. V. 1929. The mechanical theory of cometary forms. *Russ. Astron. J.*, **6**, pp. 180–186.

Orlov, S. V. 1960. *On the Nature of Comets*. (Moscow: Akad. Nauk.)

Pansecchi, L., and Fulle, M. 1990. A neck-line structure in the dust tail of the Great January Comet 1910 I. *Astron. Astrophys.*, **239**, pp. 369–374.

Pansecchi, L., Fulle, M., and Sedmak, G. 1987. The nature of two anomalous structures observed in the dust tail of comet Bennett 1970 II: a possible neck-line structure. *Astron. Astrophys.*, **176**, pp. 358–366. [Erratum: 1988, *Astron. Astrophys.*, **205**, p. 367.]

Parker, E. N. 1958. Dynamics of the interplanetary gas and magnetic fields. *Ap. J.*, **128**, pp. 664–676.

Parker, E. N. 1963. *Interplanetary Dynamical Processes*. (New York: Interscience.) 272 pp.

Pittichová, J., Sekanina, Z., Birkle, K., Boehnhardt, H., Engels, D., and Keller, P. 1999. An early investigation of the striated tail of comet Hale-Bopp (C/1995 O1). *Earth Moon Plan.*, **78**, pp. 329–338.

Prinn, R., and Fegley, B. 1989. Solar nebula chemistry: origin of planetary satellite, and cometary volatiles. In *Origin and Evolution of Planetary and Satellite Atmospheres*, eds. S. K. Atreya, J. B. Pollack, and M. S. Matthews (Tucson: Univ. of Arizona Press), pp. 78–136.

Probstein, R. F. 1969. The dusty gasdynamics of comet heads. In *Problems of Hydrodynamics and Continuum Mechanics*, eds. F. Bisshopp et al. (Philadelphia: Soc. Ind. Appl. Math.), pp. 568–583.

Rahe, J., Donn, B., and Wurm, K. 1969. *Atlas of Cometary Forms*. NASA SP-198. (Washington, D.C.: NASA.)

Renard, J.-B., Hadamcik, E., and Levasseur-Regourd, A.-C. 1996. Polarimetric CCD imaging of comet 47P/Ashbrook-Jackson and variability of polarization in the inner coma of comets. *Astron. Astrophys.*, **316**, pp. 263–269.

Renard, J.-B., Levasseur-Regourd, A.-C., and Dollfus, A. 1992. Polarimetric imaging of comet Levy (1990c). *Ann. Geophys.*, **10**, pp. 288–292.

Rettig, T. W., and Hahn, J. M. 1997. Comet Shoemaker-Levy 9: an active comet. *Planet. Space Sci.*, **45**, pp. 1271–1277.

Rettig, T. W., Tegler, S. C., Pasto, D. J., and Mumma, M. J. 1992. Comet outbursts and polymers of HCN. *Astrophys. J.*, **398**, pp. 293–298.

Reuter, D. C. 1992. The contribution of methanol to the 3.4 μm emission feature in comets. *Ap. J.*, **386**, pp. 330–335.

Richter, K., and Keller, H. U. 1987. Density and brightness distribution of cometary dust tails. *Astron. Astrophys.*, **171**, pp. 317–326.

Richter, K., and Keller, H. U. 1988. The anomalous dust tail of comet Kohoutek (1973 XII) near perihelion. *Astron. Astrophys.*, **206**, pp. 136–142.

Rieke, G. H., and Lee, T. A. 1974. Photometry of comet Kohoutek (1973f). *Nature*, **248**, pp. 737–740.

Rietmeijer, F., Mukhin, L. M., Fomenkova, M. N., and Evlanov, E. N. 1989. Layer silicate chemistry in P/comet Halley from PUMA-2 data. *Lunar Planet. Sci. Conf.*, **20**, pp. 904–905.

Rose, L. A. 1979. Laboratory simulation of infrared astrophysical features. *Astrophys. Space Sci.*, **65**, pp. 47–67.

Ryan, E. V., and Campins, H. 1991. Comet Halley spatial and temporal variability of the silicate emission feature. *Astron. J.*, **101**, pp. 695–705.

Sagdeev, R. Z., Evlanov, E. N., Fomenkova, M. N., Mukhin, L. M., Prilutskii, O. F., and Zubkov, B. V. 1987. Composition of comet Halley dust particles based on PUMA instruments measurements in zero mode. *Space Res.*, **25**, pp. 849–855.

Sagdeev, R. Z., Evlanov, E. N., Fomenkova, M. N., Prilutskii, O. F., and Zubkov, B. V. 1989. Small-size dust particles near Halley's comet. *Adv. Space Res.*, **9**, pp. (3)263–(3)267.

Samarasinha, N. H., Mueller, B. E. A., and Belton, M. J. S. 1999. Coma morphology and constraints on the rotation of comet Hale-Bopp (C/1995 O1). *Earth Moon Plan.*, **77**, pp. 189–198.

Sandford, S. A., and Walker, R. M. 1985. Laboratory infrared transmission spectra of individual interplanetary dust particles from 2.5 to 25 microns. *Ap. J.*, **291**, pp. 838–851.

Schulze, H., Kissel, J., and Jessberger, E. 1997. Chemistry and mineralogy of comet Halley's dust. In *From Stardust to Planetesimals*, Conf. Series Vol. 122, eds. Y. Pendleton and A. Tielens (San Francisco: Astron. Soc. of the Pacific Press), pp. 397–414.

Schwarzschild, K. 1901. Der Druck des Lichtes auf kleine Kugeln und die Arrhenius'sche Theorie des Cometenschweife. *Sitz. Bayer. Akad. Wiss. München 1901*, pp. 293–327.

Schwehm, G. 1976. Radiation pressure on interplanetary dust particles. In *Interplanetary Dust and Zodiacal Light*, eds. H. Elsässer and H. Fechtig (Berlin: Springer Verlag), pp. 459–463.

Schwehm, G., and Rohde, M. 1977. Dynamical effects on circumsolar dust grains. *J. Geophys.*, **42**, pp. 727–735.

Sekanina, Z. 1973. Comet Kohoutek (1973f). *IAU Circ.* No. 2580.

Sekanina, Z. 1976. Progress in our understanding of cometary dust tails. In *The Study of Comets*, NASA SP-393, eds. B. Donn, M. Mumma, W. Jackson, M. A'Hearn, and R. Harrington (Washington, D.C.: NASA), pp. 893–939.

Sekanina, Z. 1979. Fan-shaped coma, orientation of rotation axis, and surface structure of a cometary nucleus. *Icarus*, **37**, pp. 420–442.

Sekanina, Z. 1980. Physical characteristics of cometary dust from dynamical studies: a review. In *Solid Particles in the Solar System*, eds. I. Halliday and B. A. McIntosh (Dordrecht: Reidel), pp. 237–250.

Sekanina, Z. 1981a. Distribution and activity of discrete emission areas on the nucleus of periodic comet Swift-Tuttle. *Astron. J.*, **86**, pp. 1741–1773.

Sekanina, Z. 1981b. Properties of dust particles in comet Halley from observations made in 1910 during its encounter with the Earth. In *The Comet Halley Dust & Gas Environment*, ESA SP-174, eds. B. Battrick and E. Swallow (Noordwijk: ESTEC), pp. 55–65.

Sekanina, Z. 1981c. Rotation and precession of cometary nuclei. *Ann. Rev. Earth Planet Sci.*, **9**, pp. 113–145.

Sekanina, Z. 1982a. The problem of split comets in review. In *Comets*, ed. L. L. Wilkening (Tucson: Univ. of Arizona Press), pp. 251–287.

Sekanina, Z. 1982b. The path and surviving tail of a comet that fell into the Sun. *Astron. J.*, **87**, pp. 1059–1072.

Sekanina, Z. 1986. Periodic comet Halley (1982i). *IAU Circ.* No. 4187.

Sekanina, Z. 1987a. Anisotropic emission from comets: Fans versus jets. I. Concept and modeling. In *Diversity and Similarity of Comets*, ESA SP-278, eds. E. J. Rolfe and B. Battrick (Noordwijk: ESTEC), pp. 315–322.

Sekanina, Z. 1987b. Anisotropic emission from comets: Fans versus jets. II. Periodic comet Tempel 2. In *Diversity and Similarity of Comets*, ESA SP-278, eds. E. J. Rolfe and B. Battrick (Noordwijk: ESTEC), pp. 323–336.

Sekanina, Z. 1991a. Cometary activity, discrete outgassing areas, and dust-jet formation. In *Comets in the Post-Halley Era*, eds. R. L. Newburn, Jr., M. Neugebauer, and J. Rahe (Dordrecht: Kluwer Acad. Publ.), pp. 769–823.

Sekanina, Z. 1991b. Randomization of dust-ejecta motions and the observed morphology of cometary heads. *Astron. J.*, **102**, pp. 1870–1878.

Sekanina, Z. 1993. Computer simulation of the evolution of dust coma morphology in an outburst: P/Schwassmann-Wachmann 1. In *On the Activity of Distant Comets*, eds. W. F. Huebner, H. U. Keller, D. Jewitt, J. Klinger, and R. M. West (San Antonio: Southwest Research Institute), pp. 166–181.

Sekanina, Z. 1995. Evidence on sizes and fragmentation of the nuclei of comet Shoemaker-Levy 9 from Hubble Space Telescope images. *Astron. Astrophys.*, **304**, pp. 296–316.

Sekanina, Z. 1996a. Tidal breakup of the nucleus of comet Shoemaker-Levy 9. In *The Collision of Comet P/Shoemaker-Levy 9 and Jupiter*, eds. K. S. Noll, H. A. Weaver, and P. D. Feldman (Cambridge: Cambridge Univ. Press), pp. 55–80.

Sekanina, Z. 1996b. Morphology of cometary dust coma and tail. In *Physics, Chemistry, and Dynamics of Interplanetary Dust*, Conf. Series Vol. 104, eds. B. Å. S. Gustafson and M. S. Hanner (San Francisco: Astron. Soc. of the Pacific Press), pp. 377–382.

Sekanina, Z. 1996c. Activity of comet Hale-Bopp (1995 O1) beyond 6 AU from the Sun. *Astron. Astrophys.*, **314**, pp. 957–965.

Sekanina, Z. 1998. Modeling the diurnal evolution of a dust feature in comet Hale-Bopp (1995 O1). *Ap. J.*, **494**, pp. L121–L124.

Sekanina, Z. 2000. SOHO sungrazing comets with prominent tails: Evidence on dust production peculiarities. *Ap. J. (Letters)*, in press.

Sekanina, Z., and Boehnhardt, H. 1999. Dust morphology of comet Hale-Bopp (C/1995 O1). II. Introduction of a working model. *Earth Moon Plan.*, **78**, pp. 313–319.

Sekanina, Z., Boehnhardt, H., Ryan, O., and Birkle, K. 2000. A new scenario for the formation of striations in the dust tail of comet Hale-Bopp (C/1995 O1). Presented at *Dust in the Solar System and Other Planetary Systems*, IAU Colloq. 181 and COSPAR Colloq. 11, Canterbury, UK, April 2000.

Sekanina, Z., Chodas, P. W., and Yeomans, D. K. 1994. Tidal disruption and the appearance of periodic comet Shoemaker-Levy 9. *Astron. Astrophys.*, **289**, pp. 607–636.

Sekanina, Z., Chodas, P. W., and Yeomans, D. K. 1998. Secondary fragmentation of comet Shoemaker-Levy 9 and the ramifications for the progenitor's breakup in July 1992. *Planet. Space Sci.*, **46,** pp. 21–45.

Sekanina, Z., and Farrell, J. A. 1978. Comet West 1976 VI: discrete bursts of dust, split nucleus, flare-ups, and particle evaporation. *Astron. J.*, **83,** pp. 1675–1680.

Sekanina, Z., and Farrell, J. A. 1980. The striated dust tail of comet West 1976 VI as a particle fragmentation phenomenon. *Astron. J.*, **85,** pp. 1538–1554.

Sekanina, Z., and Farrell, J. A. 1982. Two dust populations of particle fragments in the striated tail of comet Mrkos 1957 V. *Astron. J.*, **87,** pp. 1836–1853.

Sekanina, Z., and Farrell, J. A. 1986. The striated dust tail of comet 1910 I. *Bull. Am. Astron. Soc.*, **18,** p. 818. (Abstract.)

Sekanina, Z., and Larson, S. M. 1984. Coma morphology and dust-emission pattern of periodic comet Halley. II. Nucleus spin vector and modeling of major dust features in 1910. *Astron. J.*, **89,** pp. 1408–1425.

Sekanina, Z., and Larson, S. M. 1986a. Coma morphology and dust-emission pattern of periodic comet Halley. IV. Spin vector refinement and map of discrete dust sources for 1910. *Astron. J.*, **92,** pp. 462–482.

Sekanina, Z., and Larson, S. M. 1986b. Dust jets in comet Halley observed by Giotto and from the ground. *Nature*, **321,** pp. 357–361.

Sekanina, Z., Larson, S. M., Hainaut, O., Smette, A., and West, R. M. 1992. Major outburst of periodic comet Halley at a heliocentric distance of 14 AU. *Astron. Astrophys.*, **263,** pp. 367–386.

Sekanina, Z., and Miller, F. D. 1973. Comet Bennett (1970 II). *Science*, **179,** pp. 565–567.

Sekanina, Z., and Pittichová, J. 1999. Distribution law for particle fragmentation times in a theory for striated tails of dust comets: application to comet Hale-Bopp (C/1995 O1). *Earth Moon Plan.*, **78,** pp. 339–346.

Sekanina, Z., and Schuster, H. E. 1978. Meteoroids from periodic comet d'Arrest. *Astron. Astrophys.*, **65,** pp. 29–35.

Simpson, J. A., Rabinowitz, D., Tuzzolino, A. J., Ksanfomality, L. V., and Sagdeev, R. Z. 1986. Halley's comet coma dust particle mass spectra, flux distributions, and jet structures derived from measurements on the Vega-1 and Vega-2 spacecraft. In *Exploration of Halley's Comet*, ESA SP-250, eds. B. Battrick, E. J. Rolfe, and R. Reinhard (Noordwijk: ESTEC), Vol.2, pp. 11–16.

Simpson, J. A., Rabinowitz, D., Tuzzolino, A. J., Ksanfomality, L. V., and Sagdeev, R. Z. 1987. The dust coma of comet P/Halley: measurements on the Vega-1 and Vega-2 spacecraft. *Astron. Astrophys.*, **187,** pp. 742–752.

Simpson, J. A., Tuzzolino, A. J., Ksanfomality, L. V., Sagdeev, R. Z., and Vaisberg, O. L. 1989. Confirmation of dust clusters in the coma of comet Halley. *Adv. Space Res.*, **9,** pp. (3)259–(3)262.

Snyder, L. E., Webber, J. C., Cruthcher, R. M., and Swenson, G. W., Jr. 1976. Radio observationss of OH in comet West 1975n. *Astrophys. J.*, **209,** pp. L49–L52.

Šolc, M., Jessberger, E. K., Hsiung, P., and Kissel, J. 1987. Halley dust composition. *Publ. Astron. Inst. Czech. Acad. Sci.*, **67,** pp. 47–50.

Šolc, M., Vanýsek, V., and Kissel, J. 1986. Carbon stable isotopes in comets after encounters with P/Halley. In *Exploration of Halley's Comet*, ESA SP-250, eds. B. Battrick, E. J. Rolfe, and R. Reinhard (Noordwijk: ESTEC), Vol.1, pp. 373–376.

Solem, J. C. 1995. Cometary breakup calculations based on a gravitationally-bound agglomeration model: the density and size of Shoemaker-Levy 9. *Astron. Astrophys.*, **302,** pp. 596–608.

Southworth, R. B. 1963. Dust in comet Arend-Roland. *Astron. J.*, **68,** pp. 293–294.

Southworth, R. B. 1964. The size distribution of the zodiacal particles. *Ann. New York Acad. Sci.*, **119,** pp. 54–67.

Stephens, J. R., and Russell, R. W. 1979. Emission and extinction of ground and vapor-condensed silicates from 4 to 14 microns and the 10 micron silicate feature. *Ap. J.*, **228,** pp. 780–786.

Sykes, M. V., Lebofsky, L. A., Hunten, D. M., and Low, F. J. 1986. The discovery of dust trails in the orbits of periodic comets. *Science*, **232,** pp. 1115–1117.

Sykes, M. V., Lien, D. J., and Walker, R. G. 1990. The Tempel 2 dust trail. *Icarus*, **86,** pp. 236–247.

Tanigawa, T., Kawakita, H., and Watanabe, J.-i. 1997. The activity of the fragmented nucleus of comet Shoemaker-Levy 9. *Planet. Space Sci.*, **45**, pp. 1417–1422.

Telesco, C. M., Decher, R., Baugher, C., Campins, H., Mozurkewich, D., Thronson, H. A., Cruikshank, D. P., Hammel, H. B., Larson, S., and Sekanina, Z. 1986. Thermal-infrared and visual imaging of comet Giacobini-Zinner. *Ap. J.*, **310**, pp. L61–L65.

Thomas, N., and Keller, H. U. 1987. Comet P/Halley's near-nucleus jet activity. In *Diversity and Similarity of Comets*, ESA SP-278, eds. E. J. Rolfe and B. Battrick (Noordwijk: ESTEC), pp. 337–342.

Tielens, A. G. G. M., and Allamandola, L. 1987. Composition, structure and chemistry of interstellar dust. In *Interstellar Processes*, eds. D. Hollenbach and H. Thronson (Dordrecht: Reidel), pp. 397–469.

Tiersch, H., and Notni, P. 1989. The electric potential on dust particles in comets and in interplanetary space. *Astron. Nachr.*, **310**, pp. 67–78.

Tokunaga, A. T., Golisch, W. F., Griep, D. M., Kaminski, C. D., and Hanner, M. S. 1986. The NASA Infrared Telescope Facility comet Halley monitoring program I. Preperihelion results. *Astron. J.*, **92**, pp. 1183–1190.

Tokunaga, A. T., Golisch, W. F., Griep, D. M., Kaminski, C. D., and Hanner, M. S. 1988. The NASA Infrared Telescope Facility comet Halley monitoring program II. Postperihelion results. *Astron. J.*, **96**, pp. 1971–1976.

Tokunaga, A. T., Hanner, M. S., Golisch, W. F., Griep, D. M., Kaminski, C. D., and Chen, H. 1992. Infrared monitoring of comet P/Tempel 2. *Astron. J.*, **104**, pp. 1611–1617.

Utterback, N. G., and Kissel, J. 1990. Attogram dust cloud a million kilometers from comet Halley. *Astron. J.*, **100**, pp. 1315–1322.

Vasundhara, R., and Chakraborty, P. 1999. Modeling of jets from comet Hale-Bopp (C/1995 O1): Observations from the Vaine Bappu Observatory. *Icarus*, **140**, pp. 221–230.

Vasundhara, R., Chakraborty, P., Hänel, A., and Heiser, E. 1999. Modeling dust jets and shells from comet Hale-Bopp. *Earth Moon Plan.*, **78**, pp. 321–328.

Walker, R. G., and Aumann, H. H. 1984. IRAS observations of cometary dust. *Adv. Space Res.*, **4**, pp. (9)197–(9)201.

Warell, J., Lagerkvist, C.-I., and Lagerros, J. S. V. 1999. Dust morphology of the inner coma of C/1995 O1 (Hale-Bopp). *Earth Moon Plan.*, **78**, pp. 197–203.

Watanabe, J.-i., and Nishioka, K. 1991. Synchronic band and its implication in the cometary dust. In *Origin and Evolution of Interplanetary Dust*, eds. A. C. Levasseur-Regourd and H. Hasegawa (Dordrecht: Kluwer Acad. Publ.), pp. 253–256.

Weaver, H. A., A'Hearn, M. F., Arpigny, C., Boice, D. C., Feldman, P. D., Larson, S. M., Lamy, P., Levy, D. H., Marsden, B. G., Meech, K. J., Noll, K. S., Scotti, J. V., Sekanina, Z., Shoemaker, C. S., Shoemaker, E. M., Smith, T. E., Stern, S. A., Storrs, A. D., Trauger, J. T., Yeomans, D. K., and Zellner, B. 1995. The Hubble Space Telescope (HST) observing campaign on comet Shoemaker-Levy 9. *Science*, **267**, pp. 1282–1288.

Weaver, H. A., Feldman, P. D., A'Hearn, M. F., Arpigny, C., Brandt, J. C., Festou, M. C., Haken, M., McPhate, J. B., Stern, S. A., and Tozzi, G. P. 1997. The activity and size of the nucleus of comet Hale-Bopp (C/1995 O1). *Science*, **275**, pp. 1900–1904.

West, R. A. 1991. Optical properties of aggregate particles whose outer diameter is comparable to the wavelength. *Appl. Optics*, **30**, pp. 5316–5324.

Whipple, F. L. 1950. A comet model. I. The acceleration of comet Encke. *Ap. J.*, **111**, pp. 375–394.

Whipple, F. L. 1951. A comet model. II. Physical relations for comets and meteors. *Astrophys. J.*, **113**, pp. 464–474.

Whipple, F. L. 1978. Rotation period of comet Donati. *Nature*, **273**, pp. 134–135.

Whipple, F. L., and Sekanina, Z. 1979. Comet Encke: Precession of the spin axis, nongravitational motion, and sublimation. *Astron. J.*, **84**, pp. 1894–1909.

Williams, D. M., Mason, C. G., Gehrz, R. D., Jones, T. J., Woodward, C. E., Harker, D. E., Hanner, M. S., Wooden, D. H., Witteborn, F. C., and Butner, H. M. 1997. Measurement of submicron grains in the coma of comet Hale-Bopp C/1995 O1 during 1997 February 15–20 UT. *Astrophys. J.*, **489**, pp. L91–L94.

Wooden, D. H., Butner, H. M., Harker, D. E., and Woodward, C. E. 2000. Mg-rich silicate crystals in comet Hale-Bopp: ISM relics or solar nebula condensates? *Icarus*, **143**, pp. 126–137.

Wooden, D. H., Harker, D. E., Woodward, C. E., Butner, H. M., Koike, C., Witteborn, F. C., and McMurtry, C. W. 1999. Silicate mineralogy of the dust in the inner coma of comet C/1995 O1 (Hale-Bopp) pre- and post-perihelion. *Ap. J.*, **517,** pp. 1034–1058.

Worms, J. C., Renard, J. B., Hadamcik, E., Levasseur-Regourd, A.-C., and Gayet, J. F. 1999. Results of the PROGRA[2] experiment: an experimental study in microgravity of scattered polarized light by dust particles with large size parameter. *Icarus*, **142,** pp. 281–297.

Xing, Z., and Hanner, M. S. 1997. Light scattering by aggregate particles. *Astron. Astrophys.*, **324,** pp. 805–820.

Yanamandra-Fisher, P., and Hanner, M. S. 1999. Optical properties of non-spherical particles of size comparable to the wavelength of light: application to comet dust. *Icarus*, **138,** pp. 107–128.

Yavnel, A. A. 1977. Chemical composition of meteors and meteoritic matter. In *Comets, Asteroids, Meteorites*, ed. A. H. Delsemme (Toledo: Univ. of Toledo Press), pp. 133–135.

Zerull, R. H., Gustafson, B. Å. S., Schultz, K., and Thiele-Corbach, E. 1993. Scattering by aggregates with and without an absorbing mantle: Microwave analog experiments. *Appl. Optics*, **32,** pp. 4088–4100.

Zinner, E., Wopenka, B., Amari, S., and Anders, E. 1990. Interstellar graphite and other carbonaceous grains from the Murchison meteorite: structure, composition and isotopes of C, N, and Ne. *Lunar Planet. Sci. Conf.*, **21,** pp. 1379–1380.

Near Earth Environment

Tony McDonnell[1], Neil McBride[1], Simon F. Green[1]
Paul R. Ratcliff[2], David J. Gardner[2], Andrew D. Griffiths[2]

[1] The Open University, Milton Keynes, UK
[2] Unit for Space Sciences & Astrophysics, University of Kent, Canterbury, UK

Abstract. Planet Earth provides an interface to the interplanetary environment; its atmosphere forms a protective shield against direct impacts and erosion and is a medium in which to observe the approach of meteoroids and even to capture intact smaller meteoroids. The Earth's gravitational well enhances the flux of interplanetary dust and modifies its velocity distribution. We consider the effect of the Earth on the dynamical properties on the interplanetary dust population, the relative contribution of sporadic meteoroids and annual streams, the efficiency of the atmosphere in capturing and fragmenting meteoroids and the effect of space debris on in situ experimental results. We review the range of modelling tools necessary to interpret the complex interaction of these populations with spacecraft, with particular emphasis on the improved calibration of impact detectors and the application of software models. Analysis of the available data from 30 years of in situ impact experiments, and more recent recovered samples reveals evidence of the relative contributions from space debris and various astrophysical sources. While temporally and spatially averaged fluxes are well represented by existing isotropic interplanetary models for meteoroids responsible for penetrating experimental foils (of thickness F_{max}) greater than approximately 30 μm, at smaller sizes a high degree of anisotropy is apparent in resolved data. An Earth apex component is observed for particles larger than a few microns in size whereas at smaller sizes, β-meteoroids from the solar direction dominate. Space debris forms an increasingly significant proportion of the LEO population at $F_{max} < 30$ μm in addition to its dominance in the centimetre size range and above.

I. INTRODUCTION

The Earth and its atmosphere provide a target for interplanetary particulates which allows us to study incoming objects directly, if they reach the surface; or indirectly, if they interact with the atmosphere or orbiting spacecraft. Although this sample is not necessarily representative of the total population of interplanetary material, it is of fundamental importance for understanding the past and present influx of material to the Earth and currently provides the *only* source of interplanetary material which can be studied in the laboratory. The relationship between these samples and their original sources depends on the physical and dynamical processes which have characterised their transport to the Earth. We observe only material from sources for which such mechanisms exist.

The identification of a source of particular samples may be based on composition or on evidence of past processes, e.g. SNC meteorites associated with Martian origin (e.g. Laul et al. 1986; Ott and Beggeman 1985) or antarctic meteorites of lunar origin (e.g. Dennison et al. 1987). We sometimes have dynamical evidence such as the collection of meteorite samples from observed fireballs for which orbits may be derived. More revealingly it can be based on a combination of both, such as the association of basaltic achondrite meteorites with the asteroid 4 Vesta. Small asteroids (i.e. large impact ejecta), of similar spectral type to Vesta (Binzel and Xu 1993) have been identified in orbits bridging the location of Vesta and the 3:1 Jovian resonance (Wisdom 1983, 1985; Yoshikawa 1990) which provides a transport mechanism to Earth. Details of the large particle influx to the Earth and associated meteor phenomena are dealt with by Ceplecha et al. (1998) and the properties of airborne microparticles which are decelerated without significant ablation are described by Jessberger et al. (this volume). We are concerned here with the near Earth particulate environment, its interaction with the Earth's atmosphere, in situ detection of sub-mm particles and associated modelling tools for deriving their properties and sources.

Orbiting spacecraft (and the Earth itself) can only sample objects with orbits which cross or closely approach the Earth's; the Earth's own gravitational influence has a modest focusing effect dependent on the relative velocity. Although the sources of these particulates may themselves be in stable orbits (planets, main belt asteroids) or distant orbits (comets), small particulates are subject to forces which significantly alter or dominate their dynamics, namely:

- perturbations due to close encounters or resonances with more massive bodies;
- non-gravitational forces due to the non-isotropic sublimation of volatiles from cometary nuclei or fragments;
- radiation pressure which exerts a radial force dependent on particle size and composition which can exceed solar gravity for sub-μm particles (β-meteoroids);
- Lorentz forces due to electrostatically charged particles moving through the interplanetary or planetary magnetic fields, which can also exceed gravity for sub-μm particles;
- the Poynting-Robertson effect (non-radial radiation pressure due to aberration of sunlight) resulting in a gradual decrease in eccentricity and semi-major axis;
- the Yarkovsky effect (non-isotropic thermal emission from an illuminated rotating object) causing a change in semi-major axis dependent on the sense of rotation.

Although purely gravitational transport mechanisms from the asteroid belt to the Earth have been identified (collisional fragments entering Jovian resonances, resulting in chaotic Mars and subsequently Earth crossing orbits; see review by Farinella et al. 1994), the Poynting-Robertson effect dominates the orbital evolution of millimetre to metre sized particles. We therefore expect particulates of asteroidal origin to approach the Earth generally with small eccentricity, low inclination orbits and therefore low relative velocity. Particles of cometary origin, on the other hand, may have a much larger variety of relative velocities reflecting the diversity of original cometary orbits and ejection locations. Particulates ejected from long period comets, which have near parabolic orbits and random inclinations, require only a small radiation pressure force to exceed the solar escape velocity. Material released from short-period comets, with more tightly bound, generally low inclination, but still highly eccentric orbits, has a more significant lifetime in the inner solar system. This material contributes to the general zodiacal background as well as meteoroid streams associated with individual comets.

In addition to possible impact ejecta from planets and their moons, interstellar particles may also reach the Earth's vicinity. Their presence, speculated by Öpik (1951), but argued strongly against by Lovell based on his Jodrell Bank radio meteoroid data, has been inferred recently (Taylor et al. 1996) from refined detection techniques for high velocity meteors (> heliocentric escape velocity at 1 AU plus Earth's orbital speed of 72 km s^{-1}) (Baggaley et al. 1994). Although of minor importance in terms of mass influx to the Earth, detection and possible capture of such particles will provide direct access to interstellar material from which planetary systems form.

In this chapter we examine the effect of the Earth on the interplanetary dust population at 1 AU, in situ data from orbiting spacecraft (concentrating on the Long Duration Exposure Facility (LDEF) and European Retrievable Carrier (EuReCa), described in section III.B) and a range of modelling tools necessary for its interpretation in terms of physical and dynamical properties of the population.

II. THE EARTH AS A TARGET

II.A. Natural Meteoroids

Dust particles are continually encountering the Earth from a variety of astrophysical sources. To model this global influx to Earth or to a satellite orbiting Earth, one must make an estimate of the mass and velocity distributions of these particles. Although visual meteor showers are prominent (we deal with these below) most of the mass influx to Earth occurs from smaller "sporadic" particles. This "sporadic background" comes mostly from prograde orbiting dust particles with relatively low eccentricity orbits and hence relative encounter velocities with the Earth which are generally quite low (a few km s^{-1}). The particles then fall into the Earth's gravitational potential well and are accelerated. This "falling in" effect means that the influx to Low Earth Orbit (LEO) appears reasonably isotropic in nature, at least at the lower end of the velocity spectrum.

The "Grün flux" (Grün et al. 1985; see section IV.A) is generally used to model the natural sporadic meteoroid influx to Earth. The flux values are given as cumulative mean values (i.e. time averaged number of particles, m^{-2} s^{-1}, of individual masses $\geq M$) for a spinning flat plate detector at 1 AU, outside the gravitational influence of the Earth (but moving in an Earth-like orbit). In the model's derivation, isotropy of the meteoroid environment at 1 AU is assumed (noting that the overestimation in particle spatial density could only be 30 % at most if in reality the meteoroid flux was highly anisotropic). In transforming the model's fluxes to LEO, maintaining the isotropy assumption is a reasonable approximation for many spacecraft which have orbital geometries which are essentially randomised over their orbital lifetimes. For a directed detector we would expect asymmetries in sampling to occur due, for example, to the spacecraft's motion around the Earth, or from sampling of specific anisotropic astrophysical sources.

II.B. Meteoroid Properties and Dynamics

1. Fluxes and Size Distributions

When detecting moving particles, we measure a *flux*: the number of particles passing through unit area per unit time (i.e. number m^{-2} s^{-1}). The total cumulative exposure to a particle flux, is described in terms of *fluence*, which is the time integrated flux (i.e. number m^{-2}). For any individual value of particle velocity (V), the flux (measured perpendicular to the velocity vector) is given by spatial density (i.e. number m^{-3}) multiplied by this velocity.

In general, we are concerned with the apparent particle flux intercepted by a detector; and so we illustrate relationships for a flat detector which is sensitive on one side only, i.e. it can be impacted from particles coming from 2π sr. Because of the cosine dependence of resolved area, we find that the flat plate detector has an *effective* sensitive solid angle of π sr (whereas a spherical detector would be sensitive to 4π sr). We need to find how the apparent flux measured by a flat plate detector depends on the detector motion and the particle environment.

If we consider a meteoroid environment which has a particle spatial density n and in which the magnitude of the particle velocities is V, the flux (F) measured by the flat plate detector can be written

$$F = \frac{n\,V}{f}, \qquad (1)$$

where f is a factor which depends on the detector motion and distribution of particle trajectories. We can highlight 3 main cases.

If we consider a unidirectional meteoroid flux (i.e. the trajectories are parallel) then: $f = 1$ for a stationary flat plate detector which is perpendicular to the meteoroid velocity vector; $f = \pi$ if the detector is spinning about an axis which is perpendicular to the meteoroid's velocity vector; $f = 4$ if the detector is randomly tumbling within the unidirectional meteoroid flux.

If we consider a "cylindrically isotropic" meteoroid environment where the trajectories are all parallel to a plane, but are randomised within that plane (i.e. velocity components V_x and V_y are randomised, but $V_z = 0$) then: $f = \pi$ for a stationary flat plate which has its normal parallel with the plane; $f = \pi$ if the detector is spinning about an axis which is perpendicular to the plane (i.e. same as stationary result); $f = 4$ if the detector is randomly tumbling.

If we consider a "spherically isotropic" (usually just referred to as isotropic) meteoroid environment where the trajectories are randomised (i.e. velocity components V_x, V_y and V_z are randomised) then $f = 4$ for all three cases of a stationary, spinning or randomly tumbling flat plate detector. Note that we sometimes refer to particle intensity I (number m^{-2} s^{-1} sr^{-1}). For the isotropic environment, the intensity I is constant over 4π sr, and so a flat-plate detector (whether stationary, spinning or randomly tumbling) would see a flux of πI.

When sampling a flux of particles, we inevitably tend to sample a large range of particle masses. This *mass distribution* may be described in a number of ways. If we assume a power law relationship between mass and flux, then the form of the *cumulative* distribution is

$$F(M) = k\, M^{-\alpha}, \tag{2}$$

where $F(M)$ is the flux of particles of individual mass greater than or equal to mass M; k is a constant and α is the *cumulative mass distribution index*.

Clearly when plotting $\log_{10} F(M)$ versus $\log_{10} M$, this relationship is linear with a gradient of $-\alpha$. We can also use the *differential mass distribution* which describes the flux of particles of individual mass M, per dM interval

$$F'(M) = \alpha\, k\, M^{-(\alpha+1)}\, \mathrm{d}M, \tag{3}$$

where $\alpha + 1$ is referred to as the *differential mass distribution index*, s (i.e. we see that $\alpha = s - 1$). Note that when $\alpha = 1$ (i.e. $s = 2$), an *equal combined mass* is contained in each logarithmic mass bin. In this case, the combined cross-sectional area of particles in each logarithmic mass bin increases with decreasing mass. We can also note that when $\alpha = 2/3$ (i.e. $s = 5/3$), an *equal combined cross sectional area* is contained in equal logarithmic mass bins.

Sometimes, the particle distribution is described in terms of the *differential size distribution*, such that the flux of particles of individual radius a, per da interval, is given by

$$F'(a) = k'\, a^{-u}\, \mathrm{d}a, \tag{4}$$

where k' is a constant, and u is the *differential size distribution index*. The *cumulative* size distribution index would thus be given by $u - 1$; it can be shown that $\alpha = (u - 1)/3$.

2. Gravitational Enhancement

When deriving fluxes near the Earth from quantitative values of the dust flux at 1 AU (given for example by the Grün flux), we must note that the meteoroid trajectories are altered in the vicinity of the Earth due to gravity. This has the effect of concentrating the particles and increasing their velocities as they fall into the gravitational well. A resulting increase (enhancement) of the "free-space" flux is obtained. If the isotropic flux of particles has a single value of velocity V_∞ in free-space, then at a given distance r from the centre of the

Earth we find that the flux, at constant particle mass, is enhanced (Öpik 1951) by the factor G, given by

$$G = 1 + \frac{V_{esc}^2}{V_\infty^2} \quad \text{which can also be written} \quad G = \frac{V_E^2}{V_E^2 - V_{esc}^2} \;, \tag{5}$$

where V_{esc} is the escape velocity at distance r, and where V_E is the enhanced meteoroid velocity at distance r given by

$$V_E = \sqrt{V_{esc}^2 + V_\infty^2} \;. \tag{6}$$

To convert from a value of the Grün flux at 1 AU to a flux in LEO, we could crudely assume a single velocity for the flux (say 20 km s^{-1}) and multiply by G. In reality however, the meteoroid flux will have a velocity distribution $n(V_\infty)$, so that G is actually a function $G(V_\infty)$, and we must integrate over all V_∞ values. For a value of the Grün flux F_G, and a velocity distribution at 1 AU normalised such that

$$\int_0^\infty n(V_\infty) \; dV_\infty = 1 \tag{7}$$

we calculate the final enhanced flux F_E by

$$F_E = \int_0^\infty F_G \; n(V_\infty) \; G(V_\infty) \; dV_\infty \tag{8}$$

giving a weighted mean enhancement $\overline{G}$ of

$$\overline{G} = \frac{\int_0^\infty F_G \; n(V_\infty) \; G(V_\infty) \; dV_\infty}{\int_0^\infty F_G \; n(V_\infty) \; dV_\infty} \;. \tag{9}$$

Note that this value of $\overline{G}$ implies a characteristic value of velocity (by putting $\overline{G}$ into Eq. (5)), but that this velocity is *not* the same as the mean velocity of the normalised velocity distribution, which is given by

$$\overline{V_\infty} = \int_0^\infty n(V_\infty) \; V_\infty \; dV_\infty \;. \tag{10}$$

For comparison, the gravitational enhancement factor for a single V_∞ value of around 20 km s^{-1} is $G \sim 1.3$ at the LDEF spacecraft's (see section IV.B) mean altitude of $\sim$470 km, whereas $\overline{G} \sim 2.0$ when a velocity distribution is applied (see below). We see that the enhancement strongly favours the slower particles.

3. *Velocity Distributions*

The velocity distribution $n(V_\infty)$ of meteoroids at 1 AU (i.e. as if viewed from a massless Earth) has generally been derived from ground based observations of photographic meteors (which can then be corrected for the effect of the

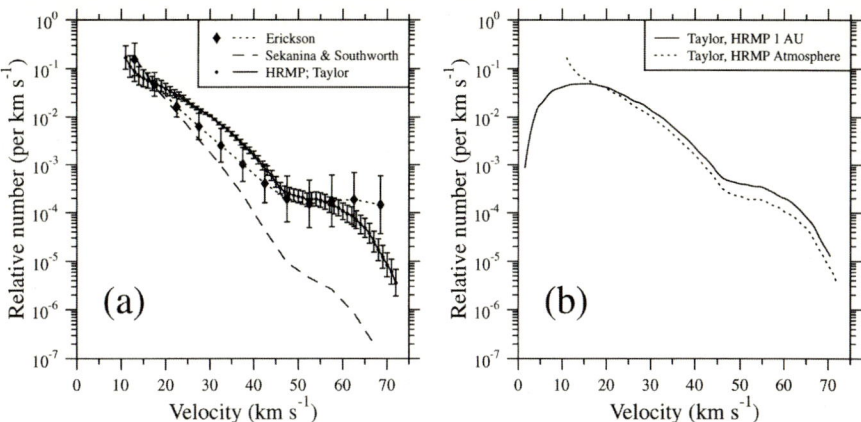

Figure 1. (a) The normalised velocity distribution of meteoroids encountering the Earth's atmosphere, following Taylor (1995b). Also shown for comparison is the Erickson (1968) distribution, and the (erroneous) Sekanina and Southworth (1975) distribution. (b) The Taylor, HRMP distribution corrected for gravitational enhancement to take the distribution to 1 AU (i.e. as seen from a massless Earth). The errors in the distribution are commensurate with those in (a). Also shown again for comparison (dotted curve) is the original distribution. Values are listed in Table 6.

Earth's gravitational acceleration). Dohnanyi (1966) obtained a distribution from 286 observations taken from Hawkins and Southworth (1958), and Erickson (1968) used the same data but attempted a more rigorous reduction to meteor number with a constant mass threshold. Kessler (1969) used 2090 sporadic meteors given by McCrosky and Posen (1961) to give a more statistically reliable distribution (see Zook 1991 for a comparison between these distributions). However, the most statistically reliable data set comes from the Harvard Radio Meteor Project (HRMP) where $\sim 20\,000$ meteor observations were taken (Southworth and Sekanina 1973; Sekanina and Southworth 1975). This data set is often used in various modelling work. Taylor (1995a) reappraised the data using an improved analysis of ionisation probability (Verniani 1973) and mass distribution index. Taylor also identified a numerical error in the original code used to reduce the data which resulted in a significant under-estimation of numbers of fast meteors (particularly 50 to 70 km s^{-1} meteors where the under-estimation is by a factor of ~ 100).

A final corrected velocity distribution of meteoroids encountering the Earth's atmosphere (Taylor 1995a, 1995b) is shown in Fig. 1a. Also shown for comparison is the Sekanina and Southworth distribution (with the under-estimation clearly apparent) and the Erickson distribution, which is in better agreement with Taylor's curve. Due to the high statistical reliability of the Taylor-HRMP data, we use this distribution in preference to others.

In order to obtain the normalised velocity distribution $n(V_E)$ shown in Fig. 1a corrected to 1 AU (i.e. converted to the $n(V_\infty)$ distribution) we must step through the $n(V_E)$ distribution and divide each bin by G (calculated with the appropriate value of V_E using the second form of Eq. (5)) to remove gravitational enhancement and then re-bin this $n(V_E)/G$ value at a velocity V_∞ (calculated from Eq. (6)). Care needs to be taken in rebinning the data as the bin widths change during the conversion. The renormalised V_∞ distribution, i.e. the meteoroid encounter velocity distribution at 1 AU (but away from the influence of Earth) is shown in Fig. 1b (values are also presented in Table 6). The mean velocity from this distribution is 17.7 km s^{-1}. In the distribution we can clearly see the contribution from prograde material (velocity up to $\sim$45 km s^{-1}) and the retrograde material (velocity greater than $\sim$45 km s^{-1}).

4. Earth Shielding

For a tumbling flat plate detector orbiting Earth the exposure to the dust influx will be reduced somewhat by the physical shielding of the Earth. This "Earth shielding factor" is generally calculated by taking the fraction of solid angle subtended by the Earth at a given distance r from the Earth's centre. A tangent to the Earth passing through a point (satellite) subtends an angle θ with the direction of the Earth's centre. Thus the shielding factor η is given by

$$\eta = \frac{1 + \cos\theta}{2}, \quad (11)$$

where $\sin\theta = (r_e/r)$ and r_e is the radius of the Earth. However, this assumes that meteoroid trajectories can be represented by a straight line. In reality the trajectories will be curved due to gravitational influence, and hence the angle θ is somewhat velocity dependent (Kessler 1972; Bandermann and Singer 1969) i.e. we are really dealing with a function, $\eta(V_\infty)$, and the angle θ is described by

$$\sin\theta = \frac{r_e\sqrt{V_\infty^2 + V_{esc}^2(r_e)}}{r\sqrt{V_\infty^2 + V_{esc}^2(r)}}, \quad (12)$$

where the total shielding factor is calculated by integrating over the $n(V_\infty)$ distribution

$$\overline{\eta} = \frac{\int_0^\infty n(V_\infty)\,\eta(V_\infty)\,dV_\infty}{\int_0^\infty n(V_\infty)\,dV_\infty}. \quad (13)$$

The value of r_e should also include the height of the atmosphere as aerocapture and ablation effectively increase the Earth's cross-section (where a true mean radius of the Earth might be taken as 6378 km and the atmosphere for this purpose might be taken as $\sim$100 km).

The error introduced by ignoring the velocity dependence, and not adding the atmosphere height, can be at most between 5 and 10 % (using the 1 AU velocity distribution given above) and so while interpretation of previous data may not have been significantly influenced by ignoring these details it is recommended that the correction is performed for subsequent analysis of high resolution flux data, such as those gathered from LDEF. Table 1 gives mean values of gravitational enhancement factor G and Earth shielding factor η for

Table 1

Mean gravitational enhancement factor, $\overline{G}$, and Earth shielding factor, $\overline{\eta}$, for a random tumbling flat plate detector, calculated using the 1 AU velocity distribution derived from the Taylor HRMP data shown in Fig. 1b (using a "working" velocity range of 0.7–71.6 km s^{-1}) and assuming an atmosphere height of 100 km. The weighted mean velocities $\overline{V}$, are also given. The mean K factor $\overline{K}$ (section III.B) is also given indicating the ratio of the flux on a circularly orbiting flat plate at altitude h, to that on a similar but stationary detector.

Altitude (km)	$\overline{G}$	$\overline{\eta}$	$\overline{V}$	$\overline{K}$
100	2.04	0.50	18.1	1.09
200	2.03	0.58	18.0	1.09
400	2.00	0.63	17.9	1.09
470 (LDEF)	1.99	0.65	17.9	1.08
800	1.94	0.70	17.8	1.08
1 000	1.92	0.72	17.8	1.08
2 000	1.81	0.79	17.6	1.08
4 000	1.65	0.87	17.3	1.07
10 000	1.41	0.95	17.1	1.05
20 000	1.26	0.98	17.1	1.04
35 790 (GEO)	1.16	0.99	17.2	1.03
100 000	1.06	1.00	17.4	1.01
At 1 AU	1.00	1.00	17.7	1.00

various altitudes using the 1 AU velocity distribution shown in Fig. 1b and assuming an atmosphere height of 100 km, as well as the mean meteoroid velocity at the particular altitude.

5. Detector Sensitivity Enhancement

In the work above we obtained a meteoroid velocity distribution, at 1 AU, *at equal particle mass*. However, when working with real particle detectors the detection threshold generally depends on the mass *and* velocity of the incoming particle. This in turn means that two "just-detectable" particles travelling at two different velocities will have different masses, and hence will lie on different parts of the mass distribution curve. These factors must be accounted for when analysing impact data from real detectors. It is especially vital for the LDEF data when we want to compare fluxes on different faces which will have different impact velocity distributions.

If the cumulative flux of particles can be described by $F(M) = kM^{-\alpha}$ (i.e. Eq. (2)) then the ratio of fluxes for masses M_1 and M_2 is given by

$$\frac{F(M_2)}{F(M_1)} = \left(\frac{M_2}{M_1}\right)^{-\alpha}. \tag{14}$$

A real sensor detects, for example, crater diameter, penetration limit, plasma production etc. which depend on mass and velocity such that

$$\text{signal} = c\, M^a\, V^b \qquad (15)$$

where c is a constant. The ratio of the number of detections for two particle velocities V_1 and V_2 will be given by

$$\frac{F(V_2)}{F(V_1)} = \left(\frac{V_2}{V_1}\right)^{(\alpha b/a)}, \qquad (16)$$

where the ratio b/a is often referred to as the factor γ.

Thus in order to convert from an *equal mass* velocity distribution to an *equal detection threshold* velocity distribution one must weight the distribution by the factor $V^{\alpha\gamma}$. The weighted mean velocity of particles at equal detection limit becomes

$$\overline{V} = \int_0^\infty n(V)\, V^{\alpha\gamma}\, dV\ . \qquad (17)$$

If we are interested in craters in metal targets, many crater depth relationships take on a form where depth is proportional to $M^{1/3}V^{2/3}$ (e.g. Shanbing et al. 1994; see section IV.A.1), so $a = 1/3$, $b = 2/3$ and hence $\gamma = 2.0$. If we are interested in the limiting foil thickness for penetration of foils (i.e. at the onset of the hole on the rear side of the foil), F_{max}, then we could use a formula where F_{max} is proportional to $M^{0.352}V^{0.806}$ (McDonnell and Sullivan 1992; see section IV.A.1) so $a = 0.352$, $b = 0.806$ and hence $\gamma = 2.29$.

For plasma production from an impact on a thick target (as in an ionisation detector) Kissel and Krueger (1987) determine a mass exponent of $a = 0.8$, although Göller and Grün (1989) show empirically that $a \sim 1.0$ is more appropriate. Using a velocity exponent of $b = 3.55$ (Dietzel 1973) and $a = 1$ gives $\gamma = 3.55$ for plasma production.

By assuming a cumulative mass index of $\alpha = 1.1$ (which was assumed by Taylor (1995a) in re-analysis of the HRMP data and is in agreement with the Grün flux in the mass range appropriate to radio meteors), we can determine the velocity distribution of particles at a given detector threshold, at a given altitude. We take the 1 AU velocity distribution, account for gravitational enhancement and Earth-shielding, and then weight the resulting distribution by $V^{\alpha\gamma}$. Fig. 2 shows the normalised geocentric velocity distributions obtained when considering detector thresholds related to equal mass, equal crater depth, equal foil penetration and equal plasma production, at an altitude appropriate to LDEF (i.e. with respect to a *stationary* detector). The weighted mean velocities for the four cases are:
 – equal mass $\overline{V} = 17.9$ km s^{-1},
 – equal crater size $\overline{V} = 24.6$ km s^{-1},
 – equal foil penetration $\overline{V} = 25.8$ km s^{-1}, and
 – equal plasma production $\overline{V} = 31.7$ km s^{-1}.

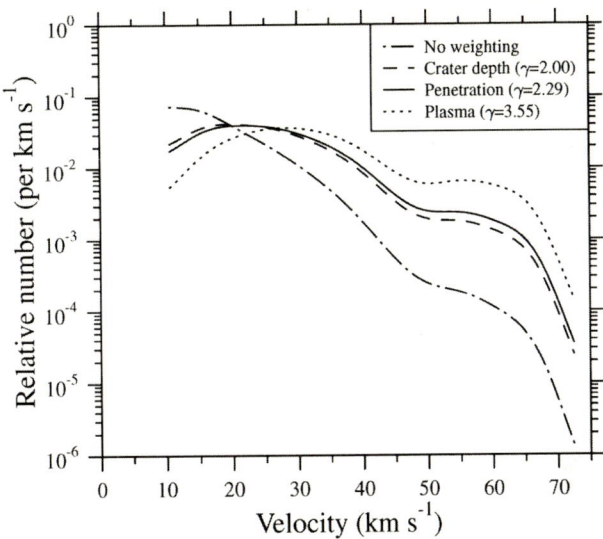

Figure 2. Normalised velocity distributions applicable to the altitude of LDEF, at equal mass, and weighted to equal crater depth, foil penetration limit F_{max} and equal plasma production. A value of $\alpha = 1.1$ has been assumed, and γ is the ratio b/a in the mass-velocity dependence $M^a V^b$ of the detector.

Because the penetration effectiveness decreases at oblique angles of incidence we are interested in the normal component of impact velocity, $V \cos A$, where A is the impact angle measured from the normal of the flat plate detector. We can, in this case, integrate over all impact angles such that the weighted mean *normal* velocity becomes $\overline{V} \cdot 2/3$ where $\overline{V}$ is given above. This is only true if there is no shielding of the spacecraft.

Finally, one should note that although the distributions in Fig. 2 have been calculated for an altitude applicable to LDEF (i.e. ~ 470 km) the orbital motion of LDEF around the Earth has not been accounted for. LDEF had a geocentric orbital velocity of ~ 7.6 km s^{-1} and so this has the effect of broadening the distributions somewhat (see section IV.B). More "extreme" impact velocities are possible depending on whether collision was "head-on" or from the rear. This can also be complicated by different faces of a satellite maintaining particular geocentric orientations (e.g. the "ram" face on LDEF), such that different velocity distributions (at equal particle mass) are obtained on different faces (see e.g. Zook 1991). To calculate the weighted mean impact velocity for a particular face we integrate over all incoming meteoroid directions and over the meteoroid velocity distribution, and vectorially add the meteoroid velocity to the spacecraft orbital velocity to calculate the impact direction and impact speed to the face. Earth shielding (which itself depends on the meteoroid velocity; see section II.B.4) must be accounted for before the vector addition and the resultant incoming meteoroid vector must be $< 90°$ from the normal to the plate (assuming a flat face).

II.C. The Sporadic Background

Considering the factors described above, and assuming an isotropic sporadic meteoroid environment with the mass distribution described by the Grün flux, we can calculate for a given face on a spacecraft, the total cumulative flux $F(M)$ to that face. This is given by

$$F_M = \int_M^\infty \int_{V_\infty=0}^\infty \int_{\phi=0}^{2\pi} \int_{\theta=0}^\pi \frac{F'_G(M)}{\pi} \, G(V_\infty) \, n(V_\infty) \, \cos A \cdot \frac{V_{\text{rel}}}{V_E} \sin\theta \, d\theta \, d\phi \, dV_\infty \, dM \qquad (18)$$

where θ, ϕ are spherical polar coordinates with the spacecraft at the origin, V_{rel} is the relative velocity of the incoming meteoroid with respect to the spacecraft, V_E is the gravitationally enhanced meteoroid velocity as given by Eq. (6) and the angle A is the instantaneous impact angle to the face (measured from the face normal). $F'_G(M) \, dM$ is the *differential* value of the Grün flux at mass M i.e. the flux within each logarithmic mass interval of the tabulated Grün flux (in section IV.A; Table 6). We must force the instantaneous contribution to be zero (e.g. by putting $F'_G = 0$) if the meteoroid cannot impact the face (i.e. if $A > \pi/2$), and also if θ and ϕ are such that the spacecraft is shielded by the Earth (noting that the Earth shielding is itself velocity dependent as described in section II.B.4).

In numerical evaluation of Eq. (18) the fluxes for given $d\theta$, $d\phi$, dV_∞ and dM intervals can be used to define a flux distribution as a function of mass as shown or as a function of a detected property such as foil penetration (F_{max}). The flux contribution is then binned at the appropriate F_{max} value for the θ, ϕ, V_∞, M element.

An enhancement of the sporadic background flux (i.e. the non-shower component) in the early hours of the morning is clearly evident when viewing visual meteors. This bias towards the Earth apex direction is also apparent in radio meteor rates (e.g. Mawrey and Broadhurst 1993) and in the radiant distribution seen in the Harvard Radio Meteor Project data (Taylor and McBride 1997). This has yet to be fully assessed, although it appears that the enhancement of *fast* (retrograde) particles is considerable (and is important for the "detectability" of the particles; see section II.B.5). Structure is also seen in the more abundant slower meteors, with enhancements in directions slightly forward of the Sun and anti-Sun direction (Taylor and McBride 1997). Seasonal variations in the sporadic background meteor rate are also apparent in meteor data. This has been linked to the broad "sporadic streams" associated with the "Taurid Complex" (Štohl 1986; McBride et al. 1995). At smaller particle sizes, anisotropy such as β-meteoroids on hyperbolic trajectories, is also seen in spacecraft data (see section V.B).

II.D. The Annual Meteor Showers

Although the assumption of isotropy in the "random tumbling plate" modelling approach is often adequate, if one wants to consider a spacecraft geometry in detail, and its exposure to anisotropic astrophysical sources (particularly at larger particle masses such as 10^{-6} kg), then the effect of annual meteoroid streams might be significant.

In terms of modelling spacecraft damage due to exposure to the interplanetary particle influx, streams can be important (particularly at times of "storms") as the detected showers tend to have higher velocities than is generally assumed for the typical sporadic background. While typical space debris velocities intercepting LEO satellites will tend to be around 8 km s^{-1} (relative to the Earth) and sporadic impacts might be typically around 20 km s^{-1}, stream meteoroids can have speeds of up to 72 km s^{-1} (and so impact speeds to LEO satellites of up to 80 km s^{-1}). Since the detection threshold of real sensors has a velocity dependence of V^γ for constant mass, where γ could be as high as ~ 3.5, a stream meteoroid might have in excess of 3 orders of magnitude more "detectability" than a piece of space debris of the same mass. Hence, the streams clearly merit investigation as a possible satellite failure mechanism (e.g. Caswell et al. 1995; Beech and Brown 1994; Beech et al. 1995).

In defining a method for calculating quantitative fluxes from meteoroid streams we can use meteor shower parameters described by various authors (e.g. Kresáková 1966; Cook 1972; Cour-Palais 1969). We will use a recent source of meteor shower parameters (Jenniskens 1994) and the method described by McBride (1997). These data were collected by observers over a 10 year period from both northern and southern hemisphere observation sites and provide an excellent database to calculate stream fluxes. Table 2 gives information, following Jenniskens (1994), for the 50 major annual showers. The radiant direction has right ascension RA$_{max}$ and declination Dec$_{max}$ at the time of shower maximum, defined by the solar longitude $\lambda_\odot^{max}$. The radiant moves ΔRA and ΔDec degrees per degree of solar longitude from $\lambda_\odot^{max}$, so the RA of the radiant for an instantaneous solar longitude $\lambda_\odot$ is given by

$$\text{RA} = \text{RA}_{max} + \Delta\text{RA}(\lambda_\odot - \lambda_\odot^{max}) \ , \tag{19}$$

and similarly for Dec. Jenniskens found that shower activity (zenithal hourly rate, ZHR) as a function of time around its maximum could be described by

$$ZHR = ZHR_{max} 10^{-B|\lambda_\odot - \lambda_\odot^{max}|} \ , \tag{20}$$

where B describes the slopes of the log-linear activity profiles. Most streams had symmetrical profiles and could be described by a single B value. Six streams did not have high enough ZHRs to produce a B value for their profiles, although for the modelling we apply a "typical" value of $B = 0.2$.

Jenniskens also found that 6 streams are best represented by the sum of 2 activity profiles. These are made from a *peak* profile, defined by ZHR_{max}^p and B^p, and a *background* profile defined by ZHR_{max}^b, with separate inward and outward slope values B^{b+} and B^{b-} respectively. The total ZHR at a given solar longitude is then found by adding $ZHR^p + ZHR^b$. The relevant parameters for these 6 streams are in Table 3.

Table 2

Meteor shower parameters following Jenniskens (1994). Solar longitude, radiant RA and Dec (all in degrees) are J2000 for the shower maximum. ΔRA and ΔDec are in degrees per degree of solar longitude. V (km s^{-1}) is the geocentric entry velocity into the Earth's atmosphere (i.e. includes gravitational enhancement). Where the activity slope B is not shown, a value of 0.2 is assumed for the modelling. Values are given for α and k calculated here for each stream, so defining each stream's mass distribution where the flux is calculated by $F(M) = kM^{-\alpha}$ where M is in kg. Note, the Bootids are also referred to as the Quadrantids.

Name	$\lambda_\odot^{max}$	RA max	Dec max	ΔRA	ΔDec	V km s^{-1}	ZHR$_{max}$	B	χ	α	k
Bootids	283.3	232	+45	+0.6	−0.3	43	133	1.84	2.5	0.92	0.84×10^{-16}
γ Velids	285.7	124	−47	+0.5	−0.2	35	2.4	0.12	3.0	1.10	0.58×10^{-18}
α Crucids	295.4	193	−63	+1.1	−0.4	50	3.0	0.11	2.9	1.06	0.19×10^{-18}
α Hydrusids	300	138	−13	+0.7	−0.3	44	< 2	–	2.8	1.03	0.34×10^{-18}
α Carinids	311.2	99	−54	+0.4	+0.0	25	2.3	0.16	2.5	0.92	0.13×10^{-16}
δ Velids	318	127	−50	+0.5	−0.3	35	< 1.3	–	3.0	1.10	0.31×10^{-18}
α Centaurids	319.4	210	−58	+1.3	−0.3	57	7.3	0.18	2.3	0.83	0.37×10^{-17}
o Centaurids	323.4	176	−55	+0.9	−0.4	51	2.2	0.15	2.8	1.03	0.19×10^{-18}
θ Centaurids	334	220	−44	+1.1	−0.4	60	< 4.5	–	2.6	0.95	0.44×10^{-18}
δ Leonids	335	169	+17	+1.0	−0.3	23	1.1	0.049	3.0	1.10	0.19×10^{-17}
Virginids	340	165	+9	+0.9	−0.2	26	< 1.5	–	3.0	1.10	0.15×10^{-17}
γ Normids	353.0	285	−56	+1.3	−0.2	56	5.8	0.19	2.4	0.87	0.19×10^{-17}
δ Pavonids	11.1	311	−63	+1.6	−0.2	60	5.3	0.075	2.6	0.95	0.51×10^{-18}
Lyrids	32.4	274	+33	+1.2	+0.2	49	12.8	0.22	2.7	0.99	0.20×10^{-17}
μ Virginids	40	230	−8	+0.5	−0.3	30	2.2	0.045	3.0	1.10	0.11×10^{-17}
η Aquarids	46.5	340	−1	+0.9	+0.3	66	36.7	0.080	2.7	0.99	0.15×10^{-17}
β Corona Australids	56	284	−40	+1.3	+0.1	45	< 3.0	–	3.1	1.13	0.15×10^{-18}
α Scorpids	55.9	252	−27	+1.1	−0.2	35	3.2	0.13	2.5	0.92	0.47×10^{-17}
ω Scorpids	72.6	241	−20	+1.0	−0.1	21	5.2	0.15	3.0	1.10	0.14×10^{-16}
daytime Arietids	77	47	+24	+0.7	+0.6	38	54	0.10	2.7	0.99	0.26×10^{-16}
γ Sagitarids	89.2	286	−25	+1.1	+0.1	29	2.4	0.037	2.9	1.06	0.19×10^{-17}
τ Cetids	95.7	24	−12	+0.9	+0.4	66	3.6	0.18	2.5	0.92	0.37×10^{-18}

Table 2 - continued

Name	$\lambda_\odot^{max}$	RA max	Dec max	ΔRA	ΔDec	V km s^{-1}	ZHR max	B	χ	α	k
θ Ophiuchids	98	292	−11	+1.1	+0.1	27	2.3	0.037	2.8	1.03	0.35×10^{-17}
τ Aquarids	98.0	342	−12	+1.0	+0.4	63	7.1	0.24	2.5	0.92	0.89×10^{-18}
υ Phoenicids	111.2	28	−40	+1.0	+0.5	48	5.0	0.25	3.0	1.10	0.26×10^{-18}
o Cygnids	116.7	305	+47	+0.6	+0.2	37	2.5	0.13	2.7	0.99	0.14×10^{-17}
Capricornids	122.4	302	−10	+0.9	+0.3	25	2.2	0.041	2.0	0.69	0.83×10^{-16}
δ Aquarids North	124.1	324	−8	+1.0	+0.2	42	1.0	0.063	3.3	1.19	0.36×10^{-19}
Pisces Australids	124.4	339	−33	+1.0	+0.4	42	2.9	0.26	3.2	1.16	0.15×10^{-18}
δ Aquarids South	125.6	340	−17	+0.8	+0.2	43	11.4	0.091	3.3	1.19	0.36×10^{-18}
ι Aquarids South	131.7	335	−15	+1.0	+0.3	36	1.5	0.070	3.3	1.19	0.12×10^{-18}
Perseids	140.2	47	+58	+1.3	+0.1	61	84	0.20	2.5	0.92	0.12×10^{-16}
κ Cygnids	146.7	290	+52	+0.6	+0.3	27	2.3	0.069	2.2	0.79	0.30×10^{-16}
ε Eridanids	153	51	−16	+0.8	+0.3	59	< 40	–	2.8	1.03	0.17×10^{-17}
γ Doradids	155.7	60	−50	+0.5	+0.2	41	4.8	0.18	2.8	1.03	0.11×10^{-17}
Aurigids	158.2	73	+43	+1.0	+0.2	69	9	0.19	2.7	0.99	0.29×10^{-18}
κ Aquarids	177.2	339	−5	+0.9	+0.4	19	2.7	0.11	2.8	1.03	0.19×10^{-16}
ε Geminids	206.7	104	+28	+0.7	+0.1	71	2.9	0.082	3.0	1.10	0.21×10^{-19}
Orionids	208.6	96	+16	+0.7	+0.1	67	25	0.12	3.1	1.13	0.16×10^{-18}
Leo Minorids	209.7	161	+38	+1.0	−0.4	61	1.9	0.14	2.7	0.99	0.11×10^{-18}
Taurids	223.6	50	+18	+0.3	+0.1	30	7.3	0.026	2.3	0.83	0.43×10^{-16}
δ Eridanids	229	54	−2	+0.9	+0.2	31	< 0.9	–	2.8	1.03	0.75×10^{-18}
ζ Puppis	232.2	117	−42	+0.7	−0.2	41	3.2	0.13	3.4	1.22	0.95×10^{-19}
Leonids	235.1	154	+22	+1.0	+0.4	71	23	0.39	3.4	1.22	0.34×10^{-19}
Puppids/Velids	252	128	−42	+0.8	−0.4	40	4.5	0.034	2.9	1.06	0.82×10^{-18}
Phoenicids	252.4	19	−58	+0.8	+0.4	18	2.8	0.30	2.8	1.03	0.25×10^{-16}
Monocerotids	260.9	100	+14	+1.0	−0.1	43	2.0	0.25	3.5	1.25	0.33×10^{-19}
Geminids	262.1	113	+32	+1.0	+0.1	36	88	0.39/.72	2.6	0.95	0.78×10^{-16}
σ Hydrusids	265.5	133	+0	+0.9	−0.3	59	2.5	0.10	3.0	1.10	0.47×10^{-19}
Ursids	271.0	224	+78	−0.2	−0.3	35	11.8	0.61	3.4	1.22	0.81×10^{-18}

Table 3

Modelling parameters for the 6 streams which are best represented by the linear addition of 2 separate *peak* and *background* activity profiles. Note that the peak profile for the Geminids has a separate inward and outward value.

Name	$\lambda_\odot^{max}$	ZHR_{max}^p	B^p	ZHR_{max}^b	B^{b+}	B^{b-}
Bootids (Quadrantids)	283.3	110	2.5	20	0.37	∼0.45
Pisces Australids	124.4	2.0	∼0.40	0.9	0.03	∼0.10
Perseids	140.2	70	0.35	23	0.050	0.092
Leonids	235.1	19	0.55	4	0.025	>0.15
Geminids	262.1	74	0.59/0.81	18	0.09	0.31
Ursids	271.0	10	0.9	2.0	0.08	0.2

To model the streams it is assumed that the cumulative flux (mass) distribution of meteoroids within a stream can be described (over a "working" range of say 10^{-9} to 10^{-3} kg) by Eq. (2), which leads to a meteor magnitude distribution of

$$n(m) = n(0)\,\chi^m \quad , \qquad (21)$$

where $n(m)$ is the flux of meteoroids (number m^{-2} s^{-1}) giving rise to meteors of magnitude between $m + 0.5$ and $m - 0.5$, $n(0)$ is this flux for $m = 0$ and χ is the "magnitude distribution index". The rate (m^{-2} s^{-1}) of zero magnitude meteors $n(0)$ is given by

$$n(0) = \frac{ZHR_{max}(0.4 + 0.6\chi)}{(\sum_m P(m)\chi^m)A_E 3600} \quad , \qquad (22)$$

where A_E is the effective meteor viewing surface area (i.e. amount of sky the observer is sampling) and $\sum_m P(m)\chi^m$ can be calculated from quantities given in Table 4. $P(m)$ is a probability function related to the observability of the meteors and this depends on their magnitudes. To calculate the effective sampling area A_E, one can simply assume that meteors with a zenith distance of $< 58°$ are visible, and so a reasonable approximation would be $A_E \approx \pi(h\tan 58°)^2$, where h is the height at which the meteor is formed (e.g. $h = 70$ km at $V = 20$ km s^{-1} and $h = 100$ km at $V = 72$ km s^{-1}).

The relationship between meteoroid mass and meteor photographic magnitude was obtained by Jacchia et al. (1967). Using their determined colour index of -1.86 (i.e. $m_{ph} = m_v - 1.86$, which is somewhat different to what

Table 4

Values of the probability function $P(m)$ as given in Jenniskens (1994), used to calculate $\sum_m P(m)\chi^m$. For example, for $\chi = 2.5$ then $\sum_m P(m)\chi^m = 15$.

m:	-2	-1	0	1	2	3	4	5	6
P(m):	0.75	0.73	0.70	0.63	0.48	0.32	0.09	0.009	0.001

Jenniskens has used) we convert to an expression for the mass of a meteoroid associated with a meteor of visual magnitude m (assuming typical observed zenith distances of 45°)

$$M_m = 10^{(3.07-0.44m-3.89\log_{10} V)} \text{ kg} , \qquad (23)$$

where V is in km s^{-1}. The 0.44 exponent leads to a conversion between α and χ which follows the relation $\alpha = 2.3 \log \chi$, and indicates that the meteor brightness is proportional to $M_m^{0.9}$, which was also found by Verniani (1973) for faint radio meteors.

We can now calculate the value of the constant k in Eq. (2) by solving for one point on the cumulative mass distribution curve. We use $N(0)$ which is the *cumulative* flux for zero magnitude meteors i.e., $N(0) = n(0) + n(-1) + n(-2)....n(-\infty)$ (and in practise using just the first few terms is adequate). Note that $N(0)$ refers to the flux of meteoroids giving rise to meteors of magnitude equal to *or brighter than* +0.5, and so the associated mass of the cumulative flux point $N(0)$ is at $M_{m=+0.5}$ (i.e. $N(0) \equiv F(M_{m=+0.5})$). From Eq. (2),

$$k = \frac{N(0)}{M_{m=+0.5}^{-\alpha}} . \qquad (24)$$

We can calculate the cumulative flux at stream maximum, $F(M)_{max}$ (i.e. at solar longitude $\lambda_\odot^{max}$) using Eq. (2). For any other solar longitude the cumulative flux is given by

$$F(M) = F(M)_{max} \frac{ZHR}{ZHR_{max}} , \qquad (25)$$

where the instantaneous ZHR at a given solar longitude is given by Eq. (20).

Table 2 includes columns of α and k, calculated here for each stream. However, spacecraft detectors can rarely measure flux directly at a particular particle mass, but rather at a given detector threshold (such as penetration limit). As an example, we calculate the flux of perforating impacts in aluminium of thickness F_{max}. To do this we use the penetration equation 1992c from McDonnell and Sullivan (1992), using the normal component of impact velocity and a meteoroid density of 2.5 g cm^{-3} in the equation. We integrate each stream contribution over its mass distribution (with lower mass limit 10^{-7} g pertaining to faint radio meteors), calculating the appropriate F_{max} for a given

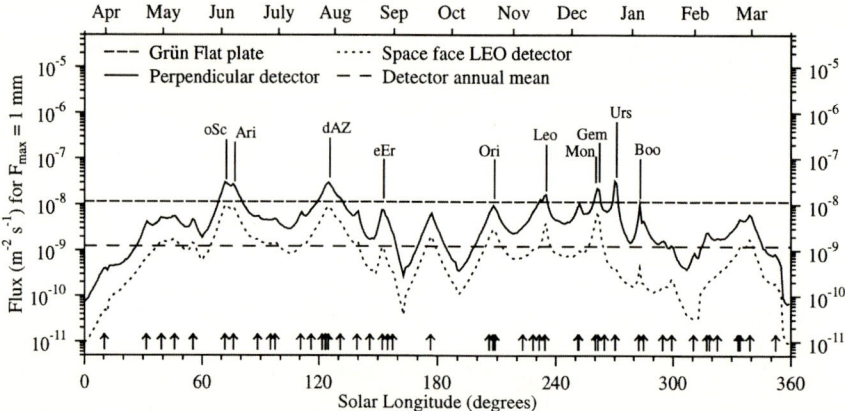

Figure 3. The instantaneous flux throughout the year, of stream meteoroids that penetrate 1 mm of aluminium foil (i.e. $F_{max} = 1$ mm). The arrows indicate the peaks of all 50 streams used, with the 10 most important streams labelled. See text for details.

particle mass and velocity and binning the flux at this F_{max} value. This is done for all streams throughout the year. Figure 3 shows the instantaneous flux as a function of solar longitude, for $F_{max} = 1$ mm i.e. this is the flux of meteoroid impacts that would penetrate 1 mm of aluminium (this being applicable to considerations of significant spacecraft damage and in a size regime where we do not have to extrapolate greatly to small meteoroid masses). The upper curve shows the results obtained using a foil detector mounted perpendicularly to all streams, and hence gives an upper level for any instantaneous exposure. The lower curve is for a foil detector mounted on the space pointing face of a gravity stabilised LEO spacecraft with its orbit parallel to the ecliptic plane (hence the contribution from very northern showers is greatly reduced). The mean value of this curve is shown, with the annual mean level also shown for comparison, as derived for the space face detector using the Grün flux (with gravitational enhancement to LEO). It is seen that at this size regime, $\sim 10\%$ of the Grün flux prediction is obtained purely from the 50 streams. It is also seen that instantaneous contributions from particular streams exceed the annual mean. The result of short exposure to the streams (e.g. a 2 week Shuttle mission) would be highly dependent on the time of year. August shows high activity whereas April is relatively benign. Note that foil detections (or penetrations to a spacecraft surface) are dependent on a combination of the deduced mass distribution and stream velocity, and so are not necessarily correlated with the ZHR of the shower. The top 10 streams from Jenniskens' (1994) list, in terms of peak flux to a foil at $F_{max} = 1$ mm are (with abbreviations used in Fig. 3): Ursids (Urs), δ Aquarids South (dAZ), ω Scorpiids (oSc), Geminids (Gem), daytime Arietids (Ari), Leonids (Leo), Bootids (Boo), Orionids (Ori), Monocerotids (Mon) and ϵ Eridanids (eEr).

II.E. Atmospheric Effects

1. Atmospheric Entry

For the dominant fraction of interplanetary meteoroids entering the atmosphere on a one way trip, we have an interesting size-dependent process. The acceleration due to the Earth's gravity causes bodies to enter the atmosphere with a velocity, V_{entry}, which must be at least equal to the escape velocity (11.1 km s^{-1} for an altitude of 100 km). Since molecular energies at the boiling point of most terrestrial materials (1 000–2 000 K) correspond to thermal velocities of < 2 km s^{-1}, the dissipation of the particulates' kinetic energy through atmospheric deceleration and heating would at first sight be expected to ablate the particle totally. However, this happens only in a fairly modest "window" of masses between 10^{-12} kg and 1 kg. For particles larger than this the atmosphere is not deep enough for complete ablation despite an exponentially increasing effect (due to increasing atmospheric density) towards the ground. Laminar flow, enhanced by forward ablation in the atmosphere, also protects and decelerates the body, often to little more than the free fall terminal velocity, so we do, fortunately, "find" meteorites. Fall rates correspond to one "football sized" particle landing on Australia every year. For increasing initial meteoroid masses retention of a greater fraction of the approach velocity occurs, and for objects above 10 m in initial size the atmosphere becomes thin. We have, therefore, a record of hypervelocity impact craters such as the Barringer "Meteor" Crater, Arizona (diameter 1.2 km) corresponding to an impacting object of order 100 m in diameter. Earlier, and larger, impactors have been significant in initiating global terrestrial changes by virtue of the impact ejecta (Alvarez et al. 1980) and residues of the impactors are found in geological formations.

The current influx of ocean and terrestrial spherules is formed by quite a different influx to the Earth. This micro-population is derived either from the ablation and fragmentation products of mid-sized meteoroids or alternatively from single interplanetary particulates; it has offered to date rich scope for sampling the interplanetary distribution in terms of morphology, mineralogy and chemistry (see Brownlee, this volume). Micrometeorites, and their capacity for survival during atmospheric entry, were first highlighted by Whipple (1950) although their existence was first detected by Murray and Renard (1883) in the Challenger expedition. The study of these "cosmic spherules" in ocean sediments is well established and advances in their extraction from a variety of other media continues (e.g. spherules and IDP's melted from antarctic ices and from Greenland ice melt pools (e.g. Maurette et al. 1987, 1991; Yamakoshi 1994)).

In his classical analysis of the formation of the meteor phenomenon, Whipple (1950) considers ablation from impact under free molecular flow in the meteoroid. The meteoroid velocity, exceeding atmospheric thermal velocities by a factor of more than 10, provides its own energy, sufficient (if the deceleration path is long enough) to melt and evaporate the meteoroid. For high velocities the meteoroid mass is reduced to zero before the particle is appreciably decelerated; few or no ablation spherules will be produced since even the micro-fragments may not be able to enter the micrometeorite survival window at the lower altitudes of meteor ablation.

In the situation where molecular impacts raise the meteoroid surface temperature to ablate surface atoms or molecules, the species find themselves travelling forward into the atmosphere at a high enough velocity to ionise the ambient atmosphere within a path length determined by the mean free path. The term "meteor" describes the light-emitting path of the body and the associated ionisation. The physics of meteoroid flight through the atmosphere has been considered, for example, by Whipple (1950, 1951), Öpik (1958), Baldwin and Sheaffer (1971), Bronshten (1983), Flynn (1989) and Love and Brownlee (1991).

We need to interpret meteor ablation to understand the incident mass distribution of meteoroids, likely sources, streams and the velocity distribution; in short, the particulate environment of near Earth space. The information complements that derived from near Earth satellites which have generally returned impact rates; Explorers 16 and 23 and the Pegasus series provided the basis for environmental definition for 30 years. Though still valuable now, their data have been superseded in the quality of definition by retrieved satellites, especially LDEF and EuReCa. Although electronic in situ detection has developed in sophistication, with few exceptions (e.g. the IDE detector on LDEF and, in more distant cis-lunar regions of space, the MDC experiment on Hiten, see section IV.B) little opportunity for LEO deployment has been offered since HEOS 2 in High Eccentricity Orbit (HEO) in 1972–74. Detection techniques are reviewed by Auer (this volume) and results of spacecraft measurements are reviewed in section IV.

The energy budget of a meteoroid in the atmosphere, and hence its temperature, is determined by the balance between the energy input by solar radiation and atmospheric drag and the energy loss by re-radiation and ablation. The rate of change of temperature of a meteoroid is then given by

$$MC\frac{dT}{dt} = \frac{1}{2}\alpha A \rho_a V^3 + AS(1-a) - B\beta\sigma T^4 + L\frac{dM}{dt} , \qquad (26)$$

where M is the meteoroid mass, C is its specific heat capacity, α is an accommodation coefficient relating the efficiency with which impinging air molecule kinetic energy is transferred to the meteoroid (~ 1 for the high relative velocities operating, i.e. molecules are effectively absorbed by the meteoroid and re-emitted at the thermal velocity corresponding to the surface temperature), A is the meteoroid cross-sectional area, ρ_a is the local atmospheric density, V is the meteoroid velocity relative to the atmosphere, S is the solar constant, a is the meteoroid albedo, B is the thermally emitting surface area, β is the emissivity, σ is the Stephan-Boltzmann constant and L is the meteoroid material

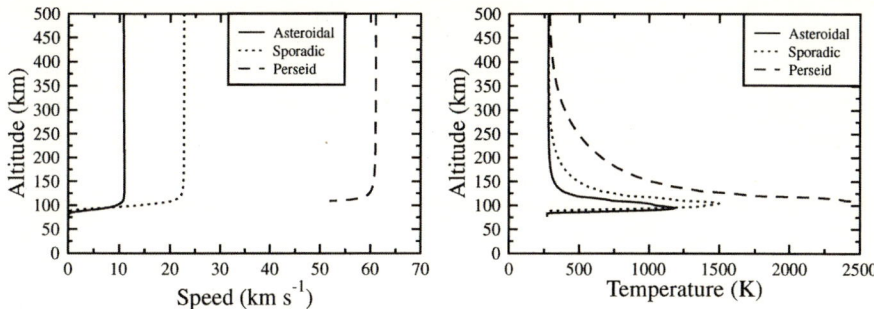

Figure 4. Speed and temperature profiles for a 10 μm diameter meteoroid entering the Earth's atmosphere with a zenith angle of 45° for $V_\infty = 0$ km s^{-1} (asteroidal meteoroid), $V_\infty = 20$ km s^{-1} (sporadic meteoroid) and $V_\infty = 60$ km s^{-1} (Perseid meteoroid). The altitude of strong deceleration and peak temperature has only a small velocity dependence. The "asteroidal" particle survives entry intact. The "sporadic" meteoroid melts, but suffers virtually no mass loss. The "Perseid" meteoroid ablates completely at 108 km.

latent heat of vaporisation. The particle acceleration is given by

$$\frac{dV}{dt} = \frac{1}{2} \frac{C_D A \rho_a V^2}{M}, \quad (27)$$

where C_D is the drag coefficient (~ 2), and the rate of change of the meteoroid mass through ablation is

$$\frac{dM}{dt} = \frac{B P_v \mu^{1/2}}{2\pi k T}, \quad (28)$$

where P_v is the vapour pressure, μ is the meteoroid mean molecular mass and k is the Boltzmann constant. These equations can be solved using explicit numerical integration to give the dynamic and thermal profile of the meteoroid flight, as shown in Fig. 4.

Whipple's analytical approach to meteoroid flight through an isothermal atmosphere (for the case where vaporisation of the meteoroid is negligible) leads to the derivation that the maximum temperature reached is given by

$$T_m^4 = T_0^4 + \frac{M \alpha V_{entry}^3 \cos Z}{3 \beta B e \sigma C_D H} \quad (29)$$

and occurs when the meteoroid velocity has fallen to

$$V_c = \frac{V_{entry}}{\exp(-1/3)} = 0.7165 V_{entry}, \quad (30)$$

where T_0 is its initial temperature, Z is the trajectory zenith angle, V_{entry} is the meteoroid atmospheric entry velocity and H is the atmospheric scale height.

Table 5

Meteoroid diameters and masses for potential ablation and melting, calculated according to Eq. (29). Velocities shown (V_∞) are the initial geocentric velocities at infinity prior to gravitational acceleration. The contrast between a low approach velocity to the Earth, possibly an asteroidal fragment of low ecliptic inclination and eccentricity, and a typical sporadic meteoroid is illustrated; the asteroidal window for survival is some 4 orders of magnitude greater.

	$V_\infty = 0$ km s^{-1} 'Asteroid' ($\rho = 3000$ kg m^{-3})	$V_\infty = 20$ km s^{-1} 'Sporadic Meteoriod' ($\rho = 3000$ kg m^{-3})
'Ablation' $T = 1900$ K	$M = 9.0 \cdot 10^{-10}$ kg $d = 83$ μm	$M = 7.4 \cdot 10^{-14}$ kg $d = 5.2$ μm
'Melting' $T = 1200$ K	$M = 4.1 \cdot 10^{-11}$ kg $d = 29$ μm	$M = 3.3 \cdot 10^{-15}$ kg $d = 1.8$ μm

A meteoroid will therefore survive atmospheric entry essentially unscathed if its thermally emitting surface area is sufficiently large compared with its mass so that it can radiate energy efficiently enough to keep the maximum temperature below a critical level. Since surface area is proportional to (dimension)^2 while mass is proportional to (dimension)^3 this effectively sets an upper limit to the size of body which survives atmospheric entry unaltered. The size limit is a strong function of the critical temperature (to the fourth power) and hence of the material of the meteoroid, and also of the initial geocentric velocity. Table 5 shows the result of these calculations (in terms of both meteoroid mass and size) for typical velocities and densities of particles.

In the size regime of stratospheric collection (diameters of 5 to 50 μm) a strong selection effect therefore operates in favour of low geocentric velocity particles (and hence by inference those of asteroidal origin) over those of higher velocity. This "filter" is critical to interpretation of the surviving fraction we collect in or below the atmosphere and to the fraction which can be captured into Earth orbit.

The altitude at which these critical phenomena occur is also a vital parameter for examining the LEO satellite impact data. Whipple's analysis leads to the derivation that the atmospheric density at which the meteoroid reaches its maximum temperature is given by

$$\rho_m = \frac{8\exp(1)\beta\sigma\left(T_m^4 - T_0^4\right)}{\alpha V_{entry}^3} . \tag{31}$$

The atmospheric density at which a typical sporadic meteoroid reaches its maximum temperature is thus ~ 80 kg m^{-3}, which corresponds to an altitude of ~ 100 km. However, from Eqs. (29) and (31) it is clear that while the peak temperature experienced by a meteoroid of a given mass is a strong function of its initial velocity, the atmospheric density (and hence altitude) at which this occurs is not. Such altitudes are sufficiently low that the flux of interplanetary particles incident upon the Earth is unmodified by the atmosphere at spacecraft orbital altitudes. However, the atmosphere does play a highly significant role in the case of Earth-orbital particles, as will be shown in the following section.

2. Aerocapture and Aerofragmentation Capture

If a particle on an initially hyperbolic trajectory with respect to the Earth penetrates deeply enough into the atmosphere, it will be absorbed. Its fate will either be total ablation or total or partial survival, as described in the previous section. If a particle only passes through the upper reaches of the atmosphere it may lose only a small proportion of its kinetic energy and remain on a hyperbolic trajectory. However, there exists a finite range of altitudes of atmospheric penetration between these two cases for which a particle is not directly absorbed by the atmosphere but loses enough kinetic energy to adopt an elliptical rather than a hyperbolic trajectory, i.e. to be captured into orbit around the Earth. The kinetic energy that must be lost is equal to the initial kinetic energy of the particle in the geocentric frame, prior to acceleration by the Earth's gravity. The process is known as aerocapture (e.g. McDonnell and Ratcliff 1992; Ratcliff and McDonnell 1992; Ratcliff et al. 1993a,b), direct evidence for which has been provided by the recovery of an 80 μm cosmic spherule displaying evidence of two distinct phases of atmospheric heating, presumably due to aerocapture into orbit followed by final atmospheric entry at the next perigee passage (Genge et al. 1996).

The kinetic energy lost by a particle passing through the atmosphere depends on the velocity of the particle and the air mass encountered. The critical altitude for aerocapture for a given particle thus depends on its initial kinetic energy (it must penetrate deeper into the atmosphere to lose more energy), its cross-sectional area (a larger particle of a given energy intercepts a greater air mass), and the atmospheric density as a function of altitude. At a given altitude this is a strong function of solar activity and time of day, as the Earth's atmosphere expands significantly at high altitudes (above ~ 140 km) under increased solar irradiation. Figure 5 shows sample aerocapture maximum altitudes as a function of initial geocentric velocity. For particles of 100 μm or smaller the capture altitudes are typically 100–250 km.

Once a particle is captured into Earth orbit kinetic energy is lost at each subsequent perigee passage, reducing the apogee and the orbital eccentricity. The perigee altitude remains almost constant. Once the orbit is circularised the particle is rapidly deorbited (typically on a timescale of minutes, or a fraction of a single orbit) as not only does the orbital altitude decay leading to greater atmospheric drag, but also the interaction with the atmosphere (and hence loss of kinetic energy) occurs continually rather than for only part of the orbit. Figure 6 shows the evolution of the orbit of an aerocaptured particle.

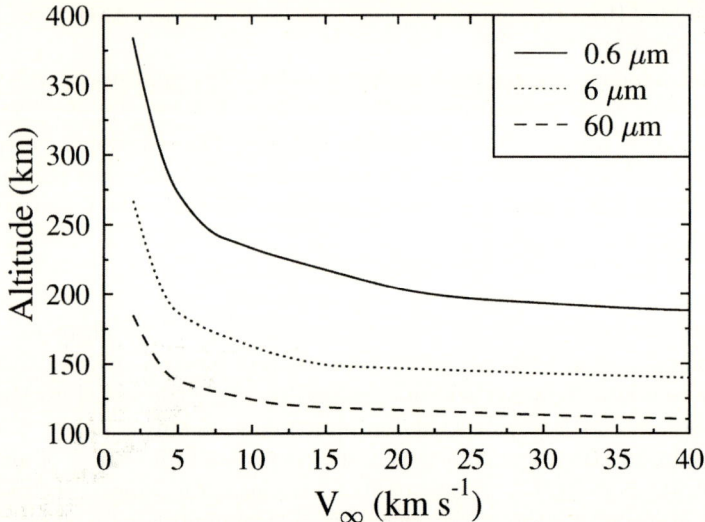

Figure 5. Maximum altitudes for aerocapture plotted against V_∞ for meteoroids of three different sizes (after Ratcliff and McDonnell 1992).

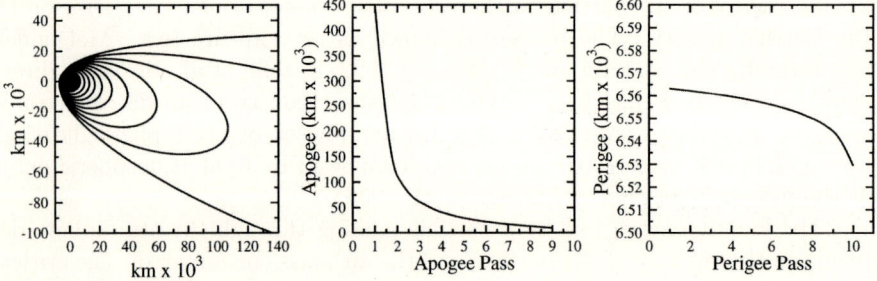

Figure 6. Numerical integration of:
(a) the path of an aerocaptured particle;
(b) the evolution of the orbit apogee;
(c) the perigee distance. The particle has $V_\infty = 2$ km s^{-1} and is 20 μm in diameter with a density of 3000 kg m^{-3}. Aerocapture (i.e. first perigee) occurs at an altitude of 192 km, the first apogee is at 447 000 km, and the particle completes 10 orbits before being absorbed by the atmosphere. The maximum number of orbits the particle can complete if the first apogee is allowed to tend to infinity is 12.

The primary factor influencing the number of orbits that a captured particle will complete is the ratio of the kinetic energy lost during capture to its initial kinetic energy. It therefore follows that the orbital lifetime of particles is a strong function of their geocentric velocity, but only a weak function of their size or mass. For a particle that is only just captured ($e \lesssim 1$) the velocity post capture is almost equal to the escape velocity, V_{esc} (in practice the eccentricity is limited to ~ 0.98 or particles would adopt heliocentric rather than geocentric orbits). The circular orbital velocity, V_c, (effectively the minimum velocity for an orbital particle) is given by $V_c = V_{esc}/\sqrt{2}$. Therefore, if the particle

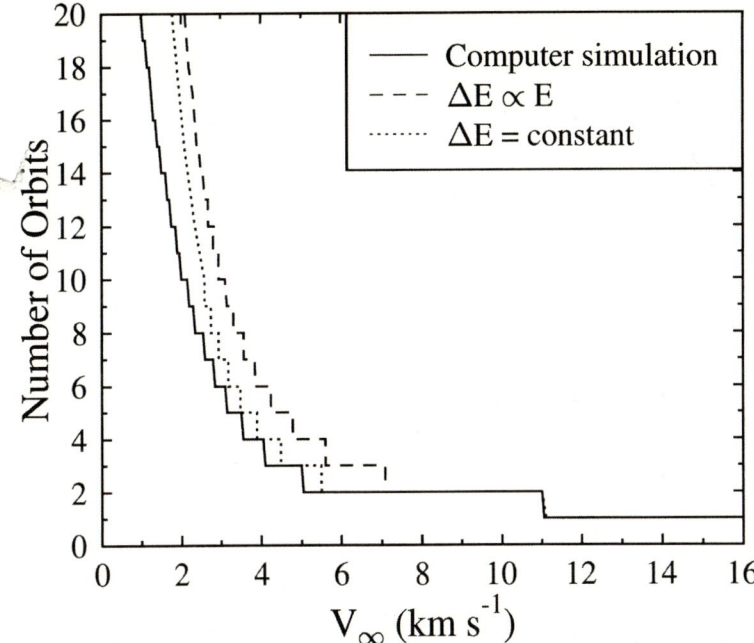

Figure 7. The maximum number of orbits that can be completed by an aerocaptured meteoroid particle as a function of initial geocentric velocity. The dashed line is the analytical result assuming constant fractional energy loss at each perigee passage. The solid line is the result obtained by computer simulation of the path of model particles, with the initial orbital eccentricity limited to 0.98 as explained in the text. The dotted line assumes a constant absolute energy loss at each perigee passage and provides a better approximation.

loses half its remaining kinetic energy at the second perigee passage then it will be absorbed. Since drag force is proportional to V^2 (and hence to the particle kinetic energy) and the energy lost is the integral of the force over distance, a particle will lose approximately the same fraction of its kinetic energy at each perigee passage. In fact it loses a slightly greater fraction each time since the interaction path length increases with increasing orbital circularity and the perigee is slightly reduced. Hence, if a particle has lost over half its kinetic energy during the initial capture process (i.e. if its atmospheric entry velocity, V_{entry}, was $> \sqrt{2} \cdot V_{esc}$, or ~ 15.6 km s^{-1}, corresponding to an initial geocentric velocity (V_∞) equal to $V_{esc} \sim 11$ km s^{-1}) it can only complete a single orbit. The number of orbits possible increases with decreasing geocentric velocity so that asteroidal particles with $V_\infty < 1$ km s^{-1} may complete 20 or more orbits before being absorbed (see Fig. 7). If the particle penetrates deeper into the atmosphere at first encounter, such that a greater fraction of its initial kinetic energy is lost, the number of orbits completed will clearly be smaller.

The above treatment assumes that the particle mass is invariant (i.e. that aerobraking is sufficiently gentle for ablation not to occur), and that the initial perigee is not modified by external forces such as solar radiation pressure or the solar-terrestrial-lunar gravitational field. If the orbital perigee is raised by one of these mechanisms into a region with significantly lower atmospheric density then orbital lifetimes can be greatly extended. Conversely the lifetime may be shortened by these mechanisms lowering the orbital perigee.

The fraction of interplanetary particles captured into orbit rather than being directly absorbed by the Earth's atmosphere also depends on their initial kinetic energy and density. This is because the range of altitudes over which capture occurs, Δh, depends on the atmospheric scale height, and hence on the altitude of capture. For particulates not matching well the Earth's heliocentric orbital elements and having an initial geocentric velocity of V_∞ the Earth presents a cross-sectional area for direct absorption of approximately

$$A_a = \pi R_c^2 (1 + V_{esc}^2/V_\infty^2) \ , \tag{32}$$

where R_c is the radius of the Earth plus the critical atmospheric altitude for particle capture. To calculate the cross-sectional area for *either* direct absorption *or* capture, R_c is replaced by R_c plus the altitude range, Δh, over which aerocapture occurs. The relative probability of capture vs. direct absorption is thus

$$P = \frac{(R_c + \Delta h)^2 - R_c^2}{R_c^2} \ . \tag{33}$$

Typical values of Δh range from several to tens of km, giving rise to capture efficiencies for particles of 100 μm diameter or smaller of $\sim 10^{-5}$ for typical sporadic meteoroid velocities, rising to 10^{-2} for asteroidal particle velocities (Ratcliff and McDonnell 1992). Although these capture probabilities are low, the dwell-time of particles in orbit (initially with a high apogee) serves to increase their population in the near Earth environment compared with the population of interplanetary particles directly entering the Earth's atmosphere.

In view of the requirements for particle capture and their subsequent orbital evolution and lifetime, a number of qualitative conclusions can be drawn:
- The population of natural interplanetary material in Earth orbit will be dominated by those particles with an initially low geocentric velocity—this implies a preponderance of asteroidal over cometary material.
- The orbital material will be predominantly in eccentric orbits since the lifetime of circular orbits is extremely small.
- The mass index, α, of the material in orbit will be greater than that in interplanetary space since the capture of smaller particles is favoured. In view of this, the relative abundance of captured orbital particles to interplanetary particles directly entering the Earth's atmosphere is mass-dependent (increasing with decreasing mass), but calculations suggest a value of $\sim 10\%$ for 1 μm particles (Ratcliff et al. 1993a).

The phenomenon of meteorite breakup in the atmosphere is well known for the relatively large bodies directly entering the atmosphere that give rise to visual and radio meteors. Breakup occurs if the aerodynamic pressure or thermal stresses experienced by the body exceed its strength. Bodies en-route for capture into orbit experience lower pressures and thermal stresses due to the longer interaction time, but particles above mm dimensions will still undergo some modification. As a minimum this will entail ablative loss of surface material, but for a structurally weak body (due to irregular structure, incipient fractures or friable material) fragmentation may also occur. The fragmentation products will then be captured into orbit. Indeed, since smaller particles are more readily captured than larger ones, the fragmentation products may even be captured into orbit if the initial body would not have been. This process is known as aerofragmentation capture (McDonnell and Ratcliff 1992).

The altitude at which fragmentation occurs is that at which entry temperatures and pressures reach their peak (i.e. of the order of 100 km) and so the lifetime of the fragmentation products is typically extremely short and may be limited to a single orbit unless an efficient perigee-raising mechanism operates. However, Fechtig (Hoffman et al. 1975b) has also considered fragmentation in the magnetosphere of fireball-type objects, which are considered to be loose and friable, as a mechanism for generating the "swarms" and "groups" seen by the HEOS 2 spacecraft in high Earth orbit (see Grün et al. this volume). Fragmentation mechanisms such as this, operating at higher altitude, result in much higher orbital lifetimes for the fragmentation products because of the potential for higher perigee altitudes and hence reduced atmospheric drag. However, for small particles radiation pressure is strong and, depending on the heliocentric longitude of the perigee, perigee may be increased or decreased rapidly.

While it is clear that fragmentation capture mechanisms can operate to enhance the population of natural material in Earth orbit, their relative contribution still remains to be proven. It is, however, an area where exploration would be profitable.

III. SPACE DEBRIS

Studies of the near Earth particulate environment must take account not only of incoming or captured interplanetary dust, but also of the continually changing population of objects introduced by human activities in space. The broad title of "space debris" spans an enormous range of size (or mass) from tens of metres for spent boosters and payloads, both active and "dead", down to sub-μm dimensions for fragmentation products and impact ejecta. Other major constituents of this population include payload separation hardware, mission related hardware such as covers, shrouds, etc., aluminium oxide spherules from upper stage motor burns and paint or fibre flakes from the degradation of spacecraft surfaces. Of this population, over 7 000 objects are regularly tracked and catalogued by US Space Command (USSpaceCom) (formerly NORAD) covering the size range above 10 cm for (LEO) or above $\sim$ 1 metre for Geostationary Equatorial Orbits (GEO), although detection of objects as small as a few mm is possible with ground-based optical and radar systems. Of this population ~ 50 % is estimated to be fragmentation products, 25 % mission related objects, 20 % non-operational payloads and only 5 % operational payloads.

Not only the present population but also its future evolution is a source of concern to the space community. Extrapolation from the present scenario is fraught with uncertainties, but under the most pessimistic assumptions a runaway growth in the debris population is seen as a possibility, resulting from collisions between members of the population, fragmentations, explosions and further normal space activities. This increase is, however, partially compensated by the reduction in the population resulting from re-entry into the Earth's atmosphere. As a result of these concerns, the Space Agencies have become aware of the potential problems and are beginning seriously to assess and quantify the situation, to define design and operational procedures to reduce the generation of debris and even to look at possible future debris removal procedures.

Although not a significant proportion of the space debris total mass, and all tracked debris objects are in the size range of a few cm and above, it is the micro-debris (sizes $\leq$ mm) which is of most relevance to studies of interplanetary dust since it is these objects which are "contaminants" for in situ detectors. In reality, the space debris population itself, and its evolution, is a target for such experiments, but the two populations must nonetheless be identified and separated. We will provide here only a brief overview of space debris properties to place the results of in situ detection (section IV) and the orbital environment models (section III) in context. An excellent overview of space debris detection techniques, modelling, shielding and mitigation may be obtained from a number of sources: USIG (1989; 1995); USNRC (1995); ESA (1988; 1993; 1997) and Johnson and McKnight (1987).

Any experiments to detect or capture interplanetary dust located on Earth-orbiting spacecraft will also be sensitive to micro-debris particles. Since the only detection mechanisms are these experiments themselves, our knowledge of the space debris population is restricted by available space platform opportunities and the limited area-time product of space detectors. All particles are characterised by short lifetimes (measured in years or less) if they are in LEO and we see the population in dynamic equilibrium between supplying sources and depleting forces, whether these sources be natural or space-age products.

The definition of space debris orbits has not generally been possible from impact experiment results since most have no discrimination between natural and space debris particles, a directional resolution of worse than a steradian, no time resolution and unknown platform pointing history. Valuable information on the relative populations of debris and interplanetary material may still be obtained from impact experiments if the spacecraft pointing history is well defined or chemical analysis of residues is possible. Results from LDEF and EuReCa are discussed in section IV. The distribution of orbits of micro-debris will be related to the orbits of their parents, the known tracked objects, despite rapid randomisation of the nodes and apsides, as well as the increased importance of atmospheric drag and radiation pressure. The distribution of inclinations of the larger satellites and debris reflects the launch site or original mission type. Strongly pronounced maxima occur (in decreasing order of significance) at 75°, $\sim 63°$ (the critical inclination for zero apsidal advance employed by Molniya satellites), $\sim 82°$, $\sim 100°$ (Sun-synchronous orbits), 90° (exact polar orbits), $\sim 0°$ (GEO's) and $\sim 28.5°$ (due East launches from Kennedy Space Center). About half the total catalogued objects down to ~ 10 cm in size have near-circular orbits with eccentricity $e < 0.01$, and 90 % with $e < 0.05$. Geostationary Transfer Orbits (GTOs) and Molniya type orbits have eccentricities ~ 0.73. Smaller debris in LEO (or with sources which have low perigees) will be subject to greater atmospheric drag effects due to their larger surface area-to-mass ratio, which would be expected to decrease the mean eccentricity with decreasing size. Around 90 % of satellites and debris lie in LEO at altitudes between 200 and 2 000 km, with accumulations at 600, 800, 1 000 and 1 500 km. Other concentrations occur for GEO at $\sim 35\,800$ km (most lying within ±500 km of GEO altitude, inclinations $\lesssim 3°$, eccentricities up to 0.01); GTO and Molniya type orbits with perigee in LEO and apogee altitudes 35 000–40 000 km; navigation satellites with near-circular 19 000 km altitude orbits.

An example of a tracked object database is ESA's Database and Information System Characterising Objects in Space (DISCOS), operated at the European Space Operations Centre at Darmstadt, Germany; it is a continuously updated collection of satellite and space debris data and software tools which may be accessed by registered users (Klinkrad 1991; Jehn et al. 1993; Klinkrad et al. 1997a). DISCOS has been developed around the ORACLE relational database management system and provides an accessible interface to widely dispersed data sources for use in characterising the space debris component of the near Earth Environment.

The key table contains, where known, for each catalogued object (identified throughout by its COSPAR-ID),
– name and USSpaceCom number,
– shape (cylinder, sphere, cone),
– dimensions (length, diameter),
– mass,
– decent date or, if still in orbit, estimated lifetime,
– country of origin,
– cross reference to a file containing additional text information.

Data are taken from the RAE Table of Earth Satellites (King-Hele et al. 1990 and updates) except country of origin and USSpaceCom number which are taken from the NASA Satellite Situation Reports (public access version, available from NASA Goddard Space Flight Centre, of the Satellite Catalogue maintained by the USSpaceCom Space Surveillance Network). Continually updated orbits are available from a modified version of the NASA Two-Line elements (TLEs) issued by USSpaceCom, (although permission from NASA is required to access these data) together with a file of elements of all objects currently in orbit, providing a snapshot of the current population. Alternatively, the Satellite Situation Reports provide more limited free-access orbital data. Information on over 100 satellite break-ups (data from Johnson and Nauer 1990) are also available. The associated software allows the usual database functions as well as access to solar and geomagnetic activity data, forward propagation of orbits, a wide variety of output formats and a file of literature sources.

Definition of the space debris environment (i.e. distribution of composition and physical properties over all possible orbits and all size ranges, and as a function of time) is severely limited by the available observational data. Debris detection techniques, either ground-based optical and radar systems or in situ impact detectors and collectors, cover only a tiny portion of the size/orbit/time space and have different selection effects which may be difficult to quantify. Routine tracking (the Space Surveillance Network: a worldwide network of radar and optical sensors operated by USSpaceCom; Chamberlain & Slauenwhite 1993) is only possible for large objects due to the limited sensitivity and range of the techniques. More sensitive optical and radar systems are being designed and employed to extend debris detection into the critical mm to cm size range (e.g. Lobb et al. 1993; Stansbery et al. 1995; Goldstein & Goldstein 1995; Girard & Worms 1997) for which serious damage or destruction of spacecraft is possible, although generally only statistical data on possible orbits can be derived. Although many thousands of detections of micro-debris particles have been made from a variety of spacecraft, the environment is only poorly defined. Only a few orbits have been sampled (mostly at around 500 km altitude) on a few occasions (or over long time averages). Moreover, the limited area-time products result in poor statistics for particle sizes greater than a few hundred μm. Such observations are used to define environment characterisation models, which may simply be tools for engineering purposes (e.g. using existing measurements and extrapolating across altitude, size and time to produce a "comprehensive" model), or predictive, accounting for sources and sinks (launches, operational debris, fragmentation events, atmospheric decay). These models use a range of catalogued, ground and space-based data, together with assumptions about objects which have not been detected (e.g. interpolating and extrapolating over size and altitude), usually assuming a steady state or steadily changing population. McKnight and Johnson (1989) highlight the errors which may be introduced simply by uncritical use of catalogues. In the micro-debris size range, we have already seen that the lifetimes in LEO may be very short, so such averaged models can only be regarded as supplying an indication of the population at any particular altitude and time. The in situ data themselves indicate the environment at the time of detection, and a more comprehensive picture of the overall population can only be obtained by continued deployment of experiments at a range of altitudes rather than the piecemeal approach which is currently possible.

IV. MODELLING TOOLS

The complexity of interplanetary dust and space debris distributions, and especially the history of the pointing direction of a particular spacecraft surface or sensor, call for sophisticated modelling in the light of the precise data now available. The use of modelling tools provides a means of comparison of modelling results with real data from different experiments and their interpretation in terms of the true debris and interplanetary dust populations.

IV.A. The Grün Interplanetary Dust Model at 1 AU

The "Grün flux" (Grün et al. 1985), introduced in section II.A, is generally used to quantify the natural sporadic meteoroid influx to Earth, and was derived taking into account in situ spacecraft experiments. The flux values are given as cumulative mean values (i.e. time averaged number of particles m^{-2} s^{-1}, of individual masses $\geq M$) for a spinning flat plate detector at 1 AU outside the gravitational influence of the Earth (but moving in an Earth-like orbit). Table 6 gives the flux values from the model. A parametric form of the cumulative flux, $F(M)$ (M in g), at 1 AU can be written (from Grün et al. 1985)

$$F(M) = (c_4 M^{\gamma_4} + c_5)^{\gamma_5} + c_6 (M + c_7 M^{\gamma_6} + c_8 M^{\gamma_7})^{\gamma_8} + c_9 (M + c_{10} M^{\gamma_9})^{\gamma_{10}} , \quad (34)$$

where $c_4 = 2.2 \cdot 10^3$, $c_5 = 15$, $c_6 = 1.3 \cdot 10^{-9}$, $c_7 = 10^{11}$, $c_8 = 10^{27}$, $c_9 = 1.3 \cdot 10^{-9}$, $c_{10} = 10^6$, $\gamma_4 = 0.306$, $\gamma_5 = -4.38$, $\gamma_6 = 2$, $\gamma_7 = 4$, $\gamma_8 = -0.36$, $\gamma_9 = 2$ and $\gamma_{10} = -0.85$. The total contribution is dominated by the first term for $M \geq 10^{-9}$ g, the second term in the $10^{-14} \leq M \leq 10^{-9}$ g range, and the third term for $M \leq 10^{-14}$ g.

The method of modelling meteoroid fluxes, gravitational enhancement and spacecraft impact damage, was described in section II.B. We can calculate the maximum meteoroid mass intercepted, by "inverting" the Grün flux distribution. This is shown in Fig. 8. We can also consider the maximum penetration in aluminium to a real space face surface orbiting in LEO or GEO, as a function of the area-time product a_t (m^2 s); shown in Fig. 8. Functions are fitted to the curves in Fig. 8 over the range shown, for the LEO surfaces:

$$M(g) = [(5.02 \cdot 10^{-15} a_t)^{-0.152} + (4.49 \cdot 10^{-8} a_t)^{-0.716}]^{-5.128} \quad (35)$$

$$F_{max}(\mu m) = [(1340 a_t)^{-0.126} + (2.02 \cdot 10^{-4} a_t)^{-0.620}]^{-2.193} , \quad (36)$$

where a_t is the area-time product (m^2 s). For GEO surfaces, the functions still apply, but instead of using a_t, we substitute $a_t/1.67$.

IV.B. Spacecraft Geometry

Considering the anisotropic distributions of meteoroids or space debris particles, one should bear in mind that some spacecraft orbital and attitude geometries directly affect the sampled flux. For example, McBride et al. (1995) investigated LDEF's orbital history in detail. The LDEF satellite maintained

Table 6

Parameters for use in a meteoroid flux model. Upper table: interplanetary flux distribution at 1 AU from Grün et al. (1985). The fluxes are cumulative (number $m^{-2}\ s^{-1} \geq M$ g) and are the annual mean to a spinning flat plate detector at 1 AU Lower table: normalised velocity distributions of meteoroids (as shown in Fig. 1b). Velocity V is in km s^{-1}. The $n(V_E)$ distribution is for the meteoroids encountering the Earth's atmosphere (Taylor 1995b) at altitude of order 100 km, whereas $n(V_\infty)$ shows the distribution corrected to 1 AU (i.e. as seen from a massless Earth). The $n(V_\infty)$ distribution is combined with flux distribution to derive fluxes to a detector as described by Eq. (18.)

M(g)	Flux (m^{-2} s^{-1})	M(g)	Flux (m^{-2} s^{-1})	M(g)	Flux (m^{-2} s^{-1})
10^{-18}	$2.6 \cdot 10^{-1}$	10^{-11}	$1.5 \cdot 10^{-5}$	10^{-4}	$3.3 \cdot 10^{-10}$
10^{-17}	$3.8 \cdot 10^{-2}$	10^{-10}	$6.4 \cdot 10^{-6}$	10^{-3}	$1.9 \cdot 10^{-11}$
10^{-16}	$5.9 \cdot 10^{-3}$	10^{-9}	$3.0 \cdot 10^{-6}$	10^{-2}	$9.7 \cdot 10^{-13}$
10^{-15}	$1.1 \cdot 10^{-3}$	10^{-8}	$1.2 \cdot 10^{-6}$	10^{-1}	$4.7 \cdot 10^{-14}$
10^{-14}	$2.5 \cdot 10^{-4}$	10^{-7}	$3.0 \cdot 10^{-7}$	10^{0}	$2.2 \cdot 10^{-15}$
10^{-13}	$8.3 \cdot 10^{-5}$	10^{-6}	$4.7 \cdot 10^{-8}$	10^{1}	$1.0 \cdot 10^{-16}$
10^{-12}	$3.4 \cdot 10^{-5}$	10^{-5}	$4.6 \cdot 10^{-9}$	10^{2}	$4.7 \cdot 10^{-18}$

V	$n(V_\infty)$ $\cdot 10^{-4}$	$n(V_E)$ $\cdot 10^{-2}$	V	$n(V_\infty)$ $\cdot 10^{-4}$	$n(V_E)$ $\cdot 10^{-4}$	V	$n(V_\infty)$ $\cdot 10^{-6}$	$n(V_E)$ $\cdot 10^{-6}$
0.5	1.97	–	25.5	234	207	50.5	398	221
1.5	8.65	–	26.5	210	175	51.5	387	207
2.5	35.0	–	27.5	196	154	52.5	367	187
3.5	88.3	–	28.5	177	138	53.5	356	191
4.5	163	–	29.5	151	120	54.5	352	194
5.5	204	–	30.5	130	107	55.5	329	184
6.5	268	–	31.5	114	90.4	56.5	297	169
7.5	333	–	32.5	99.6	74.0	57.5	268	147
8.5	377	–	33.5	84.4	64.5	58.5	236	138
9.5	403	–	34.5	70.1	55.2	59.5	211	127
10.5	431	–	35.5	58.0	44.9	60.5	195	107
11.5	461	17.0	36.5	48.3	37.3	61.5	171	95.0
12.5	478	11.1	37.5	40.3	29.9	62.5	142	87.0
13.5	483	8.81	38.5	33.1	24.3	63.5	118	71.5
14.5	486	7.12	39.5	26.3	20.1	64.5	92.6	57.0
15.5	489	6.35	40.5	20.4	16.0	65.5	72.3	48.4
16.5	487	5.72	41.5	16.2	12.6	66.5	56.8	37.4
17.5	470	4.95	42.5	13.2	9.60	67.5	38.4	25.1
18.5	444	4.60	43.5	10.4	7.57	68.5	25.5	17.2
19.5	423	4.37	44.5	7.78	5.97	69.5	18.1	12.3
20.5	401	3.89	45.5	6.04	4.42	70.5	12.8	8.44
21.5	364	3.43	46.5	5.10	3.26	71.5	1.96	6.09
22.5	325	3.04	47.5	4.67	2.72	72.5	–	3.60
23.5	297	2.79	48.5	4.41	2.51			
24.5	268	2.45	49.5	4.13	2.32			

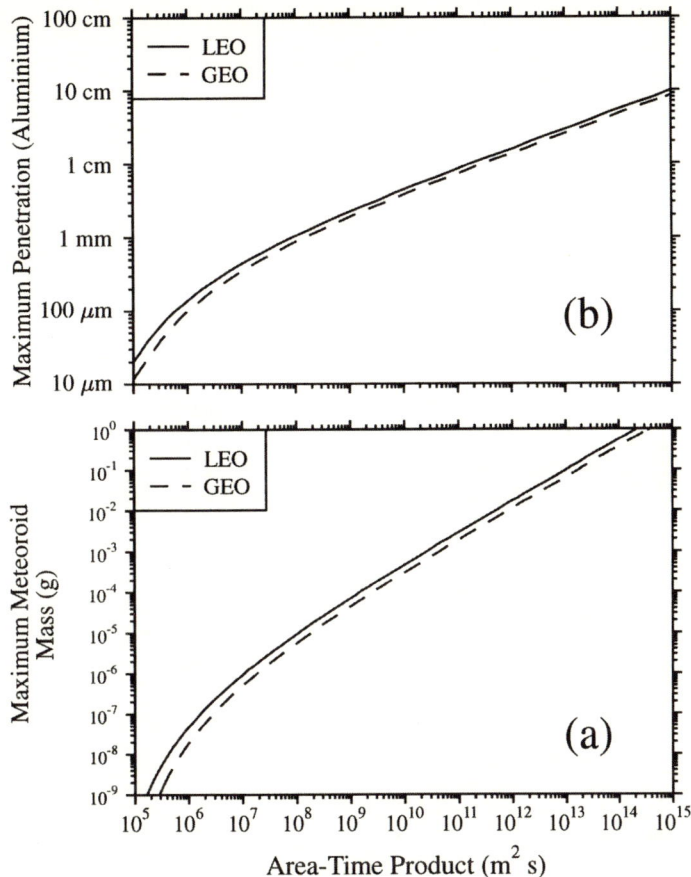

Figure 8. (a) The largest meteoroid mass encountered, and (b) the maximum aluminium penetration limit, for space pointing detectors, as a function of the exposed area-time product. For example, a typical Shuttle 2-week mission yields a maximum aluminium penetration depth of ~ 2 mm, whereas LDEF's 5.78 year exposure yields a maximum aluminium penetration depth of ~ 6 mm. (for the Shuttle, half the total surface area is assumed, 600 m^2; for LDEF an effective exposure area of 150 m^2 is assumed).

an essentially constant geocentric orientation (although its orbit precessed) and hence the faces are often referred to as North, South, East, West, space and Earth, these being the approximate facing directions with respect to the equatorial plane (the cylindrical spacecraft was actually tilted by an average of 1.1° from the geocentric radius vector and the East face was offset by 8.0° from the ram direction). The East face direction was always the "ram direction" with respect to the LDEF geocentric orbital velocity vector (see O'Neal and Lightner 1991 for an overview of the LDEF mission).

Figure 9 shows LDEF in its orbit, which is inclined at 28.5° to the Earth's equatorial plane. The orbit precesses around the North Celestial Pole (NCP) with a period of about 52 days, and hence the angle between the vector **n**, the

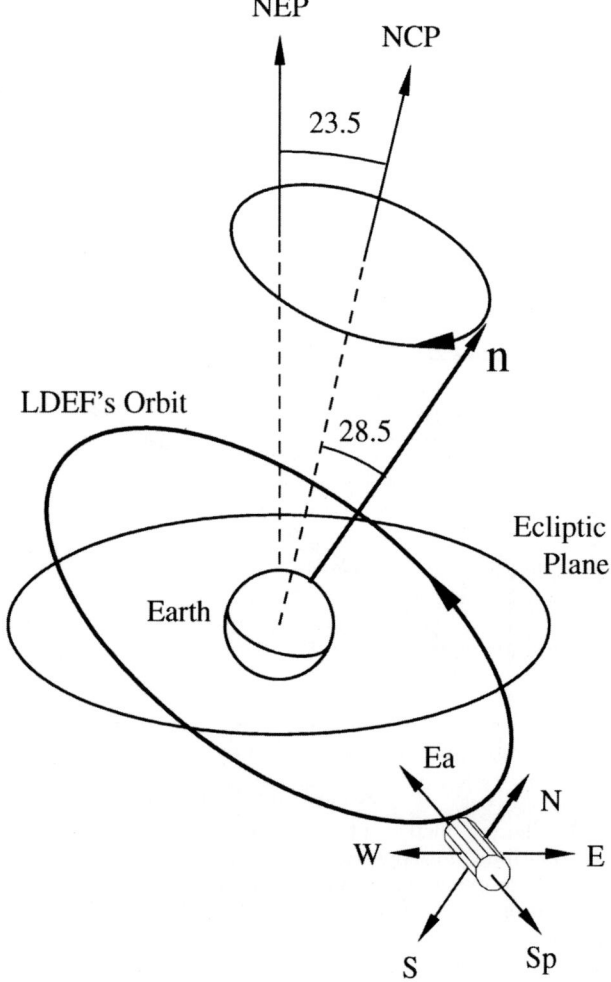

Figure 9. LDEF's orbit with respect to the Earth and the ecliptic plane. The orbit's normal vector **n** precesses clockwise around the Earth's spin axis (with a period ~ 52 days) and so the ecliptic latitude and longitude of the pointing direction of vector **n** changes throughout the year. The North face normal vector **N** is offset by $8°$ from **n** and hence essentially precesses around **n** over one geocentric orbital period.

normal to the orbit, and the ecliptic plane varies with time with a similar period. LDEF was deployed on April 7th 1984 with an initial longitude of ascending node $\Omega = 230.3°$ at a solar longitude of $17.9°$. McBride et al. (1995) present a model such that the orbital plane geometry (and hence the experimental face normal vectors) can be calculated for any time during the mission lifetime by numerically integrating the precession of the ascending node. Figure 10 shows

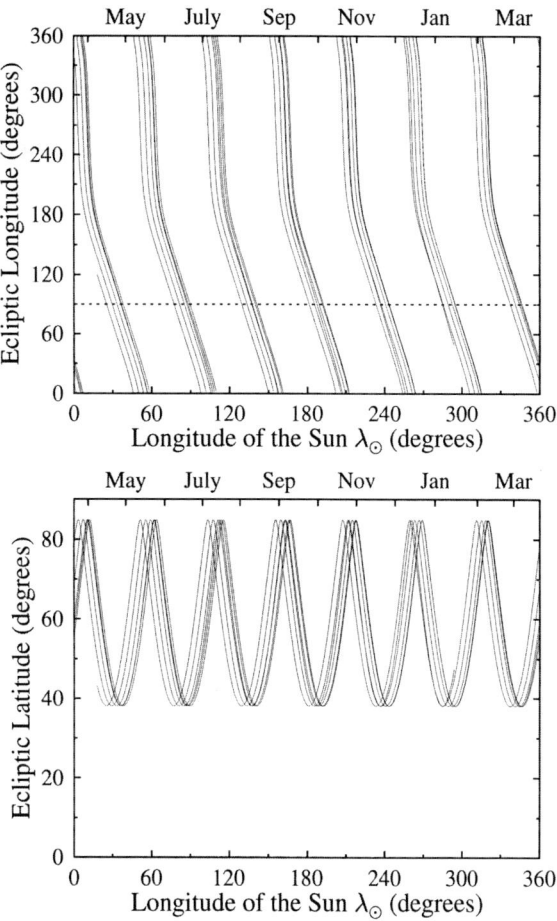

Figure 10. Ecliptic longitude and latitude of the pointing direction of the vector **n** against the longitude of the Sun over the 5.8 solar orbits that LDEF (the Earth) completes during its mission. These plots can essentially be taken as displaying the North face normal vector **N** direction (averaged over one orbital period). The dotted line in the top graph indicates the longitude of **n** when the latitude is at a minimum. These longitude values are repeatable over the lifetime and are equal to $90° \pm 1°$.

how the ecliptic longitude and latitude of the "pointing direction" of vector **n** varied with longitude of the Sun $\lambda_\odot$ (i.e. time of year). LDEF (while orbiting the Earth) made 5.8 solar orbits during its mission lifetime, and the ecliptic longitude and latitude of vector **n** has very similar values from year to year. This repeatability is a result of the precession period being approximately one seventh of a year, and also the gradual decrease in LDEF's altitude. The repeatability is vital when considering individual dust sources such as meteoroid streams or localised asymmetries in the sampling of the sporadic meteoroid

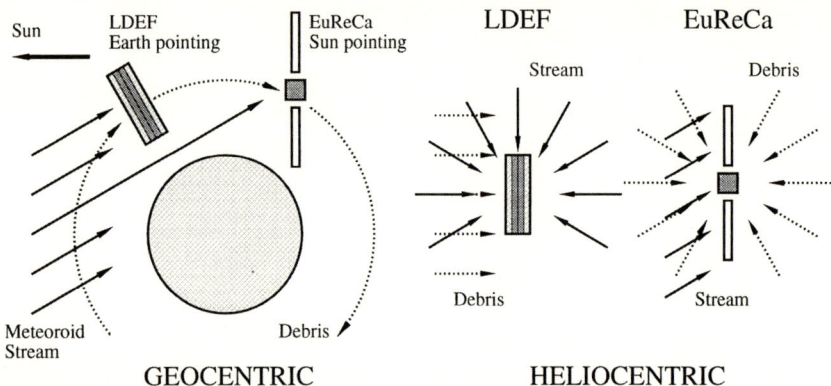

Figure 11. A comparison of the impact geometries of stream meteoroids and space debris particles, to LDEF and EuReCa. Due to the LDEF being Earth-pointing, and EuReCa being Sun-pointing, the time-integrated exposure to the faces of the spacecraft appear very different when considered in the heliocentric frame.

background. McBride et al. showed that individual stream contributions (and asymmetries in the sporadic background) could give rise to flux biases towards certain faces, and hence produce asymmetries in observed impact data.

As another example of geometry, one can consider the Sun-synchronous EuReCa satellite (see Nelleson 1995 for an overview of the EuReCa mission). Unlike LDEF, this satellite maintained an orientation such that one face always pointed to the Sun, while another face pointed approximately towards the Earth apex direction. This meant that the exposure from meteoroid streams, and from orbiting space debris, would affect EuReCa differently from LDEF. This is shown schematically in Fig. 11, where in a heliocentric frame one sees very different geometries of stream meteoroids and debris particles. For EuReCa, any meteoroid enhancement from the Earth apex direction would produce a large bias in impacts to the apex face. Also, due to Earth shielding from the apex particles when the satellite is on the anti-apex side of the Earth, one would expect an enhancement to the *Sun-pointing* face. We must clearly keep in mind the overall geometry of the spacecraft we are modelling. As an example of the influence of the spacecraft orbit and altitude on observed fluxes we determine the velocity distribution and fluxes (using Eq. (18)) for the main faces of LDEF.

The mean *normal* impact speeds ($\overline{V}_{normal}$) *at equal particle mass* on the main faces of LDEF are given in Table 7. These can be compared to the mean particle speeds ($\overline{V}$). Also given for comparison are the mean particle speeds which would be obtained if LDEF were stationary (i.e. no geocentric orbital motion). The slight enhancement to the East, North, South and West faces is a result of these faces being somewhat Earth-shielded (unlike the space face) as faster meteoroids are shielded less than slower meteoroids (section II.B.4). If we weight these velocities with $V^{\alpha\gamma}$ as described in section II.B.5, where we used $\gamma = 2.29$ to describe an equal foil penetration limit

Table 7

Mean particle impact, and normal component of impact velocity to the main faces of LDEF. Values are given for a stationary case (see text) and for impact at equal *mass* threshold (the K factor gives the ratio between the fluxes obtained compared with the stationary case). Also shown are the velocities when weighted to equal *crater depth*, equal *foil penetration thickness* and equal *plasma production*. (γ is the ratio b/a in the mass-velocity dependence $M^a V^b$ of the detector).

LDEF Face	Stationary $\overline{V}$ (km s^{-1})	Equal Mass ($\gamma = 0$)			Equal Crater ($\gamma = 2.00$)		Equal F_{max} ($\gamma = 2.29$)		Equal Plasma ($\gamma = 3.55$)	
		$\overline{V}$	$\overline{V}_n^{(1)}$	K	$\overline{V}$	$\overline{V}_n^{(1)}$	$\overline{V}$	$\overline{V}_n^{(1)}$	$\overline{V}$	$\overline{V}_n^{(1)}$
East	18.1	21.9	16.8	2.11	26.5	20.8	27.3	21.4	31.4	24.7
West	18.1	15.8	9.6	0.32	24.1	14.5	25.4	15.3	31.1	18.7
North	18.1	19.8	13.1	1.13	25.7	16.7	26.7	17.3	31.3	20.2
South	18.1	18.9	12.0	0.88	25.3	15.7	26.3	16.3	31.2	19.3
Space	17.9	19.3	12.0	1.02	25.3	15.4	26.3	16.0	31.0	18.9
Earth	19.0	19.6	3.7	0.94	27.0	4.8	28.1	5.0	32.8	5.9

$^{(1)}\overline{V}_{normal}$

(i.e. using the 1992c equation from McDonnell and Sullivan 1992) then we obtain new mean velocities weighted to equal F_{max} (in fact we have calculated these velocities by weighting with V^γ over the entire Grün mass distribution, so giving a result better reflecting the Grün flux meteoroid source). These weighted particle velocities (Table 7) show a smaller variation between the faces, which reflects the strong velocity dependence of foil penetration (i.e. we are weighting towards the faster particles and these are less affected by the spacecraft motion). The mean normal impact speeds do show a greater variation reflecting the intrinsic effects of spacecraft orientation. With impact plasma detectors the anisotropy would be even less evident.

The results of these calculations can be represented by a mean enhancement for each spacecraft surface due to the spacecraft's geocentric motion. This enhancement factor is called a "K factor". The flux on a given surface is then simply obtained by

$$F_M = F_G(M) \, \overline{G} \, \overline{\eta} \, K(\text{face}) \, , \tag{37}$$

where $\overline{G}$ and $\overline{\eta}$ are weighted mean gravitational enhancement and Earth shielding factors for a given altitude (see Table 1). The K factors associated with the LDEF faces, for fluxes *at equal particle mass*, are given in Table 7. For example, a simple prediction of the cumulative flux on the East face for $M \geq 10^{-9}$ g is given by $8.2 \cdot 10^{-6}$ m^{-2} s^{-1} (using $F_G(10^{-9}\text{g}) = 3.0 \cdot 10^{-6}$ m^{-2} s^{-1}, $\overline{G} = 1.99$, $\overline{\eta} = 0.65$, $K(\text{East}) = 2.11$) whereas doing the calculation "properly" gives the flux as $8.4 \cdot 10^{-6}$ m^{-2} s^{-1}. The K factors for the flux on a simple tumbling flat plate (i.e. spacecraft surface) in a circular orbit at any specified altitude, $\overline{K}$, were listed in Table 1.

IV.C. The NASA Orbital Debris Environment Model

The NASA orbital debris environment model was developed for use by the spacecraft community for the design and operation of spacecraft in LEO. The original version (Kessler et al. 1989) used a number of functional forms based on the USSpaceCom orbital element data set and a range of telescopic and in situ impact data to describe the cumulative particle size distribution and distribution of collision speed and direction on a randomly tumbling surface. The impacted spacecraft was assumed to be in a circular orbit with velocity 7.7 km s^{-1} and all debris orbits were circular. A debris growth rate per year and atmospheric drag depletion correction based on time-varying solar radio flux were also incorporated. This model has proved invaluable for easily calculable predictions of *average* impact rates on spacecraft for mission planning and was incorporated in the ESABASE debris module (section IV.D), but is not ideal for scientific interpretation of impacts on particular surfaces of a real spacecraft due to the circular orbit assumptions. The velocities of debris in high eccentricity orbits are different both in magnitude and direction and can therefore access spacecraft surfaces for which the model predicts no flux.

An updated version (Kessler et al. 1996), based on the latest available ground-based and in situ data, has recently been released. The new version splits the debris particles into six size dependent populations representing different debris sources (i.e. intact objects, large and small fragments, Na/K particles, paint flakes and aluminium oxide particles) in six broad orbital inclination bands. Particles follow elliptical or circular orbits in each inclination bin and have a size dependent altitude distribution. The debris spatial density as a function of size, inclination, altitude and time is converted to impact flux using the collision probability equations of Kessler 1981. Due to the much greater complexity of the 1996 revision the model is now distributed as a software program for PC compatible computers.

A range of models for environment definition, including long term growth studies and risk assessment, some of which are available to the community, have been developed at NASA's Johnson Space Center (Johnson & Christiansen 1997).

IV.D. ESABASE

ESABASE (ESTEC 1995) was developed as a systems engineering and simulation tool for spacecraft design. The Debris module predicts meteoroid and debris fluxes on a detailed geometrical model of the spacecraft; particle impact fluxes and the resulting spacecraft damage are derived. Meteoroids and debris are treated separately using the Grün et al. (1985) and Kessler et al. (1989) models respectively i.e. an isotropic meteoroid flux and a debris flux confined to a plane parallel to the Earth's surface. In other words debris are assumed to exist only in circular orbits, whereas in reality there will be an elliptical component (Flury et al. 1992).

The geometrical model is created from simple shapes composed of 3 or 4 sided planar surfaces (called "elements"). At each point on the spacecraft orbit, the surface normals of the elements are aligned according to the relevant orbital altitude and attitude information. A ray-tracing algorithm "fires" meteoroid and debris particles at the element and detects any prior intersections with other elements. This allows shielding by the Earth or by parts of the spacecraft itself to be taken into account. At present secondary cratering (by ejecta from impacts on other parts of the spacecraft) is not modelled. The properties of the impacting meteoroid (i.e. mass or diameter, speed and direction) are chosen randomly from the model distributions and the impact flux calculated. A user-selected penetration equation is used to calculate the number of "failures" (i.e. penetrations of the element wall) for a given wall thickness and construction. Currently ESABASE does not model plasma production during an impact but could be used to predict impact plasma events using mass and velocity data provided for the individual test particles which sample the debris and meteoroid populations.

A new version of ESABASE has been developed to include elliptical debris orbits, directional sources of interplanetary particles, β meteoroids, interstellar particles, annual meteoroid streams and secondary ejecta. Although space debris may be only of passing interest to interplanetary dust studies, its role is vital in modelling due to the need to remove it from the LEO data sets.

IV.E. MASTER

The ESA Meteoroid And Space debris Terrestrial Environment Reference model (MASTER; Sdunnus 1995; Klinkrad et al. 1997b) is designed to provide a semi-deterministic debris modelling technique to users with restricted CPU time and storage resources. This is achieved by using the known orbits of the trackable debris population, combined with approximately 10^{11} particles of diameter > 0.1 mm generated by a fragmentation model (Sdunnus and Klinkrad 1993) to yield a reference population of 240 000 objects which were then propagated to a 1995.0 reference epoch. The debris reference population is used to calculate the spatial density at a given point in a control volume around the Earth (divided by altitude, longitude and latitude into a total of $\sim 10^6$ "cells"). A specially developed data compression technique was required to allow the software to be distributed (along with the necessary applications software) on a standard 640 MB CD-ROM.

The user can use MASTER to predict either the flux of debris particles on a target spacecraft throughout its orbit or the number of particles observable with a ground-based radar. Unlike ESABASE there is no consideration of target geometry or micro-debris, but the larger debris population is treated in a more realistic manner. The user can extrapolate the flux generated from the reference population to a future date using altitude and traffic-dependent factors (from Kessler et al. 1989). A similar, simpler, model was developed by Green (Green and McDonnell 1992; Green et al. 1993) which included target geometry and impact equations as well as particles to sub-μm sizes.

IV.F. The Divine Interplanetary Dust Model

Although most previous work in the near Earth environment has employed the Grün interplanetary dust model (see section IV.A) at 1 AU, the Divine 5 Population Model (Divine 1993a) incorporates various spacecraft data to fit to a wide range of heliocentric distances (Divine et al. this volume). The 1 AU particulate environment as utilised in the Divine model deliberately attempts to fit to the Grün model and is therefore fully compatible at this heliocentric distance. An extension of the model to include gravitational focusing for application to LEO is given in Divine (1993b). This model offers a powerful tool to calculate quantitative fluxes and velocity distributions of particles incident on detectors. It should be noted that Divine used the Sekanina and Southworth (1975) velocity distribution (which underestimated the number of fast meteors; see Taylor 1995a) and in fact did not include any retrograde particles in the populations of importance at 1 AU. Thus, if using the model for considering fluxes on detectors that have a velocity dependence V^γ for constant mass where γ is high, the model may not adequately describe the influence of fast retrograde particles. However, inclusion of these fast particles is expected in the future.

The model has also undergone upgrades with regard to the populations which are dominant beyond 2–3 AU, based on directional data returned from the Galileo and Ulysses probes and zodiacal dust thermal measurements (Staubach and Grün 1995; Staubach et al. 1997; Grün et al. 1997). An interstellar dust component is accounted for and also out-of-ecliptic measurements are utilised.

V. MEASUREMENTS

V.A. Measurement Techniques

1. Impact Equations

Fluxes defined by environment models are normally defined in terms of the parameters of an impacting particle, whereas the damage on an exposed spacecraft surface is measured, for example, as the depth of a crater or diameter of a perforation. Comparison requires some form of cross calibration, normally determined from laboratory studies to derive some form of impact equation. The hypervelocity impact process is a complex series of events in which the material is initially shocked to pressures and temperatures only exceeded in nuclear detonation; the flow of this shocked material, combined with late-stage material failure and other factors determine the final crater morphology. It is thus unsurprising that, as yet, theoretically based approaches (Watts et al. 1993) only roughly approximate the results of experimental studies. Current best practice is therefore to use empirically derived formulae. A variety of measurements may be made of an impact feature, as shown in Fig. 12, and different authors use the same term to refer to different measurements (e.g.

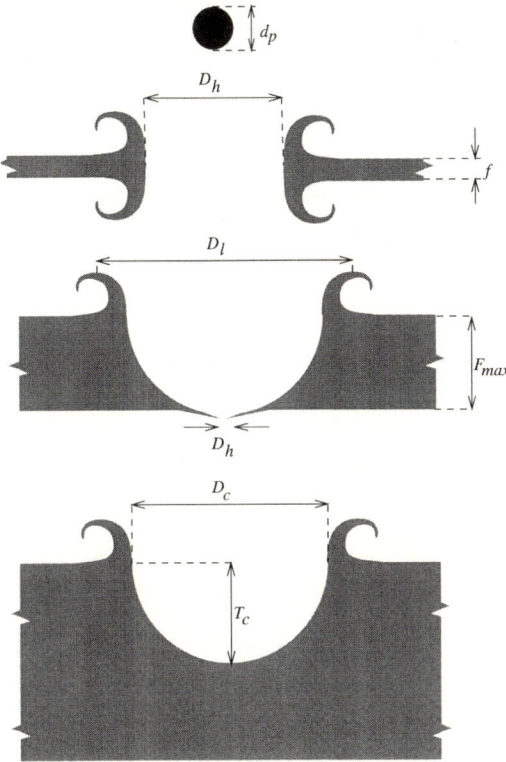

Figure 12. Definition of measurements taken from a hypervelocity impact.

crater diameter may be used to refer to the internal diameter of the crater at the original surface level, the diameter measured at the crest of the lips or even the diameter of the furthest edges of the lips).

Tables 8 to 11 list equations relevant to the decoding of impacts on exposed metallic satellite surfaces. Given here in their originally published form, it should be remembered that the equations use a variety of units. The equations are separated into three types: those that give crater depth; those that give the ballistic limit (maximum foil thickness penetrated); hole growth equations i.e. those that relate the size of a hole caused by a particle passing through a thin foil or plate.

Of the crater depth equations, that of Shanbing et al. (1994) gives reasonable results for a variety of particle materials and sizes. As with any equation obtained at lower velocities ($V \leq \sim 7$ km s^{-1}) it is likely that it will overpredict the size of the crater as impact velocities are raised to a point where a significant amount of material undergoes phase changes. It should be noted that whilst crater diameter equations exist, they tend to be less reliable than crater depth equations, due to the additional factors involved in the formation of the crater lips.

In choosing a ballistic limit equation the researcher is faced with a wide choice. As these are almost all derived empirically it is important that the chosen equation is valid in the size and velocity regime of interest and that it applies to the material under analysis. The ballistic limit equations of Gardner et al. (1997) and McDonnell and Sullivan (1992) have been derived from a wide range of μm scale metallic foils at impact velocities from 1–16 km s^{-1} and are thus suitable for decoding the impacts on space exposed foils. These equations differ principally in their definition of F_{max} (McDonnell and Sullivan use hole diameter (D_h) = foil thickness (f) whereas Gardner et al. use $D_h = 0$). The behaviour of the cratering and ballistic limit equations is compared with other similar equations, and is shown in Figs. 13, 14 and 15.

Only the hole growth equations of Carey et al. (1985) and Gardner et al. (1997) attempt to characterise the behaviour near marginal perforation, with most of the other equations only being valid for $D_h \gg f$. Even then, the work of Carey et al. does not describe perforations where $D_h < f$ and their work has also shown some anomalies compared with experimental results for different materials (Gardner et al. 1997). Sawle's equation (Sawle 1969) gives D_l, the diameter of the front surface lips, measured at their crest, thus neatly side-stepping the issue of how fast the hole at the rear opens, instead showing that the transition from cratering to penetration is a smooth one. The equation of Maiden et al. (1963) is valid only for large holes and does not address the growth phase, and the formulation of Nysmith and Denardo (1969) has problems at large particle sizes—as the size of the particle (d_p) becomes large compared with the foil thickness (f), the predicted hole becomes smaller than the particle.

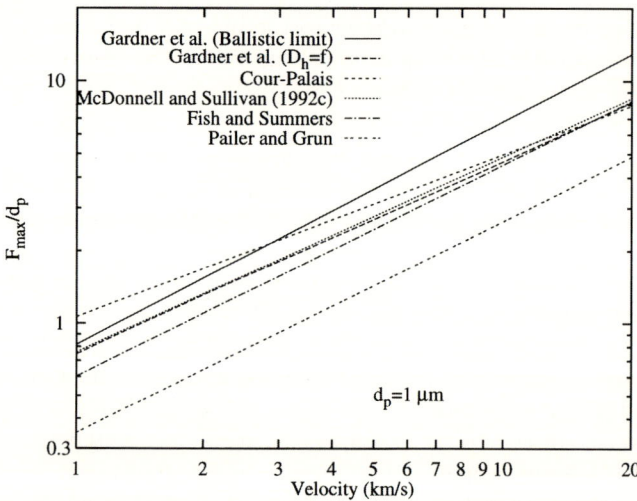

Figure 13. A comparison of ballistic limit equations at constant particle size. See Table 8 for details.

Table 8

A selection of ballistic limit equations. The equations give the thickness of foil that will just be perforated by the given particle.

$$\frac{F_{max}}{d_p} = 1.272 d_p^{0.056} \left(\frac{\rho_p}{\rho_{Fe}} \frac{\rho_{Al}}{\rho_t}\right)^{0.476} \left(\frac{\sigma_{Al}}{\sigma_t}\right)^{0.134} V^{0.806}$$

McDonnell and Sullivan (1992)[1]

$$\frac{F_{max}}{d_p} = 0.57 d_p^{0.056} \varepsilon^{-0.056} \left(\frac{\rho_p}{\rho_t}\right)^{0.5} V^{0.875}$$

Fish and Summers (1964)[1]

$$\frac{F_{max}}{d_p} = 0.772 d_p^{0.2} \varepsilon^{-0.06} \frac{\rho_p^{0.73}}{\rho_t^{0.5}} (V \cos\theta)^{0.88}$$

Pailer and Grün (1980)[1]

$$\frac{F_{max}}{d_p} = 0.653 d_p^{0.056} \rho_p^{0.52} V^{0.875}$$

Cour-Palais (1979)[1]

$$\frac{F_{max}}{d_p} = 0.129 \left(\frac{V \rho_p}{\sqrt{\sigma_t \rho_t}}\right)^{0.763} \left(\frac{\sigma_t}{\sigma_{Al}}\right)^{0.229} d_p^{0.056}$$

Gardner et al. (1997)[2]

[1] Units for this equation are: V in km s^{-1}, ρ in g cm^{-3}, d_p in cm; [2] Units for this equation are: d_p in μm with other parameters in SI.

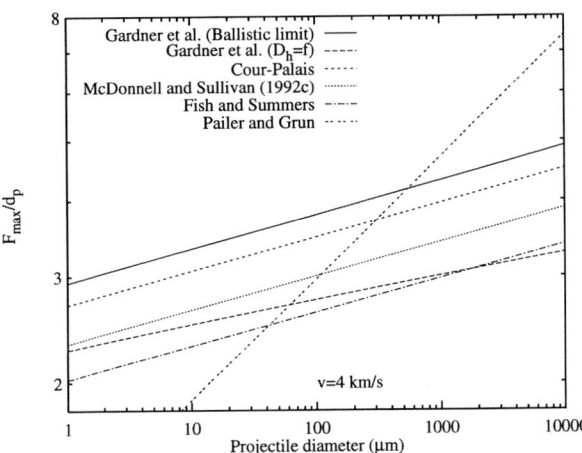

Figure 14. A comparison of ballistic limit equations at constant velocity. See Table 8 for details.

Table 9

A selection of crater depth equations. Sorensen's result assumes a crater of equal depth and radius. The equation of Watts et al. is found from a theoretical treatment, but incorporates a correction factor of 1/2 which they found necessary to match experimental results.

$$\frac{T_c}{d_p} = \frac{1}{4}\left[\frac{4}{3}\frac{\rho_p}{\rho_t}\frac{\rho_t}{Y_t}\left\{\left(c_{0t} + \frac{s(V - u_{t,\text{crit}})}{1 + (\rho_t/\rho_p)^{1/2}}\right)(V - u_{t,\text{crit}})\right\}\right]^{1/3}$$

Watts et al. (1993)[3]

$$\frac{T_c}{d_p} = 0.772 d_p^{0.2} \varepsilon^{-0.06} \frac{\rho_p^{0.73}}{\rho_t^{0.5}}(V\cos\theta)^{0.88}$$

Pailer and Grün (1980)[1]

$$\frac{T_c}{d_p} = 0.311\left(\frac{\rho_p}{\rho_t}\right)^{0.167}\left(\frac{V^2\rho_p}{s_t}\right)^{0.282}$$

Sorensen (1964)[3]

$$\frac{T_c}{d_p} = 0.27\left(\frac{\rho_p V}{\sqrt{\rho_t Y_t}}\right)^{2/3}$$

Yu Shanbing et al. (1994)[3]

[1] Units for this equation are: V in km s^{-1}, ρ in g cm^{-3}, d_p in cm; [3] Units for this equation are self consistent, e.g. SI.

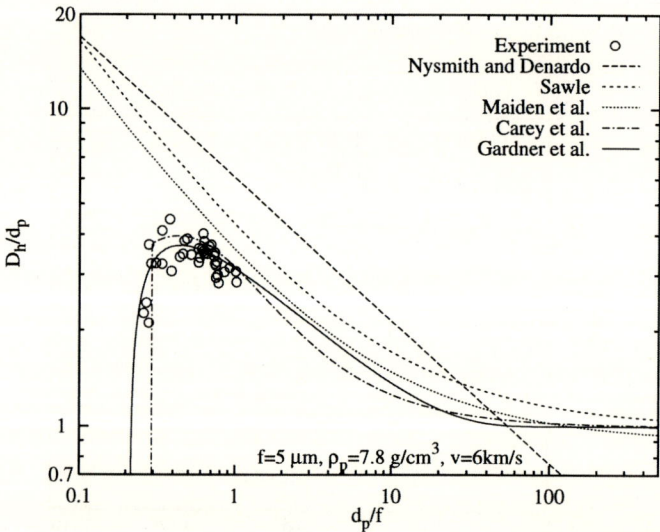

Figure 15. A comparison of hole growth equations. The problems with Nysmith and Denardo's formulation at large particle sizes is clearly seen, as are those with most of the equations at small sizes. See Table 10 for details.

Table 10

A selection of hole growth equations. Only those of Carey et al. and Gardner et al. attempt to treat near-marginal impacts.

$$\frac{D_h}{d_p} = 1 + 2.9 \left(\frac{\rho_t}{\rho_p}\right)^{0.6} \left(\frac{f}{d_p}\right) V^{0.3} \left[\frac{1}{1 + 2.9 \left(\frac{\rho_t}{\rho_p}\right) \left(\frac{f}{d_p}\right)^2 V^{-n}}\right]$$

where

$$n = \begin{cases} 2 < V < 20 \text{km/s} & : 1.02 - 4\exp\left(-0.9 V^{0.9}\right) - 0.003\,(20 - V) \\ V \geq 20 \text{km/s} & : 1.02 \end{cases}$$

Carey et al. (1985)

$$d'_p = A \left(\frac{10}{9 + e^{\frac{D'_h}{B}}}\right) + D'_h \left(1 - e^{-\frac{D'_h}{B}}\right)$$

where B is calculated from Table 11 and

$$A = 6.97 \left(\frac{V \rho_p}{\sqrt{\sigma_t \rho_t}}\right)^{-0.723} \left(\frac{\sigma_t}{\sigma_{Al}}\right)^{-0.217} f^{-0.053}$$

Gardner et al. (1997)

$$\frac{D_l}{d_p} = 2.6 \left[\left(\frac{\rho_p}{\rho_t}\right)\left(\frac{V}{c}\right)\right]^{0.2} \left(\frac{f}{d_p}\right)^{2/3} + 1$$

Sawle (1969)

$$\frac{D_h}{d_p} = 0.45 \left(\frac{f}{d_p}\right)^{2/3} V + 0.9$$

Maiden et al. (1963)

$$\frac{D_h}{d_p} = 0.88 \rho_p^{0.5} \left(\frac{f}{d_p}\right)^{0.45} V^{0.5}$$

Nysmith & Denardo (1969)

Table 11

Parameters used in the Gardner et al. (1997) hole growth equation. Parameter B is found using the above values and the relationship $B = B_1 + B_2 \cdot V$.

Target Material	Velocity km s^{-1}	Density kg m^{-3}	Yield Stress MPa	B_1	B_2 s kg^{-1}
Aluminium	1.8–5.1	2780	69	-0.684	2.12
Aluminium	5.1–8.2	2780	69	6.03	0.789
Silver	2.9–5.6	10500	150	13.7	2.88
Gold	2.1–7.5	19300	120	10.0	2.72
Beryllium-copper	3.7–6.4	8240	828	-27.7	9.96424
Copper	2.0–6.9	8950	220	5.81	1.83
Stainless steel	2.2–3.7	7840	759	-4.22	3.76
Titanium	2.3–6.6	4720	986	-0.544	2.30

2. Determining Particle Properties from Impact Data

Section II.A.3 has dealt with the velocity distributions of interplanetary dust particles in some detail, largely from radar and photographic data (typically particles of size > 0.1 mm). An additional method of obtaining a characteristic velocity for particles is to use satellite impact fluxes combined with the modelled velocity dependence implicit in K factors (see section III.B). This method considers smaller particles (down to a few μm) than are typically measured by radar. For example, comparing the fluxes actually observed on the different faces of a spacecraft such as LDEF with the expected ratios (which depend upon the mean impact velocity) allows selection between competing particle velocity distributions.

Discussion regarding the source of spacecraft impacts has continued for some time, although with the recent satellite data less scope for disagreement now exists. Analysis of the impact fluxes on different faces of a satellite, or even the impact direction determined from impact site morphology, may be used to identify the fraction of impacts from different sources. Further data may be obtained by comparing satellites with similar orbits. For example, as the LDEF and EuReCa satellites were in almost identical orbits it may be expected that any significant divergence between the data will be due to a combination of the anisotropy in the particle influx and the different pointing histories of the satellites (or possibly temporal changes).

Comparison of the impact fluxes observed on different detector surfaces, for example foils and thick targets (Gardner et al. 1996), can also prove a fruitful exercise. As damage to the different surfaces depends to a greater or lesser extent on the various impactor parameters, relationships between these parameters may be established. Such an approach may be used to obtain particle density (McDonnell and Gardner 1998)

If an ultra-low density material, such as aerogel, is used as a detector, the low impact deceleration permits relatively intact capture for future analysis. In addition to bulk parameter studies, particle shapes (lost in impacts onto thick targets) may be determined from impacts through thin foils (such that $d_p \gg f$). In this case the particle is able to "punch out" its cross-section from the foil without suffering significant morphological alteration, with the foil retaining the particle shape for later analysis.

3. Hydrocodes in the Impact Environment

While laboratory techniques, and the empirical relationships based on them, allow a wide range of impact velocities and target and projectile materials and geometries to be investigated, by no means all of the parameter space of interest can be covered. Numerical computer simulations provide a useful method of interpolating and extrapolating the parameter space. Various codes have been developed for this purpose, and are generically known as "hydrocodes" (see reviews by Anderson 1987; Zukas 1990; McGlaun and Yarrington 1993).

While various different implementation schemes are possible for hydrocodes their basic features are the same. The physical system to be modelled is represented by a discrete computational mesh where each node and/or cell has individual physical properties (e.g. volume, mass, velocity, internal energy etc.). Explicit time integration of Newton's laws of motion then allows the material to move and interact while ensuring that the conservation laws of mass, momentum and energy are satisfied. The material thermodynamic behaviour is governed by an equation of state (relating pressure, volume and energy in one or more material phases) and its mechanical behaviour by a material model (relating deformational behaviour to applied stress). A material failure model may also be included. In applying material models "hydrocodes" have outgrown their patronym and are no longer hydrodynamic. The initial and final results from a typical hydrocode calculation are shown in Fig. 16.

Hydrocodes come in a variety of forms, depending on how the material is modelled: in *Lagrangian* codes the grid defines the material, and is distorted with it; in *Eulerian* codes the grid defines the calculation area through which material moves; *SPH* is a Lagrangian model in which there is no grid linking the nodes, which are thus free to move without the imposed restrictions of preventing the grid from excessive distortions. There are a number of codes either commercially or freely available. Some of these codes specialise on one particular method (e.g. CTH, an Eulerian code), whereas others (e.g. Autodyn) present the user with the ability to use several different methods, and even to set up interactions between them (e.g. so a Lagrangian particle may be moved through a fluid modelled by an Eulerian grid). The different codes also vary in terms of ease of use and user interface, and interaction with other modelling or CAD packages.

The usefulness of hydrocode simulations depends critically on the reliability of the equations of state and material models employed. These, like most cratering and perforation equations, are also generally empirical equations and are based on experimental measurements of material response to shock loading. It is thus essential to select an equation of state and material model that is not only valid for the material in question, but also for the impact velocity regime

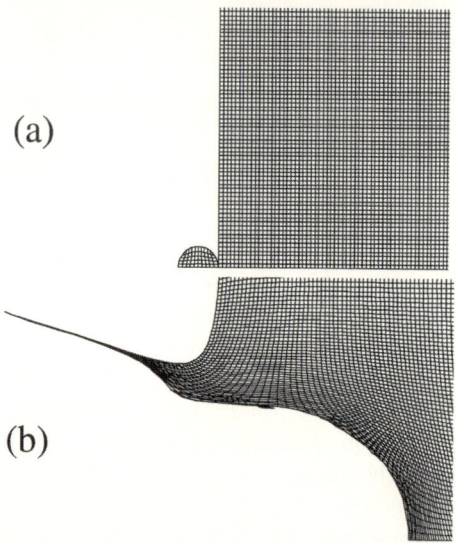

Figure 16. Typical results from a hydrocode-modelled impact event; the initial (a) and final (b) conditions are shown.

under investigation. An inappropriate choice may cause the code to crash, but worse still the simulation may run to completion and produce totally invalid results. Hydrocode validation against experimental results is thus a crucial first step, but once validated they provide an extremely powerful tool. Not only are hydrocode simulations typically cheaper and quicker to set up than experiments, but the calculation method allows far more parameters to be measured with a much higher time resolution than is possible in even the best instrumented experiments.

V.B. Results

1. *In Situ Data Sources*

Since the beginning of the space age, dust detectors of various types have sampled the near Earth environment. However, the interpretation of results has been hampered by both the limited temporal and directional resolution and the very incomplete coverage of near Earth space. The variety of techniques applied and parent spacecraft orbits has produced a heterogeneous data set which has proved difficult to interpret in terms of the contributions from different interplanetary sources and space debris. Here we briefly review the most important missions and their contribution to our current understanding of the near Earth particulate environment.

Early measurements on Explorer 8 (McCracken et al. 1961) were proved to have been influenced by the high susceptibility of piezo-electric microphone detectors in Earth orbit to thermal changes or other factors (Nilsson 1966).

Later reviews (McDonnell 1978) swept much of this early unreliable data away, leaving a core of data from Explorers 16 and 23, and Pegasus 2 and 3 whose high reliability was not "bettered" until the advent of data from recovered surfaces on the Solar Maximum Mission (SMM) and especially LDEF. Penetration and cratering data offers the most comprehensive definition of the flux rate (i.e. from the same technique) over a wide range of masses.

Explorer 16 (altitude $h = 750$–1180 km, mean $= 965$ km, inclination $i = 52°$, launch $l = 1962$ December 16, exposure $E = 0.6$ years) (Hastings 1963a, 1963b, 1963c, 1964) and Explorer 23 ($h = 464$–979 km, mean $= 722$ km, $i = 52°$, $l = 1964$ November 6, $E = 1.0$ y) (O'Neal 1965, 1968) carried experiments comprising arrays of pressurised "beer cans". A pressure-sensing switch capable of measuring a "once only' leak was activated some short time after the first perforation of any can (thickness 25–55 μm) above the ballistic limit. The data were found to be random over the exposure epoch although this did not preclude possible variations of short duration (e.g. within an orbit) or variations in flux beyond the observation period.

On Ariel 2 ($h = 289$–1358 km, mean $= 824$ km, $i = 52°$, $l = 1964$ March 27, $E = 0.2$ y), 12 and 15 μm aluminium foils were used in a series of active in situ sensors (Jennison et al. 1967). Only one penetration was detected after a total of six months of exposure in space (McDonnell 1964). Because this occurred at the edge of the detector corresponding to the region from which fresh foil was advanced during flight, it could not be ascertained if this was a true space impact perforation or an imperfection in the foil. The detection threshold for this (photometric) system is larger than that of the ballistic limit because of the need for a significant penetration area for the detection of light. The data nevertheless clearly demonstrated, by virtue of in-flight calibration, that the flux was some 3 orders of magnitude below the piezo-electric data acquired earlier by Explorer 8.

Penetration sensitive capacitance detectors were flown on three Pegasus satellites ($h \sim 580$ km, $i \sim 31°$, $l = 1965$ February 16, May 25 and July 30, $E = 0.87, 0.6, 0.42$ y) (Naumann 1965; Clifton & Naumann 1966; Dozier 1966). In space, the capacitors were charged to 40 V and penetrations were registered by discharges through the dielectric layers. Two thicknesses of 2024-T3 aluminium detectors at 203 μm and 406 μm presented reliable data. These data furnished the best assessment of the milligram meteoroid flux until the advent of LDEF and form a critical overlap, in terms of sensitivity, with faint radar meteor data (10^{-9} kg).

On the HEOS 2 spacecraft ($h = 350$–$240\,000$ km, $i \sim 90°$, $l = 1972$ January 31, $E = 2.5$ y), the eccentric orbit offered data from low perigee through to extended periods away from the space-traffic congestion region. The S215 experiment had a known pointing direction and both mass and velocity were derived from the impact plasma detected for each event. Several hundred particles were detected and categorised as (1) randomly distributed, (2) "groups" of a few particles each, interpreted as lunar ejecta, and (3) "swarms" produced in the vicinity of the Earth, of unknown origin (Hoffman et al. 1975a, 1975b).

A 1 m² array of 5 μm aluminium foils was exposed as part of NASA's OSS Pathfinder Payload (OSS-1) on STS-3 ($h = 241$ km, $i = 38°$, $l = 1982$ March 22, $E = 8$ days). Four hypervelocity perforations were detected (McDonnell et al. 1984a). Chemical analysis showed silicon-rich residues and the morphology was consistent with high velocity natural impactors. The low altitude of MFE (241 km) is of significance in the context of the poor access of microparticles which are in Earth orbit, since their lifetime in circular orbit is measured in terms of hours at this altitude. Access to unbound interplanetary particulates is, however, unabated at this altitude.

Repairs to the SMM spacecraft ($h = 533$ km, $i = 29°$, $l = 1980$ February 14, $E = 4.16$ y), recovered coincidentally at the same time as the launch of LDEF, led to the retrieval, for laboratory analysis, of multi-layer thermal insulation and aluminium louvres. These data (Laurance and Brownlee 1986) covered a range of crater diameters from sub-μm to mm dimensions. The pointing direction of the louvres was assumed to be random with respect to the Earth orbital vector because of the dedicated solar pointing direction for spacecraft observations. The flux data derived in terms of particle mass by Laurance and Brownlee was in serious error because of the use of two quite different (and mutually inconsistent) penetration formulae on the same plot; this revised, quite drastically, the interpretation of their microparticle fluxes in terms of space debris (McDonnell et al. 1992). The original source data, however, is incontrovertible in terms of crater dimension (although there is a possibility of some secondary cratering at the smallest dimensions). Analysis of the chemical residues by Laurance and Brownlee indicated a predominance of space debris, although at very small dimensions, space debris can be generated efficiently from impacts on surfaces local to the detector (e.g. from the extended SMM solar cell array which was within the acceptance angle for the louvre impacts).

LDEF's large area-time product, the wide range of materials deployed on it and its gravity gradient stabilised orbit ($h = 458$ km, $i = 28°$, $l = 1984$ April 6, $E = 5.77$ y; see section III.B) have elevated its importance to a very high level. The many investigators involved have contributed papers on the interpretation of the data obtained, in both the Proceedings of the three Post-retrieval Symposia held to date and in many other journals. The small particle penetration data was best defined by the MAP experiment (McDonnell et al. 1984b) for foil thickness and penetration in the range 2 to 30 μm, where detectors pointed in N,S,E,W and space directions. It is complemented by surveys of non-penetrating particle impacts on the smooth foils. Thick target measurements (Humes 1991) and data from the thermal control surfaces and the longerons and intercostals of the LDEF frame collated by the LDEF Meteoroid and Debris Special Investigator Group (M&D-SIG, See et al. 1991, 1993, 1995; Zolensky et al. 1996) have provided a database of impacts on spacecraft surfaces at a range of sizes. The IDE experiment (Singer et al. 1984, 1985) provided valuable time resolved fluxes for 10 months from capacitor discharge detectors with an effective F_{max} of ~ 1.5 and 3.5 μm of aluminium on the N,S,E,W, space and Earth faces (Singer et al. 1991; Mulholland et al. 1991). Intense, but short-lived, "spikes" exceeding the background by several orders of magnitude and "multiple orbit event sequences" (MOES) recurring for large numbers of spacecraft orbits were detected (Oliver et al. 1995). Detailed descriptions of all LDEF experiments may be found in the LDEF Mission 1 Experiments report edited by Clarke et al. (1984).

The Munich Dust Counter (MDC) experiment (Igenbergs et al. 1991) on the Hiten satellite of the Japanese MUSES-A mission used impact plasma detection to measure masses and velocities of particles from a few thousand km to beyond the Moon between 1991 March 3 and lunar impact on 1993 April 10. The data confirmed the three categories (random, groups and swarms) detected by HEOS 2.

The EuReCa spacecraft ($h = 508$ km, $i = 28°$, $l = 1992$ August 1, $E = 0.90$ y), provided impact data on large areas of thermal blanket, solar cell arrays and the science experiment TICCE (Timeband Capture Cell Experiment, Stevenson 1988). The impact time resolution aspect of the experiment did not function correctly due to an overload in the first exposure epoch, but it did provide flux data from 2.5–9.2 μm foil surfaces (with capture cells). EuReCa was oriented with fixed Sun and Earth-apex pointing faces. It demonstrated, like HEOS 2 and radar meteoroid data, a flux bias towards the Earth apex of motion.

The recovery of one solar panel after 3.6 years in space from the Hubble Space Telescope ($h = 614$ km, $i = 28.5°$, $l = 1990$ April 24), on its servicing mission in December 1993 has provided an additional large area of exposed material.

The MIR space station ($h = 350$ km, $i = 58°$) has also provided a platform for experiments such as Echantillons (deployed 1988 December, $E = 1.1$ y) (Mandeville 1990) with capture cells and capacitor discharge detectors, and aerogel cassettes.

Time of flight measurements through penetration of thin dielectric films have been made in GEO from the GORIZONT-41 ($h = 35\,800$ km, $E = 1993$ December–1994 October) and GORIZONT-43 ($h = 35\,800$ km, $E = 1996$ May) communication satellites (Novikov et al. 1997). Velocities were measurable for 76 impacts from particles of sizes 3–100 μm of which 80 % were inferred as natural meteoroids ($v > 12$ km s^{-1}).

The Geostationary Orbit Impact Detector (GORID), a flight-quality engineering model of the Ulysses impact ionisation detector (Grün et al. 1992) was launched with the Russian Express-2 communications satellite into GEO ($h = 35\,800$ km, $l = 1996$ September 26, $E = 5$–7 y). In the first four months of operation impacts of mass $> 10^{-17}$ kg were recorded at typically 1–5 events per day but with pronounced peaks of over 100 day^{-1}. Derived velocities indicate 20–30 % may be space debris (Drolshagen et al. 1997).

2. Flux Data

In order to compare the cratering/penetration data at a range of altitudes to search for different sources or temporal changes it is necessary either to use

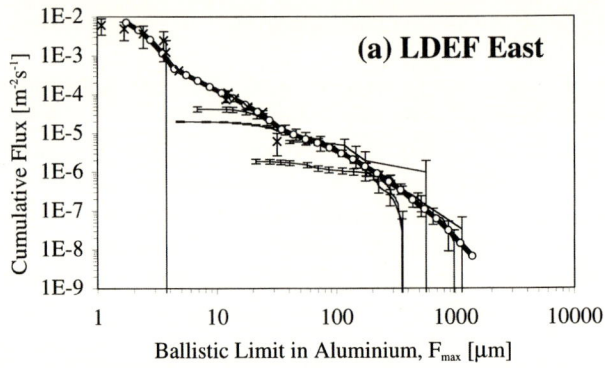

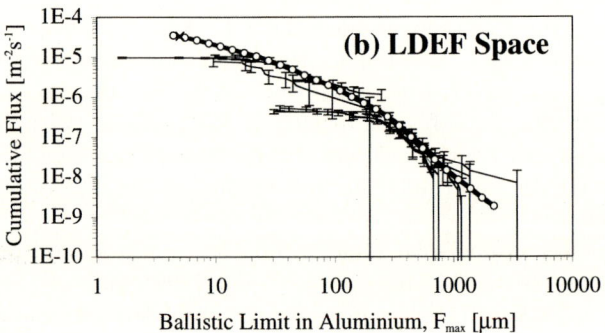

Figure 17. Impact fluxes recorded on LDEF's East (a) and space (b) faces as a function of foil penetration threshold F_{max}. Data are from a range of sources: MAP experiment foil perforations and craters (Univ. of Kent); Experiment tray clamp craters (Univ. of Kent); Frame and SDIE experiment craters (LDEF Meteoroid and Debris Special Investigator group); Thermal control panels (Love and Brownlee 1993). Error bars are based on statistics of crater counts except for the interpolated fit which also includes scatter of the different source data. Individual crater diameters D_c are converted to F_{max} using the approximate (velocity independent) relation $F_{max} = 0.68 \cdot D_c$.

appropriate cratering or penetration equations combined with environmental models or to select representative data sets obtained with the same technique or target material.

As an example of the former, we compare predictions using the Grün interplanetary flux, the HRMP velocity distribution and the McDonnell and Sullivan (1992) penetration equation, with LDEF data (as described in section III.B). Figure 17 illustrates how the source data for the East and space faces of LDEF are fitted with a mean cumulative flux as a function of F_{max}. Fig. 18 shows the comparison between the model fluxes and best fits to data from the East, West, North, South and space faces. For $F_{max} > 30$ μm the model fluxes are compatible with the predicted fluxes from the Grün interplanetary model for i) the overall flux levels and ii) the flux distribution on each individual face. In

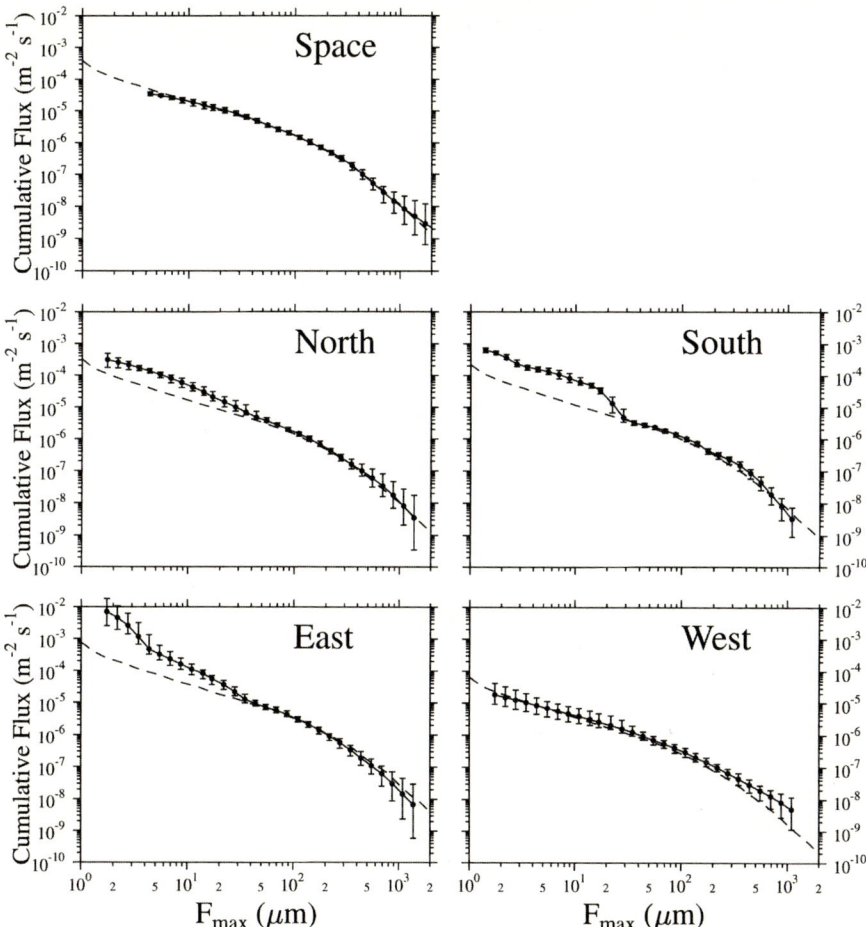

Figure 18. Fluxes on the different faces of the LDEF satellite, compared with the predictions from a meteoroid model incorporating the Grün flux and a full meteoroid velocity distribution. The meteoroid model is expected to best fit the space face as this face has no Earth shielding and very little debris. It is seen that the model fit is good. The excess flux on LDEF's North, South and East faces is attributed to an orbital debris component. However at values of $F_{max} > 30$ μm LDEF's impacts are dominated by meteoroids.

some previous analyses, an average velocity has been assigned to impacting interplanetary particles from ratios of LDEF space:West fluxes. Such a mean velocity is not directly applicable to the population of impacting particles since it is, in its impact effect, weighted by the relationship between the impact process used and the mass and velocity of individual impactors (section III.B). Table 7 lists mean impact velocities for selected faces of LDEF from the applied model.

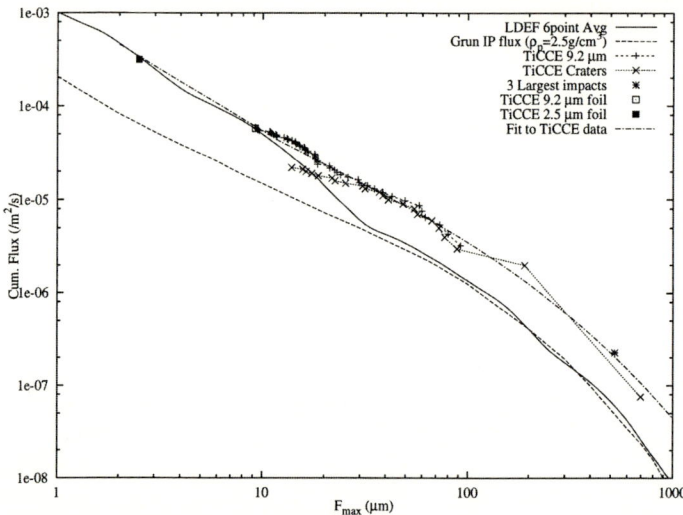

Figure 19. Impact fluxes recorded on the LDEF and EuReCa satellites. The difference between the two data sets is expected to be due to differences in the satellite pointing histories. The similarity between the fluxes below $F_{max} = 10$ μm strongly suggests that these particles are *not* from the Earth apex/Sun pointing direction and are orbital instead (i.e. space debris or aerocaptured particles).

For fluxes at $F_{max} < 30$ μm, an additional component is apparent on the East, North and South faces. This would be expected from orbital particles in near-circular orbits and has been interpreted as a space debris component based on chemical residue analysis (see below), although a small proportion could be aerocaptured natural meteoroids (see section II.E.2). At the very smallest of dimensions (0.1 μm particle size) the question of locally generated secondaries has been raised from the SMM data because the louvres subtended a finite angle with the nearby solar cell array. On MIR, certainly the local secondary particle flux has been found to be significant for crater dimensions of less than ~ 10 μm. Even allowing for these secondary craters, which may be identified by their morphology and size distribution, the MIR fluxes show an excess above the LDEF flux levels attributed to short-lived orbital debris in the manned space station environment (Mandeville and Berthoud 1995).

A comparison (shown in Fig. 19) between the average of LDEF's 6 faces (Neish 1995) and the results from the EuReCa Timeband Capture Cell Experiment (TICCE) (Gardner et al. 1996; Gardner 1995) which points towards the Earth apex/solar direction, shows a factor of ~ 2.8 difference between the flux on the two satellites for large sized particles. This suggests that the impact flux in the size range corresponding to values of F_{max} from 30 to 700 μm is dominated by a component from the Earth's apex of motion and/or the Sun (and hence clearly of natural origin). Below values of $F_{max} = 10$ μm no such difference is observed. This is consistent with the dominance at small sizes of space debris (or at least Earth-orbiting particles).

Penetration experiments generally define only the average flux, generally over a mission lifetime and with coarse pointing information from which the relative populations of orbital and interplanetary particles may be determined over a wide size range. Experiments with time resolution and in some cases velocity determination, allow a more detailed study of the distribution of particles, but often within restricted mass ranges.

Away from the Earth's gravity, we see in the interplanetary flux data from e.g. Pioneers 8 and 9 and HEOS 2 (Berg and Gerloff 1971; Hoffman et al. 1975a; Grün and Zook 1980; see also the review by McDonnell 1978) that the Earth apex flux dominates for masses $> 10^{-14}$ kg (sizes of a few μm). The distribution swings sunwards at smaller masses ($M < 10^{-16}$ kg) to a Sun-biased β-meteoroid distribution (Berg and Grün 1973; Zook and Berg 1975; Grün et al. 1980; Cooke et al. 1993). The flux distribution on HEOS 2 demonstrates a clear Earth apex bias from μm-sized interplanetary micrometeoroids approaching the Earth. Although the detection rate of particles from the apex, anti-apex, ecliptic North and South directions varied by only a factor of 2, when converted to constant mass, the flux at 10^{-15} kg was 1 to 2 orders of magnitude higher in the apex direction with mean relative velocities < 10 km s^{-1} for the apex and 20 km s^{-1} for ecliptic North and South (no velocities were measured for anti-apex particles) (Hoffman et al. 1975a, 1975b). Data from MDC on Hiten confirmed these results with particles of mass $> 10^{-17}$ kg predominantly from the apex direction and those of mass $< 10^{-17}$ kg predominantly from the Solar direction (β-meteoroids) (Iglseder et al. 1993; 1996). The apex population with low relative velocities can be explained by particles spiralling in towards the Sun being "overtaken" by the satellite, while the β-meteoroids are accelerated from the solar direction by radiation pressure. The source of the high velocity anti-apex particles detected by Hiten is not understood (high velocities will be expected from the apex direction due to particles originating from comets in high inclination or retrograde orbits).

Very large fluctuations in microparticle fluxes have been detected on timescales less than the spacecraft orbit period in HEOS 2, Hiten and LDEF IDE data. For HEOS 2 and Hiten, the "groups" with event rates from 2 per day to 4 per hour, are attributed mostly to a lunar origin with two groups possibly from the Perseid meteoroid stream. The "swarms", detected near perigee with event rates > 4 per hour could have been due to aero-fragmentation (Hoffman et al. 1975b). Although the IDE data were comparable in general terms to the penetration records of other experiments on LDEF, they show, on a shorter timescale, that the average is certainly not statistically smooth. Figure 20 illustrates the impacts plotted as a function of time and LDEF orbital longitude. "Spikes" are bursts of detections which occur only on a single orbit, whereas "MOES" (Multiple Orbit Event Sequences) consist of burst of detections which repeat for many orbits. These particles appear to be in eccentric orbits and are distributed in a ring around the orbit. Cooke et al. (1995) derive the elements for one such swarm which has low eccentricity and an inclination of 67°. Although space debris is the "prime suspect", aerocapture and fragmentation or

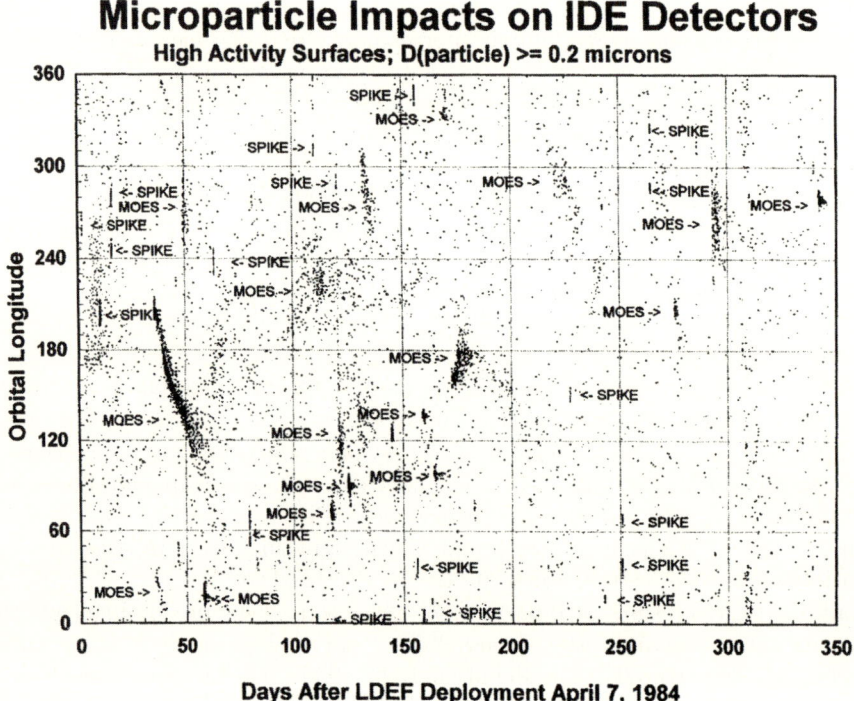

Figure 20. Time and direction resolved impacts on the LDEF IDE experiment.

the fragmented fireball hypothesis of Fechtig would result in almost identical *time sequences*. Since atmospheric drag and radiation pressure forces would be expected to dissipate such rings of μm-sized particles in shorter timescales than they appear to persist, they must be replenished by some source. For debris, we must look for direct sources such as rocket exhaust ($Al_2 O_3$ from especially GTO injection), paint flakes (having titanium pigments as a key tracer) or solar cell array impact fragmentation. It is from the chemistry that we have to pin down the sources.

3. Composition and Structure

The physical properties of interplanetary dust particles are difficult to measure, since any capture process (e.g. impact or aerocapture) will cause some degree of processing or selection effects. In situ detection gives a sample biased largely by orbital effects which may be easily determined, although the detection process itself is usually destructive. Clues to the physical properties of particles impacting a spacecraft surface may be obtained by comparing different detection surfaces, as performed by Gardner (1995). In this study a comparison of data from thick and thin targets (Gardner et al. 1996) on LDEF's space face was compared using a hole growth equation for the foil perforation which

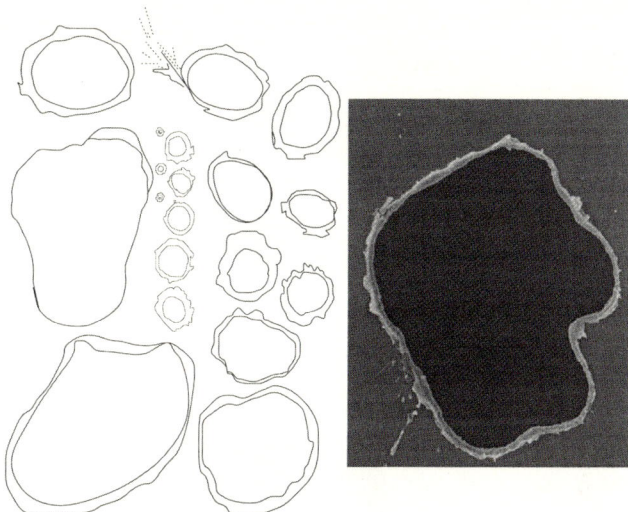

Figure 21. Perforation outlines, revealing particle cross-sections, as recorded on the EuReCa TICCE experiment.

is sensitive to impactor density (Gardner et al. 1997). The impactor density required to characterise the results on LDEF's space face and EuReCa TICCE was found to be ~ 1000 kg m^{-3}. However McDonnell and Gardner (1998) have extended this work and have shown that while this "characteristic" density is only marginally affected by impact velocity, any spread in the density of impacting particle population leads to a reduction of the characteristic density compared with the mean. They further show that for a bi-modal density distribution the low density particles have a dominant effect, even for small concentrations, reducing the characteristic density compared to the mean. They thus conclude that while for particles in the range $5 < F_{max} < 100$ μm the *characteristic* density is ~ 1000 kg m^{-3}, the *weighted mean* density for these particles is probably in the range 2000–2500 kg m^{-3}. This distinction between mean and characteristic density arises due to the general principle that for any non-linear function $\overline{f(\rho)} \neq f(\bar{\rho})$.

In the impact of a large particle on a very thin foil, such that $d_p \gg f$, the particle "punches" through the foil and is not significantly affected. The hole thus created is only marginally larger than the particle and thus serves as a measure of the cross-section of the particle. A survey of these cross-sections, as shown in Fig. 21 for perforations in a section of 2.5 μm foil from the EuReCa TICCE experiment, will thus reveal the degree of irregularity of typical impacting particles. As can be seen the typical particle is neither perfectly regular nor highly elongated, having a ratio of at most two between its long and short diameters.

From the chemical analyses of retrieved experiments we might hope to see clear evidence to pin down the relative fractions of meteoroids or space debris.

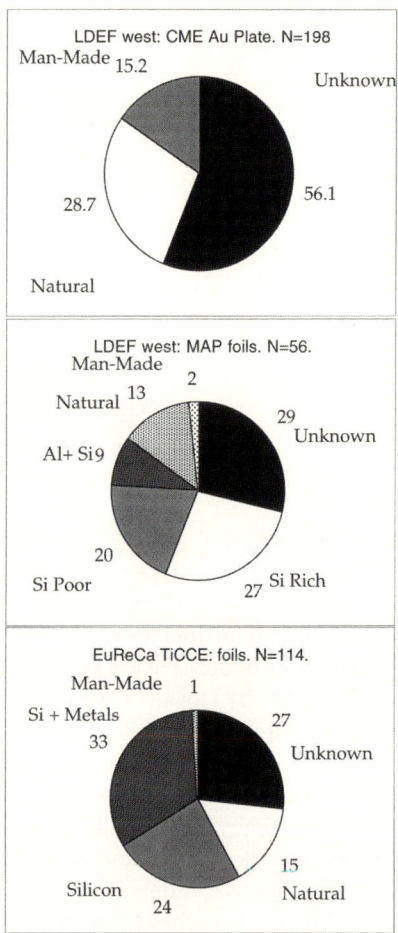

Figure 22. Results of X-ray analysis of impact residues on the LDEF and EuReCa satellites.

Residue analyses of the Chemistry of Meteoroids Experiment (CME) (Hörz et al. 1991) were performed on both the trailing face (gold surfaces) and on the leading edge (aluminium surfaces). The data yielded a highly valuable information base for attacking the problem (Bernhard et al. 1993) and results are summarised in Fig. 22 together with chemical data from the LDEF Micro Abrasion Package (MAP) experiment (Yano 1995) and the EuReCa TICCE capture cells (Gardner et al. 1996; Yano 1995; Collier 1995). The same SEM X-ray spectroscopy is used for these three sets of results.

One interesting feature of the LDEF data is the access of impactors to the trailing face which leave residues clearly corresponding to space debris materials. This calls for eccentric debris orbits (Flury et al. 1992). Despite the clear flux of debris, albeit small compared to the meteoroids, most of the craters *do not* have identified residues and (due to the preferential loss of residues by the higher velocity meteoroids) more of these "unknowns" will be meteoroids than debris. From both chemistry and flux modelling, meteoroids certainly dominate the overall scene at least for crater dimensions in the region of 30 μm to 1 mm. The LDEF CME West face debris flux corresponds to perhaps $\sim$ 15 % of the total West flux. Different data have been derived by Yano (1995) on the same LDEF West face. He was able to resolve more of the total impact chemistry from foil perforations and their captured residues and found a lower fraction of clearly identified space debris; more silicates/metals were found which could well be glassy fragments in orbit. On EuReCa TICCE (Collier 1995; Yano 1995) the debris fraction was found to be lower, but this is partly accounted for by the higher flux of meteoroids from its solar-apex attitude.

IDP's collected in the stratosphere are clearly less susceptible to impact destruction and their analyses have dramatically improved the ability to study interplanetary material in the terrestrial laboratory. We can determine readily their physical shape, study the mineralogy of sub-μm grains, their chemistry and isotopic composition. Not only can their nature be compared with cometary or asteroidal sources but evidence of their sojourn in interplanetary space be seen by, for example, the solar wind Helium excess (Rajan 1977) and by solar cosmic ray tracks (Rietmijer and Mackinnon 1987). We can also see evidence of the atmospheric entry heating (section II.B.1).

These data are reviewed by Brownlee (this volume). In this chapter we look at the evidence only in the context of clues to the sources of meteoroids detected in and above the Earth's atmosphere.

Summary statistics (Schramm et al. 1989) list "chondritic smooth" comprising 37 % of the collection, these are the non-porous micro-representatives of chondritic meteorites. Evidence in the population of lattice layer silicates (Bradley 1988) points to hydrous alteration, thus being less primitive, perhaps, than the cometary fraction. The "chondritic porous" components (45 %) are more in line with an understanding of cometary morphology; pyroxenes dominate their chemistry but the interstices are perhaps the key feature to forming the low density of impactors demanded by meteoroid studies (Verniani 1973; Babadzhanov 1994) and by LDEF's thin foil penetration measurements (Gardner 1995). "Coarse grained" particles (18 %) are also chondritic but have grains of some 3 μm or more; they are often translucent to visible light in contrast to the smaller opaque grains of the CP group. Thus the "asteroidal to cometary" ratio seen in this IDP analysis would be almost equal if, indeed, we could so clearly separate the two components; but it has to be interpreted in the light of a very strong selection bias towards the asteroidal sources in the IDP population due to the effect of gravitational enhancement. This comparability in the IDP collections could decrease away from the Earth's influence to an asteroidal fraction of only some 5 % to 10 % at 1 AU.

4. The Changing Particulate Environment

The exposure of penetration experiments over an extended time-base offers a base for examining the microparticle excess in Earth orbit, and the pene-

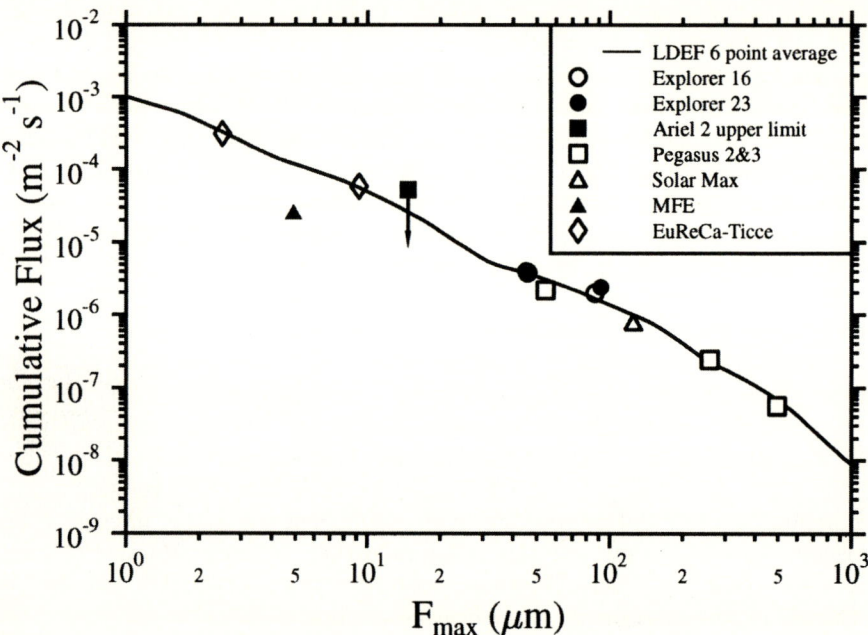

Figure 23. Impact fluxes from LEO impact detectors over the last 30 years. See text for details.

tration on a specific thickness of metal provides a readily calibrated standard (McDonnell and Baron 1995). Key exposures comprising Explorers 16 and 23, the Pegasus series, SMM, MFE (OSS-1), LDEF and EuReCa are presented in Fig. 23. Extending from the early 1960's, the flux distribution might have been expected to track the growth of space debris. But as we can see, all arguments and analyses lead to the dominance of meteoroids even now for penetration values of $F_{max} > 30$ μm Aluminium. Looking over the data for growth of the current microparticle excess, the earliest penetration results, unfortunately, were not sensitive to small impactors, except for Ariel II where μm thickness foils were exposed, but detected a maximum of only one particle in its 6 months' exposure! It provided evidence against the "Earth's Dust Belt" at the time, a hypothesis largely resulting from data from unreliable piezo-electric detectors (Nilsson et al. 1966) sensitive to larger particles (5 μm diameter). Although we now see the renewal of this concept of a debris dust belt in the new space era, it is evident only at μm particle sizes. Looking to the future, the collisional cascade from large bodies (especially in the 700–900 km altitude region) could, of course, mask the meteoroids, but the good agreement of the penetration data over 30 years of measurement reinforces the dominance of the natural population in the range of penetration thickness from upwards of some 30 μm to mm dimensions.

We do not see in the more recent impact data a significant growth in even the very small debris population i.e. looking at SMM, LDEF and EuReCa; the micro-debris orbital component is essentially constant from around 1980 onwards. It is consistent with a debris population of very short lived particles which is in equilibrium; its source should be related not to the total satellite population, but more to the injection frequency of satellites into orbit. For example GTO launches using solid rocket motors are a prime contender for the injection of aluminium oxide spherules into such orbits (although Molniya orbits may be equally viable as sources).

VI. SUMMARY

The Grün et al. (1985) interplanetary flux model appears to describe the time averaged near Earth environment well if considering randomly orientated detectors for foil penetration limits $F_{max} > 30$ μm. At smaller sizes, orbital space debris dominates in LEO (i.e. ~ 500 km), although specific detector orientation directly affects the natural to debris ratio. However, there are significant size dependent anisotropies in the natural population (β-meteoroids and apex component) which should not be ignored when modelling spacecraft flux data.

Although the mean encounter velocity of natural particles is about 18 km s^{-1}, a full velocity distribution should always be considered because the relative importance of fast particles can be greatly enhanced due to the velocity dependence of detection techniques. An additional consideration is the relatively low encounter velocities of aerocaptured natural particles and space debris.

Various methods and software exists for modelling the exposure of spacecraft to the particulate environment and while they inevitably have limitations, they are valuable tools in our endeavour to understand the near Earth environment.

Acknowledgements

We would like to thank A.D. Taylor for supplying data for inclusion in this chapter, and P. Jenniskens for helpful comments concerning the meteor shower data. We also thank J. Oliver for permission to include Fig. 20. This work would not have been possible without the financial support of UK PPARC and the flight opportunities provided by ESA and NASA.

REFERENCES

Alvarez, L. W., Alvarez, W., Asaro, F. and Michel, H. V. 1980. Extraterrestrial cause for the Cretaceous-Tertiary extinction. *Science*, **208**, pp. 1095–1108.
Anderson, B. J., and Smith, R. E. 1994. Natural orbital environment guidelines for use in aerospace vehicle development. NASA TM-4527.
Anderson, C. E. Jnr. 1987. An overview of the theory of hydrocodes. *Int. J. Impact Engng.*, **5**, pp. 33–59.
Baldwin, B., and Sheaffer, Y., 1971. Ablation and breakup of large meteoroids during atmospheric entry. *J. Geophys. Res.*, **76**, pp. 4653–4668.
Babadzhanov, P. B., 1994. Density of meteoroids and their mass influx on the Earth. In *Asteroids Comets and Meteors* 1993 IAU Symposium 160, eds. A. Milani, M. Di Martino and A. Cellino (Dordrecht: Kluwer), pp. 45–54.
Baggaley, W. J., Bennet, R. T. G., Steel, D. J. and Taylor, A. D. 1994. The advanced meteor orbit radar facility: AMOR. *Q. J. R. Astr. Soc.*, **35**, pp. 283–320.
Bandermann, L. W., and Singer, S. F. 1969. Interplanetary dust measurements near Earth. *Rev. Geophys.*, **7**, pp. 759–797.
Beech, M., and Brown, P. 1994. Space platform impact probabilities—the threat from the Leonids. *ESA J.*, **18**, pp. 63–73.
Beech, M., Brown, P., and Jones, J. 1995. The potential danger to space platforms from meteor storm activity. *Q. J. R. Astr. Soc.*, **36**, pp. 127–152.
Berg, O. E., and Gerloff, U. 1971. More than two years of micrometeorite data from two Pioneer satellites. *Space Res.*, **11**, pp. 225–235.
Berg, O. E., and Grün, E. 1973. Evidence of hyperbolic cosmic dust particles. *Space Res.*, **13**, pp. 1047–1055.
Bernhard, R. P., Hörz, F., Zolensky, M. E., See, T. H., and Barrett, R. A. 1993. Composition and frequency of impact residues detected on LDEF surfaces. In *First European Conference on Space Debris*, ed. W. Flury, ESA SD-01, pp. 189–194.
Binzel, R. P. and Xu, S. 1993. Chips off Vesta and a near resonance source for achondritic meteorites. *Science*, **260**, pp. 186–191.
Bradley, J. P. 1988. Analysis of chondritic interplanetary dust thin sections. *Geochim. Cosmochim. Acta*, **52**, pp. 889–900.
Bronshten, V. A. 1983. Physics of meteoric phenomena. (Dordrecht: Reidel).
Carey, W. C., McDonnell, J. A. M., and Dixon, D. G. 1985. Capture cells: decoding the impacting particle parameters. In *Proc. XVIth Lunar and Planetary Science Conference (Abstracts volume)*, (LPSI Houston), pp. 111–112.
Caswell, R. D., McBride N., and Taylor, A. D. 1995. Olympus end of life anomaly—a Perseid meteoroid impact event? *Int. J. Impact Engng*, **17**, pp. 139–150.
Chamberlain, S. A., and Slauenwhite, T. A. 1993. United States Space Command Space Surveillance Network overview. In *Proc. First European Conference on Space Debris*, ed. W. Flury, ESA SD-01, pp. 37–42.
Ceplecha Z., J. Borovicka, W. G. Elford, D. O. ReVelle, R. L. Hawkes, V. Porubcan, and M. Simek 1998. Meteor Phenomena and Bodies. *Space Science Reviews*, **84**, (3/4) pp. 327–471.
Clarke, L. G., Kinard, D. J., Carter, D. J., and Jones, J. L. (eds.) 1984. LDEF Mission 1 Experiments, NASA SP-473.
Clifton, S., and Naumann, R. 1966. Pegasus satellite measurements of meteoroid penetration (February 16–December 31, 1965). NASA TM X-1316.
Collier, I. 1995. Hypervelocity impact and perforation: a first examination of EuReCa TICCE. M. Sc. Thesis (University of Kent at Canterbury).
Cook, A. F. 1972. A working list of meteor streams. In *Meteor research program*, NASA CR-2109, pp. 153–166.
Cooke, W. J., Mulholland, J. D., and Oliver, J. P. 1993. IDE constraints on the beta meteoroid population. *Adv. Space Res.*, **13**, pp. 119–122.
Cooke, W. J., Oliver, J. P., and Simon, C. G. 1995. The orbital characteristics of debris particle rings as derived from IDE observations of multiple orbit intersections with LDEF. In *LDEF—69 months in space: third post-retrieval symposium*, ed. A. S. Levine,

NASA CP 3275 Part 1, pp. 361–371.

Cour-Palais, B. G. 1969. The meteoroid environment model – 1969 (Near Earth to lunar surface). NASA SP-8013.

Cour-Palais, B. G. 1979. Space vehicle meteoroid shielding design. In *Proc. comet Halley micrometeoroid hazard workshop*, ESA SP-153, pp. 85–92.

Dennison, J. E., Kaczaral, P. W., and Lipshutz, M. E. 1987. Volatile chalcophile siderophile and lithophile trace elements in lunar meteorite Yamato-82192. *Proc. Eleventh Symp. Antarctic meteorites*, pp. 89–95.

Dietzel, H., Eichorn, G., Fechtig, H., Grün, E., Hoffman, H.-J. and Kissel, J. 1973. The HEOS 2 and Helios micrometeoroid experiments. *J. Phys. E. Sci. Instrum.*, **6**, pp. 209–217.

Divine, N. 1993a. Five populations of Interplanetary Meteoroids. *J. Geophys. Res.*, **98**, pp. 17029–17048.

Divine, N. 1993b. Modelling the meteoroid distributions in interplanetary space and near Earth. In *Proc. First European Conference on Space Debris*, ed. W. Flury, ESA SD-01, pp. 245–250.

Dohnanyi, J. S. 1966. Model distribution of photographic meteors. *Bellcomm. Rep.*, TR-66-340-1 (Washington DC).

Dozier, J. B. 1966. Meteoroid data recorded on Pegasus Flights. In *The Micrometeoroid Satellite Project Pegasus*, NASA TN D-3505, Chapter V, pp. 65–76.

Drolshagen, G., Svedhem, H., Grün, E., Grafodatsky, O., Verhoturov, V., Prokopiev, U., and Gusyelnikov, V. 1997. In-situ measurement of cosmic dust and space debris in the geostationary orbit. In *Proc. Second European Conference on Space Debris*, eds. B. Kaldeich-Schürmann and B. Harris, ESA SP-393, pp. 129–134.

Erickson, J. E. 1968. Velocity distribution of sporadic photographic meteors. *J. Geophys. Res.*, **73**, pp. 3721–3762.

ESA 1988. (ESA Space Debris Working Group) Space Debris. ESA SP-1109.

ESA 1993. *Proc. First European Conference on Space Debris*, ed. W. Flury, ESA SD-01.

ESA 1997. *Proc. Second European Conference on Space Debris*, eds. B. Kaldeich-Schürmann and B. Harris, ESA SP-393.

ESTEC: Mathematics & Software Division 1995. *ESABASE Reference Manual*, ESABASE/-GEN-UM-061 Issue 2 (April 1995).

Farinella, P., Froeschlé, C., and Gonczi, R. 1994. Meteorite delivery and transport. In *Asteroids Comets and Meteors 1993* IAU Symposium 160, eds. A. Milani, M. Di Martino and A. Cellino (Dordrecht: Kluwer), pp. 205–222.

Fish, R. H., and Summers, J. L. 1965. The effect of material properties on threshold perforation. In *Proceedings of the 7th Hypervelocity Impact Symposium Volume VI—Experiments* (Florida: Orlando), pp. 1–26.

Flury, W., Janin, G., Jehn, R., and Klinkrad, H. 1992. Space debris in elliptical orbits. In *18th international symposium on space technology and science*, (Japan: Kagoshima).

Flynn, G. J. 1989. Atmospheric entry heating: a criterion to distinguish between asteroidal and cometary sources of interplanetary dust. *Icarus*, **77**, pp. 287–310.

Gardner, D. J. 1995. Hypervelocity impact morphology. Ph.D. Thesis, (University of Kent at Canterbury).

Gardner, D. J., Collier, I., Shrine, N. R. G., Griffiths, A. D., and McDonnell, J. A. M. 1996. Micro-particle impact flux on the timeband capture cell experiment of the EuReCa spacecraft. *Adv. Space Res.*, **17**, pp. 193–199.

Gardner, D. J., McDonnell, J. A. M., and Collier, I. 1997. Hole growth characterisation for hypervelocity impacts in thin targets. *Int. J. Impact Engng.*, **19**, pp. 589–602

Genge, M. J., Grady, M. M., and Hutchison, R. 1996. Evidence in a glassy cosmic spherule from Antarctica for grazing incidence encounters with the Earth's atmosphere. *Meteoritics and Planetary Science*, **31**, pp. 627–632.

Girard, O., and Worms, J. C. 1997. Microsatellite for orbital debris detection by lidar. In *Proc. Second European Conference on Space Debris*, eds. B. Kaldeich-Schürmann and B. Harris, ESA SP-393, pp. 151–153.

Goldstein, R. M., and Goldstein, S. J. 1995. Flux of millimetric space debris. *Astron. J.*, **110**, pp. 1392–1396.

Göller, J. R., and Grün, E. 1989. Calibration of the Galileo/Ulysses dust detectors with different projectile materials and at varying impact angles. *Planet. Space Sci.*, **37**, pp. 1197–1206.

Green, S. F., and McDonnell, J. A. M. 1992. A numerical model for the characterisation of the orbital debris environment. In *Hypervelocity Impacts in Space*, ed. J. A. M. McDonnell (Canterbury: University of Kent), pp. 251–256.

Green, S. F., Despande, S. P., and Mackay, N. G. 1993. A 3-D numerical model for space debris and interplanetary dust fluxes incident on LDEF. *Adv. Space Res.*, **13**, pp. 107–110.

Grün, E., Pailer, N., Fechtig, H., and Kissel, J. 1980. Orbital and physical characteristics of micrometeoroids in the inner solar system as observed by Helios 1. *Planet. Space Sci.*, **28**, pp. 333–349.

Grün, E., and Zook, H.A. 1980. Dynamics of micrometeoroids in the inner solar system. In *Solid particles in the solar system*, eds I. Halliday and B. A. McIntosh (Dordrecht: Reidel), pp. 293–298.

Grün, E., Zook, H. A., Fechtig, H., and Giese, R. H. 1985. Collisional balance of the meteoritic complex. *Icarus*, **62**, pp. 244–272.

Grün, E., Fechtig, H., Giese, R. H., Kissel, J., Linkert, D., Maas, D., McDonnell, J. A. M., Morfill, G. E., Schwehm, G., and Zook, H. A. 1992. The Ulysses dust experiment. *Astron. Astrophys. Suppl. Ser.*, **92**, pp. 411–423.

Grün, E., Staubach, P., Baguhl, M., Hamilton, D. P., Zook, H. A., Dermott, S., Gustafson, B. A., Fechtig, H., Kissel, J., Linkert, D., Linkert, G., Srama, R., Hanner, M. S., Polanskey, C., Horányi, M., Lindblad, B. A., Mann, I., McDonnell, J. A. M., Morfill, G. E., and Schwehm, G. 1997. South–North and radial traverses through the interplanetary dust cloud. *Icarus*, **129**, pp. 270–288.

Hastings Jnr., E. C. 1963a. The Explorer XVI micrometeoroid satellite—Description & Preliminary Results for the period Dec 16, through Jan 13, 1963. NASA TM X-810.

Hastings Jnr., E. C. 1963b. The Explorer XVI micrometeoroid satellite; supplement I, preliminary results for the period 14 Jan 1963 – 2 Mar 1963. NASA TM X-824.

Hastings Jnr., E. C. 1963c. The Explorer XVI micrometeoroid satellite; supplement II, preliminary results for the period 3 Mar 1963 – 26 May 1963. NASA TM X-899.

Hastings Jnr., E. C. 1964. The Explorer XVI micrometeoroid satellite; supplement III, preliminary results for the period May 27 through July 22, 1963. NASA TM X-949.

Hawkins, G. S., and Southworth, R. B. 1958. Statistics of meteors in the Earth's atmosphere. *Smithsonian Contrib. Astrophys.*, **2**, pp. 349–364.

Hoffman, H., Fechtig, H., Grün, E., and Kissel, J. 1975a. First results of the micrometeoroid experiment S215 on the HEOS 2 satellite. *Planet Space Sci.*, **23**, pp. 215–224.

Hoffman, H. J., Fechtig, H., Grün, E., and Kissel, J. 1975b. Temporal fluctuations and anisotropy of the micrometeoroid flux in the Earth-Moon system measured by HEOS 2. *Planet. Space Sci.*, **23**, pp. 985–991.

Hörz, F., Bernhard, R. P., Warren, J., See, T. H., Brownlee, D. E., Laurance, M. R., Messenger, S., and Peterson, R. B. 1991. Preliminary analysis of LDEF instrument AO187-1 "Chemistry of micrometeoroids experiment". In *LDEF 69 months in space—first LDEF Post-Retrieval Symposium*, ed. A. S. Levine, NASA CP 3134, Part 1, pp. 487–502.

Humes, D. H. 1991. Large craters on the meteoroids and space debris impact experiment. In *LDEF 69 months in space—first LDEF Post-Retrieval Symposium*, ed. A. S. Levine, NASA CP 3134, Part 1, pp. 399–418.

Igenbergs, E., Hüdepohl, A., Uesugi, K., Hayashi, T., Svedhem, H., Iglseder, H., Koller, G., Glasmachers, A., Grün, E., Schwehm, G., Mizutani, H., Yamamoto, K., and Nogami, K. 1991. The Munich Dust Counter—a cosmic dust experiment on board of the MUSES-A mission of Japan. In *Origin and evolution of interplanetary dust*, eds. Levasseur-Regourd and H. Hasegawa (Tokyo: Kluwer), pp. 45–48.

Iglseder, H., Münzenmayer, R., Svedhem, H., and Grün, E. 1993. Cosmic dust and space debris measurements with the Munich Dust Counter on board the satellites Hiten and Bremsat. *Adv. Space Res.*, **13**, pp. 129–132.

Iglseder, H., Uesugi, K., and Svedhem, H. 1996. Cosmic dust measurements in lunar orbit. *Adv. Space Res.*, **17**, pp. 177–182.

Jacchia, L.G., Verniani, F., and Briggs, R.E. 1967. Selected results from precision-reduced super-Schmidt meteors. In *Meteor Orbits and Dust*, ed. G.S. Hawkins, *Smithson. Contr. Astrophys.*, **11**, pp. 1–7.

Jehn, R., Viñals-Larruga, S., and Klinkrad, K. 1993. DISCOS—the european space debris database. In *44th Congress of the International Astronautical Federation*, paper IAF-93-742.

Jenniskens, P. 1994. Meteor stream activity. I. The annual streams. *Astron. Astrophys.*, **287**, pp. 990–1013.

Jennison, R. C., McDonnell, J. A. M., and Rodger, I. 1967. The Ariel II micrometeorite penetration measurements. *Proc. Roy. Soc. A.*, **300**, pp. 251–269.

Johnson, N., and Christiansen, E. 1997. NASA/JSC orbital debris models. In *Proc. Second European Conference on Space Debris*, eds. B. Kaldeich-Schürmann and B. Harris, ESA SP-393, pp. 225–232.

Johnson, N. L., and McKnight, D. S. 1987. *Artificial space debris*, (Malabar, Florida, USA: Orbital Book Company).

Johnson, N. L., and Nauer, D. J. 1990. *History of on-orbit satellite fragmentations*, 4th ed. (Teledyne-Brown Engineering), CS90-TR-JSC-002.

Kessler, D. J. 1969. Average relative velocity of sporadic meteoroids in interplanetary space. *AIAA*, **7**, pp. 2337–2338.

Kessler, D. J. 1972. A guide to using meteoroid-environment models for experiment and spacecraft design applications. NASA TN D-6596.

Kessler, D. J. 1981. Derivation of the collision probability between orbiting objects: the lifetimes of Jupiter's outer moons. *Icarus*, **48**, pp. 39–48.

Kessler, D. J., Reynolds, R. C., and Anz-Meador, P.D. 1989. Orbital debris environment for spacecraft designed to operate in low Earth orbit. NASA TM 100471.

Kessler, D. J., Zhang, J., Matney, M. J., Eichler, P., Reynolds, D. C., Anz-Meador, P. D., and Stansbery, E. G. 1996. A computer based orbital debris environment model for spacecraft design and observations in low earth orbit. NASA TM 104825.

King-Hele, D. G., Walker, D. M. C., Pilkington, J. A., Winterbottom, A. N., Hiller, H., and Perry, G. E. 1990. *The RAE Table of Earth Satellites 1957–1989*, 4th ed. (Surrey, England: MacMillan Publishers Ltd.).

Kissel, J., and Krueger, F. R. 1987. Ion formation by impact of fast dust particles and comparison with related techniques. *Appl. Phys. A*, **42**, pp. 69–85.

Klinkrad, H. 1991. DISCOS—ESA's Database and Information System Characterising Objects in Space. *Adv. Space Res.*, **11**, pp. 43–52.

Klinkrad, H., Tejedor, O., and Viñals, S. 1997a. The DISCOS space data publication system. In *Proc. Second European Conference on Space Debris*, eds. B. Kaldeich-Schürmann and B. Harris, ESA SP-393, pp. 367–373.

Klinkrad, H., Bendisch, J., Sdunnus, H., Wegener, P., and Westerkamp, R. 1997b. An introduction to the 1997 ESA MASTER model. In *Proc. Second European Conference on Space Debris*, eds. B. Kaldeich-Schürmann and B. Harris, ESA SP-393, pp. 217–224.

Kresáková, M. 1966. The magnitude distribution of meteors in meteor streams. *Contr. Astr. Obs. Skalnaté Pleso*, **3**, pp. 75–109.

Laul, J. C., Smith, M. R., Wänke, M., Jagontz, E., Dreibus, G., Palme, M., Spettel, B., Burghele, A., Lipschutz, M. E., and Verkouteren, R. M. 1986. Chemical systematics of the Shergotty meteorite and the composition of its parent body (Mars). *Geochem. Cosmochim. Acta*, **50**, pp. 909–926.

Laurance, M. R., and Brownlee, D. E. 1986. The flux of meteoroids and orbital space debris striking satellites in low earth orbit. *Nature*, **323**, pp. 136–138.

Lobb, D. R., Dick, J. S. B., and Green, S. F. 1993. Development of concepts for detection and characterisation of debris in Earth orbiting passive optical systems. *Adv. Space Res.*, **13**, pp. 59–63.

Love, S. G., and Brownlee, D. E. 1991. Heating and thermal transformation of micrometeoroids entering the Earth's atmosphere. *Icarus*, **89**, pp. 26–43.

Love, S. G., and Brownlee, D. E. 1993. A direct measurement of the terrestrial mass accretion rate of cosmic dust. *Science*, **262**, pp. 550–553.

McBride, N. 1997. The importance of the annual meteoroid streams to spacecraft and their detectors. *Adv. Space Res.*, **20**, pp. 1513–1516.

McBride, N., Taylor, A. D., Green, S. F., and McDonnell, J. A. M. 1995. Asymmetries in the natural meteoroid population as sampled by LDEF. *Planet. Space Sci.*, **43**, pp. 757–764.

McCracken, C. W., Alexander, W. M., Dubin, M. 1961. Direct measurements of interplanetary dust particles in the vicinity of the Earth. *Nature*, **192**, pp. 441–442.

McCrosky, R. E., and Posen, A. 1961. Orbital elements of photographic meteors. *Smithso-*

nian Contrib. Astrophys., **4**, pp. 15–84.

McDonnell, J. A. M. 1964. The Study of Micrometeorites from Rockets and Satellites. Ph.D. Thesis (Manchester, England: Victoria University).

McDonnell, J. A. M. 1978. Microparticle studies by space instrumentation. In *Cosmic Dust*, ed. J. A. M. McDonnell (Chichester: J. Wiley & Sons), pp. 337–426.

McDonnell, J. A. M., and Baron, J. M. 1995. Penetration rates over 30 years in the space age. In *LDEF—69 months in space: third post-retrieval symposium*, ed. A. S. Levine, NASA CP-3275, Part 1, pp. 337–351.

McDonnell, J. A. M., and Gardner, D. J. 1998. Meteoroid morphology and densities: decoding satellite impact data. *Icarus* In press.

McDonnell, J. A. M., and Ratcliff, P. R. 1992. The geocentric particulate distribution: cometary, asteroidal or space debris? In *Asteroids, Comets, Meteors 1991*, eds. A. W. Harris and E. Bowell (Houston: Lunar and Planetary Institute), pp. 407–411.

McDonnell, J. A. M., and Stevenson, T. J. 1991. Hypervelocity impact microfoil perforations in the LEO space environment (LDEF MAP AO023 experiment). In *LDEF 69 months in space—first LDEF Post-Retrieval Symposium*, ed. A. S. Levine, NASA CP-3134, Part 1, pp. 443–458.

McDonnell, J. A. M., and Sullivan, K. 1992. Hypervelocity impacts on space detectors: decoding the projectile parameters. In *Hypervelocity Impacts in Space*, ed. J. A. M. McDonnell (Canterbury: University of Kent), pp. 39–47.

McDonnell, J. A. M., Carey, W. C., and Dixon, D. G. 1984a. Cosmic dust collection by the capture cell technique on the Space Shuttle. *Nature*, **309**, pp. 237–240.

McDonnell, J. A. M., Ashworth, D. G., Carey, W. C., Flavill, R. P. and Jennison, R. C. 1984b. Multiple foil microabrasion package. In *The Long Duration Exposure Facility (LDEF) mission 1 experiments*, eds. L. G. Clark, W. H. Kinard, D. J. Carter and J. L. Jones, NASA SP-473, pp. 117–120.

McDonnell, J. A. M., and the Canterbury LDEF MAP Team 1992. Impact cratering from LDEF's 5.75-year exposure: decoding of the interplanetary and Earth-orbital populations. *Proc. Lunar & Planetary Science*, **22**, pp. 185–193.

McGlaun, J. M., and Yarrington, P. 1993. Large deformation wave codes. In *High pressure shock compression of solids*, ed. J. R. Asay and M. Shahinpoor (New York: Springer Verlag), pp. 323–353.

McKnight, D. S., and Johnson, N. L. 1989. Understanding the true Earth satellite population. *40th Congress of the International Astronautical Federation*, paper IAF-89-617.

Maiden, C. J., Gehring, J. W., and McMillan, A. R. 1963. Investigation of fundamental mechanism of damage to thin targets by hypervelocity projectiles. NASA TR 63-225.

Mandeville, J. C. 1990. Aragat mission dust collection experiment. *Adv. Space Res.*, **10**, pp. 397-401.

Mandeville, J. C., and Berthoud, L. B. 1995. Micrometeoroids and debris on LDEF comparison with MIR data. In *LDEF—69 months in space: third post–retrieval symposium*, ed. A. S. Levine, NASA CP-3275, Part 1, pp. 275–285.

Maurette, M., Jehanno, C., Robin, E., and Hammer, C. 1987. Characteristics and mass distribution of extra-terrestrial dust from the Greenland ice cap. *Nature*, **328**, pp. 699–702.

Maurette, M., Olinger, C., Christophe Michel-Levy, M., Kurat, G., Pourchet, M., Brandstätter, F., and Bourot-Denise, M. 1991. A collection of diverse micrometeorites retrieved from 100 tonnes of Antarctic blue ice. *Nature*, **351**, pp. 44–46.

Mawrey, R. S., and Broadhurst, A. D. 1993. Comparison of predicted and measured rates of meteor signals. *Radio Science*, **28**, pp. 415–427.

Mulholland, J. D., Singer, S. F., Oliver, J. P., Weinberg, J. L., Montague, N. L., Wortman, J. J., Kassel, P. C., and Kinard, W. H. 1991. IDE spatio-temporal impact fluxes and high time resolution studies of multi-impact events and long lived debris clouds. In *LDEF 69 months in space—first LDEF Post-Retrieval Symposium*, ed. A. S. Levine, NASA CP 3134, Part 1, pp. 517–528.

Murray, J., and Renard, A. F. 1883. On the measurement characteristics of volcanic ashes and cosmic dust and their origin in deep-sea sediment deposits. *Proc. Roy. Soc. Edinburgh*, **12**, pp. 474–495.

Naumann, R. J. 1965. Pegasus measurements of meteoroid penetrations (February 16 – July 20, 1965). NASA TM X-1192.

Neish, M. J. 1995. Particle Fluxes on the Long Duration Exposure Facility. Ph. D. Thesis (Canterbury: University of Kent).

Nelleson, W. 1995. The development of the European Retrievable Carrier "EuReCa". *Adv. Space Res.*, **16**, pp. 5–16.

Nilsson, C. 1966. Some doubts about the Earth's dust cloud. *Science*, **153**, pp. 1242–1246.

Novikov, L. S., Voronov, K. E., Semkin, N. D., Verhoturov, V. I., Grafodatsky, O. S., and Maksimov, I. A. 1997. Attempt of measurement of space debris microparticle flux in geosynchronous orbit. In *Proc. Second European Conference on Space Debris*, eds. B. Kaldeich-Schürmann and B. Harris, ESA SP-393, pp. 135–138.

Nysmith, C. R., and Denardo, B. P. 1969. Experimental investigation of the momentum transfer associated with impact into thin aluminum targets. NASA TN D-5492.

Oliver, J. P., Singer, S. F., Weinberg, J. L., Simon, C. G., and Cooke, W. J. 1995. LDEF Interplanetary Dust Experiment (IDE) results. In *LDEF—69 Months in space: third post-retrieval symposium*, ed. A. S. Levine, NASA CP-3275, Part 1, pp. 257–273.

O'Neal, R. L. 1965. The Explorer XXIII micrometeoroid satellite—description and preliminary results for the period November 6, 1964. NASA TM X-1123.

O'Neal, R. L. 1968. The Explorer XXIII micrometeoroid satellite—description and results for the period Nov 6, 1964, through Nov 5, 1965. NASA TN D-4284.

O'Neal, R. L., and Lightner, E. B. 1991. Long Duration Exposure Facility—a general overview. In *LDEF—69 months in space: first post-retrieval symposium*, ed. A. S. Levine, NASA CP-3194, Part 1, pp. 3–48.

Öpik, E. J. 1951. Collision probabilities with the planets and the distribution of interplanetary matter. *Proc. R.I.A.* **54**, pp. 165–199.

Öpik, E. J. 1958. *Physics of Meteor Flight in the Atmosphere* (New York: Wiley Interscience).

Ott, U., and Beggeman, F. 1985. Martian meteorites: are they (all) from Mars? *Nature*, **317**, pp. 509–512.

Pailer, N., and Grün, E. 1980. The penetration limit of thin films. *Planet. Space Sci.*, **28**, pp. 321–331.

Rajan, R. S., Brownlee, D. E., Tomandl, D., Hodge, P. W., Farrar, H. and Britten, R. A. 1977. Detection of ^{4}He in stratospheric particles gives evidence of extra-terrestrial origin. *Nature*, **267**, pp. 131–134.

Ratcliff, P. R., and McDonnell, J. A. M. 1992. 2-D numerical computation of the relative contributions of natural material to the orbital component of the near Earth particulate population. In *Hypervelocity Impacts in Space*, ed. J. A. M. McDonnell (Canterbury: University of Kent), pp. 115–119.

Ratcliff, P. R., Taylor A. D., and McDonnell, J. A. M. 1993a. The LEO microparticle population: computer studies of space debris drag depletion and of interplanetary capture processes. *Adv. Space Res.*, **13**, pp. 71–74.

Ratcliff, P. R., Taylor, A. D., and McDonnell, J. A. M. 1993b. The relative efficiency of aerocapture of interplanetary dust for the planets. *Planet. Space Sci.*, **41**, pp. 603–608.

Rietmijer, F. J., and Mackinnon, I. D. R. 1987. Cometary evolution: clues from chondritic interplanetary dust particles. In *Symposium on the diversity of comets*, eds. E. J. Rolfe and B. Battrick, ESA SP-278, pp. 363–367.

Sawle, D. R. 1969. Hypervelocity impact in thin sheets and semi-infinite targets at 15 km/sec. *AIAA Hypervelocity Impact Conference*, Cincinnati, Ohio, Paper No. 69 p. 378.

Schramm, L. S., Brownlee, D. E., and Wheelock, M. M. 1989. Major element composition of stratospheric micrometeorites. *Meteoritics*, **20**, pp. 99–112.

Sdunnus, H., and Klinkrad, H. 1993. An introduction to the ESA reference model for space debris and meteoroids. In *Proc. First European Conference on Space Debris*, ed. W. Flury, ESA SD-01, pp. 343–348.

Sdunnus H. 1995. Meteoroid and space debris terrestrial environment reference model—final report. Institute of Spaceflight Technology and Nuclear Reactor Technology, Technical University of Braunschweig, Germany. ESA Contract No. 10453/93/D/CS.

See, T. H., Allbrookes, M. K., Atkinson, D. R., Sapp, C. A., Simon, C. G., and Zolensky, M. E. 1991. Meteoroid and debris special investigator group data acquisition procedures. In *LDEF 69 months in space—first LDEF Post-Retrieval Symposium*, ed. A. S. Levine, NASA CP 3134, pp. 459–476.

See, T. H., Mack, K. S., Warren, J. L., Zolensky, M. E., and Zook, H. A. 1993. Continued investigation of LDEF's structural frame and thermal blankets by the Meteoroid and Debris Special Investigation Group. In *LDEF—69 Months in space: second post-retrieval symposium*, ed. A. S. Levine, NASA CP-3194, Part 2, pp. 313–324.

See, T. H., Zolensky, M. E., Bernhardt, R. P., Warren, J. L., Sapp, C. A., and Dardano, C. B. 1995. LDEF Meteoroid and Debris Special Investigator Group investigations and activities at the Johnson Space Center. In *LDEF—69 months in space: third post-retrieval symposium*, ed. A. S. Levine, NASA CP-3275, Part 1, pp. 257–273.

Sekanina, Z., and Southworth, R.B. 1975. Physical and dynamical studies of meteors: meteor fragmentation and stream distribution studies. NASA contractor report CR-2615, Smithsonian Institution, Cambridge, MA.

Yu Shanbing, Sun Gengchen and Tan Qingming 1994. Experimental laws of cratering for hypervelocity impacts of spherical projectiles into thick targets. *Int. J. Impact Engng.*, **15**, pp. 67–77.

Sorensen, N. R. 1965. Systematic investigation of crater formation. In *Proceedings of the 7th Hypervelocity Impact Symposium Volume VI—Experiments*, (Florida: Orlando), pp. 281–325.

Singer, S. F., Stanley J. E., and Kassel, P. C. 1985. The LDEF Interplanetary Dust Experiment. In *Properties and interactions of interplanetary dust*, eds. R. H. Giese and P. Lamy (Dordrecht: Reidel), pp. 117–120.

Singer, S. F., Stanley, J. E., Kassel, P. C., and Wortman, J. J. 1984. Interplanetary Dust Experiment (AO201). In *The Long Duration Exposure Facility (LDEF) mission 1 experiments* NASA SP-473.

Singer, S. F., Stanley, J. E., Kassel, P. C., Kinard, W. H., Wortman, J. J., Weinberg, J. L., Mulholland, J. D., Eichorn, G., Cooke, W. J., and Mantague, N. 1991. First spatio-temporal results from the LDEF Interplanetary Dust Experiment. *Adv. Space Res.*, **11**, pp. 115–122.

Southworth, R. B., and Sekanina, Z. 1973. Physical and dynamical studies of meteors. NASA CR-2316, Smithsonian Institution, Cambridge, MA.

Stansbery, E. G., Kessler, D. J., Tracy, T., Matney, M. J., and Stanley, J. 1995. Characterization of the orbital debris environment from Haystack radar measurements. *Adv. Space Res.*, **16**, pp. 5–16.

Staubach, P., and Grün, E. 1995. Development of an upgraded meteoroid model. *Adv. Space Res.*, **16**, pp. 103–106.

Staubach, P., Grün, E., and Jehn, R. 1997. The meteoroid environment near Earth. *Adv. Space Res.*, **19**, pp. 301–308.

Stevenson, T. J. 1988. EuReCa TICCE—a nine month survey of cosmic dust and space debris at 500 km altitude. *J. Brit. Interplan. Soc.*, **41**, pp. 429–432.

Štohl, J. 1986. The distribution of sporadic meteor radiants and orbits. In *Asteroids Comets Meteors II*, eds C.-I. Lagerkivist, B. A. Lindblad, H. Lundstedt and H. Rickman (Sweden: University of Uppsala Press), pp. 565–574.

Taylor, A. D. 1995a. The Harvard Radio Meteor Project meteor velocity distribution reappraised. *Icarus*, **116**, pp. 154–158.

Taylor, A. D. 1995b. Earth encounter velocities for interplanetary meteoroids. *Adv. Space Res.*, **17**, pp. 205–209.

Taylor, A. D., and McBride, N. 1997. A radiant-resolved meteoroid model. In *Proc. Second European Conference on Space Debris*, eds. B. Kaldeich-Schürmann and B. Harris, ESA SP-393, pp. 375-380.

Taylor, A. D., Baggaley, W. J., and Steel, D. I. 1996. Discovery of interstellar dust entering the Earth's atmosphere. *Nature*, **380**, pp. 323–325.

USIG: US Interagency Group (Space) 1989. Report on orbital debris for National Security Council. Washington DC.

USIG: US Interagency Group (Space) 1995. Report on orbital debris for National Science and Technology Council Committee on Research and Development. Washington DC.

USNRC: US National Research Council 1995. Orbital debris: a technical assessment. (National Academy Press).

Verniani, F. 1973. An analysis of the physical parameters of 5759 faint radio meteors. *J. Geophys. Res.*, **78**, pp. 8429–8462.

Watts, A., Atkinson, D., and Rieco, S. 1993. Dimensional scaling for impact cratering and perforation. Technical report (Albuquerque, New Mexico: POD Associates Inc.).

Whipple, F. L. 1950. The theory of micro-meteorites. Part I. In an Isothermal Atmosphere. *Proc. Nat. Acad. Sci.*, **36**, pp. 687–693.

Whipple, F. L. 1951. The theory of micro-meteorites. Part II. In heterothermal atmospheres. *Proc. Nat. Acad. Sci.*, **37**, pp. 19–30.

Wisdom, J. 1983. Chaotic behaviour and the origin of the 3:1 Kirkwood gap. *Icarus*, **56**, pp. 57–74.

Wisdom, J. 1985. Meteorites may follow a chaotic route to Earth. *Nature*, **315**, pp. 731–733.

Yamakoshi, K. 1994. *Extraterrestrial dust: laboratory studies of interplanetary dust* (Dordrecht: Kluwer).

Yano, H. 1995. The physics and chemistry of hypervelocity impact signatures on spacecraft: meteoroids and space debris. Ph.D. Thesis (Canterbury: University of Kent).

Yoshikawa, M. 1990. Motions of asteroids at the Kirkwood gaps, I. On the 3/1 resonance with Jupiter. *Icarus*, **87**, pp. 78–102.

Zolensky, M. E., See, T. H., Bernhardt, R. P., Barret, R., Hörz, F., Warren, J. L., Dardano, C., and Leago, K. S. 1996. Final activities of the Long Duration Exposure Facility Meteoroid and Debris Special Investigation Group. *Adv. Space Res.*, **16**, pp. 53–65.

Zook, H. A. 1991. Deriving the velocity distribution of meteoroids from the measured meteoroid impact directionality on the various LDEF surfaces. In *LDEF—69 months in space: first post-retrieval symposium*, ed. A. S. Levine, NASA CP-3134, Part 1, pp. 569–579.

Zook, H. A., and Berg, O. E. 1975. A source for hyperbolic cosmic dust particles. *Planet. Space Sci.*, **23**, pp. 183–203.

Zukas, J. A. (ed) 1990. *High velocity impact dynamics* (New York: John Wiley).

Discoveries from Observations and Modeling of the 1998/99 Leonids

Peter Jenniskens

SETI Institute, NASA Ames Research Center, Moffett Field, California, USA

Abstract. The Leonid meteor storm of November 1999 entered the history books as the second in the space age, but was the first to be well observed. This was a rare encounter with fresh comet ejecta that can be precisely dated to the 1899 return of the parent comet. Earth's atmosphere acted as a giant detector for measuring composition and morphology of large mm-cm sized grains during what was in effect a comet mission to 55P/Tempel-Tuttle. The high meteor flux and preponderance of persistent trains enabled the deployment of modern observing techniques to probe the chemistry and physics of the interaction of meteoric matter with the Earth's atmosphere, leading to many discoveries of interest to a wide array of science fields such as astrobiology, planetary astronomy, and the atmospheric sciences. Moreover, the correct timing of the 1999 Leonid storm by meteor stream modeling has ushered meteor storms into the modern era of forecasting space weather. This paper summarizes the discoveries that resulted from these observations and subsequent theoretical modeling.

I. INTRODUCTION

The return of possible Leonid storms was announced in the form of enhanced Leonid activity in 1994 (Jenniskens 1994, 1996). The Leonid parent comet 55P/Tempel-Tuttle was recovered in November of 1997, several months prior to its return to perihelion on February of 1998 (Hainaut et al. 1998). The historic importance of past Leonid storms is explained in the review paper on meteors and meteoroid studies by Ceplecha et al. (1998) and, for example, in the popular book on the Leonid showers by Littmann (1998). Ceplecha et al. also provide a summary of the state of the meteor and meteoroids field just before the new season of Leonid activity. Now, shortly after the 1999 Leonid meteor storm, we are looking back on a flurry of activity using modern observing techniques and numerical modeling.

II. METEOROID STREAMS AND METEOR STORMS

Numerical techniques to study the orbital dynamics of meteoroids and the formation of meteoroid streams came of age only in the past ten years with the advance of computing facilities. Gradually, the large parameter space for variables such as ejection processes, effects of radiation pressure, and specific aspects of orbital dynamics under planetary perturbations have been probed and many ideas are now postulated that can affect the formation and evolution of meteoroid streams. It is an ongoing struggle to identify what mechanism is responsible for what. Reviews are found in Lovell (1954), Lewin (1961), Williams (1993), Steel (1994), and Jenniskens (1998). Because meteor storms represent relatively recent ejecta, they offer our best hope of identifying the correct mechanisms of formation and evolution (Krésak 1993; Jenniskens 1995).

The Leonid storms are among the most intense in recent history. Early expectations of enhanced Leonid rates during the 1998 return of the comet were based on work by Sekanina (1975), Yeomans (1981), and Yeomans et al. (1996), who calculated the orbital dynamics of comet 55P/Tempel-Tuttle as affected by planetary perturbations and non-gravitational effects. They found that past Leonid storms occurred predominantly at times when the Earth was outside of the comet orbit and in the years following the passage of the comet by the node (Fig. 1). This situation would occur again in the autumn of 1998 and 1999. For that reason, a meteor storm was anticipated (e.g., Brown et al. 1997a,b; Jenniskens 1996, Jenniskens et al. 1998 and references therein). Much uncertainty remained, because in some suitable encounters (such as in 1932) no storm was reported. Moreover, Earth was to pass at a rather large distance from the comet orbit. Also, there was the issue that the dust trail was expected to wag in and out of Earth's orbit as a result of planetary perturbations on the individual particles, considerably complicating the picture painted by Sekanina and Yeomans (Wu and Williams 1993; Jenniskens 1997, 1998).

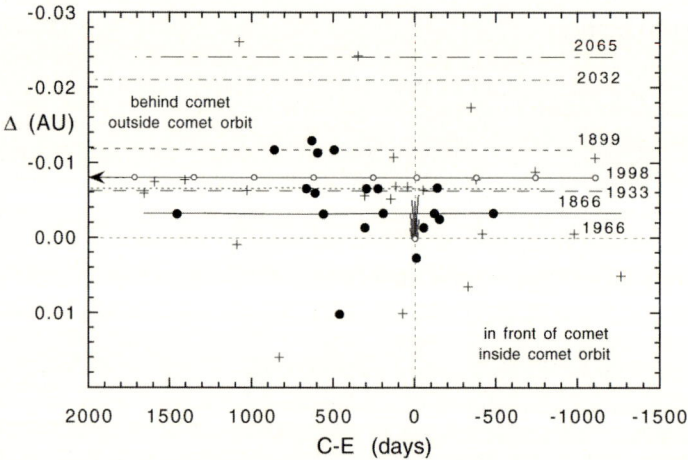

Figure 1. Detection of meteor storms for given position of Earth relative to the position of comet 55P/Tempel-Tuttle (center of figure). Δ on the vertical axis is the distance between the Earth orbit and comet orbit for the epoch of the comet's passage by the node. $C - E$ is the time in days between the passages of comet and Earth by this point. Dark circles refer to narrow Leonid outbursts, crosses are other indications of elevated Leonid rates. After: Yeomans (1981).

Following a disappointing 1998 encounter, when only a modest increase of rates was detected at the predicted time but a fireball shower half a day earlier made a spectacular display, McNaught and Asher (1999) and Lyytinen (1999) published results of dust trail models that assumed very low ejection velocities at perihelion, thus rather probing the orbital dynamics of dust near the mean motion resonance around which the Halley-type parent comet is found to librate. This approach was used before by Kondrat'eva and Reznikov (1985), based on early calculations by Kazimircak-Polonskaja et al. (1968). However, their predictions of strong Leonid meteor storms in 2001 and 2002 were so much different from those by others, that these results were not widely accepted. In fact, this approach was a step back from meteoroid stream models that attempted to describe the dust dispersion and dynamical evolution by considering the ejection process (with relatively high ejection velocities) and radiation pressure effects in some detail. Drawback of this approach was the large amount of computing time needed to handle many particles in dynamically quite different orbits, which forced the modelers to average the final meteoroid orbits over a relatively large area near the Earth's orbit (e.g., Wu and Williams 1996; Asher 1999; Brown 2000; Göckel and Jehn 2000).

The new approach was to calculate, for a small range of orbital period, where the dust from each comet return is expected to cross the ecliptic plane near Earth's orbit and combine that with observed dust dispersions from past storms. This point in the ecliptic plane changes in different years because of planetary perturbations working differently on grains at different mean anomaly from the comet, and subsequent trails are displaced from each other when the comet orbit itself is perturbed in between returns. Thus, a map is created that consists of a yearly changing cross-section of narrow dust trails (the "trailets") with Earth's path superposed (Figs. 2 and 3). The calculations included the most recent comet orbit derived from the comet's 1998 encounter and was calibrated by the timing of past meteor storms such as collected by Jenniskens (1995), Brown et al. (1997a), and Brown (1999).

This approach was validated when a meteor storm was observed in 1999, with peak rates of about Zenith Hourly Rate ZHR $= 4 \times 10^3$ per hour. Support for the premise underlying the multi-trailet model for comet dust trails came from the good timing of the 1999 Leonid meteor storm (observed at 02:02±02 UT, while predicted at 02:08±15 UT), but perhaps more so from the detection of a secondary peak in the meteor flux coincident with passing the 1866 trail (Arlt et al. 1999; Jenniskens et al. 2000a). During the 2000 return several maxima were observed that are readily identified with the passages of the 1932, 1733 and 1866 trailets, with peak rates and meteor size distributions much as predicted by Lyytinen and Van Flandern (2000). Moreover, the same tools were used to predict the return of the 2000 Ursids of Halley-type comet 8P/Tuttle, with similar success (Jenniskens and Lyytinen 2000).

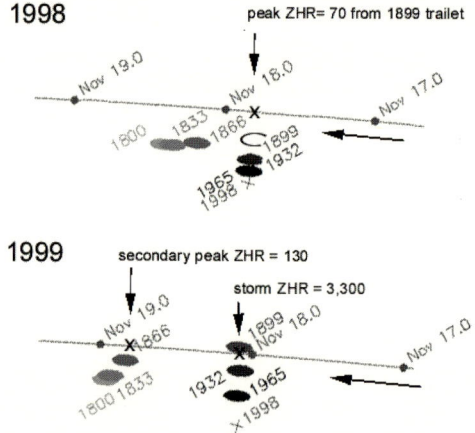

Figure 2. Pattern of dust trails intersecting with Earth's orbit in the years 1998 and 1999. The occurrence of meteor outbursts are indicated by a cross in Earth's pass (adapted from a figure by David Asher, Armagh Observatory).

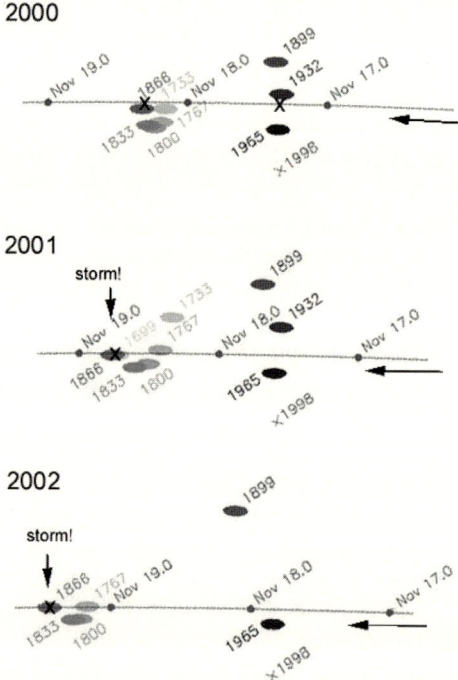

Figure 3. Same as Fig. 2, but for dust trails in 2000-2001.

An interesting case in study is the 1998 "storm peak", a maximum in activity that was observed shortly after passing the node of the comet orbit in November of 1998, shown in Fig. 4 (Arlt 1998; Jenniskens 1999a). McNaught and Asher (1999) pointed out that the Earth passed a fragment of the 1899 trail at that time, but this trail had been perturbed by Earth's encounter in 1965 and no meteoroids were expected to be present. In this light, it comes as a surprise that a peak in activity was observed at the approximate time of passing the 1899 trailet. However, recent analysis of measured radiant positions (Betlem et al. 2000), an unusual asymmetric cross section (Jenniskens 1999a), and a mass-dependence in the radiant position at that time (de Lignie et al. 2000) are evidence that the 1998 "storm component" in the shower activity curve was in fact debris from the perturbed 1899 dust trail.

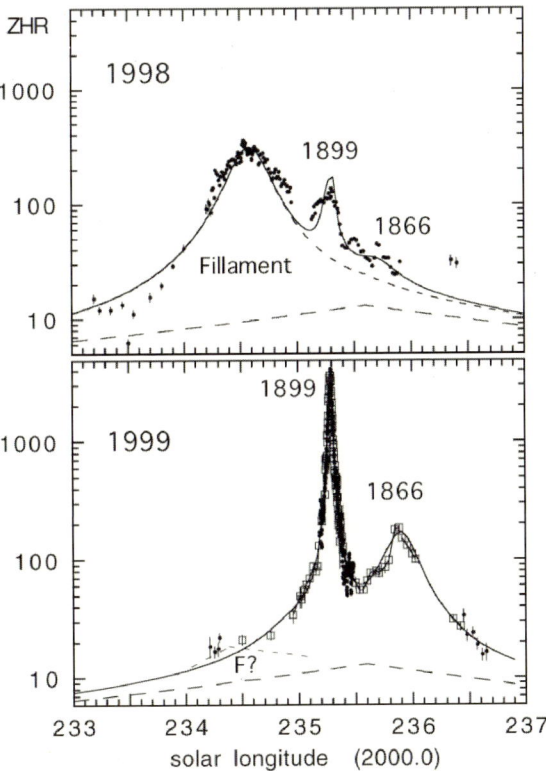

Figure 4. Zenith hourly rate of Leonid meteors with various dust components identified. This is a compilation of data from visual and video meteor observations as collected by the International Meteor Organisation (Arlt 1998; Arlt et al. 1999) and obtained during the Leonid MAC campaigns (Jenniskens 1999a; Jenniskens et al. 2000a).

The 1999 Leonid storm encounter showed the 1899 dust trailet cross section to be a smooth Lorentzian profile, much like cross sections of cometary dust trails observed at mid-IR wavelengths (Jenniskens et al. 2000a). The dispersion perpendicular to Earth's path (in the comet's orbital plane) and along the comet dust trail may be reveiled by future encounters with these dust trailets. Fortunately, future prospects for high Leonid activity are optimistic. Earth will cross the 1866 and 1767 trailets in 2001 and 2002 (Fig. 3), when similar meteor storms are predicted as seen in 1999 (McNaught and Asher 1999; Lyytinen 1999; Lyytinen and Van Flandern 2000). Those encounters will help elucidate how quickly the dust spreads along the comet orbit.

The recent Leonid showers also provided information on what happens to the comet dust trailets of Halley-type comets later in their orbital evolution. A broad meteoroid stream was detected in all years from 1994 until 1998, with a characteristic width of about a day called the Leonid "Filament". Asher et al. (1999) found that the Filament was an exceptionally strong dust trailet created in 1333, but Jenniskens and Betlem (2000) proposed that the Leonid Filament is an accumulation of dust (over many returns) in that region of mean anomaly around the comet position that is free from close encounters with Jupiter for an extended period of time. This idea followed Williams (1997), who pointed out that certain regions in the comet orbit are safe from close encounters with Uranus. However, Jupiter and Saturn play a more important role in the orbital dynamics of the meteoroids.

The Leonid Filament was the cause of the 1998 Leonid fireball shower (Fig. 4) which, like prior manifestations, was particularly lacking small meteoroids. It is expected that these smaller grains have a larger dispersion in orbital periods and thus move more rapidly to higher mean anomaly into a dangerous zone for close encounters with the main planets. Similar Filaments have now been detected in two other Halley-type comet orbits: 109P/Swift-Tuttle and 8P/Tuttle (Jenniskens et al. 1998; Jenniskens and Lyytinen 2000).

III. OBSERVING CAMPAIGNS

A collaborative effort was organised to provide an airborne platform for researchers worldwide to view the Leonid storms from the best possible observing site in good weather conditions (Jenniskens and Butow 1999; Jenniskens et al. 2000b). The Leonid Multi-Instrument Aircraft Campaign (from hereon "Leonid MAC") was sponsored by NASA and the U.S. Air Force, with additional support from NSF, ESA, ISAS, ISA and many other institutes and organisations. The 1998 mission brought an international team of 28 researchers to Okinawa, Japan, for a one-night observing run over the East China Sea. This mission included a two-beam Fe Boltzmann Lidar of the University of Illinois that was deployed onboard the NSF/NCAR "Electra" aircraft, with lasers tuned to two different iron resonant lines to measure the Boltzmann temperature from the distribution of the ground-state populations. In addition, mid- and near-IR spectrometers and imagers were operated onboard the USAF/452nd FTS

NKC-135 "FISTA" aircraft. The next 1999 mission brought 35 researchers to the Mediterranean. FISTA was now paired up with the USAF/452nd FTS B707-type aircraft "ARIA". That effort included near-real time flux measurements and live broadcast of the storm as seen from ARIA in combination with the deployment of UV, visible, near-IR and mid-IR spectrometers and imagers onboard FISTA. The meteor storm was detected just west of Greece, while the aircraft were on an east-west flight path from Israel to the Azores.

Both missions were supported by ground-based observing campaigns. At various locations in China (1998) and Spain (1999), the Dutch Meteor Society measured flux and meteoroid orbits (Betlem et al. 1999, 2000; de Lignie et al. 2000). The University of Illinois headed a Lidar experiment to measure persistent trains at the Starfire Optical Range at Kirtland AFB in New Mexico (Chu et al. 2000a; Kelley et al. 2000). In 1998, Japanese participants mounted a ground-based effort to observe the diffuse scattered sunlight from the shower in space at locations in both Japan and Hawaii (Nakamura et al. 2000). Israeli participants mounted ground-based optical, radar and ELF/VLF measurements in 1999. Other organised observing campaigns included a global ground-based effort to measure flux by radar and intensified video cameras in near-real time from 5 locations around the world. This campaign involved an international team of 41 scientists, 18 image intensified CCD systems, and two radars - one of which was positioned at high latitude for continuous coverage of the shower (Campbell et al. 2000; Brown et al. 2000). Many other ground-based efforts deserve mention here, such as the coordinated Lidar effort by the Leibniz-Institute of Atmospheric Physics in Kühlungsborn, Germany, coordinated efforts to observe impacts of Leonids on the Moon, and the efforts of many amateur meteor observers worldwide coordinated by the International Meteor Organisation. The Leonid shower was even observed from space using the Midcourse Space Experiment– MSX satellite (Jenniskens et al. 2000c).

IV. METEOROID MORPHOLOGY AND COMPOSITION

Early results from the Leonid MAC effort include the discovery that the 1998 and 1999 Leonid meteors have different light curves (Murray et al. 1999, 2000). This was interpreted in the context of the cluster model (Hawkes and Jones 1975; Ceplecha et al. 1998) to mean that, for an unknown reason, the 1999 meteoroids tended to fragment into larger pieces.

During the 1998 campaign, it was discovered that the Leonid meteor images had numerous jet-like features (LeBlanc et al. 2000). Subsequent imaging using narrow band Mg I filters onboard ARIA (Taylor et al. 2000) confirmed their existence and demonstrated that these jets are caused by meteoric matter ejected at very high speeds as far as 2 km away from the 1-cm sized object (Fig. 5). The alternating position of the jets left and right of the meteor image points at rapid spinning, at a rate of $\sim$ 200 Hz, and a jet of dust fragments ejected much like water in a lawn sprinkler system. This clearly has implications for the volume of air affected by the meteoroid's impact into the atmosphere.

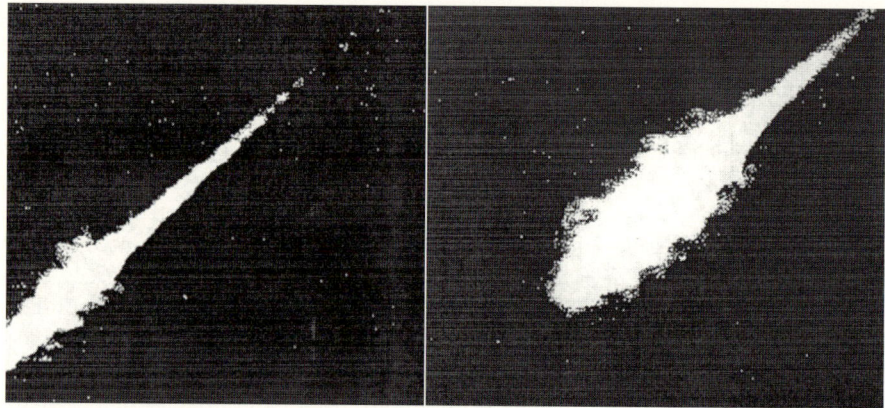

Figure 5. Two snapshots of the meteor of 03:08:48 UT (Nov. 18, 2000) in narrow band MgI images. Note the individual jets on the meteor image and the beading in the meteor wake. Courtesy: Taylor et al. (2000).

Unusual fragmentation of the meteoroids was observed also indirectly from low-resolution meteor spectroscopy. Figure 6 shows an example of an optical spectrum from a -2 magnitude Leonid as obtained by slit-less video spectroscopy. Borovička et al. (1999) found that sodium (Na) was released earlier from the meteoroids than magnesium (Mg), and lost before all magnesium was depleted. This differential ablation is expected when sodium is in a more volatile mineral than magnesium. However, it demands that all sodium-containing mineral is readily accessible for ablation. Differential ablation is not observed in most other meteors, suggesting that surface sputtering rather than evaporation from a melt is the preferred ablation pathway.

Along similar lines, Höffner et al. (1996) and Von Zahn et al. (1999) studied the relative abundance of elements K, Ca and Fe in neutral atom debris trails of meteors, using three co-aligned Lidar systems tuned at different resonant lines. They found that calcium is depleted over Iron by a factor of 11 in Leonid trails compared to chondritic materials at the altitude of the detections (97-103 km), suggesting that the calcium containing minerals survive ablation longer than most iron containing minerals. The lack of calcium rich trails in the observations suggests to me that calcium-containing minerals may even survive the ablation process.

Of special interest is also the possible fragmentation of meteoroids in the interplanetary medium. A dramatic example was filmed by Kinoshita et al. (1999), who observed a shower of meteoroid fragments at one spot in the sky during the 1997 Leonid encounter. Such showers appear to be rare. Gural and Jenniskens (2000) did an in-depth statistical analysis of the Leonid meteor's time of incidence and found no significant excess of short time intervals between 1/30th and 1 second over that expected from a Poisson distribution. Hence, no breakup of meteoroids on their final approach to Earth seems to occur. On the other hand, periodic excursions of rates and spatial correlations of meteors seem to be present in the data, which may be the result of an early breakup of large grains, for example in the comet coma during release. Those periodic variations occurred on a timescale of 2-3 minutes. In contrast, Singer et al. (2000) found periodic variations in shower activity on an approximate 8-minute timescale. This topic deserves further study, because it promises a tool for remotely probing breakup processes in the comet dust coma.

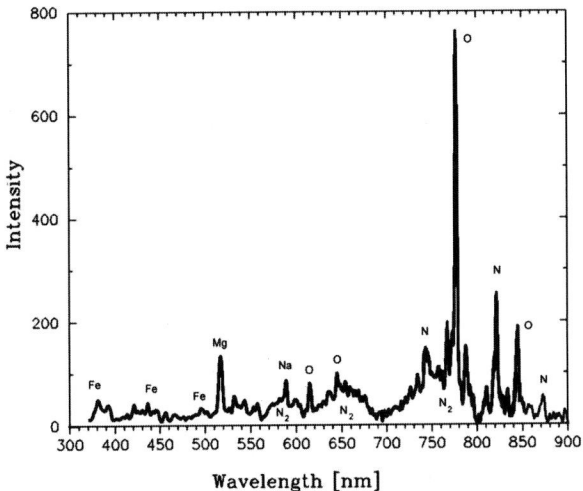

Figure 6. Typical Leonid meteor spectrum. Courtesy: Shinsuke Abe (2000).

V. THE IMPACT HAZARD

The potential danger of meteoroid impacts on satellites in orbit during a meteor storm was acknowledged by Casswell et al. (1995) and Beech et al. (1995, 1996). Until that time, the natural meteoroid complex was treated as a homogenous influx in time with directional variability only from the main dust components in the sporadic background. The influence of meteor showers was subsequently studied by McDonnell et al. (1997), Cevolani and Foschini (1998), and McBride and McDonnell (1998). See also the review by Jenniskens (1999b).

The flux anomaly during meteor storms is strongest if the satellite surface is oriented perpendicular to the radiant of the shower and speeds are unlike the typical sporadic meteoroids. Notably, the Leonid meteoroids have unusually high impact velocity (around 71 km/s). Velocity has a strong effect on plasma charge (Q) production and plasma current (I) production, which scale with velocity according to $Q \sim V^{3.48}$ and $I \sim V^{4.48}$, respectively (McDonnell et al. 1997). It was the meteoroid induced plasma discharge that was thought to be responsible for the demise of ESA's Olympus satellite (Casswell et al. 1995).

The 600 or so operational satellites weathered the 1999 Leonid storm well. No anomalies were associated with the storm. The peak flux during the storm was measured at about 1.2 ± 0.4 particles per km^2 hr^{-1} for meteoroids with visual magnitude of 6.5 or less (Arlt et al. 1999; Brown et al. 2000; Gural and Jenniskens 2000). That translates to an impact probability of only about 2 percent for the combined satellite park to be hit by such a Leonid meteoroid of sufficient mass to cause serious or disabling damage. In comparison, a Leonid storm of the intensity as that of either 1833 or 1966 would have resulted in up to 8 impacts.

Svedhem et al. (2000) reported a flurry of very small meteoroids with the impact detector GORID, on the geostationary Russian Express II telecom satellite, in roughly the direction of the Leonid radiant at a time just after the 1999 Leonid storm. Such fragments would be too small to be detected as meteors in Earth's atmosphere. Small grains have different impact effects than large grains. Hence, not only the peak dust density and spatial distribution of dust, but also the distribution of dust sizes in cometary trails, is of interest to satellite operators.

The meteor storm was found to be less rich in faint meteors than thought before. With $r = N(m+1)/N(m)$, and m a measure of the magnitude of the meteor, r values of about 2.4 seem to fit most observations in the range -4 to +9 magnitude. Earlier data for the 1966 and 1866 storms suggested a higher $r = 3.0$ (Jenniskens 1995). The Leonid Filament has a lower $r = 1.5$–2.3 (e.g., Jenniskens and Betlem 2000). As a result, the fireball tail of the magnitude distribution can be dominated by the Filament component, if present, even for relatively intense storms.

There is no obvious large size scale cut-off until at least magnitude -15 (few kg mass). Leonid masses up to 5 kg have been derived from Moon impacts, assuming a luminosity efficiency of 0.001 percent (Nemthinov et al. 1998; Bellot Rubio et al. 2000a,b; Dunham et al. 2000). Those impacts may contribute to sustain the Moon's tenuous sodium atmosphere, although no enhancement of the atmosphere in relation to the shower has yet been established (e.g., Hunten et al. 1998). During the 1998 Leonids, a diffuse glow of sodium atom emission was detected in the anti-Sun direction from the new Moon as a result of erosion of the atmosphere by the solar wind (Smith et al. 1999; Wilson et al. 1999). A spectacular result from the 1998 campaign was also the discovery of a faint glow of scattered sunlight from the comet dust trail in the direction of the true radiant of the shower (Nakamura et al. 2000). This observation provides a unique cross section of the dust trail complex and in combination with trail models could in principle provide information on the abundance of small 10-100 micron grains in the dust trail.

VI. INTERACTION OF METEOROIDS WITH THE ATMOSPHERE

Traditional arguments of conservation of energy and momentum can describe the orbit deceleration and some aspects of the meteor's ablation as a function of altitude, but they do not address the complex physical and chemical processes that lead to the observed emissions. For example, it is not understood how efficient meteor kinetic energy induces atmospheric chemistry, how much of the meteoric matter stays in solid form, and what happens to the organic matter that is present in the meteoroids. There are many open issues on how meteor processes may have affected collected samples of micrometeorites and meteorites (Rietmeijer 2000). All those questions are of interest to understanding the conditions on the early Earth at the time of the origin of life some 4 Gyr ago. The issue is complicated by the unusual rarefied flow regime under which the small meteoroids interact with the atmosphere. This is particularly relevant because small meteoroids of about 200 micron in size make up most of the present day mass influx of meteoric matter (e.g., Jenniskens et al. 2000d).

Just as with studies of meteoroid stream dynamics, the study of meteor physics traditionally has made big leaps forward during past Leonid meteor storms. Partially, because of increased interest in the topic, but also because of unusual data gathering opportunities. The storms provide not only high rates, but also feature persistent trains that allow telescopes to be pointed at the path of a meteor.

High resolution meteor spectra in the visible and optical near-Infrared obtained during the 1998 Leonid MAC mission were the first that could be directly compared with theoretical spectra of air plasmas for quantitative information on excitation temperatures and abundances. From the first positive bands of the nitrogen molecule and atomic O and N emissions (Fig. 6), a temperature of 4,300 K was measured, which is in the same range as temperatures estimated from meteoric neutral metal atom emissions (Jenniskens et al. 2000d). This work inspired the first rarefied flow models of small meteoroids, using a Direct Simulation Monte Carlo technique, which show that this temperature is characteristic for the wake of a meteor, not the head (Boyd 2000). In this light, higher excitation emissions of Ca^+, H, and Mg^+ may well originate in the ablation vapor cloud building in front of the meteoroid (Jenniskens et al. 2000d; Popova et al. 2000).

The fate of organic matter in meteors is unclear. Organic matter is abundant in cometary grains and intimately mixed with silicate grains. At first sight, one might expect that the volatile organic matter in meteoroids is lost at high altitudes, during the initial stages of heating of the grain. Indeed, Fujiwara et al. (1998) and Spurný et al. (2000a, b) established a record height for the detection of Leonids, with most bright fireballs showing first at about 196 K. However, significant ablation appears to start only below 136 km, when the first signs of (OI green line?) wake are observed. The appearance of the meteors above about 135 km is very different from the teardrop shape at lower altitudes. They look more like a bowshock or diffuse V-shaped comet coma. The cause of this emission remains unknown.

The classical ablation models, involving heating to a melt and subsequent evaporation from a molten droplet, do not well describe the loss of meteoric compounds. The more recent dust-ball model suggests an early breakup of the grains, possibly from the loss of a volatile "glue" (Hawkes and Jones 1975, Muray et al. 2000). The breakup process itself should leave a signature from the sudden release of volatile compounds. Perhaps that process is responsible for unusual mid-IR emission. Indeed, the first IR detections of meteors during the 1998 Leonid MAC imply a different light curve at mid-IR wavelengths than in the optical, with 3-5.5 micron emission peaking just before the peak of the visible light curve (Rossano et al. 2000). However, despite the slightly earlier release of sodium, the bulk of meteoric emissions at optical wavelengths appear to be in concert. Perhaps, this is because all minerals (and the organic component) are intimately mixed. This suggests that the bulk of organic matter is released not much earlier than the remainder of the meteoroid. Indeed, Russell et al. (2000) detected the C-H stretch vibration band of what appears to be complex organic matter in the meteoric debris of a bright fireball.

Another mechanism may exist as well that allows organic matter to be deposited in the atmosphere more or less intact. Carbon atoms and molecules are not strong emitters. The most likely candidate for detection is CN, which has been suspected in some fireball spectra. Until now, no quantitative information was obtained. Jenniskens et al. (2000d) searched for breakup products such as C_2 and CN in the visible and near-IR high resolution spectra of Leonid meteors, where few metal atom ablation lines overlap the bands, but without success. Upper limits were set by Rairden et al. (2000), who found CN/Fe $\leq$ 0.35. It was submitted that organic molecules survive ablation as complex compounds, in which case meteors can be an efficient vehicle for delivery of organic carbon to Earth. Many different types of future observations, numerical modeling, and laboratory experiments are needed to establish this proposed mechanism.

Meteoric solid debris has now been detected in the path of a high altitude meteor, following promising indirect evidence from an unusual rocket measurement reported by Kelley et al. (1998). During the 1999 Leonid MAC, Borovička and Jenniskens (2000) detected a red continuum in the afterglow of a bright fireball (Fig. 7) and a significant descent of part of the emitting matter, indicative of meteoric debris at a temperature of T $\sim$ 1,400 K several tenths of seconds after the fireball. This is the melting temperature of chondritic materials (Rietmeijer and Nuth 2000). In addition, the afterglow emitted strongly in neutral metal atom emission lines for about 2 seconds (Fig. 7), but not in certain air plasma lines, until the temperature inferred from the metal atom emissions decayed to below the evaporation temperature of the chondritic material. This suggests that the afterglow is in effect the result of secondary ablation.

Rietmeijer and Jenniskens (2000) identified certain spheres in the NASA Cosmic Dust Catalogs as candidates of meteoric debris from known meteor showers. Mateshvilli et al. (2000) observed what appears to be the settling of meteoric dust to lower altitudes, from photometric observations of scattered sunlight during (the altitude dependent) sunrise and sunset. The debris is not likely recondensed meteoric vapor, in light of the long lifetimes measured by Chu et al. (2000b) for meteoric neutral atom debris trails. Eighteen Fe ablation trails were observed during the 1998 Leonid MAC, with ages determined from a diffusion model of up to 28 minutes at a height of 96 km where chemistry with atmospheric compounds and dispersion by diffusion are slow.

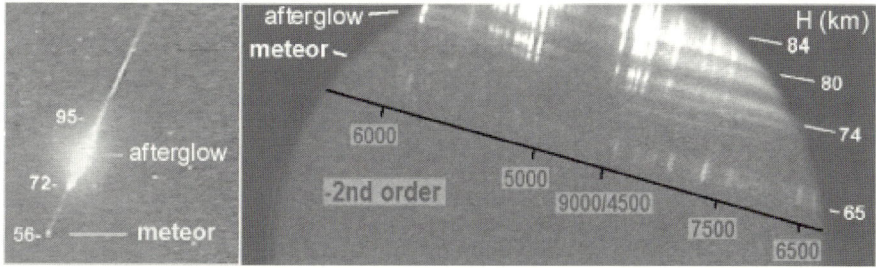

Figure 7. The phenomenon of afterglow in the 04:00:29 UT fireball (Nov. 18, 2000) as seen in single video frames from an imager (left) and a slit-less spectrograph (right). While the meteor continues to lower altitude (H), secondary ablation causes a spectrum devoid of certain air plasma emissions that are present in the meteor spectrum. Courtesy: Borovička and Jenniskens (2000).

Once the afterglow has faded, a chemiluminescence remains, which is called a persistent train. Persistent trains can be observed for minutes to hours after a bright fireball, and they come in tubular shapes with two much different amounts of billowing (Fig. 8). About 1-2 minutes after a Leonid fireball, the meteoric ablation metal Na was measured by Lidar to have a temperature of 240 and 270 K at the edges of the tubular structure, 20 - 50K above the ambient temperature (Chu et al. 2000a), consistent with a T $\sim$ 300 K temperature measured from the CO Infrared band profile in the path of a persistent train (Russell et al. 2000).

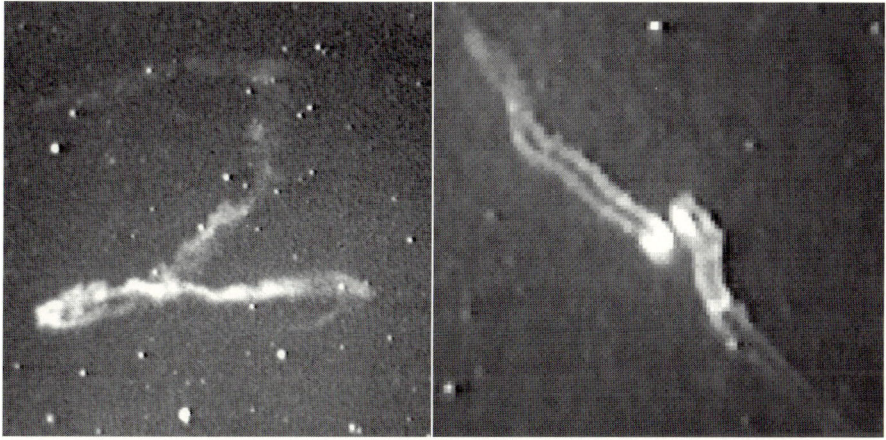

Figure 8. Leonid persistent trains of billowing (left) and tubular type (right). Images courtesy Rick Rairden and Sandy Osborough, respectively.

Zinn et al. (2000) developed a model that considered the rapid deposition of kinetic energy in air, with photodissociation and photoionization by UV photons from a T = 8,000 K shock. However, observations show that the forbidden OI line emission predicted by Zinn et al. quickly decays in a few tens of seconds. Also, earlier identifications of highly ionised O and S emissions (Borovička et al. 1996) are now thought to be in error and rather due to afterglow emissions of neutral meteoric metal atoms.

John Plane (in Jenniskens et al. 2000f) and preliminary work by Kelley et al. (2000) present dynamical models based on the Chapmann airglow mechanism that describe the recombination of internal oxygen atoms in the meteor train with ambient ozone in order to explain the tubular structure that is apparent in many trains (Fig. 8).

The models do not yet account for the absence of emission in the center of what is thought to be the tubular walls. Also, Kelley et al. (2000) pointed out that the total measured brightness is much higher than can be provided by sodium chemiluminescence alone. However, new spectroscopic measurements of persistent trains much beyond the afterglow stage, compared to laboratory measurements, identify FeO molecular emission as the dominant source of visible luminosity from persistent trains (Jenniskens et al. 2000e; Borovička and Jenniskens 2000). The FeO emission is consistent with Chapman type airglow chemistry reactions, whereby meteoric iron atoms catalyse the recombination of ozone and oxygen atoms. The dark inner region may be explained if the luminosity originates from a very thin layer between the warm inside and the ambient environment. Such thin walls are observed in high-resolution images of persistent trains and are consistent with the rocket measurement. In addition, turbulence may account for some of the unusual dynamic properties of trains (Jenniskens et al. 2000e). Indeed, many trains are dominated by billowing features (Fig. 8), sometimes only along part of the trajectory. It is not clear yet under what conditions a tubular rather than a billowing train is formed.

VII. ATMOSPHERIC PHENOMENA

The persistent trains of bright Leonids have provided a wealth of information on diffusion rates, turbulence and vertical wind profiles at altitudes of 75-100 km (Höffner et al. 1996; Grime et al. 2000; Jenniskens and Rairden 2000; Jenniskens et al. 2000f). No other shower has similarly frequent intense persistent emissions. Leonids are unique, too, in depositing matter at unusually high altitudes between 95 and 115 km. Such altitude discrimination is apparent in Lidar ranging and is also expected to selectively affect the natural airglow.

During the 1998 and 1999 Leonid showers, Lidar ranging has produced only very tentative changes in the density of the neutral atom layers. Chu et al. (2000b) report measurements for iron and Höffner et al. (2000) report measurements for potassium. Unfortunately, no measurements were obtained at the time of the 1999 storm itself and further studies are needed.

The November 1999 storm provided the first evidence of enhanced OH airglow that closely followed the meteoric influx (Kristl et al. 2000). In addition, Despois et al. (2000) reported tantalizing changes in the abundance of the upper atmosphere trace compound HCN in the night after the storm from sub-mm observations in Hawaii. No changes were found in the 1.27 μm O_2 singlet-delta emission line (Kristl et al. 2000) or the sodium airglow (Brosch and Shemmer 2000). These results need confirmation to understand the possible role of gravity and tidal waves.

A very interesting result is the detection of atmospheric trace gasses, such as CO, in the mid-IR emission spectra of persistent trains (Russell et al. 2000). The local warming of the ambient gas by the Leonid meteors may provide accurate measurements of such compounds at these altitudes for the first time. It is possible, however, that the observed emissions are the result of meteor induced aerothermochemistry.

Other interesting atmospheric phenomena may be associated with the surprisingly short ELF/VLF signals with an unusual frequency spectrum that were reported by Price and Blum (2000) and ascribed to meteors. Sprites and elves were detected at the peak of the 1999 Leonid storm (Jenniskens et al. 2000b). And, finally, Revelle et al. (2000) detected the sonic boom of a very bright Leonid fireball, despite the high altitude of its ablation, from which a luminous efficiency of 1.7 % was derived with a factor of 3 uncertainty.

Further Leonid storms are anticipated for the 2001 and 2002 encounters. The chance to follow up on the 1999 Leonid storm observations carries the promise of continued discovery and is expected to be of tremendous importance for making progress on many of the issues discussed in this chapter.

Acknowledgements

Bob Hawkes and Iwan Williams contributed to this paper by helpful peer review. The 1998 and 1999 Leonid MAC missions were made possible with financial support from NASA's Exobiology, Planetary Astronomy, and Suborbital MITM programs, NASA's Advanced Missions and Technology program for Astrobiology, NASA Ames Research Center, the National Science Foundation, and the US Air Force.

REFERENCES

Arlt, R. 1998. Bulletin 13 of the International Leonid Watch: The 1998 Leonid meteor shower. *WGN, Journal of the IMO,* **26,** pp. 239–248.
Arlt, R., Bellot Rubio, L., Brown, P., and Gyssens, M. 1999. Bulletin 15 of the International Leonid Watch: First Global Analysis of the 1999 Leonid Storm. *WGN, Journal of the IMO,* **27,** pp. 286–295.
Asher, D. J. 1999. The Leonid meteor storms of 1833 and 1966. *MNRAS,* **307,** pp. 919–924.
Asher, D. J., Bailey, M. E., and Emel'yanenko, V. V. 1999. Resonant meteoroids from Comet Tempel-Tuttle in 1933: The cause of the unexpected Leonid outbursts in 1998. *MNRAS,* **304,** pp. L53–L57.
Beech, M., Brown, P., and Jones, J. 1995. The potential danger to space platforms from meteor storm activity. *Q. Journ. Roy. Astron. Soc.*, **36,** pp. 127–152.
Beech, M., Brown, P., Jones, J., and Webster, A. R. 1996. The danger to satellites from meteor storms. *Adv. Space Res.* **20,** pp. 1509–1512.
Bellot Rubio, L. R., Ortiz, J. L., and Sada, P. V. 2000a. Luminous efficiency in Hypervelocity Impacts from the 1999 Lunar Leonids. *Ap. J.*, **542,** pp. L65–L68.
Bellot Rubio, L. R., Ortiz, J. L., and Sada, P. V. 2000b. Observation and interpretation of meteoroid impact flashes on the Moon. *Earth, Moon and Planets,* **82–83,** pp. 575–598.
Betlem, H., Jenniskens, P., Spurný, P., Docters van Leeuwen, G., Miskotte, K., Ter Kuile, C. R., Zarubin, P., and Angelos, C. 2000. Precise trajectories and orbits of meteoroids from the 1999 Leonid Meteor Storm. *Earth, Moon and Planets,* **87–88,** pp. 277–284.
Betlem, H., Jenniskens, P., Van 't Leven, J., Ter Kuile, C., Johannink, C., Zhao, H., Lei, C., Li, G., Zhu, J., Evans, S., and Spurný, P. 1999. Very precise orbits of 1998 Leonid meteors. *Meteoritics & Plan. Sci.*, **34,** pp. 979–986.
Borovička, J. and Jenniskens, P. 2000. Time resolved spectroscopy of a Leonid fireball afterglow. *Earth, Moon and Planets,* **82–83,** pp. 399–428.
Borovička J., Stork, R., and Bocek, J. 1999. First results from video spectroscopy of 1998 Leonid meteors. *Meteoritics & Plan. Sci.*, **34,** pp. 987–994.
Borovička J., Zimnikoval, P., Skvarka, J., Rajchl, J., and Spurný, P. 1996. The identification of nebular lines in the spectra of meteor trains. *Astron. Astrophys.*, **306,** pp. 995-998.
Boyd, I. D. 2000. Computation of atmospheric entry flow about a Leonid meteoroid. *Earth, Moon and Planets,* **82–83,** pp. 93–108.
Brosch, N. and Shemmer, O. 2000. Airglow and meteor rates over Israel during the 1999 Leonid shower. *Earth, Moon and Planets,* **82–83,** pp. 535–543.
Brown, P. 1999. The Leonid meteor shower: historical visual observations. *Icarus,* **138,** pp. 287–308.
Brown, P. 2000. *Evolution of Two Periodic Meteoroid Streams: the Perseids and Leonids.* Ph.D. Thesis. The University of Western Ontario (Canada), 286 pages.
Brown, P., Campbell, M. D., Ellis, K. J., Hawkes, R. L., Jones, J., Gural, P., Babcock, D., Barnbaum, C., Bartlett, R. K., Bedard, M., Bedient, J., Beech, M., Brosch, N., Clifton, S., Connors, M., Cooke, B., Goetz, P., Gaines, J. K., Gramer, L., Gray, J., Hildebrand, A. R., Jewell, D., Jones, A., Leake, M., LeBlanc, A. G., Looper, J. K., McIntosh, B. A., Montague, T., Morrow, M. J., Murray, I. S., Nikolova, S., Robichaud, J., Spondor, R., Talarico, J., Theijsmeijer, C., Tilton, B., Treu, M., Vachon, C., Webster, A. R., Weryk, R., and Worden, S. P. 2000. Global ground-based electro-optical and radar observations of the 1999 Leonid shower: First results. *Earth, Moon and Planets* **82–83,** pp. 167–190.
Brown, P., Simek, M., and Jones, J. 1997a. Radar observations of the Leonids: 1964–1995. *Astron. Astrophys.*, **322,** pp. 687–695.
Brown, P., Simek, M., Jones, J., Artl, R., Hocking, W. K., and Beech, M. 1997b. Observations of the 1996 Leonid meteor shower by radar, visual and video techniques. *MNRAS,* **300,** pp. 244-250.
Campbell, M. D., Brown, P. G., Leblanc, A. G., Hawkes, R. L., Jones, J., Worden, S. P., and Correll, R. R. 2000. Image-intensified video results from the 1998 Leonid shower. I. Atmospheric trajectories and physical structure. *Meteoritics & Plan. Sci.*, **35,** pp. 1259–1267.

Casswell, R. D., McBride, N., and Taylor, A. D. 1995. Olympus end of life anomaly – a Perseid meteoroid impact event? *Int. J. Impact Eng.*, **17**, pp. 139–150.

Ceplecha Z., Borovička J., Elford, W. G., Revelle, D. O., Hawkes, R. L., Porbucan, V., and Simek, M. 1998. Meteor phenomena and bodies. *Space Sci. Rev.*, **84**, pp. 327–471.

Cevolani, G., and Foschini, L. 1998. The effects of meteoroid stream enhanced activity on human space flight: an overview. *Planet. Space Sci.*, **46**, pp. 1597–1604.

Chambers, J. E. 1997. Shy Halley-types resonate but long-period comets don't: a dynamical distinction between short and long-period comets. *Icarus*, **125**, pp. 32–38.

Chu, X., Liu, A. Z., Papen, G., Gardner, C. S., Kelley, M., Drummond, J., and Fugate, R. 2000a. Lidar observations of elevated temperatures in bright chemiluminescent meteor trails during the 1998 leonid shower. *Geophys. Res. Lett.*, **27**, pp. 1815–1818.

Chu, X., Pan, W., Papen, G., Gardner, C. S., and Swenson, G. 2000b. Characteristics of Fe Ablation Trails Observed During the 1998 Leonid meteor shower. *Geophys. Res. Lett.*, **27**, pp. 1807–1810.

de Lignie, M. C., Langbroek, M., Betlem H., and Spurný P. 2000. Temporal variation in the orbital element distribution of the 1998 Leonid outburst. *Earth, Moon and Planets*, **82–83**, pp. 295–304.

Despois, D., Ricaud, P., Lautié, N., Schneider, N., Jacq, T., Biver, N., Lis, D. C., Chamberlin, R. A., Phillips, T. G., Miller, M., and Jenniskens, P. 2000. Search for extraterrestrial origin of atmospheric trace molecules - radio sub-mm observations during the Leonids. *Earth, Moon and Planets*, **82–83**, pp. 129–140.

Dunham, D. W., Cudnik, B., Palmer, D. M., Sada, P. V., Melosh, J., Frankenberger, M., Beech, R., Pellerin, L., Venable, R., Asher, D., Sterner, R., Gotwols, B., Wun, B., and Stockbauer, D. 2000. The first confirmed video recordings of lunar meteor impacts. *Lunar and Planetary Science XXXI, LPC 2000*, (abstract).

Fujiwara, V., Ueda, M., Shiba, Y., Sugimoto, M., Kinoshita, M., Shimoda, C., and Nakamura, T. 1998. Meteor luminosity at 160 km altitude from TV observations for bright Leonid meteors. *Geophys. Res. Lett.*, **25**, pp. 285–288.

Göckel, C. and Jehn, R. 2000. Testing cometary ejection models to fit the 1999 Leonids and to predict future showers. *MNRAS*, **317**, pp. L1–L5.

Grime, B. W., Kane, T. J., Liu, A., Papen, G., Gardner, C. S., Kelley, M. C., Kruschwitz, C., and Drummond, J. 2000. Meteor trail advection observed during the 1998 Leonid shower. *Geophys. Res. Lett.*, **27**, pp. 1819–1822.

Gural, P. and Jenniskens, P. 2000. Leonid Storm Flux Analysis from one Leonid MAC Video AL50R. *Earth, Moon and Planets*, **82–83**, pp. 221–247.

Hainaut, O. R., Meech, K. J., Boehnhardt, H., and West, R. M. 1998. Early recovery of comet 55P/Tempel-Tuttle. *Astron. Astrophys.*, **333**, pp. 746–752.

Hawkes, R. L. and Jones, J. 1975. A quantitative model for the ablation of dustball meteors. *MNRAS*, **173**, pp. 339–356.

Höffner, J., Fricke-Begemann, C., and von Zahn, U. 2000. Note on the reaction of the upper atmosphere potassium layer to the 1999 Leonid meteor storm. *Earth, Moon and Planets*, **82–83**, pp. 555–574.

Höffner, J., von Zahn, U., McNeil, W. J. and Murad, E. 1999. The 1996 Leonid shower as studied with a potassium lidar: observations and inferred meteoroid sizes. *J. Geophys. Res.*, **104**, pp. 2633–2634.

Hunten, D. M., Cremonese, G., Sprague, A. L., Hill, R. E., Verani, S., and Kozlowski, R. W. H. 1998. The Leonid Meteor Shower and the Lunar Sodium Atmosphere. *Icarus*, **136**, pp. 298–303.

Jenniskens, P. 1994. High Leonid Activity on November 17-18 and 18-19, 1994. *WGN, Journal of the IMO*, **22**, pp. 194-198.

Jenniskens, P. 1995. Meteor stream activity II. Meteor outbursts. *Astron. Astrophys.*, **295**, pp. 206–235.

Jenniskens, P. 1996. Meteoroid stream activity. III. Measurements of the first in a new series of Leonid outburst. *Meteoritics & Plan. Sci.*, **34**, pp. 177–184.

Jenniskens, P. 1997. Meteor stream activity. IV. Meteor outbursts and the reflex motion of the Sun. *Astron. Astrophys.*, **317**, pp. 953–961.

Jenniskens, P. 1998. On the dynamics of meteoroid streams. *Earth, Planets Space*, **50**, pp. 555–567.

Jenniskens, P. 1999a. Activity of the 1998 Leonid shower from video records. *Meteoritics Planet. Sci.*, **34**, pp. 959–968.

Jenniskens, P. 1999b. Update on the Leonids. *Adv. Space. Res.*, **23**, pp. 137–147.

Jenniskens, P., and Betlem, H. 2000. Massive remnant of evolved cometary dust trail detected in the orbit of Halley-type comet 55P/Tempel-Tuttle. *Ap. J.*, **531**, pp. 1161–1167.

Jenniskens P., Betlem, H., de Lignie, M., ter Kuile, C., van Vliet, M. C. A., van 't Leven, J., Koop, M., Morales, E., and Rice, T. 1998. On the unusual activity of the Perseid meteor shower (1989-96) and the dust trail of comet 109P/Swift-Tuttle. *MNRAS*, **301**, pp. 941–954.

Jenniskens, P., and Butow, S. J. 1999. The 1998 Leonid multi-instrument aircraft campaign – an early review. *Meteoritics Planet. Sci.*, **34**, pp. 933–943.

Jenniskens, P., Butow, S. J., and Fonda, M. 2000b. The 1999 Leonid Multi-Instrument Aircraft Campaign – an early review. *Earth, Moon and Planets*, **82–83**, pp. 1–26.

Jenniskens, P., Crawford, C., Butow, S. J., Nugent, D., Koop, M., Holman, D., Houston, J., Jobse, K., Kronk, G., and Beatty, K. 2000a. Lorentz shaped comet dust trail cross section from new hybrid visual and video meteor counting technique - implications for future Leonid storm encounters. *Earth, Moon and Planets*, **82–83**, pp. 191–220.

Jenniskens, P., de Lignie M., Betlem, H., Borovička, J., Laux, C. O., Packan, D., and Krueger, C. H. 1998. Preparing for the 1998/99 Leonid Storms. *Earth, Moon and Planets*, **80**, pp. 311–341.

Jenniskens, P., Lacey, M., Allan, B. J., Self, D. E., and Plane, J. M. C. 2000e. FeO "Orange Arc" emission detected in optical spectrum of Leonid persistent train. *Earth, Moon and Planets*, **82–83**, pp. 429–438.

Jenniskens, P., and Lyytinen, E. 2000. Possible outburst on December 22, 2000. *WGN, the Journal of IMO*, **28**, pp. 221–226.

Jenniskens, P., Nugent, D., and Plane, J. C. M. 2000f. The dynamical evolution of a tubular Leonid persistent train. *Earth, Moon and Planets*, **82–83**, pp. 471–488.

Jenniskens, P., Nugent, D., Tedesco E., and Murthy J. 2000c. 1997 Leonid shower from space. *Earth, Moon and Planets*, **82–83**, pp. 305–312.

Jenniskens, P., and Rairden, R. L. 2000. Buoyancy of the "Y2K" persistent train and the trajectory of the 04:00:29 UT Leonid fireball. *Earth, Moon and Planets*, **82–83**, pp. 457–470.

Jenniskens, P., Wilson, M. A., Packan, D., Laux, C. O., Krueger, C. H., Boyd, I. D., Popova, O. P., and Fonda, M. 2000d. Meteors: A delivery mechanism of organic matter to the early Earth. *Earth, Moon and Planets*, **82–83**, pp. 57–70.

Kazimircak-Polonskaja, E. I., Beljaev, N. A., Astapovic, I. S., and Terenteva, A. K. 1968. Investigation of perturbed motion of the Leonid meteor stream. In *Physics and Dynamics of Meteors, IAU Symposium 33*, eds. L. Kresak, P. M. Millman (Dordrecht: D. Reidel, pp. 449–475.

Kelley, M. C., Alcala, C., and Cho, J. Y. N. 1998. Detection of a meteor contrail and meteoric dust in the Earth's upper mesosphere. *J. Atmosph. and Solar-Terr. Phys.*, **60**, pp. 359–369.

Kelley, M. C., Gardner, C., Drummond, J., Armstrong, T., Liu, A., Chu, X., Papen, G., Kruschwitz, C., Loughmiller, P., Grime, B., and Engelman, J. 2000. First observations of long-lived meteor trains with resonant Lidar and other optical instruments. *Geophys. Research Lett.*, **27**, pp. 1811–1814.

Kinoshita, M., Maruyama, T., and Sagayama, T. 1999. Preliminary activity of Leonid meteor storm observed with a video camera in 1997. *Geophys. Res. Lett.*, **26**, pp. 41–44.

Kondrat'eva, E. D., and Reznikov, E. A. 1985. Comet Tempel-Tuttle and the Leonid Meteor Swarm. *Solar System Research*, **19**, pp. 96–101.

Krésak, L. 1993. Cometary dust trails and meteor storms. *Astron. Astrophys.*, **279**, pp. 646–660.

Kristl, J., Esplin, M., Hudson, T., and Siefring, C. L. 2000. Preliminary data on variations of OH airglow during the Leonid 1999 meteor storm. *Earth, Moon and Planets*, **82–83**, pp. 525–534.

LeBlanc, A. G., Murray, I. S., Hawkes, R. L., Worden, P., Campbell, M. D., Brown, P., Jenniskens, P., Correll, R. R., Montague, T., and Babcock, D. D. 2000. Evidence for transverse spread in Leonid meteors. *MNRAS*, **313**, pp. L9–L13.

Lewin, B. J. 1961. *Physikalische Theorie der Meteore und die Meteoritische Substanz im Sonnensystem.* Teil II, (Berlin: Akademie Verlag).

Littmann, M. 1998. *The heavens on fire: the great Leonid meteor storms.* (Cambridge: Cambridge University Press).

Lovell, A. C. B. 1954. *Meteor Astronomy.* (Oxford: Clarendon Press).

Lyytinen, E. 1999. Leonid predictions for the years 1999-2007 with the satellite model of comets. *Meta Research Bulletin*, **8**, pp. 33–40.

Lyytinen, E. J., and van Flandern, T. 2000. Predicting the strength of Leonid outbursts. *Earth, Moon and Planets*, **82–83**, pp. 149–166.

Mateshvili, N., Mateshvili, I., Mateshvili, G., Gheondjian, L., and Kapanadze, Z. 2000. Dust particles in the atmosphere during the Leonid meteor showers of 1998 and 1999. *Earth, Moon and Planets*, **82–83**, pp. 489–504.

McBride, N., and McDonnell, J. A. M. 1998. Meteoroid impacts on spacecraft: sporadics, streams, and the 1999 Leonids. *Plan. Space Sci.*, **47**, pp. 1005–1013.

McDonnell J. A. M., McBride, N., and Gardner, D. J. 1997. The Leonid meteoroid stream: Spacecraft interactions and effects. In Proceedings of the "Second European Conference on Space Debris" ESA SP-393, pp. 391–396.

McNaught, R. H., and Asher, D. J. 1999. Leonid Dust Trails and Meteor Storms. *WGN, Journal of the IMO*, **27**, pp. 85–102.

Murray, I. S., Beech, M., Taylor, M. J., Jenniskens, P., and Hawkes, R. L. 2000. Comparison of 1998 and 1999 Leonid light curve morphology and meteoroid structure. *Earth, Moon and Planets*, **82–83**, pp. 351–367.

Murray, I. S., Hawkes, R. L., and Jenniskens, P. 1999. Airborne intensified charge couple device observations of the 1998 Leonid shower. *Meteoritics & Plan. Sci.*, **34**, pp. 949–958.

Nakamura, R., Fujii Y., Ishiguro, M., Morishige, K., Yokogawa, S., Jenniskens, P., and Mukai, T. 2000. The discovery of a faint glow of scattered sunlight from the dust trail of the Leonid parent comet 55P/Tempel-Tuttle. *Ap. J.*, **540**, pp. 1172–1176.

Nemtchinov, I. V., Shuvalov, V. V., Artem'eva, N. A., Ivanov, B. A., Kosarev, I. B., and Trubetskaya, I. A. 1998. Light flashes caused by meteoroid impacts on the Lunar surface. *Solar System Research*, **32**, pp. 99–114.

Popova, O. P., Sidneva, S. N., Shuvalov, V. V., and Strelkov, A. S. 2000. Screening of meteoroids by ablation vapor in high-velocity meteors. *Earth, Moon and Planets*, **82–83**, pp. 109–128.

Price, C., and Blum, M. 2000. ELF/VLF radiation produced by the 1999 Leonid meteors. *Earth, Moon and Planets*, **82–83**, pp. 549–559.

Rairden, L. R., Jenniskens, P., and Laux, C. O. 2000. Search for organic matter in Leonid meteoroids. *Earth, Moon and Planets*, **82–83**, pp. 71–80.

Revelle, D. O., Whitaker, R. W., Armstrong, W. T., Mutschlecner, J. P., Sandoval, T. D., and Bueck, N. 1999. Infrasonic detection of a Leonid meteor. *Meteoritics & Plan. Sci.*, **34**, pp. 995–1005.

Rietmeijer, F. J. M. 2000. Interrelationships among meteoric metals, meteors, interplanetary dust, micrometeorites, and meteorites. *Meteoritics & Plan. Sci.*, **35**, pp. 1025–1041.

Rietmeijer, F. J. M., and Jenniskens, P. 2000. Recognizing Leonid meteoroids among the collected stratospheric dust. *Earth, Moon and Planets*, **82–83**, pp. 505–524.

Rietmeijer, F. J. M., and Nuth, J. A. 2000. Collected extraterrestrial materials: constraints on meteor and fireball compositions. *Earth, Moon and Planets*, **82–83**, pp. 325–350.

Rossano, G. S., Russell, R. W., Lynch, D. K., Tessensohn, T. K., Warren, D., and Jenniskens, P. 2000. Observations of Leonid meteors using a mid-wave infrared imaging spectrograph. *Earth, Moon and Planets*, **82–83**, pp. 81–92.

Russell, R. W., Rossano, G. S., Chatelain, M. A., Lynch, D. K., Tessensohn, T. K., Abendroth, E., Kim, D., and Jenniskens, P. 2000. Mid-Infrared Spectroscopy of Persistent Leonid Trains. *Earth, Moon and Planets*, **82–83**, pp. 439–456.

Sekanina, Z. 1975. Meteoric storms and formation of meteor streams. In *Asteroids, Comets, Meteoric Matter*, eds. C. Cristesu, W. J. Klepczynski, and B. Millet (Bucharest: Editura Academiei Republicii Socialiste Romania), pp. 239–267.

Singer, W., Molau, S., Rendtel, J., Asher, D. J., Mitchell, N. J., and von Zahn, U. 2000. The 1999 Leonid meteor storm: verification of rapid activity variations by observations at three sites. *MNRAS*, **318**, pp. L25–L29.

Smith, S. M., Wilson, J. K., Baumbardner, J., and Mendillo, M. 1999. Discovery of the distant lunar sodium trail and its enhancement following the Leonid meteor shower of 1998. *Geophys. Res. Lett.*, **26,** pp. 1649–1652.

Spurný P., Betlem, H., Jobse, K., Koten P., van 't Leven, J., 2000b. New type of radiation of bright Leonid meteors above 130 km. *Meteoritics & Plan. Sci.*, **35,** pp. 1109–1115.

Spurný P., Betlem, H., van 't Leven J., and Jenniskens, P. 2000a. Atmospheric behavior and extreme beginning heights of the thirteen brightest photographic Leonid meteors from the ground-based expedition to China. *Meteoritics & Plan. Sci.*, **35,** pp. 243–249.

Steel, D. 1994. Meteoroid streams. In *Asteroids, Comets, Meteors 1993*, eds. A. Milani, M. di Martino, A. Cellino (Dordrecht: Kluwer Acad. Publ.), pp. 111–126.

Svedhem, H., Drolshagen, G., and Grün, E. 2000. In IAU Coll. 181, *Dust in the Solar System and Other Planetary Systems* (abstracts), 82.

Taylor M. J., Gardner, L. C., Murray, I. S., and Jenniskens, P. 2000. Jet-like structures and wake in MgI (518 nm) images of 1999 Leonid storm meteors. *Earth, Moon and Planets*, **82–83,** pp. 379–389.

Von Zahn, U., Gerding, M., Höffner, J., McNeill, W. J., and Murad, E. 1999. Iron, calcium and potassium atom densities in the trails of Leonids and other meteors: strong evidence for differential ablation. *Meteoritics & Plan. Sci.*, **34,** pp. 1017–1027.

Williams, I. P. 1993. The dynamics of meteoroid streams. In *Meteoroids and Their Parent bodies*, eds. J. Stohl and I. P. Williams (Bratislava: Astron. Inst. Slovak Acad. Sci.), pp. 31–40.

Williams, I. P. 1997. The Leonid meteor shower: why are there storms but no regular annual activity? *MNRAS*, **292,** pp. L37–L40.

Wilson, J. K., Smith, S., Baumgardner, J., and Mendillo, M. 1999. Modeling an enhancement of the lunar sodium tail during the Leonid meteor shower of 1998. *Geophys. Res. Lett.*, **26,** pp. 1645–1648.

Wu, Z., and Williams, I. P. 1993. The Perseid meteor shower at the current time. *MNRAS*, **264,** pp. 980–990.

Wu, Z., and Williams, I. P. 1996. Leonid meteor storms. *MNRAS*, **280,** pp. 1210–1218.

Yeomans, D. K. 1981. Comet Tempel-Tuttle and the Leonid Meteors. *Icarus*, **47,** pp. 492–499.

Yeomans, D. K., Yau, K. K., and Weissmann, P. R. 1996. The impending appearance of comet Tempel-Tuttle and the Leonid meteors. *Icarus*, **124,** pp. 407–413.

Zinn, J., Wren, J., Whitaker, R., Szymanski, J., ReVelle, D. O., Priedhorsky, W., Hills, J., Gisler, G., Fletcher, S., Casperson, D., Bloch, J., Balsano, R., Armstrong, W. T., Akerlof, C., Kehoe, R., McKay, T., Lee, B., Kelley, M. C., Spalding, R. E., and Marshall, S. 2000. Coordinated observations of two large Leonid meteor fireballs over northern New Mexico, and computer model comparisons. *Meteoritics & Plan. Sci.*, **34,** pp. 1007–1015.

Properties of Interplanetary Dust: Information from Collected Samples

Elmar K. Jessberger[1], Thomas Stephan[1], Detlef Rost[1], Peter Arndt[2],
Mischa Maetz[2], Frank J. Stadermann[3], Don E. Brownlee[4], John P. Bradley[5],
Gero Kurat[6]

[1] Institut für Planetologie, Münster, Germany
[2] Max-Planck-Institut für Kernphysik, Heidelberg, Germany
[3] Washington University, St. Louis, U.S.A.
[4] University of Washington, Seattle, U.S.A.
[5] MVA. Inc. and Georgia Institute of Technology, Atlanta, U.S.A.
[6] Naturhistorisches Museum, Wien, Austria

Abstract. The properties of hundreds of interplanetary particles have been determined by direct laboratory analysis of recovered samples. The particles that span the 1 μm to 1 mm size range have been collected from the stratosphere, from polar ice, and from deep sea sediments. Typically, these particles are black, somewhat porous and have chondritic elemental compositions. They are rather complex mineral assemblages in that they are mixtures of very large numbers of sub-micrometer-sized components. While the data are not totally representative of small interplanetary meteoroids at 1 AU they provide significant insight into the common physical properties of meteoroids. These properties can be used as guidelines for analysis of spacecraft and astronomical observations and for modeling solar system dust as well as some circumstellar dust in systems around other stars.

I. INTRODUCTION

Knowledge of the properties of interplanetary dust particles (IDPs) is important for modeling the origin, evolution and nature of the meteoroid complex. It also plays an important role in understanding a variety of solar system processes including the formation and survival of the IRAS infrared dust bands, interpretation of the zodiacal light, processes in comet tails, *space weathering effects* on asteroids and alteration of surfaces of airless bodies. In a broader perspective, the properties of solar system dust also relate in important ways to circumstellar dust around other stars. This aspect has gained increasing importance with the discovery that dust systems around young and evolved stars are common and that many have dust systems analogous to the Sun's zodiacal dust complex. This is particularly true for the Vega-like systems where dust may be generated by comets in a manner similar to the Kuiper belt of the solar system (Weissman 1984). Particles are generated by larger bodies, their orbits decay by Poynting-Robertson drag, and they are also destroyed by hypervelocity collisions. These processes also dominate the evolution of interplanetary dust in the solar system and it is likely that dust properties are similar in the solar system and other circumstellar dust systems, particularly where dust may be generated by comets. At a minimum, the solar system dust properties are a reasonable analogue for the properties of dust in other circumstellar systems. In the solar system it is possible to collect dust samples and study them in detail with a wide variety of modern laboratory instrumentation. This direct information provides important insight into the limits of properties of particles in many different environments.

The idealized goal of this review would be to evaluate the various physical, chemical, isotopic, and mineralogical properties of interplanetary dust in the solar system. Due to a variety of complications, this lofty goal is impossible to accomplish with existing or even future samples. There may be several types and populations of particles that vary with size, location, orbital parameters and time. What can be done is to describe samples collected at Earth, in a limited size range, and over a limited time interval. Due to selective effects of Earth capture, of atmospheric survival, of collection, and even of personal preferences of the researcher, the collected samples are only approximately representative of particles currently colliding with the Earth.

This review contains information from two different types of collections of interplanetary particles. One is collected from the stratosphere and the other from Antarctic and Greenland ice. Although all of the particles are micrometeorites (unmelted particles) and interplanetary dust (IDP), it is common usage to refer to the stratospheric particles as IDPs and the polar ice particles as MMs (micrometeorites). Due to contamination and collection limitations, most of the particles collected in the atmosphere are limited to the 5 μm to 25 μm diameter range. Most of these small particles survive atmospheric entry without melting and they are only moderately altered by their capture from space and residence on Earth. They are probably fairly representative of the 15 μm diameter particle population at 1 AU, although gravitational focusing effects will enhance the proportion of low speed asteroid particles (Flynn 1994; Kortenkamp and Dermott 1998). Much larger extraterrestrial particles have been collected from deep sea sediments and from Greenland and Antarctic ice and they provide direct information on the meteoroids in the 20 μm to 1 mm size range (Kurat et al. 1994a; Maurette et al. 1994). The larger particles (> 50 μm) are too rare to be effectively collected in the stratosphere and they must be recovered from surface deposits where they are affected by a range of alteration processes that are not yet fully understood. These large particles are typically more strongly heated during atmospheric entry and these effects selectively alter or destroy denser, fragile, or higher velocity particles. On the ground they also require some selection processes to be distinguished from terrestrial materials. Whereas nearly all 15 μm particles survive atmospheric entry and can be collected in the stratosphere with probably only mild alteration (besides some heating) in the stratosphere (Brownlee 1981, 1985;

Warren and Zolensky 1994), the larger particles collected from the Earth's surface are a more biased sampling. Although biased, they do uniquely provide samples of larger dust particles of the 200 µm size range that actually dominate the mass of meteoroids in the interplanetary medium and the annual influx of extraterrestrial material to Earth (Kyte and Wasson 1986; Brownlee 1997).

II. ANTARCTIC AND GREENLAND MICRO-METEORITES

Large unmelted interplanetary dust particles (micrometeorites MMs) are available from Greenland ice (Maurette et al. 1986) and from Antarctica (Maurette et al. 1991, 1994, 1996). In Greenland, micrometeorites can be collected from "cryoconite", a dark sediment in melt water lakes, which consists of dust and cocoons of blue algae and siderobacteria. Cryoconite contains about 10 g/kg fine-grained sand and dust mostly of terrestrial origin and about 800 cosmic spherules and 200 unmelted to partially melted micrometeorites. In Antarctica, micrometeorites can be collected by intentionally melting blue ice. One ton of Antarctic blue ice contains about 100 cosmic spherules with diameters > 50 µm and about 500 unmelted to partially melted MMs 50–400 µm in diameter. Thus, large amounts of unmelted and almost unaltered samples of the interplanetary dust particles which contribute most to the recent accretion rate on Earth are available for study.

II.A. Mineralogy and Petrography of MMs

Many micrometeorites have experienced alteration by frictional heating in the atmosphere. They are partially or totally melted and consist of foamy glass with variable amounts of unmelted phases (scoriaceous MMs, cosmic spherules, respectively, Figs. 1a–2a). Others have been thermally metamorphosed but not melted, and surprisingly many retained their pristine mineralogy (unmelted MMs).

The pristine mineralogy of micrometeorites is remarkably simple (e.g., Maurette et al. 1991, 1993, 1994; Kurat et al. 1992, 1993, 1994a,b). Major minerals are olivine, low-Ca pyroxene, magnetite, and hydrous Mg-Fe silicates (phyllosilicates) like serpentine and saponite. The typical unmelted MMs are dense and low-porosity mixtures of anhydrous and hydrous phases in proportions ranging from all anhydrous (coarse-grained "crystalline" micrometeorites, Fig. 2b) to totally hydrous mineral assemblages (phyllosilicate micrometeorites, Fig. 3). Minor phases are Ca-rich pyroxenes, feldspars, Fe-Ni sulfides and metal, Mg-Fe hydroxides, MgAl and Fe-Cr spinels, perovskite, ilmenite, hibonite, and others (Kurat et al. 1994c; Beckerling and Bischoff 1995; Hoppe et al. 1995). A summary on the mineralogy, chemistry, and oxygen isotopic composition is given by Greshake et al. (1996). The major anhydrous silicates have variable

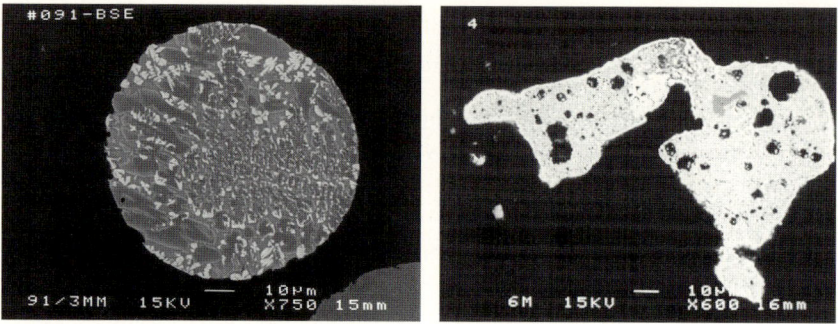

Figure 1a (left). Cosmic spherule from Antarctica, polished section, back scattered electron (BSE) scanning image, quench texture with olivine (dark gray to gray, compositionally zoned) and magnetite (white) in glassy matrix (gray). From mount 91/3 (particle 91).

Figure 1b (right). Scoriaceous micrometeorite 6M4 from Antarctica. BSE image of polished section shows magnetite-rich melt with vesicles (black) and an irregularly shaped relic olivine (gray). Note the magnetite cover.

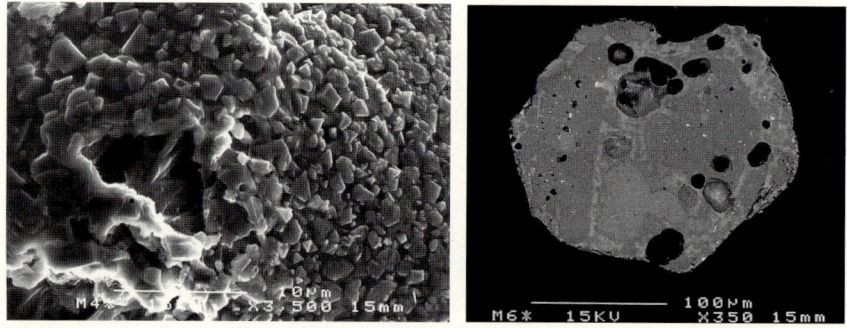

Figure 2a (left). Magnetite cover on micrometeorite M4 (Kurat et al. 1994) from Antarctica. Most thermally altered MMs are covered by magnetite.

Figure 2b (right). Crystalline micrometeorite M6 (Kurat et al. 1994a) from Antarctica. BSE image of polished section showing olivine (dark gray), pyroxene (gray) and dispersed sulfides and chromite (white). Such anhydrous mineral assemblages resemble olivine and olivine pyroxene aggregates in carbonaceous chondrites.

Fe/Mg ratios (unequilibrated mineral assemblage, Walter et al. 1994) and are usually very rich in minor elements as compared to their terrestrial counterparts and olivines from the most common meteorites, the ordinary chondrites. Ordinary chondrite matter is very rare among MMs and comprises less than 1 % by number (Walter et al. 1995). The phyllosilicates contain some elements in chondritic abundances (e.g., Ti, Al, Cr, Na, and K). Refractory minerals like Mg-Al spinel and perovskite are strongly enriched in refractory trace elements (e.g., rare earth elements, Sc, Zr, Hf, etc.; Kurat et al. 1994b, 1994d; Hoppe et al. 1995) as compared to chondritic rocks. Mineralogy, mineral chemistry, and the presence of refractory minerals in MMs are similar to those of CM-type

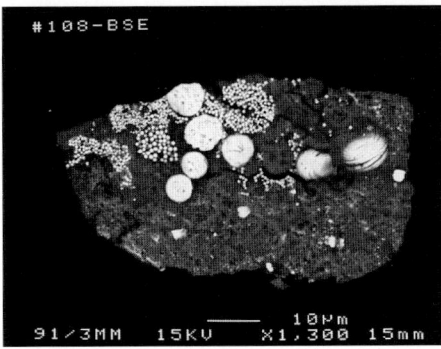

Figure 3. Phyllosilicate MM from Antarctica consisting of clay minerals (dark) and magnetite (framboidal and platy). This texture and mineral association are similar to those of CI and CM carbonaceous chondrites. BSE image. From mount 91/3 (particle 108).

(Mighei-type) and CR-type (Renazzo-type) carbonaceous chondrites. However, some differences between MMs and CM/CR chondrites are the abundance of Ca-poor pyroxene in MMs (most CM chondrites do not contain such pyroxenes) and the lack of very Fe-poor olivines with high Al and Ca contents in MMs (they are common in CM and CR chondrites).

II.B. Major, Minor and Trace Element Chemistry of MMs

Phyllosilicate-rich MMs are characterized by chondritic bulk major and minor element abundances, except for Ca, Na, Ni, and S that are depleted with respect to CI (and CM/CR) chondrites (Fig. 4a). The lithophile trace element abundances in phyllosilicate-rich MMs are similar to those in CM chondrites (that in turn are similar to those in CR chondrites), except for K which is overabundant. The abundances of siderophile elements in MMs deviate from those in CI and CM/CR chondrites. Only the refractory elements Os and Ir and the volatile Se have abundances similar to those in CI and CM/CR chondrites. The common siderophile elements Ni and Co are depleted with respect to chondritic abundances and are also fractionated from each other; the Ni/Co ratio is non-chondritic. Iron is somewhat enriched over chondritic abundances as are Au and As. The depletion of Ni, Co, and S is probably due to terrestrial leaching of Ni-bearing Mg-Fe sulfates from MMs (Presper et al. 1993). Micrometeorites do not contain sulfates, which, however, are abundant in CM and CI chondrites. Similarly, the Ca depletion of MMs compared to CM chondrites is probably due to leaching of carbonates, minerals that are common in CM chondrites but absent in MMs. The enrichments in MMs of Au, As, and K over chondritic abundances very likely are due to terrestrial contamination. These three elements are strongly enriched in the terrestrial crust as compared to chondrites.

Scoriaceous MMs have trace element abundances very similar to those of phyllosilicate MMs (Fig. 4b), except for large depletions and re-enrichments

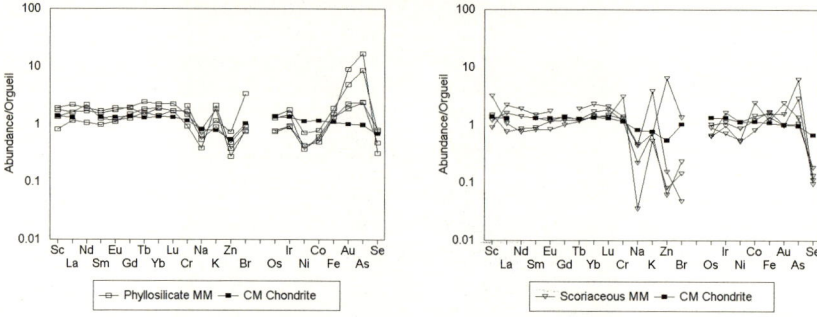

Figure 4a (left). CI-normalized selected element abundances in phyllosilicate MMs and CM chondrites (data from Kurat et al. 1994b; Anders and Grevesse 1989; Palme, pers. comm.). There is a close match between the compositions of MMs and CM chondrites for the refractory elements (Sc–Cr and Os, Ir). The siderophile element abundances are disturbed by loss of soluble minerals (sulfates containing Ni and Co) and by terrestrial contamination (Au and As).

Figure 4b (right). CI-normalized selected element abundances in scoriaceous MMs and CM chondrites (data sources as for Fig. 4a). The volatile elements (Na, K, Zn, Br, As and Se) are either depleted or enriched with respect to CI chondrites, a consequence of the heating event during atmospheric entry and subsequent contamination in the terrestrial environment.

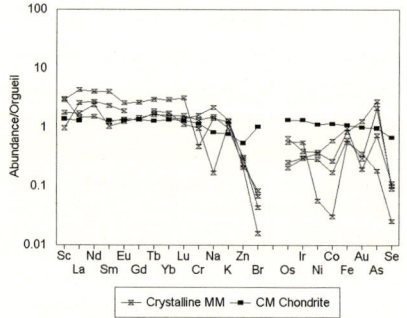

Figure 5. CI-normalized selected element abundances in coarse-grained crystalline MMs and CM chondrites (data sources as for Fig. 4a). Most fractionations with respect to CI chondrites are indigenous, except for the enrichment in As which is of terrestrial origin.

in volatile elements (Na, K, Zn, Br, As, Se). Clearly, scoriaceous MMs were formed from phyllosilicate meteoroids by heating during atmospheric entry.

Anhydrous crystalline MMs deviate in composition from chondrites, a feature typical also for anhydrous aggregates and chondrules in carbonaceous chondrites (Fig. 5). Elemental fractionations are usually not very severe for the refractory lithophile elements but are commonly strong for volatile and siderophile elements, a consequence of the conditions prevailing during aggregation.

Micrometeorites (and also IDPs) are surprisingly rich in carbon. Perreau et al. 1993 and Engrand et al. 1994 found C/O ratios that on average were higher than those in CI chondrites, the most C-rich chondrites. Micrometeorites are up to 5× richer in C than CM/CR chondrites.

II.C. Isotope Abundances in MMs

Several elements have isotopic compositions that are non-terrestrial and, in some cases, also non-solar, in both, MMs and IDPs (e.g., McKeegan 1987a, b; Stadermann 1990; Kurat et al. 1994d; Hoppe et al. 1995; Engrand et al. 1996; see also below). Specifically, anomalies in the isotopic abundances of H, C, N, and O are common, similar to carbonaceous chondrites.

II.D. Rare Gas Abundances in MMs

Interplanetary dust has been exposed to the solar wind and cosmic rays for a sufficiently long time to accumulate large amounts of solar noble gases and spallogenic isotopes. Micrometeorites have, for example, very high He (up to 10^{-1} cm^3 g^{-1} STP) and Ne contents – in excess of 10^{-5} cm^3 g^{-1} STP – comparable only to a few very gas-rich chondrites and to lunar soil (Olinger et al. 1990; Maurette et al. 1991; Nier 1994). He and Ne isotope abundances are similar to those of solar energetic particles (SEP) and thus confirm the extraterrestrial origin of IDPs and MMs (and of some cosmic spherules). A minor contribution of cosmic ray spallation Ne was also identified. Interplanetary dust (IDPs and MMs) clearly was exposed to cosmic rays and to the solar wind. The particles must have been of the size as recovered while they were exposed to the solar wind. Therefore, MMs and IDPs (Nier and Schlutter 1990) are true interplanetary dust meteoroids and cannot be atmospheric break-up products of a larger-sized meteoroid.

II.E. Conclusions from MM Studies

Micrometeorites represent the main mass of extraterrestrial matter accreting onto the Earth today. Whereas meteorites, that deliver only a few percent of the total incoming mass on Earth, are dominated by ordinary chondrites, micrometeorites bear some similarities to the rare CI/CM/CR carbonaceous chondrites. However, they also differ from them in so many aspects that they have to be considered as a solar system matter of its own. The features of MMs that are different from those of chondrites are likely to be of primordial origin. These include the mineral abundances, mineral chemistry and the bulk C content.

There is considerable overlap in mineralogical and chemical composition between MMs and IDPs (e.g., phyllosilicate dominated particles, olivine and pyroxene abundances and chemical compositions) but the fluffy, fine-grained olivine aggregates and GEMS that are abundant in IDPs are not found so far in MMs. However, the particles of sizes < 50 μm have not been investigated properly yet and that is where the crossover in mineralogical and petrographical features of MMs and IDPs can be expected. Deviations of the dust composition from that of chondrites are very probably due to extraction of water-soluble sulfates and carbonates and to contamination in the terrestrial environment.

III. STRATOSPHERIC INTERPLANETARY DUST

III.A. Shape and External Morphology

The vast majority of the typically 15 μm sized IDPs have shapes that can be modeled by rather equidimensional forms. The exteriors of most particles can be roughly approximated by ellipsoids with ratios of maximum length to minimum width that are less than two. Typical particles are aggregates of smaller components and a good model of the overall shape is a lumpy ellipsoid with second order surface modulations in the size range from 0.1 μm to a few microns. The irregularities consist of both cavities and protrusions. Highly non-regular shapes such as plates and rods are essentially nonexistent for whole particles although some subcomponents do have these shapes. Only fairly uncommon particles have wedge-shaped angular structures generally seen as fragments produced by comminution of solid rocks. Some of the collected 15 μm particles are spherical but in most cases this is thought to be due to atmospheric melting, a process experienced by many particles larger than 100 μm. Typically, the surface texture is related to the interior structure and the porosity. The most common IDPs have roughly chondritic elemental compositions and, as shown in Fig. 6, they have surface properties ranging from smooth (CS or chondritic smooth) to those that are highly rough and porous (CP or chondritic porous).

The CS particles are smooth at the micrometer scale and usually have solid non-porous interiors. Many of the CS particles are dominated by hydrated silicates. Their smooth surfaces and the presence of hydrated silicates are related, in part, to aqueous alteration and compaction processes that modified the precursor materials. Many of the CS particles are mineralogically similar to the matrix of CM and CI chondrites and phyllosilicate micrometeorites. Not all particles, however, that contain hydrated minerals have smooth exteriors. Some are porous with rough exteriors and some are rough due to the matrix construction of masses of layer silicates. Most particles with rough surfaces at the micrometer scale are of the CP type and they are usually aggregates of large numbers of sub-micrometer grains and components. Most of these are dominated by anhydrous minerals but some are composed of hydrated phases. The fundamental building blocks of the porous particles are usually grains of 0.3 μm size. They form aggregate structures that range in porosity and degree of filling materials that assist in bonding them together. The surfaces of most porous particles are highly irregular and often contain deep cavities that in some cases provide open line-of-sight pathways that pass entirely through the particle. In general, the porous particles with rough surfaces are simply aggregates of small particles clustered in random open fashion with no apparent order. There is no evidence that sub-grains were ever linked end to end forming dendritic networks as is seen in smoke particles formed by condensation.

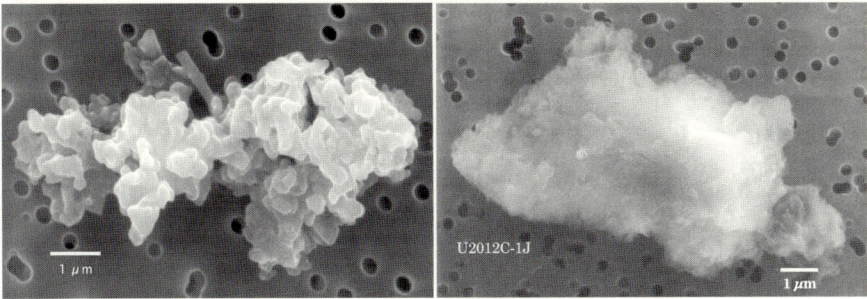

Figure 6. SEM images of (left) a porous chondritic IDP and (right) a smooth chondritic IDP.

Roughly 15 % of the IDPs are essentially single mineral grains or are simple assemblies of only a few mineral grains. Most commonly these are olivine, enstatite or pyrrhotite grains. These particles differ from the particles with chondritic composition in that they are usually more angular with broad planar surfaces. Most are anhedral although a few do show crystal faces.

III.B. Density

Particle density and area-to-mass ratio are important parameters for many processes involving interplanetary dust and it is common for calculations and models to assume a single nominal particle density. For real particles, however, there is a range of densities and even the meaning of their density is somewhat complex due to their irregular shapes. For non-spherical solid particles, the effective density will in most cases be less that the bulk density of the material that the particle is made of. For example, with higher average cross-section area, the Poynting Robertson drag force of irregular particles is higher than for spheres of the same mass and composition. The shape effects are particularly important for small particles because cavities and protrusions can make up an appreciable fraction of the particle diameter. When trying to determine the density there is often some ambiguity in what to use for its volume. In many cases, the volume can be well approximated by an imaginary rubber sheet that would surround the particle and contact its highest points.

The densities of particles can be measured in a variety of ways (Strait et al. 1995; Flynn and Sutton 1988; Fraundorf et al. 1982; Maetz et al. 1994). The most extensive study was the measurement of 150 chondritic 10 μm IDPs by Love et al. (1994). They estimated particle volumes by measuring the cross section of particles from SEM images and by measuring the heights by differential focus methods. Masses were measured by quantitative X-ray techniques that determined the total mass of elements heavier than oxygen. A weakness of the study was that C could not be measured and O had to be calculated by stoichiometry. This study found a range of 0.3 to 6.2 g cm^{-3} for chondritic composition particles with an average at 2.0 g cm^{-3} (Fig. 7). About 75 % of the measured densities were between 1 and 3 g cm^{-3}. Because the samples were collected in the stratosphere, the relative proportions of dense

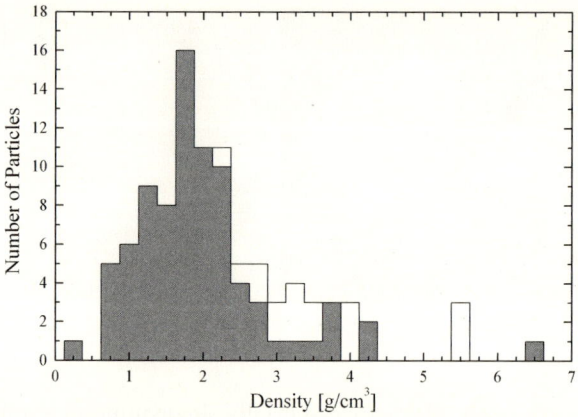

Figure 7. The density distribution of stratospheric IDPs measured by Love et al. (1994). The open bars refer to particles that are S-rich and contain abnormally large FeS grains.

and low-density particles are affected by the atmospheric fall speed. Dense particles fall at higher speeds and are relatively underrepresented. Attempting to correct for this effect gives a rough estimate of the mean density of particles impacting the top of the atmosphere of 2.8 g cm^{-3}, but the peak in the density distribution remains at 2.0 g cm^{-3} for 15 μm particles.

The Love et al. (1994) work covered only chondritic composition particles. It did not include non-chondritic particles such as those composed of single large mineral grains and it also did not include particles that fragmented during collection. The particles composed of single mineral grains are rare but they are usually non-porous and inclusion of these would shift the mean particle density to a slightly higher value. The most common dense particles are those dominated by FeS, which has a bulk density of 5 g cm^{-3}. Fragile particles that broke up during collection are perhaps the most porous IDPs and if their existence could be accurately accounted for, they would partly compensate for the also unmeasured solid mineral grain particles. It is significant that there are at least a few very highly porous IDPs in the stratospheric collections: it proves that they indeed do exist. They may be more abundant in space although it seems unlikely that they comprise a major fraction of 15 μm meteoroids.

Another method to study masses and densities of IDPs involves less assumptions, but is also more complicated, and consequently only fewer IDPs have been analyzed so far (Maetz et al. 1994; Arndt et al. 1996a): *S*canning *T*ransmission *I*on *M*icroscopy (STIM) measures the energy loss of few-MeV protons when they pass through matter. From the energy loss the mass to area ratio (area density) can directly be evaluated with only little dependence on the major element composition (Lefevre et al. 1987). By scanning a proton beam over the IDP one obtains the area density profile and thus also information on the porosity and on portions of the sample with varying density (Fig. 8a). With the very small proton beam sizes (< 250 nm) available at the

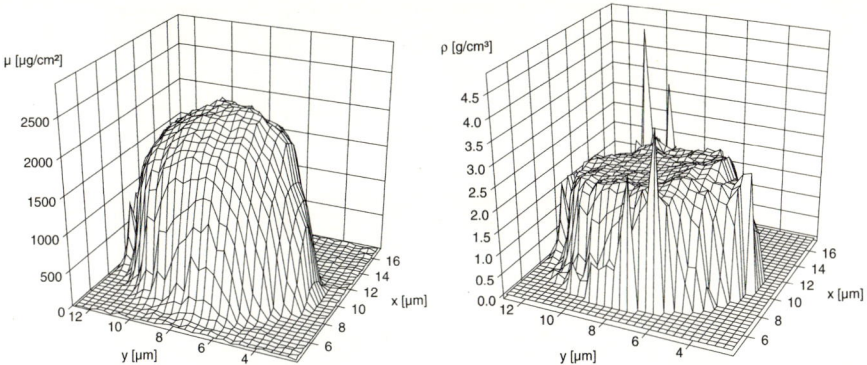

Figure 8a (left). Area density profile of IDP L2005B11 as obtained by scanning transmission ion spectroscopy.

Figure 8b (right). Density profile of IDP L2005B11 that results from combining the area density profile (Fig. 8a) and the figure of the IDP obtained from white light interferometry (resolution: 0.05 μm). The mass of the particle is 1.2 ng, its density 2.3 g cm^{-3}.

new Heidelberg Proton Microprobe a lateral resolution of 1 μm for typical particle diameter of 10–15 μm can be achieved (Arndt et al. 1997). By integration over the particle area the total IDP mass can be calculated with a detection limit of about $1-2 \cdot 10^{-12}$ g with 2 MeV protons (Maetz 1994). If the geometric surface profile and thus the volume of the IDP is known from, e.g., white light interference microscopy, both the mean density and the density profile can be calculated (Fig. 8b) (Maetz et al. 1994). Up to now this procedure resulted in the determination of four IDP masses ranging from 7.2 to 1191 pg and the densities of two IDPs, 1.0 and 2.3 g cm^{-3} (Maetz et al. 1994; Arndt et al. 1996a).

A remarkable result from particle density measurements is that common IDPs are much more porous than meteorites but they are not as highly porous as is sometimes assumed in modeling (Greenberg and Hage 1990; Greenberg and Gustafson 1981). The most common measured density is 2 g cm^{-3} consistent with moderate porosity near 40 %. Particles with a density below 1 g cm^{-3} do exist but they are rare. The scarcity of very low density particles is consistent with the basic construction of common IDPs. Most of the porous particles are simply loose aggregates of rather equidimensional sub-micrometer components. In totally random aggregates of similar sized components it is difficult to have a porosity significantly higher than 50 % because of the scarcity of contacts between grains to hold the particles together. As the density decreases the grain contacts along a transecting plane will reach zero and the particle becomes unbound. Microtome sections of very low density particles show very few grain-grain contacts in the section-plane and most of the grain-grain contacts that actually bind the particle together occur above and below the plane (Fig. 9).

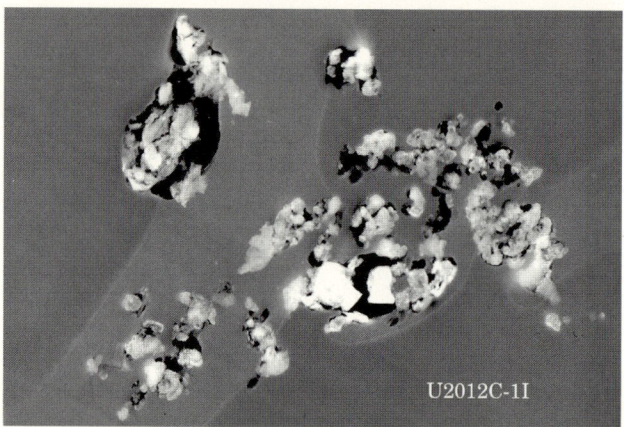

Figure 9. A TEM view of a 0.1 μm thin microtome section of a single porous IDP. The medium gray and black areas are epoxy and pore spaces, respectively, were material was torn out during cutting. The epoxy areas denote open space and show how unconnected the particle is on the section plane. Width of field is 10 μm.

Densities much lower than 1 g cm^{-3} have been suggested for meteoroids (Ceplecha 1977), but this does not seem possible for 15 μm particles based on the structures observed in collected particles. To obtain such low densities would require a level of organization not seen in the samples. For example, extreme densities would require void spaces considerably larger than the 0.3 μm size of the building blocks of the aggregates. This in turn would require either thin sheets of a binding material such as carbonaceous organic or an organizational framework where the sub-grains would be organized into non-random chains to form beads-on-a-chain structures. It is also noteworthy that the mean densities for IDPs obtained in the laboratory roughly coincides with the mean densities estimated from in-situ measurements in comet Halley's coma (Maas et al. 1989)

Although densities of particles that are much larger or smaller than 15 μm have not been widely measured, it is possible to make predictions on these based on the assumption that they are composed of the loose aggregate structures of 15 μm IDPs. The aggregate structure of IDPs suggests that as particles approach sizes smaller than a micron they will primarily be solid objects and have densities > 2 g cm^{-3}. For particles larger than 15 μm it is possible that increasingly larger void spaces will lead to increasing lower densities. If 100 μm particles are made by loose clustering of 15 μm particles, the larger particles will have lower densities. While it seems unlikely that an appreciable fraction of 15 μm IDPs can have densities < 1.5 g cm^{-3}, it is possible that larger particles could have densities < 1 g cm^{-3}. It is also possible that particles > 100 μm are different from smaller particles either because they are derived from different sources or have undergone a different set of selection processes in the interplanetary medium that allow them to survive and reach 1 AU.

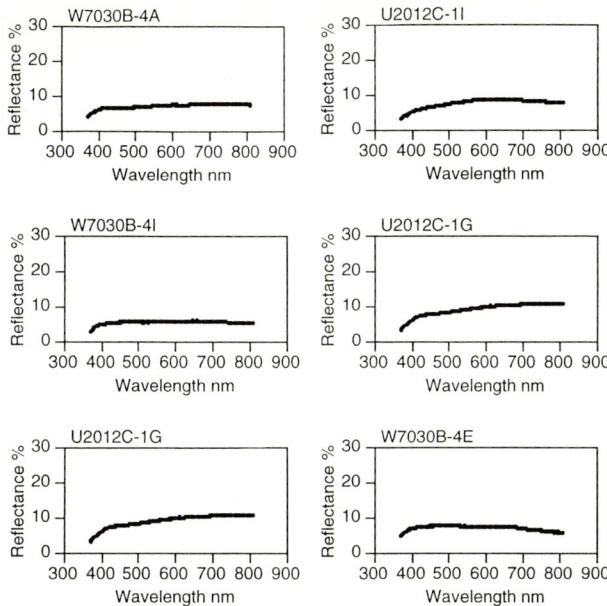

Figure 10. Reflectance spectra for individual 15 μm IDPs (Bradley et al. 1996).

At the same time, it is likewise possible that the much-larger-than-IDP particles represent the common meteorite constituents like chondrules and inclusions that are rarely found in IDPs.

III.C. Optical and Infrared Properties

In the optical microscope, most IDPs are opaque, fine-grained dark objects. Under close inspection, they are aggregate structures of mostly black irregular materials with occasional clear (silicate) grains and reflective (sulfide) grains ranging up to microns in size. Spectral reflectance measurements of individual stratospheric IDPs show that typical 15 μm IDPs have visible albedos in the range 5–15 % (Bradley et al. 1992; Bradley et al. 1996). Most have rather featureless reflectance curves except for downturns below 450 nm (Fig. 10). Most of the measured spectral reflectance curves resemble C type asteroids although some have increasing reflectance in the red, analogous to the P and D asteroid classes. The low albedo of typical IDPs is presumably due to their porous structure and the presence of strongly absorbing material. The absorbing materials include carbon, small sulfides and GEMS (sub-micrometer components composed of glass with embedded small metal and sulfide grains) (Bradley 1995). It is possible that the abundant nm-sized metal grains in GEMS are the major source of absorption in many of the IDPs. If this is the case, then the source of absorption in these particles would be very different from that in carbonaceous chondrites that are not reduced and do not contain abundant nanophase metal grains.

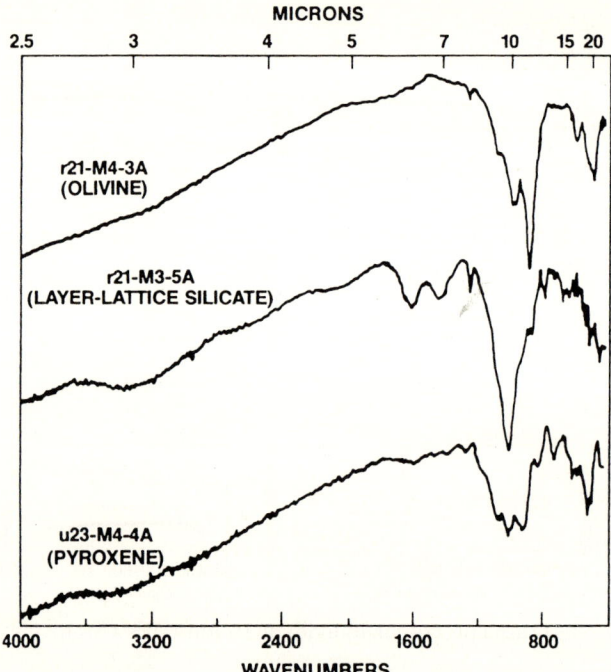

Figure 11. Infrared signatures of the three main infrared IDP groups showing differences in the 10 μm silicate feature (Sandford and Walker 1985).

The infrared properties of individual IDPs have been measured and they provide the basis of an IDP classification scheme (Sandford and Walker 1985). The shape of the 10 μm silicate absorption feature in the IR spectrum of particles varies in width, position, and fine structure (Fig. 11). These variations led to the establishment of the *olivine, pyroxene*, and *hydrated* subtypes. These classifications imply that the silicate feature of a particular IDP most closely matches the laboratory spectra of these minerals. Particles in any of the three classes can contain other minerals and although the IR classification is only a semi-quantitative assessment of a particles' bulk mineralogy, it is a very valuable classification scheme. The fine structure in the silicate structure also provides a useful means for comparing IDPs studied in the laboratory with dust in comets (cf. Sekanina et al., this volume) and dust around other stars. Fine structure producing a bump at 11.2 μm has been seen in several comets (Hanner et al. 1994) and has also been seen in dust around Beta Pictoris and related objects (Butner et al. 1994). This feature is attributed to the presence of olivine, although the best spectral match with IDPs is with the *pyroxene* class that is actually composed of glass with only minor amounts of crystalline silicate (Bradley et al. 1992). The *olivine* IDP class actually produces too much of a pronounced 11.2 μm fine structure relative to the cometary and circumstellar dust spectra. The similarity of the 11.2 μm fine structure in some IDPs, in comets, and Beta Pictoris and its apparent absence in normal interstellar "astronomical silicate" provides important clues to processing of grains in nebular or circumstellar environments.

Recent infrared observation reveal the dominance of Fe-poor, Mg-rich submicron crystalline and glassy (or amorphous) olivine and pyroxene grains as the solid dust component of comet Hale-Bopp (Hanner et al. 1998). Such a paragenesis has previously been encountered only in anhydrous chondritic aggregate IDPs strengthening the link between this type of IPDs and comets (Hanner 1999) that is further supported by prevalence of Fe-poor, Mg-rich silicates in comet Halley's dust (Jessberger 1999).

III.D. Classification and Mineralogy

Typical IDPs are fine-grained mixtures of thousands to millions of mineral grains and amorphous components. Like the most primitive meteorites, most of the IDPs are unequilibrated in the sense that a given phase in a particle can have wide ranging elemental compositions. The particles have not generally undergone metamorphic processing similar to those that equilibrated most ordinary chondrites. The particles are small, very complex and are difficult to fully characterize in a simple way. In some cases the groups may have a genetic meaning, in others, particles from different groups could be derived from common parents. Unfortunately, due to their small size, it is often difficult to classify them into such genetically meaningful groups. This can be attempted as was done with chondrites but it would be difficult using only 15 μm samples. The particles do, however, clearly fall into major categories and several investigators have suggested classification schemes (Brownlee et al. 1977; Sandford and Walker 1985; Bradley et al. 1988; Mackinnon and Rietmeijer 1987; Rietmeijer 1994). The following list is not very sophisticated but it does accommodate most unmelted particles. There are strongly heated to melted particles that are not included here because their original properties were significantly modified during atmospheric entry.

The great majority of the particles fall into the chondritic class and have approximately chondritic elemental composition as their identifying characteristic. Normally this requires that the abundant elements Mg, Al, Si, S, Ca, Cr, Mn, Fe, and Ni that can readily be detected by EDX are present within a factor of a few in CI chondritic abundance. It is often feared that this composition is used as a filter in identifying extraterrestrial particles and that thereby other particle classes are overlooked. This surely does happen, but it can only be a very minor effect. On good stratospheric collections, when the atmosphere is not contaminated by volcanic ash particles, the chondritic particles often comprise over 75 % of the total particles that are not Al-rich space debris or clear-cut contaminants from the aircraft. Most of the non-chondritic composition particles that are identified as extraterrestrial usually have some chondritic composition material attached to them. In most cases, they appear to be mineral grains or coarse grained assemblages of materials that were previously imbedded in fine-grained material similar to the chondritic groups. As such, they should not be thought of as necessarily genetically important

Table 1

Interplanetary Dust Types.

Chondritic	Coarse Grained
anhydrous (pyroxene, olivine)	sulfide
hydrated (layer silicate)	olivine
	pyroxene
	metal
	CAI
	carbonate
	phosphate

sub-classes. They are, however, legitimate interplanetary particles and their strength, density, and optical properties can differ significantly from normal chondritic particles.

Table 1 lists the coarse grained (non-chondritic) particle types in descending order of occurrence. The most common coarse grained particles are sulfides. These include particles that are single pyrrhotite crystals, polycrystalline mixtures of large sulfide grains (Zolensky and Thomas 1995), and large sulfide grains with various amounts of fine-grained chondritic material. Less common are olivine and pyroxene dominated particles (Zolensky and Barrett 1994) and the least common are particles dominated by phases that are abundant in (meteoritic) Ca-Al-rich inclusions (CAIs), large FeNi metal grains, carbonates, and phosphates.

The chondritic particle group is split into the hydrous and anhydrous classes. There is no perfect definition of the difference between the two groups. The hydrous class must have hydrated silicates but they also contain various amounts of anhydrous silicates. Even this definition is sometimes difficult to use with certain particles. These particles are typically identified by lattice fringe imaging of hydrous phases in the TEM. In some particles the "hydrous silicate" is very poorly ordered and difficult to image. This can be an original property but it can also result from thermal decomposition of layer silicates during atmospheric entry. The anhydrous particles are composed of anhydrous phases. Ideally, to be an official anhydrous IDP, it should have no hydrated phases but thermally decomposed or small amounts of hydrated phases are difficult to detect. There are also cases of what appear to be normal anhydrous IDPs that contain a 1 μm chunk of hydrated silicate (0.1 % of the IDP mass). It could be a tiny breccia fragment included into the IDP parent body, but at this level of "contamination," it also could be a tiny piece of a hydrated IDP that was picked up from the collection substrate or during handling.

Three distinct mineralogical classes of chondritic IDPs are recognized. They are referred to as the *pyroxene*, *olivine*, and *layer silicate* classes. The distinct classes were first identified using infrared (IR) spectroscopy (Sandford and Walker 1985), and later confirmed using automated X-ray point-count analyses (Bradley 1988; Germani et al. 1990). The *pyroxene* and *olivine* classes are dominated by the anhydrous silicates pyroxene, olivine, (and glass), while the *layer silicate* class contain hydrated (layer lattice) silicates. Most chondritic IDPs belong to one of the three categories, although several exceptions have been reported. Examples include IDPs with approximately equal proportions of olivine and pyroxene, and others with both anhydrous and hydrated silicates.

The *pyroxene* IDPs are the most inherently interesting and intensively studied because they are highly porous, fluffy objects unlike any other known class of meteoritic materials. Most are complex admixtures of 0.1–5 μm diameter single mineral grains, glass, carbonaceous material, GEMS, and other fine-grained matrix material. The average grain size within the matrices is $\approx$ 100 nm or less. Enstatite is the most abundant of the single mineral grains and it occurs as euhedral crystals, platelets, and sometimes as whiskers (rods and ribbons). Some of the platelets, rods, and ribbons contain crystallographic defects suggestive of nebular gas-to-solid condensation (Bradley et al. 1983). Other enstatites (and forsterites) contain elevated Mn and Cr abundances consistent with gas-to-solid condensation (Klöck et al. 1989). Fe-rich sulfides are also important single mineral constituents as well as minor amounts of forsteritic olivine. Less abundant single-mineral species include alumino-silicate glass (Bradley 1994), metal (FeNi), and FeNi carbides (Christoffersen and Buseck 1983; Bradley et al. 1984).

The carbonaceous material in *pyroxene* IDPs is predominantly disordered (as opposed to graphitic). It is found throughout the IDPs both as discrete clumps and as a matrix with embedded mineral inclusions. Carbonaceous material enriched in ^{15}N has been observed in one IDP (Messenger et al. 1996), and large D/H anomalies are believed to be associated with the carbonaceous material (Messenger and Walker 1997). PAHs have been observed in several IDPs. It is likely that *pyroxene* IDPs contain other indigenous organics but, since the particles are pulse heated during atmospheric entry (but cf. Bonny et al. 1988), collected in silicone oil, and cleaned with solvents, the issue of indigenous organics is complicated.

In the *olivine* class, the IDPs typically are coarse-grained but some of them also contain carbonaceous material, glass, and fine-grained matrix components (Bradley et al. 1989; Christoffersen and Buseck 1986). In other words, there appears to be some mineralogical overlap between the *olivine* and *pyroxene* classes of IDPs. However, a conspicuously large number of *olivine* IDPs exhibit mineralogical evidence of strong heating (e.g., Fe-sulfides with magnetite rims) and they do not contain solar flare tracks. Thus, it is likely that the *olivine* class contains members that were severely heated during atmospheric entry. Laboratory experiments have shown that *layer silicate* IDPs transform into olivine-rich IDPs with heating (Greshake et al. 1998).

The relationship between the three classes of IDPs is unclear. Since solar flare tracks have been found in IDPs from each class, the mineralogical differences are believed to be indigenous rather than the effects of strong heating during atmospheric entry. However, tracks are commonly found in *pyroxene* and *layer lattice* silicate IDPs but they are conspicuously absent in most *olivine* class IDPs. Therefore, the *olivine* class may include a significant number of IDPs that were strongly heated (above 650°C) during atmospheric entry.

It is likely that at least some of the mineralogical differences between the three mineralogical classes reflect different types of parent bodies. For example, the *layer-lattice silicate* IDPs are probably from hydrous parent bodies where there was significant aqueous alteration. The mineralogy of several *layer silicate* IDPs is similar to the mineralogy of the CI and CM chondrites whose asteroidal origins are believed. Whether these differences are reflective of distinct classes of parent bodies is unknown. High-speed IDPs thought to be of cometary origin typically belong to the *pyroxene* and *olivine* classes, whereas most low-speed IDPs belong to the *layer-lattice silicate* class.

III.E. Elemental Composition

Given the large scale compositional difference in the solar system as a whole that reflects itself in density differences from the inner to the outer planets, given the compositional heterogeneity within the planets as best known from the Earth, and finally, given the complex and manifold diversity among primitive meteorites and their constituents, the elemental composition is one essential parameter to delineate the provenance and history of solar system material, including IDPs. Because of the possible sources of IDPs – they include asteroids also from the outer belt, or comets – the understanding of which probably is crucial for the understanding of the history of the solar system as a whole, quite some experimental efforts were made and highly sophisticated techniques were developed to obtain the elemental compositions of IDPs. This is difficult because of the small size of typically 15 μm and because each IDP has to be treated as an individual entity, like a whole meteorite.

For many purposes, it is sufficient to use a model composition of interplanetary dust that fairly closely matches chondritic elemental composition. A good approximation of typical IDP compositions is nominal chondritic composition for major element ratios with a match to CI chondrites (Anders and Grevesse 1989) within a factor of two. There is particle-to-particle variation and there are notable systematic variations from chondritic composition (e.g., Ca, C, and Zn as well as other volatile minor elements) but still a fairly monotonous chondritic composition is an excellent approximation for typical cases (Brownlee 1996). This is quite remarkable because most chondritic meteorites do not typically have chondritic elemental composition at the 10 μm size scale. Most meteorites are coarser-grained and common 15 μm volumes are single mineral grains. Even the fine-grained matrix of carbonaceous chondrites is not "chondritic" in terms of elemental abundances. No mineral can approximate solar abundances for more than a few elements. The closest mineral match to chon-

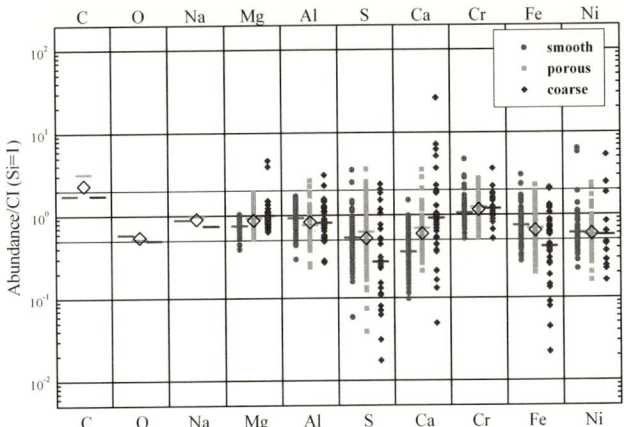

Figure 12. Bulk major element compositions of 200 (for C and O only 30) stratospheric IDPs showing a factor of two match with CI chondrites, some systematic offsets, and significant differences between smooth, porous, and coarse-grained particles. Horizontal bars for different sub-groups and diamonds for all particles are geometric means except for C, O, and Na, where only arithmetic averages were available. (Data from Schramm et al. 1989.)

dritic composition is Fo$_{50}$ olivine (Fe=Si=Mg atom fraction) that matches Fe, Mg, Si and O but severely misses for nearly all other elements. Typical IDPs are very fine-grained and even 15 μm particles are assemblages of hundreds to millions of individual mineral grains plus amorphous material. This averaging effect in particles > 2 μm results in an approximately solar composition for the condensable elements. The elemental composition of typical IDPs is remarkably simple and is likely to be the result of a simple mechanical mixing of large numbers of randomly selected tiny grains.

1. *Major Elements*

The published data on IDPs include stratospheric particles as small as 5 μm in diameter and particles as large as a millimeter collected from polar and deep sea sediments. The composition data have been obtained by a variety of techniques including the electron microprobe, X-ray fluorescence, neutron activation, and proton as well as synchrotron induced X-ray emission (PIXE and SXRF), and different types of secondary ion mass spectrometry (SIMS and TOF-SIMS). The largest data set for major elements (Schramm et al. 1989) comes from the EDX analyses of 200 stratospheric IDPs (Fig. 12). Brownlee (1997) gives electron microprobe data for 500 cosmic spherules in the 1 μm to 1 mm size range but the compositions of these melted particles only partly reflect that of their precursors because of effects related to hypervelocity entry into the atmosphere (Love and Brownlee 1991; Kornblum 1969).

The collected stratospheric IDPs appear to be a fair and probably the least biased sampling of the 15 μm size meteoroids although some particles are excluded due to break-up by the stresses of atmospheric entry and col-

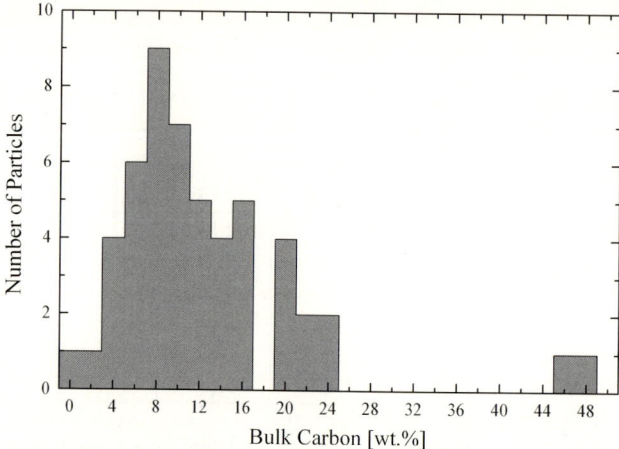

Figure 13. Bulk C concentrations of stratospheric IDPs (Keller et al. 1994). The peak of the distribution is more than twice the CI chondrites and the average is even higher.

lection. Particulate contamination problems are minimal during periods when the stratosphere is not loaded with volcanic particles. Any meteoroid type that would not be recognized or otherwise not be included in the collections probably represents less than 10 % of the 15 μm meteoroid population entering the Earth's atmosphere. There is evidence that some samples do, however, pick up contamination during their stratospheric residence (Jessberger et al. 1992; Bohsung et al. 1994, 1995a; Stephan et al. 1994a, b; Arndt et al. 1996a, b; Rost et al. 1996, 1999) but this appears to only severely affect trace elements, for example and most prominently Br (see below).

Studies of the major element abundances of 15 μm stratospheric IDPs show that most elements match the abundances of CI chondrites within a factor of two (Schramm et al. 1989). Analyses of large numbers of particles show peaking of compositions usually with only modest offsets from CI abundances. Some elements do, however, have abnormally large dispersions. Many particles show Ca and S depletions relative to CI abundances by a factor of two or more due to the high contents of phases such as FeS and carbonates in CI chondrites. Rare IDPs, however, have large *excesses* of these elements because they are largely composed of sulfide or Ca-rich minerals. If the data from many particles are combined, the depletions and excesses largely compensate for each other. C is an element that is commonly found at more than a factor of two above CI abundances (Fig. 13) (Thomas et al. 1993). C – in the sun after H, He, and O the fourth most abundant element – is highly fractionated among classes of chondrites, and it is significant that many IDPs have higher C abundances than the most C-rich meteorites. In fact, some IDPs are the most C-rich meteoritic materials and their high C abundance is strong evidence that they include samples of bodies so far not sampled by conventional meteorites.

2. Minor and Trace Element Abundances

Trace element data provide valuable information on IDPs. They may provide a means of defining sub-groups, indicate associations with chondritic groups, provide information on atmospheric heating and can act as sensitive tracers of the many potential processes that may have influenced IDPs from the time of formation to the time of sample analysis. While major elements are usually determined by standard electron probe techniques, analysis of minor and trace elements requires more elaborate methods such as PIXE (*P*roton *I*nduced *X*-ray *E*mission), SXRF (*S*ynchrotron *X-R*ay *F*luorescence), and SIMS (*S*econdary *I*on *M*ass *S*pectrometry), and this is more complex due to the more complex instrumentation and because of higher contamination hazards.

Recently a major compilation and investigation of trace elements, along with major and minor elements, from 89 IDPs has been published by Arndt et al. (1996a). They analyzed the abundances of 28 elements from Na to Zr, excluding the noble gases. The data for this enterprise mostly stem from three laboratories and – except some newly added data – had been published in the following papers: van der Stap et al. (1986); Wallenwein et al. (1987); Antz et al. (1987); Flynn and Sutton (1987, 1990, 1991, 1992a,b); Sutton and Flynn (1988); Stadermann (1990); Jessberger et al. (1992); Thomas et al. (1993, 1994); Flynn et al. (1993, 1994); Bohsung et al. (1995b); Arndt and Flynn (1995). The trace element data used by Arndt et al. (1996a) were obtained on stratospheric IDPs using PIXE, SXRF, and SIMS. Further techniques for IDP element analysis, not considered in this compilation, are: electron microprobe (Schramm et al. 1989; Thomas et al. 1993), time-of-flight (TOF) SIMS (Stephan et al. 1994a,b,c) and neutron activation (Zolensky et al. 1989; Lindstrom and Zolensky 1990). Unfortunately, a comparison of the combined data set with some of these techniques is problematic because of the very different depths (or volume) of information – many elements are rather heterogeneously distributed within the IDPs – and also because of the rather small overlap of the element list. On the other hand, a careful comparison of the PIXE, SXRF, and SIMS trace element results demonstrated (Arndt et al. 1996a) that these three techniques do not systematically differ, are coherent, and thus can be compared and combined to a single full *complete* data set.

In order to allow this comparison and the further investigation of that data set, the published material was recalculated to one common dimension: element abundance normalized to Fe and CI chondritic abundance taken from Anders and Grevesse (1989). In cosmochemistry generally Si is taken for normalization. However, Si data were published only for about 50 % of the IDPs, the quality of Si data obtained by PIXE and SXRF is rather poor due to absorption effects, and – a dilemma for all analytical techniques – because the probable contamination with silicone oil from collecting and maneuvering the IDPs.

The results of the study are shown in Fig. 14. On the average S, Ca, and Ni are depleted in IDPs relative to CI chondrites while the eleven elements

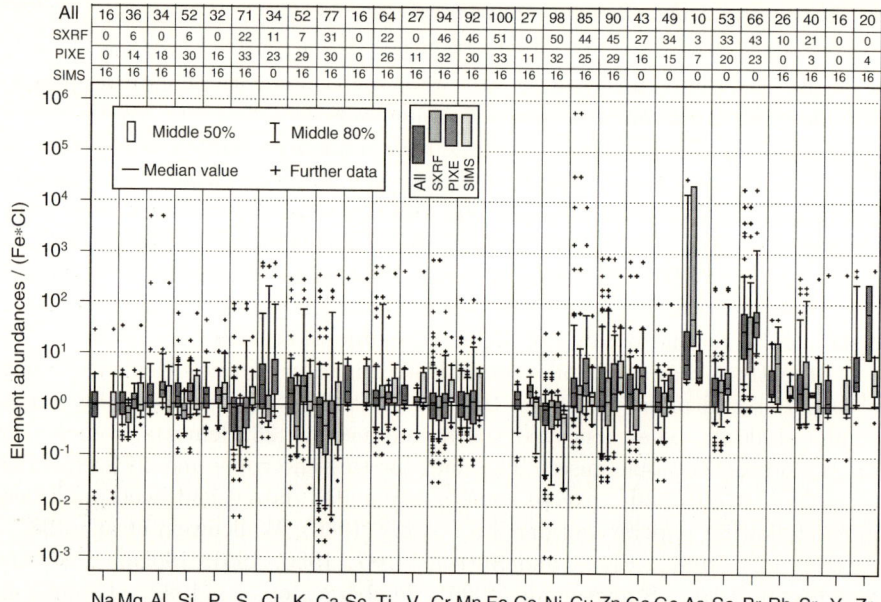

Figure 14. Element abundances of 100 bulk analyses of 89 IDPs normalized to Fe and the respective CI chondritic element/Fe ratios. The input for this plot constitutes *all* presently available major, minor, and trace element abundances in IDPs. The data set is broken down according to the three employed experimental techniques: PIXE, SXRF, and SIMS. The data are presented in a boxplot where "further data" indicates those that are outside the 80 % range. On the top line the number of available entries for the different methods are given. (From Arndt et al. 1996.)

from Cu to Zr are enriched. The abundances of the remaining elements are indistinguishable from that of CI-chondrites and thus from the mean solar system. Arndt et al. (1996a) – in order to check for bimodal or multimodal patterns – examined the distributions of the element abundances and found only unimodal distributions with the exception of Ni that has a peak near the CI-ratio and a small group with low Ni values.

The data set was also analyzed with cluster analytical techniques to search for groups and with discriminant analysis to investigate their separation (Massart and Kaufmann 1983; SAS Institute Inc. 1988). The aim of cluster analysis is to find in multidimensional space groupings within a given data set that are (a) as homogeneous as possible, (b) separated from each other as far as possible, and (c) based on as many variables as possible. The main problem in the cluster analysis of the IDP data set stems from missing data: Only *one* single element, Fe, was measured in *all* particles. In fact, 50 % of the possible element data table is empty because of insufficient sensitivities of the instruments and abundances below detection limits. As a consequence, (a) no cluster analysis could include all IDPs simultaneously, (b) what is known about the cosmochemical nature of the elements had to be ignored and (c) certain element combinations had to be chosen to encompass the maximum number of IDPs.

A cluster analysis based on the six elements Cr, Mn, Ni, Cu, Zn, and normalizing Fe, cluster 82 % of the IDPs into four groups which were named chondritic, low-Zn, low-Ni, and nonsystematic. The geometric mean abundances of all 28 elements within three groups – chondritic, low-Zn, and low-Ni – are given in Fig. 15. In the left hand panel of Fig. 15, the elements are plotted against their volatility taking as its measure the 50 % condensation temperature in a gas of solar composition as compiled by Wasson (1985). In the right hand panel, the elements are plotted against decreasing CI abundance. In order to avoid over-interpretation of less significant mean abundances, a light gray symbol is used if an element is detected only in less than half of the IDPs in the group, otherwise the symbol is black. In the following we discuss the four groups.

Chondritic Group

The abundances of many elements in the chondritic group (44 members) are rather close to CI but with frequent and major depletions of Ca and S (0.8×CI). Four elements are very close to CI: Mg, Mn, Ni, and Y. Enrichments are found at three levels: (a) most prominently (5–30×CI) for Br, As (only rarely detected), and Zr, (b) in the narrow range 2.2–2.7×CI for the elements Cl, Cu, Zn, Ga, Se, and Rb, and (c) in the equally narrow range 1.2–1.7×CI for Na, Al, Si, P, K, Sc, Ti, V, Cr, Co, Ge, and Sr. Some of the Si overabundance, however, probably stems from the unavoidable silicone oil. The fact that the latter large number of elements is tightly grouped around 1.4×CI may result from the Fe-normalization and may indicate an overall Fe-deficiency of 0.7×CI. Such a deficiency is supported by the few available absolute Fe concentrations (0.9×CI). If the Fe-deficiency of 0.7×CI is indeed true then (a) 16 elements, (Na–Sr) and (Mg–Y), in IDPs are present in solar abundance within 30 %, (b) Ca and S deficits increase, and (c) the enrichments of the second group of elements (Cl–Rb) decrease to 1.7×CI. It has clearly been demonstrated (Arndt et al. 1996a) that at least PIXE is sufficiently sensitive to detect such enrichments of Cl, Cu, Zn and Se, possibly also Ga, and marginally Rb. The elements enriched in this group are also most strongly enriched in the group of the non-chondritic particles.

Low-Zn Group

Ten of the twelve particles in that group show features that previously had been interpreted as depletions due to atmospheric entry heating. The elements detected in at least six out of the twelve IDPs can be ordered in terms of their abundance relative to Fe and CI chondrites: highly enriched is again Br; almost normal are Ga, Se and Cu; highly depleted are Ca and Zn; moderately depleted are S, Cr, Mn, Si, Ti, and Ni (0.26–0.61×CI). Assuming that Fe is enriched by a factor of 2.5 in the low-Zn group compared to the chondritic group, an excellent agreement between these two groups is found, except a

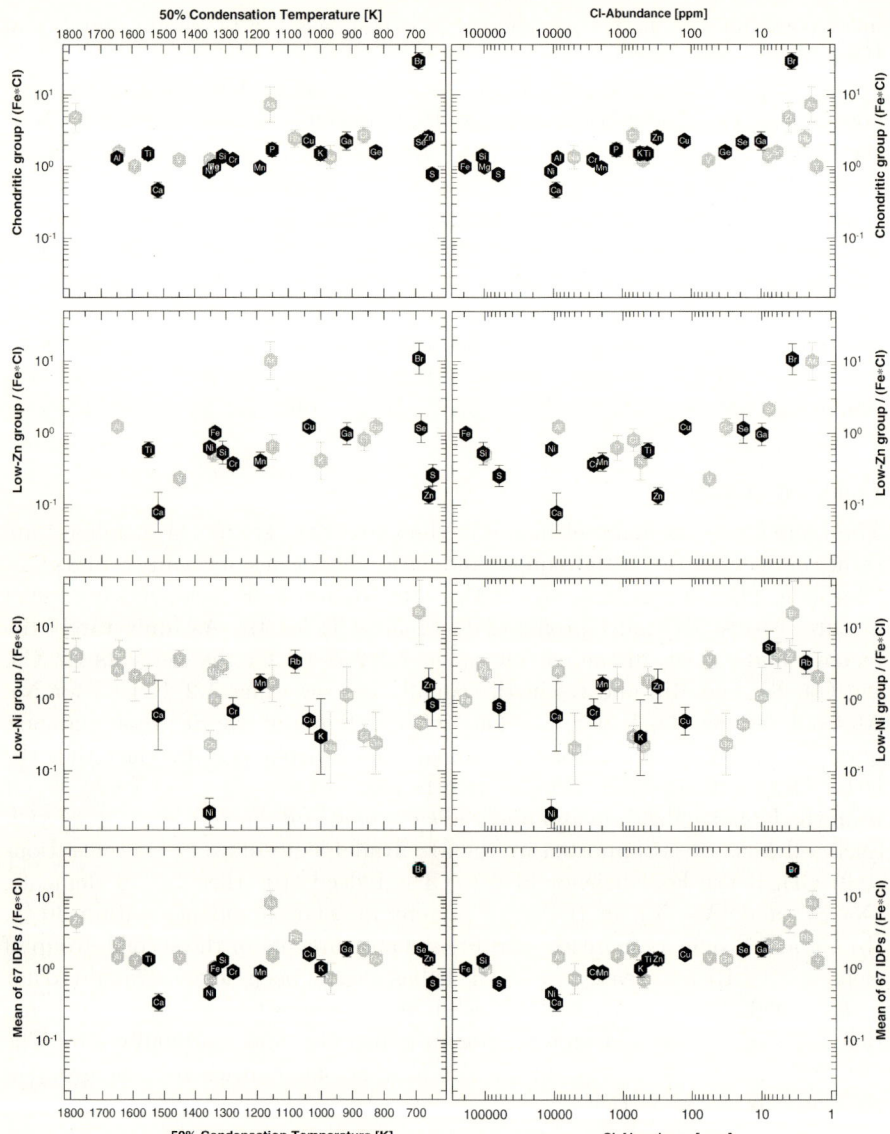

Figure 15. Geometric means and standard deviations of the element ratios within the chondritic, low-Zn, and low-Ni IDP groups (from top). The bottom panel shows the geometric means of IDPs of these three groups. In the left column the abscissa is the 50 % condensation temperature (Wasson, 1985) and in the right column the respective CI-chondrite abundance (Anders and Grevesse 1989), both decreasing to the right. Symbols of elements that are detected in < 50 % of the IDPs of the respective group are gray, the others black. (From Arndt et al. 1996a.)

probably insignificant difference in Ca. The only really significant difference between the two groups is the abundance of the volatile element Zn. Most importantly, there is no difference in two other volatile elements that have about the same condensation temperature as Zn: S is depleted in both groups by the same factor and Se in both groups is present in nearly CI abundance.

Low-Ni Group

The mean value of Ni/Fe in the low-Ni-group (11 members) is $0.03 \times$ CI. This group is quite variable with considerable spread in the element means. Some member IDPs of this group are discussed by Flynn and Sutton (1990). They describe element patterns remarkably similar to that of basalt and unless solid proof of extraterrestrial origin is available, e.g., solar flare tracks or solar wind He, some of the low-Ni particles could be terrestrial. In these particles Ca/Fe is higher than in the previous two groups and overlap the CI-value. High mean values for a number of elements indicate a generally low Fe content. This is supported by the Fe mean ($0.7 \times$ CI) based on the (few) absolute Fe concentrations. Three particles have low Zn *and* low Ni values.

Non-Systematic Group

This is the most weakly defined group with only six members that are most distinct from all the other particles. The group is characterized by enormous enrichments of the same elements that are also heavily enriched in the other groups: Br, Al, Cu, Zn, Ga, Se, Cl, and Si (in decreasing order).

In conclusion, the cluster analysis based on Cr, Mn, Ni, Cu, and Zn reveal four groups of stratospheric particles, showing some common but also some distinguishing features: The fourteen most abundant elements (Fe, Si, Mg, S, Ni, Ca, Al, Na, Cr, Mn, P, Cl, K, Co) in chondrite-like IDPs are as abundant as in CI chondrites and the solar system in general, only Ca and S are depleted. This does not hold for the low-Ni particles with Ni depletions and about normal Ca, and for the low-Zn group with high Fe. The remaining fourteen trace elements (Ti, Zn, Cu, V, Ge, Se, Ga, Sr, Sc, Zr, Br, Rb, As, Y) in *all* three groups are enriched by about the *same* factor, $2 \times$ CI, with the exception of Br with enormous enrichments of typically about $25 \times$ CI but up to $20\,000 \times$ CI.

The fourteen elements that on the average in chondrite-like IDPs are present in cosmic abundance set the baseline for enrichments of the many and depletions of the few elements. Deviations from the solar abundances may be indigenous and may be related to the origin and history of IDPs, but almost certainly they are at least in part the result of processes in the Earth's atmosphere and even on the Earth. In addition, the magnitudes of element losses and gains certainly depend, besides pre-atmospheric size and density, on entry velocity and angle, but also on the chemical compound the elements are in, all of which is unknown for individual IDPs. Some contaminating processes, however, can be studied as will be demonstrated next.

3. Contamination of IDPs

The research on the IDP contamination problem mostly rests on PIXE and TOF-SIMS analyses of minute sub-volumes of IDPs. Since TOF-SIMS, a technique primarily developed for surface analysis (Benninghoven 1994), is a rather novel system it will shortly be introduced. A flock of 500 primary ions, often Ga^+, hit with a repetition frequency of 5–10 kHz an area with ~ 0.2 μm diameter of the surface to be studied. About ten secondary ions per shot are released and, after mass separation in a drift tube equipped with a reflector, are detected with a time resolution of typically 200 ps. That together with the primary ion pulse length of $\sim$1 ns results in a mass resolution of several thousand. Spectra and secondary ion images – of positive as well as negative polarity in subsequent analyses – are generated by integrating minutes to hours of mass separated secondary ion counts. The high spatial resolution provides images of the lateral element distribution as well as quantitative measurements of element ratios on sections and surfaces of individual IDPs. Simultaneous detection of all secondary ions, sequentially with both polarities, nearly infinite mass range, an instrumental transmission of 20–80 %, and the small sample consumption that can be restricted to monolayers ensures that a maximum chemical information can be obtained without completely destroying even small samples. The lateral distribution of elements and molecules is investigated by scanning the samples with the primary ion beam (Schwieters et al. 1991). Being a surface analysis technique, the information depth is in the order of a few monolayers. Thus, TOF-SIMS is an important technique to study, among others, surface features like contamination of IDPs.

Contamination by silicone oil presents an obvious problem for *all* stratospheric IDPs since they are captured in silicone oil and are also immersed in it for manipulation.

Another most prominently enriched element in IDPs is Br. This has been noted in very early IDP trace element analyses and then has been interpreted to indicate large scale chemical heterogeneity of the early solar system (van der Stap 1986). Later, however, this contention was disputed based on indications that atmospheric contamination processes played an – up to that time underestimated – role in the trace element budget of stratospheric IDPs (Jessberger et al. 1992). From then on evidence is growing that at least the Br overabundance is due to halogen-rich aerosols although the halogen abundance in the stratosphere is not very well known (Cicerone 1981). First direct observational hints for such contamination of chondritic IDPs were Br-salt nanocrystals attached to IDP W7029E5 (Rietmeijer 1993). Stephan et al. (1994b) demonstrated by TOF-SIMS studies of a section of IDP L2006G1 that it has a halogen-rich exterior rim. The TOF-SIMS images (Plate 2) of most ions resemble a ring structure with C and H in the center. Outside the ring, Cl and F are concentrated on one side of the rim of the particle. In this rim Br is clearly identified. In a further study (Stephan et al. 1995), a set of ten stratospheric particles was selected first for bulk SEM-EDX including C measurements (Thomas

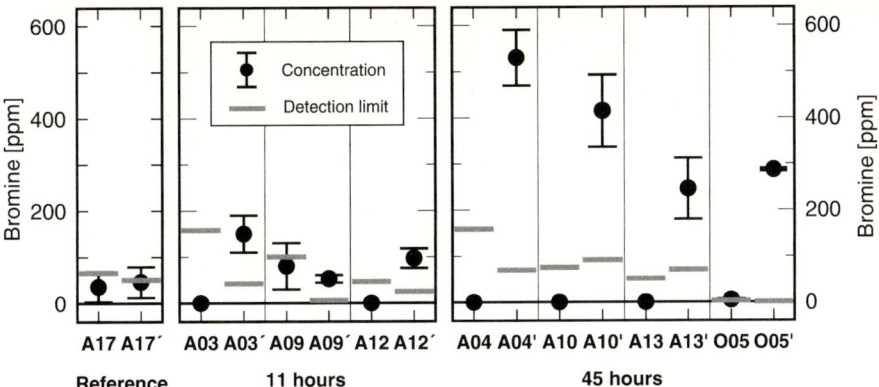

Figure 16. Br concentrations, obtained with PIXE, in seven aerogel (A##) and one Orgueil (O05) particles before (plain identification number) and after (number with ') stratospheric IDP collection flight. The limits of detection, LODs, are also indicated. They depend, among other factors, on the total proton dose and thus are variable. The duration of stratospheric exposure is given at the bottom. Aerogel particle A17 was not flown but was subject to all other transport and handling procedures. Br contamination occurred during stratospheric flight, and its magnitude is related to the flight duration. No contamination by transport and handling is observed.

et al. 1994) and then for TOF-SIMS analyses of the *very* surfaces of these IDPs (Rost et al. 1996, 1999). With such surface TOF-SIMS analyses, because of its roughness a challenge for the method, clearly the presence of F, Cl, and Br on IDP surfaces was indeed and without doubt demonstrated (Plate 3). Except for U2015G1 where Br was found in an extremely fine-grained iron oxide sub-unit (Stephan et al. 1994a), none of the particles analyzed so far with TOF-SIMS showed appreciable amounts of Br in the bulk.

Arndt et al. (1996b, 1997) imitated systematically that portion of the IDP history that can be imitated: the time on the IDP collector and the laboratory handling up the analysis. They exposed various small (10–40 μm) and porous particles – among them aerogel and CI chondrite fragments – to the stratosphere for 11 and 45 hours on an actual IDP capture flag. All particles were analyzed prior to flight with PIXE, some particles were just handled, but kept on the ground for a laboratory contamination test, and the recovered particles were again analyzed with PIXE thereafter. Br that was not detected before flight was clearly present in all particles after exposure to the atmosphere (Fig. 16). PIXE-mapping of one of these particles (Fig. 17) demonstrates that Br is exclusively confined to the particle and is absent in the surrounding material, also absent in silicone oil. This finding again corroborates the earlier contention (Jessberger et al. 1992) that contamination processes indeed play an important role although the detailed mechanisms are not yet fully understood.

Certainly, the effectivity of contaminating processes depends on a number of factors like chemical composition and state or on porosity and size of the IDP constituents (Stephan et al. 1994a) that largely remain unknown at present.

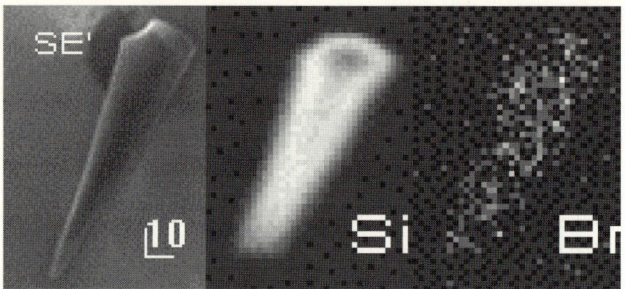

Figure 17. Secondary electron image (left) and Br distribution (right), both obtained with PIXE, of an aerogel particle and its surrounding after 45 hours of stratospheric collection flight. The scale bar length is 10 μm. The beam size for the Br-mapping was several microns. Br is exclusively found in the aerogel particle.

In addition, it must be seen what other trace elements that are also enriched in IDPs – though to a much lesser degree than Br – are affected as well (Arndt et al. 1997). Therefore, to obtain definitive information on the siting of the enriched as well as the depleted elements is a crucial goal of future IDP research that can, however, be achieved by compositional mappings with PIXE and TOF-SIMS in addition to other microanalytical techniques like ATEM.

4. Compositional Mapping

PIXE at the Heidelberg proton microprobe allows elemental mapping with a lateral resolution down to 1 μm and thus provides additional information on the siting and distribution of the elements without destroying the particle before or during the analysis (Maetz et al. 1996). An example of a Zn and Fe-rich IDP is given in Fig. 18. Here, Fe is very inhomogeneously distributed and correlated to Ni and S, but anti-correlated to Mg and Si. The Zn-enrichment ($10 \times$CI) is exclusively confined to the FeS-Ni region in the center, a possible siting for Zn in IDPs (Arndt et al. 1996a).

From imaging TOF-SIMS analyses a few element maps are presented here. Plate 4 shows secondary ion images for 18 different ion species from particle L2006E10. Here the correlation of Ca and P, both highly enriched in a 4×1 μm^2 sized region, lead to the discovery of presumably an apatite grain in a hydrated IDP (Stephan et al. 1994b).

The rather unique and long list of elements accessible with TOF-SIMS – from the light volatile to the heavy refractory elements – allows a comparison with the composition of comet Halley's particles (Jessberger et al. 1988, 1989). The composition of the only anhydrous IDP analyzed so far with TOF-SIMS (L2006B21; Stephan et al. 1993) and Halley's dust data – as far as they are available – agree relatively well at least for some elements (Fig. 19). The general trend for Halley's element/Si ratios normalized to CI chondrites, characterized by C-enrichment and depletion of elements like O, Mg, Al, Cr, Mn, Fe, and Ni, as well as almost chondritic Ti and Co, was also found for L2006B21, although to different extents. Future analyses with TOF-SIMS of anhydrous IDPs may

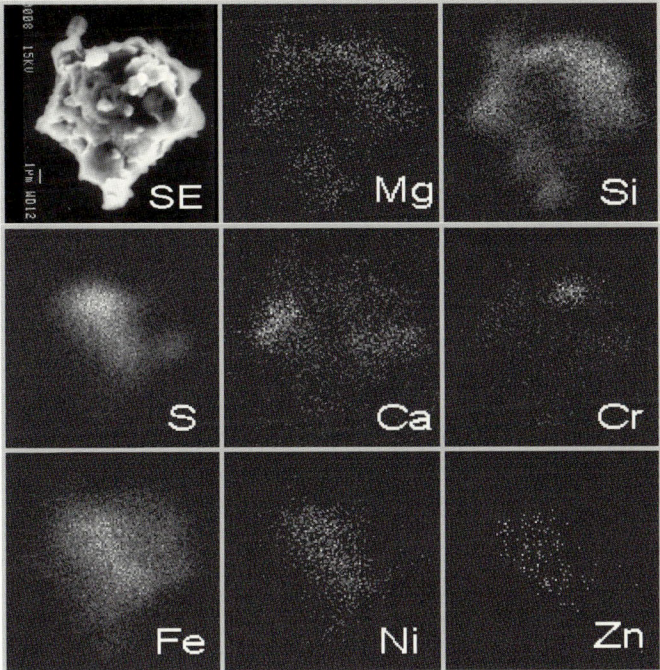

Figure 18. SEM-image (top left; W. Klöck pers. communication) and element mappings obtained by PIXE of eight elements in IDP L2005 AD14. The proton beam size was 1.5 μm, the pixel distance is 0.1 μm. Note the interior bar rich in Fe, S, Ni, and Zn.

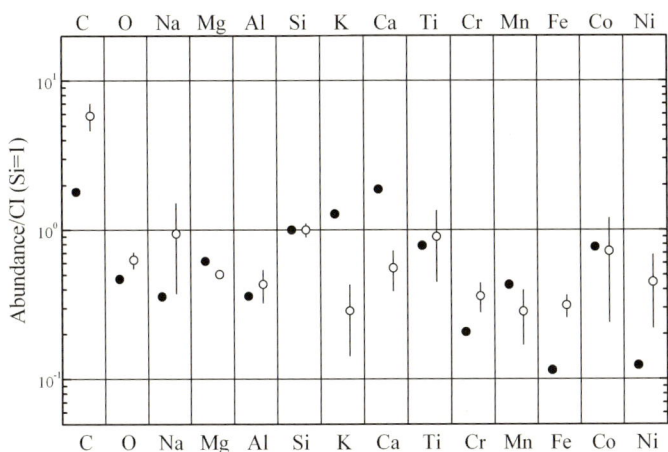

Figure 19. Comparison of element ratios – normalized to CI-chondrites and Si for anhydrous IDP L2006B21 (obtained with TOF-SIMS; filled circles) and comet Halley's dust (open circles). It should be noted that the cometary composition is uncertain by about a factor of two as discussed in detail by Jessberger et al. (1991); cf. also Sekanina et al. (1998), and that the indicated error bars reflect the variability of the cometary dust compositions.

help to solve the question whether anhydrous IDPs originate from comets, provided that more and more secure data on the chemical composition of different comets will be available. Here TOF-SIMS also may play an important role on future space missions for the *in situ* analysis of cometary matter (Beck and Kissel 1994).

III.F. Isotopic Composition

Studying the isotopic compositions of IDPs is of particular interest for two main reasons: First, isotopic signatures can prove that a given particle is actually extraterrestrial. Since the isotopic compositions of the elements in natural and most artificial terrestrial dust particles vary only within small margins – variations from radioactive decay are certainly not considered here – any composition clearly outside that range (i.e., an isotopic anomaly) represents proof of a particle's extraterrestrial origin. However, the reverse is not true and the absence of isotopic anomalies in a given particle can not establish its terrestrial origin. The possibility to positively prove the extraterrestrial origin of particles on an individual basis is particularly important because all currently available collection sites, including the stratosphere and the Low Earth Orbit (LEO), are subject to natural terrestrial and man-made contamination. The second reason for performing isotopic measurements in IDPs is to gain information on the origin and history of a particle. When the physical and chemical processes that lead to an observed isotopic anomaly are known, elements of the history of an IDP and/or its progenitor material can be inferred.

For bulk samples with macroscopic dimensions a number of mass spectrometric analytical techniques exist that allow routine precision measurements of isotopic compositions. However, when it comes to the analysis of particles with micrometer-dimensions, only a few methods offer the required sensitivity. Fortunately, many of the isotopic anomalies observed in IDPs are so pronounced that a slightly lower measurement precision than for the analysis of terrestrial samples can be accepted. The first isotopic measurements of IDPs were made with thermal ionization mass spectrometry (Esat et al. 1979; Esat and Taylor 1987). More recently, step-heating and pulse-heating techniques have been used to determine the noble gas contents of IDPs, and to some extend their noble gas isotopic compositions (Nier 1994). However, the vast majority of all isotopic measurements of IDPs has been made with double-focusing secondary ion mass spectrometry (SIMS) due to its unique ability to measure isotopes at low abundances with high spatial resolution and sufficient precision.

The distinction of whether or not an observed isotopic composition is anomalous requires the knowledge of the range of compositions that is considered *normal*. With respect to extraterrestrial materials, any isotopic composition found in terrestrial samples is seen as *normal*, i.e., the terrestrial samples with the most diverging isotopic compositions define the normal terrestrial range for that particular element. Detailed information on terrestrial isotopic ranges for a number of elements can be found in e.g., Hoefs (1980). An isotopic composition is considered anomalous, when it is separated by more than two sigma from the normal terrestrial range (McKeegan 1987a). This definition of an isotopic anomaly is distinctly different from what is generally being used in terrestrial geochemistry.

An isotopic composition is frequently expressed as a delta-value, which is the deviation of the measured isotopic ratio from a standard ratio in per mill units. Isotopic ratios are generally calculated with respect to the most abundant isotope in terrestrial standards. Thus, e.g., the expression $\delta^{13}C = 200$ ‰ indicates that in this sample the $^{13}C/^{12}C$ ratio is 200 parts per thousand (= 20 %) higher than the standard $^{13}C/^{12}C$ ratio. For many elements a standard isotopic ratio with a δ-value $\equiv 0$ ‰ has been defined, e.g., SMOW (standard mean ocean water) for H and O (Hagemann et al. 1970), PDB (Pee-Dee Belemnite) for C, and air for N. The results of isotopic measurements in IDPs are listed in the following paragraphs by element.

Hydrogen is particularly susceptible to isotopic fractionation because of the large relative mass difference of the isotopes D and H. This also increases instrumental mass fractionation effects, making the measurement of H isotopic compositions less precise than those of other elements. Zinner and McKeegan (Zinner et al. 1983; McKeegan et al. 1985, 1987) measured the H isotopic composition in 31 chondritic IDPs with SIMS. In 13 particles anomalous D/H ratios with $\delta D > 100$ ‰ and in three particles with $\delta D > 2\,000$ ‰ were determined. In five particles it could be demonstrated that D-excesses are spatially correlated with the C abundance, hinting at a carbonaceous carrier of the anomaly. Only IDPs of the hydrated silicate variety and the *pyroxene* type were found to exhibit H isotopic anomalies, but not *olivine* type IDPs. In a follow-up study, covering some of the same particles and a number of new ones, Stadermann (1990) also observed D anomalies in roughly 1/3 of the analyzed chondritic IDPs. The largest D/H-ratio found in this study was $\delta D = 1\,100$ ‰. The results of the H measurements indicate that many IDPs are isotopically very heterogeneous, with some fragments having essentially normal D/H-ratios while other fragments of the same particles were enriched in D with $\delta D > 1\,000$ ‰. An important discovery was made by Messenger et al. (1996) when he observed that H isotopic anomalies were much larger and more common in so-called "cluster IDPs" than in individual, non-cluster IDPs. These cluster IDPs are particles that fragment during impact on the stratospheric collector flags. While it is not obvious why cluster particles *per se* should exhibit larger anomalies than other IDPs, there is a hint towards a time-dependence in the data set, which could indicate a correlation with a specific dust source (Messenger and Walker 1997).

The large variations of the H isotopic composition made it interesting to study these samples with SIMS isotope imaging. In this technique, secondary ion images of the distribution of different isotopes on the sample surface, such as D and H in this case, are mathematically converted point-by-point into a distribution image of the isotopic ratios on the surface of the sample. These images can then be presented as false-color pictures, directly showing the isotopic heterogeneity of the sample. Measurements of this kind have first been made by McKeegan et al. (1987) in a chondritic IDP called Butterfly. This particle showed a local D enrichment (hotspot) with $\delta D > 9\,000$ ‰ concentrated in a spot with a diameter of 1 μm or less. Similar imaging measurements have also been made by Fleming et al. (1989) in IDPs with smaller, but significant H isotopic heterogeneity. Recent H imaging studies by Messenger (Messenger et al. 1996; Messenger and Walker 1997) identified D-hotspots with $\delta D > 50\,000$ ‰. Those are the largest D excesses ever observed in a natural sample.

Lithium is still light enough to make SIMS measurements relatively easy due to the fact that there are only few interferences at the atomic masses 6 and 7. Stadermann (1990) measured the $^6Li/^7Li$-ratios in 12 stratospheric dust particles without finding any compositions significantly different from that of the terrestrial standard. Xu et al. (1994) determined the $^6Li/^7Li$-ratios in four IDPs and found general agreement with the values in reference literature. Boron was also analyzed in that study by Xu et al. (1994) who measured the $^{10}B/^{11}B$-ratios of the same four IDPs and again found agreement with reference data.

Carbon that is found to be isotopically anomalous in presolar grains extracted from certain meteorites (Anders and Zinner 1993) was first studied in IDPs by McKeegan (1987a). He determined the $^{13}C/^{12}C$-ratios in seven IDPs and found an anomalous composition in only one case, in fine-grained chondritic material adhering to a refractory Al-oxide particle. In 124 measurements on fragments of 68 stratospheric dust particles Stadermann (1990) did not observe any C isotopic anomalies, although several of the same fragments showed large ^{15}N excesses in the same measurements. With respect to their C isotopic composition, Messenger (1997) found that cluster IDPs are normal and indistinguishable from individual IDPs.

Nitrogen isotopic compositions of IDPs were first measured by Stadermann (1990). He found significant ^{15}N enrichments with $\delta^{15}N$ values up to 442 ‰ in 12 out of 35 chondritic IDPs as well as in one overall chondritic particle that, however, was significantly enriched in Na, K, and P. Due to the nature of these SIMS measurements, where N is detected as CN^- ions, the observed anomalies only represent N in association with a C-rich phase. Messenger (1997) also observed heavy N in several IDPs with ^{15}N enrichments up to $\delta^{15}N \approx 500$ ‰. Most analyzed IDPs appear to be isotopically very heterogeneous with respect to N, similar to what was observed for H. A correlation between the isotopic anomalies of H and N, though not in their magnitude, was found by Stadermann (1990). Most particles with N anomalies also have H anomalies but the inverse is not true (Messenger et al. 1996). In cluster IDPs $\delta^{15}N$ enrichments were found to be more common than in non-cluster IDPs, but the effects were not generally larger (Messenger 1997).

Oxygen isotopic anomalies with substantial ^{16}O enrichments have been reported for a few IDPs with a refractory elemental composition. The extend of the anomalies and the composition of these IDPs strongly resemble refractory meteorite inclusions (McKeegan 1987b; Stadermann 1990). Significant O isotopic anomalies have not been observed in any chondritic IDPs, however, only very few particles have been analyzed (cf. Greshake et al. 1996).

Magnesium and *Silicon* isotopic compositions were measured in several particles by McKeegan (1987a) and Stadermann (1990). No isotopic anomalies besides some terrestrial, i.e., mass-dependent fractionation were found. Evidence of ^{26}Mg excesses at the per mill level has been reported earlier by Esat et al. (1979).

The major isotopic anomalies found in IDPs are those of H, N, and O. For both H and N, ion-molecule exchange reactions in a cold interstellar molecular cloud can explain the observed isotopic compositions (Zinner 1988; Tielens 1997). However, the same type of reactions are also expected to lead to isotopic fractionation of C that, however, has not been detected in IDPs, even though the C isotopic composition of a large number of particles has already been measured. One possible explanation for the lack of C isotopic anomalies is that C could be fractionated through several chemical routes with opposite and partially canceling effects (Tielens 1997). The evidence for molecular cloud material in IDPs and meteorites has recently been reviewed by Messenger and Walker (1997). The observation of deuterium hotspots with enrichments up to 50 000 ‰ in IDPs with the ion imaging technique has prompted similar studies in primitive meteorites. Guan (1998) indeed found comparable hotspots with δD up to 8 000 ‰ in black fragments in the matrix of the CR2 chondrite Renazzo. So far, the D carriers in meteorites and IDPs are not positively identified.

Future isotope studies of IDPs should attempt to answer the following questions: Can isotopic anomalies of other elements be found in IDPs, e.g., by extending the precision of the measurements or by extending the number of elements measured? Which are the carrier phases of the anomalies? What are the exact mechanisms leading to observed anomalies and why do certain elements show normal isotopic compositions while others in the same IDP are anomalous? At least with respect to the first two questions, improvements in micro-analytical measurement techniques will be beneficial.

IV. ORIGINS

For distinguishing different particle classes that might have genetic implications, it does not appear that major element composition is a strong discriminator. Small samples of primitive solar system materials, to first order, tend to have similar compositions with scatter due to size-scale dependent heterogeneities in their parent bodies. Thomas et al. (1995) have shown that considerable insight into these heterogeneities within IDPs can be obtained by the analysis of so-called "cluster" particles that fragment into large numbers of > 5 μm components during collection in the stratosphere. Compositional ranges in these particles are similar to the overall range of individually collected particles. The natural level of compositional variation of nanogram particles produce a "noise" background that complicates accurate comparison with meteorite groups where characteristic compositions are derived from much larger samples. Physical properties such as structure and mineralogy do provide strong clues to origin and evolutionary processing (Klöck and Stadermann 1994) but at least to first order, major element abundances do not seem to be highly diagnostic properties. This is in contrast to meteorites where elemental composition is an important criterion for classification.

The major sources of these particles most probably are asteroids and comets (Leinert and Grün 1990). Especially the latter makes IDPs exciting since they provide us with the opportunity to study cometary matter in the laboratory and hence in very much greater detail than, e.g., in previous space missions to comet Halley (Jessberger et al. 1988; Jessberger and Kissel 1991). Cometary IDPs eventually may provide information on matter from (a) the ancient solar system, (b) the outer solar system and (c) even from pre-solar epochs (Brownlee 1994). But before such information can be gathered, the fundamental, yet unsolved question in IDP research must be answered: Which IDPs are cometary and which are asteroidal?

One criterion to distinguish particles from the two sources is the velocity at atmospheric entry. Model calculations demand that cometary particles in general have higher entry velocities than asteroidal particles, and that there is only a small velocity overlap between both groups (Jackson and Zook 1992). Therefore, cometary particles are expected to be more strongly heated upon atmospheric aero-braking than asteroidal particles (Flynn 1989). Consequently, one aim in IDP research is to determine the degree of heating a particle has experienced. Maximum temperatures can be estimated if the particles still contain solar flare tracks (Bradley et al. 1984). A minimum entry temperature indicator may be the degree of loss of volatile elements. Such losses are observed for the volatile elements S, Se, Ge, and Zn in heating experiments with 100 μm fragments from the CI chondrite Orgueil taken as analogues of Antarctic micrometeorites (Greshake et al. 1998). Zn abundances in IDPs that are significantly lower than in CI chondrites have been interpreted as indicating Zn losses and thus were related to high entry temperatures (Flynn and Sutton 1992a). Also discussed by some researchers is the presence of magnetite rims as implying strong entry heating (Klöck et al. 1992) but alternative interpretations of these rims are advocated by others (Maurette et al. 1993). A possibly powerful technique to estimate minimum temperatures is provided by He stepwise degassing experiments (Nier and Schlutter 1993) that provide diffusion information – and thus an estimate on the duration and temperature at atmospheric entry. It turns out that if solar wind ^{4}He is lost, an IDP must have experienced typically at least 700°C. At such a high temperature, however, particles are conjecturally rather strongly modified especially as far as volatile or organic components are concerned. As judged at first glance from the Halley results (Jessberger et al. 1988; Krueger et al. 1991) these components probably are the ones that are most different in comets compared to meteorites. Consequently, the He method rejects IDPs as being unaltered cometary matter and does not help to identify pristine cometary particles. This applies to all methods using temperature criteria.

Then the question arises if there are chemical groups that are significantly special and that eventually may be related to cometary and asteroidal sources, respectively. One might expect that the more elements are included in tackling that problem the higher the chances are to approach an answer. A few research groups are engaged since the mid-1980s to obtain trace element concentrations in IDPs (Arndt et al. 1996a), but hitherto without plain evidence as has been detailed above.

All in all, the years of effort of quite a number of research groups did not yet result in establishing clear-cut criteria as to the source of any individual IDP or of certain groups of IDPs. Surely, one rather technical reason for this shortcoming is the tiny dimension of IDPs that makes concerted analyses with complementary methods pretty difficult. Thus they are very rare. One may expect, however, that the presently ever accelerating technologic advancement in the various required micro-analytical methods (Jessberger 1991; Walker 1991) eventually leads to better substantiated and more complete known facts on individual IDPs. A second reason is very obvious: We do not know enough about comets. Unquestionably, the experiments on the missions to comet Halley provided a wealth of unprecedented data. However, the inherent uncertainties of the elemental abundances of Halley's dust, the vague information on the molecular inventory, the very scarce knowledge of the isotopic composition (except a few high $^{12}C/^{13}C$ particles; Jessberger and Kissel 1991), and – very important – the absolute lack of structural and mineralogical information obstructs comparison with facts obtained on IDPs in sophisticated laboratories on Earth. Thus, the question of the source of an individual IDP is not answered at present, and probably must wait for the results from future space missions to comets that also include retrieval of cometary dust.

Acknowledgements

This work was supported by FWF and ÖAD in Austria, and by the Deutsche Forschungsgemeinschaft as well as by the Ministerium für Schule und Weiterbildung, Wissenschaft und Forschung des Landes Nordrhein-Westfalen in Germany.

REFERENCES

Anders, E., and Grevesse, N. 1989. Abundances of the elements: Meteoritic and solar. *Geochim. Cosmochim. Acta,* **53**, pp. 197–214.

Anders, E., and Zinner, E. 1993. Interstellar grains in primitive meteorites: Diamond, silicon carbide, and graphite. *Meteoritics,* **28**, pp. 490–514.

Antz, C., Bavdaz, M., Jessberger, E. K., Knöchel, A., and Wallenwein, R. 1987. Chemical analysis of interplanetary dust particles with synchrotron radiation. *Proc. 10th Europ. Reg. Astron. Meeting IAU,* **2**, pp. 249–252.

Arndt, P., and Flynn, G. J. 1995. On the reliability of PIXE and SXRF microanalyses of interplanetary dust particles. *Meteoritics,* **30**, p. 482.

Arndt, P., Bohsung, J., Maetz, M., and Jessberger, E. K. 1996a. The elemental abundances in interplanetary dust particles. *Meteoritics Planet. Sci.,* **31**, pp. 817–834.

Arndt, P., Jessberger, E. K., Warren, J., and Zolensky, M. 1996b. Bromine contamination of IDPs during collection. *Meteoritics Planet. Sci.,* **31**, p. A8.

Arndt, P., Jessberger, E. K., Maetz, M., Reimold, D., and Traxel, K. 1997. On the Accuracy of element and mass analyses of micron sized samples determined with the Heidelberg proton microprobe with the Heidelberg proton microprobe. *Nucl. Instr. Meth. Phys. Res. B,* **130**, pp. 192–198.

Beck, P., and Kissel, J. 1994. COMA: a cometary matter analyzer for in situ analysis with high mass resolution. *Lunar Planet. Sci.,* bf XXV, pp. 75–76.

Beckerling, W., and Bischoff, A. 1995. Occurrence and composition of relict minerals in micrometeorites from Greenland and Antarctica – implications for their origins. *Planet. Space Sci.,* **43**, pp. 435–449.

Benninghoven, A. 1994. Surface analysis by secondary ion mass spectroscopy SIMS. *Surface Sci.,* **299/300**, pp. 246–260.

Bohsung, J., Arndt, P., and Jessberger, E. K. 1994. Bromine in interplanetary dust particles. *Lunar Planet Sci.,* **XXV**, pp. 139–140.

Bohsung, J., Arndt, P., and Jessberger, E. K. 1995a. Comment on "The bromine content of micrometeorites: Arguments for stratospheric contamination" by F. J. M. Rietmeijer. *J. Geophys. Res.,* **100**, pp. 7549–7550.

Bohsung, J., Arndt, P., Jessberger, E. K., Maetz, M., Traxel, K., and Wallianos, A. 1995b. High resolution PIXE analyses of interplanetary dust particles with the New Heidelberg Proton Microprobe. *Planet. Space Sci.,* **43**, pp. 411–428.

Bonny, Ph., Balageas, D., and Maurette, M. 1988. Entry corridor of micrometeorites containing organic material. *Lunar Planet. Sci.,* **XXI**, p. 111.

Bradley, J. P. 1988. Analysis of chondritic interplanetary dust thin-sections. *Geochim. Cosmochim. Acta,* **52**, pp. 889–900.

Bradley, J. P. 1994. Mechanisms of grain formation, post-accretional alteration, and likely parent body environments of interplanetary dust particles (IDPs). In *Analysis of Interplanetary Dust,* eds. M. E. Zolensky, T. L. Wilson, F. J. M. Rietmeijer and G. J. Flynn (New York: Amer. Inst. Physics), pp. 89–104.

Bradley, J. P. 1995. GEMS and new pre-accretionally irradiated relict grains in interplanetary dust – the plot thickens. *Meteoritics,* **30**, p. 491.

Bradley, J. P., Brownlee, D. E., Fraundorf, P. 1984. Discovery of nuclear tracks in interplanetary dust. *Science,* **226**, pp. 1432–1434.

Bradley, J. P., Sandford, S. A., and Walker, R. M. 1988. Interplanetary dust particles. In *Meteorites and the Early Solar System,* eds. J. F. Kerridge and M. S. Matthews (Tucson: U. Arizona Press), pp. 861–898.

Bradley, J. P., Germani, M. S., Brownlee, D. E. 1989. Automated thin-film analyses of anhydrous interplanetary dust particles in the analytical electron microprobe. *Earth Planet. Sci. Lett.,* **93**, pp. 1–13.

Bradley, J. P., Humecki, H. J., and Germani, M. S. 1992. Combined infrared and analytical microscope studies of interplanetary dust particles. *Astrophys. J.,* **394**, pp. 643–651.

Bradley, J. P., Veblen, D. R., Brownlee, D. E. 1993. Pyroxene whiskers and platelets in interplanetary dust: evidence of vapor phase growth. *Nature,* **301**, pp. 473–477.

Bradley, J. P., Keller, L. P., Brownlee, D. E., and Thomas, K. L. 1996. Reflectance spectroscopy of interplanetary dust particles. *Meteoritics Planet. Sci.,* **31**, pp. 394–402.

Brownlee, D. E. 1981. Extraterrestrial components. In *The Sea*, **vol. 7**, (John Wiley & Sons), p. 773.
Brownlee, D. E. 1985. Cosmic dust: collection and research. *Ann. Rev. Earth Planet. Sci.*, **13**, pp. 147–173.
Brownlee, D. E. 1994. The origin and role of dust in the Early Solar System. In *Analysis of Interplanetary Dust*, eds. M. E. Zolensky, T. L. Wilson, F. J. M. Rietmeijer, and G. J. Flynn (New York: Amer. Inst. Physics), pp. 5–8.
Brownlee, D. E. 1996. The elemental composition of interplanetary dust. In *Physics, Chemistry and Dynamics of Interplanetary Dust*, eds. B. Å. S. Gustafson and M. Hanner, ASP Conf. Ser., **104**, (Astron. Soc. Pacific), pp. 261–264.
Brownlee, D. E. 1997. The elemental composition of cosmic spherules. *Meteoritics Planet. Sci.*, **32**, pp. 157–176.
Brownlee, D. E., Tomandl, D. A., and Olszewski, E. 1977. Interplanetary dust: A new source of extraterrestrial material for laboratory studies. *Proc. Lunar Sci. Conf. VIII*, pp. 149–160.
Butner, H. M., Walker, H. J., Wooden, D. H., and Witteborn, F. C. 1994. Evidence for cometary dust in the disks around beta-Pic-like stars. *Bull. American Astron. Soc.*, **187**, p. 10.
Ceplecha, Z. 1977. Meteoroid populations and orbits. In *Comets, Asteroids, Meteorites*, ed. A. H. Delsemme (University of Toledo Press), pp. 143–152.
Christoffersen, R., and Busek, P. R. 1983. Epsilon carbide: a low temperature component of interplanetary dust particles. *Science*, **222**, pp. 1327–1328.
Christoffersen, R., and Busek, P. R. 1986. Mineralogy of interplanetary dust particles from the "olivine" infrared class. *Earth Planet. Sci. Lett.*, **78**, pp. 53–66.
Cicerone, R. J. 1981. Halogens in the atmosphere. *Rev. Geophys. Space Phys.*, **19**, pp. 123–139.
Engrand, C., Christophe Michel-Levy, M., Jouret, J., Kurat G., Maurette, M., and Perreau, M. 1994. Are the most C-rich Antarctic micrometeorites exotic? *Meteoritics*, **29**, p. 464.
Engrand, C., Deloule, E., Hoppe, P., Kurat, G., Maurette, M., and Robert, F. 1996. Water contents of micrometeorites from Antarctica. *Lunar Planet. Sci.*, **XXVII**, pp. 337–338.
Esat, T. M., Brownlee, D. E., Papanastassiou, D. A., and Wasserburg, G. J. 1979. Magnesium isotopic composition of interplanetary dust particles. *Science*, **206**, pp. 190–197.
Esat, T. M., and Taylor, S. R. 1987. Mg isotopic systematics of some interplanetary dust particles. *Lunar Planet. Sci. Conf. XVIII*, pp. 269–270.
Fleming, R. H., Meeker, G. P., di Brozolo, F. R., and Blake, D. F. 1989. Isotope ratio imaging of interplanetary dust particles. In *Secondary Ion Mass Spectrometry (SIMS VII)*, eds. A. Benninghoven, C. A. Evans, K. D. McKeegan, H. A. Storms, and H. W. Werner (John Wiley & Sons), pp. 389–392.
Flynn, G. J. 1989. Atmospheric entry heating: A criterion to distinguish between asteroidal and cometary sources of interplanetary dust. *Icarus*, **77**, pp. 287–310.
Flynn, G. J. 1994. Interplanetary dust particles collected from the stratosphere: Physical, chemical, and mineralogical properties and implications for their sources. In *Analysis of Interplanetary Dust*, eds. M. E. Zolensky, T. L. Wilson, F. J. M. Rietmeijer, and G. J. Flynn (New York: Amer. Inst. Physics), pp. 127–143.
Flynn, G. J., and Sutton, S. R. 1987. First cosmic dust trace element analyses with the Synchrotron XRF microprobe. *Lunar Planet. Sci.*, **XVIII**, pp. 296–297.
Flynn, G. J., and Sutton, S. R. 1988. Cosmic dust particle densities inferred from SXRF elemental measurements. *Meteoritics*, **23**, pp. 268–269.
Flynn, G. J., and Sutton, S. R. 1990. Synchrotron X-ray fluorescence analyses of stratospheric cosmic dust: New results for chondritic and low-nickel particles. *Proc. Lunar Planet. Sci. Conf. XX*, pp. 335–342.
Flynn, G. J., and Sutton, S. R. 1991. Chemical characterization of seven large area collector particles by SXRF. *Proc. Lunar Planet. Sci. Conf. XXI*, pp. 549–556.
Flynn, G. J., and Sutton, S. R. 1992a. Trace elements in chondritic stratospheric particles: Zinc depletion as a possible indicator of atmospheric entry heating. *Proc. Lunar Planet. Sci. Conf. XXII*, pp. 171–184.
Flynn, G. J., and Sutton, S. R. 1992b. Element abundances in stratospheric cosmic dust: Indications for a new chemical type of chondritic material. *Lunar Planet. Sci.*, **XXIII**, pp. 373–374.

Flynn, G. J., Sutton, S. R., Bajt, S., Klöck, W., Thomas, K. L., and Keller, L. P. 1993. The volatile content of anhydrous interplanetary dust. *Meteoritics*, **28**, p. 349.

Flynn, G. J., Sutton, S. R., Bajt, S., Klöck, W., Thomas, K. L., and Keller, L. P. 1994. Hydrated interplanetary dust particles: Element abundances, mineralogies, and possible relationships to anhydrous IDPs. *Lunar Planet. Sci.*, **XXV**, pp. 381–382.

Fraundorf, P., Hints, O., Lowry, P., Keegan, K. D., and Sandford, S. A. 1982. Determination of the mass, surface density and volume density of individual interplanetary dust particles. *Lunar Planet. Sci.*, **XII**, pp. 225–226.

Germani, M. S., Bradley, J. P., and Brownlee, D. E. 1990. Automated thin-film analyses of hydrated interplanetary dust particles in the analytical electron microscope. *Earth Planet. Sci. Lett.*, **101**, pp. 162–179.

Greenberg, J. M., and Gustafson, B. Å. S. 1981. A comet fragment model for the Zodiacal light particles. *Astron. Astrophys.*, **93**, pp. 35–42.

Greenberg, J. M., and Hage, J. I. 1990. From interstellar dust to comets: a unification of observational constraints. *Astrophys. J.*, **361**, pp. 260–274.

Greshake, A., Hoppe, P., and Bischoff, A. 1996. Mineralogy, chemistry, and oxygen isotopes of refractory inclusions from stratospheric interplanetary dust particles and micrometeorites. *Meteoritics Planet. Sci.*, **31**, pp. 739–748.

Greshake, A., Klöck, W., Arndt, P., Maetz, M., Flynn, G. J., Bajt, S., and Bischoff, A. 1998. Heating experiments simulating atmospheric entry heating of micrometeorites: Clues to their parent body sources. *Meteoritics Planet. Sci.*, **33**, pp. 267–290.

Guan, Y. 1998. Trace and minor elements in ureilites and deuterium-enrichments in several primitive meteorites: Characteristics and geochemical implications. Ph. D. Thesis (St. Louis: Washington University).

Hagemann, R., Nief, G., and Roth, E. 1970. Absolute isotopic scale for deuterium analysis of natural waters, absolute D/H ratios for SMOW. *Tellus*, **22**, pp. 712–715.

Hanner, M. S. 1999. The silicate material in comets. *Space Sci. Rev.*, **90**, pp. 99–108.

Hanner, M. S., Gehrz, R. D., Harker, D. E., Hayward, T. L., Lynch, D. K., Mason, C. G., Russell, R. W., Wooden, D. H., and Woodward, C. E. 1998. Thermal emission from the dust coma of comet Hale-Bopp and the composition of the silicate grains. *Earth, Moon and Planets*, in press.

Hanner, M. S., Lynch, D. K., and Russell, R. W. 1994. The 8–13 micron spectra of comets and the composition of silicate grains. *Astrophys. J.*, **425**, p. 274.

Hoefs, J. 1980. Stable Isotope Geochemistry. (Heidelberg: Springer Verlag), 140 pp.

Hoppe, P., Kurat, G., Walter, J., and Maurette, M. 1995. Trace elements and oxygen isotopes in a CAI-bearing micrometeorite from Antarctica. *Lunar Planet. Sci.*, **XXVI**, pp. 623–624.

Jackson, A. A., and Zook, H. A. 1992. Orbital evolution of dust particles from comets and asteroids. *Icarus*, **97**, pp. 70–84.

Jessberger, E. K. 1991. Discussion: New techniques on the horizon for the analysis of the inorganic cometary components. In *Analysis of Samples from Solar System Bodies*, ed. E. K. Jessberger, *Space Science Reviews*, **56**, pp. 227–231.

Jessberger, E. K. 1999. Rocky cometary particulates: Their elemental, isotopic and mineralogical ingredients. *Space Sci. Rev.*, **90**, pp. 91–97.

Jessberger, E. K., and Kissel, J. 1991. Chemical properties of cometary dust and a note on carbon isotopes. In *Comets in the Post-Halley Era*, eds. R. Newburn, M. Neugebauer, and J. Rahe (Heidelberg: Springer Verlag), pp. 1075–1092.

Jessberger, E. K., Christoforidis, A., and Kissel, J. 1988. Aspects of the major element composition of Halley's dust. *Nature*, **332**, pp. 691–695.

Jessberger, E. K., Kissel, J., and Rahe, J. 1989. The composition of comets. In *Origin and Evolution of Planetary and Satellite Atmospheres*, eds. S. K. Atreya, J. B. Pollak, and M. S. Matthews (Tucson: The University of Arizona Press), pp. 167–191.

Jessberger, E. K., Bohsung, J., Chakaveh, S., and Traxel, K. 1992. The volatile element enrichment of chondritic interplanetary dust particles. *Earth Planet. Sci. Lett.*, **112**, pp. 91–99.

Keller, L. P., Thomas, K. L., and McKay, D. S. 1994. Carbon in primitive interplanetary dust particles. In *Analysis of Interplanetary Dust*, eds. M. E. Zolensky, T. L. Wilson, F. J. M. Rietmeijer, and G. J. Flynn (New York: Amer. Inst. Physics), pp. 159–164.

Klöck, W., and Stadermann, F. J. 1994. Mineralogical and chemical relationships of interplanetary dust particles, micrometeorites and meteorites. In *Analysis of Interplanetary Dust*, eds. M. E. Zolensky, T. L. Wilson, F. J. M. Rietmeijer, and G. J. Flynn (New York: Amer. Inst.), pp. 51–88.

Klöck, W., Thomas, K. L., McKay, D. S., and Palme, H. 1989. Unusual olivine and pyroxene composition in interplanetary dust and unequilibrated ordinary chondrites. *Nature*, **339**, pp. 126–128.

Klöck, W., Flynn, G. J., Sutton, S. R., and Nier, A. O. 1992. Magnetite as evidence of entry heating. *Meteoritics*, **27**, pp. 243–244.

Kornblum, J. J. 1969. Micrometeoroid interaction with the atmosphere. *J. Geophys. Res.*, **74**, pp. 1893–1906.

Kortenkamp, P., and Dermott, S. F. 1998. Accretion of interplanetary dust particles by the Earth. *Icarus*, **135**, pp. 469–495.

Krueger, F. R., Korth, A., and Kissel, J. 1991. The organic matter of comet Halley by joint gas phase and solid phase analysis. *Space Sci. Rev.*, **56**, pp. 167–175.

Kurat, G., Presper, T., Brandstätter, F., and Koeberl, C. 1992. CI-like micrometeorites from Cap Prudhomme, Antarctica. *Lunar Planet. Sci.*, **XXIII**, pp. 747–748.

Kurat, G., Brandstätter, F., Presper, T., Koeberl, C., and Maurette, M. 1993. Micrometeorites. *Russ. Geol. Geophys.*, **34**, pp. 132–147.

Kurat, G., Koeberl, C., Presper, T., Brandstaetter, F., and Maurette, M. 1994a. Antarctic micrometeorites. In *Workshop on the Analysis of Interplanetary Dust Particles*, (Lunar Planetary Inst.), p. 36.

Kurat, G., Koeberl, C., Presper, T., Brandstätter, F., and Maurette, M. 1994b. Petrology and geochemistry of Antarctic micrometeorites. *Geochim. Cosmochim. Acta*, **58**, pp. 3879–3904.

Kurat, G., Hoppe, P., Walter, J., Engrand, C., and Maurette, M. 1994c. Oxygen isotopes in spinels from Antarctic micrometeorites. *Meteoritics*, **29**, pp. 487–488.

Kurat, G., Hoppe, P., and Maurette, M. 1994d. Preliminary report on spinel-rich CAIs in an Antarctic micrometeorite. *Lunar Planet. Sci.*, **XXV**, pp. 763–764.

Kyte, F. T., and Wasson, J. T. 1986. Accretion rate of extraterrestrial matter – iridium deposited 33 to 67 million years ago. *Science*, **232**, pp. 1223–1229.

Lefevre, H. W., Schofield, R. M. S., Overley, J. C., and MacDonald, J. D. 1987. Scanning transmission ion microscopy as it complements particle induced X-ray emission microanalysis. *Scanning Microscopy*, **3**, pp. 879–889.

Leinert, C., and Grün, E., 1990. Interplanetary dust. In *Physics and Chemistry in Space*, eds. R. Schwenn and E. Marsch (Berlin:Springer), *Space and Solar Physics*, pp. 204–275.

Lindstrom, D. J., and Zolensky, M. E. 1990. INA of cosmic dust particles from large area collector. *Lunar Planet. Sci.*, **XXI**, pp. 700–701.

Love, S. G., and Brownlee, D. E. 1991. Heating and thermal transformation of micrometeorites entering the Earth's atmosphere. *Icarus*, **89**, pp. 26–43.

Love, S. G., Joswiak, D. J., and Brownlee, D. E. 1994. Densities of stratospheric micrometeorites *Icarus*, **111**, pp. 227–236.

Maas, D., Krueger, F. R., and Kissel, J. 1989. Mass and density of silicate and CHON-type dust particles released by comet p/Halley. In *Asteroids Comets Meteors III*, eds. C.-I. Lagerkvist, H. Rickmann, B. A. Lindblad, and M. Lindgren (Uppsala: Reprocentralen HSC), pp. 389–392.

Mackinnon, I. D. R., and Rietmeijer, F. J. M. 1987. Mineralogy of chondritic interplanetary dust particles. *Rev. Geophys.*, **25**, pp. 1527–1553.

Maetz, M. 1994. Scanning Transmission Ion Microscopy zur Bestimmung von Dichteprofilen von Interplanetaren Staubteilchen. Diploma Thesis (University of Heidelberg), 89 pp.

Maetz, M., Arndt, P., Bohsung J., Jessberger, E. K., and Traxel, K. 1994. Comprehensive analysis of six IDPs with the Heidelberg proton microprobe. *Meteoritics Planet. Sci.*, **29**, pp. 494–495.

Maetz, M., Arndt, P., Greshake, A., Jessberger, E. K., Klöck, W., and Traxel, K. 1996. Structural and chemical modifications of microsamples induced during PIXE analyses. *Nucl. Instr. Methods*, **B 109/110**, pp. 192–196.

Massart, D. L., and Kaufman, L. 1983. The interpretation of analytical chemical data by the use of cluster analysis. (John Wiley & Sons), 235 pp.

Maurette, M., Hammer, C., Brownlee, D. E., Reeh, N., and Thomsen, H. H. 1986. Placers of cosmic dust in the blue ice lakes of Greenland. *Science*, **233**, pp. 869–872.

Maurette, M., Olinger, C., Christophe Michel-Levy, M., Kurat, G., Pourchet, M., Brandstätter, F., and Bourot-Denise, M. 1991. A collection of diverse micrometeorites recovered from 100 tons of Antarctic blue ice. *Nature*, **351**, pp. 44–47.

Maurette, M., Kurat, G., Perreau, M., and Engrand, C. 1993. Microanalysis of Cap Prudhomme Antarctic meteorites. *Microbeam Analysis*, **2**, pp. 239–251.

Maurette, M., Immel, G., Hammer, C., Harvey, R., Kurat, G., and Taylor, S. 1994. Collection and curation of IDPs from the Greenland and Antarctic ice sheets. In *Analysis of Interplanetary Dust*, eds. M. E. Zolensky, T. L. Wilson, F. J. M. Rietmeijer and G. J. Flynn (New York: Amer. Inst. Physics), pp. 277–289.

Maurette, M., Engrand, C., and Kurat, G. 1996. Collection and Microanalysis of Antarctic Micrometeorites. In *Physics, Chemistry, and Dynamics of Interplanetary Dust*, eds. B. Å. S. Gustafson and M. S. Hanner, ASP Conf. Ser., Vol. 104, pp. 265–273.

McKeegan, K. D. 1987a. Ion microprobe measurements of H, C, O, Mg, and Si isotopic abundances in individual interplanetary dust particles. Ph. D. thesis (Washington University).

McKeegan, K. D. 1987b. Oxygen isotopes in refractory stratospheric dust particles: proof of extraterrestrial origin. *Science*, **237**, pp. 1468–1471.

McKeegan, K. D., Walker, R. M., and Zinner, E. 1985. Ion microprobe isotopic measurements of individual interplanetary dust particles. *Geochim. Cosmochim. Acta*, **49**, pp. 1971–1987.

McKeegan, K. D., Swan, P., Walker, R. M., Wopenka, B., and Zinner, E. 1987. Hydrogen isotopic variations in interplanetary dust particles. *Lunar Planet. Sci.*, **XVIII**, pp. 627–628.

Messenger, S. R. 1997. Combined molecular and isotopic analysis of circumstellar and interplanetary dust. Ph. D. thesis (Saint Louis: Washington University).

Messenger, S., and Walker, R. M. 1997. Evidence for molecular cloud material in meteorites and interplanetary dust. In *Astrophysical implications of the laboratory study of presolar materials*, eds. T. J. Bernatowicz and E. Zinner (New York: Amer. Inst. Physics), pp. 545–564.

Messenger, S., Walker, R. M., Clemett, S. J., and Zare, R. N. 1996. Deuterium enrichments in cluster IDPs. *Lunar Planet. Sci.*, **XXVII**, pp. 867–868.

Nier, A. O. 1994. Helium and neon in interplanetary dust particles. In *Analysis of Interplanetary Dust*, eds. M. E. Zolensky, T. L. Wilson, F. J. M. Rietmeijer and G. J. Flynn (New York: Amer. Inst. Physics), pp. 115–126.

Nier, A. O., and Schlutter, D. J. 1990. He and Ne isotopes in individual stratospheric particles – a further study. *Lunar Planet. Sci.*, **XXI**, pp. 883–884.

Nier, A. O., and Schlutter, D. J. 1993. The thermal history of interplanetary dust particles collected in the Earth's stratosphere. *Meteoritics*, **28**, pp. 675–681.

Olinger, C. T., Maurette, M., Walker, R. M., and Hohenberg, C. M. 1990. Neon measurements of individual Greenland sediment particles: proof of an extraterrestrial origin and comparison with EDX and morphological analyses. *Earth Planet. Sci. Lett.*, **100**, pp. 77–93.

Perreau, M., Engrand, C., Maurette, M., Kurat, G., and Presper, T. 1993. C/O atomic ratios in micrometer-sized crushed grains from Antarctic micrometeorites and two carbonaceous meteorites. *Lunar Planet. Sci.*, **XXIV**, pp. 1125–1126.

Presper, T., Kurat, G., Koeberl, C., Palme, H., and Maurette, M. 1993. Elemental depletions in Antarctic micrometeorites and Arctic cosmic spherules: comparison and relationships. *Lunar Planet. Sci.*, **XXIV**, pp. 1177–1178.

Rietmeijer, F. J. M. 1993. The bromine content of micrometeorites: Arguments for stratospheric contamination. *J. Geophys. Res.*, **98**, pp. 7409–7414.

Rietmeijer, F. J. M. 1994. A proposal for a petrological classification scheme of carbonaceous chondritic micrometeorites. In *Analysis of Interplanetary Dust*, eds. M. E. Zolensky, T. L. Wilson, F. J. M. Rietmeijer and G. J. Flynn (New York: Amer. Inst. Physics), pp. 231–240.

Rost, D., Stephan, T., and Jessberger, E. K. 1996. Surface analysis of stratospheric dust particles with TOF-SIMS: New results. *Meteoritics Planet. Sci.*, **31**, pp. A118–A119.

Rost, D., Stephan, T., and Jessberger, E. K. 1999. Surface analysis of stratospheric dust particles, *Meteoritics Planet. Sci.*, **34**, pp. 637–646.

Sandford, S. A., and Walker, R. M. 1985. Laboratory and infrared transmission spectra of

interplanetary dust particles from 2.5 to 25 microns. *Astrophys. J.*, **291**, pp. 838–851.
SAS Institute Inc. 1988 SAS/STAT User's Guide. SAS Institute Inc. Cary, NC.
Schramm, L. S., Brownlee, D. E., and Wheelock, M. M. 1989. Major element composition of stratospheric micrometeorites. *Meteoritics*, **24**, pp. 99–112.
Schwieters, J., Cramer, H.-G., Heller, T., Jürgens, U., Niehuis, E., Zehnpfenning, J., and Benninghoven, A. 1991. High mass resolution surface imaging with a time-of-flight secondary ion mass spectroscopy scanning microprobe. *J. Vac. Sci. Technol.*, **A9**, pp. 2864–2871.
Sekanina, S., Hanner, M., Jessberger, E. K., and Fomekova, M. 1998. The chemical and isotopic composition of cometary dust. In *Interplanetary Dust*, eds. E. Grün, H. Fechtig, and B. Å. S. Gustafson, this volume.
Stadermann, F. J. 1990. Messung von Isotopen und Elementhäufigkeiten in einzelnen Interplanetaren Staubteilchen mittels Sekundärionen-Massenspektrometrie. Ph. D. Thesis (Universität Heidelberg), 97 pp.
Stephan, T., Klöck, W., Jessberger, E. K., Thomas, K. L., Keller, L. P., and Behla, F. 1993. Multielement analysis of carbon-rich interplanetary dust particles with TOF-SIMS. *Meteoritics*, **28**, pp. 443–444.
Stephan, T., Jessberger, E. K., Klöck, W., Rulle, H., and Zehnpfenning, J. 1994a. TOF-SIMS analysis of interplanetary dust. *Earth Planet. Sci. Lett.*, bf 128, pp. 453–467.
Stephan, T., Jessberger, E. K., Rulle, H., Thomas, K. L., and Klöck, W. 1994b. New TOF-SIMS results on hydrated interplanetary dust particles. *Lunar Planet. Sci.*, **XXV**, pp. 1341–1342.
Stephan, T., Thomas, K. L., and Warren, J. L. 1994c. Comprehensive consortium study of stratospheric particles from one collector. *Meteoritics*, **29**, pp. 536–537.
Stephan, T., Thomas, K. L., and Warren, J. L. 1995. Particles from collection flag U2071. *Stratospheric dust catalog, vol. 1*, (Heidelberg: MPI-Kernphysik), 121 pp.
Strait, M. M., Thomas, K. L., McKay, D. S. 1995. Porosity of an anhydrous chondritic interplanetary dust particle. *Meteoritics*, **30**, pp. 583-584.
Sutton, S. R. and Flynn, G. J. 1988. Stratospheric particles: Synchrotron X-ray fluorescence determination of trace element contents. *Proc. Lunar. Planet. Sci. Conf. 18*, pp. 607–614.
Thomas, K. L., Blanford, G. E., Keller, L. P., Klöck, W., and McKay, D. S. 1993. Carbon abundance and silicate mineralogy of hydrous interplanetary dust particles. *Geochim. Cosmochim. Acta*, **57**, pp. 1551–1566.
Thomas, K. L., Keller, L. P., Blanford, G. E., and McKay, D. S. 1994. Quantitative analyses of carbon in anhydrous and hydrated interplanetary dust particles. In *Analysis of Interplanetary Dust*, eds. M. E. Zolensky, T. L. Wilson, F. J. M. Rietmeijer and G. J. Flynn (New York: Amer. Inst. Physics), pp. 165–172.
Thomas, K. L., Blanford, G. E., Clemett, S. J., Flynn, G. J., Keller, L. P., Klöck, W., Maechling, C. R., McKay, D. S., Messenger, S., Nier, A. O., Schlutter, D. J., Sutton, S. R., Warren, J. L., and Zare, R. N. 1995. An asteroidal breccia: The anatomy of a cluster IDP. *Geochim. Cosmochim. Acta*, **59**, pp. 2797–2815.
Tielens, A. G. G. M. 1997. Deuterium and interstellar chemical processes. In *Astrophysical Implications of the Laboratory Study of Presolar Materials, Vol. CP402*, eds. T. J. Bernatowicz and E. Zinner (New York: Amer. Inst. Physics), pp. 523–544.
van der Stap, C. C. A. H., Vis, R. D., and Verheul, H. 1986. Interplanetary dust: Arguments in favour of a late stage nebular origin of the chondritic aggregates. *Lunar Planet. Sci.*, **XVII**, pp. 1013–1014.
Walker, R. M. 1991. Comments on the analysis of returned cometary samples. In *Analysis of Samples from Solar System Bodies*, ed. E. K. Jessberger, *Space Science Reviews*, **56**, pp. 213–226.
Wallenwein, R., Antz, C., Jessberger, E. K., and Traxel, K. 1987. Proton microprobe analysis of interplanetary dust particles. In *Proc. 10th Europ. Reg. Astron. Meeting IAU 2*, pp. 245–248.
Walter, J., Kurat, G., Brandstätter, F., Presper, T., Koeberl, C., and Maurette, M. 1994. The chemical compositions of olivines and pyroxenes from Antarctic micrometeorites. *Meteoritics*, **29**, pp. 545–546.
Walter, J., Kurat, G., Brandstätter, F., Presper, T., Koeberl, C., and Maurette, M. 1995. The abundance of ordinary chondrite debris among Antarctic micrometeorites. *Meteoritics*,

30, pp. 592–593.

Warren, J. L., and Zolensky, M. E. 1994. Collection and curation of interplanetary dust particles recovered from the stratosphere by NASA. In *Analysis of Interplanetary Dust*, eds. M. E. Zolensky, T. L. Wilson, F. J. M. Rietmeijer, and G. J. Flynn (New York: Amer. Inst. Physics), pp. 245–254.

Wasson, J. T. 1985. Meteorites: Their record of early Solar-System history. (New York: Freeman Co.), 267 pp.

Weissman, P. R. 1984. The Vega particulate shell – comets or asteroids? *Science*, **224**, pp. 987–989.

Xu, Y., Song, L., Zhang, Y., and Fan, C. Y. (1994). ^{6}Li/^{7}Li, ^{10}B/^{11}B and ^{7}Li/^{11}B/^{28}Si in individual interplanetary dust particles. In *Analysis of Interplanetary Dust*, eds. M. E. Zolensky, T. L. Wilson, F. J. M. Rietmeijer and G. J. Flynn (New York: Amer. Inst.), pp. 211–222.

Zinner, E. 1988. Interstellar cloud material in meteorites. In *Meteorites and the Early Solar System* eds. J. F. Kerridge and M. S. Matthews (University of Arizona Press), pp. 956–983.

Zinner, E. K., McKeegan, K. D., and Walker, R. M. 1983. Laboratory measurements of D/H ratios in interplanetary dust. *Nature*, **305**, pp. 119–121.

Zolensky, M. E., and Barrett, R. 1994. Compositional variations of olivines and pyroxenes in chondritic interplanetary dust particles. In *Analysis of Interplanetary Dust*, eds. M. E. Zolensky, T. L. Wilson, F. J. M. Rietmeijer and G. J. Flynn (New York: Amer. Inst. Physics), pp. 1–90.

Zolensky, M. E., and Thomas, K. L. 1995. Iron and iron-nickel sulfides in chondritic interplanetary dust particles. *Geochim. Cosmochim. Acta*, **59**, pp. 4707–4712.

Zolensky, M. E., Lindstrom, D. J., Thomas, K. L., Lindstrom, R. M., and Lindstrom, M. M. 1989. Trace element compositions of six "chondritic" stratospheric dust particles. *Lunar Planet. Sci.*, **XIX**, pp. 1255–1256.

In Situ Measurements of Cosmic Dust

Eberhard Grün[1], Michael Baguhl[1], Håkan Svedhem[2], Herbert A. Zook[3,†]

[1] Max-Planck-Institut für Kernphysik, Heidelberg, Germany
[2] ESA, ESTEC, Noordwijk, The Netherlands
[3] NASA, Johnson Space Center, Houston, TX, U.S.A.
† deceased

Abstract. In-situ measurements of cosmic dust provide information on the spatial and orbital distributions, and on the physical and chemical properties of dust in interplanetary space. Pioneers 8 through 11, Helios, Galileo and Ulysses spaceprobes measured interplanetary dust from 0.3 to 18 AU distance from the Sun. The Earth satellites HEOS, Hiten, as well as other spacecraft, determined the flux of micrometeoroids at 1 AU. The size distribution from a few micrometer to millimeter range was also characterized by analysis of lunar microcraters and later verified by near-Earth satellites like LDEF. Distinctly different populations of dust particles exist throughout the solar system. In the inner solar system, out to about 3 AU, zodiacal dust particles are observed by in-situ detection from spaceprobes. These particles orbit the Sun on low inclination ($i < 30°$) and moderately eccentric ($e < 0.6$) orbits. Their spatial density falls off somewhat faster than the inverse of the solar distance. In addition, particles on highly eccentric orbits have been recorded in the inner solar system. β-meteoroids leave the solar system on unbound (hyperbolic) orbits because of the action of solar radiation pressure and electromagnetic forces. They have been observed by the Pioneers 8, 9, Hiten, and by the Ulysses spacecraft. Pioneers 10 and 11 showed that dust particles on high inclination or even retrograde trajectories compose much of the population of big dust particles ($> 10~\mu$m) outside about 3 AU. The dust detectors onboard the Ulysses and Galileo spaceprobes identified micrometer sized interstellar dust sweeping through the solar system. Within a distance from Jupiter of about 2 AU Ulysses and Galileo observed streams of tiny grains originating from within the jovian system.

I. INTRODUCTION

An early motivation for the study of dust in space was the risk imposed by impacts of natural meteoroids onto man-made satellites (Whipple 1958). Ground-based observations of the zodiacal light and of meteors had demonstrated that there is plenty of particulate matter in space of sizes ranging from micrometers to centimeters and bigger traveling at speeds of tens of km s^{-1}. Laboratory experiments had shown that projectiles impacting onto a metal sheet at speeds in excess of 5 km s^{-1} could penetrate it even if the sheet was 2 to 5 times thicker than the projectile dimension. Since protection of delicate space hardware is expensive (the cost of lifting one kilogram of payload into space was and still is of the order of $ 20 000) engineers implement only the minimum necessary protection. For considerations like these one had to know the total number of meteoroids and the size of the biggest one that could have an effect on any given satellite during its useful life time. For early small research satellites of short exposure times in space this issue was not as critical as for space stations and for big modern commercial satellites that are planned for long operational life times.

The relevant number to quantify the impact hazard by meteoroids is the penetration rate, i.e. the number of impacts that could penetrate a shield of a given area and thickness in a given time interval. This penetration rate has two components: the meteoroid flux and the penetration limit. The first quantity describes the meteoroid flux at a given position in space. The cumulative meteoroid flux is the number of particles of mass m or bigger that pass through a cross sectional area of one m^2 in one second. This quantity has to be provided by direct (in-situ) measurements and/or by a meteoroid model (e.g., Divine 1993 and chapter by Staubach et al.). The penetration limit describes the protective capability of a shield. It refers to the smallest particle that can penetrate the shield at a given impact speed. This second quantity is determined by laboratory experiments (Cour-Palais 1979) or by impact models (Wagner and Kreyenhagen 1979). In this chapter we are only concerned with the first aspect of the problem which we will concentrate on.

Among the first dust instruments flown were simple detectors which in some cases measured the penetration rate directly by exposing a piece of shield to the space environment and recording the number of penetrations that occurred in a given time interval (for a comprehensive review see McDonnell 1978). Detectors of this type include the "beer can" experiments that were successfully flown on several early satellites and spaceprobes (e.g., Explorer 16 and 23, Pioneers 10 and 11). These detectors consisted of a large number of pressurized cells that recorded the decrease in gas pressure that occurs when a wall was punctured by a meteoroid penetration. The walls were metal sheets (stainless steel or copper beryllium) of 25 or 50 μm thickness that had penetration limits of 10^{-9} and 10^{-8} g at 20 km s^{-1} impact speed (Humes 1974, 1980). The Pegasus detectors were large area (about 200 m^2) detectors behind a shield that recorded penetrations by the discharge of a capacitor that was triggered by the penetration itself (Naumann 1966 and Dozier 1966). Detectors like these determined the flux of meteoroids in near-Earth space in the 10 to 100 μm-size range. This range was very helpful for assessing the meteoroid hazard to typical satellites.

Detectors for μm-sized meteoroids used other phenomena, e.g., microphonics, in order to detect impacts of dust particles. Many of these first instruments were unreliable devices that responded not only to impacts but also to mechanical, thermal or electrical interferences. A dust belt around the Earth was suggested by these initial measurements (see chapter on Historical Development by Fechtig et al.) which was only dismissed years later when instruments had matured enough to reduce this noise by several orders of magnitude. Modern dust detectors are able to reliably detect a single impact of a micrometer sized meteoroid in one month or longer.

Early on it was realized (e.g. Whipple 1967) that dust lifetimes are limited by collisions, and by Poynting-Robertson and solar wind drag, much shorter than the age of the solar system. These effects along with gravitational scattering act to disperse meteoroids in space. Consequently, interplanetary dust must have presently existing sources such as meteoroids, comets and asteroids. Therefore, the motivation for the study of dust in space has shifted to astrophysical questions about the physical and chemical properties of the grains, their sources and sinks and the processes acting on them. Dust grains are now considered as test particles that probe their environment and provide information on their sources.

The importance of studying dust in the solar system has long been questioned because of a presumed weak coupling to other planetary objects and phenomena. Only in obviously dusty environments as in comets, has dust been considered important. The observation of microcraters on lunar rocks, the discovery of dust rings around most outer planets, and finally the detection of dust disks around nearby stars has awakened new interest in the study of dust. Recently, new phenomena have been discovered that show effects of dust. Numerical modeling of meteoroid orbital evolution by Jackson and Zook (1989) showed that there should be a heliocentric ring of dust gravitationally shepherded by the Earth. The predicted ring was observationally identified by Dermott et al. (1994) in the IRAS (Infra-Red Astronomy Satellite) data and confirmed by Reach et al. (1995) with COBE (Cosmic Background Explorer) data. The jovian ring has first been suspected from the drop in the energetic particle flux observed by Pioneer 11 (Fillius 1976). The composition of pick-up ions in the solar wind shows traces of interstellar dust evaporating near the Sun (Geiss et al. 1996). Signatures in the magnetic field measurements by the Phobos spacecraft near Mars (Baumgärtel et al. 1996) have been suspected to be caused by dust in the Martian environment.

Dust has been observed by astronomical means in a wide range of environments: as the F-corona near the Sun, as zodiacal light in interplanetary space, as thermal emission in the asteroid belt, it has also been detected in comae and tails of comets, in rings around planets, in interstellar space and around other stars. These remote sensing observations provide a picture of the spatial distribution of dust and information on its optical properties. Information on the size distribution of dust, its physical and chemical properties, and the trajectories the particles move on could in most cases not be obtained directly.

Addressing these questions is the domain of in-situ detectors which have been developed over the past decades. The development went from simple penetration detectors to multi-coincidence dust analyzers with capabilities to determine the mass, impact speed, trajectories, electrical charge, and even the chemical composition of dust (see the chapter on Instrumentation by Auer). Impact ionization has been found to be the most sensitive and versatile method to detect and analyze small dust particles in space. This method is based on the fact that high velocity impacts (above about 1 km s^{-1}) of dust particles on a solid target produce a small crater and vaporize and even ionize material from both the projectile and the target. The so generated ions and electrons are separated by an electric field in front of the target, collected onto electrodes and converted into electronic signals by charge sensitive amplifiers. The amplitude and rise time of these signals are functions of the mass and speed of the impacting particle and the mass spectrum of the ions relates to the chemical composition of the impacting grain.

Because of the complexity of the impact process our understanding of it is far from complete, therefore, the correspondence between the measured impact signals and the particle properties is mostly obtained by empirical calibration of the detectors with high speed simulated dust particles from dust accelerators. The accuracy of the derived meteoroid parameters depends to a large part on how well the projectiles, with which the instruments were calibrated, represent the meteoroids encountered in space. Unfortunately, the range of dust particles, in mass, speed and composition, provided by dust accelerators is rather limited. Therefore, in many cases, meteoroid parameters have to be derived by extrapolation from calibrations. In some cases analyses of the results obtained in space provide means to determine the masses and speeds of grains well beyond any previous calibration (cf. dust stream analysis by Zook et al. 1996). By combination of information from different methods, a consistent picture (model) of the whole meteoritic complex is emerging in which results from in-situ measurement play a crucial role. Related topics are discussed in the other chapters of this book. Especially closely connected to this chapter are the chapters on Instrumentation (Auer), Near-Earth Environment (McDonnell et al.), Zodiacal light (Levasseur-Regourd et al.), Physical Processes (Mukai et al.), Synthesis (Staubach et al.), and Evolution (Dermott et al.). Some overlap with these chapters is intended in order to show this relationship.

A previous review of in-situ dust measurements was given by McDonnell, (1978). Since then much progress has been made through new measurements with advanced and more sensitive detectors that provided statistically significant results on interplanetary dust in regions of space not visited before and on unexpected phenomena. However, before we discuss these results we have to understand the characteristics of detectors and missions and their constraints and biases on the results—this is done in section II. The most significant results from in-situ dust measurements are discussed in the subsequent three sections. In section III we discuss dust measurements at 1 AU especially with respect to the size distributions of the interplanetary dust flux. The radial distribution of dust and its variation within the zodiacal cloud out to about 3 AU are presented in section IV. Measurements in the outer solar system and characteristics of planetary and interstellar contributions to dust in interplanetary space are discussed in section V. In section VI we draw a picture of the characteristics of the dust complex based on observations by spacecraft detectors and we conclude with a discussion of future developments.

II. CHARACTERISTICS OF IN-SITU DUST MEASUREMENTS IN SPACE

In-situ dust measurements in space have unique characteristics that are described in this section. We start with a general overview of missions and detectors. Reliable identification of meteoroid impacts is the mandatory prerequisite of any useful dust measurement—first, we will discuss some aspects of it. Understanding the noise environment of space measurements is necessary to make meaningful dead time corrections of the impact rate. This is demonstrated in the case of Helios dust measurements. Dust measurements deal generally with small numbers of recorded particles. To make the most out of them some useful statistical methods are described. Finally, problems of dust detection geometry and of orbit determination are discussed.

II.A. Dust Missions and Detectors

The objective of dust measurements in space is the characterization of physical and chemical properties of micrometeoroids and the determination of their orbits. The result will be a multi-parameter distribution of all properties describing the meteoric complex. At present we are far from such a complete description. We have only sparse information on several of the relevant parameters and for only a few parameters (like masses and some orbital elements) can we specify their distributions.

There are obvious constraints of in-situ dust measurements set by the space mission by which the measurements were obtained. In-situ dust detectors only record and analyze meteoroids that intersect their trajectories. Therefore, several spacecraft on different trajectories are required to build up a fairly complete three-dimensional picture of the interplanetary dust complex. An underlying assumption is that the spatial dust distribution is rather smooth and that it does not vary on time scales shorter than several years. Both assumptions are supported by zodiacal light observations if one ignores localized phenomena like cometary activity or dust emissions from the planet Jupiter.

In-situ measurements of interplanetary dust have been performed in the heliocentric distance range from 0.3 AU out to 18 AU. Table 1 gives some details about the missions, spaceprobes, and dust detectors. Two types of impact detectors were used for interplanetary dust measurements: impact ionization detectors with detection thresholds ranging from 10^{-16} to 10^{-13} g and penetration detectors with detection thresholds of 10^{-9} and 10^{-8} g at impact speeds of 20 km s^{-1}. Most impact ionization detectors have sensitive areas of only 0.01 m^2, except the Galileo and Ulysses instruments (Grün et al. 1992a, b) that have ten times larger sensitive areas. The fields-of-view (FOVs) or corresponding angular sensitivities of the HEOS 2, Helios 1/2, Ulysses, and Galileo detectors are significantly less than that of a flat plate (for reference, a flat plate has π sr effective solid angle interval), therefore, they are able to provide improved directional information. For each detector a compromise has been made between the angular accuracy and the expected count rate that is determined by the width of the FOV. Only in the case of the Pioneer 8 and 9

Table 1

Characteristics of in-situ dust measurements in interplanetary space. Distance ranges are those in which dust measurements were obtained. Spacecraft spin axis directions are: N, perpendicular to ecliptic plane; S, Sun pointing; E, Earth pointing; var., variable in plane perpendicular to Earth-Sun line. Sensor orientations are given with respect to spin axis. The mass thresholds refer to 20 km s^{-1} impact speed. The effective solid angles correspond to the sensor field-of-view, and the dynamic measurement range refers to the mass determination.

Spacecraft	distance range (AU)	spin axis direction	sensor orientation (deg.)	mass threshold (g)	sensitive area (m^2)	solid angle (sr)	dynamic range
Helios 1/2	0.3–1	N	65, 134	$9 \cdot 10^{-15}$	0.012	1.23	10^4
Galileo	0.7–5.4	S, E	120	$4 \cdot 10^{-15}$	0.1	1.4	10^6
Pioneer 9	0.75–0.99	N	90	$2 \cdot 10^{-13}$	0.0074	2.9	200
Pioneer 8	0.97–1.09	N	90	$2 \cdot 10^{-13}$	0.0094	2.9	200
HEOS 2	1	var.	0	$2 \cdot 10^{-16}$	0.01	1.03	10^4
Hiten	1	N	90	$2 \cdot 10^{-15}$	0.01	1.5	$3 \cdot 10^4$
Ulysses	1–5.4	E	85	$4 \cdot 10^{-15}$	0.1	1.4	10^6
Pioneer 10	1–18	E	180	$8 \cdot 10^{-10}$	0.26[1]	2.8	1
Pioneer 11	1–10	E	180	$6 \cdot 10^{-9}$	0.56[1]	2.8	1

[1] initial area, actual area decreased as cells were punctured

detectors a higher angular accuracy was obtained for those particles for which rough trajectories were measured (Berg and Grün 1973). The penetration detectors flown onboard the Pioneer 10 and 11 spacecraft (Humes 1980) in the outer solar system have large initial sensitive areas and wide FOVs.

For most dust detectors, the signal that is used to detect a dust impact (e.g. the impact charge) depends both on the mass and speed of the impacting particle. Therefore, the sensitivity threshold, i.e. the smallest particle mass for which an impact signal can be reliably identified, is a function of the impact speed. In interplanetary space impact speeds of meteoroids vary by about a factor 20 or larger. As a consequence, signals from impacts of equal massive particles but at different impact speeds may vary over a very large range (e.g. factor 35 000 for the impact charge signal). Therefore, instruments should have the widest possible dynamic range over which impact signals can be measured. For bigger particles the impact signals may be saturated and only lower mass limits can be stated. A dynamic range of 1 for the Pioneer 10 and 11 instruments implies that only a lower mass limit for all recorded impacts can be stated.

At 1 AU dust flux measurements over a wide mass range (10^{-18} g to 10^{-3} g) were provided by several satellite detectors, the Long Duration Exposure Facility (LDEF; Humes 1993), and lunar microcrater statistics (cf. chapter on Near-Earth Environment by McDonnell et al.). As a consequence, the coverage of the meteoritic mass range is most complete at 1 AU. Both inside and outside this heliocentric distance the mass range covered is limited to the sensitive range of a few in-situ detectors. Dust has been measured near the ecliptic plane in the heliocentric distance range from 0.3 (Helios) to 18 AU (Pioneer 10). At distances between 1.3 and 5.4 AU Ulysses covered a latitude range almost from pole to pole.

As already noted only a limited number of different projectile compositions have been employed in the calibration of dust detectors. Carbon, iron and glass projectiles were used as meteoroid analogs to calibrate the Galileo and Ulysses dust sensors. Measurements of planetary, interplanetary, or interstellar particles of drastically different composition will, therefore, have systematic errors because of unknown impact ionization magnitude. Resolution of this problem may come from future instruments (Cassini, Stardust) that combine dust detectors with chemical composition analyzers.

Recent observations of dust streams from Jupiter (Zook et al. 1996) demonstrate the coupling between dust and the interplanetary magnetic field. Therefore, measurements of the electric charge become increasingly important. Early charge measurement attempts with the Helios (Grün 1981; Leinert and Grün 1990), Galileo and Ulysses (Svestka et al. 1996) detectors were not fully conclusive. The electric charge carried by the dust particle is orders of magnitude smaller than the charge signal they generate upon impact onto the detector and hence it is difficult to detect. A spherical particle of 20 μm radius and potential of +5 Volts, carries a charge of $1.1 \cdot 10^{-14}$ C which is just at the threshold of the Galileo detector before noise is considered. A new attempt to determine the electric charge on dust particles is currently being made with the Cassini Cosmic Dust Analyzer (Srama et al. 1996) which has increased charge sensitivity and an improved noise reduction scheme.

II.B. Reliability of Impact Detection and Impact Rate Measurements

Most dust detectors respond to both dust impacts and to noise events. The susceptibility to different types of noise depends on the detection principle used. Dust detectors for small dust particles are generally more susceptible to noise than detectors for bigger meteoroids. In the space environment noise sources become important, even when no noise was detected during tests carried out on the ground. Among the noise sources found in previous dust instruments were mechanical vibrations and thermal cracks onboard the spacecraft, interference from other instruments, radiation from radioactive sources, photoelectrons from solar UV, energetic particles from solar flares and planetary magnetospheres, and cosmic rays. E.g. energetic particle from solar flares and photoelectrons from solar UV were identified in the data from the Ulysses dust detector (Baguhl et al. 1993).

In such a noisy environment, modern multi-coincidence dust detectors have the advantage that they generally provide some means of identification and analysis of even unexpected noise sources. Thereby, methods can be developed that reliably distinguish noise from dust impact events. Such an analysis and the resulting learning effect allowed the Galileo and Ulysses investigators to increase the sensitivity of the instruments significantly (Baguhl et al. 1993). By reprogramming Galileo's onboard data processing computer this detector reached the same sensitivity as the Ulysses detector despite its data transmission rate to the ground was lower by more than a factor hundred (Grün et al. 1995).

Even in the case of a perfect distinction between noise and dust impacts there is still the dead time effect caused during times of high noise rates. This is due to the fact that an instrument can, at any given time, detect and analyze only a noise event or a dust impact. Each measurement takes a certain time to be performed (e.g. for the Galileo and Ulysses instruments this time is about 10 milliseconds) before the next event can be analyzed. This time is missing from the total available time and it has to be considered when a true dust impact rate is determined. Generally this effect is rather small (≤ 1 %) if the noise or impact rate is less than 1 per second (for the Galileo and Ulysses instruments) and even if the noise rate exceeds the impact rate by large factors.

The Pioneer 10 and 11 instruments that are considered noise-free also experienced the effect of dead time. They were designed to record reliably the very low impact rate of 10 μm-sized and bigger meteoroids in the asteroid belt. The implementation of this design goal caused difficulties in the more dusty planetary environments. It was only realized after Voyager's discovery of the jovian ring that Pioneer 11 several years earlier had passed through the outer fringes of this ring during its very close flyby of Jupiter. Two impacts were recorded close to the ring position. However, since the instrument had a built-in dead time of 87 minutes after each detection more penetrations may have happened during this passage. This dead time was introduced into the instrument design in order to avoid multiple counts from marginal penetrations for which the pressure release from the cell may be very slow. The same effect may have caused also a too small number of detections during the fly-through of Saturn's ring system (Humes 1980). A principal way to check this effect was to see whether the total number of recorded penetrations eventually reached the number of pressurized cells of this detector. However, this check was prevented because beyond 20 AU the detectors responded to the very low ambient temperatures and the consequently freezing of the fill gas (argon and nitrogen, Humes, personal communication). The pressure in the cells dropped and the electrical discharge (which is used to detect the pressure drop) triggered. This discharge heated the gas and the pressure increased again. After some time the whole process repeated. Because of this effect no useful data were obtained from the Pioneer 10 and 11 detectors beyond 20 AU.

Early dust detectors (before about 1980) were not controlled by microprocessors and processed dust impacts and noise events according to programs that were "hard-wired" into the instrument. These detectors could not be flexibly adjusted to their environment by reprogramming and hence they were much more affected by unexpected noise. The distinction between impacts and noise could only be made on the ground on the basis of the transmitted data. In these cases there is an additional dead time which depends on the ratio of the data transmission rate to the noise rate. When this ratio was < 1 most data received on Earth were noisy. For Helios the noise rate was highest closest to the Sun. Also the data transmission rate was a function of distance to the Earth, and the angular separation of Helios from the Sun as seen from the Earth. Therefore, for the determination of the dust impact rate all these effects had to be carefully considered which resulted in a measuring time for small impacts of only 35 % of the total available time. For bigger impacts the measuring time was about 60 %. The Helios 1 dust detector recorded 235 impacts of which information was received on the ground during the first 5 years or 10 revolutions about the Sun. Because of an increased noise rate onboard the second Helios spaceprobe, information on only 20 impacts was received on the ground from this instrument. In this situation it was impossible to estimate the dead time for this detector and hence no reliable dust flux measurement was obtained.

II.C. Small Number Statistics

Generally, results from in-situ dust detectors suffer from the relatively small number of recorded impacts and hence from limited statistical accuracy. Therefore, methods have to be employed that allow us to get the most meaningful results out of these measurements.

The number of meteoroids, X, expected to impact a space-exposed surface area, A, is proportional to the area-time product, At, and is also proportional to the meteoroid flux, F. If At is made small, so that $X \ll 1$, then X becomes the probability of an impact occurring with exposure At. These probabilities are additive and Poisson statistics can be applied (Feller 1957). For Poisson statistics, the probability of observing N events, where X is the expected event rate, is given by

$$P(N, X) = \frac{X^N}{N!} e^{-X}, \qquad (1)$$

where $X = \eta F At$, and where η is a factor that accounts for local shielding, as well as for converting a flux per unit solid angle to impacts on a flat plate, etc. The expected value, X, need not be small in this formula; however, it is then no longer the probability of the impact occurring.

A question one is often faced with is: given a number, N, of observed impacts, what are the lower and upper limits of the true meteoroid flux, F, that could give rise to this number at some assumed confidence level? As is usually known for any particular experiment, it comes down to determining the lower and upper limits of X.

At 95 % confidence, consider a true "mean value", X_u, so much larger than N that the probability of obtaining N or fewer impacts is equal to 0.025 (2.5 %). This probability is given by

$$P(\leq N, X_u) = e^{-X_u} \left(1 + X_u + \frac{X_u^2}{2!} + \frac{X_u^3}{3!} + \ldots + \frac{X_u^N}{N!}\right) = 0.025. \qquad (2)$$

Table 2

Lower and upper limits of flux values F that give rise to the observed number of impacts, N, at confidence levels 0.99, 0.95 and 0.9.

	0.99		0.95		0.90	
N	lower limit	upper limit	lower limit	upper limit	lower limit	upper limit
0	0.000	4.61	0.000	2.99	0.000	2.30
1	0.00501	7.43	0.0253	5.57	0.0513	4.74
2	0.1035	9.27	0.242	7.22	0.355	6.30
3	0.338	10.98	0.619	8.77	0.818	7.75
4	0.672	12.59	1.09	10.24	1.37	9.15
5	1.08	14.15	1.62	11.67	1.97	10.51

Likewise, consider a true "mean value", X_l, so much smaller than N that the probability of obtaining N or more impacts is similarly equal to 0.025, or

$$P(\geq N, X_l) = 1 - e^{-X_l}\left(1 + X_l + \frac{X_l^2}{2!} + \frac{X_l^3}{3!} + \ldots + \frac{X_l^{N-1}}{(N-1)!}\right) = 0.025. \quad (3)$$

X_l and X_u are, then the lower and upper limits to the "true mean value", at 95 % confidence, that corresponds to the observed N. Equations (2) and (3) are numerically solved by iteratively choosing different X_l and X_u values until $P(N, X_l)$ and $P(N, X_u)$ are both satisfactorily close to 0.025. Table 2 gives 99 %, 95 % and 90 % confidence values for N up to 5. Ricker's (1937) table goes up to $N = 50$ for 95 % and 99 % confidence limits.

Another application of small number statistics is the analysis of temporal variations of the impact rate observed by a dust instrument. A question to be answered is: "Is the observed variation of the impact rate due to statistical fluctuations or is it the signature of a dust stream?" The method described was used to identify dust streams in the Ulysses impact data observed in the vicinity of Jupiter (Baguhl et al. 1993).

Streams in dust data were identified using a method first employed by Oberst and Nakamura (1991). The authors calculate the probability for an observed sequence of events to follow Poisson statistics. A minimum in the probability function of a random occurrence of N events in the time interval Δt indicates a stream. The time intervals are formed by sliding a window with a fixed number of events, N, over the data set. The method was modified for the Ulysses data taking into account that the events in a stream are not only concentrated in time, but also in impact direction (rotation angle, ρ). For a given data set the mean rate, μ, of impacts per time and rotation angle interval was calculated. The N events correspond to an area element $\Delta F = \Delta t \cdot \Delta \rho$, where Δt and $\Delta \rho$ are the time and rotation intervals covered. The probability

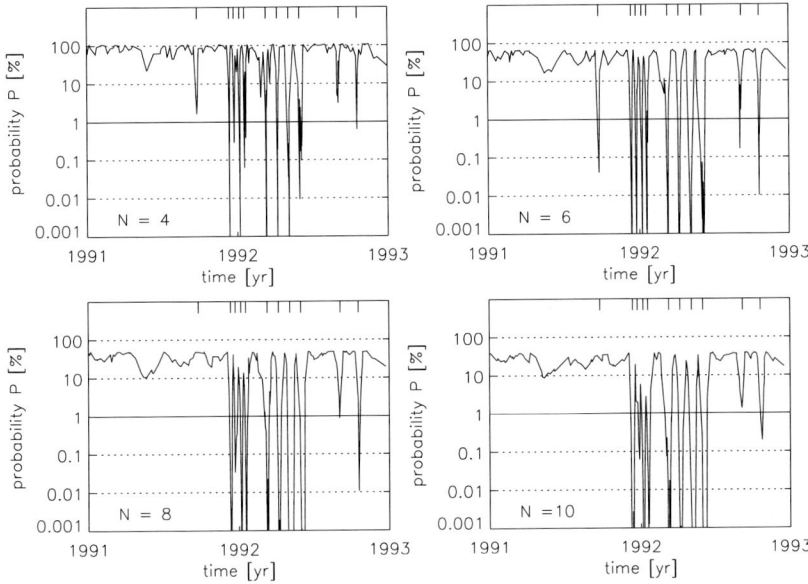

Figure 1. Stream identification in the Ulysses dust impact data. Around the time of Jupiter flyby (February 1992) the Ulysses dust detector recorded short, collimated bursts of dust impacts (cf. section V.B, below). A sliding window of variable width is applied to the data set (N denotes the number of impacts in the sliding window) and the expression for the probability P in Eq. (4) is evaluated. P denotes the probability to find a stream with N members in a random data set, i.e. for $P \ll 1$ the probability is high that the observed instantaneous impact rate represent a non-random dust stream. The identified streams are marked at the top of the diagrams.

for a random occurrence of at least N events in this area element is then calculated as:

$$P(N, \Delta F) = 1 - e^{\mu \Delta F} \sum_{m=0}^{N-1} \frac{(\mu \Delta F)^m}{m!}. \qquad (4)$$

This probability is multiplied by the total number of area elements and events, $\mu \Delta F$, in order to yield the mean number of a random occurrence in the whole data set. If this number is $< 0.1\ \%$, we identify the events as a stream. Since the stream identification works best when the fixed number window N equals the total number of events in a stream, the analysis has to be conducted with different values of N. Figure 1 shows the results of the analysis for the Ulysses dust stream data. At $N = 6$ all 11 streams cross the stream identification level.

II.D. Detection Geometry and Orbit Determination

For a single impact the maximum uncertainty of the impact direction is given by the width of the sensor field-of-view. However, the probability of an impact to occur from a certain direction is distributed like the detector angular sensitivity and hence the statistical uncertainty of the impact direction is smaller. Espe-

Table 3

Statistical uncertainty of the mean rotation angle of the Ulysses detector as function of number of events, N, recorded in a collimated stream.

N	1	2	5	10	20	50	100	200	500
1σ	30°	21°	13°	9°	6°	4°	3°	2°	1°
2σ	55°	40°	25°	18°	13°	8°	6°	4°	3°

cially, if several impacts arrive from the same direction and are recorded by a dust detector on a spinning spacecraft the statistical uncertainty of the direction determination decreases rapidly. E.g. for the dust streams observed by Ulysses, a much higher directional accuracy was obtained.

The mean value of the impact direction (rotation angle, ρ) of a dust stream is obtained by calculating the mean values x_m and y_m of the components $x = \cos\rho$ and $y = \sin\rho$ of the sensor pointing vectors of individual impacts projected into a plane perpendicular to the spin axis. Obviously the error in the determination of the mean value depends on the number N of events in the stream. We estimate the error by a numerical simulation: A set of N random impact directions within the FOV of the sensor is generated and weighted with the sensor angular sensitivity distribution. Then the mean impact direction of all N impacts is calculated as described above. This procedure is repeated 1 000 times. The resulting distribution of mean impact directions allows us to give the 1σ- and 2σ-uncertainties of individual stream directions. The values are given in Table 3.

The calculation performed corresponds to maximum uncertainties for the Ulysses sensor that is mounted almost perpendicular (85°) to the spin axis. If the stream impacts occur at angles greatly different than 90° from the spin axis, the above values are upper limits and the uncertainty of the rotation angle determination decreases. A similar analysis can be done for dust sensors mounted at different angles to the spacecraft spin axis.

In the next few paragraphs we discuss some geometric limitations of space missions for the determination of meteoroid orbits. First, we will look at the range of the orbital elements (of bound orbits about the Sun) semi-major axis, a, eccentricity, e, and inclination, i, that can be observed by an in-situ detector onboard a given spacecraft. For this study we assume random ascending nodes and random arguments of perihelion of meteoroid orbits. Both assumptions are well supported by the roughly azimuthal symmetry of zodiacal light observations.

Dust particles in space can only be detected by an in-situ detector if they reach the position of the detector and hit its sensitive area. Therefore, the range of orbital elements of observed meteoroids on bound orbits is limited. For a spacecraft at distance r from the Sun only particles with aphelia at or outside this distance and perihelia at or inside this distance can be detected, i.e. $a(1+e) \geq r \geq a(1-e)$. If the spacecraft covers a range of heliocentric distances the range of observable a- and e-values is wider. The detection probabilities of meteoroids are not uniform for orbits within the accessible a- and e-range but they generally peak for particles that are at their respective aphelion or perihelion distances because the residence times at these distances are the longest.

All particles on bound orbits about the Sun penetrate the ecliptic plane twice per orbit. Therefore, they are all observable by a detector located in the ecliptic plane. But the detection probability for orbits with high inclinations is strongly reduced in the ecliptic plane because these meteoroids spend only a short time period close to the ecliptic. Therefore, the inclination distribution of meteoroids is most easily obtained by a detector onboard a spacecraft that reaches high latitudes, like the Ulysses spacecraft. At any given spacecraft latitude, λ, meteoroids can only be observed if their orbit inclination $i \geq \lambda$.

In case a dust detector is mounted on a spinning spacecraft its FOV generally scans a larger portion of the sky than that which is determined by the detector's FOV proper. In most cases detectors are mounted close to perpendicular to the spin axis, thereby, scanning the widest range of the sky possible. In some cases a different angle with respect to the spin axis has been chosen in order to optimize the sensitivity to a special subset of the observable dust populations. In the case of the Galileo detector this is dust in orbit about Jupiter that arrives from close to the spin axis (towards the Earth) during most of Galileo's orbit about Jupiter. The Pioneer 10 and 11 detectors did not record impact directions at all because the axes of their fields-of-view were parallel to the spacecraft spin axes. The detectors recorded the dust flux from roughly the anti-Sun hemisphere. The HEOS 2 detector was mounted along the spin axis, too. However, it provided some directional information because the spin axis was turned to different directions perpendicular to the Sun direction.

In case of Helios the two sensors were mounted at different angles with respect to the spin axis in order to distinguish dust orbiting with different inclinations. The Ecliptic Sensor recorded impacts with elevations from $-45°$ to $+55°$ with respect to the ecliptic plane, whereas the South Sensor recorded only impacts with elevations from $-4°$ to $-90°$. Figure 2 shows the detection probabilities by the two Helios sensors for dust particles with different inclinations, taking into account their respective FOVs. For this example we want to describe the derivation of the detection probability in some detail.

The orbit of a dust particle is characterized by its orbital elements, semi-major axis, a (for convenience, we will use the reciprocal value $1/a$), eccentricity, e, and inclination, i. We assume here random distributions of ascending nodes and arguments of perihelion. Small dust particles are affected by radiation pressure, which effectively reduces solar gravity, i.e. $GM = 1.33 \cdot 10^{20}$ N m^2 kg^{-1} (G = gravitational constant, and M = solar mass) has to be replaced by $GM(1 - \beta)$, where $\beta = F_{\rm rad}/F_{\rm grav}$, with $F_{\rm rad}$ being the radiation pressure force and $F_{\rm grav}$ the solar gravitational force. Here, we only consider prograde orbits (i.e. $i \leq 90°$) but the extension to retrograde orbits can easily be done.

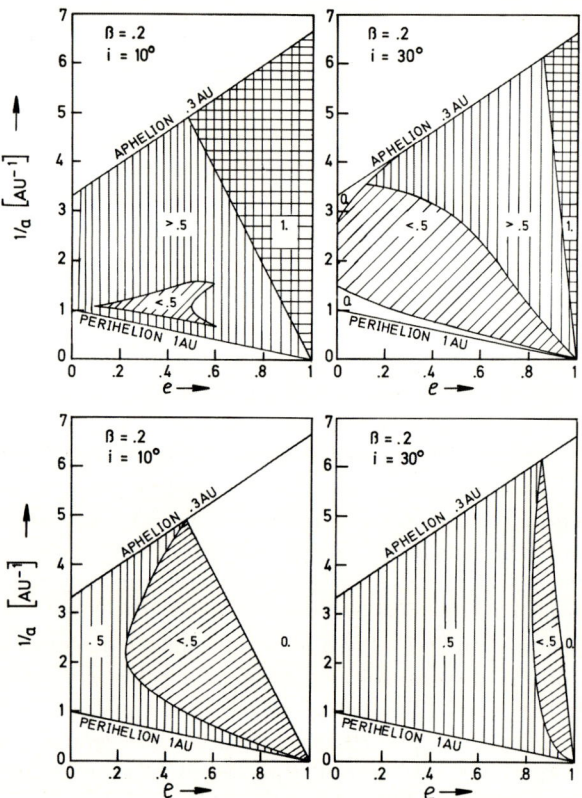

Figure 2. Micrometeoroid orbits observable by the two Helios dust sensors: Ecliptic sensor (top panel) and South sensor (bottom panel). In the reciprocal semi-major axis $(1/a)$-eccentricity (e) plane the ratio of the number of orbits that are within the field-of-view of the respective sensor and those that hit Helios from any direction is displayed by differently shaded areas. Only orbits between the lines marked aphelion at 0.3 AU and perihelion at 1 AU can reach Helios. The diagrams refer to the specific inclination, i, and radiation pressure constant, β (cf. section VI.A), assumed for the orbits.

For particles that cross the orbit of Helios ($a = 0.65$ AU $= 9.7 \cdot 10^{10}$ m, $e = 0.52$, and $i = 0°$) the relative speed vector $\mathbf{u} = \mathbf{v} - \mathbf{v}_{\text{Helios}}$ (bold letters symbolize vector quantities) between Helios and the dust particle is calculated. For a given set of dust orbital elements ($1/a, e, i,$ and β) the components (in an ecliptic RTN-system, R = radial, N = normal to the ecliptic plane, and T = perpendicular to both R and T components) of the heliocentric dust velocity vector are:

$$v_R^2 = GM(1-\beta)\left(\frac{2}{r} - \frac{1}{a} - \frac{a}{r^2}(1-e^2)\right), \tag{5}$$

$$v_T^2 = GM(1-\beta)\frac{a}{r^2}(1-e^2)\cos^2 i, \tag{6}$$

$$v_N^2 = GM(1-\beta)\frac{a}{r^2}\left(1-e^2\right)\sin^2 i. \tag{7}$$

The heliocentric speed vector of Helios is calculated correspondingly (with $\beta = 0$). In order to calculate the relative speed vector **u** at a given heliocentric radial distance of Helios one has to recognize that there are 4 different but equally probable combinations of the dust velocity possible: $u_R = \pm v_R \mp v_{R\text{Helios}}$, corresponding to inward and outward going trajectories of Helios and the dust particle, $u_T = v_T - v_{T\text{Helios}}$, and $u_N = \pm v_N$, corresponding to ascending and descending trajectories. The elevation angle, γ, of this orbit is then given by

$$\tan\gamma = \frac{u_N}{\sqrt{u_R^2 + u_T^2}}. \tag{8}$$

For assumed values of i and β the parameters $1/a$ and e were varied in small steps and the elevation angles of the dust trajectories were calculated. Then the elevation angles of the speed vectors are compared with the elevation sensitivity of the Helios dust sensors (see Grün et al. 1980). Figure 2 shows the relative number of meteoroids on the respective orbits that are detectable by the Helios sensors (compared to the total number that could have been recorded if the Helios sensors had 4π sr fields-of-view). The ecliptic detector has the highest detection probability for meteoroids with high eccentricities and small semi-major axes. These orbits can be observed both in their ascending and descending nodes. Because of its sensitivity the South Sensor can observe meteoroids only at their ascending node, therefore, the maximum detection probability is 0.5. Orbits with small eccentricities are preferably detected.

III. MEASUREMENTS AT 1 AU

III.A. Early Meteoroid Flux Measurements in the Earth-Moon System

The initial goal of dust measurements in near-Earth space was the determination of the meteoroid flux as a function of meteoroid sizes. Later the study of man made space debris also became increasingly significant (cf. chapter on Near-Earth Environment by McDonnell et al.). Early reliable measurements were obtained by the pressurized cell detector on Explorer 16 which had a total sensitive area of 1.6 m^2 (Hastings 1963, cf. chapter on Instrumentation by Auer). The detectors had 25 and 50 μm thick copper-beryllium walls corresponding to mass thresholds of about 10^{-9} and 10^{-8} g (at 20 km s^{-1} impact speed), respectively, and recorded 55 hits during a 7 months period. Later a similar experiment of 2.1 m^2 was flown on Explorer 23 which recorded 124 hits (O'Neal 1965). Similar results were reported from dust detectors on the Cosmos 135 and 163 satellites (Mazets 1971). In 1965 three Pegasus satellites were launched which consisted of huge 200 m^2 meteoroid detectors (D'Aiutolo et al. 1967; Naumann et al. 1966). The penetration detectors had sensitivity thresholds up to 10^{-6} g (corresponding to penetrations of 0.4 mm thick aluminum sheets) and recorded more than 2 000 impacts during 7 years total operational life, thereby, bridging the gap to radar meteor observations. Figure 3 shows some results from early in-situ meteoroid measurements.

Another method to determine the near Earth meteoroid environment was the study of surfaces that had been exposed to space and later returned to Earth. Microscopic examination of 14 windows (each of 0.055 m^2) from eight Gemini spacecraft revealed only a single meteoritic impact pit (of 30 μm diameter centered within a 110 μm diameter spallation zone, Zook et al. 1970). Later measurements with increased area-time products of the exposed surfaces (on Apollo, Solar Maximum Mission, LDEF and several other satellites) provided dust fluxes with improved statistical accuracy.

The interplanetary meteoroid flux is derived from near-Earth measurements by taking into account the gravitational enhancement of the dust flux near the Earth. This gravitational enhancement factor depends on the meteoroid speed. Early meteoroid measurements did not provide speed information, therefore, this information had been taken from meteor observations. Zook (1975a) obtained average impact speeds at large distances from the Earth (at the Moon) of 13 to 18 km s^{-1}. With velocity distributions corresponding to such average speeds relative to the Earth, gravitational enhancement factors of 2.4 to 1.7 are applicable to meteoroid measurements in low-Earth orbit.

Since the late sixties reliable dust detectors with sufficiently large sensitive areas to provide flux measurements of micrometer and sub-micrometer sized meteoroids were developed and flown in interplanetary space. Among the first such detectors were the dust detectors on the Pioneer 8 and 9 spaceprobes. Pioneer 8 and 9 were on low eccentric heliocentric orbits close to the Earth's orbit. These detectors had a mass threshold of 10^{-13} g at 20 km s^{-1} impact speed. Because of the spin of the spaceprobes (spin axis perpendicular to the ecliptic) the detectors were most sensitive to the flux in the ecliptic plane. During their combined operational lifetime of seven years they recorded about 800 impacts. Most of these impacts arrived from the solar direction (see below) for which Berg and Grün (1973) and Zook and Berg (1975) argued that the mass threshold was an upper limit because the impact speed could have been significantly higher than 20 km s^{-1}. It was assumed that the Pioneer 8 and 9 fluxes from the solar direction refer to particle masses of about 10^{-15} g (Fig. 3).

The HEOS 2 satellite carried an impact ionization detector with an even higher detection sensitivity than the Pioneer 8 and 9 detectors. With this detector, impact speeds and masses of impacting meteoroids could be determined. The HEOS 2 dust detector (Hoffmann et al. 1975a,b) was mounted with its axis parallel to the spin axis of the satellite and could observe dust flux only from directions perpendicular to the Earth-Sun line. The dust flux at the intersection of this plane with the ecliptic plane was determined for masses of 10^{-12} and 10^{-13} g (Fig. 3).

After the return of the first lunar samples by the Apollo astronauts in 1969 it was immediately recognized that lunar rocks were peppered by microcraters, caused by micrometeoroid impacts. Exposure ages of the rocks to this bombard-

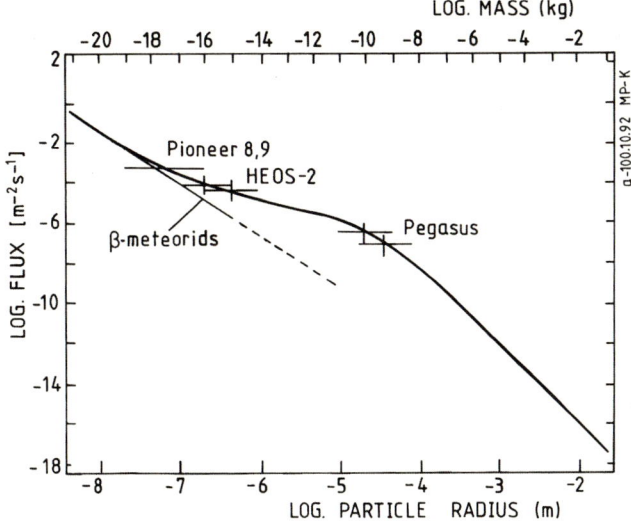

Figure 3. Cumulative flux of interplanetary meteoroids at 1 AU distance from the Sun. The flux refers to the number of meteoroids bigger than or equal to the radius or mass indicated (a density of 2.5 g cm^{-3} has been assumed to relate dust masses to sizes) that hit a flat plane detector scanning the ecliptic plane. Fluxes have been derived from lunar micro-crater analyses and measurements by in-situ space detectors (Pegasus, Pioneer 8 and 9, and HEOS). The line labeled β-meteorids has been derived from theoretical considerations (Grün et al. 1985).

ment on the order of several 10^4 years have been estimated (Hutcheon 1975). However, since the exposure time of a given surface on a lunar rock could not accurately be determined (Zook 1980) no absolute dust fluxes could be derived. Another difficulty was the calibration of microcrater dimensions with respect to the size of the impacting meteoroid. Hörz et al. (1975) reviewed all available calibration data and arrived at a size dependent crater-to-projectile diameter ratio of 2 to 9 for 10^{-18} to 1 g mass meteoroids (the values refer to the central pit of the impact crater and not to the much wider spallation zone around it). Early attempts of lunar microcrater analyses combined relative crater counts from different samples (Fechtig et al. 1974) in order to derive a combined slope of the lunar microcrater size distribution. Later it was recognized (Morrison and Zinner 1977; Morrison and Clanton 1979) that only the steepest slope measured on a single surface was least vulnerable to obstruction by thin coatings of dust on the rocks. Therefore, Grün et al. (1985) used their results and combined them with in-situ measurements in order to derive the microcrater production flux on the lunar surface. Allison and McDonnell (1981) examined the effects of secondary microcraters produced by ejecta from primary craters on lunar samples and found that the number of microcraters was significantly increased for crater diameters below about 7 μm. Zook et al. (1984, 1985) could experimentally demonstrate that the flux of hypervelocity-looking secondary impact crater due to oblique impacting primaries is significantly higher than that assumed previously. Therefore, lunar rocks give no good evidence on the interplanetary dust size distribution at micrometer sizes and below, and Grün et al. (1985) concluded that the crater-producing interplanetary flux of meteoroids $< 10^{-9}$ g is up to 2 orders of magnitude smaller than the secondary dust flux on the lunar surface.

The flux of interplanetary meteoroids is generally given in the form of the cumulative meteoroid flux $F(m)$, which is the number of meteoroids with masses bigger than or equal to mass m which impact one square meter each second. The cumulative flux is related to the differential flux $f(m)$ of particles in the mass range m to $m + \mathrm{d}m$

$$F(m') = \int_{m'}^{\infty} f(m) \mathrm{d}m,$$

or

$$f(m) = -\frac{\mathrm{d}F(m)}{\mathrm{d}m}. \tag{9}$$

Since the meteoroid flux covers many orders of magnitude in mass it is convenient to express Eq. (9) in terms of logarithmic mass interval $\mathrm{d}(\log m)$:

$$\frac{\mathrm{d}F(m)}{\mathrm{d}(\log m)} = -m \cdot \ln 10 f(m). \tag{10}$$

The relation between the cumulative flux $F(m)$ and the cumulative spatial density $N(m)$ is given by

$$F(m) = \frac{v}{k} N(m), \tag{11}$$

where v is the average impact velocity onto the sensor measuring the flux, and k is a constant: $k = 4$ in the case of an isotropic flux when $F(m)$ is measured by a one sided flat plate sensor, $k = \pi$ if the flux is concentrated in a plane and if $F(m)$ is the average flux in that plane measured by a flat plate sensor that has an effective sensitive angle of 2 rad and that is oriented perpendicular to that plane, and finally, $k = 1$ if the flux is monodirectional and measured by a flat plate sensor normal to that flux. A factor $k = \pi$ also applies to the situation when the flux is mono-directional and it is observed by a flat plate detector that spins around an axis which is perpendicular to that flux.

Important moments of the flux distribution are the distributions of the cross sectional area $\{-A(m)\mathrm{d}N(m)/\mathrm{d}(\log m)\}$ and mass densities $\{-m\mathrm{d}N(m)/\mathrm{d}(\log m)\}$ of interplanetary meteoroids (Fig. 4). Both distributions reach their maximum near the knee (10^{-5} to 10^{-6} g) of the flux curve (Fig. 3). The importance of the cross sectional area density distribution is that it represents the light scattering property of interplanetary dust that is responsible for the zodiacal light brightness (Giese and Grün 1976). The maximum in the mass distribution describes meteoroids that carry most mass (Grün et al. 1985).

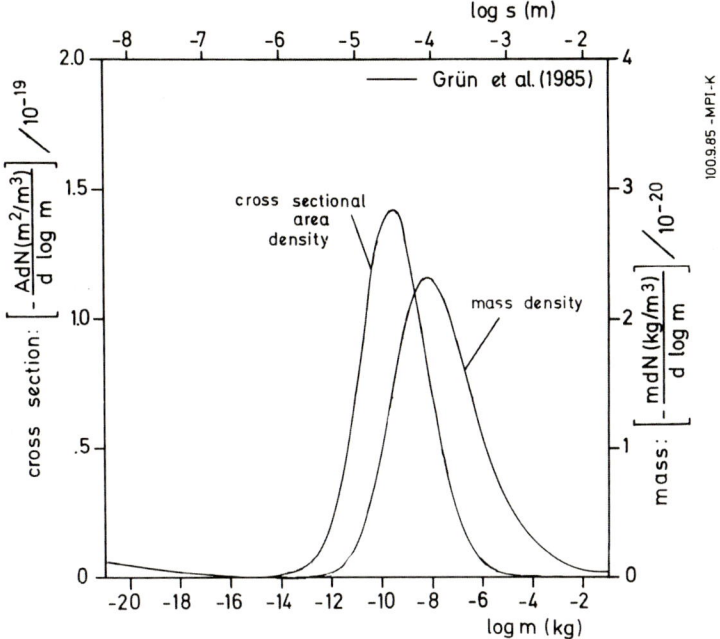

Figure 4. Distributions of cross sectional area density and mass density of interplanetary meteoroids at 1 AU (after Leinert and Grün 1990). The distributions have been derived from the flux curve shown in Fig. 3. At 1 AU the total mass density is $9.6 \cdot 10^{-17}$ g m^{-3} and the total cross sectional area density is $4.6 \cdot 10^{-19}$ m^2 m^{-3}.

III.B. HEOS 2

Although the dust detector onboard the HEOS 2 satellite was mounted along the spin axis of the satellite, directional information was obtained by pointing the spin axis into different directions in a plane perpendicular to the Earth-Sun line. Most of the time the sensor was pointing towards the Earth apex direction, thereby, determining at large Earth distances ($\approx$ 200 000 km) the dust flux from the apex direction. Part of the time the sensor was facing the ecliptic north, south, and antiapex directions. From these directions the flux was much lower and only very small (10^{-15} g) particles were recorded (Hoffmann et al. 1975a). Here, we want to ignore near-Earth effects in the HEOS measurements, such as dust swarms in the auroral zones of the Earth that had been interpreted as consequences of electrostatic break-up of bigger meteoroids (Fechtig et al. 1979), or as solid rocket motor exhaust particles (Grün and Zook 1985). A special category of dust was detected during a two month period in 1974 when the Earth and with it the HEOS satellite penetrated the orbit plane of comet Kohoutek 1973f. During this time an excess of micrometer-sized particles was recorded (Fig. 5; Hoffmann et al. 1975b) that had the correct velocity vectors and sizes for particles being emitted from this comet during its approach to

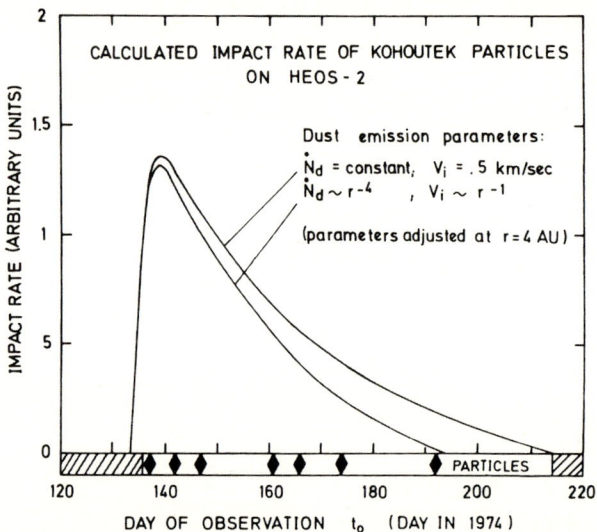

Figure 5. Model calculations of the impact rate onto the HEOS detector of dust emitted from comet C/Kohoutek (C/1973 E1) at about 4 AU distance from the Sun. Impacts were observed by the HEOS detector around the time when the HEOS satellite passed through the orbit plane of comet Kohoutek 1973f. Impacts are displayed by diamonds at the bottom of the graph. The hatched areas on the bar indicate times when the detector pointed to unfavorable directions. Two model calculations are shown with different dust emission rates, $\dot{N}_d$, and emission speeds, v_i. In one model the emission parameters were assumed to be constant, in the other they depend on the heliocentric distance, r.

the Sun (Grün et al. 1976). No other enhancement during periods of meteor streams (Hoffmann et al. 1975b) or during other cases of comet passages like the Kohoutek case have been recognized by the many dust detectors in space during the following decades (cf. Riemann and Grün 1992).

III.C. Hiten

The most recent in-situ measurements of the 1 AU-meteoroid population have been performed by the dust detector on the Japanese Hiten satellite. The satellite was on a high eccentricity orbit around the Earth with an apogee distance of up to $1.5 \cdot 10^6$ km. It spent most of its time far from the Earth and, therefore, was able to characterize the interplanetary dust population. The dust sensor was mounted perpendicular to the spin axis of the satellite and scanned the ecliptic plane like the Pioneer 8, and 9 detectors. Because of its configuration the Hiten instrument (Igenbergs et al. 1991) could determine the flight path with $\pm 75°$ accuracy for individual particles.

About 500 events that are considered reliable dust particle impacts were detected with Hiten. For a mission duration of three years and taking the sensor geometry into account this results in a flux of $1.0 \cdot 10^{-3}$ m^{-2} s^{-1} for an equivalent flat plate. This flux needs to be related to a mass threshold to enable

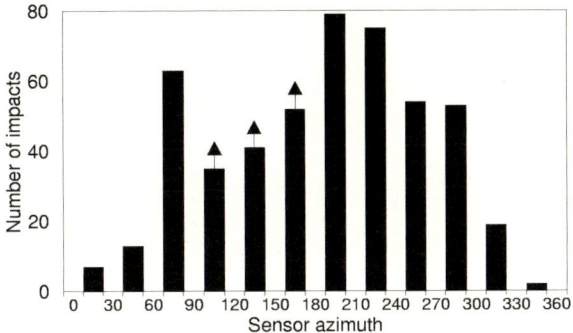

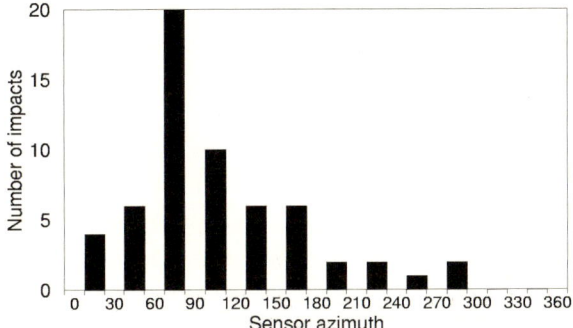

Figure 6. Impacts recorded by the Hiten detector between February 1990 and April 1993. Upper panel: The number of impacts are plotted as a function of the instrument pointing (sensor azimuth) in the ecliptic plane. The Sun direction is fixed at 90° and the Earth apex direction at 180°. The arrows indicate that these numbers of impacts are lower limits (see text). Lower panel: Number of impacts as a function of instrument pointing for impacts having an impact speed > 40 km s^{-1}. High speed β-meteoroids coming from the solar direction (90°) are clearly identified.

comparison with other experiments and models. The Hiten experiment was of the impact plasma type and thus shows a very strong velocity dependence in its mass threshold. An equivalent mass threshold of $2 \cdot 10^{-15}$ g at 20 km s^{-1} impact speed has been estimated. The interplanetary flux model (Grün et al. 1985) predicts an average flux of $7.0 \cdot 10^{-4}$ m^{-2} s^{-1} for this mass so the Hiten flux is about 50 % higher. Reasons for this discrepancy could be uncertainties in the determination of the detection threshold or a time variable flux of small meteoroids that interact with the interplanetary magnetic field.

Figure 6, upper panel, shows all impacts binned in 30° wide bins as a function of instrument pointing (sensor azimuth). The direction 90° is always pointing towards the Sun and the apex direction of Earth motion is fixed at 180°. The three arrows indicate that these bins contain fewer impacts than the experiment probably has received. This is caused by a partial blinding

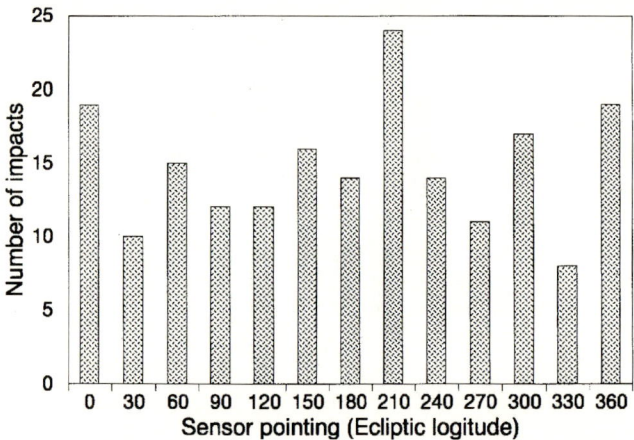

Figure 7. Impacts detected by the Hiten detector between April 1990 and April 1993 having impact velocities larger than 15 km s^{-1}, are plotted as a function of sensor pointing in ecliptic longitude. The bars correspond to average numbers in 45° bins in ecliptic longitude. This width corresponds roughly to the half width of the field-of-view of the detector. The peak at 210° may be related to impacts of interstellar particles.

effect due to solar photo electrons during the quarter of each revolution when the sensor is gradually increasing its exposure to the Sun. This is likely to be the reason why the peak in the apex direction appears to be shifted somewhat towards the anti-sun direction (270°). The directional flux in the apex direction is $9.7 \cdot 10^{-4}$ m^{-2} s^{-1}. The peak between sensor azimuth 60° and 90° in both panels of Fig. 6 is caused by β-meteoroids on prograde hyperbolic orbits. Also here, most likely a number of impacts are missing due to the blinding effect mentioned above. β-meteoroids have been identified by their much higher velocities ($v > 40$ km s^{-1}) than the background flux in Fig. 6a. A further study of the individual impacts reveals that β-meteoroids generally have lower masses than the majority of the (apex) particles. Since the impact velocity in Fig. 6, lower panel, is above 40 km s^{-1} one will find by vector addition that the heliocentric velocity of the particles from 60° to 90° is above 50 km s^{-1} which is above the escape velocity even without considering radiation pressure effects (cf. section VI). The directional flux of these particles is $2.4 \cdot 10^{-4}$ m^{-2} s^{-1}. New results from the Ulysses Dust experiment have shown strong evidence for interstellar particles in the outer solar system (outside 3 AU distance from the Sun, see below) arriving from a direction of 252° ± 20° ecliptic longitude and 5° ± 10° ecliptic latitude (Grün et al. 1993; Baguhl et al. 1996). Hiten data were used to search for these particles at 1 AU distance. In order to do this, a different coordinate system is used to present the data. Figure 7 shows the flux for three full years as a function of ecliptic longitude. A peak around 220° is clearly visible indicating an excess of particles close to the interstellar direction.

The Hiten results indicate there is indeed an enhanced flux close to the upstream direction at 220° heliocentric longitude. The absolute number is not very impressive (around 20) due to the small aperture of the instrument but it is still statistically significant thanks to the long exposure time that has resulted in a very stable background flux from other directions. If we assume the excess at 220° longitude is of interstellar origin we can calculate a minimum flux, since the number may be higher due to the elimination of particles with impact velocities below 15 km s^{-1}. Taking geometrical factors into account a flux of $1.2 \cdot 10^{-4}$ m^{-2} s^{-1} is found, indeed close to the $1.5 \cdot 10^{-4}$ m^{-2} s^{-1} as reported from Ulysses (Grün et al. 1993) at Jupiter distance. The mass thresholds for the two detectors is of the same order. The shift of the peak position from 250° to 220° can be explained by the bending of the dust trajectories by solar gravity and by the blinding effect of the instrument described above. This latter effect will reduce the number of detected interstellar particles at high longitudes during winter/spring times and thereby offsetting the peak towards lower longitudes.

IV. MEASUREMENTS WITHIN THE ZODIACAL CLOUD

In this section we discuss in-situ dust measurements that were taken within about 3 AU distance from the Sun, a region of space were sunlight is clearly observed to be scattered from interplanetary grains (cf. chapter on Zodiacal Light by Levasseur-Regourd). There are seven space missions that contributed significant dust flux measurements in this region: Helios from 0.3 to 1 AU, Galileo from 0.7 AU to Jupiter's distance ($\approx$ 5 AU), Pioneer 8 and 9 in the heliocentric distance range from 0.75 AU to 1.08 AU, and Ulysses mainly during its traverse from the ecliptic South to North pole that took place in the distance range from 1.3 AU to 2.3 AU. Pioneer 10 and 11 measurements taken outside the Earth's distance are discussed in the following section. Helios, Pioneer 8 and Pioneer 9 were spinning spacecraft with their spin axes perpendicular to ecliptic plane. The dust detectors were mounted with their axes close to perpendicular to the spin axis, therefore, during one spin revolution they scanned the dust flux in and close to the ecliptic plane. The Galileo and Ulysses spacecraft have their spin axes and antennae pointing towards the Earth (or Sun). The dust detectors again were mounted at angles close to perpendicular to the spin axes. The five dust instruments discussed here were impact ionization detectors that are sensitive to micrometer and sub micrometer-sized grains (cf. Table 1). We will analyze data from these missions for two aspects: (1) the radial profile of the dust flux, and (2) the dynamical and physical characteristics of dust in the zodiacal cloud.

IV.A. Helios

The dust detector on the Helios spacecraft consists of two sensors of 55 cm^2 and 65 cm^2 sensitive area, respectively, both in combination with a low resolution ($M/\Delta M \approx 5$ amu) time-of-flight mass spectrometer. The Ecliptic Sensor was sensitive to impacts arriving from close to the ecliptic plane. Since this sensor

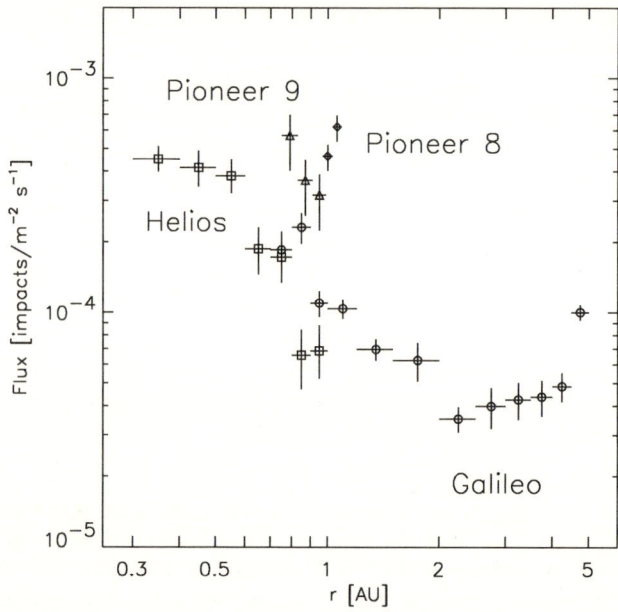

Figure 8. Fluxes of interplanetary dust particles measured by various spaceprobes as a function of heliocentric distance: Helios (squares), Pioneer 8 (diamonds), Pioneer 9 (triangles), and Galileo (circles). The fluxes refer to the detection thresholds and geometries of the individual instruments and, therefore, cannot be directly compared to each other. Although the Pioneer 8 and 9 instruments were the least sensitive (cf. Table 1) they detected the highest fluxes—mostly from the solar direction (cf. Fig. 10). Detection by Helios of dust from solar direction was impeded by a thin film in front of the sensor facing the Sun. The Galileo instrument never faced the solar direction, the fluxes shown are the highest fluxes observed in the respective distance intervals during the outward going (away from the Sun) portions of the Galileo trajectory.

viewed the Sun once per spin revolution it was covered by an aluminum coated 0.3 μm thick plastic film in order to prevent heat and solar UV radiation to enter into the sensor. This film caused a penetration cut-off for meteoroids which depended on the mass and density of impacting dust particles (Pailer and Grün 1980). The South Sensor had an open aperture that was shielded from solar radiation by the spacecraft rim and hence recorded only impacts arriving from south of the ecliptic plane. This sensor was sensitive to smaller and lower density meteoroids.

Helios measurements covered the range from 0.3 to 1 AU heliocentric distance (Fig. 8). The measured dust flux displays a steady increase towards the Sun by about a factor 10. Data from 152 impacts onto the South Sensor and 83 impacts onto the Ecliptic Sensor were received on the ground during the first 10 orbits about the Sun. Measurements by both sensors showed a comparable flux increase towards the Sun. No systematic variation of the size distribution can be derived from the data.

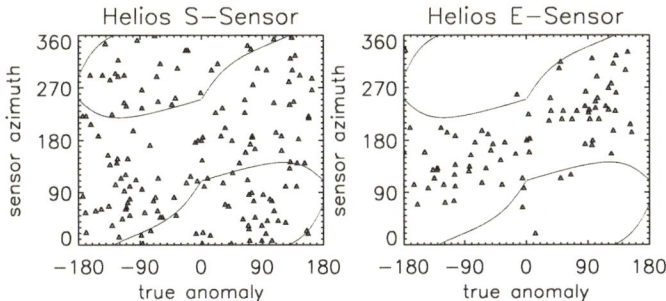

Figure 9. Impacts recorded by the Helios sensors (South sensor and Ecliptic sensor) as a function of sensor pointing (sensor azimuth) and spacecraft position (true anomaly). Apex direction is at azimuth 180° and Sun direction is at 90°, perihelion is at true anomaly 0° and aphelion is at true anomaly ±180°. Meteoroids orbiting on circular or low eccentricity orbits are recorded between the two lines running from lower left to upper right.

Figure 9 shows the impacts recorded by both sensors as a function of spin angle (sensor azimuth) and spacecraft position (true anomaly). There are significant differences between the measurements by the two sensors. The Ecliptic Sensor detected most impacts centered about the apex direction (i.e. 90° off the Sun in the direction of spacecraft motion) while the South Sensor observed particles from all around during a spin revolution with a predominance of small particles from the solar direction. There is a characteristic band within which most impacts were recorded by the ecliptic detector (Grün et al. 1980). The significance of this band is that particles moving on circular orbits around the Sun (independent of their orbit inclination or radiation pressure constant, see below, section VI) would fall fully within that band if they are detected by the Helios sensors. The width of the band is determined both by the field-of-view of the sensors (about 70°) and the impact direction. At aphelion (true anomaly = ±180°) meteoroids could be faster or slower than Helios, therefore, the band is widest there. Low eccentric orbits have maximum probability of the sensor azimuth to lie half way between the two lines.

The impacts recorded by the Ecliptic Sensor are concentrated near the center of the band indicating that these particles move on low eccentric orbits. Modeling of the Helios results by Grün et al. (1980) show that these "apex" particles have eccentricities ($e_{ave} \leq 0.6$) and semimajor axes (a_{ave} about 0.6 AU) and small inclinations ($i_{ave} < 30°$) which represent typical orbits of paticles producing the zodiacal light (Divine 1993). Since apex particles did penetrate the front film of the Ecliptic Sensor their density cannot be below about 1 g/cm^3 (Pailer and Grün 1980), at least not for the smallest particles detected. Apex particles have also been observed as a major meteoroid component by the Pioneer 8 and 9, HEOS 2 and Hiten dust detectors.

On the other hand impacts outside the band in Fig. 9 that were mostly observed by the Helios South Sensor must be from particles on orbits of higher eccentricities. Modeling shows that these "eccentric" particles have eccentricities, $e_{ave} = 0.7$, and semi major axes, a_{ave} about 0.9. and small inclinations. The inclinations ($i_{ave} < 30°$) cannot be too large because otherwise their azimuth would shift to the apex band.

Meteoroids on high eccentric orbits are principally visible also by the Ecliptic Sensor. Their absence from the data shows that they must be prevented from detection by an instrumental effect, i.e. the front film. Experimental calibration of the Helios film (Pailer and Grün 1980) indicates that at a given particle mass and impact speed, the penetration limit depends on the projectile density. By extrapolation of this dependence to the masses and speeds of the eccentric particles observed by the Helios South Sensor, Grün et al. (1980) find that at least half of the eccentric particles should have densities below 1 g cm^{-3}. A low density is also indicated by the electric charge recorded for four of the biggest particles ($m > 10^{-9}$ g) on the South Sensor. Charge values of a few times 10^{-13} C can only be explained by an extremely high surface potential of about 100 V or densities well below 1 g cm^{-3}. Svestka et al. (1996) have suggested that fluffy particles could have higher charge values than spherical particles of the same mass.

Low-resolution mass spectra obtained by the Helios detector indicate two major groups of spectra (cf. Grün 1981; Leinert and Grün 1990), those with a dominant abundance of low mass elements ($\leq$ 30 amu) and those that are dominated by higher mass elements ($>$ 30 amu). The ratio of low mass to high mass spectra by number was about 2 for the South Sensor, whereas it was about 1 for the South Sensor.

IV.B. Pioneers 8 and 9

For 20 impacts detected by both Pioneer instruments, masses and speeds could be determined separately by a time-of-flight measurement between the front film and the rear sensor. About 40 times more impacts were detected by the front sensor alone. For the latter impacts no mass and speed values can be given. Figure 8 shows the radial profiles of the spin averaged flux observed by both spacecraft. Whereas Pioneer 9 measurements showed a decrease of the dust flux in the distance range from 0.75 to 0.99 AU Pioneer 8 observed a slight increase from 0.97 to 1.09 AU. The absolute flux values measured are compatible with meteoroid masses between 10^{-15} and 10^{-14} g.

The first successful attempt to measure trajectories of micrometeoroids was made with the Pioneer 8 and 9 dust detectors. Trajectories were measured by two different segmented sensors, the front film sensor and the rear impact target, that were separated by 5 cm distance (cf. chapter by Auer). For the 20 particles that penetrated the front film, trajectories and rough orbital elements were calculated (Wolf et al. 1976; McDonnell 1978). Their average heliocentric speed was $\langle v \rangle = 29$ km s^{-1}. Half of them were more massive than 10^{-11} g and moved on prograde elliptical orbits, while four of the smaller particles moved on hyperbolic orbits away from the solar hemisphere. A similar but more pronounced, anti-sunward directed flux was found for the 800 smaller impacts that were recorded only by the front film sensor (Fig. 10). The more

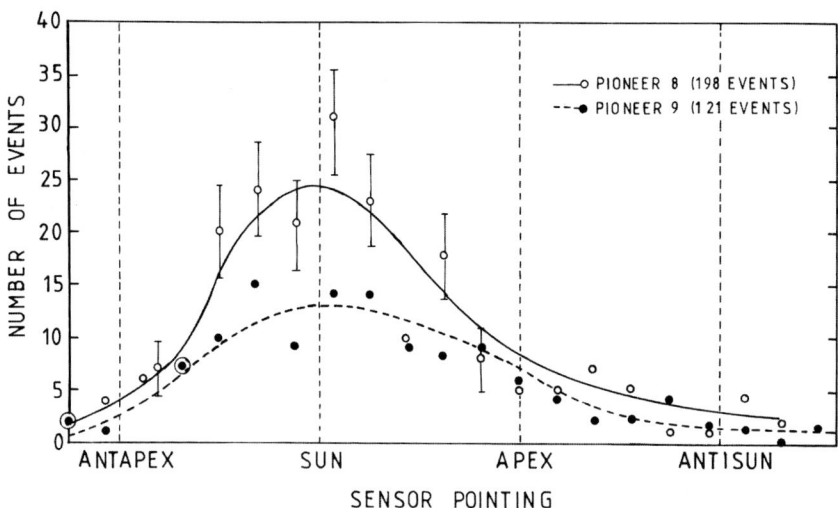

Figure 10. Sensor pointing (azimuth) distribution of impacts recorded by the front film grid detector of Pioneers 8 and 9. During one spacecraft spin revolution the detectors scan the ecliptic plane (Sun direction 90°, apex direction, 180°, after Berg and Grün 1973).

massive particles were mostly detected from the apex direction of the spacecraft motion.

The flux of micrometer to sub-micrometer sized dust grains from approximately the solar direction was identified by Berg and Grün (1973) to be meteoroids in hyperbolic orbits that are leaving the solar system. Zook and Berg (1975b) called these particles β-meteoroids and deduced that they were probably generated as fragments resulting from mutual collisions between larger meteoroids that were sunward from the Pioneer 8 and 9 sensors. Whipple (1975) concluded that most β-meteoroids, because of their directional characteristics, must have been produced outside 0.5 AU. Further analyses (Zook 1975a) showed that the flux of β-meteoroids appeared to be increasing with increasing heliocentric distance near, and just outside of, 1 AU. Almost simultaneously, however, McDonnell et al. (1975) were able to explain the same data by assuming that the production rate of β-meteoroids varied instead with heliocentric longitude. But uncertainty remains. First, zodiacal light data show that the heliocentric radial variation of meteoroids (or, more precisely, their cross-sectional area per unit volume) varies with heliocentric distance as $r^{-\alpha}$ where $\alpha = 1.3$ inside of 1 AU (Leinert et al. 1981) and $\alpha = 1.5$, or even higher, outside of 1 AU (Hanner et al. 1976). Why should the radial distribution of zodiacal particles change character right at, or near, 1 AU? Secondly, to add to the mystery, Jackson and Zook (1989) found, through numerical modeling that was later confirmed by IRAS observations (Dermott et al. 1994), that many dust grains ejected from asteroids in the main belt would be expected to be trapped into heliocentric orbital period resonances with the Earth, and would be concentrated around and external to 1 AU.

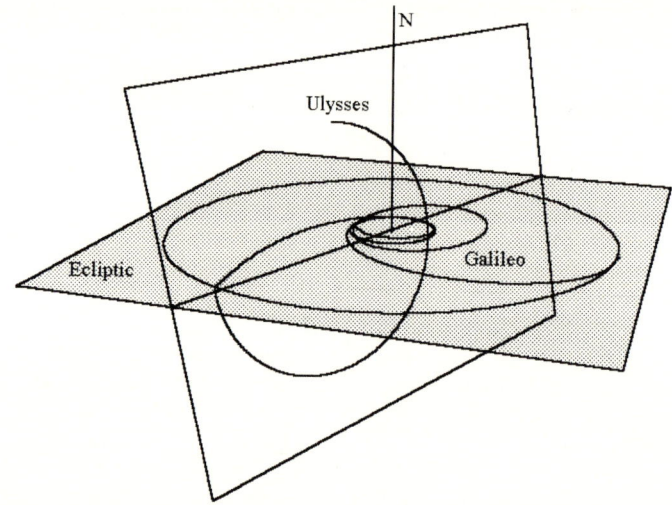

Figure 11. Trajectories of the Galileo and Ulysses spaceprobes (from Grün et al. 1997). The Sun is in the center, Earth's and Jupiter's orbits, and Galileo's trajectory are in the ecliptic plane (shaded plane). The initial trajectory of Ulysses from the Earth to Jupiter was also in the ecliptic plane. Subsequently, Ulysses was thrown by Jupiter's gravity onto an orbit plane (transparent plane) inclined 79° to the ecliptic plane. Now perihelion and aphelion of Ulysses are at 1.3 and 5.4 AU, respectively.

IV.C. Galileo

The orbit of the Galileo spacecraft is shown in Fig. 11. From launch in October 1989 until September 1993 the orbit of the Galileo spacecraft was within 3 AU distance from the Sun and close (< 5° latitude) to the ecliptic plane, i.e. in the region where interplanetary dust should be most prominent. The Galileo spacecraft had its spin axis pointing towards the Sun for thermal reasons. The dust detector was mounted at an angle of 120° from the Sun direction. This had the effect that during the outbound part of the orbit the dust detector was looking into the ram direction and consequently saw an enhanced dust flux, e.g. in the 0.8 to 0.9 AU interval the spacecraft outbound flux was a factor 4 higher than the inbound flux (Fig. 8). This difference in the dust flux during different parts of the trajectory is a consequence of the orbital distribution of interplanetary meteoroids (cf. chapter by Staubach et al.). Galileo crossed 1 AU twice on its outbound trajectory past the orbit of the Earth (at some distance from the Earth). The flux decreased monotonically and no flux enhancement was seen there within the statistical limits. However, the flux was significantly lower than the Pioneer 8 and 9 fluxes indicating that an enhancement, if any, should be in particles which were not observable by Galileo, i.e. in the small particles from the solar direction.

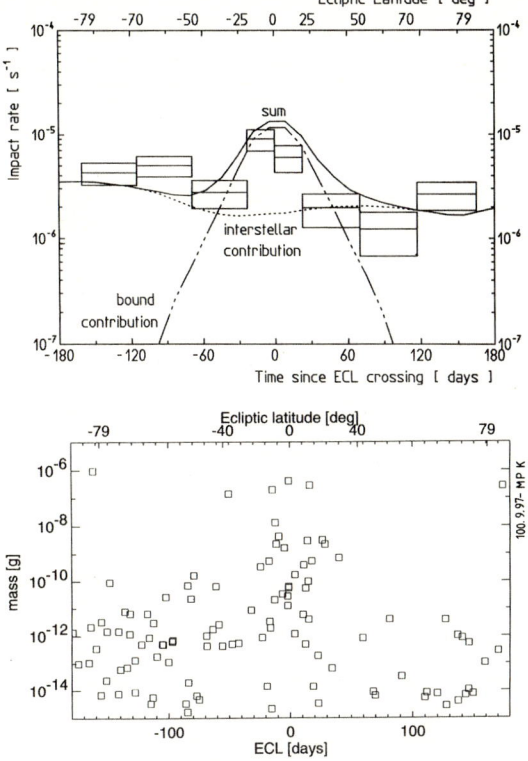

Figure 12. Dust impacts recorded by Ulysses during its passage from the ecliptic South to the ecliptic North pole (from Grün et al. 1997). Abscissas are time from ecliptic plane crossing (ECL). The top scale gives the spacecraft latitude. Upper panel: Impact rates (impact charge $> 8 \cdot 10^{-14}$ C) measured by Ulysses around the time of its ecliptic plane crossing. The boxes indicate the mean value and the standard deviation. During all that time Ulysses was inside 2.2 AU distance from the Sun. The dotted line gives the expected impact rate of interstellar particles. The dash-dotted line gives the impact rate of interplanetary meteoroids on bound orbits. Lower panel: Meteoroid masses measured by Ulysses during its S-N traverse (impact charge $> 8 \cdot 10^{-14}$ C).

IV.D. Ulysses

The Ulysses mission provided the possibility to sample nearly all latitudes during the course of its mission. This makes it possible to get the latitudinal distribution of dust. In the limited distance range from 2.3 to 1.3 AU Ulysses passed from close to the solar south pole (−79° lat.) through the ecliptic to close to the solar north pole (+79° lat.). Figure 12, upper panel, shows the impact rate during 360 days of the pole-to-pole passage (ecliptic plane crossing occurred in March 1995). A total of 117 impacts (with impact charges $> 8 \cdot 10^{-14}$ C) were recorded during this time. The passages over the south and north poles occurred 170 days before and after ecliptic plane crossing, respectively. The impact rate stayed relatively flat except for the maximum $(8 \cdot 10^{-6}$ s$^{-1})$ at ecliptic crossing.

Figure 12, lower panel, shows the masses of dust particles detected during the S-N traverse. 34 particles with masses $< 10^{-10}$ g were detected during the 400 days of the S-N traverse, 23 of these particles have been recorded during the 80 days when the spacecraft was within 30° of the ecliptic plane. This increased flux of big particles near the ecliptic is obviously due to the interplanetary dust population, while the smaller particles refer to the interstellar dust flux.

Yet another population of very small dust particles (≈ 0.1 μm) was observed at high latitudes, especially over the South pole. This population can not be explained by classical zodiacal cloud particles because of this high latitude, nor can it be the interstellar dust population that arrives from a narrowly defined direction (Grün et al. 1997a). Hamilton et al. (1996) suggest that these particles may be electromagnetically deflected β-meteoroids that have been generated at low latitudes and subsequently deflected by the contemporary interplanetary magnetic field to high latitudes. One solar cycle (11 years) later these particles should be found at low latitudes again (Morfill et al. 1986).

V. MEASUREMENTS IN THE OUTER SOLAR SYSTEM

V.A. Pioneers 10 and 11

The Pioneer 10 and 11 spacecraft were the first manmade spaceprobes to the outer solar system. After a flyby of Jupiter Pioneer 10 reached a solar system escape trajectory with an escape asymptote of 83° ecliptic longitude and +3° latitude. An objective of the Pioneer 10 and 11 missions was to explore the meteoroid environment in the asteroid belt and beyond. For this purpose it carried three instruments that were measuring dust by different means: a simple zodiacal light photometer (Hanner et al. 1976, cf. chapter on Zodiacal Light by Levasseur-Regourd et al.), a penetration detector consisting out of 234 pressurized cells mounted on the back side of the antenna (Humes et al. 1974), and the sophisticated optical Asteroid and Meteoroid Detector (AMD, Soberman et al. 1974).

The AMD instrument consisted of four telescopes that were designed to detect scattered sunlight from meteoroids passing through their field-of-view. From the timing and the amplitudes of the received light signals the trajectory of the meteoroid can be reconstructed. Because of a high noise level on individual channels no single trajectory could be uniquely identified (Auer 1974). Some coincident signal were later re-interpreted by Dubin and Soberman (1991) as light flashes from exploding meteoritic particles, which they called "cosmoids". However, since this cosmoid hypothesis is in direct conflict to zodiacal light and in-situ meteoroid measurements, it will not be considered here any further.

On Pioneer 10 only one panel (containing about half of the pressurized cells) of the meteoroid penetration detector functioned out to 18 AU distance from the Sun, the other panel failed shortly after launch. In the end 95 of the

108 pressurized cells of the panel were punctured (10 penetrations occurred in the vicinity of Jupiter) and the total area was reduced from 0.26 m² to 0.032 m². The sensitivity of the 25 μm thick cell is estimated to $8.3 \cdot 10^{-10}$ g, at 20 km s^{-1} impact speed. Figure 13, upper panel, shows the penetration flux out to 18 AU. After a sharp decline outside the Earth's orbit the flux stayed about constant out to 18 AU, except for a short peak during Jupiter flyby.

The Pioneer 11 spacecraft was deflected by Jupiter onto a trajectory which led back towards the Sun with a perihelion distance of 3.7 AU. After perihelion passage the spacecraft crossed the orbit of Jupiter again and was finally deflected by Saturn onto a solar escape trajectory. During its traverse from Jupiter to Saturn, Pioneer 11 reached a latitude of 15°. Data from the penetration detector onboard Pioneer 11 were reported out to Saturn (Humes 1980). The sensitivity of the 50 μm thick cell is $6 \cdot 10^{-9}$ g, at 20 km s^{-1} impact speed. On Pioneer 11 neither panel showed malfunctioning, but one panel recorded only about half the flux of the other panel (Fig. 13, lower panel). This discrepancy is still unexplained. Out to Saturn 58 cells were punctured on one panel and 33 on the other (2 and 3 penetrations occurred near Jupiter and 3 and 1 penetrations near Saturn, respectively). The number of counts at Saturn is a lower limit because the instrument was only operational for 30 % of the time within 3.1 R_S (R_S = Saturn radius). A similar effect may have occurred at Jupiter where Pioneer 11 flew, unknowingly at that time, through the outer fringes of the Jovian ring.

Pioneer 11 crossed the distance range from 3.7 to 4.9 AU (Jupiter's distance) three times and measured about the same flux on all three legs. Since Pioneer 11 crossed this region with different speeds and flight path angles, the measured fluxes constrain the orbit distribution of the meteoroids. During the inbound leg after Jupiter flyby no particles on prograde orbits could have reached the detector, therefore all impacts recorded during this portion were caused by meteoroids on retrograde orbits. An extensive analysis of potential bound orbit distributions by Humes (1980) showed that Pioneer 11 data is best explained by meteoroids in randomly inclined and/or highly eccentric orbits.

Meteoroid penetrations through the 25 μm thick stainless steel sensor on the Pioneer 10 spacecraft resulted in a spatial density of meteoroids that decreased from 1 to about 2 AU from the Sun, and then remained constant with increasing heliocentric distance, out to 18 AU (Humes et al. 1974; Humes 1980). This is an enigmatic result for two reasons: First, the zodiacal light photometer on the same spacecraft (Hanner et al. 1976, cf. chapter on Zodiacal Light by Levasseur-Regourd et al.) resulted in a spatial density of zodiacal meteoroids that steadily decreased from 1 to 3.3 AU, except for an approximate 30 % additional asteroid belt contribution between 2.3 and 3.3 AU, and in a spatial density below the detection limit outside 3.3 AU. More insight into this puzzle was obtained when sensors on the IRAS satellite detected bands of thermal infra-red emission parallel to the ecliptic (Low et al. 1984). Sykes and Greenberg (1986) interpreted these asteroidal bands as due to collisions within the Eos, Koronis, and Themis families in the main asteroid belt. A potential

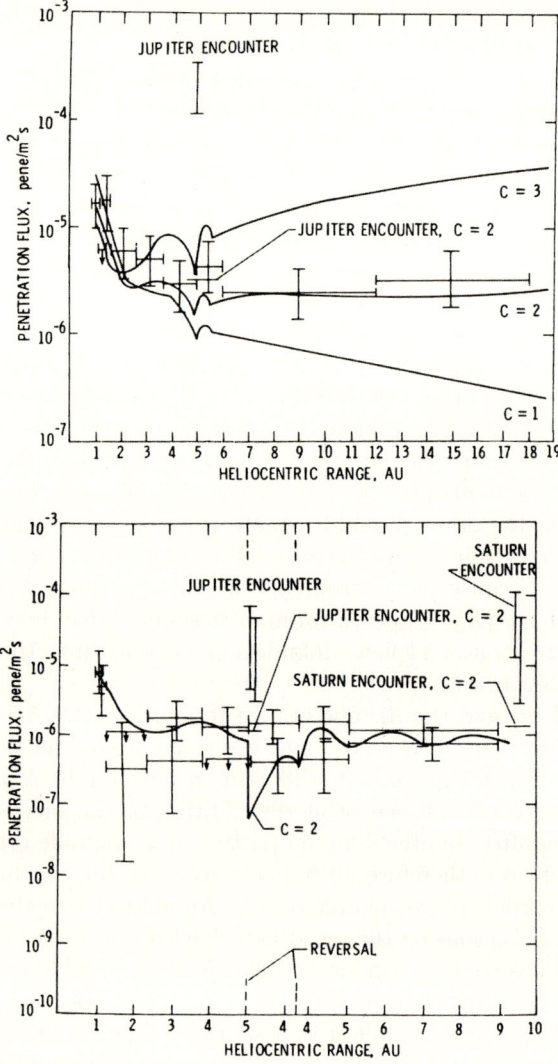

Figure 13. Meteoroid impact rates observed by the Pioneer 10 and 11 penetration detectors during their trajectories through the outer solar system (from Humes 1980). Upper panel: Pioneer 10 penetration flux of meteoroids of masses $\geq 8 \cdot 10^{-10}$ g as function of heliocentric distance. At Jupiter flyby the flux was about a factor 100 higher than in interplanetary space. Measurements are compared with model fluxes calculated for particles on randomly inclined obits with eccentricities of 0.99 and special semimajor axis distribution (C = 2 corresponds to a constant spatial dust density). Lower panel: Pioneer 11 penetration flux of masses $\geq 6 \cdot 10^9$ g. After Jupiter flyby Pioneer 11 traversed two more times the region between 3.7 and 5 AU. During Jupiter and Saturn flybys enhanced impact rates were observed.

explanation could be that main belt asteroid collisions generate mostly coarse grained fragments that are providing most of the scattering cross-section observed by the zodiacal light sensors, while comets contribute most of the smaller particles detected by the Pioneer 10 and 11 penetration sensors.

The second reason why the Pioneer 10 penetration data present an enigma, is the lack of spatial density fall-off with increasing heliocentric distance beyond 2 AU. Poynting-Robertson drag (e.g., see Burns et al. 1979) is expected to set up a spatial density of the meteoroid population that increases with decreasing heliocentric distance inside the source region of meteoroids, with a zero spatial density outside that region. This marked contradiction of theoretical expectations compared to the actual penetration data led Zook (1980) to suggest that meteoroids made of water ice were penetrating the Pioneer 10 sensor at far distances from the Sun, but were evaporating at close distances from the Sun, with only few, or none, inside 2 AU. Humes (1980) found that the meteoroid penetration data obtained between 3.5 and 5 AU from the penetration sensors on the Pioneer 11 space probe could not be explained as being due to penetrations by meteoroids in largely prograde heliocentric orbits. Both asteroids (Bender 1979) and short period comets (Porter 1963) are nearly all in prograde orbits about the Sun. So the puzzle is, what is the source of meteoroids that are in such highly inclined and highly eccentric orbits? Are they related to long period comets or are they interstellar grains? Further analysis of the Galileo and Ulysses measurements and future data from the Cassini dust detector en route to Saturn should be able to provide a great deal more information about the orbital and compositional characteristics of this family of particles, and lead us to their sources.

Although the Voyager spacecraft did not carry a dust detector, it was discovered that during the crossing of Saturn's G ring the plasma wave instrument responded to dust impacts onto the spacecraft (Gurnett et al. 1983). Subsequent flybys of Uranus and Neptune confirmed this effect and dust in the vicinity of these planets was recorded (Gurnett et al. 1987; Gurnett et al. 1991). Recent initial analysis of the Plasma Wave data indicates (Gurnett et al. 1997) that dust impacts were identified even during interplanetary cruise in the region of the Kuiper belt.

V.B. Jupiter Dust Streams

Directional information on dust impacts was obtained by the dust instruments onboard Galileo and Ulysses. These spacecraft had most of the time their antennas and hence their spin axes pointing to the Earth. The angle variable used to describe the sensor position is called rotation angle. Its zero point is taken as the sensor pointing direction that is closest to ecliptic north. The sense of rotation is chosen positive for Ulysses (with respect to the spin axis pointing towards Earth) and negative for Galileo, reflecting the opposite rotation directions of the two spacecraft.

An unexpected discovery by the Galileo and Ulysses spacecraft was the high concentration of small dust impacts in the vicinity of Jupiter now called

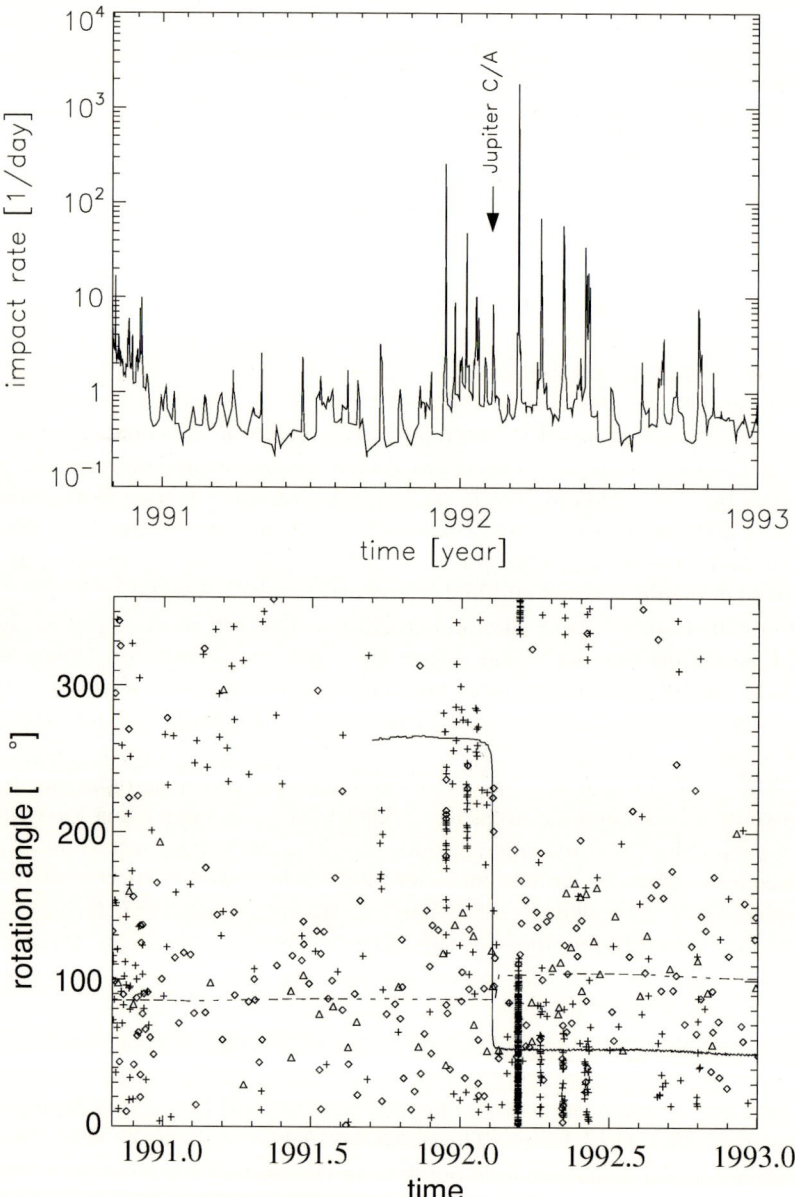

Figure 14. Dust impacts observed by Ulysses from launch to the end of 1992 (Jupiter flyby occurred on 10 February 1992). Upper panel: Impact rate of particles observed. A sliding mean over 6 impacts has been applied. Lower panel: Impact direction (rotation angle) of Ulysses dust impacts. The plus signs denote particles with impact charges $< 8 \cdot 10^{-14}$ C. The squares denote impacts with impact charges $\geq 8 \cdot 10^{-14}$ C. The solid lines show the direction of dust arriving from Jupiter, the dashed line indicates the direction from which interstellar dust would arrive.

Jupiter dust streams or even dust storms. Ulysses flew by Jupiter in February 1992 and the rate of dust impacts around the time of flyby is shown in Fig. 14, upper panel. Although the impact rates differ by several orders of magnitude between stream 10 and stream 6, the stream parameters are strikingly similar. Besides the concentration in time the distribution of the sensor pointing directions at the time of impacts is remarkable (Fig. 14, lower panel). The distribution has a maximum width of 140°. This coincides exactly with the width of the detector's angular sensitivity distribution. Within the statistical uncertainty, the distribution of rotation angles supports the assumption of a mono-directional stream of particles. From Fig. 14, lower panel, it seems quite natural that the direction the particles come from is related to Jupiter: The best argument for this assertion was the change in impact direction between the pre- and post-flyby streams. Before Jupiter flyby dust on fast trajectories from the planet is sensed from a prograde direction (270° rotation angle), whereas after the flyby dust stream particles are sensed from the opposite direction (90°). Another argument was the concentration of the streams near Jupiter. The first stream identified occurred at a distance from Jupiter of 1.1 AU and the last stream at about 2 AU distance. No dust streams were observed by Ulysses at larger distances.

The measurements by the Galileo dust instrument confirmed the Ulysses findings on approach to Jupiter and it is currently analyzing the dust streams within the jovian system (Grün et al. 1998). The first dust stream was measured in mid 1994, at 1.7 AU distance. Figure 15a shows the Galileo rate during its final orbit loop out to Jupiter. Big streams occurred about every 3 month, whereas small streams occurred irregularly in-between. There was a significant difference to the Ulysses dust stream measurements: The rates were higher, ranging up to 20 000 impacts/day for several days. On Ulysses, rates grater then 1 000 impacts/day were measured only during the peak hour of the biggest Ulysses stream. A reason for this difference is the viewing geometry of the Galileo dust sensor with respect to the Jupiter direction that was more favorable than in the Ulysses case. While comparing the rotation angle plot (Fig. 15, lower panel) with the rate plot (Fig. 15, upper panel) it has to be kept in mind that the rotation angles of only a very small portion of all recorded events could be transmitted with complete information. Nevertheless, the 4 major streams show up clearly in the data. With decreasing distance to Jupiter, the mean rotation angle is more and more aligned to the Jupiter direction. Analysis by Zook et al. (1996) showed that the Ulysses measurements can be explained as nanometer sized particles that are strongly deflected by the interplanetary magnetic field. Particles could enter the Ulysses detector only during favorable magnetic field conditions, whereas, in the Galileo case dust trajectories were always in the field of view of the sensor. Similar results were obtained for Galileo data (J.-C. Liou personal comm.).

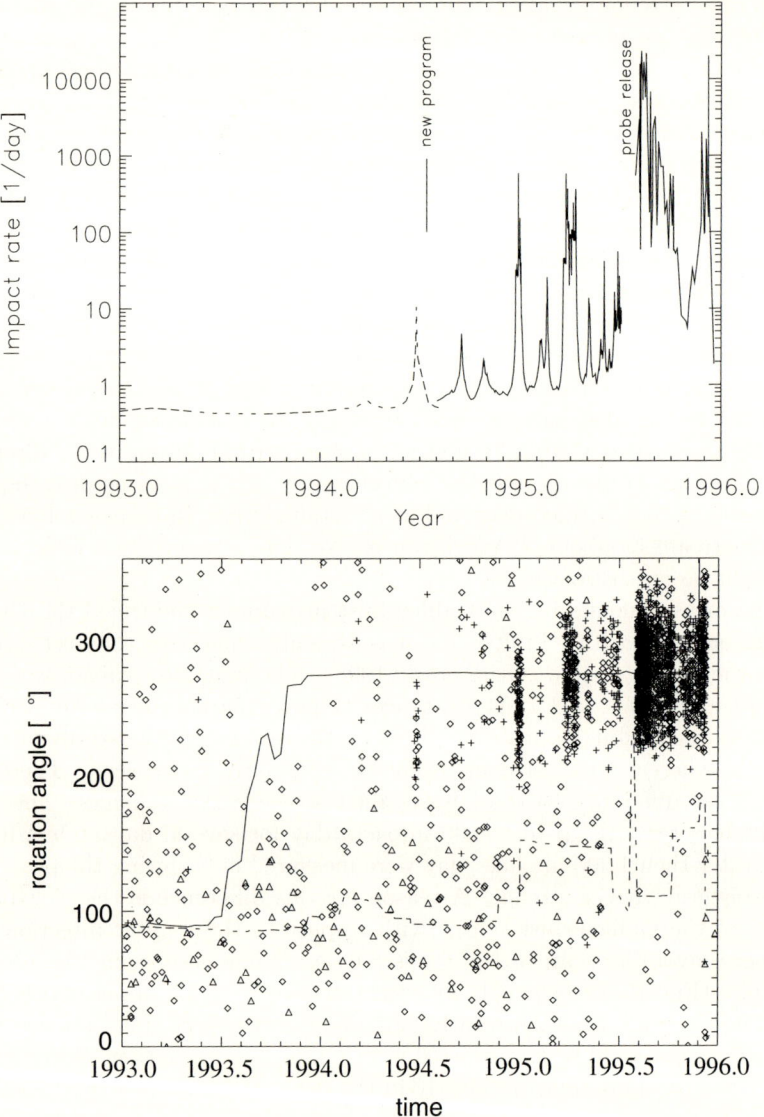

Figure 15. Dust impacts observed by Galileo during its trajectory from Earth (second flyby in December 1992) to Jupiter. Upper panel: Rate of all dust impacts recorded (solid line) and of big dust impacts (impact charge $\geq 8 \cdot 10^{-14}$ C, dashed line). Before the reprogramming of the dust instrument in July 1994 only big dust impacts could be reliably identified in the transmitted data. A data gap occurred during the release of Galileo's atmospheric probe. Lower panel: Instrument pointing directions (rotation angle) at the time of dust impacts. Plus signs denote impacts of dust particles with impact charges $< 8 \cdot 10^{-14}$ C. The squares denote particles with impact charges $\geq 8 \cdot 10^{-14}$ C. The solid lines show the direction of dust arriving from Jupiter, the dashed line indicate the direction from which interstellar dust would arrive.

V.C. Interstellar Dust

A second important discovery by Ulysses is the identification of a flow of interstellar dust particles passing through the planetary system. This finding was also confirmed by Galileo. The identification of interstellar dust is based on geometrical and dynamical arguments. It has been known that interstellar gas flows through the planetary system with a speed of 26 km s^{-1} in the direction of 73° ecliptic longitude and −5° ecliptic latitude (Lallement 1993; Witte et al. 1993). Before and after Ulysses flyby of Jupiter this direction was opposite to the prograde direction of Jupiter. The fact that the orbits of dust particles are predominantly retrograde at the Jupiter distance is very hard to explain without the assumption that the dust is from an interstellar source, since most of the heliospheric dust is on prograde orbits, like their sources, short period comets and asteroids. Although a large number of retrograde, mainly long period, comets exists, their inclinations are randomly distributed and cannot explain the near complete lack of prograde dust particle orbits at Jupiter's distance. Furthermore, a random distribution of orbital elements should result in uniform distribution of dust velocity vectors. This was not observed.

The interstellar origin of the particles should also show up as a high-velocity group in the measured impact speeds. Figure 16 shows impact speeds of non stream particles as a function of rotation angle measured during the first year after Ulysses' Jupiter flyby. The measurements are compared with lines of different β values separating bound orbits from hyperbolic ones. Assuming perfect coupling of interstellar gas and dust and a β of 1, the particles impact with a speed of about

$$v = \sqrt{v_{\text{is}}^2 + v_{\text{sc}}^2}, \tag{12}$$

where $v_{\text{is}} = 26$ km s^{-1}, the interstellar dust speed at infinity and $v_{\text{sc}} = 8$ km s^{-1}, the spacecraft speed at 5 AU, resulting in a mean impact speed of about 27 km s^{-1}, whereas the heliospheric escape speed at 5 AU is 18.8 km s^{-1} even assuming $\beta = 0$. Most measured speeds are high and seem to be centered around the expected value for interstellar grains. It has to be kept in mind, however, that the error in the speed determination is a factor of two and, therefore, speed arguments cannot be proof of the interstellar origin on their own.

In support of an interstellar origin of the dust flux observed by Ulysses and Galileo in the outer solar system is the fact that the measured dust flux and direction stays almost constant with spacecraft latitude and heliocentric distance, as can be seen in Fig. 17. The flux variations by only a factor two can be explained by the geometrical effect that the angular sensitivity for detecting interstellar grains varies with a period of approximately one year since Ulysses' spin axis follows the direction to Earth.

Galileo en route to Jupiter has also encountered interstellar dust. This can be easily seen in the data (Fig. 15b): at a distance of about 3 AU (1993.5), bigger particles are concentrated in the interstellar dust direction (rotation angle 90°). The mass distribution of interstellar grains found by Ulysses and

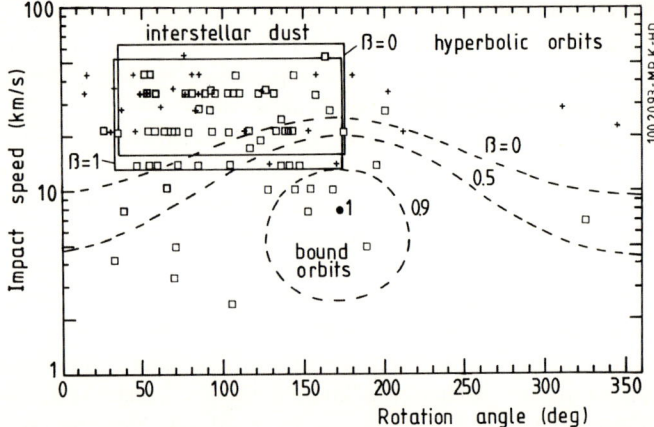

Figure 16. Impact speed versus impact direction (rotation angle) of dust impacts observed by Ulysses during a period of one year after Jupiter flyby (excluding dust stream particles, from Grün et al. 1994). Crosses denote particles with masses $\leq 2.5 \cdot 10^{-14}$ g and squares denote particles with bigger masses. Interstellar particles (arriving from rotation angle 90° at 26 km s^{-1}) should be recorded inside the boxes which take into account the uncertainty in speed and direction determination. The two boxes correspond to different assumed radiation pressure constants β (the small offset in rotation angle is introduced for better visibility). The dashed lines separate hyperbolic (above) from bound orbits (below) depending on the assumed β values.

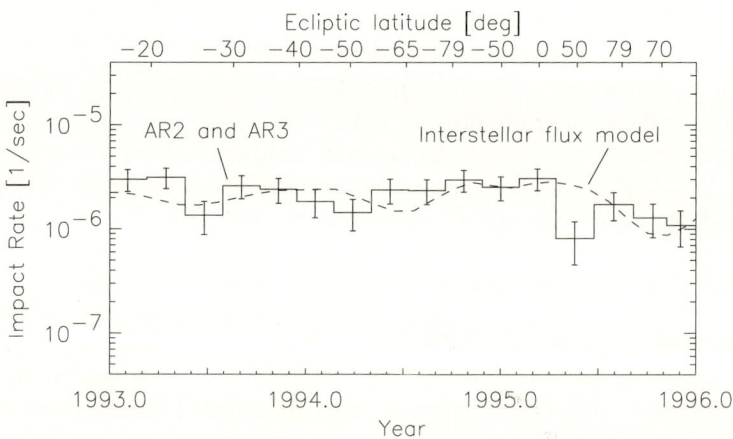

Figure 17. Flux of dust impacts observed by Ulysses during the period from Jupiter encounter to the North pole passage (solid line, from Krüger et al. 1999a). Included are only impacts with 10^{-13} C $<$ impact charges $\leq 10^{-11}$ C and dust stream particles have been excluded. A sliding mean with a width of 30 particles has been applied. The error bars correspond to the 1σ error. The broken line indicates the expected flux of interstellar dust according to a model that takes into account solar gravity and radiation pressure and the dust interaction with the interplanetary magnetic field and the viewing geometry of the dust sensor (Landgraf 2000).

Galileo has been analyzed by Baguhl et al. (1996). They found that the mass distribution only overlaps with the bigger masses of the "classical" Mathis, Rumpel, and Noordsiek (1977) distribution of astronomically observed interstellar grains but extends to much bigger particles. The total mass flux of the Ulysses particles is of the order of the mass flux expected from grains in the local interstellar medium (Frisch et al. 1999).

An attractive alternate explanation of the Pioneer 10 and 11 results in the outer solar system is the presumption that it is caused by interstellar dust as well. However, it has to be recognized that Pioneer 10 moved downstream from the interstellar flow and, hence, the interstellar dust particles that were observed by Ulysses and Galileo could not hit the detector, because it was facing the opposite hemisphere. Therefore, the big Pioneer 10 and 11 particles may be a different population of interstellar particles that do not couple as close to the interstellar gas as the smaller particles observed by Ulysses and Galileo. Meteor observations by Taylor et al. (1996) suggest that there exist a population of big ($\approx 10^{-6}$ g) interstellar meteors passing on different trajectories through the solar system.

VI. CHARACTERISTICS OF THE INTERPLANETARY DUST COMPLEX AS MEASURED BY SPACECRAFT

In this section we shall review the basic dynamic effects that act on dust particles in interplanetary space (see also chapter on Processes by Mukai et al.) and discuss whether and how, these effects are recognized in data from in-situ dust experiments. All dust particles in space feel the gravitational pull of the Sun and the planets. Table 4 gives values for the solar gravitational force on particles of different masses and compares it with other forces acting on particles in interplanetary space. For particles with masses $> 10^{-8}$ g solar gravity is by far the most dominating force. As a consequence dust grains move on Keplerian orbits that are conic sections with the Sun at one focus—other forces are only small disturbances. Certainly all observations of big particles are compatible with such orbits, although, the accuracy of in-situ measurements is not sufficient for an unique orbit determination.

VI.A. Gravity and Radiation Pressure Effects

Close to planets their gravitational effect becomes important. Inside the gravitational sphere of influence of a planet (Öpik 1951)

$$R = a_{\rm P} \left(\frac{m_{\rm P}}{2 m_\odot} \right)^{\frac{1}{3}}, \qquad (13)$$

the planetary gravitational force dominates. One effect is a gravitational flux enhancement near the planet which has been observed both near the Earth and near Jupiter, although some contribution of the particles observed there may come from dust populations orbiting these planets. Searches by HEOS 2

Table 4

Main forces on dust in interplanetary space at 5 AU, radius is for spherical, absorbing particles, density 1000 kg m^{-3}, $v_{\text{rel}} = 400$ km s^{-1}, $U = 5$ V, $B = 1$ nT, $\alpha = 80°$.

radius (μm)	0.01	0.1	1	10	100
mass (g)	$4 \cdot 10^{-17}$	$4 \cdot 10^{-14}$	$4 \cdot 10^{-11}$	$4 \cdot 10^{-8}$	$4 \cdot 10^{-5}$
F_{grav} (N)	10^{-24}	10^{-21}	10^{-18}	10^{-15}	10^{-12}
$F_{\text{rad}}/F_{\text{grav}}$	0.5	2	0.5	0.05	$5 \cdot 10^{-3}$
charge (C)	$6 \cdot 10^{-18}$	$6 \cdot 10^{-17}$	$6 \cdot 10^{-16}$	$6 \cdot 10^{-15}$	$6 \cdot 10^{-14}$
$F_{\text{L}}/F_{\text{grav}}$	2000	20	0.2	$2 \cdot 10^{-3}$	$2 \cdot 10^{-5}$

and Hiten for dust enhancements in the Lagrangian regions of the Earth-Moon system failed. The Earth shepherded ring (an orbital resonance effect with the Earth proposed by Jackson and Zook, 1989) has been found in remote IR observations by IRAS and COBE (Dermott et al. 1994, in IRAS data and Reach et al. 1995, in COBE data) but has not been uniquely identified in in-situ data. The reason for this negative results is that in-situ measurements of big particles ($m > 10^{-9}$ g) suffer from small numbers and consequently large statistical uncertainties.

Solar radiation pressure force on micrometeoroids in interplanetary space decreases with the inverse square of the distance to the Sun, i.e. it has the same radial dependence as the gravitational force. Therefore, the ratio of both forces is constant everywhere in interplanetary space and it is only dependent on material properties. This ratio is generally termed β

$$\frac{F_{\text{rad}}}{F_{\text{grav}}} = \beta = 5.7 \cdot 10^{-4} \frac{<Q_{\text{pr}}>}{\rho \cdot s}, \tag{14}$$

where $<Q_{\text{pr}}>$ is the efficiency factor for radiation pressure on the meteoroid, averaged over the solar spectrum, s is the radius of a spherical particle, and ρ is its density; all quantities are in SI units, e.g. for $s = 10^{-6}$ m (1 μm), $\rho = 1\,000$ kg m^{-3}, $\beta = 0.57$. For particles bigger than the effective wavelength of visible Sun light $<Q_{\text{pr}}> \approx 1$, depending somewhat on material properties, and it decreases for particles smaller than the wavelength. The force ratio β increases for smaller s values and reaches its maximum value between 0.1 and 1 μm. The maximum value is about 0.5 for non absorbing dielectric materials and increases with increased absorptivity; it reaches values of 3 to 10 for strongly absorbing metallic particles.

There are several consequences for the dynamics of small particles because of the radiation pressure. Firstly, from Eqs. 5 to 7 it can be seen that small particles affected by radiation pressure move slower on the same orbit than bigger particles. This is seen in in-situ data as the apex peak of dust impacts (Grün and Zook 1980): Spacecraft orbiting the Sun at about 1 AU distance (Pioneer 8, 9, HEOS and Hiten) found a maximum in the flux of micrometer-sized meteoroids arriving from the direction of spacecraft motion (apex) indicating that micrometeoroids are slower than a spacecraft on more or less circular orbits.

Secondly, small particles that are generated from big particles (e.g. particles emitted from comets or impact ejecta generated from meteoroids or asteroids) carry the kinetic energy of the bigger parent object but find themselves in a reduced effective potential field of the Sun due to radiation pressure. As a consequence these particles move on different orbits than their parent. E.g. if a dust particle is released at perihelion from a big parent object (eccentricity e_p) its orbit will have the eccentricity

$$e_d = \frac{e_p + \beta}{1 - \beta}. \tag{15}$$

It can be seen that even for a parent object on a circular orbit the ejected dust grain will move on an unbound hyperbolic orbit ($e_d \geq 1$) if its $\beta > 0.5$. We will call here particles β-meteoroids that leave the solar system on unbound orbits. Some authors used this term for all particles that are affected by radiation pressure even if they are still on elliptical bound orbits. However, this lead to some confusion since all particles in space are affected by radiation pressure, therefore, we want to restrict the term β-meteoroids to the narrower definition above.

In the data from Pioneers 8 and 9 a flow of β-meteoroids leaving the solar system on hyperbolic orbits was first detected by Berg and Grün (1973). Attempts by other spacecraft (e.g. Helios) to see them were not successful because of unfavorable detector geometries. Only recently β-meteoroids have been identified in Ulysses and Hiten data.

Besides the direct effect of radiation pressure on trajectories of small dust grains there is also a more subtle effect: the Poynting-Robertson effect. This is caused by radiation pressure force, that is not perfectly radial on a moving dust particle but has a small component opposite to the particle motion. The strength is of the order of v_t/c (v_t is the tangential particle velocity) and it leads to a loss of angular momentum and orbital energy of the orbiting particle. The effect is strongest when the particle speed is highest, i.e. close to the Sun at its perihelion. Therefore, particle orbits get slowly circularized while they spiral towards the Sun. The time, τ_{PR}, for a particle on a circular orbit to spiral to the Sun is

$$\tau_{\text{PR}} = 2.2 \cdot 10^{13} \frac{s\rho}{Q_{\text{pr}}} \left(\frac{r}{r_0}\right)^2, \tag{16}$$

with all quantities in SI units, e.g. $s = 10^{-2}$ m, $\rho = 1\,000$ kg m^{-3}, $r = r_0 = 1$ AU, $\tau_{\text{PR}} = 2.2 \cdot 10^{14}$ s $= 7 \cdot 10^6$ years. A similar but weaker effect (about 30 % of the Poynting-Robertson drag force) arises from solar wind drag (Jackson and Zook 1989; Gustafson 1994).

This spiraling of micrometeoroids to the Sun leads to a well-defined radial dependence of the spatial density inside the source region of these dust grains. The radial drift speed of a particle on circular orbit is

$$v_{\rm PR} \propto r^{-1}. \tag{17}$$

Flux conservation requires that, in steady state, the same number of particles pass through a spherical surface each second. This sets up a spatial density $n(r)$

$$n(r) v_{\rm PR} {\rm d}A = const., \quad {\rm d}A \propto r^2, \text{ and } n(r) \propto r^{-1}. \tag{18}$$

Leinert et al. (1981) show that the observed radial density $n \propto r^{-1.3}$ found from zodiacal light measurements with Helios is compatible with a distributed source of dust in the inner solar system. Other authors assume $n \propto r^{-1}$ and put the stronger intensity variation in a variation of the particle albedo or size distribution (Cook 1978; Stanley et al. 1979). In-situ measurements with Helios, Galileo, and Ulysses inside 2 AU from the Sun are compatible (Divine 1993; Grün et al. 1997) with a $r^{-1.3}$ slope of the spatial density but do not exclude a r^{-1} dependence.

For 10 to 100 μm-sized particles this circularization and inward drift of the Poynting-Robertson effect are often upset by planetary perturbations (Jackson and Zook 1989) and particles are temporarily trapped in an outer resonance with a planet. They remain trapped until their eccentricities are pumped up again such that the trap no longer holds the particles and they continue their sunward drift.

VI.B. Electromagnetic Effects

Any meteoroid in interplanetary space will be electrically charged. Several competing charging processes determine the actual charge of a meteoroid (cf. chapter by Mukai et al.). Irradiation by solar UV light frees photoelectrons which leave the grain. Electrons and ions are collected from the ambient solar wind plasma. Energetic ions and electrons cause the emission of secondary electrons. Whether electrons or ions reach or leave the grain depends on their energy and on the polarity and electric potential of the grain. Because of the predominance of the photoelectric effect in interplanetary space, meteoroids are usually charged positively at a potential of a few Volts. Only at rare times of very high solar wind densities may the electron flux to the particle dominate so that the particle is charged negatively. The final charging state is reached when all currents to and from the meteoroid cancel. The time scale for charging is seconds to hours depending on the size of the particle; small particles charge up slower. The charge q of a dust particle of mass m at a surface potential U is

$$q = \epsilon \eta U m^{\frac{1}{3}}, \tag{19}$$

where $\epsilon = 8.85 \cdot 10^{-12}$ C V^{-1} m^{-1} is the permittivity, and η is a constant describing the shape, structure, and density of the particle ($\eta = 7.8 \rho^{\frac{1}{3}}$, for a sphere of density ρ), $U \approx 5$ V due to the photo-effect. With the Cassini instrument (see chapter on Instrumentation by Auer) a new attempt is currently being made, and, indeed dust charges have been positively identified (Srama, personal communication).

The outward streaming (away from the Sun) solar wind carries a magnetic field. Due to the rotation of the Sun (with an equatorial period of 25.7 days) open magnetic field lines are drawn outward in a spiral like water from a lawn sprinkler. The polarity of the magnetic field can be either positive or negative depending on the polarity at the base of the field line in the solar corona which varies spatially and with time. A meteoroid in interplanetary space near the ecliptic plane typically sees 2 or 4 sectors of alternating magnetic field polarity per solar rotation. Above a certain heliographic latitude the field is unipolar. The components of the interplanetary magnetic field (Parker spiral) in spherical coordinates (r, ϕ, θ) are given by

$$B_r = B_{r0} \left(\frac{r_0}{r}\right)^2$$

$$B_\phi = B_{\phi 0} \left(\frac{r_0}{r}\right) \cos \theta$$

$$B_\theta = 0. \tag{20}$$

The angle, α, between the magnetic field and the radial direction is given by

$$\tan \alpha = \frac{\Omega r}{v_{\text{sw}}}, \tag{21}$$

with Sun's angular speed $\Omega = 2.866 \cdot 10^{-6}$ rad s^{-1}. The Lorentz force, F_L, on a dust grain of electric charge q is given by

$$F_L = q|\mathbf{v_{rel}} \times \mathbf{B}| = q v_{\text{rel}} B \sin \alpha. \tag{22}$$

The ratio of Lorentz force over gravity is $F_L/F_{\text{grav}} \propto s^{-2} r$, i.e. it is strongly increasing for smaller grain sizes and with larger heliocentric distances.

For a slowly moving observer or a meteoroid in interplanetary space near the ecliptic plane the magnetic field sweeps outward with the solar wind speed ($v_{\text{SW}} = 400$ to 800 km s^{-1}) with alternating polarity. In the magnetic reference frame the meteoroid moves inward at about the same speed since its orbital speed is comparatively small. The Lorentz force on a charged dust particle near the ecliptic plane is mostly either up- or downward depending on the polarity of the magnetic field. Ten-nanometer sized dust stream particles emitted from the jovian system show this effect of the local interplanetary magnetic field (Zook et al. 1996). Bigger particles do not feel such a strong electromagnetic interaction. Since the polarity varies due to the sector structure near the ecliptic plane at a frequency much faster than the orbital period of a μm-sized interplanetary dust particle the net effect of the Lorentz force is very small. Only secular effects on the bigger zodiacal particles are expected to occur that could have an effect on the symmetry plane of the zodiacal cloud close to the Sun (Morfill et al. 1979, 1986). Zodiacal light observations (Leinert et al. 1980) show such an effect on the symmetry plane but there are other explanations as well (Misconi and Weinberg 1978).

The overall polarity of the solar magnetic field changes with the solar cycle of 11 years. For one solar cycle positive magnetic polarity prevails in the northern and negative polarity in the southern solar hemisphere. As a consequence interstellar particles that enter the heliosphere are either deflected towards the solar equatorial plane (which roughly corresponds to the ecliptic plane) or away from it depending on the overall polarity of the magnetic field. Therefore, small interstellar particles are either prevented (during one solar cycle) from reaching the inner solar system or are concentrated (in the other solar cycle) in the ecliptic plane. However, one has to recognize that interstellar grains need about 20 years to traverse the distance from the heliospheric boundary (expected to be at about 100 AU from the Sun) to the inner solar system and, hence, encounter two opposing solar cycles. Focusing magnetic configuration occurred during the solar cycles from 1956 to 1967 and from 1978 to 1989. In the period from 1989 to 2000 the overall magnetic field has an unfavorable configuration, therefore, after some time lag only big (μm-sized) interstellar particles should reach the inner solar system (Landgraf 2000).

In Table 4 it is shown that for 10 nm sized particles the Lorentz force exceeds gravity by more than a factor 1 000. Therefore, the trajectories of these particles are totally dominated by the interaction with the interplanetary magnetic field. Zook et al. (1996) showed that the particles that constitute the Jupiter dust streams are of this size. The deviation of the stream direction (Fig. 15b and Fig. 18) from the Jupiter direction was correlated with the magnitude and direction of the interplanetary magnetic field (especially its tangential component). Calculations of dust trajectories backward in time by Zook et al. (1996) demonstrated the electromagnetic interaction of streams of nanometer sized particles observed by Ulysses and Galileo.

An important effect of meteoroids in interplanetary space are mutual collisions. It has been shown (Whipple 1967; Grün et al. 1985) that the meteoritic complex in the inner solar system is self destructive on a time scale of the order of 10^5 years. Grün et al. (1985) have shown that meteoroids below 10^{-6} g inside 1 AU are generated by collisions among bigger meteoroids. The life times of these bigger meteoroids are controlled by collisions rather than by the Poynting-Robertson effect. A direct consequence of collisions in space is the generation of fragments, part of which become of β-meteoroids. β-meteoroids, therefore, provide the major loss mechanism for meteoritic matter in the inner solar system. Detailed observations β-meteoroids could provide important clues to the stability of the interplanetary dust cloud and the strength of its sources.

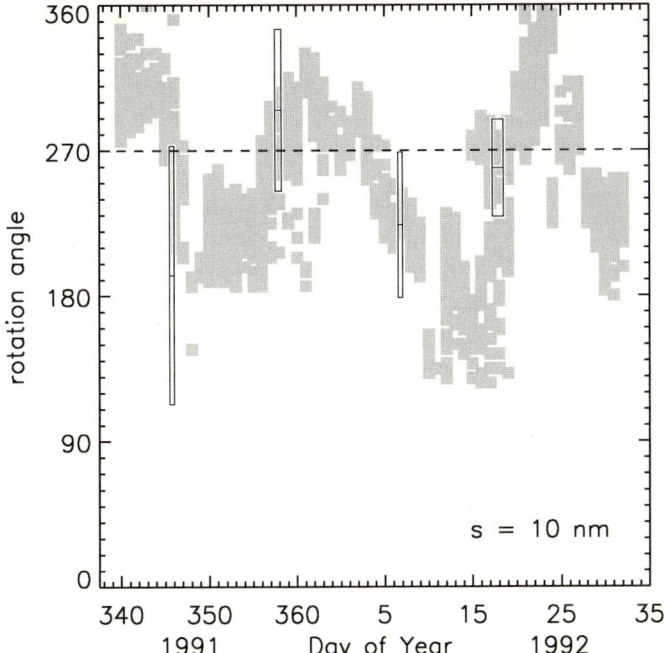

Figure 18. Calculated arrival directions (rotation angle) of dust trajectories from Jupiter to Ulysses (shaded areas) are compared with observed directions of dust streams (the 4 boxes refer to duration and range of observed impact directions). Arrival directions were calculated by tracing backwards in time many trajectories from Ulysses in various directions (Zook et al. 1996) and noting those trajectories (shaded areas) that come close to Jupiter ($<$ 100 Jupiter radii). Forces on 10 nm radius particles include solar gravity and interaction with the interplanetary magnetic field as observed by Ulysses (Balogh et al. 1993). Solar wind speed was assumed to be 400 km s^{-1} and the charge on the dust grains corresponds to $+5$ V surface charge.

VII. FUTURE DEVELOPMENTS

In this chapter, so far, we have demonstrated the outstanding capabilities of in-situ dust measurements and discussed the results obtained to date. Despite previous successes there remain a large range of questions to be addressed by future investigations. The emphasis of dust studies will shift from mere exploration and survey of the interplanetary dust inventory to detailed analyses of specific dust components and environments (planets, satellites, rings, asteroids, comets, and interstellar dust) and processes involving the formation, transport and destruction of dust grains. Chemical analyses and accurate trajectory measurements will provide important new information.

In the Earth environment the most important motivation for dust studies is the monitoring of man made space debris. Advances in instrumental techniques will provide reliable distinction between natural meteoroids and man made debris, either through orbital or compositional characterization of dust. As a side product, information on natural meteoroids will be gained. In the years 1998 to 2002 the return of the Leonid meteor shower (cf. chapter by Jenniskens) is attracting the attention of the space agencies because of the potential hazard to space missions and satellites. This meteor stream showed peak fluxes of about 3000 meteors per hour in 1999. According to Lyytinen and van Flandern (2000) the strongest meteor storms should occur in 2001 and 2002. Most in-situ debris monitors which are currently in preparation will fly in low Earth orbits (LEO) and study the submillimeter sized dust environment. Some satellites, like the Russian Express 2 telecommunication satellite (launched end of 1996) that carries a spare unit of the Ulysses dust detector, studies the dust environment in the important Geostationary Earth Orbit (GEO) regime (Drolshagen et al. 1999). When manned lunar exploration starts again the study of the lunar dust environment with instruments like the LEAM experiment (Berg et al. 1976) will become necessary.

Future studies of interplanetary dust may focus on special regions of interplanetary space where important questions involving dust are still unresolved. In the solar F-corona, i.e. inside about 20 R_S (solar radius $R_S = 7 \cdot 10^5$ km), where losses by evaporation from the zodiacal cloud occur, meteoroids of different materials are spatially separated according to their sublimation temperature. Solar probe missions exploring the solar neighborhood as close as 4 R_S are under study by space agencies. In the outer solar system the asteroidal contribution to the zodiacal cloud needs to be further characterized. Flybys of individual asteroids (Gaspra and Ida) by the Galileo spacecraft at distances of about 10^4 km did not show any enhancement of the dust flux in their immediate environment as was expected by Hamilton and Burns (1992). The search and analysis of a dust population outside the planetary region in the Kuiper belt will provide clues on primitive material from our own protoplanetary disk.

Cometary material, both gas and dust, is addressed by several space missions that are currently in preparation or in their early mission phases. NASA's Stardust (Brownlee et al. 1996) and Contour missions will analyze dust by an in-situ dust analyzer during flybys of comets. In addition, Stardust will return dust samples to Earth-based laboratories in order to be analyzed by state-of-the-art techniques. ESA's Rosetta mission will study dust collected during a rendezvous with a comet by analyzing-techniques that match most advanced dust analyzers in the laboratory of today. Compositional information on particles in the sub-micrometer size range and isotopic abundance of cosmo-chemically important elements will be gained by in-situ dust analyzers.

Circum-planetary dust in the jovian system is being studied by the Galileo mission, that reached Jupiter in late 1995. The Japanese Nozomi (Igenbergs et al. 1998) mission was launched to Mars in 1998. It carries a dust detector similar to the Hiten instrument and is taking measurements en route to and in orbit about Mars. Emissions of small dust particles may not only be restricted to the Jupiter magnetosphere but also Saturn and the other ringed planets may show such phenomena. The Cassini mission (launched in 1997) to Saturn will reach its destination in 2004 and will characterize the Saturnian dust environment during the following 4 years. The processes to be studied by all three missions (to Mars, Jupiter, and Saturn) include dust generation by impacts on satellites (Krüger et al. 1999b), by volcanoes (Io, Graps et al. 2000) and, potentially, geysers (Enceladus), and by interactions of dust with the plasmas and fields in magnetospheres (Grün et al. 1998) and in the solar wind (Zook et al. 1996). The Cassini dust instrument (Srama et al. 1996) includes a high rate detector in order to cope with the high impact rates expected while crossing the plane of Saturn's ring, an improved detector for the measurement of electric dust charges, and a medium resolution chemical analyzer. During its flyby of Jupiter in the year 2000 the Cassini spacecraft had an opportunity to look at dust streams from Jupiter with its improved instrumentation. These measurements were performed at a closest approach distance of 135 Jupiter radii while Galileo took simultaneous measurements of Jupiter dust streams at about 10 Jupiter radii.

Possibly the astrophysically most important discovery by in-situ dust detectors to date is the identification of interstellar grains sweeping through the planetary system. This species of dust could previously only be studied by extinction measurements of distant starlight and by infrared emissions with limited information content. In-situ analysis of local interstellar dust by sophisticated instrumentation provides a whole new course of scientific investigation to this important but otherwise difficult to characterize material. Only three years after the discovery of interstellar dust in the solar system, NASA approved the Stardust mission which will analyze interstellar dust (in addition to cometary dust, Landgraf et al. 1999) by an in-situ dust analyzer and collect and return dust samples to Earth-based laboratories for further analysis. Also the Cassini dust detector with its improved capabilities will contribute new information on interstellar dust and the interstellar environment of our solar system.

Acknowledgements

Support in preparation of the manuscript by O. Kress, H. Krüger, M. Landgraf, and G. Linkert is acknowledged.

REFERENCES

Allison, R. J. and McDonnell, J. A. M. 1981. Secondary cratering effects on lunar microterrain: Implications for the micrometeoroid flux. *Proc. Lunar Planet. Sci. Conf.*, **12B**, pp. 1703–1716.

Auer S. 1974. The asteroid belt: doubts about the particle concentration measured with the Asteroid/Meteoroid detector on Pioneer 10. *Science*, **186**, pp. 650–652.

Baguhl, M., Grün, E., Linkert, G., Linkert, D. and Siddique, N. 1993. Identification of 'small' dust impacts in the Ulysses dust detector data. *Planet. Space Sci.*, **41**, pp. 1085–1098. and

Baguhl, M., Grün, E. and Landgraf, M. 1996. In situ measurements of interstellar dust with the Ulysses and Galileo spaceprobes, *Space Sci. Rev.*, **78**, pp. 165–172.

Balogh, A., Erdos, G., Forsyth, R. J., Smith, E. J. 1993. The evolution of the interplanetary sector structure in *Geophys. Res. Lett.*, **20**, pp. 2331–2334.

Baumgärtel, K., Sauer, K., Bogdanov, A., Dubinin, E., Dougherty, M. 1996. Phobos events: signatures of solar wind dust interaction. *Planet. Space Sci.*, **44**, pp. 589–601.

Bender, D. F. 1979. Osculating orbital elements of the asteroids In *Asteroids*, ed. T. Gehrels (Tucson: Univ. Arizona Press), pp. 1014–1039.

Berg, O. E. and Grün, E. 1973. Evidence of hyperbolic cosmic dust particles. In *COSPAR: Space Research XIII*, pp. 1046–1055.

Berg, O. E., Wolf, H., and Rhee, J. 1976. Lunar soil movement registered by the Apollo 17 cosmic dust experiment. In *Interplanetary Dust and Zodiacal Light*, eds. H. Elsässer and H. Fechtig (Berlin: Springer Verlag), pp. 233–237.

Brownlee, D. E., Burnett, D., Clark, B., Hanner, M. S., Horz, F., Kissel, J., Newburn, R., Sandford, S., Sekanina, Z., Tsou, P., and Zolensky, M. 1996. Stardust: Comet and interstellar dust sample return mission. In *Phys. Chemistry and Dynamics of Interplanetary Dust, ASP Conf. Series*, **104**, eds B. Å. S. Gustafson and M. S. Hanner, pp. 223–226.

Burns, J. A., Lamy, Ph. L., and Soter, S. 1979. Radiation forces on small particles in the solar system. *Icarus*, **40**, pp. 1–48.

Cook, A. F. 1978. Albedos and size distribution of meteoroids from 0.3 to 4.8 AU. *Icarus*, **33**, pp. 349–360.

Cour-Palais, B. G. 1979. Space vehicle meteoroid shielding Design. In *The Comet Halley Micrometeoroid Hazard, ESA SP-153*, pp. 85–92.

D'Aiutolo, C. T., Kinard, W. H., and Naumann, R. J. 1967. Recent NASA meteoroid penetration results from satellites. In *Meteor Orbits and Dust*, ed. G. S. Hawkins, NASA SP-135, pp. 239–251.

Dermott, S. F., Jayaraman, S., Xu, Y. L., Gustafson, B. Å. S., and Liou, J. C. 1994. A circumpolar ring of asteroidal dust in resonant lock with the Earth. *Nature*, **369**, pp. 719–723.

Divine, N. 1993. Five populations of interplanetary meteoroids. *J. Geophys. Res.*, **98**, pp. 17029–17048.

Dozier, J. B. 1966. V. Meteoroid data recorded on Pegasus flights. In *The Meteoroid Satellite Project Pegasus First Summary Report*, NASA TN D-3505, pp. 65–76.

Drolshagen, G., Svedhem, H., Grün, E., Grafodatsky, O., Prokopiev, U. 1999. Microparticles in the geostationary orbit (GORID experiment). *Advances in Space Research*, **23**, pp. 123–133.

Dubin, M., and Soberman, R. K. 1991. Cosmoids: Solution to the Pioneer 10 and 11 meteoroid measurement enigma. *Planet. Space Sci.*, **39**, pp. 1573–1590.

Fechtig, H., Hartung, J. B., Nagel, K., Neukum, G., and Storzer, D. 1974. Lunar microcrater studies, derived meteoroid fluxes, and comparison with satellite-borne experiments. *Proc. 5th Lunar Sci. Conf.*, **Vol. 3**, pp. 2463–2474.

Fechtig, H., Grün, E., and Morfill, G. 1979. Micrometeoroids within Ten Earth Radii, Planet. *Space Sci.*, **27**, pp. 511–531.

Feller, W. 1957. An introduction to probability theory and its applications. 2nd Ed. (New York: Wiley).

Fillius, W. 1976. The trapped radiation belts of Jupiter. In *Jupiter*, ed. T. Gehrels (Tucson: Univ. of Arizona Press), pp. 896–927.

Frisch, P. C., Dorschner, J., Geiss, J., Greenberg, J. M., Grün, E., Landgraf, M., Hoppe, P.,

Jones, A. P., Krätschmer, W., Linde, T. J., Morfill, G. E., Reach, W. T., Slavin, J., Svestka, J., Witt, A., Zank G.P. 1999. Dust in the Local Interstellar Wind. *Astrophysical Journal*, **525**, pp. 492–516.

Geiss, J., Gloeckler, G., von Steiger, R. 1996. Origin of C+ Ions in the Heliosphere. *Space Science Reviews*, **78**, pp. 43–52.

Giese, R. H., and Grün, E. 1976. The Compatibility of Recent Micrometeoroid flux Curves with Observations and Models of the Zodiacal Light. In *Interplanetary Dust and Zodiacal Light*, eds. H. Elsässer and H. Fechtig, Lecture Notes in *Physics*, **48**, (Berlin–Heidelberg–New York: Springer Verlag), pp. 135–139.

Graps, A. L., Grün, E., Svedhem, H., Krüger, H., Horányi, M., Heck, A., Lammers, S. 2000. Io as a Source of the Jovian Dust Streams. *Nature*, **405**, pp. 48-50.

Grün, E., Kissel, J., and Hoffmann, H.-J. 1976. Dust Emission from Comet Kohoutek (1973 f) at large Distance from the Sun. In *Interplanetary Dust and Zodiacal Light*. eds. H. Elsässer and H. Fechtig, Lecture Notes in *Physics*, **48**, (Berlin–Heidelberg–New York: Springer Verlag), pp. 334–338.

Grün, E., Pailer, N., Fechtig, H., and Kissel, J. 1980. Orbital and Physical Characteristics of Micrometeoroids in the Inner Solar System as Observed by Helios 1. *Planet. Space Sci.*, **28**, pp. 333–349.

Grün, E., and Zook, H. A. 1980. Dynamics of Micrometeoroids. In *Solid Particles in the Solar System*, eds. I. Halliday and B. A. McIntosh (Dordrecht: Reidel), pp. 293–298.

Grün, E. 1981. Physikalische und Chemische Eigenschaften des interplanetaren Staubes – Messungen des Mikrometeoritenexperimentes auf Helios. Forschungsbericht, BMFT W 81-034, 194 Seiten.

Grün, E. and Zook, H. A. 1985. In-Situ Detection of Micron Sized Dust Particles in Near-Earth Space. Proc. of Orbital Debris Workshop, NASA CP- 2360, pp. 233–245.

Grün, E., Zook, H. A., Fechtig H., and Giese, R. H. 1985. Collisional balance of the meteoritic complex. *Icarus* **62**, pp. 244–272.

Grün, E., Fechtig, H., Hanner, M. S., Kissel, J., Lindblad, B. A., Linkert, D., Morfill, G. E., and Zook, H. A. 1992a. The Galileo dust detector. *Space Sci. Rev.*, **60**, pp. 317–340.

Grün, E., Zook, H. A., Baguhl, M., Balogh, A., Bame, S. J., Fechtig, H., Forsyth, R., Hanner, M. S., Horányi, M., Kissel, J., Lindblad, B.-A., Linkert, D., Linkert, G., Mann, I., McDonnell, J. A. M., Morfill, G. E., Phillips, J. L., Polanskey, C., Schwehm, G., Siddique, N., Staubach, P., Svestka, J., and Taylor, A. 1993. Discovery of jovian dust streams and interstellar grains by the Ulysses spacecraft. *Nature*, **362**, pp. 428–430.

Grün, E., Baguhl, M., Divine, N., Fechtig, H., Hanner, M. S., Kissel, J., Lindblad, B. A., Linkert, D., Linkert, G., Mann, I., McDonnell, J. A. M., Morfill, G. E., Polanskey, C., Riemann, R., Schwehm, G., Siddique, N., Staubach, P., and Zook, H. A. 1995. Three years of Galileo dust data Planet. *Space Sci*, **43**, pp. 971–999.

Grün, E., Staubach, P., Baguhl, M., Dermott, S., Fechtig, H., Gustafson, B. A., Hamilton, D. P., Hanner, M. S., Horányi, M., Kissel, J., Lindblad, B. A., Linkert, D., Linkert, G., Mann, I., McDonnell, J. A. M., Morfill, G. E., Polanskey, C., Schwehm, G., Srama, R., and Zook, H. A. 1997a. South–North and Radial Traverses Through the Zodiacal Cloud. *Icarus*, **129**, pp. 270–288.

Grün, E., Krüger, H., Graps, A. L., Hamilton, D. P., Heck, A., Linkert, G., Zook, H. A., Dermott, S., Fechtig, H., Gustafson, B. A., Hanner, M. S., Horányi, M., Kissel, J., Lindblad, B. A., Linkert, D., Mann, I., McDonnell, J. A. M., Morfill, G. E., Polanskey, C., Schwehm, G., and Srama, R. 1998. Galileo observes electromagnetically coupled dust in the jovian system. *J. Geophys. Res.*, **103**, pp. 20011–20022.

Gurnett, D. A., Grün, E., Gallagher, D., Kurth, W. S., and Scarf, F. L. 1983. Micron-Size Particles Detected Near Saturn by the Voyager Plasma Wave Instrument. *Icarus*, **53**, pp. 236–254.

Gurnett, D. A., Kurth, W. S., Scarf, F. L., Burns, J. A., J. A., Cuzzi, J. A., Grün, E., 1987. Micron-Sized Particles Impact Detected Near Uranus by the Voyager 2 Plasma Wave Instrument. *Journal of Geophys. Res.*, **Vol. 92** (No. A13), pp. 14959–14968.

Gurnett, D. A., Kurth, W. S., Granroth, L. J., and Allendorf, S. C. 1991. Micron-sized particles detected near Neptune by the Voyager 2 plasma wave instrument. *J. Geophys. Res.*, **96**, pp. 19177–19186.

Gurnett, D. A., Anher, J. A., Kurth, W. S., and Granroth, L. J. 1997. Micron-sized particles detected in the outer solar system by the Voyager 1 and 2 plasma wave instruments.

Geophys. Res. Lett., **24**, pp. 3125–3128.

Gustafson, B. Å. S. 1994. In *Physics of the zodiacal cloud*, Ann Rev. Earth Planet. Sci., **22**, pp. 553–595.

Hamilton, D. P., Grün, E., and Baguhl, M. 1996. Electromagnetic escape of dust from the solar system. In *Physics, Chemistry, and Dynamics of Interplanetary Dust, ASP Conference Series*, **Vol. 104**, eds. B. Å. S. Gustafson and M. S. Hanner, pp. 31–34.

Hamilton, D. P., and Burns, J. A. 1992. Orbital stability zones about asteroids. II - The destabilizing effects of eccentric orbits and of solar radiation. *Icarus*, **96**, pp. 43–64.

Hanner, M. S., Sparrow, J. G., Weinberg, J. L., and Beeson, D. E. 1976. Pioneer 10 observations of zodiacal light brightness near the ecliptic: Changes with heliocentric distance. In *Lecture Notes in Physics*, **48**: Interplanetary Dust and Zodiacal Light, eds. H. Elsasser and H. Fechtig (New York: Springer-Verlag), pp. 29–35.

Hastings, E. C. 1963. The Explorer XVI micrometeoroid satellite: Description and preliminary results for the period December 16, 1962 through January 13, 1963. NASA TM X-810.

Hoffmann, H.-J., Fechtig, H., Grün, E., and Kissel, J. 1975a. First results of the micrometeoroid experiment S-215 on HEOS 2 satellite. *Planet. Space Sci.* **23**, pp. 215–224.

Hoffmann, H.-J., Fechtig, H., Grün, E., and Kissel, J., 1975b. Temporal fluctuation and anisotropy of the micrometeoroid flux in the Earth-Moon system. *Plant. Space Sci.*, **23**, pp. 985–991.

Hoffmann, H.-J., Fechtig, H., Grün, E., and Kissel, J., 1976. Particles from Comet Kohoutek detected by the Micrometeoroid Experiment on HEOS 2. In *The study of Comets*, eds. B. Donn et al., NASA SP-393, pp. 949–961.

Hörz, F., Brownlee, D. E., Fechtig, H., Hartung, J. B., Morrison, D. A., Neukum, G., Schneider, E., Vedder, J. F., and Gault, D.E. 1975. Lunar microcraters: Implications for the micrometeoroid complex, *Planet. Space Sci.*, **23**, pp. 985–992.

Humes, D. H., Alvarez, J. M., O'Neal, R. L., and Kinard, W. H. 1974. The interplanetary and near-Jupiter meteoroid environments. *J. Geophys. Res.*, **79 (No. 25)**, pp. 3677–3684.

Humes, D. H. 1980. Results of Pioneer 10 and 11 meteoroid experiments: Interplanetary and near-Saturn. *Journ. Geophys. Res.*, **85**, pp. 5841–5852.

Humes, D. H. 1993. Small craters on the meteoroid and space debris impact experiment. In *LDEF—69 Months in Space, Third Post- Retrieval Symposium*, ed. A. S. Levine (NASA Conf. Publ. 3275), pp. 287–322.

Hutcheon, I .D. 1975. Micrometeorites and solar flare particles in and out of the ecliptic. *J. Geophys. Res.*, **80**, pp. 4471–4483.

Igenbergs, E., Hüdepohl, A., Uesugi, K. T., Svedhem, H., Igelseder, H., Grün, E., Mizutani, H., Yamamoto, T., Fujimura, A., and Araki, H. 1991. The Munich Dust Counter Cosmic Dust Experiment: Results of the first year of operation. 42nd Congress of the International Astronautical Federation, IAF-91-470.

Igenbergs, E., Sasaki, S., Münzenmayer, R., Ohashi, H., Farber, G., Fischer, F., Fujiwara, A., Glasmachers, A., Grün, E., Hamabe, Y., Iglseder, H., Klinge, D., Miyamoto, H., Mukai, T., Naumann, A., Nogami, K., Schwehm, G., Svedhem, H., and Yamakoshi, K. 1998. Mars dust counter, *Earth Planets Space*, **50**, pp. 241–245.

Jackson, A. A., and Zook, H. A. 1989. A solar system dust ring with the Earth as its shepherd. *Nature*, **337**, pp. 629–631.

Krüger, H., Grün, E., Landgraf, M., Baguhl, M., Dermott, S., Fechtig, H., Gustafson, B. A., Hamilton, D. P., Hanner, M. S., Horányi, M., Kissel, J., Lindblad, B.A., Linkert, D., Linkert, G., Mann, I., McDonnell, J. A. M., Morfill, G. E., Polanskey, C., Schwehm, G., Srama, R. and Zook, H. 1999a. Three years of Ulysses dust data: 1993 to 1995, *Planetary and Space Science* **47**, pp. 363–383.

Krüger, H., Krivov, A. V., Hamilton, D. P., Grün, E. 1999b. Detection of an impact-generated dust cloud around Ganymede. *Nature*, **399**, pp. 558–560.

Lallement, R. 1993. Measurements of the interstellar gas. *Adv. Space Res.* **13**, pp. (6)113–(6)120.

Landgraf, M. 2000. Modelling the motion and distribution of interstellar dust inside the Heliosphere. *J. Geophys. Res.*, **105**, pp. 10302–10316.

Landgraf, M., Müller, M., Grün, E. 1999. Prediction of the in-situ dust measurements of the Stardust mission to comet 81P/Wild 2. *Planetary and Space Science*, **47**, pp. 1029–1050.

Leinert, Ch. , Hanner, M., Richter, I., and Pitz, E. 1980. The plane of symmetry of interplanetary dust in the inner solar system. *Astron. Astrophys.*, **82**, pp. 328–336.

Leinert, C., Richter, I., Pitz, E., and Planck, B. 1981. The zodiacal light from 1.0 to 0.3 A.U. as observed by the Helios space probes. *Astron. Astrophys.*, **103**, pp. 177–188.

Leinert, Ch., and Grün, E. 1990. Interplanetary Dust. In *Physics of the Inner Heliosphere I.*, eds. R. Schwenn, and E. Marsch (Berlin: Springer Verlag), pp. 207–275.

Low, F. J. , Beintema, D. A., Gautier, T. N., Gillette, F. C., Beichman, C. A., Neugebauer, G., Young, E., Aumann, H. H., Boggess, N., Emerson, J. P., Habing, H. J., Hauser, M. G., Houck, J. R., Rowan-Robinson, M., Soifer, B. T., Walker, R. G., and Wesselius, P. R. 1984. Infrared cirrus: New components of the extended infrared emission. *Astrophys. J.*, **278**, pp. L19–L22.

Lyytinen, E. J. and van Flandern, T. 2000. Predicting the strength of Leonid outbursts. *Earth, Moon and Planets*, **82–83**, pp. 149–166.

Mathis, J. S., Rumpl, W., Nordsieck, K. H. 1977. The size distribution of interstellar grains. *Astrophys. J.*, **217**, pp. 425–433.

Mazets, E. P. 1971. Cosmic dust and meteor showers. *Space Res.*, **XI**, (Berlin: Akademie-Verlag) pp. 363-369.

McDonnell, J. A. M. 1978. Microparticle Studies for Space Instrumentation. In *Cosmic Dust*, ed. J. A. M. McDonnell, (New York: Wiley and Sons), p. 337.

McDonnell, J. A. M., Berg, O. E., and Richardson 1975. Spatial and time variations of the interplanetary microparticle flux analyzed from deep space probes Pioneers 8 and 9. *Planet. Space Sci.*, **23**, pp. 205–214.

Misconi, N. Y., and Weinberg, J. L. 1978. Is Venus concentrating interplanetary dust towards its orbital plane?, *Science*, **200**, pp. 1484–1485.

Morfill, G. E., and Grün, E. 1979. The motion of charged dust particles in interplanetary space – I. The zodiacal dust cloud. *Planet. Space Sci.*, **27**, pp. 1269–1282.

Morfill, G. E., Grün, E., and Leinert, C. 1986. The Interaction of Solid Particles with the Interplanetary Medium. In *The Sun and the Heliosphere in Three Dimensions*, ed. R. G. Marsden (Dordrecht: D. Reidel Publishing Co.), pp. 455–474.

Morrison, D. A., and Clanton, U. S. 1979. Properties of microcraters and cosmic dust of less than 1 000 A dimensions, *Proc. Lunar Planet. Conf. 10th*, pp. 1649–1663.

Morrison, D. A., and Zinner, E. 1977. 12054 and 76215: New measurements of interplanetary dust and solar flare fluxes. *Proc. Lunar Sci. Conf. 8th*, pp. 841–863.

Naumann, R. J. 1966. The near-Earth meteoroid environment. NASA TN D-3717.

Oberst, J., and Nakamura, Y. 1991. A search for clustering among the meteoroid impacts detected by the Apollo lunar seismic network. *Icarus*, **Vol. 91**

O'Neal, R. L. 1965. The Explorer XXIII micrometeoroid satellite: Description and preliminary results for the period November 6, 1964 through February 15, 1965. NASA TM X-1123.

Öpik, E. J. 1951. Collision probabilities with the Planets. *Proc. Roy. Irish Acad.*, **A54**, pp. 165–172.

Pailer, N., and Grün, E. 1980. The Penetration Limit of Thin Films. *Planet. Space Sci.*, **Vol. 28**, pp. 321–331.

Porter, J. G. 1963. The statistics of comet orbits. In *The Moon Meteorites and Comets*, eds. B. M. Middlehurst and G. P. Kuiper (Univ. Chicago Press), pp. 550–572.

Reach, W. T., Franz, B. A., Weiland, J. L., Hauser, M. G., Kelsall, T. N., Wright, E. L., Rawley, G., Stemwedel, S. W., and Spiesman, W. J. 1995. Observational confirmation of a circumsolar dust ring by the COBE satellite. *Nature*, **374**, pp. 521–523.

Ricker, W. R. 1937. The concept of confidence or fiducial limits applied to the Poisson frequency distribution. *J. American Statistical Assn.*, **32**, pp. 349–356.

Riemann, R., and Grün, E. 1992. Meteor streams, asteroids and comets near the orbits of Galileo and Ulysses. In *Hypervelocity Impacts in Space*, ed. J. A. M. McDonnell (Canterbury: Univ. of Kent), pp. 120–125.

Sobermann, R. K., Neste, S. L., and Lichtenfeld, K. 1974. Optical measurements of interplanetary particulates from Pioneer 10. *J. Geophys. Res.*, **79**, pp. 3685–3694.

Srama, R., and Grün, E., and the Cassini Dust Science Team 1996. The Cosmic Dust Analyzer for the Cassini mission to Saturn. In *Physics, Chemistry, and Dynamics of Interplanetary Dust, ASP Conference Series*, **Vol. 104**, eds. B. Å. S. Gustafson and M. S. Hanner, pp. 227–231.

Stanley, J. E., Singer, S. F., Alvarez, J. M. 1979. Interplanetary dust between 1 and 5 AU. *Icarus*, **37**, pp. 457–466.

Svestka, J., Auer, S., Baguhl, M., and Grün, E. 1996. Measurement of dust electric charges by the Ulysses and Galileo dust detectors. In *Physics, Chemistry, and Dynamics of Interplanetary Dust, ASP Conference Series*, **Vol. 104**, eds. B. Å. S. Gustafson and M. S. Hanner, pp. 481–484.

Sykes, M. V., and Greenberg, R. 1986. The formation and origin of the IRAS zodiacal dust bands as a consequence of single collisions between asteroids. *Icarus*, **65**, pp. 51–69.

Taylor, A. D., Baggaley, W. J., and Steel, D. I. 1996. Discovery of interstellar dust entering the Earth's atmosphere. *Nature*, **380**, pp. 323–325.

Wagner, M. H., and Kreyenhagen, K. N. 1979. Review of hydro-elastic-plastic code analysis as related to the hypervelocity particle impact hazard. In *The Comet Halley Micrometeoroid Hazard*, **ESA SP-153**, pp. 115–120.

Whipple, F. L. 1958. The meteoritic risk to space vehicles. In *Proc. of the VIIIth International Astronautical Congress*, (Vienna: Springer-Verlag) pp. 429–435.

Whipple, F. L. 1967. On maintaining the meteoritic complex. In *NASA SP-150: The Zodiacal Light and the Interplanetary Medium*, ed. J. L. Weinberg (Washington, D. C.: U. S. Govt. Printing Office), pp. 409–426.

Whipple, F. L. 1975. Sources of interplanetary dust. In *Lecture Notes in Physics*, **48**: Interplanetary Dust and Zodiacal Light, eds. H. Elsässer and H. Fechtig (New York: Springer-Verlag), pp. 403–415.

Witte, M., Rosenbauer, H., Banaskiewicz, M., and Fahr, H. 1993. The Ulysses neutral gas experiment: Determination of the velocity and temperature of the neutral interstellar helium. *Adv. Space Res.*, **13**, pp. (6)121–(6)130.

Wolf, H., Rhee, J., and Berg, O. E. 1976. Orbital elements of dust particles intercepted by Pioneers 8 and 9. In *Lecture Notes in Physics*, **48**: Interplanetary Dust and Zodiacal Light, eds. H. Elsässer and H. Fechtig (New York: Springer-Verlag), pp. 165–169.

Zook, H. A., Flaherty, R. E., and Kessler, D. J. 1970. Meteoroid impacts on the Gemini windows. *Planet. Space Sci.*, **18**, pp. 953–964.

Zook, H. A. 1975a. The state of meteoritic material on the Moon. *Proc. 6th Lunar Sci. Conf.*, pp. 1653–1672.

Zook, H. A. 1975b. Hyperbolic cosmic dust: its origin and its astrophysical significance. *Planet. Space Sci.*, **23**, pp. 1391–1397.

Zook, H. A. and Berg, O. E. 1975 A source for hyperbolic cosmic dust particles. *Planet. Space Sci.*, **23**, pp. 183–203.

Zook, H. A. 1980. On lunar evidence for a possible large increase in solar flare activity $2 \cdot 10^4$ years ago. *Proc. Conf. Ancient Sun*, eds. R. O. Pepin, J. A. Eddy, and R. Merril, pp. 245–266.

Zook, H. A., Lange, G., Grün, E., and Fechtig, H. 1984. Lunar primary and secondary microcraters and the micrometeoroid flux. *Lunar and Planet. Sci.*, **XV**, pp. 965–966.

Zook, H. A., Lange, G., Grün, E., and Fechtig, H. 1985. The interplanetary micrometeoroid flux and lunar primary and secondary microcraters. In *Properties and Interactions of Interplanetary Dust*, eds. R. H. Giese and P. Lamy, pp. 89–96.

Zook, H. A., Grün, E., Baguhl, M., Hamilton, D., Linkert, G., Liou, J. C., Forsyth, R., and Phillips, J. L. 1996. Solar wind magnetic field bending of jovian dust trajectories. *Science*, **274**, pp. 1501–1503.

Synthesis of Observations

Peter Staubach[1,2], Eberhard Grün[2], Mark J. Matney[3]

[1] European Space Operations Centre, Darmstadt, Germany
[2] MPI für Kernphysik, Heidelberg, Germany
[3] Lockheed Martin Engineering and Science Services,
   NASA Johnson Space Center, Houston, USA

Abstract. Based on Neil Divine's (1993) 'Five Population of Interplanetary Meteoroids'-model, a description of the interplanetary meteoroid complex is given in terms of distinct meteoroid populations. Each population has separable distributions in particle mass, orbital inclination, eccentricity and perihelion distance. The model matches particle concentrations and fluxes as derived from radar meteors and zodiacal light observations as well as from measurements by impact detectors on various spacecraft. This model has been expanded to include a comparison with directional and impact speed information obtained by the Galileo and Ulysses detectors. Small particle orbits that are affected by radiation pressure are newly included in the model. Particle populations on heliocentric hyperbolic orbits have to be considered in order to match recent spacecraft measurements.

PREAMBLE

In 1993 when this book was conceived E.G. asked Neil Divine to be principal author of this chapter. Despite his severe illness he accepted this task. Unfortunately in early 1994 Neil Divine passed away before he could lay out the plan for this chapter. Because of the close co-operation of P.S. and E.G. with him during his last years we are obligated to his memory. This chapter is dedicated to Neil Divine the modest man and his paramount work.

I. INTRODUCTION

Dust in interplanetary space can be detected and analyzed by a number of techniques. Huge surveys of radio meteor orbits observed by the Harvard-Smithsonian radar have been published by Southworth and Sekanina (1973). Recent surveys with the Adelaide radar are reported by Baggaley (1996). Interplanetary dust particles (IDPs) which are collected in the atmosphere give cosmochemical information, e.g. composition and structure (Brownlee 1996 and Bradley 1996). The size distribution of dust at 1 AU distance from the

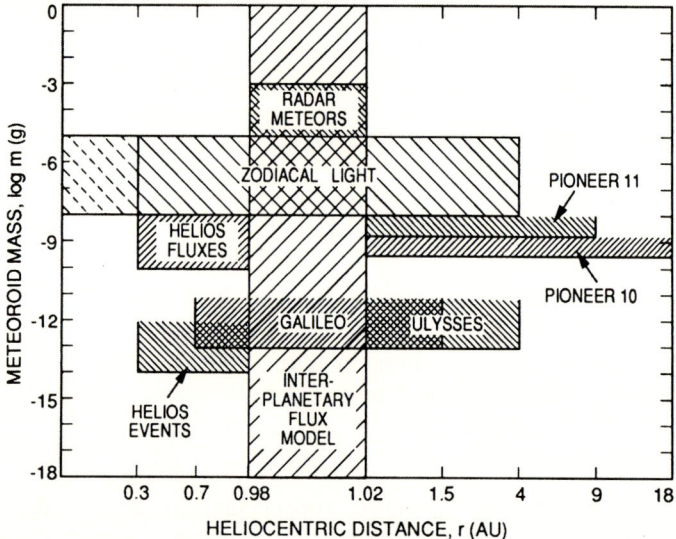

Figure 1. The mass and heliocentric distance ranges of the dust data sets which are considered in this analysis (after Divine 1993).

Sun is documented in lunar microcrater records (Grün et al. 1985). Zodiacal light observations from the Earth and from spacecraft (Leinert and Grün 1990; Levasseur-Regourd 1996) demonstrate the spatial distribution of dust in the inner solar system. Observations of the thermal emission give evidence for dust in the outer solar system (e.g. Hauser 1996, Kelsall et al. 1998). Measurements of dust particles by in-situ detectors on interplanetary spaceprobes require new populations of dust (e.g. interstellar dust and dust emitted from the Jovian system (Grün et al. 1993; Baguhl et al. 1995)) which are not easily identified at Earth's distance. Figure 1 shows the mass and heliocentric distance ranges of the dust data sets which are considered in this analysis. Note that the whole mass range is covered at 1 AU only and that a wide heliocentric distance range is only covered for masses ranging from 10^{-14} to 10^{-7} g.

All dust data obtained by various methods refer to different aspects of the interplanetary dust complex. Some data refer to micron-sized dust grains, other to hundred times bigger particles, some data refer to an ensemble of particles within the field-of-view of a telescope, other to individual particles recorded by an in-situ instrument at a specific position in space. In order to obtain a description of many aspects of the interplanetary dust complex, modeling is required. Several early attempts have been made to develop such models but with limited applications to only a few measurements. A break-through has been achieved by Divine (1993) who developed a model that derives the global dust distributions from local dust measurements and that synthesizes meteor data, measurements from in-situ dust detectors, and remote sensing data (visible zodiacal light and infrared thermal emission observations).

I.A. Physical Processes

In this section we review the basic dynamic effects which act on dust particles in interplanetary space. All dust particles in space feel the gravitational pull of the Sun. For particles with masses $m > 10^{-8}$ g solar gravitation is by far the dominant force. As a consequence they move on Keplerian orbits which are conic sections with the Sun in one focus—other forces are only small disturbances. Certainly, all observations of big particles are compatible with such orbits, although, the accuracy of the in-situ measurements is not sufficient for an unique orbit determination.

The pressure exerted on dust in interplanetary space by the solar radiation F_{rad} decreases with the inverse square of the distance to the Sun, i.e. this is the same dependence as the solar gravitational force F_{grav} (cf. Burns et al. 1979). Therefore, the ratio of both forces F_{rad}/F_{grav} is constant everywhere in interplanetary space and it is only dependent on material properties. This ratio is generally termed β

$$\frac{F_{rad}}{F_{grav}} = \beta = 5.7 \cdot 10^{-5} \frac{<Q_{pr}>}{\rho s}, \qquad (1)$$

where $<Q_{pr}>$ is the efficiency factor for radiation pressure on the meteoroid averaged over the solar spectrum, s is the radius of a spherical particle, and ρ is the density, e.g. for $s = 10^{-4}$ cm (1 micron), and $\rho = 1$ g cm^{-3} we obtain $\beta = 0.57$. For big particles $<Q_{pr}>$ is of the order of 1, depending on material properties (optical properties, shape, porosity, etc., for a full discussion of such dependencies see Gustafson 1994) but it decreases for particles smaller than the effective wavelength of Sun light (~ 0.5 μm). As a consequence β increases for smaller s values and reaches its maximum value between 0.1 and 1 microns, below which it decreases again. The maximum value is about 0.5 for non-absorbing dielectric materials and increases with increased absorptivity; it reaches maximum values of 3 to 10 for metallic spherical particles. In this study results obtained from Mie calculations for homogeneous spheres (Fig. 2) are taken from Gustafson (1994).

There are important consequences for the dynamics of small particles because of radiation pressure. Small particles which are generated from big particles (e.g. ejection from comets or impact ejecta from meteoroids or asteroids) carry kinetic energy of their parents but find themselves in a reduced potential field of the Sun. As a consequence they move on different orbits than their parents.

Photoemission of electrons by absorption of ultraviolet radiation is the main process which electrically charges interplanetary meteoroids. Therefore, the particles are influenced by the interplanetary magnetic field. For particles of less than 10^{-7} m radius or 10^{-15} g mass, the Lorentz force dominates orbital dynamics. Details are described in Leinert and Grün (1990), Morfill and Grün (1979a,b) and Morfill et al. (1986), Hamilton et al. (1996), cf. also the chapter on in-situ measurements by Grün et al. This latter effect is not yet included in the modeling described here.

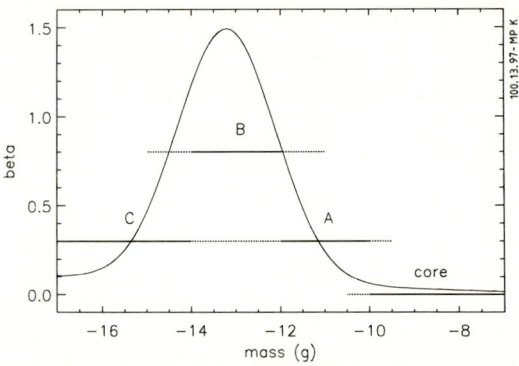

Figure 2. Ratio of radiation pressure to gravitational force β as a function of particle mass using Mie calculations for homogeneous spheres (Gustafson 1994). Mass ranges and corresponding β values for dust populations on bound orbits are indicated by solid lines (main contributions) and dashed lines (weak contributions).

I.B. Properties of Interplanetary Dust

In this section we discuss the physical properties of interplanetary meteoroids which affect their dynamics: shape, density, and efficiency for light scattering and absorption (represented by the albedo and the β value of the particle). All dynamically important compositional properties are represented by these parameters.

The albedo can be derived by comparison of the reflected and thermally emitted components of zodiacal light and leads to an average albedo of $A = 0.08 - 0.09$ at 1 AU (Leinert and Grün 1990). The shape of meteoroids is difficult to determine. Analysis of lunar microcraters lead to the assumption, that particles are roughly equidimensional (Leinert and Grün 1990).

The bulk density of meteoroid material is also uncertain. Kessler (1970) distinguished mass densities of cometary and asteroidal meteoroids. Taking into account available density data (e.g. Whipple 1963; Verniani 1964), Kessler chose 0.5 g cm^{-3} for cometary meteoroids. In a more recent paper (Kessler 1990) he modified this value to a mass-dependent density distribution with 0.5 g cm^{-3} for $m > 10^{-2}$ g, 1 g cm^{-3} for 10^{-2} g $< m < 10^{-6}$ g and 2 g cm^{-3} for masses below. For meteoroids of asteroidal origin, which have stony character, the density is assumed to be 3.5 g cm^{-3}. Based on the work of Greenberg and Hage (1990), Gustafson (1994) adopted different bulk densities for silicate (3.5 g cm^{-3}), organic-refractory (1.8 g cm^{-3}), volatile ice (1.2 g cm^{-3}) and polycyclic aromatic hydrocarbon (2 g cm^{-3}). Using literature of photographic and visual meteors, lunar microcraters and the Helios micrometeoroid experiment, Leinert and Grün (1990) reported that the average density of the majority of the meteoroids increases from about 1.5 g cm^{-3} for photographic meteor particles to 3 g cm^{-3} for micron-sized dust. In a recent paper, Drolshagen (1994) gave a distribution for meteoroid particles of all sizes: 70 % with $\rho = 0.5$ g cm^{-3}, 30 % with $\rho = 3.0$ g cm^{-3}.

Following Divine (1993) we chose a density of 2.5 g cm^{-3} for $m > 10^{-10}$ g. Since the bulk density for lower masses is probably the same or larger, 2.5 g cm^{-3} is also used for masses below 10^{-10} g (in contrast to Divine (1993) who used 0.25 g cm^{-3} for small particles).

I.C. Model Assumptions

Symmetry with respect to the ecliptic plane

From observational results (e.g. zodiacal light measurements), different symmetries of the spatial meteoroid distribution have been derived (Leinert and Grün 1990; Dumont and Levasseur-Regourd 1987). Deviations from symmetry with respect to the ecliptic plane are observable but they are small (0.01 AU offset from the Sun and 2° inclination, cf. Kelsall et al. 1998) and hence neglected here.

Rotational symmetries

In the following model we assume rotational symmetry of the interplanetary dust cloud, which is in general agreement with zodiacal measurements. A consequence of this assumption is that the orbital elements, line of node and line of apsides are randomly distributed and a meteoroid orbit can be described only by the three orbital elements: perihelion distance (r_1), or equivalently semimajor axis (a), eccentricity (e), and inclination (i).

Radial profile

Brightness measurements of the zodiacal light show a decrease of the dust concentration away from the Sun (Leinert and Grün 1990). To describe the radial distribution a power law $n(r) \sim r^\alpha$ has been assumed. Different approaches were made to determine the exponent α (e.g. -1.3 found from the *Helios* zodiacal light experiment, Leinert et al. 1981).

No time variability

In contrast to the sporadic or background meteoroid flux, meteor streams occur only at distinct times, but at those times they often exceed the sporadic meteor rate. The flux increasing factor F is defined as the ratio between the flux of streams and sporadic flux (Cour-Palais 1969) and is different for each meteor stream (e.g. Geminids ($F = 4$), Perseids (5) or Quadrantids (8)) and can vary with time, depending on the distance between the Earth and the parent comet at encounter. In this review, time variations like meteor streams are neglected and, hence, only the sporadic meteor background is modeled.

The ultimate goal of modeling the interplanetary dust complex is to provide an accurate description of each component that makes up the interplanetary dust cloud and to describe the evolution of all species from their formation to their removal from the system. The full evolutionary picture of interplanetary dust would involve a complete understanding of sources like comets, asteroids, planetary systems and interstellar dust, and processes such as dynamical evolution (gravitation, radiation pressure and electromagnetic interactions), collisions (destruction and generation of fragments), and sublimation and sputtering. We are far from a complete understanding of all these effects. One first attempt to look at combined effects of collisions and Poynting-Robertson evolution of dust from the asteroid belt was made by Gustafson et al. (1991), however much more work is needed. The models that we will present here are "snapshot" models describing the current state of the interplanetary dust cloud. They do not explain how this state is reached and where it will go.

In this chapter we will briefly mention early dynamical models but review more closely the dynamical model of dust in interplanetary space by Divine (1993) and its recent updates. The model describes the momentary state of the meteoroid complex and makes no statements concerning sources, sinks and evolution of interplanetary meteoroids other than what can be derived from the orbits directly, e.g. an interstellar origin for some grains.

II. EARLY MODELING

The objective of dynamical modeling is to describe for each position in space the spatial density of dust and the directional flux as a function of the orbital elements and their distributions for a specific dust population. Most dynamical modeling to date includes only solar gravity as the operative force. Because of the assumption of rotational symmetry of the dust cloud, the spatial dust density depends only on distance r from the Sun and latitude λ above the symmetry plane (here the ecliptic plane). Kessler (1981) gives the spatial density N of a particle with the orbital elements perihelion distance, r_1, eccentricity, e, and inclination, i:

$$N(r,\lambda) = \frac{(1-e)^{3/2}}{2\pi^3 r r_1 \sqrt{(r-r_1)((1+e)r_1 - (1-e)r)(\cos^2\lambda - \cos^2 i)}}. \qquad (2)$$

An early application of dynamical dust modeling was the interpretation of spatial dust densities which were obtained from zodiacal light observations in terms of distributions of orbital elements of interplanetary dust particles. Haug (1958) and later Banderman (1968) derived an integral which transforms distribution functions of orbital elements, $D(r_1, e, i)$, into spatial densities at any given position in space, where $D(r_1, e, i)$ is the number of meteoroids having perihelia between r_1 and $r_1 + dr_1$, eccentricities between e and de, and inclinations between i and di. If the distribution function is separable $D(r_1, e, i) = D_1(r_1) \cdot D_e(e) \cdot D_i(i)$ then the relative spatial density $n(\lambda)$ at latitude λ is given by (cf. Leinert and Grün 1990):

$$n(\lambda) = \int_{i=\lambda}^{\pi/2} \frac{D_i(i)\mathrm{d}i}{\sqrt{\sin^2 i - \sin^2 \lambda}}. \qquad (3)$$

$D_i(i) = \sin i$ results in an isotropic distribution of dust: $n(\lambda) = const.$ and hence, $D_i(i)/\sin i$ describes the deviation from isotropy.

In the past decades several models have been developed (mostly for engineering purposes) that describe the natural interplanetary meteoroid environment. In the following a summary of important meteoroid models is given and their capabilities are discussed.

II.A. Cour-Palais (1969)

A comprehensive meteoroid model was developed by Cour-Palais (1969). Different observational data sets were used to match the meteoroid environment of cometary origin in the mass range between 10^{-12} and 1 g at heliocentric distance 1 AU near the ecliptic plane. Particles of asteroidal origin were considered to be negligible at 1 AU.

To determine the meteoroid environment near Earth, both the sporadic meteoroid flux (background flux) and the flux originated from meteor streams have to be considered. Using data from photographic and radar meteor observations and results from penetration detectors aboard satellites and probes (e.g. Explorer 16, Explorer 23, Pegasus) and assuming that meteoroid velocities are independent of mass, an average atmospheric entry velocity of 20 km s^{-1} was chosen for sporadic meteors. The particle density was assumed to be 0.5 g cm^{-3} for sporadic and stream particles, both of cometary origin. To model the increase in average flux due to meteor streams, a cumulative total meteoroid flux of 1.1 times the sporadic flux at $m = 1$ g was chosen and the stream component contributed a negligible amount at $m = 10^{-6}$ g. Both gravitational focusing and shielding by the Earth have to be taken into account for calculating the meteoroid flux on a spherical, randomly oriented spacecraft with a surface area of 1 m^2. To correct for the Earth's gravitational enhancement, the interplanetary meteoroid flux must be multiplied by a focusing factor which depends on average meteoroid velocity and distance from the center of Earth (e.g. the factor varies between 1.75 at Earth's surface and 1 for deep space for an average meteoroid velocity of 20 km s^{-1}). The shielding factor ξ is defined as the ratio of shielded to unshielded flux and depends on the radius r_E of the shielding body (Earth) and the altitude H above the surface.

$$\xi = \frac{1 + \cos\theta}{2} \quad \text{where} \quad \sin\theta = \frac{r_E}{r_E + H}. \tag{4}$$

The gravitationally focused unshielded total meteoroid flux N_t on a spherical, randomly oriented spacecraft with the surface of 1 m^2 is given by:

$$10^{-12}\text{g} \leq m \leq 10^{-6}\text{g}$$

$$\log N_t = -14.339 - 1.548 \log m - 0.063 (\log m)^2 \tag{5}$$

$$10^{-6}\text{g} \leq m \leq 10^{0}\text{g}$$

$$\log N_t = -14.37 - 1.213 \log m. \tag{6}$$

To calculate the cumulative flux N_{st} of particles exceeding mass m applicable to each individual stream the geocentric velocity v_{st} of each stream in km s^{-1} and the ratio F of cumulative flux of stream to the average cumulative background flux integrated for the considered time period have to be used:

$$10^{-6} \text{g} \leq m \leq 10^0 \text{g}$$

$$\log N_{st} = -14.41 - \log m - 4.0 \log \left(\frac{v_{st}}{20}\right) + \log F. \tag{7}$$

II.B. Kessler (1970)

Kessler (1970) extended the model of Cour-Palais (1969) by distinguishing meteoroids of cometary and asteroidal origin and by calculating the flux onto a spacecraft using the relative meteoroid velocity.

The flux of cometary meteoroids exceeding particle mass m can be evaluated as a function of distance r from the Sun, heliocentric latitude λ and average velocity $\overline{u_c}$ of meteoroids relative to a spacecraft in m s^{-1}. The velocity $\overline{u_c}$ is a function of heliocentric distance r, the ratio σ of spacecraft heliocentric speed to the circular speed at distance r, and of the angle θ between the spacecraft velocity vector and the surface of an imaginary heliocentric sphere of radius r at the location of the spacecraft:

$$\overline{u_c} = r^{-0.5} \cdot 31.29 \cdot 10^3 \sqrt{(1.30 - 1.9235\sigma \cos\theta + \sigma^2)}. \tag{8}$$

The radial spatial density distribution has a dependence of $r^{-1.5}$ while the latitudinal distribution is represented by the function $e^{-2|\sin\lambda|}$. It is assumed that radial and latitudinal distributions are independent of each other. To calculate the flux F_c of cometary meteoroids onto a randomly tumbling surface, the weighted average cometary velocity is required:

$$(\overline{u_c^{-1}})^{-1} = \overline{u_c}\, \delta_c^{-1}, \tag{9}$$

where δ_c is the velocity weighting function which depends on σ and θ. Using the following equations, the flux F_c can be derived from the spatial density ρ_c assuming a mass density of 0.5 g cm^{-3}.

$$10^{-6} \text{g} \leq m \leq 10^2 \text{g}$$

$$\log \rho_c = -18.173 - 1.213 \log m - 1.5 \log r - 0.869 \,|\sin\lambda| \tag{10}$$

$$10^{-12} \text{g} \leq m \leq 10^{-6} \text{g}$$

$$\log \rho_c = -18.142 - 1.584 \log m - 0.063(\log m)^2 - 1.5 \log r - 0.869 \,|\sin\lambda| \tag{11}$$

$$F_c = \frac{1}{4} \rho_c (\overline{u_c^{-1}})^{-1}. \tag{12}$$

To calculate the flux of meteoroids of asteroidal origin, an average mass density of 3.5 g cm^{-3} and geometric albedo of 0.1 was assumed. Kessler (1970) further assumed that the spatial density of particles with masses above m varies as $m^{-0.84}$ for masses between 10^{-9} g and 10^{-19} g. The asteroidal radial distribution is a function of heliocentric distance while the latitudinal distribution is a function of heliocentric latitude. To take these effects into account for the calculation of the spatial density, two correction factors $f(r)$ and $h(\lambda)$ are added. A third factor $g(r)\cos\omega$ considers the asymmetry of the asteroid belt with heliocentric longitude ω and heliocentric distance r. The spatial density ρ_a is calculated as follows.

$$10^{-9}\text{g} \leq m \leq 10^{2}\text{g}$$

$$\log \rho_a = -15.79 - 0.84 \log m + f(r) + g(r)\cos\omega + h(\lambda) \tag{13}$$

$$10^{-12}\text{g} \leq m \leq 10^{-9}\text{g}$$

$$\log \rho_a = -8.23 + f(r) + g(r)\cos\omega + h(\lambda). \tag{14}$$

The asteroidal velocity parameter $\overline{u_a}$ of asteroidal meteoroids relative to a spacecraft is a function of r, θ and σ. The values of $\overline{u_a}$ can be calculated for three different heliocentric distances (1.7 AU, 2.5 AU, 4.0 AU) using the following equations where σ is the ratio of the heliocentric spacecraft speed to the speed of a circular orbit at the same distance from the Sun and θ is the angle mentioned above. For other heliocentric distances than those, $\overline{u_a}$ can be found by linear interpolation.

1.7 AU $\quad \overline{u_a} = 30.05 \cdot 10^3 (1.2292 - 2.1334\sigma\cos\theta + \sigma^2)^{0.5},$ \hfill (15)

2.5 AU $\quad \overline{u_a} = 29.84 \cdot 10^3 (1.0391 - 1.9887\sigma\cos\theta + \sigma^2)^{0.5},$ \hfill (16)

4.0 AU $\quad \overline{u_a} = 29.93 \cdot 10^3 (0.9593 - 1.9230\sigma\cos\theta + \sigma^2)^{0.5}.$ \hfill (17)

Using the weighted average asteroidal velocity

$$(\overline{u_a^{-1}})^{-1} = r^{-0.5}\overline{u_a}, \tag{18}$$

the asteroidal flux F_a can be calculated similar to the cometary flux F_c. The directionality of asteroidal meteoroids relative to a spacecraft is assumed to be monodirectional while cometary meteoroids are assumed to be omnidirectional.

The uncertainty of flux is a factor 28 for cometary meteoroids while for asteroidal meteoroids the uncertainty depends on particle mass. So the meteoroid population in the asteroid belt varies between factors 30 and 5 000 for $m = 10^{-6}$ g and 1 g, respectively.

II.C. Grün et al. (1985)

A comprehensive meteoroid flux model using the lunar micro crater distribution, data from in-situ spacecraft measurements, zodiacal light observations and oblique-angle hypervelocity impact studies has been developed by Grün et al. (1985). Based on earlier results of lunar crater distributions, a new distribution for smaller particles (particle mass 10^{-18} g $< m < 10^{-9}$ g) was developed taking into account recent data of the HEOS 2 satellite and Pioneer 8/9 space probes combined with dynamics of β-meteoroids. A particle density of 2.5 g cm^{-3} and an average velocity of $v_0 = 20$ km s^{-1} at $r_0 = 1$ AU was chosen. The velocity varies with heliocentric distance:

$$v(r) = v_0 \left(\frac{r}{r_0}\right)^{-0.5}. \quad (19)$$

Comparing the mass dependent collisional destruction and production rates of particles with loss rates due to Poynting-Robertson drag and radiation pressure (β-meteoroids), Grün et al. (1985) inferred that the spatial density for meteoroids in the mass range 10^{-10} g $\leq m \leq 10^{-5}$ g is presently increasing with time. They also found that the flux of small particles derived from lunar crater counts (≤ 7 μm crater diameter) is higher than the interplanetary flux, because of secondary ejecta cratering. The developed model refers only to data which were obtained far from the Earth where gravitational focusing and shielding are negligible. The cumulative flux of interplanetary meteoroids at 1 AU on a flat plate with normal vector in the ecliptic plane and spinning with the spin axis perpendicular to the ecliptic plane is given by the formula:

$$F(m) = (c_1 m^{\gamma_1} + c_2)^{\gamma_2} + c_3(m + c_4 m^{\gamma_3} + c_5 m^{\gamma_4})^{\gamma_5} + c_6(m + c_7 m^{\gamma_6})^{\gamma_7} \quad (20)$$

$c_1 = 2.2 \cdot 10^3$, $\quad c_2 = 15$, $\quad c_3 = 1.3 \cdot 10^{-9}$, $\quad c_4 = 10^{11}$,
$c_5 = 10^{27}$, $\quad c_6 = 1.3 \cdot 10^{-16}$, $\quad c_7 = 10^6$, $\quad \gamma_1 = 0.306$,
$\gamma_2 = -4.38$, $\quad \gamma_3 = 2$, $\quad \gamma_4 = 4$, $\quad \gamma_5 = -0.36$,
$\gamma_6 = 2$, $\quad \gamma_7 = -0.85$.

Mass range: 10^{-18} g $< m < 10^2$ g.

II.D. Zook (1991)

To determine the meteoroid environment for an Earth-orbiting satellite (e.g. Solar Max, LDEF) a different set of parameters has been developed by Zook (1990, 1991ab). Ignoring the spacecraft motion, it is assumed that arrival directions of meteoroids are uniform which is called the randomness assumption (Zook 1991a). Comparing fluxes and impact velocities for different surfaces (e.g. apex or anti-apex), a dependence on crater size and velocity distribution was found (Zook 1991b). Assuming constant meteoroid mass and crater size, the ratio of meteoroid impacts on the side of spacecraft motion to the opposite side is compared using different velocity distributions (Zook 1990).

II.E. Comparison

In the past various meteoroid models have been developed to pursue different physical and engineering questions. The quality and quantity of available data has increased almost as rapidly as computing speed necessary to implement the models. The models from Cour-Palais (1969), Kessler (1970) and Grün et al. (1985) have different scopes namely the near-Earth meteoroid environment and the interplanetary meteoroid environment, respectively, and are applicable only in limited regions in space. Therefore, the recent model of Divine (1993) has the advantage of describing the interplanetary meteoroid environment everywhere in the solar system.

III. BASIC FORMULATION

Keplerian dynamics (ignoring radiation pressure) of a single particle in interplanetary space is determined by its heliocentric position **r** and heliocentric velocity **v**. Assuming rotational symmetry about the ecliptic poles, the orbit distribution of an ensemble of particles can be described with just three Keplerian elements: eccentricity e, inclination i and perihelion distance r_1. Because of the symmetry assumptions, for each set of (e, i, r_1) four velocities are possible in each point in space, namely inbound, outbound, moving upward (towards ecliptic North) and moving downward (towards ecliptic South).

In order to describe not only a single particle but the whole interplanetary meteoroid complex, distributions of orbital elements and particle sizes or masses have to be determined.

III.A. Phase Space Density

Divine (1993) gives the phase space density without a derivation. Matney and Kessler (1996) on the other hand derive the six-dimensional phase space density with a rigorous definition of the distributions in orbital parameters (note that the definition of these distributions differs from the one selected by Divine—the correspondence between both definitions is given below). Both papers give the phase space density for the Keplerian case (i.e. radiation pressure constant $\beta = 0$). In order to include radiation pressure GM_0 has to be replaced by $GM_0(1 - \beta)$. Matney and Kessler start with the phase space density for a single orbiting object where δ is the Dirac delta function, **r** is the position, **v** the velocity, and $\rho(\mathbf{r})$ is the spatial density given by Eq. (2). The phase space density is given in terms of position and velocity variables. These can be transformed into orbital coordinates by using the Jacobian which is given by Divine (GM_0 is the gravitational constant) as

$$\frac{\partial(v_x, v_y, v_z)}{\partial(r_1, e, i)} = \frac{(GM_0)^{3/2} e \sin i}{2r\sqrt{r_1(r - r_1)((1 + e)r_1 - (1 - e)r)(\cos^2 \lambda - \cos^2 i)}}. \quad (21)$$

The phase space density of a single orbiting object in orbital variables is then

$$\Theta_1(\mathbf{r}, r_1, e, i) = J^{-1} \rho(\mathbf{r}) \delta(r_1 - r_1') \delta(e - e') \delta(i - i')$$

$$= \frac{(1 - e)^{3/2}}{\pi^3 e \sqrt{r_1} (GM_0)^{3/2} \sin i} \delta(r_1 - r_1') \delta(e - e') \delta(i - i'). \quad (22)$$

III.B. Orbital Parameter Distributions

Equation (22) is now integrated over "textbook distributions" of orbital parameters given by D_1, D_e, and D_i (Matney and Kessler 1996). $D_1 \, \mathrm{d}r_1$ is the number of objects having perihelion distances between r_1 and $(r_1 + \mathrm{d}r_1)$. $D_e \, \mathrm{d}e$ is the number of objects having eccentricities between e and $(e + \mathrm{d}e)$, and $D_i \, \mathrm{d}i$ is the number of objects having inclinations between i and $(i + \mathrm{d}i)$. They are normalized by

$$\int_0^\infty \mathrm{d}r_1 D_1(r_1) = 1 \, , \quad \int_0^1 \mathrm{d}e D_e(e) = 1 \, , \quad \int_0^\pi \mathrm{d}i D_i(i) = 1. \quad (23)$$

By integrating (22) over these distributions, we arrive at the correct form of the phase space density Θ_0 for a family of orbiting objects

$$\Theta_0 = \int_0^\infty \mathrm{d}r_1' D_1(r_1') \int_0^1 \mathrm{d}e' D_e(e') \int_0^\pi \mathrm{d}i' D_i(i') \Theta_1(\mathbf{r}, r_1', e', i')$$

$$= \frac{(1-e)^{3/2} D_1(r_1) D_e(e) D_i(i)}{\pi^3 e \sqrt{r_1} (GM_0)^{3/2} \sin i}. \quad (24)$$

The three distributions are normalized to unity in Eq. (23), but to represent the total number of orbiting objects in a family, a mass distribution term has to be added to Eq. (24) and the total number of objects included in it. Divine's distributions N_1, p_i, and p_e are defined differently and the relationship between Divine's distributions and the "textbook" ones are given by

$$N_1 \sim \frac{D_1}{r_1^2} \, , \quad p_i \sim \frac{D_i}{\sin i} \, , \quad p_e \sim (1-e)^{3/2} D_e. \quad (25)$$

In Fig. 3 the distributions of Divine (1993) are compared to the "textbook" distributions as given in Eq. (25). The axis definitions and units are those used by Divine. While Divine's distributions are internally consistent, any attempt to relate them to the physical distributions and numbers in space requires the use of these "textbook" distributions. Note that the "textbook" inclination distribution of the "Halo" family is a sine curve, as would be expected from a spatial density that is independent of latitude. Since all applications of Divine's method are done using Divine's distributions, we will use them from here on. However, it is suggested that any new work should use "textbook" distributions and define dust populations by simple functions in the latter representation.

III.C. Concentrations

Using the auxiliary variables $e_\chi = (1 - \sin \chi)/(1 + \sin \chi)$ and $\chi = \sin^{-1}(r_1/r)$ the concentration of particles whose masses exceed m can be calculated:

$$N_M = \frac{H_M}{\pi} \int_0^{\pi/2} \mathrm{d}\chi (\sin \chi) N_1 \int_{e_\chi}^1 \mathrm{d}e \frac{p_e}{\sqrt{e - e_\chi}}$$

$$\cdot \int_{|\lambda|}^{\pi-|\lambda|} di \frac{(\sin i)p_i}{\sqrt{(\cos \lambda)^2 - (\cos i)^2}}. \qquad (26)$$

Here N_1 has to be evaluated for perihelion distance $r_1 = (\sin \chi)r$. H_M and H_m are the cumulative and the differential mass distribution, respectively, which are given by

$$H_M = \int_m^\infty dm H_m. \qquad (27)$$

The product H_m dm represents the ratio of the numbers of particles in the mass interval m, $m+dm$ to the number of particles whose mass exceeds $m_1 = 1$ g (Divine 1993).

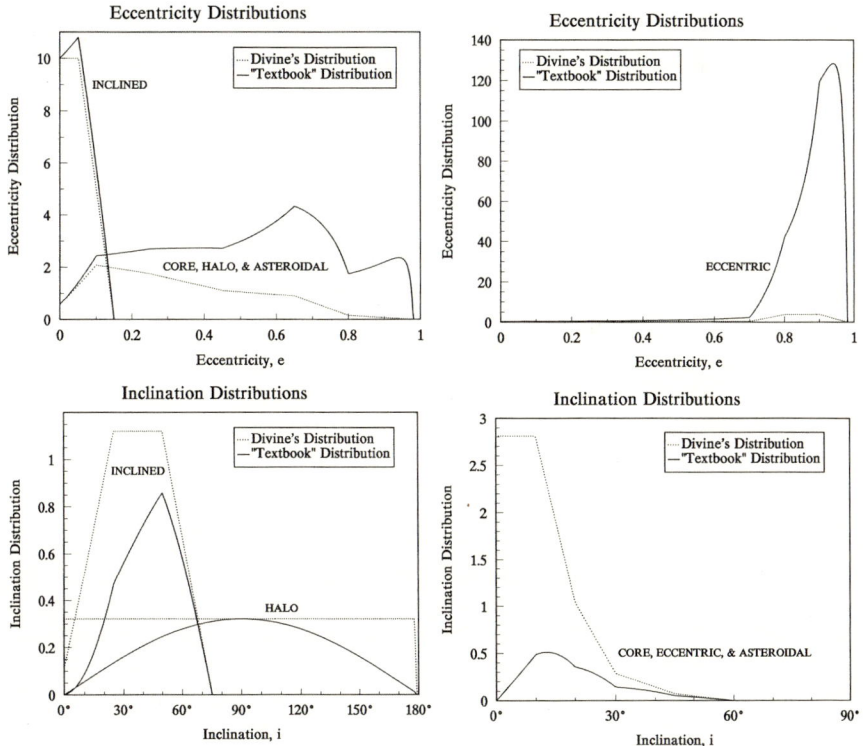

Figure 3. Divine's internal distributions from his paper (Divine 1993) are compared with the "textbook" distributions as given in Eq. (25). The axis definitions and units are those used by Divine (from Matney and Kessler 1996).

III.D. Particle Fluxes

In order to obtain particle flux onto spacecraft-borne detectors the components v_x, v_y, and v_z of particle heliocentric velocity have to be evaluated using the elements e, i, and r_1. Note, that the heliocentric particle velocity depends on the assumed radiation pressure constant β. Using intermediate quantities $\lambda = \sin^{-1}(z/r)$ and $\chi = \sin^{-1}(r_1/r)$, the following results can be obtained:

$$v_r = \pm \left[\frac{GM_0(1-\beta)}{r^2 r_1}(r-r_1)[(1+e)r_1 - (1-e)r]\right]^{1/2}$$

$$= \pm \left[\frac{GM_0(1-\beta)}{r}\frac{(\cos\chi)^2}{\sin\chi}(e-e_\chi)\right]^{1/2} \tag{28}$$

$$v_\phi = [(GM_0(1-\beta)/r)(1+e)(\sin\chi)]^{1/2} \tag{29}$$

$$v_s = \pm v_\phi[(\cos\lambda)^2 - (\cos i)^2]^{1/2} \tag{30}$$

$$v_x = \frac{x}{r}v_r - \left[\frac{ryv_\phi(\cos i) + xzv_s}{x^2+y^2}\right] \tag{31}$$

$$v_y = \frac{y}{r}v_r - \left[\frac{rxv_\phi(\cos i) - yzv_s}{x^2+y^2}\right] \tag{32}$$

$$v_z = \frac{z}{r}v_r + v_s. \tag{33}$$

The relative velocity $\mathbf{u}_D$ between a particle ($\mathbf{v}$) and a spacecraft-borne detector ($\mathbf{v}_{DB}$) is simply $\mathbf{u}_D = (\mathbf{v} - \mathbf{v}_{DB})$. The sensitivity of a detector can be expressed by its mass threshold and angular sensitivity. Let $\mathbf{r}_D$ be a unit vector which specifies the detector orientation, then the angle γ between $\mathbf{r}_D$ and the direction from which a particle arrives, is

$$\gamma = \arccos(-\mathbf{u}_D \cdot \mathbf{r}_D / u_D). \tag{34}$$

The angular sensitivity Γ, which depends on sensor geometry, is a function of the angle γ. In Fig. 4, angular sensitivities for Galileo and Ulysses are shown.

The mass threshold depends on particle mass m and particle density ρ. Using detector characteristics m_0, u_0, ρ_0, δ and α which are obtained from the calibration process (for further information see Grün et al. 1992a, b), and $u_{D'}$ is the magnitude of the impact velocity $\mathbf{u}_D$, the threshold mass m_t is calculated as follows:

$$m_t = m_0(\rho_0/\rho)^\delta (u_0/u_{D'})^\alpha. \tag{35}$$

The detector threshold m_t, angular sensitivity Γ, and a scale factor F_s to express the flux in different units are ingredients of the dimensionless weighting function η_D. The cumulative mass distribution H_M is evaluated at mass m_t.

$$\eta_D = F_s \Gamma H_M. \tag{36}$$

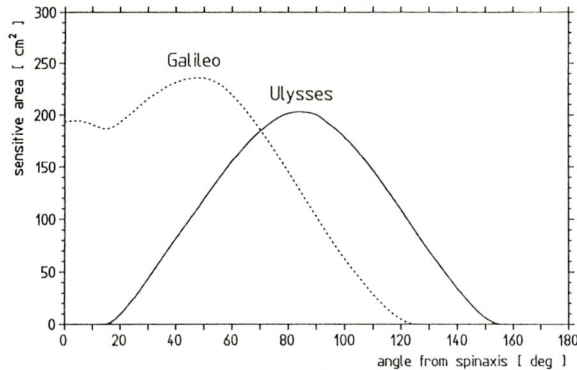

Figure 4. Average sensor areas as a function of the angle γ between impact direction and positive spin axis for Ulysses and Galileo (cf. Grün et al. 1992a).

The particle flux J_M can be obtained using the following equation (for explanation of variables see previous paragraphs):

$$J_M = \frac{1}{4\pi} \sum_{i=1}^{4} \int_0^{\pi/2} d\chi (\sin \chi) N_1 \int_{e_\chi}^{1} de \frac{p_e}{\sqrt{e - e_\chi}}$$

$$\cdot \int_{|\lambda|}^{\pi - |\lambda|} di \frac{(\sin i) p_i}{\sqrt{(\cos \lambda)^2 - (\cos i)^2}} (\eta_D u_D)_i. \quad (37)$$

III.E. Directional Flux and Impact Speed

New meteoroid data, however, which have been obtained from the dust detectors aboard the Galileo and Ulysses spacecraft, together with physical effects, which have not been taken into account by Divine, caused Staubach and Grün (1995) and Grün et al. (1997) to make major modifications to the initial model. Divine's model was based on data sets in which only spatial densities and flux values were specified. In order to include directional and velocity information the original model of Divine was extended (Staubach and Grün 1995; Grün et al. 1997).

The impact-ionization detectors aboard the Galileo and Ulysses spacecraft (Grün et al. 1992a, b; Baguhl et al. 1993) provide not only impact rates but also information about impact velocity and impact direction. To implement these data into the model, all impacts have been sorted into large bins dependent on time, impact velocity and impact direction. In Figs. 5 and 6 the trajectories of Galileo and Ulysses are shown. Each data set is divided into two periods dependent on significant points of the trajectory. Each period is divided into intervals (time bins) containing almost the same number of impacts. (Ulysses: launch to Jupiter fly-by (A1-A2) and beyond Jupiter (B1-B4); Galileo: launch to second Earth fly-by (A1-A4) and beyond second Earth fly-by (B1-B4)). The particle impact velocity which ranges from about 2 km s^{-1} to about 70 km s^{-1}

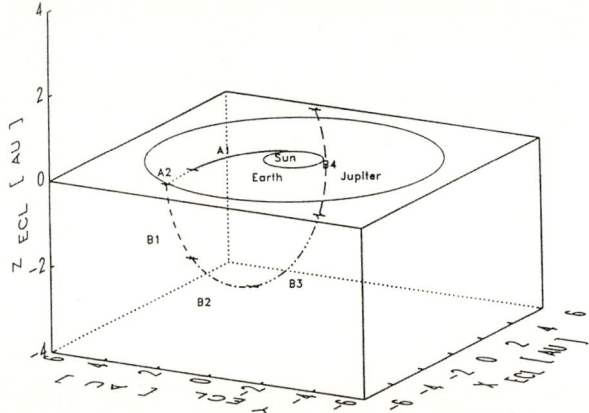

Figure 5. Trajectory and time bins for Ulysses between south and north solar pass. Vernal equinox is towards +x direction.

is divided into three intervals (2–10 km s^{-1}, 10–25 km s^{-1}, > 25 km s^{-1}). The impact direction, caused by the spin of the spacecraft, which covers the total range of 360° is divided into 4 equal intervals of 90°. To minimize statistical problems (i.e. because of the large number of bins only a few impacts are counted in each bin) a new algorithm (Staubach and Grün 1995; Grün et al. 1997) has been developed to average the flux values in the bins.

III.F. Radiation Pressure Effects and Hyperbolic Orbits

The orbits of micron and sub-micron particles are strongly affected by radiation pressure of the Sun (Eq. 1). Because of the strong influence of radiation pressure on dust orbits, the mass range is divided into 4 sections, and the heliocentric speeds (Eqs. 28–33) were calculated in each range with a different radiation pressure constant β. For particles with masses greater than 10^{-10} g the influence of radiation pressure is negligible ($\beta = 0$) and no change of the model is necessary (cf. Fig. 2). To three mass ranges non-zero radiation pressure values, β, were assigned: $\beta = 0.3$ for 10^{-12} to 10^{-10} g; $\beta = 0.8$ for $5 \cdot 10^{-15}$ to 10^{-12} g; and $\beta = 0.3$ below $5 \cdot 10^{-15}$ g.

Interstellar dust was first identified by the dust sensor aboard Ulysses and has been confirmed by Galileo dust measurements (Baguhl et al. 1995). These particles move on hyperbolic trajectories and penetrate into the solar system from a narrowly defined 'upstream' direction. In this analysis only the effects of solar gravity and solar radiation pressure (derived by D. Hamilton, private communication) are taken into account. From the analysis of impact rates, impact directions and impact speeds, determined by the Ulysses and Galileo dust instruments it was found that the speed and the direction from which interstellar particles penetrate into the Solar system are different from the measured value for interstellar gas. Interstellar gas flows through the planetary system from the direction of 253° ecliptic longitude and 5° ecliptic latitude with a speed of 26 km s^{-1} (Lallement 1993; Witte et al. 1993). Best results for the

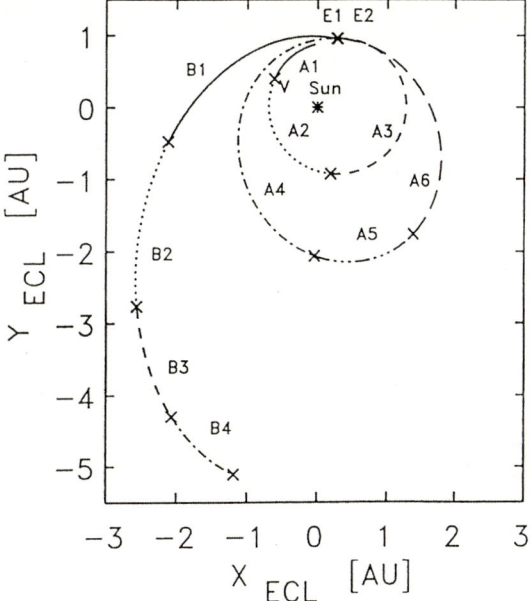

Figure 6. Trajectory and time bins for Galileo.

description of the Ulysses and Galileo measurements have been obtained using a longitude of 252° ± 20°, a latitude of 5° ± 10° and a speed of 30 km s^{-1} (Baguhl et al. 1996).

IV. METEROID DATA SETS

Data from different observational techniques have been used in Divine's model to describe the interplanetary meteoroid environment: zodiacal light, radar meteors, lunar crater counts and Earth-orbiting satellites, fluxes from impact detectors aboard the Pioneer 10 and 11, Helios 1, Galileo and Ulysses spacecraft. The size of dust particles ranges from sub-micron particles to centimeter-sized pebbles. The size distribution used in this model is derived from studies of lunar microcraters and ranges from 10^{-18} g to 1 g particle mass. In this paragraph we describe the data that were considered in the Divine model (for a more detailed description of the data sets see the chapters by Grün et al. and Levasseur-Regourd et al.)

IV.A. Meteors

A backbone of the data responsible for Divine's populations is the Harvard Radar Meteor Survey described by Sekanina and Southworth (1975). This data set of particles above 10^{-7} g mass provides distributions in perihelion distances, eccentricities and inclinations. The data has been weighted for observational selection effects.

The most sensitive detection technique of meteors is by radar. A radar beam is reflected off the ionized meteor trail allowing meteor velocity, meteor-induced ionization, and meteoroid mass to be determined. The measured mass interval ranges from 10^{-7} g to 10^{-1} g. Sekanina and Southworth (1975) divided observed radar meteors into several bins of Keplerian elements and velocity. Special weighting factors were used to account for observational selection and flux increase due to Earth's gravity.

IV.B. Lunar Microcraters

Because the Moon has no atmosphere, the lunar surface is covered by numerous impact craters which range from below one micron to hundred of km in diameter. With the aid of impact simulations in the laboratory, the particle mass can be derived from the crater diameter under assumption of impact velocity and specific particle density (Hörz et al. 1975). From the analysis of this data the meteoroid size distribution is characterized over a wide range (Grün et al. 1985).

IV.C. Zodiacal Light and Thermal Emission

In the last decades many different approaches have been used to measure both the scattered light from the incident solar radiation and the zodiacal thermal emission emitted in the infrared. Leinert et al. (1981), Levasseur-Regourd and Dumont (1980) and Hanner et al. (1974) provide summaries of zodiacal light intensities measured at different heliocentric distances from Earth, from Helios and from Pioneer 10, respectively. Shape, size and composition can be determined from zodiacal light measurements using models for polarization, brightness distribution and wavelength dependence (Leinert and Grün 1990). In the model, summaries of zodiacal light intensities seen from Helios (Leinert et at. 1981), from Earth (Levasseur-Regourd and Dumont 1980), and from Pioneer 10 (Hanner et al. 1974) are used. According to Grün et al. (1985) the mass range of meteoroids responsible for most of the scattered zodiacal light is from 10^{-8} to 10^{-5} g.

IV.D. Early Spacecraft Detectors

In this model, impact data from different kinds of spacecraft-borne detectors are used which have different mass thresholds as a function of impact velocity (Fig. 7).

Penetration detectors count particles which penetrate a thin foil. The ability to penetrate the film depends on particle density, impact velocity as well as the thickness and material composition of the foil. Examples are the detectors aboard Pioneer 10 (Humes et al. 1974) and Pegasus (Naumann et al. 1969). Puncture cells aboard Pioneer 10 and 11 provide spin averaged penetration fluxes in units of $m^{-2} s^{-1}$ averaged over unequal distance intervals between 1 and 18 AU heliocentric distance. The detectors have a field-of-view of π steradians (sr) and the angular sensitivity has the integral 2.8 sr over all directions, therefore a scale factor of $F_s = \pi/2.8 = 1.12$ is used.

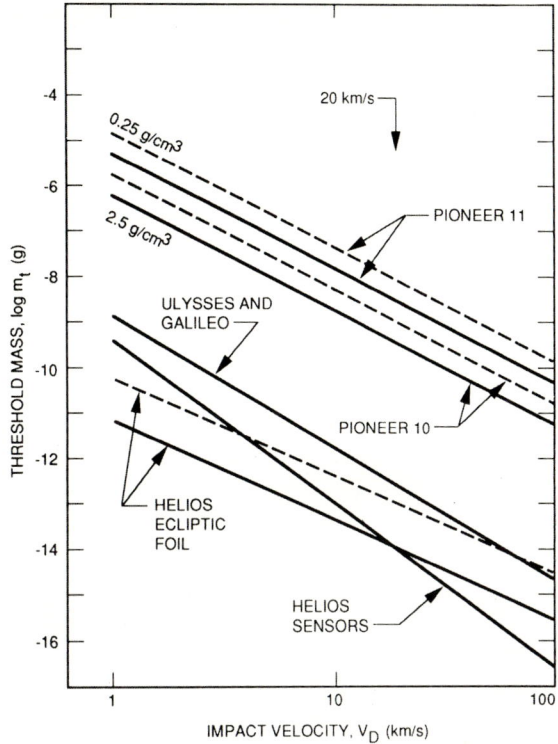

Figure 7. Threshold mass as a function of impact velocity for dust detectors on various space probes. Particle densities of $\rho = 2.5$ g cm^{-3} (solid lines) and $\rho = 0.25$ g cm^{-3} (dashed lines) are used (from Divine 1993).

The impact ionization detectors aboard Pioneer 8 and 9 and Helios have lower mass thresholds than those on Pioneers 10 and 11 so smaller particles can be detected. Aboard Helios 1, impact data between 0.31 and 0.98 AU heliocentric distance were obtained, measured with both the south sensor and the ecliptic sensor, which is covered by a foil. Spin averaged fluxes (m^{-2} s^{-1} (π sr)$^{-1}$) are obtained from these detectors.

IV.E. Ulysses

In the chapter by Grün et al., the impact rate and the directional flux information obtained from the Ulysses dust data is discussed in detail.

In order to compare this directional data with the dust model a new representation had to be found. All Ulysses dust data ($m > 6 \cdot 10^{-14}$ g) is binned into 6 time intervals, 4 rotation angle intervals and 3 speed intervals (the latter are not considered separately in the following discussion) and a smoothing routine is employed. A finer resolution in parameter space leads to poor statistics and hence was deemed impractical. Figure 8 shows the projection of the data onto the time-rotation angle plane, with the flux represented by a gray scale.

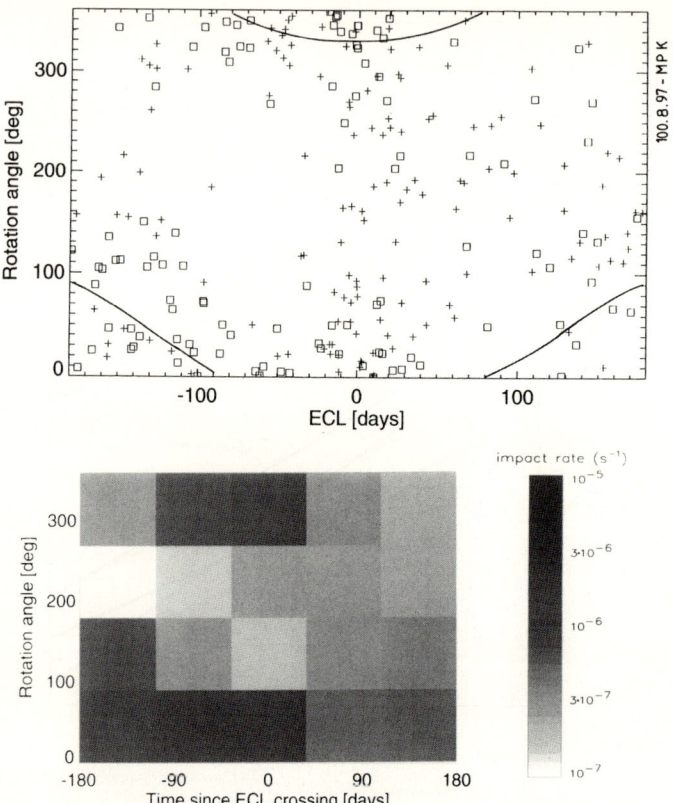

Figure 8. Directional information (rotation angle) of Ulysses dust data as function of time from ecliptic plane crossing (ECL, from Grün et al. 1997). Top panel: Traditional representation of the rotation angle of individual impacts (squares: impact charge $> 8 \cdot 10^{-14}$ C, crosses $\leq 8 \cdot 10^{-14}$ C) as a function of time. The solid line shows the direction from which interstellar grains are expected. The direction from which particles on minimum inclination ($\leq 30°$), circular and prograde orbits would arrive coincides with the interstellar direction from time ECL −40 to time ECL +40 days. Bottom panel: The projection of the data onto the time-rotation angle plane, with the flux represented by a gray scale plot.

IV.F. Galileo

The impact rate observed by Galileo during the first 4 years in orbit is shown in Fig. 9. A new presentation of rotation angles of the 437 detected impacts is also displayed whereby the fluxes in 4×9 segments of the rotation angle-time plane are represented by a gray scale. These fluxes can now be compared directly with model calculations. For most of the time, impacts were observed over the full range of rotation angles. The impact rate varies strongly with time due to varying heliocentric distance and the spacecraft motion with respect to the FOV of the dust sensor.

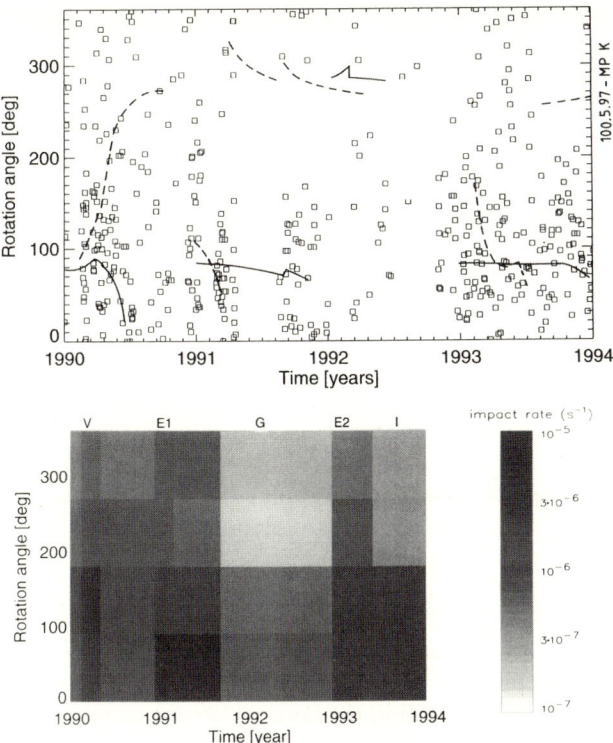

Figure 9. Rotation angles of the 437 impacts detected by Galileo during the first 4 years (from Grün et al., 1997). Fig. 9a. Individual impacts with impact charge $> 8 \cdot 10^{-14}$ C are marked by squares. Solid lines show the center direction from which interstellar grains would arrive. Dashed lines represent the center direction from which particles on minimum inclination, circular prograde orbits would arrive. These lines are only shown if the direction is visible by the detector for longer than a month. Fig. 9b shows the projection of the data onto the time-rotation angle plane, with a gray scale plot representing the directional fluxes.

V. DIVINE'S ORIGINAL MODEL POPULATIONS

Five different sets of distributions of orbital elements (populations) were required by Divine (1993) to match all of the data sets that we considered. Note, that Divine assumed radiation pressure constant $\beta = 0$ which is justified for particles with mass $m > 10^{-10}$ g. For smaller particles the distribution functions may still give an acceptable fit to the data but extrapolations to regions outside the observed spatial and mass ranges may lead to false results. Due to the sparseness of measured data, the model (which tries to minimize the difference between measured data and model predicted values) does not necessarily give a unique solution. All particles are assumed to be spherical with density $\rho = 2.5$ g cm^{-3} (except for very small particles 0.25 g cm^{-3} is used). The albedo varies between 0.02 and 0.05. Figure 3 shows the inclination and eccentricity distributions of the five original populations. For the size distributions (H_m) of the populations please see Table 4 in Divine (1993). In the following we give a short description of these populations.

1. Core Population

The core population, at mean mass 10^{-5} g, has particles on orbits of small eccentricity and inclination, and the concentrations increase inward as $r^{-1.3}$. This population contributes most to the zodiacal light, as well as making major contributions to all other data sets. The core population is the most reliable of the five populations because it is the dominant component represented in most data sets.

2. Asteroidal Population

The asteroidal population, at mean mass 10^{-3} g, also has small eccentricities and inclinations, but increases outward from 1 AU; it is developed mostly to match that part of the radar meteor data (Southworth and Sekanina 1973), which peaks in the asteroid belt. The relation of this population to asteroids is unclear.

3. Halo Population

The halo population, at mean mass 10^{-7} g, exists beyond 2.5 AU and has random orbital inclinations (including retrograde); it is needed to provide the nearly uniform fluxes detected in the outer solar system by Ulysses, Pioneer 10, and Pioneer 11.

4. Inclined Population

The inclined population, at mean mass 10^{-8} g, has particles on near-circular, moderately inclined orbits inside 1 AU; it is needed to reproduce the flux difference between the south and ecliptic sensors on Helios 1 (and has some features in common with apex particles in the literature). In addition, a more isotropic component of the zodiacal light at close solar distances supports the existence of such a population.

5. Eccentric Population

The eccentric population, having mean mass below 10^{-12} g, exists mostly inside 1 AU, and is based primarily on an attempt to match the angular distribution of individual impact events on Helios; it has some features in common with β-meteoroids but it is not well defined by the current data set. In contrast to β-meteoroid observations, the eccentric population displays no predominance of the flux from the solar direction. Berg and Grün (1973) and Zook and Berg (1975) demonstrate with Pioneer 8 and 9 data that these β-meteoroids represent a stream of small particles leaving the solar system on hyperbolic orbits due to the effect of radiation pressure. Recent measurements by the *Hiten* satellite (Svedhem 1996) support these observations.

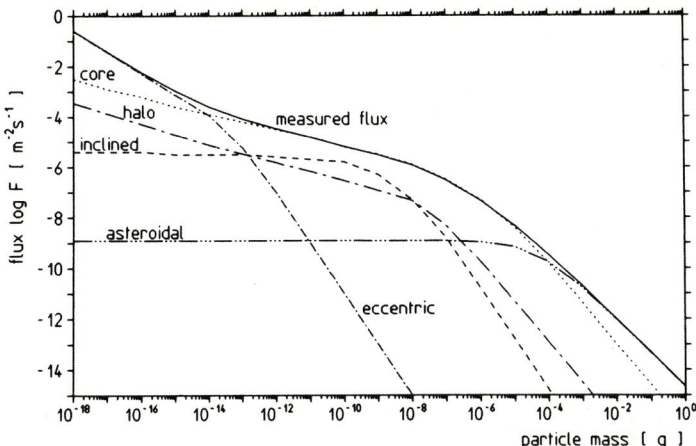

Figure 10. The mass distribution of meteoroids was obtained from the interplanetary flux model (of Grün et al. 1985) which includes lunar microcrater data and in-situ measurements from Earth satellites and spaceprobes at 1 AU distance from the Sun.

VI. COMPARISON OF DIVINE'S MODEL WITH OBSERVATIONS

In this section we demonstrate the capabilities of Divine's (1993) model by comparing various measurements with the corresponding model values.

VI.A. Interplanetary Flux Model (Size Distribution)

The mass distribution of meteoroids (Fig. 10) was obtained from the interplanetary flux model of Grün et al. (1985) which includes lunar microcrater data and in-situ measurements from Earth satellites and spaceprobes at 1 AU distance from the Sun. The core population matches the data over a wide mass range from 10^{-13} to 10^{-5} g. The asteroidal population represents bigger meteoroids at 1 AU. Smaller particles are represented by the eccentric population.

VI.B. Meteors (Radial Distribution)

Figure 11a shows the radial space density of meteor particles which was corrected for unobservable orbits (Southworth and Sekanina 1973). It displays a bimodal distribution with a minimum density near 1 AU. Divine represents this data by two populations: the core population which shows a steady increase towards the Sun and the asteroidal population peaking in the asteroid belt. Figure 11b shows the two model populations.

VI.C. Zodiacal Light and Thermal Emission

The radial concentration of zodiacal particles has been reported to scale as $r^{-1.3}$ (Leinert et al. 1981). Figure 12 shows the elongation dependence of the

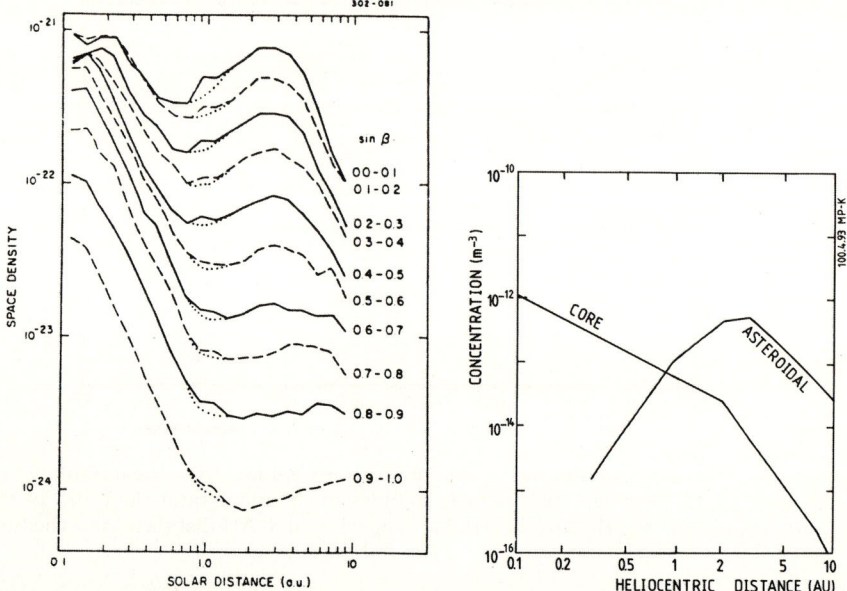

Figure 11. The radial space density of meteor particles which was corrected for unobservable orbits (Southworth and Sekanina 1973). The set of curves refers to different latitudes (represented by $\sin \beta$). The right diagram shows the two model populations (core and asteroidal).

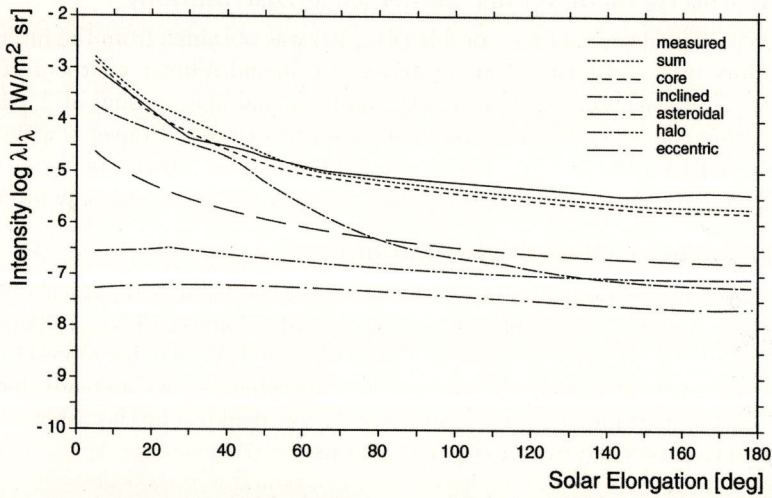

Figure 12. The elongation dependence of the zodiacal light intensity and the corresponding contributions by the model populations. The measured values refer to data of the Zodiacal Infrared Project, ZIP, at ecliptic latitude 0° and for 10.94 μm wavelength (Temi et al. 1989).

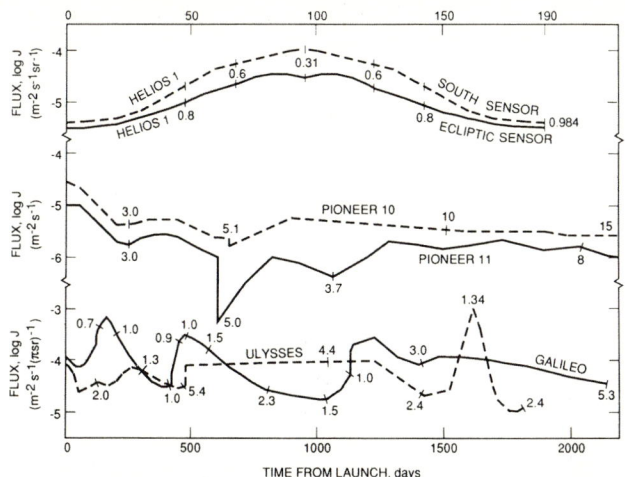

Figure 13. Contributions from Divine's dust populations to impact rate data measured by different spacecraft. The upper scale refers to time from aphelion passage for Helios, and the lower scale refers to time from launch of Pioneers 10 and 11, Galileo, and Ulysses (from Divine 1993).

zodiacal thermal emission intensity (Temi et al. 1989) and the corresponding contributions by the model populations; the zodiacal thermal emission is dominated by the core population.

One application of Divine's model is to determine meteoroid temperatures from measurements of the zodiacal thermal emission. The intensities of the zodiacal emission were measured by the IRAS satellite at four different wavelengths, at 12, 25, 60 and 100 μm (Hauser et al. 1984). Knowing the spatial meteoroid distribution and assuming an infrared emissivity and a spherical shape for the particles, the particle temperatures were evaluated as a function of particle size and heliocentric distance (Staubach et al. 1993).

VI.D. Spaceprobe Data

In Fig. 13 the total model impact rates (sum of all of Divine's populations) are displayed for interplanetary spaceprobes (Helios, Pioneers 10 and 11, Galileo and Ulysses). The core population is dominant inside 3 AU, whereas the halo population is most important at larger distances form the Sun. The model results represent well the measured impact rates.

VII. NEW RESULTS

An extension of Divine's original model is the inclusion of additional observational parameters into the model including impact direction and impact speed. A first step in this direction was taken by Divine himself by including the directionality of Helios event data. This method has been further developed by one of the authors (P.S.) and is applied to Galileo and Ulysses data. In addition new populations must be introduced (which partially replace some of the original populations) in order to match the directional flux data and to include the effects of radiation pressure.

VII.A. Interstellar Dust Population

Ulysses observations require the inclusion of an interstellar dust population in the model (Grün et al. 1994). These particles pass on hyperbolic orbits through the solar system, arriving from 252° ecliptic longitude and 5° latitude at speed about 26 km s^{-1} (Baguhl et al. 1996). For these small (10^{-13} g) particles, radiation pressure cannot be neglected which reduces the effect of solar gravity. Actually, $\beta = 1$ is used in the current model for these small particles which means that these particles move through the solar system on straight trajectories. Smaller values of β has the effect of focussing these particles near the Sun, whereas bigger values cause them to be deflected away from the Sun. Ulysses data from mid 1991 to the end of 1994 is ideally suited to characterize this new population, because in this time period interstellar dust dominates the data, except for the small particles in jovian dust streams and another small particle population over the solar poles which both can easily be separated (Hamilton et al. 1996). The flux of interstellar particles is $1.5 \cdot 10^{-4}$ m^{-2} s^{-1}; no reduction of the interstellar dust flux in the inner solar system due to sublimation has been found (Grün et al. 1997). The current model assumes no time variation of the interstellar flux, although, a 22 year variation has been predicted on theoretical grounds (Gustafson and Misconi 1979; Morfill and Grün 1979b).

The halo population which represents Pioneer 10/11 dust data do not seem to play a significant role in the Galileo and Ulysses data. Although, both fluxes can be fitted by the halo population, the directionality of the flux cannot! On the other hand, the hyperbolic interstellar dust population matches the Galileo and Ulysses data sets well. However, Pioneer 10 data can not be fit with the interstellar dust population because the flux observed by that spacecraft arrived from the opposite hemisphere than the interstellar dust seen by Ulysses. Therefore, there remains the need for the halo population in order to explain Pioneer 10 and 11 measurements. That the halo population does not prominently show up in the Galileo and Ulysses data may be due to the poor statistics of the big particles ($m > 10^{-9}$ g) to which the halo population must be restricted.

VII.B. Meteoroid Populations Affected by Radiation Pressure

Because of the strong influence of radiation pressure on small dust grains on Keplerian orbits, new populations of bound heliocentric meteoroids are defined according to most recent results of the Galileo and Ulysses dust detectors. Each population is dominant in a certain mass interval and has a constant ratio of radiation pressure to gravitational force (β). For particles with masses greater than 10^{-10} g, the influence of radiation pressure is negligible and in the Galileo and Ulysses data sets only a few events have been detected in this mass range. Therefore, Divine's core and asteroidal populations (but only for masses $m > 10^{-10}$ g) are considered which are based mainly on zodiacal light and radar meteor data. Taking into account most recent Galileo and

Ulysses dust data, three new populations are defined. The big particle population ($m > 10^{-10}$ g) has $\beta = 0$ and can be considered as part of Divine's core population. The next population (A-population) covers the mass range from 10^{-12} to 10^{-10} g and has $\beta = 0.3$. The B-population covers the range from $5 \cdot 10^{-15}$ to 10^{-12} g and has $\beta = 0.8$. For completeness, we include even smaller particles (C-population, $< 5 \cdot 10^{-15}$ g) again having $\beta = 0.3$.

The main result of the iterations is that all data sets can only be matched if we assume no depletion of interstellar dust at least down to 1.3 AU where Ulysses crossed the ecliptic plane. Any depletion of the interstellar dust flux would increase the interplanetary model flux at ecliptic plane crossing (cf. Divine's populations which do not include an interstellar population, at all) and is, therefore, unacceptable. Since we did not find any depletion of interstellar dust to that distance we assume a constant interstellar flux (beside the gravitational focusing by the Sun) over the whole range of the combined Galileo and Ulysses data, i.e. from 0.7 AU to 5.4 AU. Hiten measurements (Svedhem et al. 1996) support the assumption that interstellar dust reaches Earth's orbit.

In Fig. 14 we show the directional characteristics of the model populations, both the combined and the individual fluxes. The agreement between the general features of the measurements (Fig. 8b) and the model is satisfactory. The individual directional plots for the interstellar (Fig. 14b) and interplanetary (Fig. 14c) populations demonstrate their contributions to the fluxes observed during different mission phases.

Figure 15 shows the model rates for Galileo and the individual contributions from the interstellar and interplanetary dust populations. Until the second Earth flyby, interstellar dust contributes at most 30 % to the model flux, later this ratio increases to almost 100 %. The directional model flux matches the general features of the measurements (Fig. 9b) quite well. The individual contributions to the directional flux show the angular separation between the interstellar population and the interplanetary population over most of the time period.

Table 1 gives orbital characteristics of Divine's asteroidal and core populations and of the model populations found by this analysis. It can be seen that the inclination distributions become wider for smaller particles, an effect which may be caused by electromagnetic forces (c.f. Hamilton et al. 1996). There are significant differences from the original 5 populations determined by Divine (1993) except for the "asteroidal" population which remains unchanged since it does not contribute to the Galileo and Ulysses data sets. The "core" population is truncated below 10^{-10} g but its orbital element distributions remain unchanged. The "A", "B", and "C" populations can be regarded as the extension of the core population to small particles for which radiation pressure is important. The "interstellar" population is new. The combination of these populations matches the following data sets: radio meteors, zodiacal light, interplanetary flux at 1 AU and, of course, the Galileo and Ulysses data sets discussed here.

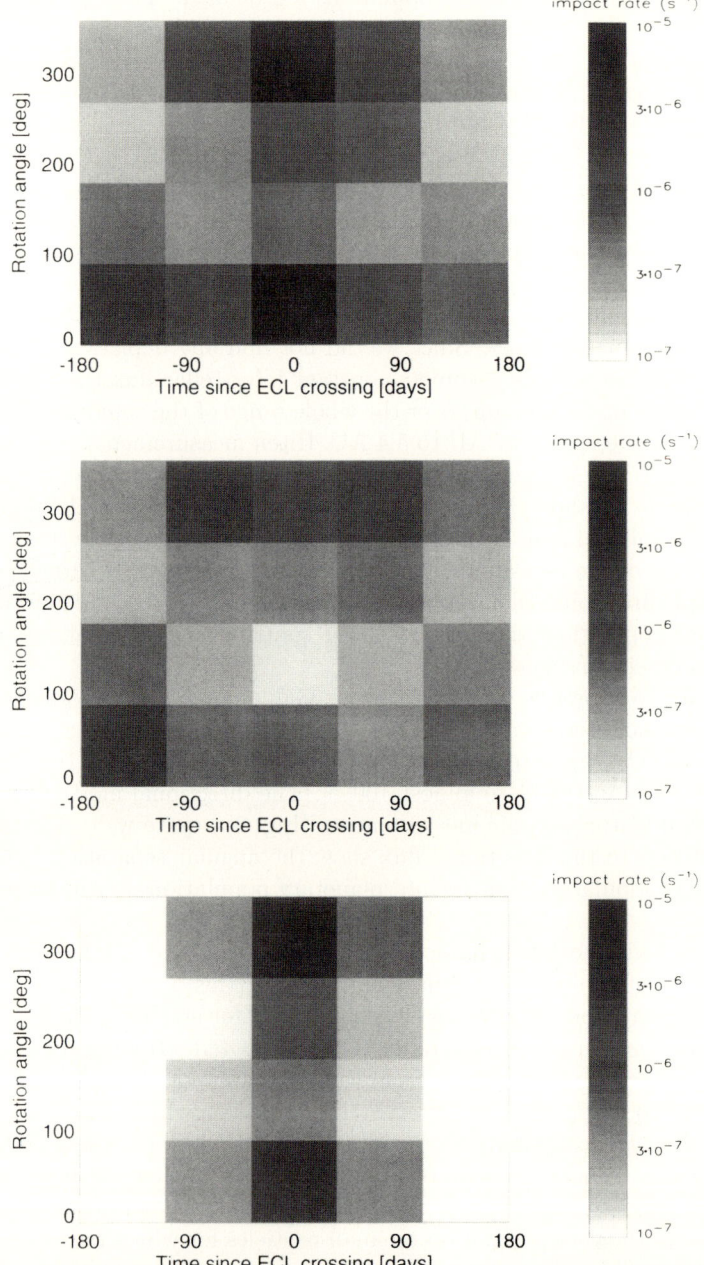

Figure 14. Directional impact rates for Ulysses as function of time from ecliptic plane crossing (ECL, from Grün et al. 1997). Top panel: Combined impact rates of all model populations. Middle panel: Interstellar dust population. Bottom panel: Sum of all bound interplanetary dust populations.

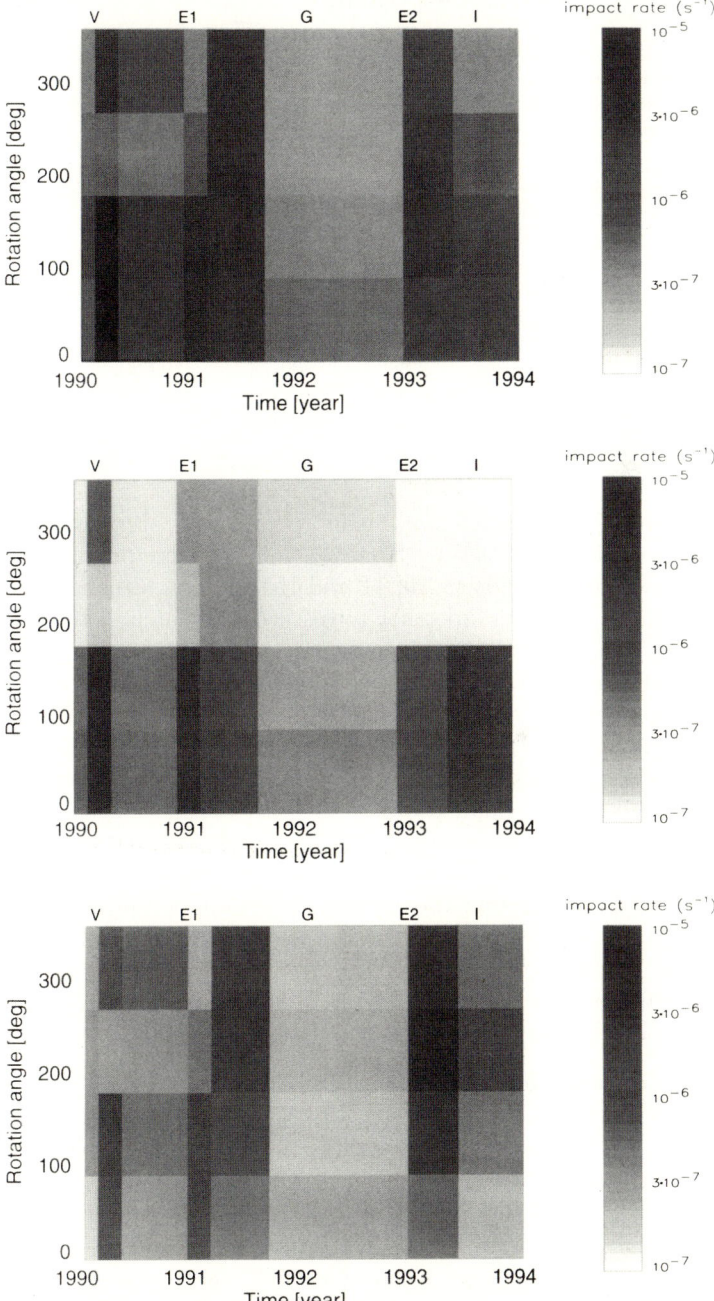

Figure 15. The directional impact rates of the model populations, both the combined (top panel) and the individual fluxes: interstellar dust (middle panel) and bound interplanetary dust (bottom panel) for Galileo.

Table 1

Comparison of the inclination distributions, p_i, of Divine's asteroidal and core populations with the new model populations of the bound dust populations needed to explain the Galileo and Ulysses observations. Below the population name the maximum of the mass distribution is given.

inclination	asteroidal 10^{-3} g	core 10^{-5} g	A 10^{-11} g	B 10^{-13} g	C $5 \cdot 10^{-15}$ g
0°	2.809	2.809	1.684	0.525	0.513
10°	2.809	2.809	2.020	2.707	1.566
20°	1.039	1.039	1.347	1.998	2.278
30°	0.286	0.286	0.673	0.608	1.109
45°	0.073	0.073	0.337	0.112	0.154
60°	0	0	0.002	0.001	0.026
90°	0	0	0	0	0.001

No "inclined", "eccentric" and "halo" populations are needed to explain the Galileo and Ulysses data sets. However, some of these populations may be required to explain the Pioneer 10, 11 and Helios data, although only with a restricted validity range in order not to disturb the match of the Galileo and Ulysses data. For example, the halo population is needed for masses above 10^{-9} g in order to describe the Pioneer 10 and 11 fluxes which were observed to arrive from directions opposite to that of the interstellar flux. The "inclined" population may still be needed in the inner solar system in order to explain some of the Helios data, although, some of that may be explained by the new interstellar, A, B, and C populations.

In summary, Divine's original five populations describe well the data sets he used. The inclusion of directional information of small particle fluxes requires the introduction of new populations affected by radiation pressure and on hyperbolic orbits. It should be noted that for small particles with $m < 10^{-14}$ g, we can draw no final conclusion on this size regime, because the relation is unclear between our B and C populations, Divine's "eccentric" population and another new hyperbolic β-meteoroid population.

VII.C. Predicted Fluxes onto the Cassini Detector

In 1997 the Cassini spacecraft was launched and it will become the first manmade satellite orbiting Saturn. After a two year trajectory through the inner solar system with flybys of Venus, and Earth, Cassini will be on its trajectory to Saturn with a further flyby of Jupiter. The Cassini spacecraft is a three-axis controlled spacecraft. For this model prediction it is assumed that it spins about its axis and it is further assumed that its axis (antenna) points to the Earth – a likely orientation for part of the mission. The dust detector on board is mounted on a turntable to the spacecraft which provides some articulation with respect to the spacecraft axis. Figure 16 shows predicted fluxes for two

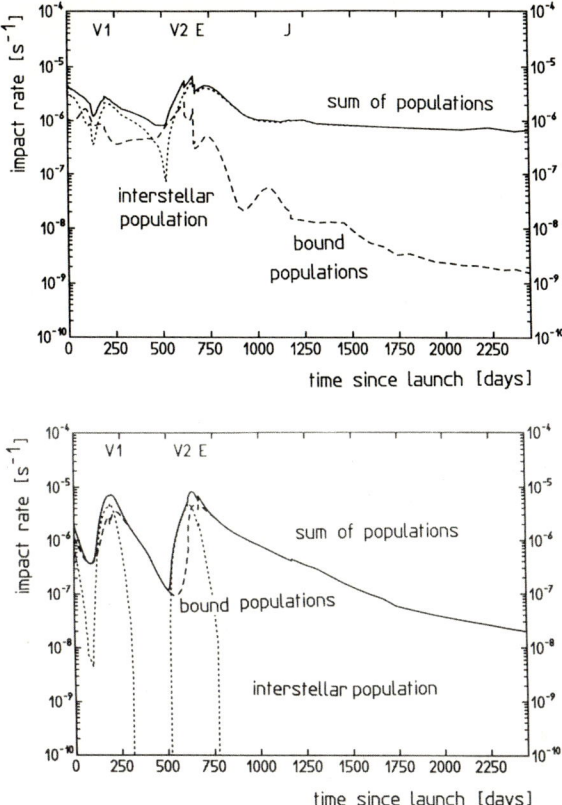

Figure 16. Predicted fluxes onto the Cassini dust detector. Fluxes for two articulation positions are shown. See text for details.

turntable positions: 1.) About at a right angle and 2.) at 125° with respect to the spin axis. These two positions separate interstellar dust from interplanetary dust on bound orbits very well. For meteoroid flux predictions for Cassini using the original Divine populations see Garret et al. (1999).

VIII. FUTURE DEVELOPMENTS

There are several areas in which new observations and new theoretical approaches will provide stimulus to improved modeling of the interplanetary meteoroid complex. Advances in meteor observations pertain to bigger meteoroids while theoretical advances are expected mostly for the dynamics of submicron sized particles.

VIII.A. New Meteor Data and Analysis

Recently, the Harvard radio meteor analysis has been re-appraised by Taylor (1995). Taylor showed that the previous analysis of the speed data was erroneous which lead to too low eccentricities of the corresponding orbits. Because of the influence of this data set on the definition of the core and asteroidal populations a redefinition of these populations has to be done as soon as a complete analysis becomes available. A redefinition of some populations will not affect model fluxes in those regions of space and in those size ranges where direct measurements are available but it will mostly affect extrapolations of the model beyond the measured ranges.

Meteor particles of interstellar origin have been found by Taylor et al. (1996). New developments in radar meteor technology provided the identification of extremely high speeds (up to several $100\,\mathrm{km\,s^{-1}}$) well in excess of the maximum speed of objects which are gravitationally bound to the solar system. Analysis of fluxes, directions and masses are forthcoming. Their existence, together with the micron sized interstellar grains observed by Ulysses and Galileo, show that interstellar space is populated by significant populations of particles which are bigger than interstellar grains observed by astronomical means (0.1 μm).

Meteor streams contribute only a fraction to the total meteoroid population. But the spectacular display of meteor storms like the Leonid meteor storm in 1998 to 2002 (see chapter by Jenniskens) and its potential hazard to space flight during this time increase the importance of modeling time variations of meteor streams.

VIII.B. Small Meteoroid Populations

Radiation pressure and interactions with the interplanetary magnetic field cause 0.1 μm or smaller particles to move on unbound trajectories through the solar system. Such small particles have been identified by a number of spaceprobe detectors (Pioneer 8, 9, Helios, Hiten and Ulysses) but detailed characteristics (including spatial distributions, orbits, speeds etc.) of this population are not yet known. In addition, theoretical understanding of their trajectories is not well advanced. Therefore, satisfactory modeling is not yet available.

VIII.C. Formulation of the Dust Environment of Earth-Orbiting Satellites

To calculate particle fluxes onto Earth orbiting satellites both planetary shielding and gravitational focusing have to be taken into account (a very similar approach can be applied to a satellite about any planet). Using Divine's original notation, focusing and shielding factors are derived and to demonstrate the capabilities of the algorithm, meteoroid fluxes onto different surfaces of an Earth-orbiting satellite are calculated (Divine 1992; Divine et al. 1993; Staubach and Grün 1995; Staubach et al. 1997)

1. Gravitational focusing

Using the heliocentric velocities of a meteoroid $\mathbf{v}$ and of the Earth $\mathbf{v}_E$, the approach velocity of a meteoroid in the absence of the gravity of the Earth is given by

$$\mathbf{w} = \mathbf{v} - \mathbf{v}_E. \tag{38}$$

Using **R** as the planetocentric position of a target point (e.g. satellite), the particle can possibly reach the target on either of two hyperbolic trajectories. Using two auxiliary velocities w_R and B the magnitude of the velocity at the target point can be calculated (Divine 1992, Divine et al. 1993):

$$w_R = (\mathbf{w} \cdot \mathbf{R}/R) \tag{39}$$

$$B = +\left[\left(\frac{w - w_R}{4}\right)\left(w - w_R + \frac{4GM_E}{Rw}\right)\right]^{1/2}. \tag{40}$$

Let $\mathbf{u_R}$ and $\mathbf{u_P}$ be unit vectors of satellite position relative to Earth and of the meteoroid angular momentum vector, another unit vector $\mathbf{u_Q}$ can be defined which lies in the orbital plane of the particle:

$$\mathbf{u_Q} = \mathbf{u_R} \times \mathbf{u_P} = \mathbf{u_R} \times \left(\frac{\mathbf{R} \times \mathbf{w}}{\sqrt{w^2 - w_R^2}}\right). \tag{41}$$

Now the meteoroid velocity vector $\mathbf{v_F}$ at the satellite (after gravitational focusing of the Earth) can be calculated:

$$v_R = \frac{1}{2}(w + w_R) \pm B \tag{42}$$

$$\mathbf{v_F} = v_R \mathbf{u_R} - \sqrt{w^2 + \frac{2GM}{R} - v_R^2}\, \mathbf{u_Q}. \tag{43}$$

The two possible trajectories differ by the sign of B. In order to calculate the meteoroid flux at the position of the satellite from the distribution functions of the populations, the focusing factor η_F has to be calculated using the following Jacobian (Divine 1992, Divine et al. 1993):

$$\eta_F = \frac{\partial(v_x, v_y, v_z)_F}{\partial(w_x, w_y, w_z)} = \left|\frac{1}{2} - \left(w - w_R + \frac{2GM_E}{Rw}\right)\left(\frac{\pm 1}{4B}\right)\right|. \tag{44}$$

2. Planetary shielding

In order to evaluate meteoroid fluxes onto Earth orbiting satellites, the planetary shielding of the Earth has to be considered. Let R_E be the radius of the Earth and r_1 be the particle's perigee, its position can be described by the true anomaly ν. If $r_1 \leq R_E$ the meteoroid can reach the target only before perigee, that means the true anomaly has to be negative. This can be described by the shielding factor η_S which can be derived from eccentricity e, velocity v_1 at perigee, and true anomaly ν:

$$e = \sqrt{1 + \left(\frac{Rw}{GM_E}\right)^2 \left(w^2 + \frac{2GM_E}{R} - \left(\frac{w + w_R}{2} \pm B\right)^2\right)} \tag{45}$$

$$r_1 = \frac{Gm_\mathrm{E}}{w^2}(e-1) \tag{46}$$

$$v_1 = \frac{R}{r_1}\sqrt{w^2 + \frac{2GM_\mathrm{E}}{R} - \left(\frac{w+w_\mathrm{R}}{2} \pm B\right)^2} \tag{47}$$

$$\sin\nu = \frac{(1+e)\left(\frac{w+w_\mathrm{R}}{2} \pm B\right)}{v_1 e}. \tag{48}$$

$$\eta_\mathrm{S} = \begin{cases} 0, & \text{for } \nu \text{ and } (R_\mathrm{P} - r_1) \text{ both positive;} \\ +1, & \text{otherwise.} \end{cases} \tag{49}$$

VIII.D. Directional Flux onto a Satellite Surface

Using both, the focusing and shielding factors, and the relative velocity $\mathbf{v}_\mathrm{D}$ which can be expressed by the meteoroid velocity $\mathbf{v}_\mathrm{F}$ at target point and the geocentric velocity $\mathbf{v}_\mathrm{DP}$ of the satellite, the meteoroid flux J onto an Earth orbiting satellite can be evaluated (Divine 1992, Divine et al. 1993):

$$\mathbf{v}_\mathrm{D} = (\mathbf{v}_\mathrm{F} - \mathbf{v}_\mathrm{DP}) \tag{50}$$

$$J = \sum \int \mathrm{d}r_1 \int \mathrm{d}e \int \mathrm{d}i \, \frac{\partial(v_x\, v_y\, v_z)}{\partial(r_1\, e\, i)} \left(\frac{r_1}{GM}\right)^{3/2} \frac{N_1 p_e p_i}{2\pi e} (\eta_\mathrm{F}\eta_\mathrm{S}\eta_\mathrm{D} v_\mathrm{D}). \tag{51}$$

Here the summation proceeds not only over several populations and over the four directions (in, out, up and down), but also over the two possible hyperbolic trajectories at the planet.

VIII.E. Meteoroid Fluxes on LDEF

Using the flux formula derived above meteoroid fluxes onto Earth-orbiting satellites have been calculated using the modified set of populations. In order to compare the model predicted fluxes with previous work and with measurements of impact craters, results for the retrieved Earth-orbiting satellite LDEF are compared with fluxes obtained from analysis of impact craters (Dauba and Drolshagen 1995).

Figure 17 shows the meteoroid flux onto the *Long Duration Exposure Facility* LDEF for distinct surface elements (leading edge and trailing edge). The flux onto the leading edge is significantly higher (approximately by a factor of 10) than onto the trailing edge. The results have been compared with fluxes obtained by Dauba and Drolshagen (1995) (ESABASE model). A comparison with measured fluxes (total flux including meteoroids and space debris) which have been obtained from impact crater analysis (published in Dauba and

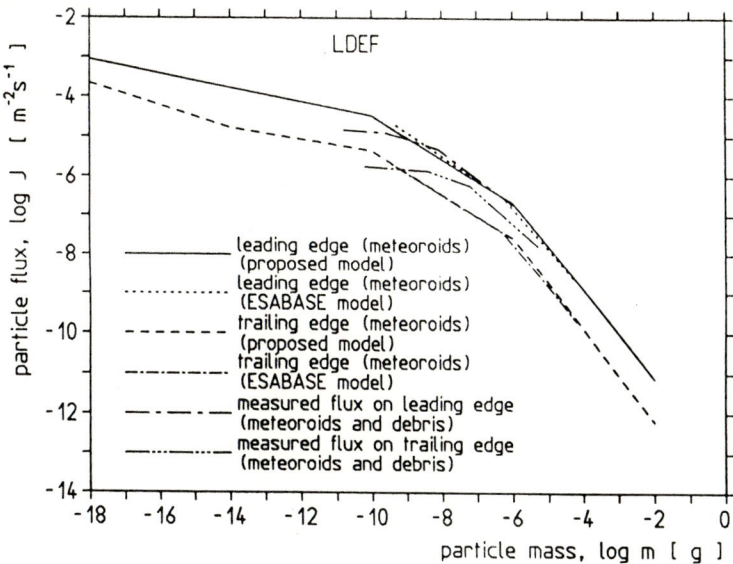

Figure 17. Fluxes onto distinct surfaces of LDEF (from Staubach et al. 1997).

Drolshagen 1995) shows a rather good conformity taking into account only a small contribution of orbital space debris for this mass range.

Acknowledgements

The authors thank Bo Å. S. Gustafson and Douglas P. Hamilton for helpful comments that improved the manuscript.

REFERENCES

Baggaley, W. J. 1996. The meteoroid orbit facility AMOR: Recent developments. In *Physics, Chemistry, and Dynamics of Interplanetary Dust, ASP Conference Series*, **104**, eds. B. Å. S. Gustafson and M. S. Hanner, pp. 65–70.

Baguhl, M., Grün, E., Hamilton, D. P., Linkert, G., Staubach, P. 1995. The flux of interstellar dust observed by Ulysses and Galileo. *Space Sci. Rev.*, **72**, pp. 471–476.

Baguhl, M., Grün, E., Linkert, G., Linkert, D., Siddique, N. 1993. Identification of 'small' dust impacts in the Ulysses dust detector data. *Planet. Space Sci.*, **41 (No. 11/12)**, pp. 1085–1098.

H.,

Baguhl, M., Grün E. and Landgraf M. 1996. In-situ measurements of interstellar dust with the Ulysses and Galileo spaceprobes. *Space Sci. Rev.*, **78**, pp. 165–172.

Banderman, L. W. 1968. Physical properties and dynamics of interplanetary dust. Thesis (Univ. Maryland).

Berg, O. E., Grün, E. 1973. Evidence of hyperbolic cosmic dust particles. *Space Res.*, **XIII**, pp. 1047–1055.

Bradley, J. P., Ireland T 1996. The search for interstellar components in interplanetary dust particles. In *Physics, Chemistry, and Dynamics of Interplanetary Dust, ASP Conference Series*, **104**, eds. B. Å. S. Gustafson and M. S. Hanner, pp. 275–282.

Brownlee, D. E. 1996. The elemental composition of interplanetary dust. In *Physics, Chemistry, and Dynamics of Interplanetary Dust, ASP Conference Series*, **104**, eds. B. Å. S. Gustafson and M. S. Hanner, pp. 261–264.

Burns J. A., Lamy, P. L., Soter, S. 1979. Radiation forces on small particles in the solar system. *Icarus*, **40**, pp. 1–48.

Cour-Palais, B. G. 1969. Meteoroid Environment Model – 1969 (Near Earth to Lunar Surface). NASA SP-8013.

Dauba, O., Drolshagen, G. 1995. Meteoroids and debris flux predictions for Eureca, the Hubble Space Telescope and LDEF. *ESA/ESTEC WP* **1874**.

Divine, N. 1992. Meteoroid focusing at a Planet. *JPL-IOM 527-92-86*.

Divine, N. 1993. Five populations of interplanetary meteoroids. *J. Geophys. Res.*, **98**, pp. 17029–17048.

Divine, N., Grün, E., Staubach, P. 1993. Modelling the Meteoroid Distributions in Interplanetary Space and near-Earth. *Proc. First Europ. Conf. on Space Debris*, **ESA SD-01**.

Drolshagen, G. 1994. Material Densities of Meteoroid and Space Debris Particles. *ESA/-ESTEC/WMA*, **WMA/94-035/GD/DENS**.

Dumont, R., Levasseur-Regourd, A. C. 1987. The symmetry plane of the zodiacal dust cloud retrieved from IRAS data. In *Interplanetary Matter, Proc. 10th European Regional Meeting of the IAU 2*, **67**, ed. Z. Ceplecha and P. Pecina (Publ. Astron. Inst. Czech. Acad. Sci.), pp. 281–284.

Garrett, H. B., Drouilhet, S. J., Oliver, J. P., Evans, R. W. 1999. Interplanetary meteoroid environment model update. *J. Spacecraft and Rockets*, **36 (No. 1)**, pp. 124–132.

Greenberg, J. M., Hage, J. I. 1990. From Interstellar dust to comets: A unification of observational constraints. *Astrophys. J.*, **361**, pp. 260–274.

Grün, E., Baguhl, M., Staubach, P., Dermott, S., Fechtig, H., Gustafson, B. Å., Hamilton, D. P., Hanner, M. S., Horányi, M., Kissel, J., Lindblad, B. A., Linkert, D., Linkert, G., Mann, I., McDonnell, J. A. M., Morfill, G. E., Polanskey, C., Schwehm, G., Srama, R., Zook, H. A. 1997. South-North and radial traverses through the zodiacal cloud. *Icarus*, **129**, pp. 270–288.

Grün, E., Fechtig, H., Giese, R. H., Kissel, J., Linkert, D., Maas, D., McDonnell, J. A. M., Morfill, G. E., Schwehm, G., Zook, H. A. 1992b. The Ulysses Dust Experiment. *Astron. Astrophys. Suppl.*, **92**, pp. 411–423.

Grün, E., Fechtig, H., Hanner, M. S., Kissel, J., Lindblad, B. A., Linkert, D., Morfill, G. E., Zook, H. A. 1992a. The Galileo dust detector. *Space Sci. Rev.*, **60**, pp. 317–340.

Grün, E., Gustafson, B., Mann, I., Baguhl, M., Morfill, G. E., Staubach, P., Taylor, A., Zook, H. A. 1994. Interstellar dust in the heliosphere. *Astron. Astrophys.*, **286**, pp. 915–924.

Grün, E., Zook, H. A., Baguhl, M., Balogh, A., Bame, S. J., Fechtig, H., Forsyth, R.,

Hanner, M. S., Horányi, M., Kissel, J., Lindblad, B.-A., Linkert, D., Linkert, G., Mann, I., McDonnell, J. A. M., Morfill, G. E., Phillips, J. L., Polanskey, C., Schwehm, G., Siddique, N., Staubach, P., Svestka, J., Taylor, A. 1993. Discovery of jovian dust streams and interstellar grains by the Ulysses spacecraft. *Nature*, **362**, pp. 428–430.

Grün, E., Zook, H. A., Fechtig, H., Giese, R. H. 1985. Collisional Balance of the Meteoritic Complex. *Icarus*, **62**, pp. 244–272.

Gustafson B. Å. S. 1994. Physics of zodiacal dust. *Ann. Rev. Earth. Planet. Sci.*, **22**, pp. 553–595.

Gustafson B. Å. S., Misconi, N. Y. 1979. Streaming of interstellar grains in the solar system. *Nature*, **282**, pp. 276–278.

Gustafson, B. Å. S., Grün, E., Dermott, S. F., Durda, D. D. 1991. Collisional and dynamic evolution of dust from the asteroid belt. In *Asteroids, Comets, Meteors*, ed. A. W. Harris and E. Bowell (Houston: Lunar and Planet. Inst.).

Hamilton, D. P., Grün, E. and Baguhl, M. 1996. Electromagnetic escape of dust from the solar system. In *Physics, Chemistry, and Dynamics of Interplanetary Dust, ASP Conference Series*, **104**, eds. B. Å. S. Gustafson and M. S. Hanner, pp. 31–34.

Hanner, M. S., Weinberg, J. L., DeShields II, L. M., Green, B. A., Toller, G. N. 1974. Zodiacal light and the asteroid belt: The view from Pioneer 10. *J. Geophys. Res.*, **79**, pp. 3671–3675.

Haug U. 1958. Über die Häufigkeitsverteilung der Bahneelemente bei den interplanetaren Staubteilchen. *Zeitschrift f. Astrophysik*, **44**, pp. 71–97.

Hauser M. G., Gillett, F. C., Low, F. J., Gautier, T. N., Beichman, C. A., Neugebauer, G., Aumann, H. H., Baud, B., Boggess, N., Emerson, J. P., Houck, J. R., Soiffer, B. T., Walker, R. G. 1984. *Astrohys. J.*, **278**, pp. L15–L18.

Hauser, M. G. 1996. COBE observations of zodiacal emission. In *Physics, Chemistry, and Dynamics of Interplanetary Dust, ASP Conference Series*, **104**, eds. B. Å. S. Gustafson and M. S. Hanner, pp. 309–314.

Hörz, F., Brownlee, D. E., Fechtig, H., Hartung, J. B., Morrison, D. A., Neukum, G., Schneider, E., Vedder, J. F., Gault, D. E. 1975. Lunar microcraters: implications for the micrometeoroid complex. *Planet. Space Sci.*, **23**, pp. 151–172.

Humes, D. H., Alvarez, J. M., O'Neal, R. L., Kinrad, W. H. 1974. The interplanetary and near Jupiter environment. *J. Geophys. Res.*, **79**, pp. 3677–3684.

Kelsall, T., Weiland, J. L., Franz, B. A., Reach, W. T., Arendt, R. G., Dwek, E., Freudenreich, H. T., Hauser, M. G., Mosley, S. H., Odegard, N. P., Silverberg, R. F., Wright, E. L. 1998. The COBE diffuse infrared background experiment search for the cosmic infrared background. II. Model of the interplanetary dust cloud. *Astrophys. J.*, **508**, pp. 44–73.

Kessler, D. J. 1970. Meteoroid Environment Model-1970. NASA SP-8038.

Kessler D. J. 1981. Derivation of collision probability between orbiting objects: The lifetimes of Jupiter's outer moons. *Icarus*, **48**, pp. 39–48.

Kessler, D. J. 1990. Update of Meteoroid and Orbital Debris Environment Definition Space Station Level II', Change Request. Marshall Space Flight Center AL, BB 000883A 1990.

Lallement, R. 1993. Measurements of the interstellar gas. *Adv. in Space Res.* **13**, pp. (6)113–120.

Leinert C. and Grün, E. 1990. Interplanetary dust. In *Physics of the Inner Heliosphere 1 Large-Scale Phenomena*, eds. R. Schwenn and E. Marsch (Heidelberg: Springer), pp. 207–275.

Leinert C., Richter, I., Pitz, E., Planck, B. 1981. The zodiacal light from 1.0 to 0.3 AU as observed by the Helios space probes. *Astron. Astrophys.*, **103**, pp. 177–188.

Levasseur-Regourd, A. C. 1996. Optical and thermal properties of zodiacal dust. In *Physics, Chemistry, and Dynamics of Interplanetary Dust, ASP Conference Series*, **104**, eds. B. Å. S. Gustafson and M. S. Hanner, pp. 301–308.

Levasseur-Regourd A. C. and Dumont, R. 1980. Absolute photometry of zodiacal light. *Astron. Astrophys.*, **84**, pp. 277–279.

Matney, M. J., Kessler, D. J. 1996. A Reformulation of Divine's Interplanetary Model. In *Physics, Chemistry, and Dynamics of Interplanetary Dust, ASP Conference Series*, **104**, eds. B. Å. S. Gustafson and M. S. Hanner, pp. 15–18.

Morfill, G. E., Grün, E. 1979a. The motion of charged dust particles in interplanetary space. I. The zodiacal dust cloud. *Planet. Space Sci.*, **27**, pp. 1269–1282.

Morfill, G. E., Grün, E. 1979b. The motion of charged dust particles in interplanetary space - 2. Interstellar grains. *Planet. Space Sci.*, **27**, pp. 1283–1292.

Morfill, G. E., Grün, E., Leinert, C. 1986. The interaction of solid particles with the interplanetary medium. The Sun and the Heliosphere in three Dimensions, ed. R. G. Marsden, D. Reidel (Dordrecht, Boston, Lancaster, Tokyo), pp. 455–474.

Naumann, R. J., Jex, D. W., Johnson, C. L. 1996. Calibration of Pegasus and Explorer XXIII Detector panels. NASA TR-R-321.

Sekanina, Z., Southworth, R. B. 1975. Physical and dynamical studies of meteors: Meteor fragmentation and stream-distribution studies. NASA CR-2615.

Southworth R. B. and and Sekanina Z. 1973. Physical and dynamical studies of meteors. NASA CR-2316.

Staubach, P., Divine, N., Grün, E. 1993. Temperatures of zodiacal dust. *Planet. Space Sci.*, **41 (No. 11/12)**, pp. 1099–1108.

Staubach, P., Grün, E. 1995. Development of an upgraded meteoroid model. *Adv. Space Res.*, **16 (No. 11)**, pp. 103–106.

Staubach, P., Grün, E., Jehn, R. 1997. The meteoroid environment near Earth. *Adv. Space Res.*, **19 (No. 2)**, pp. 301–308.

Svedhem, H., Münzenmayer, R., Iglseder, H. 1996. Detection of possible interstellar particles by the HITEN spacecraft. In *Physics, Chemistry, and Dynamics of Interplanetary Dust, ASP Conference Series*, **104**, eds. B. Å. S. Gustafson and M. S. Hanner, pp. 27–30.

Taylor, A. D. 1995. The Harvard Radio Meteor Project Meteor Velocity Distribution Reappraised. *Icarus*, **116**, pp. 154–158.

Taylor, A. D., Baggaley, W. J., Steel, D. I. 1996. Discovery of interstellar dust entering the Earth's atmosphere. *Nature*, **380**, pp. 323–325.

Temi, P., De Bernadis, B., Masi, S., Moreno, G., Salama, A. 1989. Infrared emission from interplanetary dust. *Astron. J.*, **337**, pp. 529–535.

Verniani, F. 1964. *Il nuovo cimento*, **33 (No. 4)**, pp. 4453–4464.

Whipple, F. L. 1963. On meteoroids and penetration. *J. Geophys. Res.*, **68 (No. 17)**, pp. 4929–4939.

Witte, M., Rosenbauer, H., Banaskiewicz, M., Fahr, H. 1993. The Ulysses neutral gas experiment: Determination of the velocity and temperature of the neutral interstellar helium, *Adv. in Space Res.* **13**, pp. (6)121–130.

Zook, H. A. 1990. Flux vs. direction of impacts on LDEF by meteoroids and orbital debris. *Lunar and Planet. Sci. Conf. XXI*, pp. 1385–1386.

Zook, H. A. 1991a. Deriving the velocity distribution of meteoroids from the measured meteoroid impact directionality on the various LDEF surfaces. LDEF-69 months in space, First Post Retrieval Symposium, NASA CP-3134, Part 1.

Zook, H. A. 1991b. Meteoroid directionality on LDEF and asteroidal versus cometary sources. *Lunar and Planet. Sci. Conf. XXII*, pp. 1577–1578.

Zook, H. A., Berg, O. E. 1975. A source of hyperbolic cosmic dust particles. *Planet. Space Sci.*, **23**, pp. 183–203.

Instrumentation

Siegfried Auer

Max-Planck-Institut für Kernphysik, Heidelberg, Germany
permanent address: Post Office Box 421, Basye, Virginia 22810, USA

Abstract. Information on the dynamics and properties of interplanetary dust is obtained from *in-situ* detectors on board Earth satellites and deep space probes. This chapter reviews the methods of detection and discusses their strengths and limitations. Detailed descriptions are given for those detectors which have significantly advanced the state of the art of interplanetary dust research. Also reviewed are laboratory facilities required for the calibration of the detectors with fast (1 to $100\,\mathrm{km\,s^{-1}}$) dust particles and for the simulation of electrical charging of dust in space.

I. INTRODUCTION

Ejected from celestial bodies, *interplanetary dust* particles carry along valuable information about their parent bodies, such as composition and structure. In addition, during their long travel through interstellar or interplanetary space, the particles are affected by the interplanetary medium: they are impacted by their own kind, eroded by the solar wind, charged by photoelectron emission and other effects, penetrated by energetic nuclei, and accreted by extraneous molecules. Accordingly, they are uniquely "finger-printed" as to origin and history, and planetary scientists use these "fingerprints" in continuing attempts to reveal their properties and dynamics.

These particles are collected in the Earth's atmosphere and are found deposited in Greenland and Antarctic ice and on the ocean floor. However, in order to determine their orbits in space, and in order to avoid the material effects of heating during their entry into the atmosphere or of weathering as they are exposed on the ground or ocean floor, they must be observed in space under pre-atmospheric conditions.

This chapter is devoted to instrumentation for *in-situ* measurements of interplanetary dust. It does not discuss any results of such measurements nor any methods of particle collection. Those two subjects are discussed in the following two chapters of part III of this volume and the first chapter of part V, respectively.

Section II discusses the physical interactions by which interplanetary dust particles manifest themselves as they approach the spacecraft or interact with its sensors. These interactions are the basis of the flight instruments presented in section III, which have led to our present understanding and portrait of the interplanetary dust phenomena. In section IV, the dust accelerators are discussed that generate fast $(1\ldots 100 \text{ km s}^{-1})$ particles needed to develop, test, and calibrate the flight instruments. Finally, a facility is described which simulates the electric charging of dust particles in space.

The units used throughout this chapter, especially in the equations, conform to the International System of Units (SI).

II. DETECTION AND CHARACTERIZATION OF DUST PARTICLES

In this section, the physical interactions are discussed by which dust particles are detected in space. The section has been organized to reflect the sequence in which a particle becomes detectable as it approaches the spacecraft (Fig. 1). Toward the end of the section, methods of determination of specific particle properties are discussed.

Over interplanetary distances, one can observe infrared emission and scattered sunlight from the cloud of interplanetary dust (*zodiacal light photometer*, section II.B.1). When an individual sun-illuminated particle passes by a spacecraft, it appears as a streak of light against the star background (*imaging telescope*, section II.B.2). Or it lights up while passing through an artificial *light curtain* (section II.A.3). When a particle comes close to a thin wire or other conductor, its natural charge induces a *charge* on that conductor (section II.B) which, if large enough, can reveal the particle's *velocity, trajectory, and orbit* (section II.G).

As the particle strikes a solid target at high speed (typically 10 to 70 km s^{-1}), it is destroyed producing a crater; an *impact light flash* (section II.C); and ejecta of solid and liquid particles, neutral and ionized molecules, and electrons (*impact ionization*, section II.D). In addition, the particle and impact ejecta transfer *momentum* (section II.F) to the target and generate a pulse of high pressure (up to 5 TPa) and temperature (up to 0.5 MK). The ejecta expand and strike adjacent surfaces where they may produce secondary craters, light flashes, and ejecta.

If the particle penetrates a target thinner than its diameter (*thin-foil penetration*, section II.E), its speed is reduced and its flight direction may be altered. In addition, the particle suffers ablation and, in many cases, fragmentation. Ejecta, including ions and electrons (*impact ionization*, section II.E.3), emerge from the penetration hole in both forward and backward directions. If *perforation of a pressurized cell* (section II.E.1) occurs, gas begins leaking through the hole and the loss of gas pressure actuates a pressure switch. If the foil is made of an electrically polarized material, some foil material is *depolarized* (section II.E.4). Or, if the particle strikes a charged *capacitor*, it may trigger a capacitor discharge (section II.E.2).

Impact ionization has led to the most sensitive detection methods because: the ions and electrons can be separated by an electric field into positive ions and negative ions and electrons, detected with great sensitivity, even counted individually; and different ion species can be separated in a mass spectrometer

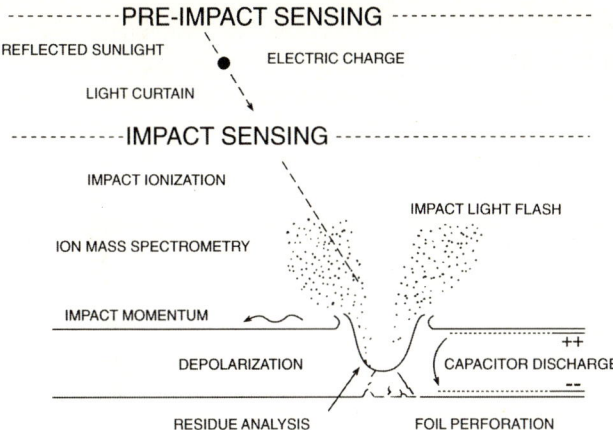

Figure 1. Simplified illustration of *in-situ* detection methods of interplanetary dust particles.

and measured quantitatively (*chemical and isotopic composition*, section II.J). The total number of electrons (or total number of positive ions) is an indication of the mass (section II.I) of the dust particle when the impact speed is known. In addition, the ratio of ions of the particle material to ions of the target material is a function of the impact speed and density of the particle; thus, with this ratio and the speed known, the *density* (section II.I) can be determined. Finally, because electron and ion signals always appear in coincidence, dust impacts can be distinguished from noise.

If only we could slow down the dust particles and place them gently under a microscope without breaking or heating or otherwise altering them! To mitigate the excessive forces and the resulting damage to a dust particle during the impact, methods of gradual *deceleration for intact capture* (section II.H) have been developed or proposed. Once captured, a particle can be imaged and analyzed on the spacecraft or it can be returned to Earth where elaborate analytical tools and facilities are available.

II.A. Detection of Scattered and Emitted Light

1. Zodiacal Light Photometer

The diffuse glow of zodiacal light provides information on properties of dust particles—spatial distribution, number density, size, shape, structure, and material—averaged over many particles and long distances. Because a spaceborne zodiacal light photometer operates in full sunlight, suppression of stray light is a major design challenge. Typically, the level of solar stray light should be below 1 % of the zodiacal light brightness. At short wavelengths (330...550 nm), this level was achieved by the Helios zodiacal light experiment (Leinert et al. 1974, 1975, 1981) which suppressed stray light by some 15 orders of magnitude! At infrared wavelengths (1...300 μm), the Cosmic Background Explorer (COBE) is an outstanding example of a design that reduced, by many orders of magnitude, the thermal radiation from the sun, Earth, and warm parts of the spacecraft (Miller et al. 1982).

Temporary interference and permanent degradation generated by cosmic rays cause additional problems. Interference is produced primarily by Čerenkov radiation in the glass envelope of the photomultiplier tube or by ionization in the semiconductor radiation detector. It tends to increase the level of the noise background and can sometimes be reduced by an electronic "despiking" circuit. Permanent darkening of ordinary glass is also caused by energetic particles and photons (starting at a dose of 10 Gy) and can be prevented (for doses up to 0.1...1 MGy) by employing radiation resistant glass or reflective optics. For methods of zodiacal light measurements, see the chapter by Levasseur-Regourd and Mann.

2. Imaging Telescope

Telescopes detecting scattered sunlight from *individual* particles are important because the sensitive area increases with the cross section of the particle and can reach many km^2 or Mm^2. The main problems are background radiation from stars and interference from energetic particles.

An attempt at optical detection has been made by Soberman et al. (1974a,b) on the Pioneer 10 and 11 missions to the outer solar system. Their instrument used a stereoscopic configuration of four spatially offset, parallel telescopes. The exact times when a particle enters and exits each field of view are significant. Because of the spatial offset, these times differ for the different fields, depending on the distance, flight direction, and speed of the particle. If at least three of the four telescopes detect the particle, the times of entry and exit and the peak brightness are recorded and are, at least in principle, sufficient for calculating the velocity, distance, and reflectance of the particle. Note, however, that the Pioneer 10 data were considered unreliable (Auer 1974).

At long wavelengths, the Infrared Astronomical Satellite (IRAS) has detected some large orbital debris (Dow et al. 1990). Unfortunately, an on-board "de-spiking" circuit, designed to suppress interference from energetic particles, also eliminated the signals from objects that passed closer than a few thousand km.

With the advent of high-speed CCD arrays and photon counting cameras, a dust particle can be distinguished more effectively from the background. The single photocathode is replaced by an array of N (e.g., $N = 1024 \times 1024$) discrete pixels. When sunlight, after reflection from a particle that moves on a straight line, falls onto the array, it strikes only pixels in the proper time sequence and on a straight line. Therefore, *all* illuminated pixels form a straight line in an x, y, t coordinate system. This line contains all photoelectrons generated by the reflected sunlight, but only a fraction N^{-1} of the photoelectrons generated by the background radiation. By selecting only this line of pixels, one can reduce the background noise $\propto N^{-0.5}$ and increase the sensitive area of the imaging telescope $\propto N^{0.67}$ (Auer 1984). For a 1024 × 1024 pixel array, this amounts to an increase of the sensitive area by a factor of about 10 000.

Also, the signatures of energetic nuclear particles can be readily recognized and eliminated by software, because their dwell times are zero.

Note, however, two unresolved problems with the imaging telescope. First, the amount of digital signal processing required to filter the data in real time is too demanding for present-time computers. This problem may disappear in the future, due to rapid advances in computer power. Second, particles passing at a close range, especially those smaller than 1 mm, are generally out of focus; thus, the method of reducing background noise loses its effectiveness with decreasing particle size.

3. Light Curtain

Dust particles passing outside a spacecraft are best detected by the sunlight they reflect. If particles passing at a close range ($\approx$ 10 cm), or falling through a sensor aperture, are illuminated by an artificial fan beam, or "light curtain", it may be possible to detect the pulse of reflected light.

An experiment applying this principle has been accepted for the comet orbiter Rosetta (Leese et al. 1996; Perruchot et al. 1996). The sensor is composed of three segmented light curtains of area 0.01 m^2 and thickness 4 mm. The first curtain operates continuously, watching for incoming particles. When a signal is detected, the two following curtains are turned on. They localize the particle and measure the time-of-flight between the curtains (to determine the speed) and the light amplitude (to determine the size). For each screen, an illumination of 5000 W m^{-2} is produced by four laser diodes ($\lambda = 830$ nm) and collimating cylinder lenses. Laser light reflected off the particle passes through a narrow-band optical filter, is collected by two facing rows of hollow concentrators (Welford and Winston 1989), and measured by photodiodes. Using a laser power of 3 mW, a glass sphere having a diameter of 100 μm and a speed of 1...3 m s^{-1} has been detected in the laboratory.

II.B. Charge

The natural electric charge of a dust particle is important, because it influences the particle's motion in an electromagnetic field. In particular, a charged particle can be levitated and transported (Berg et al. 1975) or disrupted by electrostatic stress (Fechtig et al. 1979). Two particles might become oppositely charged and, as a result, combine to form a single larger grain (Chow et al. 1993). Charges of dust particles are measured, unambiguously for the first time, by the Cassini CDA (section III.H).

Unlike reflected sunlight, charge remains with a dust particle as it enters the aperture of a sensor. The charge, if large enough, reveals the particle's presence and position by disturbing the distribution of surface charges on conductors inside the sensor as it passes by. As a result, charges are induced temporarily on the conductors, and these charge signals can be amplified and analyzed as they vary with time. The charge/velocity sensors, discussed in section II.G, and, in particular, the Cassini CDA, employ this principle.

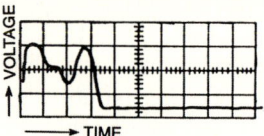

DETECTOR OUTPUT SIGNAL

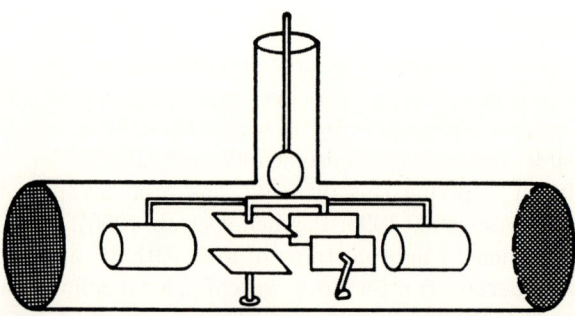

Figure 2. Detector for measuring particle speed and position, with a reproduction of the oscilloscope trace of the output signal (from Shelton et al. 1960). The two cylinders detect the charge of the dust particle, whereas the two pairs of plates measure its x, y position. The particle speed is derived from the time of flight through the detector.

Charge induction has been utilized in electrostatic dust accelerators to detect the charge, speed, position, and mass of individual particles (Shelton et al. 1960). Such a detector consists of four parts: a first tube, a pair of horizontal plates, a pair of vertical plates, and a second tube mounted within a grounded shield and connected to the input of an amplifier, as shown schematically in Fig. 2. The signal, also shown in Fig. 2, consists of four parts corresponding to the four parts of the detector. While a particle dwells inside one of the tubes, its full charge q appears as if deposited on the tube. The amplitude of the first and fourth part of the signal is, therefore, equal to q/C (where C is the known capacitance between detector and shield) and its duration τ is the time it takes the particle to drift through the tube having a known length L; thus, the particle's speed is $v = L/\tau$. Knowing v, its mass m can be calculated from the kinetic energy, $\frac{1}{2}mv^2 = qU$, where U is the high voltage of the accelerator. Finally, the x, y position is derived from the second and third parts of the signal, as the induced charge is directly proportional to the position of the particle within a pair of plates.

In interplanetary space, the *surface potential* of a dust particle is established by a balance between various charging currents, which depend on the particle's environment and on properties of the particle, such as material composition, size, and shape. The currents are caused primarily by photoelectrons,

Table 1

Characteristics[1] of charge detectors actually flown or intended for flight.

Spacecraft	path length in sensor [m]	dwell time in sensor [μs]	number of signal samples	detection threshold [fC]	minimum particle diameter [μm] when $\Phi = 10$ V
Helios, Galileo, Ulysses	0.03	3	1[2]	10	15
Cassini	0.20	20	120	2	3
CDCF[3]	0.30	30	180	0.04	0.07

[1] data are for $v = 10$ km s^{-1}
[2] peak value
[3] trajectory sensor developed for the Cosmic Dust Collection Facility (see section II.G)

secondary electrons, and solar wind particles. Potentials can range from about -10 kV in planetary magnetospheres to some $+10$ V in interplanetary space (see, e.g., Whipple 1981), depending on size and material. The charging of particles has been investigated in the laboratory by Švestka et al. (1993) and is further discussed in section IV.B and chapter 16. The charge of a spherical particle as a function of its diameter d and surface potential Φ is:

$$q = 2\pi d \epsilon_0 \Phi , \qquad (1)$$

where ϵ_0 is the permittivity of free space. Table 1 lists sensitivities of different charge detectors.

The detection of fast moving charges as small as 0.04 fC (≈ 250 electrons) is limited by amplifier noise and interference from electric fields, ultraviolet radiation, plasma, energetic particles, and mechanical vibrations. The internal noise of a charge-sensitive amplifier at room temperature can be as low as 4 aC rms (= 24 electrons rms, equivalent to 200 eV (Si) fwhm, according to Kern and McKenzie (1970)). However, external interference often limits the detection of small charges more severely than amplifier noise and must be controlled carefully: electric fields can be attenuated by grounded grids across the sensor aperture; ultraviolet radiation, especially from the sun, must be blocked and the materials of exposed sensor surfaces should be selected for low photoelectron emission yield; solar wind and ionospheric plasma can be deflected out of the sensor by electric fields (the dust detectors on HEOS 2 and Helios included ion and electron repelling grids); and mechanical vibrations can be attenuated by mounting the detector on special shock absorbers. In addition to these preventive measures, the recording of entire signals, first implemented on Hiten (see section III.G), has proven to be an effective tool for confirming true and eliminating false impacts.

The charge can also be utilized to decelerate particles and to measure their masses, as is further discussed in sections II.H and II.I, respectively.

II.C. Impact Light Flash

As can be seen from meteor phenomena, particle impacts at cosmic velocities produce light. The light flash from impact on a solid target was utilized for dust detection by Berg and Meredith (1956) on a sounding rocket and by Kissel (1986) on the VeGa 1 and 2 and Giotto spacecraft to comet Halley. It was studied in the laboratory by Jean and Rollins (1970), Eichhorn (1975, 1976, 1978a,b), and Burchell et al. (1996).

Eichhorn (1975) found that the light flash, in the case of iron particles impacting a gold target, consists of two parts. The first part has a rise time shorter than 100 ns and is the light produced by the primary impact. When ejecta from the primary impact strike adjacent surfaces, the second part of the light flash is produced. The rise time of the second part depends on the speed of the impacting particle and can be 200 μs at low impact speed.

Eichhorn (1976) measured the spectrum of light in the wavelength range 300...700 nm and determined an effective plasma temperature of about 3000 K for a micrometer-sized iron particle impacting tungsten at a speed of 10 km s^{-1}. Figure 3 shows the total light energy normalized to the projectile mass and plotted as a function of the impact speed for four different projectile materials. The light energy increases with power 3.0...3.6 of the impact speed for aluminum, iron, and tungsten projectiles. Eichhorn attributed the different velocity dependence for carbon to its high melting temperature and extremely high heat of fusion.

Burchell et al. (1996), firing iron particles onto molybdenum, determined that the bulk of the light flash cannot come from either the target heated by the impact or from recombination in the plasma and considered the ejecta a more plausible source. They observed only a single light flash, with a short duration (0.3...6 μs) and well-defined ($\pm$10 ns) start. Since the start of the light flash defines the time of impact, it served as one of three trigger signals for the ion time-of-flight mass spectrometers of the PUMA and PIA experiments (see section III.E).

Ang (1990) explained the flash as caused by emissions from a jet of shocked material formed through a shock interaction analogous to the process which occurs in a shaped charge. According to Ang, the jet is initiated from either the target or the particle, depending on the shock properties of both target and particle materials and the impact velocity. For carbon striking gold, most of the material in the jet would originate from the gold target at low impact speeds (< 5 km s^{-1}) and from the carbon particle at higher impact speeds, whereas for W, Fe, and Al striking gold, the jet would originate from the particle material at any speed. This model could explain Eichhorn's data in Fig. 3 including the curve for carbon striking gold.

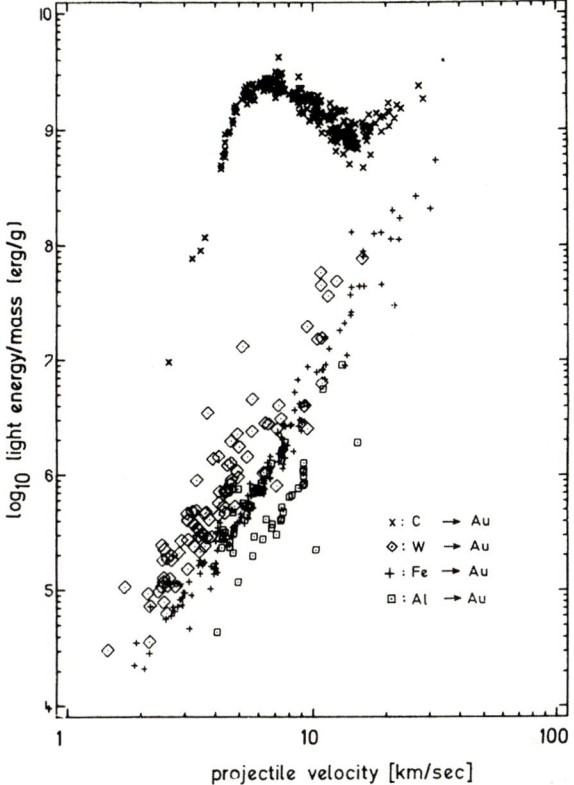

Figure 3. Total light energy normalized to the particle mass, versus projectile velocity for four different projectile materials. The target material in each case was gold. Reprinted from Planet. Space Sci., vol. 24, Eichhorn, G., 1976, copyright, with permission from Elsevier Science.

II.D. Impact Ionization

1. Detection of Impact Ionization

When a dust particle strikes a solid target at high speed, it produces a crater in the target and ejecta of both particle and target material. The ejecta consist of: neutral or charged debris, neutral molecules, positive and negative ions, and electrons. Because of its high internal pressure (up to 5 TPa), the cloud of ejecta then expands into the surrounding vacuum. As the ejecta strike sensor walls, grid wires, and other surfaces, they produce secondary ions, electrons, and debris which, in turn, can strike more surfaces and produce additional ions and electrons. This production of charged particles is generally known as impact ionization. Its detection yields the greatest sensitivity (into the attogram region), and information on the speed and chemical composition of the particle can also be obtained. It has been employed in numerous space instruments (see Table 5 in section III).

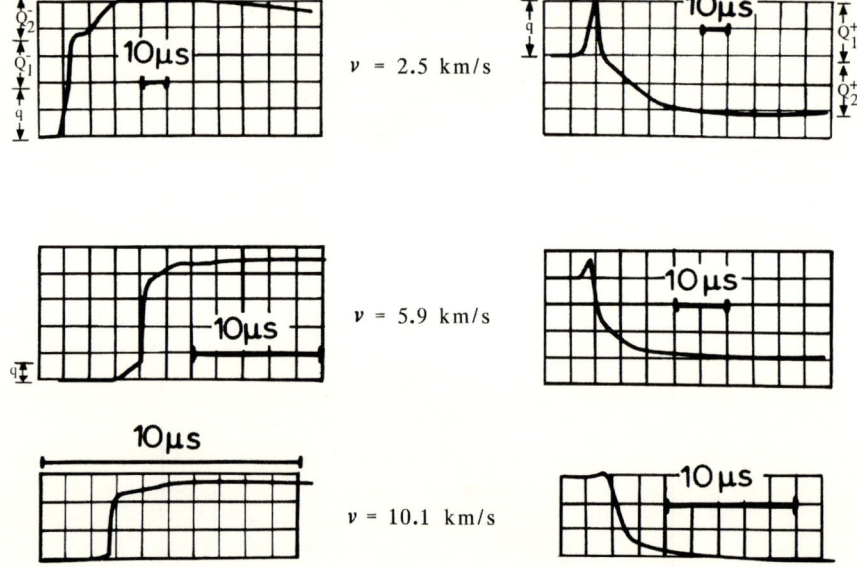

Figure 4. Oscilloscope traces of signals from impact ionization (Fe striking a plane target of W) at different impact speeds. The first, rising part of each signal is charge induced by the charge of the projectile as it approaches the target. The polarity of the second part of each signal depends on the direction of the electric field in front of the target: on the left panel, Fig. 4a, electrons and negative ions leaving the target cause the signal to swing further positive, whereas on the right panel, Fig. 4b, positive ions leaving the target cause it to swing negative. Note how the rise time decreases with increasing speed.

Impact ionization has been studied over a wide range of impact speeds from 50 m s^{-1} to 90 km s^{-1}. Figure 12 schematically shows the HEOS 2 dust sensor as an example of an impact ionization detector. In a typical laboratory set-up, positively charged microspheres from an electrostatic dust accelerator strike a target. By applying an electric field, positive and negative charge carriers are separated: electrons and negative ions are collected on the target and positive ions are drawn to a negatively biased ion collector. Typical signals from the target are shown in Fig. 4b (right panel of Fig. 4). When the bias voltage is reversed, one obtains signals as shown in Fig. 4a (left panel of Fig. 4).
Each signal consists of three distinct parts:

1. An initial positive-going linear ramp that starts when the particle passes through the entrance grid. It indicates the charge induced on the target by the positive charge, q, of the approaching dust particle. The impact occurs at the break in Fig. 4a and at the positive peak in Fig. 4b. At the time of impact, the charge flows onto the target.

2. A pulse with a short rise time $(1\ldots 2\ \mu s)$ immediately following the impact. It is positive-going for negative charges moving away from the target

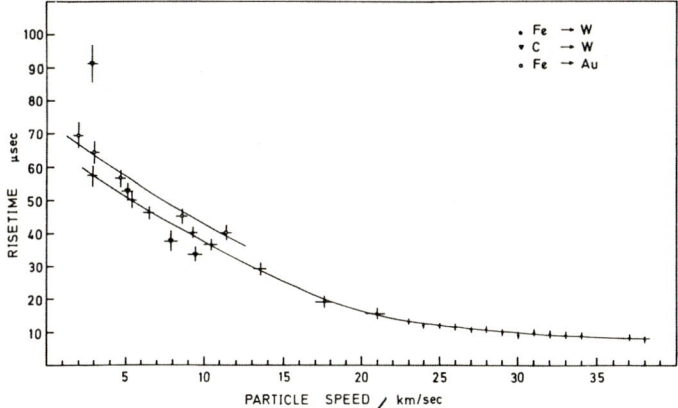

Figure 5. Rise time of the ion pulse of the HEOS 2 sensor versus particle speed (from Dietzel et al. 1973). Note: a different geometrical configuration may produce a different relation.

(Fig. 4a) and negative-going for positive charges (Fig. 4b). This first part of the pulse, Q_1, is caused by charges emitted from the *primary* impact crater.

3. Usually, there follows a pulse having a longer rise time (3...200 μs, depending on the impact speed) and the same polarity as Q_1. This second part of the pulse, Q_2, is caused by charges emitted from *secondary* (*tertiary*, ...) impact craters. For any given impact speed v, the amplitude (Q_2) and rise time of the third part are functions of the geometry and material of adjacent surfaces and the applied electric field: the more confined the space around the target, the shorter is the rise time and the larger the amplitude Q_2; vice versa, if the closest surface is at a large distance from the target, the pulse rise time is long and the amplitude small. Significantly, for any given experiment configuration the rise time depends on the impact speed: if v is high, the rise time is short; if v is low, the rise time is long. The same holds for the combined detector signal, $Q = Q_1 + Q_2$. This relationship is the result of two facts: (1) the speed of the debris increases with v and (2) the ratio of primary to secondary ionization also increases with v (Eichhorn 1978b; Ratcliff et al. 1996).

Thus, the rise time $\tau = \tau(v)$ of the detector signal is a function of the impact speed v, independent of the mass of the dust particle, and can be used to determine v directly from τ (Auer and Sitte 1968). Because it depends on the instrument's geometry, $\tau(v)$ must be determined for each setup. The particular function $\tau(v)$ for the HEOS 2 dust experiment is shown in Fig. 5.

The total emitted charge, Q, is generally expressed as a power law:

$$Q = |Q^-| \approx Q^+ = K m^\alpha v^\beta \ . \tag{2}$$

Table 2

Negative charge Q^- from impact ionization on different targets at normal incidence.

Target	Q^- [C]	v-range [km s^{-1}]	$Q^{-(1)}$ [fC]	References
W	$1.6 \cdot 10^{-11} mv^{3.5}$	0.05...40	1600	Dietzel et al. (1973)
Au	$1.0 \cdot 10^{-20} mv^{5.6}$	9...51	260	Grün (1984)
Al	$5.5 \cdot 10^{-18} mv^{4.8}$	8...46	90	Grün (1984)
PCB-Z (white paint)	$2.4 \cdot 10^{-15} mv^{4.1}$	3...36	60	Grün (1984)
H$_2$O ice	$1.6 \cdot 10^{-12} m^{0.8} v^{2.5}$	8...60	10	Timmermann et al.(1991)
100 nm thick Al foil	$1.3 \cdot 10^{-24} mv^{6.2}$	10...51	8	Auer (1994a)
Si aerogel (8 kg m^{-3})	$8.5 \cdot 10^{-23} mv^{5.5}$	27...80	0.8$^{(2)}$	Auer (1998)

$^{(1)}$ for iron projectiles having $m = 1$ pg and $v = 10$ km s^{-1}
$^{(2)}$ extrapolated to $v = 10$ km s^{-1}

Negative (Q^-) and positive (Q^+) charges are roughly equal in magnitude, although $|Q^-|$ often exceeds Q^+ by a factor of up to 3. With Q and v known, one can determined m from Eq. (2). A mass exponent $0.7 < \alpha < 1$ has been reported by some researchers (e.g., Auer and Sitte 1968; Dietzel et al. 1972; Timmermann and Grün 1991), and also the desorption ionization model (Eqs. (3) and (4)) leads to $0.5 < \alpha < 0.93$, but usually $\alpha = 1$ has been chosen as an adequate approximation. The charge yields of diverse target materials, when struck by spherical iron particles from the Heidelberg electrostatic dust accelerator, are listed in Table 2.

While Table 2 is for normal incidence only, Dietzel et al. (1972) reported that Q^- strongly increases with increasing angle of incidence (measured from the surface normal), reaching a maximum at 0.79 rad, with $Q^-(0.79 \text{ rad}) = 4.1\ Q^-(0 \text{ rad})$, then decreasing to $Q^-(1.20 \text{ rad}) = 1.6\ Q^-(0 \text{ rad})$. Svedhem et al. (1992) reported a similar increase of Q^- with the angle of incidence. This angular dependence must be taken into account when calibrating a flight sensor and when reducing flight data of impacts from unknown directions.

An entire spacecraft surface can occasionally serve as a large impact ionization target with an electric field antenna acting as a charge detector. This method has been discussed in detail by Oberc (1996). It enabled, for example, Gurnett et al. (1983, 1987) to estimate the mass and size distributions of dust particles near Saturn and Uranus.

2. Models of Impact Ionization

The processes involved in impact ionization are rather complex and not yet well understood. Not all details are accessible to experimental investigations. Laboratory facilities are severely limited in simulating the wide ranges of mass, speed, density, shape, and composition of interplanetary dust particles. In order to explain the experimental observations and to predict the response of a specific detector, even when not all impact conditions can be achieved in the laboratory, various assumptions have been made and models of ion formation have been developed.

The Plasma Ionization Model. When a dust particle impacts on the surface of a solid target with supersonic speed, shock waves propagate into both particle and target, thereby compressing some material to several times its normal density and converting impact energy into heat. When the shock front arrives at the opposite free surface of the particle (or of the target if it is a thin foil), rarefaction waves propagate back into the heated material. Thus, the internal energy produced by the shock is converted to expansion energy and the material expands adiabatically into the vacuum.

Since the dissipated energy per molecule may exceed many eV per degree of freedom and since the collision frequency is high because of pressures in the TPa region, one often describes this material as a very hot, dense plasma of ions, electrons, and neutral atoms populating quasi-equilibrium states. The degree of ionization of each atomic species can be described by a Saha-equation, assuming detailed balance for equilibrium. The temperature is the main parameter controlling the distributions of various ion types, independent of the heating mechanism. The plasma is assumed to expand adiabatically and to recombine in part during the expansion. When the pressure of the expanding material is insufficient to maintain the high collision rates of the quasi-equilibrium, the state of ionization is frozen (Drapatz and Michel 1974).

The plasma model can describe many features of the ion formation, but not all. Ion mass spectra predicted by this model contain species with multiple charges and high kinetic energies (up to 1 keV), especially at higher impact speeds (above $40\ldots60$ km s^{-1}). However, in experiments on the Heidelberg dust accelerator, with impact speeds ranging up to 64 km s^{-1}, no ions with multiple charges or with energies higher than about 100 eV were detected. On the other hand, a number of cluster ions were present in all mass spectra generated by Fe particles impacting on Ag, such as H_2^+, CH^+, $C_2H_3^+$, Fe_2^+, $FeAg^+$, Ag_2^+ and Ag_3^+ ions or negative H^- and CN^- ions, which the plasma ionization model does not predict.

The Desorption Ionization Model. The apparent similarity of the mass spectra obtained from dust impacts and non-equilibrium processes leading to methods such as laser-induced desorption, ion bombardment, or electric pulse induced desorption led Knabe and Krueger (1982), Kissel and Krueger (1987), and Krueger (1996) to develop a semi-empirical approach toward an understanding of impact ionization.

Desorption ionization is understood as a fast, i.e., far from equilibrium, dissipation of energy on the *surface* of a solid that leads to the release of preformed ions from the surface. The abundance of ion species is nearly independent of the method of primary excitation. A non-equilibrium desorption, i.e., solid/vacuum phase transition, may occur. If large surface areas are affected, because of a large primary interaction radius or very high energy density, the desorbed or otherwise produced ions appear as if they were formed under quasi-equilibrium conditions.

The strength of the excitation determines only the yield of ion formation but not the composition of the mass spectra. The mass spectra depend only on the chemical and physical conditions of the surfaces of target and particle. Varying the excitation parameters, e.g., the energy deposition density, will hardly change the ion types—even large molecules can appear—but determine the quantity of ions released.

Krueger (1996) developed semi-empirical formulae for the charge of ions of the particle material (iron):

$$Q_p = c_{Fe} C_p m_p^{0.93} \left(\frac{\rho_t}{\rho_p}\right)^2 F_1(V); \text{ with } c_{Fe} C_p = 10^{(7.0\pm0.5)} \text{ ions}, \quad (3)$$

for the charge of ions of the target material (in the PUMA instrument only):

$$Q_t = C_t m_p^{0.5} F_2(V); \text{ with } C_t = 10^{(7.5\pm0.5)} \text{ ions}, \quad (4)$$

where the particle mass m_p is in [pg], the reduced shock speed V is approximated by $V = v(\rho_p/\rho_t)^{0.5}$, ρ_p and ρ_t are the densities of projectile and target materials, respectively, and $F_1(V)$ and $F_2(V)$ are given in Fig. 6. The shape of $F_1(V)$ reflects the limited number of atoms in the projectile and the shape of $F_2(V)$ the quasi-unlimited number of atoms in the target. When the particle consists of several elements, the charge of ions of the particle material is:

$$Q_p = \sum Q_{pi} = \left(\sum s_i c_i\right) C_p m_p^{0.93} \left(\frac{\rho_t}{\rho_p}\right)^2 F_1(V), \quad (5)$$

where the summation is taken over all elements, and s_i is the mole fraction ($\sum s_i = 1$) and c_i the relative ion yield of the i-th element, respectively (as tabulated by Krueger, 1996). The particle's mean density and mass are obtained from:

$$\rho_p = C_1 \rho_t \left[\frac{Q_t^{1.55}}{\sum (Q_{pi}/c_i)}\right]^{0.25}, \quad (6)$$

$$m_p = C_2 \frac{Q_t^{1.7}}{\rho_p^{2.5}}, \quad (7)$$

where $C_{1,2}$ can be calculated from C_p and C_t. The total ion charge is $Q = Q_p + Q_t$.

The Volume Ionization Model. At higher speed, the surface mechanisms are still present, but now the particle completely vaporizes and also (for $v > 50$ km s^{-1}) ionizes throughout its entire volume because of the large dissipation of energy during the impact. This *volume* ionization then exceeds the surface effects if the particle is not too small ($m > 1$ fg). Hornung and Kissel (1994) and Hornung et al. (1996) developed a model that explains impact ionization hydrodynamically and mainly uses the Thomas-Fermi equation of state for the

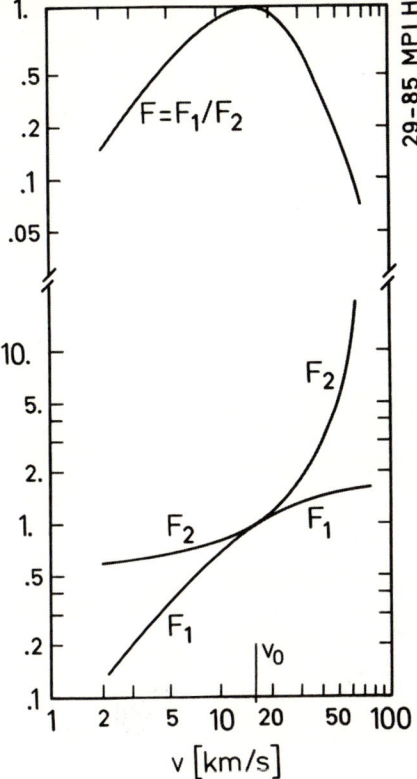

Figure 6. The functions $F(V)$, $F_1(V)$, and $F_2(V)$ used in the desorption ionization model, versus the reduced shock speed $V = v(\rho_p/\rho_t)^{0.5}$. The shape of $F_1(V)$ reflects the limited number of atoms in the projectile and the shape of $F_2(V)$ the quasi-unlimited number of atoms in the target (from Kissel and Krueger, 1987).

high temperature (0.1...1 MK) high pressure (0.1...10 TPa) states in the dust and in the target as well as for the expansion isentrope. The most important feature of the volume ionization model is that there are no unique values for the ionization efficiencies, E_i, of the elements in a dust particle of given mass and density since each E_i also depends on the concentrations, s_i, as well as on the ionization behavior of all chemical species: $E_i = E_i(s_1, s_2, \ldots, s_k, m_p, \rho_p)$. The $s_1, s_2, \ldots, s_k$, however, are also unknown. Therefore, one starts an iterative computational loop with crude estimates of E_i and s_i and varies the concentrations s_i of all chemical elements until all measured and theoretical numbers of ions coincide.

By applying this model to flight data of the PUMA 1 instrument, Hornung and Kissel (1994) showed that it gives a meaningful interpretation of the elemental composition of the dust particles, which is in accord with previous data reduction schemes based on the desorption model.

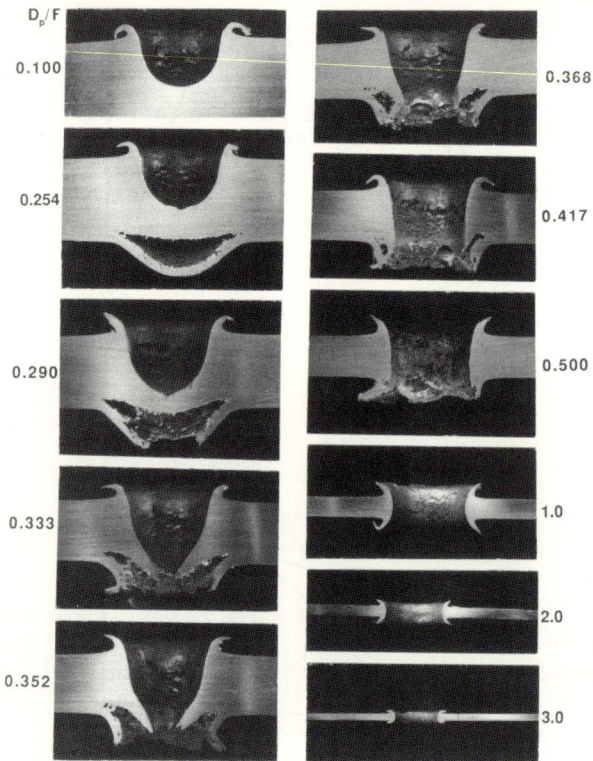

Figure 7. Cross sections of craters and penetration holes made in aluminum 1100 plates of thickness f by soda-lime glass projectiles having speed $v = 6$ km s^{-1} and diameter $d_p = 3.2$ mm, from $d_p/f = 0.1$ (thick target) to $d_p/f = 3$ (thin foil). Note the marginal penetration at $d_p/f = 0.33$ and the decreasing hole size for $d_p/f > 0.5$ (photograph provided by courtesy of F. Hörz).

II.E. Thin-Foil Penetration

When a fast particle strikes a thin foil (foil thickness f < particle diameter d_p), both particle and foil materials are compressed and shock waves propagate away from the moving contact surface. As the shock front reaches the rear surface of the foil, a rarefaction wave, if strong enough, relieves the pressure by bulging the rear surface or blowing off spall. Figure 7 (from Hörz et al. 1994, 1995) shows partly to fully penetrated foils. The foils shown are aluminum, cross-sectioned after being struck by glass spheres. At $d_p/f = 0.1$, a crater is formed that looks like a standard impact crater in a semi-infinite target; at $d_p/f = 0.25$, the opposite surface buckles; at $d_p/f = 0.29$, spall is blown off the buckled surface but the foil remains gas-tight; at $d_p/f = 0.33$, the ballistic limit is exceeded and the gas seal broken; as d_p/f increases further to $d_p/f = 0.5$, the penetration hole opens up and its back looks increasingly like its front including

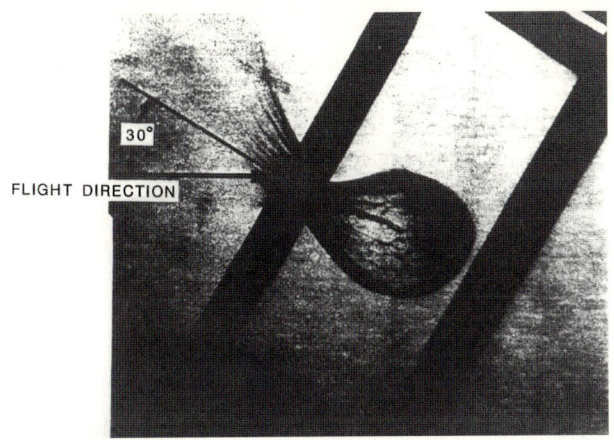

Figure 8. Flash x-ray shadowgraph of an oblique (0.52 rad from surface normal) impact at 6.4 km s^{-1} of a 3.2 mm Cd sphere on a sheet of 1.2 mm Cd, taken 7 μs after the impact. Note the change in the direction of the propagating mass after penetration (from McMillan, 1968).

the characteristic lip on the rim of the crater; at $d_p/f > 1$, the hole diameter gradually decreases until it is almost equal to d_p at $d_p/f \approx 100$ (virtually independent of the impact speed, according to Hörz et al. 1995). If the impact occurs at an angle to the foil normal, the bulk of ejecta emerges from the opposite surface, generally not in the original flight direction but in a direction somewhere (depending on d_p/f) between the impact direction and the surface normal, as illustrated in Fig. 8 (from McMillan 1968).

All phenomena known from the impact on a semi-infinite target also occur during thin-foil penetration, namely: damage to the particle; removal of foil material; ejection of solid, liquid, and/or gaseous material; secondary (and tertiary) cratering on nearby surfaces; ionization; light flash; and momentum transfer. In addition, after having penetrated the foil, the particle propagates at a reduced speed, possibly in a different direction than before, and material, including plasma, is ejected from both front and rear faces of the foil.

Before considering individual detection methods that use thin foils, the reduction of speed will be briefly discussed, because it is undesirable for thin-foil velocity detectors and desirable for the intact capture of particles by a stack of foils. Also, the condition for total particle destruction by a foil is presented, because the foil can serve as a "bumper shield" to protect a space structure from damage by meteoroid impacts.

Reduction of speed. During penetration, the particle is subjected to a brief (≈ 1 ns), yet very strong, deceleration ($10^{11} \ldots 10^{14}$ m s^{-2}). As its mass m_p picks up foil material m_f (= mass of foil intercepted by the geometrical cross section of the particle), the principle of momentum conservation, $m_p v_p = (m_p + m_f)(v_p - \Delta v)$, leads to:

$$m_p(v_p - \Delta v) = 1 + \frac{3f\rho_f}{2d_p\rho_p}, \qquad (8)$$

where v_p and $(v_p - \Delta v)$ are the particle's speed before and after the penetration, respectively, ρ_p is its density, d_p its diameter, and ρ_f and f are the density and thickness of the foil, respectively. Experimentally, Grün and Rauser (1969) found the speed loss Δv in Al foils to be proportional to v_p over the velocity range from 2 to 12 km s^{-1} and the relative speed loss $\Delta v/v_p$ to depend only on the particle mass, in agreement with Eq. (8). Note, however, that the above relations are invalid for some foil materials; for example, the deceleration in Au foils is greater than one would expect on the basis of momentum conservation (Grün and Rauser 1969).

Total Destruction of the Projectile. Whipple (1947) described an arrangement called "bumper shield", which can protect a spacecraft from damage by dust particles, e.g., the Giotto spacecraft encountering the dusty environment of comet Halley. This "bumper shield" is a thin foil placed at a distance from the main structure. The principle is, first, the destruction of the dust particle and, second, the spreading of the impact debris (momentum, energy) over a wide area of the spacecraft structure. According to Swift et al. (1983), a hypervelocity (> 3 km s^{-1}) dust particle is totally destroyed and vaporized by a bumper foil having a thickness f that satisfies the formula:

$$\frac{3f\rho_f}{2d_p\rho_p} > 0.25\ldots0.33 \qquad (9)$$

The design of bumper shields has been discussed in more detail by Cour-Palais and Crews (1990), Christiansen (1993), and Christiansen and Kerr (1993).

1. Perforation of Pressurized Cell

One of the first dust detectors, yielding reliable near-Earth dust fluxes, was the thin-walled pressurized "beer-can" cell on the Explorer 16 satellite, flown in 1962. A decade later, a similar detector was put on the Pioneer 10 and 11 spacecraft to Jupiter and beyond. This detector type registers an impact when the cell's gas pressure (monitored by a pressure switch) drops, because gas leaks into space through a penetration hole in the cell wall. The objective was to determine the nature of the particles that were able to penetrate such walls. By employing walls with increasing thicknesses and assuming particle speeds and densities, one estimated the cumulative fluxes of particles with increasing masses.

Calibrations, using the dust accelerators described in section IV.A and extrapolations from a limited set of data, led to empirical relations between the properties of a foil (thickness f, density ρ_f, and tensile strength σ_f) and the properties of the particle (diameter d_p, density ρ_p, speed v, and tensile strength σ_p) that is marginally able to perforate the foil. Results by McDonnell (1970) are depicted in Fig. 9 for iron microspheres striking various materials. According to McDonnell and Sullivan (1992), the following formula applies over a wide range of velocities ($4\ldots16$ km s^{-1}) and dimensions (μm to cm):

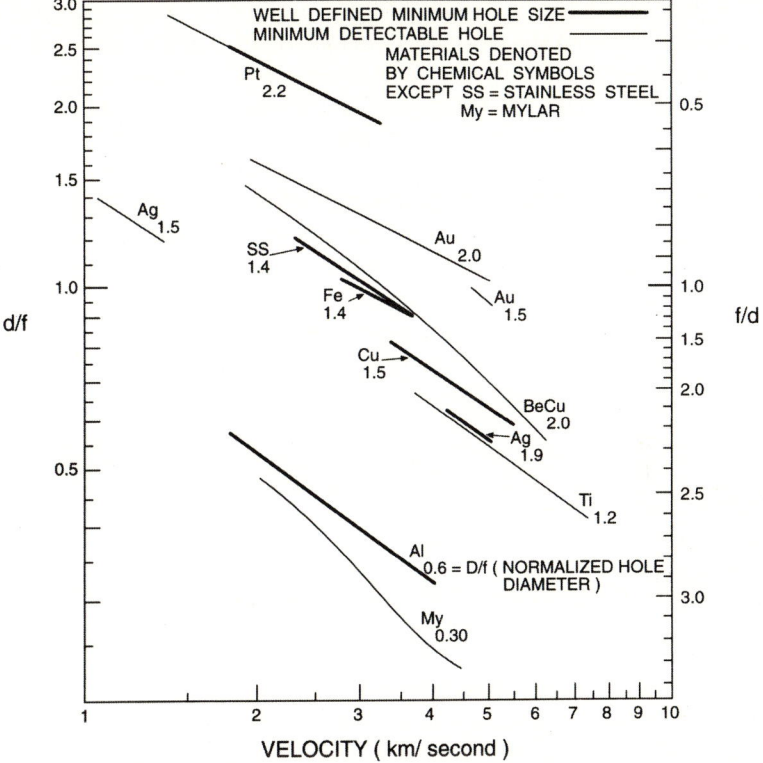

Figure 9. Marginal penetration of Fe microspheres on metal foils of different thicknesses: relative projectile diameter d/f as a function of impact velocity. D indicates the smallest detectable hole (modified from McDonnell, 1970).

$$f = 0.0063 d_p^{1.056} \left(\frac{\rho_p}{\rho_{Fe}}\right)^{0.476} \left(\frac{\rho_{Al}}{\rho_f}\right)^{0.476} \left(\frac{\sigma_{Al}}{\sigma_f}\right)^{0.134} (v \cos \alpha)^{0.806} , \quad (10)$$

with $\sigma_{Al} = 80$ MPa and $\alpha =$ angle between impact direction and surface normal.

2. *Capacitor Discharge*

When a fast dust particle penetrates the surface electrode of a charged capacitor and strikes the dielectric, the dielectric behind the shock wave front is compressed and may become conductive because of the high pressure (Zel'dovich 1968), high temperature (impact ionization), or locally increased electric field strength. As a result, a conductive path through the dielectric may form through which the capacitor is discharged. This capacitor discharge has been the basic mechanism of the thin-foil detectors flown on the Pegasus spacecraft (Naumann et al. 1969; see section III.A) and of the Metal-Oxide-Silicon (MOS) detectors (Kassel 1973) flown on Explorer 46 and the Long Duration Exposure Facility (LDEF) (Singer et al. 1985).

Each Pegasus detector was shielded by an Al plate. Only dust particles or spall which penetrated the shield were detected. In contrast, an MOS detector is unshielded and the surface electrode is made as thin as possible so that very small particles ($d_p \sim 0.2$ μm) can penetrate. It consists of a thin (0.4...1 μm) SiO_2 dielectric on a silicon substrate and, on top of the dielectric, a 0.1 μm thick aluminum film that acts as the outer electrode. A bias voltage of at least 30 V generates a high electric field (≈ 100 MV m^{-1}) across the dielectric. If an impacting particle triggers a discharge, a voltage drop across the capacitor occurs within nanoseconds, which can be readily detected. The discharge is accompanied by an audible crack and a visible flash. While the discharge current flows, aluminum from the outer electrode is removed in an area (depending on the stored energy) usually much larger than the area of the actual impact crater. As a result, the conductive path is cleared and the detector re-conditioned.

According to Kassel (1973), the sensitivity of an MOS detector with a 0.4 μm thick dielectric is given by the relationship:

$$d_p v^2 = 2.8 \, , \qquad (11)$$

where d_p is the minimum diameter of an iron test particle capable of triggering a discharge and v ($= 1.1...2$ km s^{-1}) is the component of the particle velocity normal to the detector surface. The sensitivity is inversely proportional to the combined thickness of the dielectric and outer electrode.

An MOS detector vaporizes an impacting dust particle, and some detector material, and ionizes the vapor with an efficiency that may be close to 100 %. The high ionization efficiency is a result of the large amount of energy that is stored on the detector (approximately 10^{-7} J) and discharged through the compressed material, adding about 10 keV/atom of the impacting particle to the available energy (the kinetic energy is typically of the order of 10 eV/atom). Utilizing this high ionization efficiency, Auer and Berg (1975) combined an MOS detector with a double-focusing ion time-of-flight mass spectrometer to analyze the composition of dust particles.

Because an MOS detector contains Al, O, and Si, large quantities of these elements are always present in an MOS ion mass spectrum and cannot be distinguished from Al, O, or Si of the particle material. In order to quantitatively measure all elements of the particle, a detector would have to be developed consisting exclusively of elements uncommon in interplanetary dust, such as a large-area, reverse-biased germanium diode. Because of the high ionization efficiency, such a detector might enable one to quantitatively measure all elements of the particle material and thus its mass, independent of other particle parameters such as impact speed, density, or composition.

3. Impact Ionization

When a particle penetrates a thin foil ($f < d_p$), ions and electrons are emitted from both ends of the penetration hole and from the particle. Also, neutral or charged spallation products may be ejected.

By utilizing the impact ionization from a thin foil, Berg and Richardson (1969) detected dust particles on Pioneer 8 (see section III.B and Fig. 11). Positive charges were collected on the negatively biased (-3.5 V) foil and negative charges on the positively biased ($+24$ V) grids of the front sensor array. A pair of additional grids, more negatively biased (-7 V) than the foil, prevented negative charges from escaping the system.

For the quantity of negative charges produced by impacts on a 100 nm thick Al foil see Table 2.

4. Depolarization

A depolarization sensor detects an impacting high-velocity dust particle by means of local removal of polarized material of the foil volume. A typical sensor consists of a thin (≥ 1.5 μm) foil of permanently polarized polyvinylidene fluoride (PVDF) or vinylidene fluoride/trifluoro-ethylene (PVDF copolymer). A high-velocity ($v \gtrsim 1$ km s^{-1}) dust particle, impacting the sensor, produces very rapid removal of dipoles from within and near the PVDF volume destroyed (impact crater or penetration hole) by the particle, as shown in Fig. 15. This local volume depolarization results in a fast current pulse (ns range) in the external circuit. The theory and details of PVDF and PVDF copolymer dust sensor operation have been described by Simpson and Tuzzolino (1985).

A PVDF based sensor has several advantages over other impact detectors: it requires no bias voltage, is simple, inexpensive, reliable, electrically and thermally stable, mechanically rugged, radiation resistant up to 100 kGy, does not respond to high fluxes of charged nuclei, has a large dynamic range ($1 : 10^6$) and fast response (up to 10 000 random impacts/s with no corrections and 100 000/s with known corrections). Special sensors have been developed which measure x, y coordinates of particle impact ($\pm 1 \ldots 3$ mm); and two planar sensor arrays permit measurements of particle velocity and trajectory (Simpson and Tuzzolino, 1989). PVDF sensors operated successfully on the VeGa 1 and 2 missions to comet Halley (DUCMA, see section III.E.2) and on the ERIS mission to study the effects of an energetic collision in space and are also a subsystem of the Cassini Cosmic Dust Analyzer (section III.H).

The output signal Q of a sensor having a volume polarization P (typically $P = 0.047$ C m^{-2} for PVDF and $P = 0.061$ C m^{-2} for PVDF copolymer) is approximately:

$$Q = PD^2 \frac{\epsilon p/f}{1 + (\epsilon - 1)p/f} \frac{\pi}{4}, \qquad (12)$$

where p is the depth of the crater formed, D its diameter, f the foil thickness, and $\epsilon \approx 12$, the dielectric constant of PVDF. When the particle penetrates the foil ($p = f$), Eq. (12) becomes:

$$Q = PD^2 \frac{\pi}{4}. \qquad (13)$$

Specifically, from experiments at the Heidelberg dust accelerator with non-penetrating iron particles having mass m and speed v ($2 \ldots 11$ km s^{-1}), Simpson and Tuzzolino (1985) found the response of a 28 μm thick PVDF detector to be:

$$Q = 4.8 \cdot 10^{-7} m^{1.3} v^{3.0}, \qquad (14)$$

whereas from experiments at the Munich dust accelerator with penetrating glass particles (Simpson et al. 1989), the response of the same detector and in the same speed range was:

$$Q = 3.9 \cdot 10^{-5} m^{0.90} v^{1.05}. \qquad (15)$$

Results obtained with 1.5...3.5 μm thick PVDF copolymer detectors were reported by Tuzzolino (1992).

II.F. Momentum

The very first rocket-borne dust detector (Bohn and Nadig 1950) was a microphone, or "momentum sensor". Based on its data, very high near-Earth dust fluxes were derived, which caused much concern about the safety of satellites and led to the hypothesis of a "dust belt around the Earth". Later it was found, however, that microphones also respond to thermal gradients in the instrument itself and to solar protons. Therefore, data from these sensors, by themselves, are considered unreliable (Berg and Grün 1973).

The strength of a momentum detector lies in its mechanical simplicity and low weight. As part of the Dust Impact Detection System (DIDSY), several piezoelectric crystals were attached to the dust shields of the Giotto spacecraft to comet Halley (McDonnell et al. 1986 a). Detectors with sensitive areas as large as the spacecraft were, in effect, made by means of crystals weighing only a few grams.

Three types of momentum detectors have been used on spacecraft: thick piezoelectric crystals, thin (typically 0.1 mm) piezoelectric diaphragms, and capacitor microphones. Detection thresholds vary from $3 \cdot 10^{-12}$ kg m s^{-1} (McDonnell and Abellanas 1972) to 10^{-3} kg m s^{-1} (McDonnell et al. 1986b), depending on the specific configuration of the experiment.

Only at low speed ($v < 0.55...1$ km s^{-1}) is the response of a momentum detector proportional to the momentum, mv, of the dust particle. At higher speed, the recoil from gaseous, liquid, or solid ejecta produces additional momentum beyond that carried by the incoming particle. As a result, the momentum imparted to the target is enhanced by a factor, E, that increases with v. Based on empirical data at $v = 1...8$ km s^{-1}, McDonnell et al. (1984) expressed the enhancement factor as:

$$\text{(momentum transferred)}/(mv) = E(v) = 1 + (0.0005v)^2. \qquad (16)$$

Stradling et al. (1990) determined E for speeds $v = 5...21$ km s^{-1} using the Los Alamos 6 MV electrostatic dust accelerator. Based on their data, Beard (1991) modified Eq. (16):

$$E(v) = 0.96 + 0.00015v. \qquad (17)$$

Weishaupt (1987) developed a method to determine both the diameter and momentum of a particle from the first rise time and amplitude of the signal from a piezoelectric plate. This rise time increases with the travel time of the pressure waves through the projectile and, thus, with the projectile's diameter. Results of laboratory tests with both slow (60 m s^{-1}) and fast (6.6 km s^{-1}) particles agreed well with his calculations.

When dust is encountered near a comet or in a planetary ring, the momentum transfer may noticeably decelerate an entire spacecraft. By analyzing the Doppler frequency and amplitude of the radio signal from the Giotto spacecraft, Edenhofer et al. (1986) detected the impact of a single large particle ($m = 40$ mg) as well as the total mass of all particles that hit the spacecraft during its encounter with comet Halley.

II.G. Velocity, Trajectory, and Orbit

A dust particle's *velocity*, relative to the spacecraft, is needed in order to calculate its mass and most recent orbit. Its *trajectory* is needed in order to determine the coordinates of the point of impact on a capture medium.

The impact speed, $v = |\mathbf{v}|$, can be obtained from the rise times of the charge signals of an impact ionization detector, as was discussed in section II.D.1 and shown in Fig. 5. This indirect method, used on HEOS 2, Helios, Galileo, Ulysses, and Hiten, is accurate within a factor of about 2. Another indirect method (Grün 1981) derives v from the ratio of negative-to-positive charge amplitudes as $v = 2400(Q^-/Q^+)^{0.59}$. Using these indirect methods, however, the direction of $\mathbf{v}$ remains rather uncertain, only known to be within the field of view of the instrument (typically $\approx \pm 1$ rad).

For a more accurate determination of the *velocity* vector $\mathbf{v}$, the coordinates of two points on the particle's trajectory and the time of flight between those points are to be measured. Any such measurement requires some kind of interaction with the particle. At least the first of the two interactions should be so weak as to not noticeably alter the particle's velocity that is to be measured. Interactions with light (see section II.A) or by charge induction (see section II.B) fulfill this requirement and are discussed in more detail below. However, a mechanical interaction with a thin foil, as discussed in section II.E, reduces the velocity and may alter the trajectory of the penetrating particle (Auer 1972; Peterson 1994; and Auer 1994b) and thus tends to defeat the objective, the measurement of velocity and trajectory.

The Pioneer 8 cosmic dust sensor by Berg and Richardson (1969) (see section III.B) did use a thin foil in front of a thick target. Both foil and target were, in effect, coarse (4×4) x, y position detectors of the impact ionization type, the combination being a coarse velocity/trajectory sensor (± 10 % in speed and ± 0.47 rad in angle). Changes in magnitude and direction of $\mathbf{v}$, which the mechanical interaction with the front foil may have caused, were small enough to be ignored. However, many small particles were unable to penetrate the front foil, thus, their speeds could not be determined at all.

A *non-destructive charge/speed/position detector* was introduced by Shelton et al. (1960) for positioning the beam of charged dust particles from an electrostatic dust accelerator. It consists of a pair of parallel plates as was discussed in section II.B (Fig. 2).

Three *non-destructive charge/velocity/trajectory sensors* were described by Auer (1975) and Auer and von Bun (1994). Each employs four one-dimensional position-sensitive detectors in series, two for the x-coordinates and two for the y-coordinates, of points on the trajectory, of which each consists of a plane grid of metal wires. The first method utilizes a charge-division technique; the second a time-of-flight technique; and the third an array of charge-division detectors, each similar in essence to Shelton's with the plates reduced to thin wires.

A velocity sensor of the *time-of-flight* type, measuring v_y and v_z (but not v_x), is included in the Cassini CDA (section III.H) and is schematically shown as part of Fig. 18. The two outer grids and the walls are grounded whereas the two inner, inclined grids are together connected to a charge-sensitive amplifier. The charge pulse has two peaks, the amplitudes of which indicate the charge q of the particle. v_z, the component parallel to the sensor axis, can be calculated from the pulse duration (i.e., time of flight between the grounded grids), whereas the angle to the axis, $\arctan(v_y/v_z)$, can be calculated from the ratio of: the time interval between the peaks (i.e., time of flight between the inclined grids) and the pulse duration. Also, the distance of the trajectory from the sensor wall can be derived from the sag between the two peaks: the deeper the sag, the closer is the trajectory to the wall.

The third version, a *velocity sensor array*, was developed for the Cosmic Dust Collection Facility (CDCF). Mechanically, it is similar to the charge-division type except that multi-strand miniature spring-loaded wires are used instead of single wires (for reasons of reliability). The main difference is electrical: each wire is connected (in principle at least) to a separate amplifier; thus two adjacent wires act as one charge/speed/position sensor element, similar to Shelton's parallel-plate sensor. A schematic cross section is depicted in Fig. 10. The charge induced on the i-th wire is:

$$q_i \approx \frac{q/r_i}{\sum(1/r_j)}, \qquad (18)$$

where q is the particle's charge, r_i the perpendicular distance between the particle and the i-th wire, and the sum is taken over all j wires of the sensor (Auer and von Bun 1994). The sum of all induced signals, $\sum q_i$, approaches the charge q as the particle moves away from grounded walls and grids and approaches a wire. Using Eq. (18), the charge induced on any wire as a function of time can thus be calculated if the particle's charge, velocity, and trajectory are known. By an inverse procedure, an unknown particle's charge, velocity, and trajectory can be derived from the sensor signals, provided they are recorded in their entirety (detailed shape and amplitude) using high-speed transient recorders. The measurement accuracy at high signal-to-noise ratios was demonstrated to be at least 0.1 % in speed and 1.7 mrad in angle (Auer 1996). Even for a low

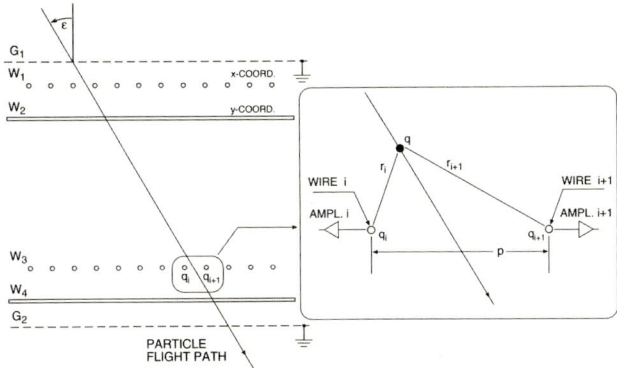

Figure 10. Schematic cross section of the charge/velocity/trajectory sensor array developed for the CDCF. It consists of two electrical shielding grids, G_1 and G_2, and four arrays W_k ($k = 1 \ldots 4$) of charge sensing wires. The charge q_i on each wire i of each array W_k is individually amplified and recorded as a function of time. Wires i and $i+1$, between which a particle passes (see insert), sense the strongest signals, q_i and q_{i+1}, and, thus, identify the particle's *coarse* position. The *fine* position can be derived from the ratio of the amplitudes, q_i/q_{i+1} ($\approx r_{i+1}/r_i$ from Eq. (18)). Signals from wires of W_1 and W_3 are analyzed to yield the x-coordinates of the particle's position along the trajectory as a function of time and, thus, its velocity component v_x. Similarly, y-coordinates and velocity component v_y are derived from the signals detected at W_2 and W_4 (from Auer, 1996).

signal-to-noise ratio of only 2.5:1 (particle diameter of $x \approx 70$ nm, see Table 1), an accuracy of about 0.3 % in speed and 3.4 mrad in angle has been estimated, based on a computer simulation (Auer and von Bun 1994).

Another non-destructive but coarse velocity sensor consists of arrays of light curtains and is to fly on the Rosetta mission (Perruchot et al. 1996; see section II.A.3). Also as a velocity sensor, two arrays of 4×4 depolarization detector elements have been produced for flight (Tuzzolino 1991, 1996). Both resemble the Pioneer 8 instrument (impact ionization detectors) but use different detection principles.

An impact position sensor consisting of a diaphragm with three or four piezoelectric detectors bonded to its corners was described by McDonnell and Abellanas (1972) and Igenbergs et al. (1987). The authors used the arrival times of the bending waves at the detectors to locate the impact within ± 1 mm.

A particle's *orbit* in a field of known forces can be calculated from its position, $\mathbf{r_p}(t)$, and velocity, $\mathbf{v_p}(t)$, e.g., in heliocentric coordinates. When the particle is detected at time t_0, the orbits of the spacecraft and the particle intersect, i.e., their positions coincide and the particle's position $\mathbf{r_p}(t_0)$ is:

$$\mathbf{r_p}(t_0) = \mathbf{r_s}(t_0). \tag{19}$$

In addition, the particle's velocity, $\mathbf{v_p}(t_0)$, can be written as:

$$\mathbf{v_p}(t_0) = \mathbf{v_s}(t_0) + \mathbf{v_m}(t_0) , \tag{20}$$

Table 3

Ranges of decelerations produced by various media.

deceleration medium	break length	typical deceleration [m s^{-2}] for $v_p = 10$ km/s	$v_p = 30$ km/s	references
atmosphere	5...50 km	$10^3...10^4$	$10^4...10^5$	chapters II and V
electrostatic field	5...50 m	$10^6...10^7$	$10^7...10^8$	Wolfe et al.(1986)
aerogel	0.5...500 mm	$10^8...10^{11}$	$10^9...10^{12}$	Zolensky et al.(1994)
thin foil[1]	0.05...5 μm	$10^{11}...10^{13}$	$10^{12}...10^{14}$	Capaccione et al. (1986)

[1] from Eq. (8), for penetrating particle, $d_p = 10$ μm, $\rho_p = \rho_f = 1$Mg m^{-3}, $f = 0.05...5$ μm

where $\mathbf{r_s}(t_0)$ and $\mathbf{v_s}(t_0)$ are the position and velocity vectors of the spacecraft, respectively, and $\mathbf{v_m}(t_0)$ is the measured velocity vector of the particle relative to the spacecraft. Because $\mathbf{r_s}(t_0)$ and $\mathbf{v_s}(t_0)$ are accurately known from spacecraft tracking data, only the impact velocity, $\mathbf{v_m}(t_0)$, is required for the calculation of the particle's orbit. While the calculation of a purely Keplerian orbit around a single central body from $\mathbf{r_s}$, $\mathbf{v_s}$, and $\mathbf{v_m}$ is straightforward, it becomes more complex when gravitation from several bodies and other forces, such as solar radiation pressure or the Lorentz force from a magnetic field, act on the particle. For more details, see the chapters on particle dynamics in this book.

II.H. Deceleration for Intact Capture

During a hypervelocity impact on a solid target, usually the structure of a dust particle is destroyed and often little of the dust material is left in the crater. In order to study the undamaged dust particle, its shape, morphology, and mineralogical composition, one has to capture it intact. The problem is how to deal with the particle's high velocity. Table 3 lists deceleration media (in use or proposed) and typical decelerations.

Over the past decade, much effort has gone into the utilization of low density silica aerogel as a capture medium. Water-clear aerogels with densities of only 10 to 50 kg m^{-3} have been used in space collections and tested on particle accelerators (e.g., Zolensky et al. 1994). Laboratory tests have shown that the lowest density aerogels can capture solid particles with diameters $d_p = 10...100$ μm at $v = 6$ km s^{-1}. Particles have been recovered from the aerogel, and they retained their fundamental physical properties such as mineralogical composition, morphology, and solar flare tracks. Most of the impacting particles formed narrow cone-shaped tracks roughly 100d_p long. In each track, the particle was found at the tip of the hollow cone. In some cases, projectiles fragmented into components, each producing a track. In a few impacts the projectile did not survive and produced more of a crater than a deep cone-shaped track.

Impact ionization from aerogel was found to be very weak (see Table 3), suggesting that ejecta are mostly captured beneath the surface of the aerogel. The near-absence of ejecta has potential applications for aerogel as the equivalent of a black body "absorbing" dust particles and emitting very little material.

II.I. Mass, Density, and Diameter

The *mass*, m, of a dust particle is usually determined from the amplitudes of the charge signals obtained from impact ionization, using Eq. (2), when the impact speed, v, is known. Since these amplitudes also depend on the particle's composition, density, and angle of incidence, which are generally unknown, the accuracy is only a factor of about 6...10. Perhaps a slightly better accuracy can be obtained when the composition is also measured (Eq. (7)). Similarly, m can be determined from the impact light flash, thin-foil depolarization, or momentum transfer signals, using the empirical relations given in sections II.C, II.E.4, and II.F, respectively, provided that v is known. The MOS/time-of-flight technique mentioned in section II.E.2 is potentially capable of obtaining an ion mass spectrum of nearly 100 % of the particle material and, from it, the mass of the dust particle. Also, Auer (1982) proposed to determine non-destructively the mass of a charged dust particle from the bending of its trajectory in a strong electric field, using high-resolution velocity/trajectory sensors.

The *density*, ρ_p, of a dust particle strongly affects its ability to penetrate a thin foil (Eq. (10)). This effect was used to obtain statistical information on particle densities from the Helios twin detectors (Grün 1981). Also, the density of an individual dust particle can be calculated from the ratio of target and particle material derived from the ion mass spectrum (Eq. (6)) if the impact speed is known. Krueger (1996) estimated this method to be accurate within a factor of 1.5.

The *diameter*, d, of course, is related to the mass and density discussed above. Independently, it can be estimated from the particle's charge if the surface potential is known (Eq. (1)). Grün (1981) interpreted the high charges of some dust particles found with the Helios experiment as an indication of large diameters. Also, optical detectors (section II.A.2 and II.A.3) measure the product of cross-sectional area and surface reflectivity of individual particles. For a given reflectivity, the diameter can be estimated. Finally, the size distribution of interplanetary dust has been derived from the zodiacal light, as discussed in part II of this book.

II.J. Chemical and Isotopic Composition

The greatest sensitivity is offered by an impact ionization detector combined with an ion time-of-flight (TOF) mass spectrometer, as employed on the Helios, VeGa, Giotto, and Cassini dust analyzers. Crude spectra of the positive ions were first observed by Auer and Sitte (1968) and Hansen (1968). The mass lines most abundant at low speed and easily identified were those of alkaline elements (especially Na, K). They originated primarily from impurities on the surfaces of the particle and target. Dietzel et al. (1973) took great care in procuring for the Helios instrument a target without impurities in the material or on its surface.

A TOF mass spectrometer is particularly well suited because (1) the time of ion formation is very short ($\lesssim 1$ ns), (2) ions are registered sequentially by a single ion collector or electron multiplier and, therefore, a complete spectrum is obtained for each impact of a dust particle, (3) the ion transmission can be high (up to 50 percent), (4) the mechanical design is relatively simple, and (5) mass and power consumption can be low. In operation, ions of both particle and target materials, produced by impact ionization, are drawn into the mass spectrometer by a potential difference U applied between the target and an acceleration grid. Once the ions have passed the grid, their kinetic energies are: $e_i(U + U_0) = \frac{1}{2} m_i v_i^2$, where $e_i U_0$ is the initial kinetic energy and e_i, m_i, and v_i are the charge, mass, and speed of the i-th ion species, respectively. Each species travels through the TOF drift space of length s with its characteristic speed, $v_i \propto (e_i/m_i)^{0.5}$, and arrives after a time, $t_i = s/v_i$, at the ion collector. The mass-to-charge ratio is:

$$\frac{m_i}{e_i} = \frac{2t_i^2}{(U + U_0)s^2}. \tag{21}$$

The initial energy, eU_0, tends to broaden the mass lines and can be counteracted by energy focusing. While the mass resolution of the first composition analyzer on Helios had been $m/\Delta m = 5\ldots 15$, it was increased to > 150 for the Giotto and VeGa missions and 10 000 for Rosetta.

Information on the composition of interplanetary dust can also be obtained from the spectrum and polarization of zodiacal light; possibly from a spectrum of the impact light flash; from the mass spectrum of ions produced by an MOS capacitor discharge; and from any of the analytical tools that are known in the laboratory and applied to captured particles, such as: energy-dispersive x-ray spectrometry (EDS), secondary ion mass spectrometry (SIMS), or laser microprobe mass analysis (LAMMA).

III. FLIGHT INSTRUMENTATION

A few dust instruments have significantly advanced the state of the art and are briefly discussed in the following sections. Tables 4 and 5 give an overview of these instruments with essential characteristics.

III.A. Explorer 16, Pegasus, and Pioneer 10: Large-Area Penetration Detectors

Not relying on the "ping-ping" detected by momentum sensors but on actual material perforation, these experiments established the near-Earth dust flux at levels still considered reasonable today and some 6 orders of magnitude below the levels of the early "dust belt around the Earth".

Table 4

Characteristics of penetration detectors.

Mission	Explorer 16		Pegasus I			Pioneer 10
Detection Technique	pressurized cell		capacitor discharge			pressurized cell
see section	II.E.1, III.A		II.E.2, III.A			II.E.1, III.A
Spacecraft Launch	Dec. 1962		Feb. 1965			Mar. 1972
Type of Orbit	geocentric		geocentric			heliocentric
Wall Material	copper-beryllium		aluminum			stainless steel
Sensitive Area [m^2]	1	0.4	7.5	16	171	0.26
Wall Thickness [μm]	25	51	38	200	410	25
Mass Threshold[1] [ng]	8	30	1	200	1000	8

[1] for $v = 20$ km s^{-1}, $\rho = 0.5$ Mg m^{-3}

The Explorer 16 satellite carried several penetration detectors, the most successful type being the "beer-can" pressurized cell. As discussed in section II.E.1, this cell detects the loss of gas pressure after perforation of its wall by a fast dust particle. Also utilizing the penetrating ability of dust, the Pegasus satellites employed capacitor discharge detectors (see section II.E.2). Named for the winged horse of Greek mythology, Pegasus was one of the largest spacecraft, visible from Earth. Its 29 m long and 4.3 m wide wing carried 208 panels of aluminum sheets of various thicknesses, with a thin foil of polymer plastic inside each aluminum sheet and a copper coating on the other side of the plastic. In Table 4 the wall (or shield) thickness, penetration threshold, and sensitive area are compiled for the Explorer XVI and Pegasus detectors and also for the Pioneer 10 "beer-cans" (which operated on the way to Jupiter and beyond). The dust detectors on Explorer have been described by Hastings (1964), those on Pegasus by Naumann et al. (1969), and those on Pioneer 10 by Meshejian et al. (1970) and Humes et al. (1974).

III.B. Pioneer 8: Reliable Coincidence Detector

Pioneer 8 was launched 13 December 1967 into a heliocentric orbit with perihelion and aphelion distances of 0.99 and 1.088 AU, respectively. It was spin-stabilized with the spin axis normal to the ecliptic plane and a spin period of one second. The cosmic dust experiment was mounted in the equator of the spacecraft with the axis of its field of view scanning the ecliptic plane. Thanks to a

Table 5

Characteristics of impact ionization and depolarization detectors.

Mission	Pioneer 8	HEOS 2	Helios 1	VeGa, Giotto	VeGa	Galileo	Hiten	Cassini
Instrument				PUMA, PIA	DUCMA	DDS	MDC	CDA
see section	II.E.3, III.B	II.D, III.C	III.D	III.E.1	II.E.4, III.E.2	III.F	III.G	III.H
Time of Operation from	Dec. 67	Feb. 72	Dec. 74	Dec. 84	Dec. 84	Dec. 89	Jan. 90	Oct. 97
to	70	Jan. 74	84	Mar. 86	Jan. 87	2001	Jan 92	June 08
Type of Orbit	heliocentric	geocentric polar	heliocentric	Halley fly-by	Halley fly-by	Jupiter orbiter	geocentric	Saturn orbiter
Heliocentric Distance [AU]	0.97 – 1.09	1	0.3 – 1	0.7 – 1.1	0.7 – 1.1	0.7 – 5.2	1	0.7 – 9.5
Sensitive Area [m^2]	0.01	0.01	2 · 0.012	0.0005	0.0075	0.1	0.01	0.1
Viewing Angle [rad]	2.1	2.1	1.20	—	—	2.4	2.6	1.7
Eff. Solid Angle [sr]	2.9	1.03	1.23	—	—	1.45	1.5	1.06
Mass Threshold [fg] at								
5 km/s	7000	14	14	—	—	120	180	120
10 km/s	1000	1.2	1.2	—	—	15	20	15
20 km/s	200	0.1	0.1	—	—	2	0.5	2
40 km/s	40	0.01	0.01	—	—	0.13	0.02	0.13
80 km/s	—	—	—	0.3	150	—	—	—
Dynamic Mass Range	230	10 k	10 k	100 k	100 k	1 M	10 k	1 M
Charge Threshold ± [fC]	—	—	10	—	—	10	100	1
Mass Spectrum [u]	—	—	16…69	1…110	—	—	—	1…120
Resolution	—	—	5…15	> 150	—	—	—	20…50

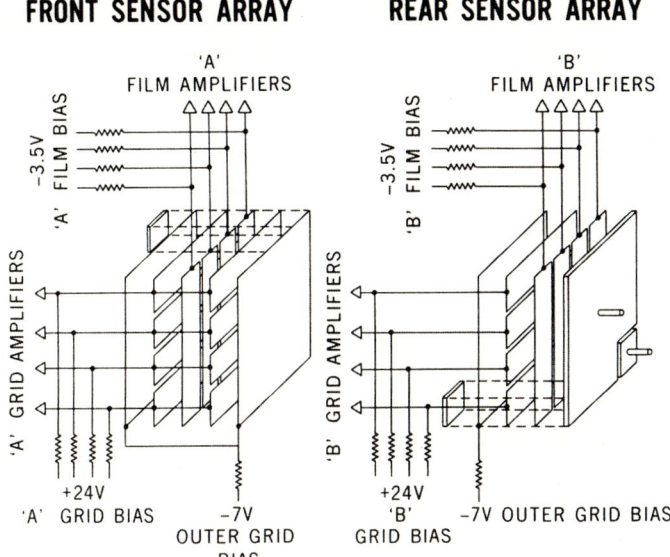

Figure 11. Schematic view of the Pioneer 8 dust sensor. The front sensor array consists of four thin-foil impact ionization detector strips and two sets of four grid strips, one in front and another behind the strips of foil. The rear sensor array consists of four thick-foil sensor strips, a set of four grid strips, and two target plates, each with an acoustic detector. A high reliability of detection results from the coincidence of impact ionization signals. Speed and directional information is obtained from the particle time-of-flight between front and rear array and the two impact locations. (from Berg and Richardson, 1969)

coincidence technique, this instrument generated the first reliable data of dust fluxes in deep space. It has been described by Berg and Richardson (1969), Berg and Grün (1973), and Grün et al. (1973). A similar experiment flew on Pioneer 9 since November 1968 and another one was deployed in December 1972 on the lunar surface by Apollo 17 astronauts (lunar ejecta and meteorites, or LEAM, experiment, see Berg et al. (1973)).

Figure 11 shows the schematics of the sensor. It consists of a front thin-foil impact ionization detector array and a rear solid-target impact ionization detector array. The arrays are 10 cm square and separated by 5 cm from each other in order to determine, with a time-of-flight technique, the speed of an impacting particle that has been registered by both detector arrays. The front detector array employs a 580 nm thick Parylene-C/metal composite foil. Positive charges produced upon impact or penetration are collected by the negatively (-3.5 V) biased foil and negative charges by positively ($+24$ V) biased grids in front of, and behind, the foil. Two additional outer grids are negatively (-7 V) biased to prevent the electrons from escaping the detection system. The rear detector consists of a 60 μm thick molybdenum foil cemented to a plate with a quartz acoustical sensor (microphone). Positive charges from the impact ionization are collected by the molybdenum foil and negative charges by a grid in front of the foil. Both front and rear detector arrays consist of 16 square segments (2.5 cm × 2.5 cm) formed by crossing four foil strips (each 2.5 cm × 10 cm) with four grid strips of the same size. Each of the foil and grid strips is connected to a separate charge sensitive amplifier in order to uniquely identify which of the 16 front segments and 16 rear segments was affected by the impact. The positive pulse from the foil is pulse-height analyzed in eight logarithmic steps covering a dynamic range of 1:230. The negative pulse at the grid is used as a coincidence signal.

The experiment generated four classes of events: time-of-flight (TOF) events, registered by both front and rear detectors (fourfold coincidence); front foil-grid (FFG) events (double coincidence), detected only by the front detector; microphone events; and front foil-only events. The TOF events represent the most reliable data obtained from the experiment because they require a fourfold coincidence within a limited time window and in the proper sequence. They yield the speed, direction, and mass of a particle. From these data the orbital elements of dust particles were derived. The FFG events are regarded as impacts of particles on the front foil which did not penetrate it. The microphone events were found to be unreliable (see section II.F), as were the front foil-only events.

III.C. HEOS 2: The First Speed-and-Mass Sensor for Small Dust Particles

While the FFG events of the Pioneer 8 sensor were impacts of small particles, including β-meteoroids, which did not penetrate the front foil, the HEOS 2 sensor avoided such cutoff, by determining the particle speed from the rise time of the impact ionization signal.

The Highly-Eccentric-Orbit Satellite (HEOS) was launched 31 January 1972 into a polar orbit with a perigee varying from 350 to 3000 km and an apogee near 240 Mm. It operated until 3 January 1974. The dust experiment was mounted in a cylindrical cavity at the bottom of the spin stabilized spacecraft, viewing along the spacecraft spin axis. By turning the spin axis, any viewing direction perpendicular to the Earth-sun line could be established.

The experiment has been described by Dietzel et al. (1973) and Hoffmann et al. (1975a,b). Its sensitive area is 0.01 m^2 and its field of view a cone with an angle of 2.1 rad (effective solid angle of 1.0 sr). It consists of a hemispherical impact ionization target, in the center of which an ion collector is located, as schematically shown in Fig. 12. An ion yield independent of the chemical composition of the particle is obtained since a small amount ($\approx$ 1 %) of sodium was admixed to the gold of the target. A radial electric field separates the charges from the impact ionization: negative charges return to the target, which is biased at 0 V, and positive charges are pulled to the ion collector which is biased at -350 V. The two charge pulses are coincident and of similar shape but opposite polarity. An impact is recorded if both pulse amplitudes, Q^+ and Q^-, exceed detection thresholds at 2 fC and if the pulses

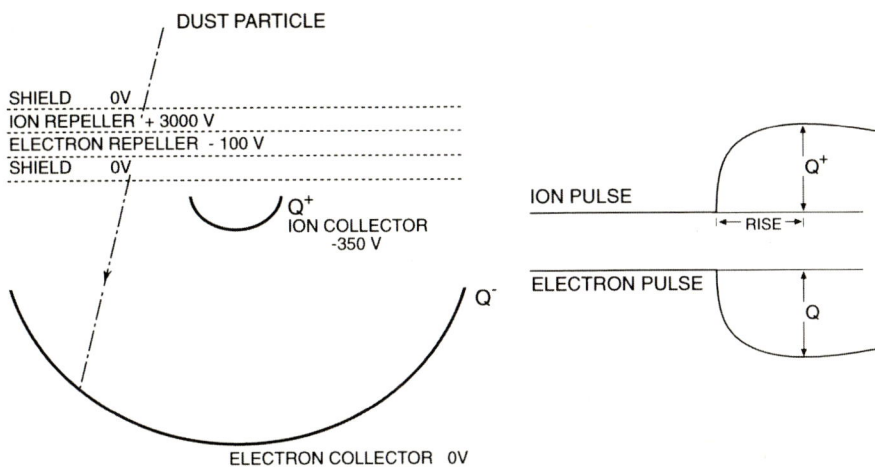

Figure 12. Schematic diagram of the HEOS 2 dust detector and its two output signals. A dust particle strikes the hemispherical target. Electrons are collected on the target and positive ions on the central ion collector. A plasma shield rejects both ions and electrons approaching from space but does not affect dust particles.

show the proper polarities and coincide within 20 μs. In addition, if an event appears as a sequence of pulses, such as one caused by mechanical vibrations or electrical ringing, it is marked as an "oscillation" and considered unreliable.

The detection threshold varies from $m = 10$ ag at $v = 40$ km s^{-1} to 160 fg at $v = 2.5$ km s^{-1}. The particle's speed v relative to the sensor is determined from the rise time of the ion pulse (Fig. 5) if the amplitude is $Q^+ \geq 50$ fC. In this case, the particle's mass m can be calculated from the geometric mean, $(Q^+Q^-)^{0.5} = Q$, of the two amplitudes and v, according to $m \propto Qv^{-3.5}$ (Eq. (2)). If a charge pulse reaches 20 pC, the amplifier saturates and neither rise time nor amplitude can be measured; therefore, v and m are determined only for the range $Q = 50$ fC ... 20 pC.

The dust sensor was shielded from direct sunlight and from electrons and ions of the solar wind. It was shielded from sunlight by virtue of its orientation perpendicular to the Earth-sun line. The shield from solar wind particles consisted of a combination of four parallel grids, placed across the sensor aperture: a grid biased at $+3$ kV (to repel ions) and a grid biased at -100 V (to repel electrons), both enclosed by grounded grids to ensure proper electric field conditions for both the sensor and the outside of the spacecraft. Without these shields, solar photons producing photoelectrons, or charged particles of the solar wind, could have reached the charge collectors and upset the measurement of very small charges from particle impacts.

III.D. Helios: The First Dust Composition Analyzer

Helios carried into space the first analyzer of the elemental composition, and also the first detector of the natural charge, of interplanetary dust particles.

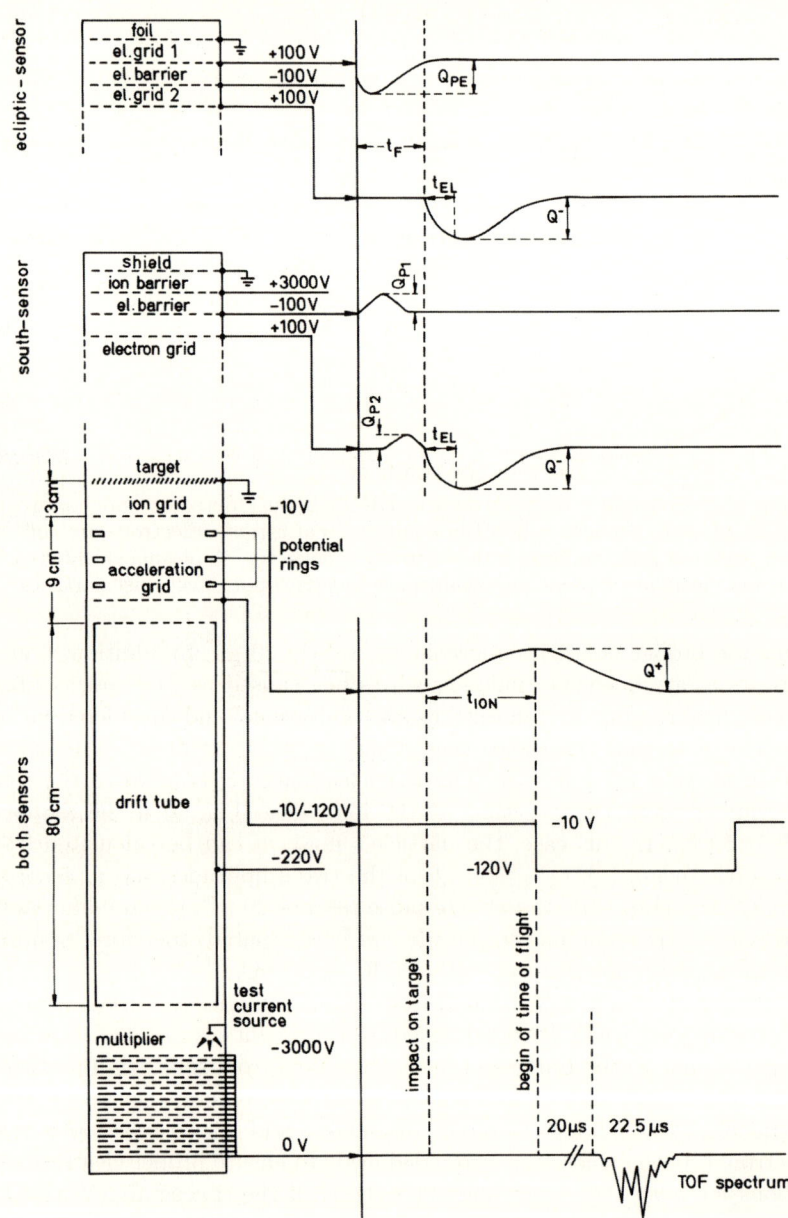

Figure 13. Schematic diagram of both Helios dust sensors with output signals. Ions from the impact ionization on the target are electrically drawn into the drift tube. Different species drift through the tube at different speeds $v_i \propto m_i^{-0.5}$, according to their atomic mass numbers m_i. The ecliptic sensor employs a thin foil as a shield from both solar wind plasma and sunlight, whereas the south-sensor, not exposed to sunlight, uses an electrostatic plasma shield similar to the one on HEOS 2. Note that the south-sensor is sensitive to particle charges (signals Q_{P1} and Q_{P2}).

The Helios 1 spacecraft was launched 10 December 1974 into a heliocentric orbit with perihelion and aphelion distances of 0.3 and 1.0 AU, respectively. It was spin-stabilized with the spin axis normal to the ecliptic plane and a spin period of one second, similar to Pioneer 8. It carried two dust experiments, the ecliptic-sensor, which was exposed to sunlight, and the south-sensor, which was shielded from direct sunlight by the spacecraft edge. The experiment has been described by Dietzel et al. (1973) and Grün et al. (1975, 1979). A twin experiment flew on Helios 2 one year after Helios 1.

Each sensor combined an impact ionization detector with a time-of-flight mass spectrometer. A schematic diagram of both versions is shown in Fig. 13. A solar wind protection system was placed in front of the impact ionization detector. It consisted, in the ecliptic-sensor, of a grounded thin foil (300 nm Parylene coated with 70 nm Al) that reflected sunlight and stopped the solar wind particles and, in the south-sensor, of a grounded grid, an ion barrier grid (at +3 kV), and an electron barrier grid (at −100 V), similar to the grids of the HEOS 2 dust experiment. The detection system consisted of a Venetian-blind type target with a surface of pure gold and two grids, the electron grid in front (at +100 V), that detected negative charges, and the ion grid behind the target (at −10 V), that detected positive ions (Q^+). The electron barrier grid and the electron grid of the south-sensor also detected the natural charges (Q_{P1} and Q_{P2}) of dust particles. The TOF mass spectrometer consisted of a time focusing section, a drift tube, and an open electron multiplier. The focusing section was designed to increase the mass spectrometric resolution by ion bunching, similar to the time-lag focusing technique described by Wiley and McLaren (1955). Potential rings were installed to homogenize the electric field between the ion grid and acceleration grid. The potentials of the ion grid and the focusing section were identical until an impact was detected. As soon as most of the ions had passed the ion grid, a negative pulse was applied to the acceleration grid. Ions lagging behind on their way to the drift tube sustained a stronger push by the electric field than ions running ahead. As a result, ions of the same mass but with different kinetic energies arrived near-simultaneously at the multiplier. This way, an ion mass resolution of $m/\Delta m = 5\ldots 10$ in the mass range $15\ldots 70$ atomic mass units was achieved.

A measurement cycle was initiated when a pulse from the positive charge grid exceeded a fixed threshold of 7 fC. The peak amplitudes and rise times of both positive and negative charge pulses were recorded. The speed of a dust particle was derived from the rise times of both charge pulses while its mass was determined from the pulse amplitudes and the speed, as in the HEOS 2 instrument. Calibration data were presented by Dalmann et al. (1977) who also showed that for particles with similar composition the functions $Q^+/m(v)$ are similar. Using the information of the measured spectrum, the particle mass could be determined within a factor of 3.

The thin foil of the ecliptic sensor acted similar to the thin foil of the Pioneer 8 sensor. Electrons from the impact ionization were detected, providing an independent charge pulse (Q_{PE}), and the time of flight between the foil and

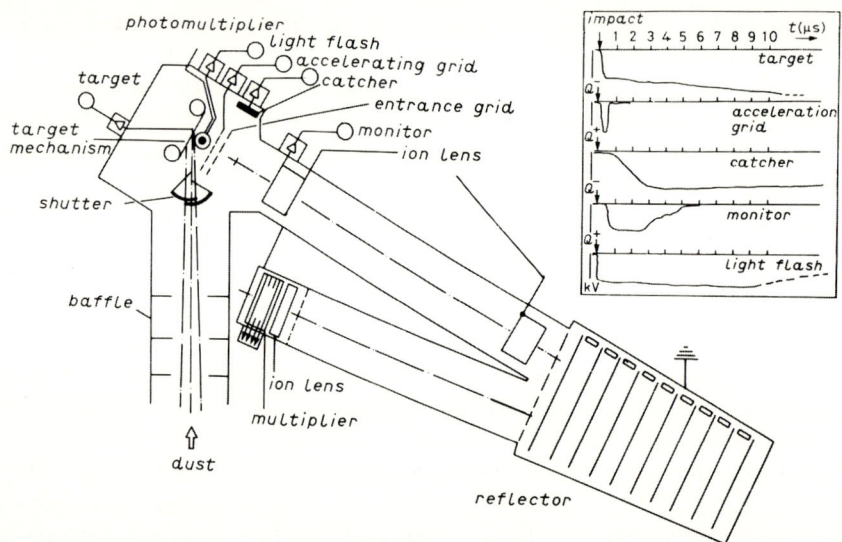

Figure 14. Schematic cross section of the Giotto Particulate Impact Analyser (PIA) with signals indicated. Dust particles enter through the light baffle and shutter and strike the target. Ions from the impact ionization are drawn into the first leg of the drift tube, reflected into the second leg, and detected by the multiplier. The reflector also produced first-order energy focusing of the drifting ions, resulting in a high mass spectroscopic resolution ($m/\Delta m > 150$). Courtesy of Jochen Kissel.

target (t_F). On the other hand, particles having a small diameter and/or low material density may have been stopped by the foil and not reached the target (Grün, 1981).

III.E. VeGa 1/2 and Giotto to Comet Halley

The VeGa 1 and 2 and the Giotto spacecraft were launched in December 1984 and July 1985, respectively, and encountered comet Halley in March 1986 at a distance of 1 AU from Earth. Giotto carried the Particulate Impact Analyzer (PIA), while VeGa 1 and 2 carried experiments (PUMA 1 and 2) almost identical to PIA, as well as the Dust Counter and Mass Analyzer (DUCMA).

1. High-Resolution Dust Composition Analyzer (PIA, PUMA)

Like the Helios dust analyzer, the PIA and PUMA instruments employed an impact ionization detector and a time-of-flight (TOF) ion mass spectrometer. The PIA has been described by Kissel (1986). A schematic cross section is shown in Fig. 14. Five significant changes from the Helios instrument were made: (1) the electric field strength above the target was increased by about a factor of 100 to accelerate the ions quickly and to eliminate the influence of secondary impacts; (2) an electrostatic ion reflector with a folded drift tube and first-order energy focusing (Mamyrin et al. 1973) was introduced; (3) the impact light flash was detected by a photomultiplier tube and served as one of three start signals for the ion TOF measurement; and (4) a microprocessor was used for adapting the instrument to the rapidly varying cometary environment and for data selection. As a result of (1)–(3), the resolution of the mass spectrometer was significantly increased ($m/\Delta m > 150$).

The fast Halley fly-bys provided a unique application of this type of instrument since: (1) the flux near the comet was several orders of magnitude higher than in interplanetary space; (2) the flight direction of the particles was known; and (3) the impact speed was also known and very high (69/80 km s^{-1}), yielding a high mass sensitivity. Because of the *high flux*, the target area was small (5 cm^2). The small area facilitated surveillance by a photomultiplier for impact light flashes and simplified the design of the energy-focusing spectrometer. Since the *direction* of the dust particles was known, the photomultiplier could be shielded from external light by a narrow, baffled aperture. In addition, the high impact speed guaranteed a very high degree of ionization of the particle material (see section II.D.2 for models of impact ionization).

Based on the PIA design, the more advanced Cometary and Interstellar Dust Analyzer (CIDA) was launched on the Stardust spacecraft to comet Wild 2 in February 1999. The CIDA exposes a large target area (110 cm^2) to a large field of view. In addition, its mass spectrometric resolution is increased to $m/\Delta m = 250$ at 100 atomic mass units and, for the first time, it is possible to measure negative ions from the impact ionization.

2. Dust Counter and Mass Analyzer (DUCMA)

DUCMA was the first depolarization-type dust sensor in space. It used a 28 μm thick PVDF foil with a sensitive area of 75 cm^2. It has been described in detail by Simpson et al. (1987). Its operation is discussed in section II.E.4 and schematically shown in Fig. 15.

III.F. Galileo/Ulysses: Large-Area Multi-Coincidence Dust Detector System (DDS)

The large sensitive area of 0.1 m^2 provided a comfortable detection rate of about 1 impact per day, leading to the discovery of new phenomena, such as jovian dust streams and interstellar grains (Grün et al. 1993). The Galileo spacecraft was launched 18 October 1989 from the space shuttle Atlantis. It swung by Venus and Earth, then passed by the asteroid Gaspra, made another swing by Earth, passed by the asteroid Ida, released a probe that entered Jupiter's atmosphere, and finally entered into an orbit around Jupiter on 7 December 1995, where it investigates the planet's magnetosphere, atmosphere, and moons. The main body of Galileo that carries the DDS is slowly spinning (period = 19 s) about an axis that usually points in the anti-sun direction. The DDS is oriented with its axis at 0.96 rad from the spacecraft spin axis. It has been described by Grün et al. (1992a). A twin instrument, operating on the Ulysses spacecraft since October 1990, has been described by Grün et al. (1992b). Both instruments were calibrated by Göller and Grün (1989). Also, the DDS-spare instrument was named GORID and placed on the Russian Express 2 spacecraft into geosynchronous orbit in mid-1996.

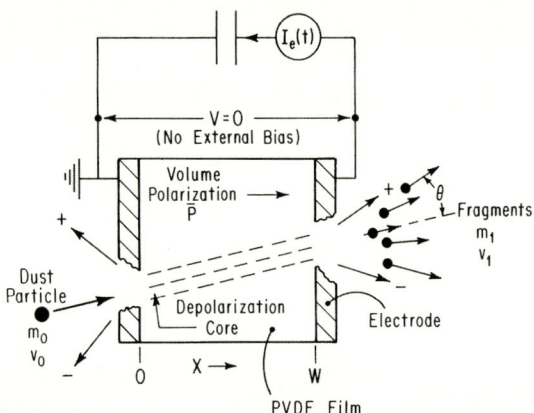

Figure 15. Schematic drawing of a polarized PVDF sample with conducting contact electrodes. The sample has a volume polarization directed along the x-axis. There is no applied bias voltage. An incident high-velocity dust particle penetrates into the sample, resulting in complete depolarization within and near the crater formed. The impact generates a fast current pulse $I(t)$ in the external circuit, and neutral or charged spallation products may be ejected from the sample (from Simpson and Tuzzolino, 1985, with permission from Elsevier Science).

The DDS is an impact ionization detector, with a design resembling the HEOS 2 instrument but a 10 times larger sensitive area (0.1 m^2). Generally in the shadow, it does not need a solar wind protection system but has additional features as schematically shown in Fig. 16. An incoming dust particle passes through a grid triplet consisting of two electric shields and, between them, a charge sensing grid. The charge amplitude, Q_P, and travel time, t_{PE}, from the charge sensing grid to the target, are measured as well as the rise time of the electron pulse, t_{RE}, and the time lag of the ion pulse, t_{EI}, provided that $|Q_P| \geq 10$ fC. Also different from HEOS 2 is the ion collector, consisting of another grid triplet. The center grid detects only collected ions and no charge is induced by ions outside the shielding grids. Thus, Q_I, t_{RI}, and t_{EI} are additional, independent parameters. Some ions pass through the ion collector grids and are detected by a channel electron multiplier. This multiplier produces an additional coincident signal, Q_C, that is important because it is not affected by electric or acoustic interference. A solar wind protection system was deemed unnecessary since the DDS is generally directed 2.2 rad away from the sun.

Note that a positively charged dust particle induces a positive charge pulse, not only on the charge sensing grid, but also on the target, as it approaches the target and before it impacts. The induced charge (Q_P) drives the target signal (Q_E) positive at first, i.e., into the "wrong" direction, and tends to neutralize the charge of electrons from the impact ionization, arriving later. As a result, the target signal may never cross the negative detection threshold or only after a delay that can be long enough for t_{EI} to appear negative (i.e., Q_I detected before Q_E). This condition has been observed by the DDS with slow (few

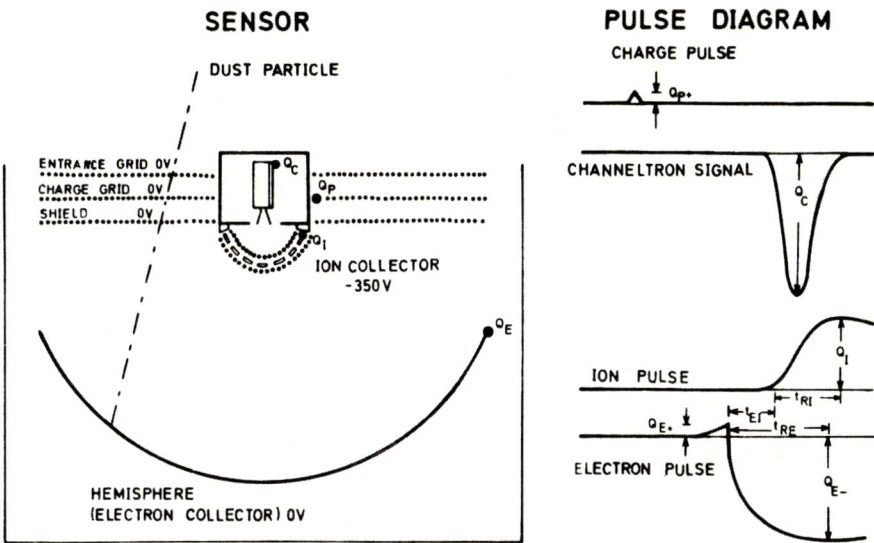

Figure 16. Schematic cross section of the Galileo Dust Detector System (DDS) with signals indicated. Basically an enlarged version of the HEOS 2 sensor (with 10 times the sensitive area), the DDS uses a hemispherical target and central ion collector. In place of the plasma shield, it employs a charge sensing grid. Also, a channeltron behind the partially transparent ion collector produces an additional signal. (From Grün, E., et al. 1992a, Fig. 1)

km s^{-1}) particles, both on the electrostatic dust accelerator and in flight, and is considered an indication of positive particle charges.

III.G. Hiten: Dust Counter (MDC) with a Transient Recorder

Sampling the entire shape of a detector signal from some time before until some time after the impact has enabled the MDC to extract a wealth of information about each event that previous experiments had missed; such as the precise time of impact, location of impact, data quality, and interfering signals.

The Japanese satellite Hiten (Muses A) was launched 24 January 1990 into a highly elliptical Earth orbit extending to 1.5 Gm and including the vicinity of the moon, the Earth's geomagnetic tail, and the Lagrangian points $L4$ and $L5$ of the Earth-moon system. The cylindrical body of the spacecraft was spin stabilized with its axis perpendicular to the plane of the ecliptic and a spin period of 3 s. The Munich Dust Counter (MDC) was mounted on the perimeter of the spacecraft, scanning the ecliptic plane. It has been described by Igenbergs et al. (1991). A second MDC, on Bremsat, was placed in a low Earth orbit in February 1994 and a third MDC is part of the Japanese Planet-B Mars Orbiter mission that was launched in July 1998.

The MDC is an impact ionization detector designed to measure the mass, speed, and crude flight direction of interplanetary dust particles and space de-

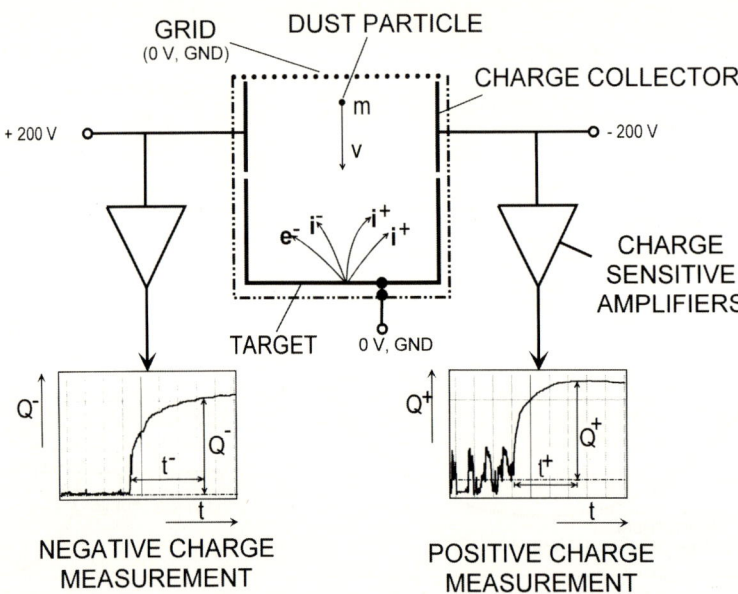

Figure 17. Schematic diagram of the Hiten MDC. The sensor consists of a rectangular box with two charge collectors on opposite walls. Unlike previous detectors, the MDC does not draw a signal from the impact target. The signals shown have been logarithmically compressed and are reproductions of actual flight recordings. The periodic noise on Q^+ is from a known internal source. (Illustration provided by courtesy of R. Münzenmayer)

bris. It consists of a rectangular box with a square aperture of 0.01 m^2 and a field of view of 2.6 rad. It includes a gold target and two charge collector plates: one for the negative charge and one for the positive charge, as schematically shown in Fig. 17. Both charge signals are amplified; logarithmically compressed over a range from 5 fC to 50 pC; sampled once every 200 ns; digitized to 8 bits/sample; and stored in a 512-sample (102 µs) continuous-loop memory. The memory, being of the first-in-first-out (FIFO) type, loads the newest sample while it deletes the oldest sample, similar in operation to a digital storage oscilloscope, or transient recorder. When an impact is detected, recording continues until 51 µs (256 samples) after the impact, then it stops, and the entire contents of the FIFO (512 samples or 102 µs total) are read out. Thus, the data include signal samples from 51 µs before until 51 µs after the impact.

From the recorded data not only the speed and mass of a particle are derived but also the precise time of impact, the impact location, and the charge (if sufficiently large). In addition, interfering signals can usually be recognized from their shape. In fact, in a few cases the pre-impact portion of the signal seemed to indicate a particle charge but, after careful examination, was identified as electric interference (Münzenmayer 1995). The signal samples shown in Fig. 17 were actually received from space.

III.H. Cassini: Multi-Parameter Cosmic Dust Analyzer (CDA)

The Cassini spacecraft was launched 15 October 1997. It swung by Venus (twice), Earth, and Jupiter; will reach Saturn in July 2004; release a probe (named Huygens) which descends on Titan; and, during the following four years, make over 50 specifically designed orbits around Saturn, investigating its magnetosphere, atmosphere, rings, and moons. Cassini is three-axis stabilized with its prime antenna pointing toward the Earth. The Cosmic Dust Analyzer (CDA) is mounted on its perimeter, pointing away from the sun.

The CDA combines, for the first time, capabilities to investigate simultaneously the velocity, mass, chemical composition, electric charge, and direction of individual dust particles, and to reliably detect dust fluxes from 10 impacts/(m^2 month) to 10 million impacts/(m^2 s). It consists of a large (0.1 m^2) hemispherical impact ionization detector with a central ion collector, similar to the Galileo DDS but augmented by a charge/velocity sensor in the aperture and a chemical analyzer target (CAT, area 0.02 m^2) in the middle of the impact target. Also, two depolarization (PVDF) detectors are mounted outside the hemisphere, capable of recording very high dust fluxes, particularly in the environment of Saturn's rings. Transient recorders store the waveforms of several signals. All operations are controlled by a microprocessor that can be re-programmed during flight. The detectors and electronics are mounted on a dedicated single-axis turntable for viewing interesting regions within a 2π sr area of the sky. The CDA has been described in detail by Ratcliff et al. (1992), Srama et al. (1996), and Srama et al. (2001), and the PVDF subsystem more specifically by Tuzzolino (1996).

Figure 18 shows schematically a cross section of the large detector and shapes of the signals in the chemical analyzer mode. A dust particle enters through the four grids of the charge/velocity sensor where it induces a signal of trapezoidal shape with a sag-shaped plateau. From this signal, the particle's electric charge q; two velocity components v_y and v_z; the y-coordinates of the points of entry and exit; and the approximate distance of the trajectory from the sensor wall can be determined, provided that $|q| \gtrsim 2$ fC. While the particle continues its flight, the flight time until impact on the target is recorded, serving as another coarse indicator of the particle's velocity. As the particle approaches the hemisphere, its charge induces ramp-shaped charge signals on both the main target and grid in front of the CAT, the amplitudes and shapes of which provide information on its trajectory.

When the particle strikes the main target surrounding the CAT, the impact ionization signals produced are similar to those of the Galileo DDS, discussed in section III.F. In addition, an ion mass spectrum is produced when the particle strikes the CAT. The main difference between impacts on the main target (0 V potential) and the CAT (+1 kV potential) is in the electric field strength. At the main target the strength is about 100 V m^{-1} whereas between the CAT and the grid above it the strength is about 300 kV m^{-1}. As a result, the separation of impact-generated electrons and ions at the CAT occurs much faster than at the main target, leading to faster rise times. At the CAT, the ions immediately

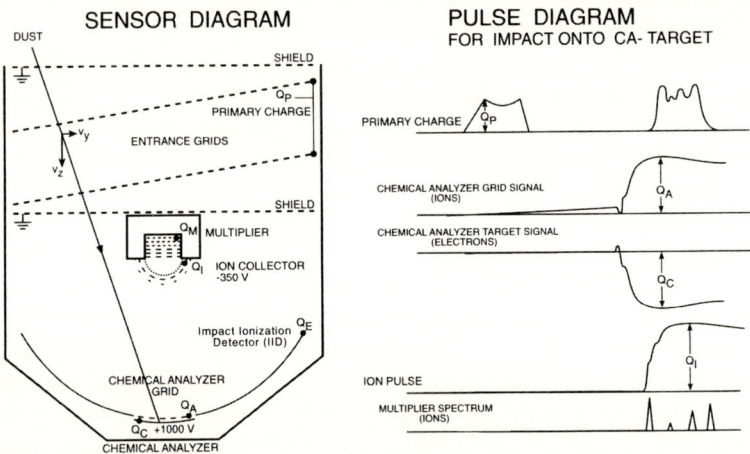

Figure 18. The Cassini Cosmic Dust Analyzer (PVDF subsystem not shown) uses a large hemispherical target and central ion collector, like the Galileo and Ulysses instruments. It is shown in the chemical analyzer mode with signals indicated. A high electric field strength at the central portion of the target (chemical analyzer target, CAT) is critical for achieving an ion time-of-flight mass spectrum. The primary charge (Q_p) grids at the entrance are not only sensitive to the charge carried by the dust particle but also to ions from the impact ionization leaving the sensor.

reach energies of 1 keV, and different ion species start separating according to their masses, $v \propto m^{-0.5}$. The ions are further accelerated, and their paths focused, by the potential (-350 V) of the ion collector. The ion collector consists of three hemispherical grids, similar to the Galileo-DDS. Some ions pass through these grids; are post-accelerated by the high potential (-3 kV) of the first dynode of the multiplier; and produce secondary electrons which then undergo fast amplification and detection by the multiplier. Thus, the combination of CAT, ion collector/multiplier, and fixed flight distance (0.23 m) between them is utilized as a time-of-flight ion mass spectrometer. The multiplier output represents the mass spectrum of the impact-generated ions, indicating the composition of the particle. Srama et al. (1996) reported a mass resolution in the range between $m/\Delta m = 20$ for H^+ and 50 for Rh^+. While the Galileo DDS used a channel electron multiplier for simplicity, the CDA, requiring a greater dynamic range and high time resolution, employs a 20-stage focused mesh multiplier, similar to the composition analyzers on Helios, VeGa, and Giotto.

The recording of entire signals, as known from the MDC on Hiten, enables the CDA to extract data not only on the speed, mass, and composition of each particle but also on its charge, trajectory, location and time of impact, and helps to discriminate real signals from noise. Also, some ions from the impact ionization, while leaving the sensor, are detected at the entrance grids by the charge/velocity detector.

A 16-bit microprocessor controls all operations of the instrument and turntable and the flow of data and commands; sets event priorities, detection thresholds, high voltage levels, time stamps, and error flags; extracts essential parameters (138 bytes) from the raw data (4608 bytes) of each event and thereby compresses the data volume by a factor of 30. During low dust activity, both raw and compressed data are transmitted to Earth. When the activity increases, raw data are deleted so as to make best use of the available data transmission rate.

The PVDF subsystem operates independently under control of its own microprocessor. Two PVDF copolymer thin-foil sensors (discussed in section II.E.4) are mounted side by side; a larger one (50 cm^2) being 28 μm thick and a smaller one (10 cm^2) being 6 μm thick. During ring crossings, the particle speed is known, so the mass can be determined from the signal amplitude. Impact rates up to 10 000/s can be measured with a time resolution as high as 0.1 s.

III.I. Very-High-Resolution Cometary Dust Composition Analyzer (COSIMA)

From the Halley results one generally expects atomic mass numbers of the cometary organic molecules up to many thousands. To distinguish elemental ions from molecules of the same integer mass number also requires a mass resolution of many thousands. COSIMA covers both requirements, mass range and mass resolution, and does so for minute dust samples.

The Cometary Secondary Ion Mass Analyzer (COSIMA), accepted for the comet orbiter Rosetta, was originally developed for the Comet Rendezvous and Asteroid Flyby (CRAF) mission and has been described by Zscheeg et al. (1992). This instrument utilizes ion bombardment of particles already collected, instead of impact ionization, to generate the ion mass spectra. It includes a dust collector, an optical microscope for target characterization, a primary ion gun, and a double focusing time-of-flight ion mass spectrometer. The dust collector contains up to 25 targets of 4 cm^2 substrates each in a storage wheel. After the target has been exposed to cometary dust, the respective lot is moved in front of the microscope and scanned. On-board image evaluation software detects the presence and location of dust particles larger than a few μm in diameter. Once the presence of cometary particles is established, the target is moved in front of the mass spectrometer. Pulses of 3 ns duration of a beam, about 10 μm in diameter, of isotopically pure ^{115}In-ions at 10 keV from the primary ion gun strike the selected dust and produce ions from the cometary matter. These secondary ions are extracted by an electric field; travel through a field-free section; pass a two-stage electrostatic reflector; and return through the drift section as schematically shown in Fig. 19. A single-stage reflector, the ion-mirror, sends the ions back into the drift section, again through the two-stage reflector and a fourth time through the drift section, the total effective path length being about 3.5 m. Finally, they impact onto the ion detector, made of a two-stage microchannel plate, where the arrival time of each ion

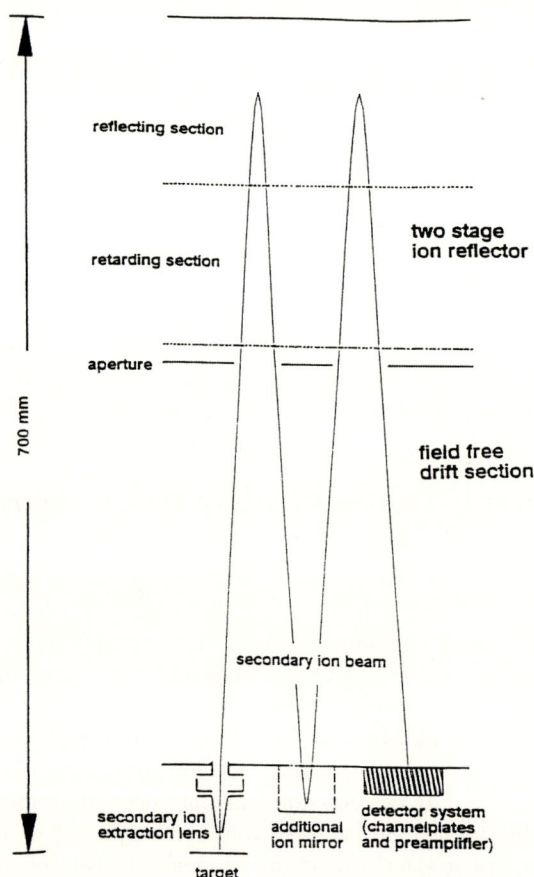

Figure 19. Schematic cross section of the COSIMA time-of-flight spectrometer. In this instrument, the ions are produced by ion bombardment. The ion drift path is folded three times and passes twice through a double focusing ion reflector, before the ions are individually detected.

is measured. The performance of COSIMA includes: (1) a high sensitivity that allows the analysis of few-picogram dust samples; (2) a wide mass range of $1 \leq M \leq 3500$ atomic mass units; and (3) a mass resolution of up to $M/\Delta M = 10\,000$ as required to distinguish organic from inorganic molecules.

IV. LABORATORY SIMULATION

Before an instrument is carried into space, it must be designed and tested on the ground to verify and calibrate its response. Because the most striking characteristic of an interplanetary dust particle is the high speed of typically $1\ldots 100$ km s^{-1}, one has to accelerate microparticles to speeds of similar magnitude in the laboratory. Acceleration techniques have been reviewed by Cable

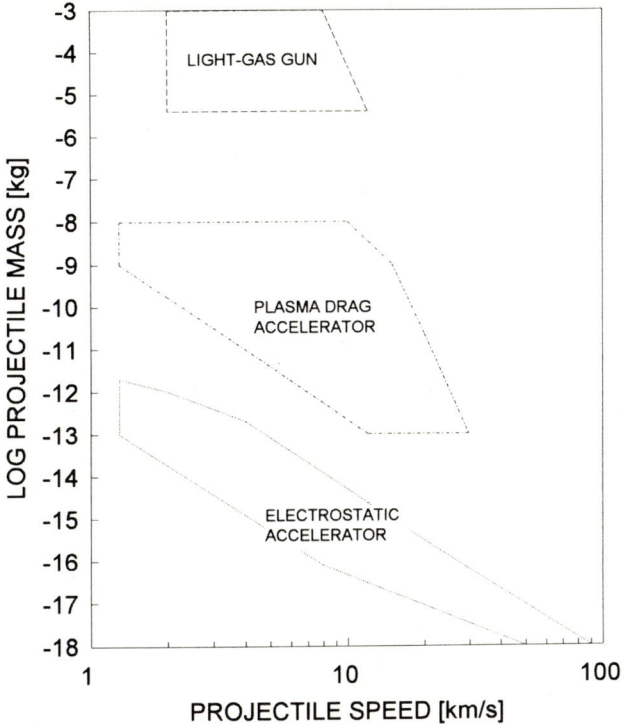

Figure 20. The mass versus speed ranges of the three common types of microparticle accelerators.

(1970) and Martelli and Cerroni (1983). Basically three accelerator types are in use to produce test particles with these high speeds. Their ranges of operation are compared in Fig. 20. The electrostatic dust accelerator (section IV.B) covers the entire velocity range but only for very small particles (0.1...10 μm). The plasma drag accelerator (section IV.C) and the light gas gun (section IV.A) produce projectiles having sizes more like micrometeoroids (10...100 μm) and meteoroids (1...10 mm), respectively, but at lower speeds.

While many materials can be accelerated, low-density or porous substances are often too fragile to survive the mechanical forces involved. Also, velocities > 30 km s^{-1} are simply not attainable except for the smallest particles ($\approx$ 0.1 μm). Understanding the physics of hypervelocity impact phenomena (e.g., the impact ionization models mentioned in section II.D.2) can help to predict the response of a given detector to particles having certain properties (mass, velocity, composition, density) which are not available from existing acceleration facilities. For some applications, laser pulses are considered an adequate and more convenient substitute for real dust particles. According to Tuzzolino (1983), depolarization detectors can be tested with a pulsed laser beam. Similarly, a laser has been used by Burton (1983) for simulating particle impacts onto a momentum sensor for the Giotto mission. Also, Kissel and

Krueger (1987) showed that the impact of a hypervelocity dust particle and a laser pulse produce similar ionization effects.

For accelerated particles to be meaningful test objects, one needs to know their velocities, masses, state of matter, and trajectories after acceleration. Isbell (1987) has reviewed the diagnostic technology in general; high-speed imagers in particular have been presented by Swift (1987); and some individual methods are discussed below.

Particle charging in space is an important aspect that is also being investigated in the laboratory. Model calculations, especially of charges on very small dust particles, are rather unreliable and sometimes differ by more than one order of magnitude. Therefore, an experimental program was started to determine, in the laboratory, charges and equilibrium potentials of particles of various sizes and materials during exposure to beams of electrons and ions having specific energies.

Repulsive electrostatic forces within a charged dust particle produce mechanical stress proportional to the square of the electric field strength on the particle's surface. Electrostatic disruption (fragmentation) of the particle should occur when this stress exceeds the tensile strength of the particle material. Several phenomena observed within the solar system can be explained by the electric charging and subsequent electrostatic fragmentation of dust particles (see, e.g., Fechtig et al. 1979; Grün et al. 1984; or Boehnhardt and Fechtig 1987). Tensile strengths of interplanetary dust particles, however, have only been estimated because they depend on the size and shape of the particle, its mineralogic composition, its irradiation history, etc. (for references see Draine and Salpeter 1979). Therefore, the experimental program was also designed to determine the tensile strength of a particle from the electric field strength at the moment of disruption.

The significance of experimental studies became apparent when photoelectric yields of very small particles were found to be up to a factor of about 100 greater than the bulk yield. These measurements were started in 1980 (for the latest results see Schleicher et al. 1994).

IV.A. Acceleration of Dust Particles

1. Light-Gas Gun

This widespread technique allows a good choice of projectile materials having well defined masses and shapes. Typically, projectiles of $0.01\ldots 1$ g are accelerated to about 7 km s^{-1}. Light-gas guns are relatively easy to operate, and they are clean and finely controllable. However, the highest speed that can be achieved is intrinsically limited to about $10\ldots 14$ km s^{-1}.

In a first stage, conventional gun powder drives a solid piston along a pump tube filled with a light gas (H_2 or He) towards a diaphragm which ruptures at a preset pressure of some GPa. This pressure, considerably higher than the pressure in the powder gas, is achieved by the deceleration of the rather massive piston ($0.1\ldots$ several kg). The light gas then flows into a second gun

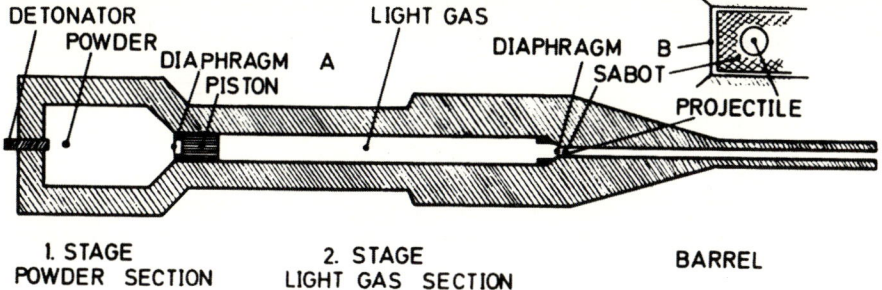

Figure 21. Simplified sketch of a two-stage light-gas gun.

barrel having a smaller bore and accelerates the projectile which is usually located in a protective sabot. Upon leaving the gun, the sabot is stopped either mechanically or by aerodynamic drag before entering the experimental chamber. Figure 21 shows a simplified sketch of the two-stage light-gas gun first built and described by Crozier and Hume (1957).

The theoretical maximum speed for a projectile of vanishing density is $2/(\gamma - 1)$ times the speed of sound (Seigel 1979), where γ ($= 1.40$ for H_2) is the ratio of the specific heats of the gas. Since the speed of sound, being $v_s = (\gamma p/\rho)^{0.5}$, increases with the gas pressure p and the inverse of the gas density ρ, the lightest gas at the highest possible pressure is required.

Of particular note is the NASA Ames vertical gun facility, which offers the possibility of impacting horizontal targets at angles varying from vertical to almost horizontal. One of the fastest guns now in operation is based at the Fraunhofer-Institut für Kurzzeitdynamik in Freiburg, Germany. It has routinely achieved speeds of about 10 km s^{-1} with aluminum projectiles of 1.5 mm diameter (Stilp 1987).

The particle speed can be measured, with an accuracy of < 1 %, by arrays of photodiodes that sense either the occultation of laser beams in the free-flight chamber or the impact flashes at the sabot separator and the target (Stilp 1987).

2. *Electrostatic Accelerator*

This technique is based on the acquisition of kinetic energy by a particle of mass m and positive charge q falling through a potential difference U, thus $\frac{1}{2}mv^2 = qU$, where v is the terminal speed of the particle. It was first described by Früchtenicht (1962, 1964) and Vedder (1963). If the particles are spheres of diameter d, the charge is related to the surface field E_s by

$$q = \pi\epsilon_0 d^2 E_s. \tag{22}$$

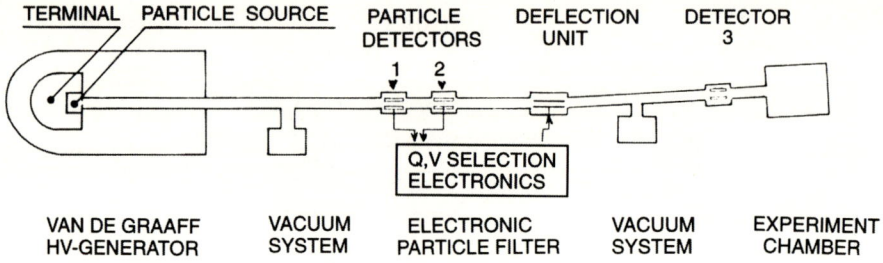

Figure 22. Schematic sketch of an electrostatic dust accelerator facility.

For negatively charged particles, the values of E_s are limited by electron field emission to ≈ 1 GV m^{-1} and, for positively charged particles, to ≈ 10 GV m^{-1} by the tensile strength of the material. Thus the terminal speed of a particle of a given mass cannot be increased beyond a certain limit by increasing the ratio q/m, since $q \propto d^2$ and $m \propto d^3$. For a given accelerating voltage it follows that $m \propto v^{-6}$ and $q \propto v^{-4}$, and since they have the higher q/m ratio, the smallest particles are the fastest.

Figure 22 is a schematic sketch of the electrostatic dust accelerator facility at the Max-Planck-Institut Heidelberg using a van de Graaff high-voltage generator. Dust particles are charged in the source located in the generator terminal which is at the positive potential U with respect to ground. Leaving the accelerator, the particles pass through two q, v detectors as depicted in Fig. 2. With q and v known, the mass m of the particle is calculated as $m = 2qU/v^2$. The experimenter can program an electronic system, first described by Rudolph (1966), to admit to the experiment chamber only particles within specific speed and/or charge windows. All particles outside the selected windows are deflected electrostatically out of the beam.

Obviously, for producing high speeds, both the charge-to-mass ratio (q/m) of the projectile and the accelerator voltage should be as high as possible. The voltage routinely achieved by the accelerator at Heidelberg is 2 MV, whereas the accelerator at Los Alamos produces 6 MV (Keaton et al. 1990). Slattery et al. (1973) described a linear accelerator of the Sloan Lawrence type, consisting of a 1.6 MV van de Graaff machine and 92 stages of 0.1 MV each, resulting in a total accelerating voltage of 10.8 MV.

In order to put high charges on the particles, a dust source as described by Shelton et al. (1960) and schematically shown in Fig. 23 is used. Here the particles are raised from the reservoir by high-voltage electric pulses and subsequently charged through contact with a very thin tungsten needle. Once they are charged, they leave the source immediately. This system produces a beam with a broad statistical distribution of particle masses and speeds. Empirically, Eichhorn (1974) found the maximum field strength at the particle's

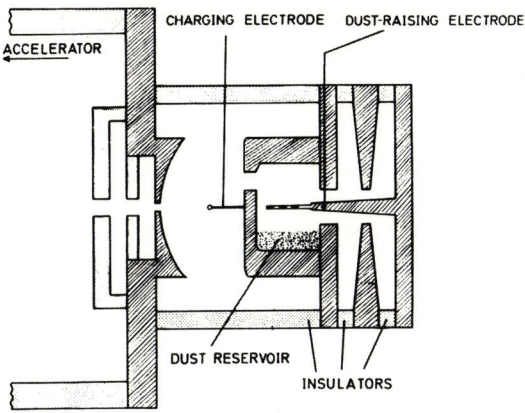

Figure 23. Schematic sketch of a dust particle charging system according to Shelton et al. (1960) that generates a statistical beam of particles.

surface to increase with decreasing particle size as:

$$E_{max} \approx 2.8 \cdot 10^6 d^{-0.5}, \qquad (23)$$

thus $q_{max} \approx 7.9 \cdot 10^{-5} d^{1.5}$. While Shelton's dust source features high repetition rates, it is limited to conductive dust materials. Carbonyl iron powder works best, while aluminum, carbon, cobalt, nickel, tungsten, boron carbide, silver coated glass, and polypyrrole colloid have also been used. Non-conductive particles have been made conductive, e.g., silicate coated with aluminum or nickel. Even hollow microspheres have been produced in order to increase the q/m-ratio.

Vedder (1963) described another type of source similar to the one shown in Fig. 25 that charges conductive as well as non-conductive dust particles (e.g., quartz, glass, Al_2O_3, kaolin, polystyrene). During the charging process a single particle is suspended in an electric quadrupole field and observed through a microscope. During the charging process, which takes up to 1 hour, all edges of the particle are smoothed by ion field emission. When the particle's charge approaches the limit beyond which electrostatic self-destruction occurs, the operator applies an electric field which kicks the particle out of the quadrupole and into the accelerator. While field strengths obtained with Shelton's source are typically $E_s = 1 \ldots 2.5$ GV m^{-1}, Vedder's source reaches the limit of 10 GV m^{-1} mentioned above.

3. Plasma Drag Accelerator

Another approach to overcome the speed limit of the light-gas gun is to substantially raise the temperature of the accelerating gas. This cannot be accomplished using chemical energy. Instead, electric energy stored on a high-voltage capacitor is discharged into either a volume of hydrogen or helium or a thin metal wire or disc which explodes into a hot plasma. The speed of the plasma generated in this process reaches up to 100 km s^{-1}. To obtain similar thermal-expansion speeds of a gas, temperatures above 0.5 MK would be required!

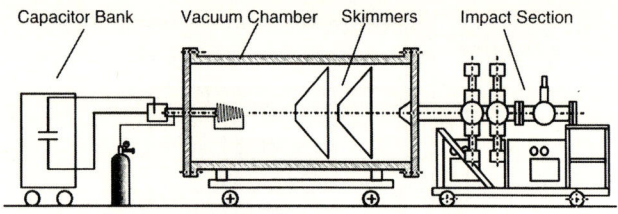
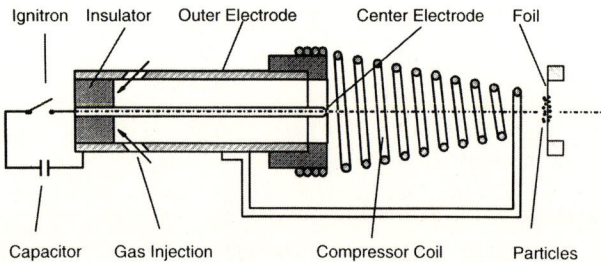

Figure 24. Schematic sketch of a plasma drag accelerator (upper portion) and details of the source (lower portion). (Illustration provided by courtesy of R. Münzenmayer)

Hüdepohl et al. (1989) described a combination of a coaxial plasma accelerator and a plasma compressor coil as schematically shown in Fig. 24. Electric energy from a capacitor (350 μF at 16 kV) is discharged into the coaxial part of the accelerator which ionizes helium gas, or vaporizes and ionizes a thin aluminum disc, between the center electrode and the outer electrode. The plasma is then accelerated by the Lorentz force that is produced by the radial current from the center electrode to the cylindrical outer electrode and the azimuthal magnetic field of the center electrode. As the plasma enters the compressor coil, a current (up to 50 kA) starts to flow from the tip of the center electrode to the windings of the coil. The current generates a strong magnetic field (flux density up to 1 T) in the coil which compresses the plasma (pressure up to ≈ 1 GPa). The high-density, high-speed plasma ruptures a thin foil that carries the particles and drags the particles along.

Since acceleration is caused by aerodynamic drag, the intense contact between projectile and plasma requires special precautions. The selection of projectile materials is limited to those which have the lowest possible absorption coefficient for visible and ultraviolet light and an emissivity in the infrared as high as possible. These two conditions together with a small particle size help to keep the projectile temperature below the evaporation point. Since the acceleration by drag increases only in proportion to the cross-section of the projectile, whereas mass increases in proportion to the volume, the highest speeds are again expected for small projectiles. The projectiles are usually soda-lime glass beads having masses between 100 pg and 100 μg. They are accelerated to speeds of up to about 15 km s^{-1} (see summary Fig. 20), highly eroded, partially melted, and sometimes fused by the plasma.

Because the plasma expands into a wide angle, the particles are not well focused. Usually, many particles are launched simultaneously and only a few reach the target. A particle's speed can be calculated by dividing the known flight distance, from dust source to target, by the time of flight from firing of the source to detection of the impact (error < 1 %). Diameter, shape, and x, y position are derived from the hole which the particle punches in an ultra-thin (< 200 nm) nitrocellulose foil mounted in front of the target. The foil is inspected under the microscope after each shot and replaced. In addition, when four piezoelectric detectors are mounted at the corners of the target plate, the x, y position of the impact can be determined, to an accuracy of < 1 mm, from the arrival times of the four detector signals (Igenbergs et al. 1987).

IV.B. Dust Charging in an Electrodynamic Quadrupole

1. Laboratory Set-Up

Laboratory studies of dust electric charging by electrons and ions were started in 1986 at the Max-Planck-Institut für Kernphysik in Heidelberg. In the present set-up, a dust particle is suspended in an electrodynamic quadrupole inside an ultra-high vacuum (pressure $\approx$ 100 nPa) chamber and exposed to a beam of electrons or ions (He^+, Ar^+, H_2^+) having energies of up to 20 keV or 5 keV, respectively. While glass, carbon, aluminum, iron, MgO, and loosely bound Al_2O_3 particles with radii $d = 0.4 \ldots 60$ μm have been used, most experiments were performed with spherical glass particles ($d = 2 \ldots 10$ μm). The charge-to-mass ratio q/m of a particle is determined from the frequency f_z of its motion in the vertical direction and from the amplitude V and frequency f of the alternating voltage applied to the middle electrode:

$$\frac{q}{m} = \frac{\pi^2 r_0^2 f f_z}{V}, \qquad (24)$$

where r_0 is the inner radius of the middle electrode (Fig. 25). While the electrodynamic quadrupole had been developed originally for applications in mass spectrometers and, later, dust accelerators (Vedder 1963), the objective here is to deposit as much charge on a dust particle as possible.

A particle suspended in the quadrupole and illuminated by a He-Ne laser can be observed through a long-distance microscope. Its image is also projected, via lens and image intensifier, on a two-dimensional position sensitive detector which puts out two coordinates of the particle position. For a vertical cross-section of the suspension system see Fig. 25. The frequency f_z and amplitude of the particle motion are measured and fed into a control computer. The computer damps the amplitude of the particle motion, via damping wires mounted on the quadrupole electrodes, and thereby improves the stability of the motion; it calculates Q/m using Eq. (24); and continuously predicts and adjusts f as q/m increases. As a result, the stability of the particle motion during the experiment is guaranteed even in extreme situations.

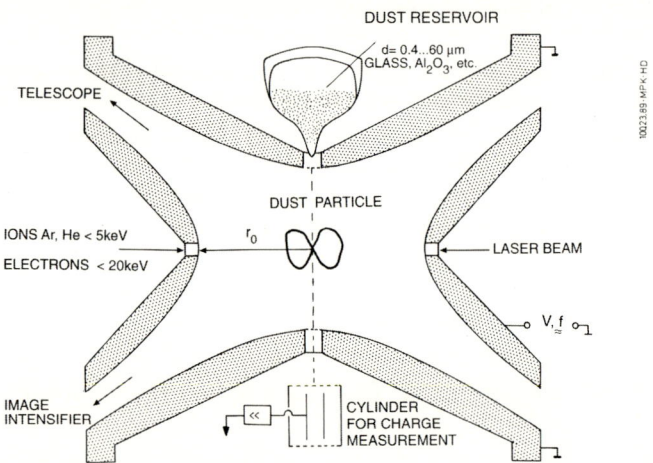

Figure 25. Schematic sketch of a dust particle charging system, that uses an electric quadrupole for suspending a single particle, and a beam of ions, electrons, or ultraviolet radiation for charging the particle.

When the operator wants to terminate the charging process, he applies a voltage pulse to the upper electrode which kicks the particle out of the quadrupole and through a cylinder that measures the particle's charge q (as discussed in section II.B). If the particle is spherical, its diameter d, surface potential Φ, and surface electric field strength E_s can be calculated from q, q/m, and the density of the particle material, using Eqs. (1) and (22). It can also be determined by a dynamic method, observing i/m, the temporal derivative of q/m, in the following manner. A highly positively charged particle is exposed to a beam of ions having a low energy $E_i < e_0 \Phi$ (e_0 is the ion charge). Initially, no ion can reach the particle surface because of electrostatic repulsion. However, some ions strike the electrodes of the quadrupole producing secondary electrons. Some of these electrons, in turn, strike the particle and slowly discharge it, thereby decreasing Φ. When $\Phi = E_i/e_0$ is reached, a few ions can strike the particle, the discharging process is slowing down, a characteristic break in the q/m versus i/m plot is observed, and the particle diameter can be calculated from q/m and E_i. A third method of determining the particle diameter has been developed during the study of electrostatic fragmentation (Švestka and Grün 1992a). It is based on the characteristic time of damping of the particle movement caused by collisions of the particle with the residual gas, the pressure of which is known.

2. Charging by Electrons

Glass particles with diameter $d = 0.4\ldots 4$ μm, being charged by electrons having energies $E_i = 1\ldots 20$ keV, always ended up with positive potentials. This is compatible with results of recent calculations (see the section on charging in chapter 16) for small particles consisting of non-conducting materials. It can happen only to micron and sub-micron sized particles having a secondary electron emission yield greater than one. On the other hand, the potentials of larger particles can be negative (Grün et al. 1984) and approximately equal to E_i/e_0. This ambiguous behavior has implications for particles (including ices) in planetary magnetospheres where electrons in this energy range are present: particles of the same material but with charges of opposite polarity can coexist in the same magnetospheric environment. Opposite polarities may contribute to the coagulation of particles. For details see Švestka and Grün (1991) and Švestka et al. (1993).

The charging experiments also showed that the q/m versus i/m characteristic is practically linear when using either electron or ion beam, as long as the electric field strength is not too high. This means that the sticking coefficient of ions and electrons on surfaces of charged particles is constant, equal to one, and independent of both the impact energy of the ions or electrons and the electric field strength at the particle surface.

3. Charging by Ions

When glass particles were charged by ions it was found that strong discharging starts at certain field strengths, independent of the vacuum pressure. The discharging current was analyzed by the dynamic method described above and ascribed to field emission of ions from the particle surface. The threshold for this process is about 500 MV m^{-1}, depends on the history of the particle surface, and is significantly lower than the expected field strength of about 10 GV m^{-1}. The large discrepancy is probably attributable to the fact that the theoretical calculations were based on electrically conductive particles. In the case of dielectric materials, the electric field penetrates into the particle and substantially alters the conditions on the particle surface. For the same reason, the field emission of electrons from semiconductors starts at field strengths that are about one order of magnitude lower in comparison with metals, as has been shown experimentally.

Field emission prevents compact dielectric particles from electrostatic fragmentation. The highest attainable field strength of about 500 MV m^{-1} produces an "electrostatic stress" of about 1 MPa which is insufficient to fragment compact particles. A further consequence of the lower field emission threshold is that every positively charged particle with a radius of curvature < 10 nm is discharged at +5 V, a typical surface potential of interplanetary particles. As a result, particles made of the same material but of different sizes and shapes can be charged to different surface potentials. For more details about recent measurements see Čermák et al. (1995).

Electrostatic fragmentation of loosely bound particles having diameters of 2...20 μm and made of Al_2O_3 (also some of glass or tungsten) was also studied. Their tensile strengths were found (from q/m and d at the time of disruption) to range from 1 to 100 kPa, compatible with theoretical estimates, and to be proportional to the inverse square of the particle diameter. If this relation holds true, the onset of fragmentation depends on the electrostatic potential Φ

rather than the field strength E_s. In addition, if Φ is practically independent of particle size, as is expected in many environments for particles within the size range studied, the onset of fragmentation is also independent of particle size. For details see Švestka and Grün (1992a,b).

Acknowledgements

During the preparation of this manuscript, I was a guest at the Max-Planck-Institut für Kernphysik, Heidelberg, Germany. J. Švestka contributed to section IV.B. My special thanks go to O.E. Berg, J. Kissel, J.N. Swab, and A.J. Tuzzolino for helpful suggestions to improve the manuscript.

REFERENCES

Ang, J. A. 1990. Impact flash jet initiation phenomenology. *Int. J. Impact Eng.*, **10**, pp. 23–33.
Auer, S. 1972. Cosmic dust impact location detector. U.S. Patent no. 3,694,655.
Auer, S. 1974. The asteroid belt: Doubts about the particle concentration measured with the Asteroid/Meteoroid Detector on Pioneer 10. *Science*, **186**, pp. 650–652.
Auer, S. 1975. Two high resolution velocity vector analyzers for cosmic dust particles. *Rev. Sci. Instrum.*, **46**, pp. 127–135.
Auer, S. 1982. Imaging by dust rays: a dust ray camera. *Optica Acta*, **29/10**, pp. 1421–1426.
Auer, S. 1984. Space debris monitor. Feasibility study. Final Report to NASA Johnson Space Center under contract NAS9–17028.
Auer, S. 1994 a. Plasma produced by impacts of fast dust particles on a thin film. *LPI Tech. Rpt.* **94-05**, pp. 21–25.
Auer, S. 1994b. CDCF trajectory sensor development and calibration. Final Report for NASA Johnson Space Center under purchase order no. T-2899T.
Auer, S. 1996. Accuracy of a velocity/trajectory sensor for charged dust particles. In *Physics, Chemistry, and Dynamics of Interplanetary Dust*, eds. B. Å. S. Gustafson and M. S. Hanner (Provo, Utah: Astronomical Soc. of the Pacific Press), pp. 251–254.
Auer, S. 1998. Impact ionization from silica aerogel. *Int. J. Impact Eng.*, **21**, pp. 89–95.
Auer, S. and Berg, O.E. 1975. Composition analyzer for microparticles using a spark ion source. *Rev. Sci. Instrum.*, **46**, pp. 1530–1534.
Auer, S., and Sitte, K. 1968. Detection technique for micrometeoroids using impact ionization. *Earth Planet. Sci. Letters*, **4**, pp. 178–183.
Auer, S., and von Bun, F. O. 1994. Highly transparent and rugged sensor for velocity determinations of cosmic dust particles. *LPI Tech. Rpt.*, **94-05**, pp. 25–29.
Beard, R. 1991. Impacts on the meteoroid and rear shields of the Giotto spacecraft at the GEM encounter with Grigg-Skjellerup. In *Hypervelocity Impacts in Space*, ed. J. A. M. McDonnell (Canterbury: Univ. of Kent), pp. 94–99.
Berg, O. E., and Grün, E. 1973. Evidence of hyperbolic cosmic dust particles. *Space Research*, **XIII**, (Berlin: Akademie-Verlag), pp. 1047–1055.
Berg, O. E., and Meredith, L. H. 1956. Meteorite impacts to altitude of 103 kilometers. *J. Geophys. Res.*, **61**, pp. 751–754.
Berg, O. E., and Richardson, F. F. 1969. The Pioneer 8 Cosmic Dust Experiment. *Rev. Sci. Instrum.*, **40**, pp. 1333–1337.
Berg, O. E., Richardson, F. F., and Burton, H. 1973. Lunar ejecta and meteorites experiment. Apollo 17 Prelim. Science Report, NASA SP-330, 16–1.
Berg, O. E., Wolf, H., and Rhee, J. 1975. Lunar soil movement registered by the Apollo cosmic dust experiment. *Proc. IAU Colloq.*, **31**, pp. 233–238.
Bohn, J. L., and Nadig, F.H. 1950. Researches in the physical properties of the upper atmosphere with special emphasis on acoustical studies with V-2 rockets. Report No.8 (Research Institute of Temple University), pp. 1–26.
Boehnhardt, H., and Fechtig, H. 1987. Electrostatic charging and fragmentation of dust near P/Giacobini-Zinner and P/Halley. *Astron. Astrophys.*, **187**, pp. 824–828.
Burchell, M. J., Kay, L., and Ratcliff, P. R. 1996. Use of combined light flash and plasma measurements to study hypervelocity impact processes. *Adv. Space Res.*, **17/12**, pp. (12)141–(12)145.
Burton, W. M. 1983. Cometary particle impact simulation using pulsed lasers. *Adv. Space Res.*, **2/12**, pp. 255–258.
Cable, A. J. 1970. Hypervelocity accelerators. In *High-velocity impact phenomena*, ed. Ray Kinslow (New York, London: Academic Press), pp. 1–21.
Capaccioni, F., and McDonnell, J. A. M. 1986. Experimental measurement of particle deceleration and survival in multiple thin foil targets. *Adv. Space Res.*, **6/7**, pp. 17–20.
Čermák, I., Grün, E., and Švestka, J. 1995. New results in studies of electric charging of dust particles. *Adv. Space Res.*, **15**, pp. (10)59–(10)64.
Chow, V. W., Mendis, D. A., and Rosenberg, M. 1993. Role of grain size and particle velocity distribution in secondary electron emission in space plasmas. *J. Geophys. Res.*, **98**, pp. 19,065–19,076.

Christiansen, E. L. 1993. Design and performance equations for advanced meteoroid and debris shields. *Int. J. Impact Eng.*, **14**, pp. 145–156.

Christiansen, E. L., and Kerr, J. H. 1993. Mesh double-bumper shield: a low-weight alternative for spacecraft meteoroid and orbital debris protection. *Int. J. Impact Eng.*, **14**, pp. 169–180.

Crozier, W. D., and Hume, W. 1957. High-velocity, light-gas gun. *J. Appl. Phys.*, **28**, pp. 892–894.

Cour-Palais, B. G., and Crews, J. L. 1990. A multi-shock concept for spacecraft shielding. *Int. J. Impact Eng.*, **10**, pp. 135–146.

Dalmann, B.-K., Grün, E., Kissel, J., and Dietzel, H. 1977. The ion composition of the plasma produced by impact of fast particles. *Planet. Space Sci.*, **25**, pp. 135–147.

Dietzel, H., Neukum, G., and Rauser, P. 1972. Micrometeoroid simulation studies on metal targets. *J. Geophys. Res.*, **77**, pp. 1375–1395.

Dietzel, H., Eichhorn, G., Fechtig, H., Grün, E., Hoffmann, H.-J., and Kissel, J. 1973. The HEOS 2 and Helios micrometeoroid experiments. *J. Phys.(E) Scientific Instrum.*, **6**, pp. 209–217.

Dow, K. L., Sykes, M. V., Low, F. J., and Vilas, F. 1990. The detection of Earth orbiting objects by IRAS. *Adv. Space Res.*, **10**, pp. (3)381–(3)384.

Draine, B. T., and Salpeter, E. E. 1979. On the physics of dust grains in a hot gas. *Astrophys. J.*, **231**, pp. 77–94.

Drapatz, G., and Michel, K. W. 1974. Theory of shock-wave ionization upon high-velocity impact of micrometeorites. *Z. Naturforsch.*, **29 a**, pp. 870–879.

Edenhofer, P., Bird, M. K., Brenkle, J. P., Buschert, H., Esposito, P. B., Porsche, H., and Volland, H. 1986. First results from the Giotto radio-science experiment. *Nature*, **321**, pp. 355–357.

Eichhorn, G. 1974. Untersuchung der Lichtemission bei Hochgeschwindigkeitseinschlägen. Dissertation (University of Heidelberg, Germany).

Eichhorn, G. 1975. Measurements of the light flash produced by high velocity particle impact. *Planet. Space Sci.*, **23**, pp. 1519–1525.

Eichhorn, G. 1976. Analysis of the hypervelocity impact process from impact flash measurements. *Planet. Space Sci.*, **24**, pp. 771–781.

Eichhorn, G. 1978a. Heating and vaporization during hypervelocity particle impact. *Planet. Space Sci.*, **26**, pp. 463–467.

Eichhorn, G. 1978 b. Primary velocity dependence of impact ejecta parameters. *Planet. Space Sci.*, **26**, pp. 469–471.

Fechtig, H., Grün, E., and Morfill, G. 1979. Micrometeoroids within ten Earth radii. *Planet. Space Sci.*, **27**, pp. 511–531.

Friichtenicht, J. F. 1962. Two-million-Volt electrostatic accelerator for hypervelocity research. *Rev. Sci. Instrum.*, **34**, pp. 209–212.

Friichtenicht, J. F. 1964. Micrometeoroid simulation using nuclear accelerator techniques. *Nucl. Instrum. Meth.*, **28**, pp. 70–78.

Göller, J. R., and Grün, E. 1989. Calibration of the Galileo/Ulysses dust detectors with different projectile materials and at varying impact angles. *Planet. Space Sci.*, **37**, pp. 1197–1206.

Grün, E. 1981. Physikalische und chemische Eigenschaften des interplanetaren Staubes – Messungen des Mikrometeoritenexperimentes auf Helios. Bundesministerium für Forschung und Technologie, Report BMFT-FB-W 81-034.

Grün, E. 1984. Impact ionization from gold, aluminium and PCB-Z. In *The Giotto Spacecraft*, eds. E. Wolfe and B. Battrick, ESA SP-224, pp. 39–41.

Grün, E., Berg, O. E., and Dohnanyi, J. S. 1973. Reliability of cosmic dust data from Pioneers 8 and 9. *Space Research*, **XIII**, (Akademia-Verlag), pp. 1057–1062.

Grün, E., Fechtig, H., Gammelin, P., and Kissel, J. 1975. Das Staubexperiment auf Helios (E10). *Raumfahrtforschung*, **19**, pp. 268–269.

Grün, E., Fechtig, H., Gammelin, P., Kissel, J., Auer, S., Braun, G., Dalman, B.-K., Dietzel, H., and Hoffmann, H.-J. 1979. Das Helios-Mikrometeoritenexperiment (Sonnensonde Helios A und B – Experiment 10). Bundesministerium für Forschung und Technologie, Report BMFT-FB-W 79-09.

Grün, E., Fechtig, H., Hanner, M. S., Kissel, J., Lindblad, B.-A., Linkert, D., Maas, D., Morfill, G. E. and Zook, H. A. 1992 a. The Galileo dust detector. *Space Sci. Rev.*, **60**, pp. 317–340.

Grün, E., Fechtig, H., Giese, R. H., Kissel, J., Linkert, D., Maas, D., McDonnell, J. A. M., Morfill, G. E., Schwehm, G. and Zook, H. A. 1992b. The Ulysses dust experiment. *Astron. Astrophys. Suppl. Ser.*, **92**, pp. 411–423.

Grün, E., Morfill, G. E., and Mendis, D. A. 1984. Dust-magnetosphere interactions. In *Planetary Rings*, eds. R. Greenberg and A. Brahic (Tucson: Univ. of Arizona Press), pp. 275–332.

Grün, E. and Rauser, P. 1969. Penetration studies of iron dust particles in thin foils. *Space Research*, **IX**, eds. K.S.W. Champion et al., pp. 147–154.

Grün, E. and 22 co-authors 1993. Discovery of jovian dust streams and interstellar grains by the Ulysses spacecraft. *Nature*, **362**, pp. 428–430.

Gurnett, D. A., Grün, E., Gallagher, D., Kurth, W. S., and Scarf, F. L. 1983. Micron-sized particles detected near Saturn by the Voyager plasma wave instrument. *Icarus*, **53**, pp. 236–254.

Gurnett, D. A., Kurth, W. S., Scarf, F. L., Burns, J. A., Cuzzi, J. N., and Grün, E. 1987. Micron-sized particle impacts detected near Saturn by the Voyager 2 plasma wave instrument. *J. Geophys. Res.*, **92**, pp. 14,959–14,968.

Hansen, D. O. 1968. Mass analysis of ions produced by hypervelocity impact. *Appl. Phys. Letters*, **13**, pp. 89–91.

Hastings, E. C. 1964. The Explorer XVI micrometeoroid satellite. Supplement III, preliminary results for period May 27, 1963 through July 22, 1963. NASA TM X-949.

Hoffmann, H.-J., Fechtig, H., Grün, E. and Kissel, J. 1975a. First results of the micrometeoroid experiment S 215 on the HEOS 2 satellite. *Planet. Space Sci.*, **23**, pp. 215–224.

Hoffmann, H.-J., Fechtig, H., Grün, E., and Kissel, J. 1975b. Temporal fluctuations and anisotropy of the micrometeoroid flux in the Earth-moon system measured by HEOS 2. *Planet. Space Sci.*, **23**, pp. 985–991.

Hornung, K., and Kissel, J. 1994. On shock wave impact ionization of dust particles. *Astron. Astrophys.*, **291**, pp. 324–336.

Hornung, K., Malama, Yu. G., and Thomas, K. 1996. Modeling of the very high velocity impact process with respect to in-situ ionization measurements. *Adv. Space Res.*, **17/12**, pp. (12)77–(12)86.

Hörz, F., Cintala, M. J., Bernhard, R. P. and See, T. H. 1994. Dimensionally scaled penetration experiments: aluminum targets and glass projectiles 50 μm to 3.175 mm in diameter. *Int. J. Impact Eng.*, **15**, pp. 257–280.

Hörz, F., Cintala, M. J., Bernhard, R. P., Cardenas, F., Davidson, W. E., Haynes, G., See, T. H., and Winkler, J. L. 1995. Penetration experiments in aluminium 1100 targets using soda-lime glass projectiles. NASA Technical Memorandum 104813.

Humes, D. H., Alvarez, J. M., O'Neal, R. L., and Kinard, W. H. 1974. The interplanetary and near-Jupiter meteoroid environments. *J. Geophys. Res.*, **79,25**, pp. 3677–3684.

Hüdepohl, A., Rott, M., and Igenbergs, E. 1989. Coaxial plasma accelerator with compression coil and radial gas injection. *IEEE Trans. Magnetics*, **25**, pp. 232–237.

Igenbergs, E., Aigner, S., Hüdepohl, A., Iglseder, H., Kuczera, H., Rott, M., and Weishaupt, U. 1987. Launcher technology, in-flight velocity measurement and impact diagnostics at the TUM/LRT. *Int. J. Impact Eng.*, **5**, pp. 371–380.

Igenbergs, E., Hüdepohl, A., Uesugi, K., Hayashi, T., Svedhem, H., Iglseder, H., Koller, G., Glasmachers, A., Grün, E., Schwehm, G., Mizutani, H., Yamamoto, T., Fujimura, A., Ishii, N., Araki, H., Yamakoshi, K. and Nogami, K. 1991. The Munich dust counter—A cosmic dust experiment on board of the MUSES—A mission of Japan. In *Origin and evolution of interplanetary dust*, eds. A. C. Levasseur-Regourd et al., (Kluwer Academic Publishers), pp. 45–48.

Isbell, W. M. 1987. Historical overview of hypervelocity impact diagnostic technology. *Int. J. Impact Eng.*, **5**, pp. 389–410.

Jean, B. and Rollins, T. L. 1970. Radiation from hypervelocity impact generated plasma. *AIAA Journal*, **8**, pp. 1742–1748.

Kassel, P. C., Jr. 1973. Characteristics of capacitor-type micrometeoroid flux detectors when impacted with simulated micrometeoroids. Technical Note D-7359, NASA, Washington.

Keaton, P. W., Idzorek, G. C., Rowton Sr., L. J., Seagrave, J. D., Stradling, G. L., Bergeson, S. D., Collopy, M. T., Curling Jr., H. L., McColl, D. B., and Smith, J. D. 1990. A hypervelocity-microparticle-impacts laboratory with 100 km/s projectiles. *Int. J.*

Impact Eng., **10**, pp. 295–308.

Kern, H. E., and McKenzie, J. M. 1970. Noise studies of ceramic encapsulated junction field effect transistors (JFETs). *IEEE Trans. Nucl. Sci.*, **17/3**, pp. 425–432.

Kissel, J. 1986. The Giotto particulate impact analyser. ESA **SP-1077**, pp. 67–83.

Kissel, J., and Krueger, F. R. 1987. Ion formation by impact of fast dust particles and comparison with related techniques. *Appl. Phys. A*, **42**, pp. 69–85.

Knabe and Krueger 1982. Ion formation from alkali iodide solids by swift dust particle impact. *Z. Naturforsch.*, **37a**, pp. 1335–1340.

Krueger, F. R. 1996. Ion formation by high- and medium-velocities dust impacts from laboratory measurements and Halley results. *Adv. Space Res.*, **17/12**, pp. (12)71–(12)75.

Leese, M. R., McDonnell, J. A. M., Green, S. F., Busoletti, E., Clark, B. C., Colangeli, L., Crifo, J. F., Eberhardt, P., Giovane, F., Grün, E., Gustafson, B., Hughes, D. W., Jackson, D., Lamy, P., Langevin, Y., Mann, I., McKenna-Lawlor, S., Tanner, W. G., Weissman, P. R., and Zarnecki, J. C. 1996. Dust flux analyser experiment for the Rosetta mission. *Adv. Space Res.*, **17/12**, pp. 137–140.

Leinert, C., and Klüppelberg, D. 1974. Stray light suppression in optical space experiments. *Applied Optics*, **13**, pp. 556–564.

Leinert, C., Link, H., Pitz, E., Salm, N., and Klüppelberg, D. 1975. The Helios zodiacal light experiment (E9). *Raumfahrtforschung*, **19/5**, pp. 264–267.

Leinert, C., Pitz, E., Link, H., and Salm, N. 1981. Calibration and in-flight performance of the zodiacal light experiment on Helios. *Space Science Instrumentation*, **5**, pp. 257–270.

Mamyrin, B. A., Karataev, V. I., Shmikk, D. V., and Zagulin, V. A. 1973. The mass-reflectron, a new non-magnetic time-of-flight mass spectrometer with high resolution. *Zh. Eksp. Teor. Fiz.*, **64**, pp. 82–89, and *Sov. Phys.-JETP*, **37**, pp. 45–48 (in English).

Martelli, G., and Cerroni, P. 1983. Hypervelocity acceleration techniques: a review of existing capabilities and prospects for future developments. *Adv. Space Res.*, **2**, pp. 259–268.

McDonnell, J. A. M. 1970. Factors affecting the choice of foils for penetration experiments in space. *Space Research*, **X**, (North Holland), pp. 314–325.

McDonnell, J. A. M., and Abellanas, C. 1972. A technique for position sensing and improved momentum evaluation of microparticle impacts in space. *Rev. Sci. Instrum.*, **43**, pp. 1214–1216.

McDonnell, J. A. M., Alexander, M., Lyons, D., Tanner, W., Anz, P., Hyde, T., Chen, A.-L., Stevenson, T. J., and Evans, S. T. 1984. The impact of dust grains on fast fly-by spacecraft: momentum multiplication, measurements and theory. *Adv. Space Res.*, **4/9**, pp. 297–301.

McDonnell, J. A. M., and 24 co-authors 1986a. The Giotto dust impact detection system. ESA **SP-1077**, pp. 85–107.

McDonnell, J. A. M. and 27 co-authors 1986b. Dust density and mass distribution near comet Halley from Giotto observations. *Nature*, **321**, pp. 338–341.

McDonnell, J. A. M., and Sullivan, K. 1992. Hypervelocity impacts on space detectors: decoding the projectile parameters. *Proc. Hypervelocity Impacts in Space*, ed. J. A. M. McDonnell (Canterbury: Univ. of Kent, 1–5 July 1991), pp. 39–47.

McMillan, A. R. 1968. Experimental investigations of simulated meteoroid damage to various spacecraft structures. Contractor Report, contract no. NAS9-3081, NASA CR-915, p. 89.

Meshejian, W. K., Ramamurti, K., Trower, W. P., and Wollan, D. S. 1970. A gas density detector for use in space. *J. Spacecr. Rockets*, **7**, pp. 1228–1233.

Miller, M. S., Evans, D. C., Moseley, H., and Ludwig, U. W., 1982. Optical design of the Diffuse Infrared Background Experiment for NASA's Cosmic Background Explorer. *SPIE 331 Instrumentation in Astronomy*, **IV**, pp. 483–489.

Münzenmayer, R. 1995. Beiträge zur experimentellen Erforschung des Staubes im Weltall. Ph. D. Thesis, (München: Technische Universität).

Naumann, R. J., Jex, D. W., and Johnson, C. L. 1969. Calibration of Pegasus and Explorer XXIII detector panels. NASA Technical Report R-321.

Oberc, P. 1996. Electric antenna as a dust detector. *Adv. Space Res.*, **17/12**, pp. 105–110.

Perruchot, S., Lamy, P. L., Giovane, F., and Gustafson, B. Å. S. 1996. Concepts for dust velocity measurements on a cometary orbiter. Proceedings, IAU Colloquium 150.

Peterson, R. 1994. Charge collection during hypervelocity penetrations of thin foils. *LPI*

Tech. Rpt. **94-05**. pp. 64–76.
Ratcliff, P. R., McDonnell, J. A. M., Firth, J. G., and Grün, E. 1992. The cosmic dust analyser. *J. Brit. Interplan. Soc.*, **45**, pp. 375–380.
Ratcliff, P. R., Gogu, F., Grün, E., and Srama, R. 1996. Plasma produced by secondary impacts: implications for velocity measurements by in-situ dust detectors. *Adv. Space Res.*, **17/12**, pp. (12)111–(12)115.
Rudolph, V. 1966. Massen-Geschwindigkeitsfilter für künstlich beschleunigten Staub. *Z. Naturforsch.*, **21a**, pp. 1993–1996.
Schleicher, B., Burtscher, H., and Siegmann, H.C. 1994. Photoelectric quantum yield of nanometer metal particles. *Applied Phys. Letters*, **63(9)**, p. 1191.
Seigel, A. E. 1979. Theory of high-muzzle-velocity guns. In *Interior ballistics of guns*, eds. H. Krier and M. Summerfield, *Progress in Astronautics and Aeronautics*, vol. **66**, (published by the AIAA), pp. 135–175 (Eq. 23).
Shelton, H., Hendricks Jr., C. D., and Wuerker, R. F. 1960. Electrostatic acceleration of microparticles to hypervelocities. *J. Appl. Phys.*, **31**, pp. 1243–1246.
Simpson, J. A., Rabinowitz, D., Tuzzolino, A. J., Ksanfomality, L. V., and Sagdeev, R. Z. 1987. The dust coma of comet P/Halley: measurements on the VeGa-1 and VeGa-2 spacecraft. *Astron. Astrophys.*, **187**, pp. 742–752.
Simpson, J. A., Rabinowitz, D., and Tuzzolino, A. J. 1989. Cosmic dust investigations I. PVDF detector signal dependence on mass and velocity for penetrating particles. *Nucl. Instr. and Meth.*, **A279**, pp. 611–624.
Simpson, J. A., and Tuzzolino, A. J. 1985. Polarized polymer films as electronic pulse detectors of cosmic dust particles. *Nucl. Instr. and Meth.*, **A236**, pp. 187–202.
Simpson, J. A., and Tuzzolino, A. J. 1989. Cosmic dust investigations II. Instruments for measurement of particle trajectory, velocity and mass. *Nucl. Instr. and Meth.*, **A279**, pp. 625–639.
Singer, S. F., Stanley, J. E., and Kassel, P. C. 1985. The LDEF interplanetary dust experiment. In *Properties and Interactions of Interplanetary Dust*, eds. R. H. Giese and P. Lamy (Dordrecht: Reidel), pp. 117–120.
Slattery, J. C., Becker, D. G., Hammermesh, B., and Roy, N. L. 1973. A linear accelerator for simulated micrometeors. *Rev. Sci. Instrum.*, **44**, pp. 755–762.
Soberman, R. K., Neste, S. L., and Lichtenfeld, K. 1974a. Particle concentration in the asteroidal belt from Pioneer 10. *Science*, **183**, pp. 320–321.
Soberman, R. K., Neste, S. L., and Lichtenfeld, K. 1974b. Optical measurement of interplanetary particulates from Pioneer 10. *J. Geophys. Res.*, **79, 25**, pp. 3685–3694.
Srama, R., Grün, E., and the Cassini Dust Science Team 1996. The cosmic dust analyzer for the Cassini mission to Saturn. In *Physics, Chemistry, and Dynamics of Interplanetary Dust*, eds. B. Å. S. Gustafson and M. S. Hanner (Provo, Utah: Astronomical Soc. of the Pacific Press), pp. 227 – 231.
Srama, R., Bradley, J. G., Grün, E., Ahrens, T. J., Auer, S., Cruise, M., Fechtig, H., Graps, A., Havnes, O., Heck, A., Helfert, S., Igenbergs, E., Jessberger, E. K., Johnson, T. V., Kempf, S., Krüger, H., Lamy, P., Landgraf, M., Linkert, D., Lura, F., McDonnell, J. A. M., Möhlmann, D., Morfill, G. E., Schwehm, G. H., Stübig, M., Švestka, J., Tuzzolino, A. J., Wäsch, R., and Zook, H. A. 2001. The Cassini Cosmic Dust Analyser. *Space Science Reviews*, special issue on Cassini, submitted.
Stilp, A. 1987. Review of modern hypervelocity impact facilities. *Int. J. Impact Eng.*, **5**, pp. 613–621.
Stradling, G. L., Idzorek, G. C., Keaton, P. W., Studebaker, J. K., Blossom, A. A. H., Collopy, M. T., Curling Jr. H. L., and Bergeson, S. D. 1990. Searching for momentum enhancement in hypervelocity impacts. *Int. J. Impact Eng.*, **10**, pp. 555–570.
Svedhem, H., and Pedersen, A. 1992. Behaviour of ejecta particles and generated plasma at hypervelocity impact. In *Hypervelocity Impacts in Space*, ed. J. A. M. McDonnell (Canterbury: University of Kent), pp. 72–77.
Švestka, J., and Grün, E. 1991. Methods, difficulties and first results in laboratory simulation of cosmic dust electric charging. In *Origin and Evolution of Interplanetary Dust*, eds. A. C. Levasseur-Regourd and H. Hasegawa (Dordrecht: Kluwer Acad. Publ.), pp. 367–370.
Švestka, J., and Grün, E. 1992a. Electrostatic fragmentation of dust particles. In *Hypervelocity Impacts in Space*, ed. J. A. M. McDonnell (Canterbury: Univ. of Kent), pp. 139–143.

Švestka, J., and Grün, E. 1992b. Electrostatic fragmentation of dust particles in laboratory. In *Astrochemistry of Cosmic Phenomena*, ed. P. D. Singh (Dordrecht: Kluwer Acad. Publ.), pp. 17–18.

Švestka, J., Čermák, I., and Grün, E. 1993. Electric charging and electrostatic fragmentation of dust particles in laboratory. *Adv. Space Res.*, **13**, pp. (10)199–(10)202.

Swift, H. F. 1987. High-speed image-forming instrumentation for hypervelocity impact studies. *Int. J. Impact Eng.*, **5**, pp. 623–634.

Swift, H. F., Bamford, R., and Chen, R. 1983. Designing space vehicle shields for meteoroid protection: a new analysis. *Adv. Space Res.*, **2/12**, pp. 219–234.

Timmermann, R. and Grün, E. 1991. Plasma emission from high velocity impacts of microparticles onto water ice. In *Origin and Evolution of Interplanetary Dust*, eds. A. C. Levasseur-Regourd et al., (Kluwer), pp. 375–378.

Tuzzolino, A. J. 1983. Pulse amplitude method for determining the pyroelectric coefficient of pyroelectric materials. *Nucl. Instr. Meth.*, **212**, pp. 505–516.

Tuzzolino, A. J. 1991. Two-dimensional position-sensing PVDF dust detectors for measurement of dust particle trajectory, velocity, and mass. *Nucl. Instr. Meth.*, **A301**, pp. 558–567.

Tuzzolino, A. J. 1992. PVDF copolymer dust detectors: particle response and penetration characteristics. *Nucl. Instr. and Meth.*, **A316**, pp. 223–237.

Tuzzolino, A. J. 1996. Applications of PVDF dust sensor systems in space. *Adv. Space Res.*, **17/12**, pp. (12)123–(12)132.

Vedder, J. F. 1963. Charging and acceleration of microparticles. *Rev. Sci. Instrum.*, **34**, pp. 1175–1183.

Weishaupt, U. 1987. Hypervelocity impact of small masses on large surfaces of piezoelectric ceramics. *Int. J. Impact Eng.*, **5**, pp. 663–670.

Welford, W. T., and Winston, R. 1989. High collection nonimaging optics. (San Diego: Academic Press).

Whipple, E. C. 1981. Potentials of surfaces in space. *Rep. Prog. Phys.*, **44**, pp. 1197–1250.

Whipple, F. L. 1947. Meteorites and space travel. *Astron. J.*, **52/1161**, p. 131.

Wiley, W. C., and McLaren, T. H. 1955. Time-of-flight mass spectrometer with improved resolution. *Rev. Sci. Instrum.*, **26**, pp. 1150–1157.

Wolfe, J. H., Ballard, R. W., Carle, G. C., and Bunch, T. E. 1986. A micrometeoroid deceleration and capture experiment: conceptual experiment design description. In *Trajectory Determinations and Collection of Micrometeoroids on the Space Station*, LPI Tech. Rpt. **86-05** (Houston TX), pp. 91–93.

Zel'dovich, Ya. B. 1968. EMF produced by a shock wave moving in a dielectric. *Soviet Physics JETP*, **26**, pp. 159–162. (Russian original in: *Zh. Eksp. Teor. Fiz.*, **53**, 1967, pp. 237–243).

Zolensky, M. E., Barrett, R. A., and Hörz, F. 1994. The use of silica aerogel to collect interplanetary dust in space. *LPI Tech. Rpt.* **94-05**, pp. 94–98.

Zscheeg, H., Kissel, J., Natour, Gh., and Vollmer, E. 1992. COMA - advanced space experiment for in situ analysis of cometary matter. *Astrophysics and Space Sciences*, **195**, pp. 447–461.

Physical Processes on Interplanetary Dust

Tadashi Mukai[1], Jürgen Blum[2], Akiko M. Nakamura[1],
Robert E. Johnson[3], Ove Havnes[4]

[1] Kobe University, Kobe, Japan
[2] University of Jena, Jena, Germany
[3] University of Virginia, Charlottesville, Virginia, USA
[4] University of Tromsø, Tromsø, Norway

Abstract. This chapter discusses physical processes affecting interplanetary dust grains, including processes determining dust formation, growth, disruption, and alteration. Computer simulations and laboratory studies of coagulation and aggregation show that mutual collisions between solid grains determine the growth of solid aggregates in the early solar system when the impact energies are too low to destroy the colliding grains. On the other hand, fragmentation occurs at the high impact energies that currently prevail, generating dust from meteoroids, asteroid, and satellite surfaces as well as from sublimating comets. Such impact processes are discussed based on a compilation of laboratory measurements (e.g., size, shape, velocity and spin distributions of fragments). Gradual alteration of the dust grains occurs in the present solar system due to solar radiation and energetic particle impact. Sublimation, sputtering and charging can alter the nature of interplanetary dust grains. Dust grain temperatures, erosion rates due to solar-wind-induced sputtering, and surface charges are studied. The evidence for alteration of physical/chemical/mineralogical properties of interplanetary dust grains is still rather poor. Therefore, we discuss the expected changes, referring to those physical processes that can be simulated in laboratory experiments, and to their analogues obtained through theoretical modeling.

I. INTRODUCTION

Small solid particles with typical sizes of 0.01-1 μm stick and form porous aggregates as they collide in the primordial solar nebula (e.g., Weidenschilling and Cuzzi 1997; Weidenschilling 1997; Wurm and Blum 1998). These aggregates grow to the order of km in diameter by further mutual collisions during settlement toward the central plane of the protoplanetary nebula (Weidenschilling and Cuzzi 1997). These km-sized planetesimals eventually become proto-planets by mutual gravitational interaction. This is the standard scenario describing the earliest history of the solar system. Recently, several improvements on the physical mechanisms in this model have been discussed (e.g., Weidenschilling and Cuzzi 1997; Weidenschilling 1997). For example, the coagulation of solid particles into fluffy or fractal aggregates is an important step leading to larger preplanetary dust grains. Very fluffy dust aggregates are found to be dissipative so that collisions among them lead to further particle growth. On the other hand, fragmentation occurs when the collision velocity among massive particles exceeds a critical value (Blum and Münch 1993; Blum and Wurm 2000).

These ancient solid aggregates are now unfortunately removed from interplanetary space and cannot be found and examined; even if they had survived the T-Tauri winds, the Poynting-Robertson effect combined with sublimation and other processes prevents dust and meteoroids from surviving on Keplerian orbits for 4.6 billion years. Exposure to interplanetary space also leads to disruption by mutual collisions with alteration through sublimation and sputtering that changes the grain properties.

If all ancient aggregates now have disappeared from the Solar System is still an open question. It is likely that some primitive bodies, such as comets and asteroids, may preserve them. It is widely accepted that IDPs, interplanetary dust particles, collected in the Earth's upper atmosphere are samples of primitive materials (see, e.g., Gibson 1992 or the chapter by Jessberger et al. this volume), because they are supplied primarily from the primitive bodies. However, even as these samples were stored inside the larger bodies, their physical/chemical/mineralogical properties might have been altered. For example, aqueous alteration has been proposed for hydrated interplanetary dust grains (see, e.g., Tomeoka 1991). Consequently, studies of such alteration processes can potentially reveal the "original" nature of interplanetary dust grains.

The short lifetime of interplanetary dust, compared to 4.6 billion years, suggests that solid particles observed in interplanetary space are supplied continuously. Comets, as well as asteroids, are possible candidates for such sources. Although the grains released from parent bodies are newly born members in the Solar System, they may have experienced some alteration after leaving their parent bodies. Nuclear tracks generated by solar high-energy particles, found by Bradley et al. (1984), suggest that grain properties are altered during the "short" stay in interplanetary space, of roughly 10^4 years (e.g., Johnson R. E. 1985). In addition, Bradley (1994) identified certain chemical anomalies that accumulate in grains exposed to ionizing radiation.

We need to examine the physical processes that determine their alteration in order to understand the original interplanetary dust grains. While little is known about the "real" alteration processes affecting the nature of interplanetary dust particles, research into the signatures of these phenomena has proceeded. For example, Pinho and Duley (1994) estimated, based on laboratory data, the chemical change in carbon species following irradiation with high-energy particles, i.e., CH_4 changes to CH_n ($n < 4$) after irradiation (Lanzerotti et al. 1987; Gustafson et al. in this volume). The alteration of the nature of dust grains by the energy deposited in it by plasma-ion and electron impact has been reviewed (e.g., Johnson 1990). Using theoretical predictions based on simulations in the laboratory, we can try to understand the spatial/temporal variations of dust grains in the interplanetary space, such as the dust free zone close to the sun. However, high quality observations are needed to confirm the alteration of dust grains due to solar radiation and/or solar energetic particles.

Recently, the Ulysses spacecraft found interstellar dust grains penetrating into the solar system in the vicinity of Jupiter (Grün et al. 1993) and to a heliocentric distance of at least 2.2 AU at high ecliptic latitudes (Baguhl et al. 1995). It is expected, therefore, that the interstellar dust component may be significant in the populations of interplanetary dust grains. Before the findings by Ulysses, it was believed that the interplanetary magnetic field, accompanying the solar wind, prevents charged, interstellar dust from penetrating the inner Solar System (e.g., Levy and Jokipii 1976), although Gustafson and Misconi (1979) showed that this depends on the solar cycle. In addition, solar radiation removes the volatile component of dust by sublimation. How could interstellar dust be in the inner Solar System? To study this, the relevant physical processes (e.g., charging of the grains and their interaction with the magnetic field, as well as the mass loss due to sublimation/sputtering) need to be examined in detail.

We focus on fundamental processes that influence the physical properties of dust particles in the Solar System, determining the origin and evolution of interplanetary dust grains. The collisional growth of solid particles is reviewed in section II while fragmentation during collisions is addressed in section III, sublimation in section IV, sputtering in section V, charging in section VI, and grain lifetimes in section VII.

II. COLLISIONAL GROWTH OF SOLID PARTICLES

Mutual collisions between solid bodies play an important role in many astrophysical and cosmophysical scenarios. Collision velocities range from below $10^{-3}\,\mathrm{m\,s^{-1}}$ for collisions between the ring particles around the giant planets to well above $10^4\,\mathrm{m\,s^{-1}}$ for collisional interactions between asteroids. Thus, the specific impact energies (kinetic impact energy per unit mass) can vary by more than 14 orders of magnitude. A whole variety of physical processes occur when the collision velocity ranges from extremely low to hypervelocity values. In this section, we concentrate on collision processes in which the specific impact energies are low enough so that the solid material of the colliding bodies is not destroyed. According to Johnson K. L. (1985), the physics of the impact regime is well described by the parameter

$$Z = \frac{\rho v^2}{Y_d}, \qquad (1)$$

where ρ is the density of the colliding solid material, v is the collision velocity, and Y_d denotes the dynamic yield strength of the material. For $Z < 10^{-6}$ (corresponding to collision velocities $v < \sim 0.1\,\mathrm{m\,s^{-1}}$), impacts are fully elastic, in the region $\sim 10^{-6} < Z < \sim 10^{-3}$ ($\sim 0.1 < v < \sim 5\,\mathrm{m\,s^{-1}}$), the transition from fully elastic to fully plastic deformation occurs, and the limit of the shallow indentation theory is exceeded when $Z > \sim 0.1$ ($v > \sim 100\,\mathrm{m\,s^{-1}}$). For the latter case, heating effects are no longer negligible and the hydrodynamic interaction of the colliding solid particles must be taken into account. Material fragmentation dominates at even higher velocities, which is discussed in section III.

The collisional growth of solid particles is in many cases of astrophysical interest due to weak surface- or volume-proportional attractive forces. In the following, we discuss solid particles for which low-velocity collisions lead to sticking (coagulation). The above-mentioned limitations are valid in these collisional interactions so that $v < {\sim}100\,\mathrm{m\,s^{-1}}$ which will be assumed in the following discussions.

The interesting cosmophysical applications of low velocity grain-grain collisions with surface force-induced coagulation are mainly the origin of the solid bodies in the early solar system, i.e., the growth of planetesimals in the solar nebula, aerosol aggregation processes in planetary atmospheres, and the mutual interactions between the particles in the ring systems of the giant planets. The collision velocities of interest range from below $10^{-3}\,\mathrm{m\,s^{-1}}$ to approximately 10 $\mathrm{m\,s^{-1}}$, the particle sizes are between μm or even sub-μm dust particles in the pre-planetary nebula and decimeter-sized boulders in planetary rings.

II.A. Two-particle Collisions

Hertz (1882) was first to develop a formulation for the collision of two solid particles. Hertz described the frictionless contact problem of two perfectly elastic solids with radii of curvature R_1 and R_2 for small deformations. In the simplified case of spherical bodies, the contact area between the two spheres is circular of radius a. Hertz proposed a pressure distribution inside this contact area of the form

$$p = p_0 \left[1 - \left(\frac{r'}{a}\right)^2\right]^{1/2}, \tag{2}$$

in which p_0 is the maximum local pressure and r' is the radial distance from the midpoint in the contact circle. Integrating the pressure distribution over the contact area gives the total load (i.e., the total force):

$$P = \frac{2}{3} p_0 \pi a^2. \tag{3}$$

Summing all material parameters in the reduced modulus of elasticity

$$E^* = \left(\frac{1 - \nu_1^2}{E_1} + \frac{1 - \nu_2^2}{E_2}\right)^{-1}, \tag{4}$$

with E_1 and E_2 being the Young's moduli and ν_1 and ν_2 being the Poisson's ratios of the materials of the two contacting bodies, a relation between the total load P, and the size of the contact area can be formulated

$$a = \left(\frac{3PR}{4E^*}\right)^{1/3}. \tag{5}$$

Here, $R = (1/R_1 + 1/R_2)^{-1}$ is the reduced radius of the two spherical bodies. In practical problems, the total displacement δ of the centers of the two spheres, given by the sum of the individual total compressions, $\delta = \delta_1 + \delta_2$, is a value of interest and is given by

$$\delta = \frac{a^2}{R} = \left(\frac{9P^2}{16RE^{*2}}\right)^{1/3}. \tag{6}$$

The Hertz theory is valid as long as no attractive or repulsive forces other than the elastic reaction of the contacting bodies are present. The onset of plastic yield is another limitation to the theory so that the maximum local pressure inside the contact area, p_0, must not exceed the critical yield stress for compression Y.

In the quasi-static limit, the Hertz theory may also be adopted for the description of the central (head-on) collisions between two perfectly elastic bodies. In the case of spherical dust particles with masses m_1 and m_2 and reduced mass $\mu = (1/m_1 + 1/m_2)^{-1}$, the equation of motion along the axis connecting the centers of the two spheres is:

$$\mu \frac{d^2\delta}{dt^2} + P = 0. \tag{7}$$

Using the relation between the contact force P and the displacement δ in Eq. (6), we get:

$$\mu \frac{d^2\delta}{dt^2} + \frac{4}{3}R^{1/2}E^*\delta^{3/2} = 0. \tag{8}$$

For the determination of the displacement versus time and force versus time dependence, Eq. (8) must be solved numerically. Analytic solutions for the maximum compression δ^* and for the total collision time T_c,

$$\delta^* = \left(\frac{15\mu v^2}{16R^{1/2}E^*}\right)^{2/5} \tag{9}$$

and

$$T_c = 2.87 \left(\frac{\mu^2}{RE^{*2}v}\right)^{1/5}, \tag{10}$$

give first estimates of the collisional properties of "real" dust grains.

1. Dynamic Interactions

Such a quasi-static Hertzian approach cannot account for physical processes on short time scales. In general, it does not describe energy loss processes in mutual dust particle collisions that are in particular responsible for the phenomena of particle sticking and subsequent dust aggregation.

A variety of energy loss processes can be responsible for the occurrence of particle sticking. Among those, hysteresis-type deformations, excitation of surface vibrations (Chokshi et al. 1993), plasticity (see, e.g., Johnson K. L. 1985 and references therein), impact heating (Leliwa-Kopystynski et al. 1984), tangential surface friction, and, strongly related to the latter, the transfer between translational and rotational degrees of freedom (Blum and Münch 1993) are the most important in collisional interactions between electrically neutral dust particles.

A phenomenological description of the outcome of mutual dust grain collisions can be given by the so-called coefficient of restitution ϵ which is defined by the ratio between the relative velocity of rebound v' and the relative velocity of approach v

$$\epsilon = v'/v. \tag{11}$$

In this context, $1 - \epsilon^2$ gives the fractional energy loss in such a collision and the critical velocity below which sticking occurs is given by (Dahneke 1971)

$$v_s = \left[\frac{2}{\mu \epsilon^2}(U_r - \epsilon^2 U_i)\right]^{1/2}. \tag{12}$$

Here, U_i and U_r are the contact potentials of the two colliding particles during incidence and at rebound. For symmetric potentials ($U_i = U_r \equiv U$), Eq. (12) reduces to

$$v_s = \left(\frac{2U}{\mu}\frac{1-\epsilon^2}{\epsilon^2}\right)^{1/2}. \tag{13}$$

Perfectly elastic central collisions with Hertzian forces only, give $\epsilon = 1$ and, hence, $v_s = 0$. For elastic-plastic central impacts at moderate speeds, a more general discussion of plasticity leads to

$$\epsilon \approx 3.8 \left(\frac{Y_d}{E^*}\right)^{1/2} Z^{-1/8}, \tag{14}$$

with Z taken from Eq. (1) (see, e.g., Johnson K. L. 1985). In the case of non-sliding surfaces (e.g., in rough aggregate-aggregate collisions), grazing (tangential) collisions between non-rotating spherical objects lead to an effective coefficient of restitution of

$$\epsilon = 5/7 \tag{15}$$

(Blum and Münch 1993) that is independent of the two spheres' radii.

2. Particle Sticking

However, the coagulation of dust particles requires, besides an energy loss during the collision, attractive forces that act against the repulsion of the grains in the elastic rebound phase. If we assume that the dust grains are smooth, spherical, and consist of a material for which the elastic continuum theory is valid at the collision velocities of interest, a quasi-static approach based on the contact mechanics by Johnson et al. (1971) predicts that upon impact a neck forms due to the long-range van der Waals attraction of the two bodies. The formation process of this contact neck happens on a time scale much shorter than the collisional time scale. Since static elastic Hertz theory is applied, the description of the contact potential and contact force is straightforward. Under the assumption that the attractive surface potential can be formulated by

$$U_s = -2\pi a^2 \gamma, \tag{16}$$

in which a denotes the radius of the common contact circle and γ is the material-dependent specific surface energy, the interaction force $F \geq -F_c$ is given by

$$\frac{F}{F_c} = 4\left(\frac{a}{a_0}\right)^3 - 4\left(\frac{a}{a_0}\right)^{3/2} \tag{17}$$

and the total interaction potential, the sum of elastic and surface potential, can be expressed as

$$\frac{U_T}{F_c \delta_c} = 2 \cdot 6^{1/3} \left[\frac{8}{5}\left(\frac{a}{a_0}\right)^5 - \frac{8}{3}\left(\frac{a}{a_0}\right)^{7/2} + \frac{2}{3}\left(\frac{a}{a_0}\right)^2 \right]. \tag{18}$$

The normalization factors in Eqs. (17) and (18) are the separation force

$$F_c = 3\pi \gamma R, \tag{18a}$$

the radius of the equilibrium contact area

$$a_0 = \left(\frac{9\pi \gamma R^2}{E^*}\right)^{1/3}, \tag{18b}$$

and the tear-off separation distance

$$\delta_c = \frac{1}{2} \frac{a_0^2}{6^{1/3} R}. \tag{18c}$$

The relation between the displacement δ and the radius of the contact area is given by

$$\frac{\delta}{\delta_c} = 6^{1/3} \left[2\left(\frac{a}{a_0}\right)^2 - \frac{4}{3}\left(\frac{a}{a_0}\right)^{1/2} \right]. \tag{19}$$

A static scenario for zero impact velocity collision behavior of spherical dust grains can be derived using the formulation of Eqs. (16) through (19). Upon impact, an immediate formation of a connecting neck between the dust particles is assumed ($\delta/\delta_c = 0$). As this neck is formed at sound speed and as the collision velocity is much smaller than that, a jump in the force (F/F_c) and potential [$U/(F_c\delta_c)$] curves results. The gain of potential energy at the point of first contact, $U/(F_c\delta_c)(\delta = 0) = -8 \cdot 4^{2/3}/15$, is converted into kinetic energy that pulls the particles together. Compression stops for $U = 0$ at the instance of maximum compression, $\delta/\delta_c = 2.794$, and is reversed into expansion for the quasi-Hertzian case. At $\delta = 0$, the point of initial contact, the symmetry between the incoming and outgoing dust particles is broken due to the still existing contact neck between the grains. As there is no energy loss in a quasi-Hertzian contact, the dust grains separate until $U = 0$. This is reached for a separation $\delta/\delta_c = -0.882$ at which the interaction force (Eq. (17)) is still attractive, $F/F_c(U = 0) = -0.850$ so that the particles do not separate. Hence, the two-sphere system oscillates about the point of minimum total energy, i.e., $F/F_c = 0$, until the excess energy is dissipated. In the case of non-zero collision velocities, the maximum separation is larger until tear-off is reached at $\delta/\delta_c = -1$. The condition for tear-off is the limit of the stability of the connecting neck between the contacting spheres $\partial F/\partial a \to \infty$, similar to the break-up of liquid bridges. The excess energy needed for the separation of two spheres is $U/(F_c\delta_c)(\delta = -\delta_c) = 4/45$ which leads to the formulation of the absolute minimum of the critical sticking velocity v_s for central collisions

$$v_s = \left(\frac{8}{45}\frac{F_c\delta_c}{\mu}\right)^{1/2}. \qquad (20)$$

The above-described collisional energy loss due to a hysteresis-type deformation asymmetry is not the only irreversible energy dissipation channel. Chokshi et al. also investigated the excitation of elastic waves in low velocity dust particle collisions (see Chokshi et al. 1993 and references therein). An analysis of this energy loss mechanism and corrections to the formalism by Dominik and Tielens (1997) lead to a somewhat larger threshold velocity for sticking for centrally colliding spherical dust grains

$$v_s \approx 1.07 \frac{\gamma^{5/6}}{E^{*1/3}R^{5/6}\rho^{1/2}}. \qquad (21)$$

Recent experimental work by Poppe and Blum (1997) and Poppe et al. (1999) in which micron-size spherical SiO_2 dust grains impacted a flat SiO_2 target in vacuum showed that for dust particles with these perfect geometries a well-defined threshold velocity for sticking exists at much larger velocities than predicted by theory. The experimental and theoretical values for the critical sticking velocity v_s differ by more than an order of magnitude. Obviously, the energy dissipation mode used in the theoretical model (i.e., elastic surface waves) is not the only relevant physical dissipation process and there is now strong experimental evidence that micron-sized silicate grains coagulate when colliding at velocities below ~ 1 m/s.

II.B. Aggregation Phenomena

When the collision velocities in dust particle impacts are sufficiently small so that inter-particle sticking is a frequent process, a sequence of such coagulation processes leads to the aggregation of the dust grains into larger dust clusters. A variety of scientific publications deal with such aggregation processes and the morphologies and structures of the resulting agglomerates.

In the "classical" approach to the aggregation problem, Monte Carlo techniques are used to simulate the aggregation of dust particles and lots of simplifications have been introduced to generalize the results. Smirnov (1990) and Meakin (1991) have written recent reviews on aggregation phenomena. These, and references therein may lead the interested reader to the details of the aggregation simulations.

In general, the growth of a specific dust particle cluster (target cluster) is followed through a sequence of randomly oriented and directed mutual collision processes that may result in sticking. For simplicity, the target cluster as well as the projectile consists of the same type of particle, usually spherical dust grains of a single size. Depending on whether the projectile is a single dust grain or a cluster of size comparable to the target cluster, the aggregation process is denoted as *particle-cluster aggregation* (PCA) or *cluster-cluster aggregation* (CCA), respectively. Moreover, the structure of growing dust aggregates is also dependent on the transport processes responsible for the collision of the dust grains. If the motion of the colliding dust aggregates or dust particles is diffusive with mean free paths smaller than the typical grain or cluster size (e.g., dust-gas interactions at high gas densities), the so-called *diffusion-limited aggregation* (DLA) model is applied. On the other hand, in the case of transport with large portions of linear trajectories or even a pure ballistic transport (e.g., transport without dust-gas interactions), the aggregation process is called *ballistic aggregation*.

For most astrophysical applications, clustering effects may be well described by the latter process while atmospheric aggregation phenomena are mostly in the DLA regime due to the large atmospheric gas pressure. The general description of large aggregates is done by the relation between the number of equal constituent grains N and a typical radius a_a of the aggregate:

$$N \approx \alpha \left(\frac{a_a}{a_i} \right)^D . \qquad (22)$$

Here, D is the so-called fractal dimension which characterizes the specific growth process involved, $D \leq 3$ for aggregation processes in three-dimensional space, α denotes a scaling parameter, and a_i is the radius of the constituent particles. The total mass of the aggregate m is related to the number of constituent particles through

$$m = m_0 N, \qquad (23)$$

where m_0 stands for the mass of a constituent grain. There have been many proposals for the definition of the typical size a_a of the aggregates. Among those, the maximum radius a_m and the radius of gyration

$$a_g = \left[\frac{1}{2N^2}\sum_{i=1}^{N}\sum_{j=1}^{N}(\mathbf{r_i}-\mathbf{r_j})^2\right]^{1/2}, \qquad (24)$$

where $\mathbf{r_i}$ and $\mathbf{r_j}$ are the position vectors of the i-th and j-th constituent particle inside the aggregate are widely used. We usually describe the typical size of aggregates by their "characteristic" radius

$$a_c = (5/3)^{1/2}\, a_g. \qquad (25)$$

This characteristic radius is the radius of a homogeneous sphere with the same moment of inertia as the aggregate. Other definitions of the aggregate radius have been proposed but will not be addressed here for conciseness.

In general, ballistic PCA simulations yield fractal dimensions of $D \approx 3$ for the growing clusters such as the one illustrated to the left in Fig. 1. The influence of the actual aggregation scenario on the structure of the aggregates was found to be in the internal density or filling factor. Large filling factors of 0.3 were found for low sticking probabilities and extremely low densities down to 0.05 resulted for projectiles of elongated shape. Aggregates produced in the ballistic CCA process have a fractal dimension of $D = 1.95$, see the aggregate to the right in Fig. 1. Possible restructuring processes may increase their fractal dimension slightly (Meakin and Jullien 1988).

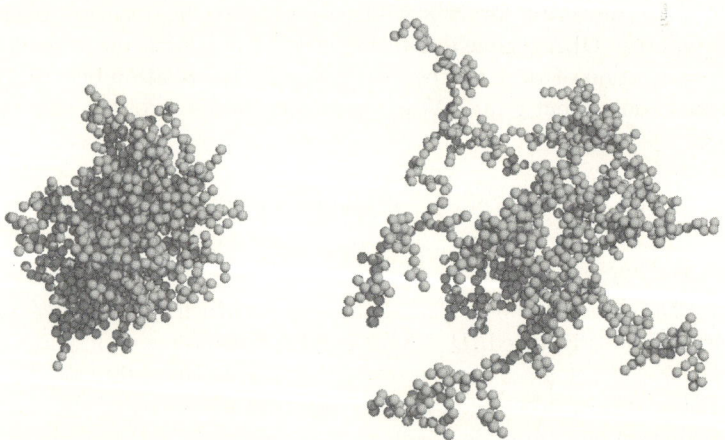

Figure 1. Examples of Monte Carlo simulations of ballistic aggregates. The left-side cluster has been computed along the ballistic *particle-cluster aggregation* model, the cluster to the right is a ballistic *cluster-cluster aggregation* product. Both aggregates consist of the same number of constituent particles (1024) and are shown on the same scale. In these simulations, sticking probabilities of unity were assumed.

To clarify which type of aggregation process actually occurs; we must again consider the problem of sticking efficiencies. We have seen that for a two-particle collision the threshold velocity for sticking depends on material properties as well as on the size of the particles (see Eqs. (20) and (21)). The situation is somewhat different for the collision between two aggregates. Let us, for the moment, assume that the aggregates themselves behave like rigid bodies. In this case, collisions between two aggregates can be treated as collisions between pairs of constituent grains and we can apply the above-derived formulation with corrections due to an increased mass of the colliding grains. This simple consideration shows that for collisions between clusters of similar masses, the critical sticking velocity will scale with the inverse square root of the cluster mass. This means that for CCA processes or aggregation of pairs of PCA clusters, the threshold velocity for sticking decreases considerably with increasing aggregate mass. On the other hand, collisions between single dust grains and macroscopic aggregates (PCA case) are also quantitatively very similar to the collisional interaction between two isolated dust grains so that the description of the critical sticking velocity in mutual dust particle interactions is valid with only minor corrections.

In practice, subsequent collisions and related aggregation processes are much more complex. Dust aggregates are neither rigid nor are collisions always central. To cover these circumstances, the "modern" approach to the aggregation problem is to simulate this complicated entanglement of microscopic (solid state and surface properties of dust grains) and macroscopic (collisions and motions of dust grains) behavior using computer simulations. The first publications of n-particle interactions for the simulation of cosmic aggregation processes are relatively recent. Sablotny et al. (1995) and Dominik and Tielens (1997) independently developed similar approaches to a realistic aggregation simulation. Both groups use for their microscopic particle-particle interactions a formulation derived by Johnson et al. (1971). In combination with energy loss mechanisms, their mutual particle interactions are very similar to those described above. The solution of the equations of motion of all interacting dust grains in the two colliding aggregates using such force terms and taking into account energy losses gives the appropriate description for the collisional behavior of dust aggregates. Dominik and Tielens (1997) demonstrate the influence of the collision velocity on the outcome of an impact between two loosely bond aggregates taking also into account the frictional behavior of the spherical dust surfaces. In their two-dimensional model, low velocity collisions ($v \ll v_s$) lead to the sticking of the two aggregates without any deformation. With increasing velocity, both aggregates are compacted more and more until they fragment for velocities well above the critical sticking velocity for single grain collisions (v_s). This is in agreement with the experimental results obtained in aggregate-aggregate collisions (Blum and Münch 1993; Blum and Wurm 2000). At even higher velocities ($v \gg v_s$) both aggregates completely disintegrate into their constituent dust grains.

Recent numerical studies by Kempf et al. (1999) who used an alternative n-particle code show that a single fractal dimension cannot describe aggregates, which form due to Brownian motion in a rarefied gas environment. However, their best mathematical characterization is given by a bell-shaped distribution function around a mean fractal dimension $\overline{D} = 1.85$ with a full width at half maximum of $\Delta D = 0.36$. The low mean fractal dimension suggests that the predominant growth is by sticking collisions between dust clusters of similar sizes.

II.C. Coagulation and Aggregation Studies in the Laboratory

Aggregation in a cloud of dust particles is a generally observed phenomenon in flame synthesis (e.g., soot formation experiments), in condensation experiments in which individual grains are condensed from a supersaturated vapor, and in liquid solutions. Subsequent collisions between these dust particles lead to coagulation and, hence, to the formation of dust aggregates. However, there are problems of astrophysical relevance and applicability: (1) often, important physical parameters are not accessible (e.g., particle and cluster velocities, number densities of dust grains) so that collision rates cannot be estimated; (2) dust particle and dust aggregate characterizations are often very poor so that, for example, size distributions are not available which makes the physical modeling of the aggregation scenario almost impossible. Many of these studies (see the review by Smirnov 1990 for detailed references) were performed either with silica (amorphous SiO_2) particles in the size range 2-10 nm or with metal (Au, Fe, Zn) grains of similar sizes. The general result of these investigations is that aggregates are fractal particles, and their fractal dimensions range from $D \approx 1.5$ to $D \approx 2.2$. These results show that the dominating mechanism responsible for the aggregation is cluster-cluster coagulation (see section II.B). However, almost all experiments were reported to be in the DLA regime or in the so-called *reaction-limited aggregation* regime which is a DLA process with a very low sticking efficiency.

Wurm and Blum (1998) performed laboratory aggregation experiments in an astrophysically relevant regime (i.e., for micron-sized silica particles) in a rarefied gas environment, and with ballistic collisions. In their experimental setup, a cloud of de-agglomerated dust grains (using the turbomolecular pump (TMP) technique described by Blum et al. 1996) embedded in a rarefied, turbulent gas, is produced. The gas pressure inside the device was kept so low (about 1 mbar) that the Knudsen number Kn (i.e., the ratio of the mean free path of the gas molecules to the particle size) was always Kn$\gg$ 1 so that free molecular flow dynamics was present. Due to the turbulent motion of the gas in the vicinity of the TMP rotor blades, frequent collisions among the dispersed dust grains with velocities of typically 0.2 m s^{-1} lead to grain sticking and hence to a rapid growth of dust aggregates. Through a small outlet in the device, samples of the aggregates escaped into a diagnostics chamber in which the aggregate morphologies and the mass distributions were analyzed using long-distance microscopy and flash illumination techniques.

The extracted aggregates have fractal structures with a $D \approx 1.9$. The observation of a relatively narrow aggregate size distribution having a typical width of a factor of 3 (10% and 90% thresholds of the mass distribution function) supports the theoretical find that a fractal dimension below $D \approx 2$ is characteristic of the cluster-cluster aggregation process in which predominantly aggregates of comparable size (or mass). The low fractal dimension of the aggregates produced in the turbulent gas environment shows that no aggregate restructuring takes place during the individual collisions since this would increase D to values around $D \approx 2.1$ (Meakin and Jullien 1988). From the temporal evolution of the mass distribution function of the growing aggregates, starting with single grains of 0.95 μm radius and ranging to aggregates consisting of several hundred constituent particles, Wurm and Blum (1998) deduce that the sticking probability for this aggregation scenario is unity throughout the entire size range of the aggregates.

The experiments described above belong to the category of cloud experiments and are thus lacking direct information on each individual aggregate-aggregate collision. With a small modification of the experimental setup, Wurm and Blum (1998) circumvented this disadvantage: after the extraction of the fractal aggregates from the turbulent TMP environment, the sedimentation of the grains in a tubular vacuum chamber was observed. Typical sedimentation velocities in a mbar gas environment range around 5 cm s^{-1} so that collisions between differentially settling aggregates occur at velocities between about 1 mm s^{-1} to 1 cm s^{-1}. The slightly different sedimentation velocities originate from slightly different ratios of mass to surface area of the individual clusters. In total, 28 individual collisions between aggregates consisting of 1 to approximately 150 constituent particles were observed with a random distribution of impact parameters. All 28 collisions resulted in sticking between the aggregates, which agrees with the predictions by Chokshi et al. (1993). Restructuring of the aggregates during the impacts was not detected which is in agreement with the low fractal dimension of the clusters found in the TMP experiments which were aggregated in collisions of more than one order of magnitude higher velocities. An example of such a collision between two aggregates consisting of monodisperse SiO$_2$ grains with 1.9μm diameter is shown in Fig. 2.

In a modified experiment, Blum and Wurm (2000) directed a jet of fractal SiO$_2$ dust agglomerates onto a thin solid Si$_3$N$_4$ target at various velocities and under microgravity and vacuum conditions. The authors observed the subsequent impacts with a long distance microscope and found with increasing impact velocity a hit-and-stick behavior (impact velocity $v_c \leq 0.2$ m s^{-1}), impact restructuring (0.2 m s^{-1} $< v_c \leq 0.65$ m s^{-1}), compact growth (0.65 m s^{-1} $< v_c \leq 1.2$ m s^{-1}), loss of monomers (1.2 m s^{-1} $< v_c \leq 1.9$ m s^{-1}), and fragmentation ($v_c > 1.9$ m s^{-1}).

Figure 2. A collision between two fluffy dust aggregates at relative velocity of $3 \cdot 10^{-3} \text{m s}^{-1}$. Each aggregate consists of monodisperse SiO_2 grains of 1.9 μm diameter. Subsequent images from left to right were taken every 6.5 ms.

Individual collisions between macroscopic dust aggregates were experimentally studied by Blum and Münch (1993) who found a transition between restitution (i.e., rebound of the colliding particles) and fragmentation at a critical fragmentation velocity v_f. They collided mm-sized aggregates consisting of μm-sized $ZrSiO_4$ dust grains under micro-gravity conditions (free-fall experiment) and in a vacuum with arbitrary impact parameter in the velocity regime from 0.15m s^{-1} to 3.9m s^{-1}. Onset of fragmentation was found at $v_f \approx 1 \text{m s}^{-1}$ for central collisions and $v_f \approx 3 \text{m s}^{-1}$ for grazing impacts. Their aggregates were PCA type clusters with fractal dimensions $D = 3$ and mean relative densities of $\alpha = 0.26$. Their results compare well with earlier experiments on different types of aggregate-solid collision experiments performed under normal gravity (see references in Blum and Münch 1993 and in the review by Blum 1995).

These experiments show that besides coagulation and aggregation processes, fragmentation events (see section III for the fragmentation of solid bodies) may also be important even at relatively low collision velocities as proposed by Dominik and Tielens (1997) (see section II.B.). This demonstrates the need for more experimental work on both fields; the physics of aggregation and fragmentation of dust aggregates as well as detailed theoretical investigations of these processes.

III. COLLISIONAL FRAGMENTATION

III.A. Impact Processes

Impact processes are ubiquitous in the Solar System. Fragmentation caused by impacts is undoubtedly one source of interplanetary dust particles. For example, the IRAS dust bands were found in the direction of the orbits of major asteroid families and could have been supplied from asteroid-asteroid collisions (Dermott et al. in this volume). The amount of ejectae from impact fragmentation, their size, shape, velocity, and spin distributions, and the degree of shock metamorphism are of interest since fragmentation is one source of interplanetary dust particles.

The controlling parameters of the collisional outcome are collision velocity, size of the colliding bodies, material property, and impact geometry. Here, we concentrate on collisional fragmentation of rigid bodies. Fragmentation of either fluffy dust aggregates or gravitationally accumulated planetesimals is not included (see, e.g., Ryan et al. 1991; section II). We will also not discuss impact into pulverized bodies although this is an important source mechanism for interplanetary dust (see recent experimental works on the impact cratering into regolith-like surfaces at various impact velocities, e.g., Yamamoto and Nakamura 1997; Colwell and Taylor 1999). Typical collision velocities range from several hundreds of m s^{-1} to several tens of km s^{-1}; a typical collision velocity among main-belt asteroids is estimated to be several km s^{-1} and that among Edgeworth-Kuiper Belt Objects is about a factor of 10 lower than for the asteroids (Farinella and Davis 1996). The shock wave generated by such an impact typically produces a pressure of ~100 GPa when the impact velocity is ~10 km s^{-1}. The shock wave decays into plastic deformation and finally into an elastic stress wave as it spreads. The stress level at which the behavior of rocks turns from elastic to plastic is the Hugoniot elastic limit (HEL), which is typically a few GPa. Fracturing of brittle material by tensile stress can extend far from the impact point, this is because the tensile strengths for intact rocks are ~10 to 100 MPa. If the colliding bodies are small enough, the outcome is overall disruption, not just cratering.

Laboratory studies on impact cratering and disruption have primarily been performed using powder-guns and light-gas-guns that can launch projectiles $10^{-3} \sim 10^{-2}$ m in diameter at velocities up to several km s^{-1}. An electrostatic dust accelerator or a plasma gun can achieve higher velocity, though only for smaller particles. Target materials are usually rocks and H$_2$O ices, but natural meteorites are also impacted. Projectile materials are mostly metals and plastics, but icy or rocky projectiles are also accelerated. Because laboratory studies are limited in the collisional velocity and in the size of colliding bodies, the laboratory results need to be extrapolated to higher velocities and to larger scale-phenomena by using both theoretical insights and numerical simulations.

In the following, we briefly outline the generation and propagation of a shock wave in the planar impact approximation (see, e.g., Melosh 1989). At the instant of the first contact of the two bodies, shock waves are generated and propagate into both bodies. The shock pressure P in a coordinate frame in which unshocked material is at rest, is related to the particle velocity u_p and initial density ρ_{0i},

$$P = \rho_{0i} U_{si} u_{pi} \ (i = 1, 2), \tag{31}$$

where the subscripts *1* and *2* denote the materials. This is one of the Hugoniot relations. It is also known that the empirical relation between the shock-wave velocity U_s and particle velocity u_p

$$U_{si} = C_{Bi} + S_i u_{pi}, \tag{32}$$

holds for many materials. Here C_B and S are the bulk sound velocity and a dimensionless constant which is related to the Gruneisen parameter, respectively. The pressure P is derived in terms of the impact velocity V, using the

relation
$$V = u_{p1} + u_{p2}. \tag{33}$$

When the two colliding bodies are made of basalt from Yakuno, for example, then $\rho_0 = 2650$ kg m^{-3}, $C_B = 3.17$ km s^{-1}, and $S = 1.25$ (Mizutani et al. 1990). The shock pressure is estimated to be 42 GPa and 125 GPa in the basalt-basalt collisions of velocity of 5 km s^{-1} and 10 km s^{-1}, respectively.

Shock compression deposits internal energy as it passes through the solid. The onset pressures for shock-induced melting and vaporization of rocks are typically several tens of GPa and $\sim$100 GPa, respectively. The shock-compressed material is unloaded by rarefaction waves, also called release waves or tensile waves. Rarefaction waves generate and propagate inward when the compression wave reaches a free surface, such as the projectile's backside.

III.B. Fragmentation and Strength

The compression wave eventually also reaches the free surface of the larger body (target) where it is reflected as a rarefaction wave, which can cause spallation from the target surface when the tensile strength is exceeded.

The interference of compression and rarefaction waves near the surface leads to velocity doubling in dense rocks that can cause ejecta to be thrown out at nearly two times the velocity of the compression wave. Such spall is responsible for the ejection of some solid target material at velocity that can exceed planetary escape velocity (see, section III.E.).

The final amount of material removed (excavated) by an impact is controlled by the strength (called strength regime) and/or the gravity (called gravity regime) of the bodies. A dimensional analysis can be used to predict the crater radius R_c in the strength regime as (Housen et al. 1983),

$$\frac{R_c}{s_1} = k_1 \left(\frac{Y_2}{\rho V^2}\right)^{\frac{\alpha}{\alpha-3}}, \tag{34}$$

where s_1, Y_2, ρ, k_1, and α are impactor radius, target strength, density of the colliding bodies, and two dimensionless constants, respectively (see, also Eq. (1)). For example, α was found to be 3/4 from experimental data of basalt impacts. This value of α indicates that crater volumes are proportional to the energy of the impactors. Similar scaling laws are proposed for disruptions. For example, the mass of the largest fragment M_l from the disruption of a target body of mass M_2 and radius s_2 by a relatively small projectile is given as (Mizutani et al. 1990),

$$\frac{M_l}{M_2} = k_2 \left[\frac{Y_2}{P}\left(\frac{s_2}{s_1}\right)^3\right]^{\beta}, \tag{35}$$

where P and k_2 are the initial pressure in Eq. (31) and dimensionless constant, respectively. The index β was determined to be $\sim$1 from experimental results on the destruction of basalt targets. Because strength plays a crucial role in disruptions, and because the strength of a rock varies with impact conditions, such as the strain-rate and the size of bodies, laboratory results cannot be directly applied to collisional fragmentation of asteroidal bodies (Holsapple 1994; Benz et al. 1994). In the following, the variation of the strength with strain-rate and the size of colliding bodies is introduced.

In contrast to a fracture under low strain-rate, in which case the strength is independent of the loading rate, the tensile strength of a dynamically fragmented material increases with increasing strain-rate. A model of crack growth and coalescence in brittle fracture has been used to explain this (Grady and Kipp 1980). Higher loading rates activate a larger number of flaws, even more resistant flaws, in order to accommodate the growing stresses. The peak failure stress becomes higher, and the fragment sizes smaller. Dynamic tensile strength Y_t has been measured for rocks (Grady and Lipkin 1980) and ice (Lange and Ahrens 1983). The power-law relation,

$$Y_t \propto \dot{\epsilon}^\eta \tag{36}$$

was fitted to the data of rocks and ice, where $\dot{\epsilon}$ denotes strain-rate. The power indices $\eta = 1/3$ and $1/4$ were obtained for granite and ice, respectively.

Grady and Kipp (1980) constructed a continuum dynamic fragmentation model. The average effect of many individual fractures is incorporated in a scalar parameter called "damage", D' ($0 \leq D' \leq 1$), which regulates the elastic modulus when the material is in tension. The intact material before any fragmentation has taken place, corresponds to $D'=0$, whereas at $D'=1$, material is completely fragmented and offers no resistance to further extensional strain. By introducing this parameter, the fragmentation process, although discontinuous in nature, is treated by continuum equations. The value of D' is defined to be the superposition of the spherical volumes of all the cracks which have been activated. Starting from a flaw distribution,

$$n = k\epsilon^m, \tag{37}$$

where n is the number density of active flaws which can be activated at or below a tensile strain level of ϵ, k and m are material constants, it is possible to derive the strain-rate dependent strength

$$Y_t \propto \dot{\epsilon}^{\frac{3}{m+3}} \tag{38}$$

and the strain-rate dependent most probable fragment size

$$s_M \propto \dot{\epsilon}^{-\frac{m}{m+3}}. \tag{39}$$

The fracture strength of rocks in static tests decreases as the size of the rock sample increases due to the greater likelihood of finding a weaker flaw. The weakest flaw corresponds to the largest flaw. According to the classical model by Griffith (1920), critical stress above which crack begins to grow is proportional to square root of the crack length (see, e.g., Lawn 1993). However, it is known that static strengths of rocks are constant when the size of rocks is larger than a few meters. A very large body, on the contrary, is strengthened statically by the gravitational force (Davis et al. 1985). Under static loading, rocks subjected to a confining pressure are stronger. Housen et al. (1991) conducted fragmentation experiments at elevated external pressure using explosive charges to fragment targets. The overpressure used in the experiment corresponded to the volume-averaged lithostatic stress of the interior of bodies up to 460 km in diameter. The resultant decrease of the degree of fragmentation, i.e., increase of the largest fragment mass M_l (see, Eq. (35)), was interpreted as due to the effective impact strength of the target Y_e being increased by external pressure P_{ext}, i.e.,

$$Y_e = Y_t + P_{ext}. \qquad (40)$$

No thermal effects were found in impact experiments performed on warm ($\sim$298 K) and cold ($\sim$100 K) rock targets (Smrekar et al. 1986), although rocks under static loading are strengthen when temperature is low. A temperature effect was reported in impact cratering of iron meteorites (Matsui and Schultz 1984). Brittle behavior was observed at low temperatures (<200 K).

III.C. Size Distribution of Fragments

The size (mass) distribution of collisional fragments has been extensively investigated by laboratory experiments (e.g., Fujiwara et al. 1989). Figure 3 shows cumulative mass of fragments of diameter smaller than s, obtained by cratering events and disruptions, which were performed under similar conditions except for the size of targets (Fujiwara et al. 1977). The amount of fine fragments decreases when the size of target increases, approaching the value of the cratering asymptotically. The propagation geometries of the stress wave in finite targets (i.e., in disruptions) are more complicated than that in the infinite half-plane. Reflection of the stress wave from the surface of finite targets leads to a complex stress history at most points in the targets. Consequently, lots of fine fragments originate from regions other than the vicinity of the impact site.

The power index γ of size distribution density or differential size distribution $n(s)$ for smaller fragments are found to be similar over a variety of impact conditions,

$$n(s) = n_0 s^{-\gamma}, \qquad (41)$$

where n_0 changes with impact conditions. The index γ was found to be $\sim$3 to 3.5 for basalt and pyrophillite (Fujiwara et al. 1977; Takagi et al. 1984), and 2.9 to 3.7 for 255 K and 77 K ice (Kato et al. 1992, 1995). The size distribution of finer fragments (s =1 to 70 μm) was studied for a basalt block, in which the grain sizes of the constituent minerals ranged from 30 to 300 μm, with the average size about 80 μm (Asada 1985). The power index γ was found to be 3.5 and 3.0 for $10° < \theta < 55°$, and $\theta > 55°$, respectively, where θ was the ejec-

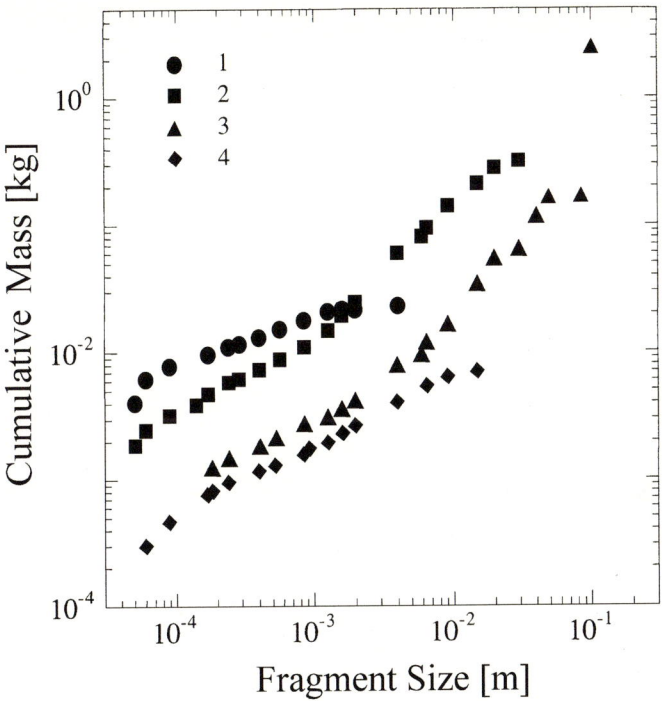

Figure 3. Size distribution of fragments from disruptions of finite targets of 2.1 cm in diameter (1), 5.0 cm in diameter (2), and 10.0 cm in diameter (3), and from a cratering of an infinite target (4) (from Fujiwara et al. 1977).

tion angle of fragments measured from the target surface. A similar index, 3.3 to 3.6 is proposed for the size distribution of interstellar grains (MRN-distribution).

The size distribution of fragments has been explained in many ways. For example, according to the Grady-Kipp model, in which, fragment size (distribution) is given as a function of strain-rate (Eq. (39)), an overall fragment size distribution can be derived if a strain-rate distribution is given. This was done analytically by assuming a power-law decay of strain-rate from the impact site (Asphaug et al. 1992),

$$\dot{\epsilon}(r) \approx \frac{V}{2s_1}\left(\frac{s_1}{r}\right)^n, \qquad (42)$$

where r is the distance from the impact point. The size distribution was incorporated in numerical simulations of impact fragmentation (Melosh et al. 1992). Other examples are geometrical statistics (i.e., random partitioning of lines, area or volumes, e.g., Gilvarry 1961; Grady and Kipp 1985), and entropy maximization (Englman et al. 1988). In the latter, the probability of having $n(s)$ fragments of size s is looked for under the constraint of mass and energy conservation. Through the expression of total energy of fragments of size s, the physics of fragmentation is incorporated in the model.

III.D. Shape Distribution of Fragments

The shape of fragments affects optical properties including radiation pressure. Moreover, shape can be an indicator or a clue to the physical process that produced the fragments.

Detailed quantitative data on the shape of impact fragments are sparse. Surface roughness of basalt and dunite fragments was investigated by fractal analysis (Fujimura et al. 1986). Fractal dimension of the cross section boundaries was obtained both by the area-perimeter relationship and by the perimeter-opening of the dividers (used to measure the perimeter) relationship. The values were centered about 1.1, which corresponds to the fractal dimension ~2.1 of the fragment surface.

The shape of fragments is more often described by the axial ratios, B/A and C/A ($C \leq B \leq A$). One method starts by measuring the largest dimension of the fragment, A (Capaccioni et al. 1984, 1986; Giblin et al. 1994), and the other method starts by measuring the smallest dimension of the fragment, C (Fujiwara et al. 1978). These two main methods of measuring the axes can lead to different results (Verlicchi et al. 1994).

Larger fragments from cratering events are presumably spall fragments and have plate-like shapes (Melosh 1984). It was reported that most of the fragments (masses $> 10^{-4}$ kg) from craters excavated in Gabbro had B/A values greater than 0.6 and C/A values less than 0.25. In two shots reported by Lange et al. (1984), the means were 0.77±0.13 and 0.73±0.11 for B/A. For C/A they were 0.21±0.07 and 0.16±0.06.

Fragments from catastrophic disruptions rarely have B/A or C/A ratios below 0.3 and 0.2. The mean values center around 0.7 and 0.5 over widely different experimental conditions (see, e.g., Fujiwara et al. 1989). Smaller mean values, 0.6 for B/A and 0.4 for C/A, were obtained both for ice fragmentation at 257 K (Lange and Ahrens 1981), and for experiments on artificial rock in open air under 1 atm pressure conditions, where ejectae are considered to be free from artificial disruption caused by their collision with the walls of the laboratory chamber (Giblin et al. 1994).

III.E. Velocity and Spin Distribution of Fragments

In general, fragment velocity and rotational frequency are the highest near the impact point and decrease with increasing distance from this site. Surface fragments tend to have higher velocities than fragments from the interior of the target. Anisotropies ("collimated jets") are observed in the azimuthal distribution of ejectae (e.g., Fujiwara et al. 1989; Martelli et al. 1994).

The cumulative mass of fragments with velocities greater than a given value can be used to estimate the fraction of material that can "escape" in asteroidal collisions. Figure 4 shows results from cratering and disruption events. The upper and the lower limits to the cumulative mass are shown for some of the disruptions since not all fragments were examined. The dimensional analysis which proposes Eq. (34) predicts the total volume of ejectae with velocity exceeding V_{eject} to be,

$$\frac{V_e(>V_{eject})}{R_c^3} = k_3 \left[V_{eject}\left(\frac{\rho}{Y_2}\right)^{\frac{1}{2}}\right]^{\frac{6\alpha}{\alpha-3}}, \qquad (43)$$

where $V_e(>V_{eject})$ is the total volume of ejectae having velocity higher than V_{eject} and k_3 is a dimensionless constant. The value of α obtained from the crater-size data, 3/4, leads to a relation,

$$V_e(>V_{eject}) \propto V_{eject}^{-2}, \qquad (44)$$

which fits the data of basalt cratering in Fig. 4.

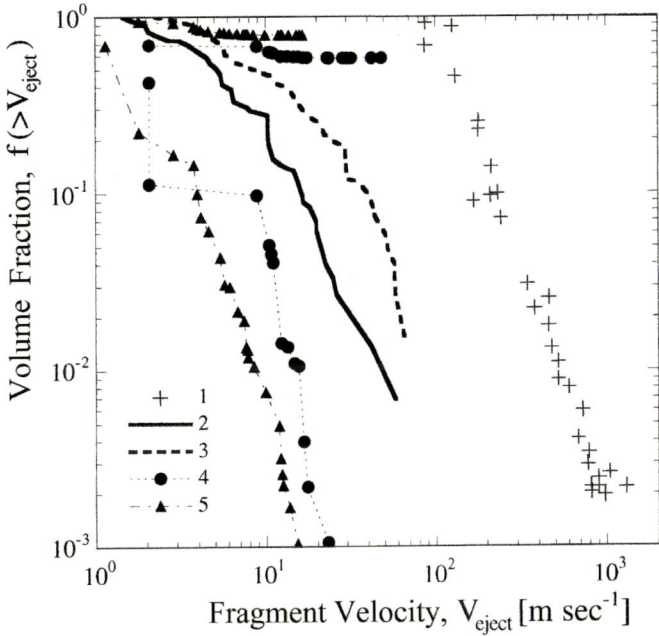

Figure 4. Volume fraction of fragments with velocity exceeding V_{eject}. The volume of fragments is normalized to the total volume of ejectae in cases (1, 2, and 3) and by the total volume of the targets for disruptions (4 and 5). Targets are basalt (1, from Gault and Heitowit 1963; 4, from Nakamura et al. 1992), gypsum (5 from Nakamura et al. 1992), and a mixture of 50% basalt fragments, 20% iron grit, 24% fly ash, and 6% of water (2 and 3, from Housen 1992). Upper and lower limits are shown for the cases of disruptions (4 and 5).

Ejection of material from a planetary surface has been investigated assuming that there is no severe loading and a triagonal-shaped stress wave (Melosh 1984, 1987). The stress wave generated by an impact of an asteroidal body of radius a_1 at velocity V is assumed to rise in the time

$$\tau_r \simeq a_1/V \qquad (45)$$

required for the impactor to bury itself in the target. The decay time τ_d is roughly the interval of the rarefaction wave from the rear surface of the projectile to propagate into the maximum pressure region. If the impactor and the target are made of the same material, this simplifies to

$$\tau_d \simeq \frac{2a_1}{C_L}, \qquad (46)$$

where it is assumed that the rarefaction wave travels at the P-wave velocity C_L, and that the peak stress decays as the stress pulse propagates in the target, while the rise time and the decay time remain constant. Points below the target surface experience compressive pressure followed by tensile pressure due to reflected waves at the surface. If the tension exceeds the material tensile strength, Y_t, it causes spallation. The surface material is accelerated both by the compressive and the reflected tensile waves, but it experiences diluted pressure. Thus an approximate, but important relation between the spall thickness Z_s, and the ejection velocity is obtained for regions far from the impact point:

$$Z_s \simeq \left(\frac{Y_t}{\rho C_L}\right)\left(\frac{2a_1}{V_{eject}}\right). \qquad (47)$$

The spall thickness is proportional to the radius of the impactor and to the tensile strength of the target and inversely proportional to the ejection velocity. We assume that the spall thickness is directly related to the size (diameter) to compare to experiments. This is supported by experimental results by Polanskey and Ahrens (1990) who studied the relation between the maximum length and the thickness of spall fragments. Figure 5 shows fragment diameter versus velocity data from basalt impact experiments. Antipodal fragments (i.e., spalls from the backside of targets) seem to obey the above relation as well.

Experiments have shown that the velocity of the largest fragments (measured in the center of mass system) is about 1 to 10 m s^{-1}, and that the power index of the upper bound of the velocity distribution versus fragment size (diameter) is approximately $-1/2$ to -1 (Nakamura 1993; Giblin 1998, and references therein). There is larger velocity dispersion at smaller sizes (Giblin et al. 1994). Similar to the velocity distribution, the dispersion of rotational frequency is large at small sizes, whereas large bodies have spin rates that are fairly close to each other. Overall size-velocity, size-spin relations for fragments, including their dispersion, are not revealed yet and further investigation is needed.

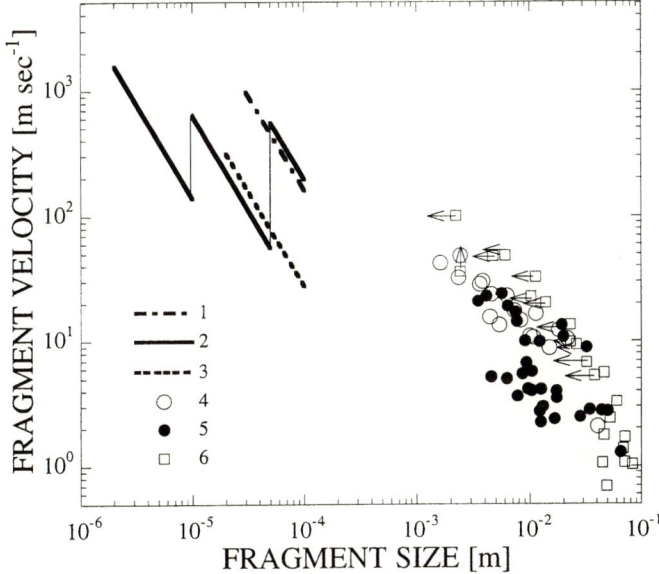

Figure 5. Fragment size (diameter) versus velocity relations (Figure 5. from Nakamura et al. 1994). Lower limits to the velocity of the fastest fine fragments were obtained for an oblique impact (1); and for normal impacts (2 and 3, corresponding to different shots). Data obtained for individual larger fragments are also shown (4, 5, and 6, each cited from different data sources).

IV. SUBLIMATION

Sublimation is a process to convert material in the solid state to vapor. We prefer the more specific term of sublimation to the word "evaporation" that also refers to the transformation from liquid to vapor. Sublimation may allow the material to condense in a purified form and can lead to material transport, crystalline, as well as chemical modifications.

The simplest way to cause sublimation is to heat the solid by one of several processes that can heat cosmic dust grains. Absorption of solar radiation dominates heating of dust in interplanetary space, while impacts between grains or by solar energetic particles, such as proton and alpha-particles, also contribute. Eisenhour et al. (1994) proposed that electromagnetic energy released from lightning and solar magnetic reconnection flares have played an important role in the formation of chondrules in transient heating events in the early solar nebula. Collisions with gas molecules heats grains intercepted by a planet during their atmospheric entry and is an important factor in the study of collected Interplanetary Dust Particles (IDPs) in the Earth's atmosphere (e.g., Love and Brownlee 1991; Svetsov et al. 1995). Chemical reactions are possible heat sources in the very low temperature regions of interstellar space such as in dark clouds (e.g., Draine 1985). A phase change from the amorphous to crystalline state of water-ice can also produce heat (e.g., Smoluchowski 1985).

Mass loss through desorption and sputtering can also occur when energy is deposited at a sufficiently high rate (e.g., Kissel 1991; Roessler 1991). These catastrophic events are discussed in section V while this section concentrates on the relatively slow process of sublimation. Several important topics related to sublimation have been omitted, e.g., chemical alteration of grain materials consisting of mixed components with unequal sublimation temperatures and sublimation of organic minerals found in the cometary coma (such as $"CHON"$ in comet P/Halley, see the chapter by Sekanina et al.). Sublimation of multi-components ices and the catastrophic events related to sublimation are also neglected in this text.

IV.A. Equilibrium

The condition when the sublimation rate equals the condensation rate is known as the dynamic equilibrium between sublimation and condensation in which the number of molecules leaving a unit surface area of the solid per unit time equals the number of molecules striking the unit surface area and sticking to it per unit time. When the molecules of interest have Maxwellian velocity distribution, the number of molecules striking a unit area per unit time is $\zeta = N_m[kT/(2\pi m_{gas})]^{1/2}$, where N_m is the number density of molecules of mass m_{gas}, and k is Boltzman's constant. The vapor pressure in an ideal gas is $p = N_m kT$ so that we may approximate $\zeta = p(2\pi kT m_{gas})^{-1/2}$.

In interplanetary space, however, the system is far from being in equilibrium. Molecules sublime into a practically perfect vacuum so that no condensation occurs, except in very special environments such as close to a comet nucleus (Yamamoto and Ashihara 1985; Crifo 1987) and in the cold primitive solar nebula (Lunine et al. 1991). It is in general assumed that the sublimation rate into vacuum is the same as it would be at equilibrium. As a result, a mass loss rate that Patashnick and Rupprecht (1975) called a "theoretical sublimation rate" is defined based on the mass of imagined gas molecules striking the unit surface per unit time. Using the approximation for an ideal gas, this rate is

$$\Pi = m_g \zeta = \left(\frac{m_g}{2\pi kT}\right)^{1/2}, \qquad (48)$$

where the temperature T of the imagined gas is the same as that of the solid grain. We therefore need to know the vapor pressure $p(T)$ and the grain temperature.

IV.B. Vapor Pressure Versus Temperature

The vapor pressure of solid material $p(T)$ has been studied using two basic experimental methods: the Langmuir rate method which is sometimes called the surface method and the Knudsen effusion method or equilibrium method. Neither method involves a direct measure of the pressure; instead the rate of escape either from a surface or from a cell is measured (e.g., Palmer and Shelef 1968).

In a phase transition from solid to vapor, the Clausius-Clapeyron equation gives the relation

$$\frac{dp}{dT} = \left(\frac{L}{T}\right)\left(\frac{1}{\delta V}\right), \qquad (49)$$

where L is the heat of sublimation and δV is the change in volume. We can substitute $\delta V = RT/p$ where R denotes a gas constant based on the implicit assumption of ideal gas behavior and neglecting the difference in molar volume of the solid and the vapor. One should note that this relation applies to single species vapor. For the case of multiple species, it is necessary to have either a vapor pressure function for each species and information on the number density relationship between the various species, or to have a single function for the total vapor pressure. For example, carbon vapor is in general a mixture of species. The dominant species in the 2400-2700 K temperature range are a monomer C_1, a dimer C_2 and a trimer C_3 of carbon atoms. It is reported that at 2500 K, the vapor composition is $C_1/C_2/C_3 = 1:0.1:0.2$ (Palmer and Shelef 1968). In another example, noted by Lamy (1974), SiO_2 molecules do not sublimate directly from silicate minerals, they first dissociate into SiO and $(1/2)O_2$. This suggests that the vapor pressure curve for SiO, not for SiO_2, applies to the sublimation of silicates.

A simple integration of Eq. (49) gives the relation

$$\log_{10} p(T) = -\frac{A}{T} + B \qquad (50)$$

between p and T, where the constants A and B depend on the molecular species, and A is explicitly related to L. The values of A and B for silicate and iron (cited by Lamy 1974), for carbon (cited by Mukai and Mukai 1973) and for $CO_2, HCN, CH_3CN, C_2H_2$ and C_2H_4 (cited by Yamamoto et al. 1983) are listed in Table 1, where p and T are in units of N m^{-2} and Kelvin, respectively.

The sublimation process of volatile materials has also been treated as a molecular desorption rate (e.g., Draine 1985; Sandford and Allamandola 1993). We assume that the sticking efficiency of molecules striking the surface equals unity and that the escape from the surface takes place at a rate

$$\nu = \nu_0 \exp\left(-\frac{\theta}{T}\right), \qquad (51)$$

where ν_0 denotes the lattice vibration frequency of the molecule on its surface matrix site, and $k\theta = L$ is the binding energy per molecule on the surface. Consequently, we can derive the relation between p and ν:

$$n^{2/3}\nu = \frac{p}{(2\pi m_g kT)^{1/2}} = \zeta, \qquad (52)$$

where n is the number density of the species. This relation is sometimes used to interpret laboratory data on the sublimation properties of ice mixtures, such as H_2O:CO ices (e.g., Kouchi 1990 and those compiled by Sandford and Allamandola 1993). It is noteworthy that the sublimation rate of molecules depends on the surface material through the surface binding energy. For example, CO molecules on CO-ice easily sublime at lower temperature than CO on H_2O-ice (e.g., Sandford and Allamandola 1988).

Table 1.
Vapor pressure parameters: $log_{10}p = -A/T + B$, where p [N m^{-2}] and T [Kelvin]

Substance	A	B
Silicate[1]	26335	13.57
Iron[1]	21080	19.01 - 2.14 $log_{10}(T)$
Carbon[2]	39039	13.24
CO_2[3]	1367	12.03
HCN[3]	1942	11.74
CH_3CN[3]	1789	10.07
C_2H_2[3]	1145	11.06
C_2H_4[3]	789.7	9.70

after 1) Lamy (1974), 2) Mukai and Mukai (1973), 3) Yamamoto et al. (1983)

1. One Example: Water-ice

It should be noted that the accuracy of the relationship between p and T given in Eq. (50) is limited by the implicit assumption of ideal gas behavior and by neglect of the molar volume of vapor atoms or molecules. As an example of a more detailed pressure curve, we discuss water-ice, the most important species in the interplanetary space.

Kelley's formula has been cited widely, especially for $T \geq 100$ K (e.g., Lamy 1974; Mukai and Schwehm 1981; Patashnick and Rupprecht 1975). That is,

$$log_{10}p(T) = -\frac{2461}{T} + 3.857 log_{10}T + 3.41 \times 10^{-3}T + 4.875 \times 10^{-8}T^2 + 3.332, \quad (53)$$

where the pressure p is in units of N m^{-2} and T is in Kelvin.

On the other hand, the International Critical Tables (vol. IV and V, 1929) gives a formula, which is valid in the range of T from 173 K to 273 K:

$$log_{10}p(T) = -\frac{2445.5646}{T} + 8.2312 log_{10}T - 0.01677006\,T + 1.20514 \times 10^{-5}T^2 - 4.63227. \quad (54)$$

This formula, abbreviated ICT, is sometimes known as Washburn's formula (Jancso et al. 1970).

Both formulae are derived from experimental results for crystalline water-ice. However, Léger et al. (1979) argued that interstellar ice formed in the cold region with $T \leq 30$ K is probably amorphous. Kouchi et al. (1994) showed that amorphous ice is preserved in the region of the primordial solar nebula where $r \geq 12$ AU. Higher sublimation rates are widely observed for fresh water-ice films condensed at low temperatures. These films presumably consist of amorphous water-ice (Kouchi 1987; Sack and Baragiola 1993). It is therefore likely that amorphous water-ice grains experience higher sublimation rates than crystalline water-ice. In fact, Kouchi (1987) reported, based on laboratory measurements, that the vapor pressure of amorphous water-ice is one or two orders of magnitude larger than that of crystalline water-ice, and depends greatly on the condensation temperature and on the rate of condensation. Mukai (1986) derived the relation

$$log_{10}p(T) = -\frac{2391}{T} + 4log_{10}T - 5.065 \times 10^{-4} T^{1.4} + 3.286 \quad (55)$$

for amorphous water-ice based on data by Léger et al. (1983).

Wagner et al. (1994) compiled formulae for the vapor pressure of crystalline water-ice accounting for experimental errors, especially in the low sublimation region. They yield the following equation, which is valid for the temperature range 190 K $\leq T \leq$ 273.16 K, i.e.,

$$log_{10}p(T) = -6.0489 \left[1 - \left(\frac{T}{273.16}\right)^{-1.5}\right]$$
$$+ 15.073 \left[1 - \left(\frac{T}{273.16}\right)^{-1.25}\right] + 2.7865. \quad (56)$$

Sandford and Allamandola (1988) derived the surface binding energy of a H_2O molecule on the surface of H_2O-ice from experimental studies of condensation and sublimation. In unit of Kelvin, $\theta = 4815 \pm 15$ K for unannealed ice and $\theta = 5070 \pm 50$ K for annealed ice. Applying the values of $\nu_0 = 1.3 \times 10^{16}$ s^{-1} and $n = 3.3 \times 10^{28}$ m^{-3} found in Draine (1985), the vapor pressure curve for water-ice can now be deduced from Eqs. (51) and (52), i.e.,

$$log_{10}p(T) = -\frac{2091}{T} + \frac{1}{2}log_{10}T + 11.33, \quad (57)$$

for unannealed water-ice, and

$$log_{10}p(T) = -\frac{2202}{T} + \frac{1}{2}log_{10}T + 11.33, \quad (58)$$

for annealed water-ice.

Figure 6 shows the values of p for water-ice as a function of T according to the formulae presented above. It is noteworthy that beyond $T =150$ K (where sublimation is extremely effective at determining the equilibrium temperature of water ice in interplanetary space, e.g., Lamy 1974; Mukai and Schwehm 1981), the difference in p among the cases of $p(T)$ considered here appears to result in significantly different temperatures. The difference in p is within one order of magnitude, except as given by Eqs. (57) and (58) (SA-cases). The dependence of p on T seems to be less strong in the SA-cases than indicated by the other formulae. This comes from the fact that the heat of sublimation derived in the SA-cases is significantly lower than in the other formulae, which in turn is due to the difference in definition of the sublimation process. CO contamination in the H_2O ice used in the experiments by Sandford and Allamandola (1988) might also contribute to the difference in temperature dependence. On the other hand, the differences in $p(T)$ discussed here may to a large extent be due to extrapolation of $p(T)$ from the range of T used in the experiment. In choosing the best representation of $p(T)$ for use in studies of sublimation, one should take into consideration the type of water-ice and the temperature range of interest.

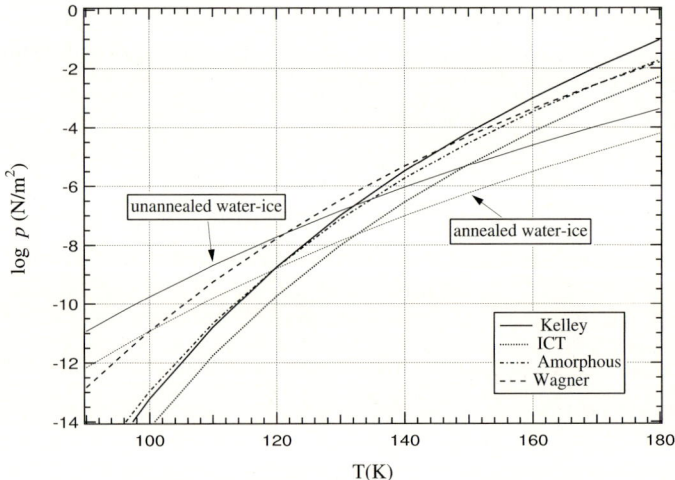

Figure 6. Vapor pressure curve p(T) for water-ice. See the text for references to the abbreviations.

IV.C. Sublimation Rate

We have seen that the mass loss rate dM/dt from an interplanetary dust grain of mass M due to sublimation is calculated using the total mass of molecules needed to strike the unit surface per unit time Π to exert a pressure equal to the vapor pressure. However, experiments suggest that some molecules striking the surface are reflected and do not condense onto the solid. Therefore, a factor of α_a has been introduced to fit experimental data using the equation

$$\frac{dM}{dt} = \alpha_a S \Pi, \tag{59}$$

where S denotes the total surface area of the dust grain. The quantity α_a is a probability, sometimes called the "evaporation coefficient" or "vaporization coefficient", it is also referred to as the "condensation coefficient" or "sticking coefficient". We refer to α_a as the "accommodation coefficient" following Lamy (1974) and Patashnick and Rupprecht (1975).

In general, the accommodation coefficient depends on both the temperature of the solid surface and on the species of molecules. Ignorance of α_a has been a continuing source of uncertainty and controversy in studies off sublimation. Lamy (1974) used $\alpha_a=0.7$ while Patashnick and Rupprecht (1975) set $\alpha_a=1$ for all cases that they considered. Reliable experimental data on α_a has been derived for carbon by comparing the flux in the Langmuir method with that obtained in the Knudsen method as discussed in detail in Palmer and Shelef (1968). They suggested that α_a for C_1 must be between 0.1 and 1 and that it probably is of the same general magnitude for C_2 and C_3. Since these species are major components of sublimating carbon, the accommodation coefficient can be of the order of one tenth. Unfortunately, reliable experimentally determined α_a values as a function of T are rare. As a result, the uncertainty in the mass loss rate due to sublimation remains large. However, since p depends strongly on T, the accuracy of dM/dt is mainly governed by the formula used to represent $p(T)$.

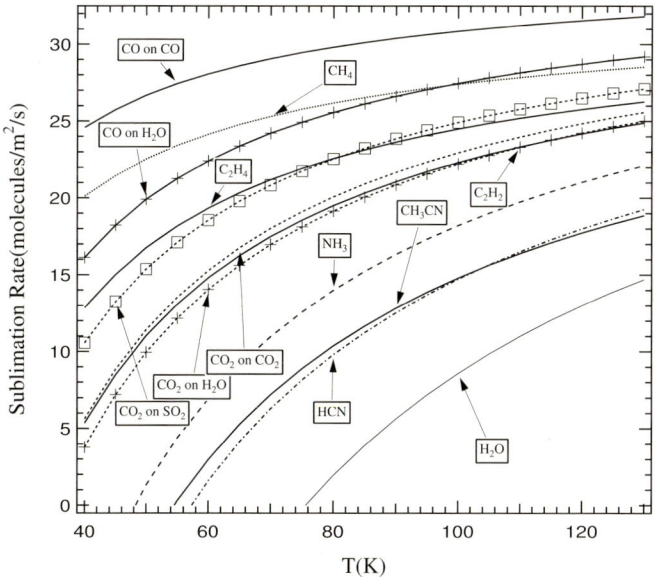

Figure 7. Sublimation rate for volatile molecules. See references for each of molecule species in the text and in Tables 1 and 2.

Figure 7 shows the number of molecules sublimating per unit surface and unit time ζ in $m^{-2}s^{-1}$ when $\alpha_a=1$ is assumed. This is intended to give a rough estimate of what molecular species easily sublimate at a certain temperature.

These relations have been derived from the values of the desorption parameters defined in Eqs. (51) and (52), i.e., ν_0, n and θ listed in Draine(1985) and the values of θ reported for several molecules in Sandford and Allamandola (1993). Using the relation between ζ and p given in Eq. (52), we can also estimate $\zeta(T)$ for the substance listed in Table 1. It is clear that the mass loss rate due to sublimation strongly depends on the grain temperature. We therefore need to know the temperature of an interplanetary dust grain as a function of the grain composition, geometry, and the heliocentric distance to study where in the solar system volatile ices sublimate. Table 2 gives the parameters X and Y used to calculate ζ from $log_{10}\zeta = -X/T + Y$.

Table 2.

Sublimation rate parameters: $log_{10}\zeta = -X/T + Y$, where ζ [m^{-2} sec^{-1}] and T [Kelvin]

Substance	X	Y
CO on CO ice[1,2]	416.9	35.02
CO on H$_2$O ice[1,2]	755.7	35.02
CO$_2$ on CO$_2$ ice[1,3]	1168	35.63 − 0.5 $log_{10}T$
CO$_2$ on H$_2$O ice[1,3]	1242	35.63 − 0.5 $log_{10}T$
CO$_2$ on SO$_2$ ice[1,3]	970.6	35.63 − 0.5 $log_{10}T$
CH$_4$[2]	486.4	32.27
NH$_3$[2]	1691	35.12
HCN[2]	1968	34.40
H$_2$O[2]	2655	35.13

after 1) Sandford and Allamandola (1993), 2) Draine (1985), 3) Yamamoto et al. (1983)

IV.D. Interplanetary Dust Grain Temperatures

The grain temperature is in general calculated from a balance between the energy input and output. The major energy gain is through absorption of solar photons E_a, while the major energy loss process is thermal emission, E_r. When we consider volatile grain materials, such as water-ice, the energy loss though sublimation E_s should also be taken into account. Of course, sublimation plays an important role in determining the equilibrium temperature of any material when it is sufficiently close to the Sun (e.g., Mann et al. 1994).

Other sources of heat can also be important, usually over short time spans. An example is energy released in phase transition from amorphous to cubic ice structure (section IV.A). Based on careful measurements of structural transitions in amorphous water ice, Jenniskens and Blake (1994) reported that transient heating events from the high-density amorphous form to the low-density form occur gradually over the 38 to 68 K range in addition to the generation of heat during the crystallization of cubic ice at 148 K.

The heating of interplanetary dust by impinging solar wind particles is known to be insignificant when compared with E_a (e.g., Mukai and Schwehm 1981). We neglect any heating by phase transition of water-ice and by collisions with solar wind particles in this section for simplicity.

The definitions of E_a, E_r and E_s for a grain with the geometric cross section G and a total surface area S at a heliocentric distance r are:

$$E_a = G \left(\frac{R_0}{r}\right)^2 \int_0^\infty B_\odot(\lambda) Q_{abs}(\lambda, m^*) d\lambda, \qquad (60)$$

$$E_r = S \int_0^\infty B(\lambda, T_d) Q_{abs}(\lambda, m^*) d\lambda, \qquad (61)$$

$$E_s = \left(\frac{dM}{dt}\right) L, \qquad (62)$$

where $Q_{abs}(\lambda, m^*)$ is the efficiency factor for absorption, which due to Kirchhoff's law equals that for emission at the same wavelength λ. R_0 and $B_\odot(\lambda)$ are the radius and energy spectrum of the Sun, respectively and $B(\lambda, T_d)$ is the Planck function at the grain temperature T_d.

Many works have derived the temperature of interplanetary dust grain as functions of the heliocentric distance and the radius s of spherical grains taking into account the wavelength dependence of the refractive indices m^* of the grain material and that of Q_{abs}, thereby postulating $G = \pi s^2$ and $S = 4\pi s^2$. For example, temperatures for graphite, silicate (quartz), and for water-ice were shown by Mukai and Mukai (1973), and for quartz, obsidian, andesite, water-ice and iron by Lamy (1974). Patashnick and Rupprecht (1975) gave results for water-ice. Hanner (1985) showed results representing a Titan tholin, which is an organic residue formed during electric discharges in a CH_4 and nitrogen atmosphere obtained by Khare et al. (1984).

To study the temperature of grains that are not directly illuminated by the Sun, for example close to a comet nucleus, we need to account for the attenuation of the incident sunlight in the cometary coma (e.g., Crifo 1995). The reduction in energy input from direct sunlight is in part compensated for by scattered light from other grains and due to their thermal emission that should also be accounted for in calculating the energy input E_a.

The complex refractive index m^* of candidate materials must be known across a wide wavelength interval before we can estimate the grain equilibrium temperature T_d. The material's m^* must be known to $\lambda = 0.02$ μm to account for the UV absorption peak of dielectric materials such as silicate and water-ice. This is especially important for submicron grains for which the UV peak's contribution to E_a is especially strong while the longer wavelengths are the most important when we estimate low grain temperatures far from the Sun. The dependence m^* beyond $\lambda = 100$ μm usually cannot be neglected in the outer solar system. Data on m^* and Q_{abs} that depends on the particle geometry as well as on m^* are increasingly important as research is concerned not only with the grain temperature but also with the spectroscopic analysis of thermal emission by the grains. The reader can find references to data on m^* in the Chapter by Gustafson et al. in this volume.

It has recently become more popular to represent interplanetary dust using irregularly shaped particle models. How can we deal with the effect of particle shape on the grain temperature? Greenberg and Shah (1971) showed that the shape effect on the temperature of interstellar dust grains is negligibly small. However, a significant difference between the temperatures of porous aggregates and of a sphere with an equivalent mass has been revealed (e.g., Greenberg and Hage 1990; Kozasa et al. 1992; Okamoto et al. 1994; Mann et al. 1994).

Three major shape dependent effects are considered: (1) changes in the ratio of the geometric cross section G to the total surface area S, (2) the Q_{abs} values used to calculate E_a and E_r, and (3) the sublimation process affecting E_s.

It is known (van de Hulst 1957) that the average geometric cross section of a convex particle over random orientation is one-fourth of its surface area: i.e., $G/S=1/4$ is the same as for a sphere. Kitada et al. (1993) studied a correlation between G and S for fluffy aggregates represented by a fractal analogue model where the computer simulations for collisions of incident particles colliding with and building up an aggregate have been performed to estimate the values of G and S in the manner initially used by Meakin and Donn (1988). They found a relation of $G/S \approx 1/4$, although the number of constituent particles in their test aggregate (≤ 8192) is insufficient to represent a typical size interplanetary dust grain: i.e., 1 to 200 μm. We never the less conclude that the ratio G/S for a broad range of non-spherical shapes is likely to be similar to that of a sphere.

A proposed simple way to calculate Q_{abs} for irregularly shaped particles is to use Mie theory in combination with the Maxwell Garnett mixing theory that we may refer to as MG-Mie. The Maxwell Garnett effective medium theory is used to obtain average optical constants for mixed materials (e.g., Hage and Greenberg 1990; Lien 1991; Mukai et al. 1992, and Gustafson et al. in this volume). Perrin and Lamy (1990) expose and stress the weak points with the representation. On the other hand, Hage and Greenberg (1990) and Kozasa et al. (1992) found that MG-Mie is in good agreement with more sophisticated methods to derive Q_{abs}, more so at the longer wavelengths that dominate thermal emission than at the peak of the solar spectrum where most of the energy is absorbed. In general, neither a phase function for the angular scattering nor a remarkable feature in the spectrum of thermal emission can be obtained using the MG-Mie method. However, it seems that an integrated value over phase angles and over a broad spectrum, such as the average Q_{abs} values needed in radiative energy transfer can be accurately calculated using this method.

The sublimation process is, in general, independent of the shape of a dust aggregate, since it is a pure physical process between the solid on atomic scales and the vapor. However, in the irregularly shaped grain it becomes important to consider re-condensation of vaporized molecules on the surface of the solid aggregate. This is analogous to condensation and sublimation of gas molecules in the interior of a porous comet nucleus (e.g., Keller 1990), although the scale differs. It is expected that the sublimation process for a porous grain might be modified by the reduced vapor pressure due to re-condensation of subliming molecules on concave portions of the rough surface. We conclude that grain temperatures are not likely to depend very strongly on the grain geometry except for porous structures such as aggregates whose Q_{abs} values can be calculated to a reasonable approximation using the MG-Mie method.

IV.E. Comets

Observations of activity in a few distant comets suggests that some molecular species sublime far from the Sun. Comet Bowell (or 1980b), which is believed to be a new comet from the Oort cloud based on its dynamical behavior, released grains already between 11 and 12 AU from the Sun (Sekanina 1982). Sublimation of extremely volatile ices could generate activity in such distant comets. Senay and Jewitt (1994) have recently reported the detection of submillimeter emission of carbon monoxide from periodic comet Schwassmann - Wachmann 1, which has perihelion and aphelion distances 5.77 AU and 6.28 AU, respectively. This suggests that the sublimation of volatile ices containing carbon monoxide may be the driver of episodic outbursts of dust observed from this comet. Sandford and Allamandola (1993) studied the residence times t_r of molecules on ices, where t_r is defined by ν given in Eq. (51) as $t_r = 1/\nu$. They showed that the residence time of CO molecules on CO ice is about 10 years at 20 K, and 30 s at 30 K. On the other hand, the residence time of CO on water-ice is 2×10^5 years at 30 K, and 0.1 years at 40 K. It is expected that the temperature of a freely sublimating CO surface at 6 AU is nearly 26 K (Senay and Jewitt 1994). However, the temperature of CO ice in comets may be higher than 30 K at 6 AU because of the contamination of absorbing materials. Consequently, it is expected that CO molecules sublime significantly already at about 6 AU from the Sun even though they are stored in water-ice matrices.

IV.F. Reaction Force

Although an interplanetary dust grain is very small in size, there exists a temperature difference δT_d between the area facing the Sun and that on the dark side. From a simple derivation, referring to Burns et al. (1979),

$$\delta T_d \sim \frac{(1-g)sB}{2\kappa}, \qquad (63)$$

where g is the albedo of the grain, s is the grain radius, B denotes the solar flux at a heliocentric distance r, and κ is the thermal conductivity of the grain material. For $B = 1.36 \times 10^3$ J m^{-2} s^{-1} at $r=1$ AU and $s = 10^{-4}$ m, Hasegawa et al. (1977) found that δT_d is 0.04 K for a silicate glass grain with $\kappa = 1.4$ J m^{-1}s^{-1}K^{-1} and $g=0.1$, and $\delta T_d \sim 0.0009$ K for an iron grain with $\kappa = 76$ J m^{-1}s^{-1}K^{-1} and $g=0.05$.

The temperature difference δT_d leads to anisotropic thermal emission from the grain and therefore to a finite net reaction force. When the grain is spinning and its rotation is retrograde, the grain's leading hemisphere has a larger force on it than the trailing side. As a result, the grain will experience a pseudo drag force. On the other hand, if the rotation is prograde, the grain experiences acceleration. This is the Yarkovsky force, which was quantitatively examined by Burns et al. (1979), and studied by Hasegawa et al. (1977). Gustafson (1994) estimated the range of possible values of the ratio of the Yarkovsky force to Poynting-Robertson drag force as a function of grain radius in a recent review.

Just as the Yarkovsky force results from the anisotropic emission of thermal radiation, it is natural to assume that a grain having non-uniform surface temperature will also experience a reactive force from anisotropic mass flow from its surface due to sublimation. Consequently, the effect on the orbit should also be similar to the Yarkovsky effect. This is similar to the well-known effect on periodic comets that tend to change their orbital elements after perihelion passage. In improving the ephemeris prediction accuracies for short period comets, substantial perturbations due to such reactive forces acting upon the cometary nucleus have been introduced (e.g., Yeomans 1991). Direct measurements of jet-like features of gas and dust streams from the cometary nucleus in comet P/Halley (Keller et al. 1987) have confirmed the existence of anisotropic mass flow from the nucleus.

Mukai and Fechtig (1983) proposed that this process also produces a variation in porosity of an interplanetary dust grain by reducing the packing. Consider aggregates consisting of many individual grains with volatile matrix and refractory inclusions. The reactive force generated by sublimation acts in a direction from the illuminated side facing the incident light to the dark side facing the inside of aggregate. This produces a net force that drives each constituent particle toward the inner part of the aggregate. It is therefore expected that, when anisotropic mass ejection due to sublimation occurs, reaction forces cause a restructuring of the aggregates grains as well as perturbation on the orbital motion.

V. SPUTTERING

Plasma ions and electrons in addition to UV photons bombard grains throughout all regions of space. This can alter a grain chemically and structurally and can also cause erosion referred to as sputtering or desorption (Johnson 1990, Strazzulla and Johnson 1991) that cannot be separated from the chemical alterations. Erosion by ions, electrons and UV photons competes with sublimation and accretion (adsorption) in determining the size and composition of cosmic grains. The surface composition and, hence, the sputtering rate for a grain can vary depending on the exposure age since sputtering preferentially depletes the surface region of the most volatile species and causes implantation of ions that can change the chemistry in the surface region (Johnson and Baragiola 1991).

V.A. Plasma Parameters

Typically plasma ion sputtering is characterized by the impact energy of the ions (or the electrons). Both the relative flow speed between the plasma and the grain *and* the plasma temperature determine the distribution of impact energies. Plasma parameters (density, flow speed and temperature) can vary enormously throughout the various regions of space in which grains reside from the relatively low temperature plasmas in interstellar shocks, primarily responsible for erosion of interstellar grains, to galactic cosmic ray particles (e.g., Table 3) that permeate all regions of space and are characterized by a very high temperature. Plasma parameters in various regions are summarized below. However, in addition to plasma flux and temperature, the residence time of the grain in the plasma is of critical importance.

1. The Interplanetary Medium

Objects in interplanetary space are exposed to the continuous expansion of the solar corona, referred to as the solar wind (e.g., Lanzerotti 1987) which is of the order of 1 keV/u (~ 400 km s^{-1}) at 1 AU. The solar wind flux (see Table 3 for constituents and abundances) is of the order of $2 \cdot 10^{12}$ particles m^{-2} s^{-1} at ~ 1 AU, corresponding to a number density of ions (and electrons) of the order of $(4 \text{ to } 5) \times 10^6$ particles m^{-3} that forms on the average a neutral plasma. The flux decreases, roughly, as if from a point source, i.e., $\sim 1/r^2$, where r is the solar distance. The plasma has a temperature of $\sim 10^5$ K at 1 AU that decreases with increasing distance from the Sun, so the energy connected with the expansion (flow) determines the bombardment energy of the ions.

Table 3.
Plasma Characteristics

Plasma	Some species		Characteristics	Typical penetration depths [g cm^{-2}]
galactic cosmic rays	H$^+$ He^{++} C^{+6} O^{+8} Si^{+14} Fe^{+26}	0.87 0.12 10^{-3} 10^{-3} 10^{-4} 10^{-4}	$\phi_i(E) \approx k(E + m_p c^2)^{-2.5}$ $k = 0.8$ [cm^2 sec sr GeV / nucleus]$^{-1}$ E in GeV $m_p c^2 = 0.938$ GeV	$\sim (10$ to $10^{-2})$
solar co-rotating	H$^+$ He$^+$ others	0.95 0.05 <0.001	$\phi_i \approx E^{-4}$ [cm^2 sec sr MeV]$^{-1}$ E in MeV for $E > 0.01$ MeV	$\sim (10^{-2}$ to $10^{-3})$
typical large solar flare	H$^+$ He^{++} others	0.95 0.05 <0.001	$\phi_i \approx 3 \cdot 10^2 E^{-2}$ [cm^2 sec]$^{-1}$ E in MeV (for $E > 10$ keV)	$\sim (10^{-4}$ to $10^{-5})$
solar wind	H$^+$ He^{++} C^{+6} O^{+6+8} Ne Si Fe Au	0.96 0.04 $4 \cdot 10^{-4}$ $5 \cdot 10^{-4}$ $7 \cdot 10^{-5}$ $8 \cdot 10^{-5}$ $5 \cdot 10^{-5}$ $2 \cdot 10^{-6}$	$n_i \sim 4.5$ cm^{-3} (at 1 AU) $v_i \sim 400$ km sec^{-1} $T_i \sim 7$eV (at 1 AU)	$\sim 10^{-6}$

Superimposed on this flux are higher-energy particles associated with active solar regions, very energetic particles produced in solar flares, and galactic cosmic-ray particles. A typical large flare event has a total fluence of protons with energies greater than 1 MeV of the order of 10^{14} particles m^{-2}. The energy spectrum in these events typically follows $\propto E^{-2}$, where E is the proton energy. The average flux of such particles is of the order of 10^4 below that of the solar wind flux, but with a very different energy spectrum. Co-rotating events associated with active solar regions have much steeper energy spectra ($\sim E^{-4}$) with an average flux of 1 MeV protons of $\sim 10^4$ particles m^{-2}s^{-1}sr^{-1}MeV^{-1}. If the occurrence rate is of the order of two events per solar rotation ($\sim$ 27days) with interplanetary space filled with such particles 1/3 of the time and the spectra extrapolated to 10 keV, an average flux of the order of 10^2 smaller than the solar-wind flux is obtained, but again involving more energetic particles (Geiss 1982).

The description above approximates the present conditions of a quiescent Sun. Solar type stars may however have passed through phases in which the particle radiation was much more intense. In a T-Tauri phase the solar-flare rate is much greater than at the present Sun and the particles are more energetic. Worden et al. (1981) give an average flux from these flares that is roughly equivalent to the solar-wind flux, $2 \cdot 10^{12}$ particles m^{-2}s^{-1} at $\sim$ 1 AU, but for particles of energies greater than 10 MeV. As the Sun may have gone through such a phase, Feigelson (1982) deduced an average proton flux for the young Sun of $\sim 4 \cdot 10^{10}$ ions m^{-2}s^{-1} at $\sim$ 1 AU for energies greater than 6 MeV which could have lasted anywhere from 10^4 to 10^7 years.

Besides particles ejected from our Sun, the interplanetary medium and the interstellar medium are filled with a background of galactic (cosmic-ray) particles (Simpson 1983). The energy in the cosmic-ray background is primarily contained in the energetic ions with smaller fractions carried by γ-rays and electrons. The decrease seen at small energies is due to solar modulation. Table 3 gives a reasonable approximation to the proton spectrum in the absence of any solar modulation.

The UV and EUV radiation from the Sun produces effects comparable to that of the incident charged particles. The solar Lyman-flux, shown in Fig. 8, is of the order of 10^3 times the solar-wind flux, but the total *energy flux* is only about a factor of 10 higher. In this regard it is worth noting that at Neptune's and Triton's distance from the Sun ($\sim$ 30 AU) the total UV energy available with photon energy above that for dissociating CH$_4$ ($\sim$ 155 nm, $\sim$ 8 eV) is of the same order as the galactic cosmic-ray *energy flux* (Delitsky and Thompson 1987).

2. The Interstellar Medium

Grains in the interstellar medium are exposed to the background cosmic ray radiation described above, without the solar modulation. In addition, there are many regions in which shock-heated plasmas or hot gases propagate through regions containing grains and neutral gases (Draine et al. 1983). Plasmas can be generated by supernova remnants and temperatures can vary enormously depending on the nature of the shock. Typical densities and temperatures (H, H$^+$, He, He$^+$...) range from 10^6 m^{-3} at 10^5 K to 10^3 m^{-3} at 10^8 K. These

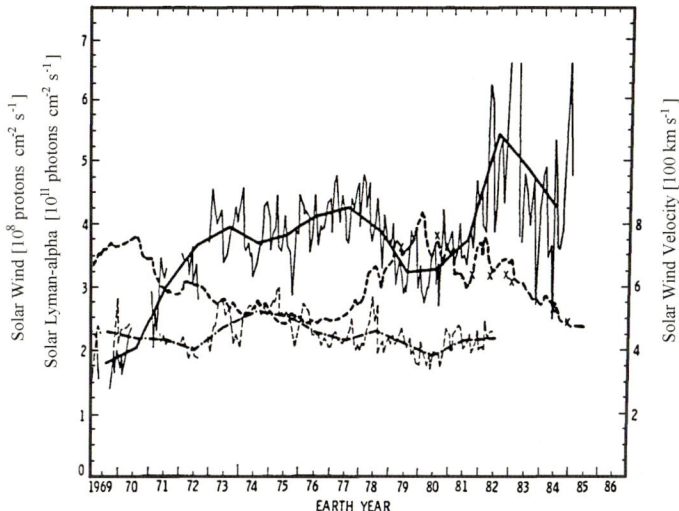

Figure 8. Monthly averages of the solar wind and Lyman-α fluxes. Solid lines solar-wind monthly and yearly averages; dashed: Lyman-α; dash-dot solar wind velocities monthly and yearly.

shock-heated gases (plasmas) can interact with material in low density (diffuse) interstellar clouds and at the edges of as well as in the interiors of high density (molecular) clouds. In addition, molecular clouds are permeated by the galactic cosmic ray ion flux. This flux can affect grains directly or indirectly by way of the Lyman-α photons produced as the ions transverse the cloud's hydrogen gas (Sternberg et al. 1987).

3. Cometary Comae

Dust grains ejected from comets first pass through the plasma in a cometary coma before being exposed to the interplanetary medium. The plasma in the inner coma, which is inside the 'contact surface' defining the inner cavity, consists of heavy ions ($\sim 10^{15}$ m^{-3}. They are typically H_3O^+) that are relatively cool (< 500 K). Beyond this the temperature rapidly increases through a region inside the cometopause, the stagnation region, where the plasma flow is slow but the temperatures high (~ 1000 K). The interaction of the ionized gas with the solar-wind fields accelerates freshly formed ions (pick-up ions) in this region and beyond the bow shock. The pick-up ions of interest are dominated by heavy ions originating from cometary material (e.g., O^+). These are efficient sputtering agents with energies varying from ~ 1 keV/u to ~ 1 MeV/u, depending on whether they are formed outside or inside the bow shock (Cravens 1991).

4. Planetary Magnetospheres

The Earth and the outer planets possess magnetospheres that can trap solar particles for considerable periods (Lanzerotti 1987). Plasma is also supplied by ionization of atoms coming from sources within the magnetospheres, such as planetary atmospheres, the surfaces of their moons, and ring particles. In turn, these objects are subject to bombardment by the trapped plasma, resulting in an interesting feedback process (e.g., Johnson 1990). Whereas an interplanetary grain traverses such regions rapidly and, hence, experiences little erosion, the objects embedded in planetary magnetospheres can themselves be sources of grains. It was typically assumed that such grains were relevant only to the regions close to the planet but streams of small dust particles ejected from the Jovian magnetosphere have now been detected at large distances from Jupiter (Grün et al. 1993).

We show the heavy ion (O^+) plasma flux experienced by grains in the inner magnetosphere of Saturn near the orbit of the moon Dione in Fig. 9 for comparison with the above sources. The compositions of both the Jovian and Saturnian plasmas suggest that the satellites and grains are sources of the heavy ions in the plasma (Cheng and Johnson 1989). The densities achieved, particularly near Io, can be many orders of magnitude above the local solar-wind density, but the flux decreases rapidly away from the satellite's orbital plane. The ions also have an overall motion with respect to objects in orbit since the magnetosphere rotates with the planet so that a plasma stream flows onto the grains in addition to random bombardment (e.g., Johnson 1990).

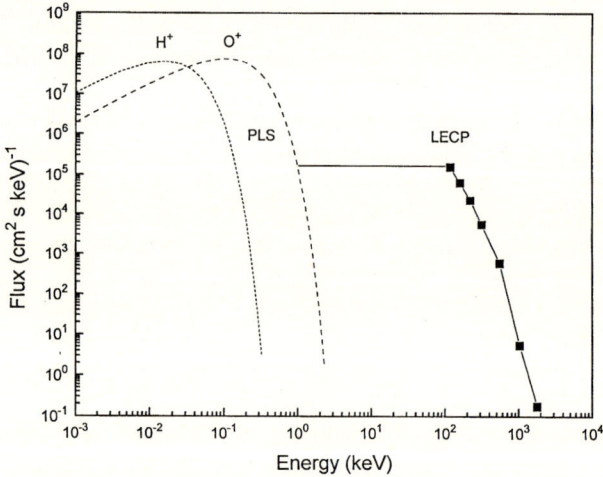

Figure 9. Model fluxes at Dione extrapolated from the Voyager data obtained by the PLS and LECP instruments. Solid line is predominantly O^+ (from Shi et al. 1995).

V.B. Materials

Grain materials throughout space are predominantly dielectrics, but these insulating solids can differ enormously in composition, volatility, and size. Individual grains are often heterogeneous, but a simplified picture is sometime used in which a refractory core grows an organic mantle and then an ice mantle before, for instance, aggregation and incorporation into a comet. The most volatile species such as molecular ices may condense into mantles on grains in molecular clouds, in the outer solar system prior to aggregation in comet nuclei, and in cometary comas. In addition, ice grains permeate the inner Saturnian magnetosphere.

Organics are the next most volatile material and, in fact, can be formed by irradiation processing of ices, as discussed by Gustafson et al. in this volume. Finally, interstellar grains, cometary grains, and dust from the lunar surface, asteroids, and, possibly, Io contain silicates with varying amounts of metals. Graphite or other carbonaceous grains also exist in the interstellar medium.

Refractory cosmic dust particles collected at Earth are found to be predominantly porous aggregates of small, heterogeneous grains with dark coatings of processed organics (Bradley et al. 1984). They have chondritic composition with metallic inclusions, glasses, single crystal minerals, and polyphase submicron grains. A particular glass of recent interest called GEMS (glasses with embedded metals and sulfides) is found to have nanoscale compositional gradients thought to be indicative of radiation processing. Dust grains vary considerably in size: 5 to 30 μm GEMS; ~ 1 μm for Saturn's E-ring grains; 0.1 μm to 0.02 μm in diffusive clouds; and nanometer grains formed in cool stellar atmospheres (e.g., Bradley 1994).

V.C. UV Irradiation

The Lyman-α energy flux exceeds the solar-wind ion energy flux as seen in Fig. 8 and the galactic cosmic-ray ions in dark clouds produce Lyman-α photons. The UV energy deposition rate in a grain therefore dominates over the ion energy deposition rate, especially if lower energy UV photons are included. Some regions in outer planet magnetospheres are the only exception. Whereas the UV flux only heats refractory silicates, it can chemically alter the ices and organics and produce desorption.

Desorption and chemical alterations of ices and organics are interrelated. That is, bond breaking produces radicals and, depending on the background temperature, these radicals can react when their density is sufficiently high. Volatile species are formed such as H_2 (from H_2O, NH_3, CH_4, and organics), O_2 (from H_2O), N_2 (from NH_3), and CO and O_2 (from CO_2). These very volatile species accumulate in voids and can diffuse to the surface and desorb depending on the ambient temperature, thereby permanently altering the grain structure and composition. Relatively volatile organic solids become carbonized, refractory materials in this manner. Radical reactions and volatile loss can be suppressed at very low temperatures (e.g., in molecular clouds). Rapid heating of a grain with a significant density of stored radicals and volatiles (e.g., by collision with another grain or a heavy cosmic-ray ion) may result in explosive desorption of materials (Greenberg 1983).

Photo-desorption of H_2 from H_2O may account for the remarkable recent finding that very low-temperature, fresh ice, which does not lose H_2O efficiently to photo-sputtering, does so after a radiation dose. The 'incubation' dose required depends on the ambient temperature (Westley et al. 1995). The efficiency of desorption after the 'incubation' dose is high so that this process can account for rapid loss of an icy mantle when a grain is ejected from a molecular cloud. Photo-desorption of H_2O ice can exceed solar wind sputtering and exceeds sublimation beyond 5 AU from the Sun. Note that absorption of a single photon can initiate reactions producing a transient 'heat pulse' causing loss of very volatile species (e.g., CO) from extremely small grains (~ 1 nm).

V.D. Plasma-Induced Sputtering and Alteration

Like for UV irradiation, the energy deposited in a grain by plasma-ion and electron impact both alters the material and causes erosion that is known as sputtering in this case. The interaction can occur due to momentum transfer collisions of the ions with atoms in the target or by the electronic excitation of atoms and molecules by the incident plasma ions and electrons (Johnson 1990).

Plasma-ion and electron impact excitations can act in the same manner as UV-induced excitations, producing alterations and desorption. However, energetic impacts can also cause effects that differ in that a density of 'excitations' is produced by a *single particle*. This is critical for two reasons. Closely-spaced electronic excitations can interact and lead to alternate chemical pathways, e.g., incident ions more efficiently produce H_2 and O_2 in H_2O (Brown et al. 1984) than do UV photons (Westley et al. 1995). Also, at the low fluxes typical in space, thermally-induced recombination can suppress radical accumulation at low temperatures since radical densities must be *accumulated* during UV irradiation, i.e., the rate of radical production is important. This needs to be accounted for when applying laboratory data on radiation effects to astronomical scenarios and is particularly difficult to test in the laboratory where the photon or charged particle fluxes are typically five orders of magnitude higher than in space. Finally, impacting ions can alter the materials in a grain simply by implantation (e.g., Roth 1983; Strazzulla 1999). For example, a proton plasma impacting carbon can form CH_4 by this process after a large fluence and implanted H can compete for O in silicates to form hydroxyl (-OH) or H_2O. Implanted C in a silicate can form CO.

1. Defects, Amorphization and Tracks

Bond breaking reactions and momentum transfer collisions cause displacements and heat the lattice locally. This typically produces amorphous regions in crystalline materials, e.g., the amorphous rims on lunar grains (Taylor 1982) and some grains in IDPs (Bradley 1994) damaged by solar wind ions. In addition, the transient thermal spike produced by a very heavy incident ion may be sustained long enough so that small crystalline regions can form in an otherwise amorphous material, along with regions dominated by defects and vacancies.

Displacements and agitation can lead to 'track' formation along the path of an ion. Tracks formed by solar flare ions are used to determine exposure ages of collected IDPs. The agitation along the particle path also produces ion-enhanced diffusion. Therefore, volatile species can be brought to the surface where they are lost (Johnson 1990, 1996). Loss of H_2 and O_2 is seen from low temperature H_2O irradiated by heavy ions. Laboratory measurements show that loss occurs from the full penetration depth of the ion (Reimann et al. 1984; Benit and Brown 1990). The agitation initiated by a fast ion also causes overall mixing and can cause local segregation of species (e.g., Si inclusions in SiO; Fe in iron oxides).

Vacancies conglomerate in materials after prolonged exposure to radiation, forming voids in which newly formed or implanted gases are trapped (Roth 1983) (e.g., O_2 is formed and 'trapped' in voids produced in solid H_2O while implanted He is trapped in silicates). The trapped gas acts to increase void (bubble) size. If voids interconnect, new pathways to the surface can form, resulting in porous solids. That this occurs in organics is dramatically demonstrated through studies of irradiation of low temperature CH_4 (Lanzerotti et al. 1987). It was found that when H_2 is formed and lost, cross-linking occurs between carbon atoms which changes the material structure. The loss of H_2 is observed to be low for either energetic H^+ or He^+ irradiation of a fresh sample since H_2 diffusion is inefficient in low temperature CH_4. However, a large increase in H_2 loss occurs at a particular fluence, with the amount depending on the ion penetration depth. This suggests that H_2 formed at depth could escape by a newly formed 'percolation' pathway to the surface (Lanzerotti et al. 1987). This change is permanent since the large loss rates were found to persist until the hydrogen was mostly depleted, leaving behind a porous carbonized material. A related phenomenon was seen at ≤ 15 K in H_2O, but it was reversible (Moore and Hudson 1992). Although such processes are most easily observed and studied in molecular ices, they also occur in more refractory materials.

2. Ion Mixing and "Welding"

The agitation causing tracks and other defects can also cause mixing across interfaces when energetic ions penetrate a heterogeneous mix of grains or a mantle on a grain. This process is well studied in layered materials (e.g., Benit and Brown 1990). Collisional mixing and new bond formation at the interface can change the effective binding properties between materials that are in contact. It was therefore pointed out that aggregates of grains, even with icy mantles, are not necessarily destroyed by ion bombardment (unless charging is excessive). Rather, the loss of newly formed volatile species is a non-disruptive process. In this manner, long-term exposure to penetrating ions can cause non-disruptive loss and carbonization of volatile mantles, as well as enhanced adhesion ('welding') of refractory grains whose organic upper layers are carbonized (Johnson R. E. 1985; Johnson and Lanzerotti 1986).

3. Sputtering

Agitation, defect formation, and ion-enhanced diffusion of volatiles leads to material loss to the gas phase known as sputtering when near the vacuum interface. Chemical effects leading to material loss like the ones discussed above (e.g., H_2 from organics; CH_4 when it is implanted into carbon for example), are referred to as chemical sputtering. Roth (1983) reviewed these effects in refractories while Johnson and Schou (1993) and Johnson (1998) reviewed the sputtering of volatile and refractory species by electronic energy deposition (electronic sputtering). Earlier, Betz and Wehner (1983) reviewed sputtering of refractory insulators, primarily by momentum transfer.

Although sputtering is a relatively old area of study and there is a large amount of data (e.g., Andersen and Bay 1981; Sigmund 1993), there is still a shortage of data for those ion energies and materials of interest in many regions of space. This is particularly so for low energy ions or atoms (≤ 0.5 keV) incident on refractory insulators. Therefore, models constructed in the important 'threshold' region for sputtering have only been tested for metallic solids. The scarcity of laboratory data for the sputtering of insulators by low energy ions is due to problems associated with surface charging and the fact that yields are small. Monte Carlo particle tracking programs have been developed more recently but these generally do not address chemical alterations. Care must be taken when modeling the erosion of small grains whose size is comparable to or smaller than the ion range. For example, the physical sputtering yield in a 1 μm carbon grain bombarded by 10 keV O^+ is $\sim 30\%$ larger than for a flat carbon surface. Further enhancements occur due to nonlinear sputtering when the energy deposited by a single ion raises the grain temperature sufficiently and sublimation can also contribute to the loss from very small grains.

Solar-wind-induced sputtering of silicates causes an erosion rate of $\sim 0.01-0.02$ nm yr^{-1}. The most volatile species that can form in the transiently agitated lattice are lost first. The material is modified to the depth of ion penetration after long-term exposure ($\sim 10^4$ years for solar wind ions) and the sputtering can then approach stoichometry. Johnson and Baragiola (1991) expected depletion of O in the surface regions based on their experiments with non-reactive, heavy ions incident on silicates. However, Bradley (1994) appears to have found O-excesses in the irradiated surface regions of IDP's caused by -OH formation when protons are incident. Mg and Ca appear to be depleted in irradiated rims of IDPs whereas these regions are enriched in O, S, Fe and Ni (Bradley 1994). This has not yet been confirmed in a laboratory study.

As we have seen, sputtering of organics by light fast ions like H^+ and He^+ leads to depletion of volatiles from organics and gradually produces a carbonaceous residue. Similarly, sulfuric residue also accumulates when sulfur is present. On the other hand, large organic molecules can be ejected intact when heavy ions irradiate a grain with an organic mantle rather than light ions, a process suggested to be important in the ISM (Johnson 1991). We also note that electronic processes are extremely efficient in the sputtering of organics.

Electronic sputtering has been shown to dominate the sputtering of ices and is very efficient except for low energy ions (≤ 1 keV/u), Fig. 10. Many aspects of the laboratory data have been described using the so-called thermal spike model for scattering (e.g., Johnson 1990). The conversion factor used in these models to describe electronic energy that is rapidly converted into lattice motion and can lead to sputtering, is ~ 0.2, while the rest of the energy dissi-

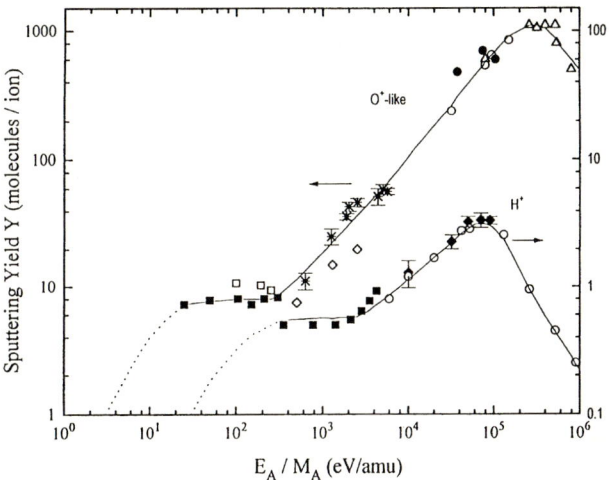

Figure 10. Compilation of data for the sputtering yield (molecules removed per incident ion) of water ice for temperatures below 80 K (from Shi et al. 1995).

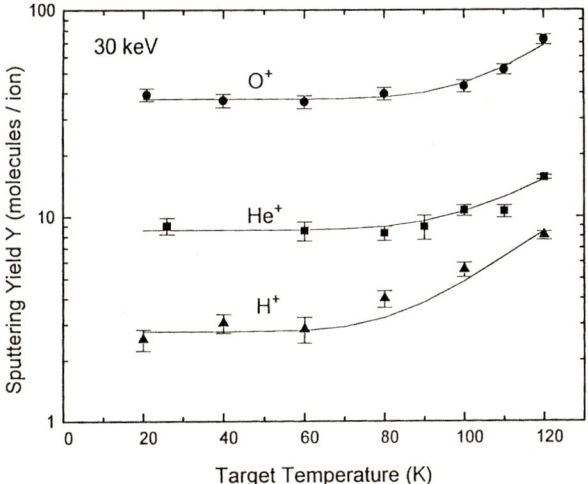

Figure 11. Temperature dependence of the sputtering yield when ions are incident on water ice (from Shi et al. 1995).

pates more slowly. In very small, nanometer-sized, grains the full electronic energy deposited may act due to confinement. The concept of hot-spots or cylindrically-heated regions around the track of the ion have frequently been used (e.g., Léger et al. 1985), but laboratory data has often been ignored. However, there is now considerable data on the ices, and extrapolations using the thermal spike model have been shown to be valid as long as semi-empirical parameters are used to roughly account for the associated chemical and physical alterations (Johnson 1990). The sputtering yield of water ice also exhibits temperature dependence, Fig. 11, which is determined by the reactivity of the radical species produced in ice, as discussed above. Such a temperature dependence is also seen in photo-stimulated desorption of H_2O (Westley et al. 1995) and in the sputtering of other molecular ices (Johnson 1990). The temperature dependent component of the yield for water ice has a slower dependence on the energy density deposited than does the temperature independent component in Fig. 10. Activation energy of the order of 0.04-0.07 eV is associated with the increase of the yield with temperature. The data in Figs. 10 and 11 can therefore be directly applied to the sputtering of icy grains.

VI. CHARGING

VI.A. Charging of Single Isolated Dust Particles

A considerable number of mechanisms can contribute to the charging of dust grains. The time evolution of the dust charge q_d can be represented as

$$\frac{dq_d}{dt} = \sum_k I_k. \tag{64}$$

if each such mechanism produces a current I_k across the grain boundary. The currents I_k are functions of the grain properties, the plasma conditions including the dust density and the radiation field. The important currents in Solar System space conditions are those due to: plasma electron (I_e) or ion (I_i) attachment, secondary electron production (I_{sec}) if the plasma impact energies are sufficiently high, photoelectric effect (I_{ph}) and the thermoionic effect (I_{th}) if the dust grain is hot. Field emission prevents further charging of a dust particle once the dust grain possesses a sufficiently strong electric surface potential (Muller and Tsong 1969; Draine and Salpeter 1979; Whipple 1981). For a negatively charged dust grain with a radius of s_μ (in unit of μm), the limiting surface potential for conductors is

$$U \approx -10^3 \, s_\mu, \quad \text{Volt} \tag{65}$$

and for positively charged dust

$$U \approx +3 \cdot 10^4 \, s_\mu, \quad \text{Volt.} \tag{66}$$

The limiting surface potentials on dielectric dust particles should be much lower. Svestka et al. (1993) find a limiting field of $+5 \cdot 10^8 \, \text{V m}^{-1}$ for micronsized glass particles as compared to the $+3 \cdot 10^{10} \, \text{V m}^{-1}$ given by Eq. (66). The limiting surface potentials may be reached by very small dust grains of radius $s_\mu \leq 10^{-2} \, \mu m$ but is less likely reached by positively charged dust.

Observations by Ulysses (Svestka et al. 1996) may indicate the existence of a few times 10^{-18} kg mass dust grains that have extremely high surface potentials near 10^4 V. Field emission according to Eq. (66) should have limited the charge to $\lesssim 2\cdot 10^3$ Volt if the dust material density is $\sim 10^3$ kg m^{-3}. Svestka et al. (1996) state, however, that the measurement of dust charges by the Ulysses dust detector is subject to large uncertainties.

In addition to the above dust currents, charging can also result from dust fragmentation and dust–dust collisions with charge exchange. These two mechanisms may be important in thunderclouds (e.g. Chiu 1978).

1. Ion and Electron Attachment

The theory for grain charging is only well developed for spherical grains and we will in the following use this as our model. In various papers it has been shown that the cross section for a plasma particle j ($= e$ or i) of charge $Z_j e$, mass m_j and velocity w to collide with a spherical grain of radius s_d and surface potential U is

$$\sigma_j(w) = \pi s_d^2 \left(1 - \frac{2Z_j eU}{m_j w^2}\right). \tag{67}$$

The total current caused by plasma particles j with a polar velocity distribution $f_j(w,\varphi,\theta)$ that is symmetric in θ is given by

$$I_j = 2\pi Z_j e \int_{w_1}^{\infty} \int_0^{\pi/2} S_j \cdot \sigma_j(w) f_j(w,\varphi) w^3 \sin\varphi \, d\varphi \, dw. \tag{68}$$

We adopt a sticking coefficient $S_j = 1$ for all (w,φ) although others (e.g., Draine and Salpeter 1979; Mukai 1981) have included the effects of secondary electrons in calculating the sticking coefficients. The lower limit in particle velocity is $w_1 = 0$ if particles are attracted and $w_1 = (2Z_j eU/m_j)^{1/2}$ if they are repelled. The simplest case is when f_j is Maxwellian (e.g., Whipple 1981). For the plasma particles that are repelled ($\chi_j = Z_j eU/kT_j \geq 0$) we find

$$I_j = n_j \pi s_d^2 Z_j e \left(\frac{8kT_j}{\pi m_j}\right)^{1/2} \exp(-\chi_j) \tag{69}$$

and for attracted particles ($\chi_j < 0$)

$$I_j = n_j \cdot \pi s_d^2 Z_j e \left(\frac{8kT_j}{\pi m_j}\right)^{1/2} (1 - \chi_j). \tag{70}$$

A velocity distribution which is shifted by a velocity V_d of the dust particle relative to the plasma particles of type j is

$$f_j(w,\varphi,V_d) = n_j \left(\frac{m_j}{2\pi kT_j}\right)^{3/2} \exp\left[-\frac{m_j}{2kT_j}(V_d^2 + w^2 - 2V_d w \cos\varphi)\right]. \tag{71}$$

The currents now become for $\chi_j \geq 0$ (Whipple 1981; Havnes et al. 1987)

$$I_j = n_j \pi s_d^2 Z_j e \frac{c_j}{2M_j} \left\{ \frac{1}{2} \left[1 + 2M_j^2 - 2\chi_j\right] \cdot \right.$$
$$\left[\text{erf}\left(M_j + \chi_j^{1/2}\right) + \text{erf}\left(M_j - \chi_j^{1/2}\right)\right]$$
$$+ \frac{1}{\sqrt{\pi}} \left(M_j + \chi_j^{1/2}\right) \exp\left[-\left(M_j - \chi_j^{1/2}\right)^2\right]$$
$$\left. + \frac{1}{\sqrt{\pi}} \left(M_j - \chi_j^{1/2}\right) \exp\left[-M_j + \chi_j^{1/2}\right]^2 \right\} \quad (72)$$

and for $\chi_j \leq 0$

$$I_j = n_j \pi s_d^2 Z_j e \cdot \frac{c_j}{2M_j} \left\{ \left[(1 + 2M_j^2 - 2\chi_j) \text{erf}(M_j)\right] \right.$$
$$\left. + \frac{2M_j}{\pi^{1/2}} \cdot \exp(-M_j^2) \right\}. \quad (73)$$

Here $c_j \equiv (2kT_j/m_j)^{1/2}$ and the Mach number $M_j = V_d/c_j$. The dust drift velocity V_d is often small compared to c_e and $M_e \ll 1$. One can use the appropriate of Eqs. (72) or (73) for the electron currents in such cases. The plasma currents must be found by appropriate integrations of Eq. (71) when the velocity distributions are non-Maxwellian (e.g. Havnes et al. 1987).

2. Secondary Electron Emission

Plasma electrons hitting a dust grain may ionize the dust material thereby ejecting (secondary) electrons and producing a current $I_s \geq 0$. The yield δ_y, which is the ratio of emitted to incoming electrons, is a function of grain material and size and the kinetic energy E of the primary electrons in the inertial frame of the dust particle. It is often approximated by

$$\delta_y(E) = 7.4 \, \delta_M \left(\frac{E}{E_M}\right) \exp\left[-2\left(\frac{E}{E_M}\right)^{1/2}\right], \quad (74)$$

when the grain size is large compared to the stopping length of the incident electrons in the dust material (Sternglass 1954). The yield has a maximum of $\delta_y = \delta_M$ (from 1 to 10) at incident energy E_M (~ 100 to ~ 1000 eV).

One finds $kT_s \approx 1$ to 5 eV if the velocity distribution of the secondary electrons is approximated by a Maxwellian distribution. Meyer-Vernet (1982) calculated the resulting secondary current if the primary electrons also have a Maxwellian distribution with temperature T_e. The currents become for negatively charged dust where all secondary electrons escape

$$I_{\text{sec}}(U \leq 0) = 3.7 \, \delta_M n_e \left(\frac{kT_e}{2\pi m_e}\right)^{1/2} \exp(-\chi_e) \cdot F_5\left(\frac{E_M}{4kT_e}\right), \quad (75)$$

where

$$F_5(x) = \chi^2 \int_0^\infty u^5 \exp[-(xu^2 + u)]du. \tag{76}$$

For positively charged dust

$$I_{\text{sec}}(U \geq 0) = 3.7\, \delta_M n_e \left(\frac{kT_e}{2\pi m_e}\right)^{1/2} (1 - \chi_s)$$

$$\cdot \exp(\chi_s - \chi_e) F_{5,B}\left(\frac{E_M}{4kT_e}\right), \tag{77}$$

where $\chi_s = -eU/kT_s$ and

$$F_{5,B}(\chi) = \chi^2 \int_B^\infty u^5 \exp[-(xu^2 + u)]du \tag{78}$$

with

$$B = \left(-\frac{\chi_e}{E_M/4kT_e}\right)^{1/2}. \tag{79}$$

The energy dependence is in part due to the distance at which secondary electrons are produced to the surface. The proximity to the surface can also strongly enhance secondary electron emission from dust grains over that from macroscopic bodies while primary electrons may also pass through sufficiently small dust particles without producing many secondary electrons. Draine and Salpeter (1979) give corrections to account for dust sizes. More recent work by Chow et al. (1993, 1994) includes detailed models for the corrections due to secondary electron production from dust of a variety of sizes and compositions that are in good agreement with experimental results by Svestka et al. (1993).

Illustrating the dependence on energy and size, Fig. 12 shows the equilibrium potential of a dust particle as a function of the incoming electron energy (Svestka et al. 1993). The initial decrease in potential is due to secondary electron production deeper into the grain but eventually the production is closer to the exit side where the-mainly-forward produced secondary electrons can escape more easily. These size-dependent effects shift to electron energies of several keV in micron-sized dust particles. For large dust particles the secondary emission becomes important at primary electron energies from ~ 40 eV for glass, ~ 50 eV for silicon, ~ 60 eV for copper and ~ 90 eV for graphite (Walch et al. 1995).

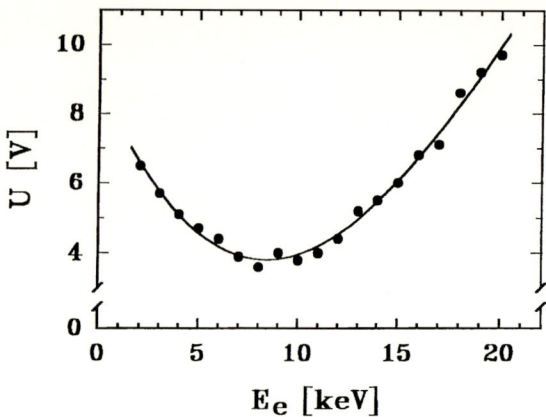

Figure 12. Typical dependence of particle equilibrium surface potentials U on the energy of energetic electrons. The example is for glass particles of radius 1.12 μm, at $p = 10^{-6} mbar$.

3. Photoemission

When dust particles are exposed to photons of energy exceeding the work function of the dust material ε_w, the minimum energy required to knock off an electron, the result is a photocurrent across the surface. The current across the surface of a spherical grain of radius s_d at heliocentric distance r can be written

$$I_{\rm ph} = 4\pi s_d^2 \left(\frac{R_0}{r}\right)^2 \int_{h\nu=\varepsilon_w+\varepsilon_{\rm MIN}}^{\infty} \frac{\mathcal{F}_*}{h\nu} Q_{\rm abs} Y_m \cdot \int_{\varepsilon_{\rm MIN}}^{\varepsilon_{\rm MAX}} f(\varepsilon) d\varepsilon \, d(h\nu), \qquad (80)$$

where R_0 is the solar radius and $(\mathcal{F}_*/h\nu)$ m^{-2} s^{-1} Hz^{-1} the photon flux. The size dependent efficiency factor (ratio between the effective and geometric cross section) for absorption of photons $Q_{\rm abs}$ can be calculated using Mie-theory and the refractive index of the material. The number of electrons released per absorbed photon, the yield function Y_m, was predicted to also be a function of dust size (e.g., Gail and Sedlmayer 1980). Analogous to the case of secondary electron emission, the yield from small particles of dielectrics should be greatly enhanced compared to those from macroscopic materials due to the proximity from the site of electron emission to the grain surface. Gail and Sedlmayer found that the yield is close to unity already at photon energies close to the workfunction for the material. This should apply to dust sizes ≤ 1 μm. For conductors the yield function is close to that for bulk materials (flat half space surfaces) except for very small dust particles of radius of only a few nanometers. The yield increases up to a factor of several hundred in such cases (Müller et al. 1988; Schleicher et al. 1993). The energy distribution of released photoelectrons is $f(\varepsilon)$ and has a cut-off at a maximum energy $\varepsilon_{\max}$ equal to the photon energy

$h\nu$ minus the work function. The minimum energy required for a released photoelectron to escape is $\varepsilon_{\min} = 0$ for $U \leq 0$ and $\varepsilon_{\min} = eU$ for $U \geq 0$.

Eq. (80) is difficult to determine accurately since the solar UV spectrum can vary with time and since the absorption efficiency Q_{abs} depends on the not fully known material and geometric properties of the grains. See the chapter by Gustafson et al. and articles by Lou and Charalampopoulos (1994), McGuire and Hapke (1995), and Hirst et al. (1994). Impurities can also seriously affect the value of the photoelectric workfunction ε_w. As an extreme example, Qiu et al. (1989) found that if ammonia (NH_3) is co-deposited with metals like Na, Cs, Yb, that can be solvated in NH_3, workfunctions as low as 0.9 (± 0.1) eV could be obtained. We list only an approximate expression (Whipple 1981; Wallis and Hassan 1983) due to the uncertainties in evaluating (Eq. (80))

$$I_{\rm ph} = \pi s_d^2 e\, 2.5 \cdot 10^{10} \frac{\chi}{r^2} \exp\left(-\frac{\max[eU, 0]}{kT_\nu}\right), \tag{81}$$

where the efficiency factor χ is close to 1 for conductors and close to 0.1 for dielectrics. Here, the heliocentric distance r is in astronomical units and kT_ν ($\sim$ 1 to 3 eV) is the average energy of the photoelectrons.

4. Thermoionic Effect

At extremely high temperatures T_d, electrons may reach a sufficient energy to escape from the dust grain. This effect is normally only important considerably above 1000 K and therefore is relevant only on highly refractive materials like carbon and silicates close to the Sun. The thermoionic current can be written (e.g., Evans 1994)

$$I_{\rm th} = A T_d^2 \exp\left(-\frac{\varepsilon_w}{kT_d}\right), \tag{82}$$

where the coefficient A is $\sim 4 \times 10^4$ ampere $m^{-2} K^{-2}$ for pure carbon. The value can depend on the geometry as well as on possible impurities in the material.

5. Grain Charging in the Solar System

Charges on single dust particles in the solar system can vary between comparatively wide limits depending on the conditions. A comparison between I_e (Eqs. (72) or (73)) and $I_{\rm ph}$ (Eq. (81)) reveals that these currents are often of the same order. With $U=0$ we find

$$\frac{I_{\rm ph}}{I_e} = \frac{3.7 \cdot 10^2 \chi}{n_e T_e^{1/2} ({\rm eV}) r^2}. \tag{83}$$

At interplanetary conditions with $T_e \sim 10$ eV and $n_e \sim 10^6$ m^{-3}, Eq. (83) indicates that the photoelectric effect dominates to considerable distances from the Sun. The exceptions are low yield dust and when $n_e \gtrsim 10^7$ m^{-3}, when Eq. (83) implies that $I_{ph} \sim I_e$ for $r \gtrsim 1$ AU. Charging by plasma particle impacts (I_e, I_i, $I_{\rm sec}$) normally dominates at magnetospheric conditions where n_e often is in the range $10^7 - 10^8$ m^{-3} (e.g., Hartquist et al. 1992).

The surface potential U of a spherical grain is related to the dust charge number Z_d by

$$Z_d = (4\pi \epsilon_0 \, s/e) \, U \text{ (Volt)} \approx 700 U \text{ (Volt)} \, s_\mu, \tag{84}$$

so that U normally falls in the range from minus several times ten Volts to plus 10 Volts. The largest negative potentials are achieved at moderately high plasma temperatures so that $I_e \gg I_{sec}$ and when n_e is sufficiently high so that $I_e \gg I_{ph}$. The equilibrium surface potential, U_{eq}, when the charging is by plasma attachment only, is found from Eq. (64) with $dq_d/dt = 0$. This now reads $I_e + I_i = 0$, giving

$$U_{eq} = -\beta \frac{kT}{e} \tag{85}$$

where $\beta = 2.5$, 3.6, 3.9 and 3.4 for plasmas with ion types H^+, O^+, S^+ and S^{++}, respectively (e.g., Havnes et al. 1990). The effect of secondary electrons, that become increasingly important with increasing plasma temperature, is to make the surface potential less negative and eventually positive. Since ejected secondary electrons have approximately Maxwellian velocity distributions with fairly low (1 to 5 eV) temperatures, they cannot overcome high positive surface potentials and the positive potentials will normally stay below $\approx +10$ Volt. Since photoelectrons also have low temperature (1 to 3 eV) energy distributions, they cannot produce high positive dust surface potentials either. The large dust surface potentials of up to plus several hundred Volt indicated by Ulysses and Galileo for interplanetary dust (Svestka et al. 1996) is therefore a serious challenge if the values are real.

VI.B. Collective Effects on Dust Charging

Equal electron and ion charge densities $n_e = n_i$ is normally assumed when calculating the charges on single isolated dust particles. Implicitly postulated is then that the charge density carried by dust is negligible compared to that of electrons, or that $n_e \gg N_d Z_d$. While this is true of isolated dust grains, it is usually not the case in a cloud of dust grains, neither in space nor under laboratory conditions. The charge $N_d Z_d e$ must be accounted for as part of the space charge. Space charge quasi-neutrality in a cloud of identical dust particles requires that

$$e n_e - Z_i e n_i - U s_d N_d = 0 \tag{86}$$

when the cloud is of large dimensions L_d compared to the plasma Debye length $\lambda_D^2 = \epsilon_0 \, kT/(n_e e^2)$, which can be expressed as

$$L_d = \frac{N_d}{\nabla N_d} \gg \lambda_D. \tag{87}$$

Equation (86) must be replaced by the full Poisson equation to apply to smaller clouds, such as for dust levitation near solid surfaces in space as well as in laboratories. Nitter et al. (1998) investigated this much more difficult problem in the context of dust levitating above the lunar surface or above asteroid surfaces. They found that the sheath electric field near the solid surface (Langmuir et al. 1924; Chen 1984) can be severely modified by the presence of levitated dust.

A dust cloud permeated by a plasma will also be partly charged (Havnes and Morfill 1984) and will develop a local potential V. The dust charge affects the local plasma densities, which often can be described by the Boltzmann relations

$$n_\alpha = n_{\alpha 0} \exp\left(-\frac{Z_d eV}{kT_\alpha}\right), \quad \alpha = i, e. \tag{88}$$

In Eq. (88), the electron density of the ambient plasma is $n_{e0} = Z_i n_{i0}$ and T_α is the plasma temperature which we assume to be constant throughout the cloud. Using Eq. (88) in Eq. (86) and the current equilibrium equation $I_e + I_i + I_{ph} = 0$, we have two equations that can be used to determine the dust surface potential U and the local plasma potential V. The quasineutrality equation becomes

$$\exp(\hat{V}) - \exp(-Z_i X \hat{V}) - P \cdot \hat{U} = 0 \tag{89}$$

where

$$P = \frac{4\pi\epsilon_0 k T_e s_d N_d}{e^2 n_{e0}} = 695 \frac{T_e(eV) s_\mu N_d}{n_{e0}}, \tag{90}$$

and the current equilibrium for $R \leq 1$

$$X^{1/2} \left(\frac{m_i}{m_e}\right)^{1/2} \exp(\hat{U} + \hat{V}) - (1 - Z_i X \hat{U}) \exp(-Z_i X \hat{V}) - R = 0. \tag{91}$$

The ratio between the photoelectric current and plasma current is

$$R = \frac{I_{ph}}{\pi s_d^2 n_{e0} \left(\frac{8kT_e}{\pi m_e}\right)^{1/2}}. \tag{92}$$

Here, we used the abbreviations $\hat{U} = eU/kT_e$, $\hat{V} = eV/kT_e$ and $X = T_e/T_i$. The photocurrent I_{ph} in Eq. (92) can be approximated by Eq. (81).

The parameter P (Eqs. (90) and (91)), which is a measure of the dust collective effect, is modified to

$$P = \frac{4\pi\epsilon_0 kT_e}{e^2 n_0} \int_{s_d(\min)}^{s_d(\max)} s_d dN_d \tag{93}$$

when the cloud contains a distribution of dust sizes (Havnes et al. 1990). The collective effect of the dust on the dust charging is negligible when $P \ll 1$(tenuous clouds) but increases in importance as P increases. The dust surface potentials approach zero when $P \gg 1$ while the local plasma potential V is maximized. Figures 13 and 14 show that the surface potential U of the dust and the plasma potential V are functions of P only for a given plasma condition. Havnes et al. (1990) give numerical solutions to Eqs. (91) and (92) and rational functions to approximate the solutions to eU/kT_e and eV/kT_e as a function of P. They treat a range of values of $R < 1$ for plasmas with ions of types H^+ and O^+, S^+ and S^{++} with X=1. Havnes (1984) and Wilson (1991) also considered cases where $R > 1$, so that $U > 0$.

If we consider the case where $R \ll 1$ in Eq. (91) (i.e., when the photoelectric effect can be neglected), we see from Eq. (91) with $\hat{U} = 0$ that

$$\hat{V}_{\text{MAX}} = -\frac{1}{2} \frac{\ln\left(\frac{m_i}{m_e} \cdot X\right)}{(1 + Z_i X)}. \tag{94}$$

This is the maximum negative value for $\hat{V}$ when $P \gg 1$. If the velocities of ions and electrons are Boltzmann distributions, we find from Eqs. (88) and (94) that we have $Z_i n_i / n_e = (X m_i/m_e)^{1/2}$ for $P \gg 1$. The reason is that the dust cloud charges negatively and attracts ions while it repels electrons. Ions scatter into the negative potential well of the dust cloud where its density exceeds the ambient ion density. The opposite applies to the electrons and their number densities are lower inside the dust cloud than outside. The effect on ions increases when $X \gg 1$, while the electron density inside and outside the clouds is practically the same.

Figure 14 shows how eU/kT_e and eV/kT_e vary as a function of P (which is proportional to the dust density, see Eq. (90) for values of $X = 1$ to 40 in the case of hydrogen ions. We see how eV_{MAX}/kT_e decreases following Eq. (94) as X increases. In a cloud in quasineutrality, practically all of the negative charge is on the dust and the dust space charge density is practically identical to that for ions when $P \gg 1$. The approximate average charge number on dust for $P \gg 1$ can therefore be obtained by solving $\overline{Z}_d N_d \approx -n_{e0} \exp(-Z_i X \hat{V}_{\text{MAX}})$ which leads to

$$\overline{Z}_d \approx -\frac{n_{e0}}{N_d} \left(\frac{m_i}{m_e} X\right)^{\frac{Z_i X}{2(1+Z_i X)}}. \tag{95}$$

This shows that $\overline{Z}_d$ increases with increasing X unless, of course, this implies that the electron temperature increases to the extent that secondary electron production becomes important. If $|\overline{Z}_d| \lesssim 1$ we should no longer use the average dust charge but consider the distribution of charges in discrete charge states (e.g., Gail and Sedlmayr 1975; Umebayashi and Nakano 1990). Collective effects on dust charging have been verified experimentally (e.g., Xu et al. 1993; Barkan et al. 1994).

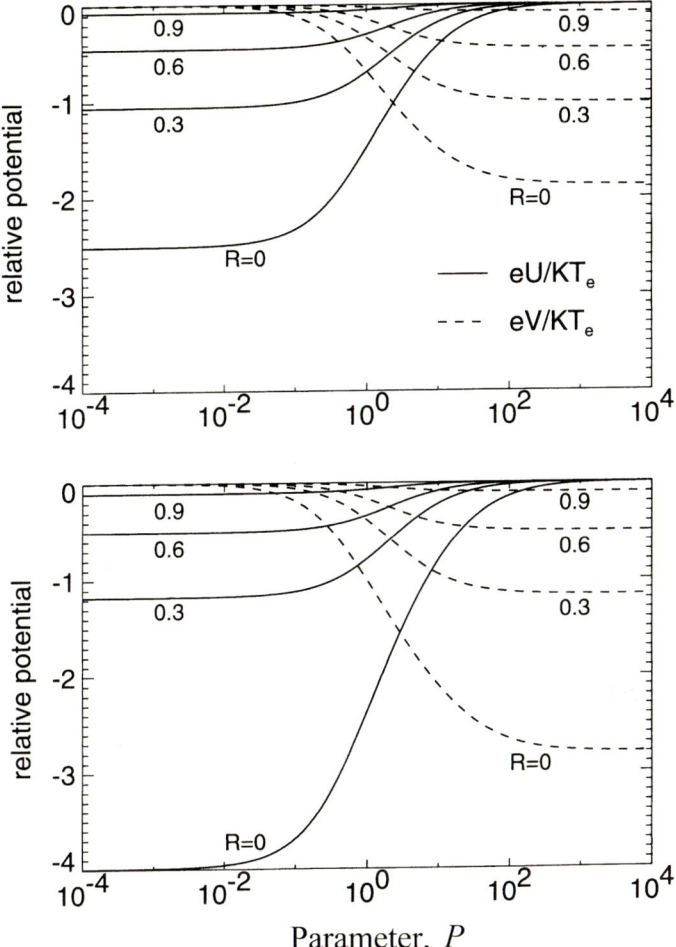

Figure 13. The local dust plasma potential eV/kT_e and dust surface potential eU/kT_e as a function of P for different ratios R between the photoelectron flux and electron flux (Eq. (94)). Results are for protons (upper figure) and singly ionized oxygen (lower figure). In both cases $X = 1$.

VII. LIFETIMES

Dohnanyi (1978) showed that the mean survival time of interplanetary grains of mass $\geq 10^{-12}$ kg before catastrophic collision with another grain, t_c, is $\sim 10^4$ years. Leinert et al. (1983) concluded that t_c is shorter than the time required to dissipate orbital energy by way of the Poynting-Robertson effect, t_{PR}, as long as the grains exceed 80 to 100 μm in radii. Later, using empirical dust flux data, Grün et al. (1985) also found that t_c at 1 AU is shortest ($\sim 10^4$ years) for meteoroids of 10^{-7} to 10^{-3} kg mass, and t_c is shorter than t_{PR} for grains of mass $\geq 10^{-8}$ kg.

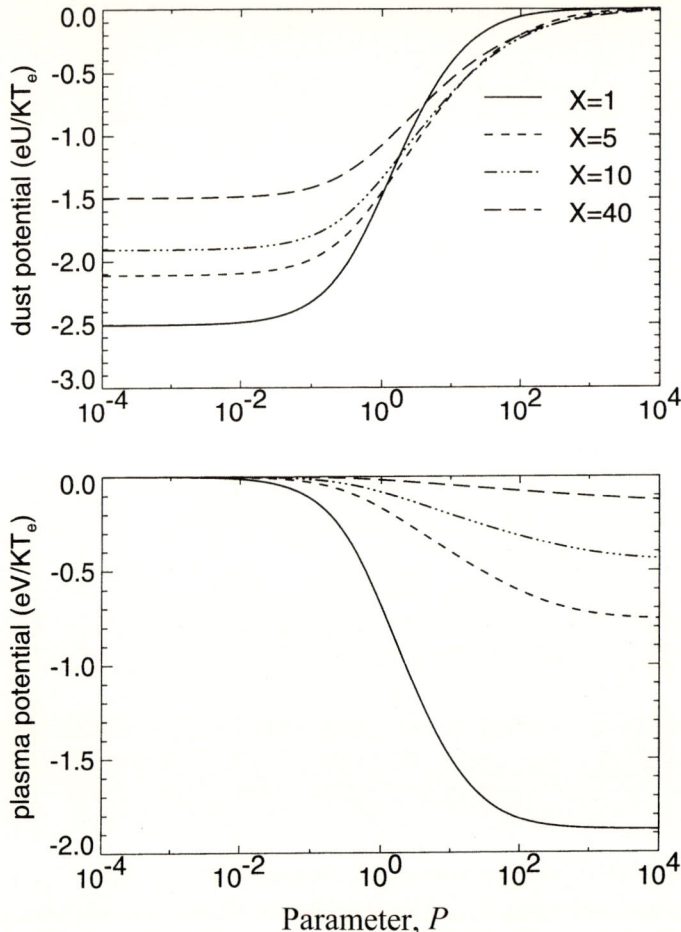

Figure 14. The potentials eV/kT_e and eU/kT_e as a function of P for a range of values of $X = T_e/T_i$ when $R = 0$. This case is with protons as ions.

It is hard to define a critical grain mass or size beyond which the lifetime is limited by catastrophic collisions rather than by the Poynting-Robertson effect. This is since t_c depends in a complicated way on the grain size as well as on other physical grain properties and on the heliocentric distance. However, we can make a rough estimate; when the grains are larger than ~ 100 μm, they suffer catastrophic collisions with other grains before their orbital elements change significantly as a result of the Poynting-Robertson effect. By contrast, the Poynting-Robertson effect controls the lifetime of grains ≤ 100 μm.

Mass loss through sublimation can also limit the lifetime, especially for volatile grains or for volatile mantles on grains. While the dynamical lifetimes t_c and t_{PR} depend mostly on the grain size, the sublimation lifetime, t_m (defined as $M/(dM/dt)$) is a strong function of the heliocentric distance. For comparison, the lifetime of a grain suffering mass loss by sputtering (section V) is proportional to the square of the heliocentric distance, r, since the solar wind flux has a r^{-2}-dependence (e.g., Mukai and Schwehm 1981). The sputtering process therefore dominates over sublimation beyond some critical heliocentric distance r_c.

Size dependence cannot be neglected although the size dependence of the sublimation lifetime is not as strong as the dependence on heliocentric distance. Since there is a size dependence of the grain temperature, we may estimate the value of r_c as a function of a grain radius. For pure water-ice, r_c becomes about 1 AU for radii $s = 10$ μm spheres and r_c=2.8 AU for $s = 100$ μm and ≤ 1 μm. The mimimum in mass loss rate of pure water-ice due to sublimation at about $s = 10$ μm was already pointed out by Patashnick and Rupprecht (1975). This is due to a resonance effect leading to a peak in the thermal emission and therefore lower temperature when the grain radius is close to the wavelength of thermal emission, i.e., about 10 μm (e.g., Mukai 1986).

It is widely believed that no pure water-ice grains exist in interplanetary space. Instead, icy mantles accrue on grains and consequently several material types are mixed together in the grain. Absorbing inclusions in water-ice enhance the absorption properties of visible light. Consequently, they attain higher temperatures for sublimation far from the Sun than those expected for pure water-ice grain. Mukai (1986) deduced a temperature for a dirty water-ice grain with a radius of 500 μm, composed of a water-ice matrix with magnetite (volume fraction of 10%) and silicate (volume fraction of 10 %), i.e., T_d=290 $r^{-0.48}$, where T_d and r, respectively, are in units of Kelvin and AU. This implies that dirty water-ice grains lose significant mass to sublimation already at $\sim 4 - 5$ AU from the sun. It is interesting to compare these values with $\sim 1 - 2.8$ AU for pure water-ice.

A comparison of lifetimes, such as t_c, t_{PR}, t_m, has been discussed in previous works (e.g., Dohnanyi 1978; Mukai and Schwehm 1981; Leinert et al. 1983; Mukai and Giese 1984; Grün et al. 1985).

Acknowledgements

We thank Jiri Svestka of Prague Observatory for his assistance in making Fig. 12, and for his useful comments concerning the experimental works on electric charging of dust grains.

REFERENCES

Andersen, H. H., and Bay, H. L. 1981. Sputtering yield measurements. In *Sputtering by particle bombardment I*, ed. R. Berisch (Berlin: Springer Verlag), pp. 145–218.

Asada, N. 1985. Fine fragments in high-velocity impact experiments. *J. Geophys. Res.*, **90:12,** pp. 445–453.

Asphaug, E., Melosh, H. J., and Ryan, E. 1992. Theoretical predictions for fragment size distributions. (abstract) *Lunar Planet Sci. Conf.*, XXIII pp. 45–46.

Baguhl, M., Hamilton, D. P., Grün, E., Dermott, S. F., Fechtig, H., Hanner, M. S., Kissel, J., Lindblad, B.-A., Linkert, D., Linkert, G., Mann, I., McDonnell, J. A. M., Morfill, G. E., Polanskey, C., Riemann, R., Schwehm, G., Staubach, P., and Zook, H. A. 1995. Dust measurements at high ecliptic latitudes. *Science*, **268,** pp. 1016–1019.

Barkan, A., D'Angelo, N. D., and Merlino, R. L. 1994. Charging of dust grains in a plasma. *Phys. Rev. Letters*, **72,** pp. 3093–3096.

Benit, J., and Brown, W. L. 1990. Electronic sputtering of oxygen and water molecules from thin films of water ice bombarded by MeV ions. *Nucl. Inst. Methods/*, **B46,** pp. 448–451.

Benz, W., Asphaug, E., and Ryan, E. V. 1994. Numerical simulations of catastrophic disruption: recent results. *Planet. Space Sci.*, **42,** pp. 1053–1066.

Betz, G., and Wehner, G. K. 1983. Sputtering of multicomponent materials. In *Sputtering by particle bombardment II*, ed. R. Berisch (Berlin: Springer Verlag) pp. 11–90.

Blum, J. 1995. Laboratory and space experiments to study pre-planetary growth. *Adv. Space Res.*, Vol.15, **10,** pp. 39–54.

Blum, J., and Münch, M. 1993. Experimental investigations on aggregate-aggregate collisions in the early solar nebula. *Icarus*, **106,** pp. 151–167.

Blum, J., Schnaiter, M., Wurm, G., and Rott, M. 1996. The De-Agglomeration and Dispersion of Small Dust Particles - Principles and Applications. *Rev. Sci. Instrum.*, **67,** pp. 589–595.

Blum, J., and Wurm, G. 2000. Experiments on Sticking, Restructuring and Fragmentation of Preplanetary Dust Aggregates. *Icarus*, bf 143, pp. 138–146.

Bradley, J. P. 1994. Chemically anomalous, preaccretionally irradiated grains in interplanetary dust from comets. *Science*, **265,** pp. 925–929.

Bradley, J. P., Brownlee, D. E., and Fraundorf, P. 1984. Discovery of nuclear tracks in interplanetary dust. *Science*, **226,** pp. 1432–1434.

Brown, W. L., Auqutyniak, W. M., Maucoutonio, K. J., Simmous, E. H., Boring, J. W., Johnson, R. E., and Reimana, I. T. 1984. Electronic sputtering of low temperature molecular solids. *Nucl. Inst. Methods/*, **B1,** pp. 307–314.

Burns, J. A., Lamy, P. L., and Soter, S. 1979. Radiation forces on small particles in the solar system. *Icarus*, **40,** pp. 1–48.

Capaccioni, F., Cerroni, P., Coradini, M., Di Martino, M., Farinella, P., Flamini, E., Martelli, G., Paolicchi, P., Smith, P. N., Woodward, A., and Zappala, V. 1986. Asteroidal catastrophic collisions simulated by hypervelocity impact experiments. *Icarus*, **66,** pp. 487–514.

Capaccioni, F., Cerroni, P., Coradini, M., Farinella, P., Flamini, E., Martelli, G., Paolicchi, P., Smith, P. N., and Zappala, V. 1984. Shapes of asteroids compared with fragments from hypervelocity impact experiments. *Nature*, **308,** pp. 832–834.

Chen, F. F. 1984. *Introduction to Plasma Physics and Controlled Fusion.*, Vol. I, (New York: Plenum Press)

Cheng, A. F., and Johnson, R. E. 1989. Effects of magnetospheric interactions on the origin and evolution of atmospheres. In *Origin and evolution of atmospheres*, eds. S. K. Atreya and J. B. Pollack (Tucson: Univ. of Arizona Press) pp. 682–722.

Chiu, C.-S. 1978. Numerical study of cloud electrification in an axisymmetric, time-dependent cloud model. *J. Geophys. Res.*, **83,** pp. 5025–5049.

Chokshi, A., Tielens, A. G. G. M., and Hollenbach, D. 1993. Dust coagulation. *Ap. J.*, **407,** pp. 806–819.

Chow, V. W., Mendis, D. A., and Rosenberg, M. 1993. Role of grain size and particle velocity distribution in secondary electron emission in space plasmas. *J. Geophys. Res.*, **98**, pp. 19065–19076.

Chow, V. W., Mendis, D. A., and Rosenberg, M. 1994. Secondary emission from small dust grains at high electron densities. *IEEE Trans. Plasma Sci.*, **22**, pp. 179–186.

Colwell, J. E., and Taylor, M. 1999. Low-velocity microgravity impact experiments into simulated regolith. *Icarus*, **138**, pp. 241–248.

Cravens, T. E. 1991. Plasma processes in the inner coma. In *Comets in the Post-Halley Era*, eds. R. L. Newburn et al. (Amsterdam: Kluwer Acad. Publ.) pp. 1211–1258.

Crifo, J. F. 1987. Improved gas kinetic treatment of cometary water sublimation and recondensation, application to comet P/Halley. *Astron. Astrophys.* **187**, pp. 438–450.

Crifo, J. F. 1995. A general physicochemical model of the inner coma of active comets: I. Implications of a spatially distributed gas and dust production. *Ap. J.*, bf 445, pp. 470–488.

Dahneke, B. 1971. The capture of aerosol particles by surfaces. *J. Colloid and Interface Sci.*, **37**, No.2, pp. 342–353.

Davis, D. R., Chapman, C. R., Weidenschiling, S. J., and Greenberg, R. 1985. Collisional history of asteroids: Evidence from Vesta and the Hirayama families. *Icarus*, **62**, pp. 30–53.

Delitsky, M. L., and Thompson, W. R. 1987. Chemical processes in Triton's atmosphere and surface. *Icarus*, **70**, pp. 354–365.

Dohnanyi, J. S. 1978. Particle Dynamics. In *Cosmic Dust*, ed. J. A. M. McDonnell (Chichester: Wiley-Interscience Publ.) pp. 527–605.

Dominik, C., and Tielens, A. G. G. M. 1997. The Physics of Dust Coagulation and the Structure of Dust Aggregates in Space. *Ap. J.* **480**, pp. 647–673.

Draine, B. T. 1985. Grain evolution in dark clouds. In *Protostars and Planets II*, eds. D. C. Black and M. D. Matthews, (Tucson: Univ. of Arizona Press), pp. 621–640.

Draine, B. T., Roberg, W. G., and Dalgarno, A. 1983. Magnetospheric hydrodynamic shock waves in molecular clouds. *Ap. J.*, bf 264, pp. 485–507.

Draine, B. T., and Salpeter E. E. 1979. Destruction mechanisms for interstellar dust. *Ap. J.*, **231**, pp. 438–455.

Eisenhour, D. D., Daulton, T. L., and Buseck, P. R. 1994. Electromagnetic heating in the early solar nebula and the formation of chondrules. *Science*, **265**, pp. 1067–1070.

Englman, R., Rivier, N., and Jaeger, Z. 1988. Size-distribution in sudden breakage by the use of entropy maximization. *J. Appl. Phys.*, **63**, pp. 4766–4768.

Evans, A. 1994. *The Dusty Universe.*(New York: Wiley & Sons.)

Farinella, P., and Davis, D. R. 1996. Short-period comets: Primordial bodies or collisional fragments? *Science*, **273**, pp. 938–941.

Feigleson, E. D. 1982. X-ray emission from young stars and the implications for the early solar system. *Icarus*, **51**, pp. 155–163.

Fujimura, A., Takagi, Y., Furumoto, M., and Mizutani, H. 1986. Fractal dimensions of fracture surfaces of rock fragments. *Mem. Natl. Inst. Polar Res.*, **41**, pp. 348–357.

Fujiwara, A., Cerroni, P., Davis, D., Ryan, E., Di Martino, M., Holsapple, K., and Housen, K. 1989. Experiments and scaling laws for catastrophic collisions. In *Asteroids II*, eds. R. P. Binzel, T. Gehrels, M. S. Matthews, (Tucson: Univ. of Arizona Press), pp. 240–269..

Fujiwara, A., Kamimoto, G., and Tsukamoto, A. 1977. Destruction of basaltic bodies by high-velocity impact. *Icarus*, **31**, pp. 277–288.

Fujiwara, A., Kamimoto, G., and Tsukamoto, A. 1978. Expected shape distribution of asteroids obtained from laboratory impact experiments. *Nature*, **272**, pp. 602–603.

Gail, H.-P., and Sedlmayr, E. 1975. On the charge distribution of interstellar dust grains. *Astron. Astrophys.*, **41**, pp. 359–366.

Gail, H.-P., and Sedlmayr, E. 1980. On the photoelectric yield of insulating dust grains. *Astron. Astrophys.*, **86**, pp. 380–385.

Gault, D. E., and Heitowit, E. D. 1963. The partition of energy for hypervelocity impact craters formed in rock. *Proc. 6th Hypervelocity Impact Symposium*, pp. 419–456.

Geiss, J. 1982. Processes affecting abundances in the solar wind. *Space Sci. Rev.*, **33**, pp. 201–217.

Giblin, I. 1998. New data on the velocity-mass relation in catastrophic disruption. *Planet. Space Sci.*, **46**, pp. 921–928.

Giblin, I., Martelli, G., Smith, P. N., Cellino, A., Di Martino, M., Zappala, V., Farinella, P., and Paollichi, P. 1994. Field fragmentation of macroscopic targets simulating asteroidal catastrophic collisions. *Icarus*, **110**, pp. 203–224.

Gibson, Jr. E. K. 1992. Volatiles in interplanetary dust particles: A review. *J. Geophys. Res.*, **97**, pp. 3865–3875.

Gilvarry, J. J. 1961. Fracture of brittle solids. I. Distribution function for fragment size in single fracture (theoretical). *J. Appl. Phys.*, **32**, pp. 391–399.

Grady, D. E., and Kipp, M. E. 1980. Continuum modeling of explosive fracture in Oil Shale. *Int. J. Rock Mech. Min. Sci. & Geomech. Abstr.*, **17**, pp. 147–157.

Grady, D. E., and Kipp, M. E. 1985. Mechanisms of dynamic fragmentation: Factors governing fragment size. *Mechanics of Materials*, **4**, pp. 311–320.

Grady, D. E., and Lipkin, J. 1980. Criteria for impulsive rock fracture. *Geophys. Res. Letters*, **7**, pp. 255–258.

Greenberg, J. M. 1983. The largest molecules in space - Interstellar dust. In *Cosmochemistry and the origin of life*, (Dordrecht: D. Reidel Publ. Co.), pp. 71–112.

Greenberg, J. M., and Hage, J. I. 1990. From interstellar dust to comets: A unification of observational constraints. *Ap. J.*, **361**, pp. 260–274.

Greenberg, J. M., and Shah, G. A. 1971. Interstellar grain temperatures. Effects of Shape. *Astron. Astrophys.*, **12**, pp. 250–257.

Griffith, A. A. 1920. The phenomena of rupture and flow in solids. *Philosophical Transactions of Royal Society*, **221**, pp. 163–198.

Grün, E., Zook, H. A., Baguhl, M., Balogh, A., Bame, S. J., Fechtig, H., Forsyth, R., Hanner, M. S., Horanyi, M., Kissel, J., Lindblad, B.-A., Linkert, D., Linkert, G., Mann, I., McDonnell, J. A., Morfill, G. E., Phillips, J. L., Polansky, C., Schwehm, G., Siddique, N., Staubach, P., Svestka, J., and Taylor, A. 1993. Discovery of Jovian dust streams and interstellar grains by the Ulysses spacecraft. *Nature*, **362**, pp. 428–430.

Grün, E., Zook, H. A., Fechtig, H., and Giese, R. H. 1985. Collisional balance of the meteoritic complex. *Icarus*, **62**, pp. 244–272.

Gustafson B. Å. S. 1994. Physics of zodiacal dust. *Annu. Rev. Earth Planet. Sci.*, **22**, pp. 553–595.

Gustafson, B. Å. S., and Misconi, N. Y. 1979. Streaming of interstellar grains in the solar system. *Nature*, **282**, pp. 276–278.

Hage, J. I., and Greenberg, J. M. 1990. A model for the optical properties of porous grains. *Ap. J.*, **361**, pp. 251–259.

Hanner, M. S. 1985. A preliminary look at the dust in comet Halley. *Adv. Space Res.*, **5**, pp. 325–334.

Hartquist, T. W., Havnes, O., and Morfill, G. E. 1992. The effects of dust on the dynamics of astronomical and space plasmas. *Fund. Cosm. Phys.*, **15**, pp. 107–142.

Hasegawa, H., Fujiware, A., Koike, C., and Mukai, T. 1977. Effect of the spin of an interplanetary dust on its motion. *Memoirs of the Faculty of Science, Kyoto Univ. Series A of Physics, Astrophysics, Geophysics and Chemistry*, **35**, pp. 131–139.

Havnes, O. 1984. Charges on dust particles. *Adv. in Space Res.*, **4**, pp. 75–83.

Havnes, O., Aanesen, T. K., and Melandsø, F. 1990. On dust charges and plasma potentials in a dusty plasma with dust size distribution. *J. Geophys. Res.*, **95**, pp. 6581–6585.

Havnes, O., Goertz, C. K., Morfill, E., Grün, E., and Ip, W.-H. 1987. Dust charges, cloud potential and instabilities in a dust cloud embedded in a plasma. *J. Geophys. Res.*, **92**, pp. 2281–2287.

Havnes, O., and Morfill, G. E. 1984. Effects of electrostatic forces on the vertical structure of planetary rings. *Adv. Space Res.*, **4**, pp. 85–90.

Hertz, H. 1882. Über die Berührung fester elastischer Körper (on the contact of elastic solids). *J. Reine und Angewandte Mathematik*, **92**, pp. 156–171.

Hirst, E, Kaye, P. H., and Guppy, J. R. 1994. Light scattering from nonspherical airborne particles: experimental and theoretical comparisons. *Appl. Optics*, **33**, pp. 7180–7186.

Holsapple, K. A. 1994. Catastrophic disruptions and cratering of solar system bodies. *Planet Space Sci.*, **42**, pp. 1067–1078.

Housen, K. R. 1992. Crater ejecta velocities for impacts on rocky bodies (abstract). *Lunar Planet. Sci. Conf. XXIII*, pp. 555–556.

Housen, K. R., Schmidt, R. M., and Holsapple, K. A. 1983. Crater ejecta scaling laws: Fundamental forms based on dimensional analysis. *J. Geophys. Res.*, **88**, pp. 2485–2499.

Housen, K. R., Schmidt, R. M., and Holsapple, K. A. 1991. Laboratory simulations of large scale fragmentation events. *Icarus*, **94**, pp. 180–190.

Jancso, G., Pupezin, J., and van Hook, W. A. 1970. The vapor pressure of ice between $+10^{-2}$ and $-10^{+2°}$. *J. of Physical Chem.*, **74**, pp. 2984–2989.

Jenniskens, P., and Blake, D. F. 1994. Structural transitions in amorphous water ice and astrophysical implications. *Science*, **265**, pp. 753–756.

Johnson, K. L. 1985. *Contact Mechanics*, (Cambridge: Cambridge Univ. Press).

Johnson, K. L., Kendall, K., and Roberts, A. B. 1971. Surface energy and the contact of elastic solids. *Proc. R. Soc. Lond. A*, **324**, pp. 301–313.

Johnson, R. E. 1985. Comment on the evolution of interplanetary grains. In *Ices in the Solar System*, eds. Klinger, et al. (Dordrecht: D. Reidel Publ. Co.) pp. 337–339.

Johnson, R. E. 1990. *Energetic charged particle interaction with atmospheres and surfaces*, (Berlin: Springer Verlag)

Johnson, R. E. 1991. Irradiation of solids: theory. In *Solid-State Astrophysics*, eds. E. Bussoletti and G. Strazzulla (Amsterdam: North Holland), pp. 129–168.

Johnson, R. E. 1996. Sputtering of ices in the outer solar system. *Rev. Modern Phys.*, **68**, pp. 305–312.

Johnson, R. E. 1998. Sputtering and desorption from icy surfaces. In *Solar System Ices* eds. B. Schmitt, C. de Bergh and M. Festou (Dordrecht: Kluwer Acad. Publ.) pp. 303–334.

Johnson, R. E., and Baragiola, R. A. 1991. Lunar surface: sputtering and secondary ion mass spectrometry. *Geophys. Res. Lett.*, **18**, pp. 2169–2175.

Johnson, R. E., and Lanzerotti, L. J. 1986. Ion bombardment of interplanetary dust. *Icarus*, **66**, pp. 619–624.

Johnson, R. E., and Schou, J. 1993. Sputtering of inorganic insulators. In *Fundamental processes in the sputtering of atoms and molecules*. ed. P. Sigmund, (Copenhagen: Roy. Dan. Acad.), pp. 403–493.

Kato, M., Iijima, Y., Arakawa, M., Okimura, Y., Fujimura, A., Maeno, N., and Mizutani, H. 1995. Ice-on-ice impact experiments. *Icarus*, **113**, pp. 423–441.

Kato, M., Iijima, Y., Okimura, Y., Arakawa, M., Maeno, N., Fujimura, A., and Mizutani H. 1992. Impact experiments on low temperature H_2O ice. In *Physics and Chemistry of Ice*, eds. N. Maeno and T. Hondo (Sapporo: Hokkaido Univ. Press), pp. 237–244.

Keller, H. U. 1990. The nucleus. In *Physics and chemistry of comets*, ed. W. F. Huebner, (Heidelberg: Springer-Verlag), pp. 13–68.

Keller, H. U., Delamere, W. A., Huebner, W. F., Reitsema, H. J., Schmidt, H. U., Whipple, F. L., Wilhelm, K., Curdt, W., Kramm, R., Thomas, N., Arpigny, C., Barbieri, C., Bonnet, R. M., Cazes, S., Coradini, M., Cosmovici, C. B., Hughes, D. W., Jamar, C., Malaise, D., Schmidt, K., Schmidt, W. K. H., and Seige, P. 1987. Comet P/Halley's nucleus and its activity. *Astron. Astrophys.*, **187**, pp. 807–823.

Kempf, S., Pfalzner, S., and Henning, Th. K. 1999. N-particle- simulations of dust growth. *Icarus*, **141**, pp. 388–398.

Khare, B. N., Sagan, C., Arakawa, E. T., Suits, F., Callcott, T. A., and Williams, M. W. 1984. Optical constants of organic tholins produced in a simulated Titanian atmosphere - From soft X-ray to microwave frequencies. *Icarus*, **60**, pp. 127–137.

Kissel, J. 1991. Mass-spectrometric in situ analysis of solid-state extraterrestrial samples. *Solid-State Astrophysics*, eds. E. Bussoletti and G. Strazzulla, pp. 169–195.

Kitada, Y., Nakamura, R., and Mukai, T. 1993. Correlation between cross section and surface area of irregularly shaped particle. *Proc. of 3rd International Congress on Optical Particle Sizing*, ed. K. Takahashi, pp. 121–125.

Kouchi, A. 1987. Vapour pressure of amorphous $H2O$ ice and its astrophysical implications. *Nature*, **330**, pp. 550–552.

Kouchi, A. 1990. Evaporation of H_2O-CO ice and its astrophysical implications. *J. Crystal Growth* **99**, pp. 1220–1226.

Kouchi, A., Yamamoto, T., Kozasa, T., Kuroda, T., and Greenberg, J. M. 1994. Conditions for condensation and preservation of amorphous ice and crystallinity of astrophysical ices. *Astron. Astrophys.*, **290**, pp. 1009–1018.

Kozasa, T., Blum, J., and Mukai, T. 1992. Optical properties of dust aggregates I. Wavelength dependence. *Astron. Astrophys.*, **263**, pp. 315–320.

Lamy, P. L. 1974. Interaction of interplanetary dust grains with the solar radiation field. *Astron. Astrophys.*, **35**, pp. 197–207.

Lange, M. A., and Ahrens, T. J. 1981. Fragmentation of ice by low velocity impact. *Proc. Lunar Planet Sci. Conf. XII* pp. 1667–1687.

Lange, M. A., and Ahrens, T. J. 1983. The dynamic tensile strength of ice and ice-silicate mixtures. *J. Geophys. Res.*, **88**, pp. 1197–1208.

Lange, M. A., Ahrens, T. J., and Boslough, M. B. 1984. Impact cratering and spall failure of Gabbro. *Icarus*, **58**, pp. 383–395.

Langmuir, I., Found, C. G., and Dittmer, H. F. 1924. A new type of electric discharges: The streamer discharge. *Science*, **60**, pp. 392–394.

Lanzerotti, L. J. 1987. Solar-terrestrial physics. In *Encyclopedia of physical science and technology*, (SanDiego: Academic Press) **12**, pp. 833–843.

Lanzerotti, L. J., Brown, W. L., and Marcantonio, K. J. 1987. Experimental study of erosion of methane ice by energetic ions and some considerations for Astrophysics. *Ap. J.*, **313**, pp. 910–919.

Lawn, B. 1993. *Fracture of brittle solids*, 2nd ed. (Cambridge: Cambridge Univ. Press)

Léger, A., Gauthier, S., Defourneau, D., and Rouan, D. 1983. Properties of amorphous H_2O ice and origin of the 3.1-micron absorption. *Astron. Astrophys.*, **117**, pp. 164–169.

Léger, A., Jura, M., and Omont, A. 1985. Desorption from interstellar grains. *Astron. Astrophys.*, **144**, pp. 147–160.

Léger, A., Klein, J., de Cheveigne, S., Guinet, C., Defourneau, D., and Belin, M. 1979. The 3.1 μm absorption in molecular clouds is probably due to amorphous H_2O ice. *Astron. Astrophys.*, **79**, pp. 256–259.

Leinert, C., Röser, S., and Buitrago, J. 1983. How to maintain the spatial distribution of interplanetary dust. *Astron. Astrophys.*, **118**, pp. 345–357.

Leliwa-Kopystynski, J., Taniguchi, T., Kondo, K., and Sawaoka, A. 1984. Sticking in moderate velocity oblique impact - Application to planetology. *Icarus*, **67**, pp. 280–293.

Levy, E. H., and Jokipii, J. R. 1976. Penetration of interstellar dust into the solar system. *Nature*, **264**, pp. 423–424.

Lien, D. J. 1991. Optical properties of cometary dust. In *Comets in the Post-Halley Era*, eds. R. L. Newburn, Jr., M. Neugebauer, and J. Rahe, (Dordrecht: Kluwer Acad. Publ.) Vol. 2: pp. 1005–1041.

Lou, W., and Charalampopoulos, T. T. 1994. On the electromagnetic scattering and absorption of agglomerated small spherical particles. *J. Phys. D: Appl. Physics*, **27**, pp. 2258–2270.

Love, S. G., and Brownlee, D. E. 1991. Heating and thermal transformation of micrometeoroids entering the earth's atmosphere. *Icarus*, **89**, pp. 26–43.

Lunine, J. I., Engel, S., Rizk, B., and Horanyi, M. 1991. Sublimation and reformation of icy grains in the primitive solar nebula. *Icarus*, **94**, pp. 333–344.

Mann, I., Okamoto, H., Mukai, T., Kimura, H., and Kitada, Y. 1994. Fractal aggregate analogues for near solar dust properties. *Astron. Astrophys.*, **291**, pp. 1011–1018.

Martelli, G., Ryan, E. V., Nakamura, A. M., and Giblin, I. 1994. Catastrophic disruption experiments: recent results. *Planet. Space Sci.*, **42**, pp. 1013–1026.

Matsui, T., and Schultz, P. H. 1984. On the brittle-ductile behavior of iron meteorites: New experimental constraints. *J. Geophys. Res.*, **89**, pp. 323–328.

McGuire, A. F., and Hapke, B. W. 1995. An experimental study of light scattering by large, irregular particles. *Icarus*, **113**, pp. 134–155.

Meakin, P. 1991. Fractal aggregates in geophysics. *Rev. of Geophys.*, **29(3)**, pp. 317–354.

Meakin, P., and Donn, B. 1988. Aerodynamic properties of fractal grains; Implications for the primordial solar nebula. *Ap. J.*, **329**, pp. L39–L41.

Meakin, P., and Jullien, R. 1988. The Effects of Restructuring on the Geometry of Clusters formed by Diffusion-limited, Ballistic and Reaction-limited Cluster-Cluster Aggregation. *J. Chem. Phys.*, **89**, pp. 246–250.

Melosh, H. J. 1984. Impact ejection, spallation, and the origin of meteorites. *Icarus*, **59**, pp. 234–260.

Melosh, H. J. 1987. High-velocity solid ejecta fragments from hypervelocity impacts. *Int. J. Impact Engng.*, **5**, pp. 483–492.

Melosh, H. J. 1989. *Impact Cratering: A Geologic Process*, (New York: Oxford Univ. Press).
Melosh, H. J., Ryan, E. V., and Asphaug, E. 1992. Dynamic fragmentation in impacts: Hydrocode simulation of laboratory impacts. *J. Geophys. Res.*, **97**, pp. 14735–14759.
Meyer-Vernet, N. 1982. "Flip-flop" of electric potential of dust grains in space. *Astron. Astrophys.*, **105**, pp. 98–106.
Mizutani, H., Takagi, Y., and Kawakami, S. 1990. New scaling laws on impact fragmentation. *Icarus*, **87**, pp. 307–326.
Moore, M. H., and Hudson, R. 1992. Far-infrared spectral studies of phase changes in water ice by photon irradiation. *Ap. J.*, **401**, pp. 353–360.
Mukai, T. 1981. On the charge distribution of interplanetary grains. *Astron. Astrophys.*, **99**, pp. 1–6.
Mukai, T. 1986. Analysis of a dirty water-ice model for cometary dust. *Astron. Astrophys.*, **164**, pp. 397–407.
Mukai, T., and Fechtig, H. 1983. Packing effect of fluffy particles. *Planet. Space Sci.*, **31**, pp. 655–658.
Mukai, T., and Giese, R. H. 1984. Modification of the spatial distribution of interplanetary dust grains by Lorentz forces. *Astron. Astrophys.*, **131**, pp. 355–363.
Mukai, T., Ishimoto, H., Kozasa, T., Blum, J., and Greenberg, J. M. 1992. Radiation pressure forces of fluffy porous grains. *Astron. Astrophys.*, **262**, pp. 315–320.
Mukai, T., and Mukai, S. 1973. Temperature and motion of the grains in interplanetary space. *Pub. Astron. Soc. Japan*, **25**, pp. 481–488.
Mukai, T., and Schwehm, G. 1981. Interaction of grains with the solar energetic particles. *Astron. Astrophys.*, **95**, pp. 373–382.
Muller, E. W., and Tsong, T. T. 1969. *Field Ion Microscopy*, (New York: Am. Elsevier Press).
Müller, U., Schmidt-Ott, A., and Burtscher, H. 1988. Photoelectric quantum yield of free silver particles near threshold. *Z. Phys. B – Cond. Matter*, **73**, pp. 103–106.
Nakamura, A. M. 1993. Laboratory simulation on the velocity of fragments from impact disruptions. *Institute of Space and Astronautical Science, (Kanagawa, Japan) Report* 651.
Nakamura, A. M., Fujiwara, A., and Kadono, T. 1994. Velocity of finer fragments from impact. *Planet. Space Sci.*, **42**, pp. 1043–1052.
Nakamura, A., Suguiyama, K., and Fujiwara, A. 1992. Velocity and spin of fragments from impact disruptions: an experimental approach to a general law between mass and velocity. *Icarus*, **100**, pp. 127–135.
Nitter, T., Havnes, O., and Melandsø, F. 1998. Levitation and dynamics of charged dust in the photoelectron sheath above surfaces in space. *J. Geophys. Res.*, **103**, pp. 6605–6620.
Okamoto, H., Mukai, T., and Kozasa, T. 1994. The $10\mu m$-feature of aggregates in comets. *Planet. & Space Sci.*, **42**, pp. 643–649.
Palmer, H. B., and Shelef, M. 1968. Vaporization of carbon. *Chem. & Phys. of Carbon*, **4**, pp. 85–135.
Patashnick, H., and Rupprecht, G. 1975. The size dependence of sublimation rates for interplanetary ice particles. *Ap. J.*, **197**, pp. L79–L82.
Perrin, J.-M., and Lamy, P. L. 1990. On the validity of effective-medium theories in the case of light extinction by inhomogeneous dust particles. *Ap. J.*, **364**, pp. 146–151.
Pinho, G. P., and Duley, W. W. 1994. Effect of variable graphitic and diamond-like content on the temperature of carbonaceous dust. *Mon. Not. R. Astron. Soc.*, **269**, pp. 121–126.
Polansky, C. A., and Ahrens, T. J. 1990. Impact spallation experiments: Fracture patterns and spall velocities. *Icarus*, **87**, pp. 140–155.
Poppe, T., and Blum, J. 1997. Experiments on pre-planetary grain growth. *Adv. Space Res.*, **20**, pp. (8)1595–(8)1604.
Poppe, T., Blum, J., and Henning Th. 1999. New experiments on collisions of solid grains related to the preplanetary dust aggregation. *Adv. Space Res.*, **23**, pp. (7)1197–(7)1200.
Qiu, S. L., Lin, C. L., Jiang, L. Q., and Strongin, M. 1989. Photoemission studies of the metal-nonmetal transition of sodium on solid ammonia. *Phys. Rev. B*, **39**, pp. 1958–1961.
Reimann, C. T., Boring, J. W., Johnson, R. E., Garrett, J. W., Farmer, K. R., and Brow, W. L. 1984. Ioninduced molecular ejection from D_2O ice. *Surf. Sci.*, **147**, pp. 227–240.

Roessler, K. 1991. Suprathermal chemistry in space. In *Solid-State Astrophysics*, eds. E. Bussoletti and G. Strazzulla, (Dordrecht: North-Holland, Elsevier Sci. Publ.), pp. 197–266.

Roth, J. 1983. Chemical sputtering. In *Sputtering by particle bombardment II*, ed. R. Berisch, (Berlin: Springer Verlag), pp. 91–146.

Ryan, E. V., Hartmann, W., and Davis, D. R. 1991. Impact experiments 3: Catastrophic fragmentation of aggregate targets and relation to asteroids. *Icarus*, **94**, pp. 283–298.

Sablotny, R. M., Kempf, S., Blum, J., and Henning, Th. 1995. Coagulation simulations for interstellar dust grains using an n-particle code. *Adv. Space Res.*, **15**, pp. (10)55–(10)58.

Sack, N. J., and Baragiola, R. A. 1993. Sublimation of vapor-deposited water ice below 170 K, and its dependence on growth conditions. *Phys. Rev. B*, **48**, pp. 9973–9978.

Sandford, S. A., and Allamandola, L. J. 1988. The condensation and vaporization behavior of $H_2O{:}CO$ ices and implications for interstellar grains and cometary activity. *Icarus*, **76**, pp. 201–224.

Sandford, S. A., and Allamandola, L. J. 1993. Condensation and vaporization studies of CH_3OH and NH_3 ices: Major implications for astrochemistry. *Ap. J.*, **417**, pp. 815–825.

Schleicher, B., Burtscher, H., and Siegmann, H. C. 1993. Photoelectric quantum yield of nanometer metal particles. *Appl. Phys. Lett.*, **63**, pp. 1191–1193. pp. 190–204.

Sekanina, Z. 1982. Comet Bowell /1980b/ - an active-looking dormant object. *Astron. J.*, **87**, pp. 161–169.

Senay, M. C., and Jewitt, D. 1994. Coma formation driven by carbon monoxide release from comet Schwassmann-Wachmann 1. *Nature*, **371**, pp. 229–231.

Shi, M., Baragiola, R. A., Grosjean, D. E., Johnson, R. E., Jurac, S., and Schou, J. 1995. Sputtering of water ice surfaces and the production of extended neutral atmospheres. *J. Geophys. Res.*, **100**, pp. 26387–26396

Sigmund, P. 1993. *Fundamental processes in the sputtering of atoms and molecules*, (Copenhagen: Roy. Dan. Acad. of Sci.)

Simpson, J. A. 1983. Introduction to the galactic cosmic radiation. In *Composition and origin of cosmic rays*, ed. M. M. Shapiro, (Amsterdam: Reidel), pp. 1–24.

Smirnov, B. M. 1990. The properties of fractal clusters. *Physics Reports*, **188**, pp. 1–78.

Smoluchowski R. 1985. Amorphous and porous ices in cometary nuclei. In *"Ices in the Solar System"*, eds. J. Klinger, D. Benest, A. Dollfus and R. Smoluchowski, (Dordrecht: D. Reidel Publishing Co.), pp. 397–406.

Smrekar, S., Cintala, M. J., and Hörz, F. 1986. Small-scale impacts into rock: An evaluation of the effects of target temperature on experimental results. *Geophys. Res. Lett.*, **13**, pp. 745–748.

Sternberg, A., Dalgarno, A., and Lepp, S. 1987. Cosmic-ray induced photodestruction of interstellar molecules. *Ap. J.*, **320**, pp. 676–682.

Sternglass, E. J. 1954. *Sci. Paper 1772*, (Pittsburgh: Westinghouse Res. Lab.)

Strazzulla, G. 1999. Ion Irradiation and the Origin of Cometary Materials. *Space Sci. Reviews*, **90**, pp. 269–274.

Strazzulla, G., and Johnson, R. E. 1991. Irradiation effects on comets and cometary debris. In *Comets in the Post-Halley Era*, eds. R. L. Newburn et al. (Amsterdam: Kluwer Acad. Publ.), pp. 243–275.

Svestka, J., Auer, S., Baguhl, M., and Grün, E. 1996. Measurement of dust electric charges by the Ulysses and Galileo dust detectors. In *Physics, Chemistry and Dynamics of Interplanetary Dust*, Conf. Series Vol. 104, eds. B. Å. S. Gustafson and M. S. Hanner (San Francisco: Astron. Soc. of the Pacific Press), pp. 481–484.

Svestka, J., Cermak, I., and Grün, E. 1993. Electric charging and electrostatic fragmentation of dust particles in laboratory. *Adv. Space Res.*, **13**, pp. (10)199–(10)202.

Svestsov, V. V., Nemtchinov, I. V., and Teterev, A. V. 1995. Disintegration of large meteoroids in Earth's atmosphere: Theoretical models.*Icarus*, **116**, pp. 131–153.

Takagi, Y., Kawakami, S., and Mizutani, H. 1984. Impact fragmentation experiments of basalts and pyrophyllites. *Icarus*, **59**, pp. 462–477.

Taylor, S. R. 1982. *Planetary science: A lunar perspective* (Houston: Lunar and Planet. Inst.) Chap.4.

Tomeoka, K. 1991. Aqueous alteration in hydrated interplanetary dust particles. In *Origin and Evolution of Interplanetary Dust*, eds. A. C. Levasseur-Regourd and H. Hasegawa, (Dordrecht: Kluwer Acad. Publ.), pp. 71–78.

Umebayashi, T., and Nakano, T. 1990. Magnetic flux loss from interstellar clouds. *Mon. Not. R. Astr. Soc.*, **243,** pp. 103–113.

van de Hulst, H. C. 1957. *Light Scattering by Small Particles*, (New York: Wiley [Also New York: Dover 1981]) 8.41.

Verlicchi, A., La Spina, A., Paolicchi, P., and Cellino, A. 1994. The interpretation of laboratory experiments in the framework of an improved semi-empirical model. *Planet. Space Sci.*, **42,** pp. 1031–1042.

Wagner, W., Saul, A., and Pruss, A. 1994. International equations for the pressure along the melting and along the sublimation curve of ordinary water substance. *J. Phys. Chem. Ref. Data*, **23,** pp. 515–527.

Walch, B., Horányi, M., and Robertson, S. 1995. Charging of dust grains in plasma with energetic electrons. *Phys. Rev. Letters*, **75,** pp. 838–841.

Wallis, M. K., and Hassan, M. H. A. 1983. Electrodynamics of submicron dust in the cometary coma. *Astron. Astrophys.*, **121,** pp. 10–14.

Weidenschilling, S. J. 1997. The origin of comets in the solar nebula: A unified model. *Icarus*, **127,** pp. 290–306.

Weidenschilling, S. J., and Cuzzi, J. N. 1997. In *Protostars and Planets III*, eds. E. H. Levy and J. I. Lunine, (Tucson: Univ. of Arizona Press) pp. 1031–1060.

Westley, M. A., Baragiola, R. A., Johnson, R. E., and Barratta, G. A. 1995. Photo desorption from low temperature water ice in interstellar and circumstellar grains. *Nature*, **373,** pp. 405–407.

Whipple, E. C. 1981. Potentials of surfaces in space. *Rep. Prog. Phys.*, **44,** pp. 1197–1250.

Wilson, G. R. 1991. The plasma environment, charge state, and currents of Saturn's C and D rings. *J. Geophys. Res.*, **96,** pp. 9689–9701.

Worden, S. P., Schneeberger, T. J., Kuhn, J. R., and Africano, J. L. 1981. Flare activity on T-tauri stars. *Ap. J.*, **244,** pp. 520–524.

Wurm, G., and Blum, J. 1998. Experiments on preplanetary dust aggregation. *Icarus*, **132,** pp. 125–136.

Xu, W., D'Angelo, N., and Merlino, R. L. 1993. Dusty plasmas: The effect of closely packed grains. *J. Geophys. Res.*, **98,** pp. 7843–7847.

Yamamoto, S., and Nakamura, A. M. 1997. Velocity measurements of impact ejecta from regolith targets. *Icarus*, **128,** pp. 160–170.

Yamamoto, T., and Ashihara, O. 1985. Condensation of ice particles in the vicinity of a cometary nucleus. *Astron. Astrophys.*, **152,** pp. L17–L20.

Yamamoto, T., Nakagawa, N., and Fukui, Y. 1983. The chemical composition and thermal history of the ice of a cometary nucleus. *Astron. Astrophys.*, **122,** pp. 171–176.

Yeomans, D. K. 1991. Cometary orbital dynamics and astrometry. In *Comets in the Post-Halley Era*, eds. R. L. Newburns Jr., M. Neugebauer and J. Rahe (Amsterdam: Kluwer Acad. Publ.), pp. 3–17.

Interactions with Electromagnetic Radiation: Theory and Laboratory Simulations

Bo Å. S. Gustafson[1], J. Mayo Greenberg[2], Ludmilla Kolokolova[1], Yu-lin Xu[1], Ralf Stognienko[3]

[1] University of Florida, Gainesville, Florida, USA
[2] Leiden University, Leiden, Netherlands
[3] Jena-Optronik GmbH, Jena, Germany

Abstract. At the time of the most recent major book on interplanetary dust (McDonnell 1978), most theoretical studies were confined to homogeneous and spherical dust models. Only microwave analogue experiments were used to explore the scattering by more realistic structures representing either the first stratospheric collections of interplanetary dust particles or complex aggregates of interstellar grains proposed on theoretical grounds. Advances in computing power, light scattering theory, and in experimental capabilities have since allowed the implementation of powerful and flexible theoretical solutions, more sophisticated solutions for the calculation of scattering by interacting particles based on classical methods, and have led to multi-wavelength microwave analogue investigations of structures in the size range of interplanetary dust. We attempt to summarize these advances, place them into context, and suggest a framework for models of interplanetary dust that facilitates systematic studies. The electromagnetic scattering sections apply to a broad range of natural and artificial terrestrial and cosmic dust or aerosols as well as to interplanetary dust.

I. INTRODUCTION

Studies described in this book point to a complex interplanetary dust system where fragments from comets and asteroids are believed to dominate over impact ejecta and volcanic cinders originating from planetary satellites. These dust populations mix with smaller amounts of ejectae from planetary surfaces and with interstellar grains that are now also known to permeate interplanetary space.

Modern grain models are based on a combination of properties derived from in-situ and remote measurements, collected samples, theoretical and laboratory models all of which are plagued by uncertainties and biases. There is however no doubt that they reveal that at least some of the particles are aggregates. The realization that we are not faced with dust of a single and simple morphology but a range in particle properties including some highly complex structures has become widely accepted only gradually. At the time of the most recent major book on interplanetary dust (McDonnell 1978), most works on light scattering by interplanetary dust were still based on a homogeneous grain material with a spherical boundary. The "frontier" of light scattering research was for the most part in exploration of the effects of non-spherical particle shapes. While some groups also worked on the development of theoretical solutions for scattering by porous or aggregate particle structures, only microwave analogue experiments were used to explore the scattering by "Fluffy" structures representing the first stratospheric collections (Giese et al. 1978) and complex "Bird's-Nest" aggregates of interstellar grains proposed on theoretical grounds (Greenberg and Gustafson 1981). Complex aggregate structures have increasingly become the focus of light scattering theory and experimental studies as both theory and experiments have evolved in sophistication and scope.

We may observe these dust grains with the unaided eye as they collectively scatter sunlight to produce the zodiacal light at visible wavelengths and it is now possible to record the thermal emission from the dust cloud in the infrared part of the spectrum. Definite relations and similarities exist in the physics of light scattering, radiative heating, and thermal emission. We therefore follow common practice and refer to the common term of "light scattering" for these and for the evaluation of the resulting radiation forces that affect the motion of the grains. Besides the prediction of observable quantities based on a given grain model, the direct problem, astronomers and physicists are interested in the inverse problem of derivation of grain physical properties from zodiacal light observations, thermal emission recordings, and dynamics.

Interactions between dust grains and electromagnetic radiation result in complex transformations that depend on the material, shape, size, internal structure and orientation of a grain. Some of these dependencies may average out when we observe averages from a large number of particles in a cloud but the distributions of properties must be known for the proper averaging to be made. It is often assumed that the dust particles in a cloud are randomly oriented. Although this is not necessarily justified for particles that are subject to orbital dynamics in a central force system, randomness is likely to be a good approximation for interplanetary dust (Gustafson 1994). Even so, a method to deal with the number of free parameters to represent possible distributions of interplanetary particle types, their size distributions and mixtures is probably our biggest challenge and we therefore address this need in section II where we propose a systematic framework based on the "Bird's- Nest" dust model.

In section III we discuss the optical constants that define the relevant properties of a particle's material. This will help us identify possible evolutionary sequences for dust materials as well as provide reasonable starting points for systematic investigations. Section IV addresses the problem of solving for the scattering once all relevant parameters are known. While it is not possible to fit an exhaustive discussion of known solutions, this section introduces some of the many methods used in practical solutions. Here, we include some less often used methods that we feel may deserve another look as advancing computer technology continually changes the desirability of specific computing techniques.

We discover in section V that a systematic picture of the scattering properties of aggregate dust grains is now emerging. Theoretical advances generally confirm results from experiment-based studies that continue to probe for scattering properties by ever larger and more complex structures.

II. A PHYSICAL DUST MODEL

Following Greenberg and Gustafson (1981) we find it useful to base models of interplanetary dust on the properties of their assumed interstellar precursors. This leads to a working description of interplanetary dust as grain aggregates defined by the

- individual grain size, shape, and refractive index
- packing factor
- overall aggregate size.

The starting point in parameter space is a representation of the refractory residues left behind as ices sublimate away from aggregated pristine interstellar grains. It must be consistent with cosmic aboundances while sizes, shapes, and refractive indexes are constrained by interstellar extinction data. Packing is close to 10% to account for grain accumulation in the absence of a compacting force and subsequent loss of material to ice sublimation. This packing is close to the lowest possible in an aggregate where all grains are interconnected randomly oriented 2:1 to 3:1 prolate shapes. Only the third parameter, the overall aggregate size remains to be fixed, presumably as a size distribution. The parameters can naturally be extended to encompass

- core/mantle or other individual grain geometries
- grain alignment
- overall shape of an aggregate
- fractal dimension of aggregate.

This framework is useful since there is considerable evidence in favor of the basic evolutionary model of interplanetary dust that relates the grains to their interstellar precursors as was originally proposed by (Greenberg and Gustafson 1981, and by Greenberg 1982, 1983) who advanced their "Bird's-Nest" model for comet grains and zodiacal dust of cometary origin. We now seek to define a starting point to describe the interplanetary dust population using this set of parameters. First, we review evidence for the base model – the original "Bird's-Nests".

We learned from the chapter by Jessberger et al. that at least part of some interplanetary particles derive from interstellar grains that have undergone only minor processing over the 4.5 Gyrs since the solar system formed. These materials are preserved in comets and even in asteroids and the dust they produce. Specifically, Bradley (1999) show that chondritic porous particles of interplanetary origin collected in the Earth's atmosphere are physically, chemically, and isotopically primitive objects in which large D/H and ^{15}N isotopic anomalies are evidence for presolar interstellar components. While these grains and D-rich IDPs analyzed by Messenger et al. (1995) and Messenger (2000) are often attributed to comets, meteoritic organics attributed to asteroids have also been found to carry deuterium enrichments and there are further indications of the interstellar origin of parts of the Murchison, Murray, (e.g., Kerridge 1999) and other meteorites (Lawless 1980; Cronin et al. 1988; Sephton and Gilmour 2000). Clearly, there are at least parts of both comets and asteroids that have undergone little evolution during the formation of the solar system or since then.

Greenberg and Gustafson (1981) and Greenberg and Hage (1990) point out that there is no compelling process to fuse silicates while such primitive attributes are preserved. The silicate found in comets and possibly in some D- and C-type asteroids as well as the dust grains they produce should therefore bear some resemblance to the silicate cores of classical interstellar grains from which they are believed to originate. This means dimensions of the order of $0.1\,\mu$m and prolate elongation ratios of 2:1 to 3:1. They argue that mantles incorporating smaller grains and organic molecules condense on the silicates. Ice-mantles that are more volatile also grow before grains collide and aggregate in the outer parts of the solar nebula. These most volatile components have either sublimated away or are chemically processed and incorporated with the refractory organic mantles by the time the dust reemerges from comets and primitive asteroids to be observed in the inner solar system.

Greenberg and Hage (1990) and Greenberg and Li (1999a, b) estimated that the mass in organic refractory material is comparable to that in silicates based on cosmic abundances. The combination of loose packing (there is over 45% void even in a tight cubic lattice packing of spheres) and voids left behind from sublimating volatiles should produce loosely packed aggregates of submicron elongated silicate grains coated by an organic tar-like material. In the absence of a compacting process, the expected packing factor (ratio of the volume occupied by the material to the total volume) is near 10% so that the volume of the interstices between constituent particles in an aggregate occupies $\sim 90\%$ of the total volume.

We note that transport of refractory material by the gas and therefore repacking can take place at least in the upper layers of comets. Repeated collisions also should lead to some compaction so that a fragment from a comet or from an undifferentiated asteroid could be more compact depending on its history. Sirono and Greenberg (2000) show that substantial compaction can take place in comets as they repeatedly collide with each other in the Kuiper belt and in the Oort cloud but the compaction is limited to the boundary layers between large-scale components making up the comets.

Physical processes of the formation of dust particles, their physical alteration upon irradiation, ion bombardment, heating, freezing and collisional breakup discussed in the chapter by Mukai et al. have also been studied in the laboratory, although possibly less extensively. Ibadinov (1989) and Stephens and Gustafson (1991) studied the structure of dirty-ice surfaces in simulated space environments and the formation of dust particles. An ambitious study of the evolution of icy materials, the KOSI experiments, were undertaken by Grün et al. (1991; Kochan et al. 1998; Sears et al. 1999). The process of dust agglomeration, discussed in the chapter by Mukai et al., was studied experimentally by Blum et al. (Wurm and Blum 1998; Poppe et al. 2000) and theoretically by for example Ossenkopf (1993), Weidenschilling (1997) and Dominik (1999).

More may now be known of dust from comets or dust presumed to be from comets than from asteroids. Comet grain temperatures in combination with the color and polarization in scattered light indicate porous structures (e.g., Giese et al. 1978; Greenberg and Hage 1990; Li and Greenberg 1998; Greenberg and Li 1999a; Levasseur-Regourd 1999). While this was known before the recent perihelion passage of comet Hale-Bopp, the best evidence of the nature of comet dust may come from studies of this comet. These include near-infrared, far-infrared and visual, ground-based and remote space-based observations that rule out many comet materials (e.g., the chapter by Sekanina et al.). Li and Greenberg (1998) show that no matter what the size distribution of the aggregates, the constituent units of the aggregates are like classical size ($0.1\,\mu$m) interstellar grains. Gustafson and Kolokolova (1999) showed that color and wavelength dependence of polarization in the visual also supports this view and that the dependence on distance from the nucleus indicates the sublimation of some organic compound while aggregates recede from the nucleus. This leaves a sparse matrix of refractory material behind to make up comet debris not unlike the original BN-model by Greenberg and Gustafson (1981).

In models, we may represent each 2:1 elongation ratio constituent particle in an aggregate by a bisphere for simplicity. This differs from the case of aggregated spheres only in that the packing is lower except in the extreme case of close packing. Comet dust densities of 200 to $300\,\mathrm{kg\,m^{-3}}$ can be estimated based on interstellar grain properties and from solar system abundances (Greenberg and Hage 1990). The strength of the silicate emission features indicate low densities perhaps even as low as $100\,\mathrm{kg\,m^{-3}}$ (Rickman 1989; Li and Greenberg 1998), see also insitu mass spectrometer data by Kissel (1999). Adopting $200\,\mathrm{kg\,m^{-3}}$ as a nominal value, and equal mass refractory and volatile components with roughly a 2:1 elongation ratio $0.1\,\mu$m size constituent grains, we find 10% packing as one extreme for interplanetary dust. Close packing near 70% is the other extreme that may represent fragments from some D-type asteroids and possibly also from C-types. This asteroid dust packing is suggested based on the $2600 \pm 500\,\mathrm{kg\,m^{-3}}$ dynamically determined density of S-type asteroid Ida (Thomas et al. 1996) and the interstellar material densities above. For asteroid densities and their possible evolution, see also Wilson et al. (1999). Many asteroids are likely to be less porous or even compact on the scales of interest to optical properties, e.g., Flynn et al. (1999), Moore et al. (1999), Britt and Consolmagno (2000), and we may represent their debris as low porosity angular aggregates or they may be fitted into our general framework as single particle "aggregates".

The chondritic porous subset of deuterium rich Interplanetary Dust Particles (or IDPs) collected in the Earth's stratosphere with inclusions traced to presolar materials (Bradley 1999), have size distributions of inclusions that also include micron size and larger structures, while some appear to feature a narrow size distribution peaked near $0.1\,\mu$m (Rietmeijer 1998) as expected from classical interstellar grains. Overall, collected IDPs are close in both composition and structure to the expectation for interplanetary dust. Although there remains no doubt that IDPs are examples of interplanetary dust, selection bias may be severe. The structures that we seek to model therefore include but are not restricted to IDPs.

Compacted or not, modified or not, this extended "Bird's-Nest", or "BN", grain model represents dust as fragments of comets and asteroids that themselves are much larger aggregates. The fragments are therefore also aggregates and are neither self-similar on all scales nor do they posses a radial density gradient. "BNs" are therefore fluffy aggregates but they are not formally fractal grains or grains of fractal dimension (see the glossary for definitions).

III. OPTICAL CONSTANTS

The electromagnetic scattering properties of a dust particle are usually evaluated based on the grain materials, which are defined using so called optical constants, and on the geometry of the boundary to the surrounding space. Geometric influences on the light scattering properties are the topic of section IV while this section addresses the optical properties of the grain material. While the separation is conceptually straightforward in the case of idealized grains such as the classical homogeneous, compact and isotropic sphere, it is not as clear if the material is porous. Such particles may either be considered as simple geometries consisting of porous materials (if a suitable refractive index can be found) or as compact materials of complex geometries. We first deal with electromagnetic properties on atomic, molecular, and lattice scale before we discuss larger scales, the use of effective medium theories and the issue of their applicability.

Bohren and Huffman (1983, Chapter 9) and most basic textbooks describe electrodynamic properties of bulk material based on interacting oscillators that dissipate some, usually small, fraction of the energy. This leads to a phenomenological interpretation of the interaction of matter with electromagnetic radiation and to theoretical expressions for the macroscopic optical constants based on atomic level properties, see for example the excellent description by Scharf (1994, Chapter 4). For our purpose, it is sufficient to notice that charges interact in what can be viewed as electromagnetic coupling or alternatively as mutual illumination (compare with the coupled dipole approximation in section IV.D). Because the speed of light is finite, coupling or higher order illumination acts as a retardation of the response to the incident field in addition to the energy loss. Optical constants are therefore complex numbers. The real part n of the refractive index, $m = n+in'$, is inversely proportional to the phase velocity of an electromagnetic wave inside the material in units of the speed of light in empty space c, while the imaginary part n' is a measure of the attenuation. Both the refractive index, and the dielectric constant $\varepsilon = n^2 - n'^2 +$ i $2nn'$ for non-magnetic materials, are usually referred to as optical constants although they vary with the wavelength λ (or frequency ω). We recall (e.g., Bohren and Huffman 1983) that stretches of chemical bonds affect m the most in the infrared (2.5 to 25 μm) part of the spectrum while coupling to lattice vibrations dominate in the thermal infrared. Frequency dependencies of the real and imaginary parts are related through the Kramers-Kronig relations (e.g., Scharf 1994).

The infrared refractive index of a crystalline material is expected to change and the lattice resonance frequencies shift as crystals are damaged under exposure to solar UV-photons. Visual refractive indices change primarily as chemical bonds in exposed crystalline or amorphous organic materials break. This leads to the loss of H, O, and N, thus leaving a larger fraction of C behind which in general increases the imaginary component. The broken bonds allow new bonds to form and often lead to highly refractive organics with large molecular weight. Since collisions increase with the temperature, the imaginary part grows with temperature in a manner that is composition and crystalline dependent (Agladze et al. 1996; Brucato et al. 1999). Space exposure is therefore expected to alter n, increase n' and alter the temperature dependence of n'. The aging of materials under simulated space conditions has been studied in several laboratories and the refractive indices measured. A few examples are:

- Allamandola, Sandford et al. (NASA AMES); ices and organics
- Bussoletti, Colangeli et al. (Napoli); silicates; carbon, and carbon compounds
- Dorschner, Mutschke, Begemann, Henning et al. (Jena); silicates and analogs of natural minerals, e.g., FeO, FeS, SiS_2
- Greenberg, Ehrenfreund, Schutte et al. (Leiden); mixtures of ices, e.g., H_2O ice with N_2, O_2, SO_2, CO, CO_2 and simple organics
- Khare et al. (NASA AMES, Cornell); organics

We do not list the many and extensive studies of chemistry in planetary interiors and on the surface of planets and satellites and in their atmospheres, although ejectae from these bodies are expected to supply a minor fraction of interplanetary grains. Laboratory simulations of planetary chemistry has been reviewed by Raulin (1990), Khanna (1995), Gazeau et al. (2000).

III.A. Bulk Materials

Interstellar grains incorporate silicates and organic compounds with mixtures of ices whose amount and composition depend strongly on the environment. All components are modified to some extent by cosmic (including solar) irradiation, interaction with solar-wind particles and cosmic rays. Thus, to understand the composition of interplanetary dust and build on the framework of the "BN" model, we need to consider the following steps:

- Creation and processing of a broad range of compounds in molecular clouds
- Partial destruction during repeated cycling into the diffuse medium
- The incorporation of these compounds in their various stages of processing in early solar materials

Systematic laboratory studies of ultraviolet processing and cosmochemistry in molecular clouds go back to the 1970s (e.g., Greenberg et al. 1972; Smith and Adams 1977). These were extended by Greenberg and his group in Leiden (e.g., Greenberg 1983; Schutte 1995, 1999) and by several other groups including Allamandola's group at NASA AMES (Allamandola et al. 1988; Schutte et al. 1992, 1993; Bernstein et al. 1997). They produce complex organic compounds starting with the mixture of simple molecules such as H_2O, CO, CH_4, NH_3, CH_3OH, CO_2, etc.

The end-products can be separated into at least four groups. (1) Very complex polymers of at least 400 AMU that have not yet been identified. (2) A large hydrocarbon part that includes minor oxygen and nitrogen containing materials in the region of 100 to 250 AMU that are in the process of being fully characterized, (3) specific compounds in the region of less than 100 AMU, already confirmed to be real through isotope labeling and (4) PAH's.

The lower molecular weight compounds were studied in detail by Briggs et al. (1992) using gas chromatography and liquid chromatography to separate compounds and use of a mass spectrometer for their identification (GC/MS). Briggs et al. found C_2-C_3 hydroxy acids and hydroxy amides, glycerol, urea, hexamethylene tetramine, formamidine, ethanolamine, glycerine and other amino acids. Many heterocyclic aromatic molecules seem also to form.

Glycine, glycolic acid and oxalic acid that were identified in the organic refractory have also been found in the Murchison meteorite (Lawless 1980; Peltzer et al. 1984; Cronin et al. 1988). The middle molecular weight fraction of the photo-produced laboratory residues has been analyzed using several types of mass spectrometry (e.g., Mendoza-Gómez et al. 1995). These indicate the presence of highly unsaturated aromatic hydrocarbons, which agrees qualitatively with finds in meteoritic material (Grady et al. 1983; Gilmour and Pillinger 1992; Allamandola et al. 1999; Sephton and Gilmour 2000) and in cometary material (Kissel and Krueger 1987; Mukhin et al. 1989; Moreels et al. 1994; Jochims et al. 1999).

Li and Greenberg (1997) considered the residues obtained in the experiments described above as "first generation organic refractories". This means organic compounds emerging from molecular clouds as mantles on classical size silicate dust cores. Further ultraviolet irradiation should be applied to laboratory organics to simulate dust mantles on the presumably aged grains that went into making the solar system. While this is necessary to reproduce infrared properties, the optical constants can already be used to represent interstellar dust organic refractories in the UV and visual region (e.g., Jenniskens 1993). Greenberg et al. (1995, 2000) flew laboratory created residues on the ERA (Exobiology Radiation Assembly) platform of the EURECA satellite for six months to obtain a simulated "second generation" organic refractory. The exposure to interplanetary space, while not ideal, is the closest analogue so far to ultraviolet processing of dust mantles in diffuse clouds after leaving molecular clouds. This more processed organic refractory is representative of the dust mantles that have undergone a single evolutionary cycle. Meanwhile, the organic component from the Murchison meteorite mentioned above can be considered as the other extreme consisting of greatly processed material. Li and Greenberg (1997) obtained a fit to the interstellar infrared silicate polarization feature at 10 and 20 μm as well as the 3.4 μm diffuse cloud feature using a model of interstellar dust as 0.1 μm prolate silicate grains covered by organics with optical constants intermediate between the EURECA samples and the organic component of the Murchison meteorite.

The infrared spectral region is especially sensitive to molecular bonds and lattice vibrations. Identification of thermal emission features directly translates to chemical components in the dust material. For example the 10 and 20 μm "silicate" features indicate the presence of Si-O bonds. In the solar-system, the silicate feature was first discovered in comets (Ney 1974) and in the solar corona (Lena et al. 1974) but had previously been seen in many galactic sources. Lattice vibrations can modify the strength and shape of the spectral feature. Identification of the specific silicates in cosmic dust is complicated by the unknown size and geometry of the dust that also cause distortions of the spectrum. The recent ISO data in the infrared beyond 20 μm are not subject to this problem.

The first 10 μm observations showed a smooth band whose resemblance to the spectrum of a synthetic protosilicate, as well as to the spectrum of primitive carbonaceous chondrites, led to the suggestion (Day 1974) that cosmic silicates are amorphous. However, later observations revealed structure spanning from 8 to 13 μm with two main features at 9.7 and 11.3 μm. According to Knacke (1978), amorphous olivine or a phyllosilicate seem to be responsible for the broad shape and 9.7 μm peak. The 11.3 μm peak is more difficult to identify. Laboratory spectra (Koike et al. 1981) as well as spectra from collected stratospheric interplanetary dust or IDPs show the 11.3 μm feature when crystalline silicates are present (Sandford and Walker 1985). This led Campins and Ryan (1989) to suggest the presence of crystalline silicates in comets. Colangeli et al. (1993) reached similar conclusions based on laboratory spectra. As an alternative, Orofino et al. (1994) showed that laboratory spectra of silicone carbide SiC also produce a good fit to the observed feature, especially when mixed with amorphous carbon (Blanco et al. 1994). Evolution of the silicate feature under cosmic conditions is studied by Hallenbeck et al. (2000).

The silicate band at 20 μm, as well as weaker features observed in the 8 to 13 μm range, result from the O-Si-O bending mode and have been simulated in the laboratory, e.g., in Jena (Dorschner, Henning, Mutschke et al.) and in Japan (Koike et al.). A number of silicate-containing minerals were suggested as candidates for cometary and interplanetary dust components as a result of the simulations, among them glassy and crystalline olivine and pyroxene.

Another important feature near 3.4 μm was discovered in thermal emission from comet Halley dust (Hanner 1985). The band resembled but was not identical to interstellar absorption near 3.4 μm and has been linked to the C-H stretching bond (Knacke et al. 1986). This feature was soon reproduced through laboratory synthesis of solid organic residue from irradiated low-occupancy methane ice clathrate (Chyba and Sagan 1987) and in spectra from hydrogenated amorphous carbon (Blanco et al. 1987; Scott et al. 1997). Exactly what material produces the 3.4 μm feature is not known. Although it was considered strong evidence for the presence of some organic refractory material (Greenberg 1999; and the chapter by Sekanina et al.), it is now thought to be largely produced by CH_3OH (DiSanti et al. 1995).

Transformations of the laboratory infrared spectral features under the influence of admixtures, temperature changes, UV, ion, and cosmic-rays irradiation were also studied (e.g., Henning et al. 1995; Colangeli et al. 1999). This allows the study of not only composition but also evolution of cosmic grains from remotely observable IR spectra.

We base the refractive index of silicates and organics in comets, asteroids and IDPs to a large extent on refractive indices for interstellar matter. Some laboratory and astronomical data have been joined into a consistent set. Draine and Lee (1984) introduced the concept of "Astronomical Silicates" that they presented in the form, not of a chemical compound, but through the wavelength dependent optical constants obtained through a fit to astronomical and laboratory data related to interstellar dust. "Astronomical Silicate" has optical properties satisfying the Kramers-Kronig relations but the set does not necessarily have a material counterpart, instead the set is intended to represent average effective optical interstellar grain properties. The term "silicate" is due to prominent 10 and 20 μm features in the $m(\lambda)$ curve corresponding to the silicate feature discussed above. Actual interstellar materials that are represented by these sets of optical parameters may well contain significant amounts of non-silicate materials.

Later, Li and Greenberg (1997) inverted modern laboratory data and used a more realistic elongated core-mantle interstellar particle model. Their fit reproduces all prominent interstellar extinction data including polarization. We therefore now have a consistent set of refractive indices to approximate interstellar silicates (the cores in Li and Greenberg's model) separate from the organics (the mantles in the same model). "Astronomical Silicate" by Draine and Lee also appears to serve as a good approximation when a single effective refractive index is needed, as we shall see in sections IV and V.

III.B. Aggregates and Other Inhomogeneous Materials

In this subsection, we ask how the refractive index changes as interstellar matter condenses, aggregates, and partially sublimates, and if the complex material can indeed be described using a refractive index. A refractive index is intended to represent materials on scales where the grain is homogeneous and isotropic. Specifically, it is required and sufficient that both the phase velocity and attenuation rate of an electromagnetic wave propagating through the grain is constant and independent of direction. This allows Maxwell's equations to be integrated across an electromagnetically homogeneous and isotropic volume. The scale of these volumes depends both on the grain geometry and on the wavelength.

Two distinct physical scales are relevant according to the framework presented in Sec. II; small and large scales in comparison to the dimensions of classic interstellar grains. The averaging to produce an effective refractive index can often be made on small scales in comparison to interstellar grain dimensions at optical wavelengths while averaging across grain interstices can simplify thermal emission calculations.

Starting with the smaller scale compared to interstellar grains, we note that since classical interstellar grain sizes are submicron, any randomly distributed inhomogeneities must also be small. It is therefore reasonable to invoke the small-scale asymptotic approximation for these structures while mantle condensation and other non-random variations usually cannot be accounted for by using effective medium theory alone.

The energy scattered by internal inhomogeneities is accounted for through an increase of the imaginary part n'. Although the scattering indeed causes a loss of energy propagating in the original direction, the energy removed by the scattered wave is not removed from the scattering problem. Since internal scattering has the same frequency as the original incident wave, it adds vectorially to the other fields. The effect in any specific point may be either an increase or a decrease in the field amplitude depending on the phase relations of the interfering waves. This is analogous to defects in bulk materials that also cause scattering.

A pioneering effort to describe optical constants of inhomogeneous material using an effective refractive index is due to Garnett (1904), who proposed the so-called Maxwell Garnett rule. Also successful, and often predicting similar results, is the solution due to Bruggeman (1935). The Maxwell Garnett and the Bruggeman rules are both based on the assumption that the inhomogeneities are small compared to the wavelength. This is the Rayleigh or static limit approximation. The difference is in the topology of the mixtures. The Maxwell Garnett rule assumes a separated grain topology in which inclusions are embedded in a matrix material. That is why the effective refractive index depends on which material is taken as the matrix. The Bruggeman rule avoids this problem by considering a set of randomly distributed cells of different materials, called aggregated grain topology. These classic mixing rules have been generalized to multiphase mixtures, including magnetic, anisotropic and chiral components, polydisperse inhomogeneities, non-spherical (ellipsoidal, cylindrical, cubic, flakes and rods) and stratified inclusions, etc.. There are also mixing rules developed for more complex, fractal particle topologies including percolated topology or topologies with a radially varying filling factor. An extensive review of mixing rules can be found in the book by Sihvola (1999).

For larger scales, Stroud and Pan (1978) proposed a rule that allows inhomogeneities to be comparable in size to the wavelength. They considered spherical inclusions but the method can be applied to any shape inclusions as long as an exact solution to Maxwell equations is known for their geometry. Chýlek et al. (2000) discuss some modifications to this type of the effective medium theories, known as extended effective medium theories.

In all effective medium theories, the task is to replace a function $\varepsilon(\mathbf{r})$ used to represent the dielectric constant that may be an arbitrary function of position $\mathbf{r}$ by an effective dielectric constant ε_{eff} that is independent of $\mathbf{r}$. We consider a two-phase composite, where the first phase with fillingfactor $f(\mathbf{r})$ has the dielectric constant ε_1 of the bulk material and the second phase is empty space with $\varepsilon_2 = 1$ to simulate porous structures and aggregates. An analytic expression for the quantity ε_{eff} of the two-phase composite in an arbitrary (inhomogeneous and irregular) scattering volume is given by Bergman's theorem (Bergman 1978) and can be written

$$\varepsilon_{\text{eff}} = 1 - f_{\text{av}} \int_0^1 \frac{G(u, f_{\text{av}})}{T - u} \, du. \tag{1}$$

Here $f_{\text{av}} = \frac{1}{V} \int f(r) \, dV$ is the average filling factor inside the smallest sphere enclosing the volume V. All dielectric properties are contained in $T = 1/(1 - \varepsilon_1)$, and the integration variable u can be interpreted as a "depolarization factor". The spectral function $G(u, f_{\text{av}})$ is the distribution of depolarization factors and includes only geometrical information on the scattering volume. *The geometric properties are thus separated from the dielectric properties, which facilitates systematic analysis as well as computations considerably.*

An interpretation of Eq. (1) is that all possible geometric resonances in a two-phase composite occur when the complex quantity T assumes only real values in the interval $[0, 1]$. The integration over u probes all possible resonances. Whether a resonance occurs or not is determined by the spectral function. While, this function is not always known, solutions have been found for some topologies and the spectral function can in some cases be obtained from laboratory measurements (e.g., Sturm et al. 1991).

Some simple mixing rules and their corresponding spectral functions are compiled in Table 1. Generalizations of the Maxwell Garnett and the Bruggeman mixing rules require additional parameters to account for the size and/or shape distribution of the particles forming the inhomogeneous composite (e.g., Bohren and Huffman 1983; Ossenkopf 1991). In Ossenkopf (1991) and in Henning and Stognienko (1996) these parameters are adjusted to model the spectra of aggregates that are either in the range, $f_{\text{av}} \approx 0.1 - 0.4$, or very porous, $f_{\text{av}} < 0.05$. Spectral functions for these generalized rules can hardly be determined analytically. However, it is possible to calculate the spectral functions associated with empirical data through numerical inversion of Eq. (1).

Although the spectral representation does not always feature a direct solution to the scattering problem, it is a useful tool to analyze the physical background of rivaling effective medium theories or mixing rules. A database of spectral functions for samples with well-defined topologies would be valuable. Knowledge of the spectral function for a given topology is particularly useful in the inverse problem, i.e., the derivation of bulk optical constants from measured effective constants (Rouleau and Martin 1991).

TABLE 1

Mixing rules and corresponding spectral functions

Mixing rules	Spectral function
Garnett (1904) $$\frac{\epsilon_{\text{eff}} - \epsilon_2}{\epsilon_{\text{eff}} + 2\epsilon_2} = f_{\text{av}} \frac{\epsilon_1 - \epsilon_2}{\epsilon_1 + 2\epsilon_2}$$	Sturm et al. (1991) $$\delta\left(u - \frac{1 - f_{\text{av}}}{3}\right)$$
Bruggeman (1935) $$f_{\text{av}} \frac{\epsilon_1 - \epsilon_{\text{eff}}}{\epsilon_1 + 2\epsilon_{\text{eff}}} + (1 - f_{\text{av}}) \frac{\epsilon_1 - \epsilon_{\text{eff}}}{\epsilon_1 + 2\epsilon_{\text{eff}}} = 0$$	Sturm et al. (1991) $$\frac{3f_{\text{av}} - 1}{2f_{\text{av}}} \delta(u)\Theta(3f_{\text{av}} - 1) +$$ $$\frac{3}{4\pi f_{\text{av}} u} \sqrt{(u - u_-)(u_+ - u)} \Theta(u - u_-)\Theta(u_+ - u)$$ $$u_{+/-} = \tfrac{1}{3}\left(1 + f_{\text{av}} \pm 2\sqrt{2f_{\text{av}} - 2f_{\text{av}}^2}\right)$$
Looyenga (1965) $$\epsilon_{\text{eff}}^{1/3} = f_{\text{av}} \epsilon_1^{1/3} + (1 - f_{\text{av}})\epsilon_2^{1/3}$$	Sturm et al. (1991) $$f_{\text{av}}^2 \delta(u) +$$ $$\frac{3\sqrt{3}}{2\pi}\left[(1 - f_{\text{av}})^2 \left|\frac{u-1}{u}\right|^{\tfrac{1}{3}} + (1 - f_{\text{av}})f_{\text{av}} \left|\frac{u-1}{u}\right|^{\tfrac{2}{3}}\right]$$
Monecke (1989) $$\epsilon_{\text{eff}} = \frac{2(f_{\text{av}}\epsilon_1 + (1 - f_{\text{av}})\epsilon_2)^2 + \epsilon_1 \epsilon_2}{(1 + f_{\text{av}})\epsilon_1 + (2 - f_{\text{av}})\epsilon_2}$$	Monecke (1994) $$\frac{2f_{\text{av}}}{1 + f_{\text{av}}} \delta(u) + \frac{1 - f_{\text{av}}}{1 + f_{\text{av}}} \delta\left(u - \frac{1 + f_{\text{av}}}{3}\right)$$

IV. SCATTERING SOLUTIONS

An incident electromagnetic wave causes a perturbation of the internal and surface fields of the particle. By symmetry, a perturbation in the field that can be the above mentioned induced perturbation, a result of thermal motion, or any phenomenon leading to acceleration of a charge, would cause the emission of an electromagnetic wave. This exchange is completely symmetric with respect to time. The relation between direction of incidence and emission are governed by probabilities in a quantum mechanical view and scattering functions in classical electrodynamics. Both can be viewed as interference phenomena.

This disturbance can be viewed as an induced field added vectorially to the undisturbed field. In general, the corresponding energy, which is proportional to the square of the field amplitudes, is partially absorbed and therefore only a fraction of the energy is the source of the electromagnetic wave spreading from the particle, the scattered wave. Expressed differently, charges accelerate in response to the induced field and therefore radiate. Charges keep accelerating and therefore keep radiating as long as the incident field is periodic. The motion, an oscillation that may be viewed as an A/C current, does not occur without energy loss to heat through resistance, i.e., through collisions. There is therefore in general a finite absorption as well as scattering. The scattered wave has the same frequency as the incident radiation and therefore the same wavelength outside the particle. This distinguishes it from other emitted waves. Sometimes, the term elastic scattering is used to distinguish same frequency scattering from Raman scattering, thermal emission and other special circumstances that can cause a frequency change. In this section, we seek a relation between the incident light and the light observed at a large distance due to elastic scattering. This relation is often used to predict the scattering from a cloud of particles or to derive particle properties from the remotely observable scattered light.

Energy conservation dictates that the scattered wave from a finite particle approaches a spherical geometry at sufficiently large distances. This distance should be large compared to all particle dimensions and the wavelength so that phase differences between contributions from different parts of the particle vanish. Following common practice, the spherical wave can then locally be approximated by a plane-wave and it is this plane wave that we seek to describe.

Arbitrary electromagnetic plane-waves propagating through empty space can be fully defined through the direction of propagation and the orthogonal amplitude components of the electric field

$$E_1 = A\cos(\tau + \delta_1),$$
$$E_2 = B\cos(\tau + \delta_2). \qquad (2)$$

The variable part of the phase factors is $\tau = \omega t - ks$, where t is the time, s allows for the phase shift along the direction of propagation and $k = 2\pi\omega/c = 2\pi/\lambda$ is the propagation constant or wavenumber. The constant parts are the phases δ_1 and δ_2. The corresponding Stokes vector is

$$\begin{aligned} I &= A^2 + B^2, \\ Q &= A^2 - B^2, \\ U &= 2AB\cos(\delta_1 - \delta_2), \\ V &= 2AB\sin(\delta_1 - \delta_2). \end{aligned} \qquad (3)$$

Here, the degree of linear polarization $P = \sqrt{Q^2 + U^2}/I$ which is often encountered in astronomy is sometimes substituted for Q in an alternate Stokes vector. The squared amplitudes are directly measurable intensities and the phases can also be derived from measurable parameters (e.g. Gustafson, 1996a). We note that there are only three independent parameters describing the time-averaged wave; the two intensities and the phase difference $\delta_1 - \delta_2$. There exists the relation $I^2 = Q^2 + U^2 + V^2$ for a strictly monochromatic wave so that the Stokes vector also has only three independent parameters.

To the extent that the incident as well as the scattered radiation can be locally described as plane waves, these are fully defined by their Stokes parameters. The scattering can then be described by a linear transformation, mathematically the 4×4 Müller matrix $\mathbf{F}$;

$$(I, Q, U, V) = \frac{\mathbf{F}}{k^2 r^2}(I_0, Q_0, U_0, V_0), \tag{4}$$

between two Stokes vectors where subscripts denote the incident beam parameters and r is the distance from the scatterer. The 16 element $\mathbf{F}$-matrix contains only seven independent parameters. Eight parameters are found in the intensity and phase of the orthogonal components that can be experimentally determined (Gustafson 1996a, 2000), or equivalently in the corresponding complex amplitudes. Of these eight, only the phase differences, not the absolute phases, are relevant which reduces the number to seven.

The principle of optical equivalence (e.g., Bohren and Huffman 1983) states that the Stokes parameters contain the complete set of quantities to characterize the intensity and state of polarization of a beam of light, *in practical analysis* since optical measurements involve linear transformations only. It follows that the transformation matrix $\mathbf{F}$ between the incident and scattered Stokes vectors fully describe the scattering process to the same level of detail.

The matrix $\mathbf{F}_\lambda(\theta, \phi)$ is a dimensionless function of the scattering angle, $0 \leq \theta \leq 180°$, between the direction of propagation before and after the scattering process and of the azimuth angle $0 \leq \phi < 360°$. The energy scattered into a unit solid angle $(\sin(\theta)\,d\theta\,d\phi)$ around the direction (θ, ϕ), equals that of light incident on the area $\mathbf{F}_\lambda(\theta, \phi)/k^2$. This is the cross section, $C_{\theta,\phi}$, for scattering in the direction (θ, ϕ). Integrating, we obtain the area intercepting the total scattered energy $C_{\text{sca}} = \iint C_{\theta,\phi}\,d\theta\,d\phi$. Similar cross sections can be defined for the absorbed energy, C_{abs}, and for extinction – the total energy removed from the beam $C_{\text{ext}} = C_{\text{sca}} + C_{\text{abs}}$. Known as the fundamental extinction formula, the relation $C_{\text{ext}} = \frac{4\pi}{k^2}\,\text{Re}\{S(0)\}$ where $S(0)$ is the forward scattering amplitude, directly expresses a removal of energy from the forward direction in the form of interference between the incident and scattered waves. This formula simplifies the extraction of cross sections from laboratory measurements and can serve as a check on theoretical calculations. The fraction of energy transferred to the scattered field is the single scattering albedo which equals the ratio of the scattering to extinction cross sections.

We represent the solar spectrum using the Planck function $B(T)$ at the solar temperature T and normalize to the geometric cross section C_g to generate average efficiency factors for scattering

$$Q_{\text{sca}} = \frac{\int B(T)\, C_{\text{sca}}(\lambda)\, d\lambda}{C_g \int B(T)\, d\lambda} \tag{5a}$$

with a similar formula for absorption

$$Q_{\text{abs}} = \frac{\int B(T)\, C_{\text{abs}}(\lambda)\, d\lambda}{C_g \int B(T)\, d\lambda}. \tag{5b}$$

Dependencies on shape and morphology when particles are of a given material (refractive index) are minimized in efficiencies versus mass plots. Figure 1a illustrates three main light scattering domains using efficiencies for scattering versus mass and Fig. 1b shows absorption versus mass. All data are for averages over uniform orientations to simulate randomly oriented grains in a cloud. The solid curves represent efficiencies for homogeneous spheres calculated using the refractive index of "Astronomical Silicate" by Draine and Lee assuming a bulk density of 2500 kg m^{-3}. Circles correspond to single classical-size interstellar grains. Small asterixes are for sets of 2, 3, 5, 10, 30, and 50 spheres where sets of 2 to 3 spheres represent classical interstellar grains. The calculations were made using the T-matrix formulation code by Mackowski and Mishchenko (1996). Larger numbers in the aggregates could not be reached in the calculations but the data were extended using results from the microwave laboratory. The large asterix is for a 1450 sphere model representing 725 aggregated interstellar grains shown in Fig. 2a by Gustafson et al. (1999), this aggregate is similar but not identical to target 1 in Table 2, see Fig. 5. The ellipse shaped symbol is for target 9 representing microwave analogue results for tightly packed interstellar grains in an elongated compact aggregate. Squares are for microwave analogue results for silicate cubes that are also represented in Fig. 5.

While all calculations are for the wavelength dependent refractive index of "Astronomical Silicate" with the data averaged from the infrared to the far UV, the 500 wavelength laboratory data translate to the 0.44 to 0.65 μm interval, which is only as broad as the sensitivity of the human eye. To the accuracy of measurements, a single refractive index is used across the waveband so that the 1450 ± 20 ensemble of 0.63 ± 0.35 mm diameter polystyrene spheres has $m \approx 1.615 + i0.03$ at the microwave frequencies. This makes the polystyrene plastic equivalent to a silicate at the optical wavelengths it is used to simulate. The compact aggregate (target 9) consists of ~ 3000 nylon cylinders $m \approx 1.74 + i0.005$, each representing the silicate core of a classic-size interstellar grain. The cores are embedded in a $m \approx 1.8 + i0.1$ compound representing an organic refractory of volume approximately 1.5 times that of all cores added together. The acrylic cubes representing silicates have refractive indices closer to $\sim 1.605 + i0.003$ so that while the refractive indices differ from model to model they are in the range of silicates and the organic refractory materials.

The Mie-solution using the refractive index of "Astronomical Silicate" by Draine and Lee is found to be a good approximation to the absorption efficiencies by BN-structures *of the same mass*. This is practically independent of the particle shape and structure. Scattering is more sensitive to particle type above $\sim 10^{-15}$ kg where the efficiencies for the BN-type structures and the cubes fall significantly below those for "Astronomical Silicate" spheres. While the approximation to the scattering cross sections is good in the lower mass ranges, the angular dependence of the scattering and polarization is more sensitive and dependence on the refractive index is much greater than at large sizes.

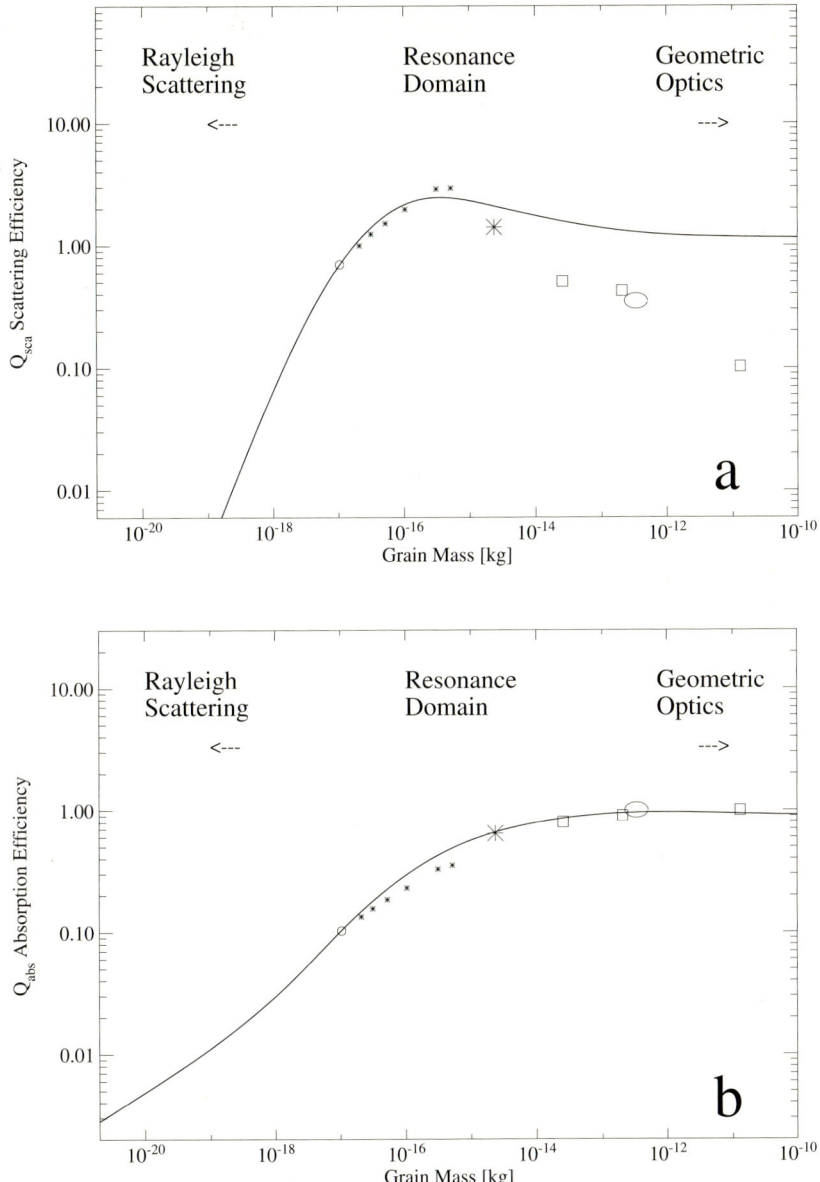

Figure 1. Efficiencies for scattering Q_{sca} and absorption Q_{abs} averaged over the solar spectrum are shown as a function of the particle mass for "Astronomical Silicate" spheres (solid curve). This is a good approximation to scattering by small BN-type aggregates and to the absorption at all sizes tested when averaged over orientation. Circles are for a single classical interstellar grain. Asterixes are for aggregates of the BN-type representing interplanetary dust as aggregates of interstellar grains with the large asterix representing an aggregate of 725. The ellipse shaped symbol is for tightly packed interstellar grains of the BN-composition and squares denote cubes.

For example a 10^{-20} kg organic refractory grain of $m = 1.7 + i0.15$ has the absorption efficiency $Q_{abs} = 0.03$, a factor 10 larger than for "Astronomical Silicate", while the scattering efficiency is a few magnitudes lower than this for both.

Significant simplifications take place when a particle is either sufficiently small, or sufficiently large in comparison to the wavelength, which are in the approximate regions marked "Rayleigh Scattering" and "Geometric Optics". The incident field is the same everywhere across a sufficiently small particle, which is the condition for Rayleigh scattering. Scattering in the Rayleigh region is weak for particles with refractive indexes in this range so that absorption tends to dominate over scattering, which is clearly seen in the figures. At the other extreme, sections of the incident wavefront can be treated as units as long as the particle is uniform over sufficiently large dimensions compared to the wavelength. This is the condition for geometric optics. At larger masses than shown in the figure, the cubes asymptotically approach the scattering by spheres of the same material (refractive index) when averaged over random orientations. The angular distribution of scattering as well as the Q_{sca} is strictly the same for very large randomly oriented convex particles (e.g., Gustafson, 1994) and $Q_{ext} = Q_{sca} + Q_{abs} = 2$ for all particles so that these large cubes necessarily also have the same Q_{abs} as the spheres.

Scattering is more complex in the intervening region known as the resonance domain ranging from approximately 0.1 to 100 μm across for solar illumination and more when particles have heterogeneities on smaller scales as BN-structures do. /looseness-1 Since interaction per surface area is weak in the Rayleigh size range, most of the scattering tends to occur in the resonance region and most of the absorption in the geometric optics region. The dust size distribution strongly affects what range particles contribute the most and thus the type of interaction that dominates in a cloud as a whole. Figure 2 illustrates this for two size distributions of Zodiacal dust cloud particles using the "Astronomical Silicate" sphereres represented by solid curves in Fig. 1. Figure 2a shows the geometric area per unit of volume, or the "specific geometric area" following the Grün et al. (1985) mass distribution and, for comparison, the Dohnanyi (1969) distribution expected as particles of equal strength break-up in mutual collisions. Following common practice, the cumulative area to a given grain mass is plotted on a log-log scale starting from the most massive grains. This shows that most of the area in the Grün et al. distribution is in the 10^{-9} kg range corresponding to 100 μm diameter while it is always in the smallest particles in the Dohnanyi distribution. The total scattering from the Dohnanyi distribution peaks around 10^{-16} kg (0.2 or 0.3 μm), see Figure 2b while the peak contribution shifts depending on the scattering angle to near 10^{-14} kg (1 μm) in backscattering (Figure 2c). While absorption is less peaked it also is mostly from the 10^{-14} kg range (Figure 2d). From this, we see that the range of particles in a Dohnanyi distribution contributing to most of the interaction depends on the type of interaction while the Grün et al. distribution is sufficiently peaked so that particles in the 10^{-9} kg (100 μm) range always dominate. This probably holds for BN-type particles as well as for the spheres used in our example, but this needs to be confirmed. We conclude that depending primarily on the mass (or size) distribution, most of the interaction takes place somewhere in the resonance scattering region. We therefore discuss scattering in this scattering range.

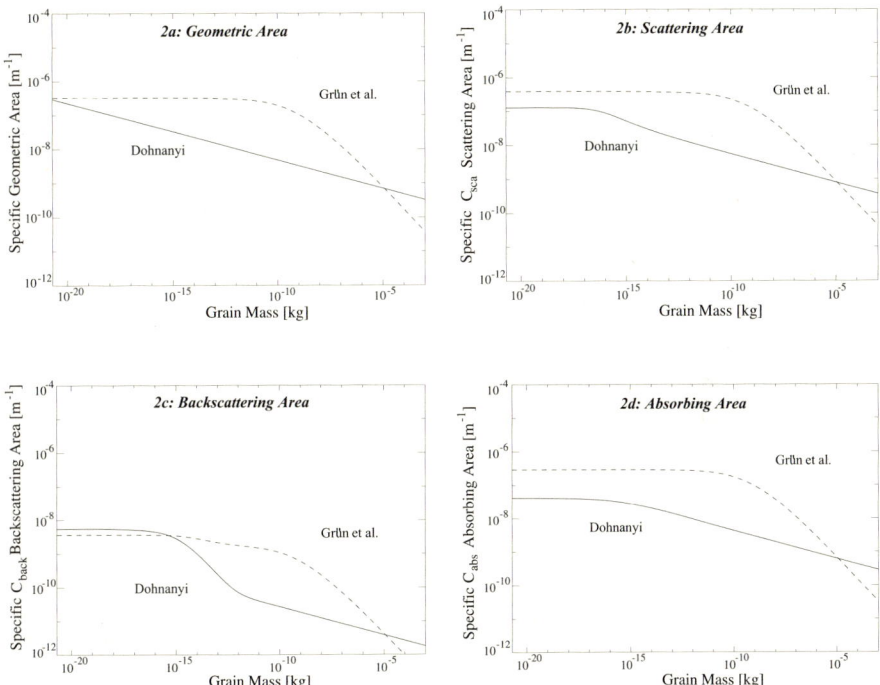

Figure 2. Accumulative areas for "Astronomical Silicate" spheres down to a given mass as a function of the particle mass in the Dohnanyi (1969) size distribution and in the Grün et al. (1985) distribution, represented by solid and dashed lines respectively.

Even given a specific refractive index or a dielectric constant to represent the material(s) a particle is made from, no general solution to theoretically solve the scattering problem in the resonance region exists. Instead, we have to choose from a range of specialized methods that may be exact or approximate, theoretical or empirical. The best choice depends on the particle size, geometry, type of function $\varepsilon(\mathbf{r})$, and on available facilities. To date, many cases can still not be reliably solved on a computer. In a review, van de Hulst (1993) concludes that calculations of scattering by arbitrary particles using current numerical techniques and computers is limited to circumference to wavelength ratios $x < 15$. Even then, there are sometimes still no à priori knowledge of when the solution is correct and when it breaks down (see, Xu and Gustafson 1999). The selection of a method of calculation is therefore not obvious. While this section is not intended as a general review of scattering solutions it may serve as a starting point in selecting a method to solve for light or other electromagnetic interaction in the resonance range. Four main methods are widely recognized and applied to theoretical studies of cosmic dust at present. Perhaps the simplest to use and numerically most economic is the Mie-solution for a perfect homogeneous sphere, an example of the exact boundary condition family of solutions. For non-spherical particles, the T-matrix formulation is convenient when we seek averages over a broad range of orientations. It can be obtained using either analytic or numerical methods depending on the geometry. The interacting spheres (N-sphere) solutions are rigorous extensions to Mie theory and are based on solutions to the boundary condition on the particle's surface. By contrast, the highly flexible and often used coupled-dipole approximation is based on a discretized representation of the internal field. All have limitations and work well only for some particle types.

The boundary condition family of solutions takes advantage of the fact that charges cannot be created or removed inside, outside, or on the boundary of the particle. These elegant solutions use Maxwell's equations to describe the electromagnetic radiation outside the particle, an analogue set that involves the refractive index to describe the internal field, and a boundary condition to assure that charges are not created or removed outside the particle, inside, or at the interface. Maxwell's equations can be derived from the conservation of charge condition using the Lorentz force law to deduce the basic electromagnetic field tensor and use of the Lorentz transformation of Special Relativity to translate these equations to a moving reference system (e.g., Scharf 1994, Chapter 2). Charge conservation is therefore guaranteed inside and outside of the particle using rather simple expressions and the solution to the scattering problem is obtained by making sure of charge conservation at the boundary as well. The surface integral over the boundary conditions has been exactly and analytically solved for a few particle geometries that include single and multiple spherical particles, spheroids, and infinite cylinders. The best-known example is the famous Mie-solution to the scattering by an arbitrary single homogeneous sphere, a solution used in a broad range of scientific disciplines.

IV.A. Mie Theory and Related Boundary Solutions

A century ago Lorenz (1898), Mie (1908), Debye (1909) and others independently solved the problem of light scattering by a homogeneous and isotropic sphere of arbitrary size and arbitrary homogenous and isotropic refractive index referred to as Mie-theory (see Logan 1965). In Mie-theory, the plane incident electromagnetic field, the scattered electromagnetic field and the internal electromagnetic field are expanded in infinite series of vector spherical harmonics. The major task in Mie calculations is to find the partial wave coefficients (a_n, b_n) through the use of rigorous analytical expressions in terms of only size and refractive index of the sphere. These are obtained from the solution of the standard boundary conditions at the spherical surface. The analytical expressions for a_n and b_n can be found, for example, in the books by van de Hulst (1957), Kerker (1969), or Bohren and Huffman (1983). Bohren and Huffman (1983) include an efficient computer source code for Mie scattering calculations.

Wang and van de Hulst (1991) improved the numerics of Mie-calculations using their ratio method to compute the Riccati-Bessel Neumann and Hankel functions. This allows numerically accurate Mie-calculations well into the geometric optics region (up to circumference to wavelength ratio $x \approx 5 \cdot 10^4$), corresponding to a few millimeters size in visual light.

A solution to the problem of scattering by a single homogeneous sphere coated with a homogeneous concentric spherical layer was obtained by Aden and Kerker (1951) while Toon and Ackerman (1981) give a modern solution. The major complication is the inclusion of the additional boundary conditions inside the scatterer. Kaiser and Schweiger (1993) obtained a two-layer solution. The solution for coated spheres can also be generalized to a radially and continuously stratified sphere (Bhandari 1985; Kai and Massoli 1994). Maxwell's equations are also separable in cylindrical coordinates (e.g., Wait 1955) and in spheroidal coordinates (e.g., Asano and Yamamoto 1975; Voshchinnikov and Farafonov 1993) so that the boundary conditions have been rigorously solved for these particle shapes as well. A core-mantle spheroid solution was obtained by Cooray and Ciric (1991) and independently by Farafonov et al. (Farafonov et al. 1996). In astronomy, these solutions are primarily useful when estimating scattering based on classical interstellar homogeneous or core mantle grain models and can also serve as a starting point for the scattering by BN-aggregates of such grains.

IV.B. Extension of Boundary Conditions to N-Spheres

Many solids that fall in the resonance region size range at visual wavelengths are aggregates, often of low packing. This includes at least some major class of interplanetary dust particles that are often referred to as fluffy. The resulting cooperative scattering includes two main effects: (1) interaction between particles, i.e., the scattered radiation by each particle illuminates the other particles and is again scattered, and (2) systematic far-field interference between the waves scattered by different particles owing to phase difference. Compared to isolated particles, these two effects alter the scattering from an aggregate of particles as a whole. The first effect becomes increasingly important when the number and the sizes of the individual particles are not small and when the refractive index is large or particles are closely packed and vanishes when they are sufficiently dispersed.

A simple approximation to scattering by a small and fluffy aggregate therefore is to solve for (2), the far-field interference between the waves scattered by constituent particles in an aggregate. This is the so-called coherent scattering solution that approximates the incident field on each constituent particle by the undisturbed incident field that would be there in the absence of the particle. Consequently, each particle scatters as though no other particles were present. The *difference* in optical path length from each particle to any distant point depends on the location in the aggregate only and leads to systematic interference. This solution closely follows the Rayleigh-Gans scattering formulation (van de Hulst 1957) but uses summations over discrete scattering centers instead of integrals over a continuous medium. The coherent scattering solution is a fast and effective means to estimate light scattering by tenuous aggregates up to several hundred classical size interstellar grains (Zerull et al. 1993).

The first attempt to a rigorous solution appears to be due to Trinks (1933) who considered a pair of identical spheres of rather small size so that they scatter within the Rayleigh regime. In general, accounting for interactions is increasingly elusive as the aggregate grows in size and as the interactions grow in importance. Mutual interaction between scattered waves from individual particles and use of multiple reference systems are the main reasons. To set up the standard electromagnetic boundary conditions on each of the spherical surfaces as they appear in Mie-theory, one is faced with a large number of translations. The scattered waves from each sphere is expanded in a reference system with origin at the center of the sphere and must be translated into the coordinate systems centered on its neighboring sphere where it becomes part of the incident wave. This is repeated for every pair of spheres in the aggregate using addition theorems for vector spherical harmonics. The more complex the scattered field and the larger the displacement, the larger the numerical demands of these translations are in traditional solutions. When applied to BN-type interplanetary dust models, this usually restricts use of the rigorous solutions to small aggregates. Ironically, this is where interactions can be neglected anyway. We therefore do not discuss the extensive body of work by many researchers who have spent considerable effort to investigate the dependent scattering by an arbitrary aggregate of spheres, cylinders and spheroids of small size analytically.

Xu and Gustafson (2001) showed that the size limitation can be circumvented in a fully rigorous and efficient multisphere-scattering theory (Xu 1995, 1997; Xu et al. 1999). Xu's generalized multiparticle Mie-solution is based on a fully rederived framework and has been thoroughly tested using modern microwave techniques (Xu and Gustafson 1996, 1997, 1999). The formulation includes the following major components:

(1) rigorous expressions and efficient numerical techniques in practical computations for the vector translation coefficients introduced by the addition theorems for vector spherical wave functions,

(2) the solution of partial (sometimes called differential) scattering coefficients of all individual spheres from an extended formulation of Mie-theory to allow incident spherical waves,

(3) an asymptotic form of the vector addition theorems valid in the far zone and an asymptotic single-field representation of the total scattered far-field consisting of all the partial scattered fields expanded in respective sphere-centered reference systems, and

(4) rigorous analytical expressions for all scattering properties of an aggregate of spheres derived from the far-field solution, including the four elements of the amplitude scattering matrix and thus all sixteen elements of the Müller matrix, the efficiencies for extinction, scattering, absorption, and radiation pressure, as well as the asymmetry parameter.

The generalized multiparticle Mie-solution is limited only by numerical computing constraints and provides exact results within the formalism of classical electrodynamics. The CPU time increases as the fourth power of the size parameter of the constituent particles and with the square of the number of constituent particles in an aggregate. For example, calculations for an aggregate of 400 spheres of size parameter $x = 10$ require about 2 hours on a DEC alpha-station equipped with 4 Gb RAM. This is for a single wavelength and a single orientation of the aggregate. The method is therefore still not practical for calculation of colors of large aggregates but will likely become so as computing power increases. A way to increase the efficiency of averaging over orientations is to express the solution in the form of a T-matrix.

IV.C. T-Matrix Solutions

The T-matrix method (Waterman 1965, 1973; Varadan and Varadan 1980; Mishchenko et al. 1996) is also known as the null field method and as the extended boundary condition method (EBCM). The incident, transmitted, and scattered fields are all expanded into spherical vector wave functions. The expansion coefficients of the scattered field are then related to the coefficients of the incident field by the T- (transmission-) matrix. The task is to obtain the matrix elements through numerical integration over or near the scattering body's surface. This means that the integral needs to satisfy a set of boundary conditions that are a direct consequence of Maxwell's equations thereby simply assuring that there is no net flow of charges in or out of the closed volume encompassing the scatterer. It is useful to note that the T-matrix formulation can be used with a variety of integration schemes for the boundary conditions. When an analytic expression for the integration is found, such as for the sphere, the solution is equivalent to Mie theory. Originally, Waterman applied the method to particles for which a solution could be found near but not necessarily on the surface and the analytic expression for the integral could be replaced by numeric integration. The T-matrix formulation can therefore be a rigorous solution in the classical electrodynamics representation of a particle and its material, or it may be an approximation.

Practical T-matrix implementations are usually for particles with a high degree of symmetry to reduce the computational burden, e.g., spheroids and circular cylinders. The method is demanding on computer RAM so that most existing computers cannot process T-matrix codes for particles with surface area equal to a sphere of size parameter exceeding ~ 50 although Mishchenko and Travis (1998) reached the order of 100 which corresponds to dimensions of tens of microns across in the visual. In addition, the larger the particle, the smaller the range of refractive indices that can be used. The axial ratio of elongated particles also rapidly becomes more restricted. The codes are relatively easy to use despite numerical problems. The greatest advantage may be that scattering can be calculated for any incident direction and expressions can be derived for averages over particle orientations once the T-matrix is known.

T-matrix codes have also been developed for core-mantle spheres and spheroids (e.g., Peterson and Ström 1974), and for the multi-spheres case (Mackowski and Mishchenko 1996). The T-matrix formulation is in some cases only a restatement of a solution that my have been obtained using other methods. The difference between these and their analytic boundary condition counterparts discussed in the previous section are therefore in some cases negligible. The major advantage is that the T-matrix formulation allows direct averaging over orientation. The disadvantage is that T-matrix codes are very memory-demanding so that the overall size of the particle is restricted when compared for example to the generalized multiparticle Mie-solution. The tractability of this formulation is in part a function of the ratio of computer RAM to MIPS costs.

IV.D. Internal Field Solutions

Maxwell's equations for dielectric matter (i.e., assuming a magnetic permeability of unity) can be expressed as the simple condition

$$\nabla \times \nabla \times \mathbf{E}(\mathbf{r}) - k^2 \epsilon(\mathbf{r}) \mathbf{E}(\mathbf{r}) = 0. \tag{6}$$

A direct solution of this partial differential equation using finite element methods is complicated by the boundary conditions at the surface of the scatterer. Although there has been some success (e.g., Peterson 1991), most numerical methods start from the volume integral equation

$$\mathbf{E}(\mathbf{r}) = \mathbf{E}_0(\mathbf{r}) + \frac{k^2}{4\pi} \int_V d^3\mathbf{r}' \, \widehat{G}(\mathbf{r}, \mathbf{r}') \left[\epsilon(\mathbf{r}') - 1\right] \mathbf{E}(\mathbf{r}'), \tag{7}$$

which may be derived from Eq. (6) and that contains the boundary conditions implicitly. The integration is over the volume of the scatterer where $\mathbf{E}_0$ is the incident electric field and $\widehat{G}(\mathbf{r}, \mathbf{r}')$ is the free space Green dyadic function. After discretization, the method of moments (Harrington 1982) or the finite element method (Silvester and Ferrari 1990) may be used to numerically solve the integral equation. However, the coupled dipole approximation or the *discrete dipole approximation* (DDA) is more often used in the astrophysical literature. While Purcell and Pennypacker (1973) originally proposed this method without reference to the integral equation, several authors showed that it may be derived directly from Eq. (7) (e.g., Hage and Greenberg 1990). Lakhtakia (1992) discussed the relations between commonly used algorithms. We describe the coupled dipole approximation and its practical implementations in some detail since it is widely applied to cosmic dust.

Purcell and Pennypacker (1973) originally used the coupled dipole formulation to approximate the scattering by interstellar grains. They first divided the particle into sufficiently small volumes so that each unit can be represented by a single dipole. The polarizability α may be a complex tensor, but in the common case of a homogeneous and isotropic material it may reduce to a (complex) scalar depending on the positioning of the dipoles. The electric field at the distance r from an oscillating dipole of polarizability α and electric dipole moment $\mathbf{p} = \alpha \mathbf{E_0}$ where $\mathbf{E_0}$ is the incident field, can be written

$$\mathbf{E} = \frac{k^2}{4\pi\varepsilon}(\mathbf{n}\times\mathbf{p})\times\mathbf{n}\frac{e^{ikr}}{r} + \frac{[3\mathbf{n}(\mathbf{n}\cdot\mathbf{p})-\mathbf{p}]}{4\pi\varepsilon}\left(\frac{1}{r^3}-\frac{ik}{r^2}\right)e^{ikr}, \qquad (8)$$

where $\mathbf{n}$ is the unit vector in the direction of propagation. The electric field at the i:th dipole in a system of N dipoles then becomes the sum of the incident field, and the scattered field from all the other dipoles:

$$\mathbf{E_i} = \sum_{j\neq i}^{N}[a_{ij}\mathbf{p_j} - b_{ij}(\mathbf{p_j}\cdot\mathbf{n_{ij}})\mathbf{n_{ij}}] + \mathbf{E_0}e^{ikr_i}, \qquad (9)$$

where

$$a_{ij} = \frac{e^{ikr_{ij}}}{4\pi\varepsilon r_{ij}}\left(k^2 + \frac{ik}{r_{ij}} - \frac{1}{r_{ij}^2}\right),$$
$$b_{ij} = 3a_{ij} - 2k^2\frac{e^{ikr_{ij}}}{4\pi\varepsilon r_{ij}} \qquad (10)$$

and r_{ij} is the distance between the i:th and the j:th dipole. This can be written as the matrix operation

$$A\mathbf{x} = \mathbf{b}, \qquad (11)$$

where A is the $3N \times 3N$ complex interaction matrix, $\mathbf{x}$ is a $3N$ vector containing the unknown field at the dipoles and $\mathbf{b}$ is a $3N$ vector containing the incident field components. The task is to solve the system of equations for $\mathbf{E_i}$. The scattered field at any point can then be calculated by adding the contributions from individual dipoles accounting for the phase. Since there are three field components, there are $3N$ linear equations to solve to a given accuracy, in addition, multipole expansion methods has also been used on small aggregates. Several numerical methods have been proposed.

The scattering order iterative method may have been first used by Chiapetta (1980) and later by Singham and Bohren (1988). They noted that Eq 9 can be written as

$$\mathbf{E_i} = \mathbf{E_0}e^{ikr_i} + \sum_{j\neq i}^{N} C_{ij}\mathbf{E_j}, \qquad (12)$$

where

$$C_{ij} = \alpha_j \begin{pmatrix} a_{ij} + b_{ij}(n_{ij}^x)^2 & b_{ij}n_{ji}^x n_{ji}^y & b_{ij}n_{ji}^x n_{ji}^z \\ b_{ij}n_{ji}^y n_{ji}^x & a_{ij} + b_{ij}(n_{ji}^y)^2 & b_{ij}n_{ji}^y n_{ji}^z \\ b_{ij}n_{ji}^z n_{ji}^x & b_{ij}n_{ji}^z n_{ji}^y & a_{ij} + b_{ij}(n_{ji}^z)^2 \end{pmatrix}. \qquad (13)$$

Equation (13) can be expanded in an infinite series in terms of C_{ij} through successive substitutions of the electric field on the right-hand side of the equation

$$\mathbf{E_i} = \mathbf{E_0}e^{ikr_i} + \sum_{j\neq i}^{N} C_{ij}\mathbf{E_0}e^{ikr_j} + \sum_{j\neq i}^{N} C_{ij}\sum_{k\neq j}^{N} C_{ij}e^{ikr_k} + \ldots \qquad (14)$$

The physical meaning of the terms in the series has been interpreted in terms of scattering orders, which gave the method its name. The first term represents the incident field on the i:th dipole. The second term contains the correction to the field due to scattering of the incident field by all the other dipoles. The third term is the correction due to the scattering from all other dipoles corrected for the previous (second) term and so on.

Each successive term in Eq (14) entails an additional level of summation and it appears that the computing effort soon becomes prohibitive. However, the previous scattering order field at each dipole is known at each step and can be used to calculate a corrected field at any dipole. Each term in the scattering order is used to calculate the next order term.

Convergence depends on the strength of dipole interactions. If the aggregate is optically thin, interaction is weak and each successive term is smaller than the previous one so that the series converges quickly. If interactions are strong, the series diverge. The fact that this iteration scheme places no constraint on the placement of dipoles preserves the inherent flexibility of the Coupled Dipole Approximation that makes it so attractive. Another advantage is that RAM requirements are minimized and partial results can be stored on disk when necessary. The number of computer operations needed to calculate each term in the series is proportional to N^2. When the series converge, it is usually after much fewer than N terms so that the number of computer operations grows slower than N^3. The convergence properties are probably the largest limitation of this method, which is better suited for classical size interstellar grains than for interplanetary dust and similar size and refractive index scattering systems.

The Conjugate Gradient Method (used in the DDA) is an iterative approach used by Draine (1988) and Draine and Flatau (1994, 1995) to solve the general matrix-equation Eq. (11) based on the conjugate method (Petravic and Kuo-Petravic 1979). They approximate the electric field and the Greens function in each element by its value in the center and confine the dipoles to a cubic lattice to exploit symmetry relations. An initial guess **x** is successively improved through iteration. Most computing cycles are in the matrix-vector multiplications and in two vector multiplications which makes each iteration use approximately two times as many computer cycles as a scattering order iteration (requiring only one matrix-vector multiplication). However, each step in the conjugate gradient method approaches the solution monotonically so that the series converges given sufficient numerical precision unless the interaction-matrix is ill-conditioned. The coupled dipole approximation as implemented by Draine and Flatau (1994, 1995) and Draine (2000) is the most commonly used although, or maybe because, they traded some of the flexibility for computational ease and speed.

The interaction matrix in Eq. (11) may also be inverted directly in computer RAM when the number N of dipoles is sufficiently small. A big advantage is that the inverted matrix can be used to calculate the scattering for any particle orientation or to derive averages over orientation. The matrix uses $144N^2$

bytes so that a 1000 dipole matrix requires 144 Mb. This is approximately the practical limit for matrix-inversion in 512 Mb RAM when using standard library routines since some manipulation is required. Practical applications usually involve the inversion of larger interaction matrixes. For example, the approximate number of dipoles in a 10 μm particle interacting with visual light is in the 10^6 or 10^7 range or more which is why iterative methods are used. However as computer hardware improves, direct inversion becomes more attractive and may presently be a convenient method for small systems including common interstellar grain models.

IV.E. Experiments

Experiments are often the only means of obtaining the required scattering data, especially from large and complex structures. It remains the more expedient method in many more cases. Another essential role for laboratory investigations is verification of theoretical solutions. We distinguish scaled microwave experiments from the direct method and show that each has its own advantages and that the methods are complementary.

1. Direct method

By levitating particles in an electrostatic field, Weiss-Wrana (1983) obtained angular dependency of intensity and polarization of laser light ($\lambda = 0.633\,\mu$m) scattered by single particles. However, the particles were hard to characterize and the scattering from single particles is usually too faint to measure accurately given the quality of the background, which is plagued by straylight.

Less systematic, but easier is the investigation of scattering by clouds of particles rather than single grains. Interest in this type of measurements was rekindled mainly as a result of new techniques to produce stable particle clouds or jets in a laser beam so that the average, or "volume" scattering functions can be measured. The main advantage is the possibility to study polydisperse particle samples in a single measurement as long as the supply of particles is abundant. While this technique is suited to broad surveys, it is less well suited to the testing of theories, for systematic exploration, or for investigation of scattering by "BN"-type grains. The difficulty in obtaining samples and limited knowledge of the particle parameters as well as poor particle control are among the reasons. Laboratory samples of particles whose size is in the range of cosmic grains (0.1-100 μm) usually are prepared from particulates of terrestrial rocks and minerals or from meteorite samples that are hard to characterize. Artificial grains with relatively well-known shape and refractive index can be prepared. However, they are available in a limited range of usually regular shapes and refractive indices that often do not correspond to the refractive index of cosmic grains. Deformed samples can be seen among the depicted Latex spheres used for test and calibration purposes in a recent study (Combet et al. 2001).

Measurement of the F-matrix components involving the phase shift suffered by a scattered wave is more difficult and requires special techniques. The large contribution to the phase shift due to each particle's location in the cloud (except in forward scattering) makes phase information difficult to extract even when phase relations can be measured. However, Bottiger et al. (1980) and Hovenier (2000) obtained all elements of the scattering matrix using polarization modulation in combination with lock-in detection. Bottiger et al. measured scattering matrix elements for sets of two, three, and four spheres. Kuik et al. (1991) obtained data on a broad selection of compact, porous, or aggregate particle ensembles of sizes ranging from a few microns to 100 micron in diameter using the laboratory by Hovenier. Samples include Saharan sand, volcanic dust, loess and red clay feldspar.

The accuracy of the direct method is limited by straylight and beam distortions that are especially significant at small scattering angles that can lead to direct illumination of the detector. Forward scattering amplitude and phase needed to deduce the forward cross sections are currently not accessible although improvement in laser and beam handling technology may soon make this possible.

The capability to measure characteristics of the scattered light averaged over an ensemble of particles is an important time saving feature. Exploration is fast but systematic investigations are difficult since there is limited control over the particles. The lack of control prevents rigorous tests of theoretical predictions. This has often led to questionable claims of experimental confirmation of theoretical values where an impressive agreement can also be the result of numerical fitting of a large number of unconstrained and therefore adjustable parameters. The direct method is well suited to investigate average optical properties from polydisperse particle clouds where a broad parameter range can be covered with relative ease and to the search for empirical rules.

2. Microwave analog method

The main limitations of the direct method can be routinely overcome using the microwave analogue technique that takes advantage of the ability to scale the light-scattering problem up or down to any convenient dimension. This feature is inherent in the form of Maxwell's equations where dimensions are only encountered as a ratio to the wavelength. This simplification is also widely used by theoreticians who as a rule report scattering dimension in terms of a dimensionless size parameter such as $x = 2\pi a/\lambda$, (where a is a typical dimension, e.g., the radius of a sphere). For example, radar cross sections for a tree or bush may fall in the same resonance scattering regime (between Rayleigh scattering and Geometric Optics, depicted in Figure 1) as the cross sections of interplanetary dust scattering sunlight. Depending on the detailed assumptions made, the two solutions may even be identical, or the radar scattering may shift into either region depending on the specific radar wavelength or size of the dust grain.

The first generation analogue to light scattering laboratory by Greenberg et al. (1961) used X-band equipment, a mature and widely available technology following WW-II. The 3.18 cm wavelength, when used to study scattering by classical size interstellar grains ($\sim 0.1\,\mu$m) at visual wavelengths ($\lambda \approx 0.5\,\mu$m), placed the models in the convenient centimeter size-range depending on the interstellar grain model and the visual wavelength of study. Since scattering and other electro-magnetic interactions depend on the refractive index as well as on the particle geometry and size relative to the wavelength, care must be taken so that the refractive index of the scale model at the laboratory wavelength is representative of the material to be studied at the simulated wavelength. For example common acrylic plastics have refractive indices with real parts near 1.6 and imaginary parts in the 0.01 to 0.001 range throughout the microwave range of frequencies which makes it a convenient, easily machinable, analog to silicate materials at visual wavelengths.

Interplanetary dust, which range into larger dimensions than the classical interstellar grains, scale to meter size models at these wavelengths. The modern microwave analogue to light scattering facility at the University of Florida (Gustafson 1996a, 2000) operates in the w-band (λ= 2.7 to 4 mm) which brings the models back to the decimeter scale or less. The facility covers the size range from near Rayleigh scattering to $\sim$70 wavelengths across where the upper limit is set by the beam dimensions. And it uses the full bandwidth to cover a wavelength interval comparable to the visual range in width. A now defunct facility at the Ruhr University that produced much scattering data in support of astrophysics operated at an intermediate wavelength near 8 mm in the ka-band (Zerull 1985; Zerull et al. 1993).

The use of millimeter or centimeter wavelengths in lieu of optical 0.4 – 0.7 μm light allows the simulation of light scattering by micron-sized grains with conveniently sized model particles. Also, we may take advantage of commercially available laboratory components developed for radar and telecommunication. These are supplemented with custom components in the University of Florida facility including antennae to achieve a controlled and flat wavefront. These advantages over the direct method allows:

- Cancellation of the direct illumination of the receiver antenna at any scattering angle including small angles so that forward scattering quantities can be measured
- Extraction of phase shift in addition to intensities from measured quantities so that **all** elements of the scattering **F**-matrix are obtained, and therefore all parameters of the scattering process
- Sufficient amplitude and phase stability to allow unattended operation for weeks at the time
- Scattering measurements for single and interacting particles of precisely controlled and specified size, shape, and orientation. This allows systematic investigations and is ideal for theory tests
- Repeatability of specific measurements where specific particle models can be retained and subject to control measurements and additional scattering investigations at any time
- Geometry dependent color effects can be separated from material dependencies since many materials are known to have a wavelength independent refractive index across the w-microwave band (e.g., Gustafson 2000)

- Near infinite angular resolution is possible using custom antennae. This is achieved through a highly controlled and parallel wavefront near the target (Gustafson 2000).

Practically all refractive indices found in the optical range can be reproduced using an analogue material (Gustafson 1996a, 2000). Expanded plastics can be used to conveniently represent ices and other difficult to control substances.

The University of Florida facility has an angular coverage from the forward direction to near backscattering, $\theta = 0°$ to $165°$, leaving the $165°$ to $180°$ range without coverage. This angular interval is of great significance in astronomy and is the range of radar studies. The limitation could be greatly reduced or even eliminated. Size coverage is limited by sensitivity and is therefore refractive index dependent at the small particle end. It extends from $\sim 1\lambda$ across for most materials which is near Rayleigh scattering to $\sim 70\lambda$ across near the Geometric Optics regime set by the beam width. The automated facility and its performance have been described in detail elsewhere (Gustafson 1996a, 2000).

V. RESULTS

V.A. Theory Based Studies

Surprisingly, although evidence now exists that many cosmic grains are irregular in shape and fluffy or likely aggregated in structure, Mie-theory for homogeneous spheres remains a popular tool in the interpretation of zodiacal (e.g., Berriman et al. 1994), solar corona (e.g., Macqueen and Greeley 1995), and cometary (e.g., Lisse et al. 1998) observational data. This may be since it is often possible to fit the scattering by complex particles using scattering theory for any particle geometry as long as the observations refer to broad averages (e.g., Fig. 1) or as long as the particle parameters are not independently constrained. Oscillations in the wavelength and angular dependencies of intensity and polarization that are seldom observed in natural systems but is predicted by Mie-theory as specific resonances are stimulated, fade when averaged across a waveband and can be completely removed if a particle size distribution is assumed, thereby removing a key indicator of spherical geometry.

The discrepancy in shape and structure between real and model particles does not necessarily mean that the idealized spherical model is always irrelevant. However, interpretation requires considerable care. For example, Mie-calculations fit very well the angular dependence of the forward scattering peak by more complex particles, zone 1, in the article by Zerull et al. (1993), see their Figure 14. However the fit usually deteriorates rapidly beyond $\approx 90°/(x+1)$, the start of zone II that is the most sensitive to particle parameters. Beyond $\approx 360°/(x+1)$ is zone III where multiple or dependent scattering is the strongest and the possibility to fit Mie-scattering also is reasonably good. Oscillations in the scattering intensity versus scattering angle decrease toward these, the highest scattering angles. Even when it is possible to find a composition and size distribution of spherical particles that fit observational data across a specific range of scattering angles or across a wavelength interval, the same set of parameters usually do not fit observational data at other wavelengths or at other scattering angles (e.g., Sen et al. 1991). This is illustrated by Mukai et al. (1987) who show that an unrealistic composition of comet Halley dust is found from a fit to Mie-theory. They computed polarization for spheres using a wide range of input parameters and concluded that the only refractive indices fitting the comet dust polarization were close to those of slightly dirty ice. However, spheres do not provide a unique fit and this is not the cometary material in the coma near the sun.

Thus, Mie theory cannot be recommended for quantitative solutions to the inverse problem for cosmic dust, i.e., for the estimate of physical dust properties based on observed light-scattering properties. Mie-theory can however be a useful tool for qualitative comparative analysis of sets of polydisperse particles, e.g., dust in different regions of cometary coma or in different comets (Hoban et al. 1989; Kolokolova and Jockers 1998). It also can be used in studies of qualitative regularities, such as correlations between color and polarization (Kolokolova et al. 1997). See also the examples by Eaton (1984).

The presence of a strong brightness increase in backscattering and of negative polarization at large scattering angles found in interplanetary and cometary observations make the light-scattering properties similar to those of atmosphereless planets, satellites and asteroids. This fact led to models where dust grains were believed to be spheres. But even small deviations from a perfect spherical shape drastically reduce the backscattering. Diffraction from the particle as a whole, described using Mie-theory was combined with geometric-optics concepts of multiple light-scattering on rough surfaces (Perrin and Lamy 1985; Mukai et al. 1982). These computations were able to reproduce the observed intensity and polarization of scattered light using a set of free, and partially empirical, input parameters. The true significance of the fit however, is not obvious.

As we have seen in Fig. 1, the absorption efficiency is less sensitive to the details of the morphology than is the scattering. This hints at the legitimacy of the widespread application of Mie-theory to extract dust temperatures from infrared observations and to represent emission efficiencies. The dust material is usually simulated as a mixture of several substances (silicates, carbonaceous compounds, ice) and empty space using usually the Maxwell Garnett or Bruggeman rules. The first successful attempt based on this approach was probably that by Hage and Greenberg (1990). They applied the integral equation formulation on porous structures, that they also use to represent aggregates and showed that the extinction and absorption cross sections, which depend directly on the internal field, can be rather well approximated using effective medium theories. This is sometimes but not always true of the angular scattering (see Bohren 1986; Chýlek et al. 1988; Comberg and Wriedt 1998). Hage and Greenberg could now reduce the computing demand using the Maxwell Garnett mixing rule in combination with Mie-theory. However, this approximation set a low limit on the largest dimensions of particles in an aggregate relative to the wavelength. Using this approximation, Hage and Greenberg (1990) covered aggregates of up to 8000 absorbing (complex refractive index $m = 1.33(1.7) + i0.5(0.25)$) constituent spheres of size parameter $x = 0.19$. The main result was the dependence of extinction and absorption cross section on porosity (or on the packing factor = 1 - porosity) in an aggregate. Especially, the shape of the $10\,\mu$m silicate absorption feature was shown to depend strongly on the porosity.

Doubts are sometimes expressed about the ability to represent irregular core-mantle particles using effective-medium theory in conjunction with Mie-theory. To check the validity we made a set of calculations related to the interstellar dust model by Greenberg and Hage (1990). The interstellar model may be relevant to comets where interstellar grain derivatives are released. We use the effective medium theories by Maxwell Garnett and by Bruggeman and applied these to various particle geometries. For spheres we use Mie calculations and compare to exact boundary condition solutions for core-mantle spheres (based on Toon and Ackerman 1981) and for spheroids (based on Farafonov et al. 1996). In addition, we present T-matrix calculations using the code by Mishchenko and Travis (1998). The calculations show (Fig. 3) that the Maxwell Garnett theory provides the best fit to exact core-mantle solutions as long as the cores are taken to be inclusions while the Bruggeman theory works only slightly less well. However, the Maxwell Garnett solution taking the mantle to be the inclusion is not a good approximation even when the volume of the mantle is very small in comparison with the volume of the core. Prolate spheroids usually have slightly larger absorption cross sections but they reproduce the same shape of the silicate feature as the spherical particles. It thus appears that the particle geometry does not influence the shape of the silicate feature appreciably while it affects the strength of the feature per unit volume (or mass) of grain material much more. This is true for particles with size parameter less than one while the shape dependencies and differences develop between exact core-mantle and effective medium calculations at larger sizes. The deviations increase with increasing particle size, even though the Bruggeman and the Maxwell Garnett solution with the core as inclusion still produce similar results.

It thus appears that the method based on the combination of Maxwell Garnett theory and Mie-calculations is reliable for particles in the Rayleigh region and slightly larger so that it can be used for interpretation of infrared data using the grain model of Greenberg and Hage (1990). Li and Greenberg (1998) used the method to simulate the $3-20\,\mu$m emission from comet Hale-Bopp. They showed that several reasonable sets of cometary dust properties could fit the spectrum obtained by Williams et al. (1997). For example, a good fit was obtained for particles with porosity of 0.975 when using a power 1.375 size distribution with maximum grain mass 10^{-10} kg and an organic to silicate mass ratio of 0.5.

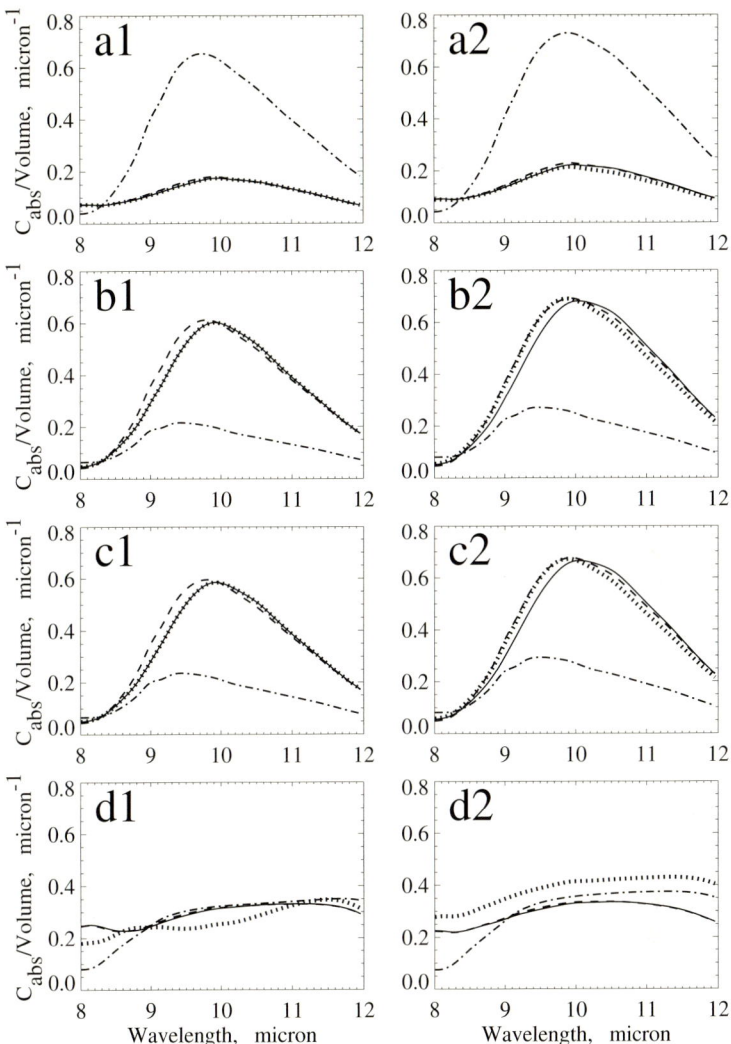

Figure 3. The 10-micron silicate feature calculated for interstellar grains of physical parameters comparable to those given by Greenberg and Hage (1990): 1:3 prolate spheroids (set 2, right panel) and equal-volume spheres (set 1, left panel) made of silicate core and organic mantle (refractive indices from Li and Greenberg 1998). Panels a are for the minor axis of the core equal to 0.2 μm and 0.33 μm for the mantle (as in Greenberg and Hage 1990). Set b of panels show the case of a larger core with the same outer mantle dimensions (all dimensions of the core are 1.5 times larger than in panels a). Panels c show the same volume ratio of the core and mantle as in panels a but for particles of linear dimensions reduced to 1/3 of the classical interstellar grains. Panels d are for 15 times larger linear dimension particles. The dotted line shows the exact core-mantle solutions. The solid line gives the Maxwell Garnett solution with the core taken to be the inclusion while the dash-dotted line is for the mantle as the inclusion. Dashed lines are for the Bruggeman theory.

Mukai et al. (1992) used this approach to estimate radiation pressure on porous grains which requires the constituent particles to be much smaller than interstellar grains. Okamoto et al. (1994) used the same method to study efficiencies of cometary grains in the infrared. This approach works better the smaller the size parameter of the particles themselves as well as the smaller the inclusions, and the closer the refractive index is to unity. More generally, the approximation works better the closer the ratio to the ambient value of m is to unity. This is since these conditions approach the static limit where effective medium theories are strictly valid and since Rayleigh-Gans and eventually Rayleigh scattering is also approached so that both the inhomogeneties and the shape become less important. For example, Wolff et al. (1994) showed that the method is successful on Rayleigh-sized inclusions but show significant deviations for larger inhomogeneities.

Related to this is that the observed infrared spectral energy distribution is often used to obtain the color-temperature of the grains. Equating the grain temperature with the color temperature, the ratio of the grain absorption and emission cross sections is obtained by way of the energy equilibrium equation (see e.g., the chapter by Sekanina et al. or that by Mukai et al.). The cross sections are averages over the whole spectrum dominated respectively by the visual part of the spectrum where the absorption of solar radiation peaks and by infrared wavelengths where the particle energy emission peaks. The emission efficiency at a given wavelength is usually denoted by the same symbol as the absorption efficiency $Q_{abs}(\lambda)$ at the same wavelength since these are numerically identical. This is a consequence of Kirchhoff's law and can also easily be realized from time reversal symmetry arguments. Mie-theory is used to calculate $Q_{abs}(\lambda)$ and thereby the size and the refractive index of the dust material that produces the efficiency ratios corresponding to the temperature determination. The method is likely to work well as long as major emission features are not present. Especially the 10 and 20 μm emission features can be very strong depending on the particle size and porosity (e.g., Figure 9 by Gustafson 1994), and as their relative strength varies, they can strongly perturb the temperature determination.

Increasing computer power in the 1990s made the Coupled Dipole Approximation usable for modeling cosmic dust grains. Perrin and Sivan (1991) simulated varying levels of porosity by removing dipoles representing matter from the surface and interior of an initially homogeneous body. They showed that for an initially Rayleigh sphere, $x = 0.286$, as well as for larger, $x = 1.43$ but transparent (obsidian) spheres, porosity does not affect the angular dependence of polarization $P(\theta)$ or the shape of angular intensity distribution $I(\theta)$. However, the light-scattering properties of large ($x = 1.43$ and $x = 1.96$) absorbing (glassy carbon) spheres approach the Rayleigh polarization $P(\theta) = (1 - cos^2\theta)/(1 + cos^2\theta)$ with increasing porosity.

West (1991), using the code by Draine (1988), may have carried out the first systematic theoretical study of light scattering by aggregates. He considered two kinds of aggregates of spheres: rather low packing assemblies, consisting of

170 constituent particles (CP) of size parameter in the range [0.19 - 0.57] and a compact aggregate, consisting of eight CP of size parameter in the range [0.6 - 1.8]. He reported scattering angle dependencies of intensity and polarization for a silicate-like material of refractive index $m = 1.7 + i0.029$ averaged over six orientations. The intensity of light scattered by an aggregate in the forward domain approaches the diffraction by a sphere of equal cross section to the projected-area, whereas the dependence of polarization on the scattering angle mimics the dependence for the constituent particles. The last conclusion is the strongest for loosely packed aggregates (low packing factor): all computed $P(\theta)$-values are close to the polarization from Rayleigh particles.

Kozasa et al. (1993) conducted an independent study of angular scattering by aggregates, also based on Draine's (1988) code. These authors used a 3-dimensional Monte-Carlo simulation for the construction of large, very porous (BCCA) and relatively compact (BPCA) structures of silicate and/or magnetite. However, they could compute the scattering from small Rayleigh-like constituent particles ($x = 0.1$) only. Although the authors considered a much larger number N of constituent particles [256, 4096], they obtained a similar result as West (1991), namely, $P(\theta)$ for all aggregates was found to be of the Rayleigh-type. Only very absorbing compact aggregates of $N = 4096$ showed a slightly lower polarization maximum, shifted to larger scattering angles. The forward-scattering intensity peak increases with the number of constituent particles and is more pronounced for compact, BPCA aggregates, than for the BCCA aggregates.

Using improved efficiency algorithms by Draine and Flatau (1994) and by Lumme and Rahola (1994), Lumme et al. (1997) investigated rather large (N up to 200) aggregates of non-Rayleigh ($x = 1.2$ to 1.9) constituent particles. They reported all 16 elements of the scattering matrix averaged over approximately 500 orientations of the aggregate. The main purpose was the validation of empirical formulas and of some simplified theoretical methods to calculate light-scattering, not a systematic study of the scattering by aggregates. In particular, they found that scattering in the forward domain could be calculated using the first-order scattering approximation (also known as coherent scattering) with the same accuracy as using the DDA. Thus, interaction between constituent particles is not manifest at small scattering angles. One can see from this paper that $P(\theta)$ differs more from that for individual CP the larger the size parameter and the refractive index of the CPs are, as long as the CP are outside the Rayleigh region.

Xing and Hanner (1997) undertook a consistent study of light-scattering properties of aggregates. Using the Draine and Flatau (1994) version of the DDA, they calculated cross sections, intensity and polarization for aggregates of four and ten spherical and tetrahedral CPs of size parameter $x > 1$. They calculated $I(\theta)$ and $P(\theta)$ for spherical and tetrahedral CPs of size parameter $x = 2.64$ (in some cases $x = 5.24$) and for the refractive index $m = 1.88 + i0.71$, corresponding to a glassy carbon (in some cases $m = 1.65 + i0.01$, corresponding more closely to silicate) and for four types of aggregate structure that they call "totally separated", "separated", "touching", and "overlapping". All their aggregate structures are described in terms of the size of an equivalent-volume sphere. The results for the forward-scattering intensities confirm results in earlier papers that scattering at small angles can be represented using an equal-cross-section sphere and that it does not depend significantly on the properties of the CPs. By contrast, the size and refractive index of the CPs determine most of the polarization. The non-spherical shape of a CP, the larger number of CPs per aggregate and "touching" aggregate structures generate scattering effects of higher order. When averaged over orientation, these fill in resonant oscillations that are typical of particles in the size range $1 < x < 10$.

The plots of cross sections per unit volume versus wavelength were obtained in the wavelength range [0.1-100] μm for aggregates made of 10 spherical or tetrahedral glassy-carbon CP. For a single sphere and a single tetrahedron the cross sections are rather different, but are more alike when aggregated. This is independent of their packing density within the investigated range. For aggregates, the change of the cross sections with wavelength tends to have less steep slopes towards the longer wavelength than a single CP, and typically $C_{abs} \sim 1/\lambda$ for $\lambda = [10 - 100]\,\mu m$.

Yanamandra-Fisher and Hanner (1999) used the DDA to study optical properties of solid and porous particles of varying shapes (sphere, brick, cube, cylinder, hexagon and tetrahedron) and composition (silicate, carbon, and their mixture). The effect of packing which, in the same way as in the study by Hage and Greenberg (1990), simulates aggregates, was studied for tetrahedral and spherical shapes of the size parameter $x = 2.5$. Increasing porosity enhances the forward-scattering peak and depresses side and back-scattering (compare Kozasa et al. 1993). Influence of porosity on polarization depends on the shape of the particle. For example, decreasing the packing in tetrahedron shaped structures makes $P(\theta)$ more Rayleigh-like. For a given volume of material, the same porosity in the shape of a sphere does not give Rayleigh-like polarization. The reason is that for this an even higher porosity (lower packing) is required than investigated by Yanamandra-Fisher and Hanner. Our interpretation is that Rayleigh-like polarization occurs whenever scattering volumes that are small compared to the wavelength do not interact significantly, i.e., when the first-order scattering approximation or coherent scattering applies. Since the average distance between volume elements is at a minimum when the scattering body is spherical, this shape requires a lower packing than any other geometry before the polarization becomes Rayleigh-like. Thus, a porous structure or low packing aggregate with small CPs produces a non-Rayleigh intensity distribution with Rayleigh-like polarization.

The Coupled Dipole Approximation is so far the most versatile and the most often used theoretical tool to study light-scattering properties of complex shapes and of aggregates. It becomes increasingly attractive as computer power grows.

V.B. Experiment Based Studies

Weiss-Wrana (1983) carried out an extensive and consistent study of the angular dependency of intensity and polarization of laser light ($\lambda=0.633\,\mu$m) scattered by single particles encompassing a broad range of terrestrial minerals (including crystalline and amorphous, black, colored and transparent materials) and fragments from meteoritic particles. The study showed that light-scattering properties of real particles could not be reproduced using Mie theory and she concluded that light-scattering properties of dark opaque particles with very rough surface or with fluffy structure resemble the characteristic features of the empirical scattering function derived from measurements of the zodiacal light.

More recently, Worms et al. (1999) compared scattering by levitated particles in microgravity conditions during aircraft flights with the same sample packed by gravity. For now, only the results of calibrations and preliminary results of their study have been reported. The only reported effect of levitation is an increase in the maximum polarization. The reason of this observation is not yet clear. It can be, for example, a result of particle sorting by area to mass ratio under gravity, i.e., by size.

Combet et al. (2001) used a continuously flowing gas jet loaded with dust to obtain light-scattering properties from 16 samples of terrestrial particulates with an additional two control and calibration samples. They systematically studied deviations of their results from Mie-theory. Using separate lasers, the measurements were repeated at a second wavelength so that color effects in both intensity and polarization of scattered light were obtained although only color was reported. These laboratory data can be used as a reference in the interpretation of color and polarimetric color of cometary dust and in the zodiacal light.

The microwave analog technique that can be used with very complicated particles of controlled size, shape, structure and composition manufactured to simulate natural complex particles including aggregates, has produced more usable results. The microwave analog technique was used to study light-scattering properties of "fluffy" dust models as far back as the 1970s. Giese et al. (1978) summarized the results of a study of scattering by particles made to represent some of the first interplanetary dust particles collected in the upper stratosphere. Unfortunately, the model particles did not have completely controlled physical characteristics. Usually, only the overall size parameter of the aggregate ($x \approx 15-30$) and the approximate refractive index of the material were given, whereas the size and shape of CPs were unspecified and varied within an aggregate. The packing factor was also unspecified with the aggregates broadly characterized as "loose", "fluffy", or "compact". Photographs show that the aggregates, apparently, consisted of from a single to some hundreds of CP.

However, even these first experiments showed the main fact, already mentioned, namely; (1) intensity in the forward-scattering domain does not depend on the properties of constituent particles, it depends only on the size of the aggregate itself; (2) the values of polarization and the shape of $P(\theta)$ depend very much on the composition (refractive index) of CPs but not much on the packing factor of the aggregate. We recall that these results are for aggregates that are about ten wavelengths across or less.

The first microwave measurements with completely controlled aggregates were reported by Gustafson (1980) and by Greenberg and Gustafson (1981). Data was obtained using the facility built by Giese et al. in addition to that built by Greenberg. Greenberg and Gustafson made aggregates of acrylic ($m = 1.6 + i0.03$ at microwave frequencies) cylindrical CPs of size parameter 0.47 (in the Greenberg facility using $\lambda = 3.18\,\text{cm}$) and 1.88 (in the Giese facility, $\lambda = 0.8\,\text{cm}$). Here the size parameter is defined as the cylinder circumference to wavelength ratio, in analogy to the size parameter for spheres. The centimeter size acrylic models are the equivalent of micron size-range silicate grains at optical wavelengths and are made to represent the cores of classical interstellar grains. Two hundred fifty identical 2:1 aspect ratio prolate cylinders were randomly positioned with random orientation within an aggregate at a packing factor near 0.1 to simulate the expected aggregation process thereby forming the "BN" structure by Greenberg and Gustafson (1981). For comparison, 125 prolate cylinders of elongation 4:1 and 500 1:1 elongation cylinders were also used to form aggregates while maintaining a constant amount of scattering material and packing in all three BN-aggregates. Besides, systematically investigating the effect of particle elongation, they also made comparisons to the scattering by aggregates in which cylinders were aligned with their axis of symmetry in parallel. The investigation also extended to aggregated core-mantle cylinders of size $x = 0.78$ and $x = 0.6$ cores (both at the Greenberg facility) simulating silicate ($m = 1.6 + i0.03$). These were covered by a mantle of simulated water ice ($m = 1.3$) so that the overall size of a core-mantle CP was $x = 2.45$ and $x = 1.98$, respectively. The models were based on an evolutionary model of dust growth in space. The main findings are:

- Polarization and intensity per unit volume of CP of $x = 0.47$ size do not depend on the elongation of the cylinders and show Rayleigh-like dependencies on the scattering angle (cf., West 1991, Lumme et al. 1997).
- The measurements show a dramatic change of $P(\theta)$ as CPs exceed $x = 1$. The smooth change of polarization with scattering angle that is typical for aggregates of constituent particles of $x < 1$, now oscillates around zero for aggregates of CP of $x = 1.88$ (cf., Xing and Hanner 1997).
- Core-mantle cylinder aggregates exhibited smaller values of polarization and shifted maximum (cf., Kozasa et al. 1993). This behavior is intermediary to the two cases above; it thus appears that the size is a major factor in decreasing the polarization. What the effect of the core-mantle structure of CP is, if any, is unclear.

Thus, all main optical properties of aggregates now known from theoretical calculations were obtained as much as 20 years ago from microwave modeling. Moreover, they were studied for non-spherical particles, which still present almost insurmountable difficulties in numerical computations.

While both the facility originally built by Greenberg and that built by Giese et al. are now defunct, microwave measurements have progressed as well as has theory. Ironically, the closing of the Greenberg and Giese microwave laboratories was preceded by massive improvements of both facilities. In a final collaboration between the facilities, Gustafson expanded his experimental investigation of BN-type structures in what became the only study in the significantly improved Giese et al. facility renovated by Zerull. Besides the quality of measurements, the main improvement was in the models built by Gustafson. Larger overall size aggregates this time were assembled from sets of two to three spherical particles representing silicate ($m = 1.735 + i0.007$) cores of classical interstellar grains and measurements were made with and without $m = 1.86 + i0.12$ mantles. Mantles this time represented an organic refractory residue left behind after water ice sublimation rather than the water-ice mantles investigated by Greenberg and Gustafson (1981). Zerull et al. (1993) reported dependencies of intensity and polarization on scattering angle for six aggregates of size parameter $x = 0.58$ and 0.73 CPs and from 10 mantled aggregate models. The mantles were obtained by repeatedly dipping the six original models in a carbon-based paint and hollow microsphere mixture. The main conclusions are:

- The number of constituent particles, at least for the aggregates of 250 to 500 CPs, does not affect the polarization and side-scattering intensity. However, the forward scattering peak is higher for larger number of CP.
- The size of CPs, as long as they remain Rayleigh particles, does not affect either the intensity or the Rayleigh-like polarization of scattering from the aggregates.
- More compact aggregates have smaller values of polarization and higher values of forward-scattering and back-scattering intensity.
- The thicker an absorbing mantle on CPs is, the higher forward-scattering intensity and the lower the back-scattering intensity.
- The polarization of light, scattered by mantled aggregates, decreases in comparison with mantle-free aggregates. However, polarization increases again when the mantle becomes very thick.

The most recent microwave data were obtained using a new generation of microwave facilities designed and built in the Laboratory for Astrophysics at the University of Florida. It is based on the same microwave analogue principle that scales micron-sized particles in the visual to centimeter-sized particles in the microwave region, however this facility works across a waveband from 2.7 to 4 mm. The broadband property allows the simulation of a fairly wide region of the visual from 0.4 to 0.65 μm using a single model (exact limits depend on the adopted scaling factor) and, thus, to study not only angular but also wavelength dependencies of light-scattering, in particular, colors (Figure 4).

The refractive index of most materials does not change with wavelength in the microwave region so that m is constant (Bohren and Huffman 1983;

Gustafson 2000). This allows us to separate the effect of size and structure of our particle models from the spectral change of the refractive index. We use analogue materials whose complex refractive index in the microwave region equals the refractive index of the simulated material in the visual. E.g., polystyrene $m = 1.6 + i0.003$ is an analog for silicate, while nylon $m = 1.74 + i0.005$ represents organic materials expected in some cometary and other cosmic and natural terrestrial dust. Mixtures of expanded plastic, plaster, and iron oxide was used to simulate organic refractory material with $m \approx 1.7 + i0.2$ and $1.8 + i0.1$.

Intensities and polarizations shown in Figure 4 were obtained in the scattering angle range $[0 - 165]°$, and in the wavelength intervals blue [2.7-3] and red [3.5 − 4] mm. These wavelength intervals simulate the cometary continuum filters centered at 0.443 and 0.642 μm (as used by Chernova et al. 1993 and Jockers et al. 1997) after scaling. Table 2 gives the physical parameters: size, shape and number of constituent particles in each aggregate, the refractive index of their material and the structure of aggregates as given by the packing factor, i.e., the ratio of the volume of material to the total volume of the aggregate. All size parameters are calculated for the middle wavelength in the red $\lambda = 3.75$ mm and blue $\lambda = 2.83$ mm wavebands. The size parameter for nonspherical CPs is given in terms of an equal-volume sphere. Figure 5 illustrates some of these particles and particles used to generate the data in Fig. 1. Aggregate 1 has parameters that allow it to represent the aggregated silicate cores of approximately 825 classical interstellar grains. In making Aggregate 1, we assumed that interstellar grains are stripped of their organic mantles as well as any ice mantle to make this aggregate comparable to the second "BN" in Table 1 by Greenberg and Gustafson (1981). Differences are the number of interstellar grain cores (825 versus 240), the representation of individual cores using sets of two spheres as opposed to 2:1 elongation cylinders, and the irregular overall shape of our aggregate as compared to the overall approximately spherical shape of the original BN-aggregates. We note that the irregular shape makes the packing factor ill defined but it is of the same order. The remaining models extend the parameter space considerably and in a systematic way. Our Aggregate 3 has larger CPs of a more absorbing material corresponding to an organic but otherwise resembles Aggregate 1. The effect of considerably tighter packing is investigated using Aggregate 4. Lower packing models are more difficult to produce and we therefore use a loose tangle of string as model 6. In Aggregate 2, cylinders replaced spherical CPs to separate out the effect of the shape of CPs. Aggregate 5 is similar to Aggregate 3 except for the lower number of aggregated particles. Table 2 also lists aggregates 7 and 8 made of larger CPs arranged in the nominal lose packing (~ 0.1) and tightly packed, respectively. Aggregate 9 is a compact aggregate representing silicate interstellar dust cores emedded in organics (Gustafson, 1996b). Aggregate 10 has intermediate size CPs of highly absorbing organic material.

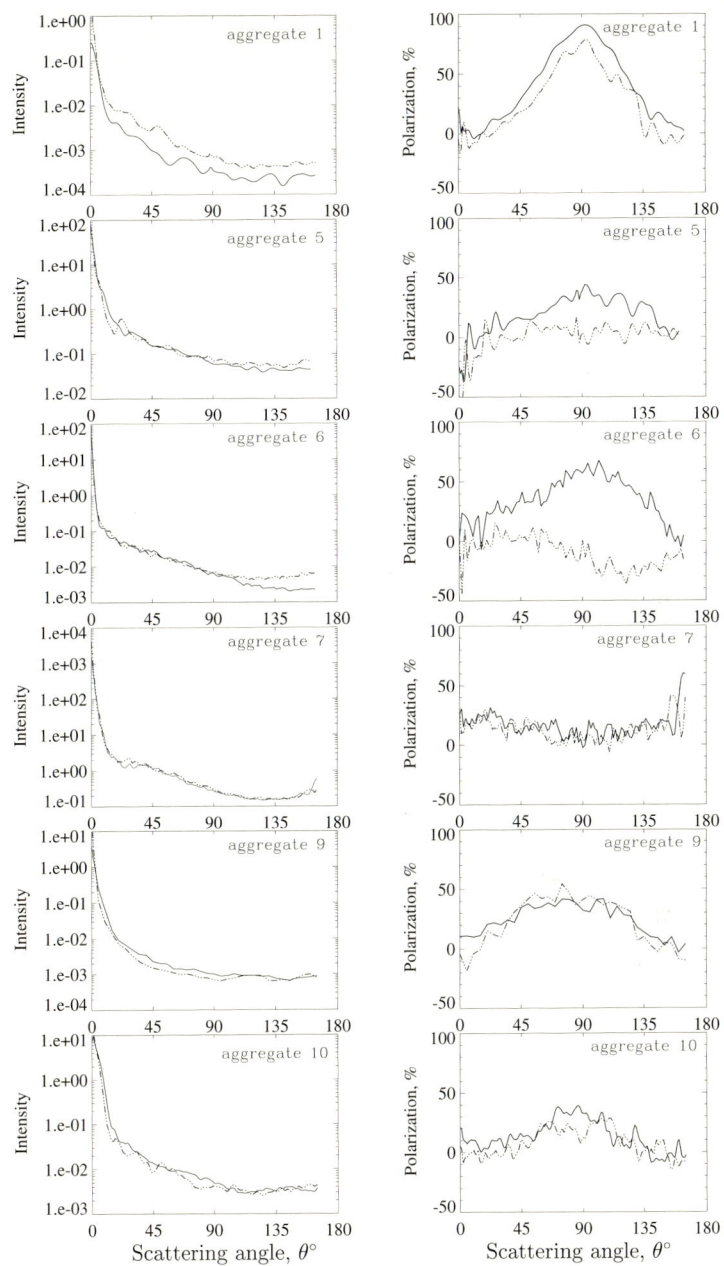

Figure 4. Intensity and polarization in the blue (dashes) and red (solid) wavebands obtained as averages over 12 to 36 orientations. This simulates a cloud of randomly oriented aggregate particles sufficiently well to determine broad trends in average scattering properties, specifically the sign of color and of polarimetric color. Remaining artifacts of insufficient averaging show up in the form of fine structure in the angular distributions.

TABLE 2
Physical Characteristics of Aggregates

	Characteristics of Constituent Particles						Packing Factor (Approximate)
	∅, Diameter, mm	$x = 2\pi r/\lambda$		Shape	Number	Refractive Index	
		Red	Blue				
1	0.65	0.54	0.72	sphere	∼1650	1.615+i0.03	0.1
2	1.25∅ × 2	1.40	1.85	cylinder	∼1500	1.74+i0.005	0.1
3	1.58	1.32	1.75	sphere	∼1650	1.74+i0.005	0.1
4	1.58	1.32	1.75	sphere	∼1600	1.74+i0.005	0.45
5	1.58	1.32	1.75	sphere	∼500	1.74+i0.005	0.1
6	loosely wound ball of string of 1.25∅					1.74+i0.005	<0.05
7	19	15.9	21.1	sphere	43	1.74+i0.005	0.1
8	19	15.9	21.1	sphere	37	1.74+i0.005	0.55
9	1.25∅ × 2.5 embedded in a matrix of organic refractory material	1.51	2.00	cylinder	∼3000	1.74+i0.005 1.8+i0.1	∼1
10	∼4	3.35	4.44	spheroid	88	1.7+i0.2	0.2

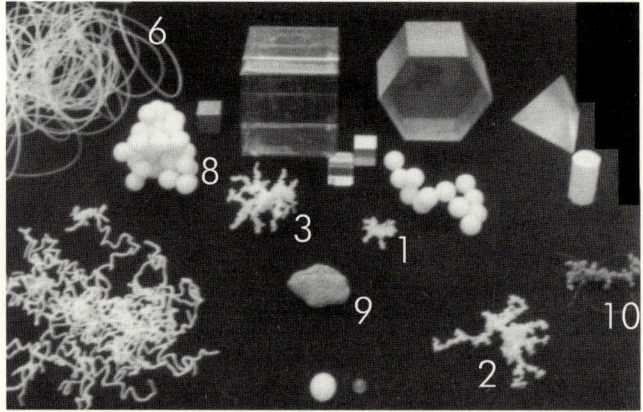

Figure 5. Some targets used in the microwave analogue measurements.

We draw the following conclusions about angular dependencies of intensity and polarization on the physical properties of aggregates based on the combined results cited above:

1. **Intensity in the forward-scattering domain** does not depend directly on the properties of CPs and their packing in an aggregate. Instead, it depends on the size of an aggregate or, more specifically, on the projected area of an aggregate particle. Our interpretation is that forward scattering directions are dominated by Fraunhofer diffraction that is partially canceled by transmitted light interfering (nearly) destructively and thereby decreasing forward scattering.

2. **Intensity at large scattering angles**. Scattering at the largest scattering angles is more sensitive to properties of individual particles in an aggregate, especially to the absorption. However, the microwave data reach

only to 165° and available information is insufficient to draw definite conclusions.

3. **Polarization in the forward-scattering domain.** As expected based on (1), there is no change in polarization with changing properties of aggregates at these scattering angles. This is, in our interpretation, since Fraunhofer diffraction is insensitive to polarization.

4. **Polarization at large scattering angles** is sensitive to the size, packing and imaginary (absorption) index of CPs and becomes negative when the size parameter of CPs exceeds 1. The negative polarization branch is more pronounced for more compact and for more absorbing particles. The negative polarization near backscattering angles that has been observed in the scattering from many astronomical objects including comet dust occurs mostly in the range of angles not covered by the microwave data.

5. **Polarization in the side-scattering domain,** $P(\theta)$, does not depend on the number of CPs in an aggregate as long as $N < 1650$. However, both the shape of $P(\theta)$ and the amount of polarization depend strongly on the size parameter of CPs. They change from a Rayleigh-like curve for $x < 1$ to curves of the same shape but of lower polarization and with the maximum shifted toward larger scattering angles for $x > 1$. Oscillations about small polarization values set in as CPs grow larger. This is especially evident for transparent materials and for loose aggregates. Oscillations become smoother for more compact aggregates and for aggregates consisting of absorbing materials. For very compact and absorbing aggregates $P(\theta)$ approaches a shape that is reminiscent of Rayleigh-like polarization but with a lower maximum that is also shifted to smaller scattering angles.

Recent data (Gustafson and Kolokolova 1999) also allows conclusions concerning the wavelength dependencies of intensity and polarization of scattered light on an aggregate's physical parameters:

1. Color and polarimetric color depend mostly on the size and composition of the constituent particles, specifically;

    an increase in the size of constituent particles results in:
    - larger (more red) color,
    - smaller (more blue) polarimetric color,

    an increase in the imaginary part of the refractive index results in:
    - larger (more red) color,
    - smaller (more blue) polarimetric color.

2. The spectral properties are not severely dependent on the number of constituent particles, N, at least for $N < 1500$.

3. Increasing compactness of an aggregate makes the color larger, and the polarimetric color smaller.

4. The shape of constituent particles could not be seen to influence the aggregates' color. However, it slightly changes the position and the amount of maximum polarization. It, apparently, also affects the polarimetric color.

We formulate the following set of provisional rules for the interpretation of remote sensing data for aggregate dust that were also given by Gustafson and Kolokolova (1999):

1. High values of polarization at $\theta \approx 90°$ indicate that the constituent particles are small ($x < 1$).
2. A weak wavelength dependence of the polarization and of the angular dependence of polarization, $P(\theta)$, indicates the presence of either small ($x \ll 1$) or large ($x \gg 1$) constituent particles. We found that strong dependencies occur at CP sizes comparable to the wavelength.
3. A combination of red polarimetric color ($P_{red} - P_{blue} > 0$) and blue color ($I_{red} - I_{blue} < 0$) is typical of aggregates of non-absorbing particles that are comparable and larger in size than the wavelength ($0.5 < x < 10$).
4. Red color and red polarimetric color are indicative of dark, absorbing aggregates.

V.C. Radiation Pressure

The momentum carried by the absorbed photons ($=$ energy$/c$) transfers to the dust particle alongside the energy heating it. In addition, there is a reaction force due to the change in momentum experienced by the scattered light. Following common practice, we quantify the momentum transferred to the particle as that equal to the momentum carried by light incident on the surface C_{pr} in analogy to the cross sections for absorption and scattering. The component in the direction of the incident radiation is

$$C_{pr} = C_{abs} + \frac{1}{k^2} \iint (1 - \cos(\theta)) F(\theta, \phi) \sin(\theta) \mathrm{d}\theta \mathrm{d}\phi \tag{16}$$

with the perpendicular components

$$-\frac{1}{k^2} \iint \sin^2(\theta) \cos(\phi) F(\theta, \phi) \mathrm{d}\theta \mathrm{d}\phi, \tag{17a}$$

and

$$-\frac{1}{k^2} \iint \sin^2(\theta) \sin(\phi) F(\theta, \phi) \mathrm{d}\theta \mathrm{d}\phi. \tag{17b}$$

These components need to be averaged over orientation and over the incident spectrum. While the non-radial components average out for a cloud of particles in random orientation, individual grains are expected to spin so that a net non-radial component results. Gustafson (1994) argues that this is most likely to oppose the effect known as PR-drag associated with radiation pressure. However, the extent of this effect is unknown so, for simplicity, we consider averages over uniform distribution of orientation. This average results in a radiation pressure force that varies with the heliocentric distance r as the flux density of sunlight, which like gravity has an inverse-square dependence. It is therefore customary to write the radiation pressure force in units of the local solar gravity and thus make it independent of heliocentric distance;

$$\beta = \frac{(\text{Radiation flux}/r^2)\frac{C_{pr}}{\text{Grain mass}}}{GM/r^2} = C_r \frac{C_{pr}}{\text{Grain mass}}. \qquad (18)$$

The values of C_r and C_{pr} and therefore β depend on the mass (M), luminosity, and spectrum of the star. Here, G is the gravitational constant. For the Sun we have $C_r = 7.6 \cdot 10^{-4}$ kg m^{-2}.

The solid curve extending across the mass range in Figure 6 shows the β values for the "Astronomical Silicate" spheres used in Fig. 1 with assumed bulk density of 2500 kg m^{-3}. The curves covering the 10^{-19} to 10^{-15} kg range (calculated using the T-matrix formulation code by Mishchenko and Travis 1998) show that 2 : 1 aspect ratio oblate spheroids (dotted curve) as well as cylinders (dashes) of "Astronomical Silicate" in random orientations have almost the same β-values as spheres of the same mass. The same example particles as in Figures 1a and 1b (same symbols) are used to show the likely range for BN-structures. Data for porous and compact aggregates and for cubes show that any shape and structure BN-material particle of the types investigated have β-values close to that of an "Astronomical Silicate" sphere of the same mass. This is a long sought after conclusion that greatly simplifies the calculation of interplanetary dust dynamics for this specific type of particles. However, the relative insensitivity to particle morphology found for the "Bird's-Nest"-type structures does not apply to all particle types and in all environments. See Gustafson et al. (2001) for a discussion. Even for "B-N" structures, the result needs to be checked using averages over the spectrum of the central star before it is applied to exosolar dust grains.

The solar radiation pressure force peaks near 10^{-16} kg where it is comparable to the solar gravity. At β of unity, marked by the dashed line, no net radial force is experienced by the grains and they can to first approximation float freely around the solar system. All particles above this line are repelled when exposed to sunlight and cannot remain in bound orbit. All grains between the dashed lines, $0.5 < \beta < 1$, are still not on bound trajectories if they are released from a parent body in circular orbit. This is since they "inherit" kinetic energy from the parent body (e.g., Gustafson 1994). Grains separating from a parent around perihelion, as many small comet grains are expected to, will be ejected even when their β-values are below 0.5 and loosely bound dust are easily ejected by planetary perturbations (Gustafson et al. 1987).

Significantly higher β-values reaching the 2 to 3 range corresponding to repulsion were obtained in microwave measurements carried out in the Bochum single frequency facility and reported by Gustafson (1994). Such values are possible for some natural materials. However, the data obtained at the Florida wideband microwave laboratory indicate that they probably are not reached by BN-structures. The difference between the Bochum and Florida results are due to averaging. Because the Bochum facility operated at a single frequency, the

C_{pr} could not be integrated over the solar spectrum. The value near 0.55 μm where the solar luminosity peaks was assumed to be representative and this value was used in place of an average. Using the University of Florida facility that operates across a waveband corresponding to 0.44 μm to 0.66 μm, this was found to be a poor approximation. Here, eighty-five discrete frequencies were used to generate the average.

In conclusion, the mass of a grain of BN-type material and structure has a larger effect on β and therefore the dynamics of the dust grain than other grain parameters including the linear dimensions or "size". "Astronomical Silicate" spheres are therefore likely to yield a good approximation to the true β-values of most interplanetary dust grains. We emphasize that this result is for materials expected from the BN-model. The exception is probably any large and highly elongated particles for which Gustafson (1994) used the geometric optics approximation (see Figure 3 in that paper).

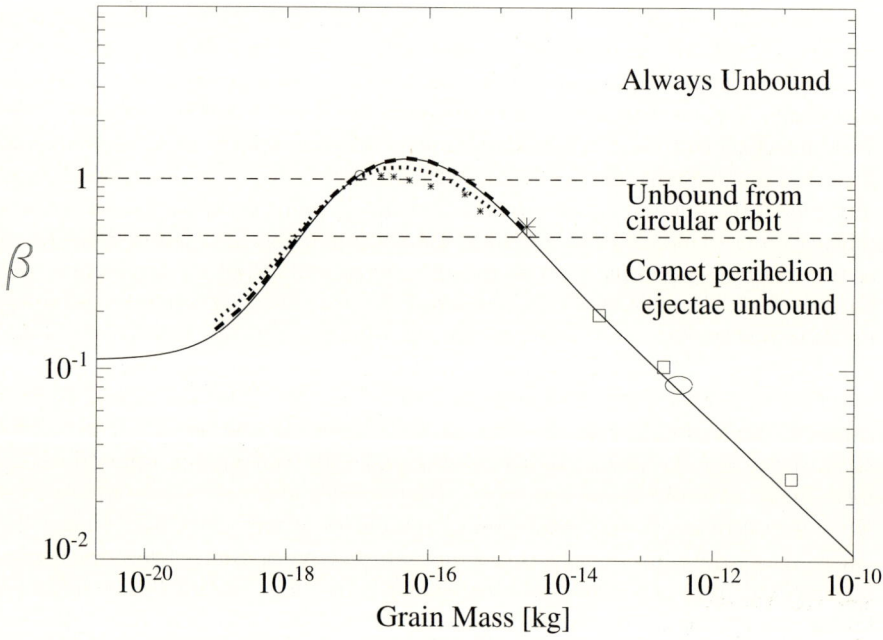

Figure 6. Radiation to gravity force ratio β as a function of the particle mass for "Astronomical Silicate" spheres (solid curve), circle represents a single classical interstellar grain. Results for 2:1 prolate cylinders (dashes) and 2:1 oblate spheroids (dots) were obtained using the code by Mishchenko and Travis (1998). Results for small BN-type aggregates (small asterixes) were calculated using the code by Mackowski and Mishchenko (1996). All other data are from microwave laboratory measurements at the University of Florida. The large asterix is for a BN-type aggregate of several hundred interstellar grains, the ellipse shaped symbol for a compact aggregate representing tightly packed interstellar grains of the BN-composition, and squares denote silicate cubes. All data are for averages over random orientations. We see that the mass is more important in determining β than shape or structure.

VI. CLOSING REMARKS

We reviewed evidence that interplanetary dust includes silicates and an organic refractory material. These materials, especially the organics, evolve under space exposure. The material becomes inhomogeneous as chemical bonds break and new bonds form even if the material was uniform to start with, which is unlikely. The materials are therefore almost certainly inhomogenous and very complex. Since the concept of a refractive index formally applies to perfectly homogeneous materials, it is not à priori obvious that these materials can be represented using a refractive index. This is fundamental since most light scattering theories, based on classic electrodynamics, start from the assumption that all particles are uniquely described by the size and shape (geometry) and refractive index. They make use of the complex refractive index to assure charge conservation inside the particle using Maxwell's equations. So, when do these classical theories apply?

Using effective medium theory we see that some refractive index can be used as long as the scale of inhomogeneities is sufficiently small compared to the wavelength. It does not matter what the inhomogeneities are as long as they are either random or can be described using some regular function. This is in analogy with even the most perfect material, since all matter is inhomogeneous on the atomic distance scale which leads to the well known breakdown of the refractive index representation at x-rays below ~ 0.1 nm used in crystal diffraction studies. The issue now is if complex aggregate structures such as IDPs can be adequately represented using Mie-theory and a single refractive index obtained through one of the several effective medium formulae that have been proposed.

As a systematic framework and to fix the scale of the inhomogeneities, we consider the "BN" model of interplanetary grains. These are aggregates of classic size interstellar grains each with dimensions near $0.1\,\mu$m. We show that the absorption, which is a direct consequence of the internal field, can be rather well described in this way. This is especially true at infrared wavelengths where the scale is more favorable than in the optical regime. Scattering efficiencies that depend on the external field are more difficult to reproduce. It may at first not be fully clear that this can be seen in Figures 1a and 1b, since the refractive indices we used were not referred to as "effective" indices. However, the set of wavelength dependent refractive indices known as the "Astronomical Silicates" by Draine and Lee (1984) are derived from astronomical observations and from laboratory measurements of presumably inhomogeneous materials. As such, they correspond to effective refractive indices. We find (Figures 1a, 1b and 6) that using these refractive indices with Mie-theory, we reproduce the absorption and radiation force suffered by the much more complex BN-type aggregates. Significantly, the figures also show that the correct interpretation is not a vindication of the "Astronomical Silicate" but rather that these *averaged properties* do not depend strongly on the particle parameters other than the mass. They were reproduced by all models we tested; loose aggregates, a compact aggregate, and silicate cubes. Using specific mixing rules we show how the emission features are more sensitive to the geometry.

Inhomogeneities or any structures such as individual particles in an aggregate that cannot be accounted for using an effective refractive index should be accounted for through formulation of the condition for charge conservation on the boundary. This means that they are treated as individual, although dependent, particles by way of simultaneous solution of the boundary condition on all surfaces. Scattering theories (Xu 1995, 1997) can do this and therefore account for the interactions. We see that it now may become possible to quantify rules for the limitations of effective medium theory and criteria for when a structure must be treated as part of the particle's shape. We suggest that the derivation of such rules, whether theoretical or empirical, should be encouraged. We can anticipate that the rules depend on what specific quantity is of interest.

We summarized the principles on which the most relevant light scattering methods are based and we see that several solutions to the scattering by interacting particles are now becoming available in principle. However, the computational limitations are such that usually only smaller particles than classic size interstellar grains can make up aggregates. Alternatively, the number of aggregated particles must be small, typically less than 100 at optical wavelengths. This leads to only weak interactions between grains when they are loosely packed such as in the BN-model and in many IDPs. Coherent scattering approximations that account for interference at large distances but neglect internal interactions usually apply in these circumstances. Simply stated, numerics limit calculations with sophisticated scattering codes to conditions where simple approximations apply. The number of computer cycles increase rapidly with the number of interacting bodies or with increasing size to wavelength ratio. A major advance in computer performance is therefore needed before the realm of strong interactions can be fully penetrated unless the numerical methods are improved (e.g., Xu 1997). However, interaction is stronger for more tightly packed structures that may also be abundant in interplanetary space or for aggregates with higher refractive index materials that are likely less common in space.

Possibly the most widely used method to solve the scattering problem in astronomy is the discrete dipole approximation. It differs from the boundary condition methods in that it evaluates the (discretized) internal field instead of the boundary conditions. The material is represented by a set of polarizability tensors and there is no reliance on the refractive index to assure charge conservation. This circumvents the problem with the applicability of a refractive index although this fact is seldom taken advantage of in practice. Comparisons to microwave measurements and exact calculations indicate that discretization of the internal field can lead to unexpectedly large errors when it is used to calculate the scattered field (Xu and Gustafson 1999). This is presumably since steep gradients can develop in the internal field as the material becomes strongly inhomogeneous, or equivalently, as particles interact.

That radiation pressure efficiencies as well as absorption are likely to be reasonably well represented using "Astronomical Silicates" with Mie-theory when averaged over orientation and over the solar spectrum does not apply to the smallest particles (mass below 10^{-18} kg) if the imaginary part of their refractive index differs significantly from that of "Astronomical Silicate". The approximation also does not apply to the radiation pressure efficiency of large grains (mass above 10^{-12} kg) if they are highly elongated. The Rayleigh solution and geometric optics apply in these cases, respectively, so that accurate values can easily be calculated using standard and numerically economic techniques even in these cases. The radiation force is proportional to the volume in the Rayleigh region and to the surface area in geometric optics.

The scattering cross section of grains exceeding 10^{-15} kg may be significantly overestimated using "Astronomical Silicate" spheres. All scattering quantities including absorption and radiation pressure efficiencies can also be significantly different from the values for "Astronomical Silicate" spheres when referring to either non-random orientations, to specific wavelengths, or to narrow wavebands. This is especially true in or near emission features that depend strongly on the material properties and packing of the material. It is for example risky to estimate the grain temperature based on brightnesses near the strong silicate bands at 10 and 20 μm where many observations are made. The grain temperature, color, and polarization as well as wavelength dependence of polarization are diagnostic of particle properties and indicate that a large fraction of the comet dust particles are likely to be of the BN-aggregate type.

A comprehensive collection of references and links as well as tables and plots showing refractive indices of a variety of materials, as well as links to light scattering codes can be found at the Jena - St. Petersburg Database of Optical Constants (http://www.astro.spbu.ru/JPDOC).

We have seen that studies of light scattering and other electromagnetic interactions with interplanetary dust have changed dramatically since the book edited by McDonnell (1978). This is largely, but probably not only, because of broad-front and dramatic advances in theory, numerical computations, and in laboratory facilities. We have seen a shift in the models of study guided to a large extent by collected samples, paralleling the evolutionary guiding principles adopted by Greenberg and Gustafson (1981). The internal structure of particles, not only the size, shape, and bulk materials is now at the focus of research. The added set of parameters needed to describe a scattering problem has made the need for a framework to help make our work systematic and focus the effort even greater. The proposed use of the BN-type particles as a framework is intended to focus research efforts by relating the dust models under study to an evolutionary sequence and therefore reduces the number of free parameters. This facilitates the incorporation of knowledge gained in studies of interstellar matter. The next important step is the incorporation of knowledge obtained from dust dynamics. This will help us better understand evolution and therefore the relation between particles in different parts of the zodiacal cloud. It will also help us better take advantage of the ever-changing observing geometries and to better understand observational biases. The challenge is to not let the effort to reduce the number of free parameters unduly restrict our work or to misguide our focus.

Acknowledgements

This work was in part supported by NASA through grants NAG5-6378 and NAG5-8944.

REFERENCES

Aden, A. L., and Kerker, M. 1951. Scattering of electromagnetic waves from two concentric spheres. *J. Appl. Phys.*, **22**, pp. 1242–1246.

Agladze, N. I., Sievers, A. J., Jones, S. A., Burlitch, J. M., and Beckwith, S. V. W. 1996. Laboratory results on millimeter-wave absorption in silicate grain materials at cryogenic temperatures. *Ap. J.*, **462**, pp. 1026–1040.

Allamandola, L. J., Bernstein, M. P., Sandford, S. A., and Walker, R. L. 1999. Evolution of interstellar ices. *Sp. Sci. Rev.*, **90**, pp. 219–232.

Allamandola, L. J., Sandford, S. A., and Valero, G. J. 1988. Photochemical and Thermal Evolution of Interstellar/Pre-Cometary Ice Analogs. *Icarus*, **76**, pp. 225–252.

Asano, S., and Yamamoto, G. 1975. Light scattering by a spheroidal particle. *Appl. Opt.*, **14**, pp. 29–49.

Bergman, D. J. 1978. The dielectric constant of a composite material – a problem in classical physics. *Phys. Rep.*, **43**, pp. 377–407.

Bernstein, M. P., Allamandola, L. J., and Sandford, S. A. 1997. Complex organics in laboratory simulations of interstellar/cometary ices. *Adv. in Sp. Res.*, **19**, pp. 991–998.

Berriman, G. B., Boggess, N. W., Hauser, M. G., Kelsall, T., Lisse, C. M., Moseley, S. H., Reach, W. T., and Silverberg, R. F. 1994. COBE DIRBE near-infrared polarimetry of the zodiacal light: Initial results. *Ap. J. Let.*, **431**, pp. L63–L66.

Bhandari, R. 1985. Scattering coefficients for a multilayered sphere: Analytic expressions and algorithms. *Appl. Opt.*, **224**, pp. 1960–1967.

Blanco, A., Borghesi, A., Fonti, S., and Orofino, V. 1994. Amorphous carbon and silicon carbide grain mixtures in the envelopes of carbon stars. *Astron. Astrophys.*, **283**, pp. 561–566.

Blanco, A., Borghesi, A., Fonti, S., Orofino, V., Bussoletti, E., and Colangeli, L. 1987. Laboratory amorphous carbon: A possible analog of cometary dust. In *Proceedings of the International Symposium on the Diversity and Similarity of Comets*, eds. J. Rolfe and B. Battrick (Paris: ESA Pub. Div. SP-278), pp. 677–679.

Bohren, C. F. 1986. Applicability of effective-medium theories to problems of scattering and absorption by nonhomogeneous atmospheric particles. *J. of the Atmospheric Sciences*, **43**, pp. 468–475.

Bohren, C. F., and Huffman, D. R. 1983. *Absorption and scattering of light by small particles.* (New York: John Wiley & Sons).

Bottiger, J. R., Fry, E. S., and Thompson, R. C. 1980. Phase matrix measurements for electromagnetic scattering by sphere aggregates. In *Light scattering by irregularly shaped particles*, (New York: Plenum Press), pp. 283–290.

Bradley, J. P. 1999. Interstellar dust - evidence from interplanetary dust particles. In *Formation and Evolution of Solids in Space*, eds. J. M. Greenberg and A. Li (Dordrecht: Kluwer Academic Publishers), pp. 485–503.

Briggs, R., Ertem, G., Ferris, J. P., Greenberg, J. M., McCain, P. J., Mendoza-Gómez, C. X., and Schutte, W. A. 1992. Comet Halley as an aggregate of interstellar dust and further evidence for the photochemical formation of organics in the interstellar medium. In *Origins of Life and Evolution of the Biosphere*, **22**, pp. 287–307.

Britt, D. T., and Consolmagno, G. J. 2000. The porosity of dark meteorites and the structure of low-albedo asteroids. *Icarus*, **146**, pp. 213–219.

Brucato, J. R., Mennella, V., Colangeli, L., and Bussoletti, E. 1999. Temperature dependence of the FIR absorption coefficient for carbon and silicate grains. In *Formation and Evolution of Solids in Space*, eds. J. M. Greenberg and A. Li (Dordrecht: Kluwer Academic Publishers), pp. 291–296.

Bruggeman, D. A. G. 1935. Berechnung verschiedener physikalischer Konstanten von heterogenen Substanzen. *Ann. Phys. (Leipzig)*, **24**, pp. 636–679.

Campins, H., and Ryan, E. V. 1989. The identification of crystalline olivine in cometary silicates. *Ap. J.*, **341**, pp. 1059–1066.

Chernova, G., Kiselev, N., and Jockers, K. 1993. Polarimetric characteristic of dust particles as observed in 13 comets: comparison with asteroids. *Icarus*, **103**, pp. 144–158.

Chiapetta, P. 1980. Multiple scattering approach to light scattering by arbitrarily shaped particles. *J. Phys. A*, **13**, pp. 2101–2108.

Chyba, C., and Sagan, C. 1987. Infrared emission by organic grains in the coma of comet Halley. *Nature*, **330**, pp. 350–353.

Chýlek, P., Srivastava, V., Pinnick, R. G., and Wang, R. T. 1988. Scattering of electromagnetic waves by composite spherical particles: Experiment and effective medium approximations. *Appl. Opt.*, **27**, pp. 2396–2404.

Chýlek, P., Videen, G., Geldart, D., Dobbie, J., and Tso, H. W. 2000. Effective medium approximations for heterogeneous particles. In *Light Scattering by Nonspherical Particles: Theory, Measurements, and Geophysical Applications*, eds. M. I. Mishchenko, J. W. Hovenier and L. D. Travis (New York: Academic Press), pp. 274–308.

Colangeli, L., Mennella, V., Busoletti, E., Merluzzi, P., Rotundi, A., Palumbo, P., and di Marino, C. 1993. Similarities between cometary, meteoritic and laboratory analog dust: hints from the attribution of the 10-micrometer band. *Meteoritics*, **28**, p. 338.

Colangeli, L., Mennella, V., Palumbo, P., and Rotundi, A. 1999. Cosmic dust and laboratory simulation: Wishes, results and open problems. In *Formation and Evolution of Solids in Space*, eds. J. M. Greenberg and A. Li (Dordrecht: Kluwer Academic Publishers), pp. 203–228.

Comberg, U., and Wriedt, T. 1998. Scattering by inhomogeneous particles, exact theories and effective-medium theories. In *Electromagnetic and light scattering - theory and applications III*, eds. T. Wriedt and Y. Eremin, (Bremen: Bremen University Press), pp. 43–50.

Combet, P., Lamy, P. L., and Loesel, J. 2001. Light scattering by complex dust particles. I. Laboratory measurements on ensembles of particles. *Astron. Astrophys.*, submitted.

Cooray, M. F. R., and Ciric, I. R. 1991. Scattering of electromagnetic waves by a coated dielectric spheroid. *J. Electromagnetic Waves and Applications*, **11**, pp. 1491–1507.

Cronin, J. R., Pizzarello, S., and Cruikshank, D. P. 1988. Organic matter in carbonaceous chondrites, planetary satellites, asteroids and comets. In *Meteorites and the early solar system*, (Tucson, AZ: University of Arizona Press), pp. 819–857.

Day, K. L. 1974. A possible identification of the 10-micron "silicate" feature. *Ap. J.*, **192**, pp. L15–L17.

Debye, P. 1909. Der lichtdruck auf kugeln von beliebigem material. *Ann. Phys.*, **30**, pp. 57–136.

Disanti, M. A., Mumma, M. J., Geballe, T. R., and Davies, J. K. 1995. Systematic observations of methanol and other organics in Comet P/Swift-Tuttle: Discovery of new spectral structure at 3.42 micron. *Icarus*, **116**, pp. 1–17.

Dohnanyi, J. W. 1969. Collisional models of asteroids and their debris. *J. Geophys. Res.*, **74**, pp. 2431–2554.

Dominik, C. 1999. Dust coagulation and the structure of dust aggregates in space. In *Formation and Evolution of Solids in Space*, eds. J. M. Greenberg and A. Li (Dordrecht: Kluwer Academic Publishers), pp. 377–387.

Draine, B. T. 1988. The discrete-dipole approximation and its application to interstellar graphite grains. *Ap. J.*, **333**, pp. 848–872.

Draine, B. 2000. The discrete dipole approximation for light scattering by irregular targets. In *Light Scattering by Nonspherical Particles: Theory, Measurements, and Geophysical Applications*, eds. M. I. Mishchenko, J. W. Hovenier and L. D. Travis (New York: Academic Press), pp. 131–145.

Draine, B. T., and Flatau, P. J. 1994. Discrete-dipole approximation of scattering calculations. *J. Opt. Soc. Amer. A*, **11**, pp. 1491–1499.

Draine, B. T., and Flatau, P. J. 1995. DDSCAT.4b.1 FORTRAN program. Available via anonymous ftp from astro.princeton.edu.

Draine, B. T., and Lee, H. M. 1984. Optical properties of interstellar graphite and silicate grains. *Ap. J.*, **285**, pp. 89–108.

Eaton, N. 1984. Comet dust – Application of Mie scattering. *Vistas in Astronomy*, **27**, pp. 111–129.

Farafonov, V. G., Voshchinnikov, N. V., and Somsikov, V. V. 1996. Light scattering by a core-mantle spheroidal particle. *Appl. Opt.*, **35**, pp. 5412–5426.

Flynn, G. J., Moore, L. B., and Klöck, W. 1999. Density and porosity of stone meteorites: implications for the density, porosity, cratering, and collisional disruption of asteroids. *Icarus*, **142**, pp. 97–105.

Garnett, J. C. M. 1904. Colours in metal glasses and in metallic films, *Phil. Trans. R. Soc. Lond.*, **203**, pp. 385–420.

Gazeau, M.-C., Cottin, H., Vuitton, V., Smith, N., and Raulin, F. 2000. Experimental and theoretical photochemistry: application to the cometary environment and Titan's atmosphere. *Planet. Sp. Sci.*, **48**, pp. 437–445.

Giese, R. H., Weiss, K., Zerull, R. H., and Ono, T. 1978. Large fluffy particles: a possible explanation of the optical properties of interplanetary dust. *Astron. Astroph.*, **65**, pp. 265–272.

Gilmour, I., and Pillinger, C. 1992. Isotopic Differences Between PAH Isomers in Murchison. *Meteoritics*, **27**, pp. 224–225.

Grady, M. M., Wright, I. P., Fallick, A. E., and Pillinger, C. T. 1983. The stable isotopic composition of carbon, nitrogen and hydrogen in some Yamato meteorites. In *National Institute of Polar Research, Symposium on Antarctic Meteorites,* Nat. Inst. of Polar Res. Memoirs, Special Issue, no. 30, pp. 292–305.

Greenberg, J. M. 1982. What are comets made of? A model based on interstellar dust. In *Comets*, ed. L. L. Wilkening (Tucson: Univ. of Arizona Press), pp. 131–163.

Greenberg, J. M. 1983. The largest molecules in space – Interstellar dust. In *Cosmochemistry and the origin of life*, (Dordrecht: D. Reidel Publishing Co.), pp. 71–112.

Greenberg, J. M. 1999. Tracking the organic refractory component of interstellar dust. In *Formation and Evolution of Solids in Space*, eds. J. M. Greenberg and A. Li (Dordrecht: Kluwer Academic Publishers), pp. 53–76.

Greenberg, J. M., Gillette, J. S., Muñoz Caro, G. M., Mahajan, T. B., Zare, R. N., Li, A., Schutte, W. A., de Groot, M., and Mendoza-Gómez, C. 2000. Ultraviolet photoprocessing of interstellar dust mantles as a source of polycyclic aromatic hydrocarbons and other conjugated molecules. *Ap. J.*, **531**, pp. L71–L73.

Greenberg., J. M., and Gustafson, B. Å. S. 1981. A comet fragment model of Zodiacal light particles. *Astron. Astrophys.*, **93**, pp. 35–42.

Greenberg, J. M., and Hage, J. I. 1990. From interstellar dust to comets - A unification of observational constraints. *Astrophys. J.*, Part 1, **361**, pp. 260–274.

Greenberg, J. M., and Li, A. 1999a. Morphological structure and chemical composition of cometary nuclei and dust. *Sp. Sc. Rev.*, **90**, pp. 149–161.

Greenberg, J. M., and Li, A. 1999b. Tracking the organic refractory component from interstellar dust to comets. *Adv. Sp. Res.*, **24**, pp. 497–504.

Greenberg, J. M., Li, A., Mendoza-Gómez, C., Schutte, W. A., Gerakines, P. A., and de Groot, M. 1995. Approaching the interstellar grain organic refractory component. *Ap. J. Let.*, **455**, pp. L177–L180.

Greenberg, J. M., Pedersen, N. E., and Pedersen, J. C. 1961. Microwave analog to the scattering of light by nonspherical particles. *J. Appl. Phys.*, **32**, No. 2, pp. 233–242.

Greenberg, J. M., Yencha, A. J., Corbett, J. W., and Frisch, H. L. 1972. Ultraviolet effects on the chemical composition and optical properties of interstellar grains. *Mém. Soc. Roy. Sci. Liège*, **6(III)**, pp. 425–436.

Grün, E., Bar-Nun, A., Benkhoff, J., Bischoff, A., Dueren, H., Hellmann, H., Hesselbarth, P., Hsiung, P., Keller, H. U., and Klinger, J. 1991. Laboratory simulation of cometary processes - Results from first KOSI experiments. In *Comets in the post-Halley era*, Vol. 1, eds. R. L. Newburn, Jr., M. Neugebauer, J. Rahe (Dordrecht: Kluwer Academic Publishers), pp. 277–297.

Grün, E., Zook, H. A., Fechtig, H., and Giese, R. H. 1985. Collisional balance of the meteoritic complex. *Icarus*, **62**, pp. 244–272.

Gustafson, B. Å. S. 1980. Scattering by ensembles of small particles - experiment theory and application. In *Reports Observ. of Lund*, **17** (Lund: Lund University Press).

Gustafson, B. Å. S. 1994. Physics of Zodiacal dust. *Ann. Rev. of Earth and Planetary Sciences*, **22**, pp. 550–592.

Gustafson, B. Å. S. 1996a. Microwave analog to light scattering measurements: A modern implementation of a proven method to achieve precise control. *J. Quant. Spect. Rad. Transf.*, **55**, pp. 663–672.

Gustafson, B. Å. S. 1996b. Optical Properties of Dust from Laboratory Scattering Measurements. In *Physics, Chemistry, and Dynamics of Interplanetary Dust*, eds. B. Å. S. Gustafson and M. S. Hanner (Provo, Utah: Astronomical Soc. of the Pacific Press), pp. 401–408.

Gustafson, B. Å. S. 2000. Microwave analog to light scattering measurements. In *Light Scattering by Nonspherical Particles: Theory, Measurements, and Geophysical Applications*, eds. M. I. Mishchenko, J. W. Hovenier and L. D. Travis (New York: Academic Press), pp. 367–390.

Gustafson, B. Å. S., and Kolokolova, L. 1999. A systematic study of light scattering by aggregate particles using the microwave analog technique: Angular and wavelength dependence of intensity and polarization. *J. Geophys. Res.*, **104**, pp. 31711–31720.

Gustafson, B. Å. S., Kolokolova, L., Thomas-Osip, J. E., Waldemarsson, K. W. T., Loesel, J., and Xu, Y.-l. 1999. Scattering by simple and complex systems II: Microwave measurements. In *Formation and Evolution of Solids in Space*, eds. J. M. Greenberg and A. Li (Dordrecht: Kluwer Academic Publishers), pp. 549–564.

Gustafson, B. Å. S., Kolokolova, L., and Xu, Y.-l. 2001. Radiation pressure on cosmic dust. *Ap. J.*, submitted.

Gustafson, B. Å. S., Misconi, N. Y., and Rusk, E. T. 1987. Interplanetary dust dynamics III. Dust released from P/Encke: Distribution with respect to the Zodiacal cloud. *Icarus*, **72**, pp. 582–592.

Hage, J. I., and Greenberg, J. M. 1990. A model for the optical properties of porous grains. *Ap. J.*, **361**, pp. 251–259.

Hallenbeck, S., Nuth, J., and Nelson, R. N. 2000. Evolving optical properties of annealing silicate grains: from amorphous condensate to crystalline mineral. *Ap. J.*, **535**, pp. 247–255.

Hanner, M. S. 1985. A preliminary look at the dust in Comet Halley. *Adv. Space Res.*, **5**, pp. 325–334.

Harrington, R. F. 1982. *Field Computation by Moment Methods*, (Malabar: Krieger).

Henning, Th., Begemann, B., Mutschke, H., and Dorschner, J. 1995. Optical properties of oxide dust grains. *Astron. Astrophys. Suppl. Ser.*, **112**, pp. 143–149.

Henning, Th., and Stognienko, R. 1996. Dust opacities for protoplanetary accretion disks: influence of dust aggregates. *Astron. Astrophys.*, **311**, pp. 291–303.

Hoban, S., A'Hearn, M. F., Birch, P. V., and Martin, R. 1989. Spatial structure in the color of the dust coma of Comet P/Halley. *Icarus*, **79**, pp. 145–158.

Hovenier, J. 2000. Measuring Scattering Matrices of small particles at optical wavelengths. In *Light Scattering by Nonspherical Particles: Theory, Measurements, and Geophysical Applications*, eds. M. I. Mishchenko, J. W. Hovenier and L. D. Travis (New York: Academic Press), pp. 355–365.

Ibadinov, Kh. I. 1989. Laboratory investigation of the sublimation of comet nucleus models. *Adv. Space Res.*, **9**, pp. 97–112.

Jenniskens, P. 1993. Optical constants of organic refractory residue, *Astron. Astrophys.*, **274**, pp. 653–661.

Jochims, H. W., Baumgärtel, H., and Leach, S. 1999. Structure-dependent photostability of polycyclic aromatic hydrocarbon cations: Laboratory study and astrophysical implications. *Ap. J.*, **512**, pp. 500–510.

Jockers, K., Rosenbush, V. K., Bonev, T., and Credner, T. 1997. Images of polarization and colour in the inner coma of comet Hale-Bopp. *Earth, Moon and Planets*, **78**, pp. 373-379.

Kai, L. I., and Massoli, P. 1994. Scattering of electromagnetic-plane waves by stratified sphere model. *Appl. Opt.*, **33**, pp. 501–511.

Kaiser, T., and Schweiger, G. 1993. Stable algorithm for the computation of Mie coefficients for scattered and transmitted fields of a coated sphere. *Comput. Phys.*, **7(6)**, pp. 682–686.

Kerker, M. 1969. *The scattering of light, and other electromagnetic radiation*, (New York: Academic Press).

Kerridge, J. F. 1999. Interstellar material in meteorites. In *Formation and Evolution of Solids in Space*, eds. J. M. Greenberg and A. Li (Dordrecht: Kluwer Academic Publishers), pp. 447–484.

Khanna, R. K. 1995. Infrared spectroscopy of organics of planetological interest at low temperatures. *Adv. in Space Res.*, **16**, pp. 109–118.

Kissel, J. 1999. In situ measurements of evolved solids in space with emphasis on cometary particles. In *Formation and Evolution of Solids in Space*, eds. J. M. Greenberg and A. Li (Dordrecht: Kluwer Academic Publishers), pp. 427–445.

Kissel, J., and Krueger, F. R. 1987. The organic component in dust from comet Halley as measured by the Puma mass spectrometer on board Vega 1. *Nature*, **326**, pp. 755–760.

Knacke, R. F. 1978. Mineralogical similarities between interstellar dust and primitive solar system material. In *Protostars and Planets*, ed. T. Gehrels (Tucson: University of Arizona Press), pp. 112–133.

Knacke, R. K., Brooke, T. Y., and Joyce, R. R. 1986. The 3.2-3.6 micron emission features in Comet Halley. Comparison with interstellar and laboratory spectra. In *20th ESLAB Symposium on the Exploration of Halley's Comet. Volume 2: Dust and Nucleus*, eds. B. Battrick, E. J. Rolfe, and R. Reinhard (Paris: ESA Pub. Div.), pp. 95–99.

Kochan, H. W., Huebner, W. F., and Sears, D. W. G. 1998. Simulation experiments with cometary analogous material. *Earth, Moon and Planets*, **80**, pp. 369–411.

Koike, C., Hasegawa, H., Asada, N., and Hattori, T. 1981. The extinction coefficients in mid- and far-infrared of silicate and iron-oxide minerals of interest for astronomical observations. *Astrophys. Space Sci.*, **79**, pp. 77–85.

Kolokolova, L., and Jockers, K. 1998. Composition of cometary dust from polarization spectra. *Planet. Space Sci.*, **45**, pp. 1543–1550.

Kolokolova, L., Jockers, K., Chernova, G., and Kiselev, N. 1997. Properties of cometary dust from the color and polarization. *Icarus*, **126**, pp. 351–361.

Kozasa T., Blum, J., Okamoto, H., and Mukai, T. 1993. Optical properties of dust aggregates. II. Angular dependence of scattered light. *Astron. Astrophys.*, **276**, pp. 278–288.

Kuik, F., Hovenier, J. W., and Stammes, P. 1991. Experimental determination of scattering matrices of water droplets and quartz particles. *Appl. Opt.*, **30**, pp. 4872–4881.

Lakhtakia, A. 1992. Strong and weak forms of the method of moments and the coupled dipole method for scattering of time-harmonic electromagnetic fields. *Intern. J. Mod. Phys. C* **3**, pp. 583–603.

Lawless, J. G. 1980. Organic compounds in meteorites. In *Life sciences and space research XVIII*, eds. Oxford and Elmsford (New York: Pergamon Press), pp. 19–27.

Léna, P., Viala, Y., Hall, D., and Soufflot, A. 1974. The thermal emission of the dust corona, during the eclipse of June 30, 1973. II - Photometric and spectral observations. *Astron. Astrophys.*, **37**, pp. 81–86.

Levasseur-Regourd, A.-C. 1999. Polarization of light scattered by cometary dust particles: Observations and tentative interpretations. *Space Sci. Rev.*, **90**, pp. 163–168

Li, A., and Greenberg, J. M. 1997. A unified model of interstellar dust. *Astron. Astrophys.*, **323**, pp. 566–584.

Li, A., and Greenberg, J. M. 1998. From Interstellar Dust to Comets: Infrared Emission from Comet Hale-Bopp (C/1995 O1). *Ap. J. Let.*, **498**, pp. L83–L87.

Lisse, C. M., A'Hearn, M., Hauser, M., Kelsall, T., Lien, D., Moseley, S., Reach, W., and Silverberg, R. F. 1998. Infrared observations of comets with COBE. *Ap. J.*, **496**, pp. 971–991.

Logan, N. A. 1965. Survey of some early studies of the scattering of plane waves by a sphere. *Proc. IEEE*, **53**, pp. 773–785.

Lorenz, L. V. 1898. Sur la lumière reflechie et refractée par une sphere transparente. In *Oeuvres Scientifiques de L. Lorentz*. (Copenhagen: Librairie Lehman et Stage), pp. 405–529.

Lumme, K., and Rahola, J. 1994. Light scattering by porous dust particles in the discrete-dipole approximation. *Ap. J.*, **425**, pp. 653–667.

Lumme, K., Rahola, J., and Hovenier, J. 1997. Light scattering by dense clusters of spheres, *Icarus*, **126**, pp. 455–469.

Mackowski, D. W., and Mishchenko, M. I. 1996. Calculation of the T matrix and the scattering matrix for ensembles of spheres. *J. Opt. Soc. Amer. A*, **13**, pp. 2266–2278.

Macqueen, R. M., and Greeley, B. W. 1995. Solar coronal dust scattering in the infrared. *Ap. J.*, **440**, pp. 361–369.

McDonnell, J. A. M. (editor) 1978. *Cosmic dust*. (Chichester: John Wiley and Sons).

Mendoza-Gómez, C. X., de Groot, M. S., and Greenberg, J. M. 1995. The fate of polycyclic aromatic material in space. *Astron. Astroph.*, **295**, pp. 479–486.

Messenger, S. 2000. Identification of molecular-cloud material in interplanetary dust particles. *Nature*, **404**, pp. 968–971.

Messenger, S., Clemett, S. J., Keller, L. P., Thomas, K. L., Chillier, X. D. F., and Zare, R. N. 1995. Chemical and mineralogical studies of an extremely Deuterium-rich IDP. *Meteoritics*, **30**, no. 5, pp. 546–547.

Mie, G. 1908. Beiträge zur optik trüber medien, speziell kolloidaler metallösungen. *Ann. Phys.*, **25**, pp. 377–445.

Mishchenko, M. I., and Travis, L. D. 1998. Capabilities and limitations of a current FORTRAN implementation of the T-matrix method for randomly oriented, rotationally symmetric scatterers. *J. Quant. Spectrosc. Radiat. Transfer*, **60**, pp. 309–324.

Mishchenko, M. I., Travis, L. D., and Mackowski, D. W. 1996. T-matrix computations of light scattering by nonspherical particles: A review. *J. Quant. Spectrosc. Radiat. Transfer*, **55**, pp. 535–575.

Moore, L. B., Flynn, G. J., and Klöck, W. 1999. Density and porosity measurements on meteorites: Implications for the porosities of asteroids. In *30th Annual Lunar and Planetary Science Conference*, March 15-29, 1999, Houston, TX, abstract no. 1128.

Moreels, G., Clairemidi, J., Hermine, P., Brechignac, P., and Rousselot, P. 1994. Detection of a polycyclic aromatic molecule in comet P/Halley. *Astron. Astroph.*, **282**, pp. 643–656.

Mukai, S., Mukai, T., Giese, R. H., Weiss, K., and Zerull, R. H. 1982. Scattering of radiation by a large particle with a random rough surface. *Moon and the Planets*, **26**, pp. 197–208.

Mukai, T., Ishimoto, H., Kozasa, T., Blum, J., and Greenberg, J. M. 1992. Radiation pressure forces of fluffy porous grains. *Astron. Astrophys.*, **262**, pp. 315–320.

Mukai, T., Mukai, S., and Kikuchi, S. 1987. Complex refractive index of grain material deduced from the visible polarimetry of comet P/Halley. *Astron. Astrophys.*, **187**, pp. 650–652.

Mukhin, L. M., Grechinsky, A. D., and Ruzmaikina, T. V. 1989. On the origin of P/Halley dust component. *Adv. in Sp. Res.*, **9**, pp. 23–27.

Ney, E. P. 1974. Multiband photometry of Comets Kohoutek, Bennett, Bradfield, and Encke. *Icarus*, **23**, pp. 551–560.

Okamoto, H., Mukai, T., and Kozasa, T. 1994. The 10 micron feature of aggregates in comets. *Planet. Sp. Sci.*, **42**, pp. 643–649.

Orofino, V., Blanco, A., and Fonti, S. 1994. Silicon carbide: A possible component of the cometary dust. *Astron. Astrophys.*, **282**, pp. 657–662.

Ossenkopf, V. 1991. Effective-medium theories for cosmic dust grains. *Astron. Astrophys.*, **251**, pp. 210–219.

Ossenkopf, V. 1993. Dust coagulation in dense molecular clouds: the formation of fluffy aggregates. *Astron. Astrophys.*, **280**, pp. 617–646.

Peltzer, E. T., Bada, J. L., Schlesinger, G., and Miller, S. L. 1984. The chemical conditions on the parent body of the Murchison meteorite: Some conclusions based on amino, hydroxy, and dicarboxylic acids. *Adv. Space Res.*, **4**, pp. 69–74.

Perrin, J. M., and Lamy, P. L. 1985. Optical properties of rough grains – A theoretical study. In *Properties and interactions of interplanetary dust*, eds. P. L. Lamy and R. H. Giese (Dordrecht: D. Reidel Publishing Co.), pp. 245–248.

Perrin, J. M., and Sivan, J. P. 1991. Scattering and polarization of light by rough and porous interstellar grains. *Astron. Astrophys.*, **247**, pp. 497–504.

Peterson, A. F. 1991. Analysis of heterogeneous electromagnetic scatterers: Research progress of the past decade. *Proc. IEEE*, **79**, pp. 1431–1441.

Peterson, B., and Ström, S. 1974. T-matrix formulation of electromagnetic scattering from multilayered scatterers. *Phys. Rev. D*, **10**, pp. 2670–2684.

Petravic, M., and Kuo-Petravic, G. 1979. An ILUGC algorithm which minimizes the euclidean norm. *J. Comput. Phys.*, **32**, pp. 263–269.

Poppe, T., Blum, J., and Henning, Th. 2000. Analogous experiments on the stickiness of micron-sized preplanetary dust. *Ap. J.*, **533**, pp. 454–471.

Purcell, E. M., and Pennypacker, C. R. 1973. Scattering and absorption of light by nonspherical dielectric grains. *Ap. J.*, **186**, pp. 705–714.

Raulin, F. 1990. Prebiotic chemistry in the solar system. In *ESA, Formation of Stars and Planets, and the Evolution of the Solar System*, pp. 151–157.

Rickman, H. 1989. The nucleus of Comet Halley - Surface structure, mean density, gas and dust production. *Adv. in Sp. Res.*, **9**, pp. 59–71.

Rietmeijer, F. J. M. 1998. Interplanetary dust particles. In *Planetary Materials*, ed. J. J. Papike (Washington: Mineralogical Soc. of Am.), pp. (2)1-(2)95.

Rouleau, F., and Martin, P. G. 1991. Shape and clustering effects on the optical properties of amorphous carbon. *Ap. J.*, **377,** pp. 526–540.

Sandford, S. A., and Walker, R. M. 1985. Laboratory infrared transmission spectra of individual interplanetary dust particles from 2.5 to 25 micron. *Ap. J.*, **291,** pp. 838–851.

Scharf G. 1994. *From Electrostatics to Optics, a Concise Electrodynamics Course.* (Berlin: Springer-Verlag).

Schutte, W. A. 1995. The formation of organic molecules in astronomical ices. *Adv. in Sp. Res.*, **16,** pp. (2)53-(2)60.

Schutte, W. A. 1999. Laboratory simulation of processes in interstellar ices. In *Formation and Evolution of Solids in Space*, eds. J. M. Greenberg and A. Li (Dordrecht: Kluwer Acad. Publ.), pp. 177–201.

Schutte, W. A., Allamandola, L. J., and Sandford, S. A. 1992. Laboratory simulation of the photoprocessing and warm-up of cometary and pre-cometary ices - Production and analysis of complex organic molecules. *Adv. in Sp. Res.*, **12,** pp. 47–51.

Schutte, W. A., Allamandola, L. J., and Sandford, S. A. 1993. An experimental study of the organic molecules produced in cometary and interstellar ice analogs by thermal formaldehyde reactions. *Icarus*, **104,** pp. 118–137.

Scott, A. D., Duley, W. W., and Jahani, H. R. 1997. Infrared emission spectra from hydrogenated amorphous carbon. *Ap. J.*, **490,** pp. L175–L177.

Sears, D. W. G., Kochan, H. W., and Huebner, W. F. 1999. Invited Review: Laboratory simulation of the physical processes occurring on and near the surfaces of comet nuclei. *Meteoritics and Plan. Sci.*, **34,** pp. 497–525.

Sen, A. K., Deshpande, M. R., Joshi, U. C., Rao, N. K., and Raveendran, A. V. 1991. Polarimetry of Comet P/Halley: properties of dust. *Astron. Astrophys.*, **242,** pp. 496–502.

Sephton, M. A., and Gilmour, I. 2000. Aromatic moieties in meteorites: Relics of interstellar grain processes? *Ap. J.*, **540,** pp. 588–591.

Sihvola, A. 1999. *Electromagnetic mixing formulas and applications.* (London: IEE Publishing).

Silvester, P., and Ferrari, R. L. 1990. *Finite elements for electrical engineers*, 2nd ed. (Cambridge: Cambridge Univ. Press).

Singham, S. B., and Bohren, C. F. 1988. Light scattering by an arbitrary particle: the scattering-order formulation of the coupled-dipole method. *J. Opt. Soc. Am. A*, **5,** pp. 1867–1872.

Sirono, S.-i., and Greenberg, J. M. 2000. Do Cometesimal collisions lead to bound rubble piles or to aggregates held together by gravity? *Icarus*, **145,** pp. 230–238.

Smith, D., and Adams, N. G. 1977. Molecular synthesis in interstellar clouds: Some relevant laboratory measurements. *Ap. J.*, **217,** pp. 741–748.

Stephens, J. R., and Gustafson, B. Å. S. 1991. Laboratory reflectance measurements of analogues to 'dirty' ice surfaces on atmosphereless solar system bodies. *Icarus*, **94,** pp. 209–217.

Stroud, D., and Pan, F. P. 1978. Self-consistent approach to electromagnetic wave propagation in composite media: Application to model granular metals. *Phys. Rev. B*, **17,** pp. 1602–1610.

Sturm, J., Grosse, P., and Theiß, W. 1991. Effective dielectric function of alkali halide composites and their spectral representation. *Z. Phys. B – Condensed Matter*, **83,** pp. 361–365.

Thomas, P. C., Belton, M. J. S., Carcich, B., Chapman, C. R., Davies, M. E., Sullivan, R., and Veverka, J. 1996. The shape of Ida. *Icarus*, **120,** pp. 20–32.

Toon, O. B., and Ackerman, T. P. 1981. Algorithms for the calculation of scattering by stratified spheres. *Appl. Opt.*, **20,** pp. 3657-3660.

Trinks, W. 1933. Zur vielfachstreuung an kleinen kugeln. *Ann. Phys.*, **22,** pp. 561–590.

van de Hulst, H. C. 1957. *Light Scattering by Small Particles.* (New York: John Wiley & Sons, Inc.)

van de Hulst, H. C. 1993. Light scattering by seed particles - a review. In *Proceedings of the SPIE - the International Society for Optical Engineering*, **2052,** pp. 3–14.

Varadan, V. K., and Varadan, V. V. 1980. Recent Developments in Classical Wave Scattering: Focus on the T-matrix. (Oxford: Pergamon Press).

Voshchinnikov, N. V., and Farafonov, V. G. 1993. Optical properties of spheroidal particles. *Astrophys. Space Sci.*, **204**, pp. 19–86.

Wait, J. R. 1955. Scattering of a plane wave from a circular dielectric cylinder at oblique incidence. *Can. J. Phys.*, **33**, pp. 189–195.

Wang, R. T., and van de Hulst, H. C. 1991. Rainbows: Mie computation and the Airy approximation. *Appl. Opt.*, **30**, pp. 106–114.

Waterman, P. C. 1965. Matrix formulation of electromagnetic scattering. *Proc. IEEE*, **53**, pp. 805–812.

Waterman, P. C. 1973. Numerical solution of electromagnetic scattering problems. *Phys. Rev. D*, **8**, pp. 3661–3678.

Weidenschilling, S. J. 1997. The origin of comets in the solar nebula: A unified model. *Icarus*, **127**, pp. 290–306.

Weiss-Wrana, K. 1983. Optical properties of interplanetary dust: comparison with light scattering by larger meteoritic and terrestrial grains. *Astron. Astrophys.*, **126**, pp. 240–250.

West, R. 1991. Optical properties of aggregate particles whose outer diameter is comparable to the wavelength. *Appl. Optics*, **30**, pp. 5316–5324.

Williams, D. M., Mason, C. G., Gehrz, R. D., Jones, T. J., Woodward, C. E., Harker, D. E., Hanner, M. S., Wooden, D. H., Witteborn, F. C., and Butner, H. M. 1997. Measurement of submicron grains in the coma of comet Hale-Bopp C/1995 01 during 1997 February 15–20 UT. *Ap. J. Let.*, **489**, pp. L91–L94.

Wilson, L., Keil, K., and Love, S. J. 1999. The internal structures and densities of asteroids. *Meteoritics & Planetary Science*, **34**, pp. 479–483.

Wolff, M. J., Clayton, G. C., Martin, P. G., and Schulte-Ladbeck, R. E. 1994. Modeling composite and fluffy grains: The effects of porosity. *Ap. J.*, **423**, pp. 412–425.

Worms, J.-C., Renard, J.-B., Levasseur-Regourd, A.-Ch., and Hadamcik, E. 1999. Light scattering by dust particles in microgravity: The PROGRA^2 achievements and results. *Adv. Space Res.*, **23**, pp. 1257–1266.

Wurm, G., and Blum, J. 1998. Experiments on preplanetary dust aggregation. *Icarus*, **132**, pp. 125–136.

Xing, Z., and Hanner, M. 1997. Light scattering by aggregate particles. *Astron. Astrophys.*, **324**, pp. 805–820.

Xu, Y.-l. 1995. Electomagnetic scattering by an aggregate of spheres. *Appl. Optics*, **34**, pp. 4573–4588.

Xu, Y.-l. 1997. Electromagnetic scattering by an aggregate of spheres: Far field. *Appl. Optics*, **36**, pp. 9496–9508.

Xu, Y.-l., and Gustafson, B. Å. S. 1996. A complete and efficient multisphere scattering theory for modeling the optical properties of interplanetary dust. In *Physics, Chemistry, and Dynamics of Interplanetary Dust*, eds. B. Å. S. Gustafson and M. S. Hanner (Provo, Utah: Astronomical Soc. of the Pacific Press) pp. 417–420.

Xu, Y.-l., and Gustafson, B. Å. S. 1997. Experimental and theoretical results of light scattering by aggregates of spheres. *Ap. Optics*, **36**, No. 30, pp. 8026–8030.

Xu, Y.-l., and Gustafson, B. Å. S. 1999. Comparison between multisphere light-scattering calculations: (I) Rigorous solution and (II) DDA. *Ap. J.*, **513**, pp. 896–911.

Xu Y.-l., and Gustafson, B. Å. S. 2001. A generalized multiparticle Mie-solution: Further experimental verification. *J. Quant. Spectrosc. Radiat. Transfer*, in press.

Xu, Y.-l., Gustafson, B. Å. S., Giovane, F., Blum, J., and Tehranian, S. 1999. Calculation of the heat-source function in photophoresis of aggregated spheres. *Phys. Rev. E*, **60**, pp. 2347–2365.

Yanamandra-Fisher, P. A., and Hanner, M. S. 1999. Optical properties of non-spherical particles of size comparable to the wavelength of light: application to comet dust. *Icarus*, **138**, pp. 107–128.

Zerull, R. H. 1985. Laboratory investigations and optical properties of grains. In *Properties and interactions of interplanetary dust*, eds. P. L. Lamy and R. H. Giese (Dordrecht: D. Reidel Publishing Co.), pp. 197–206.

Zerull, R. H., Gustafson, B. Å. S., Schultz, K., and Thiele-Corbach, E. 1993. Scattering by aggregates with and without an absorbing mantle: microwave analog experiments. *Appl. Optics*, **32,** pp. 4088–4100.

Orbital Evolution of Interplanetary Dust

Stanley F. Dermott[1], Keith Grogan[2], Daniel D. Durda[3], Sumita Jayaraman[4], Thomas J. J. Kehoe[1], Stephen J. Kortenkamp[5,6], Mark C. Wyatt[7]

[1] University of Florida, Gainesville, Florida, USA
[2] NASA Goddard Space Flight Center, Greenbelt, Maryland, USA
[3] Southwest Research Institute, Boulder, Colorado, USA
[4] Vanguard Research Inc., Scotts Valley, California, USA
[5] University of Maryland, College Park, Maryland, USA
[6] Carnegie Institution of Washington, Washington, District of Columbia, USA
[7] Uk Astronomy Technology Centre, Royal Observatory, Edinburgh, UK

Abstract. The two most important dynamical features of the zodiacal cloud are: (i) the dust bands associated with the major Hirayama asteroid families, and (ii) the circumsolar ring of dust particles in resonant lock with the Earth. Other important dynamical features include the offset of the center of symmetry of the cloud from the Sun, the radial gradient of the ecliptic polar brightness at the Earth, and the warp of the cloud. The dust bands provide the strongest evidence that a substantial and possibly dominant fraction of the cloud originates from asteroids. However, the characteristic diameter of these asteroidal particles is probably several hundred microns and the migration of these large particles towards the inner Solar System due to Poynting-Robertson light drag and their slow passage through secular resonances at the inner edge of the asteroid belt results in large increases in their eccentricities and inclinations. Because of these orbital changes, the dividing line between asteroidal and cometary-type orbits in the inner Solar System is probably not sharp, and it may be difficult to distinguish clearly between asteroidal and cometary particles on dynamical grounds alone.

I. INTRODUCTION

Advances in infrared astronomy have revealed that the structure of the zodiacal cloud is complex and substantially different from the smooth, rotationally symmetric cloud assumed prior to the launch of the IRAS (Infrared Astronomical Satellite) and COBE (Cosmic Background Explorer) spacecraft (Giese et al. 1986). We now know that the Sun is not at the center of symmetry of the cloud (Kelsall et al. 1998; Dermott et al. 1999; Wyatt et al. 1999b), that the cloud contains dust bands originating from the disintegration of asteroids (Low et al. 1984; Dermott et al. 1984), dust trails derived from known comets (Sykes and Walker 1992), and clouds of dust associated with a circumsolar ring of dust particles trapped in resonant lock with the Earth (Dermott et al. 1994a; Reach et al. 1995). These features pose challenging dynamical problems. However, they are also our best source of information on the sources that supply the cloud. The plan of this review is as follows.

In Section II, we discuss the LDEF (Long Duration Exposure Facility) results on the sizes of the particles accreted by the Earth. These results show very clearly that, in the vicinity of the Earth, the dominant particles in the cloud have characteristic diameters $\sim 10^2\,\mu$m (Love and Brownlee 1993). A similar conclusion was reached by Grün et al. (1985), who reviewed evidence from a variety of other sources including the lunar microcratering record as well as spacecraft micrometeoroid detectors. These particles are not primordial (their lifetimes are $\sim 10^6$ yr) but are replenished from sources that must be either asteroidal or cometary. However, which of these sources is dominant is still a matter of debate. The main purpose of this review is to analyze the constraints on the origins of the particles that are imposed by the dynamics. Section II also contains a brief discussion of the collisional evolution of the particle sizes and of the radiation forces that act on the particle orbits. In Section III, we discuss in detail the variations of the orbital elements of asteroidal and cometary particles due to Poynting-Robertson light drag. Some of these variations arise from gravitational interactions with the planets and these changes determine many of the dynamical features of the cloud. Of particular importance are the long term changes in the eccentricities and inclinations of the particle orbits due to secular perturbations, and the dependence of these changes on particle size.

The most important source of information on the asteroidal contribution to the zodiacal cloud are the IRAS multi-waveband observations of the dust bands. In Section IV, we use these IRAS data to quantify the magnitude of the asteroidal source and to determine the size-frequency distribution of the particles. We do not have any useful way of assessing the direct contribution of cometary material and our only recourses are either to estimate the fraction of the cloud that is asteroidal and ascribe the remainder to comets, or to examine whether there are any features of the cloud that dictate that there must be a significant cometary contribution. COBE provided particularly useful information on the structure of the broad-scale, background cloud. In Section V, we discuss the shape of the cloud, the offset of the center of symmetry from the Sun, the plane of symmetry, and the variation of the ecliptic polar brightness with heliocentric distance. One of the most striking features of the zodiacal cloud revealed by IRAS and confirmed by COBE is the trailing/leading asymmetry which has been accounted for by resonant trapping of dust particles by the Earth (Dermott et al. 1994a). In Section VI, we discuss the origin of this asymmetry and the dynamics of the Earth's resonant ring.

The problem of the origin of the IDPs (Interplanetary Dust Particles) collected in the Earth's atmosphere is obviously related to the origin of the particles in the cloud but there are two other dynamical factors that influence the rate of accretion. In Section VII, we discuss the role of gravitational focusing in the accretion of low-eccentricity asteroidal particles. We also discuss the possible role of the Earth's resonant ring in the accretion process. Finally, we discuss the variation of the accretion rate with time due to the variation of the Earth's orbital elements.

II. FORCES AND COLLISIONS

A wide range of gravitational and non-gravitational forces act to change the orbits of particles in the zodiacal cloud. At the same time, the particle size distribution evolves due to mutual collisions. Of particular importance for our discussion is the critical diameter, D_{crit}, that divides the larger particles in the cloud, those that are collisionally evolved and have lifetimes determined by mutual collisions, from the smaller particles that have lifetimes determined by drag and radiation forces. The summary that follows is concerned largely with the evolution of asteroidal particles and is based on the account given recently by Wyatt et al. (1999b). For a more complete discussion of the collisional evolution of asteroidal and cometary dust see the paper by Grün et al. (1985).

A typical asteroidal dust particle is created by the breakup of a larger "parent" body. This parent body could have been created by the breakup of an even larger body, and the particle itself will most likely end up as a parent body for particles smaller than itself. This "collisional cascade" spans the complete size range of disk material, and the particles that share a common ancestor are said to constitute a "family" of particles. The size distribution that results from this collisional cascade is given by

$$N(D) = \frac{1}{3(3q-1)} \left(\frac{D_0}{D}\right)^{3(q-1)}, \tag{1}$$

where $N(D)$ is the number of asteroids with diameter $> D$, D_0 is a constant, and q is the power law index (Dohnanyi 1969). If $5/3 < q < 2$, then the total area, A, in the cascade population is dominated by contributions from the smallest particles of diameter $\sim D_{\min}$, where $D_{\min}$ is the lower cutoff of the size distribution, whereas the total volume, V, of the source population is dominated by the contributions from the larger fragments. If we write V in terms of the equivalent diameter, D_e, that is, $V = (\pi/6)D_e^3$, then

$$A = \frac{(\pi/4)D_{\min}^2}{(3q-5)} \left(\frac{D_0}{D_{\min}}\right)^{3(q-1)} \tag{2}$$

and

$$D_0 = D_e \left[3(2-q)\right]^{1/3} \left(\frac{D_0}{D_{\max}}\right)^{2-q}, \tag{3}$$

where $D_{\max}$ is the diameter of the largest family fragment in the cascade. For the three major Hirayama families, Eos, Themis and Koronis (Hirayama 1918), $D_e/D_0 \simeq 1.2$ (Dermott et al. 1984).

Dohnanyi's derivation of the above power law assumed an infinite range of sizes. However, in the asteroid belt, the smallest particles are removed by drag forces and light pressure (as discussed in the following two subsections). Given that these small particles are the bullets that destroy bodies a magnitude larger, their removal could have a profound effect on the size distribution of the larger bodies. Numerical experiments (Davis et al. 1993; Durda and Dermott 1997) have shown that the rapid removal of the smallest particles could result in strong departures from Dohnanyi's simple power law. However, Durda and Dermott (1997) argue that, in the asteroid belt, Poynting-Robertson light drag does not in fact remove the smaller particles on a fast enough timescale to affect the size distribution.

Dohnanyi (1969) also assumed that all particles in the asteroid collision cascade have the same, size-independent impact strength, in which case it can be shown analytically that $q = 1.83$. However, strain-rate effects (Housen and Holsapple 1990) and gravitational overburden of large asteroids (Davis et al. 1985) lead to size-dependent strengths among real asteroids. Durda and Dermott (1997) and Durda et al. (1998) have shown that size-dependent impact strengths can lead to evolved size distributions that deviate from the classic Dohnanyi power law distribution. Observation of the small ($D < 30$ km) asteroids in the inner region of the main belt shows that $q = 1.78 \pm 0.02$, indicating some dependence of the strength of asteroids on their size (Durda and Dermott 1997). The distribution of the very largest ($D > 30$ km) asteroids also deviates from the Dohnanyi distribution because of the transition from strength-scaling to gravity-scaling for asteroids larger than ~ 150 m (Durda et al. 1998), and also because some of the larger asteroids are original and not part of the collisional cascade. Overall, however, the basic collision cascade theory is well-supported by evidence from the larger, observable members of the main belt asteroid population.

The size distribution of the zodiacal cloud's medium-sized (1 mm $< D <$ 3 km) members is also expected to follow Eq. (1), but there is no observational proof of this, since these members are too faint to be seen. However, the proof that the zodiacal cloud's collisional cascade extends from its largest members down to its smallest dust particles is, of course, provided by the existence of the Solar System dust bands that we know are derived from asteroidal collisions (Dermott et al. 1984; Grogan et al. 2001).

Mutual collisions among asteroidal particles are typically not catastrophic: $N(D)$ increases rapidly with decreasing D and so a particle in the main belt is most likely to be broken up by a projectile that has just enough mass (and hence energy) to do so. This in turn means that the collisional fragments have velocities, and hence orbital elements, that are almost identical to those of the original particle; that is, in the absence of other forces, all members of the same family have near-identical orbits. This, of course, is the reason for the close grouping of the orbital elements of the Hirayama family members. It is also the reason why dust bands actually exist and are observable.

II.A. Radiation Forces

In this chapter, we are concerned only with the dynamical effects of solar radiation on micron-sized dust particles, but given the current interest in exozodiacal clouds it is useful to follow Wyatt et al. (1999b) and to generalize all of our equations to any star. In this context, radiation forces are caused by the absorption, scattering, and re-emission of incident photons by the dust particle (refer to Burns et al. 1979, for a thorough description). Radiation pressure is the component of the radiation force that points radially away from the star. It is defined for different size particles by its ratio to the gravitational force, and is denoted by the symbol β. We can write (Gustafson 1994):

$$\beta(D) = F_{\text{rad}}/F_{\text{grav}} = C_r(\sigma/m)\langle Q_{\text{pr}}\rangle_{T_*}(L_*/L_\odot)(M_*/M_\odot), \quad (4)$$

where $C_r = 7.65 \times 10^{-4}$ kg m^{-2}; σ/m is the ratio of the particle's cross-sectional area to its mass (for example, $\sigma/m = 1.5/\rho D$ for spherical particles of density ρ);

$$\langle Q_{\text{pr}}\rangle_{T_*} = \frac{\int Q_{\text{pr}}(D,\lambda) F_\lambda d\lambda}{\int F_\lambda d\lambda} \quad (5)$$

is the particle's radiation pressure efficiency averaged over the stellar spectrum, F_λ; T_* is the star's effective temperature; M_* and L_* are the mass and luminosity of the star; and $M_\odot$ and $L_\odot$ are the mass and luminosity of the Sun.

A useful approximation for large particles is that $\langle Q_{\text{pr}}\rangle_{T_*} \approx 1$; in which case

$$\beta(D) = (1150/\rho D)(L_*/L_\odot)(M_*/M_\odot), \quad (6)$$

where ρ is measured in kg m^{-3}, and D is in μm. This approximation is valid for astronomical silicate particles in the Solar System with $D \gtrsim 1\,\mu$m (see Fig. 1).

The effect of radiation pressure is equivalent to reducing the mass of the star by a factor $1 - \beta$. This means that small daughter fragments created by the breakup of a parent body move on orbits that can be substantially different from that of the parent. The reason for this is that while the positions and velocities of a parent and its daughter fragments are the same at the moment of breakup (apart from a small velocity dispersion), their β are different, and so the daughter fragments move in effective potentials that are different from that in which the parent moves. Daughter fragments created in the breakup of a parent body with $\beta = 0$, and for which the orbital elements at the time of the breakup were semi-major axis a, eccentricity e, inclination I, longitude of ascending node Ω, longitude of pericenter ϖ, and true anomaly f, move in the same orbital plane as the parent, $I' = I$ and $\Omega' = \Omega$, but on orbits with semi-major axes a', eccentricities e', and pericenter orientations ϖ', given by (Burns et al. 1979; Kortenkamp and Dermott 1998a; Wyatt et al. 1999b)

$$a' = a(1-\beta)/[1 - 2\beta(1 + e\cos f)/(1 - e^2)], \quad (7)$$

$$e' = (1-\beta)^{-1}\sqrt{(e^2 + 2\beta e\cos f + \beta^2)}, \quad (8)$$

$$\varpi' - \varpi = f - f' = \arctan[\beta \sin f/(\beta \cos f + e)]. \quad (9)$$

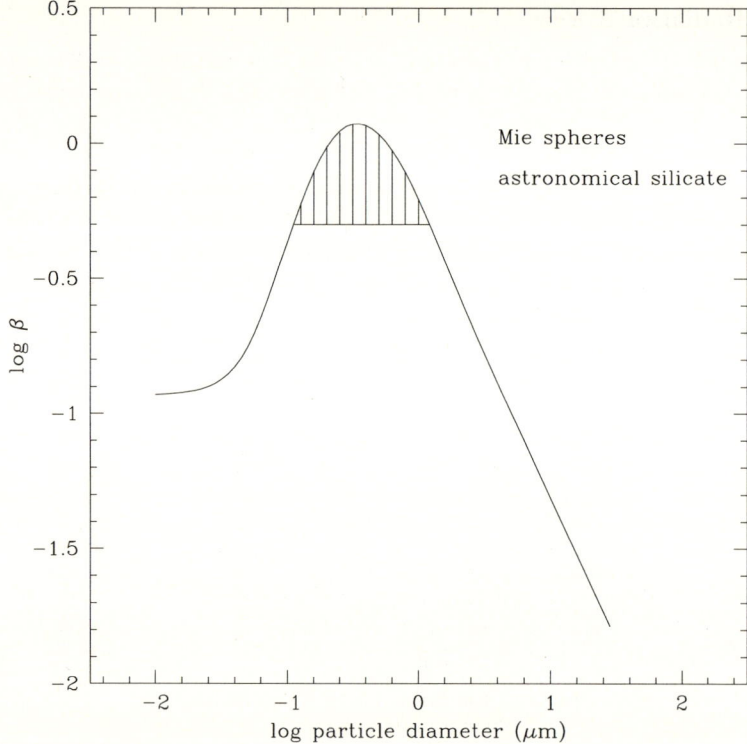

Figure 1. Variation with particle diameter D of the ratio, β, of the forces due to solar radiation and gravity acting on a spherical particle of density $2{,}500\,\mathrm{kg\,m^{-3}}$ with the optical properties of astronomical silicate calculated using Mie theory. Radiation pressure acts on particles with $\beta > 0.5$ (shown hatched) derived from larger particles moving on near-circular orbits, to blow these small particles out of the Solar System. For $D(\mu\mathrm{m}) \gtrsim 1$, $\beta \approx 1/(2.2D)$.

Analysis of these equations shows that the orbits of the largest fragments, those for which $\beta < 0.1$, are similar to that of the parent. On the other hand, the smallest fragments, those for which $\beta > 0.5(1-e^2)/(1+e\cos f)$, have hyperbolic orbits ($e' > 1$) and are known as "β meteoroids" (Zook and Berg 1975). Since β meteoroids are blown out of the system on the timescale of the orbital period of the parent, the diameter of the particle for which $\beta > 0.5$ essentially defines the lower limit of the collisional cascade. However, as we have assumed here that the parent body is on a nearly circular orbit, this result does not apply to bodies on comet-like orbits. Figure 1 shows that there may also be a population of submicron particles that have $\beta < 0.5$ (Gustafson 1994).

II.B. Poynting-Robertson (P-R) Light Drag

The component of the radiation force tangential to a particle's orbit is called the P-R drag force. This force is also proportional to β. It results in an evolutionary decrease in both the semi-major axis and the osculating eccentricity of the particle's orbit (Wyatt and Whipple 1950):

$$\dot{a}_{\rm PR} = -(\alpha/a)(2 + 3e^2)/(1-e^2)^{3/2} = -2\alpha/a + O(e^2), \qquad (10)$$
$$\dot{e}_{\rm PR} = -2.5(\alpha/a^2)e/(1-e^2)^{1/2} = -2.5\alpha e/a^2 + O(e^3), \qquad (11)$$

where $\alpha = 6.24 \times 10^{-4}\beta(M_*/M_\odot)\,{\rm AU}^2\,{\rm yr}^{-1}$. P-R drag therefore causes the orbit of the particle to spiral in towards the star. However, it does not change the plane of the particle's orbit, $\dot{I}_{\rm PR} = \dot{\Omega}_{\rm PR} = 0$; neither does it affect the orientation of the particle's pericenter, $\dot{\varpi}_{\rm PR} = 0$.

For a particle with zero eccentricity, Eq. (10) can be solved to find the time it takes for the particle to spiral in from an astrocentric radial distance r_1 to r_2:

$$t_{\rm PR} = 400(M_\odot/M_*)\left[(r_1/a_\oplus)^2 - (r_2/a_\oplus)^2\right]/\beta, \qquad (12)$$

where $t_{\rm PR}$ is given in years and $a_\oplus = 1$ AU is the semi-major axis of the Earth's orbit. The time taken for an asteroidal particle to migrate from its source region to the Earth is $\propto D$ (Eq. 6, large particle approximation). For a particle of diameter $D = 100\,\mu{\rm m}$, released at a radial distance $r_1 = 3$ AU, $t_{\rm PR} \sim 7 \times 10^5$ yr. It follows from Eq. (10) that the P-R decay timescale, $t_{\rm PR}$, decreases as the eccentricity of the particle increases. Dividing Eq. (10) by Eq. (11) gives da/de, the rate of change of the semi-major axis of a particle with eccentricity, which is clearly independent of β and hence the size of the particle. This leads to the interesting result that the eccentricity distribution of a wave of particles spiraling towards a star, is only dependent, at any given semi-major axis, on the initial distribution of semi-major axes and eccentricities of the particles. Although the time taken for any particular particle to reach the given semi-major axis is dependent on its size.

The effects of the average force due to scattering of incident protons in a stellar wind, which acts in the same way as P-R light drag, should also be added to Eqs. (10)–(12), Gustafson (1994). In the Solar System, this force, known as solar wind drag, is usually taken to be 30% of the P-R light drag force, varying over the 11-year solar cycle from 20% to 40%, thus reducing the timescale in Eq. (12) by about 30%.

II.C. Collisions

The importance of collisions in determining a particle's evolution depends on its collisional lifetime. The collisional lifetime of the particles of diameter $D_{\rm typ}$ that constitute most of a disk's cross-sectional area (that is, those particles that are expected to characterize the disk's mid-IR emission), can be approximated by

$$t_{\rm coll}(D_{\rm typ}, r) = \frac{t_{\rm per}}{4\pi\tau_{\rm eff}(r)}, \qquad (13)$$

where r is the astrocentric radial distance and $\tau_{\rm eff}(r)$ is the disk's effective face-on optical depth, which would be equal to the disk's true optical depth if its particles had unity extinction efficiency (Artymowicz 1997; Wyatt et al. 1999b). The orbital period of a particle in years is given by $t_{\text{per}} = \sqrt{(a/a_\oplus)^3 (M_\odot/M_*)}$.

Consider the fragments created in the breakup of an asteroid at a heliocentric distance r. The largest fragments, with $D > D_{\text{crit}}$, are broken up by collisions before their orbits have suffered any significant P-R drag evolution, while the smaller fragments, with $D < D_{\text{crit}}$, for which the P-R drag evolution is faster, can reach the Sun without a catastrophic collision. By equating the collisional and P-R drag lifetimes given by Eqs. (12) and (13), and using the large particle approximation for β, Eq. (6), Wyatt et al. (1999b) estimate that

$$D_{\text{crit}} = \frac{0.23}{\rho \tau_{\text{eff}}(r)} \left(\frac{L_*}{L_\odot}\right) \sqrt{\left(\frac{M_\odot}{M_*}\right)\left(\frac{a_\oplus}{r}\right)}, \qquad (14)$$

where ρ is measured in kg m^{-3}, and D_{crit} in μm.

Consider the daughter fragments created in the breakup of an "endless" supply of asteroids on orbits with semi-major axis a_s that flow towards the Sun due to P-R drag. If we ignore any further disintegrations of the particles that are involved in the flow, then the orbits of all the particles in a given size range will be distributed between the source and the Sun according to

$$N(a) \propto 1/\dot{a}_{\text{PR}} \propto a, \qquad (15)$$

where $N(a)da$ is the number of orbits with semi-major axes in the range a to $a + da$. Thus, the spacing of the orbits increases as the particles approach the Sun and this fact tends to decrease the number density of the particles, defined as the number of particles per unit volume. But given that both the circumferences of the orbits and the vertical extent of the particle distributions also decrease proportionally with decreasing a, it follows that, for particles in near-circular orbits, the number density of these particles, regardless of their size, will increase inversely with heliocentric distance.

However, because the flow rate of the particles is inversely proportional to their diameter, that is, because $\dot{a}_{\text{PR}} \propto 1/D$ (see Eqs.(6) and (10)), it follows that the size distribution of the particles in the flow region interior to the asteroid belt, must be quite different from that in the source region.

If the collisional processes leading to the size distribution of the large parent bodies, $N_s(D)$, still holds for the production of the P-R drag affected particles, then the size distribution in the flow region is given by:

$$N(D) \propto N_s(D)/\dot{a}_{\text{PR}} \propto N_s(D)D. \qquad (16)$$

Thus, if $N_s(D)$ is given by Eq. (1) with $q = 11/6$, then the cross-sectional area of a disk's smaller, P-R drag affected particles is concentrated in the largest of these small particles, while the cross-sectional area of the particles that are large enough to be unaffected by P-R drag ($D > D_{\text{crit}}$) is concentrated in the smallest of these larger particles. The result is that most of a disk's cross-sectional area is expected to be concentrated in particles with $D_{\text{typ}} \approx D_{\text{crit}}$, justifying the use of Eq. (13) for the collisional lifetime of these particles.

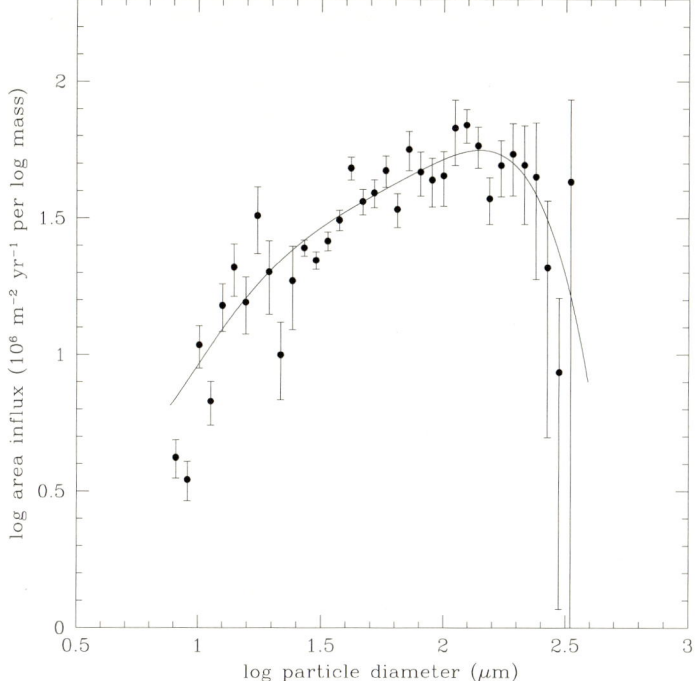

Figure 2. Incremental area of particles accreted annually by the Earth as a function of particle diameter. The bold line is the polynomial derived by Love and Brownlee (1993) from the cratering record on LDEF (Long Duration Exposure Facility). The total area of the accreted particles is dominated by particles with diameters between 60 μm and 200 μm.

Observations of the mean polar brightness of the zodiacal cloud at 1 AU (see Fig. 19) can be used to estimate (the results are somewhat model dependent) that, near the Earth, the effective, normal optical depth is $\sim 5 \times 10^{-8}$. If, as we believe, these zodiacal particles originated in the asteroid belt and migrated to 1 AU due to the P-R drag, then the zodiacal cloud's volume density should vary $\sim 1/r$ and its effective optical depth in the asteroid belt should be similar to that at 1 AU. Assuming the zodiacal cloud particles to have a density 2,500 kg m^{-3}, the cross-sectional area of material in the asteroid belt should be concentrated in particles with $D_{\text{typ}} \sim 10^3$ μm for which the collisional lifetime, and the P-R drag lifetime, are $\sim 10^7$ yr.

However, because the collisional and P-R lifetimes are similar, we must expect many of these large particles to be broken up by collisions before they reach the inner Solar System; in which case we must expect the cross-sectional area of material at 1 AU to be concentrated in particles smaller than that in the asteroid belt. This is in agreement with the LDEF observations of Love and Brownlee (1993) and other evidence (see the review by Grün et al. 1985) that shows the cross-sectional area distribution at 1 AU to peak for particles with $D \sim 10^2$ μm (see Fig. 2).

The analysis of the collisional lifetimes of Wyatt et al. (1999b) also predicts that the collisional lifetime of the large bodies in the asteroid belt should be $\sim 10^9 \sqrt{D}$ yr, where D is in km. Since the Solar System is 4.5×10^9 years old, this implies that the population of asteroids larger than ~ 20 km should be progressively dominated by primordial objects; this is in agreement with the observed size distribution of these asteroids – see Fig. 15.

Analysis of the collision rates of objects in the Kuiper Belt (Stern 1995) shows that a collisional cascade should exist there too; there is also evidence to suggest that the Kuiper Belt was once more massive than it is today (Jewitt 1999), meaning that in the past collisions would have played a much larger role in determining its structure than they do today, maybe even causing the supposed mass loss (Stern and Colwell 1997). The size distribution of the observed Kuiper Belt objects appears to be slightly steeper than that in the inner Solar System ($q > 11/6$, Jewitt 1999), while observations have been unable, as yet, to determine its dust distribution (Backman et al. 1995; Gurnett et al. 1997).

The migration of small dust grains from the Kuiper Belt has been investigated by Liou et al. (1996), who discovered that only grains with diameter $\sim 1\,\mu$m survived collisions with interstellar dust particles to reach the inner Solar System. As the P-R drag lifetime for dust grains with diameter $\sim 10^2\,\mu$m is significantly longer, it is questionable whether such large grains could reach the inner Solar System without suffering some disruption due to mutual collisions with other Kuiper Belt dust grains or with interstellar dust particles. An investigation by Grogan et al. (1996) concluded that interstellar dust particles themselves provide only a minor contribution to the zodiacal cloud. In this chapter, we therefore focus on the orbital evolution of asteroidal and cometary dust particles with diameters in the range 1 to $10^3\,\mu$m. The dominant non-gravitational forces acting on these particles in the inner Solar System are radiation pressure, P-R drag and solar wind drag. Particles smaller than this are affected by the Lorentz force, which can also become important for larger dust particles in the outer Solar System (Leinert and Grün 1990). Whereas meter-sized bodies are acted upon by the Yarkovsky effect which, although not directly relevant to the dynamical behaviour of micron-sized dust particles, can lead to significant changes in the orbital distribution of their source bodies (Bottke et al. 2000). These effects are beyond the scope of the discussion presented here.

III. ORBITAL EVOLUTION

The orbital evolution of a given dust particle in the cloud will also be affected by both secular (long-period), resonant, and short-period gravitational perturbations imposed by the planets. Secular perturbations give rise to long term variations in a particle's orbit (as discussed below) and short-period perturbations can lead to gravitational scattering. A resonant perturbation arises when two periods or frequencies are in a simple numerical ratio and can lead to large changes in a particle's orbit. A mean motion resonance occurs when such a relationship exists between the orbital periods of two bodies (this type of resonance is discussed further in Section VI). In the case of secular resonance, the relevant frequencies are the rates of change of the proper longitude of pericenter ($\dot{\varpi}_p = A$) or proper longitude of ascending node ($\dot{\Omega}_p = -A$) of the particle, and one of the eigenfrequencies of the planetary system (g_k or f_k respectively, see Eqs. 27 and 28 below). In order to determine the secular evolution of asteroidal dust particles (i.e., particles in low eccentricity and inclination orbits), low order secular perturbation theory may be employed (e.g., Brouwer and Clemence 1961; Dermott and Nicholson 1986; Murray and Dermott 1999; Wyatt et al. 1999b); as the secular evolution of the particle's complex eccentricity, z, and complex inclination, y, are decoupled (see Eqs. 21 and 22). This low order theory is not, however, suitable for determining the secular evolution of cometary dust particle orbits with moderate to high eccentricities or inclinations; as above a certain threshold value the secular evolution of the particle's complex eccentricity and inclination are no longer decoupled. These threshold values for eccentricity, e, and inclination, I, have not yet been accurately determined, but our investigations indicate that a maximum e and $\sin I$ of ~ 0.2 may be adopted.

The gravitational perturbations imposed on a particle's orbit by a planetary system containing N_{pl} bodies with masses M_j, orbiting a central star of mass $M_\star$, can be described by the particle's disturbing function, R. This function can be decomposed into the sum of many terms, and those which do not depend on the mean longitudes of either the particle or the planets (i.e., long-period terms) can be identified as contributing to the secular perturbations, R_{sec}. To second order in the eccentricities, e, and inclinations, I,

$$R_{sec} = na^2 \left[\frac{1}{2} A \left(e^2 - I^2 \right) + \sum_{j=1}^{N_{pl}} A_j e e_j \cos \left(\varpi - \varpi_j \right) \right. \\ \left. + \sum_{j=1}^{N_{pl}} B_j I I_j \cos \left(\Omega - \Omega_j \right) \right], \quad (17)$$

where ϖ is the longitude of pericenter, Ω is the longitude of ascending node, n is the mean motion of the particle (accounting for the effect of radiation pressure), and

$$A = +\frac{n}{4(1-\beta)} \sum_{j=1}^{N_{pl}} \left(\frac{M_j}{M_\star} \right) \alpha_j \bar{\alpha}_j b_{3/2}^{(1)}(\alpha_j), \quad (18)$$

$$A_j = -\frac{n}{4(1-\beta)} \left(\frac{M_j}{M_\star} \right) \alpha_j \bar{\alpha}_j b_{3/2}^{(2)}(\alpha_j), \quad (19)$$

$$B_j = +\frac{n}{4(1-\beta)} \left(\frac{M_j}{M_\star} \right) \alpha_j \bar{\alpha}_j b_{3/2}^{(1)}(\alpha_j), \quad (20)$$

where $\alpha_j = a_j/a$ and $\bar{\alpha}_j = 1$ for $a_j < a$; $\alpha_j = \bar{\alpha}_j = a/a_j$ for $a_j > a$; a is the semi-major axis, and $b_{3/2}^{(s)}(\alpha_j)$ are the Laplace coefficients ($s = 1, 2$). A, A_j and B_j are in units of radian s^{-1}, R_{sec} is in units of m^2 s^{-2}; quantities subscripted j refer to the jth planet and unsubscripted quantities refer to the particle.

The effects of these secular perturbations are such that the semi-major axis remains constant while the eccentricity and inclination vary in a manner coupled with the variations of its longitude of pericenter and ascending node, described by the complex eccentricity z and complex inclination y,

$$z = e \exp i\varpi, \tag{21}$$

$$y = I \exp i\Omega. \tag{22}$$

Lagrange's planetary equations then give the eccentricity and inclination variations due to secular perturbations as

$$\dot{z}_{\text{sec}} = +iAz + i\sum_{j=1}^{N_{\text{pl}}} A_j z_j, \tag{23}$$

$$\dot{y}_{\text{sec}} = -iAy + i\sum_{j=1}^{N_{\text{pl}}} B_j y_j, \tag{24}$$

where z_j and y_j are the complex eccentricities and inclinations of the perturbers, which vary over time, t, according to

$$z_j(t) = \sum_{k=1}^{N_{\text{pl}}} e_{jk} \exp i(g_k t + \zeta_k), \tag{25}$$

$$y_j(t) = \sum_{k=1}^{N_{\text{pl}}} I_{jk} \exp i(f_k t + \gamma_k), \tag{26}$$

where g_k and f_k are the eigenfrequencies of the perturber system, the coefficients e_{jk} and I_{jk} are the corresponding eigenvectors, and ζ_k and γ_k are constants dependent on the initial conditions of the perturber system.

The solution of Eqs. (23) and (24), giving the secular evolution of the particle's instantaneous complex eccentricity and inclination (the osculating elements) can be decomposed into two distinct time-varying elements, the forced elements and the proper elements, to be added vectorially in the complex plane:

$$z(t) = z_{\text{f}}(t) + z_{\text{p}}(t)$$
$$= \sum_{k=1}^{N_{\text{pl}}} \left[\frac{\sum_{j=1}^{N_{\text{pl}}} A_j e_{jk}}{g_k - A} \right] \exp i(g_k t + \zeta_k) + e_p \exp i(+At + \zeta_0), \tag{27}$$

$$y(t) = y_{\text{f}}(t) + y_{\text{p}}(t)$$
$$= \sum_{k=1}^{N_{\text{pl}}} \left[\frac{\sum_{j=1}^{N_{\text{pl}}} B_j I_{jk}}{f_k + A} \right] \exp i(f_k t + \gamma_k) + I_p \exp i(-At + \gamma_0), \tag{28}$$

where e_p, ζ_0 and I_p, γ_0 are determined by the initial conditions of the particle.

These equations have simple physical and geometrical interpretations. A particle's forced elements, z_f and y_f, depend only on the orbits of the perturbers in the system (that have a slow secular evolution, Eqs. 25 and 26), as well as on the particle's semi-major axis (which has no secular evolution). Thus, at a time t_0, a particle that is on an orbit with a semi-major axis a, has forced elements imposed on its orbit by the perturbers in the system that are defined by $z_f(a, t_0)$ and $y_f(a, t_0)$. The contribution of the particle's proper elements to its osculating elements, $z(t_0)$ and $y(t_0)$, is then given by: $z_p(t_0) = z(t_0) - z_f(a, t_0)$ and $y_p(t_0) = y(t_0) - y_f(a, t_0)$; thus defining the particle's proper eccentricity, e_p, and proper inclination, I_p, which are its fundamental orbital elements (i.e., those that the particle would have if there were no perturbers in the system), as well as the orientation parameters ζ_0 and γ_0. Since both the forced elements, and the osculating elements, of collisional fragments are similar to those of their parent (apart from fragments with $\beta > 0.1$, see Eqs. 7–9), particles from the same family have almost the same proper elements, e_p and I_p.

The evolution of a particle's proper elements is straight-forward — they precess around circles of fixed radius, e_p and I_p, at a constant rate, A, counterclockwise for z_p, clockwise for y_p. The secular precession timescale depends only on the semi-major axis of the particle's orbit:

$$t_{\text{sec}} = 2\pi/At_{\text{year}}, \tag{29}$$

where t_{sec} is given in years, t_{year} is one year measured in seconds, and A is given by Eq. (18); secular perturbations produce long period variations in a particle's orbital elements (e.g., $t_{\text{sec}} \sim 10^5$ yr in the asteroid belt). The centers of the circles that the proper elements precess around are the forced elements (Fig. 3).

Now consider the family of collisional fragments originating from a primordial body, the orbital elements of which were described by a, e_p, and I_p. Here we consider only fragments that are unaffected by Poynting-Robertson drag, i.e. the largest particles in the distribution. The orbital elements of the largest fragments, those with $\beta < 0.1$, created in the breakup of the primordial body are initially very close to those of the primordial body; they do not have identical orbits due to the velocity dispersion imparted to the fragments in the collision. The forced elements imposed on the orbits of all of these collisional fragments are similar to those imposed on the primordial body (as their semi-major axes are almost the same). The secular evolution of their proper eccentricities and inclinations is to precess about the forced elements (which are also varying with time), but at slightly different rates (due to their slightly different semi-major axes). A similar argument applies for all particles created by the collisional breakup of these fragments. Thus, after a few precession timescales, the complex eccentricities and complex inclinations of the collisional fragments of this family lie evenly distributed around circles that are centered on $z_f(a, t)$ and $y_f(a, t)$, and that have radii of e_p and I_p (e.g., their complex eccentricities lie on the circle shown in Fig. 21 (left)), while their semi-major axes are all still

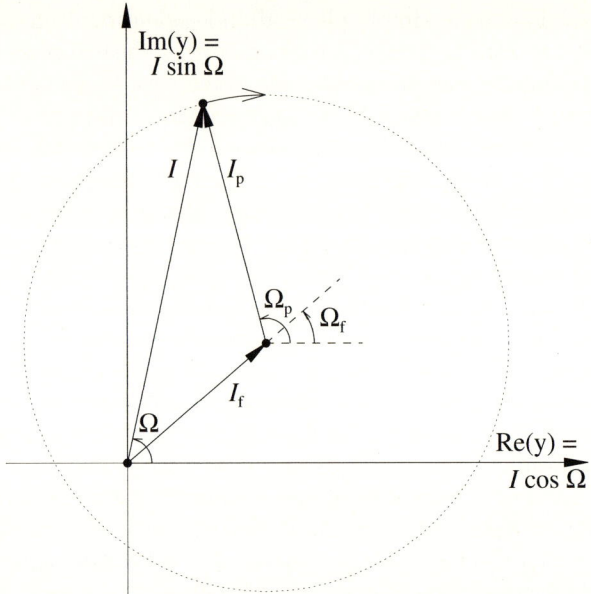

Figure 3. Schematic showing the vectorial combination of the proper and forced inclinations (I_p, I_f) and their ascending nodes (Ω_p, Ω_f) to give the osculating inclination and node (I, Ω). Reprinted with permission from Grogan et al. (2001).

close to a. This is seen to be the case in the asteroid belt: there are families of asteroids that have similar a, e_p, and I_p, that are the collisional fragments resulting from the breakup of a much larger asteroid (Hirayama 1918).

III.A. P-R Drag Affected Orbits

The solution given by Eqs. (27) and (28) accounts for the fact that small particles see a less massive star due to the action of radiation pressure, but not for the P-R drag evolution of their orbits. To find the secular evolution of the orbital elements of a particle that is affected by P-R drag, the equations governing the evolution of its complex eccentricity and inclination (Eqs. (23) and (24)), must both be solved in conjunction with the P-R drag evolution of its semi-major axis and eccentricity (Eqs. (10) and (11)). While the solution given by Eqs. (27) and (28) is no longer applicable, the decomposition of the particle's complex eccentricity and complex inclination into forced and proper elements, and the physical meaning of these elements, is still valid; however, each of these elements now depends on the particle's dynamical history.

Consider the P-R drag affected particles created by the breakup of an asteroid family group. Immediately after they are created, the osculating orbital elements of these particles are similar to those of the rest of the family; i.e., they have similar semi-major axes, and complex eccentricities and complex inclinations that are uniformly distributed in these planes around circles of radii approximately equal to the proper elements of the family, e_p and I_p. The dynamical evolution of a wave of these particles, i.e., those that were created at the same time, can be followed by numerical integration to ascertain how the orbital elements of the particles in the wave vary as their semi-major axes decrease due to P-R drag; this is the "particle in a circle" method (Dermott et al. 1992). The complex eccentricities and complex inclinations of a wave of particles originating in the asteroid belt remain on circles, and the wave's semi-major axis, a_{wave}, decreases: its effective proper eccentricity (the radius of the wave's circle in the complex eccentricity plane) decreases $\propto e_{\text{p}}(a_{\text{wave}}/a)^{5/4}$; its effective proper inclination (the radius of the wave's circle in the complex inclination plane) remains constant at I_{p}; the distributions of the particles' ϖ_{p} and Ω_{p} remain random; while its effective forced elements (the centers of the circles in the complex eccentricity and complex inclination planes) have a more complicated variation. Figure 4 shows this behavior in the variation of the inclination distribution of a wave of 40 μm diameter dust particles derived from the Eos family. The radius of the circle represents the proper inclination I_{p} which remains unchanged until the mean semi-major axis of the wave approaches Earth where many of the particles are gravitationally scattered.

Thus, the orbital element distributions, $n(z)$ and $n(y)$, of P-R drag affected particles are like that of the large particles, in that they are the vector addition of forced elements, z_{f} and y_{f}, to symmetrical proper element distributions; however, their forced and proper elements are different for particles from different families, as well as being different for particles of different sizes and with different orbital semi-major axes. The presence of forced elements, and their variation, can lead to asymmetries in the large-scale distribution of dust particle orbits that can be detected observationally (see Sections IV and V for details) to provide information about the perturbers in the system, as well as the properties of the dust particles themselves.

III.B. Numerical Simulations

We have developed a unique integration code specifically designed for evolving the orbits of large populations of dust particles under the effects of radiation pressure, P-R drag and solar wind drag, as well as point-mass gravitational forces. To achieve this, we have applied the dissipative mapping technique (Malhotra 1994) to the specific problem of deriving a MVS (Mixed Variable Symplectic) type integration code (Wisdom and Holman 1991) that also incorporates the effects of these non-gravitational forces (Burns et al. 1979). The development and testing of this dissipative code, significantly faster than more conventional integration techniques, is described in detail elsewhere (Kehoe 1999; Kehoe et al. 2001). We have employed the code to evolve representative samples of asteroidal dust particles forward in time to the present epoch, along with the planets Jupiter, Saturn, Uranus, and Neptune, from a number of different epochs in the past (Kehoe et al. 2002). As the timescale for a dust particle orbit to decay under the effect of P-R and solar wind drag is dependent on the particle size (Eqs. (6) and (12)), each set of past epochs chosen was dependent on the size of the dust particles considered, and a separate set

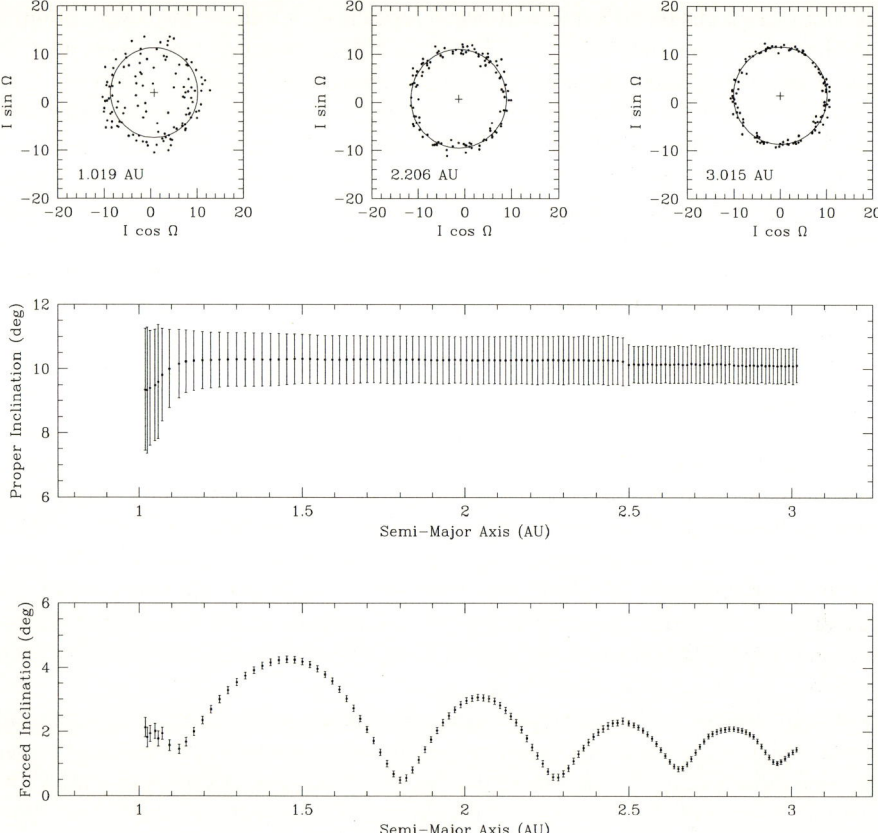

Figure 4. Variation of the inclination distribution of a wave of 40 μm diameter dust particles derived from the Eos family. The particles were released 2×10^5 years ago with a mean semi-major axis of $a = 3.015$ AU and migrated from the asteroid belt toward the Sun due to the action of drag forces. The mean semi-major axis of the particles in the wave reached $a = 1$ AU at the present time. The variation in the "center of mass" (marked by a cross in the upper three panels) corresponds to the variations of the forced elements, I_f and Ω_f, of the particles. The radius of the circle represents the proper inclination, I_p which remains unchanged until the mean semi-major axis of the wave approaches Earth where many of the particles are gravitationally scattered. Note that the forced elements are calculated at different times in the past (Grogan et al. 2001).

of integrations had to be carried out for each different particle size. Up to 80 past epochs were selected for each particle size, in order to provide a comprehensive picture of the forced element distribution of asteroidal dust particles across a wide range of semi-major axis values in the inner Solar System at the present time. Here we consider asteroidal dust particles (originating in this case from the Eos family, although this is not critical) composed of astronomical silicate of density 2,500 kg m^{-3} with diameters 10, 100 and 200 μm, for which we calculated β values (the ratio of radiation pressure to solar gravity, Eq. 4) of 0.04871, 0.00446 and 0.00221 respectively, using Mie theory. The longest integrations performed for the 10, 100 and 200 µm diameter dust particles were for timescales of 0.06, 0.6 and 1.2 Myr, respectively.

To obtain initial orbital element distributions for our forward integrations we first employed a standard MVS integration code (incorporating point-mass gravitational forces only) to evolve Eos family asteroids, along with the gas giant planets, backwards in time from the present. Initial osculating orbital elements for 444 Eos family asteroids were obtained from The Asteroid Orbital Element Database (Bowell 1997) for the epoch of Julian Date 2450700.5, using the family classification of Zappalà et al. (1995). Osculating orbital elements for the planets were obtained for the same epoch using the data from Standish et al. (1992). Using the particle on a circle method we then generated initial osculating orbital elements for 124 dust particles, representative of the whole Eos asteroid family, at each of the past epochs required.

The results of the integrations presented in Fig. 5 represent a total of over 4 months CPU time running on a variety of Pentium processors. All orbital elements are heliocentric and given with respect to the mean ecliptic and equinox of the standard J2000 reference frame. In the region of the main asteroid belt (between 2.5 and 3 AU), the forced elements of the large particles display similar behaviour to that of the small particles. That is, their forced elements are locked onto Jupiter's osculating elements such that

$$z_\mathrm{f} \approx \left[b_{3/2}^{(2)}(\alpha_j)/b_{3/2}^{(1)}(\alpha_j)\right] e_j \exp i\tilde{\omega}_j, \qquad (30)$$

$$y_\mathrm{f} \approx I_j \exp i\Omega_j. \qquad (31)$$

The low dispersion of the inclinations and nodes in this region of the main belt, regardless of particle size, is the fundamental reason why dust bands are observed at these heliocentric distances. However, as the large dust particles encounter the ν_{16} secular resonance at the inner edge of the asteroid belt (at about 2 AU), the effect of the resonance disperses their forced inclinations and nodes, diffusing the dust band particles into the broad-scale zodiacal background. The ν_6 secular resonance (also at about 2 AU) produces analogous behaviour in the forced eccentricities and longitudes of pericenter of the dust particles. The effects of these secular resonances are more pronounced for the large dust particles because they are acted on by the resonances for longer periods of time. The orbital element distributions of large asteroidal dust particles produced by intra-family collisional attrition therefore lose their characteristic family signatures in the inner region of the main belt and become indistinguishable from the general background cloud of zodiacal dust. We also expect that as the dust particles spiral in further towards the Sun, gravitational scattering by the terrestrial planets will act to disperse the particles even more. This effect should be particularly marked for the more slowly evolving large dust particles as the probability of a close planetary encounter is greater. However,

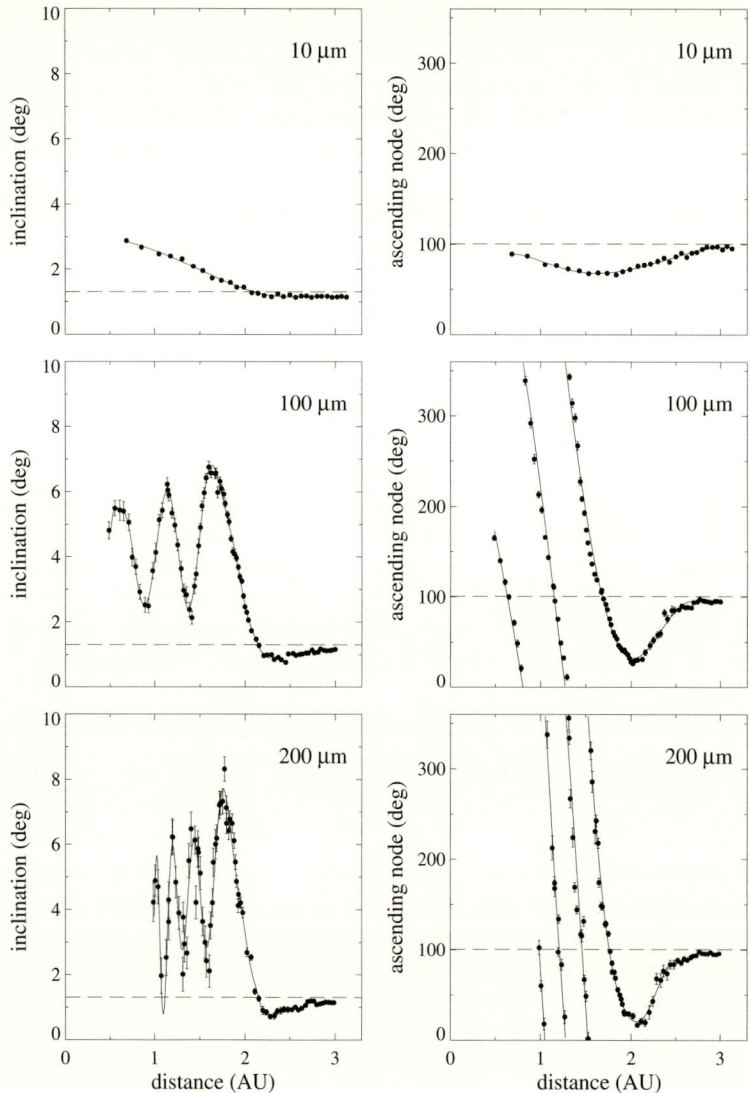

Figure 5(a). Variation of the forced inclination I_f (left), and the forced longitude of ascending node Ω_f (right), with heliocentric distance for Eos family dust particles at the present epoch (Julian Date 2450700.5). The dashed lines show the present osculating inclination (left) and osculating longitude of ascending node (right) for Jupiter (Kehoe et al. 2002).

this result can not be ascertained from the integrations described above as the terrestrial planets have not been included here, but will be added in future models. The action of secular resonances also means that large asteroidal dust particles in the inner Solar System may have orbits with significant eccentricities and inclinations, comparable to some cometary orbits.

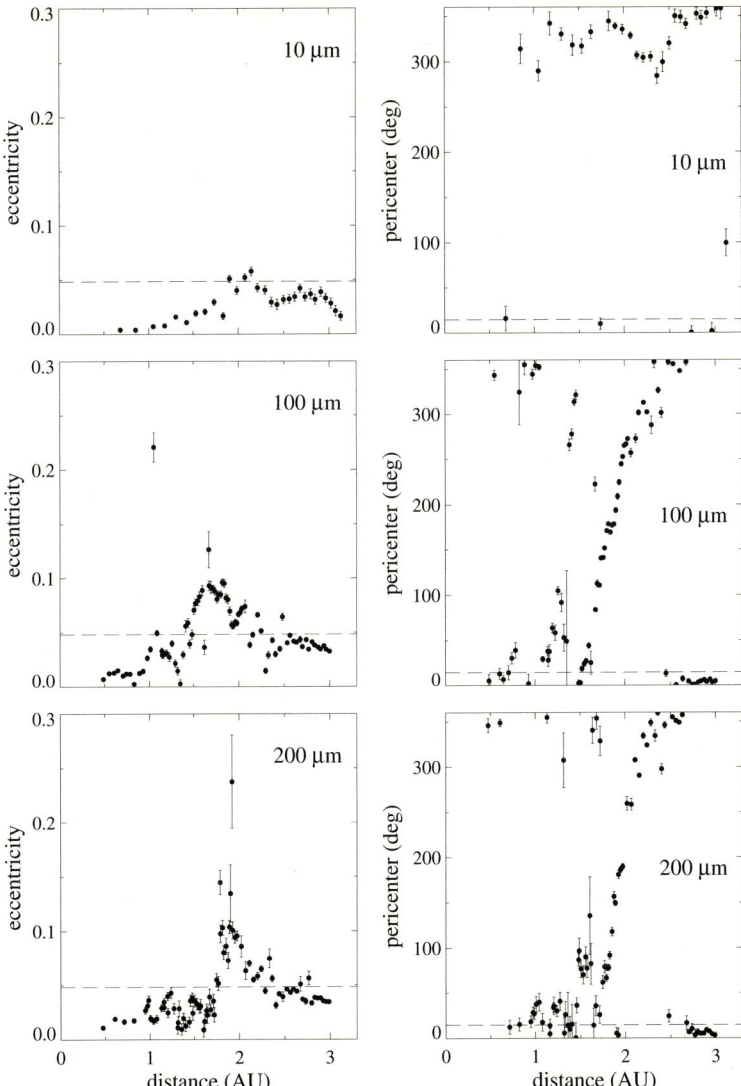

Figure 5(b). Variation of the forced eccentricity e_f (left), and the forced longitude of pericenter ϖ_f (right), with heliocentric distance for Eos family dust particles at the present epoch (Julian Date 2450700.5). The dashed lines show the present osculating eccentricity (left) and osculating longitude of pericenter (right) for Jupiter.

In the future we will extend our knowledge of asteroidal dust particle dynamics to a wider range of particle sizes, and address the main belt (non-family) contribution as well as the dust band (family) component on the way to our ultimate goal of providing a physically motivated, global model for the zodiacal emission. The dynamics of cometary material is a separate issue, we discuss this in Section III.D.

III.C. SIMUL - Visualizing the Orbital Distribution

The knowledge of the orbital distribution of a population of interplanetary dust particles demands some mechanism by which these orbits can be visualized, so that they can be compared to observational data. We have developed a FORTRAN algorithm, SIMUL, for this purpose (Dermott et al. 1988a).

The basic ideas and assumptions behind SIMUL are as follows:

1. A cloud is represented by a large number of dust particle orbits. The total cross-sectional area of the cloud is divided equally among all the orbits.
2. The orbital elements of the dust particle orbits in the cloud can be decomposed into proper and forced vectorial components. When inclination and eccentricity are low, as is typically the case for asteroidal type orbits, at any given time the forced elements are independent of the proper elements and depend only on the semi-major axis and the particle size.
3. As a first approximation, the dust particles in the cloud produced by asteroid families have the same mean proper elements as those of the parent bodies, although the gaussian distribution of these elements is a free parameter.
4. The forced elements as a function of semi-major axis and size are calculated using secular perturbation theory via direct numerical integrations, as outlined above.
5. The semi-major axis of each orbit is chosen randomly from a given radial distribution, and the remainder of the orbital elements are sampled from the distributions found from the numerical integrations.
6. Along each of the orbits, particles are distributed according to Kepler's Law. Once the spatial distribution of the orbits is specified, space is divided into a sufficiently large number of ordered cells and then every orbit is investigated for all the possible cross-sectional area contributions to each of the space cells. The model generates a large three-dimensional array which serves to describe the spatial distribution of the effective cross-sectional area.
7. The viewing geometry of any telescope can be reproduced exactly by calculating the Sun-Earth distance and ecliptic longitude of Earth at the observing time and setting up appropriate coordinate systems. In this way, IRAS-type brightness profiles can be created and compared with the observed profiles.

Although waves of only $\sim 10^2$ individual dust particles have been evolved to produce the forced elements shown in Fig. 5, the distributions are used to populate SIMUL models with $\sim 10^8$ individual orbits. These smooth models of individual components of the cloud for each particle size (which we can then weight according to any given form of size distribution), can then be compared directly with observational data. In this way we ensure that the effects of secular perturbations are fully incorporated into the models. We will discuss how we have produced accurate models for the Solar System dust bands, and how we have used these models to predict the asteroidal contribution to the zodiacal cloud in Section IV.

III.D. Cometary Particles

The high eccentricity of typical cometary orbits renders them unsuitable for study using the particle in a circle method. Following earlier work by Liou et al. (1995), we have therefore performed direct numerical integrations of a representative sample of cometary orbits in order to determine the orbital evolution of the cometary population of the zodiacal cloud (Kortenkamp and Dermott 1998a). The initial distribution of 10 μm cometary dust particles is generated from a set of orbital elements for 175 short-period Jupiter family comets. This is a subset of the MPC catalog (Marsden 1995). As of December 1995, this set represented most of the known short-period Jupiter family comets with established orbits. Because we have fewer cometary than asteroidal parent bodies we use Eqs. (7) and (8) to generate a set of 50 dust particle orbits from each cometary orbit. We randomly distributed ϖ, Ω, and the mean anomaly, λ, between 0° and 360°. Because most of these cometary dust particles are Jupiter-crossing one suspects that their orbital evolution is strongly influenced by Jupiter, at least initially. Liou and Zook (1996) have shown that some dust particles originating from the short-period comet Temple 2 (one of our 175 comets) can be injected directly into the 1:2 interior mean motion resonance with Jupiter near 3.2 AU. Some of the particles in their study remained trapped in this resonance for thousands of years and had their eccentricities significantly reduced by the resonant forces. Our primary concern in the handling of the cometary dust particles was to accurately account for this effect.

We used the RADAU fifteenth-order integrator program of Everhart (1985) with variable time steps taken at Gauss-Radau spacing to investigate the dynamical evolution of cometary particles. All of our numerical simulations with RADAU include gravitational interactions with seven planets (Mercury and Pluto are excluded) and include the effects of radiation pressure, P-R drag, and solar wind drag. We numerically evolved the cometary dust particles with RADAU until their orbits had decayed into the Sun or were ejected from the Solar System. Some dust particles decayed into the Sun in less than 30,000 years. Other dust particles that were trapped in the Jovian mean motion resonances (trapping for some lasted as long as 100,000 years) required nearly 150,000 years of integration before decaying into the Sun. Figure 6 gives ten examples of the wide range of evolutionary paths we found for these cometary dust particles. In the course of our study we scrutinized the evolution of 7,114 cometary dust particles with initial semi-major axis $a \leq 6$ AU. We found that 1,330 particles ($\sim 20\%$) were trapped in various mean motion resonances with Jupiter. The remaining 5,784 particles evolved into the inner Solar System without having been previously trapped in mean motion resonances with Jupiter. Typically this trapping leads to a decrease in the dust particle's orbital eccentricity (see discussion following Eq. (34)). Eventually the trapped dust particles escape from the resonances and P-R light drag and solar wind drag then cause their orbits

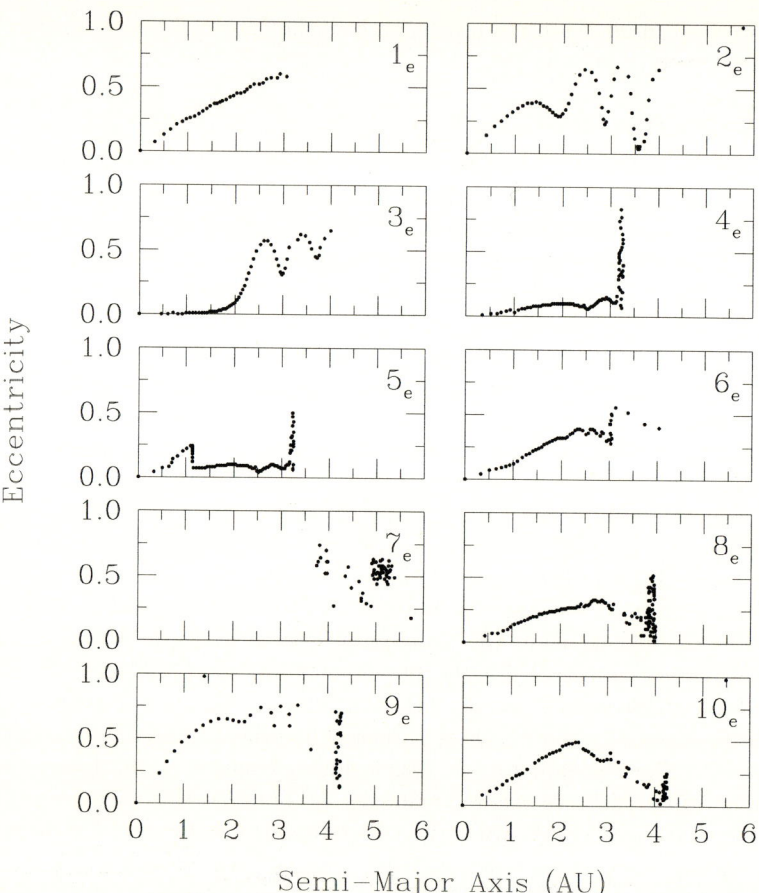

Figure 6. Ten examples showing the diversity of the evolutionary paths followed by cometary dust particles. Points are plotted every 1,000 years and the evolution typically proceeds towards smaller semi-major axes, with the exception of Example 7 (Kortenkamp and Dermott 1998a).

to decay through the inner Solar System. Figure 7 shows the Earth-crossing eccentricity and inclination (top two panels) of both the previously trapped set, and the set that were not previously trapped. The significantly lower eccentricity and slightly lower inclination of the previously trapped set results in lower atmospheric entry velocities (third panel) and a higher average spatial density at 1 AU, which directly translates into a higher capture rate (bottom panel). The cometary dust particles that were previously trapped are about ten times more likely to be captured by Earth than those that were not previously trapped. Because 20% of the total population was trapped this indicates that the ratio of previously trapped to untrapped cometary IDPs in the atmosphere may be as high as 2/1. This is of particular importance when one considers the practice of classifying collected IDPs as asteroidal or cometary based upon

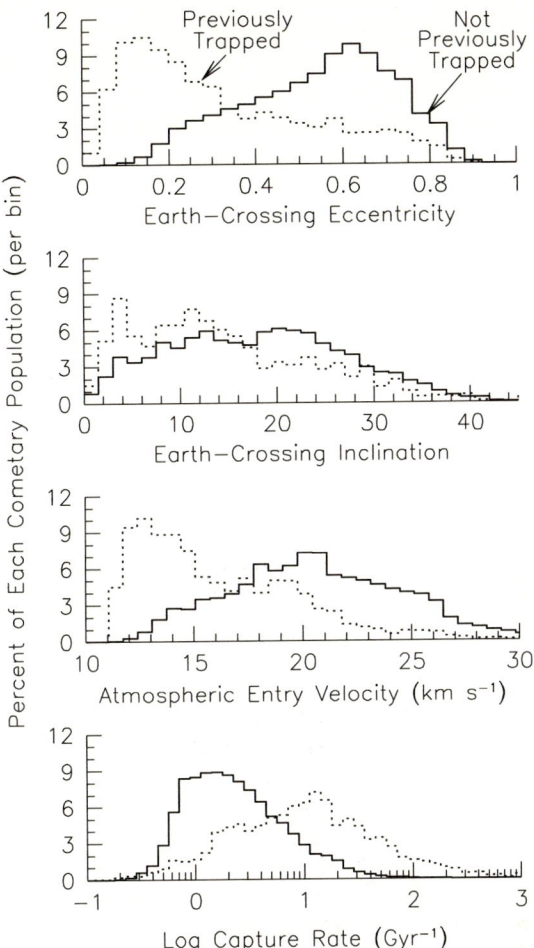

Figure 7. Comparison of two populations of Earth-crossing cometary dust particles — those that were previously trapped in mean motion resonances with Jupiter (dashed lines) and those not previously trapped (solid lines). The top two panels show the Earth-crossing eccentricity and inclination. The third panel shows the atmospheric entry velocity ($v_{\text{atm}}^2 = v_e^2 + v_0^2$, where v_e is the escape velocity of Earth and v_0 is the encounter velocity). Dust particles in the previously trapped set have low entry velocities that are indistinguishable from entry velocities of typical asteroidal dust particles. The bottom panel shows the capture rate as given by Eq. (38). Dust particles that were previously trapped have a capture rate about 10 times higher than dust particles that were not previously trapped. Reprinted from Kortenkamp et al. (2001) with kind permission from Kluwer Academic Publishers.

their inferred atmospheric entry velocities (Flynn 1989). In fact, cometary dust particles that were previously trapped in mean motion resonances with Jupiter will have atmospheric entry velocities indistinguishable from typical asteroidal dust particles.

Once we have obtained a satisfactory cometary particle distribution we will generate a SIMUL model and use it to constrain the relative contribution of cometary and asteroidal material in the cloud. Preliminary work on this problem was discussed by Liou et al. (1995). Much work remains to be done: we must investigate a much wider range of cometary particle sizes, and then face the prospect of combining the results into some form of size distribution. Until we complete this work we must continue to extract as much information from the more tractable dynamics of the asteroidal material, and attribute the remainder of the zodiacal signal by default to cometary emission.

IV. DUST BANDS

One of the most basic of all zodiacal cloud questions, namely: "What is the relative contribution of asteroidal and cometary material?", remains perhaps the most difficult to answer. The problem is that we have little information on the dust production rates from the various sources, and even armed with this knowledge we would still be faced with unraveling the complex dynamical processes to which the particles are subject over their lifetime. Mapping of the zodiacal brightness distribution is simply not enough to constrain the problem. The Solar System dust bands (Low et al. 1984) are of fundamental importance in this regard because they are discrete features which have been unambiguously related to the breakup of asteroidal material. In particular, we argue that they are associated with the collisional debris of the Hirayama asteroid families (Dermott et al. 1984; Sykes and Greenberg 1986; Grogan et al. 1997; Reach et al. 1997) and as such they represent a unique observational constraint on the contribution of asteroidal material to the cloud.

Figure 8 shows an IRAS brightness profile of the cloud, along with the results of passing the profile through a fast Fourier filter to isolate the near-ecliptic dust band features. They appear as "shoulders" superimposed on the background emission at $\pm 10°$, and a "cap" near the ecliptic plane. A dust band is a toroidal distribution of asteroidal dust particles with both common proper inclinations and common forced inclinations and nodes. The particles' common proper inclination derives from their common source in a given asteroid family, and their common forced inclinations and nodes result from the dominant perturbing force of Jupiter in the asteroid belt (as discussed in Section III). After a collisional event within an asteroid family, secular perturbations act to distribute the proper longitudes of ascending node of the particles around the sky on a timescale of order 10^5 years (Eq. (29)). Since particles in inclined orbits spend a disproportionate amount of time at the extremes of their vertical harmonic oscillations, a set of such orbits with randomly distributed proper nodes will give rise to two apparent bands of particles symmetrically placed above and below the mean plane of the system (Neugebauer et al. 1984). This gives a natural explanation for the shoulders on the IRAS profiles at approximately $\pm 10°$. Similarly, the central cap may be simply explained as a low inclination dust band. Any dispersion in the proper inclinations of the dust particles will

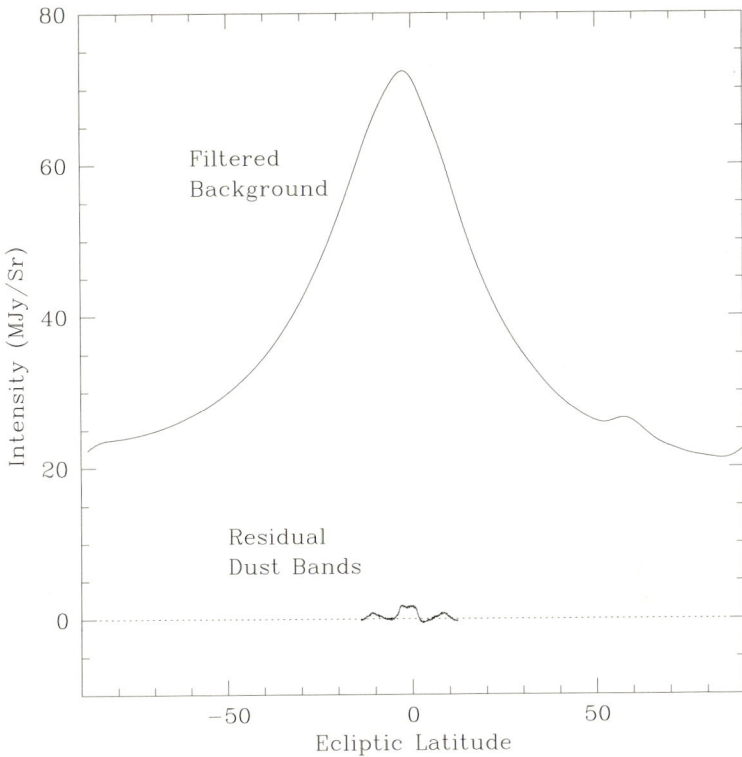

Figure 8. IRAS observation of the zodiacal cloud in the 25 μm infrared wave band. This observation was made at 90° solar elongation (the angle between the telescope pointing direction and the Earth-Sun line) in the direction leading Earth in its orbit when the planet was at an ecliptic longitude of 293°. The dust bands can be seen as projecting "shoulders" near latitudes of ±10° and 0°. The structure around 60° latitude is due to dust in the plane of the Galaxy. By applying a Fourier filter to the IRAS observation, a smooth background profile is separated from the high frequency dust band profile (shown at bottom). This filtered high frequency dust band profile is merely a residual representing the "tip of the iceberg" in terms of the dust band material in the zodiacal cloud (Dermott et al. 1994b; Grogan et al. 1997; Kortenkamp and Dermott 1998a).

lead to the dust band profile appearing broader, with the peak intensity shifted to a lower latitude (Dermott et al. 1990; Grogan et al. 1997).

Particles in cometary type orbits have high orbital eccentricities, and secular gravitational perturbations due to the planets produce large variations in these eccentricities, which are in turn coupled to variations in their inclinations (Liou et al. 1995). Therefore even if a group of cometary type orbits initially had identical inclinations, secular perturbations would disperse those inclinations over a wide range on a timescale of a few precession periods, showing that it is impossible for a comet to produce a well defined dust band.

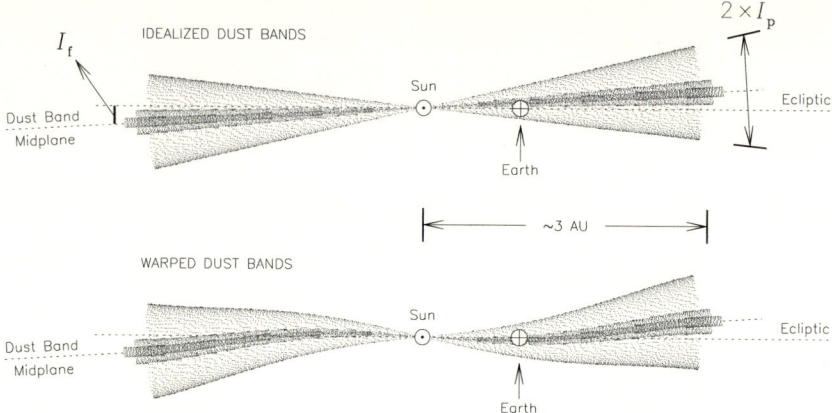

Figure 9. (Top) Cross-sections of *idealized* dust bands. Dust particles are produced by the gradual comminution of asteroid families and decay toward the Sun under the influence of drag forces. The narrow Themis and Koronis dust bands are embedded within the wider Eos dust band. The angular width of each dust band is twice the proper inclination (I_p) of its parental asteroid family. The midplane of the dust bands is inclined to the ecliptic by the forced inclination (I_f). Earth ($\oplus$) orbits the Sun within the dust bands. The spatial density of dust particles is enhanced near their extremes in latitude, which results in the "bands" of emission that were observed by the Infrared Astronomical Satellite (see Fig. 8). (Bottom) The real dust bands have warped midplanes due to variation in the forced inclination (I_f). The forced inclination is dependent on time, the semi-major axis of the decaying dust particle orbits, and on the diameter of the dust particle (see Fig. 5a). For clarity this diagram illustrates the warp obtained for a single size particle distribution at a fixed epoch. Reprinted from Kortenkamp et al. (2001) with kind permission from Kluwer Academic Publishers.

Figure 9 shows a schematic diagram of a cross-section through a dust band. The idealized dust band in the top panel has a constant forced inclination which dictates the inclination of the dust band to the ecliptic. In reality the dust bands have warped midplanes due to the variation in the forced inclination as a function of heliocentric distance, which we sketch in the lower panel. As we will discuss later, even this is a simplified picture as the forced inclinations are also a function of particle size and time, and interior to 2 AU the dispersion in the forced elements is large enough to completely degrade the integrity of the dust band structure. Figure 10 suggests the association of the dust bands with the major Hirayama asteroid families, showing the number of asteroids as a function of proper inclination with absolute visual magnitude $H < 11$. The family asteroid members are dominated by groups near 2° (Themis and Koronis) and 10° (Eos).

Dust band structures are not observed independently from the rest of the zodiacal cloud. The IRAS observations consist of a series of line of sight

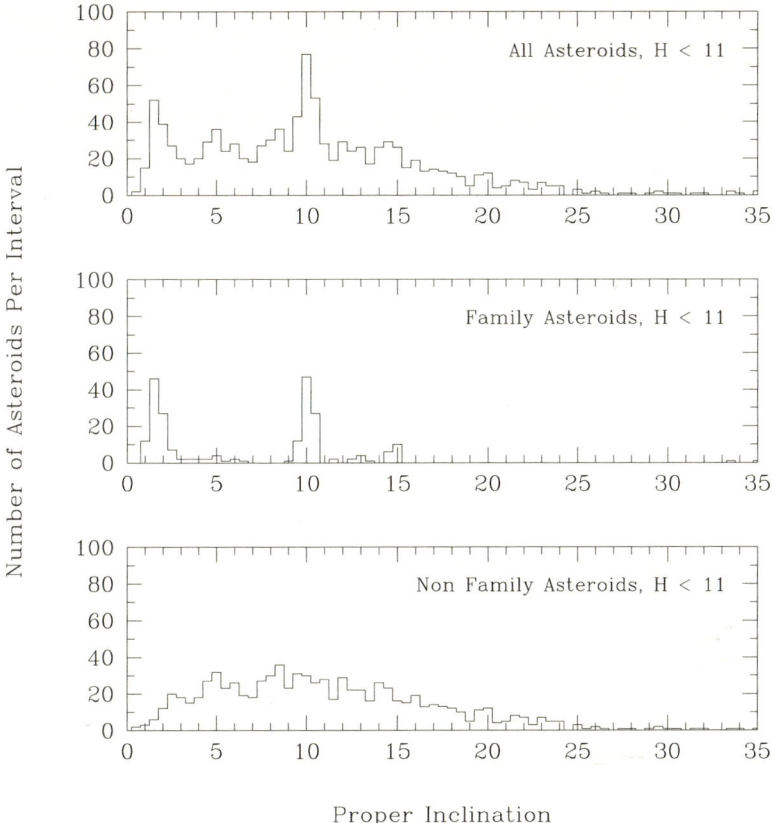

Figure 10. Histogram showing the number of asteroids versus proper inclination, I_p, for all asteroids with absolute visual magnitude $H < 11$. (Top) All 1,053 asteroids in the set. (Middle and Bottom) Asteroids associated with families and not associated with families, respectively. (Middle) The enhancement near 2° is due to the Themis and Koronis families and the enhancement near 10° is due to the Eos family. The next largest family, Maria, can be seen near 15° (Kortenkamp and Dermott 1998a).

brightness profiles taken through the zodiacal cloud as a whole and to study the bands they must somehow be isolated from the remainder of the cloud. Various techniques have been employed in the literature for this purpose: box-car averaging (Sykes 1990), background subtraction (Reach 1992; Jones and Rowan-Robinson 1993) and Fourier analysis (Dermott et al. 1986; Sykes 1988; Grogan et al. 1997; Reach et al. 1997). The important point is that isolating the dust bands is an arbitrary process. Two different techniques will produce two different sets of dust band residuals. Making the assumption that the residuals obtained from any of these processes gives the complete dust band structures is simply incorrect, because in any filtering process, the low-frequency component of the dust band structures will be indistinguishable from the low-frequency background zodiacal cloud, and the high frequency residuals will be merely the "tip of the iceberg". We have developed an iterative process (Dermott et al. 1994b; Grogan et al. 1997) to estimate the low-frequency component of the dust band using a combination of the observations and our dust band models. By using the same filter in the modeling process that we use to define the observed dust bands, and iterating, we are able to bypass the arbitrary divide associated with the filter, and extract the underlying low-frequency component of the dust bands which other techniques are unable to retrieve. This is essential in revealing the true extent to which asteroidal dust contributes to the cloud.

IV.A. IRAS Observations

The viewing geometry of the IRAS spacecraft was ideal for the study of the zodiacal cloud. The Medium Resolution (2′ in scan) Zodiacal Observational History File (ZOHF) consists of 5,757 sky brightness profiles, each providing a detailed view of the pole-to-pole cloud structure in a given line of sight defined by the ecliptic longitude of Earth, $\lambda_\oplus$, and the solar elongation angle, with most scans being taken at around 90° solar elongation. The changes in shape and amplitude of the dust band residuals from profile to profile are caused by a combination of the complex three-dimensional structure of the dust bands themselves and also the observing geometry of the IRAS satellite. The two primary causes for a change in the line of sight are: (i) the solar elongation angle, and (ii) the longitude of Earth. The changes due to these two parameters can be taken as independent to first order, allowing a quantitative parameter to be associated with each. Changes in elongation angle produce a parallax effect: there is a change in the effective distance to the bands, and hence in their observed peak latitude. For small changes in elongation angle the effect can be assumed to be linear. Characterizing the manner in which the brightness profiles change with elongation angle and longitude of Earth enables the thousands of individual IRAS scans to be reduced to a few representative profiles normalized to 90° solar elongation spaced around the sky in both trailing and leading directions, with an enhancement in the signal to noise ratio of more than an order of magnitude. Figure 11 plots the mean North/South peak latitude of the "ten degree" band for the normalized 25 μm scans. The sinusoidal variation indicates that the plane of symmetry of the bands, the plane about which on average the proper inclinations of the particles precess, is inclined to the ecliptic. This tilt of the plane of symmetry is due to the secular perturbations of the planets (discussed in Section III), and its orientation depends on the forced elements imposed on the dust particles. When viewed from Earth such a plane would appear as a sine curve, its amplitude equal to the inclination, I_f, of the plane with respect to the ecliptic (see Fig. 9). Also, the displacement from the ecliptic will be equal in the trailing and leading directions at the ascending and descending nodes. These values are taken from Fig. 11 and listed in Table 1. Notice that for the dust band particles, the forced inclination and node are close to Jupiter's inclination and node and quite different from that of the zodiacal background cloud, strong evidence for the fact that this material is located in the asteroid belt and dominated by Jupiter's influence.

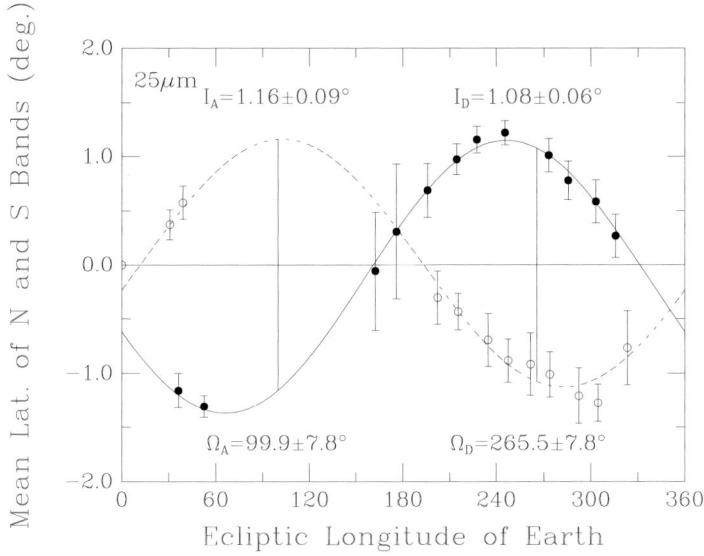

Figure 11. Variation of the latitudes of the median points separating the North and South "ten degree" bands seen by IRAS at an elongation angle of 90° in the leading (open circles) and trailing (filled circles) directions. The vertical lines, where the North and South latitudes of the midpoints are equal and opposite, are associated with the ascending and descending nodes with respect to the ecliptic (Grogan et al. 2001).

IV.B. Modeling the Dust Bands

Our approach to providing a physical model for the various components of the zodiacal cloud, including the dust bands, is essentially a two-step process. (1) Given a postulated source of particles, we describe the orbital evolution of these particles, due to Poynting-Robertson and solar wind drag, using equations of motion that also include the effects of radiation pressure and planetary gravitational perturbations. (2) Once the dust particle orbits have been specified along these lines, the distribution is visualized in three dimensions via the FORTRAN code SIMUL (Dermott et al. 1988a; Grogan et al. 1997), taking into account the thermal and optical properties of the particles and their variation with particle size. The viewing geometry of any telescope can be reproduced exactly by calculating the Sun-Earth distance and ecliptic longitude of Earth at the observing time and setting up appropriate coordinate systems. In this way, IRAS-type brightness profiles can be created and compared with the observed profiles.

This approach differs from that found elsewhere in the literature. Taking the analysis of the dust bands as an example, other authors (Sykes and Greenberg 1986; Sykes 1990; Reach 1992; Reach et al. 1997) employ simple empirical formulations for their three-dimensional structure. Interpretation of the dust band observations relies on the assumption that the spatial distribution

TABLE 1.

Orientation of planes of symmetry with respect to the ecliptic (25 μm waveband).

Structural Feature	Ascending Node		Descending Node	
	Inclination	Node	Inclination	Node
Jupiter	1.31°	100.0°	1.31°	280.0°
Eos family	1.19°	97.1°	1.19°	277.1°
Themis family	1.22°	97.8°	1.22°	277.8°
Ten degree bands	1.16±0.09°	99.9±7.8°	1.08±0.06°	265.5±7.8°
Zodiacal Cloud (Ecliptic)	1.49±0.07°	58.4±2.3°	1.59±0.07°	232.8±2.3°

of material can be explained by various combinations of gaussians and power laws in which particles "migrate" into the inner Solar System from the source regions as expected by P-R drag (orbital inclinations remain constant). However no actual orbital evolution is performed, and the effects of secular perturbations on the dust particles are therefore ignored. The results of our numerical integrations show that such secular perturbations are highly significant, particularly for the larger particles (see Section III).

The availability of cheap, fast processors has recently allowed us to extend our numerical investigation of the dynamical history of dust particles to much larger sizes and enhance our previous models of the dust bands (Grogan et al. 1997) to include a size-frequency distribution (Grogan et al. 2001), rather than being restricted to particles of a single size. This is critical in our efforts to provide a model of the dust bands that can match the IRAS observations in multiple wavebands. Particles ranging in size from 1 to 100 μm are included, each of which we assume to be a Mie sphere composed of astronomical silicate (Draine and Lee 1984). We realize that particles larger than 100 μm in diameter will exist in the zodiacal cloud, but we have yet to obtain the complete dynamical history of these particles. We continue to work on the dynamics of particles up to and beyond 500 μm (some results are shown in Figs. 4 and 5), but in this regime we will have to start incorporating the effects of particle-particle collisions as the P-R drag timescales become longer than the collisional lifetimes. In addition, the nature of the size distribution will be a complex function of dust production rates, P-R drag rates, collisional lifetimes and the nature of particle-particle collisions and will, presumably, be some function of heliocentric distance.

The situation is further complicated by the fact that even if the debris of an asteroidal collision could be described by a power law, the size-frequency index q will reflect the characteristics of the parent. The equilibrium size distribution of the collisional cascade originating from a single asteroid has been shown to be a function of the impact strength of that asteroid (Durda and Dermott 1997). Thus, it is possible for the value of q associated with a given family to

TABLE 2.

Dust band model parameters — proper elements and cross-sectional areas. The material originating from each family is distributed into the inner Solar System as far as 2 AU according to a $1/r$ P-R drag distribution (Grogan et al. 2001).

Asteroid family	$a, \Delta a$ (AU)	$e, \Delta e$	$I, \Delta I$ (°)	Area (10^9 km^2)
Eos	3.015, 0.012	0.076, 0.009	9.35, 1.5	4.0
Themis	3.148, 0.035	0.155, 0.013	1.43, 0.32	0.35
Koronis	2.876, 0.026	0.047, 0.006	2.11, 0.09	0.35

be different from that of other families and different from the value for the background cloud. In the case of a "rubble-pile" (Davis et al. 1989), the value of q associated with the initial disruption may be significantly higher than that associated with the disruption of a solid, coherent asteroid. This provides us with further motivation to relate the dust bands to given parent bodies in the main belt. However, as a first step in answering the fundamental question of the extent to which large and small particles contribute to the dust band emission, we model the size-frequency distribution as a single power law. We will refine this assumption in the future when we have a better understanding of the role of the complicating factors outlined above.

The dust band model parameters input to SIMUL are listed in Table 2. For a given size-frequency index q, the total surface area of material associated with the model bands is adjusted until the amplitudes of the 25 μm model dust bands matches the 25 μm observations; q can then be varied until a single model provides a match in amplitude to the 12, 25 and 60 μm observations simultaneously. Figure 12 shows the best results of our modeling, comparing the dust band observations (solid curves) to the dust band models (dotted curves) in the 12, 25 and 60 μm wavebands. The models were constructed as described above, and have a size-frequency index q equal to 1.43. Large particles dominate this distribution. The amplitudes in all wavebands are well matched, and the shapes of the dust band models describe the variation in shape of the observations around the sky very well.

In essence, the wavebands act as filters through which different particle sizes in the cloud are seen. For a distribution in which the small particles dominate the total surface area ($q > 5/3$, Dohnanyi 1969), the 12 μm waveband preferentially detects emission from the smaller particles, and the 60 μm waveband preferentially detects emission from the larger particles. When q is too high, too many small particles are included in the model, and the amplitudes of the 12 μm models are too large. In addition, too few large particles are included and the amplitudes of the 60 μm models are too small. For a distribution in which large particles dominate ($q < 5/3$), it becomes difficult to discriminate the exact value of q since the particles in the distribution radiate like grey bodies (Gustafson 1994).

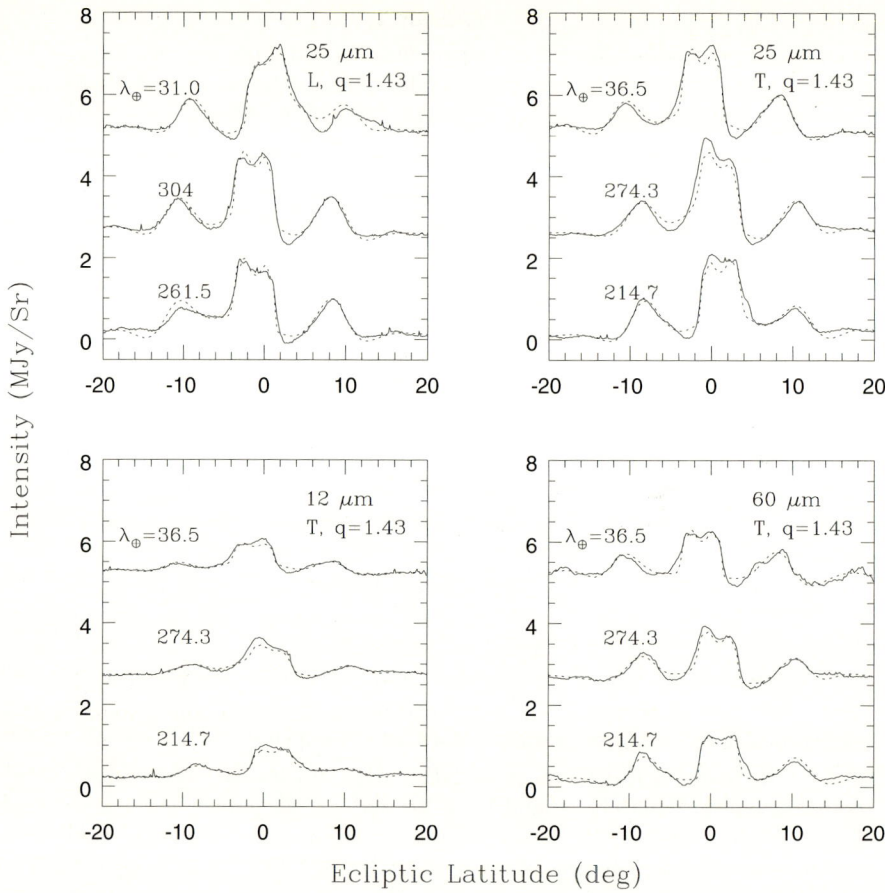

Figure 12. Filtered IRAS dust band profiles (solid lines) in three wavebands are compared with models with a size-frequency distribution, $q = 1.43$. All profiles were made at 90° solar elongation angle in a direction either leading (L) or trailing (T) the Earth in its orbit. The model dust bands (dashed curves) were constructed using particles from the Themis and Koronis families for the central band pair and from the Eos family for the 10° band pair. A dispersion of 1.5° was imposed on the proper inclination of the Eos material in this model and all of the material was confined to the asteroid belt ($2.0\,\text{AU} < a < 3.1\,\text{AU}$). The low value of $q(< 1.66)$ implies that the dominant particles are large (diameters $\lesssim 10^3\,\mu$m). Adapted with permission from Grogan et al. (2001).

A clear result from our modeling is that a high size-frequency index q, in which small particles dominate, fails to account for the observations of the Solar System dust bands. This index has to be reduced to the point where large particles dominate the distribution, and we place an upper limit of $q = 1.4$. This is consistent with the cratering record on the LDEF satellite (Love and Brownlee 1993) which suggests a q of approximately 1.15 at Earth and a peak in the particle diameter at around 100–200 μm. Since the Fourier filter preferentially isolates material exterior to the 2 AU secular resonance (in the inner Solar System the dust band material is dispersed into the background cloud due to the action of secular resonances), our results are more indicative of the size-frequency index of dust in the asteroid belt.

IV.C. The Importance of Secular Perturbations

The origin of the large dispersion in proper inclination (1.5°) required to successfully model the ten degree band (see Table 2), in rough agreement with the 1.4° found by Sykes (1990) and the 2° found by Reach et al. (1997), remains unclear, although the most likely source of the dispersion is simply the action of the secular resonance at 2 AU. However, this leaves open the question of why a large dispersion is required to model the ten degree band, and only the small dispersion of the Themis and Koronis families is required to successfully reproduce the central band observations. One answer may be that the emission associated with the central band is due to relatively recent collisions within these families. Figure 13 shows the variation with time of the total cross-sectional area associated with the main belt and describes the stochastic breakup of asteroidal fragments. This numerical approach to describing the collisional evolution of the asteroid belt is detailed by Durda and Dermott (1997). The initial main belt mass is taken to be approximately three times greater than the present mass (Durda et al. 1998); this population evolves after 4.5 Gyr to resemble the current main belt. The calculation is performed for particles from $100 \, \mu m$ through the largest asteroidal sizes, with a fragmentation index $q = 1.90$. The dust production rate in the main asteroid belt becomes more stochastic with time following a relatively smooth decrease in area as the small particles created directly from the breakup of the parent body are destroyed. The "spikes" in the dust production are due to the breakup of small to intermediate size asteroids. Therefore while the observable volume of a family may decay at a fairly constant and well-defined rate, the total area of dust associated with the family during that time may fluctuate by an order of magnitude or more.

As the particles move out of the asteroid belt the action of secular resonance disperses them into the background cloud. The integrity of the dust band is lost as the forced inclinations increase and the forced nodes are spread around the sky (Fig. 5a), an effect which is more marked as the particle size increases. We find that models confining the material to the asteroid belt (exterior to 2 AU) match the observations very well. In the future, our models will populate the inner Solar System as well as the main belt region, and the dust bands will disperse naturally into the background cloud. To do this properly we will have to: (i) investigate the dynamical history of a much greater range of particle sizes than we have considered so far in order to properly account for their behavior at the 2 AU secular resonance; and (ii) take into account collisional processes, as larger particles will have shorter collisional lifetimes compared to their P-R drag lifetimes and will therefore not penetrate as far into the inner Solar System.

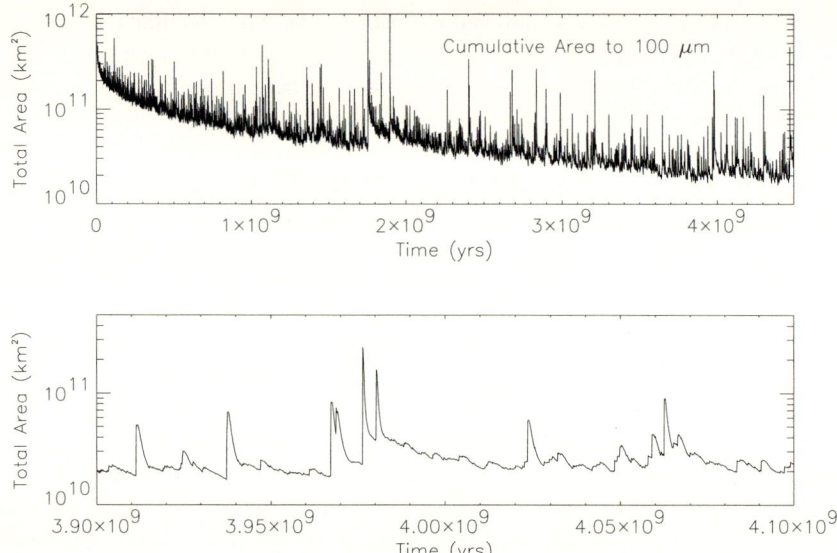

Figure 13. Variation with time of the total cross-sectional area of dust associated with the breakup of an asteroid that was big enough to supply all the observed collision products of the belt. A size-frequency distribution q of 1.9 was assumed for the initial breakup of each asteroid and this accounts for the heights of the "spikes" in the plots (Grogan et al. 2001; adapted from Durda and Dermott 1997).

However, we can obtain an estimate for the dust band contribution to the zodiacal cloud as a whole by simply extending our best fit dust band models to populate the inner Solar System. The distribution of orbits obtained in this manner will not be exactly correct, due to our insufficient treatment of the secular resonance, but will still be reasonably accurate in terms of the total surface area associated with the dust bands. Figure 14 compares the thermal emission obtained from this raw dust band model to the corresponding IRAS profile in the $25\,\mu$m waveband. The result is shown for inner Solar System distributions of material corresponding to $1/r^\gamma$ where $\gamma = 1.0$, as expected for a system evolved by P-R drag, and $\gamma = 1.3$ as predicted in parametric models of the zodiacal cloud, most recently by Kelsall et al. (1998). The dust bands appear to contribute approximately 30% to the total thermal emission. Also shown is the amplitude of the dust band material confined to the main belt (exterior to 2 AU), which represents the component of the dust band material isolated by the fast Fourier filter. This indicates that approximately 4% of the in-ecliptic infrared emission from the zodiacal cloud is produced by dust band particles that orbit exterior to 2 AU, and also clearly demonstrates the extent to which the dust band contribution is underestimated if it is assumed that the filtered dust band observations represent the entirety of the dust band component of the cloud.

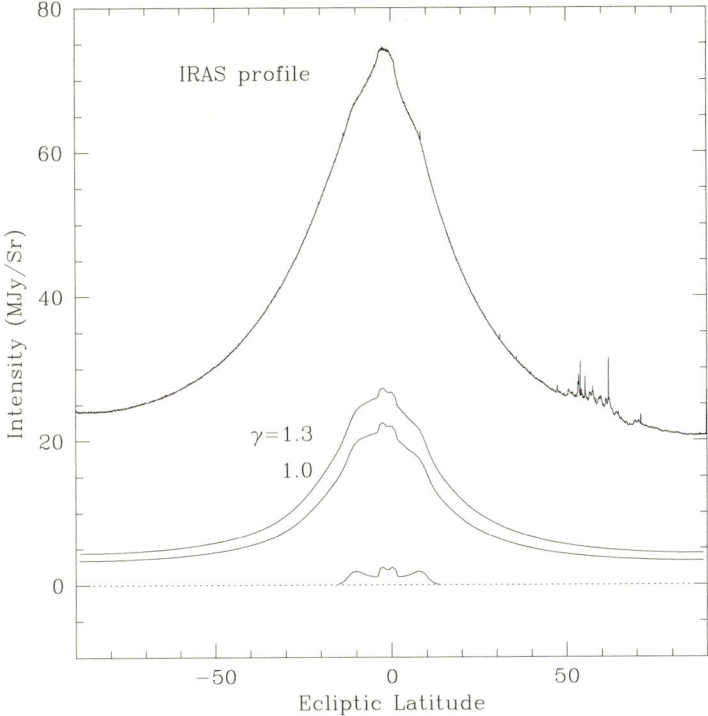

Figure 14. Contribution to the unfiltered IRAS 25 µm wave band observation shown at the top due to three simulated unfiltered profiles taken through three different dust band models. Each of the models was constructed using particles from only three asteroid families: Eos, Themis and Koronis. The model with the smallest amplitude refers to material confined to the asteroid belt. The other two models (labeled $\gamma = 1.0$ and $\gamma = 1.3$) contain, in addition, asteroidal material that has migrated towards the Sun due to Poynting-Robertson light drag (Grogan et al. 2001).

Figure 15 shows the ratio of areas of material associated with the entire main belt asteroid population and all families, for asteroid diameters greater than 1 km. The best fit lines have a slope corresponding to a size-frequency index $q = 1.795$. This diagram can be used to estimate the total contribution of main belt asteroid collisions to the dust in the zodiacal cloud, by extrapolating the observed size distributions of larger asteroids in both populations assuming a collisional equilibrium power law size distribution. The result is that the main belt asteroid population contributes approximately three times the dust area of the Hirayama families alone, and the total asteroidal contribution to the zodiacal cloud could account for almost the entirety of the interplanetary dust complex (Grogan et al. 2001). In reality, evolved size distributions are more complex than simple power laws (Durda et al. 1998) and the size distribution of individual asteroid families likely preserve some signatures of the original fragmentation events from which they were formed. However, small dust-size

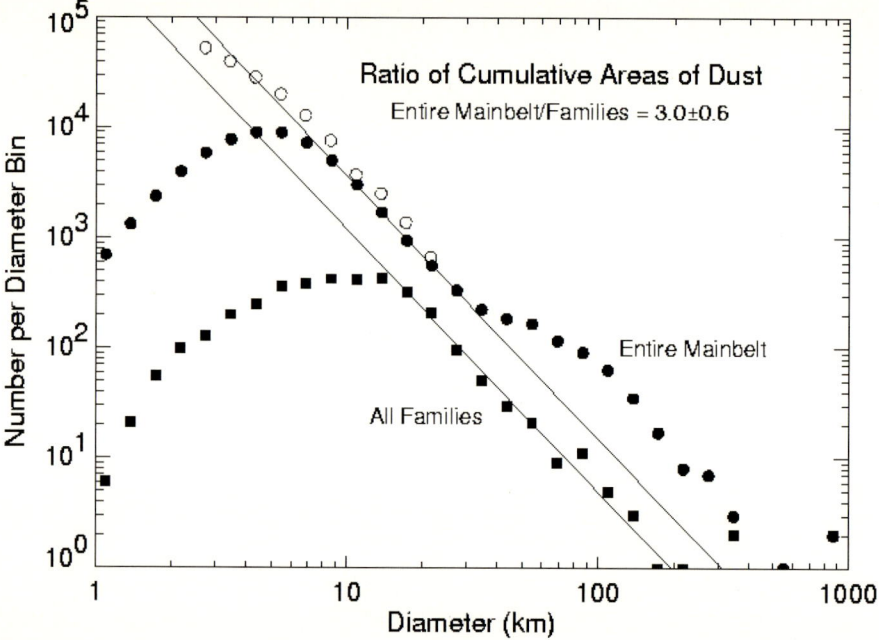

Figure 15. Diameter-frequency diagram for the main belt asteroid population, obtained by combining data from the catalogued population and McDonald/Palomar-Leiden surveys (MDS/PLS). Open points represent counts for which the PLS data had to be corrected for incompleteness. These were not included in the least-squares fits to the linear portion of the distribution. The ratio of the area of dust associated with the entire main belt asteroid population to that of the asteroid families alone was calculated to be 3.0 ± 0.6 (Grogan et al. 2001; updated from Durda and Dermott 1997).

particles and their immediate parent bodies have collisional lifetimes in the main belt that are considerably shorter than the age of the Solar System or the major asteroid families. Thus the dust size distributions associated with both the background main belt and family asteroids may well be considered to have achieved an equilibrium state, with total areas related to the equivalent volumes of the original source bodies in each population. However, to arrive at a more quantitative solution we need to apply our methods to the main belt asteroid population in the same way we have investigated the dust bands. This is an ongoing investigation.

IV.D. Equilibrium vs. Non-Equilibrium

Figure 16 shows the members of the Eos asteroid family in (e, I) space as determined by the hierarchical clustering method (Zappalà et al. 1995). Shown on this diagram is the position of the mean proper inclination of the ten degree band model. The consequence is that the ten degree dust band material is not tracing the orbital element space of the Eos family as a whole, as would perhaps be expected from the equilibrium model (Dermott et al. 1984) in which

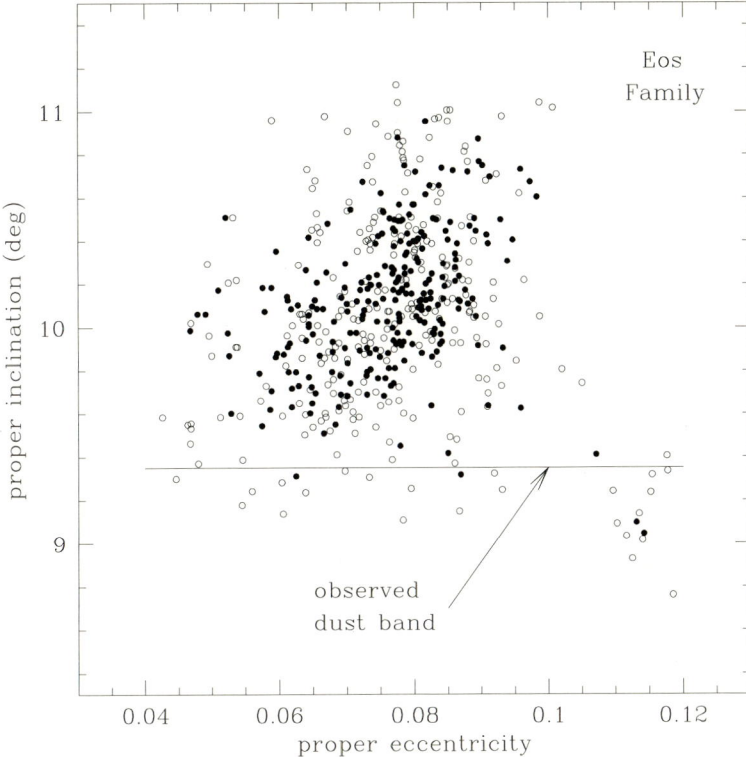

Figure 16. Proper inclination, I_p, and proper eccentricity, e_p, of asteroids in the Eos family. Asteroids with diameters $> 15\,\mathrm{km}$ are shown with filled circles. Our model dust band that best fits the observed "ten degree" band has a proper inclination of $9.35°$, well below the mean of the Eos family (Grogan et al. 2001).

the dust bands represent the continual grinding down of family asteroids. On the surface this appears to be evidence for a catastrophic (non-equilibrium) origin for the band (Sykes and Greenberg 1986), the dust bands being produced from the disruption of random main belt asteroids. But our results have shown that the structure of a dust band, shaped from the dynamical history of its constituent particles of varying sizes, is much more complex than previously thought. Two important points should be kept in mind here: (i) the action of the secular resonances serves to disperse the inclinations and nodes of the particles (as demonstrated in Section III); and (ii) the grinding down of an asteroid family, as we have modeled and shown in Fig. 13, is a stochastic process. With regard to the first point, the observed latitude of the dust band will decrease as the dispersion of the particle's inclinations and nodes increases. This effect has not yet been fully characterized. The second point leads us to conclude that spikes in the dust production rate correspond to the breakup of individual asteroids, and can therefore originate from any asteroid within the family. The dust band associated with the family, produced by the fresh injection of material from the most recent fragmentation, may therefore shift in latitude over time to reflect the orbital characteristics of its parent.

Our modeling predicts the amount of cross-sectional area required to produce the dust bands. Can this amount be provided by a relatively small (15 km) asteroid? Figure 17 shows the cumulative surface area as a function of different size-frequency distribution indices for the Eos, Themis and Koronis families and also a single 15 km diameter asteroid. At first this appears to contradict our result that a low q of at most 1.4 is needed to model the dust bands. However, the diagram as set up is indicative of the disruption process, such that the size-frequency distribution is constant from the source body all the way down to the smallest particles. Our result reflects the size distribution imposed by the combination of the dust production, P-R drag flow, and particle-particle collisions. The diagram does suggest that for a single asteroid to be responsible for the ten degree dust band, the size-frequency index of the collisional debris would initially have needed to be extremely high, well over 1.90, to produce the surface area required to match the observations. Perhaps if the initial parent body was a rubble-pile (Davis et al. 1989) rather than a single body, more small material and therefore a high q distribution would be produced. Again, the question is difficult to answer because the problem is poorly constrained, in this case because we have little information on the disruption process.

V. BACKGROUND CLOUD

We showed in the previous section how using a Fourier filter can separate the high-frequency dust band residuals from the smooth, low-frequency "background cloud" (Fig. 8). The dust bands are of great interest because they are features that have been attributed with confidence to main belt asteroids. But it is clear that the background cloud provides the majority of the zodiacal emission (albeit including significant low-frequency components from the dust bands and Earth's resonant ring), and much can be learnt from a study of its structure, especially since the observing geometry changes throughout the year as the Earth moves around its elliptical orbit. Mid-IR geocentric satellite observations (such as the IRAS and COBE observations) clearly show that the cloud is inclined to the ecliptic, that its axis of rotational symmetry is offset from the Sun, and that it is warped. Here we will discuss these observations and their dynamical consequences.

V.A. Tilt, Warp and Offset

Figure 18 shows how the plane of symmetry of the background cloud may be found. The two curves represent the variation of the latitude of peak brightness of the background cloud with ecliptic longitude of Earth, in directions both trailing and leading the Earth at 90° solar elongation. If the plane of symmetry of the cloud was the ecliptic, the latitude of peak brightness would remain constant and in the ecliptic plane. However, the observations give a sinusoidal variation. The points at which the latitudes of peak intensity are equal and

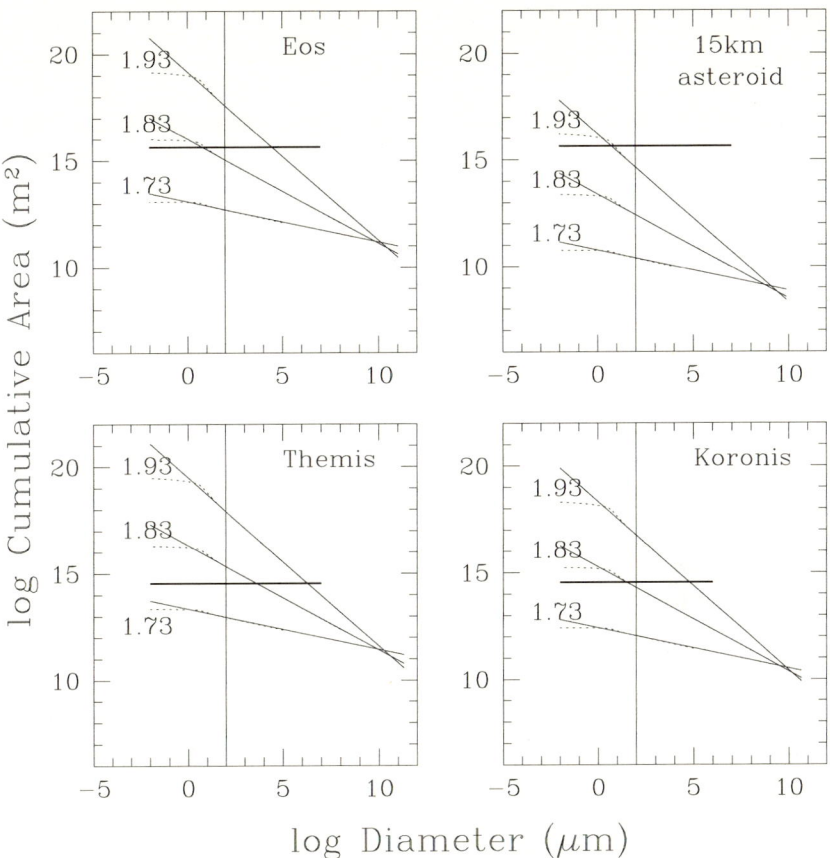

Figure 17. Cumulative area of dust associated with the Zappalà et al. (1995) Eos, Koronis and Themis families for three values of the size-frequency index q. The solid lines represent the geometrical area, while the broken lines are the cross-sectional area of emission calculated for spheres of astronomical silicate of density 2,500 kg m^{-3} calculated using Mie theory. In each panel, the cumulative area of dust from each family (Table 2) needed to model the IRAS observations of the dust bands is represented by a heavy, horizontal line. A vertical line corresponding to a diameter of 100 μm is shown for reference. We also show the area of dust associated with the breakup of a 15 km diameter asteroid. If the "ten degree" band was formed from the disruption of such an asteroid, the size-frequency index q of the collisional debris would have had to be very high, well over 1.90, to account for the amplitude of the band in the IRAS observations (Grogan et al. 2001).

opposite in the leading and trailing directions give the nodes of the cloud; the magnitude of the latitude at these points gives the inclination of the cloud with respect to the ecliptic. The diagram suggests that the inclination of the cloud to the ecliptic is equal to 1.49 ± 0.07°, and the longitude of ascending node is 58.4 ± 2.3° (Dermott et al. 1996a).

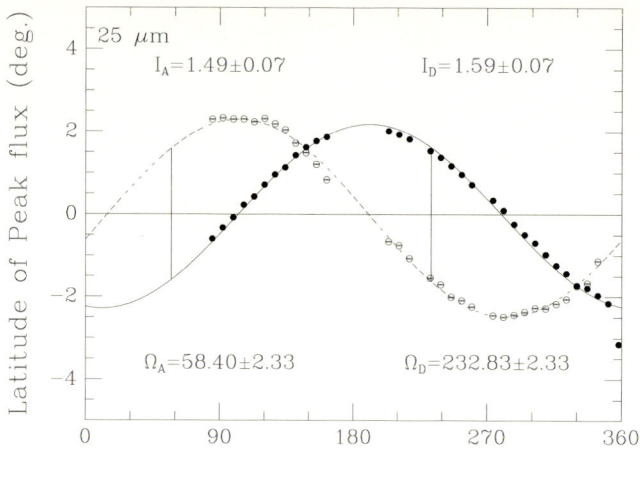

Figure 18. Variation of the latitude of peak flux of the zodiacal cloud with ecliptic longitude of Earth in the trailing (solid curve) and leading (dotted curve) directions derived from COBE observations in the 25 μm wave band. Reprinted with permission from Dermott et al. (1996a). Copyright 1996, American Institute of Physics.

Figure 19 (left) shows the variation of the brightnesses of the ecliptic poles with ecliptic longitude of the Earth (Dermott et al. 1999). The North and South polar brightnesses are equal when the Earth is at either the ascending ($70.7 \pm 0.4°$) or descending node of the local plane of symmetry (at 1 AU) of the cloud. This is in contrast to the result found earlier from the latitudes of peak intensity of the cloud in the trailing and leading directions, that gave a result of 58.4°. Since the latter observations sample the cloud external to 1 AU, this implies that the plane of symmetry of the zodiacal cloud varies with heliocentric distance, i.e., that the zodiacal cloud is warped (shown schematically in Fig. 9).

Figure 19 (right) shows an attempt to model the variation in polar brightness using a purely asteroidal model for the background cloud (Dermott et al. 1996b). The source population for the model particles are the main belt asteroids, and the material is distributed through the inner Solar System as expected by Poynting-Robertson drag. The total surface area in the model is scaled such that the peak ecliptic brightness of the model matches the peak ecliptic brightness of the observations. It is evident that this model fails to account for the flux observed at 1 AU and falls short by a factor of ~ 2. This is another route (in addition to the dust bands) by which we can answer the question of the relative contribution of asteroidal and cometary material: by combining such asteroidal and cometary models we can discover which combination provides the best fit to the observations. Early attempts have been published (Liou et al. 1995) which suggest a cometary:asteroidal ratio of about 3:1, but these results suffer from the fact that the asteroidal and cometary models were extremely

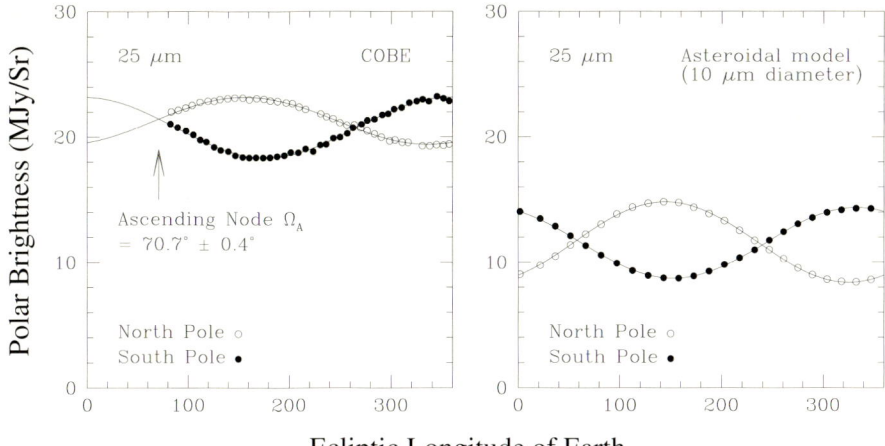

Figure 19. (Left) COBE observations of the North (open circles) and South (filled circles) polar fluxes obtained from observations in the $25\,\mu m$ wave band. Reprinted from Dermott et al. (1999) with kind permission from Kluwer Academic Publishers. (Right) Polar flux predicted by an asteroidal model of the background zodiacal cloud based on single size astronomical silicate particles of diameter $10\,\mu m$. The model flux has been normalized to match the peak ecliptic flux observed in the leading direction with a solar elongation angle of $90°$. Adapted from Dermott et al. (1996b) by kind permission of the Astronomical Society of the Pacific.

limited. The asteroidal model employed by Liou et al. (reproduced in Fig. 19), was composed only of $10\,\mu m$ diameter particles, and we have shown that the distribution of the various size particles in the cloud varies significantly due to the secular perturbations (see Fig. 5). The amplitudes of the sinusoidal variations are also too large (once the absolute fluxes are scaled to the observations), suggesting that the $10\,\mu m$ asteroidal model gives the incorrect mean forced inclination at 1 AU (the inclination of the plane of symmetry of the cloud with respect to the ecliptic). The distribution of material may also vary from the assumed $1/r$, for example, if additional surface area is produced by collisions in the inner Solar System then the asteroidal model would better describe the observations. In addition the sole source for the cometary model of Liou et al. was Comet P/Encke. We are working on producing a more complete description of the material originating from both asteroids (by increasing the range of particle sizes included) and from comets (by adopting a much more complete inventory of the cometary population).

In Table 1 we have summarized the results for the planes of symmetry with respect to the ecliptic for several features of the Solar System. In addition to the conclusions drawn about the structure of the background cloud from these results, the data provide support for the localized nature of the dust band material separated from the background by the Fourier filter. The observed

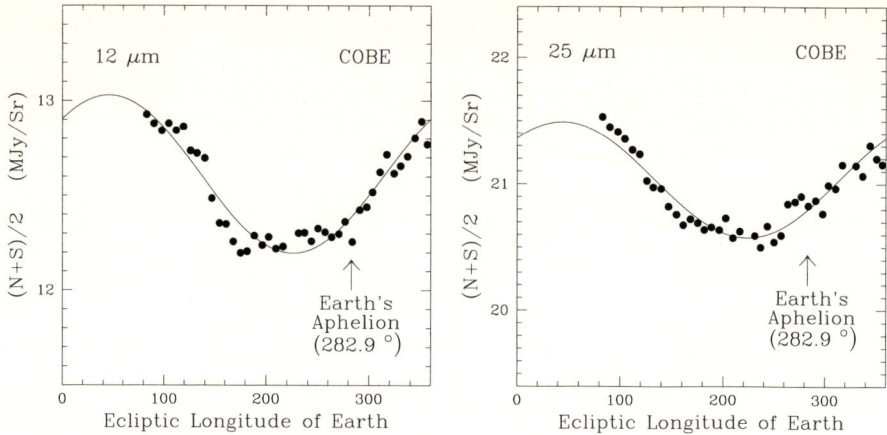

Figure 20. Variation of the mean polar brightness, (North + South)/2, as a function of ecliptic longitude of Earth, in both the 12 μm (left) and the 25 μm (right) COBE wave bands. Clearly, the minimum in the mean polar brightness does not occur at Earth's apocenter, indicating that the center of symmetry of the zodiacal cloud is offset from the Sun (Dermott et al. 1998, 1999; Holmes et al. 1998). Reprinted from Dermott et al. (1999) with kind permission from Kluwer Academic Publishers.

plane of symmetry of the ten degree band is quite distinct from that of the background cloud, but is closely similar to that of the asteroid families in the outer part of the main belt.

Figure 20 shows COBE observations of the sum of the brightnesses in the 12 and 25 μm wavebands at the North and South ecliptic poles, $(N + S)/2$ (Dermott et al. 1998, 1999; Holmes et al. 1998), where there is no contamination from the galactic plane. If the zodiacal cloud was rotationally symmetric with the Sun at the center, then the cross-sectional area density of particles in the near Earth region would vary according to $\sigma(r, \theta, \phi) \propto r^{-\gamma} f(\phi)$, where r is radial distance from the Sun, θ is azimuth, ϕ is latitude, and γ is a constant. Because the Earth's orbit is eccentric, geocentric observations sample the zodiacal cloud at different radial distances from the Sun. Thus, the minimum of the $(N + S)/2$ observation is expected to occur either at the Earth's aphelion, $\lambda_\oplus = 282.9°$, or perihelion, $\lambda_\oplus = 102.9°$, depending on whether $\gamma > 1$ or $\gamma < 1$, which is determined by the collisional evolution of particles in the near-Earth region (e.g., Leinert and Grün 1990, discuss the observational evidence and conclude that $\gamma \approx 1.3$ as found by the Helios zodiacal light experiment). However, the minimum in the 25 μm waveband observations occurs at $\lambda_\oplus = 224°$, and a similar result is found in the 12 μm waveband (the close similarity of these two curves is further evidence that the dominant particle diameter is large). This is expected only if the Sun is not at the center of symmetry of the zodiacal cloud. Parametric models of the zodiacal cloud have also shown the need for an offset to explain the observations (e.g., Kelsall et al. 1998).

Changes in the heliocentric distance of the Earth over the course of the year (due to its eccentric orbit) also provides an opportunity to compare the observed variation of the polar flux with heliocentric distance to theoretical values appropriate for asteroidal and cometary distributions (the polar flux gradient is a strong function of the eccentricity of the dust particle population at 1 AU, and can be observationally estimated from Fig. 20 by dividing the ratio of the amplitude to the mean of the sine curve by the Earth's eccentricity). Preliminary work (Dermott et al. 1999) showed that the observed polar flux gradient is indicative of a predominantly asteroidal distribution, but we have already shown that secular perturbations will strongly affect the distributions of both asteroidal (see Section III) and cometary material (see discussion in Section IV) at 1 AU. A more accurate qualitative result will follow from our ongoing investigation of the dynamical history of both asteroidal and cometary dust particles.

V.B. Physical Understanding of the Asymmetries

The effect of secular perturbations on the structure of a disk can be understood by considering the effect of the secular evolution of the constituent particles' orbits on the distribution of their orbital elements. The discussion in Section III shows that secular perturbations affect only the distribution of disk particles' complex eccentricities, $n(z)$, and complex inclinations, $n(y)$, while having no effect on their size distribution. However, secular changes in the eccentricity do affect the orbital decay rate (see Eq. 11) and this in turn affects the radial distribution of material.

In Section III, it was shown that the P-R drag affected particles in the zodiacal cloud have a distribution of complex eccentricities, $n(z)$, that lie on a circle. This means that the distribution of pericenters is biased towards the orientation in the disk that is defined by ϖ_f. The consequence of this biased orbital element distribution on the spatial distribution of this family material is best described with the help of Fig. 21 (right). This shows a face-on view (i.e., perpendicular to the plane of symmetry), of the family material in orbit around a star S. The resulting disk is made up of particles on orbits that have the same a, e_f, ϖ_f, and e_p, but random ϖ_p. The contribution of each particle to the spatial distribution of material in the disk can be described by an elliptical ring of material coincident with the particle's orbit. These elliptical rings have been represented by uniform circles of radius a, with centers that are offset by ae in a direction opposite to the pericenter direction, ϖ (this is a valid approximation to first order in the particles' eccentricities); a heavy line is used to highlight the orbital ring with a pericenter located at P, and a displaced circle center located at D, where $DP = a$. The vector SD can be decomposed into its forced and proper components; this is shown by the triangle SCD, where $SD = ae$, $SC = ae_f$, and $CD = ae_p$ (there is a similar triangle in Fig. 21 (left)). Given that the distribution of ϖ_p is random, it follows that the distribution of the rings' centers, D, for the family disk are distributed on a circle of radius ae_p and center C. Thus, the family forms a uniform torus of inner radius $a(1-e_p)$

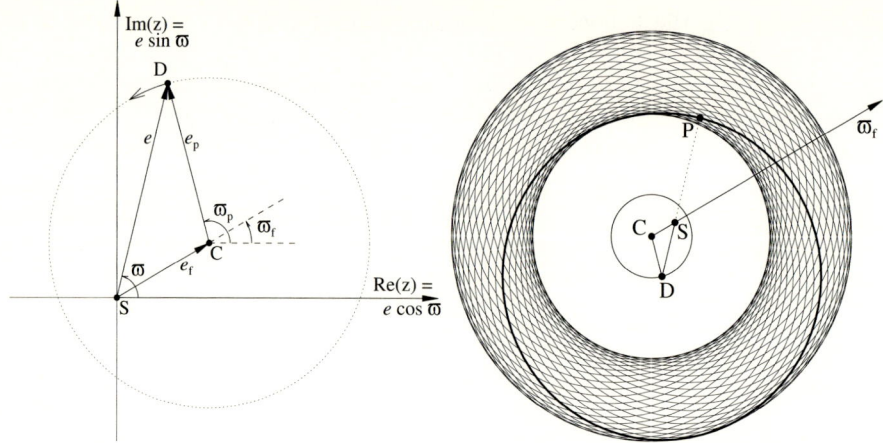

Figure 21. (Left) The osculating (instantaneous) eccentricity, e, of the orbit of a dust particle can be resolved vectorially into two components: a forced eccentricity, e_f, imposed by the perturbers in the system; and a proper eccentricity, e_p, that is determined by initial conditions. The corresponding osculating, forced and proper longitudes of pericenter, ϖ, ϖ_f and ϖ_p define the orientation of the orbit with respect to an arbitrary fixed direction. (Right) Shown here is an idealized disk of dust around a star S. The disk particles all have the same a, e_f, ϖ_f, e_p but random ϖ_p leading to variable osculating e, ϖ. To first order, the elliptical orbits can be represented by circles of radius a whose centers are offset by ae in a direction opposite to the pericenter direction, ϖ. A heavy line is used to highlight one orbit with pericenter P and displaced circle center D, where $DP = a$. The location of the center D can be decomposed into its two components shown by the triangle SCD with sides $SD = ae$, $SC = ae_f$ and $CD = ae_p$ correspond to a similar triangle in the panel on the left. Given that the distribution of ϖ_p is random, it follows that the points D for all the dust particles in this idealized disk will be distributed on a circle of radius $CD = ae_p$ and center C. Thus, the disk forms a uniform torus of inner radius $a(1 - e_p)$ and outer radius $a(1 + e_p)$ centered on a point C displaced from the star S by a distance ae_f in a direction away from the forced pericenter, ϖ_f (Wyatt et al. 1999b). Adapted from Dermott et al. (1985) with kind permission from Kluwer Academic Publishers.

and outer radius $a(1+e_p)$ centered on a point C displaced from the star S by a distance ae_f in a direction away from the forced pericenter, ϖ_f (Dermott et al. 1985; Dermott et al. 1998).

The distribution of the complex inclinations, $n(y)$, of these particles, is also the distribution of their orbital planes. Changing the reference plane relative to which the particles' orbital inclinations are defined to that described by y_f, shows that the secular complex inclination distribution of this family material leads to a disk that is symmetrical about the y_f plane; the opening angle of this disk is described by I_p. If the y_f plane varies with heliocentric distance, as indeed it was shown to do in Section III, the resulting disk appears warped (see Fig. 9). The distribution of nodes is biased towards the orientation in the disk that is defined by Ω_f.

V.C. Application to Circumstellar Disks

Observations of the disk of dust around HR4796A (Telesco et al. 2000) show a double-lobed feature, consistent with observations of a nearly "edge-on" disk of dust with a central clearing hole that is almost completely devoid of dust. The images obtained also show that the NE lobe is 5% brighter than the SW lobe; an asymmetry that could be due to gravitational perturbations by seen or unseen companions. Our modeling of the disk (Wyatt et al. 1999b), as well as accounting for the large-scale symmetrical structure, shows how a forced eccentricity imposed on all disk particles would be expected to produce the lobe brightness asymmetry. As outlined above, the center of symmetry of particles at the inner edge of the disk (those that contribute most to the flux) is offset, so that particles in the forced pericenter direction are closer to the star than those at the forced apocenter, so these particles are hotter and emit more flux. This is the so-called "pericenter glow" phenomenon, and reveals the presence of a perturbing body in the system. HR4796 is a binary system, but the orbit of the companion star is poorly constrained (the orbital period of the star is estimated at 7,000 years). Therefore we cannot definitively say whether the observed asymmetry is due to the companion or alternatively a planet or planets embedded in the disk. If the perturber in the system is a planet then depending on its location it would give rise to a detectable pericenter glow for even low eccentricity orbits. The only constraint on the planetary mass, $M_{\rm pl}$, is that the disk must be older than the time it takes for the secular perturbations to build up, an approximation for which is the precession period, $t_{\rm prec} = 2\pi/A \propto 1/M_{\rm pl}$. In this system we find that for secular perturbations to have built up at the edge of the disk, $M_{\rm pl} > 10\,M_\oplus$, where the mass of the Earth $M_\oplus = 3 \times 10^{-6} M_\odot$, but for older systems this mass limit would be lower. If there are two or more perturbers in the system then the forced element variation with semi-major axis will depend on both the mass of the perturbers and the orientation of their orbits. Such a variation could cause the disk to be warped, as discussed above.

The brightness asymmetry in HR4796A does not have high statistical significance, but we have shown here that if moderately sized planets in moderately eccentric orbits are present, then such asymmetries are to be expected. HR4796A was observed with the infrared imager OSCIR for only one hour on Keck II. Given that the significance level of any asymmetry will increase at a rate $\propto \sqrt{t}$, one good night on a 10 m telescope would settle the question of whether this asymmetry is real, and if so set some constraints on the planets that may be causing it. The measurement of such asymmetries and the detection of the presence of planets, even small planets, in circumstellar disks is clearly now within our reach. For a review of circumstellar disk structure see, for example, Beckwith (1999).

VI. RESONANT RING

Over the past 30 years or more, the possibility of resonant trapping of dust particles by the Earth and other planets has been discussed by several authors, including Schmidt (1967), Gold (1975) and Jackson and Zook (1989), but no observational evidence was found to support these discussions. In 1983, the IRAS spacecraft gave us our highest resolution data on the structure of the zodiacal cloud. Dermott et al. (1988a) analyzed these data and pointed out a marked but peculiar asymmetry, namely that the peak brightness of the cloud in the trailing direction, opposite to the Earth's orbital motion, is consistently greater than that in the leading direction. This finding was later confirmed by Reach (1991) who concluded that either there is a calibration inconsistency between the leading and trailing IRAS scans, or there is an enhancement in dust density that follows the Earth around the Sun, and that this could be related to the resonant trapping described by Jackson and Zook (1989), or could be due to gravitational focusing. In 1994, we showed that if asteroidal collisions are a significant source of zodiacal dust, then a trailing/leading asymmetry of the zodiacal cloud due to resonant trapping is to be expected, and that the asymmetry observed by IRAS is quantitatively consistent with predictions based on our numerical investigation of the resonant trapping of asteroidal dust particles, that is, particles with low orbital eccentricities (Dermott et al. 1994a).

Figure 22 shows the more complete COBE observations (in four infrared wavebands) of the peak, near-ecliptic flux of the smooth (filtered) zodiacal background at a constant solar elongation of 90°, as function of the ecliptic longitude of the Earth. There are at least three reasons why the peak brightness of the zodiacal cloud should vary with ecliptic longitude when viewed at a constant elongation angle. Variations are expected to arise from: the forced eccentricities of the dust particle orbits which produce a displacement of the Sun from the center of rotational symmetry (Fig. 20); from the inclination of the plane of symmetry of the cloud with respect to the ecliptic which produces a double sine variation (see Dermott et al. 1996a); and from the Earth's orbital eccentricity (Fig. 20). The curves fitted to the data in Fig. 22 are a combination of sine and double sine components only (Dermott et al. 1996a; Jayaraman and Dermott 1996a), and the mean difference between the two curves for the leading and trailing directions is a measure of the flux asymmetry. This asymmetry is present in all four wavebands, but is seen most clearly in the 25 μm waveband. In the 12 μm waveband, the peak-to-peak difference between the trailing and leading curves is as high as $\sim 10\%$. This fact, coupled with the large differences in the shapes of the curves seen in the four wavebands, makes it difficult to confirm the existence of a mean trailing brightness excess of $\sim 3\%$ by subtracting a simple model of the background cloud from the COBE data, as attempted by Reach et al. (1995). It is our opinion that, at present, we do not have a model of the background cloud in the various wavebands that is good at the $\sim 3\%$ level.

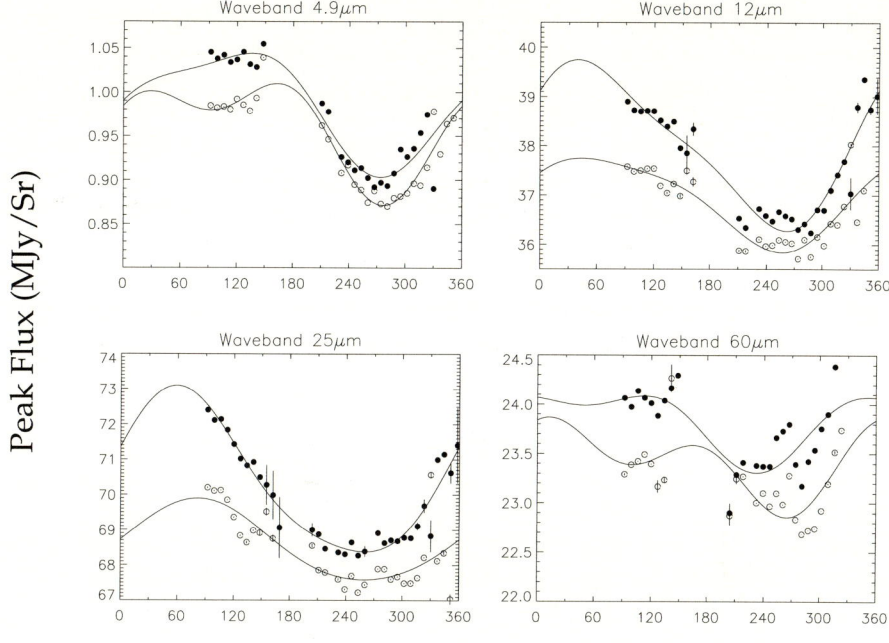

Figure 22. COBE observations of the peak flux of the smooth (filtered) zodiacal background as function of the ecliptic longitude of Earth at a solar elongation of 90° in four infrared wavebands. The flux in the Earth's trailing (filled circles) and leading (open circles) directions are fitted with independent sine curves. These curves are a combination of sine and double sine components only (Dermott et al. 1996a; Jayaraman and Dermott 1996a), and the mean difference between the two curves for the leading and trailing direction is a measure of the flux asymmetry (Jayaraman and Dermott 2001). Adapted with permission from Jayaraman and Dermott (1996a). Copyright 1996, American Institute of Physics.

Mean motion resonances occur at those heliocentric distances for which the ratio of the orbital periods of a particle and the planet are the ratio of two small integers, $p + q : p$. If we retain only the leading term in the disturbing function of the particle, then the equation of motion of the resonant argument, ϕ, defined by (Dermott et al. 1988b)

$$\phi = p\lambda - (p+q)\lambda' + q\varpi', \qquad (32)$$

is that of a damped harmonic oscillator and the acceleration of ϕ, $\ddot{\phi}$, is given by

$$\ddot{\phi} = -(\mathcal{G}M_\oplus/a')f(\alpha)e'^q \sin\phi - (p+q)\dot{n}'_{\mathrm{PR}}, \qquad (33)$$

where p and q are integers, λ is the mean longitude, ϖ is the longitude of perihelion, $\mathcal{G}M_\oplus$ is the gravitational mass of the Earth, a is the semi-major axis, e the eccentricity, $\alpha = a/a'$, $f(\alpha)$ is a function of Laplace coefficients that increases markedly with increasing p, $\dot{n}'_{\text{PR}}$ is the rate of change of the mean motion of the dust particle due to the action of Poynting-Robertson and solar wind drag, unprimed quantities refer to the orbit of the Earth and primed quantities refer to a dust particle on an orbit exterior to that of the Earth.

A particle is said to be "trapped" in a resonance with the Earth while the effect of P-R and solar wind drag acting on the particle is counterbalanced by resonant gravitational perturbations due to the Earth. In this case, we have

$$\frac{\langle \dot{e}' \rangle}{e'} \simeq -\frac{q}{2(p+q)e'^2} \left(\frac{\dot{a}'}{a'}\right)_{\text{PR}}, \tag{34}$$

where $\langle \dot{e}' \rangle$ represents the rate of change of the dust particle's eccentricity averaged over the librational period of the resonant argument. Given that $\dot{a}' < 0$ and the orbit of the dust particle is *converging* on the orbit of the Earth, trapping into resonance results in a rapid increase in the particle's orbital eccentricity and a strengthening of the resonance (Dermott et al. 1988b). Paradoxically, this leads to resonance disruption on timescales $\sim 10^4$ yr — see Fig. 23. If the eccentricities are large, then the libration widths associated with the resonances are also large, and the dynamics may break down and become chaotic due to resonance overlap (Wisdom 1980). However, Marzari and Vanzani (1994) used a numerical approach to map the capture probability in the e–ϖ (eccentricity and longitude of perihelion) phase space of the particle. They found that the increase in the orbital eccentricity of the particles after resonant capture leads to Earth-crossing orbits and the particles eventually escape out of resonance due to close encounters with the Earth. Therefore, close encounters with the planet probably play a critical role in curtailing the lifetime of a particle in resonance. These encounters are not included in any theory of orbital evolution, although Beaugé and Ferraz-Mello (1994) have made an attempt in that direction.

We should note here that Eq. (34) also applies to the orbital evolution of cometary particles trapped in resonances interior to the orbit of Jupiter. But in this case, since the resonances are interior, the equation for $\dot{e}$ contains a sign change and because $\dot{a} < 0$ and the orbits are *diverging* from Jupiter, orbital evolution results in a decrease in the orbital eccentricity (Dermott et al. 1988b) — see Fig. 6 (Examples 4 and 5).

The paths of particles librating in the 4:3, 5:4, 6:5, and 7:6 mean motion resonances with the Earth are shown in Fig. 24. Without drag, these paths would have mirror symmetry about the Earth-Sun line. But drag introduces a phase lag into the equations of motion, with the result that the paths are asymmetric; as the particles pass through perihelion, they approach the Earth closer in the trailing direction (behind the Earth in its orbit) than in the leading direction. By setting the left-hand side of Eq. (33) to zero, we find that the phase lag is a maximum ($|\sin \phi| \approx 1$) when

$$|(\mathcal{G} M_\oplus/a') f(\alpha) e'^q| = |(p+q)\dot{n}'_{\text{PR}}|, \tag{35}$$

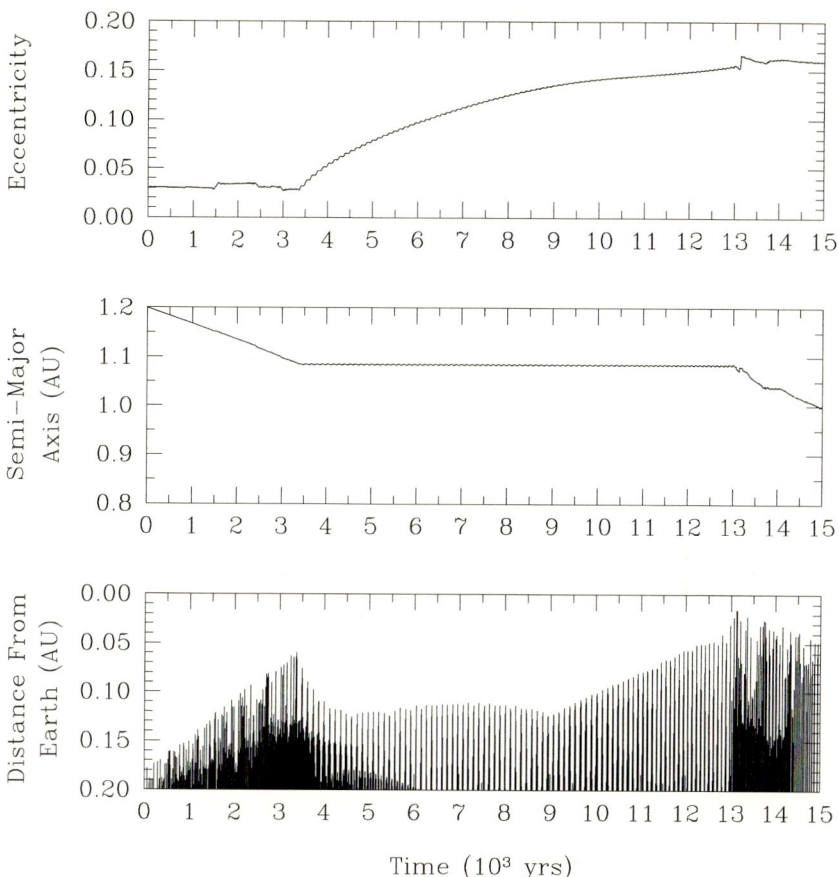

Figure 23. Orbital evolution of a 20 μm diameter dust particle. From left to right we see the semi-major axis (middle panel) of the particle orbit decay until, after about 3,500 years, it becomes trapped in a first-order exterior mean motion resonance with Earth. At this point the mean semi-major axis remains constant while the eccentricity (top panel) begins to increase. While the orbit is decaying the distance of closest approach of the particle from Earth (bottom panel) decreases. Upon capture into resonance the particle is initially kept away from Earth by the resonance. However, as the eccentricity increases the minimum gap between Earth and particle narrows until eventually the particle is released from the resonance, after being trapped for about 10,000 years (Kortenkamp and Dermott 1998a).

and the strength of the resonance is just sufficient to counteract the effect of drag on the particle's semi-major axis. Given that the P-R drag rate increases with decreasing diameter (see Eqs. (6) and (10)), it follows that there is a lower limit to the size of the particles that can be trapped in a particular resonance. We have determined the capture probabilities of asteroidal particles composed of astronomical silicate with density 2,500 kg m^{-3}, and found numerically that the lower cut-off for resonance trapping with the Earth is $D \sim 5\,\mu$m (see

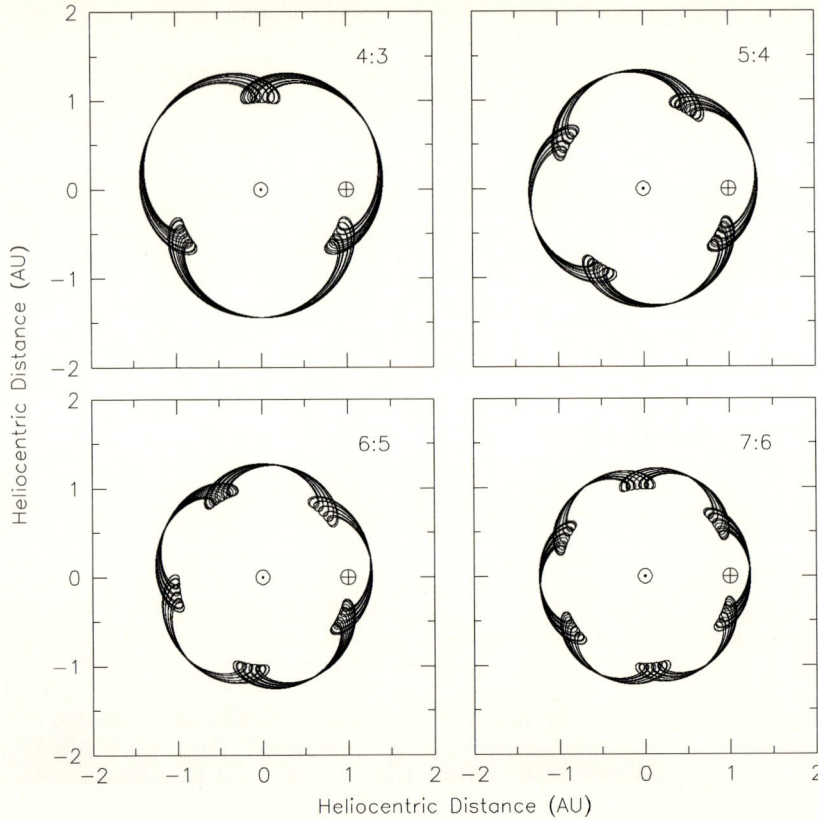

Figure 24. Illustrations showing the orbital paths traced out by asteroidal dust particles trapped in four different external mean motion resonances with Earth. The resonances occur in regions where the ratio of the orbital periods of the dust particle and the planet can be expressed as the ratio of two small integers, like the 4:3, 5:4, 6:5, and 7:6 resonances shown here. These paths are shown in a Sun-centered reference frame that is corotating with Earth's mean orbital motion. Earth ($\oplus$) is nearly stationary in this reference frame. Over many years the orbit of the dust particle librates about a quasi-stable resonant configuration. Because of the drag forces the paths of the dust particles are not symmetric about the Earth-Sun line. Different resonances cause a similar effect, but with differing numbers of lobes in the paths. Two common features of all resonant orbital paths are the cavity that Earth sits in and the proximity of the dust particles trailing behind the planet. The super-position of numerous resonances results in a circumsolar ring of asteroidal dust with Earth embedded in a cavity and followed in its orbit by a cloud of trailing dust particles. Adapted from Kortenkamp et al. (2001) with kind permission from Kluwer Academic Publishers.

Fig. 25). Thus, the observed trailing/leading asymmetry of the zodiacal cloud must be produced by particles larger than $5\,\mu$m (unless, of course, they have lower densities, the lower limit is more correctly a lower limit on β – see Eqs. (6) and (10)).

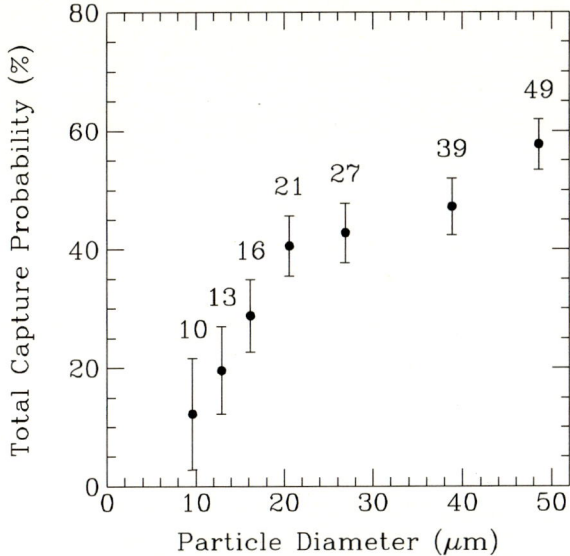

Figure 25. Capture probabilities of asteroidal dust particles into all first-order mean motion resonances of Earth between 2:3 and 15:16 as a function of particle diameter. For particle diameters less than $\sim 5\,\mu$m (which corresponds to $\beta = 0.1$ for a density of $2{,}500\,\text{kg m}^{-3}$) the capture probability is zero, setting a lower limit to the diameters of particles trapped in the ring (Jayaraman and Dermott 2001). Adapted with permission from Jayaraman and Dermott (1996a). Copyright 1996, American Institute of Physics.

The structures of the rings produced by resonant trapping have been estimated from numerical integrations. In these investigations, we only considered main belt asteroidal particles, because the high orbital eccentricities of both cometary particles and particles originating from the disruption of near-Earth asteroids make resonant trapping improbable (Dermott et al. 1988b; Gomes 1995). The particles in the Earth's resonant ring must originate from a low eccentricity source in the zodiacal cloud, and this is most likely the main asteroid belt.

To obtain a detailed description of the structure of a resonant ring, the dynamics of a wide range of particle sizes needs to be considered. However, for heuristic purposes, it is instructive to compare the results of modeling rings of single-size particles with density $2{,}500\,\text{kg m}^{-3}$ and diameters $13\,\mu$m and $39\,\mu$m, respectively. These ring structures were found by determining the fraction of particles trapped in each resonance (Fig. 25) and the average time spent by the particles in these resonances (Fig. 26). The trapping times decrease sharply with increasing p as the locations of the resonances get closer to the Earth and the probability of close encounters increases. The overall distribution of particles in a ring at any given time is a convolution of the quantities given in Figs. 25 and 26 with the distributions of the positions of the particles in the

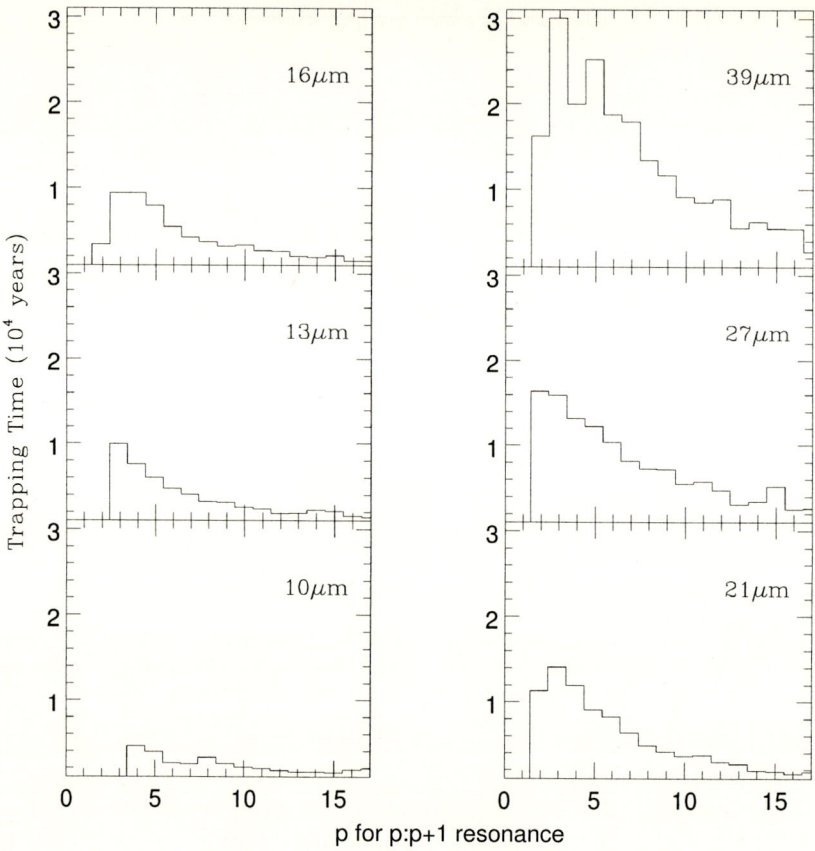

Figure 26. Average trapping lifetimes of particles in first order mean motion resonances between 2:3 and 15:16, determined empirically from numerical integrations for various particle diameters. The trapping lifetimes are longer for the larger particles and for those particles in resonances most distant from Earth's orbit (Jayaraman and Dermott 2001).

various resonances, as tracked in a reference frame centered on the Sun and rotating with the Earth from the time of capture to the typical time of release. Our final models are simulated "images" binned in pixels of 0.04×0.04 AU (Jayaraman and Dermott 2001).

The 6:5 resonance shown in Fig. 24 has five lobes, with the Earth residing asymmetrically in one of those lobes. All the other possible resonances have similar structures, but different numbers of lobes. The resultant image obtained by the super-position of the various paths, weighted according to the probability of capture and the trapping times shows that, in a rotating reference frame, the trapped particles form a near-uniform ring around the Sun that corotates in inertial space with the Earth. The rings have two notable features. First, there is a cavity in the ring at the location of the Earth. Figure 27 also shows that apart from an asymmetry in the mean position of the Earth in the ring's

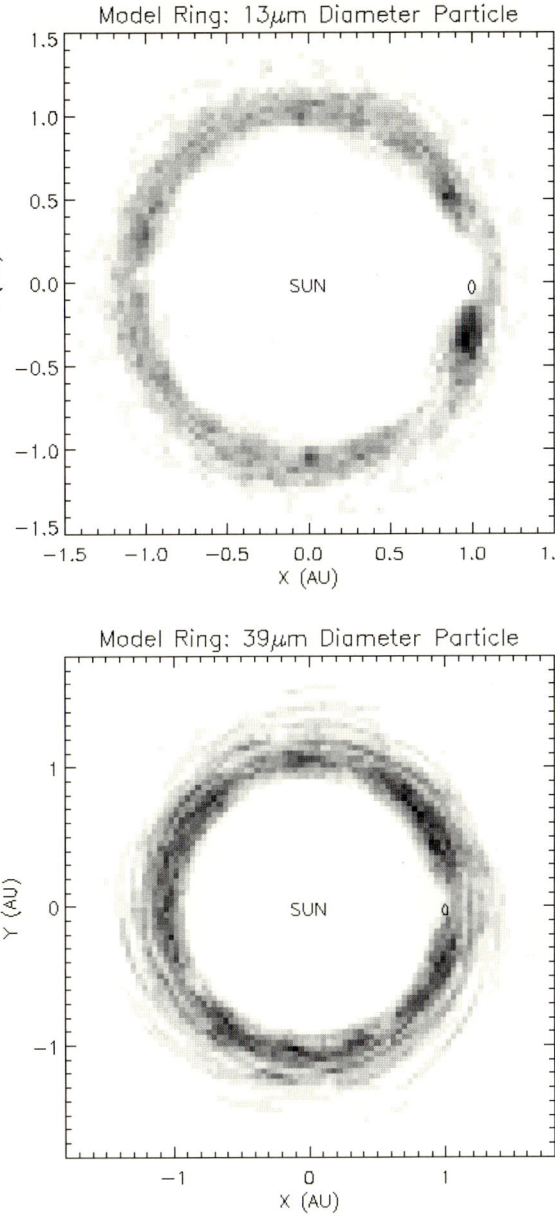

Figure 27. Simulated images of the variations in the particle number density in rings of asteroidal dust particles of diameters $13\,\mu$m (top) and $39\,\mu$m (bottom). The images are plotted in a rotating reference frame and have resolutions of 0.04×0.04 AU (Jayaraman and Dermott 2001). Adapted with permission from Jayaraman and Dermott (1996a). Copyright 1996, American Institute of Physics.

cavity, for the smaller particles alone there is also a marked asymmetry in the longitudinal variation of the particle number density. The increase in resonance strength, and the corresponding decrease in phase shift with increasing p value, disperses the longitudes of the perihelia of the particle paths in the rotating frame in the leading direction, while concentrating the longitudes of the perihelia in the trailing direction. For a ring of small particles that just satisfy the trapping criterion (Eq. 35), this results in a marked enhancement of particle number density behind the Earth in its orbit, as if the Earth had a trailing cloud of dust permanently in its wake. This trailing cloud is absent from the ring of larger particles. The probability of capture into resonance increases with increasing diameter (see Fig. 25), reaching a maximum that is determined by orbital eccentricity (Dermott et al. 1988b). However, simply because these particles are larger, their drag rates are smaller as are the phase shifts and asymmetries associated with resonant trapping. Thus, although large particles form resonant rings more easily than small particles, provided the trapping criterion is satisfied, resonant rings associated with large particles do not produce a trailing/leading asymmetry. While we expect resonant rings to be common in our Solar System, and in exosolar systems, trailing clouds are likely to be uncommon, only occurring where there is a near critical balance between the resonance force and the drag force. For example, almost all of the material that migrates out of the Kuiper Belt is expected to become trapped in resonances with Neptune forming a large irregular ring, but without a trailing cloud as the drag rates this far from the Sun are very small and the resonant phase lags are negligible. However, due to the multiple-lobed structure of resonant orbits detectable brightness asymmetries should still exist in Neptune's resonant ring (Liou and Zook 1999). Liou et al. (2000) argue that such a signature can be interpreted as evidence for a planetary system in the dust disk of ε Eridani.

Most of our work on the structures of resonant rings has been concerned with the dynamics of particles just larger than the lower limit of $D \sim 5\,\mu$m dictated by the trapping criterion (Eq. (35)). It is a fact that only particles close to this limit can contribute to a marked trailing/leading asymmetry (Jayaraman and Dermott 1996a, 2001). However, we have argued here that the diameters of most of the particles that migrate from the asteroid belt to the Earth are probably $\sim 10^2\,\mu$m and much larger than this lower limit. This paradox may be resolved by allowing for the disintegration of large particles *while* they are trapped in resonance with the Earth. More work is needed on the dynamics of this problem. We also need to complete a more detailed analysis of the COBE data. Figure 28 shows a geocentric view of the trailing dust cloud obtained from line of sight integrations through a three dimensional model of the ring. The four lower panels show images of the trailing cloud as seen in the model at ecliptic longitudes $\lambda_\oplus = 0°, 90°, 180°, 270°$. The plane of symmetry of this model is described by $I_\mathrm{f} = 3°$, $\Omega_\mathrm{f} = 50°$, which causes an apparent latitudinal oscillation of the trailing cloud. Although these oscillations are expected, they have yet to be detected in the COBE data and they will probably remain undetected until we have models accurate at the 1% level for both the background

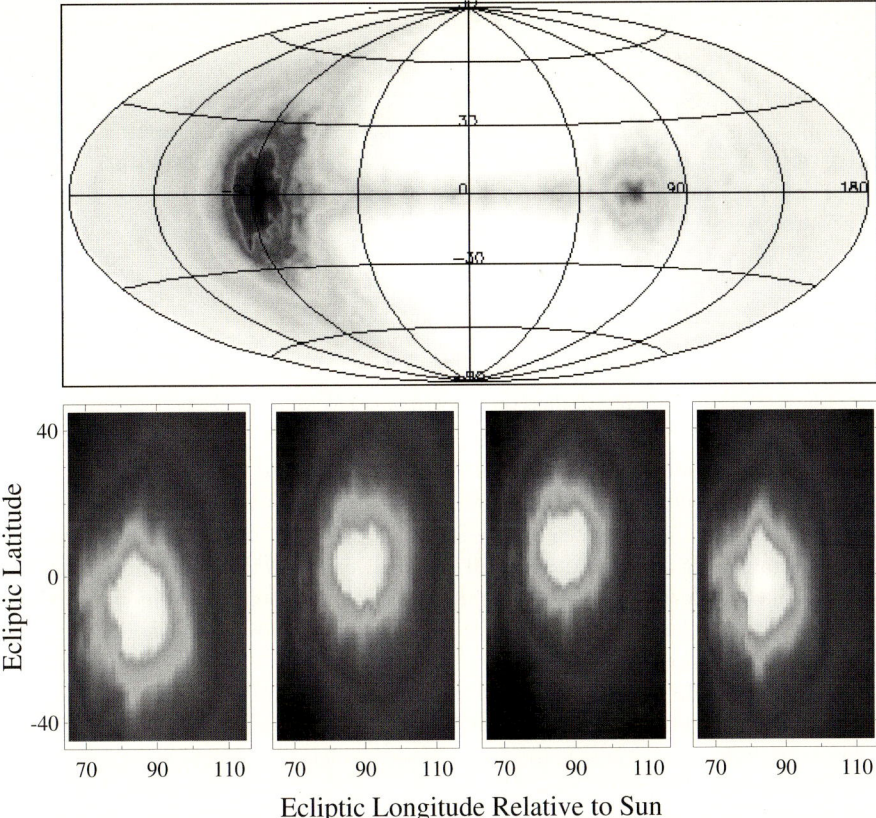

Figure 28. (Top) Geocentric view of the trailing dust cloud obtained from line of sight integrations through a three dimensional model of the ring. The longitude is defined to be zero in the direction opposite to the Sun and is measured in the clockwise direction as viewed from the North. This view of a ring of 13 μm diameter particles shows that the brightness of the trailing dust cloud (on the left) is much stronger than the leading cloud (on the right). Reprinted from Dermott et al. (1999) with kind permission from Kluwer Academic Publishers. (Bottom) Four panels showing images of the trailing cloud as seen in the model at ecliptic longitudes $\lambda_\oplus = 0°$, $90°$, $180°$, $270°$. The plane of symmetry of this model is described by $I_f = 3°$, $\Omega_f = 50°$, which causes an apparent latitudinal oscillation of the trailing cloud. Reprinted from Wyatt et al. (1999a) by kind permission of the Astronomical Society of the Pacific.

cloud and the dust bands. Once we have these models to subtract from the COBE observations, the time-varying structure of the resonant ring should be revealed. The forthcoming launch of SIRTF (Space Infrared Telescope Facility) should also provide us with a wealth of data on the structure of the Earth's resonant ring as it drifts through the trailing cloud of dust in the wake of the Earth's orbit (Jayaraman and Dermott 1996b; Wyatt et al. 1999a).

VII. ACCRETION OF IDPs

The rate at which Earth captures dust particles from different sources is dependent on the average spatial density of particles near Earth and the geocentric encounter velocity of the particles. As particles approach Earth their heliocentric trajectories are deflected toward the planet by its gravitational field. Öpik (1951) showed that this deflection results in an effective "gravitational" capture cross-section of Earth given by

$$\sigma_c = \sigma_\oplus \left(1 + \frac{v_e^2}{v_0^2}\right), \tag{36}$$

where $\sigma_\oplus$ is the geometric cross-section of Earth (taken out to 100 km altitude), v_e is the escape velocity at the surface of the planet (at 100 km altitude, $v_e \simeq 11.1 \,\mathrm{km\,s^{-1}}$) and v_0 is the geocentric encounter velocity of the particle prior to its acceleration by the planet. To illustrate the importance of this gravitational focusing effect consider that a dust particle approaching Earth with $v_0 \approx 1\,\mathrm{km\,s^{-1}}$ will "see" the effective cross-sectional area of the planet as ~ 100 times greater than its actual physical cross-section. Wetherill and Cox (1985) have shown numerically that even for extremely low encounter velocities ($v_0/v_e \leq 0.02$) Eq. (36) is still valid – albeit in a statistical sense – but approaches an upper limit of $\sigma_c/\sigma_\oplus \simeq 3{,}000$.

Earth-crossing orbits of typical asteroidal and cometary dust particles have similar distributions in ecliptic inclination. The drag forces acting on dust particles that cause their semi-major axes to decay towards the Sun also act to reduce their eccentricities (Eqs. 10 and 11). Earth-crossing asteroidal dust particles have low orbital eccentricities, typically $e < 0.1$, while Earth-crossing cometary dust particles generally have considerably higher orbital eccentricities, reaching as high as $e \simeq 1$ (Flynn 1989; Jackson and Zook 1992; Kortenkamp and Dermott 1998a). The disparity in eccentricities results in lower geocentric encounter velocities for asteroidal dust particles and thus larger gravitational capture cross-sections of Earth compared to cometary particles. Flynn (1990) noted that this would result in a near-Earth enhancement of asteroidal over cometary dust. Flynn also showed that any low inclination asteroidal dust particles would encounter a gravitational capture cross-section as much as two orders of magnitude larger than the actual geometric cross-section of Earth. It was later suggested that two of the largest families in the asteroid belt — Themis and Koronis — which have low ecliptic inclinations and are known dust producers, may contribute significantly to the IDPs (Love and Brownlee 1992).

Kortenkamp and Dermott (1998a) studied the orbital evolution of $10\,\mu\mathrm{m}$ diameter dust particles released from: 797 members of the Eos, Themis, and Koronis asteroid families (orbital elements were obtained from the PDS Small Bodies Node at http://pdssbn.astro.umd.edu); 830 non-family asteroids (a bias-free set was obtained from the Small Bodies Node by selecting bright main belt asteroids with absolute visual magnitudes $H < 11$); and 175 short-period

Jupiter family comets (a subset of the MPC catalog, Marsden 1995). Their evolution was followed until the dust particles had decayed through the entire Earth-crossing region. The cometary dust particles were divided into two distinct populations — those that were temporarily trapped in Jovian mean motion resonances and those that were not, in order to properly take into account the effect of resonant capture on the dust particles. Distributions of semi-major axis (a), eccentricity (e), and inclination (I) were determined for the Earth-crossing orbits in each population. Using these orbital elements the average spatial density (Kessler 1981) at heliocentric distance R and latitude l, is given by

$$S(R,l) = \frac{1}{2\pi^3 Ra \left[(\sin^2 I - \sin^2 l)(R-q)(Q-R)\right]^{1/2}}, \qquad (37)$$

where $q = a(1-e)$, $Q = a(1+e)$, $q < R < Q$ and $0 < l < I$. [$S(R,l)$ is used to create the schematic diagrams of the dust bands shown in Fig. 9]. If Earth is at position (R,l) then the fraction of dust particles captured per unit time (i.e., the capture rate for a given a, e, I) is

$$P = v_0 \sigma_c S(R,l). \qquad (38)$$

Typical capture rates are quite low and usually expressed as the fraction of particles with the given a, e, I captured per 10^9 years (dimensionally, Gyr^{-1}). [For details on the implementation of $S(R,l)$ for each population see Kortenkamp and Dermott (1998a).]

Figure 29 shows the mean geocentric encounter velocity (v_0) plotted against mean values of σ_c, $v_0\sigma_c$, and $v_0\sigma_c S(R,l)$ for each of the six Earth-crossing populations. The mean σ_c values (top plot) for the low eccentricity, low inclination Themis and Koronis dust particles (labeled T and K) are about a factor of ten higher than for the cometary population not previously trapped in mean motion resonances with Jupiter (solid C) and five times the previously trapped cometary population (open C). The Eos (E) and non-family asteroidal (A) populations have similar mean σ_c values. In terms of the effective volume of each population swept out by Earth per unit time ($v_0\sigma_c$; middle plot), the enhancements of the Themis and Koronis populations are reduced to less than a factor of two over the cometary populations. The enhanced capture cross-sections associated with the low velocity encounters are significantly offset by the greater volume swept up in the high velocity encounters. However, in the total capture rate (bottom plot) the differences in σ_c and $v_0\sigma_c$ are dwarfed by the remarkable range in the average spatial densities of the six populations, which spans about two orders of magnitude.

Combining estimates of the contribution of each source to the zodiacal cloud with the average capture rates for each population yields the contribution of each source to the atmospheric IDPs. A wide range of these estimates can be accommodated by Fig. 30. The right-hand panel is for the case where only 5% of the population of all Earth-crossing dust particles is due to Eos, Themis and Koronis, a conservative estimate. Here cometary dust could possibly account for 0% to 95% with the complement of this range (95% to 0%) from other

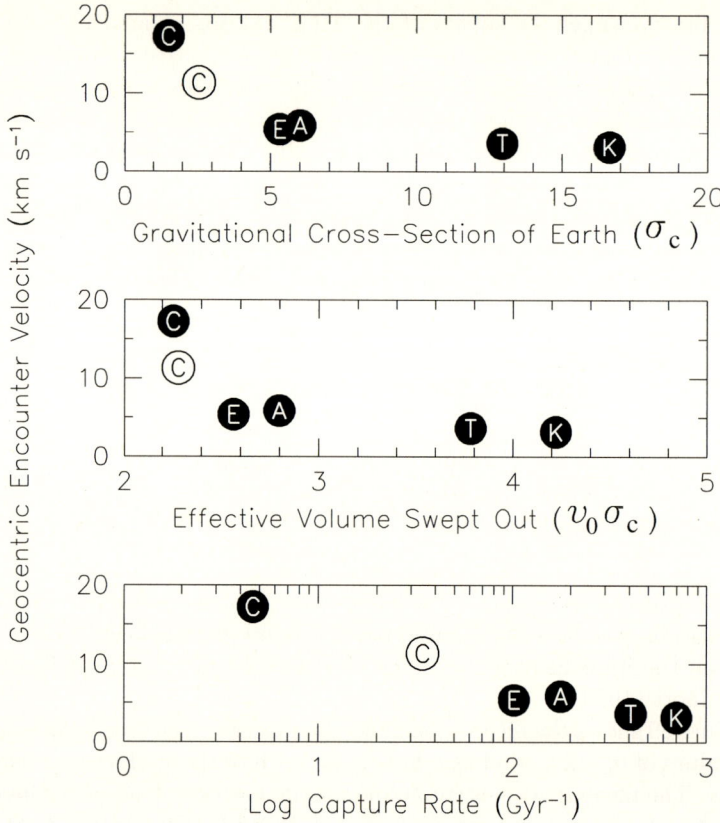

Figure 29. Mean geocentric encounter velocities (v_0 in text) for 10 μm diameter dust particles are plotted against mean gravitational capture cross-section of Earth (σ_c; top), mean effective volume of each population swept up by Earth each second ($v_0\sigma_c$; middle), and log of the mean capture rate ($v_0\sigma_c S(R,l)$; bottom). E, T, and K labels indicate the Eos, Themis and Koronis populations while A indicates other non-family asteroidal particles. Open points labeled C indicate cometary particles that were previously trapped in Jovian mean motion resonances. Solid points labeled C indicate cometary particles that were not previously trapped in Jovian mean motion resonances. Reprinted from Kortenkamp et al. (2001) with kind permission from Kluwer Academic Publishers.

asteroids. The left-hand panel is for the case where 25% of the population of all Earth-crossing dust particles is due to Eos, Themis and Koronis, a result of the dust band modeling presented earlier (Fig. 14). Cometary dust can then range from 0% to 75%, with the complement of this range (75% to 0%) being due to other asteroids. As an example, it has been suggested that the ratio of dust production by asteroid families to dust production by all asteroids is about 1:3 (Grogan et al. 2001). In this example, if dust from asteroid families contributes 25% to the population of all Earth-crossing dust particles then

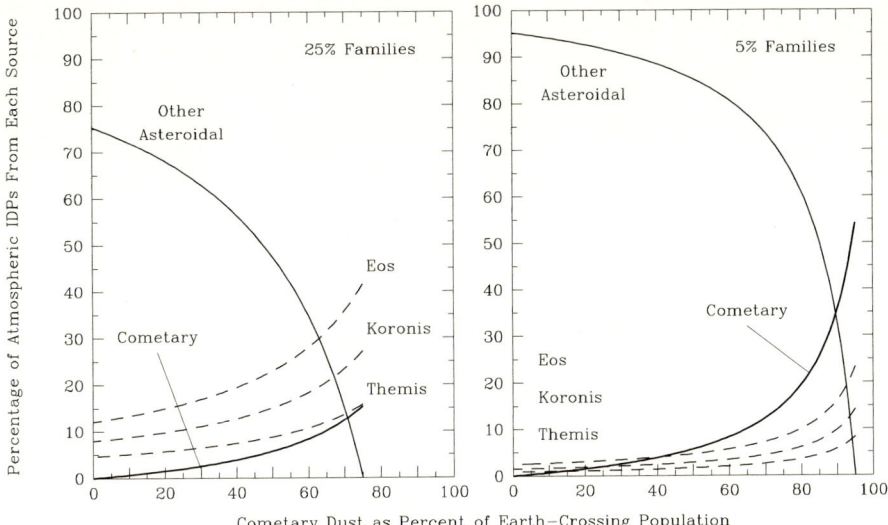

Figure 30. Based on the two dust band models of Grogan et al. (1997) these plots assume that the broad range of 5 to 25% offers a good estimate of the probable contribution of asteroid families to the population of all Earth-crossing dust particles. Using this range, we calculate the percentage of IDPs deposited in the atmosphere from each source as a function of the cometary contribution to the population of all Earth-crossing dust particles. The left-hand panel is for the case in which 25% of the dust comes from the Eos, Themis, and Koronis families. The cometary contribution can then range from 0 to 75%, with the complement of this range (75 to 0%) being due to other asteroidal dust. The three dashed lines represent Eos, Koronis, and Themis, respectively, in order of decreasing atmospheric percentages. The right-hand panel is for the case in which 5% of the dust comes from the Eos, Themis, and Koronis families. The cometary contribution can then range from 0 to 95%, again with the complement of this range (95 to 0%) being due to other asteroidal dust. Under most conditions shown in these two panels, asteroidal dust dominates the atmospheric IDPs. Only when cometary dust represents nearly 95% of the Earth-crossing population (far right, right-hand panel) does cometary dust represent more than half of the IDPs deposited in the atmosphere (adapted from Kortenkamp and Dermott 1998a).

all asteroids (non-family and family) contribute 75%. This leaves 25% to be supplied by comets. In the left-hand panel of Fig. 30 with cometary dust as 25% of the Earth-crossing population we see that Eos contributes about 17% to the IDPs, Themis and Koronis combine to provide another 18%, and nearly all of the remaining IDPs are due to other asteroids. In this example, cometary particles represent only about 2% of all IDPs.

From this figure certain conclusions are almost unavoidable, regardless of the cometary contribution to the zodiacal cloud. The existence of the dust bands and the evidence that asteroidal dust is transported to the inner Solar System (Dermott et al. 1994a) implies that most IDPs are asteroidal. Also, it shows that even though the Themis and Koronis families may contribute only between 1% and 4% to the population of all Earth-crossing dust particles, these two families make a very significant contribution to the population of IDPs. What may be most striking is that only under the most extreme of circumstances does cometary dust represent more than 50% of the IDPs. A conservative estimate based on Fig. 30 and using a value between 5% and 25% dust from families is that probably fewer than 25% of the dust particles entering the atmosphere are cometary. Losses due to melting and vaporization during atmospheric entry of the higher velocity cometary particles will further reduce their contribution to collections of intact IDPs retrieved from the stratosphere. The conclusions from our work so far are clear, the Earth predominantly accretes low inclination and eccentricity particles, and a large — perhaps dominant — fraction of this dust comes from the three asteroid families Eos, Themis and Koronis. Samples of these known K-type (Eos), S-type (Koronis) and C-type (Themis) asteroids may already exist in our IDP collections. However, we have shown in Section III that the distributions of large asteroidal particles in the cloud are different to the distributions of smaller particles, and more work needs to be done to characterize the accretion rates of this larger size asteroidal dust population. We have also discussed (see Section III; Liou and Zook 1996) how cometary particles trapped into interior mean motion resonances with Jupiter can evolve onto lower eccentricity orbits and hence increase their rate of capture by the Earth.

VII.A. Long Term Variations

Muller and MacDonald (1997) have suggested that the accretion rate of IDPs might be variable and could be responsible for driving Earth's 100,000 year climate cycle (Rial 1999, raises serious questions about the validity of this hypothesis). They proposed that the accretion rate was linked to the varying inclination of Earth's orbit with respect to the invariable plane of the Solar System. This hypothesis has been tested for accretion of asteroidal IDPs from the dust bands and, in a preliminary fashion, from Earth's circumsolar resonant dust ring.

1 Accretion from the dust bands

A raw number representing the inclination of an orbital plane is meaningless if it is not accompanied by a description of the reference plane. Traditionally in Solar System astronomy this reference plane is the ecliptic (the plane of the Earth's orbit) at a specific date. Some astronomers prefer to use the invariable plane of the Solar System as a reference. Two different reference planes are used to determine the proper and forced inclination of the dust particle orbits in a dust band. Proper inclination is referred to the midplane of the dust band and is therefore a measure of the angular half-width of the dust band. The forced inclination indicates the tilt of the dust band midplane with respect to a designated reference plane. The first step in studying the accretion rate of dust band material is the recognition that we must use Earth's orbital inclination measured with respect to the midplane of the dust bands at 1 AU, not with respect to the invariable plane.

As dust particles are produced by the gradual comminution of asteroid family members the resulting dust band takes on the orbital characteristics of the family. As the orbits of dust particles decay toward the Sun under the effect of Poynting-Robertson drag, secular gravitational perturbations due to the planets cause a variation in the forced inclination, essentially warping the dust band midplane (see Fig. 9). Furthermore, because the orbital decay rate (Eq. (12)) is dependent, through β, on the physical properties of the dust particles (Eq. (4)), dust bands composed of different sized dust particles will exhibit different warping (Fig. 5a).

Kortenkamp and Dermott (1998b) reconstructed the orientation of the Earth-crossing portion of the dust bands back to 1.2 million years ago. The simulations included gravitational interactions with seven planets (Mercury and Pluto were excluded), radiation pressure, P-R drag and solar wind drag. A wave of 10 μm diameter dust particles was released from the Eos asteroid family every 20,000 years and their orbits were allowed to decay into the inner Solar System. When each wave of dust particles became Earth-crossing (about 50,000 years after release) the eccentricity and inclination of Earth's orbit were determined (A and B of Fig. 31). The inclination was calculated with respect to the midplane of the Earth-crossing portion of the dust bands. The capture rate was then calculated using Eq. 38) and the average capture rates at each time step were normalized over the 1.2 million year period (solid lines in C and D of Fig. 31).

Several aspects of Fig. 31 deserve closer attention. First, from B, Earth's orbital inclination with respect to the Earth-crossing dust bands does not vary in a smooth periodic fashion. This fact alone argues against the idea of an inclination driven 100,000 year periodicity in the dust accretion rate, as suggested by Muller and MacDonald (1997), Farley and Patterson (1995), and Patterson and Farley (1998). Second, the accretion rate of Eos, Themis, and Koronis dust particles is strongly anti-correlated with Earth's orbital eccentricity. Third, when Earth's orbital inclination reaches or exceeds the mean proper inclination of the Themis and Koronis dust bands (dashed line in B) the planet is actually outside most of the dust band material for some period of time each year. During these years (e.g., near 0.5×10^5 and 9×10^5 years ago in C) the capture rate of Themis and Koronis material falls despite the fact that Earth's eccentricity may be near its minimum value. The higher proper inclination of the Eos dust band ensures that Earth is always deeply embedded within that dust band and so the capture rate of Eos dust particles does not show a correlation with Earth's orbital inclination (D). The transition from an inclination-driven to an eccentricity-driven accretion rate is complete for proper inclinations above $4°$ (Kortenkamp and Dermott 1998a). The mean proper inclination for the asteroid belt is about equal to that for the Eos family (about $10°$), so variations in the accretion rate of most asteroidal dust particles will

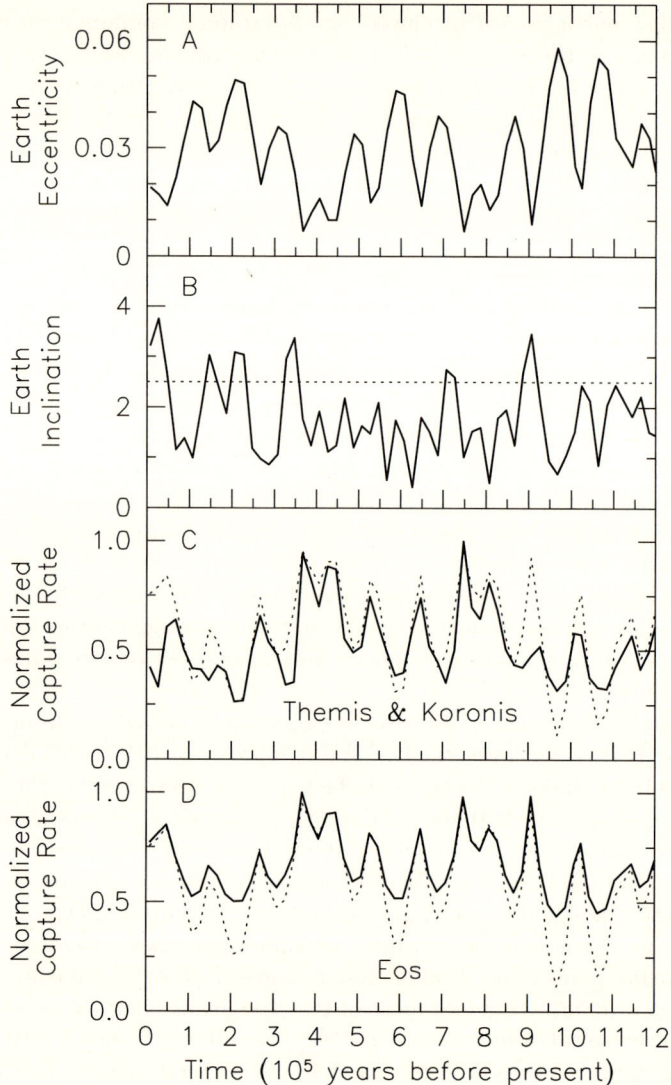

Figure 31. (A) Earth's orbital eccentricity. (B) Earth's orbit inclination with respect to the midplane of the Earth-crossing dust bands. The dashed line marks the mean proper inclination of Earth-crossing dust particles in the Themis and Koronis dust bands, which are indistinguishable at 1 AU. (C and D) Normalized average capture rate (solid lines) for $10\,\mu$m diameter dust particles from the Themis and Koronis families and the Eos family. In C and D the dashed line is Earth's orbital eccentricity plotted with the y-axis flipped from that in A. Reprinted from Kortenkamp et al. (2001) with kind permission from Kluwer Academic Publishers. [The capture rates shown in C and D are slightly different from those reported in Kortenkamp and Dermott (1998b) after correcting for a size-dependent term in the heliocentric velocity of the dust particles.]

also be anti-correlated with Earth's orbital eccentricity. Furthermore, because Earth's orbital eccentricity is not controlled by the dynamics of interplanetary dust particles, it follows that variations in the accretion rate of most asteroidal dust particles will be independent of the size of the particles. The long term accretion of asteroidal dust of all sizes should be anti-correlated with Earth's orbital eccentricity and varying with a 100,000 year periodicity.

The abundance of extraterrestrial ^{3}He in deep-sea sediments is indicative of the flux of IDPs to the sea floor, although opinions differ as to whether this amounts to a tracer of the actual accretion rate of IDPs (see Marcantonio et al. 1996, Farley and Patterson 1995). Analyses of sediment core samples do show a 100,000 year periodicity in the concentration of ^{3}He. However, the sea-floor ^{3}He periodicity is 50,000 years out of phase with the expected IDP accretion rate. Accretion from Earth's resonant ring of asteroidal dust is being investigated as a possible explanation of this 50,000 year phase lag.

2 Accretion from the resonant ring

We have numerically integrated the orbits of 40 μm diameter asteroidal dust particles from the main belt to the Earth, and divided them into those orbits that were temporarily trapped in mean motion resonances with the Earth and those that were not (Kortenkamp et al. 2001). The dust particles were assumed to be derived from the Eos asteroid family and were composed of astronomical silicate with density 2,500 kg m^{-3}. We then examined the total distribution of the minimum distances of these two populations from Earth, and also their relative velocities at close encounter, and found that the resonant particles (those that were trapped in resonance with Earth) do not have closer encounters than the non-resonant particles. However, they do tend to have higher geocentric encounter velocities (v_0) and so their capture probability is likely to be lower (see Eq. 38). There is one other consideration: when a particle escapes from resonance it is still in a near-resonant orbit and so may suffer further close encounters. The resonant population of particles may therefore undergo more close encounters with the Earth than the non-resonant particles, although our work to date indicates that this is not, in fact, the case. Hence, capture into resonance may ultimately hinder rather than aid accretion. These results are based upon integrations involving only $\sim 10^2$ dust particles of a single size, which is not sufficient to allow us to reach firm conclusions. We now need to repeat these simulations over a range of particle sizes and with a population of $\sim 10^5$ particles or more, to better determine their dynamical behaviour.

VIII. CONCLUSIONS

The motive for most of the work described in this review arises from the problem of determining the relative contributions of asteroidal and cometary material to the zodiacal cloud. Our approach to this problem is narrow, in that it is purely dynamical. We do not, for example, consider the physical properties of captured IDPs or their chemistry, even though these data may ultimately prove to be pivotal. We choose instead to focus on a small number of well-determined features of the cloud that clearly demand a dynamical explanation. We consider that a detailed analysis of these features may also produce a solution to the question of origins; a question that otherwise may remain unresolved. However, our work on this subset of dynamical problems is far from complete and, at present, it is biased towards the dynamics of asteroidal rather than cometary dust. This bias exists partly because the dynamics of particles in low eccentricity and inclination orbits are more tractable, largely because the secular variations of the orbital eccentricities and inclinations are decoupled and thus easier to describe in a full and systematic manner. Thus, we study the dynamics of asteroidal dust because we are actually able to solve some of the problems presented by these particles. However, our bias also exists for a more positive reason, namely that there is one important feature of the zodiacal cloud that can only be accounted for in terms of asteroidal dynamics and that feature is the Solar System dust bands discovered by IRAS.

Our modeling and analysis of the collisional evolution of the dust bands demonstrates that large particles with diameters between 10^2 and 10^3 μm dominate the dust band structures. Numerical investigation of their dynamical history demonstrates that these bands have a natural inner edge just exterior to 2 AU, because the action of secular resonances at the inner edge of the main asteroid belt disperses the inclinations and nodes of the large dust band particles into the background cloud. This leads us to estimate that approximately 4% of the in-ecliptic infrared emission from the zodiacal cloud is produced by dust band particles that orbit exterior to 2 AU. This is clear, unambiguous evidence that asteroidal dust is a significant component of the zodiacal cloud, at least at the 4% level.

If we now make the assumption that some of these asteroidal particles migrate to the inner Solar System, where they must be both hotter and closer to Earth, then we conclude that infrared emission from the asteroidal dust associated with the dust bands alone is likely to be much greater than 4% of the total emission. If these asteroidal particles migrate to the inner Solar System without further breakup, we calculate that the contribution is 30%. If the particles are broken up and blown out of the Solar System before reaching the inner Solar System, then our estimate would, of course, remain closer to 4%. However, it is certainly possible and perhaps even more likely that particle breakup leads to an increase in the effective area of the dust in which case our estimate would be greater than 30%. Consider, for example, that if a 10^2 μm particle breaks up to produce 10^3 particles of diameter 10μm, then the total effective surface area of the dust would increase by a factor of 10. Whether, the effective area of the dust actually increases or decreases is not known and thus 30% is, at present, our best estimate of the contribution of asteroidal dust, from the dust bands alone, to the zodiacal cloud.

But 30% must be an underestimate of the total asteroidal contribution. From a separate discussion of the ratio of the average rate of dust production in the asteroid belt as a whole to that due to those asteroids in the Eos, Themis and Koronis families, which we estimate to be 3:1, we are led to conclude that asteroidal dust may constitute 90% of the zodiacal cloud. There are other large uncertainties here because of the stochastic nature of asteroidal collisions, and because we do not have a good estimate of the quantity of dust that would be liberated by the destruction of, for example, a single 15 km diameter rubble-pile.

Thus, we have strong, quantitative evidence that asteroidal dust exists in the asteroid belt, but this does not prove that asteroidal dust is transported to the inner Solar System. However, there are other, equally well-defined features of the zodiacal cloud that require a dynamical explanation. These include: the inclination and orientation of the plane of symmetry, as seen in a given waveband, and the variation of these parameters with waveband and heliocentric distance; the offset of the center of symmetry from the Sun; and the trailing/leading asymmetry. Given that the origin of the particles in the cloud is either asteroidal or cometary, the aim of the Florida group is to follow the orbital evolution of these two sets of particles from source to sink, and to compare the structures of the various possible clouds, which could be purely asteroidal, purely cometary, or some combination of the two, with the range of observations listed above.

We have accounted for the trailing/leading brightness asymmetry of the zodiacal cloud in terms of the resonant trapping of asteroidal particles with diameters between approximately 5 and 30 μm. We consider that the particles are asteroidal because the probability of capture into resonance decreases markedly with increasing orbital eccentricity and thus it is less likely that the dominant particles in the resonant ring are cometary. This may be evidence that asteroidal particles are actually transported from the asteroid belt to Earth. That the particles must be small is apparently at odds with other considerations that indicate that the asteroidal particles that migrate out of the asteroid belt are large, with diameters between 10^2 and 10^3 μm. However, these small particles may be formed by collisional processes within the resonant ring. Dust particles are also expected to be trapped in resonances external to the orbits of other planets, including those embedded in circumstellar disks. These rings will always have some structure due to the geometry of particles in resonant orbits, but trailing clouds of dust may be rare.

Using the numerical and analytical methods described in this review, it will be possible for us to address the other dynamical features of the cloud, namely, the parameters describing the planes of symmetry and the offset of the center of symmetry from the Sun. These features may well provide a clear answer to the problem of the asteroidal or cometary origin of the cloud. However, our work on these models is still incomplete. The asteroidal models need to be extended to include much larger diameter particles than those that have been included to date. The dynamics of these larger asteroidal particles will need to take account of particle breakup as these particles migrate to the inner Solar System. Their dynamics will also be strongly influenced by resonant trapping, certainly with Earth and Mars, but possibly also with Venus, and by point-mass gravitational scattering by these terrestrial planets.

The cometary models must also be extended to include a range of particle sizes. It may well be that we need a significant cometary component of the cloud to account for the dynamical features listed above, but, at this stage, we have no clear reason to invoke a cometary model to account for any of these observations. However, we have also pointed out that the distinction between the orbital elements of cometary and asteroidal dust particles may not be as sharp as that displayed by the orbits of their parent bodies. Cometary particles that are trapped in resonance with Jupiter can have their orbital eccentricities decreased to asteroidal values. On the other hand, the large asteroidal particles that migrate slowly through the secular resonances at the inner edge of the main belt can have their orbital elements increased to the low end of the cometary range. Caution may be in order when classifying orbits as asteroidal or cometary without an appreciation of their possible orbital history. What is clear is that accretion of particles by the Earth is strongly biased towards particles in low eccentricity and inclination orbits. If our arguments with respect to the dust bands are correct, and asteroidal particles constitute a significant, and possibly dominant fraction of the particles in the broad-scale, background zodiacal cloud, then it is unlikely that cometary particles constitute a significant fraction of the IDPs that are archived in our collections.

This surprising conclusion has now been presented, by us and by others, at several meetings including the recent meeting in Canterbury (IAU Colloquium 181) without gaining, shall we say, wide acceptance. Our final comment in this regard is that it will not be possible to settle the question of particle origin by considering the rates of supply from the two sources. We now know the numbers and orbits of nearly all the asteroids in the main belt with diameters greater than about 20 km. Thus, we have a very good estimate of the collision rates of asteroidal bodies. However, we do not have good estimates of the strengths of these bodies, we do not have good estimates of the quantity of dust that would be liberated from the disruption of a rubble-pile, and we do not have a good description of the size-frequency distribution of asteroidal dust and its variation with heliocentric distance. Thus, even in the case of an asteroidal source, although we can show that asteroids could be an adequate source, we cannot claim using this supply rate argument alone that asteroids must be the dominant source.

However, the supply argument for a cometary source is markedly less useful. We know very little about either the size-frequency distribution of comets or their strength. Cometary dust trails may be clear evidence that cometary dust is deposited in the zodiacal cloud, but these trails have never been used to make a useful, quantitative assessment of the overall contribution of comets to the broad-scale, background zodiacal cloud.

Acknowledgements

The other members of the Florida Solar System Dynamics Group whose work has contributed to this paper and who we would like to thank are J.-C. Liou, E.K. Holmes and D. Fogle.

REFERENCES

Artymowicz, P. 1997. Beta Pictoris: an early Solar System? *Ann. Rev. Earth Planet. Sci.* **25**, pp. 175–219.

Backman, D. E., Dasgupta, A., and Stencel, R. E. 1995. Model of a Kuiper Belt small grain population and resulting far-infrared emission. *Astrophys. J.* **450**, pp. L35–L38.

Beaugé, C., and Ferraz-Mello, S. 1994. Capture in exterior mean-motion resonances due to Poynting-Robertson drag. *Icarus* **110**, pp. 239–260.

Beckwith, S. V. W. 1999. Circumstellar Disks. In *The Origin of Stars and Planetary Systems*, eds. C. J. Lada and N. D. Kylafis (Dordrecht: Kluwer Acad. Publ.), pp. 579–612.

Bowell, E. L. G. 1997. The Asteroid Orbital Elements Database. Lowell Observatory, ftp://ftp.lowell.edu/pub/elgb/astorb.html.

Bottke, W. F., Jr., Rubincam, D. P., and Burns, J. A. 2000. Dynamical evolution of main belt meteoroids: Numerical simulations incorporating planetary perturbations and Yarkovsky thermal forces. *Icarus* **145**, pp. 301–331.

Brouwer, D., and Clemence, G. M. 1961. *Methods of Celestial Mechanics*. (New York: Academic Press).

Burns, J. A., Lamy, P. L., and Soter, S. 1979. Radiation forces on small particles in the Solar System. *Icarus* **40**, pp. 1–48.

Davis, D. R., Chapman, C. R., Weidenschilling, S. J., and Greenberg, R. 1985. Collisional history of asteroids: Evidence from Vesta and the Hirayama families. *Icarus* **62**, pp. 30–53.

Davis, D. R., Weidenschilling, S. J., Farinella, P., Paolicchi, P., and Binzel, R. P. 1989. Asteroid collisional history: Effects on sizes and spins. In *Asteroids II*, eds. R. P. Binzel, T. Gehrels and M. S. Matthews (Tucson: Univ. of Arizona Press), pp. 805–826.

Davis, D. R., Farinella, P., Paolicchi, P., and Bagatin, A. C. 1993. Deviations from the straight line: Bumps (and grinds) in the collisionally evolved size distribution of asteroids. *Lunar Planet. Sci.* **24**, pp. 377–378.

Dermott, S. F., Nicholson, P. D., Burns, J. A., and Houck, J. R. 1984. Origin of the Solar System dust bands discovered by IRAS. *Nature* **312**, pp. 505–509.

Dermott, S. F., Nicholson, P. D., Burns, J. A., and Houck, J. R. 1985. An analysis of IRAS' Solar System dust bands. In *Properties and Interactions of Interplanetary Dust*, ASSL Proc. **119**, eds. R. H. Giese and P. Lamy (Dordrecht: D. Reidel Publ. Co.), pp. 395–410.

Dermott, S. F., and Nicholson, P. D. 1986. Masses of the satellites of Uranus. *Nature* **319**, pp. 115–120.

Dermott, S. F., Nicholson, P. D., and Wolven, B. 1986. Preliminary analysis of the IRAS Solar System dust data. In *Asteroids, Comets, Meteors II*, eds. C.-I. Lagerkvist, B. A. Lindblad, H. Lundstedt and H. Rickman (Uppsala: Reprocentralen HSC), pp. 583–594.

Dermott, S. F., Nicholson, P. D., Kim, Y., Wolven, B., and Tedesco, E. F. 1988a. The impact of IRAS on asteroidal science. In *Comets to Cosmology*, ed. A. Lawrence, (Berlin: Springer-Verlag), pp. 3–18.

Dermott, S. F., Malhotra, R., and Murray, C. D. 1988b. Dynamics of the Uranian and Saturnian satellite systems: A chaotic route to melting Miranda? *Icarus* **76**, pp. 295–334.

Dermott, S. F., Nicholson, P. D., Gomes, R. S., and Malhotra, R. 1990. Modeling the IRAS Solar System dust bands. *Adv. Space Res.* **10**, pp. 171–180.

Dermott, S. F., Gomes, R. S., Durda, D. D., Gustafson, B. Å. S., Jayaraman, S., Xu, Y.-L., and Nicholson, P. D. 1992. Dynamics of the zodiacal cloud. In *Chaos, Resonance and Collective Dynamical Phenomena in the Solar System*, ed. S. Ferraz-Mello, (Dordrecht: Kluwer Acad. Publ.), pp. 333–347.

Dermott, S. F., Jayaraman, S., Xu, Y.-L., Gustafson, B. Å. S., and Liou, J.-C. 1994a. A circumsolar ring of asteroidal dust in resonant lock with the Earth. *Nature* **369**, pp. 719–723.

Dermott, S. F., Durda, D. D., Gustafson, B. Å. S., Jayaraman, S., Liou, J.-C., and Xu, Y.-L. 1994b. Zodiacal dust bands. In *Asteroids, Comets and Meteors 1993*, eds. A. Milani, M. Martini and A. Cellino (Dordrecht: Kluwer Acad. Publ.), pp. 127–142.

Dermott, S. F., Jayaraman, S., Xu, Y.-L., Grogan, K., and Gustafson, B. Å. S. 1996a. The origin and dynamics of the interplanetary dust cloud. In *Unveiling the Cosmic Infrared Background*, AIP Conference Proc. **348**, ed. E. Dwek, (New York: Woodbury), pp. 25–36.

Dermott, S. F., Grogan, K., Gustafson, B. Å. S., Jayaraman, S., Kortenkamp, S. J., and Xu, Y.-L. 1996b. Sources of interplanetary dust. In *Physics, Chemistry and Dynamics of Interplanetary Dust*, ASP Conference Series **104**, eds. B. Å. S. Gustafson and M. S. Hanner (San Francisco: Astron. Soc. of the Pacific Press), pp. 143–153.

Dermott, S. F., Grogan, K., Holmes, E. K., and Wyatt, M. C. 1998. Signatures of planets. In *Exozodiacal Dust Workshop Conference Proceedings*, eds. D. E. Backman, L. J. Caroff, S. A. Sanford and D. H. Woodford (Washington DC: NASA CP-1998-10155), pp. 59–84.

Dermott, S. F., Grogan, K., Holmes, E. K., and Kortenkamp, S. J. 1999. Dynamical structure of the zodiacal cloud. In *Formation and Evolution of Solids in Space*, eds. J. M. Greenberg and A. Li (Dordrecht: Kluwer Acad. Publ.), pp. 565–582.

Dohnanyi, J. S. 1969. Collisional model of asteroids and their debris. *J. Geophys. Res.* **74**, pp. 2531–2554.

Draine, B. T., and Lee, H. M. 1984. Optical properties of interstellar graphite and silicate grains. *Astrophys. J.* **285**, pp. 89–108.

Durda, D. D., and Dermott, S. F. 1997. The collisional evolution of the asteroid belt and its contribution to the zodiacal cloud. *Icarus* **130**, pp. 140–164.

Durda, D. D., Greenberg, R., and Jedicke, R. 1998. Collisional models and scaling laws: A new interpretation of the shape of the main-belt asteroid size distribution. *Icarus* **135**, pp. 431–440.

Everhart, E. 1985. An efficient integrator that uses Gauss-Radau spacings. In *Dynamics of Comets: Their origin and evolution*, eds. A. Carusi and G. B. Valsecchi (Dordrecht: D. Reidel Publ. Co.), pp. 185–202.

Farley, K. A., and Patterson, D. B. 1995. A 100-kyr periodicity in the flux of extraterrestrial ^{3}He to the sea floor. *Nature* **378**, pp. 600–603.

Flynn, G. J. 1989. Atmospheric entry heating: A criterion to distinguish between asteroidal and cometary sources of interplanetary dust. *Icarus* **77**, pp. 287–310.

Flynn, G. J. 1990. The near-Earth enhancement of asteroidal over cometary dust. In *Proceedings of the 20th Lunar and Planetary Science Conference*, (Houston: Lunar and Planet. Inst.), pp. 363–371.

Giese, R. H., Kneissel, B., and Rittich, U. 1986. Three-dimensional models of the zodiacal dust cloud — A comparative study. *Icarus* **68**, pp. 395–411.

Gold, T. 1975. Resonant orbits of grains and the formation of satellites. *Icarus* **25**, pp. 489–491.

Gomes, R. S. 1995. Resonance trapping and evolution of particles subject to Poynting-Robertson drag: Adiabatic and non-adiabatic approaches. *Celest. Mech. Dynam. Astron.* **61**, pp. 97–113.

Grogan, K., Dermott, S. F., and Gustafson, B. Å. S. 1996. An estimation of the interstellar contribution to the zodiacal thermal emission. *Astrophys. J.* **472**, pp. 812–817.

Grogan, K., Dermott, S. F., Jayaraman, S., and Xu, Y.-L. 1997. Origin of the ten degree dust bands. *Planet. Space Sci.* **45**, pp. 1657–1665.

Grogan, K., Dermott, S. F., and Durda, D. D. 2001. The size-frequency distribution of the zodiacal cloud: Evidence from the Solar System dust bands. *Icarus*, in press.

Grün, E., Zook, H. A., Fechtig, H., and Giese, R. H. 1985. Collisional balance of the meteoritic complex. *Icarus* **62**, pp. 244–272.

Gurnett, D. A., Ansher, J. A., Kurth, W. S., and Granroth, L. J. 1997. Micron-sized dust particles detected in the outer Solar System by Voyager 1 and 2 plasma wave instruments. *Geophys. Res. Lett.* **24**, pp. 3125–3128.

Gustafson, B. Å. S. 1994. Physics of zodiacal dust. *Ann. Rev. Earth Planet. Sci.* **22**, pp. 553–595.

Hirayama, K. 1918. Groups of asteroids probably of common origin. *Astron. J.* **31**, pp. 185–188.

Holland, W. S., Greaves, J. S., Zuckerman, B., Webb, R. A., McCarthy, C., Coulson, I. M., Walther, D. M., Dent, W. R. F., Gear, W. K., and Robson, I. 1998. Submillimetre images of dusty debris around nearby stars. *Nature* **392**, pp. 788–790.

Holmes, E. K., Dermott, S. F., Xu, Y.-L., Wyatt, M. C., and Jayaraman, S. 1998. Modeling the effects of an offset of the center of symmetry in the zodiacal cloud. In *Exozodiacal Dust Workshop Conference Proceedings*, eds. D. E. Backman, L. J. Caroff, S. A. Sanford and D. H. Woodford (Washington DC: NASA CP-1998-10155), pp. 272–273.

Housen, K. R., and Holsapple, K. A. 1990. On the fragmentation of asteroids and planetary satellites. *Icarus* **84**, pp. 226–253.

Jackson, A. A., and Zook, H. A. 1989. A Solar System dust ring with earth as its shepherd. *Nature* **337**, pp. 629–631.

Jackson, A. A., and Zook, H. A. 1992. Orbital evolution of dust particles from comets and asteroids. *Icarus* **97**, pp. 70–84.

Jayaraman, S., and Dermott, S. F. 1996a. COBE-DIRBE observations of the Earth's resonant ring. In *Unveiling the Cosmic Infrared Background*, AIP Conference Proc. **348**, ed. E. Dwek, (New York: Woodbury), pp. 47–52.

Jayaraman, S., and Dermott, S. F. 1996b. SIRTF: A unique opportunity for probing the zodiacal cloud. In *Physics, Chemistry and Dynamics of Interplanetary Dust*, ASP Conference Series **104**, eds. B. Å. S. Gustafson and M. S. Hanner (San Francisco: Astron. Soc. of the Pacific Press), pp. 159–162.

Jayaraman, S., and Dermott, S. F. 2001. Formation, structure and observations of the Earth's resonant ring. *Icarus*, submitted.

Jewitt, D. 1999. Kuiper Belt objects. *Ann. Rev. Earth Planet. Sci.* **27**, pp. 287–312.

Jones, M. H., and Rowan-Robinson, M. 1993. A physical model for the IRAS zodiacal dust bands. *Mon. Not. R. Astron. Soc.* **264**, pp. 237–247.

Kehoe, T. J. J. 1999. Long term dynamics of small bodies in the Solar System using mapping techniques. Ph.D. thesis, University of London.

Kehoe, T. J. J., Murray, C. D., and Porco, C. C. 2001a. A dissipative mapping technique for the N-body problem incorporating radiation pressure, Poynting-Robertson drag and solar-wind drag. *Astron. J.*, submitted.

Kehoe, T. J. J., Dermott, S. F., and Grogan, K. 2002. A dissipative mapping technique for integrating interplanetary dust particle orbits. In *Dust in the Solar System and Other Planetary Systems*, Proceedings of IAU Colloquium 181/COSPAR Colloquium 11, eds. J. A. M. McDonnell et al. (Amsterdam: Elsevier), submitted.

Kelsall, T., Weiland, J. L., Franz, B. A., Reach, W. T., Arendt, R. G., Dwek, E., Freudenreich, H. T., Hauser, M. G., Moseley, S. H., Odegard, N. P., Silverberg, R. F., and Wright E. L. 1998. The COBE diffuse infrared background experiment search for the cosmic infrared background. II. Model of the interplanetary dust cloud. *Astrophys. J.* **508**, pp. 44–73.

Kessler, D. J. 1981. Derivation of the collision probability between orbiting objects: The lifetime of Jupiter's outer moons. *Icarus* **48**, pp. 39–48.

Kortenkamp, S. J., and Dermott, S. F. 1998a. Accretion of interplanetary dust particles by the Earth. *Icarus* **135**, pp. 469–495.

Kortenkamp, S. J., and Dermott, S. F. 1998b. A 100,000 year periodicity in the accretion rate of interplanetary dust. *Science* **280**, pp. 874–876.

Kortenkamp, S. J., Dermott, S. F., Fogle, D., and Grogan, K. 2001. Sources and orbital evolution of interplanetary dust accreted by Earth. In *Accretion of Extraterrestrial Matter Throughout Earth's History*, eds. B. Peucker-Ehrenbrink and B. Schmitz (Dordrecht: Kluwer Acad. Publ.), pp. 13–30.

Leinert, C., and Grün, E. 1990. Interplanetary dust. In *Physics and Chemistry in Space: Physics of the Inner Heliosphere I*, Space and Solar Physics **20**, eds. R. Schween and E. Marsch (Berlin: Springer-Verlag), pp. 207–275.

Liou, J.-C., Dermott, S. F., and Xu, Y.-L. 1995. The contribution of cometary dust to the zodiacal cloud. *Planet. Space Sci.* **43**, pp. 717–722.

Liou, J.-C., and Zook, H. A. 1996. Comets as a source of low eccentricity and low inclination interplanetary dust particles. *Icarus* **123**, pp. 491–502.

Liou, J.-C., Zook, H. A., and Dermott, S. F. 1996. Kuiper Belt dust grains as a source of interplanetary dust particles. *Icarus* **124**, pp. 429–440.

Liou, J.-C., and Zook, H. A. 1999. Signatures of the giant planets imprinted on the Edgeworth-Kuiper Belt dust disk. *Astron. J.* **118**, pp. 580–590.

Liou, J.-C., Zook, H. A., Greaves, J. S., and Holland W. S. 2000. Does planet exist in ε Eridani? A comparison between obervations and numerical simulations. In *31st*

Annual Lunar and Planetary Science Conference, (Houston: Lunar and Planet. Inst.), abstract no. 1416.

Love, S. G., and Brownlee, D. E. 1992. The IRAS dust band contribution to the interplanetary dust complex: Evidence seen at 60 and 100 microns. Astrophys. J. **104**, pp. 2236–2242.

Love, S. G., and Brownlee, D. E. 1993. A direct measurement of the terrestrial mass accretion rate of cosmic dust. Science **262**, pp. 550–553.

Low, F. J., Beintema, D. A., Gautier, T. N., Gillet, F. C., Beichmann, C. A., Neugebauer, G., Young, E., Aumann, H. H., Boggess, N., Emerson, J. P., Habing, H. J., Hauser, M. G., Houck, J. R., Rowan-Robinson, M., Soifer, B. T., Walker, R. G., and Wesselius, P. R. 1984. Infrared cirrus: New components of the extended infrared emission. Astrophys. J. **278**, pp. L19–L22.

Malhotra, R. 1994. A mapping method for the gravitational few-body problem with dissipation. Celest. Mech. Dynam. Astron. **60**, pp. 373–385.

Marcantonio, F., Anderson, R. F., Stute, M. S., Kumar, N., Schlosser, P., and Mix, A. 1996. Extraterrestrial ^{3}He as a tracer of marine sediment transport and accumulation. Nature **383**, pp. 705–707.

Marsden, B. 1995. Catalogue of cometary orbits. Minor Planet Center, http://cfa-www.harvard.edu/iau/mpc.html.

Marzari, F., and Vanzani, V. 1994. Orbital evolution of dust particles near mean motion resonances with the Earth. Planet. Space Sci. **42**, pp. 101–107.

Muller, R. A., and MacDonald, G. J. 1997. Glacial cycles and astronomical forcing. Science **277**, pp. 215–218.

Murray, C. D., and Dermott, S. F. 1999. Solar System Dynamics. (Cambridge: Cambridge Univ. Press).

Neugebauer, G., Beichman, C. A., Soifer, B. T., Aumann, H. H., Chester, T. J., Gautier, T. N., Gillett, F. C., Hauser, M. G., Houck, J. R., Lonsdale, C. J., Low, F. J., and Young, E. T. 1984. Early results from the infrared astronomical satellite. Science **224**, pp. 14–21.

Öpik, E. J. 1951. Collision probabilities with the planets and the distribution of interplanetary matter. Proc. R. Irish Acad. **54**, pp. 165–199.

Patterson, D. B., and Farley, K. A. 1998. Extraterrestrial ^{3}He in sea floor sediments: Evidence for correlated 100 kyr periodicity in the accretion rate of interplanetary dust, orbital parameters, and Quaternary climate. Geochim. Cosmochim. Acta. **62**, pp. 3669–3682.

Reach, W. T. 1991. Zodiacal emission. II — Dust near ecliptic. Astrophys. J. **369**, pp. 529–543.

Reach, W. T. 1992. Zodiacal emission. III — Dust near the asteroid belt. Astrophys. J. **392**, pp. 289–299.

Reach, W. T., Franz, B. A., Weiland, J. L., Hauser, M. G., Kelsall, T. N., Wright, E. L., Rawley, G., Stemwedel, S. W., and Spiesman, W. J. 1995. Observational confirmation of a circumsolar dust ring by the COBE satellite. Nature **374**, pp. 521–523.

Reach, W. T., Franz, B. A., and Weiland, J. L. 1997. The three-dimensional structure of the zodiacal dust bands. Icarus **127**, pp. 461–484.

Rial, J. A. 1999. Pacemaking the ice ages by frequency modulation of Earth's orbital eccentricity. Science **285**, pp. 564–568.

Schmidt, H. 1967. The possibility of dust concentration near the Earth. In The Zodiacal Light and the Interplanetary Medium, ed. J. L. Weinberg, (Washington DC: NASA SP-150), pp. 333–336.

Standish, E. M., Newhall, X. X., Williams, J. G., and Yeomans, D. K. 1992. Orbital ephemerides of the Sun, Moon, and Planets. In Explanatory Supplement to the Astronomical Almanac, ed. P. K. Seidelmann, (Mill Valley: University Science Books), pp. 279–323.

Stern, S. A. 1995. Collisional time scales in the Kuiper Disk and their implications. Astron. J. **110**, pp. 856–868.

Stern, S. A., and Colwell, J. E. 1997. Collisional erosion in the primordial Edgeworth-Kuiper Belt and the generation of the 30–50 AU Kuiper Gap. Astrophys. J. **490**, pp. 879–882.

Sykes, M. V. 1988. IRAS observations of extended zodiacal structures. Astrophys. J. **334**, pp. L55–L58.

Sykes, M. V. 1990. Zodiacal dust bands: Their relation to asteroid families. *Icarus* **84**, pp. 267–289.

Sykes, M. V., and Greenberg, R. 1986. The formation and origin of the IRAS zodiacal dust bands as a consequence of single collisions between asteroids. *Icarus* **65**, pp. 51–69.

Sykes, M. V., and Walker, R. G. 1992. Cometary dust trails. I — Survey. *Icarus* **95**, pp. 180–210.

Telesco, C. M., Fisher, R. S., Piña, R. K., Knacke, R. F., Dermott, S. F., Wyatt, M. C., Grogan, K., Holmes, E. K., Ghez, A. M., Prato, L., Hartmann, L. W., and Jayawardhana, R. 2000. Deep 10 and 18 micron imaging of the HR 4796A circumstellar disk: Transient dust particles and tentative evidence for a brightness asymmetry. *Astrophys. J.* **530**, pp. 329–341.

Wetherill, G. W., and Cox, L. P. 1985. The range of validity of the two-body approximation in models of terrestrial planet accumulation II: Gravitational cross-sections and runaway accretion. *Icarus* **63**, pp. 290–303.

Wisdom, J. 1980. The resonance overlap criterion and the onset of stochastic behavior in the restricted three-body problem. *Astron. J.* **85**, pp. 1122–1133.

Wisdom, J., and Holman, M. 1991. Symplectic maps for the n-body problem. *Astron. J.* **102**, pp. 1528–1538.

Wyatt, S. P., Jr., and Whipple F. L. 1950. The Poynting-Robertson effect on meteor orbits. *Astrophys. J.* **111**, pp. 134–141.

Wyatt, M. C., Dermott, S. F., Grogan, K., and Jayaraman, S. 1999a. A unique view through the Earth's resonant ring. In *Astrophysics with Infrared Surveys: A prelude to SIRTF*, ASP Conference Series **177**, eds. M. D. Bicay, C. A. Beichman, R. M. Cutri and B. F. Madore (San Francisco: Astron. Soc. of the Pacific Press), pp. 374–380.

Wyatt, M. C., Dermott, S. F., Telesco, C. M., Fisher, R. S., Grogan, K., Holmes, E. K., and Piña, R. K. 1999b. How observations of circumstellar disk asymmetries can reveal hidden planets: Pericenter glow and its application to the HR 4796 disk. *Astrophys. J.* **527**, pp. 918–944.

Zappalà, V., Bendjoya, P. H., Cellino, A., Farinella, P., and Froeschle, C. 1995. Asteroid families: Search of a 12,487 asteroid sample using two different clustering techniques. *Icarus* **116**, pp. 291–314.

Zook, H. A., and Berg, O. E. 1975. A source for hyperbolic cosmic dust particles. *Planet. Space Sci.* **23**, pp. 183–203.

Dusty Rings and Circumplanetary Dust: Observations and Simple Physics

Joseph A. Burns[1], Douglas P. Hamilton[2], Mark R. Showalter[3]

[1] Cornell University, Ithaka, New York, USA
[2] University of Maryland, College Park, Maryland, USA
[3] Stanford University, Stanford, California, USA

Abstract. Each giant planet is encircled by planetary rings, usually composed of particles centimeters to meters in radius, but each system also contains regions where much smaller dust grains predominate. This chapter summarizes the techniques used to determine the properties of circumplanetary material, and then gives a precis of the known characteristics of circumplanetary rings (with emphasis on tenuous structures) and dust grains, before describing some of the physics and orbital dynamics relevant to them. *Jupiter*'s dusty rings (as discovered by the Voyager and Galileo spacecraft) have three components: i) a radially confined and vertically extended *halo* which rises abruptly, probably due to an electromagnetic resonance; ii) a 6500-km-wide flattened *main ring* that shows patchiness and whose outer edge is bounded by the orbit of the satellite Adrastea; and iii) a pair of exterior *gossamer rings* that seem to be derived from the satellites Amalthea and Thebe whose orbits circumscribe these rings. In addition, small particles are strewn throughout the inner Jovian magnetosphere, especially near the paths of the Galilean moons, and the jovian system seems to eject very tiny particles at hypervelocities to interplanetary space. *Saturn*'s circumplanetary dust is unusual in the size distribution of its various rings: the broad and diffuse E ring seems to be mainly 1-micron grains whereas the narrow F and G rings have quite steep size distributions, indicating the predominance of very small grains. Surprisingly little dust resides in the main Saturnian rings, except in the localized spokes. Dust is interspersed between the narrow classical *Uranian* rings, forming a sheet that is punctuated by narrow bands and gaps. *Neptune*'s system contains at least some grains that lie well off the planet's equatorial plane, perhaps as a result of Neptune's highly tilted and offset magnetic field. The debris lost off the small moons Phobos and Deimos is believed to produce very tenuous *dust tori around Mars*. Complex orbital histories for circumplanetary grains result from conservative and non-conservative forces (gravity, radiation pressure and electromagnetism); the latter become most important for smaller particles and may even lead to ejection or planetary impact. Orbital resonance phenomena, several of which are unique to circumplanetary dust, seem to govern the distribution of grains orbiting planets. Circumplanetary dust is short-lived in a cosmic sense, owing to erosion through sputtering by the surrounding magnetospheric plasma and orbital loss due to various evolution mechanisms. These brief lifetimes imply continual regeneration to supply new material. Circumplanetary dust is often found in intimate relation with embedded small moonlets since it can be generated through energetic impacts into such bodies but is also absorbed by them.

I. INTRODUCTION

Innumerable dust grains circle each of the giant planets, not only interspersed among the macroscopic bodies that comprise the familiar opaque ring systems, but also elsewhere forming tenuous structures of their own. The original detections of dust clustered near the equatorial planes of Jupiter and Saturn were accomplished a quarter-century ago by pressurized "beer-can" experiments aboard the Pioneer spacecraft (Humes 1976, 1980). The distribution and properties of these tiny motes in the neighborhoods of all the giant planets were more thoroughly explored in Voyager images. In addition sensitive plasma detectors aboard Voyager found dust strewn throughout planetary systems in quantities too faint to be visible. Most recently circumjovian dust has been studied by Ulysses and Galileo instruments. From the ground, circumsaturnian and circumjovian dust rings were viewed during the 1995-96 and 1997 ring-plane crossings. Cassini's scientific payload, including imaging systems extending from the UV to the IR, a sophisticated dust detector and plasma instruments, is capable of revolutionizing our understanding of Jupiter's dust rings (during flyby at the end of 2000) and Saturn's complement of dust (throughout the 4-year tour, commencing in 2004).

Circumterrestrial dust is, of course, well known historically with the most diagnostic information coming from LDEF (see chapter by McDonnell et al.). Because the circumterrestrial data differ so markedly from those about the other planets, we will not consider the Earth's particles at all although our dynamical modeling is of course relevant. Dust has not been unequivocally found around any other terrestrial planet, although Dubinin et al. (1990) claim to have detected some material about Mars; the Japanese Planet-B (Nozomi) spacecraft is carrying an ionization dust detector that should map the dust distribution in this system. In explanation of the Soviet observation and in preparation for this latter mission, a dozen or so papers have been written concerning Martian dust since the late 1980s. This interest is appropriate since Soter's prescient (1971) report first explored the modeling of circumplanetary dust in connection with this system.

Several motivations provoke interest in circumplanetary dust. First, ring dynamicists are challenged by the wide range of forces to which these grains are subject and by the counter-intuitive behavior of some of this material. Second, circumplanetary grains intimately interact with the surrounding magnetospheric plasma and with neighboring satellites, in some cases being derived therefrom and in others modifying those surfaces. Thirdly, dust probes conditions in the surrounding magnetospheric plasma and, through its response, calibrates the nature of those fields. Finally, Cassini mission designers are justifiably concerned about sizes and realms of circumsaturnian dust since Cassini will pass continually (and usually at high relative speed) through the faint rings; optical surfaces can be scoured by the impacts of small grains, while larger collisions can destroy other components or even the entire spacecraft. In an earlier case Galileo's probe was redirected to avoid Jupiter's gossamer ring following its discovery (Showalter et al. 1985).

The applicable physics and dynamics acting on circumplanetary dust are distinct from those pertinent to interplanetary dust, because the dust orbits through a magnetosphere and about a central mass other than the radiation source. But circumplanetary dust particles are not classical ring particles either. In the latter case, collisions dominate and the resultant structures can be studied with the tools of fluid dynamics and kinetic theory. For faint rings, collisions among ring particles are rare, and each particle behaves as a miniature independent satellite circling its primary; thus the methods of single-particle dynamics may be applied. However, as constituent particles are generally tiny, non-gravitational forces (electromagnetism, radiation and drag) must be included.

Previous overviews of circumplanetary dust have been written mainly in the context of planetary rings (e.g., Burns et al. 1984; but see Burns 1991). Reviews by Mendis et al. (1984) and Grün et al. (1984) emphasized electrodynamic processes in rings, whereas Mignard (1984) was concerned with the role of radiation pressure; the dynamics of circumplanetary dust were described by Burns et al. (1979) and Hamilton (1993) among others. Goertz (1989), Northrop (1992), Hartquist et al. (1992), Mendis and Rosenberg (1994) and Horányi (1996) have provided the most recent reviews of dusty plasmas in an astrophysical context. Much activity, seeking explanations for various puzzling phenomena found by spacecraft, has been carried out in the last decade but, to date, has not yet been summarized. The only available text on the subject is Bliokh et al. (1995).

This section will be organized as follows. After describing the techniques used to characterize circumplanetary dust, we will outline our knowledge of the fine material strewn around the giant planets sequentially in distance from the Sun. We will then discuss the forces that act on circumplanetary dust and the relevant physical processes before considering celestial mechanics. Finally we will suggest future studies. As mentioned above, we do not discuss circumterrestrial dust; we also do not consider interplanetary particles found to be streaming away from Jupiter by the Ulysses and Galileo dust detectors and thought to have originated somewhere in the bowels of the planet's magnetosphere. We do not discuss interplanetary and interstellar dust particles, although they penetrate planetary magnetospheres, except insofar as they may generate circumplanetary dust through impacts onto satellites and other orbiting bodies.

The sorts of questions that we will address in this chapter concern the sources and fates of circumplanetary dust. How do faint rings evolve? How old are planetary rings? What causes the tenuous ring systems to differ so much? Why are some faint rings confined while others are vertically or radially extended? Which of the phenomena displayed by our relatively simple dynamical systems are relevant to the collisionally dominated classical ring systems?

II. DESCRIPTION

In this section we review how ring properties are determined and then we summarize the properties of the known dusty rings. Detailed imaging observations of a planetary ring, at multiple wavelengths and phase angles, provide a wealth of information on a ring's global characteristics including its radial and vertical structure, particle size distribution, and normal optical depth. In addition, particle detectors on spacecraft can provide information on local conditions within diffuse dusty rings. Pioneer 10 and 11's dust detectors returned limited data on $\sim$ 5–10 μm particles from the near-Jupiter and near-Saturn environments, while Voyager's PWS and PRA experiments proved to be sensitive to micron-sized dust around all of the giant planets. The dust detectors aboard Ulysses, and especially Galileo, provide the best calibrated and most useful dust detections in the jovian system. As described below, both imaging and *in situ* observations allow us to determine – or at least put useful constraints on – the gross properties of a planetary ring

II.A. Physical Models

For our purposes, we define a ring to be any ensemble of particles orbiting a planet. In general, most rings are circular, vertically thin, equatorial and axisymmetric, although specific rings exist that violate each of these generalizations (Burns 1999). The challenge to astronomers is, from remote measurements of a ring, to infer the physical and orbital properties of the constituent particles. In this section we present a brief overview of the methods used to model the physical properties of a ring.

Typical properties that one wishes to learn about a planetary ring are its radial distribution, particle sizes and composition. At any location in a ring, particle sizes can be described by a differential size distribution $n(s)$, defined such that $n(s)\,\mathrm{d}s$ is the number of particles per unit ring *area* (i.e., integrated normal to the ring plane) in the radius range s to $s+\mathrm{d}s$. The particles that comprise the known rings extend in size from smaller than a micron to $\sim$ 10 m.

One of the most fundamental properties of a ring is its *normal optical depth* τ, which is the quantity directly probed by ring occultation experiments. It is related to the local size distribution via

$$\tau = \int \pi s^2 Q_{\mathrm{ext}}(s) n(s) \mathrm{d}s \ . \tag{1}$$

Here the extinction efficiency Q_{ext} is the dimensionless ratio of a particle's extinction cross-section to its physical cross-section; it describes the fraction of light impinging upon a particle that is either absorbed or scattered into a different direction while the remainder of the light continues unimpeded. If Q_{ext} were to equal 1, τ would be the fractional area of a ring filled by particles (at least for $\tau \ll 1$; more precisely, the filling fraction is $1 - e^{-\tau}$ if particles are positioned randomly).

However, this simple interpretation of τ is actually only appropriate for particles much larger than the wavelength of light. For tiny dust grains, Mie theory is typically employed to derive Q_{ext} as a function of radius s (van de Hulst 1981). Mie theory assumes the grains are homogeneous spheres; variant formulations (e.g., Pollack and Cuzzi 1980) can be used to model more irregular shapes. Within these theories, the two key free parameters are the refractive index, which depends on the composition, and the particle size. Particle sizes are best described by a dimensionless "size parameter" X, defined as $2\pi s/\lambda$, where λ is the wavelength of light. In general, Q_{ext} is of order unity for X of order unity (cf. Fig. A2 of Cuzzi et al. 1984); it decreases rapidly (typically $\propto X^4$) for smaller X, in the Rayleigh scattering limit. For this reason, measurements of a ring's optical depth are generally insensitive to particles much smaller than the wavelength. Accordingly, a ring's optical depth generally decreases with increasing λ, and this decrease can be used to constrain the particle sizes. For larger X, Q_{ext} rapidly levels out to a value of two; this difference from our expected value of unity will be discussed further below.

In practice, the optical depth measured in a ring occultation experiment is not the normal optical depth τ, but the larger value τ/μ, where μ is the cosine of the emission angle (measured from the ring normal vector to the line of sight). This μ factor corrects for the increased line of sight when the ring (assumed to be a flat slab) is not observed pole-on. Because the value of μ is known in any given experiment, however, recovery of the ring's normal optical depth is straightforward.

Rings are detected usually through the light that they absorb or reflect. For a ring in which τ/μ is small, a simple relationship exists between the ring's properties and the intensity of light I (power per area per wavelength interval per steradian) reflected:

$$\frac{I}{F} = \frac{\tau \varpi_0 P(\alpha)}{4\mu}. \qquad (2)$$

Here I is expressed as a dimensionless ratio relative to F, where πF is the incoming solar flux density (power per area per wavelength interval). By this definition, I/F removes the effects of the Sun's spectrum and its distance from the ring, and it equals unity for a perfectly diffusing "Lambert" surface illuminated at normal incidence. In addition to τ, the key ring properties here are the single scattering albedo ϖ_0 and phase function $P(\alpha)$, where α is the phase angle or Sun-ring-observer angle. Both quantities can be derived from Mie theory (for example) and represent averages over the size distribution. Single-scattering albedo ϖ_0 describes the fraction of impinging light not absorbed by the particle, and always ranges between zero and one. The phase function describes the fraction of light scattered into various directions; it is normalized to an average value of unity when integrated over all solid angles.

The phase function is extremely sensitive to particle size. In the Rayleigh-scattering limit ($X \ll 1$) the phase function becomes isotropic to within a factor of two (see Fig. A2 in Cuzzi et al. 1984). As X increases to order unity, the phase function becomes predominantly forward-scattering due to diffraction, with a large peak at $\alpha = 180°$. It also retains a shallower peak near backscatter ($\alpha = 0°$) and a minimum at intermediate α. As X increases further, half of the energy is diffracted into an ever-narrower forward-scattering peak, of angular width $\sim \pi/X$, while the remainder of the phase function is predominantly backscattering.

The mysterious result that $Q_{\text{ext}} \to 2$ for large X, called Babinet's principle and mentioned above, is closely related to this narrow diffraction spike in forward-scatter. For X extremely large (as is the case for centimeter and larger objects under visible light) it becomes impractical to distinguish the narrow diffraction peak from un-scattered light rays. One therefore tends to eliminate this component of the phase function and simultaneously halve Q_{ext} to its expected value of unity. For this reason, phase functions of macroscopic bodies (such as moons) never include the diffraction spike. However, it should be noted that, under special circumstances, this spike cannot be neglected. For example, radio occultations of planetary rings record the phase of the signal in addition to its amplitude; the forward-diffracted signal undergoes a phase shift relative to the direct signal so the two components can be distinguished. For this reason, the definition of τ, as given in (1), for a radio occultation experiment typically differs from its visual counterpart by a factor of two. Cuzzi (1985; see also Porco et al. 1995) discusses other circumstances in which this pitfall arises.

In Eq. (2) above we made the assumption that the ring was optically thin. This is not the case for some rings, so the relation breaks down. For larger τ it becomes probable that some ring particles will shadow or block others, in which circumstance (2) becomes a significant overestimate of I/F. Furthermore, multiple scattering of light among particles leads to a more isotropic phase function. Under these circumstances, more accurate "radiative-transfer" calculations are employed, analogous to those used in atmospheric sciences. The "doubling" algorithm is most general, in which the theoretical I/F pattern for a ring is built up by successive summings of lower-τ layers (Hansen 1969). In these approaches, it is possible and often necessary to include the secondary illumination from the planet in addition to sunlight (Cooke 1991; Dones et al. 1993; Showalter 1996); for rings close to a planet, this secondary illumination can be quite significant.

Nevertheless, as will be shown below, usually the dusty planetary rings are extremely optically thin, so that (2) can be used directly. In these cases it is convenient to introduce the concept of the "normal I/F," $\equiv \mu I/F$, meaning not the value measured, but instead the value that would have been measured if the ring were viewed normally. By applying this μ factor, one can trivially compensate for a ring's intensity variations with emission angle, and this can simplify data analysis.

Of course, astronomers are confronted with the inverse problem to that described above—one is not given a ring and asked to deduce its scattering behavior; instead one acquires a set of measurements and wishes to infer the ring's properties. Then the size distribution $n(r)$ has no unique solution, and one can only consider a restricted set of models. For tiny dust, the most common distribution considered is a power law, of the form

$$n(s) = Cs^{-q}, \qquad (3)$$

where q is called the power-law index, with larger q implying a steeper size distribution, and C a normalization constant. These distributions have the advantage of simplicity, with only two free parameters. In addition, power laws are widely observed in astrophysical and geological systems (Dohnanyi 1972; also see below). In practice, one needs to specify lower and upper limits to the size distribution in order to calculate the integrals above. However, in most cases the precise limits are not important. For $q < 7$, the lower limit is unimportant as long as it falls well into the Rayleigh scattering limit; in practice, distributions steeper than $q = 7$ have never been encountered. For $q > 3$, the number density drops off fast enough with s to make the precise upper limit irrelevant. Even for flatter distributions, the upper limit rarely plays a major role in the scattering properties, although one should always perform tests to verify this.

In addition to the dust, photometric models usually include a population of larger bodies. As we will see below, the lifetime of dust in planetary rings can be quite short, so these larger bodies are needed to serve as "parents" for the visible dust (Burns et al. 1980). In this regime, very little can be inferred about actual sizes of the parent bodies, because anything larger than ~ 1 cm scatters light indistinguishably. In this situation (2) can be simplified somewhat by using the geometric albedo $k \equiv \varpi_o P(0°)/4$, yielding

$$I/F = (k\tau/\mu)P(\alpha)/P(0°). \qquad (4)$$

The unknowns k and $P(\alpha)$ are then based on values inferred for nearby or analogous moons. For this reason, the best one can generally hope for is a constraint on the ring's total optical depth in larger bodies, not their sizes. Sometimes, however, occultation data at radio wavelengths are available to better constrain this size regime.

These two populations, "small" and "large" bodies, are well distinguished by their scattering properties, because dust tends to be highly forward-scattering whereas the parent bodies are mostly backscattering. In practice, one usually first lets measurements at higher phase angles constrain the dust distribution. Then one uses photometric models to predict the ring's brightness in backscatter, and any shortfall of the model relative to the measurements serves as a constraint on the parent bodies. The ratio of dust to parent bodies in a ring is typically characterized by a dust fraction f, equal to the dust optical depth divided by the ring's total.

II.B. Observational Methods

Most of our knowledge about the diverse family of planetary rings comes from the reconnaissance of the outer planets by Voyagers 1 and 2. These two spacecraft encountered Jupiter in 1979 and Saturn in 1980 and 1981, while Voyager 2 proceeded to Uranus in 1986 and to Neptune in 1989. However, Pioneer 11 actually provided the first closeup data from Saturn a year before Voyager, in 1979. Although its imaging capabilities were inferior, its trajectory sampled very different regions of the Saturn system and provided complementary data. In recent years the capabilities of Earth-orbiting instruments such as Hubble Space Telescope (HST) and large ground-based, infrared optimized instruments like Keck's 10-m telescopes, have improved tremendously, and these data sets, especially those taken during ring-plane crossings, provide valuable complementary information. Ongoing observations by Galileo at Jupiter and future ones by Cassini at Jupiter, and especially at Saturn, will likely revolutionize the field of ring studies in much the way that the Voyager encounters did in previous decades.

1. Images

By far the largest body of information we have about planetary rings comes from images. A single image can record a ring system's I/F as a function of radius and longitude, unlike other data sets which typically only constrain a single location at any one time. During the Voyager encounters, the ring systems were imaged by wide- and narrow-angle cameras through a variety of phase angles and emission angles over periods of weeks to months. Spatial resolution was as fine as a few km. Wavelength coverage ran through the visual band using several broadband filters (Smith et al. 1979a, 1979b, 1981, 1982, 1986, 1989). Galileo's images of Jupiter's ring, which were primarily taken through the clear filter (0.6 μm), had much improved signal to noise (Ockert-Bell et al. 1999; Burns et al. 1999); a sequence of infrared images (0.9–5.2 μm) was also obtained (McMuldroch et al. 2000).

As we will see below, many of the known dusty rings are extremely faint, some with $\tau < 10^{-6}$. Such rings required long exposure-times and this seriously reduced the number of useful images acquired. Indeed, some of the rings discussed below are only known because of a handful of Voyager observations, or just a single detection. Color information was also often severely limited; the Voyager and Galileo clear filters (with pass-bands centered near 0.5 μm and 0.6 μm, respectively) had much greater transmissivity than the typical narrowband filters, and so fainter rings were usually imaged through the clear filters. On the other hand, because an image comprises many pixels, it is often possible to make suitable pixel-averages to improve significantly the detectability of faint rings that are not obvious to the eye.

Compared to spacecraft, Earth-based observatories are capable of observing rings over much longer time periods and through a much broader range of wavelengths and ring opening angles, although they are restricted to small phase angles. A few Earth-based images have begun to rival the quality of some Voyager data; the Planetary Camera aboard HST can image Saturn's ring system with a resolution of 300 km per pixel. HST and other Earth-based observatories were used widely in 1995-96 to observe Saturn during the Earth and Sun's passages through the ring plane (Nicholson et al. 1996). Jupiter's ring-plane crossings in 1997 have afforded comparable viewing opportunities (de Pater et al. 1999). These rare edge-on viewing geometries, which in Saturn's case only come every 15 years, make it possible to detect faint rings and small moons that are normally lost in the glare of the main rings. This geometry also maximizes the line-of-sight optical depth of a ring, because the factor $1/\mu$ becomes very large.

In addition to the terminology introduced above, a few additional concepts are valuable when working with ring images. In particular, many of the rings of interest are quite narrow, often unresolved in an image. Under such circumstances, it makes no sense to talk about the ring's peak I/F because that value varies inversely with image resolution. To compensate, one introduces the "equivalent width," $\int (I(a)/F) da$, where a is the projected distance from the planet's center in the ring plane. By converting the image to a radial profile $I(a)/F$ and then integrating under the curve, the effect of variations in resolution, width, and smear can be eliminated. For rings that are both optically thin and narrow, the "normal equivalent width," scaled by μ, removes the emission angle dependence as well.

2. Spectra

In much of planetary astronomy, spectral measurements provide our most direct information about the composition of surfaces and atmospheres. The same is true for Saturn's rings, where absorption bands in the infrared indicate that water ice is the major component (see Cuzzi et al. 1984; Esposito et al. 1984 and references therein). However, beyond the recognizable absorption bands, particle composition is difficult to infer because spectra from laboratory samples do not duplicate the multiple scattering prevalent within denser rings.

Multiple scattering is not an issue in faint and dusty rings, but other more substantial difficulties arise. First, these rings are especially difficult to detect unless one averages over broad swaths of wavelength (eliminating spectral information), or else uses exceedingly long exposures. For this reason, even broadband spectrophotometry has rarely been acquired for most optically thin rings. Second, absorption bands are a phenomenon uniquely associated with macroscopic bodies, arising because the material rapidly dissipates light energy at specific wavelengths. For particles comparable in size to the wavelength, the path length of a light ray through the substance is too small for this absorption to become significant. Composition, therefore, is never measured directly for dusty rings; it is usually assumed, based on the composition of the nearby moons and denser rings.

This is not to say, however, that color measurements of a ring are not useful. Although they do not constrain composition, they provide valuable information about the dust particle sizes. The reason is that the wavelength defines the "yardstick" by which the size distribution is measured, so the phase function and optical depth can be strong functions of wavelength. For example, very steep distributions are dominated by tiny Rayleigh-scatterers and so tend to appear blue, for the same reason that the sky is blue. Hence, the color of a ring can provide very useful size information as a complement to phase angle coverage.

3. Occultation Profiles

Occultation experiments fall into two general categories—stellar and radio. In the former case, a ring passes in front of a star as seen from the observer, who measures the star's brightness as a function of time. Upon reconstructing the projected path of the star, one derives a profile of the ring's optical depth as a function of radial position and longitude. Stellar occultation experiments have been performed from both spacecraft and Earth-based observatories. The rings of Uranus and Neptune were, in fact, discovered in this manner (Elliot et al. 1977; Hubbard et al. 1986). Radio occultation experiments consist of a spacecraft transmitting continuous-wave radio signals through a ring and back to Earth. In principle, "uplink" radio experiments are possible, in which an Earth-based transmitter sends the data to a spacecraft, but this has never been attempted.

Occultation experiments are capable of obtaining much finer spatial resolution than images, although only along a one-dimensional track. The photopolarimeter (PPS) aboard Voyager performed stellar occultation experiments at Saturn, Uranus and Neptune, acquiring profiles with spatial resolution of 10–100 m at $\lambda = 0.265$ μm (Lane et al. 1982, 1986, 1989); the precise resolution depends on the available signal-to-noise (Colwell et al. 1990). Voyager's ultraviolet spectrometer (UVS) performed simultaneous observations at 0.11 μm, but these profiles are generally lower in resolution and signal-to-noise ratio (Sandel et al. 1982; Broadfoot et al. 1986, 1989).

Occultation profiles of Saturn's and Uranus' rings using the Voyager radio science subsystem (RSS) have resolutions comparable to the PPS experiments (Tyler et al. 1981a, 1986; Marouf et al. 1986). The RSS transmitter operated at two wavelengths simultaneously, 3.6 and 13 cm; comparison of the results from the two wavelengths provides our best constraints on the upper end of the particle size distribution (Marouf et al. 1983; Tyler et al. 1983; Zebker et al. 1985).

Earth-based occultation observations, now spanning over more than two decades, provide a useful complement to these data sets. The detailed shape models for Uranus' rings (French et al. 1991) have only been possible because of regular occultation observations of these rings. On July 3, 1989 the bright star 28 Sgr was occulted by Saturn and its ring system, and the resultant ring profiles (French et al. 1993; Harrington et al. 1993; Hubbard et al. 1993) have resolutions approaching that attained by Voyager. An occultation experiment using the high-speed photometer aboard HST provided similar resolution (Elliot et al. 1993). Bosh et al. (2000) have used the occultation of GSC 5249-01240 to understand the F ring's kinematics and photometry.

However, it should be remembered that most dusty rings are extremely optically thin, and in this case occultation observations become problematic (Tyler et al. 1981b). No occultation experiment has ever attained a sensitivity to $\tau \lesssim 10^{-3}$, so dusty rings cannot be detected in this manner. French et al. (1996) have summed many occultation profiles to bring out Uranus' λ ring. In the general situation, τ can only be modeled, along with the phase function, from image photometry.

By analogy to the equivalent width defined above for images, the "equivalent depth" D of a ring is defined as $\int \tau(a) \mathrm{d}a$. This quantity is again useful when rings are only marginally resolved, or when occultation measurements are to be compared to results from images. We note in passing that a slightly different quantity also called "equivalent width" is sometimes used in occultation studies (Porco et al. 1995; see also references therein). It is defined as the radially integrated fraction of light removed from the stellar beam, or approximately $W_{occ} = \mu \int (1 - e^{-\tau(a)/\mu}) \mathrm{d}a$. This is similar to the equivalent depth for small τ, but is a more directly measured quantity and can differ significantly from D when τ is large. However, in this paper we reserve the term "equivalent width" for the radially integrated I/F, as discussed above.

4. Charged Particle Absorption Signatures

All the ring observation methods discussed above rely on a ring's interaction with light. The ring's interaction with the charged particle environment of a planet provides an additional probe of the ring material. Pioneer 11, both Voyagers and Galileo's atmospheric probe, all of which passed near rings, carried arrays of instruments to study the electrons/ions trapped in the planetary magnetospheres they traversed. Some of these electrons/ions are absorbed when they intercept rings and moons, so the presence of rings can be inferred by a decrease in the phase space density of ions/electrons as the spacecraft crosses the ring's magnetic-field L shell. At both Jupiter and Saturn, Pioneer 11 returned data sets in some ways superior to those of the Voyagers because this spacecraft passed much closer to the planets and therefore probed more L shells containing ring material. The Voyagers did not cross any L shells that intercepted rings at Jupiter or Uranus. As Galileo's probe plunged at 6° out of the equatorial plane towards its destination in the atmosphere, its high-energy electron detector measured clean absorption signatures (Fisher et al. 1996). Specific experiments and their implications will be discussed below in the context of the particular rings to which they relate.

Unfortunately no instrument yet flown can characterize the full six-dimensional phase space of particles, so investigators must make assumptions to fill in the missing information; these assumptions are often subject to criticism. The distinction is widely made between a "macrosignature" caused by an axisymmetric ring versus a longitudinally incomplete "microsignature" trailing behind a moon or ring clump (Van Allen 1982; Cuzzi and Burns 1988). However, even this distinction is often subject to dispute; one observer's macrosignature may be another's microsignature. These uncertainties often hinder our inferences about ring properties from charged-particle-absorption data sets.

5. Direct Sampling of Ring Dust

The final way to probe a planetary ring is the (possibly perilous!) method of directly sampling the material using a dust impact-detector. On Pioneers 10 and 11, pressurized "beer can" dust detectors recorded occasional impacts by circumplanetary, interplanetary and interstellar dust particles, from which their density in space was inferred (Humes et al. 1974; Humes 1976, 1980). It is telling that these instruments recorded substantial increases in impacts during the crossings of Saturn's (and, to a lesser extent, Jupiter's) ring planes. Unfortunately, the design of the Pioneer instruments was such that impacts separated by upwards of 80 minutes were recorded as a single event, so one can only place a rough lower limit on the number of dust particles in dense regions. The instruments were sensitive to a limited range of particle energies, corresponding to $s \gtrsim 5$ μm for Pioneer 10 and twice as large for Pioneer 11.

The Voyagers did not carry any dust detectors *per se*, but nevertheless acquired some useful information on ring dust. The Plasma Wave Spectrometer (PWS) and Planetary Radio Astronomy (PRA) instruments recorded substantial increases in impulsive noise during the ring-plane crossings at Saturn, Uranus and Neptune (Aubier et al. 1983; Gurnett et al. 1983, 1987, 1991; Meyer-Vernet et al. 1986, 1996, 1998; Pedersen et al. 1991; Oberc 1994; Tsintikidis et al. 1994, 1995). Plasma-wave data from Jupiter's flyby have also been reanalyzed (Tsintikidis et al. 1996). This noise is now recognized as having been caused by dust impacts into the spacecraft. Despite the challenges associated with interpreting data for which the instruments were never calibrated, these measurements have been used to constrain particle sizes, densities and vertical distributions.

The next generation of dust detectors is flying on the Galileo and Ulysses spacecrafts (Grün et al. 1992c) and an improved version is aboard Cassini. These sophisticated instruments obtain information about the number, mass, vector velocity and electric charge over a much broader range of particle sizes (Grün et al. 1992b, 1996b). The Cassini detector even will provide some compositional information. These devices have detected periodic bursts of sub-micron grains streaming away from Jupiter (Grün et al. 1992b, 1993, 1996a,b; Zook et al. 1996; Krüger et al. 1999a; Graps et al. 2000), presumably accelerated outward by the magnetospheric corotational electric field after originating at Io (Horányi et al. 1993) or Jupiter's rings (Hamilton and Burns 1993a). Although safety constraints prevent these spacecraft from actually crossing through most known rings, the dust distribution in other regions of circumplanetary system, plus measurements of the influx of interplanetary and interstellar dust, are being characterized.

II.C. Physical Properties of the Dusty Rings

In this section we briefly summarize the properties of each of the dusty rings. The overall architecture of the known planetary ring systems is shown in Fig. 1 which illustrates the intimate association of rings and small ring-moons. Various general characteristics of the dusty rings are listed in Table 1.

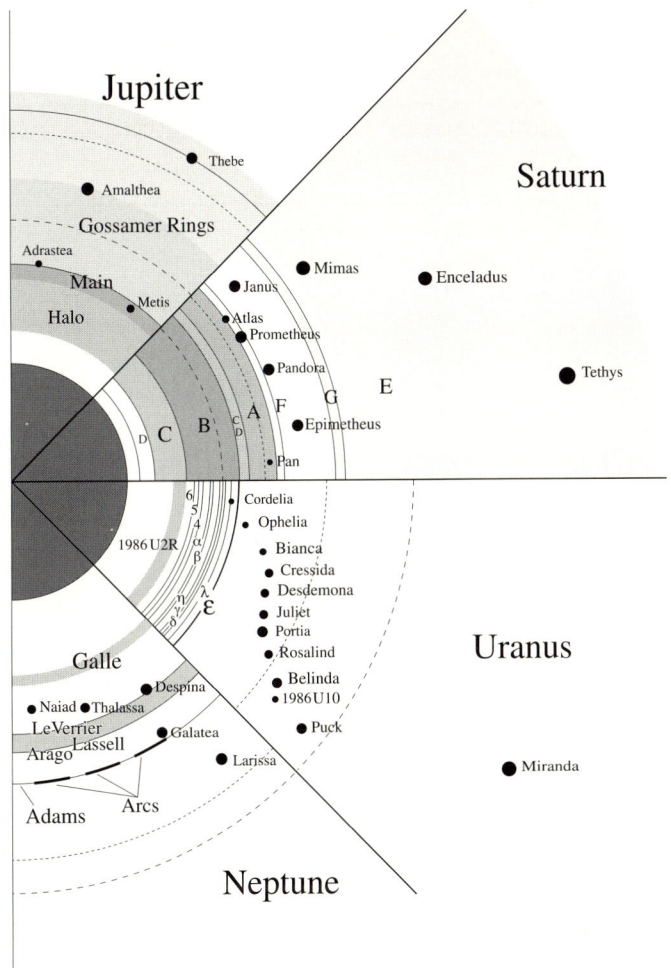

Figure 1. A schematic that compares the four known planetary ring systems (with distinct boundaries shown as solid arcs; radial limits that are poorly defined are not plotted), scaled to a common planetary radius (plotted as the solid central circle); the intimate association with ring-moons (small filled circles) is apparent. Within any planet's ring system, stippling suggests the relative optical depths of different ring components. All the ring systems lie within the Roche zones (tidal break-up, shown as dotted line for $\rho_{sat} = 1$ g cm^{-3}) of their planets, and all are fairly near the synchronous orbit position (plotted as a dashed line). Courtesy of Judith K. Burns.

1. Jupiter's Ring System

Jupiter's ring system was discovered in a single, very-long-exposure (672 sec) image taken during Voyager 1's approach in 1979, and was subsequently imaged in greater detail by Voyager 2 and Galileo. The ring was vastly brighter in the high-phase images from Voyager 2, indicating diffraction by a large population of fine dust. As a result, it was the very first dusty ring to be identified from space. Earlier hints that this faint ring might exist came from a dip in

Table 1

Characteristics of Dusty Planetary Rings

PLANET	Location[1]; W[2]	Optical Depth τ; D[5]	Dust Fraction f	q[3] ; s[4]	Comments
JUPITER					
Halo	92 000– 122 500	10^{-6}	100	?	12 500 km thick
Main Ring	122 000– 128 980	$3 \cdot 10^{-6}$	~50?	$q = 2.5 \pm 0.5$	Brightness dip near Metis' orbit; bounded by Adrastea
Amalthea Gossamer	$\leq$ 129 000– 182 000	10^{-7}	100?	?	2 600 km thick; bounded by Amalthea
Thebe Gossamer	$\leq$ 129 000– 226 000	$3 \cdot 10^{-8}$	100?	?	4400 km thick; bounded by Thebe
SATURN					
D Ring	66 000– 74 500	10^{-3}	50–100	?	Internal structure as fine as 100 km
D72 Ringlet	71 710 ($W \leq 40$)	$D = 0.01$	89 ± 4	$q \leq 2.8$ or $s \geq 10$	
Main (C, B, A) Rings	74 500– 136 800	0.05–2.5	< 3	$q = 2.7 - 3.1$	Much variation
B Ring Spokes	100 000– 116 000	~0.03	100	$s = 0.6 \pm 0.2$	Minimum width at R_{syn}
Encke Ringlet	133 580 ($W \simeq 10$)	$D = 1$	100?	?	Shares orbit with Pan
F Ring Inner Sheet	136 780– 140 200	$(1-2) \cdot 10^{-5}$	100?	?	Bounded by F Ring
F Ring	140 200 ($W \simeq 50$)	$D = 5$	≥ 98	$q = 4.6 \pm 0.5$	Shepherded by Pandora and Prometheus
G Ring	166 000– 173 000	10^{-6}	> 99	$q = 1.5$–3.5	
E Ring	180 000– 450 000	peak $(1-2) \cdot 10^{-5}$	100	$s = 1.0 \pm 0.3$	Peak near orbit of Enceladus; thickness 8 000–18 000 km, increasing with radius

Characteristics of Dusty Planetary Rings (continued)

PLANET	Location[1]; W[2]	Optical Depth τ; D[5]	Dust Fraction f	q[3] ; s[4]	Comments
URANUS					
1986U2R	37 000–39 500	10^{-4}–10^{-3}	?	?	
Dust Belts	41 000–50 000	10^{-5}	?	?	Fine internal structure down to < 100 km
Lambda Ring	50 024 ($W \simeq 2$)	10^{-3}	> 95	?	Adjacent to orbit of Cordelia; 5-cycle brightness pattern
NEPTUNE					
Galle Ring	41 000–43 000	4–$10 \cdot 10^{-5}$	?	?	
LeVerrier Ring	53 000 ($W \simeq 10$)	10^{-2}	40–70	?	Adjacent to orbit of Despina
Lassell Ring	53 000–58 000	1–$3 \cdot 10^{-4}$	?	?	Bounded at inner edge by Le Verrier
Adams Ring	62 930 ($W \simeq 50$)	10^{-2}	20–50	?	Adjacent to orbit of Galatea
Adams Arcs	62 930 ($W \simeq 10$)	10^{-1}	40–70	?	Adjacent to orbit of Galatea

[1] Radial distance from planet's center; in km, [2] W = Radial width; in km, [3] q = dust power-law index, [4] s = particle radius; in microns, [5] D = Normal equivalent depth; in km

the density of charged particles near Pioneer 11's closest approach to Jupiter (Fillius et al. 1975; Acuña and Ness 1976), as well as impact events recorded by the Pioneer dust detectors (Humes 1976; Fig. 9.1 in Elliot and Kerr 1985). Multiple explanations for the absorption signature were available at the time, however, although Acuña and Ness did propose a faint ring as a possible cause.

Dust particles, some prograde and some retrograde have been found by Galileo's dust detector to orbit Jupiter (Grün et al. 1997). Clouds of grains are noted near the Galilean moons (Grün et al. 1998; Krüger et al. 1999b), presumably ejected during impacts. Horányi (1994) has suggested that some debris might be captured to form a ring and Colwell et al. (1998) have shown how a similar process could lead to Galileo's retrograde grains. Direct evidence of the charged dust interacting with the Jovian magnetosphere is available in these data (Horányi et al. 1997; Grün et al. 1997; Graps et al. 2000), which exhibit 5- and 10-hr periodicities.

The most complete, post-Voyager, physical characterization of this ring system was performed by Showalter et al. (1985, 1987; Showalter 1989). Earth-based images of very high quality have been acquired by Nicholson and Matthews (1991) at Palomar's 5-m telescope and by de Pater et al. (1999) at Keck's 10-m telescope, the latter during the 1997 ring-plane crossings. These images rival Voyager's in clarity but were acquired at unique phase angles and wavelengths. However, a thorough analysis of them is not yet complete. Near-IR images (Meier et al. 1999) and polarimetric data (J. Gougen, private communication, 1998) were obtained by HST. The Galileo orbiter has returned high-resolution images of Jupiter's ring system mainly in forward-scattered light; these images are comparable in number to Voyager's, but the resolution is twice as good and the signal-to-noise and smear characteristics are much superior. A preliminary assessment of the Galileo images (Ockert-Bell et al. 1999) taken during Galileo's nominal mission for the most part confirms and refines Voyager results.

The current analysis of the Jovian ring system (Ockert-Bell et al. 1999) is based on measurements obtained during four orbits of Galileo's nominal mission, when 25 clear-filter images of the rings were taken as planned at spatial resolutions of 23 to 134 km/pixel; the ring appeared fortuitously in an additional 11 images taken during two of the same orbits plus a third. In confirmation of the previous interpretations from Voyager data, the tenuous Jovian rings can be considered to have three components: the main ring, halo, and gossamer ring. The first two of these components have typical normal optical depths of a few times 10^{-6}, while the third's is one-twentieth or less that amount; all contain large fractions of micron-sized dust. The much better quality of the Galileo images has allowed the nature of these components to be refined and has revealed hints of interesting fine structure.

The innermost component, a toroidal halo (Figs. 1 and 2c), extends radially from approximately 92 000 km to about 122 500 km (near the 3:2 Lorentz resonance, which is described in Sec. IV.B) and has a full-width, half-maximum thickness of 12 500 km; its brightness decreases with height off the equatorial plane and decreases as the planet is approached. The main ring reaches from the halo's outer boundary across about 6 400 km to 128 940 km, just interior to Adrastea's orbit (128 980 km); at its outer edge, the main ring takes nearly 1 000 km to develop its full brightness. The ring's brightness noticeably decreases around 127 850 km in the vicinity of Metis' semimajor axis (127 980 km) (Fig. 2b). The main ring has a faint, vertically extended component that thickens as the halo region is approached. Brightness variations of ± 10 % are visible in the central main ring and may be due to vertical corrugations, density clumps or "spokes" in the ring. Unexplained differences between the near- and far-arm brightnesses are visible. Lying exterior to the main ring, the gossamer ring has two primary components (see Fig. 3), each of which is fairly uniform: one originates just interior to Amalthea's orbit (181 000 km) while the other is situated radially interior to Thebe's orbit (222 000 km). Very faint material continues far past Thebe, blending into the background perhaps as far as 250 000 km

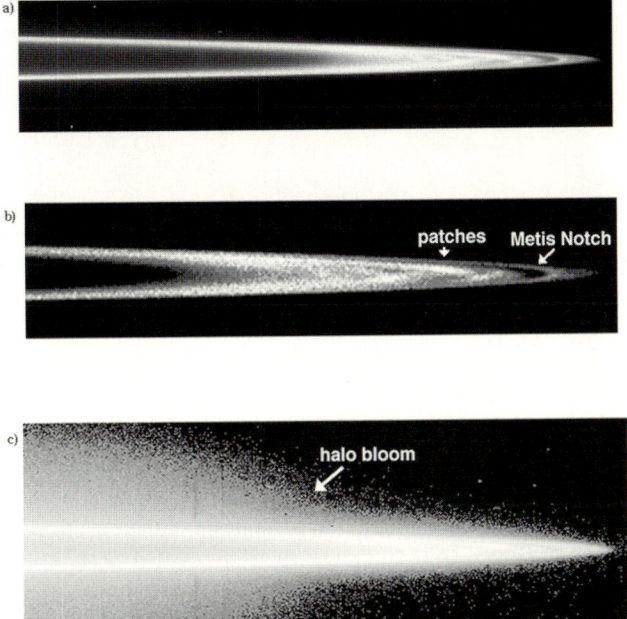

Figure 2. A Galileo view of the Jovian ring's west ansa, showing both the main ring and the halo's outer parts, processed in three different ways to highlight various features. (a) Stretched to contrast the main ring's diffuse inner periphery versus its much crisper outer boundary. (b) A stretch that emphasizes the patchy nature of the main ring's central region and the brightness dip associated with Metis' orbit. (c) A stretch to reveal the halo's development at the main ring's inner edge; it appears that the main ring itself has a faint cloud of material above and below. From Ockert-Bell et al. (1999).

out. The gossamer rings have thicknesses that match quite well the maximum elevations of these satellites off Jupiter's equatorial plane; from Galileo's nearly equatorial view, the gossamer rings have greater intensities along their top and bottom surfaces. The rings seem to be derived from the satellites (Burns et al. 1999).

Earth-based images taken at 2.27 μm during ring-plane crossing (de Pater et al. 1999) are generally consistent with this interpretation. These images and those from HST (Meier et al. 1999) show the vertically extended halo in backscattered light, and the halo is substantially fainter relative to the main ring than seen in the Voyager high-phase images. These differences may be attributed as due to dust which dominates the backscattering by the halo (since–in the nominal model–only dust, responding to non-gravitational forces, can be lifted out of the ring plane). However, further dynamical and photometric modeling (cf. Horányi and Cravens 1996), combined with the better data coming from Galileo, Cassini and Earth-based observatories, will be needed to settle this issue.

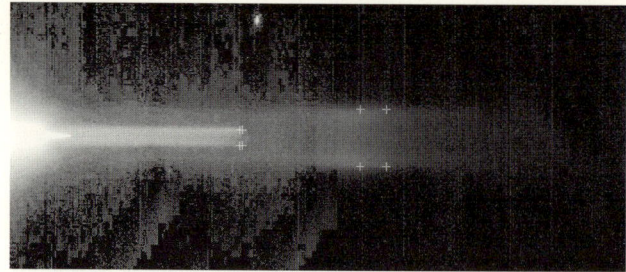

Figure 3. This mosaic of four Galileo images (416088922, 8968, 9022 and 9045), taken through the clear filter (0.611 μm) at an elevation of 0.15°, shows the edge-on gossamer rings of Jupiter across phase angles of 177–179°. The main ring and halo appear at left in white. The Amalthea ring is the narrower and brighter band extending to the right; the Thebe ring is the thicker and fainter band extending further. Each ring is bounded radially by the orbit of the corresponding moon; the crosses show the extreme positions (up/down; in/out) of the two satellites' orbits, which are eccentric and inclined. This image has been enhanced logarithmically to show all the ring components; in reality the Amalthea ring is ten times fainter than the main ring and halo, and the Thebe ring is ten times fainter than the Amalthea ring. The image, with a reprojected radial resolution of 400 km, has been expanded vertically by a factor of two to better show rings' vertical structure.

The photometry to date is based on Voyager measurements only (Showalter et al. 1987). The phase curve of Jupiter's main ring at high phase angles is compatible with a power-law size distribution of dust with index $q = 2.5 \pm 0.5$. The ring is distinctly red at these high phase angles, and its color implies the same index. The RSS atmospheric-occultation track crossed the main ring; since Tyler et al. (1981b) were unsuccessful in detecting the ring, their upper limit on the ring's optical depth was $5 \cdot 10^{-4}$, well above the actual value.

The Jovian ring's phase function is not sufficiently well known to place firm constraints on the population of larger bodies. Showalter et al. (1987) believed that $\lesssim 30\%$ of the ring's I/F in backscatter could be ascribed to the dust population, and that the remainder was caused by macroscopic bodies. However, this could not be proven rigorously. The interpretation that the backscatter is dominated by parent bodies is supported by the fact that the ring's color is very red, virtually identical to that of Adrastea (Meier et al. 1999) and Amalthea (Thomas et al. 1998; Simonelli et al. 2000). Gradie et al. (1980) have proposed that sulfur contamination from Io, combined with charged particles in the Jovian magnetosphere and impacts from micrometeoroids, act to darken and redden Amalthea's surface; if this is so, it stands to reason that parent bodies in the Jovian ring have been reddened similarly.

The analysis to date is inadequate to place severe constraints on the size distribution in the halo and gossamer ring, although the brightening toward high phase and the halo's possibly blue color relative to the ring (Meier et al. 1999) both imply that dust is the major constituent. If one assumes that the dust distribution matches that in the main ring, then the halo has a normal optical depth very similar to that of the main ring, whereas the gossamer ring's $\tau \sim 10^{-7}$ (Showalter et al. 1985, 1987). Showalter et al. (1987) have estimated the exponent in the power-law of the size distribution to be -2.5, but McMuldroch et al. (2000), from an analysis of a single forward-scattered sequence of infrared data, argue for a steeper size distribution.

2. Dust in Saturn's Main Rings and Spokes

Saturn's is the most expansive and most diverse of the planetary ring systems. The two most prominent rings, designated A and B, were discovered by G. Galileo using his first crude telescope in 1612; the inner C or crepe ring was not identified as a separate entity until 1850. The close scrutiny by the Voyagers in 1980 and 1981 revealed a system with remarkable amounts of structure; the closer one looks at Saturn's rings, the more detail one sees. The three main rings have optical depths in the range ~ 0.1–5 (Fig. 1). Photometric models for these rings indicate that all are extremely depleted in dust, with limits on the dust fraction in the A, B and C rings each at $\lesssim 3$ % (Cooke 1991; Dones et al. 1993; Doyle et al. 1989). Spectra reveal that water ice is the predominant constituent of Saturn's main rings and inner moons (see Cuzzi et al. 1984 and references therein); this is assumed to be the composition of the tinier motes as well.

The so-called "spokes" in Saturn's B ring are the primary sites of dust within the main rings (Fig. 4). As Voyager 1 approached Saturn, its cameras detected dark, wedge-shaped radial markings rotating in the B ring, predominantly on the "morning ansa" where the ring material first emerges from Saturn's shadow (Smith et al. 1981; Grün et al. 1983, 1992a; Porco and Danielson 1982). In forward-scattered light the spokes become brighter than the surrounding B ring, thereby exhibiting the characteristic phase function of dust. Because the spokes are so variable in time, their photometric behavior is rather difficult to model photometrically. Nevertheless, Doyle and Grün (1990) used nearly simultaneous images through multiple filters to measure the spokes' colors. They found that the optical depth increases with wavelength, contrary to what one observes in typical dust size-distributions. They concluded that spoke particles need to be rather narrowly confined to sizes $s = 0.6(\pm 0.2)$ μm to be compatible with these measurements. Similar techniques are being applied to HST observations of the spokes (French et al. 1998).

The dynamical properties of the spokes are also quite peculiar. The spokes occupy the radial zone 100 000–116 000 km, straddling the synchronous orbit location at 112 500 km (Smith et al. 1981, 1982; Grün et al. 1983, 1992a). New spokes seem to develop in periods as brief as a few minutes, and are initially radial. Afterward, most spokes propagate at the local Keplerian velocity, so they tilt away from the radial direction as the inner endpoint orbits faster than the outer. However, portions of some spokes have been found to corotate with the planet, while other spokes have been seen to switch from one orbital rate to the other (Grün et al. 1983, 1992a; Eplee and Smith 1984, 1985).

Porco and Danielson (1982) found that spoke activity has a periodicity matching that of the planet's rotation; in addition, more spokes are created

Figure 4. This Voyager 2 image (43643.34) of Saturn's A and B rings has been strongly enhanced in contrast to show spoke activity across the B ring. The phase angle is 7.5°. The Cassini Division is the sharply defined band between the rings.

when the magnetic sector associated with Saturn's kilometric radiation (SKR) is aligned with the morning ansa. The spoke particles are widely believed to be lifted out of the ring plane (or at least photometrically separated from parent bodies) to make them so visible; however, Grün et al. (1983) place an upper limit of 80 km on this vertical distance. In most of these dynamical properties, the spokes of Saturn's B ring appear to be unlike any other phenomenon observed in any planetary ring system.

3. *Saturn's D Ring*

Saturn's innermost (or D) ring has a peculiar history. The experienced observer Guérin (1973) reported first sighting material interior to Saturn's main rings. For the remainder of the 1970's, numerous astronomers attempted to confirm the D ring's existence, with inconsistent results. In 1979, the imaging photopolarimeter on Pioneer 11 did not detect this ring at a level ten times fainter than the Earth-based reports (Gehrels et al. 1980). Shortly thereafter Voyagers 1 and 2, with their more sensitive cameras, did in fact detect a small amount of material interior to the C ring (Fig. 5; Smith et al. 1981, 1982). Thus, although the earlier reports of the D ring were erroneous, this exceedingly faint region has continued to be known as the D ring. It is too faint to be detected by any occultation experiments or Earth-based images, even with our latest technology.

Showalter (1996) recently completed the only detailed analysis of the entire D ring data set. The ring is composed of two major narrow ringlets and a set

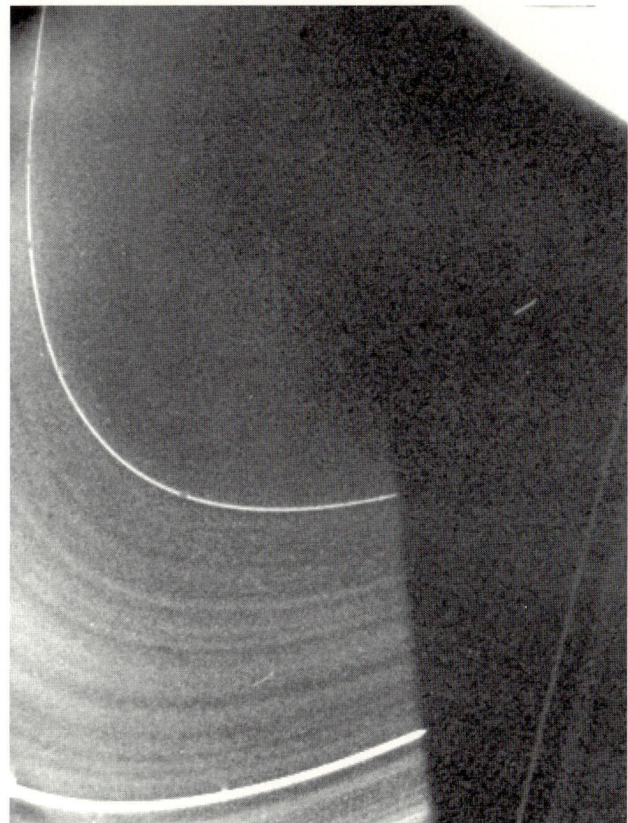

Figure 5. This Voyager 2 image (44007.53) shows the D Ring at a phase angle of 164°. The ring becomes invisible abruptly once it enters Saturn's shadow. Two narrow ringlets, identified as D68 and D72, are surrounded by fainter material which itself contains a great deal of internal structure.

of fainter wave-like structures surrounding them. The narrow rings are radially unresolved (widths $\lesssim$ 40 km) and are centered at 67 580 km and 71 710 km. The ring vanishes from sight at an inner radius $\sim$ 66 000 km; at its outer edge it merely merges into the much brighter C ring.

The brightest of the two ringlets could be detected at a broad range of phase angles, so its size distribution was readily modeled by Showalter (1996). The forward-scattering phase function is characteristic of fine dust and is compatible with a relatively flat size distribution ($q \lesssim 2.8$) or else with particles generally larger than $\sim$ 10 μm. Furthermore, an excess of light is definitely detected in backscatter, indicating that macroscopic bodies are present: $f = 89\ \% \pm 4\ \%$. The ring has a normal equivalent width of 15 m, so a radial width of $\sim$ 15 km would imply $\tau \approx 10^{-3}$. The D ring's other components were only detected in two images at especially high phase angles (156° and 164°), so their populations are poorly constrained. Surprisingly, however, the ratio of each component's intensity between these two phase angles is highly variable. Although this ratio has no unique interpretation in terms of the dust size distribution, its variations indicate that the dust populations are extremely changeable throughout the D ring. The power-law index q may well range between 2 and 6 within the system.

4. Saturn's F Ring and the Encke Gap Ringlets

Traveling outward, the next major ring to show a preponderance of dust is the narrow, slightly elliptical F ring (Fig. 6), just outside ring A, which is composed of at least four strands (Murray et al. 1997). The ring has an inclination of $(0.0067 \pm 0.0012)°$ (Bosh et al. 2001) and a physical thickness of 21 ± 4 km (Poulet et al. 2000b). Gehrels et al. (1980) discovered this ring using the imaging photopolarimeter aboard Pioneer 11, and also noted that the ring was clumpy. The clumpiness was confirmed and imaged in much greater detail by Voyager's cameras a few years later (Smith et al. 1981, 1982); the results from various Voyager instruments are reviewed by Burns et al. (1984) and Mendis et al. (1984). It was, at the time at least, the archetypical "shepherded" ring, since it appears to be confined by the two nearby moons Pandora and Prometheus (cf. Goldreich and Tremaine 1979); however, the torques on the ring do not seem to balance as they should (Showalter and Burns 1982). Furthermore, as measured on HST images taken during ring plane crossings (Nicholson et al. 1996; McGhee et al. 2000) and subsequently (French et al. 1999), the orbits of these shepherds are changing in unexpected ways. High-resolution images from Voyager 1 showed the ring to hold a variety of surprising internal structures, including strands, kinks, and the so-called "braids" in addition to the clumps. The braids themselves were of course illusory, resulting from radial "wiggles" that caused separate strands to appear to intertwine. Nevertheless, the observed structures continue to challenge dynamicists almost two decades after the F ring's discovery; unseen, embedded moonlets may play a pivotal role in the dynamics and variability of the ring (see Cuzzi and Burns 1988; Murray et al. 1997). Kolvoord et al. (1990) used Fourier analysis to identify underlying periodicities that are present in the F ring's longitudinal profile and connected the periodicities with Prometheus but also a smaller unseen moon. When a moon perturbs a nearby ring gravitationally, clumps are produced with a characteristic longitudinal spacing of $3\pi\Delta a$ (Dermott 1981; Showalter and Burns, 1982; Hänninen 1993), where Δa is the difference between the semimajor axes of the ring and moon.

Showalter (1994, 1998) recently undertook the tedious process of tracking individual clumps in the F ring to see how they evolved over the nine months between the two Voyager encounters. Initial results indicate that the F ring is the most dynamic ring in the Solar System. Clumps appear abruptly, perhaps being produced by impacts of cm-sized interplanetary meteoroids into unseen parent bodies, and disappear over times as brief as days; typical clumps survive for weeks to months. None of the most prominent clumps persisted between the two Voyager encounters. The individual clumps orbit with mean motions

Figure 6. This Voyager 2 image (44006.49) shows the F ring's ansa and its inner dust sheet at a very shallow opening angle of 1°. Saturn lies just off the left edge of the frame. Several of the F Ring's internal ringlets are also visible. The diagonal stripe crossing the ring is an artifact of an earlier image.

that differ by $\sim 0.5°$/day, corresponding to a radial width of 50–100 km. This is comparable to the dimension of the band visible in the most sensitive images and in the PPS profile (Lane et al. 1982).

Observations of Saturn's edge-on rings using HST in May 1995 led to reports of previously undetected moons orbiting just outside the A ring (Bosh and Rivkin 1996). These reports were initially plausible–after all, 13 of Saturn's 18 known moons were discovered during previous ring plane crossings–but the brightnesses indicated that the objects should have been visible to Voyager. The mystery was compounded a few months later when Nicholson et al. (1996) detected additional moon candidates but their positions did not coincide with the May objects. In retrospect, now it is clear that these objects were simply temporary clumps of material in the F ring masquerading as moons. Observations during the Sun crossing in November 1995 settled the issue; under this unique lighting, the F ring was the brightest of Saturn's rings, and numerous clumps could be seen within it (Nicholson et al. 1996; cf. Roddier et al. 2000, Poulet et al. 2000a). The longitudinal distribution of clumps is very reminiscent of that seen 15 years earlier by Voyager; at any given time the ring seems to have 2 or 3 especially bright features plus numerous smaller and fainter ones (Showalter 1998).

Furthermore, Cuzzi and Burns (1988) have studied a set of charged particle absorption signatures near the F ring detected by Pioneer 11, some of which cannot be attributed to known bodies. They infer that a belt of smaller (0.1–10 km) moonlets fills a 2000-km band surrounding the F ring; the additional absorptions can then be explained by microsignatures from temporary clumps of faint debris, arising from collisions within the belt. However, the inferred optical depth of this belt is quite large by the standard of faint rings, $\tau \sim 10^{-4}$–10^{-3}. Since meteoroid impacts into these bodies and collisions between them will raise clouds of dust continuously, it seems surprising that no broader, fainter band of dust is visible in the Voyager images (see below and Showalter et al. 1998). Some available Voyager images show the much fainter G ring ($\tau \sim 10^{-6}$; see below) but not the dust ring that one would expect from this hypothesized belt.

Photometry by Showalter et al. (1992) indicates that the F ring has a narrow core of relatively high optical depth, surrounded by a skirt composed primarily of dust. This interpretation was based on large-scale longitudinal averaging of the ring images in an attempt to smooth over the variations represented by the brightness clumps. It also made use of F ring detections by the Voyager PPS and RSS occultation experiments; more recent detections during the 28 Sgr stellar occultation (French et al. 1993; Harrington et al. 1993; Hubbard et al. 1993) have not yet been incorporated in ring photometry, but have been used to define the orbital shape (see Murray et al. 1997). Using an occultation observed by HST's Faint Object Spectrometer, Bosh et al. (2001) conclude that the F ring's equivalent depth has no significant dependence on wavelength (0.25–0.74 μm), implying ring particles are greater than s ~ 10 μm. In general terms, the size distribution does not seem to vary radically between the F ring's brightest clumps and its fainter regions; however, more subtle variations may well have been overlooked in the analyses performed so far. The dust appears to obey a rather steep power-law distribution, with $q = 4.6 \pm 0.5$. Interestingly, this phase function matches the ring in backscatter as well as forward-scatter, which severely limits the macroscopic population to $1 - f \lesssim 2$ %. However, a very narrow feature found in the RSS occultation (Tyler et al. 1983) presumably indicates a direct detection of a core of parent bodies. At visual wavelengths, the ring's normal equivalent width is 5.0 ($\pm$ 0.3) km. Poulet et al.'s (2000b) best model to explain thickness measurements during ring-plane crossing has a radial $\tau \sim 0.20$ and a dust fraction $f > 0.80$.

The F ring's color in HST images is neutral or slightly blue, like its inner shepherd, but unlike Pandora whose color is red, like the main rings (Poulet et al. 1999). Several long exposures at high phase angles reveal a faint inward extension to the F ring (Fig. 6), which is present throughout the entire gap down to the A ring. Burns et al. (1983; cf. Fig. 1 of Murray et al. 1997; Nicholson et al. 2000; Showalter et al. 1998) estimate $\tau \sim (1–2) \cdot 10^{-4}$, with the large uncertainty caused by the dearth of data. Careful analysis of the Voyager PPS data set showed some baseline variations near the F ring, which Graps et al. (1984, see also Graps and Lane 1986) attributed to fainter ring material. However, the inferred τ of this material is $\sim 10^{-2}$, which is much too large to be compatible with the images. It appears, in retrospect, that these features arose from background variations in the PPS instrument rather than from actual "unseen" ring material; nevertheless, after checking the instrumental light variations, Graps (private communication, 1995) maintains her original interpretation.

The nearest dynamical and physical analog in Saturn's rings to the F ring is a pair of narrow ringlets (see Fig. 19 in Cuzzi et al. 1984) orbiting within the Encke Gap, an opening 320 km wide in the outer third of Saturn's A ring. The gap is apparently shepherded open by the small moon Pan (Showalter 1991), orbiting at 133 583 km near the gap's middle. Two variable ringlets can be seen in the images, one near the inner edge and the other near the center; this latter ringlet appears to occupy the same orbit as Pan itself. The ringlets also appear to be incomplete, since each one is prominent in some images but absent in others, even when the phase angle and spatial resolution are unchanged.

Photometry of these ringlets is extremely difficult owing to their marked variations. Nevertheless, the ringlets brighten significantly at high phase angles implying a prevalence of dust (Showalter 1991). Neither ringlet appears in the RSS data, but the middle one is visible in the PPS profile. It has a peak $\tau \approx 0.1$ and a radial width of ~ 20 km; its equivalent width is ~ 1 km. In these properties it is quite comparable to the F ring although, given the ring's marked variations, one cannot know if this section of the ring is atypical. The ringlets have not been detected in Earth-based occultations.

5. Saturn's G Ring

Continuing outward, Saturn has two other very faint rings, designated G and E (Fig. 7). Interest in these two rings has been rekindled recently because of the potential hazard they may pose to the Cassini orbiter, scheduled to arrive at Saturn in 2004 (Cuzzi and Rappaport 1996). Cassini will need to pass repeatedly through, or very close to, both of these rings during its four-year tour. Significant new Earth-based observations of these outlying rings were acquired during the Saturn ring-plane crossings of 1995-96 (Bauer et al. 1997; de Pater et al. 1996; Nicholson et al. 1996).

The G ring is tenuous ($\tau \approx 10^{-6}$), relatively narrow and centered on orbital radius 168 000 km, far from any other known ring or moon. In Voyager observations its radial width is $\sim 7 000$ km, between 166 000 and 173 000 km, with no apparent internal structure beyond an inverted "V" profile, in which the ring brightness decays linearly both inward and outward from a central peak (Showalter and Cuzzi 1993; Throop and Esposito 1998). In HST observations (Nicholson et al. 1996) the G ring is roughly uniform in brightness, with a full radial width of 8 000 ($\pm 2 000$) km, with half-flux points at 166 000 and 170 000 ($\pm 1 000$) km. Its thickness is $< 1 300$ km. The earliest evidence for this ring was acquired by Pioneer 11 in 1979, when it detected a high-energy charged particle absorption signature in the region, that was variously attributed to a new satellite or perhaps to Janus (Simpson et al. 1980; Van Allen et al. 1980). It was later recognized as a ring in only two Voyager images, both at the same high phase angle $\alpha \simeq 160°$. Showalter and Cuzzi (1993) managed to identify it marginally at three other phase angles, yielding a crude phase function. From the four-point phase curve, Showalter and Cuzzi (1993) inferred that the dust size distribution is surprisingly steep, with $q \approx 6.0 \pm 0.2$. This means that the ring's light-scattering is dominated by extremely tiny particles, $s \sim 0.03$ μm.

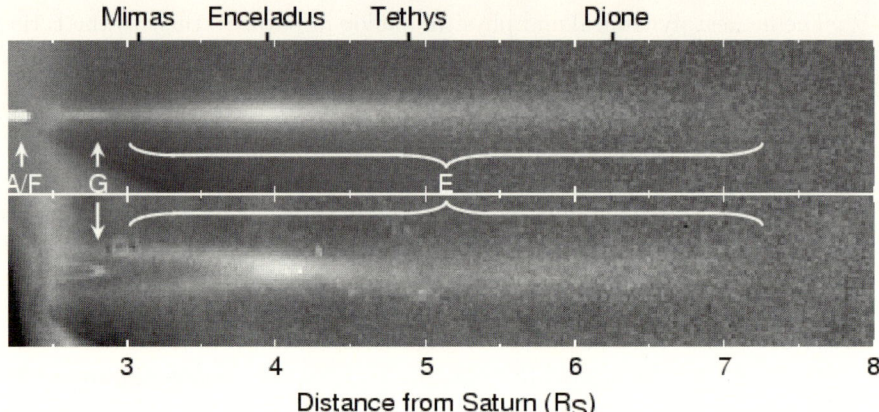

Figure 7. Two HST images that depict Saturn's outer rings: i) when the rings were nearly edge-on (top, taken on 9 August 1995) and ii) viewed from an elevation of about 2° (bottom, taken on 28 November 1995). At the left of the top image, the outermost A/F rings are over-exposed. The G ring is seen to be narrowly confined whereas the E ring, peaking at Enceladus' orbit, extends across many satellite paths. The orbital positions of several satellites, as measured at each frame's center, are shown at top. In order to bring out the G ring from the glare of the main rings, a smooth background has been subtracted. Follows Nicholson et al. (1996).

The G ring was also detected by Voyager's unintended "dust detectors," the PWS and PRA instruments. Showalter and Cuzzi (1993) found that Voyager 2's crossing radius was 172 124 km plus or minus a few km, near but distinctly inside the G ring's outer boundary. Thus, the PWS and PRA noise can be interpreted as direct, *in situ* measurements of the G ring's particles (Aubier et al. 1983; Gurnett et al. 1983; Tsintikidis et al. 1994; Meyer-Vernet et al. 1998). The final interpretation of these data indicates $\tau \approx 10^{-6}$, particles of radii a few microns, $q < 3.5$ and a ring thickness of ~ 1000 km. The issue of q is not resolved since Showalter and Cuzzi's steep distribution was supported by modeling of the PWS data set by Gurnett et al. (1983; cf. Tsintikidis et al. 1994), although these data are only sensitive to grains in the size range 0.5–5 μm, but not by Meyer-Vernet et al.'s (1998) PWS interpretation.

On the other hand, recent Earth-based images (Nicholson et al. 1996; de Pater et al. 1996; Bauer et al. 1997) reveal the G ring to be neutral to slightly red in color, whereas the color of Showalter and Cuzzi's steep distribution is expected to be blue in backscatter. This seeming contradiction has not yet been reconciled. Throop and Esposito (1998) present a range of particle size distributions in an attempt to satisfy this discrepancy, and favor a power-law q between 1.5 and 3.5. Both Canup and Esposito (1997) and Throop and Esposito (1998) demonstrate that such a size distribution may develop following the disruption of a progenitor satellite.

Regarding the larger particles from which the dust is derived, Van Allen (1983, 1987) modeled the Pioneer 11 absorption signature to demonstrate that the G ring comprises a total cross-section of (10–40) km^2 in bodies larger than

25 cm, the distance needed to stop a high-energy proton Thus, the absorption signature is likely to serve as a direct detection of the parent population within the ring. However, this population corresponds to $\tau \sim 10^{-8}$, much too small to be visible in the images; the G ring's dust therefore dominates its light-scattering at all phase angles. This result is consistent with Hood's (1989) modeling of the absorption signature, which suggests that the parent bodies are confined to a band $\leq 1\,000$ km wide, much narrower than the visible ring.

6. Saturn's E Ring

The outermost of Saturn's bands, the E ring (Fig. 7), spanning the orbits of Mimas, Enceladus, Tethys and Dione, encompasses an area and a volume larger than those of all other planetary rings put together. This ring was discovered during the 1966 crossing of Saturn's ring-plane (Feibelman 1967) and observed again from Earth during the 1979-1980 and 1995-96 crossings, with substantial improvements of data quality on each occasion. In addition, both Voyagers imaged the E ring, although the few available images are confined to three narrow ranges of phase angle. Two dust impacts were recorded in this region by the Pioneer 11 detector (Humes 1980).

The properties of Saturn's E ring were reviewed by Showalter et al. (1991) based on the Voyager and pre-1995 Earth-based data. The ring has a distinct peak in brightness near the orbit of Enceladus (Fig. 7), which likely serves as a source for the ring dust in some manner (Hamilton and Burns 1994). The ring is vertically thick compared to most rings, ranging from $\sim 6\,000$ km near its inner limit to $\sim 20\,000$ km near its outer edge; however, it appears to narrow significantly about the orbit of Enceladus. Roddier et al. (1998) claim to have seen a short-lived arc within the E ring, near one of Enceladus' triangular Lagrange points.

The ring is exceedingly faint, with a peak $\tau \sim 10^{-5}$. Several of its photometric properties, including a peculiar blue color in backscattered light (Larson 1984) and a sharp diffraction peak near forward-scatter, led Showalter et al. (1991) to the unexpected conclusion that the ring dust does not obey a power-law size distribution. They found that a narrow size range centered near $s \approx 1$ μm provides the best fit to the available data.

Testing this surprising conclusion was a high priority for the 1995-96 E ring observations from Earth. In general, the old model of the ring's properties held up quite well after the influx of new, higher-quality data. The ring's expanding thickness with radius, plus its local thinning near the orbit of Enceladus, appeared quite clearly in images from HST (Fig. 7; Nicholson et al. 1996). Images spanning wavelengths from the infrared to the ultraviolet confirm the ring's distinctive blue color (Nicholson et al. 1996; de Pater et al. 1996; Bauer et al. 1997). This color, when combined with the Voyager measurements, remains incompatible with any power-law size distribution. Although further refinements to the photometric models are needed, it is clear that the size distribution in the E ring is unlike that in other known rings.

Tsintikidis et al. (1995) and Meyer-Vernet et al. (1996) have re-examined PWS and PRA measurements from Voyager 1, which passed through the E ring at $a \sim 368\,000$ km, near the orbit of Dione. Like their Voyager 2 counterparts (Tsintikidis et al. 1994; Oberc 1994), both instruments detected impulsive noise near the ring plane, which can now be attributed to impacts with E ring dust. Meyer-Vernet et al. inferred that the particle size distribution is narrowly peaked near $s \approx 1$ μm, in close agreement with Showalter et al. (1991). However, Tsintikidis et al. inferred $s \approx 5$ μm, a significant discrepancy that has not yet been explained (see Cuzzi and Rappaport 1996). Meyer-Vernet et al. measured the ring's thickness as 12 000 km, comparable to the value in the images. However, they reported a southward offset to the density peak by 5 000 km, something not observed by others.

7. Uranus' Ring System

The nine main Uranian rings are designated, in order of increasing radius, rings 6, 5, 4, α, β, η, γ, δ, and ϵ. All are very narrow; ring ϵ is the broadest with a width ranging between 20–100 km, while every other one is $\lesssim 10$ km wide. Optical depths are generally of order 0.1-1.0. As with Saturn's denser rings, these rings show very low dust content; Ockert et al. (1987) inferred $f \lesssim 0.2$ %. French et al. (1991) and Esposito et al. (1991) review these rings' properties.

One additional narrow ring, λ (1986U1R), was discovered in images as Voyager approached Uranus (Smith et al. 1986). Unlike the other narrow rings, this one brightened substantially in forward-scatter, indicating the presence of dust. In a single image (Fig. 8) acquired at a very high phase angle ($\alpha \sim 173°$), it is the brightest of Uranus' rings. Photometry by Ockert et al. (1987) revealed this ring to have significant longitudinal variability. Showalter (1995) used additional images to determine that these variations are generally periodic, so that the ring could be better viewed as a set of five uniformly-spaced arcs. This makes the λ ring one of the few rings with known longitudinal variations, and the only one in which a single periodicity dominates.

By accounting for these variations in his modeling, Showalter (1993) attempted to infer that ring's size distribution and dust fraction. Using four phase angles and also τ measurements at four wavelengths (UVS, PPS, RSS 3.6 cm, plus an image measurement where the ring was backlit by the planet), he arrived at an interesting contradiction. A strong wavelength dependence to τ required a steep power law, with $q > 3.5$. On the other hand, the shape of the diffraction peak near forward-scatter required a flatter distribution, with $q < 3$. This led him to the conclusion that at least three different populations are necessary to model the size distribution: a steep "submicron" population to account for the τ dependence, a flatter "micron" population for the phase curve, and finally the usual macroscopic population to explain the backscatter. The best-fit model involved $f \approx 75$ %, with the dust evenly split between the two populations.

Aside from ring λ, the Uranian system contains additional dust distributed widely in radius. At high phase (our Fig. 8, and Figs. 11 and 12 in Esposito et

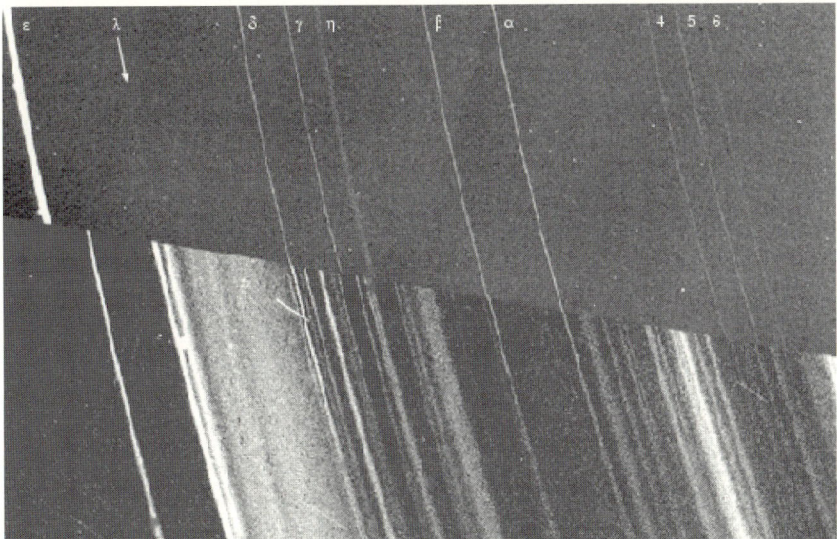

Figure 8. Compare these images of the Uranian ring system. Voyager 2 acquired the upper one the day before it passed Uranus in January 1986; at the time sunlight striking the ring particles was reflected back toward the camera (phase angle of 18°). In the lower image, taken somewhat later, the spacecraft was looking almost directly back toward the Sun (phase angle of 172.5°). This backlighting of the rings dramatically enhanced the visibility of any micrometer-sized dust particles that the rings contain. The nine narrow rings discovered in ground-based occultations (note the poor segment match for the markedly eccentric ϵ ring) can be readily discerned; they are surrounded by dust belts containing a great deal of internal structure. The λ ring, so bright in forward-scattered light, is only visible in the top frame if you put your eye close to the bottom and look toward the arrow. The short linear streaks in the bottom frame are background stars that smeared during the 96-sec exposure; even the rings themselves are smeared somewhat, especially near the bottom of the frame. Courtesy of J. Kelly Beatty.

al. 1991) the Uranian system is seen to contain an extraordinary family of dust belts and gaps around and among all the better known narrow rings. Radial structure is visible at a variety of scales from 50 km to $> 1\,000$ km, with the lower limit set by the image's resolution. Unfortunately, these dust belts only appear in this single image, so detailed inferences about their optical depths and particle sizes are not possible. Murray and Thompson (1990) have attempted to connect the orbital distribution of this material to unseen shepherd satellites.

One additional broad ring, designated 1986U2R, is visible in a lone Voyager image at a 90° phase angle. This ring is interior to all of the structure discussed above, with a peak at a radius of 38 000 km and a radial width of $\sim 5\,000$ km. Once again, very little can be determined about the particle properties of this ring from a single view, although a predominance of dust is strongly suspected.

Although no other rings were seen by the Voyager cameras, the PWS and PRA instruments again detected dust impacts during the ring plane crossing at $a \sim 118\,000$ km, which is inside the orbit of Miranda but well outside the orbits of the Voyager-discovered moons and rings (Meyer-Vernet et al. 1986; Gurnett et al. 1987). Particles are inferred to be microns in size and have total $\tau \sim 10^{-8}$–10^{-7}. Such a faint ring would not have been visible to the cameras. The data seem to disagree about the thickness of this unseen ring; Meyer-Vernet et al. infer a thickness of 150 km in the PRA data, whereas Gurnett et al. believe 3 500 km from the PWS. A difference in particle-size sensitivity seems unlikely to explain fully this discrepancy.

8. Neptune's Ring System

After the 1977 discovery of the Uranian rings by stellar occultations, Neptune's environs were searched for analogous narrow rings. Initial results were negative, except for a very fortuitous detection of the tiny moon Larissa by Reitsema et al. (1982). However, in 1984 Hubbard et al. (1986) detected an occultation event on one side of Neptune that was not repeated on the other, suggesting that incomplete arcs might be orbiting the planet. Subsequent occultation observations reported occasional events, but it took the closeup images from Voyager 2 in 1989 to settle the issue: 3–5 slender and discontinuous ring arcs are embedded within a narrow, fainter Neptunian ring (Smith et al. 1989).

Based on the Voyager images (Fig. 9), Neptune's ring system is now known to comprise several distinct components (Fig. 1). The most prominent narrow ring is Adams, at $a \approx 62930$ km, with a radial width of ~ 20 km. A 40° segment of this ring contains the aforementioned arcs, now designated Courage, Liberté, Egalité 1 and 2, and Fraternité (Fig. 10). In the images with finest resolution ($\sim$ a few km per pixel) the arcs are composed of much smaller discrete clumps. By comparing these visible arcs with the longitudes of Earth-based detections, Sicardy et al. (1991) and Nicholson et al. (1995) established that these arcs have been stable from the time of their discovery in 1984. Outside the arc region, Adams varies rather little in brightness; however, Showalter and Cuzzi (1992) found that it is faintest at the longitude opposite to the arc region and brightens as the arc region is approached in either direction. At its minimum, the Adams ring is only 10 % as bright as it is in the arcs.

A second narrow ring, LeVerrier, falls at $a \approx 53\,200$ km. It is in all respects similar to Adams except for the absence of arc-like structures. A fainter, uniform ring named Lassell extends outward from LeVerrier about half-way to Adams, ending at $a \approx 58\,000$ km. A slight increase in Lassell's brightness at its outer boundary has, perhaps unwisely, been given a different name, Arago. Finally, a much broader ring named Galle has a peak near $a \approx 42\,000$ km and probably reaches all the way down to the planet (Showalter and Cuzzi 1992). Porco et al. (1995) provides a much more detailed review of these rings.

Interspersed among these rings are a number of small moons (see Fig. 1). Galatea orbits just inside the Adams ring and its gravitational perturbations are likely to play a significant role in the arcs' confinement (Porco 1991; Horányi and Porco 1993; but see Sicardy et al. 1999; Dumas et al. 1999). Despina traverses just inward from the LeVerrier ring, while Thalassa and Naiad occupy the gap between the LeVerrier and Galle rings. Interestingly, Galatea shares

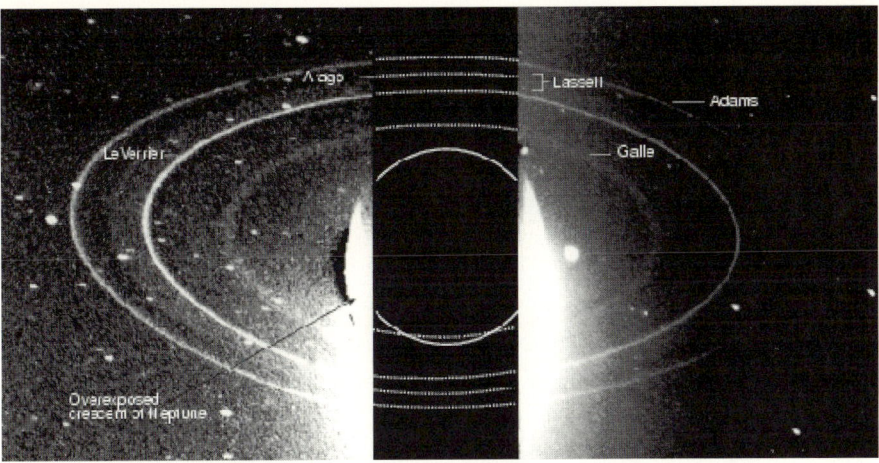

Figure 9. All of Neptune's major rings are visible in this outbound image from Voyager 2 (11446.21) at a phase angle of 134°. The outermost ring is Adams, whose arcs are not visible here. LeVerrier is the other bright ring, and Lassell is the faint band extending half way out to Adams. The faint inner Galle Ring is also visible. The many bright spots are background stars.

its orbit with a faint and possibly incomplete dusty ringlet (recall the Encke ringlet and Pan), but the other moons do not.

Voyager imaged the rings at only four narrow ranges of phase angle near 8°, 15°, 135° and 155°; furthermore, the longitudinal coverage at the two extremes is very limited. Nevertheless, all of Neptune's rings have been found to brighten at higher phase angles, indicating that they are dusty. Additional detections of the arcs in PPS and UVS stellar occultations, plus Earth-based occultations at $\lambda \approx 2.2$ μm, serve as direct probes of τ; arc equivalent widths are typically $\sim$ 1 km; given the radial widths of $\sim$ 20 km, this yields τ values of several percent (Horn et al. 1990). Unfortunately, the RSS occultation profile missed the arc region of the Adams ring and nothing was sensed (Tyler et al. 1989). In fact, except for the arcs, the Adams ring has never been detected via occultation, and only very marginal detections of the LeVerrier ring have been reported (Horn et al. 1990, Nicholson et al. 1995). The best Earth-based data sets have placed rather strict upper limits on the equivalent widths of these narrow rings (Sicardy et al. 1991; Nicholson et al. 1995).

Owing to the very limited data, inferences about the composition and sizes of Neptune's ring particles are subject to major uncertainties. Nevertheless it is clear that Neptune's rings are significantly dustier than the main rings of Saturn and Uranus. Showalter and Cuzzi (1992) and Ferrari and Brahic (1994) have performed independent analyses; Porco et al. (1995) thoroughly summarize and tabulate the dust observations. Measurements at the two high phase angles show that dust properties vary among the rings, and possibly also among the different arcs. Power–law indices $q \approx 2$–4 are generally implied. Although

Figure 10. A long-exposure, forward-scattered image (phase angle of 134°) of Neptune's outer two rings shows that the ring arcs are in fact clumps of material within the outermost (Adams) ring, which is less than 50 km wide; the LeVerrier band is about 110 km across. In this image, the direction of orbital motion is clockwise, with the longest arc (Fraternité) trailing. Neptune's over-exposed crescent is at lower right.

no ring images were acquired through Voyager's color filters, Showalter and Cuzzi inferred that the Adams ring is very red, using the slightly different passbands of the clear filters on the wide- and narrow-angle cameras. This red color would tend to favor $q < 3$. All photometric models reveal that the dust distribution is inadequate to account for the rings' backscatter I/F, indicating a significant presence of parent bodies. Accepting a broad sweep of assumptions, $0.2 \leq f \leq 0.7$ in the Adams and LeVerrier rings (Smith et al. 1989).

Ferrari and Brahic (1994) suggest that a less dissipative, "dirty ice" composition provides the best fit to the ring photometry. Nonetheless, others have assumed amorphous carbon because this is more consistent with the exceedingly dark surfaces of Neptune's inner moons. Porco et al. (1995) present a method to infer the geometric albedos of the ring particles and find values of 5–7 %, comparable to those of the inner moons (Thomas et al. 1995).

As at Uranus, dust impacts were recorded by the PWS and PRA instruments, revealing the presence of a far more widely separated dust population well outside the rings (Gurnett et al. 1991; Pedersen et al. 1991). Ring-plane crossings were at 85 290 km and at 103 950 km. Inferences from the PWS and PRA data sets are not in perfect agreement, as demonstrated by Table III of Porco et al. (1995). Nevertheless, it is reasonable to conclude that this faint cloud consists of dust a few μm in size with $\tau \sim$ a few $\times\ 10^{-6}$. Furthermore, the cloud is hundreds of km thick and is not centered on the equatorial plane; measured offsets were $\sim$ 150 km northward on the first crossing and

~ 800 km southward on the second; these perhaps merely reflect the position of the warped Laplace plane (Porco et al. 1995). Among the more interesting circumplanetary dust findings are reports by both the PWS and PRA teams of the presence of considerable dust at high latitudes.

Charged–particle absorption signatures have been identified in data from Voyager's low-energy charged particle experiment (Mauk et al. 1991, Paranicas and Cheng 1991). However, Neptune's substantially inclined magnetic field caused Voyager to cross in and out of L shells repeatedly, hindering a straightforward interpretation of these results.

III. PHYSICAL AND DYNAMICAL PROCESSES ACTING ON CIRCUMPLANETARY DUST

As a prelude to looking at the forces that act on the constituents of dusty planetary rings, we first mention that dust motes will be electrically charged. Then we consider some of the forces that act on circumplanetary particles, including electromagnetic forces, radiation pressure, gravity, and various drag forces that cause orbital evolution; typical accelerations due to these forces acting in various ring systems are tabulated. In the middle of this section we describe various processes, including the generation and destruction of grains, interactions with parent bodies, likely size distributions and resonance processes; it lists the resultant timescales. The end part of this section discusses the dynamics of particles that are subject to such forces.

III.A. Electrical Charging

Because objects in space are bathed in plasma and struck by ultraviolet photons, they acquire net electrical charges. The precise charging history for any grain depends on the grain's properties (composition, size and surface character) and on the ambient environment (plasma composition, number density and temperature, plus radiation flux); it also varies according to the grain's past since charging by any process does not occur instantaneously and since, in rare instances, more than one equilibrium state exists (Meyer-Vernet 1982).

To compute the electrical charge q, one tallies the current flows to the grain which, for Maxwellian distributions, are given by Goertz (1989), Northrop (1992), Mendis and Rosenberg (1994), Schaffer and Burns (1995), and Horányi (1996). Usually the grain's charge q is taken to be its equilibrium value, i.e., the charge at which, on average, no net current flows to the grain.

Two limiting cases of electrical charging are instructive. First, consider an isolated grain in vacuum; it will expel photoelectrons when exposed to energetic ultraviolet photons. The loss of these electrons means that the grain becomes positively charged; subsequently, the only photoelectrons that are able to escape are those whose kinetic energies are sufficient to overcome the surface electric potential. Of course, whenever a larger charge develops, a smaller fraction of the ejected photoelectrons have sufficient energy to fully escape the surface. Since photoelectrons exit with energies similar to the work function of the surface material, the equilibrium potential is roughly that work function, or a few volts positive. This limiting case applies to interplanetary dust and circumterrestrial grains where photoelectron currents dominate charging; in such a circumstance, an immediate corollary is that passage through the planetary shadow, where the primary current is absent, will have measurable consequences (cf. Horányi and Burns 1991). However, because the solar flux density decays quadratically with heliocentric distance, photo-charging is less relevant in the outer solar system. Instead, plasmas play the major role in the charging of grains. Each of the giant planets is shielded from the onslaught of the solar wind by a substantial planetary magnetic field that, in the planet's neighborhood, traps magnetospheric plasma.

The second heuristic example considers just ion and electron thermal currents to be present. If a neutral grain is placed in a plasma where, owing to energy equipartition, both components have the same kinetic temperature T, the particle at first will be struck primarily by the less-massive, swifter-moving electrons. As a result, a negative charge will build until the particle's electric potential becomes sufficient to attract enough positive ions to counterbalance the somewhat-repelled electrons. Thus, a particle embedded in a plasma should develop an electric potential ϕ (in eV) like the plasma temperature. In a classical result, Spitzer (1962) found that

$$\phi_{eq} = -\frac{bkT}{e}, \tag{5}$$

where k is the Boltzmann constant, e is the magnitude of the electron charge, and b is a constant that depends on the ion species; typical values are $2-4$. We note that this result, like that for purely photoelectric charging, is independent of particle size.

This treatment is inadequate once the grain moves appreciably with respect to the mean plasma, in which case the currents need modification. What are typical speeds? Slightly charged dust circles the giant planets at nearly the Keplerian speeds, of order tens of km s^{-1}. To a very good approximation, magnetospheric plasma near the planet is tied to (and co-rotates with) the planet's magnetic field (Stern 1976); it therefore orbits at speeds ranging from tens to hundreds of km s^{-1}, varying linearly with radial distance to the planet's center. Synchronous orbit is located at the radius r^* where a particle's circular orbital period matches the planet's spin period, or

$$r^* = \left(\frac{4\pi G \rho}{3\omega^2}\right)^{\frac{1}{3}} R_p, \tag{6}$$

where G is the gravitational constant, ρ is the planet's density, ω its spin rate, and R_p its mean radius. At other radii, uncharged grains drift relative to the plasma; their speeds vary linearly with $r^* - r$ at small and very large $r^* - r$ with typical values ranging up to many tens of km s^{-1}. Because electron thermal speeds (for plasma temperatures in typical magnetospheres of ~ 10 eV) are

$\sim 10^3$ km s^{-1}, the relative drift scarcely alters the electron flux; in contrast, ion thermal speeds are ~ 10 km s^{-1}, comparable to the plasma drift, and thus ion fluences may become quite anisotropic; the effect of a grain's charge on current fluences is given by Northrop and Birmingham (1990). At low relative speeds, the positive ion flux is reduced (because Coulomb attraction is less effective at focusing ions), allowing the equilibrium charge to become more negative (cf. Fig. 1 of Burns and Schaffer 1989); eventually at higher relative speeds, ions arc rammed onto the grain and the grain's charge grows less negative, linearly with relative speed.

The above calculations are idealized in many regards: i) they assume spherical dust grains despite the expectation that circumplanetary grains will be ejecta fragments or convoluted aggregates, like captured interplanetary dust particles; ii) they consider that all impacting ions/electrons stick on hitting the target, yet some will actually pass totally through and others will be scattered; iii) they take the plasma to contain a single species and to be Maxwellian, contrary to evidence from space probes (Mendis and Rosenberg 1994 also discuss a Lorentzian distribution); iv) they neglect changes to ion and electron currents imposed by local B–fields; and v) finally, secondary electron emission, most important at higher electron energies (> 50eV), is ignored. In the latter process, bombarding electrons penetrate the target, ionizing some material, after which the liberated electrons diffuse out to the surface; as emphasized by Horányi (1996), the ratio of emitted secondary electrons to incident ones varies sharply with the target's properties and the primary electron's energy, reaching a maximum (a factor of several) when the primary electrons penetrate to depths that are comparable to the target's size.

According to the above discussion, a grain's charge will vary as a grain moves about a planet owing to passage into/out of planetary shadows, non-uniform plasma conditions, and relative velocity changes due, for example, to a noncircular or nonequatorial orbit. Therefore, all three of these effects cause the particle's charge to oscillate with the orbital period; thus to some degree all electromagnetic forces, by their very nature, resonate with the orbital motion and accordingly can profoundly alter the orbit. These evolutionary changes are discussed in the Resonance Section IV.B. below.

It is instructive to realize that the number of excess charges corresponding to typical equilibrium potentials on micron-sized grains is not very large: since the potential $\phi = q/s$, the number of extra charges on an isolated spherical grain can be expressed as

$$N \simeq 700 \left(\frac{\phi}{\text{Volt}}\right)\left(\frac{s}{\mu\text{m}}\right). \tag{7}$$

Because individual plasma particles carry quantized charges and arrive at random but specific times, the charge carried by circumplanetary dust fluctuates stochastically. Schaffer and Burns (1995), by modeling grain-charging as a Markov process, compute the distribution functions to have half-widths of $0.5 N^{1/2}$. Thus, the typical fractional charge variation for our nominal 1 μm grain in a hydrogen plasma is about 1 %. Such a slight jittering of charge, and indeed relatively much larger variations, have minimal effects on orbital resonances (Schaffer and Burns 1995).

Somewhat counter-intuitively, smaller grains take longer than large to achieve equilibrium charges: this occurs because, from (7), the equilibrium charge $q \sim s$, while the currents that lead to this charge are proportional to the grain's surface area or $\sim s^2$; thus the charging time $\sim s^{-1}$. For grains of 1 μm radius and typical plasma conditions in the outer solar system, the time to reach equilibrium is 10 to 10^3 sec, much less than the orbital periods of 10^4 to 10^5 sec (Schaffer and Burns 1995). Once charging times become comparable to dynamical timescales, say for 0.1 μm grains or smaller, significant energy and angular momentum can be exchanged between the grain and the magnetosphere. In turn these changes may produce profound orbital modifications (Burns and Schaffer 1989; Northrop et al. 1989; Colwell et al. 1998).

An important issue related to charging is whether the grains may be considered to be "isolated" individuals or whether they form an organized ensemble to repel the bombarding plasma. In the latter case, charges will be significantly reduced (Grün et al. 1984; Mendis et al. 1984; Goertz 1989). Collective effects must be considered if the typical spacing between particles $d \ll \lambda_D$, the Debye length, which is the distance over which the charge in a plasma is effectively shielded. Goertz (1989) used nominal values of the mean particle size and optical depth to argue that charges are significantly overestimated for the F ring, most Neptunian ring arcs and the spokes. Hartquist et al. (1994) disagree. Goertz et al. (1988) have suggested that electrostatic repulsion may inflate ring thicknesses and, under some circumstances, may lead to a Coulomb lattice in which the grains form a regular array held apart electrostatically.

The general subject of collective effects in dusty plasmas – e.g., waves, instabilities, and wave scattering – has a rapidly expanding literature, much of it not concerning solar system or astrophysical studies; the reviews by Northrop (1992), Hartquist et al. (1994), Mendis and Rosenberg (1994), and Bliokh et al. (1995) provide introductions to the field. Havnes, Morfill, Goertz and their co-workers have proposed that several periodic features in the rings are associated with various instabilities. Goertz et al. (1986) have maintained that submicron dust particles sporadically elevated above Saturn's rings induce angular momentum exchange between the ring and the planet. Goertz and Morfill (1988) argue that this process leads to a radial instability in the ring.

The shape and size distribution of dust grains may be affected at high enough charges. Particles will be disrupted once their strength is unable to withstand electrostatic repulsive stresses; since the electrostatic tension is inversely proportional to the local radius of curvature, surface asperities will become rounded (Burns et al. 1980, 1984, Grün et al. 1984, Mendis et al. 1984, Mendis and Rosenberg 1994). Since most grains have like (negative) charges, the processes of nucleation and coagulation are generally inhibited by Coulomb repulsion (Mendis and Rosenberg 1994). However, small grains occasionally achieve positive potentials (cf. Meyer-Vernet 1982), in which case growth rates accelerate.

Most studies of the electrical charging of circumplanetary dust consider, as we have just discussed, currents to arrive from magnetospheric plasma or photoelectrons. Horányi and Cravens (1996) suggest that jovian ionospheric plasma may play an important role in the charging of members of Jupiter's rings, vastly accelerating the evolution of individual grains.

III.B. Forces

Since, by definition, circumplanetary dust is small and therefore has a large area-to-mass ratio, it is significantly affected by some forces that are much less important for bigger bodies. We first consider individual ring particles larger than a micron or so in size, typical of circumplanetary dust. By far the dominant force acting on such grains arises from the point-source gravity of the central planet, but several other forces perturb the orbits of these particles; in contrast to the planet's gravity, the latter may change the particle's energy and/or angular momentum and thereby modify the orbital character (Burns 1976).

The strongest gravitational perturbation is from the planet's non-spherical shape, since most rings are located within a few planetary radii. This perturbation, due primarily to the "oblateness" represented by the axially symmetric quadrupole (J_2) component in a multipole expansion of gravity (see, e.g., Danby 1988), is typically 0.1 % to 1 % the strength of the point-source or monopole term (Fig. 11). Other gravitational perturbations include: i) solar gravity, which is weak close to the planet, but dominates oblateness beyond several tens of planetary radii (exact values depend on the particular planet); ii) gravity from the other planets which is always small compared to solar gravity; and iii) the attractions of planetary satellites, which are only important in narrow zones surrounding the satellites and at associated resonant locations (see the treatments of satellite interactions and of resonances in the next section). Several non-gravitational forces also perturb particles in planetary rings. Principal among them are electromagnetic (Lorentz) forces and radiation pressure (Burns et al. 1979). For micron-sized grains, these forces range from 0.01 % to 1 % the strength of planetary gravity (Fig. 11). For smaller submicron-sized grains, however, the strength of these non-gravitational forces may equal or even exceed that of gravity. The various drag forces—plasma drag, Poynting-Robertson drag, and resonant charge variations—are much weaker ($\sim 10^{-6}$ times gravity), but because they alter orbital energy, often dominate the long-term dynamics (see section on timescales).

1. Planetary Gravity

Working in a planet-centered reference frame rotating at the planet's spin rate Ω_p, the gravitational potential Φ outside an arbitrarily shaped body can be shown to satisfy Laplace's equation, $\nabla^2 \Phi = 0$ (Danby 1988). Solving Laplace's equation in spherical coordinates yields the standard spherical harmonic ex-

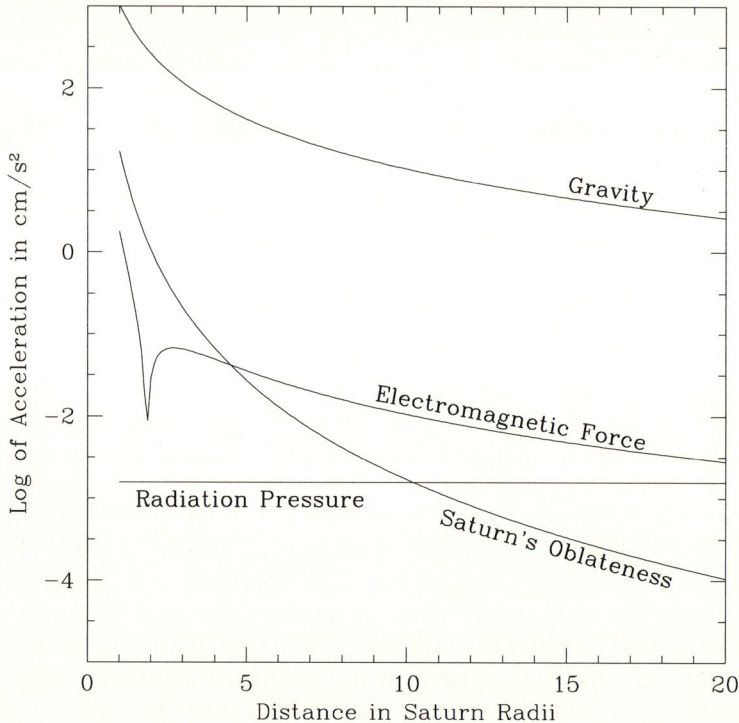

Figure 11. The strength of some perturbation forces around Saturn for a totally absorbing ($\beta = 1$) 1 μm dust grain charged to -5V. The electromagnetic force vanishes at synchronous orbit where the grain's velocity relative to the magnetic field is zero.

pansion of the gravitational potential:

$$\Phi = -\frac{GM_p}{R_p} \sum_{j=0}^{\infty} \left(\frac{R_p}{r}\right)^{j+1} \sum_{k=0}^{j} [C_{j,k} \cos(k\phi_R) + S_{j,k} \sin(k\phi_R)] P_j^k(\cos\theta) , \quad (8)$$

where G is the gravitational constant, M_p and R_p are the planetary mass and radius, and r, θ, ϕ_R are the usual spherical coordinates defined in a frame rotating with the planet. The $P_j^k(x)$ are associated Legendre polynomials (Kaula 1966; Schaffer and Burns 1992) and the coefficients $C_{j,k}$ and $S_{j,k}$ are dimensionless quantities whose values are set by the mass distribution within the planet.

The non-axisymmetric terms in a planet's gravitational potential (those with $k \neq 0$) are typically small since the primary result of planetary rotation is an axisymmetric equatorial bulge. Furthermore, away from narrow resonance zones, the non-axisymmetric terms lead to small short-period oscillations that average to zero over several orbital periods. Hence, we can often approximate

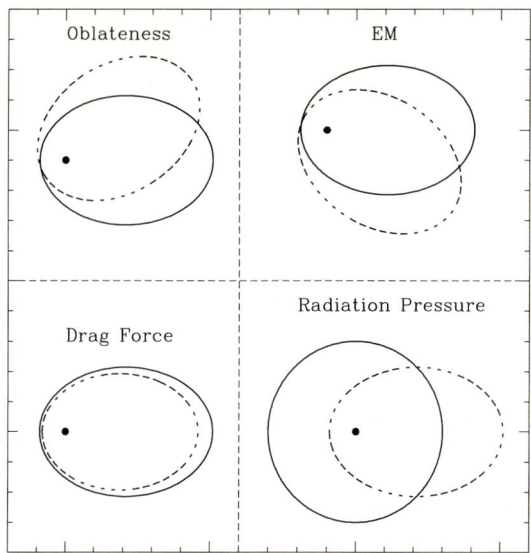

Figure 12. Sketches of orbital evolution under four different secular perturbations: planetary oblateness, the electromagnetic force from a dipolar magnetic field, radiation pressure, and drag. In the EM case, the drift direction varies with the sign of the particle's charge

(8) by

$$\Phi = -\frac{GM_p}{r}\left(1 + \sum_{j=2}^{\infty} J_j \left(\frac{R_p}{r}\right)^j P_j(\cos\theta)\right), \quad (9)$$

where P_j are the Legendre polynomials and only even terms are needed for the fluid giant planets. Choosing the origin of our coordinate system to correspond to the planet's center of mass yields $J_1 = 0$, so there is no gravitational dipole term. For all of the planets except slowly rotating Mercury and Venus, $J_2 \gg J_j$ (with $j > 2$), so in practice we can represent planetary gravitational potentials reasonably accurately by the first two terms (the monopole and quadrupole terms) in Eq. (9). Orbit-averaged equations of motion for the J_2 perturbation are given by Danby (1988) whereas the general problem is discussed more fully by Kozai (1959) and Kaula (1966). The solution to these equations is an ellipse that precesses slowly in space while retaining its size, shape and inclination (see Fig. 12).

2. Radiation Forces

Because photons carry momentum, they impart small forces when they are absorbed, emitted, or scattered by matter. The reaction of a ring particle to these forces causes orbital evolution, but this evolution is usually significant only for small dust particles (tenths to tens of micrometers) which have large surface-area-to-mass ratios. Radiation pressure forces for spheres of various compositions subject to the solar insolation are given by Burns et al. (1979) and Gustafson (1994), among others.

For a spherical grain, the ratio of the acceleration due to solar radiation pressure vs. that due to solar gravity is usually given by

$$\beta = \frac{5.7 \cdot 10^{-5} \rho s}{Q_{pr}}, \tag{10}$$

where Q_{pr} (of order 1) is the radiation pressure coefficient averaged over the solar spectrum, and ρ and s are in cgs units (Burns et al. 1979). Grains of a micron to tenths of microns (where geometric optics fails and Q_{pr} drops) are most affected by solar light. For dust grains orbiting the Sun, the main consequence of radiation pressure is simple to calculate since the force applied to a dust grain is primarily radial and the intensity of radiation emanating from the Sun falls off quadratically with distance, just like solar gravity. Thus the effective gravity felt by an interplanetary dust grain is reduced and these motes take slightly longer to orbit the Sun than they would under gravity alone. For heliocentric particles the only important dynamical outcomes are the arc of a cometary dust tail and the ejection (from the solar system) of small grains, the so-called β-meteoroids, that are released from larger parent bodies near perihelion.

The effects of solar radiation pressure on a dust grain orbiting a planet, however, are much more complex. In the planet's vicinity, a dust grain experiences a solar radiation force that is nearly constant in magnitude, but whose inertial direction varies slowly as the planet orbits the Sun. Since the orbital period of the planet is much longer than that of the ring particle, we usually make the approximation that the solar direction remains constant over one circuit of the ring particle around the planet. Orbit-averaged equations of motion for a circumplanetary particle influenced by radiation pressure were first obtained by Burns et al. (1979) and Chamberlain (1979) by averaging the effects of the perturbing force over one Keplerian orbit. Mignard (1982; Mignard and Henon 1984) later derived equivalent expressions by considering the perturbing potential due to radiation pressure. Hamilton (1993) extended this work to cover the general case of a tilted planet on an eccentric orbit about the Sun, working in equatorial coordinates so that the perturbations due to radiation pressure and planetary oblateness could be combined. Smyth and Marconi (1993), and Ishimoto (1996) give alternate forms of these equations. Solutions to the orbit-averaged equations (Richter and Keller 1995) show that radiation pressure primarily affects an orbit's eccentricity and orientation (see Fig. 12).

When particles pass into the planet's shadow, radiation forces cease. Mignard (1984), however, has demonstrated that the orbital consequences of this periodic interruption are slight.

3. Electromagnetic Forces

A charged dust grain orbiting through a planetary magnetic field experiences a Lorentz force. Close to the planet, the magnetic field **B** rotates at the planet's constant spin rate $\mathbf{\Omega}_p$ and, in a frame rotating at this rate, the Lorentz force is given by:

$$\mathbf{F}_{EM} = \frac{q}{c}(\mathbf{v}_{rel} \times \mathbf{B}) ,\qquad(11)$$

where q is the charge on the grain, c is the speed of light, and

$$\mathbf{v}_{rel} = \mathbf{v} - (\mathbf{\Omega}_p \times \mathbf{r})\qquad(12)$$

is the velocity relative to the magnetic field, with $\mathbf{v}$ the inertial velocity.

When describing a magnetic field evaluated in a current-free region ($\mathbf{J} \sim \nabla \times \mathbf{B} = 0$), the only remaining constraint that must be satisfied is Maxwell's equation $\nabla \cdot \mathbf{B} = 0$ (Stern 1976). Thus, defining $\mathbf{B} = -\nabla \Phi_{mag}$, $\nabla \times \mathbf{B} = 0$ is automatically satisfied, and we obtain Laplace's equation $\nabla^2 \Phi_{mag} = 0$ with solutions of the same form as (9) above. Hence

$$\Phi_{mag} = R_p \sum_{j=1}^{\infty} \left(\frac{R_p}{r}\right)^{j+1} \sum_{k=0}^{j} [g_{j,k}\cos(k\phi_R) + h_{j,k}\sin(k\phi_R)] P_j^k(\cos\theta) ,\qquad(13)$$

where the $g_{j,k}$ and $h_{j,k}$ are planetary magnetic field coefficients with units of gauss [Schaffer and Burns (1992) tabulate values for the giant planets and give additional references; see also Connerney (1993) and Connerney et al. (1996)]. As in the case of planetary gravity, we can capture the dominant secular effects with the lowest order $k = 0$ terms.

We now absorb several of the parameters from Eqs. (11)–(13) into a single dimensionless constant representing the ratio of the Lorentz force to planetary gravity (Schaffer and Burns 1987; Hamilton 1994). We calculate the Lorentz force due to an aligned dipolar magnetic field on a motionless grain in the equatorial plane (i.e., $\mathbf{v} = 0$ and $\theta = 90°$); for this choice the force ratio is independent of the distance from the planet. We define

$$L \equiv \frac{F_{EM}}{F_{GR}} = \frac{qg_{1,0}R_p^3\Omega_p}{cGM_p m_g};\qquad(14)$$

note that the ratio depends both on properties of the grain (the charge-to-mass ratio q/m_g) and properties of the environment (planetary mass, radius, spin rate, and dipole strength).

The Lorentz force can be treated as a perturbation to gravity for grains satisfying $L \ll 1$; thus the particles follow basically Keplerian orbits (with the usual 2:1 epicyclic motions) whose elements gradually evolve. Assuming typical grain potentials of a few volts (see Mendis et al. 1984; Horányi et al. 1992; Horányi 1996), this inequality translates to circumplanetary grains larger than several tenths of a micron in radius. For many applications, including the Jovian ring (Showalter et al. 1987) and the Saturnian E ring (Showalter et al. 1991), dust grains are inferred to be micron-sized and gravitationally dominated; hence their orbital evolution can be handled as a perturbation to a Keplerian orbit. Dust fluences onto the Galileo instrument have been found to exhibit periodicities of 5 hr and 10 hr, demonstrating the important role of Jupiter's magnetic field (rotation period of ~ 10 hr; Horányi et al. 1997; Grün et al. 1998, Graps et al. 2000).

On the other hand, when $L \gg 1$, these highly charged grains are effectively plasma particles that gyrate about the planet's magnetic field lines (the particles move on 1:1 epicycles; Mendis et al. 1984; Northrop et al. 1989; Horányi 1996); in this case the guiding centers of the paths gradually drift under the perturbing action of gravity.

The precession of orbital pericenter for a charged dust grain on an uninclined orbit in an aligned dipolar magnetic field was first calculated by Horányi and Burns (1991; see Fig. 12). Hamilton (1993) provides orbit-averaged equations for fully three-dimensional orbits in a magnetic field with aligned dipole and quadrupole components.

For values of $|L| \gg 1$ and of the appropriate sign, ejection from the system may occur. Horányi et al. (1993) have suggested that the Jovian dust streams detected in interplanetary space by the Ulysses spacecraft (Grün et al. 1993, 1996a) are composed of particles born deep in the Jovian magnetosphere and flung outwards along the corotation electric field by this process once charges become positive through secondary electron emission. Since all forces are conservative, the ejection speed (Hamilton and Burns 1993a) can be readily computed as

$$v_\infty = [2(GM/R)(L - 1/2)]^{1/2} ; \qquad (15)$$

this yields speeds that range from tens of km s^{-1} to hundreds of km s^{-1}, depending on L's precise value and how deep in the potential well the particles originate.

4. Summary of Forces

The nature of the various perturbation forces as a function of orbital radius is plotted in Fig. 11. In order to scale this plot to the several ring systems, we have tabulated estimates of the various forces discussed above. Table 2 gives typical accelerations and their dependence on orbital distance r and particle radius s, as caused by various forces, including the Lorentz force; these apply to 1-micrometer grains placed in the equatorial planes of the giant planets at $r = 1.8\ R_p$, where R_p is the planet's radius; most entries are from Burns (1991) and Esposito et al. (1991) assuming $Q_{pr} \sim 1$ and $\phi = 1$ V.

Two points must be made. First, accelerations provide at best only a rough idea of the long-term significance of any force since, as described elsewhere in the next section, orbital evolution depends on the phasing of the perturbations (Burns 1976). Second, it is valuable to note how these accelerations vary with particle size. Any forces, such as drags, that depend on particle cross-sectional area πs^2 have accelerations varying as s^{-1}, whereas electromagnetic accelerations differ as s^{-2} since surface potentials are generally constant for a specific plasma. This of course means that nongravitational accelerations, especially

Table 2

Orbital Accelerations and Timescales for 1 Micrometer Dust at 1.8 R_p

	Jupiter	Saturn	Uranus	Neptune
Accelerations (in cm s^{-2})				
Gravity ($\sim r^{-2}$)	800	350	275	350
Oblateness ($\sim r^{-4}$)	5	3	0.5	0.6
Radiation Pressure ($\sim s^{-1}$)	0.1	0.03	0.01	0.003
Electromagnetic ($\sim r^{-2}s^{-2}$)	1	0.05	0.03	0.02
Lifetimes (in years)				
Orbital Evolution				
Poynting-Robertson ($\sim s$)	$\lesssim 10^5$	10^5	$\lesssim 10^6$	10^6
Plasma Drag ($\sim s$)	[$> 10^5$]	$10^{5\pm1}$	$10^{5\pm1}$	$10^{7\pm2}$
Atmospheric Drag ($\sim s$)$^{(1)}$	-	-	$10^{2\pm1}$	$10^{5\pm2}$
Resonant Charge Variations	?	?	?	?
Destruction				
Sublimation of pure H_2O ice ($\sim s$)	$10^{3\pm1}$	$10^{5\pm1}$	$\leq \infty$	$\leq \infty$
Sputtering ($\sim s$)	$10^{3\pm1}$	$10^{3\pm1}$	$10^{5\pm2}$	$10^{7\pm2}$
Micrometeoroid Shattering ($\sim s^{-2}$)	$10^{5\pm1}$	$10^{6\pm1}$	$10^{6\pm2}$	$10^{6\pm2}$

$^{(1)}$ Also depends strongly on r (Herbert et al. 1987)

electromagnetic ones, become more important on small grains. It is for this reason that tiny motes, for which electromagnetic forces overwhelm planetary gravity, can be ejected from circumplanetary space if they are charged properly. Horányi et al. (1993) and Hamilton and Burns (1993a) have claimed that this effect can eject dust from the Jupiter system. When grains become very small, the charge-to-mass ratio can be large enough that they no longer follow nearly Keplerian paths but instead can be treated as massive plasma particles that spiral around magnetic field lines (Northrop 1992; Horányi 1996).

5. Drag Forces

In addition to the perturbation forces described above, relatively weak drag forces also act on dust grains around planets. Although they are usually feeble compared to other perturbations, drags have paramount importance because they, unlike the stronger perturbation forces, can cause secular (monotonic) changes in the orbital energy and angular momentum, and hence in the orbital semimajor axes, eccentricities, and inclinations of dust grains (Burns 1976); in turn, these changes allow grains access to resonant positions (see below) where particles can become trapped and eccentricities/inclinations can be dramatically altered. Over long periods of time, the effects of drag forces are cumulative and noticeable (Fig. 12).

The strongest drag force operating on distant circumplanetary dust is *Poynting-Robertson drag*, which arises from the transfer of momentum from solar radiation to the dust. The effects of Poynting-Robertson drag on dust grains in interplanetary space have been well characterized (Wyatt and Whipple 1950; Burns et al. 1979; Mignard 1984; Gustafson 1994). For circumplanetary particles, however, the effect is more complicated since the primary photon flux does not originate from the body at the orbit's focus. Nevertheless, orbital energy is still dissipated and the semimajor axis of the circumplanetary orbit shrinks with time. Burns et al. (1979) have shown that orbits collapse under Poynting-Robertson drag in approximately the time it takes for a particle (whether in circumplanetary or circumsolar orbit) to absorb the equivalent of its own mass in radiation:

$$T_{PR} \approx 10^3 \left(\frac{a_P}{AU}\right)^2 \left(\frac{\rho}{\text{g}\cdot\text{cm}^{-3}}\right) \left(\frac{s}{\mu\text{m}}\right) Q_{pr}^{-1} \text{yr} , \qquad (16)$$

where a_P is the planet's orbital radius and Q_{pr} is the non-dimensional radiation pressure coefficient (of order 1); this expression is written such that each term in parentheses is non-dimensional. Mignard (1984) has derived expressions for how circumplanetary eccentricities and inclinations evolve due to Poynting-Robertson drag.

Plasma drag and *atmospheric drag* are thought to account for most energy dissipation very close to the giant planets (see two paragraphs below, however); these drags arise due to momentum transfer during the collisions of dust grains with orbiting charged particles and neutral molecules/atoms, respectively. Both of these forces damp orbital eccentricities and inclinations, causing orbits to eventually adopt the motion of the impacting species, i.e., to move on circular and uninclined paths. Dust orbiting within the synchronous distance moves faster than the plasma, loses energy in collision with the plasma, and spirals in toward the planet. In contrast, dust outside the synchronous distance orbits more slowly than the plasma, and hence gains energy, requiring that it drifts away from the planet. Thorough treatments of plasma drag may be found in Grün et al. (1984) and Mendis et al. (1984); Northrop and Birmingham (1990) improve previous treatments of this problem. The orbital evolution timescale for plasma drag is

$$T_{PD} \approx \frac{2s\rho\zeta}{3\rho_p v} , \qquad (17)$$

where ρ_p is the mass density of the thermal plasma, v is the grain's speed relative to the plasma; the constant ζ is about 1 for effectively uncharged grains but is much lower, 10^{-2}, once the flow is subsonic and Coulomb attraction must be included.

Broadfoot et al. (1986) have noticed that atmospheric drag, caused by an extended planetary exosphere, will lead to rapid orbital evolution for dust at the inner edge of the Uranian ring system. We express the timescale for loss by atmospheric drag as

$$T_{atm} \approx \frac{\rho s v_{orb}}{\rho_{atm} v^2} , \qquad (18)$$

where ρ_{atm} is the atmosphere's mass density, v_{orb} is the particle's orbital speed and v is its speed relative to the atmosphere.

Atmospheric drag is concentrated near the planet where it may be most influential in causing dust orbits to collapse. Because of the surprisingly high exospheric temperatures in the outer atmospheres of most giant planets (Broadfoot et al. 1986), atmospheric gases extend to significant altitudes, implying that gas drag may be relevant for the inner parts of many ring systems. Broadfoot et al. (1986) and Herbert et al. (1987) assert that the Uranian exosphere sweeps the inner Uranian ring region clear of material in quite short times. Even with such a rapid exospheric drag, Uranian dust can be replaced by sufficiently active collisions (Colwell and Esposito 1990a).

Even though atmospheric drag and plasma drag have been suspected to determine evolution timescales for circumplanetary dust, neither is well-characterized because the mass densities of the impacting species are basically not known in the region of interest. This has become particularly clear following Galileo's observations of Jupiter's gossamer ring components (Ockert-Bell et al. 1999; Burns et al. 1999; see discussion below) which lie across a region where the orbital evolution due to plasma drag is outward whereas Poynting-Robertson drag leads to orbital collapse. The images strongly suggest that the rings are composed of inward-evolving ejecta from the satellites Amalthea and Thebe; hence P-R drag must dominate. Yet, earlier calculations (Burns et al. 1984; see Table 2) estimate that plasma drag should be orders of magnitude more effective. The implication is that the putative values for plasma density must be wrong by orders of magnitude, which is not implausible in this poorly studied region. This is but one case in which circumplanetary dust may provide a visible probe of magnetospheric conditions; expelled dust from the jovian system also highlights those magnetospheric conditions (Horányi et al. 1993).

Just as collisions with circumplanetary material might account for momentum transfer to circumplanetary grains, the interaction of orbiting material with *mass from outside* the system may also lead to orbital collapse since it brings mass, but no net angular momentum to the grains. This mass drag has been estimated most recently by Stevenson et al. (1986; cf. Cuzzi and Estrada 1998) and is insignificant for our purposes.

Drag resulting from *resonant charge variations* (Burns and Schaffer 1989; Northrop et al. 1989), a process even less quantifiable as to its efficacy than the drags described above, arises because the charge imposed on a grain lags the local equilibrium grain potential, meaning that work can be done on the orbit. Paths evolve toward or away from synchronous orbit at rates that can be rapid. This process is implicit in the remarkably rapid evolution of Jovian ring material identified by Horányi and Cravens (1996). This same energy-transfer is responsible for the magnetospheric capture of interplanetary particles, turning hyperbolic orbits into elliptical ones (Horányi 1994; Colwell et al. 1998).

Timescales for the processes described above are contained in Table 2. The plasma drag timescale, listed as $2 \cdot 10^{2\pm 1}$ yrs by Burns et al. (1984) using a nominal plasma density, has been raised substantially to agree with the absence of features in Jupiter's gossamer ring (Burns et al. 1999; see below). The sublimation timescale is very short ($< 10^4$ yrs) for "dirty" and absorbing water-ice particles at Saturn but much longer for pure ice and refractory materials (10^3–10^5 yrs) (J. Colwell, personal communication, 1999). Refractory material, more likely relevant to circumplanetary dust, evaporates very slowly.

III.C. Size Distributions

Populations of natural objects differ in their numbers as a function of radius, depending upon their properties, modes of origin and evolution. Terrestrial rocks, asteroids and interplanetary meteoroids, as well as impact craters, are represented usually by exponential power laws. The exponents will vary, according to whether these laws are written in mass or radius, and whether they are expressed differentially or cumulatively (Fujiwara et al. 1989 gives the algebraic relations between the exponents for these various expressions.). In particular, catastrophic fragmentations have a differential power-law index of -3.5 for the radius (see Eq. (3); Dohnanyi 1972; cf. Durda and Dermott 1997).

Laboratory experiments indicate that the exponent deviates somewhat depending on the conditions of fragmentation; with extensive grinding, small sizes become more common and the slope steepens (Hartmann 1969). However, due to practical constraints, experiments are usually conducted at speeds and with sizes that differ vastly from those appropriate to many space collisions; thus somewhat-speculative scalings are used to extrapolate results (Fujiwara et al. 1989; Housen and Holsapple 1990). In other natural systems, nucleation or condensation products are often generated near a single size.

Size distributions for circumplanetary dust are generally inferred by trying to match, for assumed optical properties and functional dependences (e.g., power laws in s), ring brightnesses obtained at a discrete number of phase angles; they are thus non-unique. Because the circumplanetary particles that are discussed in this chapter are usually widely separated, these rings are intrinsically very faint. They become most visible when the signal is primarily diffracted light (for example Jupiter's ring is about ten to twenty times brighter in forward-scattered light than back-scattered). Most images of circumplanetary dust were taken when the spacecraft was in the planet's shadow with phase angles of 170–179°. The particle sizes that most effectively diffract visible light into such angles are microns. Thus most faint rings seem to be composed primarily of micron-sized grains (Burns et al. 1984). Size distributions can also be estimated by interpreting the data from dust detectors (Grün et al. 1992c) and from plasma measurements (Meyer-Vernet et al. 1996, 1998; Oberc 1994).

As listed in the penultimate column of Table 1, particles in the various ethereal rings seem to exhibit great variety in their size distributions – some apparently collisionally derived, others monodisperse and yet others with steep power-laws. The typical view is that dust particles are derived from larger bodies by collisional fragmentation, and particles are eventually lost from the system at the small end. Electrostatic bursting has also been suggested to shatter particles and smooth off asperities (Mendis et al. 1984; Burns et al.

1980). Models for the evolution of the size distribution do not include aggregation because collisions occur at high speeds. Observed size distributions are not those that particles are born with whenever the lifetime of particles differs with particle size; for example, if small particles are rapidly eliminated, the observed size distribution will be flatter than that introduced (the power-law index is lowered, Burns et al. 1984). And, as described in the earlier sections on orbital evolution and grain destruction, lifetimes of circumplanetary grains do change with particle sizes; not surprisingly, smaller grains survive less time in the system.

Few attempts have been made to explain the observed size distributions. The Jovian main ring and Saturn's D ring, with exponents somewhat shallower than the expected fragmentation result of -3.5 may be collisional debris that has evolved; the shallow slope of Jupiter's size distribution, -2.5 ($\pm$0.5), may indicate orbital evolution that is faster in inverse proportion to particle radius (Burns et al. 1984). The steep values for Saturn's F ring (Showalter et al. 1992) and perhaps its G ring (Showalter and Cuzzi 1993, cf. Throop and Esposito 1998) are not understood. The monodisperse sizes of Saturn's spokes (Doyle and Grün 1990) and its E ring (Showalter et al. 1991) speak to unique origins. Goertz et al. (1983), Mendis et al. (1984), Grün et al. (1984) and Tagger et al. (1991) argue that spokes form when sub-micron grains are ejected by electrostatic forces from larger grains. Hamilton and Burns (1994) aver that Saturn's E ring grains are naturally selected as those having sizes that allow a resonance condition to be satisfied (Horányi et al. 1992; see next section); others believe that this condition incriminates vapor condensation near Enceladus. The most thorough treatments of the evolution of the size distribution of dusty rings have been carried out numerically at the University of Colorado, where the dust is viewed as the end product of collisional processes in ring-moon systems. Colwell and Esposito (1990a,b) have combined calculations of dust supply with Markov chain models for dust transport between the Uranian/Neptunian rings and their moons to generate radial profiles of dust optical depth (see concluding section). The most recent applications of these simulations are contained in review chapters about the Uranian (Esposito et al. 1991) and Neptunian (Porco et al. 1995) rings. These find that the extant systems can be matched fairly well with reasonable parameter choices.

III.D. Destruction and Generation of Grains

Circumplanetary grains are swept out of systems by the orbital evolution described above (see Table 2), and simultaneously are destroyed by the fierce environments in which most reside. Depending on their composition, some particles may sublimate away but, more likely, they will vanish because of the mass flux striking them, either being shattered apart in hypervelocity collisions with bombarding interplanetary micrometeoroids (Cuzzi and Estrada 1998) or sputtered away by the impinging magnetospheric flux, many of the same particles that account for the electric charges that the grains acquire (Johnson 1990; Johnson et al. 1993; Jurac et al. 1995). We will learn below that lifetimes of extant grains are remarkably short, albeit not very well constrained. In turn, since faint rings are visible about all the giant planets, grains must also be born so that the ethereal rings are continually resupplied: the play remains the same even as its characters change. Most authors consider that tiny grains are born in collisions, the same violent events as those that ultimately account for their demise. Fine debris may be sloughed off moonlets (Cuzzi and Burns 1988; Colwell and Esposito 1990a,b; Esposito et al. 1991) following mutual collisions. Gentle versions of these same occurrences may temporarily remove grains from independent orbit.

Sputtering by energetic ions and electrons is surprisingly capable of destroying ring particles because the ejection process is efficient, and because large fluxes of energetic particles populate planetary magnetospheres. This process has substantial variations and uncertainties, owing to our incomplete knowledge of the impinging particle fluxes and to large differences in the sputtering yield depending upon target properties and impacting particle energy (Johnson 1990; Jurac et al. 1998). The sputtering lifetime T_s of a grain can be estimated by simply dividing its mass M by the mass loss rate ($\dot{M}$): thus

$$T_s \sim \frac{M}{\dot{M}} \sim \frac{Ns}{(\sum_i F_i \zeta_i)} , \qquad (19)$$

where N is the number density of molecules in the grain and a sum is taken over all ions present of the product F_i (flux of the ith ion) times ζ_i (the sputtering yield per impact of the ith ion). Obviously these sputtering lifetimes depend on knowing yields accurately from laboratory experiments (Johnson 1990), where it may be difficult to duplicate conditions in space, and on having comprehensive information available as to the bombarding magnetospheric fluxes. The latter are generally based on very limited measurements (often merely a single spacecraft passage at best). Nevertheless the lifetimes, given in Table 2 and likely to be accurate to perhaps no more than an order of magnitude or so, are meaningful insofar as they are very brief on cosmic timescales. Also the fluences of damaging ions may depend sharply on location, so that lifetimes may vary substantially across a broad ring, especially if the ring absorbs the magnetospheric particles.

Collisions with micrometeoroids (Cuzzi and Estrada 1998) destroy circumplanetary dust effectively because interplanetary material acquires very high speeds as it is gravitationally accelerated towards the giant planets. If there is a lower cut-off in the sizes of dust grains that can penetrate a magnetosphere (see Colwell and Horányi 1996), then catastrophic fragmentation, in which a grain is shattered by a single impact, dominates over progressive erosion (see Burns et al. 1980). This implies that *number* fluxes (and not *mass* fluxes) of projectiles govern grain lifetimes and, depending on the size of the lower cut-off, that small particles might survive longer than large ones. Typical estimates (see Burns et al. 1984) are collisional lifetimes of $T_c \approx 10^5 (1 \ \mu m \ s)^2$ yrs around all the giant planets. The identification of some interstellar projectiles within the impinging flux (Grün et al. 1993; Baguhl et al. 1994) will not alter this timescale much.

Other processes have been invoked to generate some ring material. At one time volcanic dust from Io was considered a viable source for the jovian ring; even today it is a primary contender to produce the dust that forms the Jovian dust streams (Horányi et al. 1993; Grün et al. 1996a; Graps et al. 2000) and some other material that inhabits the Galilean satellite region (Grün et al. 1997, 1998a; Krüger et al. 1999b). A small fraction of interplanetary and interstellar dust may be captured by the giant planets, and could account for the few retrograde particles discovered by Galileo (Colwell et al. 1998). Minute grains of a narrow size distribution, such as those in Saturn's E ring, could have been launched by volcanoes or geysers. Particles may even condense directly from the local gas (Johnson et al. 1989).

Small ring-moons are almost always intimately intermingled with ethereal rings (see Fig. 1). In particular, all the faint rings that are clumpy or time-variable suffer significant satellite perturbations. This association is evident in most of the ring systems: Saturn's Pan (Showalter 1991) was discovered within the Encke gap where an incomplete arc of filamentary material is located; Cordelia abuts Uranus' dust-laden λ ring; Galatea, which was found skimming along the inner edge of Neptune's Adams ring and may be kinematically connected to its arcs (Porco et al. 1995), resides in an unnamed dusty (perhaps discontinuous) ringlet (Ferrari and Brahic 1994); the activity in the F ring's environs as seen during Saturn's 1995-96 ring-plane crossings (Nicholson et al. 1996) may be related to perturbations by the shepherd satellites; other clumping is due to systematic tugs by the satellites (Showalter and Burns 1982; Kolvoord et al. 1990) and micrometeoroid impacts (Showalter 1998); and Jupiter's gossamer rings are clearly derived from its heavily bombarded satellites Amalthea and Thebe (Ockert-Bell et al. 1999, Burns et al. 1999).

Even though not directly visible, parent bodies–from which the rings are derived–are assumed to be present in most faint rings. Modelers have inferred their presence from the short lives of individual dusty ring particles; more direct evidence is contained in the back-scattering opacity of rings and in the ability of diaphanous rings to absorb energetic charged particles (Van Allen 1982, 1983). In particular, Showalter et al. (1987) argue, from the Jovian ring's phase function, that the parent bodies have an opacity like that of the visible ring; a preliminary comparison (Burns et al. 1999) of back-scattered Keck images (de Pater et al. 1999) with forward-scattered Galileo images (Ockert-Bell et al. 1999) also finds that the Jovian ring contains "large" bodies, in agreement with charged particle absorptions measured in this neighborhood by Pioneer 11 (Fillius et al. 1975) and the Galileo entry probe (Fischer et al. 1996). The presence of parent bodies has been inferred for Saturn's G ring (Van Allen 1983, 1987). Saturn's F ring must contain some large objects to have affected the Voyager radio signal (Tyler et al. 1983b). Pioneer 11 absorptions in the neighborhood of Saturn's F ring have been interpreted to be localized "microsignatures", caused by clouds of ejecta from mutual collisions of parent bodies (Cuzzi and Burns 1988).

The destruction section above mentions that most circumplanetary dust is probably *generated through meteoroid impacts* with visible or unseen parent bodies. This process has been observed by the Galileo dust detectors as the spacecraft flew past various Galilean satellites (Krüger et al. 1999b), and was inferred from brightness bursts measured in Saturn's F ring (Showalter 1998). The current ring mass in small grains, M_{ring}, will be produced by a meteoroid mass flux density Φ in a time T_Φ according to

$$M_{ring} = Y \Phi A_p T_\Phi f \; , \qquad (20)$$

where Y is the yield (i.e., the mass excavated/impacting mass) from a typical hypervelocity impact, A_p is the cross-sectional area of the parent bodies, and f is the mass fraction of ejecta that are micron-sized. This assumes that all ejecta escape from the parent bodies. In the case of impact ejecta leaving the larger of the ring-moons, this may not be a good assumption because typical launch speeds from hypervelocity events are comparable to the escape speeds from these bodies (cf. Fig. 17 of Burns et al. 1984; Table 1 in Burns et al. 1999). Then the nature of the surface (which determines the launch characteristics) becomes important as does the size, shape and orbital location of the ring moon (which affect the escape speed from various locales of the moons).

Source satellites of a specific size produce the most impact ejecta (Burns et al. 1984, 1999): satellites need to be as big as possible up until they become so large that they are able to retain ejecta. The latter size for an isolated satellite is that at which the moon's escape speed matches the slowest speed at which ejecta leave a hypervelocity impact site. For a soft regolith, this corresponds to an object about 5–10 km in radius, i.e., the size of the smallest of the known ring-moons (e.g., Adrastea). Furthermore escape from the ring-moon closest to the planet is generally favored because, due to tides and non-spherical shapes, escape over much of its surface requires no impulse whatsoever (Burns et al. 1999); loss of all surface debris will inevitably lead to a "hard" regolith.

III.E. Interactions with Nearby Satellites

As the jovian gossamer ring's connection to its nearby moons demonstrates, the erosion of satellites through impacts with interplanetary projectiles may be the principal supplier of ring material. In this section we discuss additional effects of satellites on rings. Once released, the dust that comprises the ethereal rings continues to interact significantly with the largest of the parent bodies: through long-range gravity, ring-particle orbits are perturbed and these tugs will shift individual orbits but can also shepherd rings in ensemble. Grains whose orbits cross the satellites will eventually collide with these targets: some grains may be absorbed, but the most energetic impacts will generate new ring debris.

What are the initial paths of ejecta launched from ring-moons? The gross nature of these orbits depends solely upon the velocities of the sources and the relative velocities at which the ejecta leave the sources.

When considering the speed distribution of the ejecta, and averaging over it, typical departure speeds from small, isolated spherical objects turn out to be comparable to the classical escape speed, which as a rough rule of thumb is that the escape speed (in m s^{-1}) is roughly equal to the object's radius in kilometers. Once shape and tidal effects are included, escape generally only becomes easier. Thus most escaping debris from a typical ring moon, even when the causative impact took place at many tens of km s^{-1}, happens at 10–10^2 m s^{-1}. Since orbital speeds are many tens of km s^{-1}, after departing the satellite, typical orbits of ejecta have very small eccentricities and inclinations, $\sim v_{esc}/na$. That is to say, the ejecta will form a narrow tube about the source satellite. This tube will precess and smear out under the action of oblateness as discussed below.

The departure trajectories along which ejecta leave moons in the Roche zone (Fig. 1) are quite convoluted by tidal forces, the influence of aspherical moons and non-inertial effects. Burns et al. (1980; their Fig. 4) show a jovian example, while the complex interplay of a stream of particles shearing past a moon embedded in the Saturn's F ring is animated by Weidenschilling et al. (1984; their flip-chart figures running between pp. 379-413).

Since the dust grain originated at the satellite (of radius R_{moon}), its orbit (ignoring perturbations) will forevermore cross the moon's (taken to be a low-eccentricity orbit at radial distance a_{moon}). Using a particle-in-a-box formalism, the e-folding collisional timescale is

$$T_{col} \sim \pi \left((sin\, i_{dust})^2 + (sin\, i_{moon})^2 \right)^{\frac{1}{2}} \left(\frac{a_{moon}}{R_{moon}} \right)^2 \left(\frac{U_r}{U} \right) T_{orb},$$

where $T_{orb} = 2\pi a_{dust}/v_{dust}$ is the dust grain's orbital period with a_{dust} its semimajor axis and v_{dust} its orbital speed; U is the relative velocity between the moon and the dust grain; U_r is its radial component; and the orbital inclinations of the dust and ring-moon are measured relative to the planet's equatorial plane (Soter 1971; Hamilton and Burns 1994). The ratio U_r/U is nearly independent of e_{dust} and, to within < 20 %, equals one.

Grain-moon collisions occur on a very rapid timescale, typically a few years to a decade for impacts into ring-moons by zero-inclination particles. Once inclinations make the problem three-dimensional, the collisional lifetimes of grains become much longer; e.g., in the Jovian case, grains crossing the orbits of large Thebe and Amalthea (whose inclinations are 1.09° and 0.37°, respectively) will strike the satellites in $\sim 10^2$ to 10^3 yrs; collisions with the much smaller ring-moons happen in $\sim 10^5$ yrs. A totally different and more sophisticated approach (Canup and Esposito 1995) to recollision finds similar times.

The relative speed between a particle traveling on a low-inclination, arbitrarily sized eccentric orbit and a moon moving along a circular, nearly equatorial path is about

$$v_{col} \sim e v_{moon},$$

a result accurate to about 10 % for particle orbits of all sizes and shapes, as long as the collision does not occur near an orbital turning point (Hamilton and Burns 1994). Since orbital speeds are several tens of km/sec, impact speeds for orbits with $e > 0.01$ generally vastly exceed the escape speed of the ring-moon or parent body, and thus impact trajectories are little modified by the "massive" target.

Impacts become hypervelocity (exceed the speed of sound) once orbital eccentricities reach modest values. Yields, $Y \sim 5v^2$ with v in km s^{-1} (Grün et al. 1984), achieve values $\gg 1$ (i.e., impacts generate much more ejecta than the impactor's mass) at speeds above a km s^{-1}. With such values, systems may be self-sustaining.

The few collisions that occur at gentle speeds do not necessarily lead to absorption since these systems usually lie within or near the Roche zone (Fig. 1). Re-accretion is quite complicated and depends on the surface properties, spin rate, orbital location and morphology of the target, and on the relative masses of the target/projectile (Colwell and Esposito 1992).

Even if grains do not collide directly with the source moons, those that pass nearby will be scattered during their close flybys. Of course, these gravity-assists occur on shorter timescales than direct collisions but are less potent too. As a rough rule of thumb, maximal scatterings are like the local v_{esc}, or at most tens of m s^{-1}. Thus the eccentricities and inclinations of ring particles should random-walk with typical stepsizes like v_{esc}/v_{orb}, and so produce hardly any long-range effects, although they do mean that ring-moons scatter particles to a few times their own size. This should lead to diffuse ring edges (see Burns and Gladman 1998).

The accelerations due to ring-moons drop off like d^{-2} and so are relevant only within a narrow region surrounding the moon (Goldreich and Tremaine 1982). However, as the next section mentions, even small forces, if they resonate with a ring particle's preferred motions, can have important consequences.

As described above, ejecta from continual collisions with a moon will form a debris tube surrounding the orbital path of the source moon. Because of the planet's oblateness and electromagnetic effects, orbital planes precess swiftly (a few months to a few years; see Fig. 12), whereas inclinations are preserved. Thus, after differential precession, grains launched in a single event soon lie on a hoop of height $2a[sin\ i + (v_{esc}/v_{orb})]$ and width $2a[e + (v_{esc}/v_{orb})]$. This hoop will be brighter along its top and bottom edges, as well as the inner and outer radial boundaries, because the individual particles making up the distribution undergo epicyclic motions, spending more time in these regions. Burns et al. (1999) argue that this process accounts for the form of the Jovian gossamer ring (Fig. 3).

The global dynamics of tenuous rings have been investigated by Lissauer and Espresate (1998) who show that low-opacity rings can be shepherded much like their denser compatriots as long as constituents can be properly mixed before their next encounter with the satellite. For this reason, when the Jovian ring moons were located precisely at the ring edge, questions were raised whether these objects were sources or shepherds of ring material (Showalter et al. 1987; Showalter 1989). In the Jovian case, because of the close similarity of the gossamer ring thickness with the elevations of the satellites off the equatorial plane, Burns et al. (1999) are confident that the ring is derived from the moons, and, by Occam's Razor, they argue that the main ring is similarly born from Adrastea, the tiniest of the known Jovian inner satellites.

IV. CELESTIAL MECHANICS AND ORBITAL EVOLUTION

IV.A. Introduction

Recall our working definition of a ring as an ensemble of individual particles orbiting a planet. Ring particles are created, their orbits evolve, and they are destroyed, often on very short timescales. To understand the structure and dynamics of a ring as a whole, it is necessary to first appreciate the orbital evolution of the individual ring particles. We then build detailed models from realistic ensembles of individual orbits. The quality of a ring model is judged by two criteria: i) how well the model's predictions match actual observations and ii) how plausible the model's physical assumptions are.

In this section, we focus on faint dusty rings for which the methods of single-particle dynamics may be applied. First we introduce a simple way to describe an orbit.

If a ring particle orbits a spherical planet (point-mass gravity), it follows a bound Keplerian ellipse around the central planet (Danby 1988). Knowledge of the six quantities that specify a ring particle's position and velocity in space at a given time is sufficient to determine uniquely its Keplerian path. In other words, there exists a one-to-one correspondence between phase space ($x, y, z, \dot{x}, \dot{y}$, and $\dot{z}$, where the dot signifies differentiation with respect to time) and *orbital elements* (geometric quantities that specify the size, shape, and orientation of - as well as the particle's position along - an elliptic orbit). Many choices of orbital elements are possible; the ones that we use here are a, e, i, Ω, ω, and ν. The semimajor axis a is one-half the largest dimension of an orbit, the eccentricity e characterizes the orbit's ellipticity, the inclination i indicates the tilt of the orbit relative to the equatorial plane, the longitude of the ascending node Ω describes the orientation of the orbital plane, the argument of pericenter ω determines the angular position of the ellipse's shortest radius within the orbital plane, and the true anomaly ν identifies where the ring particle is located along its elliptic path. The particle's mean motion, or average angular speed,

$$n = \left(\frac{GM_p}{a^3}\right)^{1/2} \tag{21}$$

and its orbital period, $\tau_{Kep} = 2\pi/n$, are determined solely by the planetary mass M_p, the orbital size a, and the gravitational constant G. Further description of the orbital elements is given in basic celestial mechanics sources (e.g., Burns 1976; Danby 1988) and by Hamilton (1993, his Fig. 1).

The use of orbital elements offers several advantage over working with coordinates. For an unperturbed elliptical orbit, all coordinates and velocities are functions of time $[x(t), y(t), z(t), \dot{x}(t), \dot{y}(t), \text{ and } \dot{z}(t)]$, while for orbital elements all of the time dependence is contained in $\nu(t)$. The other elements a, e, i, Ω, and ω are simple constants which determine the invariant size, shape, and three-dimensional orientation of the elliptical orbit in space. Thus a description of an orbit in terms of the orbital elements allows one to visualize how the orbit actually appears much better than does a description in terms of time-dependent coordinates.

Given the physical properties of a dust grain (its size, shape, charge, and light-scattering properties), the details of the various forces that act upon it, and its initial conditions, we can use Newton's second law, $\mathbf{F} = m\mathbf{a}$, to track the particle's position at all times. The resulting second-order differential equation cannot usually be solved analytically, and is often difficult to solve numerically because of the large differences in the strengths of the strongest and weakest forces.

For large ensembles of ring particles, however, we are generally not interested in the details of how a single ring particle's position and velocity change, but only in the general character of its orbit. Accordingly, we can, in many cases, make use of an orbit-averaging procedure which utilizes the geometrical orbital elements. Orbit-averaging is a perturbation technique that requires that the planet's gravity be the dominant force acting on the particle. There are three advantages to this approach: i) orbital elements give a geometric picture of how a ring particle's orbit evolves; ii) the orbit-averaged equations of motion are 500–1 000 times faster to integrate than the original equations (Hamilton 1993); and iii) in many cases, approximate analytic solutions to interesting problems can be found.

Before discussing orbit-averaged equations of motion, we first detour to describe resonances, which dominate the dynamics in certain ring regions.

IV.B. Resonances

Dusty rings at Jupiter, Saturn, and Uranus cover broad radial swaths that often include one or more major resonances, those narrow regions of circumplanetary space where the frequencies of perturbations are commensurate with orbital frequencies (see Murray and Dermott 1999). In these locales, orbit-averaging, which is described below as a generally effective method for treating small perturbations, misses virtually all of the dynamics. Because of the relative scarcity of resonances, it is tempting to dismiss them as unimportant in determining the structure of planetary rings - this however is an erroneous conclusion! Moreover, drag forces, which secularly alter orbital semimajor axes (Table 3), are effective at transporting material from non-resonant zones into resonant locations.

Adjacent to resonant locations, small periodic perturbations can produce large effects by adding together in phase. Typically, one or more slowly-varying "resonant arguments"–in addition to the secular perturbations discussed above–can have important dynamical consequences. The most well-studied resonant perturbations in celestial mechanics arise from the gravitational perturbations between two orbiting bodies (see, e.g., Murray and Dermott 1999).

Resonances, however, need not be gravitational; they arise any time a periodic perturbation force drives an orbit at one of its natural frequencies. The periodically varying electromagnetic force on an orbiting dust grain is one such example (Burns et al. 1985); resonances with electromagnetic forces have been dubbed *Lorentz resonances* and for micron-sized dust in the jovian ring these resonances are far stronger than their gravitational cousins. Resonant charge variations (Burns and Schaffer 1989; Northrop et al. 1989) and shadow passage (Horányi and Burns 1991) also cause resonant perturbations.

Hamilton (1994) showed that the same mathematical formalism governs all types of resonances—gravitational and non-gravitational—and used this fact to derive resonant strengths for the most important electromagnetic resonances. In this section, we show how a generic resonance affects the orbital properties of a dust grain.

An orbiting dust grain is near a resonance when $d\Psi/dt \approx 0$ with the general resonant argument given by

$$\Psi = A\phi + B\phi' + C\varpi + D\Omega . \qquad (22)$$

Here ϕ is the mean longitude of the particle, ϖ and Ω are the particle's longitudes of pericenter and the ascending node, respectively, ϕ' is the mean longitude of the perturber, and A, B, C, D are integer constants. This resonant argument applies equally well to many different types of resonances including those due to the gravitational perturbations from a satellite on a circular orbit as well as those due to electromagnetic perturbations. Valid resonances must satisfy the relation $A+B+C+D = 0$, because the choice of direction for the zero point of longitude (from which all longitudes are measured) is arbitrary (Hamilton 1994). The strongest resonances are typically the first-order resonances, whether in inclination $(C = 0, D = \pm 1)$ or eccentricity $(C = \pm 1, D = 0)$. These resonances occur at places given roughly by:

$$\frac{n}{n'} = \frac{A}{A \pm 1} , \qquad (23)$$

where the mean motion n is expressed by Eq. (21) above and n' is the frequency of the perturbation, be it from another satellite or a rotating planet's gravity and magnetic fields. The locations of first-order Lorentz resonances in the Jovian ring system are displayed in Fig. 13. We focus here on the 3:2 inclination resonance that Burns et al. (1985) believe causes the transition between Jupiter's main ring and its vertically extended interior halo. The appropriate resonant argument is

$$\Psi_{3:2} = 2\phi - 3\phi' + \Omega . \qquad (24)$$

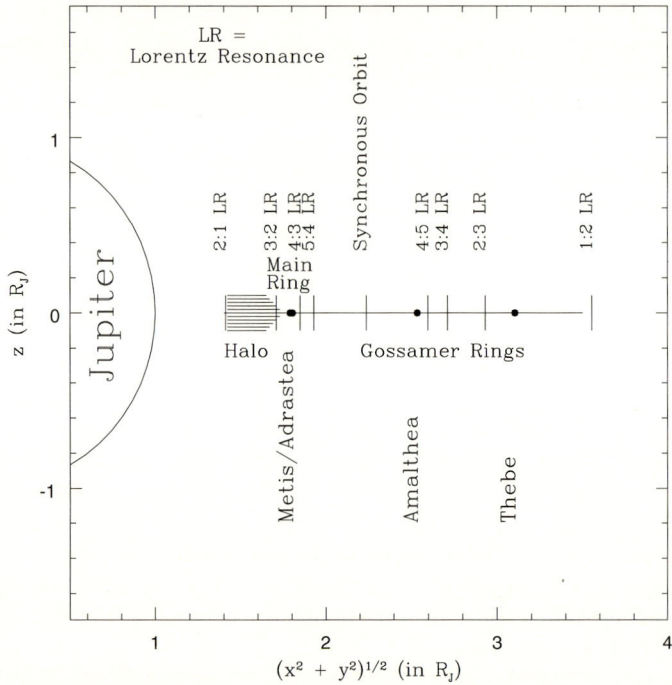

Figure 13. Location of the Lorentz resonances in the Jovian system. For evolutions that are slow enough, resonant trapping will occur when grains move toward synchronous orbit; resonant jumps in eccentricity and inclination are expected when particles leave the vicinity of synchronous orbit.

The orbital elements most strongly affected by an inclination resonance are a, i, and Ω. Hamilton (1994) derives the following expressions for the effects of the resonance:

$$\frac{da}{dt} = 2nai\beta \cos \Psi_{3:2} + \dot{a}_{drag} \tag{25a}$$

$$\frac{di}{dt} = -\frac{n\beta}{2} \cos \Psi_{3:2} \tag{25b}$$

$$\frac{d\Omega}{dt} = \frac{n\beta}{2i} \sin \Psi_{3:2} . \tag{25c}$$

Here t is time and β, the resonance strength, is nearly constant across the width of the resonance.

We include the drag term in the first of the above equations because i) orbital evolution due to drag brings dust grains to resonant locations, and ii) the drag term primarily affects the semimajor axis (Table 3) which is important in determining the outcome of a resonant interaction. Drag effects on the inclination and other orbital elements are less important and usually lead only to periodic wiggles; these are ignored here for simplicity.

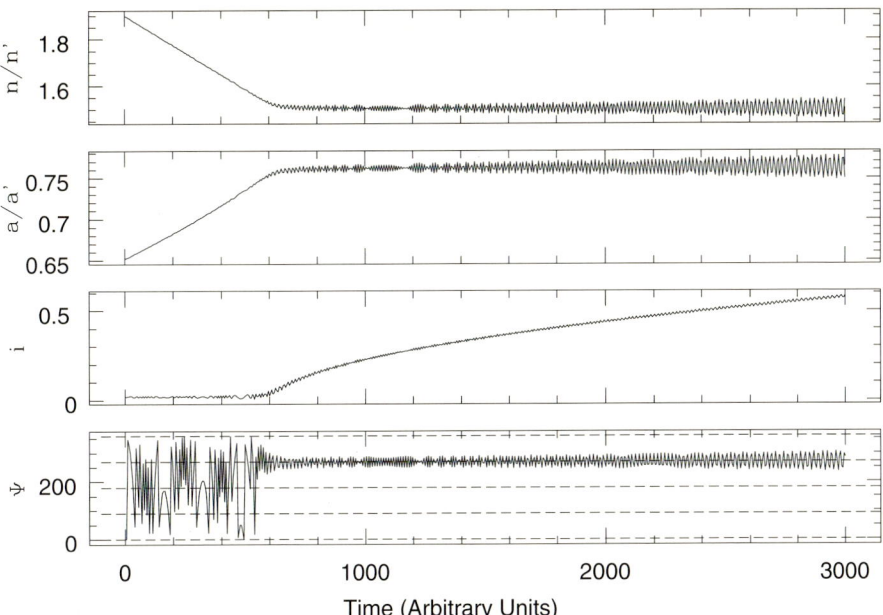

Figure 14. The trapping of a hypothetical ring particle that drifts outward through the 3:2 vertical Lorentz resonance; the orbit becomes trapped when the ratio between the particle's mean motion n and the planet's spin rate n' is $3/2$. During trapping, the evolution of a halts, while the inclination i grows at a rate that is independent of the resonant strength. Until the capture takes place, the resonant argument Ψ takes on all values, but once trapping occurs, it oscillates around $270°$. Compare to Fig. 15.

Depending on the sign of the drag term, either of two outcomes can occur at a resonance: resonant trapping or a resonant jump. In the first case, if evolution due to the drag is slow enough, then the two terms in the da/dt equation can balance and the evolution of the semimajor axis due to the drag term can be halted; this is *resonant trapping*, which occurs only when the evolution is toward the synchronous location (Fig. 13). In this case, we can take $da/dt = 0$ in Eq. (25a), solve for $\cos \Psi_{3:2}$, and eliminate this variable from Eq. (25b). After integration (assuming an initially uninclined orbit) we have:

$$i = \sqrt{\frac{\dot{a}_{drag} t}{2a}}. \qquad (26)$$

The evolution of the semimajor axis is arrested by the resonance, but the inclination grows as the square root of time to compensate (Fig. 14). This is a general feature of resonant trapping and follows from the constancy of the Jacobi integral (Hamilton 1994). Note that the resonant strength does not enter into Eq. (26).

Resonant jumps occur when the evolution is away from the synchronous location (Fig. 15). Small resonant jumps also occur anytime the resonant strength is slight compared to the drag term, because then the resonant term in Eq. (25a)

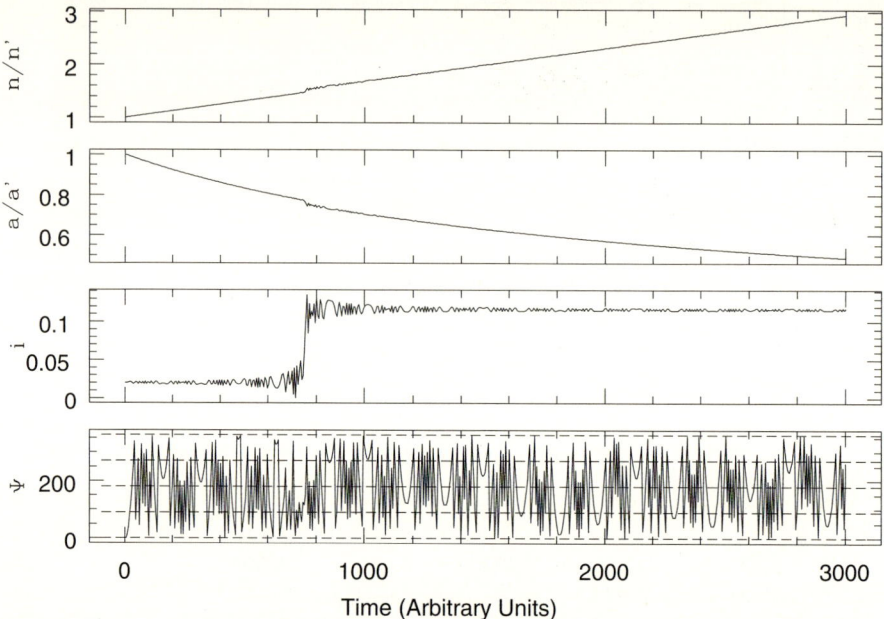

Figure 15. The history of a particle that evolves through the same 3:2 resonance as shown above, but now in the inward direction expected for jovian ring particles. When the ratio n/n' reaches 3/2, the orbital elements a and i experience sharp kicks. The size of the kick depends on the strength of the Lorentz resonance which, in turn, depends on particle size. Compare to Fig. 14.

cannot balance the drag term for any value of $\Psi_{3:2}$. These particles do not become trapped into resonance, but rather are "kicked" across the resonance. Both the semimajor axis and inclination experience jumps, whose overall amplitudes are determined by the resonance strength (Fig. 15). The relative jumps in a and i, however, are related by the Jacobi constant.

Both types of behavior are expected to be important in the jovian ring (Fig. 2). Drag evolution in the ring is dominated by the Poynting-Robertson effect (Burns et al. 1999) which acts inward toward the planet. Inside synchronous orbit, this moves dust away from synchronous orbit and, accordingly, resonant jumps are to be expected. The vertically extended jovian halo appears downstream from the 3:2 Lorentz resonance (Fig. 15), and it disappears (perhaps widened to the point of invisibility) near the 2:1 Lorentz resonance (Fig. 2; Schaffer and Burns 1992; Hamilton 1994; cf. Horányi and Cravens 1996). See also Fig. 20 later.

Beyond synchronous orbit, in the faint gossamer rings, orbital evolution is primarily toward synchronous orbit, so resonant trapping is expected for particles smaller than about 2–5 microns. Larger grains, for which electromagnetic forces are weak, will pass through the resonances relatively unscathed. This is probably what is happening in the gossamer rings where Galileo seems to have observed the large particles that evolve inward but not the small particles that are spread into a diffuse background halo by resonant trapping. In numerical simulations of the evolution of charged dust through this region (Burns et al. 1999), grains smaller than tens of micrometers become trapped.

IV.C. Orbit-Averaged Equations of Motion

We now consider how an orbit, described in terms of its orbital elements (see above), changes when non-resonant perturbation forces act. If these perturbation forces are small compared to the planet's point-source gravity, the first five orbital elements will change but only slowly. All will vary on characteristic timescales τ_{pert} which are long compared to the unperturbed Kepler period ($\tau_{pert} \gg \tau_{Kep}$). Thus we can separate timescales: over short times ($t \sim \tau_{Kep}$) a ring particle follows a nearly Keplerian path, the properties of which slowly vary over the longer timescale ($t \sim \tau_{pert}$).

The perturbations that act on circumplanetary dust grains larger than about a micron in size are much weaker than the gravity of the central planet; thus, as just discussed, over short intervals the actual orbit does not differ greatly from a Keplerian ellipse. We can take advantage of this fact by averaging the perturbations to an orbit over one Kepler period, making the explicit assumption that the particle follows an *unperturbed* Kepler ellipse during this time. This technique, which is accurate to first order in the small parameters that determine the perturbations, is called orbit-averaging. Orbit-averaged equations of motion tell us how the orbital elements change in time over intervals much longer that the orbital period.

1. Coupled Equations of Motion

In circumplanetary environments, the various perturbation forces (Sec. III) operate simultaneously. In this section we present a coupled set of orbit-averaged equations that describe the orbital evolution of a dust grain under the simultaneous influences of planetary oblateness, radiation pressure, the electromagnetic force, and an unspecified drag force. In order to simplify the equations, we assume that the orbital inclination i and the planetary obliquity γ are both less than about 30°. Ring inclinations all satisfy the first assumption, and only Uranus and Pluto do not satisfy the second. In addition, since rings are composed of a great ensemble of particles, we are usually only interested in the gross character of how these orbits change with time, not in the details of where the particles are along their paths. Accordingly, we consider changes to a, e, i, Ω and ω but not those of the sixth orbital element ν. We work in a non-rotating equatorial coordinate system and, after lengthy orbit-averaging calculations (see Hamilton 1993), we find:

$$\left\langle \frac{da}{dt} \right\rangle_{Total} = 0 + \dot{a}_{drag} , \qquad (27)$$

$$\left\langle \frac{de}{dt} \right\rangle_{Total} = \alpha(1-e^2)^{1/2} \sin\phi_\odot + \dot{e}_{drag} , \qquad (28)$$

$$\left\langle \frac{di}{dt} \right\rangle_{Total} = Z\cos\omega, \tag{29}$$

$$\left\langle \frac{d\Omega}{dt} \right\rangle_{Total} = Z\frac{\sin\omega}{\sin i} + \dot{\Omega}_{xy}, \tag{30}$$

and

$$\left\langle \frac{d\omega}{dt} \right\rangle_{Total} = -Z\frac{\sin\omega}{\sin i} + \frac{\alpha(1-e^2)^{1/2}\cos\phi_\odot}{e} + \dot{\omega}_{xy}, \tag{31}$$

where perturbations due to drag forces have been left unspecified. Here α is a frequency that parameterizes the strength of radiation pressure; the dimensionless ratio of solar radiation pressure to planetary gravity is $2\alpha/(3n)$ (Hamilton 1993) with the mean motion n given by Eq. (21). For problems in which radiation pressure is important, it is useful to define the solar angle:

$$\phi_\odot = \Omega + \omega - n_\odot t - \delta, \tag{32}$$

where $n_\odot$ is the planet's mean motion, t is time, and δ is a constant (Hamilton 1993). For orbits with $i \lesssim 30°$, $\phi_\odot$ is roughly the angle between orbital pericenter and the Sun as seen from the planet. The frequencies $\dot{\Omega}_{xy}$ and $\dot{\omega}_{xy}$ are the rates at which the node and pericenter change due to oblateness and the assumed dipolar magnetic field. They are given by

$$\dot{\Omega}_{xy} = -\frac{3nJ_2 R_p^2}{2a^2(1-e^2)^2} + \frac{nL}{(1-e^2)^{3/2}}\left(1 - e^2 - \frac{n}{\Omega_p}\right) \tag{33}$$

and

$$\dot{\omega}_{xy} = \frac{3nJ_2 R_p^2}{a^2(1-e^2)^2} - \frac{nL}{(1-e^2)^{3/2}}\left(1 - e^2 - \frac{3n}{\Omega_p}\right). \tag{34}$$

with L from Eq. (14). Combining Eqs. (30–34) yields the rate at which the solar angle $\phi_\odot$ changes:

$$\left\langle \frac{d\phi_\odot}{dt} \right\rangle_{Total} = \frac{\alpha(1-e^2)^{1/2}\cos\phi_\odot}{e} + \frac{3nJ_2 R_p^2}{2a^2(1-e^2)^2}$$

$$+ \frac{2nL}{(1-e^2)^{3/2}}\left(\frac{n}{\Omega_p}\right) - n_\odot. \tag{35}$$

Finally, Z is a frequency determined by the sum of the vertical components of radiation pressure (first term in the expression below) plus the electromagnetic force (second term):

$$Z = \frac{\alpha e[\sin\gamma \sin(n_\odot t + \delta) + \sin i \sin(\Omega - n_\odot t - \delta)]}{(1-e^2)^{1/2}} +$$

$$\frac{3nLe}{2(1-e^2)^{5/2}}\left(\frac{g_{2,0}}{g_{1,0}}\right)\left(\frac{R_p}{a}\right)\left(\frac{n}{\Omega_p}\right), \tag{36}$$

Table 3

Dynamical Outcomes from Orbital Perturbations[1] in Planetary Rings

Orbital Perturbation Force	Semimajor Axis	Eccentricity and Inclination	Node and Pericenter
Planetary Oblateness	No	No	MD[2]:Strong
Solar Gravity	No	Weak	MD:Weak
Lorentz Force away from Resonance	No	Weak	MD:Strong
Lorentz Force at Resonance	Weak	Strong	Strong
Radiation Pressure w/o shadowing	No	Strong	Strong
Radiation Pressure w/ shadowing	Weak	Strong	Strong
All Drags (Plasma, Poynting-Robertson, Neutral)	MD:Weak	MD:Weak	MD:Weak

[1] Assumes micrometer-sized grain with -5V potential [2] MD = monodirectional. This enhances the effect over time.

where $g_{1,0}$ and $g_{2,0}$ are the aligned dipolar and quadrupolar components of the planetary magnetic field.

Despite the simplifying assumptions made at the beginning of this section, the above set of equations is imposing. Most of the complexity, however, is contained in the frequencies $\dot{\Omega}_{xy}$, $\dot{\omega}_{xy}$, and Z. The contributions of the oblateness force, radiation pressure, and the electromagnetic force can be readily identified by the terms containing J_2, α, and L, respectively. For problems in which one of these forces is known to be unimportant, the appropriate terms can be simply set to zero.

With the approximations $i \lesssim 30°$ and $\gamma \lesssim 30°$, the changes in the orbital elements that determine azimuthal structure $(e, \phi_\odot)$ are independent of those that determine vertical structure (i, Ω). We will invoke this property to make simple analytic approximations in the next section. First, however, we combine the results of our orbit-averaged derivations with typical strengths of the perturbation forces to determine how strongly the various forces influence each of the orbital elements. The results are summarized in Table 3. If the perturbation force does not affect a given orbital element, we put a "No" in the appropriate bin of Table 3; otherwise we classify the effect as "Weak", or "Strong".

In addition, we note when a force causes an orbital element to increase or decrease secularly instead of simply oscillating by putting "MD" for monodirectional in the appropriate column. So, for example, in the orbit-averaged approximation, the oblateness force does not affect the semimajor axis (Eq. (27)), the eccentricity (Eq. (28)), or the inclination (Eq. (29)), but causes monodirectional changes in Ω (Eq. (33)) and ω (Eq. (34)). Under the influence of planetary oblateness, the longitude of pericenter $\Omega + \omega$ precesses (increases with time) for low inclinations (Fig. 12). Monodirectional changes in the orbital elements are important because even small variations will build up to have noticeable effects over time. The consequences of periodic perturbations, by contrast, tend to average out in time. Thus, not only are the strongest perturbations to each orbital element important, but the strongest *monodirectional* perturbations are often relevant too.

Most entries in Table 3 are taken from the orbit-averaged rates in the perturbation equations (27)–(31) above. The effects of planetary oblateness, radiation pressure (without shadowing by the planet), and the two drag forces are inferred directly from these equations (the effects of drag on the orbital node and pericenter are neglected in Eqs. (30)–(31) since other forces produce much stronger monodirectional outcomes). The fully three-dimensional, orbit-averaged equations for the Lorentz force away from resonance predict small oscillations in eccentricity and inclination that we neglected in the low-inclination approximation made above. Perturbations from solar gravity can also be orbit-averaged (Hamilton and Krivov 1996, 1997), but because the resulting expressions are long, we did not reproduce them here. Nevertheless, we show in Table 3 that the primary result of solar gravity on ring particles is to cause a slow orbital precession. We also include some effects that cannot be orbit-averaged: shadowing by the planet for radiation pressure (Mignard 1984) and Lorentz resonances (Hamilton 1994). The main consequence of shadowing is to produce small periodic changes in the semimajor axis (Horányi and Burns 1991), while Lorentz resonances strongly alter eccentricities, inclinations, and semimajor axes (see earlier discussion and Figs. 14 and 15).

2. *Visualizing Orbital Evolution*

The time histories of the orbital elements come from integrating Eqs. (27)–(36) numerically. In this section, we describe how plots of these time histories can be employed to visualize orbital evolution. We use dust around Mars as our example. The small satellites of Mars, Phobos and Deimos, are continually bombarded by interplanetary grains. Most micron-sized ejecta from these microcollisions is expelled from the satellites and goes into circumplanetary orbit around Mars, forming a very faint—and as yet undetected—ring system (Soter 1971; Horányi et al. 1990, 1991; Ip and Banaszkiewicz 1990; Juhász et al. 1993; Juhász and Horányi 1995; Hamilton 1996; Krivov and Hamilton 1997). Although Martian rings are currently theoretical constructs, the predicted evolution of dust from Deimos provides a simple example of circumplanetary dust dynamics. The strongest two perturbation forces affecting Deimos dust are Mars' oblateness and solar radiation pressure. Mars' magnetic field is weak enough that electromagnetic effects may be safely ignored and drag forces, primarily Poynting-Robertson drag, are too weak to cause significant consequences over timescales of a few tens or hundreds of years.

Figure 16, obtained by numerical integration of Eqs. (27)–(36), shows the evolution of a 20 μm particle launched from Deimos. When we integrate $\mathbf{F} = m\mathbf{a}$ in coordinates, and translate into orbital elements, we obtain a nearly identical orbit, thereby validating the averaging approximation; for a detailed comparison of these two approaches, see Hamilton (1993). The orbital semi-

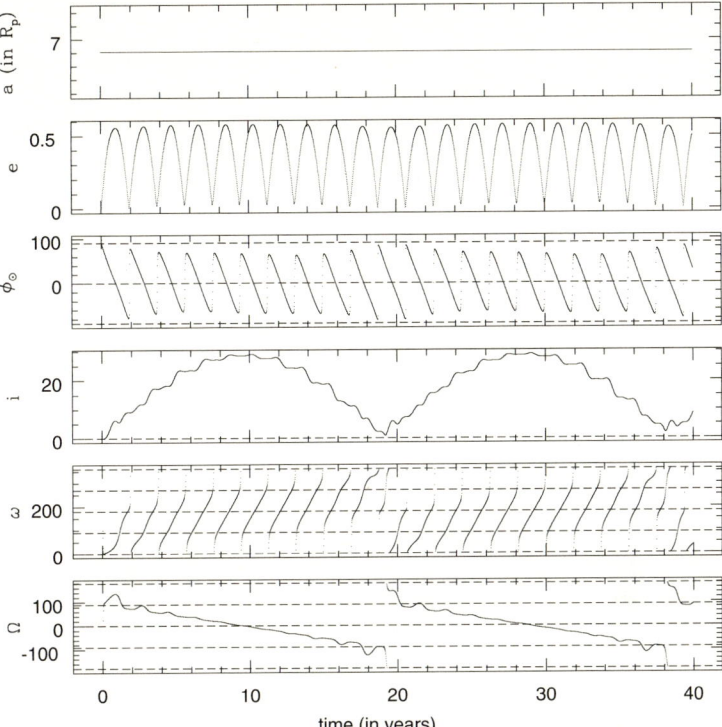

Figure 16. The evolution of the orbital elements of a 20 μm particle launched from Mars' moon Deimos onto an initially circular, uninclined orbit about Mars (J_2 = 0.001960). The active perturbation forces are planetary oblateness and solar radiation pressure. For this initial condition, we have $\alpha = 1.078$ rad yr^{-1}, $L = 0$, and $\gamma = 25.2°$. For $e = 0$, we calculate $\dot{\varpi}_T = -3.228$ rad yr^{-1}, $\dot{\omega}_{xy} = -2\dot{\Omega}_{xy} = 0.2239$ rad yr^{-1}, and $\dot{\Omega}_T = 0.2921$ rad yr^{-1}.

major axis, a, remains constant in agreement with Eq. (27) and Table 3. The orbital eccentricity oscillates sinusoidally with a period near Mars' orbital period of 1.88 years, while the solar angle $\phi_\odot$ regresses nearly linearly from roughly 90° to −90° over one eccentricity oscillation. Since the perturbation forces are small, the orbit can be well approximated by a Keplerian ellipse at each instant in time. At any moment we can visualize the full orbit of Fig. 16 by using the instantaneous values of the slowly-varying orbital elements. We notice that the points in Fig. 16 where the solar angle is zero are also the positions where the eccentricity is maximum. Since a is constant, these are also the locations where the orbital pericenters and apocenters of the instantaneous Kepler orbit are at their closest and furthest distances from Mars, respectively. Now, for low inclinations, the solar angle, $\phi_\odot$, is nearly the angle between the Sun and pericenter as seen from Mars. So the closest material to Mars should be found in the solar direction while the most distant material should be located in the anti-solar direction. To check this assertion, we transform the orbit elements of

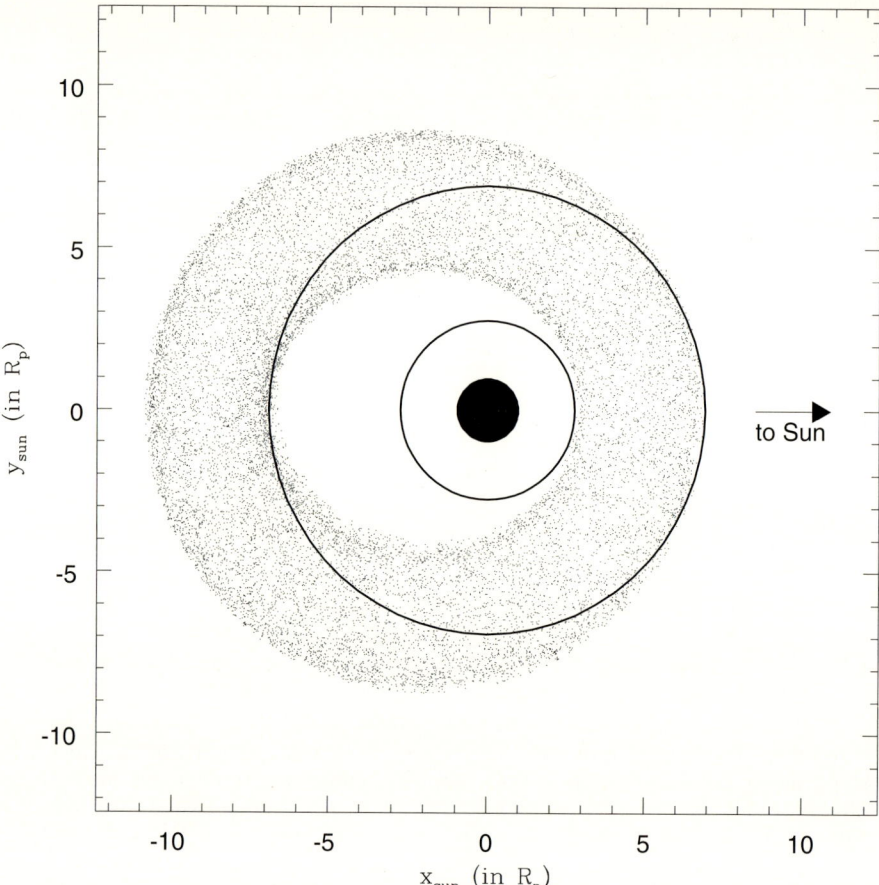

Figure 17. The positions of a particle whose orbital history is shown in Fig. 16 projected onto rotating xy coordinates. Mars (solid circle), the orbit of Phobos (inner ring), and the orbit of Deimos (outer ring) are shown for comparison. Note that the distribution of the dust grains forms an annular region that is offset in the antisolar direction.

Fig. 16 into cartesian coordinates and project them into the rotating xy plane; the results are shown in Fig. 17. The points form an annular region which is offset in the antisolar direction from Mars, and the closest and furthest material from Mars is found in the expected directions.

We can also use the orbital elements to help picture the vertical structure of the Deimos ring. The inclination in Fig. 16 varies roughly sinusoidally over a 20-year period, and the longitude of the ascending node Ω regresses linearly from 90° to −90° over one inclination oscillation. The maximum inclination out of Mars' equatorial plane occurs when $\Omega = 0$ (this is the ascending node of the ecliptic plane on the equatorial plane). If we view the dust distribution from along the line where $\Omega = 0$, to the right we should find material most elevated out of the equatorial plane (since these dust grains have just passed the ascending node). Similarly, to the left, ring material should be depressed (since the dust grains are just approaching the ascending node). More generally, the distribution of dust should be tilted so that most material to the right is elevated while that to the left is depressed. Figures 17 and 18 show the results of transforming and projecting the data in Fig. 16; the resulting distribution of dust is significantly tilted as expected. The perturbations of planetary oblateness and radiation pressure cause Deimos' dusty ring to be offset away from the Sun and tilted out of the equatorial plane. This example shows how the time histories of the orbital elements can be used to visualize the actual orbit.

IV.D. Approximate Analytic Solutions

One of the major advantages of the orbit-averaged equations discussed in section IV.C is that, in many situations, approximate analytic solutions can be found. For example, the nearly sinusoidal oscillations of e and i, and the almost linear behavior of $\phi_\odot$ and Ω in Fig. 16 hint at underlying analytic solutions. In this section, we discuss several simple analytic solutions to approximations of the full equations of motion, Eqs. (27–36).

1. Azimuthal Structure; Forced Eccentricity

We begin the search for analytic solutions by taking advantage of the fact that Eqs. (28) and (35) decouple from the other equations for small inclinations. In addition, we make the further approximation that e remains small. Following Hamilton (1996; see Horányi and Burns 1991) we find that, with the initial condition $e = 0$ at $t = 0$, Eqs. (28) and (35) have the solution:

$$e = \frac{2\alpha}{\dot{\varpi}_T} \sin \frac{\dot{\varpi}_T t'}{2} \tag{37}$$

and

$$\phi_\odot = \frac{\dot{\varpi}_T t'}{2} + \frac{\pi}{2} \tag{38}$$

with

$$\dot{\varpi}_T = \dot{\Omega}_{xy} + \dot{\omega}_{xy} - n_\odot. \tag{39}$$

Here $\dot{\varpi}_T t' = (\dot{\varpi}_T t \text{ modulo } 2\pi)$ and t is time. The solution states that during one sinusoidal oscillation of e, $\phi_\odot$ will change linearly from $-90°$ to $90°$ crossing through either $\phi_\odot = 0°$ or $\phi_\odot = 180°$ when $e = e_{max}$, depending on the sign of $\dot{\varpi}_T$. As the eccentricity returns to zero, the solar angle "jumps" back to $90°$ to restart the cycle. The jump in $\phi_\odot$ is physical (since $e \sim 0$) and arises from radiation pressure (the first term on the right-hand side of Eq. (35)) which rapidly drives $\phi_\odot$ to $90°$. Figure 16 provides a stringent test of our analytic solution, since the values of e and i are not always low as we assumed (in Fig. 16, $i_{max} \sim 30°$ and $e_{max} \sim 0.56$). Qualitatively, the approximate analytic solution agrees reasonably well with the behaviors of the eccentricity and solar

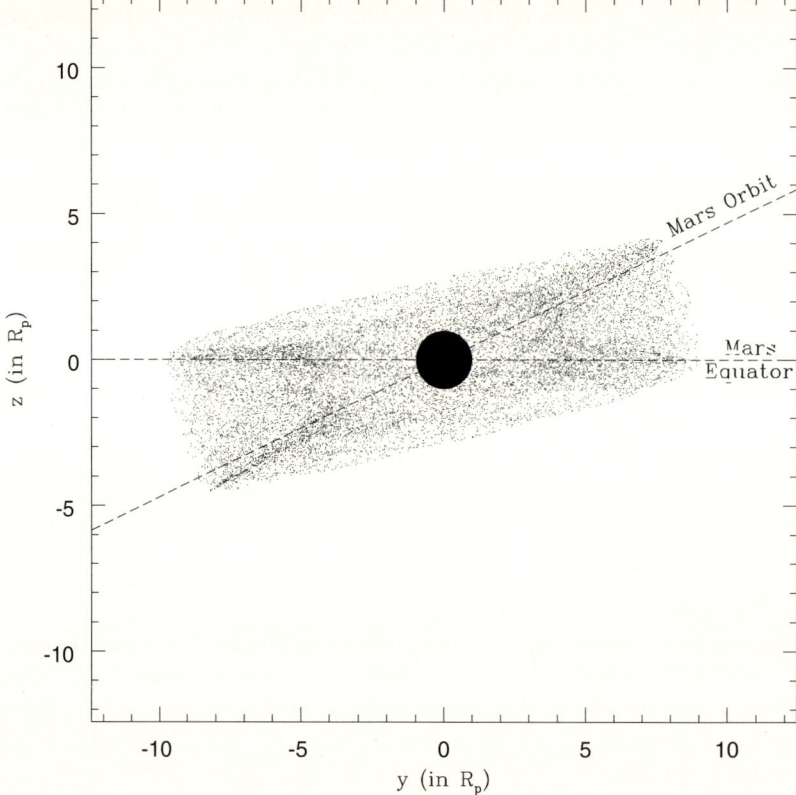

Figure 18. The positions of a particle whose orbital history is shown in Fig. 16 projected onto non-rotating yz coordinates viewed from along the positive x axis (the intersection of Mars' equatorial and orbital planes). Mars is shown to scale as a solid circle. Note that the distribution of dust grains in this projection is rectangularly shaped and tilted out of Mars' equatorial plane. The amount of tilt depends on the particle size; larger particles form a narrower distribution which is tilted less steeply from Mars' equation.

angle in Fig. 16, although some deviations can be seen. For instance, the angle $\phi_\odot$ and, to a lesser extent, e are modulated at the 20-year oscillation period of the orbital inclination. These effects arise because we neglected the inclination dependence of $\phi_\odot$ in our solution; the fact that the deviations are greatest at points in Fig. 16 where the inclination is largest supports this claim.

We measure $e_{max} = 0.56$ off Fig. 16. Using the numerical values given in Fig. 16's caption, Eqns. (37) and (38) predict an oscillation period $2\pi/|\dot{\varpi}_T|$ of 1.95 years and $e_{max} = 2\alpha/\dot{\varpi}_T = 0.67$. Our simple theory does well at predicting the period, but overestimates e_{max} by $\sim 15\ \%$ larger than this value; this discrepancy arises from the fact that e is not always small in Fig. 16, as was assumed in the derivation of the equations.

The analytic solution given by Eqs. (37)–(38) can be understood geometrically in terms of free and forced eccentricities (see Horányi and Burns 1991,

Fig. 2). The forced eccentricity is simply the eccentricity at the stationary point of Eqs. (28) and (35) which depends solely on the strengths of the perturbation forces. It can be found numerically for arbitrary eccentricity and reduces to

$$e_{forced} = \frac{\alpha}{|\dot{\varpi}_T|} \tag{40}$$

for $e \ll 1$. The free eccentricity is the amplitude of the oscillation about this value and is determined by starting conditions. With our choice of initial condition ($e = 0$ at $t = 0$), $e_{free} = e_{forced} = e_{max}/2$. The forced eccentricity causes the ring of dust in Fig. 17 to be centered about a point located a distance ae_{forced} away from Mars in the antisolar direction (Hamilton 1996).

2. Vertical Structure; Forced Inclination

The inclination and node time histories of Fig. 16 appear to be longer-period and noisier versions of the eccentricity and solar angle traces. Accordingly, we look for a solution like Eqs. (37)–(38) and ignore the short-period jitter. Mathematically, we use the solution for e and $\phi_\odot$, and average Eqs. (29)–(30) over the eccentricity-oscillation period. Following Hamilton (1996), we find:

$$i_{long} = -\frac{\alpha^2 \sin\gamma}{\dot{\Omega}_T \dot{\varpi}_T} \sin\frac{\dot{\Omega}_T t'}{2} \tag{41}$$

and

$$\Omega_{long} = \frac{\dot{\Omega}_T t'}{2} + \frac{\pi}{2} \tag{42}$$

with

$$\dot{\Omega}_T = \frac{\alpha^2}{2\dot{\varpi}_T} + \dot{\Omega}_{xy}. \tag{43}$$

As expected, the solution is of the same form as that for the azimuthal structure with i taking the role of e, and Ω acting like $\phi_\odot$. Quantitatively, Eqs. (41)–(43) predict a maximum inclination of 30° and an oscillation period of 21.5 years, values in reasonable agreement with Fig. 16.

By analogy to the forced eccentricity, we define a forced inclination:

$$i_{forced} = \frac{\alpha^2 \sin\gamma}{2|\dot{\Omega}_T \dot{\varpi}_T|}. \tag{44}$$

This has a direct interpretation in terms of the Laplace plane (see Hamilton 1996). For orbits that begin with $i = 0$, we have $i_{free} = i_{forced} = i_{max}/2$.

3. Vertical Structure; Locked Pericenter

Figure 19 shows the evolution of a 1.2-μm E ring particle originally started on a circular orbit at Enceladus' distance (Fig. 7). Over the fifty years displayed in the plot, the dominant perturbations are Saturn's oblateness, solar radiation

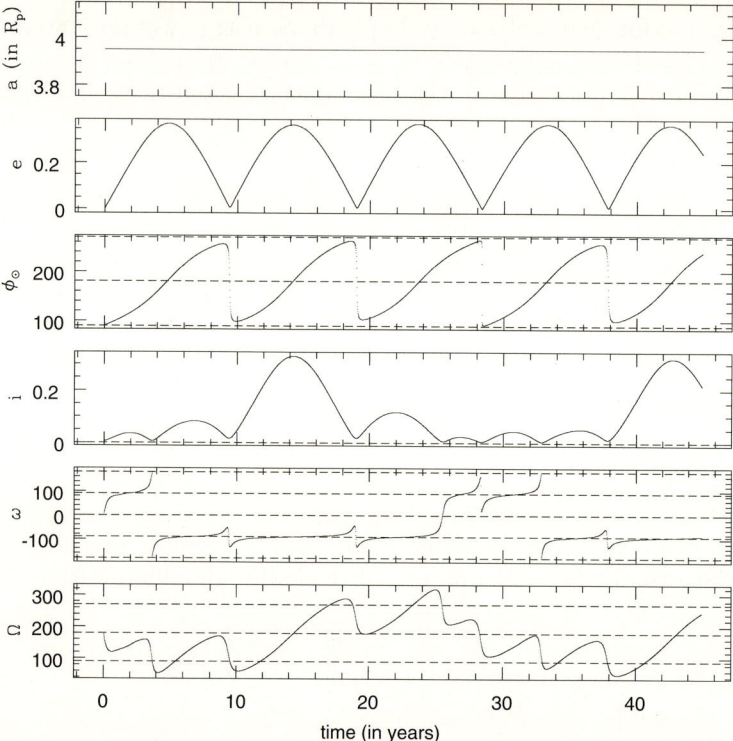

Figure 19. The orbital evolution of a 1.2 μm particle launched from Saturn's moon Enceladus onto an initially circular uninclined orbit. The perturbation forces are solar radiation pressure, planetary oblateness, and electromagnetism. For this initial condition, we have: $J_2 = 0.01630$, $\alpha = 0.1169$ rad yr^{-1}, $L = -0.001875$, and $\gamma = 26.7°$. For a characteristic $e = 0.3$, we calculate $\dot{\varpi}_T = 0.6093$ rad yr^{-1} and $\dot{\omega}_{xy} = 5.329$ rad yr^{-1}.

pressure, and the electromagnetic force (Fig. 11). The eccentricity and solar-angle time-histories are similar to those of Fig. 16 for Deimos dust, but here the solar angle precesses rather than regresses which follows from the fact that $\dot{\varpi}_T$ is positive in (43). In addition, the $n_\odot$ term doesn't dominate Eq. (39) here as it did for Deimos; instead the other two terms are large and nearly equal in magnitude. This causes $\dot{\varpi}_T$ to be very sensitive to e. Using $e \sim 0.3$ as a representative value to calculate $\dot{\varpi}_T$, we predict the amplitude and period of the eccentricity oscillation for Fig. 19: $e_{max} \sim 0.38$ and $\tau \sim 10.3$ years, respectively.

The out-of-plane history of this E ring grain is quite different than that of the Deimos grain (Fig. 16). The most striking feature is that the argument of pericenter appears to try to lock to $\pm 90°$ at all times when $e \neq 0$ and $i \neq 0$. Although the inclination remains small ($i \lesssim 0.4°$), this behavior predicts a distinctive vertical structure for the ring. For most of the time, the argument of pericenter is locked to $-90°$, which means that the location of pericenter is as far below Saturn's equatorial plane as possible. So at times when $\omega = -90°$, the inner parts of the ring are depressed below the equatorial plane while the outer parts of the ring are elevated above the plane. At instants when $\omega = 90°$ the opposite occurs, although since the inclination is very small at those times, the effect is less pronounced.

The forced inclination solution discussed above does not apply to the dust grain of Fig. 19 because here we are interested in explaining changes that occur on timescales comparable to the eccentricity-oscillation period. Accordingly, we cannot average over one eccentricity-oscillation as we did in the previous section. Instead, we find an approximate analytic solution for pericenter-locking by assuming $\omega = \pm 90°$, solving for the other orbital elements, and showing that the solution is stable against small perturbations. Following Hamilton (1993), we find

$$\sin i_{eq} = \left| \frac{Z}{\dot{\omega}_{xy}} \right|, \tag{45}$$

$$\left| \frac{d\Omega}{dt} \right|_{eq} = \dot{\Omega}_{xy} + \dot{\omega}_{xy}, \tag{46}$$

and

$$\omega_{eq} = \frac{\pi}{2}\left(\mathrm{sign}\left(\frac{\dot{\omega}_{xy}}{Z}\right)\right). \tag{47}$$

Not surprisingly, the vertical structure of the E ring is governed by the term Z, which contains the effects of the vertical forces in this problem. Since $\dot{\omega}_{xy}$ is positive for the parameters of Fig. 19, Eq. (47) implies that the sign of ω_{eq} follows that of Z. The term Z is dominated by the negative contribution from the aligned magnetic quadrupole. As the Sun moves above and below the ring plane, its contribution to Z changes sign (Eq. (36)), and occasionally dominates the electromagnetic contribution. The inclination is not free; Eq. (45) shows that it is linearly proportional to e for small e, a feature that can be clearly seen in Fig. 19.

V. PUTTING IT TOGETHER

This chapter began by recalling the techniques used to observe planetary rings and then summarized our knowledge about the dust that circles the giant planets. These sections were followed by a review of the physical and dynamical processes that act on isolated circumplanetary grains, and a survey of some relevant celestial mechanics. Now we connect all of these topics by explaining briefly how the physics that we've discussed might lead to the observed character of three dust-rich systems: the Jovian rings, Saturn's E ring and the dust bands of Uranus and Neptune. As we will see, these three systems exemplify three different modes of origin: Jupiter's rings are likely to be debris from impacts into the moons by interplanetary meteoroids, while the E ring may be self-sustained as high-speed impacts of ring dust generate new material, and the dust in the Uranian/Neptunian rings may be merely the small-size tail of a continually evolving mix of satellites and ring material.

V.A. Jovian Rings

The several distinctive structures that are visible in Jupiter's ring system provide strong circumstantial clues as to how this system works. In particular, observations by the Galileo spacecraft show that Jupiter's thickened gossamer bands are precisely circumscribed by the orbits of the ring-moons Amalthea and Thebe (Fig. 3; Ockert-Bell et al. 1999; de Pater et al. 1999; Burns et al. 1999). Voyager images had earlier hinted that very tiny and equatorial Adrastea skims the flattened main ring's periphery and that the Jovian ring's halo arises at the 3:2 Lorentz resonance (Figs. 2, 13 and 15); higher-quality Galileo frames have now confirmed and refined these findings. The gossamer rings' unique morphology—especially the rectangular end profiles at the satellites' orbits with half-thicknesses that match the satellites' excursions above/below the planet's equator and the enhanced brightnesses along the top/bottom ring edges (see Fig. 3)—can be understood if all gossamer ring particles are ejecta lost from the satellites when high-speed meteoroids strike them (see section on Generation of Grains; Eq. (20)) and this debris then evolves inward toward Jupiter (Burns et al. 1999). Apparently this orbital drift is due to Poynting-Robertson drag (Eq. (16)), rather than caused by plasma drag (Eq. (17)) and/or resonant charge variations, because the rings appear uniform across synchronous orbit, where these latter drags would change sign (see Table 2). Following impacts into tiny satellites within the Roche limit, all impact ejecta will escape. Hence, despite equatorial Adrastea being a much smaller target than Thebe and Amalthea, which lie beyond the Roche limit and whose gravities are strong enough to retain substantial amounts of ejecta, the inner moonlet might be a more prolific supplier of collisional debris than the others. This suggests that the thin main ring is composed primarily of Adrastea's detritus (Burns et al. 1999). Once again, as a result of Poynting-Robertson drag, the orbits of this material decay inward, across the main ring where some of it will be absorbed by Metis (see the section on interactions with satellites); the remainder will eventually reach the 3:2 Lorentz resonance (Eq. (23); Schaffer and Burns 1992; Hamilton 1993; cf. Horányi and Cravens 1996; see Fig. 2) where its inclination will be raised substantially (see Fig. 15), thereby accounting for the ring's expansion into the vertically extended halo (Fig. 20). Since grains with large charge-to-mass will be most affected in such a manner, this scenario is supported by the fact that halo particles are smaller than those in the main ring (de Pater et al. 1999). By the time that particles drift in as far as the 2:1 resonance, the orbits of many of them will have become highly elongated and/or inclined, meaning that some are lost to the planet's atmosphere. Although many of the Jovian rings' unique features can now be understood, questions remain – e.g., the nature of the main ring's patchiness, the cause of the exterior gossamer material and whether the known ring-moons are the primary parent bodies. Some of these may be resolved by Cassini's observations during its flyby of Jupiter at the close of 2000.

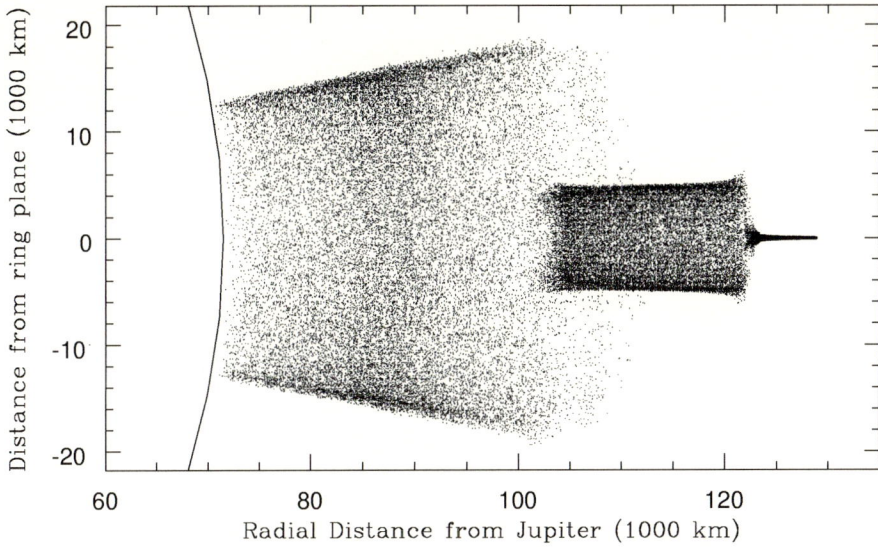

Figure 20. Orbital evolution of a 2 micron dust grain released from the small jovian satellite Metis ($a=128\,000$ km) on an initially circular uninclined orbit. The grain evolves under the combined forces of gravity, radiation pressure, the Lorentz force ($\phi=+5$ V, $L=0.002735$), and an artificially enhanced drag force. Using an enhanced drag force speeds up the numerical integrations without introducing spurious effects; a plot made with much slower Poynting-Robertson drag would look much the same as this one. The drag force pulls the dust grain in toward Jupiter across the strong 3:2 and 2:1 Lorentz resonances, where the inclination and eccentricities of the dust grain receive strong kicks (see Fig. 15). The thin main ring (between 123 000 km and 129 000 km) and the diffuse halo (between 100 000 km and 123 000 km) are clearly visible in spacecraft images (Fig. 2). The extremely diffuse "second halo" interior to the first has not yet been observed; it is a prediction of this model.

V.B. Saturn's E Ring

This diaphanous entity (Fig. 7) has two distinguishing features: first, nearly all of its particles are roughly the same size, about a micrometer in radius; and, second, the system's brightness peaks sharply at the satellite Enceladus' orbit, just where this vertically thickened ring is its narrowest. A dynamical simulation by Horányi et al. (1992) hinted at a possible connection between these facts, finding that micron-sized grains launched from Enceladus satisfy an orbital resonance condition. For a simple 2-D model (circular orbits, dipolar magnetic field and constant grain charge determined by the expected magnetospheric conditions) the precession of the orbit driven by Saturn's oblateness (Fig. 12) roughly matches—and thereby approximately cancels—the orbit's regression caused by the corotational electric field (Fig. 12); see also Eq. (35). Thus the $\phi_\odot$ term varies slowly, allowing radiation pressure to drive orbits to large eccentricities (Fig. 12, Eq. (28)); the variation in eccentricity is periodic

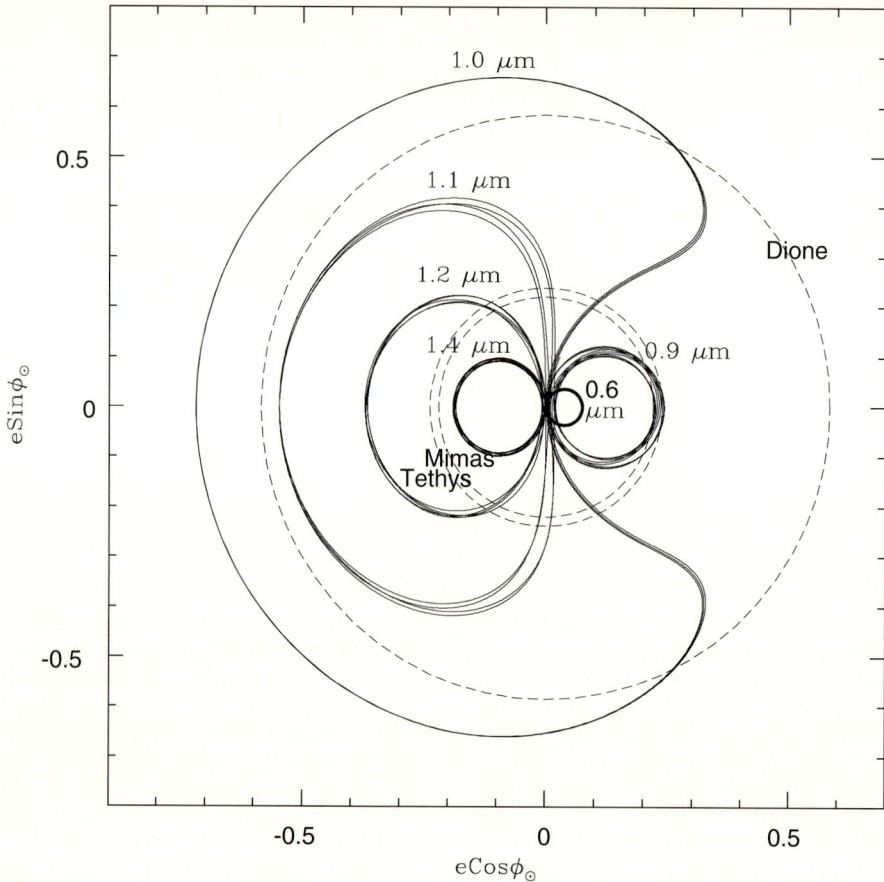

Figure 21. Evolution of the orbital eccentricity and solar angle for different-sized E-ring dust grains. Particles are launched from Enceladus, and assumed to be charged to a potential of -5 V. Note the large differences in orbital eccentricities for similar grain sizes (the radial distance of a point from (0,0) gives its eccentricity); this behavior is caused by a secular resonance that occurs when the sum of the three terms on the right-hand side of (39) is near zero. Eccentricities of 0.25 are sufficient for crossing orbits with the nearby satellites Mimas and Tethys; these eccentricities are attained by particles with sizes between 0.9 and 1.4 microns. Particles of 1 μm reach Dione. Large particles are offset away from the Sun (toward $\phi_\odot = 180°$) as was the case for dust from Deimos (see Figs. 16 and 17); the orbital histories of very large grains are nested inside the 1.4 μm curve. The pericenters of smaller grains regress due to strong electromagnetic forces (see Fig. 12), which causes the distribution of these grains to be offset toward the Sun ($\phi_\odot = 0$). The cluster of curves shown for a few of the particle sizes depict the results for different orbital inclinations. The regular appearance of the plot is due to an underlying integral of the motion (Hamilton and Krivov 1996).

on longer timescales (Figs. 19 and 21). Once the orbit achieves moderate eccentricities, particles travel along paths that cross the orbits of other Saturnian satellites, resulting in occasional collisions. Since these impacts occur typically at several km s^{-1}, they generate significant debris (see Eq. (20)); indeed enough may be kicked off for the E ring to be self-sustained at its current optical depth. With several reasonable assumptions, this model can be shown to naturally select Enceladus as the primary source of ring material and to favor micrometer-sized grains (Hamilton and Burns 1994). The prevalence of highly eccentric orbits and the consequent energetic collisions may explain the generation of the unusual amount of submicrometer dust in the neighboring F and G rings, the excess of OH molecules observed within the E ring (Hamilton and Burns 1993b) and the orbital brightness variations of nearby satellites (Buratti et al. 1998).

V.C. The Dust Bands of Uranus and Neptune

The Uranian system, with its dust sheet punctuated by bright strands and gaps (Fig. 8), and the Neptunian rings, with its unusually large fraction of dust (Fig. 9), are good testing grounds for ideas on the operation of dusty rings. As Fig. 1 shows, each of the ring systems has several embedded and nearby ring-moons. Because small moons near the giant planets are shattered by cometary impacts in cosmogonically short times (Smith et al. 1989) and because the rings themselves seem to have lifetimes much less than the solar system's age (Goldreich and Tremaine 1982; Esposito et al. 1991; Porco et al. 1995), one must ask whether the intimate mix of small satellites and dusty rings that we observe today is being continually destroyed but then recreated. Cuzzi and Burns (1988) developed such a scenario to generate hypothetical debris clouds, which they claim populate the region interior to Saturn's F ring, through the sloughing off of regoliths when moonlets collide.

The most detailed collisional histories of the small Uranian and Neptunian satellites and their associated dusty ring systems have been carried out by Colwell and Esposito (1990a, b, 1992, 1993; Esposito et al. 1991; Porco et al. 1995; Canup and Esposito 1995). These authors use both Monte Carlo and Markov approaches to follow the process of satellite disruption from an initial distribution through subsequent fragmentations. The Colorado team has modeled the creation of dust in these ring systems from macroscopic bodies suffering mutual collisions and being impacted by extrinsic meteoroids. Their models include the loss of dust due to sweep-up onto the parent bodies as well as drag (exospheric and Poynting-Robertson). Using believable parameter choices, their simulations reproduce the dust contents of the Uranian and Neptunian rings as well as the impact rates at ring-plane crossings. Snapshots giving typical radial profiles of the dust optical depth are generated by combining steady-state calculations of dust within the rings with a Markov chain model for transport between the rings. These models demonstrate that vigorous interparticle collisions are an essential ingredient to produce the high dust content, especially in light of the rapid evolution caused by the Uranian exosphere. Similar mechanisms seem likely to dominate Saturn's F ring and Encke ringlet. These results imply that all dusty ring systems must continually change their appearance.

VI. EXPECTED ADVANCES

Our knowledge about circumplanetary dust and its dynamical attributes has expanded remarkably in the last two decades. Space missions, notably Voyager to the outer solar system, first brought dusty planetary rings to the attention of scientists. But many Voyager observations of dust (imaging and plasma detections) occurred by chance, when the imaging field-of-view caught part of the "blank" background in the planet's neighborhood or when "noise" in the plasma-wave/radio data was especially prevalent as ring planes were pierced. A much more systematic survey of Jupiter's environs was planned for Galileo to carry out but it has been severely hampered by the spacecraft's damaged antenna.

The next opportunity for significant advances in our knowledge of circumplanetary dust will be in late December 2000 during Cassini's flyby of Jupiter. This mission, which carries an excellent complement of instruments (ultraviolet through infrared imaging, a sophisticated dust detector and magnetospheric imaging), is expected to unravel much about magnetospheric interactions, the nature of the Jovian ring particles and whether these diaphanous rings show any time-variability. On Cassini's arrival at Saturn, a system that displays the widest (both geometrically and phenomenologically) array of circumplanetary dust, its observations should clarify the symbiosis believed to exist between circumplanetary dust and the other inhabitants of the magnetosphere: plasma, satellites and main rings.

Perhaps the greatest progress in the next decade will be made through terrestrial laboratory and theoretical studies on the behavior of dusty plasmas. This may be driven by an entirely different motivation: the increasing commercial interest in charged dust, owing to its role in the manufacture of microchips. Insofar as theoretical studies of circumplanetary dust, fresh numerical schemes, often incorporating non-gravitational forces into symplectic orbital integrations, are starting to be developed (Hamilton et al. 1999) and will prove useful in following the complex physics of circumplanetary dust. These improvements will be necessary to reap the harvest of puzzling data likely to be revealed during the Cassini years.

Acknowledgements

Over the years, as the field of circumplanetary dust was born and developed, we have benefited from many collaborations, notably with Mihály Horányi, François Mignard, Les Schaffer, Philippe Lamy, James Pollack, Steven Soter, Eberhard Grün, Philip Nicholson and Imke de Pater, whom we thank for their insights and encouragement. We recognize the editors for their forbearance with the lateness of this manuscript. We appreciate careful reviews by Joshua Colwell, Amara Graps and an anonymous party. Our research has been supported by NASA.

REFERENCES

Acuña, M. H., and Ness, N. F. 1976. The complex main magnetic field of Jupiter. *J. Geophys. Res.*, **81**, pp. 2917–2922.

Aubier, M. G., Meyer-Vernet, N., and Pedersen, B. M. 1983. Shot noise from grain and particle impacts in Saturn's ring plane. *Geophys. Res. Lett.*, **10**, pp. 5–8.

Baguhl, M., Grün, E., Linkert, G. Linkert, D., and Siddique, N. 1993. Identification of "small" dust impacts in the Ulysses dust detector data. *Planet. Space Sci.*, **41**, pp. 1085–1098.

Baguhl, M., Grün, E., Hamilton, D. P., Linkert, G., Riemann, R., Staubach, P., and Zook, H. A. 1994. Galileo dust. *Space Sci. Rev.*, **72**, pp. 471–476.

Baguhl, M., Hamilton, D. P., Grün, E., Dermott, S. F., Fechtig, H., Hanner, M. S., Kissel, J., Lindblad, B.-A., Linkert, D., Linkert, G., Mann, I., McDonnell, J. A. M., Morfill, G. E., Polanskey, C., Riemann, R., Schwehm, G., Staubach, P., and Zook, H. A. 1995. Dust measurements at high ecliptic latitudes. *Science*, **268**, pp. 1016–1019.

Banaszkiewicz, M., and Krivov, A. V. 1997. Hyperion as a dust source in the Saturn system. *Icarus*, **129**, pp. 289–303.

Bauer, J., Lissauer, J. J., and Simon, M. 1997. Edge-on observations of Saturn's E and G Rings in the near-IR. *Icarus*, **125**, pp. 440–445.

Bliokh, P. V., Sinitsin, V., and Yaroshenko, V. 1995. Dusty and Self-Gravitational Plasmas in Space. (Dordrecht, Boston).

Bosh, A. S., and Rivkin, A. S. 1996. Observations of Saturn's inner satellites during the May 1995 ring-plane crossing. *Science*, **272**, pp. 518–521.

Bosh, A. S., Olkin, C. B., French, R. G., and Nicholson, P. D. 2001. Saturn's F ring: Kinematics and particle sizes from stellar occultation studies. *Icarus*, submitted.

Broadfoot, A. L., Herbert, F., Holberg, J. B., Hunten, D. M., Kumar, S., B., Sandel, R., Shemansky, D. E., Smith, G. R., Yelle, R. V., Strobel, D. F., Moos, H. W., Donahue, T. M., Atreya, S. K., Bertaux, J. L., Blamont, J. E., McConnell, J. C., Dessler, A. J., Linick, S., and Springer, R. 1986. Ultraviolet Spectrometer observations of Uranus. *Science*, **233**, pp. 74–79.

Broadfoot, A. L., Atreya, S. K., Bertaux, J. L., Blamont, J. E., Dessler, A. J., Donahue, T. M., Forrester, W. T., Hall, D. T., Herbert, F., Holberg, J. B., Hunten, D. M., Krasnopolsky, V. A., Linick, S., Lunine, J. I., McConnell, J. C., Moos, H. W., Sandel, B. R., Schneider, N. M., Shemansky, D. E., Smith, G. R., Strobel, D. F., and Yelle, R. V. 1989. Ultraviolet Spectrometer observations of Neptune and Triton. *Science*, **246**, pp. 1459–1466.

Buratti, B. J., Mosher, J. A., Nicholson, P. D., McGhee C. A., and French, R. G. 1998. Near-infrared photometry of the Saturnian satellites during ring plane crossing. *Icarus*, **136**, pp. 223–231.

Burns, J. A. 1976. An elementary derivation of the perturbation equations of celestial mechanics. *Am. Jnl. Phys.*, **44**, pp. 944–949 (Erratum: 45, 1230).

Burns, J. A. 1991. Physical processes on circumplanetary dust. In *The Origin and Evolution of Interplanetary Dust*, eds. A.-C. Levasseur-Regourd and H. Hasegawa (Dordrecht: Kluwer Academic Publisher), pp. 341–348.

Burns, J. A. 1999. Planetary rings. In *The New Solar System*, 4th ed., eds. J. K. Beatty, C. C. Petersen and A. Chaikin (Cambridge MA: Sky Publishing), pp. 221–240.

Burns, J. A., and Gladman, B. J. 1998. Dynamically depleted zones near Saturn for Cassini's safe passage. *Planet. Space Sci.*, **46**, pp. 1401–1407.

Burns, J. A., and Schaffer, L. 1989. Orbital evolution of circumplanetary dust by resonant charge variations. *Nature*, **337**, pp. 340–343.

Burns, J. A., Lamy, P. L., and Soter, S. 1979. Radiation forces on small particles in the solar system. *Icarus*, **40**, pp. 1–48.

Burns, J. A., Showalter, M. R., Cuzzi, J. N., and Pollack, J. B. 1980. Physical processes in Jupiter's ring: Clues for an origin by Jove! *Icarus*, **44**, pp. 339–360.

Burns, J. A., Cuzzi, J. N., and Showalter, M. R. 1983. Discovery of gossamer rings. *Bull. Amer. Astron. Soc.*, **15**, pp. 1013–1014.

Burns, J. A., Showalter, M. R., and Morfill, G. 1984. The ethereal rings of Jupiter and Saturn. In *Planetary Rings*, eds. R. Greenberg and A. Brahic (Tucson: University of

Arizona Press), pp. 200-272.

Burns, J. A., Schaffer, L., Greenberg, R. J., and Showalter, M. R. 1985. Lorentz resonances and the structure of Jupiter's rings. *Nature*, **316**, pp. 115–119

Burns, J. A., Hamilton, D. P., Mignard, F., and Soter, S. 1996. The contamination of Iapetus by Phoebe dust. In *Physics, Chemistry and Dynamics of Interplanetary Dust*, eds. B. Å. S. Gustafson and M. S. Hanner (Dordrecht: Kluwer), pp. 179–182.

Burns, J. A., Showalter, M. R., Hamilton, D. P., Nicholson, P. D., de Pater, I., Ockert-Bell, M., and Thomas, P. C. 1999. The formation of Jupiter's faint rings. *Science*, **284**, pp. 1146–1150.

Canup, R. M., and Esposito, L. W. 1995. Accretion in the Roche zone. Coexistence of rings and ring moons. *Icarus*, **113**, pp. 331–352.

Canup, R. M., and Esposito, L. W. 1997. Evolution of the G ring and the population of macroscopic ring particles. *Icarus*, **126**, pp. 28–41.

Chamberlain, J. W. 1979. Depletion of satellite atoms in a collisionless exosphere by radiation pressure. *Icarus*, **39**, pp. 286–294.

Cheng, A. F., Haff, P. K., Johnson, R. E., and Lanzerotti, L. J. 1986. Interactions of planetary magnetospheres with icy satellite surfaces. In *Planetary Satellites*, eds. J. A. Burns and M. S. Matthews (Tucson: University of Arizona Press), pp. 403–430.

Colwell, J. E., and Esposito, L. W. 1990a. A numerical model of the Uranian dust rings. *Icarus*, **86**, pp. 530–560.

Colwell, J. E., and Esposito, L. W. 1990b. A model of dust production in the Neptune ring system. *GRL.*, **17**, pp. 1741–1744.

Colwell, J. E., and Esposito, L. W. 1992. Origin of the rings of Uranus and Neptune. 1. Statistics of satellite disruptions. *JGR*, **97**, pp. 10,227–10,241.

Colwell, J. E., and Esposito, L. W. 1993. Origin of the rings of Uranus and Neptune. 2. Initial distribution of disrupted satellite fragments. *JGR*, **98**, pp. 7387–7401.

Colwell, J. E., and Horányi, M. 1996. Magnetospheric effects on micrometeoroid fluxes. *JGR-Planets*, **101**, pp. 2169–2175.

Colwell, J. E., and 12 colleagues 1990. Voyager photopolarimeter observations of Uranian ring occultations. *Icarus*, **83**, pp. 102–125.

Colwell, J. E., Horányi, M., and Grün, E. 1998. Capture of interplanetary and interstellar dust by the Jovian magnetosphere. *Science*, **280**, pp. 88–91.

Connerney, J. E. P. 1993. Magnetic fields of the outer planets. *J. Geophys. Res.*, **98**, pp. 18,659–18,679.

Connerney, J. E. P., Acuna, M. H., and Ness, N. F. 1996. Octupole model of Jupiter's magnetic field from Ulysses observations. *J. Geophys. Res.*, **101**, pp. 27,453–27,458.

Cooke, M. L. 1991. Saturn's Rings: Radial Variation in the Keeler Gap and C Ring Photometry. Ph. D. dissertation (Cornell University), xii + 206 pp.

Cuzzi, J. N. 1985. Rings of Uranus: Not so thick, not so black. *Icarus*, **63**, pp. 312–316.

Cuzzi, J. N., and Burns, J. A. 1988. Charged particle depletion surrounding Saturn's F Ring: Evidence for a moonlet belt? *Icarus*, **74**, pp. 284–324.

Cuzzi, J. N., and Estrada, P. R. 1998. Compositional evolution of Saturn's rings due to meteoroid bombardment. *Icarus*, **132**, pp. 1–35.

Cuzzi, J. N., and Rappaport, N. 1996. Report to Cassini Project on Possible Ring Hazard. JPL internal document.

Cuzzi, J. N., Lissauer, J. J., Esposito, L. W., Holberg, J. B., Marouf, E. A., Tyler, G. L., and Boischot, A. 1984. Saturn's rings: Properties and processes. In *Planetary Rings*, eds. R. J. Greenberg and A. Brahic (Tucson: University of Arizona Press), pp. 73–199.

Danby, J. M. A. 1988. *Fundamentals of Celestial Mechanics*, (2nd ed.), (Richmond, VA: Willmann-Bell).

de Pater, I., Showalter, M. R., Lissauer, J. J., and Graham, J. R. 1996. Keck infrared observations of Saturn's E and G Rings during Earth's 1995 ring plane crossings. *Icarus*, **121**, pp. 195–198.

de Pater, I., Showalter, M. R., Burns, J. A., Nicholson, P. D., Liu, M., Hamilton, D. P., and Graham, J. R. 1999. Keck infrared observations of Jupiter's ring system near Earth's 1997 ring-plane crossing. *Icarus*, **138**, pp. 214–223.

Dermott, S. F. 1981. The braided F ring of Saturn. *Nature*, **290**, pp. 454–457.

Dohnanyi, J. S. 1972. Interplanetary objects in review: Statistics of their masses and dynamics. *Icarus*, **17**, pp. 1–48.

Dones, L., Cuzzi, J. N., and Showalter, M. R. 1993. Voyager photometry of Saturn's A Ring. *Icarus*, **105**, pp. 184–215.
Doyle, L. R., and Grün, E. 1990. Radiative transfer modeling constraints on the size of Saturn's spoke particles. *Icarus*, **85**, pp. 168–190.
Doyle, L. R., Dones, L., and Cuzzi, J. N. 1989. Radiative transfer modeling of Saturn's outer B Ring. *Icarus*, **80**, pp. 104–135.
Dubinin, E. M., Lundin, R., Pissarenko, N. F., Barabash, S. V., Zakaharov, A. V., Koskinen, H., Schwingenshuh, K., and Yeroshenko, Ye. G. 1990. Indirect evidence for a dust/gas torus along the Phobos orbit. *GRL*, **17**, pp. 861–864.
Dumas, C., Terrile, R. J., Smith, B. A., Schneider, G., and Becklin, E. E. 1999. Stability of Neptune's ring arcs in question. *Nature*, **400**, pp. 733–735.
Durda, D. D., and Dermott, S. F. 1997. The collisional evolution of the asteroid belt and its contribution to the zodiacal cloud. *Icarus*, **130**, pp. 140–164.
Elliot, J., and Kerr, R. 1985. *Rings: Discoveries from Galileo to Voyager.* (Cambridge, Mass.: MIT Press).
Elliot, J. L., Dunham, E. W., and Mink, D. J. 1977. The rings of Uranus. *Nature*, **267**, pp. 328–330.
Elliot, J. L., Bosh, A. S., Cooke, M. L., Bless, R. C., Nelson, M. J., Percival, J. W., Taylor, M. J., Dolan, J. F., Robinson, E. L., and van Citters, G. W. 1993. An occultation of Saturn's rings on 1991 October 2–3 observed with the Hubble Space Telescope. *Astron. J.*, **106**, pp. 2544–2572, and p. 2598.
Eplee, R. E., Jr., and Smith, B. A. 1984. Spokes in Saturn's rings: Dynamical and reflectance properties. *Icarus*, **59**, pp. 188–198.
Eplee, R. E., Jr., and Smith, B. A. 1985. Radial growth of an extended spoke in Saturn's B Ring. *Icarus*, **63**, pp. 304–311.
Esposito, L. W., Cuzzi, J. N., Holberg, J. B., Marouf, E. A., Tyler, G. L. and Porco, C. 1984. Saturn's rings: Structure, dynamics and particle properties. In Saturn, eds. T. Gehrels and M. S. Matthews (Tucson: University of Arizona Press), pp. 463–545.
Esposito, L. W., Brahic, A., Burns, J. A., and Marouf, E. A. 1991. Particle properties and processes in Uranus' rings. In *Uranus*, eds. J. T. Bergstralh, E. D. Miner, and M. S. Matthews (Tucson: University of Arizona Press), pp. 410–465.
Farinella, P., Gonczi, R., Froeschlé, Ch., and Froeschlé, C. 1993. The injection of asteroid fragments into resonances. *Icarus*, **101**, pp. 174–187.
Feibelman, W. A. 1967. Concerning the "D" Ring of Saturn. *Nature*, **214**, pp. 793–794.
Ferrari, C., and Brahic, A. 1994. Azimuthal brightness asymmetries in planetary rings I. Neptune's arcs and narrow rings. *Icarus*, **111**, pp. 193–210.
Fillius, R. W., McIlwain, C. E., and Mogro-Campero, A. 1975. Radiation belts of Jupiter: A second look. *Science*, **188**, pp. 465–467.
Fischer, H. M., Pehlke, E., Wibberenz, G., Lanzerotti, L. J., and Mihalov, J. D. 1996. High-energy charged particles in the innermost jovian magnetosphere. *Science*, **272**, pp. 856–858.
French, R. G., Nicholson, P. D., Porco, C. C., and Marouf, E. A. 1991. Dynamics and structure of the Uranian rings. In *Uranus*, eds. J. T. Bergstralh, E. D. Miner, and M. S. Matthews (Tucson: University of Arizona Press), pp. 327–409.
French, R. G., Nicholson, P. D., Cooke, M. L., Elliot, J. L., Matthews, K., Perkovic, O., Tollestrup, E., Harvey, P., Chanover, N. J., Clark, M. A., Dunham, E. W., Forrest, W., Harrington, J., Pipher, J., Brahic, A., Grenier, I., Roques, F., and Arndt, M. 1993. Geometry of the Saturn system from the 3 July 1989 occultation of 28 Sgr and Voyager observations. *Icarus*, **103**, pp. 163–214.
French, R. G., Roques, F., Nicholson, P. D., Mc Ghee, C. A., Bouchet, P., Maene, S. A., Mason, E. C., Matthews, K., and Mosqueira, I. 1996. Earth-based detection of Uranus' lambda ring. *Icarus*, **119**, pp. 269–284.
French, R. G., Cuzzi, J., Danos, R., Dones, L., and Lissauer, J. 1998. Hubble Space Telescope observations of spokes in Saturn's rings. Abstract from the International Symposium: "The Jovian system after Galileo. The Saturnian system before Cassini-Huygens", Nantes, France, 11-15 May 1998. p. 36.
French, R. G., McGhee, C. A., Nicholson, P. D., Dones, L., and Lissauer, J. 1999. Saturn's wayward shepherds: Pandora and Prometheus. *BAAS*, **31**, p. 1228
Fujiwara, A., Cerroni, P., Davis, D., Ryan, E., diMartino, M., Holsapple, K., and Housen,

K. 1989. Experiments and scaling laws for catastrophic collisions. In *Asteroids II*, eds. R. P. Binzel, T. Gehrels and M. S. Matthews (Univ. Arizona Press), pp. 240–265.

Gehrels, T., Baker, R. L., Beshore, E., Blenman, C., Burke, J. J., Castillo, N. D., DaCosta, B., Degewij, J., Doose, L. R., Fountain, J. W., Gotobed, J., KenKnight, C. E., Kingston, R., McLaughlin, G., McMillan, R., Murphy, R., Smith, P. H., Stoll, C. P., Strickland, R. N., Tomasko, M. G., Wijesinghe, M. P., Coffeen, D. L., and Esposito, L. 1980. Imaging Photopolarimeter on Pioneer Saturn. *Science*, **207**, pp. 434–439.

Goertz, C. K. 1989. Dusty plasmas in the solar system. *Rev. Geophys.*, **27**, pp. 271–292.

Goertz, C. K., and Morfill, G. E. 1983. A model for the formation of spokes in Saturn's ring. *Icarus*, **53**, pp. 219–229.

Goertz, C. K., and Morfill, G. E. 1988. A new instability of Saturn's ring. *Icarus*, **74**, pp. 325–330.

Goertz, C. K., Morfill, G. E., Ip, W.-H., Grün, E., and Havnes, O. 1986. Electromagnetic angular momentum transport in Saturn's ring. *Nature*, **320**, pp. 141–143.

Goertz, C. K., Shan, L., and Havnes, O. 1988. Electrostatic forces in planetary rings. *Geophys. Res. Ltrs.*, **15**, pp. 84–87.

Goldreich, P., and Tremaine, S. 1979. Towards a theory for the Uranian rings. *Nature*, **277**, pp. 97–99.

Goldreich, P., and Tremaine, S. 1982. The dynamics of planetary rings. *Ann. Rev. Astron. Astrophys.*, **20**, pp. 249–283.

Gradie, J., Thomas, P., and Veverka, J. 1980. The surface composition of Amalthea. *Icarus*, **44**, pp. 373–387.

Graps, A. L., and Lane, A. L. 1986. Voyager 2 photopolarimeter experiment: Evidence for tenuous outer ring material at Saturn. *Icarus*, **67**, pp. 205–210.

Graps, A. L., Lane, A. L., Horn, L. J., and Simmons, K. E. 1984. Evidence for material between Saturn's A and F Rings from the Voyager 2 photopolarimeter experiment. *Icarus*, **60**, pp. 409–415.

Graps, A. L., Grün, E., Svedhem, H., Krüger, H., Horányi, M., Heck, A., and Lammers, S. 2000. Io as a source of the jovian dust streams. *Nature*, **405**, pp. 48–50.

Grün, E., Morfill, G. E., Terrile, R. J., Johnson, T. V., and Schwehm, G. 1983. The evolution of spokes in Saturn's B Ring. *Icarus*, **54**, pp. 227–252.

Grün, E., Morfill, G. E., and Mendis, D. A. 1984. Dust-magnetosphere interactions. In *Planetary Rings*, eds. R. J. Greenberg and A. Brahic (Tucson: University of Arizona Press), pp. 275–332.

Grün, E., Goertz, C. K., Morfill G. E., and Havnes, O. 1992a. Statistics of Saturn's spokes. *Icarus*, **99**, pp. 191–201.

Grün, E., Baguhl, M., Fechtig, H., Hanner, M. S., Kissel, J., Lindblad, B. A., Linkert, D., Linkert, G., Mann, I. B., McDonnell, J. A. M., Morfill, G. E., Polanskey, C., Riemann, R., Schwehm, G., Siddique, N., and Zook, H. A. 1992b. Galileo and Ulysses dust measurements: From Venus to Jupiter. *Geophys. Res. Let.*, **19**, pp. 1311–1314.

Grün, E., Fechtig, H., Hanner, M. S., Kissel, J., Lindblad, B.-A., Linkert, D., Maas, D., Morfill, G. E., and Zook, H. A. 1992c. The Galileo dust detector. *Space Sci. Rev.*, **60**, pp. 317–340.

Grün, E., Zook, H. A., Baguhl, M., Balogh, A., Bame, S. J., Fechtig, H., Forsyth, R., Hanner, M. S., Horányi, M., Kissel, J., Lindblad, B.-A., Linkert, D., Linkert, G., Mann, I., McDonnell, J. A. M., Morfill, G. E., Phillips, J. L., Polanskey, C., Schwehm, G., Siddique, N., Staubach, P., Svestka, J., and Taylor, A. 1993. Discovery of jovian dust streams and interstellar grains by the Ulysses spacecraft. *Nature*, **362**, pp. 428–430.

Grün, E., Baguhl, M., Hamilton, D. P., Riemann, R., Zook, H. A., Dermott, S. F., Fechtig, H., Gustafson, B. A., Hanner, M. S., Horányi, M., Khurana, K., Kissel, J., Kivelson, M., Lindblad, B.-A., Linkert, D., Linkert, G., Mann, I., McDonnell, J. A. M., Morfill, G. E., Polanskey, C., Schwehm, G., and Srama, R. 1996a. Constraints from Galileo observations on the origin of jovian dust streams. *Nature*, **381**, pp. 395–398.

Grün, E., Hamilton, D. P., Riemann, R., Dermott, S. F., Fechtig, H., Gustafson, B. A., Hanner, M. S., Heck, A., Horányi, M., Kissel, J., Krüger, H., Lindblad, B.-A., Linkert, D., Linkert, G., Mann, I., McDonnell, J. A. M., Morfill, G. E., Polanskey, C., Schwehm, G., Srama, R., and Zook, H. A. 1996b. Dust measurements during Galileo's approach to Jupiter and Io encounter. *Science*, **274**, pp. 399–401.

Grün, E., Krüger, H., Dermott, S., Fechtig, H., Graps, A. L., Gustafson, B. A., Hamilton,

D. P., Hanner, M. S., Heck, A., Horányi, M., Kissel, J., Lindblad, B.A., Linkert, D., Linkert, G., Mann, I., McDonnell, J. A. M., Morfill, G. E., Polanskey, C., Schwehm, G., Srama, R., and Zook, H. A. 1997. Dust measurements in the Jovian magnetosphere. *GRL*, **24**, pp. 2171–2174.

Grün, E., Krüger, H., Graps, A. L., Hamilton, D. P., Heck, A., Linkert, G., Zook, H. A., Dermott, S., Fechtig, H., Gustafson, B. A., Hanner, M. S., Horányi, M., Kissel, J., Lindblad, B. A., Linkert, D., Mann, I., McDonnell, J. A. M., Morfill, G. E., Polanskey, C., Schwehm G., and Srama, R. 1998. Galileo observes electromagnetically coupled dust in the jovian magnetosphere. *JGR*, **103**, pp. 20011–20022.

Guérin, P. 1973. Les anneaux de Saturne en 1969. Etude morphologique et photométrique I. Obtention et dépouillement des photographies. *Icarus*, **19**, pp. 202–211.

Gurnett, D. A., Grün, E., Gallagher, D., Kurth, W. S., and Scarf, F. L. 1983. Micron-sized particles detected near Saturn by the Voyager Plasma Wave instrument. *Icarus*, **53**, pp. 236–254.

Gurnett, D. A., Kurth, W. S., Scarf, F. L., Burns, J. A., Cuzzi, J. N., and Grün, E. 1987. Micron-sized particle impacts detected near Uranus by the Voyager 2 Plasma Wave instrument. *J. Geophys. Res.*, **92**, pp. 14,959–14,968.

Gurnett, D. A., Kurth, W. S., Granroth, L. J., Allendorf, S. C., and Poynter, R. L. 1991. Micron-sized particles detected near Neptune by the Voyager 2 plasma wave instrument. *J. Geophys. Res.*, **96**, pp. 19,177–19,186.

Gustafson, B. A. S. 1994. Physics of zodiacal dust. *Ann. Rev. Earth Planet. Sci.*, **22**, pp. 553–595.

Hamilton, D. P. 1993. Motion of dust in a planetary magnetosphere: Orbit-averaged equations for oblateness, electromagnetic, and radiation forces with application to Saturn's E ring. *Icarus*, **101**, pp. 244–264. Erratum: *Icarus*, **103**, p. 161.

Hamilton, D. P. 1994. A comparison of Lorentz, planetary gravitational, and satellite gravitational resonances. *Icarus*, **109**, pp. 221–240.

Hamilton, D. P. 1996. The asymmetric time-variable rings of Mars. *Icarus*, **119**, pp. 153–172.

Hamilton, D. P., and Burns, J. A. 1993a. The ejection of dust from Jupiter's gossamer ring. *Nature*, **364**, pp. 695–699.

Hamilton, D. P., and Burns, J. A. 1993b. OH from Saturn's rings. *Nature*, **365**, p. 498.

Hamilton, D. P., and Burns, J. A. 1994. The origin of Saturn's E ring: Self-sustained, naturally. *Science*, **264**, pp. 550–553.

Hamilton, D. P., and Krivov, A. V. 1996. Circumplanetary dust dynamics: Effects of solar gravity, radiation pressure, planetary oblateness, and electromagnetism. *Icarus*, **123**, pp. 503–523.

Hamilton, D. P., and Krivov, A. V. 1997. Dynamics of distant moons of asteroids. *Icarus*, **128**, pp. 141–149.

Hamilton, D. P., Rauch, K., and Burns, J. A. 1999. Electromagnetic resonances in Jupiter's rings. *BAAS*, **31**, p. 1223.

Hänninen, J. 1993. Numerical simulations of moon-ringlet interaction. *Icarus*, **103**, pp. 104–123.

Hansen, J. E. 1969. Radiative transfer by doubling very thin layers. *Astrophys. J.*, **155**, pp. 565–573.

Harrington, J., Cooke, M. L., Forrest, W. J., Pipher, J. L., Dunham, E. W., and Elliot, J. L. 1993. IRTF observations of the occultation of 28 Sgr by Saturn. *Icarus*, **103**, pp. 235–252.

Hartmann, W. K. 1969. Terrestrial, lunar and interplanetary rock fragmentation. *Icarus*, **10**, pp. 201–213.

Hartquist, T. N., Havnes, O., and Morfill, G. E. 1992. The effects of dust on the dynamics of astronomical and space plasmas. *Fund. Cosmic Physics*, **15**, pp. 107–142.

Havnes, O., Morfill, G. E., and Meland, F. 1992. Effects of electromagnetic and plasma drag forces on the orbit evolution of dust in planetary magnetospheres. *Icarus*, **98**, pp. 141–150.

Herbert, F., Sandel, B. R., Yelle, R. V. 1987. The upper atmosphere of Uranus- EUV Occultations observed by Voyager 2. *JGR*, **92**, pp. 15093–15109.

Hood, L. L. 1989. Investigation of the Saturn dust environment from the analysis of energetic charged particle measurements. JPL PD 699-11, Vol. XIII.

Horányi, M. 1994. New Jovian ring. *GRL*, **21**, pp. 1039–1042.

Horányi, M. 1996. Charged dust dynamics in the solar system. *Ann. Rev. Astron. Astrophys.*, **34**, pp. 383–418.
Horányi, M., and Burns, J. A. 1991. Charged dust dynamics: Orbital resonance due to planetary shadows. *J. Geophys. Res.*, **96**, pp. 19,283–19,289.
Horányi, M., and Cravens, T. E. 1996. Structure and dynamics of Jupiter's ring. *Nature*, **381**, pp. 293–295.
Horányi, M., and Porco, C. C. 1993. Where exactly are the arcs of Neptune? *Icarus*, **106**, pp. 525–535.
Horányi, M., Burns, J. A., Tátrallyay, M., and Luhmann, J. G. 1990. On the fate of dust lost from the Martian satellites. *GRL*, **17**, pp. 853–856.
Horányi, M., Tátrallyay, M., Juhász, A., and Luhmann, J. G. 1991. The dynamics of submicron dust lost from Phobos. *Jnl. Geophys. Res.*, **96**, pp. 11,283–11,290.
Horányi, M., Burns, J. A., and Hamilton, D. P. 1992. The dynamics of Saturn's E ring particles. *Icarus*, **97**, pp. 248–259.
Horányi, M., Morfill, G., and Grün, E. 1993. Mechanism for the acceleration and ejection of dust grains from Jupiter's magnetosphere. *Nature*, **363**, pp. 144–146.
Horányi, M., Grün, E., and Heck, A. 1997. Modeling the Galileo dust measurements at Jupiter. *GRL*, **24**, pp. 2175–2178.
Horn, L. J., Hui, J., Lane, A. L., and Colwell, J. E. 1990. Observations of Neptunian rings by the Voyager photopolarimeter experiment. *GRL.*, **17**, pp. 1745–1748.
Housen, K. R. and Holsapple, K. A. 1990. On the fragmentation of asteroids and planetary satellites. *Icarus*, **84**, 226–253.
Hubbard, W. B., Brahic, A., Sicardy, B., Elicer, L.-R., Roques, F., and Vilas, F. 1986. Occultation detection of a Neptunian ring-like arc. *Nature*, **319**, pp. 636–640.
Hubbard, W. B., Porco, C. C., Hunten, D. M., Rieke, G. H., Rieke, M. J., McCarthy, D. W., Haemmerle, V., Clark, R., Turtle, E. P., Haller, J., McLeod, B., Lebofsky, L. A., Marcialis, R., Holberg, J. B., Landau, R., Carrasco, L., Elias, J., Buie, M. W., Persson, S. E., Boroson, T., West, S., and Mink, D. J. 1993. The occultation of 28 Sgr by Saturn: Saturn pole position and astrometry. *Icarus*, **103**, pp. 215–234.
Humes D. H. 1976. The Jovian meteoroid environment. In *Jupiter*, ed. T. Gehrels (Tucson: Univ. Arizona Press), pp. 1052–1067.
Humes, D. H. 1980. Results of Pioneer 10 and 11 meteoroid experiments: Interplanetary and near-Saturn. *J. Geophys. Res.*, **85**, pp. 5841–5852.
Humes, D. H., Alvarez, J. M., O'Neal, R. L., and Kinard, W. H. 1974. The interplanetary and near-Jupiter meteoroid environments. *J. Geophys. Res.*, **79**, pp. 3677–3684.
Ip, W.-H. 1995a. The exospheric system of Saturn's rings. *Icarus*, **115**, pp. 295–303.
Ip, W.-H. 1995b. Implications of meteoroid-ring interaction for observations of the 1995 Saturn ring-plane crossing. *Icarus*, **117**, pp. 212–215.
Ip, W.-H., and Banaszkiewicz, M. 1990. On the dust-gas tori of Phobos and Deimos. *Geophys. Res. Ltrs.*, **17**, pp. 857–860.
Ishimoto, H. 1996. Formation of Phobos/Deimos dust rings. *Icarus*, **122**, pp. 153–165.
Johnson, R. E. 1990. *Energetic Charged-Particle Interactions with Atmospheres and Surfaces.* (New York: Springer-Verlag), 232 pp.
Johnson, R. E., Pospieszalska, M. K., Sitter, E. G., Cheng, A. F., Lanzerotti, L. J., and Sieveka, E. M. 1989. The neutral cloud and heavy ion inner torus at Saturn. *Icarus*, **77**, pp. 311–329.
Johnson, R. E., Grosjean, D. E., Jurac, S., and Baragiola, R. A. 1993. Sputtering, still the dominant source of plasma at Dione? *EOS*, **74**, p. 569, pp. 572–73.
Juhász, A., and Horányi, M. 1995. Dust torus around Mars. *Jnl. Geophys. Res.*, **100**, pp. 3277–3284.
Juhász, A., Tátrallyay, M., Gévai, G., and Horányi, M. 1993. On the density of the dust halo around Mars. *Jnl. Geophys. Res.*, **98**, pp. 1205–1211.
Jurac, S., Baragiola, A., Johnson, R. E., and Sittler Jr., E. C. 1995. Charging of ice grains by low-energy plasmas-application to Saturn's E ring. *Jnl. Geophys. Res.*, **100**, pp. 14,821–14,835.
Jurac, S., Johnson, R., and Donn, B. 1998. Monte Carlo calculations of the sputtering of grains: Enhanced sputtering of small grains. *Ap. J.*, **503**, pp. 247–252.
Kaula, W. M. 1966. *Theory of Satellite Geodesy.* (Waltham, MA: Blaisdell Publishing Co.).
Kolvoord, R. A., Burns, J. A., and Showalter, M. R. 1990. Periodic features in Saturn's

F ring. *Nature*, **345**, pp. 695–697.

Kozai, Y. 1959. The motion of a close Earth satellite. *Astron. J.*, **64**, pp. 367–377.

Krivov, A. V., and Hamilton, D. P. 1997. Martian dust belts: Waiting for discovery. *Icarus*, **128**, pp. 335–353.

Krüger, H., Grün, E., Hamilton, D. P., Baguhl, M., Dermott, S., Fechtig, H., Gustafson, B. A., Hanner, M. S., Heck, A., Horányi, M., Kissel, J., Lindblad, B. A., Linkert, D., Linkert, G., Mann, I., McDonnell, J. A. M., Morfill, G. E., Polanskey, C., Riemann, R., Schwehm, G., Srama, R., and Zook, H. A. 1999a. Three years of Galileo dust data: II. 1993 to 1995. *Planet. Space Sci.*, **47**, pp. 85–106.

Krüger, H., Krivov, A. V., Hamilton, D. P., and Grün, E. 1999b. Detection of an impact-generated dust cloud around Ganymede. *Nature*, **399**, pp. 558–560.

Lane, A. L., Hord, C. W., West, R. A., Esposito, L. W., Coffeen, D. L., Sato, M., Simmons, K. E., Pomphrey, R. B., and Morris, R. B. 1982. Photopolarimetry from Voyager 2: Preliminary results on Saturn, Titan and the rings. *Science*, **215**, pp. 537–543.

Lane, A. L., Hord, C. W., West, R. A., Esposito, L. W., Simmons, K. E., Nelson, R. M., Wallis, B. D., Buratti, B. J., Horn, L. J., Graps, A. L., and Pryor, W. R. 1986. Photopolarimetery from Voyager 2: Initial results from the Uranian atmosphere, satellites, and rings. *Science*, **233**, pp. 65–70.

Lane, A. L., West, R. A., Hord, C. W., Nelson, R. M., Simmons, K. E., Pryor, W. R., Esposito, L. W., Horn, L. J., Wallis, B. D., Buratti, B. J., Brophy, T. G., Yanamandra-Fisher, P., Colwell, J. E., Bliss, D. A., Mayo, M. J., and Smythe, W. D. 1989. Photopolarimetery from Voyager 2: Initial results from the Neptunian atmosphere, satellites, and rings. *Science*, **246**, pp. 1450–1454.

Larson, S. 1984. Summary of optical groundbased E Ring observations at the University of Arizona. In *Anneaux des Planètes/Planetary Rings*, ed. A. Brahic (Toulouse, France: Cepadues-Editions), pp. 111–113.

Lissauer, J. J., and Espresate, J. 1998. Resonant satellite torques on low optical depth particulate disks. I. Analytic development. *Icarus*, **134**, pp. 155–162.

Maravilla, D., Flammer, K. R., and Mendis, D. A. 1995. On the injection of fine dust from the Jovian magnetosphere. *Astrophys. Jnl.*, **438**, pp. 968–974.

Marouf, E. A., Tyler, G. L., Zebker, H. A., and Eshleman, V. R. 1983. Particle size distributions in Saturn's rings from Voyager 1 radio occultation. *Icarus*, **54**, pp. 189–211.

Marouf, E. A., Tyler, G. L., and Rosen, P. M. 1986. Profiling Saturn's rings by radio occultation. *Icarus*, **68**, pp. 120–166.

Mauk, B. H., Keath, E. P., Kane, M., Krimigis, S. M., Cheng, A. F., Acuña, M. H., Armstrong, T. P., and Ness, N. F. 1991. The magnetosphere of Neptune: Hot plasmas and energetic particles. *J. Geophys. Res.*, **96**, pp. 19,061–19,084.

McGhee, C. A., Nicholson, P. D., French, R. G., and Hall, K. J. 2000. HST Observations of Saturnian satellites during the 1995 ring plane crossings. *Icarus*, submitted.

McMuldroch, S. , Pilorz, S. H., Danielson, G. E., and the NIMS science team 2000. Galileo NIMS near-infrared observations of Jupiter's ring system. *Icarus*, **146**, pp. 1–11.

Meier, R., Smith, B. A., Owen, T. C., Becklin, E. E., and Terrile, R. J. 1999. Near infrared photometry of the jovian ring and Amalthea. *Icarus*, **141**, pp. 253–262.

Mendis, D. A., and Rosenberg, M. 1994. Cosmic dusty plasma. *Ann. Rev. Astron. Astrophys.*, **32**, pp. 419–463.

Mendis, D. A., Hill, J. R., Ip, W.-H., Goertz, C. K., and Grün, E. 1984. Electrodynamic processes in the ring system of Saturn. In *Saturn*, eds. T. Gehrels and M. S. Matthews (Tucson: University of Arizona Press), pp. 546–589.

Meyer-Vernet, N. 1982. Flip-flop of electric potential of dust grains in space. *Astron. Astrophys.*, **105**, pp. 98–106.

Meyer-Vernet, N., Aubier, M. G., and Pedersen, B. M. 1986. Voyager 2 at Uranus: Grain impacts in the ring plane. *GRL.*, **13**, pp. 617–620.

Meyer-Vernet, N., Lecacheux, A., and Pedersen, B. M. 1996. Constraints on Saturn's E ring from the Voyager 1 radio astronomy experiment. *Icarus*, **123**, pp. 113–128.

Meyer-Vernet, N., Lecacheux, A., and Pedersen, B. M. 1998. Constraints on Saturn's G ring from the Voyager radio astronomy instrument. *Icarus*, **132**, pp. 311–320.

Mignard, F. 1982. Radiation pressure and dust particle dynamics. *Icarus*, **49**, pp. 347–366.

Mignard, F. 1984. Effects of radiation forces on dust particles in planetary rings. In *Planetary Rings*, eds. R. Greenberg and A. Brahic (Tucson: Univ. of Arizona Press), pp. 333–366.

Mignard, F., and Hénon, M. 1984. About an unsuspected integrable problem. *Cel. Mech.*, **33**, pp. 239–250.

Morfill, G. E., Grün, E., and Johnson, T. V. 1980. Dust in Jupiter's magnetosphere: Physical processes. *Planet. Space Sci.*, **28**, p. 1087.

Morfill, G. E., Grün, E., Johnson, T. V., and Goertz, C. K. 1983. On the evolution of Saturn's spokes: Theory. *Icarus*, **53**, pp. 230–235.

Murray, C. D., and Dermott, S. F. 1999. Solar System Dynamics. (Cambridge University Press).

Murray, C. D., and Thompson, R. P. 1990. Orbits of shepherd satellites deduced from the structure of the rings of Uranus. *Nature*, **348**, pp. 499–502 (Erratum: **350**, p. 90).

Murray, C. D., Gordon, M., and Giulatti-Winter, S. M. 1997. Unraveling the strands of Saturn's F ring. *Icarus*, **129**, pp. 304–306.

Nicholson, P. D., and Matthews, K. 1991. Near-infrared observations of the Jovian ring and small satellites. *Icarus*, **93**, pp. 331–346.

Nicholson, P. D., Mosqueira, I., and Matthews, K. 1995. Stellar occultation observations of Neptune's rings: 1984–1988. *Icarus*, **113**, pp. 295–330.

Nicholson, P. D., Showalter, M. R., Dones, L., French, R. G., Larson, S. M., Lissauer, J. J., McGhee, C. A., Seitzer, P., Sicardy, B., and Danielson, G. E. 1996. Observations of Saturn's ring plane crossings in August and November 1995. *Science*, **272**, pp. 509–515.

Northrop, T. G. 1992. Dusty plasmas. *Physica Scripta*, **45**, pp. 475–490.

Northrop, T. G., and Birmingham, T. J. 1990. Plasma drag on a dust grain due to Coulomb collision. *Planet. Space Sci.*, **38**, pp. 319–326.

Northrop, T. G., Mendis, D. A., and Schaffer, L. 1989. Gyrophase drift and the orbital evolution of dust at Jupiter's ring. *Icarus*, **79**, pp. 101–115.

Oberc, P. 1994. Dust impacts detected by Voyager 2 at Saturn and Uranus: A post-Halley view. *Icarus*, **111**, pp. 211–226.

Ockert, M. E., Cuzzi, J. N., Porco, C. C., and Johnson, T. V. 1987. Uranian ring photometry: Results from Voyager 2. *J. Geophys. Res.*, **92**, pp. 14,969–14,978.

Ockert-Bell, M. E., Burns, J. A., Daubar, I. J., Thomas, P. C., Veverka, J., Belton, M. J. S., and Klaasen, K. P. 1999. The structure of Jupiter's ring system as revealed by the Galileo imaging system. *Icarus*, **138**, pp. 188–213.

Paranicas, C. P., and Cheng, A. F. 1991. Theory of ring sweeping of energetic particles. *J. Geophys. Res.*, **96**, pp. 19,123–19,129.

Pedersen, B. M., Meyer-Vernet, N., Aubier, M. G., and Zarka, P. 1991. Dust distribution around Neptune: Grain impacts near the ring plane measured by the Voyager Planetary Radio Astronomy experiment. *J. Geophys. Res.*, **96**, pp. 19,187–19,196.

Pollack, J. B., and Cuzzi, J. N. 1980. Scattering by nonspherical particles of size comparable to a wavelength: A new semi-empirical theory and its application to tropospheric aerosols. *J. Atmos. Sci.*, **37**, pp. 868–881.

Porco, C. C. 1991. An explanation for Neptune's ring arcs. *Science*, **253**, pp. 995–1001.

Porco, C. C., and Danielson, G. E. 1982. The periodic variation of spokes in Saturn's rings. *Astron. J.*, **87**, pp. 826–833.

Porco, C. C., Nicholson, P. D., Cuzzi, J. N., Lissauer, J. J., and Esposito, L. W. 1995. Neptune's ring system. In *Neptune and Triton* ed. D. P. Cruikshank (Tucson: University of Arizona Press), pp. 703–804.

Poulet, F., Karkoschka, E., and Sicardy, B. 1999. Spectrophotometry of Saturn's small satellites and rings from Hubble Space Telescope images. *JGR*, **104**, pp. 24095–24110.

Poulet, F., Sicardy, B., Nicholson, P. D., Karkoschka, E., and Caldwell, J. 2000a. Saturn's ring-plane crossings of August and November 1995: A model for the new F-ring objects. *Icarus*, **144**, pp. 135–148.

Poulet, F., Sicardy, B., Dumas, C., Jorda, L., and Tiphéne, D. 2000b. The crossings of Saturn's ring-plane by the Earth in 1995: Ring thickness. *Icarus*, **145**, pp. 147–165.

Reitsema, H. J., Hubbard, W. B., Lebofsky, L. A., and Tholen, D. J. 1982. Occultation by a possible third satellite of Neptune. *Science*, **215**, pp. 289–291.

Richter, K., and Keller, H. U. 1995. On the stability of dust particle orbits around cometary nuclei. *Icarus*, **114**, pp. 355–371.

Roddier, C., Roddier, F., Graves, J. E., and Northcott, M. J. 1998. Discovery of an arc of particles near Enceladus' orbit: A possible key to the origin of the E ring. *Icarus*, **136**, pp. 50–59.

Roddier, F., Roddier, C., Brahic, A., Dumas, C., Graves, J. E., Northcott, M. J., and Owen, T. C. 2000. Adaptive optics observations of Saturn's ring-plane crossing in August 1995. *Icarus*, **143**, pp. 299–307.

Sandel, B. R., Shemansky, D. E., Broadfoot, A. L., Holberg, J. B., Smith, G. R., McConnell, J. C., Strobel, D. F., Atreya, S. K., Donahue, T. M., Moos, H. W., Hunten, D. M., Pomphrey, R. B., and Linick, S. 1982. Extreme ultraviolet observations from the Voyager 2 encounter with Saturn. *Science*, **215**, pp. 548–553.

Schaffer, L., and Burns, J. A. 1987. The dynamics of weakly charged dust: Motion through Jupiter's gravitational and magnetic fields. *Jnl. Geophys. Res.*, **92**, pp. 2264–2280.

Schaffer, L. E., and Burns, J. A. 1992. Lorentz resonances and the vertical structure of dusty rings: Analytical and numerical results. *Icarus*, **96**, pp. 65–84.

Schaffer, L. E., and Burns, J. A. 1994. Charged dust in planetary magnetospheres: Hamiltonian dynamics and numerical simulations for highly charged grains. *Jnl. Geophys. Res.*, **99**, pp. 17,211–17,223.

Schaffer, L. E., and Burns, J. A. 1995. Stochastic charging of dust grains in planetary rings: Diffusion rates and their effect on Lorentz resonances. *Jnl. Geophys. Res.*, **100**, pp. 213–234.

Showalter, M. R. 1989. Anticipated time variations in (our understanding of) Jupiter's ring system. In *Time-Variable Phenomena in the Jovian System*, eds. M. J. S. Belton, R. A. West, and J. Rahe, NASA-SP 494, pp. 116–125.

Showalter, M. R. 1991. Visual detection of 1981S13, Saturn's eighteenth satellite, and its role in the Encke Gap. *Nature*, **351**, pp. 709–713.

Showalter, M. R. 1993. Longitudinal variations in the Uranian lambda Ring. *Bull. Amer. Astron. Soc.*, **25**, p. 1109.

Showalter, M. R. 1994. Tracking clumps in Saturn's F Ring. *Bull. Amer. Astron. Soc.*, **26**, pp. 1150–1151.

Showalter, M. R. 1995. Arcs and clumps in the Uranian λ Ring. *Science*, **267**, pp. 490–493.

Showalter, M. R. 1996. Saturn's D Ring in the Voyager images. *Icarus*, **124**, pp. 677–689.

Showalter, M. R. 1998. Detection of centimeter-sized meteoroid impact events in Saturn's F Ring. *Science*, **282**, pp. 1099–1102.

Showalter, M. R., and Burns, J. A. 1982. A numerical study of Saturn's F ring. *Icarus*, **52**, pp. 526–544.

Showalter, M. R., and Cuzzi, J. N. 1992. Physical properties of Neptune's ring system. *Bull. Amer. Astron. Soc.*, **24**, p. 1029.

Showalter, M. R., and Cuzzi, J. N. 1993. Seeing ghosts: Photometry of Saturn's G Ring. *Icarus*, **103**, pp. 124–143.

Showalter, M. R., Burns, J. A., Cuzzi, J. N., and Pollack, J. B. 1985. The discovery of Jupiter's 'gossamer' ring. *Nature*, **316**, pp. 115–119.

Showalter, M. R., Burns, J. A., Cuzzi, J. N., and Pollack, J. B. 1987. Jupiter's ring system: New results on structure and particle properties. *Icarus*, **69**, pp. 458–498.

Showalter, M. R., Cuzzi, J. N., and Larson, S. M. 1991. Structure and particle properties of Saturn's E Ring. *Icarus*, **94**, pp. 451–473.

Showalter, M. R., Pollack, J. B., Ockert, M. E., Doyle, L., and Dalton, J. B. 1992. A photometric study of Saturn's F Ring. *Icarus*, **100**, pp. 394–411.

Showalter, M. R., Burns, J. A., and Hamilton, D. P. 1998. Saturn's "gossamer" ring: The F ring's inner sheet and its interaction with Prometheus *BAAS*, **30**, p. 1044.

Sicardy, B., Roques, F., and Brahic, A. 1991. Neptune's rings 1983–1989 Ground-based stellar occultation observations I. Ring-like arc detections. *Icarus*, **89**, pp. 220–243.

Sicardy, B., Roddier, F., Roddier, C., Perozzi, E., Graves, J. E., Guyon, O., and Northcott, M. J. 1999. Images of Neptune's ring arcs obtained by a ground-based telescope. *Nature*, **400**, pp. 731–733.

Simonelli, D. P., Rossier, L., Thomas, P. C., Veverka, J., Burns, J. A., and Belton, M. J. S. 2000. Leading-trailing albedo asymmetries of Thebe, Amalthea and Metis. *Icarus*, **147**, 353–365.

Simpson, J. A., Bastian, T. S., Chenette, D. L., McKibben, R. B., and Pyle, K. R. 1980. The trapped radiations of Saturn and their absorption by satellites and rings. *J. Geophys. Res.*, **85**, pp. 5731–5762.

Smith, B. A., Soderblom, L. A., Johnson, T. V., Ingersoll, A. P., Collins, S. A., Shoemaker, E. M., Hunt, G. E., Masursky, H., Carr, M. H., Davies, M. E., Cook, A. F., II, Boyce,

J., Danielson, G. E., Owen, T., Sagan, C., Beebe, R. F., Veverka, J., Strom, R. G., McCauley, J. F., Morrison, D., Briggs, G. A., and Suomi, V. E. 1979a. The Jupiter system through the eyes of Voyager 1. *Science*, **204**, pp. 951–972.

Smith, B. A., Soderblom, L. A., Beebe, R., Boyce, J., Briggs, G., Carr, M., Collins, S. A., Cook, A. F., II, Danielson, G. E., Davies, M. E., Hunt, G. E., Ingersoll, A., Johnson, T. V., Masursky, H., McCauley, J., Morrison, D., Owen, T., Sagan, C., Shoemaker, E. M., Strom, R., Suomi, V. E., and Veverka, J. 1979b. The Galilean satellites and Jupiter: Voyager 2 Imaging Science results. *Science*, **206**, pp. 927–950.

Smith, B. A., Soderblom, L., Beebe, R., Boyce, J., Briggs, G., Bunker, A., Collins, S. A., Hansen, C. J., Johnson, T. V., Mitchell, J. L., Terrile, R. J., Carr, M., Cook, A. F., II, Cuzzi, J., Pollack, J. B., Danielson, G. E., Ingersoll, A., Davies, M. E., Hunt, G. E., Masursky, H., Shoemaker, E., Morrison, D., Owen, T., Sagan, C., Veverka, J., Strom, R., and Suomi, V. E. 1981. Encounter with Saturn: Voyager 1 Imaging Science results. *Science*, **212**, pp. 163–191.

Smith, B. A., Soderblom, L., Batson, R., Bridges, P., Inge, J., Masursky, H., Shoemaker, E., Beebe, R., Boyce, J., Briggs, G., Bunker, A., Collins, S. A., Hansen, C. J., Johnson, T. V., Mitchell, J. L., Terrile, R. J., Cook, A. F., II, Cuzzi, J., Pollack, J. B., Danielson, G. E., Ingersoll, A. P., Davies, M. E., Hunt, G. E., Morrison, D., Owen, T., Sagan, C., Veverka, J., Strom, R., and Suomi, V. E. 1982. A new look at the Saturn system: The Voyager 2 images. *Science*, **215**, pp. 504–537.

Smith, B. A., Soderblom, L. A., Beebe, R., Bliss, D., Boyce, J. M., Brahic, A., Briggs, G. A., Brown, R. H., Collins, S. A., Cook, A. F., II, Croft, S. K., Cuzzi, J. N., Danielson, G. E., Davies, M. E., Dowling, T. E., Godfrey, D., Hansen, C. J., Harris, C., Hunt, G. E., Ingersoll, A. P., Johnson, T. V., Krauss, R. J., Masursky, H., Morrison, D., Owen, T., Plescia, J. B., Pollack, J. B., Porco, C. C., Rages, K., Sagan, C., Shoemaker, E. M., Sromovsky, L. A., Stoker, C., Strom, R. G., Suomi, V. E., Synnott, S. P., Terrile, R. J., Thomas, P., Thompson, W. R., and Veverka, J. 1986. Voyager 2 in the Uranian system: Imaging Science results. *Science*, **233**, pp. 43–64.

Smith, B. A., Soderblom, L. A., Banfield, D., Barnet, C., Basilevsky, A. T., Beebe, R. F., Bollinger, K., Boyce, J. M., Brahic, A., Briggs, G. A., Brown, R. H., Chyba, C., Collins, S. A., Colvin, T., Cook, A. F., II, Crisp, D., Croft, S. K., Cruikshank, D., Cuzzi, J. N., Danielson, G. E., Davies, M. E., De Jong, E., Dones, L., Godfrey, D., Goguen, J., Grenier, I., Haemmerle, V. R., Hammel, H., Hansen, C. J., Helfenstein, P., Howell, C., Hunt, G. E., Ingersoll, A. P., Johnson, T. V., Kargel, J., Kirk, R., Kuehn, D. I., Limaye, S., Masursky, H., McEwen, A., Morrison, D., Owen, T., Owen, W., Pollack, J. B., Porco, C. C., Rages, K., Rogers, P., Rudy, D., Sagan, C., Schwartz, J., Shoemaker, E. M., Showalter, M., Sicardy, B., Simonelli, D., Spencer, J., Sromovsky, L. A., Stoker, C., Strom, R. G., Suomi, V. E., Synnott, S. P., Terrile, R. J., Thomas, P., Thompson, W. R., Verbiscer, A., and Veverka, J. 1989. Voyager 2 at Neptune: Imaging Science results. *Science*, **246**, pp. 1422–1449.

Smyth, W. H., and Marconi, M. L. 1993. The nature of the hydrogen tori of Titan and Triton. *Icarus*, **101**, pp. 18–32.

Soter, S. 1971. The dust belts of Mars. Cornell CRSR Report 472.

Spitzer, L. 1962. Physics of Fully Ionized Gases(2nd ed.), (NY: Interscience), 190 pp.

Stern, D.P. 1976. Representation of magnetic fields in space. *Rev. Geophys. Space Phys.*, **14**, pp. 199–214.

Stevenson, D. J., Harris, A. W., and Lunine, J. I. 1986. Origins of satellites. In *Satellites*, eds. J. A. Burns and M. S. Matthews (Tucson: Arizona Press), pp. 39–88.

Synnott, S. P., Terrile, R. J., Jacobson, R. A., and Smith, B. A. 1983. Orbits of Saturn's F ring and its shepherding satellites. *Icarus*, **53**, pp. 156–158.

Tagger, M., Henricksen, R. N., and Pellat, R. 1991. On the nature of the spokes in Saturn's rings. *Icarus*, **91**, pp. 297–314.

Thomas, P. C., Veverka, J., and Helfenstein, P. 1995. Neptune's small satellites. In *Neptune and Triton*, ed. D. P. Cruikshank (Tucson: University of Arizona Press), pp. 685–699.

Thomas, P. C., Burns, J. A., Rossier, L., Simonelli, D., Veverka, J., Chapman, C. R., Klaasen, K., Johnson, T. V., and Belton, M. J. S. 1998. The small inner satellites of Jupiter. *Icarus*, **135**, pp. 360–371.

Throop, H. B., and Esposito, L. W. 1998. G ring particle sizes derived from ring plane crossing observations. *Icarus*, **131**, pp. 152–166.

Tsintikidis, D., Gurnett, D., Granroth, L. J., Allendorf, S. C., and Kurth, W. S. 1994. A revised analysis of micron-sized particles detected near Saturn by the Voyager 2 Plasma Wave instrument. *J. Geophys. Res.*, **99**, pp. 2261–2270.

Tsintikidis, D., Kurth, W. S., Gurnett, D. A., and Barbosa, D. A. 1995. Study of dust in the vicinity of Dione using the Voyager 1 plasma wave instrument. *J. Geophys. Res.*, **99**, pp. 2261–2270.

Tsintikidis, D., Gurnett, D., Kurth, W. S., and Granroth, L. J. 1996. Micron-sized particles discovered in the vicinity of Jupiter by the Voyager plasma wave instruments. *GRL*, **23**, pp. 997–1000.

Tyler, G. L., Eshleman, V. R., Anderson, J. D., Levy, G. S., Lindal, G. F., Wood, G. E., and Croft, T. A. 1981a. Radio Science investigations of the Saturn system with Voyager 1: Preliminary results. *Science*, **212**, pp. 201–206.

Tyler, G. L., Marouf, E. A., and Wood, G. E. 1981b. Radio occultation of Jupiter's ring: Bounds on optical depth and particle size, and a comparison with infrared and optical results. *J. Geophys. Res.*, **86**, pp. 8699–8703.

Tyler, G. L., Marouf, R. A., Simpson, R. A., Zebker, H. A., and Eshleman, V. R. 1983. The microwave opacity of Saturn's rings at wavelengths of 3.6 and 13 cm from Voyager 1 radio occultation. *Icarus*, **54**, pp. 160–188.

Tyler, G. L., Sweetnam, D. N., Anderson, J. D., Campbell, J. K., Eshleman, V. R., Hinson, D. P., Levy, G. S., Lindal, G. F., Marouf, E. A., and Simpson, R. A. 1986. Voyager 2 Radio Science observations of the Uranian system: Atmosphere, rings and satellites. *Science*, **233**, pp. 79–84.

Tyler, G. L., Sweetnam, D. N., Anderson, J. D., Borutzki, S. E., Campbell, J. K., Eshleman, V. R., Gresh, D. L., Gurrola, E. M., Hinson, D. P., Kawashima, N., Kursinski, E. R., Levy, G. S., Lindal, G. F., Lyons, J. R., Marouf, E. A., Rosen, P. A., Simpson, R. A., and Wood, G. E. 1989. Voyager Radio Science observations of Neptune and Triton. *Science*, **246**, pp. 1466–1473.

Van Allen, J. A. 1982. Findings on rings and inner satellites of Saturn by Pioneer 11. *Icarus*, **51**, pp. 509–527.

Van Allen, J. A. 1983. Absorption of energetic protons by Saturn's Ring G. *J. Geophys. Res.*, **88**, pp. 6911–6918.

Van Allen, J. A. 1987. An upper limit on the sizes of shepherding satellites at Saturn's Ring G. *J. Geophys. Res.*, **92**, pp. 1153–1159.

Van Allen, J. A., Randall, B. A., and Thomsen, M. F. 1980. Sources and sinks of energetic electrons and protons in Saturn's magnetosphere. *J. Geophys. Res.*, **85**, pp. 5679–5694.

van de Hulst, H. C. 1981. *Light Scattering by Small Particles*. (New York: Dover Publications).

Weidenschilling, S. J., Chapman, C. R., Davis, D. R. and Greenberg, R. 1982. In Planetary Rings, eds. R. Greenberg and A. Brahic (Tucson: University of Arizona Press). pp. 367–415.

Wyatt, S. P., and Whipple, F. L. 1950. The Poynting-Robertson effect on meteor orbits. *Ap. J.*, **111**, pp. 134–141.

Zebker, H. A., Marouf, E. A., and Tyler, G. L. 1985. Saturn's rings: Particle size distributions for thin layer models. *Icarus*, **64**, pp. 531–548.

Zook, H. A., Grün, E., Baguhl, M., Hamilton, D. P., Linkert, G., Liou, J.-C., Forsyth, R., and Phillips, J. L. 1996. Solar magnetic field bending of jovian dust trajectories. *Science*, **274**, pp. 1501–1503.

Interstellar Dust and Circumstellar Dust Disks

Johann Dorschner

Astrophysikalisches Institut und Universitäts-Sternwarte Jena, Germany

Abstract. Interstellar dust research belongs to the young branches of astrophysics. With the establishment of sensitive observational techniques in the astronomical infrared spectroscopy in the 1960s, diagnostic circumstellar and interstellar dust bands were detected and induced an explosive development of the whole field. In this context, the branch of solid-state astrophysics (synonymous to laboratory astrophysics) was founded. Special dust populations attributed to characteristic phases of the interstellar medium and to special circumstellar environments could be defined. Dust turned out to be the key to the understanding of the evolution of the interstellar medium and, closely connected with this, to the early and the late stages of stellar evolution. In this review, the population scheme is used for the division of the text to the main chapters. Evolutionary dust characteristics on the galactic scale and in the context of the formation of stars and planetary systems are stressed. Relations between interstellar and interplanetary solids are pointed out.

I. LANDMARKS IN INTERSTELLAR DUST RESEARCH

I.A. From Early Conjectures to a Physical Theory

Interstellar dust started its career as an astronomical research topic only around 1930, when the existence of a general interstellar extinction weakening and reddening starlight became evident (Schalén 1929, Trümpler 1930). This belated start belongs to the curiosities in the history of astrophysics. As a matter of fact, a light-absorbing medium was assumed much earlier (Loys de Chéseaux 1744, Olbers 1823). It was introduced in order to solve the paradox of the dark night sky (Olbers' paradox), which had occupied astronomers since Kepler's time. However, Olbers' absorbing medium was resembling a kind of ether rather than dust grains distributed between the stars.

At the end of the 19th century, E.E. Barnard at the Lick Observatory and, independently, Max Wolf in Heidelberg applied the new method of photography to systematic studies of dark nebulae in the Milky Way. However, impressed by W. Herschel's "hole in the sky"-statement, they were not aware of the real character of this phenomenon. Barnard used the designation "black holes" for small dense dark clouds in the Milky Way. Later he called them cautiously "dark markings" avoiding the term "nebula". Wolf used the phrase "dunkle Höhlen" (dark caves). Although Wolf proposed an excellent statistical method to evaluate the amount of light absorption, the physical nature of the absorber remained unclear to him (for historical details see Lynds 1968; Verschuur 1989).

It was the merit of the investigators of the interstellar extinction in the early 1930s that they did not only describe the observational phenomenon, but also found the adequate physical tool for its understanding, the application of Mie's scattering theory (Öpik 1931, Schalén 1934, Schoenberg and Jung 1934). Since, at that time, iron was considered the most abundant element in the meteorites, interstellar dust was proposed to consist of submicrometer-sized iron grains. From meteor statistics some experts drew the conclusion that a significant part of the particles entering the earth's atmosphere must have hyperbolic velocities (Hoffmeister 1929). This seemed to prove a connection between meteorites and interstellar dust. The debate on the interstellar origin of the meteors was occasionally very brisk and controversial (see Öpik 1929, Hoffmeister 1931), but finally ceased with the rejection of the interstellar hypothesis. For more than three decades, analogy considerations between interstellar and solar system solids were out of the scope.

I.B. The Classical Dust Model

The first dust model with a thoroughly astrophysical foundation was proposed by Oort and van de Hulst (1946) and van de Hulst (1949). The authors came to the conclusion that the interstellar grains should consist of an ice conglomerate. Its ingredients, mainly hydrogen-saturated compounds like H_2O, CH_4, and NH_3, are formed because interstellar O-, C-, and N-atoms that condense onto the cold grain surfaces are hydrogenated there by interstellar H atoms permanently hitting the grains.

The continuous growth of a grain ends abruptly by the collision with another grain leading to the total evaporation of both. Collisions of grains must occur during interstellar cloud collisions. From the assumption of a steady state between grain growth and grain evaporation, the authors computed an average grain size distribution, the shape of which was consistent with the observed wavelength dependence of the interstellar extinction as determined by Stebbins and Whitford (1943). This obviously very successful dust model got the name "classical" dust model.

I.C. Interstellar Polarization

The discovery of the interstellar polarization by Hiltner (1949) and Hall (1949) opened a new observational access to the interstellar dust. The light of reddened stars often shows a weak linear polarization, i.e. the dust extinction coefficient has different values for different vibrational planes of the passing light waves. This effect revealed two new aspects:

1. The dust grains must be either optically or morphologically anisotropic. The latter means that they are elongated rather than spherical.
2. The grains must have a preferential orientation in space: they must be aligned.

A successful mechanism explaining the grain alignment by paramagnetic relaxation of elongated grains in the galactic magnetic field was first proposed by Davis and Greenstein (1951). The connection between polarization vector and magnetic field became a valuable tool to trace the large-scale structure of the magnetic field in the Galaxy. In addition to the extinction curve, the determination of the wavelength dependence of the interstellar polarization (polarization curve) offered an additional criterion for the selection among theoretical dust models.

I.D. Refractory Dust Grains

In the 1960s dust models based on refractory dust sorts of stellar and circumstellar origin came into use: graphite grains (Hoyle and Wickramasinghe 1962), SiO_2 grains (Kamijo 1963), grains composed of meteoritic silicates (Dorschner 1967), and silicon carbide grains (Friedemann 1969). Condensation of refractory solids in stellar atmospheres was proposed much earlier (Wildt 1933), but only in connection with stellar opacity considerations; it was not imaginable that these grains could be conveyed to the interstellar space.

UV-observations by rockets (Stecher 1965) and later by satellite observatories demonstrated that the interstellar extinction in the UV was much stronger than predicted by the classical model. In addition, the UV observations failed to detect the ice absorption edge at 160 nm in the spectra of heavily reddened OB stars, but instead demonstrated the ubiquitous presence of a strong solid-state absorption band at 217.5 nm (the famous extinction "bump"), which superposed the continuous rise of the extinction curve. For decades, this band was assigned to small (nm-sized) graphite grains (Bless and Savage 1972, cf. models in Sec. III.B.3. Table 2). Today this strict association with the mineral graphite is no longer maintained. There is, however, a consent that the band is due to the $\pi-\pi^*$ transition of aromatically bonded carbon atoms of graphitized soot grains.

The studies of refractory dust models suggested to pay greater attention to the analogy between primitive solar system solids and interstellar dust (Dorschner 1967); however, it lasted about two decades until interplanetary dust was generally accepted as guide to interstellar dust (Jones and Williams 1987, McDonnell 1988). The ways of these fields, diverging since the 1930s, began to approach again.

I.E. Diagnostic Dust Bands and Laboratory Astrophysics

Enormous progress in dust research was achieved in the late 1960s by the detection of identifiable (diagnostic) IR bands. In 1968 and 1969 a broad emission band at about 10 μm was detected in the spectra of evolved stars and in the Orion Trapezium nebula, respectively. It was assigned to stretching vibrations of Si–O bonds in SiO_4 tetrahedra of silicate grains. In the following years, this silicate feature was found in emission as well as in absorption. The assignment was confirmed by the detection of a second silicate band due to the O–Si–O bending vibrations at about 18 μm (for details of the silicate story see Gürtler et al. 1989).

A band at 11.3 μm found in carbon stars was assigned to silicon carbide (SiC). Today a large number of C-rich sources showing this band in emission and some sources with SiC absorption are well-known (for the SiC story see Speck et al. 1997). The SiC band is analogously characteristic of C-rich evolved stars as are the silicate features of O-rich stars. Both solids have been the first representatives of circumstellar dust species later designated as "stardust" (Ney 1977). In 1983, the Infrared Astronomical Satellite (IRAS) yielded several thousand stardust spectra in the range 8–25 μm (Olnon and Raimond 1986).

In addition to the ubiquitous interstellar 217.5 nm band (Sec. I.D.) a number of IR features was found, which confirmed the dominant role of carbonaceous solids in the interstellar dust chemistry. In C-rich circumstellar environments, reflection nebulae, planetary nebulae, HII regions, young stellar objects, Vega-excess stars, and in the large-scale galactic emission strong emission bands at 3.3, 6.2, 7.7, 8.6, 11.3 and 12.7 μm and additional weaker features related to them have been detected. As a rule, they are called Unidentified IR Bands (UIR bands, sometimes also UIBs) because their origin is not yet entirely understood. Following a suggestion by Donn (1968), they have been assigned to polycyclic aromatic hydrocarbons (PAHs, see Sec. V.B.3.). Because of the attribution to aromatic carbon some authors prefer using the term Aromatic Interstellar Bands (AIBs).

Apart from the evidence for aromatically bonded carbon, IR spectroscopy has also revealed the occurrence of aliphatic material in interstellar dust. The first indication was an absorption feature at 3.4 μm found in the spectrum of the Galactic Center (Butchart et al. 1986), which is due to C–H stretching vibrations of hydrocarbons. This carbonaceous solid, which is typical of the diffuse interstellar medium, clearly differs from the UIR band carrier. Meanwhile, the list of interstellar dust absorption features has considerably grown (see Table 1 in Sec. III.).

In addition to refractory solids, volatiles ("ices") have been found in dense interstellar clouds. As expected, H_2O ice was the first ice species to be detected (Gillett and Forrest 1973). After a decade of unsuccessful search the breakthrough came with the discovery of the absorption band of solid CO at 4.67 μm (Lacy et al. 1984). The list of frozen volatiles in molecular cloud sources has permanently grown (see Table 3 in Sec. IV.).

The identification of IR bands has put the field of the interstellar dust to a new empirical basis. The concept of the multi-component dust was born, and distinct dust populations typical of special cosmic environments were defined. The adequate interpretation of the observed spectra required laboratory simulation experiments with dust analog materials. This led to the foundation of an experimental branch of astrophysics called laboratory or solid-state astrophysics. Today, laboratory astrophysics facilities are working in the USA, in the Netherlands, in France, Italy, Japan and Germany.

The laboratory work does not only include measurements of imitated dust materials. One of the most fascinating discoveries of the last decade has been the detection and isolation of resistant presolar dust grains preserved in primitive meteorites (Anders and Zinner 1993, Zinner 1997, Hoppe and Zinner 2000). Another weighty discovery has been the detection of interstellar dust grains in the solar system. Sufficiently large dust particles of the local interstellar medium deeply penetrate into the heliosphere and can be subject of in-situ studies by spacecraft (Grün et al. 1994, Frisch et al. 1999, Grün et al. 2000). This has opened the fantastic possibility of collecting and recovering interstellar dust and, thus, establishing galactic mineralogy and material science. These discoveries and also the detection of micrometeoroids of probably interstellar origin (Taylor et al. 1996, Baggaley 2000) prove that close interstellar/interplanetary connections really exist which were once categorically rejected (see Sec. I.A).

II. DUST AND GALACTIC EVOLUTION

II.A. The Multi-Phase Interstellar Medium

The interstellar medium (ISM) in the Galaxy is permanently changing its spatial distribution, dynamics, and chemical composition by the exchange of matter and energy with the galactic stellar content. Large amounts of mechanical energy are transmitted to the ISM from the violent blasts of supernova explosions and by stellar winds. The resulting shocks heat the gas, produce density gradients, destroy molecules, and ionize atomic species. The ejected stellar matter gradually enriches the ISM with heavy elements. The energy input by photons from hot stars produces ISM regions, in which the thermal and chemical balance is dominated by photo-processes (photon-dominated or photodissociation regions: PDR). Cool dense clouds of the ISM become gravitationally unstable and dissolve by star formation. Massive new stars ionize the gas and evaporate the volatile dust grains and, thus, drastically change the environmental conditions in the star-forming regions.

The modern large-scale modelling of the ISM rests on the definition of several distinct phases characterized by special values of temperature, density, volume filling factor and dust content. The phases result from the aforementioned interactions. The standard ISM model originally developed by McKee and Ostriker (1977) was repeatedly modified (for the present state see, e.g. Tielens 1995, Spaans and Ehrenfreund 1999). The diffuse ISM has been assumed to consist of spherical dusty gas clouds (radius $\approx$ 2 pc, mainly neutral atomic hydrogen: H I region) embedded in thin hot ($n_\mathrm{g} = 0.003$ cm^{-3}, $T_\mathrm{g} \approx 10^6$ K) ionized matter coming from supernova remnants and being called "coronal gas". The clouds are assumed to consist of cold dense cores ($n_\mathrm{g} = 50$ cm^{-3}, $T_\mathrm{g} = 80$ K) and less dense, warm envelopes ($n_\mathrm{g} = 0.8$ cm^{-3}, $T_\mathrm{g} = 8000$ K) of neutral gas which is ionized at the outer edge. These ISM phases are in pressure equilibrium. Atoms with low ionization thresholds in the H I gas of the clouds are ionized by the interstellar radiation field.

Dust contained in the diffuse clouds with a dust-to-gas mass ratio of about 0.01 causes a mean rate of visual extinction amounting to $\langle A_V/l \rangle = 1.8$ mag per kpc near the galactic plane. From this value an average dust density $\overline{\rho_\mathrm{d}} =$

$1.8 \cdot 10^{-26}$ g cm^{-3} can be derived (Whittet 1992). Since the stellar UV radiation penetrating the average clouds dissociates molecules and erodes away volatile solids, only refractory dust components are observed in the diffuse ISM.

II.B. Molecular Clouds and Star-Forming Regions

There are two additional ISM phases that are not in pressure balance with the other phases: molecular clouds and HII regions around young stars (Blitz 1993, Lada et al. 1993, Tielens 1995). Molecular clouds are roughly by a factor of 10^2–10^3 denser than the clouds of the diffuse ISM. Giant molecular clouds with diameters exceeding 100 pc and masses of the order 10^6 $M_\odot$ are the biggest individual objects in the Galaxy. Numerous well-known dark clouds turned out to be identical with molecular clouds in the mass range 10 to 10^3 $M_\odot$. The smallest molecular clouds are the globules.

Since the large dust content prevents stellar radiation from penetrating into the interior, the molecular cloud gas is extremely cold (down to 10 K) and undissociated, H$_2$ and CO being the most abundant molecules. Although the volume filling factor of the molecular gas near the galactic plane is less than 0.01, its contribution to the total ISM mass is considerable. Estimates provide total masses of $4.6 \cdot 10^9$ $M_\odot$ for the diffuse ISM, $3.5 \cdot 10^9$ $M_\odot$ for the molecular clouds and $0.1 \cdot 10^9$ $M_\odot$ for the HII regions (Tielens 1995).

Molecular clouds exhibit internal structure. Cold cores with diameters of 0.5-3 pc and densities of 10^4–10^6 cm^{-3} are the preferential places where star formation sets in. In some cores, molecular line observations have presented evidence for gravitational contraction. Because of their high gas densities and low temperatures such cores are ideal places for the accretion of molecules onto the grains, forming ice mantles. The formation of massive stars in molecular cloud cores causes significant heating of the gas due to their output of ionizing radiation and of high-velocity stellar wind. This way, originally cold cores are transformed to "hot cores". When the ionization of the gas breaks through the molecular cloud, the star-forming region becomes visible as an emission nebula (HII region) with embedded aggregates of very young luminous stars and young stellar objects surrounded by dust envelopes. Depending on the local gas density, the ionization front gains ground with different velocities towards different directions. This way, bizarre dark dust structures, filaments, "elephant trunks" (compare the famous Hubble Space Telescope images of the central part of the Eagle Nebula, M16) and globules are formed and can be seen against the bright background of the HII regions. The coexistence of aggregates of extremely young massive stars, hot ionized nebular gas, and conspicuous extinction structures is typical of a few visible star birthplaces, e.g. Orion, Rosette, and Eagle nebulae. However, in most cases star-forming regions are deeply embedded in the parent molecular cloud that is transparent only for far infrared and millimeter-wave radiation.

II.C. Dust Populations and the Lifecycle of Dust

The dust research of the last decade has led to the definition of "dust populations". A dust population is a *multi-component dust mixture typical of a special interstellar or circumstellar environment, in which the grains are formed and/or are decisively modified* (Dorschner and Henning 1995). Because of its basic significance, the population concept is used as the main division scheme of this review.

Interstellar dust populations in a strict sense are:

1. dust of diffuse ISM clouds (Section III) and
2. dust of molecular clouds and star-forming regions (Section IV).

There are at least two additional dust populations closely connected with stellar evolutionary states:

3. dust in stellar outflows and explosion shells of evolved stars ("stardust", Section V) and
4. circumstellar dust in young stellar objects ("YSO dust") and in disks around main-sequence stars ("Vega-phenomenon dust", Section VI).

These main populations represent a rough classification scheme, and the introduction of subdivisions appears necessary. For example, stardust of oxygen-rich sources is chemically completely different from that of carbon-rich sources. The dust of a carbon-rich Mira star differs from that of a carbon-rich planetary nebula or a WC star.

The investigation of the interaction between stars and ISM, the definition of ISM phases and of dust populations have been the obstetricians of an evolutionary dust concept which describes individual grain evolution (Greenberg 1984) as well as the general role of dust as an evolving medium within the galactic evolution ("galactic dust metabolism", Dorschner 1992; the "lifecycle of dust", Jones 1997). The evolutionary concept is based on four typical chains of phenomena and processes:

1. Evolved stars inject virgin stardust in the interstellar space, where the fresh grains are mixed with old, heavily processed diffuse ISM dust. Calculations show that the present interstellar dust could be replenished by stardust on a time-scale of $2.5 \cdot 10^9$ years (Jones and Tielens 1994).

2. In the diffuse ISM, the grains are subjected to UV irradiation, supernova shocks, and cosmic rays. These environmental factors considerably modify the material properties and limit the grain lifetime to less than 10^9 years (Jones et al. 1996). If these theoretical estimates are realistic, additional contributions are necessary to compensate the destruction losses. Interstellar recondensation and phase transitions between diffuse and molecular clouds have been proposed. However, these processes are insufficiently understood.

3. In molecular clouds the grain lifetime is even more reduced. After 10^5–10^7 years (Blitz 1993) a cloud dissolves by star formation. Dust is incorporated in new stars, and the radiation output of a newly formed massive star destroys volatile grain components in a large volume. Only a small portion of the carbon-bearing mantle ices is transformed to organic refractories due to photolytic

processing by the stellar UV radiation. They form thin coatings around the refractory grain cores, e.g. silicate grains, the remainders of the core-mantle grains after the heating and evaporating of ices in star-forming regions. These processed refractory remainders of the molecular cloud dust are swept into the diffuse ISM. Many dust grains in the diffuse ISM could have passed the molecular cloud stage and the photolytic mantle processing several times and, therefore, should have relatively thick organic layers (Greenberg and Li 1995).

4. Star formation, which destroys large quantities of molecular cloud dust, is a sink for interstellar dust. However, the protostellar disks and the envelopes around newly formed stars are a breeding ground of solids on all size scales, from dust grains up to planets. Disintegration processes of larger bodies in planetary systems permanently produce small grains driven out of the system by stellar radiation and stellar wind.

III. DUST IN DIFFUSE INTERSTELLAR CLOUDS

III.A. Basic Observational Phenomena

1. Interstellar Extinction

Apart from dark nebulae that strikingly demonstrate the presence of single heavily absorbing clouds between the stars, there is a general dust extinction by the diffuse ISM providing dimming and reddening of starlight. Comparing the spectrum of a dimmed star with that of a star having the same spectral type and not being influenced by extinction allows to derive the wavelength dependence of the interstellar extinction, the so called extinction curve. Usually, it is given by the normalized amount of extinction A (in mag) plotted vs. the wavenumber λ^{-1} (in μm^{-1}). $A(\lambda^{-1})$ is connected with the optical grain properties expressed by the extinction efficiency Q_{ext} and the grain radii distribution function $n(a)$, and the distance Δl to the star by the integral relation

$$A(\lambda^{-1}) = 1.086 \Delta l \int_0^\infty Q_{\text{ext}} \pi a^2 n(a) \, da \, . \tag{1}$$

The efficiency factor for extinction, Q_{ext}, depends on wavelength, grain radius, and dielectric function of the grain material.

Now as ever, extinction studies are considered the most important observational access to the interstellar dust. Therefore, this subject has been thoroughly reviewed in the last decade (Mathis 1993, Cardelli 1994). From extinction measurements we have learnt that the diffuse ISM in the Galaxy is characterized by a nearly uniform gas-to-dust mass ratio of about 90 and that the average dust density amounts to about $\overline{\rho_d} = 1.8 \cdot 10^{-26}$ g cm^{-3} (Whittet 1992). From the shape of the extinction curve and its variations important conclusions on the interstellar dust have been drawn.

For small values of λ^{-1} the extinction curve is rather uniform in the Galaxy. In the range 1–5 μm, Martin and Whittet (1990) derived the "universal" law

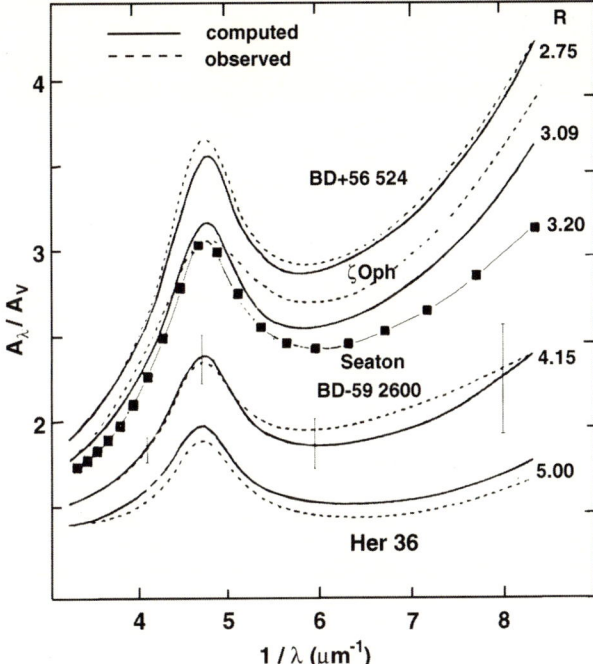

Figure 1. Normalized observed UV extinction curves (dashed lines) compared with parameterized representations (solid lines). With increasing parameter R the curves become flatter. The representation with $R = 3.20$ reproduces the mean extinction curve (squares) labeled by "Seaton". The error bars are the general r.m.s. deviations from the parameterization ansatz. After Cardelli et al. (1988).

$A(\lambda) \propto \lambda^{-1.8}$. In the visual part, i.e. for wavenumbers between 1.5 and 3 μm^{-1}, the shape is approximately linear ("$1/\lambda$ law"). A basic observational parameter measuring the slope in this range is $R = A_V/E(B-V)$ (A_V: total extinction in the visual passband V; $E(B-V)$ color excess in the UBV photometric system). For the diffuse ISM, R takes the rather uniform value $R = 3.1$, whereas dense clouds and star-forming regions show values $3.0 < R \leq 6.0$, i.e. they have flatter extinction curves in the visual and UV ranges (cf. Fig. 1).

Extraterrestrial observations have extended the extinction curve into the far UV up to $\lambda^{-1} \approx 10$ μm^{-1}. At $\lambda^{-1} \approx 4.6$ the monotonous extinction rise is interrupted by the "UV bump", a superimposed prominent solid-state band at 217.5 nm. After removing the continuous extinction background, the remaining band excess can be easily fitted by a dispersion profile (Lorentz profile; see, e.g. Dorschner et al. 1984). Some authors have used the denotation "Drude profile" (Fitzpatrick and Massa 1986, 1990) which is mostly used in the current literature. The profile shape proves that the particles responsible for this band must be very small compared with the wavelength. In this case, the extinction is pure absorption, scattering does not play a role.

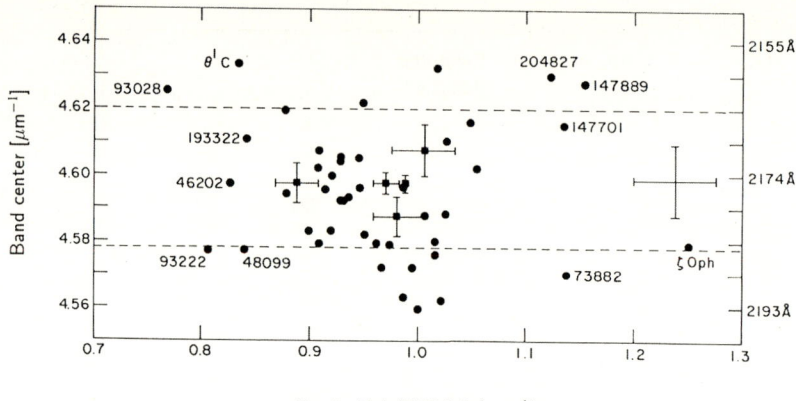

Figure 2. Peak position of the 217.5 nm band vs. band width (full width of half maximum) for interstellar extinction curves toward 45 reddened Milky Way stars measured by the IUE satellite. Squares with error bars are mean values of star clusters. The error bars at the right give the 1σ uncertainty for an individual measurements of a star with $E(B-V) = 0.55$ mag. Points representing extinction curves with most extreme band widths are labeled by the stars' HD numbers or names. The dashed lines limit the $\pm 2\sigma$ departures from the mean peak position $\lambda_0^{-1} = 4.599$ μm^{-1}. There is no correlation between positions and widths. The scatter of the peak positions is much smaller than that of the widths. After Fitzpatrick and Massa (1986).

The most noteworthy property of the 217.5 nm band is its nearly invariant position (λ_0=217.5 ± 1 nm) in spite of the large scatter of the width (FWHM = 48 ± 12 nm) along various lines of sight (Fig. 2). There is a general consent that this band is due to the π–π^* transition in nm-sized particles of graphitized carbon. The best experimental reproduction of the profile was reached by matrix-isolated nm-sized hydrogenated soot particles (Schnaiter et al. 1998; see Sec. III.B.4).

Beyond the bump, the extinction curve rises non-linearly. Its reproduction requires very small grains. According to a proposal by Greenberg (1973), the whole extinction curve is usually decomposed in three components: the visual part caused by the extinction of "big" grains (size in the order of 0.1 μm), the bump part with the Lorentz profile, and the steeply rising far-UV part. Several authors derived mathematical expressions for the whole curve, which contained observational parameters only. Cardelli et al. (1988) obtained a parameterized representation, in which, apart from the 217.5 nm band profile parameters, the quantity R alone determines the shape of the whole extinction curve (Fig. 1).

Much observational work has been devoted to the search for environmental influences on the shape of the extinction curve and its parameters. The increase of the extinction parameter R flattens the extinction curve in dense regions, this does not necessarily mean that the strength of the 217.5 nm feature is also reduced. However, in HII regions its strength decreases with increasing radiation field intensity. No correlations of the large far UV extinction variations with environmental parameters were found (Jenniskens and Greenberg 1993).

2. Interstellar Polarization

In addition to dimming and reddening, interstellar dust grains also polarize starlight. Extinction is a necessary, but not a sufficient condition for the occurrence of polarization. Of decisive importance is the alignment of the dust grains. A concise description of the alignment mechanism by magnetic fields can be found, e.g. in the monograph by Whittet (1992). An overview over the problems of polarization of starlight by interstellar dust from the standpoint of an observer was given by the same author (Whittet 1996).

The wavelength dependence of the polarization $P(\lambda)$ has a broad maximum in the visual range with its peak at $\lambda_{\max}$ varying from 350 to 900 nm; the typical value is $\lambda_{\max} = 550$ nm. A well-studied parameterized representation of the polarization curve was derived by Serkowski (1973):

$$\frac{P(\lambda)}{P_{\max}} = \exp\left[-K \ln^2(\frac{\lambda_{\max}}{\lambda})\right]. \qquad (2)$$

The parameter K turned out to be weakly dependent on $\lambda_{\max}$: $K = (0.01 \pm 0.05) + (1.66 \pm 0.09) \cdot \lambda_{\max}$ (Whittet et al. 1992). Between the basic polarization and extinction parameters the simple relation $R = 5.7\, \lambda_{\max}$ was found, wherein $\lambda_{\max}$ is measured in μm (Vrba et al. 1981).

In the UV, spectropolarimetric data are still rare. Significant UV polarization along several lines of sight was measured with the Wisconsin Ultraviolet Photo-Polarimeter Experiment (WUPPE) during the Astro-1 and Astro-2 missions (Anderson et al. 1996, Wolff et al. 1997). The data indicate that in some cases stars follow the extrapolation of Eq. (2) in the UV, whereas in others excess polarization in comparison to Eq. (2) is indicated. Clear evidence of polarization within the 217.5 nm band has only been found for the stars HD 147933-4 and HD 197770. The 217.5 nm carrier seems to be distinct from the dust component responsible for the interstellar continuous polarization.

In addition to the linear interstellar polarization a weak circular polarization has been observed (Martin 1989). It is caused by the birefringence of the interstellar space due to the aligned elongated dust grains that produce the linear polarization. The circular polarization changes its sign at the wavelength $\lambda_{\max}$.

3. Diagnostic Absorption Features

Up to now, about 200 diffuse interstellar bands (DIBs) in the wavelength range 400–1300 nm are known (Jenniskens and Désert 1993, Herbig 1995). The strongest band is that at 443 nm with FWHM = 2 nm. These features have resisted identification so far; they represent the biggest problem in astronomical spectroscopy and in laboratory astrophysics. Whilst the DIBs formerly were considered a dust phenomenon, at present, there is a preference for explaining them by transitions of carbon-bearing molecules (Freivogel et al. 1994). A critical review of the different carrier hypotheses was given by Herbig (1995).

Table 1

Vibrational dust bands detected in diffuse ISM clouds. λ: wavelength; certainty (c: certain, t: tentative assignment); S: band's strength (s: strong, m: medium, w: weak). Abbreviations: DC diffuse cloud, GC Galactic Center, MC molecular cloud, RO refractory organics, SIL silicate. Compiled from data by Tielens et al. (1996), Lutz et al. (1996), Whittet et al. (1997), Schutte et al. (1998), Chiar et al. 2000.

λ μm	Identification (Species; certainty)	S	Comment
2.75	O–H stretch (OH in hydrated SIL; t)	w	Cyg OB2 No.12 only
3.0	O–H stretch (OH, H_2O in hydrated SIL or H_2O ice[1]; t)	s	GC sources only [1] MC instead DC dust?
3.4-3.5	C–H stretch of -CH_3 and -CH_2 (aliphatic RO; c)	m	several bands many sources
5.5	C=O stretch (aliphatic RO; t)	w	GC sources only
5.9	C=O stretch (metal carbonyles; t)	w	GC sources only
6.1	O–H bend (same species as λ3.0; t)	w	GC sources only
6.2	C=C stretch (aromatic RO; t)	w	GC sources only
6.8	C–H bend (asym.) (aliphatic RO; c)	m	
7.3	C–H bend (sym.) (aliphatic RO; c)	w	GC sources only
9.7	Si–O stretch (amorphous SIL; c)	ss	many sources
18.7	O–Si–O bend (amorphous SIL; c)	s	many sources

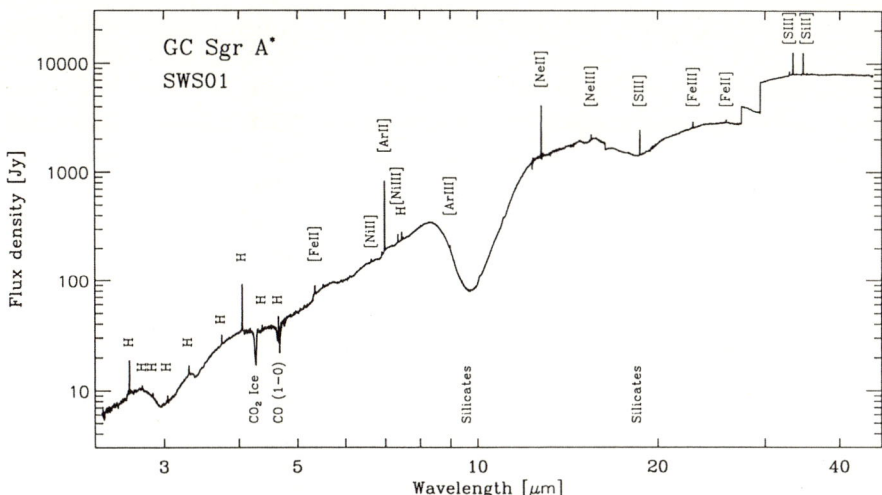

Figure 3. ISO-SWS spectrum of the Galactic Center source Sgr A*. The solid-state features (in μm) are: 3.0 (O–H stretch), 3.4/3.5 (aliphatic hydrocarbons), 4.26 (CO_2 ice), 6.1 (O–H bend), 6.8 (hydrocarbon, methanol?), 7.3 (hydrocarbons), 7.7 (CH_4 ice), 9.7 (silicate), 15.2 (CO_2 ice), 18 (silicate). The gas lines at $\approx$ 4.66 and 4.77 are the fundamentals of ^{12}CO and ^{13}CO, resp., of cold foreground gas. A band at 34.6 μm is due to the OH molecule. The hydrogen and other gas lines come from HII regions. After Lutz et al. 1996.

In the spectra of some heavily obscured strong IR sources (WC stars, the hypergiant Cyg OB2 No. 12, GC sources) a growing number of vibrational bands has been detected which are due to refractory dust components present in the diffuse ISM along the line of sight towards these sources (cf. Fig. 3). Most of these features listed in Table 1 are positioned in the wavelength range 5–8 μm that, up to now, was accessible only by the Kuiper Airborne Observatory (KAO) and more recently by the Infrared Space Observatory (ISO) spectrometers.

Most of the bands in Table 1 are due to refractory organics in carbonaceous grain material. The precise chemical composition is unknown, however, the bands allow identifying chemical groups. In most cases they indicate aliphatic compounds, but aromatics cannot be excluded. There is convincing evidence that these refractory organics can be explained as the result of UV-photoprocessing of interstellar ices in the frame of dust evolution (Greenberg 1984; see Sec. II.C.). During their life-cycle (Jones 1997) interstellar dust grains repeatedly pass the molecular-cloud stage, at the end of which a small part of the carbon-bearing mantle ices (cf. Sec. IV.A.II) is transformed to refractory organics by the UV radiation output of newly formed massive stars. This way, carbonaceous material gradually concentrates in the refractory dust grains of the diffuse ISM. Striking resemblance of the spectra of experimentally produced photolytic residues and of carbonaceous components of primitive meteorites with the interstellar bands supports this picture (Sandford et al. 1995, Pendleton and Chiar 1997).

The silicate bands are the strongest among the interstellar absorption features (cf. Fig. 3). In the astrophysical context, the term "silicate" only means that in this solid Si–O bonds are present in a tetrahedral arrangement allowing both Si–O stretching vibrations (10-μm band) and O–Si–O bending vibrations (18 μm band). In contrast to terrestrial silicate minerals, the interstellar silicates show broad and almost structureless band profiles that prove lacking long-range order. Usually, these amorphous cosmic silicates are compared with silicate glasses as laboratory analogs (Jäger et al. 1994, Dorschner et al. 1995). Single glass species in most cases fail to reproduce the large widths of the observed interstellar silicate bands. However, mixtures of different glasses give satisfactory fittings. Furthermore, the large bandwidths can be caused by disordering effects typical of the interstellar environment, e.g. long-time irradiation by cosmic rays.

Both the strengths of the interstellar Si–O stretching bands and the C–H stretching bands correlate well with the visual extinction. The relationship is not strictly linear, it is, however, of the same type for both dust components (Sandford et al. 1995). This suggests some kind of association of these components, e.g. by a core-mantle structure (Greenberg and Li 1995), but it does not cogently mean that the band carriers and the components providing the main part of the visual extinction are identical with each other.

4. Scattering, Fluorescence, and Luminescence

If light of a bright star is scattered by nearby dust grains the phenomenon of a reflection nebula occurs. The total of light scattered by the dust near the galactic plane produces the diffuse galactic light. Bright rims sometimes observed with dark clouds are due to the scattered interstellar radiation field.

From observations of the scattered starlight the grain albedo $\gamma(\lambda)$ and the asymmetry factor g of the scattering function have been derived. They are defined by

$$\gamma(\lambda) = \frac{Q_{\rm sca}(\lambda)}{Q_{\rm ext}(\lambda)} \qquad \text{and} \qquad g = \langle \cos \alpha \rangle, \qquad (3)$$

wherein $Q_{\rm sca}$ and $Q_{\rm ext}$ are the efficiencies for scattering and extinction, respectively, and $\langle \cos \alpha \rangle$ is the weighted mean of the cosine of the scattering angle α with the scattering function $\varphi(\alpha)$ as the weighting function. Isotropic scattering means $g = 0$, forward-scattering $g = 1$. Representative values in the visual are $\gamma = 0.6 \pm 0.1$ and $g = 0.6 \pm 0.2$ (Witt 1989). This points to dielectric grains with sizes comparable to those of the big grain mode in the extinction models.

Early UV data of the diffuse galactic light measured by the OAO-2 satellite showed that the dust albedo is strongly reduced within the range of the 217.5 nm extinction feature (Lillie and Witt 1976). This clearly points to the absorption nature of this band. Observations of the best-studied reflection nebula, NGC 7023, with the Ultraviolet Imaging Telescope of the Astro-1 mission (Witt et al. 1992) confirmed this finding. The data also show that the UV scattering functions (up to 140 nm) must be strongly forward directed ($g \approx 0.75$) with an albedo of $\gamma \approx 0.65$.

The IR spectra of reflection nebulae in the range 1–25 μm show emission surprisingly far in excess over the scattered light and the thermal dust emission. Independent of the distance from the star, the color temperature of this IR continuum ($\approx$ 1000 K) was the same in most nebulae. Two mechanisms have been proposed for the explanation of this excess: non-equilibrium thermal emission of nanometer-sized grains (Sellgren 1989) and fluorescence radiation of large PAH molecules (Puget and Léger 1989). In the first mechanism the absorption of single UV photons produces discrete temperature spikes of the grains far above the equilibrium temperature and in the second the stellar UV radiation excites the molecules, and the deexcitation is performed by the vibrations of the chemical groups present in these molecules.

A phenomenon closely connected with the non-equilibrium IR continuum seems to be the UIR bands. As mentioned in Sec. I.E., their most intensive representatives are positioned at the wavelengths 3.3, 6.2, 7.7, 8.6, 11.3, and 12.7 μm. They have been first observed in reflection and planetary nebulae and later on in a great variety of dusty environments, e.g. novae, WC stars, H II regions, young stellar objects, Vega-excess stars (Sellgren 1994, Geballe 1997). Most of the strong UIR bands have also been detected in the large-scale galactic emission, the so-called IR cirrus (Giard et al. 1994, Mattila et al. 1996).

The UIR bands are surely due to aromatically bonded carbon. There is, however, a basic controversy on the nature of their carriers: Are they large two-dimensional clusters, i.e. polycyclic aromatic hydrocarbon molecules (PAHs), Allamandola et al. 1987, Puget and Léger 1989, d'Hendecourt 1997), or are they very small, but three-dimensional grains of carbonaceous solids. Estimates of the size of the molecules responsible for the strong UIR bands resulted in numbers between 50 and 300 C-atoms. From the observations of the 60 μm IRAS cirrus (see next paragraph), it was concluded that the carrier should consist of $5 \cdot 10^4$ C-atoms, which corresponds to nanograins of about 5 nm diameter (Tielens 1995). The multitude of different experimental approaches to solve the identification problem of the UIR bands has produces a correspondingly large number of laboratory analogs for these solids (see the review by Sellgren 1994).

Different dusty environments of the Galaxy show photoluminescence. It was detected as a band-like excess radiation in the range 540–900 nm and got the name Extended Red Emission (ERE). For many years the Red Rectangle was the only source (Witt 1989), then many other objects with the ERE feature were found: reflection nebulae, dark nebulae, galactic cirrus clouds, H II regions, planetary nebulae, finally ERE was detected in the diffuse galactic light (Gordon et al. 1998). The observations suggest that ERE requires excitation by UV radiation, and the band strength is proportional to the incident flux. The peak of the wide band (FWHM $\approx$100 nm) shifts with increasing radiation field density from 610 to 820 nm. Reviews of the observed ERE characteristics can be found in Gordon et al. (1998) and Witt et al. (1998). The interpretation of the ERE photoluminescence was based for a long time on the supposed carbonaceous nature of the carrier (for a literature review on such hypotheses see Seahra and Duley 1999). Recently, alternative interpretations

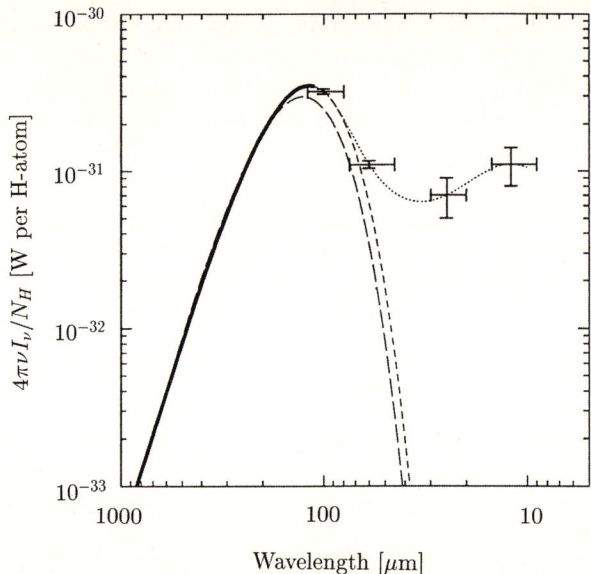

Figure 4. Thermal dust emission spectrum after COBE FIRAS data (Wright et al. 1991) compared with the normalized fluxes in IRAS passbands at 100, 60, 25, and 12 μm. The COBE data (solid line) were normalized to join the IRAS 100 μm flux. The dashed lines are two extrapolations to shorter wavelengths. For details see Draine (1994).

to the carbon-based hypotheses have been proposed that discuss the ERE as being due to silicon nanograins (Ledoux et al. 1998, Witt et al. 1998). In some sources exhibiting strong ERE some sharp features have been detected which could possibly be somehow connected with the ERE carrier (Duley 1988, Gordon et al. 2000).

5. Thermal Dust Emission

Starlight is both scattered and absorbed by the dust near the galactic plane. The energy absorbed by the grains is transformed to heat and reemitted as thermal dust radiation supplying for most of the galactic FIR background. Although discovered by rockets, the full extent of the thermal radiation output of the Galaxy became visible only after the IRAS observations (Cox and Mezger 1989, Sodroski et al. 1989).

In its passbands at 60 and 100 μm, IRAS discovered a filamentarily distributed emission called IRAS cirrus, which accounts for most of the galactic IR background. The cirrus intensity at 100 μm correlates well with many interstellar gas and dust parameters (Boulanger 1994). Whilst the FIR cirrus emission can be explained by dust grains in thermal equilibrium with the interstellar radiation field, Boulanger et al. (1985) also found cirrus emission in the MIR IRAS passbands at 12 and 25 μm far in excess of equilibrium emission of dust. These observations confirmed the occurrence of nm-sized grains that are stochastically heated by the absorption of single stellar UV photons. This

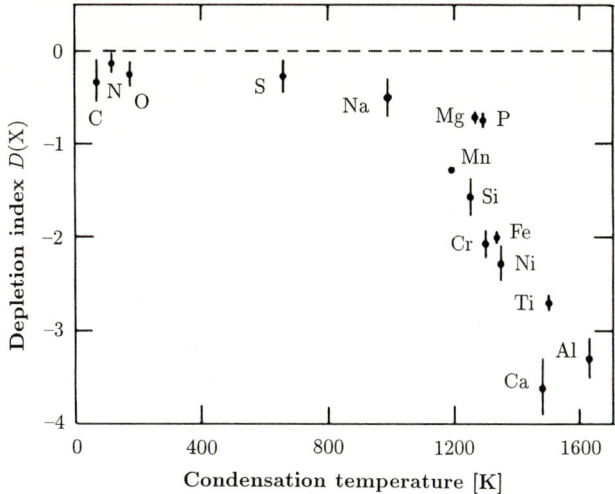

Figure 5. Depletion index $D(X)$ (for definition see Eq. (4)) vs. condensation temperature of various elements. Note the almost total depletion of the refractory metals. After Whittet (1992).

non-equilibrium emission is confined not only to regions with enhanced UV radiation, e.g. reflection nebulae and HII regions, but also occurs on the whole diffuse ISM scale. It is correlated with extinction characteristics (Boulanger et al. 1994).

More recent FIR results obtained by the COBE satellite (Wright et al. 1991, Sodroski et al. 1994) confirmed the IRAS results. From the Far InfraRed Absolute Spectrophotometer (FIRAS) measurements which show the FIR emission peak at about 140 μm, Draine (1994) estimated $T = 18 \pm 2$ K as a "typical" temperature of dust grains in the diffuse clouds of the solar neighborhood (Fig. 4).

6. Interstellar Elemental Depletion

Investigations of interstellar absorption lines show that dust-forming elements are conspicuously underabundant in the interstellar gas (Fig. 5). This depletion is considered as evidence for interstellar in-situ dust formation (Draine 1994) which, on the other side, has been required for compensating the destruction losses (cf. Sec. II.C.)

Depletion refers to the underabundance of a gas phase element in comparison with a standard. Usually the solar system abundances are taken as the standard for the interstellar gas, a somewhat arbitrary assumption. The fractional depletion $\delta(X)$ of the element X is given by

$$\delta(X) = \frac{(\frac{N_X}{N_H})_\odot - \frac{N_X}{N_H}}{(\frac{N_X}{N_H})_\odot} = 1 - 10^{D(X)}, \qquad (4)$$

wherein N_X is the number of atoms of the element X, $D(X)$ is the depletion index (logarithm of the ratio of the interstellar to the solar abundance), and $\odot$ labels the solar standard. Values of $D(X)$ can be derived with high accuracy from interstellar line observations. They show conclusive correlations with condensation temperatures of the elements, with gas densities, and with cloud velocities. For discussion of these connections see, e.g. Whittet (1992).

Depletion data have been used for estimates of the average grain composition in the diffuse ISM (Whittet 1984). More recent depletion determinations (Sofia et al. 1994) of O, C, N, Mg, Si, Fe, and S with the Goddard High Resolution Spectrograph aboard the Hubble Space Telescope, in which apart from solar also B star abundances have been used as references, allowed the conclusion that in contrast to former assumptions Mg and Fe atoms cannot be contained exclusively in silicates. This points to oxides as possible additional grain ingredients.

III.B. Dust Models

1. Constructing Dust Models

Based on the observations discussed in Sec. III.A., dust models have been proposed. Constructing a dust model requires the following steps:

1. Make observationally justified assumptions on chemical composition, shape, and size of the grains.
2. Obtain realistic data of the optical properties of the proposed dust materials, e.g. from laboratory measurements.
3. Calculate absorption and scattering efficiencies for each grain type, considering the individual grain model.
4. Construct appropriate mean values of observable quantities and compare them with the observations.

Since ices are not stable in the diffuse ISM, the dust models consist of mixtures of refractory grain components (Table 2). The grain models cover a wide range of structure types from the simplest case of compact naked or core/mantle-grains to coagulated grains built up from regular (spherical) subgrains. Exact solutions for the interaction with electromagnetic radiation only exist for spheres, ellipsoids and infinite cylinders (Bohren and Huffman 1983). The cross-sections of inhomogeneous and irregularly shaped grain aggregates can be calculated by applying the mixing rules of the effective-medium theory and approximative methods, e.g. discrete dipole approximation (Gustafson et al., this book).

Natural "dust analogs" such as IDPs and presolar meteoritic grains and laboratory experiments simulating particle formation in non-equilibrium conditions suggest that most of the current dust models are likely based on too simplified assumptions, e.g. monomineralic components. In this context, Nuth et al. (1999) proposed a petrological analogy. The refractory dust grains are considered as rock-like chemical systems containing on a microscale a greater number of chemical components, which are the fingerprints of the formation and evolution conditions of the grains. Gaining experimental experience on this "petrological way" could be the tool for substantial improvement of the dust models.

2. Grain Size Distribution

In expressions relating theoretical and observational quantities, e.g. Eq. (1), the radii distribution function $n(a)$ of the interstellar dust grains appears under the integral sign, i.e. in a mathematical structure preventing the direct determination of $n(a)$. The current practice to solve this inverse problem is to calculate the cross-sections of the interaction with electromagnetic radiation for the grain species present in the model and to find astrophysically reasonable functions $n(a)$ that allow reproducing the observed absorption bands and the wavelength dependences of observable continuous quantities, e.g. extinction curve.

Greenberg (1973) proposed a decomposition of the extinction curve in three contributions: a big mode ($\approx 0.1\ \mu$m) of dielectric grains (silicates) responsible for the visual and NIR extinction, a very small, almost monodisperse "bump" mode for the 217.5-nm band carrier (commonly graphite grains), and a very small mode for the representation of the FUV extinction slope (silicates). The size distribution of the big grain mode was derived from the assumption of a steady-state between collisional grain destruction and continuous grain growth by gas accretion (Hong and Greenberg 1978). The result was

$$n(a) \propto \exp(-Ca^3), \qquad (5)$$

wherein C is a parameter depending on the probability for collisional destruction and on the growth rate of the grains.

In contrast to this function being based on a theoretical grain evolution concept, Mathis et al. (1977) could show that a power law of the type

$$n(a) \propto a^{-3.5} \qquad (6)$$

gives a satisfactory reproduction of the extinction curve in the wavenumber range 1–9 μm^{-1} if spherical graphite and silicate grains with a between the limits $a_-=0.005$ μm and $a_+=0.25$ μm are admitted. The function in Eq. (6) is frequently used as a standard size distribution (abbreviated MRN-distribution) for the dust in the diffuse ISM. The empirical approach was improved by Kim et al. (1994) and Kim and Martin (1995), who extracted size distributions from extinction and polarization curves, respectively, by adopting the MRN model, but applying the more sensitive maximum entropy method. The results are shown in Fig. 6.

Power laws like Eq. (6) reflect a basic distribution principle of nature, the scale invariance. Collisional fragmentation is a special mechanism generating a scale-invariant size distribution (Henning et al. 1989). It was the success of the MRN size distribution, which brought collisional fragmentation of interstellar and circumstellar grains permanently in the discussion (Biermann

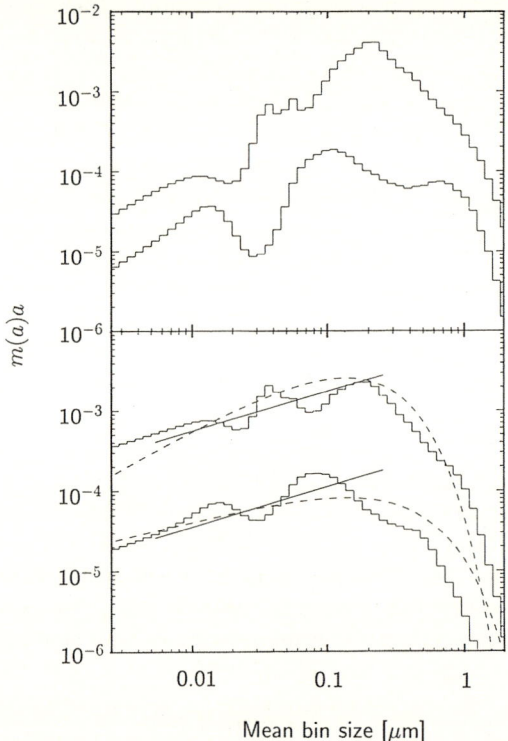

Figure 6. Distribution of grain masses derived from parameterized extinction curves according to Cardelli et al. (1988) for bare silicates (upper histograms in each graph) and for bare graphite grains (lower histograms, for clarity scaled down by a factor of 10). The ordinate is a modified mass distribution, $m(a)a$, because the application of the Maximum Entropy Method requires the use of distributions as flat as possible. Bottom: Dust in diffuse clouds (parameter $R = 3.1$); straight lines are the MRN power law, dashed lines are power laws with exponential decay. Top: Dust in dense clouds ($R = 5.3$). After Kim et al. (1994).

and Harwitt 1980, Dorschner 1982). Extrapolation of the MRN distribution runs into the size range of the IRAS-cirrus nanograins and of the PAH clusters (see Sec. III.A.4 and 5). Supernova shocks could provide the mechanism that produces the nanograins by collisional shattering (Jones et al. 1996).

There are, however, some arguments speaking against the universality of the MRN distribution. The mass determination of the dust particles of the local ISM, which were detected on their passage through the solar system by the spacecraft Ulysses and Galileo (Grün et al. 1994, Frisch et al. 1999) revealed striking deviations from the MRN distribution: There is a conspicuous deficiency of grains with masses smaller than 10^{-13} g (corresponding to diameters smaller than about 0.4 μm) and, further, the cut-off at diameters of 0.5 μm is not observed. There are, of course, some uncertainties in this detection techniques and in the assumptions of the dynamical model necessary for the mass derivation, however, the overabundance of the big particles seems to be a trustworthy fact (Landgraf et al. 2000). Using the new data, estimates of the mean dust density and the gas-to-dust ratio of the local interstellar cloud are compatible with the mean interstellar values (Frisch et al. 1999, Landgraf et al. 2000). The observed deficiency of the small grains finds a simple explanation by the electrodynamic filtration process in the heliosphere due to the Lorentz force on charged grains (Linde and Gombosi 2000).

Recent radar meteor studies have shown that micrometeoroids with hyperbolic velocities indicating interstellar origin exist down to masses of 10^{-7} g, corresponding to grain sizes of some 10 μm radius (Taylor et al. 1996, Baggaley 2000). If confirmed this would mean that the interstellar dust also contains a population of grains with sizes comparable to those in the circumstellar dust disks of the Vega phenomenon (see. Sec. VI.B.).

3. Synopsis of Current Dust Models

In Table 2 the parameters of some important dust models are listed. Most of them have been proposed in order to explain quantitatively the average extinction curve, prominent IR absorption features, and the galactic FIR emission. The "unified model" by Li and Greenberg (1997) also includes the interstellar polarization curve (it has been a strong point of Greenberg's strategy for decades always to include polarization arguments in his "unified models"). All models in Table 2 use silicates for the explanation of the observed bands at 9.7 and 18.7 μm. Carbonaceous phases are also a necessary ingredient in order to explain the C–H absorption bands at 3.4–3.5 μm and the 217.5 nm feature. The latter is mostly reproduced by cross-section calculations based on laboratory data of graphite. Désert et al. (1990) used simply the average observed profile of the 217.5 nm band and supposed that it is due to very small grains of a carbonaceous solid. An exception is the model by Williams (1989) who explained the 217.5-nm feature by a surface effect of silicate grains. In the last years, the revised data of elemental abundances have confronted the interstellar dust modelling with tight constraints (Mathis 1996, 1998).

Table 2

Dust models for the diffuse ISM. Grain types: c-m core-mantle, co composite, ml multi-layered, b bare. Species: AC amorphous carbon, HAC hydrogenated amorphous carbon, CAR unspecified hypothetical carbon, GRA graphite, PAH polycyclic aromatic hydrocarbons, RO refractory organics, MOX metal oxides, SIL silicates, SI silicon, FE iron, VAC vacuum, WI water ice. Size distribution function: d discrete size or narrow interval, exp exponential law (Eq. (5)), g giant grains ($\geq 10\ \mu$m), p power law (Eq. (6)), vs very small grains.

Authors (year)	Grain type (Species)	Size distr. function	217.5 nm carrier
Draine & Lee (1984)	b (SIL, GRA)	p	GRA
Chlewicki & Laureijs (1988)	c-m (c: SIL, m: RO), b (GRA, PAH, FE)	exp, d	GRA
Greenberg (1989)	c-m (c: SIL, m: RO), b (GRA)	exp, d	GRA
Williams (1989)	c-m (c: SIL, m: HAC), b (SIL),	p, vs	SIL
Mathis & Whiffen (1989)	co (SIL, GRA, HAC, VAC), b (GRA)	p	GRA
Désert et al. (1990)	c-m (c: SIL, m: RO), b (AC, PAH)	p, vs	CAR
Sorrell (1990)	b (SIL, AC, GRA)	d	GRA
Rowan-Robinson (1992)	b (SIL, AC, GRA)	d, g	GRA
Siebenmorgen & Krügel (1992)	b (SIL, AC, GRA, PAH)	p	GRA
Mathis (1996)	co (AC, HAC, GRA, SIL, MOX, VAC), b (GRA, SIL, MOX)	p	GRA
Li & Greenberg (1997)	c-m (c: SIL, m: RO), b (CAR, PAH)	exp, p	CAR
Zubko et al. (1999)	c-m (c: SIL, m: RO), ml (SIL, RO, WI), b (GRA, SI)	d	GRA

4. The Enigmatic Interstellar Carbon Dust

The spectroscopic investigation of solid circumstellar and interstellar phases dominated by carbon (we use the summarizing term carbonaceous dust) presents a lot of problems. In cosmic environments, different carbonaceous solids occur (for review see Henning and Salama 1998), the structures of which are only qualitatively understood. In laboratory experiments carbon dust analogs have been produced along very different experimental ways (combustion of hydrocarbons, carbon vaporization in arcs or by laser pulses, recondensation of sputtered carbon, laser pyrolysis of organic compounds etc.). The result was a confusingly great manifold of laboratory products with different (in some cases insufficiently characterized) carbon microstructure, which have been used for the explanation of the observations. We point to important experimental studies by Duley (1993), Koike et al. (1995a, b), Jäger et al. (1998a), Colangeli et al. (1999), Henning and Schnaiter (1999).

The experimental approach to the identification of the 217.5 nm band carrier is mainly based on "amorphous carbon" (a-C) or "hydrogenated amorphous carbon" (a-C:H, HAC), i.e. heterogeneous soot-like solids containing sp^3- as well as sp^2-hybridized carbon in different proportions. The designation "amorphous" is somewhat misleading because this material contains microcrystalline units of graphitic structure consisting of staples of plane or slightly bent graphene layers (planar networks of aromatically bonded carbon). The aromatic bonds rest on sp^2 hybridized orbitals and delocalized π-electrons. The plasmon resonance of the latter is considered the cause of the 217.5 nm band (π–π^*-transition).

The possible significance of HAC for the explanation of the interstellar 217.5 nm feature was demonstrated by Mennella et al. (1995a, b), who gradually dehydrogenated a-C:H and could show that the hydrogen content is of great influence on the structure of the solid and on the wavelength position of the π–π^*-band. They succeeded in correctly reproducing the observed position, however, the band width in the laboratory spectra was considerably greater than in the observations. The experiments could also explain why the UV carbon absorption band in hydrogen-deficient C-rich stars (see Sec. V.B.1.) appears at much longer wavelengths than in interstellar space.

Meanwhile, more sophisticated laboratory experiments by Schnaiter et al. (1998) have considerably improved the understanding of the carbon structure responsible for the interstellar UV band. The method started with carbon condensation in a hydrogen atmosphere and applied molecular beam techniques to form a beam of nanometer-sized hydrogen-bearing soot particles. The particles were frozen in a noble gas matrix to prevent them from forming clumps. Via this matrix-isolation spectroscopy the up to now best reproduction of the interstellar band profile was reached. However, the peak absorption coefficient of the laboratory grains was not big enough to be consistent with the new constraints of the cosmic carbon abundance (Sec. III.B.3). Interstellar grains of pure graphite would overcome this problem, but they do not appear to be

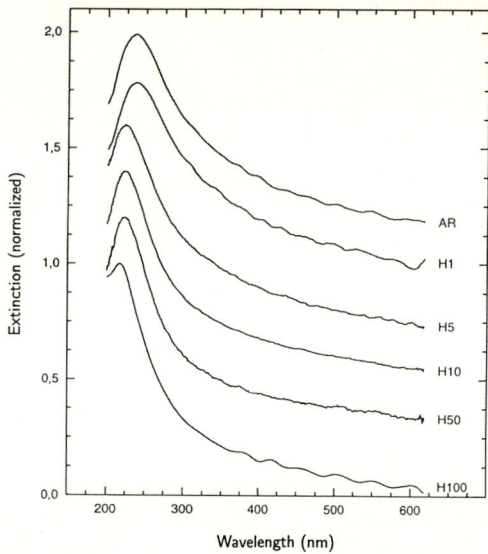

Figure 7. Laboratory spectra of nanometer-sized particles of hydrogenated carbon elucidating the observed properties of the interstellar 217.5 nm band assigned to the electronic π–π^* transition of carbon. The spectra are normalized and for clarity shifted by amounts of 0.2 to 0.25. The soot particles condensed in a 10 mbar quenching Ar/H_2 atmosphere of varying composition: pure Ar (AR), volume ratios of H_2:Ar=i:100 with i=1, 5, 10, and 50 (H1, H5, H10, H50), and pure hydrogen (H100). Hydrogen incorporation shifts the band peak toward shorter wavelengths and diminishes the width. H50 represents the best experimental reproduction of the interstellar 217.5 nm band profile, so far achieved. After Schneider et al. (1998).

very realistic, even if dust model designers (see Table 2) like to resort to this possibility. However, the abundance problem is not the only one of interstellar carbon solids. We do not know why the UV band is absent in circumstellar environments of C-type stars and in solar system solids. A further problem is the compensation of the high destruction rate of carbon grains in the diffuse ISM.

Another ubiquitous form of interstellar carbonaceous material containing aromatical carbon bonds are the carriers of the UIR bands (see Sec. I.E. and III.A.4.). PAH molecules are certainly present in C-rich stars and form the molecular precursors of graphitic carbon. PAHs or nm-sized clusters of them would also result from the destruction of soot grains in the interstellar environment. Probably such particles form the small-size tail of the MNR distribution.

However, there is also clear evidence for the occurrence of aliphatically bonded carbon (sp^3-hybridized orbitals pointing toward the corners of a regular tetrahedron). The spectral fingerprint of this refractory organic material are the bands at 3.4–3.5 μm and 6.8 μm (see Sec. III.A.3). Possibly, this refractory organic dust forms a more abundant fraction of the interstellar dust than the 217.5 nm band carrier. Laboratory experiments have suggested that this carbonaceous dust can be understood as the result of photolytic processing of organic ices contained in the mantles of molecular cloud dust (see Sec. III.A.3. and Sec. IV.B.4.). Probably, this material forms thin coatings on all kinds of dust particles in the diffuse ISM (Li and Greenberg 1997).

The understanding of the cosmic carbon dust has been promoted by the detection of presolar grains in primitive meteorites that have survived the formation of the solar system largely unaltered. This most abundant presolar species present in all classes of unmetamorphosed chondrites with an average content of 500 ppm, at least, are crystallites of face-centered cubic (fcc) diamond with diameters of about 2 nm ("nanodiamonds"). Their density is about 30 percent lower than that of terrestrial diamonds. They contain 10–40 atom percent hydrogen with a D/H ratio comparable with that of interstellar molecules. They further contain 0.3 to 1.2 mass percent of nitrogen. The carbon isotopes show no significant deviation from the solar system ratio, whereas the nitrogen isotopes are clearly anomalous. A unique isotopical anomaly of the diamonds is that of the heavy and the light xenon isotopes (Xe-HL anomaly), which points to a connection with nucleosynthetically processed gas typical of supernovae of type II (Anders and Zinner 1993, Zinner 1997). Thus, the diamonds have an interstellar as well as a stardust touch. The formation mechanism of the diamonds is unknown. Chemical vapor deposition (CVD) in a carbon-rich environment commonly suggested as low-pressure formation mechanism of diamonds meets the problem that it requires a substrate, but is apparently preferred to shock metamorphism of colliding interstellar carbon grains (Zinner 1997). Recently, new possibilities, e.g. homogeneous nucleation (Bürki 1996) and formation within carbon onions (Banhart and Ajayan 1996), have come within scope and deserve further attention. The IR spectrum of the nanodiamonds shows prominent absorption bands due to chemical groups associated with the highly reactive surfaces. There is, however, clear evidence that, at least, some of these bands are artifacts of the chemical isolation procedure (Mutschke et al. 1995, Braatz et al. 1998, 2000).

There have been some attempts to identify spectral features observed in circumstellar or interstellar environments with those found in laboratory measurements on diamonds. Immediately after the isolation of presolar diamonds from meteorites Duley (1988) explained sharp red emission lines in the Red Rectangle (around HD 44179) with the zero-phonon lines accompanying the broad band photoluminescence of diamond (cf. Sec. III.A.4). Some authors identified the observed 21 μm band of evolved stars (see Sec. V.C.2) with a feature found in laboratory studies of diamonds (Koike et al. 1995, Hill et al. 1998). Indirect indication for the presence of nanodiamonds in circumstellar environment have been discussed by Guillois et al. (1999) who suggested that two of the sharp UIR bands in the 3.4–3.5 μm range of the two Herbig Ae/Be stars Elias 1 and HD 97048 could be C–H vibrational modes of the hydrogen-terminated facets of diamond crystals.

A second form of presolar carbon solids in meteorites is graphite grains showing spherical arrangements (onions) as well as polyhedral graphite grains

(Zinner 1997, Hoppe and Zinner 2000). Their mass fraction in the meteorites amounts to about 1 ppm. The graphite onions are much larger (0.8 – 20 μm) than the onions produced in laboratory experiments (see, e.g. de Heer and Ugarte 1993, Banhart and Ajayan 1996, Jäger et al. 1998a), and they are much less regularly structured than the latter. Isotope measurements of single grains prove their origin in evolved stars (Zinner et al. 1995, Hoppe and Zinner 2000). Therefore, they must be considered as stardust components ejected into the parent cloud from which the solar system formed.

IV. DUST IN MOLECULAR CLOUDS AND STAR-FORMING REGIONS

IV.A. Basic Observational Phenomena

1. Dust Extinction and Mean Grain Size

Because of the large dust extinction (see Sec. II.B.), it is very difficult to get direct information on dust properties in the dense cold cores of a molecular cloud. Our knowledge mainly rests on observations of well-studied nearby dark cloud complexes, e.g. the Taurus-Auriga and Ophiuchus complexes (Lada et al. 1993). At present, only a few examples of cold cores are known which are apparently undisturbed by star formation ("quiescent cores").

At the edges of the cold parts of a molecular cloud and in star-forming regions that have broken through the cloud, extinction and polarization curves can be determined. They are characterized by values of R and λ_{max} considerably larger than in the diffuse clouds. For example, in the ρ Ophiuchi dark cloud the mean value $R = 3.99 \pm 0.18$ was found (Vrba et al. 1993). The increase of R and λ_{max} proves the presence of grains larger than the big grain mode in the diffuse ISM. Dust grains can grow by accretion of gas and by agglomeration. In the dust literature the last process is mostly denoted by the term coagulation, which, however, is not quite adequate. In this review, we prefer "agglomeration" to "coagulation".

It has been estimated that in the short lifetime of a molecular cloud ($\leq 10^7$, see Sec. II.C.) gas accretion is irrelevant. The adequate mechanism for an effective grain growth seems to be agglomeration (Ossenkopf 1993).

2. Diagnostic Ice Bands

With increasing resolution and sensitivity of IR spectroscopy, a growing number of volatile solids (ices) has been detected in the IR spectra of stars situated in young stellar objects deeply embedded in molecular clouds (see the reviews by Whittet 1993, Dorschner and Henning 1995; Fig. 8).

The water ice band at 3.08 μm being due to O–H stretching vibrations provided the first definitive evidence for interstellar ice at all (Gillett and Forrest 1973). Meanwhile, this band has been detected in a large number of sources.

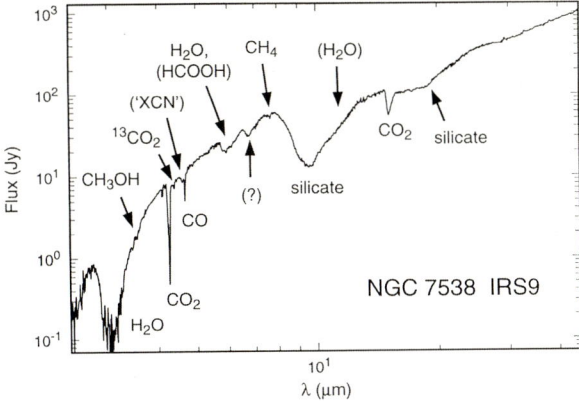

Figure 8. ISO-SWS spectrum of the deeply embedded protostellar object NGC 7538 IRS9. In addition to the silicate absorption bands, the following grain mantle ices have been detected (in parenthesis wavelengths in μm): H_2O (3.0 and 6.0), CH_3OH (3.53; 6.85?), CH_4 (7.68), CO in polar/unpolar matrix (4.67), CO_2 (15.2), XCN (4.62), and HCOOH (5.83). After Whittet et al. (1996).

In some of them the weaker band at 6.0 μm due to H–O–H bending vibrations also occurs. Some sources with strong H_2O bands, e.g. AFGL 961 (Cox 1989), show an additional band at 13.6 μm which has been interpreted as being due to hindered torsional vibrations of the H_2O molecules (libration band). For pure crystalline H_2O ice, the libration band appears at 11 μm, for amorphous ice at 12.5 μm. Admixtures of other ice sorts shift it towards longer wavelengths. Transverse optical lattice vibrations of ice have been detected in the Kleinmann-Low nebula in the Orion molecular cloud at about 45 μm (Erickson et al. 1981). The 3.08 μm feature appears only if the interstellar extinction exceeds a certain threshold value which amounts to about 3 mag in the Taurus dark cloud (Whittet et al. 1988, Smith et al. 1993) and to 10–15 mag in the ρ Ophiuchi dark cloud (Tanaka et al. 1990). As a rule, the 3.08 μm band profile is broader than that expected for pure ice absorption. This points to additional absorption contributors. They extend the "red" wing of the band up to about 3.6 μm in a typical way (see Fig. 9) and are attributed to refractory carbonaceous solids.

The "blue" wing of the water ice band has been of interest in connection with attempts to prove the presence of ammonia ice (NH_3) in the grain mantles. However, the intimate blend of the strong H_2O band and the much weaker NH_3 band prevented a clear decision. The shoulder visible in some sources in the range 2.9 to 3.0 μm around the expected position of the NH_3 band could be explained without assuming ammonia absorption (see Fig. 9). Observations of the bending mode range of both ices around 6 μm did not give additional evidence in favor of ammonia ice. The break-through leading to the definitive detection of NH_3 ice (Lacy et al. 1998) was achieved by observations of the inversion band at 9.01 μm which, for a long time, escaped the discovery because of its position near the center of the strong silicate absorption band. This

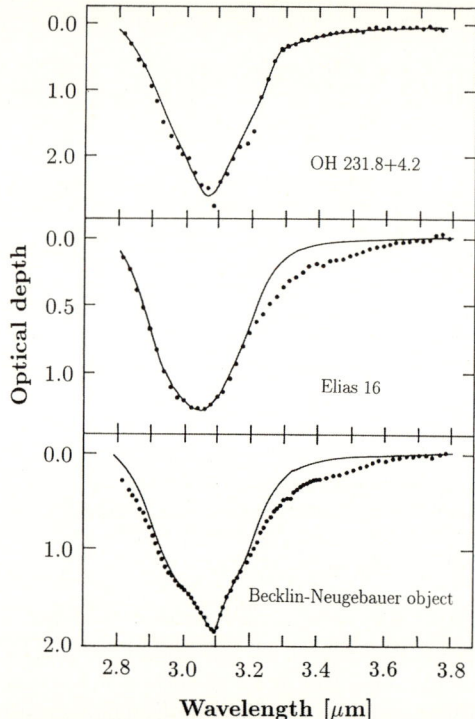

Figure 9. The profile of the 3.08 μm H_2O ice band of the Becklin-Neugebauer object (deeply embedded object in the Orion molecular cloud) and Elias 16 (field star behind Taurus dark cloud). For comparison the circumstellar ice absorption band of the OH-IR star OH 231.8 +4.2 is given. The solid curve is a profile calculated for amorphous water ice deposited on silicate cores with the radius $a_c = 0.1$ μm, having a MRN size distribution for $a \geq a_c$. The fit for BN is optimized for a mixture of ices of different temperatures. The molecular cloud objects show the typical red-wing anomaly. After Whittet (1993).

case demonstrates that the profile of the silicate feature could be significantly modified by mantle-ice absorption.

A band at 4.67 μm found in the spectra of embedded sources (Lacy et al. 1984) was assigned to C–O stretching vibrations of solid carbon monoxide (CO). In all sources exhibiting the CO-ice band, also the 3.08 μm H_2O band was detected, whereas by far not all sources with the H_2O band show the CO feature. This finding can be explained by the different sublimation temperatures (pure CO: ≈ 20 K, H_2O: ≈ 100 K): Many sources contain dust sufficiently cool for H_2O ice, but too warm for the condensation of CO.

In most cases, the observed CO bands are superpositions of two different profiles: a narrow feature at 4.675 μm (FWHM ≈ 0.011 μm) and a broader feature at 4.681 μm (FWHM ≈ 0.022 μm). Laboratory experiments (Sandford et al. 1988) suggest that the difference is due to a matrix effect. The narrow CO band is produced in an ice matrix of predominantly non-polar molecules, e.g.

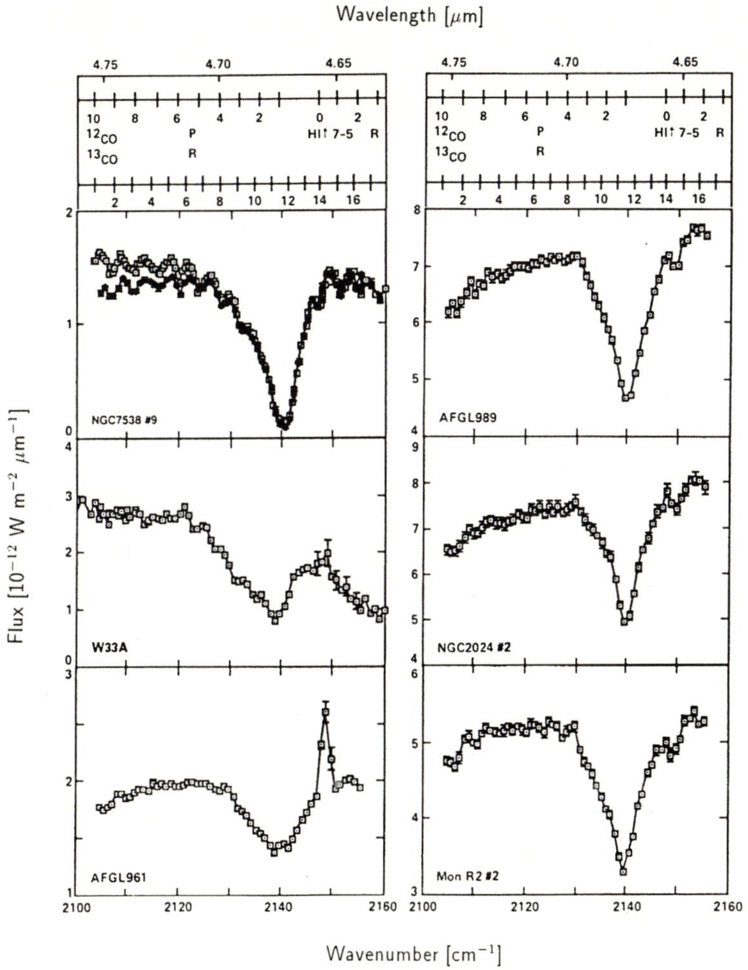

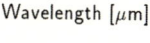

Figure 10. Solid CO features in six typical molecular cloud sources. For comparison, in the top panel the expected positions of the gas lines of the P and R branch of ^{12}CO, the R branch of ^{13}CO, and the HI 7-5 recombination line are indicated. After Tielens et al. (1991).

N_2, O_2, CO_2, whereas the broader band stems from CO embedded in a matrix of polar molecules, e.g. H_2O (Tielens et al. 1991). Figure 10 shows some typical spectra.

In contrast to solid CO, the C–O stretching band of CO_2 ice is not positioned within an atmospheric window and, thus, not accessible by ground-based observations. However, d'Hendecourt and Jourdain de Muizon (1989) detected the O–C–O bending vibration at 15.2 µm in IRAS LRS spectra. The detection of the C–O stretching band at 4.27 µm was accomplished in 1996 by the ISO satellite (Gürtler et al. 1996, de Graauw et al. 1996). The CO_2 profile also shows a matrix effect.

A typical molecular cloud ice is solid methanol (CH_3OH). The first assignment of a band at 3.54 μm to the C–H stretches of solid methanol (Baas et al. 1988, Grim et al. 1991) has been confirmed by extensive laboratory work on CH_3OH/H_2O mixtures and is now considered certain. The identification got strong support by the detection of additional vibrational modes at the characteristic wavelength positions of this solid (see Table 3). There are, however, some problems since the ratios of the band strengths are not in agreement with the expectations. This left some doubt whether the assignment, e.g. of the band at 6.85 μm to the C–H bends of methanol, is correct (Schutte et al. 1996). For many sources only upper limits for the abundance of methanol ice could be derived. In some protostellar sources CH_3OH is the most abundant ice species behind H_2O (Dartois et al. 1999).

There is evidence for the presence of ices containing CN-groups. A strong band at 4.62 μm in the deeply embedded source W33 A, which is weakly indicated in sources with strong CO features, has been tentatively attributed to C–N stretching vibrations of a CN-group attached to an unknown organic residue ("X–CN"). Isonitrile (CH_3NC) and the cyanate ion (OCN^-) have also been discussed (see Larson et al. 1985, Grim and Greenberg 1987). The longstanding proposal of a sulfur-bearing species responsible for weak bands at 4.9 μm and 3.94 μm also observed in W33A (Larson et al. 1985; Geballe et al. 1985) got some support by extensive laboratory work. Palumbo et al. (1995) found convincing evidence for the attribution of the 4.9 μm feature to C=O stretching vibrations of carbonyl sulfide (OCS). The best fitting was reached for OCS in a methanol matrix.

Other important mantle ice components are expected to be O_2 and N_2. However, these unpolar ice species have very weak vibrational bands which are overlapped by much stronger features of other components and, therefore, could not yet be detected.

Recently, detections with the ISO Short Wavelength Spectrometer have continued the list of interstellar ices. There is definitive evidence for the occurrence of the methane (CH_4) band at 7.68 μm (Boogert et al. 1996) and the band of formic acid (HCOOH) at 5.83 μm (Whittet et al. 1996, Schutte et al. 1996). In Table 3 the spectral features and abundances of the most relevant interstellar ice components are listed.

3. Absorption Bands due to Refractory Dust

Molecular cloud dust contains refractory components forming the cores of the grains. The common dust models are based on core-mantle grains consisting of silicate and/or carbon (or carbonaceous) cores surrounded by mantles of ices (see Sec. IV.A.2 and IV.B.). Silicate absorption bands at 9.7 and 18 μm have been detected in the IR spectra of very young stars deeply embedded in the molecular clouds from which they originated. According to the prototype, the Becklin-Neugebauer object in the Orion Molecular Cloud (Gillett and Forrest 1973), such objects have been called BN-objects. A great number of spectra of deeply embedded IR sources showing strong silicate

Table 3

Ice species detected in molecular clouds and deeply embedded protostars. "t" behind the wavelength indicates tentative assignment to the species. "Abundance" means column density of molecules toward the source in units of that of H_2O ice. Data from Schutte (1999) or the references of the comments.

Species	Feature(s) (μm)	Abundance ($H_2O=1$)	Comments
H_2O	3.08, 6.0, 13.3[1] 45[2]	1	[1] Cox (1989) [2] Erickson et al. (1981)
HDO	4.1 t	0.001...0.02	Teixeira et al. (1999)
H_2O_2	3.5 t	< 0.05	upper limit
CO	4.67	0.01...0.40	matrix and isotopic shift
CO_2	4.27, 15.2	0.12...0.16	matrix and isotopic shift
CH_3OH	3.54, 3.93 6.85 t, 9.75	0.02...0.30[1]	values > 0.1 only in protostars
H_2CO	3.47 t, 5.81	0.02...0.06	
HCOOH	5.85 t	0.03	X–COOH? X:organic residue
CH_4	7.68	0.01...0.02	ISO
XCN	4.62	0.001...0.008	X:organic residue (OCN^-?)
NH_3	2.97 t, 9.01	0.1	Lacy et al. (1998) Gürtler et al. (1999) Gibb et al. (2000)
XCS	4.91 t	0.002	X:organic residue OCS? Palumbo et al. (1995)

absorption is contained in the IRAS LRS spectral catalog (Olnon and Raimond 1986).

In contrast to other dust populations the silicate bands of the molecular dust are extremely wide and without any diagnostic profile structure or indications of crystallinity. Since the "ISO revolution" has presented strong arguments in favor of the existence of crystalline silicates among the dust grains injected into interstellar medium by evolved stars (see Sec. V.A.1) the total lack of such indications in the silicate band profiles of IR-sources in molecular clouds points to amorphization during the stay of the grains in the interstellar medium. A possible mechanism could be the long-time irradiation by cosmic rays.

Much less unambiguous than the case for silicate core material is that for the carbon cores. Up to now, no clear indications of the 3.4–3.5 μm bands typical of the refractory carbonaceous dust in the diffuse ISM (see Sec. III.A.3) have been found in deeply embedded objects. However, it cannot be entirely excluded that bands due to C–H bonds could be hidden in the "red" wing of the water ice band which extends up to 3.6 μm. The discovery of presolar diamonds in primitive meteorites has suggested to search for spectral features due to C–H vibrations of tertiary carbon atoms at the surface of diamond particles ($\equiv$CH). A weak band at 3.47 μm found in high-resolution spectra of the deeply embedded sources NGC 7538 IRS9, W33A, W3 IRS5, S140 IRS1, and Mon R2 IRS3, where strong C–H bands of primary (–CH$_3$) and secondary (–CH$_2$–) carbon atoms are absent, has tentatively been attributed to C–H vibrations of tertiary carbon atoms (Allamandola et al. 1993; Sellgren et al. 1994). A narrow absorption feature at 3.25 μm was assigned to aromatic C–H stretching vibrations (Sellgren et al. 1995). However, aromatically bonded carbon must be present in molecular cloud dust because photodissociation regions, the interfaces between molecular cloud gas and HII regions around newly formed stars in star-forming regions, show strong UIR band emission (see Sec. IV.A.4).

4. Dust Emission Bands in HII Regions

Newly formed massive stars in star-forming regions of molecular clouds ionize the gas and create emission nebulae. Apart from the gas lines, these nebulae also show emission bands of refractory dust that can survive the energetic UV radiation field. The first and best known case is the Trapezium region of the Orion Nebula, where the silicate emission band at 9.7 μm was detected already in the early days of IR dust spectroscopy (Stein and Gillett 1969). ISOCAM observations by Cesarsky et al. (2000) have confirmed that silicate emission is present over the whole extent of the nebula. The authors further found in ISO SWS spectra additional emission peaks in the spectral ranges 15–20, 20–28 and $>$ 32 μm, that roughly coincided with the peaks of crystalline Mg-silicate dust (cf. the forsterite spectrum in Fig. 11). They interpreted this finding and the observed structure of the 9.7 μm band profile as indications of the presence of crystalline silicates. Since molecular cloud dust does not show any evidence for crystalline silicates, the inevitable conclusion is that there has to be a mechanism that transforms amorphous to crystalline silicates during the star formation process in the molecular cloud.

5. Properties of Hot Molecular Cores

Star formation processes transform cold quiescent molecular cloud cores in hot cores. Here, gas and dust are subjected to the radiation output and the strong stellar winds of extremely young OB stars and to the shocks associated with the stellar outflows and the expansion of HII regions. The hot cores have diameters of $\approx$ 0.1 pc, masses of a few solar masses, and temperatures of $\geq$ 100 K.

Examples are the Orion Hot Core, the Orion Compact Ridge, and NGC 7538 (Walmsley and Schilke 1993).

In hot cores, close to the newly formed stars, mantle ices evaporate and cause local overabundances of single molecular species in the gas phase. In the Orion Hot Core, NH_3 is such a mantle ice "fossil". Only 5″ to the south of this region, in the Orion Compact Ridge CH_3OH and O-bearing molecules are enriched. This points to local variations in the mantle composition of the grains. The observed deuterium enrichment in the gas is also considered a "fossil" of the fractionation in the grain mantles throughout the cloud lifetime (van Dishoeck 1999).

IV.B. Processes in Molecular Clouds and Star-Forming Regions

In molecular clouds dust grains are involved in a great manifold of physical and chemical processes (Dorschner and Henning 1995, Schutte 1996, van Dishoeck 1999). In the following, only a short summary of the most important dust-relevant processes is given.

1. Molecular Accretion and Ice Formation

At the low temperatures of quiescent cores impinging gas molecules should easily stick at grain surfaces. The accretion time-scale for dust grains in molecular clouds is shorter than or comparable with both the time-scales for reactions between ions and molecules (10^7 yrs) and the average lifetime of molecular clouds (10^5–10^7 yrs).

At least, some of the accreted species will be sufficiently mobile at the grain surfaces, so that chemical reactions should take place. It must be expected that molecular hydrogen is efficiently produced on the grain surfaces because at the densities in question the H_2 formation in the gas phase is ineffective. In contrast to heavier molecules, H_2 can easily be desorbed from the grain surfaces. Of great importance for surface reactions is the fractional abundance of atomic hydrogen, H/H_2, in the gas. If its value exceeds about 10^{-3}, O-, C-, and N-atoms will be hydrogenated to H_2O, CH_4, and NH_3, CO to H_2CO and CH_3OH, and O_2 to H_2O_2 and H_2O. Otherwise CO, N_2, and O_2 should dominate the grain mantle ices; unfortunately N_2 and O_2 are not directly observable. At high abundance of atomic oxygen, CO is oxidized to CO_2.

2. Grain Agglomeration

Cohesive collisions of dust grains in cold cores can produce grain aggregates. Agglomeration dominates only if the collisional velocity $v \leq 10$ m s^{-1}, for $v \geq 1$ km s^{-1} shattering will occur. For the description of grain collisions experimental experience is urgently needed. It is one of the main goals of the laboratory astrophysics research unit in Jena, to make available such data (space project CODAG, Wurm and Blum 1998, Blum and Wurm 2000, Blum et al. 2000).

Estimates show that the main agglomeration processes in dense regions are driven by turbulence at gas densities below 10^8 H-atoms cm^{-3} and by Brownian motion at higher densities. Brownian motion at a temperature of 20 K produces relative velocities below the critical velocity for sticking. However, the growth to large aggregates is too slow because of the low collisional frequency. The lower size limit of grains which can gain turbulent relative velocities was estimated to be 10 nm for the density 10^4 H-atoms cm^{-3} and 250 nm for densities of 10^9 H-atoms cm^{-3}.

The properties of aggregates produced by agglomeration depend on the size and structure of the colliding particles. Very small grains (≤ 5 nm) should form compact aggregates. For larger particles fluffy structure is expected. The agglomeration can be modeled as a ballistic cluster growth process and studied by computer simulation. There are two limiting cases: particle-cluster aggregation (PCA, addition of single grains) and cluster-cluster aggregation (CCA, addition of subclusters of identical size). The main result of such calculations (see Dorschner and Henning 1995) is that, as a rule, PCA particles have a well-defined center of growth and are of nearly spherical shape, whereas CCA clusters show an irregular structure. The results have been applied to study numerically the evolution of the grain size distribution in cold cores and the evolution of the dust opacity in protostellar objects. The methods available for computing optical properties of such particle aggregates have been reviewed by Gustafson et al. (this book).

3. Molecular Desorption

Thermal evaporation of grain atoms, molecules, and radicals is only possible at the temperatures of hot cores. The different sublimation temperatures of the main mantle ices (CO, CO_2, and H_2O) cause local differences in the grain composition. Desorption by spot heating caused by cosmic ray nuclei, e.g. Fe, is limited to the most volatile components (CO). Effective chemical desorption could be connected with explosions of stored radicals which can be triggered by cosmic ray bombardment or by grain collisions. This mechanism requires that a sufficiently high radical concentration is reached.

4. Grain Mantle Processing and Dust Recycling

Ultraviolet radiation in the less opaque parts of molecular clouds and in the environment of young massive stars can considerably change the structure of the grain mantles. Experiments with UV irradiation of ice mixtures containing carbon-bearing species have shown that refractory organic material is formed this way. The UV photolysis produces reactive radicals which can diffuse and react after heating of the mantle to above 30 K, forming more complex organic molecules. After further heating, the volatile components of the irradiated sample evaporate and, finally, a small amount of an organic refractory residue remains. The IR spectra of such residues resemble those observed in the diffuse ISM dust and those extracted from carbonaceous chondrites (Sec. III.A.3.).

Energetic ions of cosmic rays can also modify material properties. They deposit energy in the target material by collisions with its atoms. This can lead to sputtering, bond-breaking and new association, and implantation effects.

Ion bombardment experiments with mixtures of water ice with carbon-bearing molecules simulate such effects and show the production of complex organic molecules. High doses of ion irradiation lead to complete carbonization of carbon-containing ices.

V. DUST IN STELLAR OUTFLOWS

V.A. Oxidic Stardust

Although astrophysical conjectures on the condensation of solids in cool stellar atmospheres existed already in the 1930s (Wildt 1933), the connection of these stellar solids with interstellar dust was not seen before 1962. Theoretical papers in the 1960s, which studied mechanisms of stellar origin of interstellar grains, got strong support by the detection of circumstellar IR dust bands and the investigation of stellar mass loss phenomena. These new observations allowed a first understanding of the chemical nature and the transport mechanism of the stardust to the diffuse ISM. Among the cosmic dust populations, stardust exhibits the greatest mineralogical diversity (Dorschner 1999).

1. Silicates

Silicate bands presented the first evidence of stardust (cf. Sec. I.E.). The broad and almost structureless features bear witness of the disordered structure of these solids. The most comprehensive collection of stardust silicate spectra can be found in the IRAS LRS catalog (Olnon and Raimond 1986). It contains 1808 objects with the 10- and 18-μm bands in emission and 297 objects with the 10-μm band in absorption. Most of the silicate emission sources are dusty envelopes around late-type giants, some are disks connected with young stars (see Sec. VI).

From the observed amorphous silicate bands, no direct information on the average number of bridging oxygen atoms per SiO_4-tetrahedron, which connect neighboring tetrahedra, and on the cations incorporated in the silicate structure can be obtained. We do not even know if all Si ions are completely coordinated by four oxygen ions and how many of the central tetrahedral ions are replaced by Al or Fe^{3+}. As laboratory analogs mainly vitreous Mg-Fe silicates have been studied so far (Jäger et al. 1994, Dorschner et al. 1995, Mutschke et al. 1998).

The IRAS LRS spectra and the growing number of sufficiently resolved ground-based spectra of the 8–13 μm range exhibited significant variety in the silicate band profiles and stimulated classification work (see Sloan and Price 1995, 1998, Speck et al. 2000). The observed differences in the profile shapes, in particular indications of different fine-structure peaks, contradict the theoretical concept of a unique "astronomical silicate" as well as that of "structureless" band profiles pointing to completely amorphous stardust silicates.

The more realistic concept of some mineralogical diversity was strongly supported by the "ISO revolution". The ISO spectrometers with their unprecedented wide wavelength coverage, high spectral resolution and large sensitivity

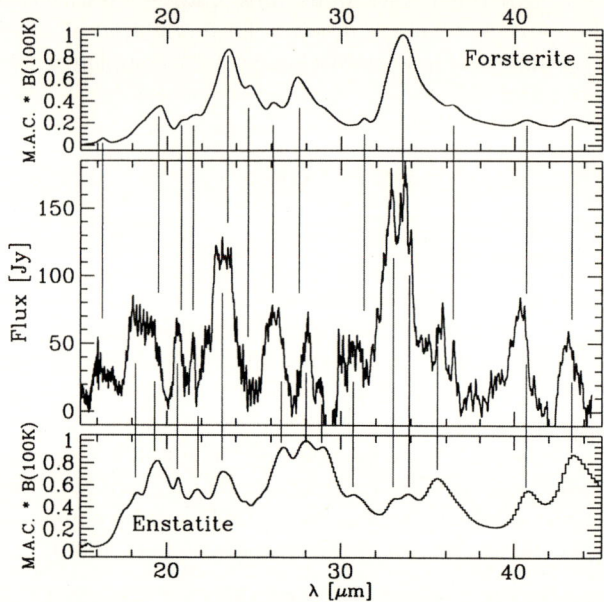

Figure 11. ISO MIR spectra of the post-AGB star AFGL 4106 show many emission bands that can be assigned to crystalline silicate grains. The continuum-subtracted spectrum (center) is compared with normalized laboratory-based spectra of forsterite (Mg_2SiO_4) and enstatite ($MgSiO_3$). M.A.C. is the mass absorption coefficient of Rayleigh spheres, B(100 K) the Planck function for the adopted dust temperature of 100 K. Coincidences of stellar with laboratory bands are indicated. Additional bands not reproduced by the laboratory spectra could be due to other silicates or oxides. After Jäger et al. (1998).

presented clear evidence for the occurrence of crystalline silicates in circumstellar environments around evolved (Waters et al. 1996, 1998a,b) as well as very young stars and remnants of protoplanetary disks (Malfait et al. 1998b, 1999), altogether in about 50 sources. The strongest of the relatively sharp MIR bands (20–50 μm) are at wavelength positions that can be attributed to Mg-rich (Fe-poor), cold olivine and pyroxene grains (see Fig. 11, Jäger et al. 1998b, Molster et al. 1999a, Molster 2000). There are, however, some bands that could not yet certainly be identified. With the successful mineralogical diagnosis via IR spectroscopy stardust mineralogy (Dorschner 1999) or astromineralogy (Molster 2000) have established themselves as new research fields.

In contrast to the wavelength positions of the prominent MIR emission bands that can be satisfactorily well reproduced by laboratory analogs, there are problems in understanding band widths and relative intensities. In some cases, the observed bands are sharper than those of the laboratory analogs. If the model of non-equilibrium condensation of the primary silicate particles that are thought to be amorphous and chemically inhomogeneous (Nuth 1996, Nuth et al. 1999, Dorschner 1999) is realistic, mechanisms have to be found that transform the dirty and amorphous silicates to clean, monomineralic, and crystalline species. In agreement with these differences between amorphous and crystalline silicates is the finding that the former are warmer than the latter. The cause could be a higher opacity of the amorphous silicates, possibly due to higher Fe content (Molster 2000).

First estimates indicate that in most of the studied sources only a small percentage of the silicates is crystallized. Molster (2000) found that the percentage of crystalline silicates is significantly higher in disk objects – extremely young stars with accretion disks as well as evolved disk objects (probably binaries) – than in normal outflow sources and interpreted this as a sign of different crystallization processes in the sources. However, the mechanisms discussed so far, do not offer satisfactory explanations. Thermal annealing (Hallenbeck et al. 1998, Fabian et al. 2000) has the handicap that the temperatures at which crystallization takes place in the laboratory experiments are significantly higher than the silicate dust temperatures estimated by theorists (see, e.g. Hron et al. 1997). Therefore, low-temperature crystallization (Molster et al. 1999b) deserves particular attention.

2. Oxides

Apart from silicates, in O-rich stars metal oxide grains can be expected. Henning et al. (1995) discussed MgO, FeO, $Mg_xFe_{1-x}O$ as possible stardust components and recommended to search for sources with weak or lacking 10 μm bands and excess emission in the 17–25 μm range. The problem is that low dust temperatures can mask this effect.

The cumulative effect of many oxides in outflow sources could raise the dust opacity between the silicate bands at 10 and 18 μm and, thus, solve the trough opacity problem. All laboratory analogs studied so far failed to reproduce the relative high absorption coefficient observed in the "trough" between the main Si–O vibrational bands. Candidates would be oxides of Al and several 3d-elements (Cr, Fe, Mn, Ti): Al_2O_3 (Begemann et al. 1997, Mutschke et al. 1998), $MgAl_2O_4$ (Posch 1999, Fabian et al. 2001), Cr_2O_3, Fe_2O_3, $FeCr_2O_4$, MnO_2, TiO_2, $FeTiO_3$ (Dorschner 1999).

The only direct spectroscopic evidence proving the occurrence of a stardust oxide is a weak but distinct band at 13 μm, which was detected in IRAS-LRS spectra of Mira stars and semiregular variables and tentatively assigned to aluminum oxide grains (Onaka et al. 1989). The detection of presolar corundum (α-Al_2O_3) grains in primitive meteorites (Nittler 1997) lent this identification some support.

Begemann et al. (1997) derived the average 13 μm-band profile from IRAS-LRS spectra and compared it with laboratory data. Posch et al. (1999) obtained more accurate profiles from ISO-SWS spectra and critically discussed the identification problem with the result that the Al–O vibrations of spinel ($MgAl_2O_4$) give better reproduction of the 13 μm profile then those of corundum. New experiments (Fabian et al. 2001) and the detection of presolar spinel grains in meteorites (Choi et al. 1998) support this conclusion. Highly refractory oxides like Al_2O_3, $MgAl_2O_4$, and also TiO_2 are discussed as potential condensation seeds for silicates in evolved O-rich stars (Gail and Sedlmayr 1998).

In the cold outer regions of optically thick envelopes volatile oxides, too, can condense. In such sources emission bands in the range 40–70 μm, which are due to lattice vibrations of H_2O ice, have been observed (Omont et al. 1990). The detection has been confirmed by ISO spectra of both evolved (Molster et al. 1999a) and also very young stars (Malfait et al. 1998b, 1999).

V.B. Carbonaceous Stardust

1. Amorphous and Hydrogenated Amorphous Carbon

The presence of carbon grains in the envelopes of C-rich stars is a well-established claim, even if there is only marginal spectral evidence. Molecules like acetylene (C_2H_2) and polyynes (H–C≡C–...–C≡C–H) are assumed to be parent molecules for the nucleation of some sorts of amorphous carbon (a-C) grains (Tielens 1990, Sedlmayr 1994). In this process, polycyclic aromatic hydrocarbons (PAH) play an important role as intermediate links (for PAHs see section V.B.3.).

Soot-like carbon grains can be expected in hydrogen deficient environments, e.g. in WC and R CrB stars. An emission band at 7.7 μm in WC stars (Cohen et al. 1989) has been assigned to aromatic C-C stretching vibrations in soot grains. In IUE spectra of variables of the R CrB type a wide UV absorption band at 240–250 nm was detected and assigned to soot grains (Hecht 1991) Attempts to detect the regular crystalline carbon modification graphite in stellar spectra failed (Glasse et al. 1986). In the H-rich carbon stars hydrogenated amorphous carbon forms (a-C:H, HAC) should occur rather than pure carbon solids.

The basic problem of the investigation of carbon dust via laboratory simulation is the great number of possible carbon solids with different micro-structure and, therefore, different spectral appearance (Henning and Salama 1998). Extended laboratory work has been carried out in order to find the right carbon solids that could explain the observations (Mennella 1995a,b; Schnaiter et al. 1996, 1998; Jäger et al. 1998a; see also Sec. III.B.4, Fig. 7). The experiments by Schnaiter et al. (1998), who applied molecular beam techniques to nm-sized soot particles (pure and hydrogenated carbon) that were frozen in a nobel-gas matrix in order to obtain UV spectra of the isolated grains, represented the break-through to the satisfactory reproduction of the R-CrB-type UV band profiles as well as of the interstellar carbon dust band at 217.5 nm.

2. Fossil Carbon Stardust in Meteorites

The discussion about the carbonaceous stardust got a new impact by the detection of presolar diamond and graphite grains in meteorites. The diamonds have been already discussed in Sec. III.B.4. The graphite grains, present in meteorites with a mass fraction of about 1 ppm (Hoppe and Zinner 2000), occur in various species differing in morphology, structure, density and isotopic patterns. In many cases, they are μm-sized onion-like spheres composed of bent graphite layers in a discontinuous concentric arrangement. In other cases, they are well-crystallized μm-sized graphite grains and cauliflower-like aggregates of smaller carbon particles (Zinner et al. 1995). The isotopic patterns suggest different stardust sources: AGB stars, WC stars, novae, and supernovae. Many of the graphite onions contain nm-sized crystallites of metal carbides (TiC, ZrC, MoC) between the bent graphite layers. These additional stardust species and the lack of silicon carbide (SiC) reveal important details on the sequence of the grain formation in the sources (Bernatowicz et al. 1996).

3. Polycyclic Aromatic Hydrocarbons

In the discussion of carbonaceous stardust polycyclic aromatic hydrocarbons (PAHs) have attracted great attention (Allamandola et al. 1987, Puget and Léger 1989, Cherchneff and Barker 1992, Geballe 1997, d'Hendecourt 1997). PAHs are large planar molecules consisting of polymerized benzene rings with hydrogen atoms at their outer rim. The strongest impact to their investigation was given by the detection of the UIR bands. Sellgren (1994) distinguished three groups of these features:
1. main features at 3.3, 6.2, 7.7, 8.6, and 11.3 μm;
2. weak features at 3.40, 3.46, 3.52, 3.57, 5.2, 5.7, 6.9, 11.9, and 12.7 μm;
3. broad features or plateaus at 3.2–3.6, 6–9, 11–13 μm.

UIR bands have been detected in objects with enhanced carbon abundance and UV radiation intensity, but very different evolutionary status (see Sec. III.A.4). The basic question that still remains is whether the UIR band carriers are large molecules or small solids. Spectral features of PAH molecules and their ions have been found at the positions of all observed UIR bands. However, most of the observed profiles are much wider than the molecular profiles. A satisfactory reproduction of the whole observed UIR band spectrum by laboratory data was not accomplished so far. The detection of the overtone band (2-0 vibrational transition) of the 3.3 μm band at 1.67 μm in the planetary nebula IRAS 21282 +5050, a strong UIR band emitter, points to a molecular carrier rather than to a solid (Geballe et al. 1994, d'Hendecourt 1997). From the intensity ratio 2-0 band/1-0 band, it was derived that the carrier should contain about 60 carbon atoms. On the other hand, there are numerous experimental approaches to explain the UIR bands by carbonaceous solids that can reproduce satisfactorily the strong features, at least (see Sellgren 1994).

V.C. Other Stardust Components

1. Silicon Carbide

Many carbon-rich stars show a strong emission band at 11.3 μm, the strength of which is tightly correlated with the star's mass loss rate (Skinner and Whitmore 1988). Based on laboratory spectroscopy (cf. the review by Dorschner 1999), it has been assigned to Si–C vibrations of silicon carbide grains condensed in the stellar wind. The long-standing efforts to find out by comparison of the observations with laboratory measurements which modification (α- or β-SiC) is the prevailing one did not result in getting convincing conclusions. The reason for this is that the profile shape of the 11.3 μm very sensitively depends on size and morphology and on the impurity content, e.g. nitrogen (Mutschke et al. 1999). On the other hand, variations of these grain parameters can explain the observed variety of the spectra (cf. Speck et al. 1997). Experimental simulation by Mutschke et al. (1999) could rule out that the carrier of the 11.3 μm band is amorphous SiC.

Up to now, more than 600 stars are known showing the 11.3 μm band in emission. There are, however, only few optically thick carbon-rich shells which show the SiC band in absorption (Speck et al. 1997). In contrast to silicates, interstellar SiC absorption has never been detected, in spite of the fact that the parent cloud from which the solar system originated must have contained SiC grains. They have been found as "fossils" with an abundance of about 5 ppm among the presolar grains in primitive meteorites. The presolar SiC grains populate the size range from 0.05 to 20 μm and appear as euhedral crystallites of β SiC (cubic) as well as aggregates of smaller subgrains containing also α SiC. Because of the large grain sizes and the relative high abundance, SiC is the best studied presolar mineral phase in meteorites. Isotopic studies confirmed the existence of different SiC sources: AGB and post-AGB stars as main sources (main stream SiC), supernovae, novae, and WC stars (Hoppe and Ott 1997). First IR spectra of isolated presolar SiC grains have been investigated by Andersen et al. (1999).

2. Nitrides, Sulfides and Silicides

The reducing conditions present in carbon stars should also be favorable for the condensation of other oxygen-free solids, e.g. nitrides and sulfides. Actually, in meteorites presolar silicon nitride (Si_3N_4) grains have been found. The isotopic pattern points to the origin in supernovae ejecta (Nittler et al. 1995). However, clear spectral evidence of the presence of Si_3N_4 in C-rich stars is still lacking. It cannot be excluded that in some of the rare SiC absorption profiles (Speck et al. 1997) a weak Si_3N_4 absorption could be hidden.

In contrast to nitrides, some types of carbon-rich evolved stars show spectral evidence for the occurrence of sulfides. An extremely strong band at 30 μm in carbon-rich AGB stars, protoplanetary nebulae and planetary nebulae (Cox 1993, Omont et al. 1995) has been tentatively attributed to the strong vibrational band of magnesium sulfide (Goebel and Moseley 1985, Nuth et al. 1985). Begemann et al. (1994) have confirmed this interpretation by laboratory measurements. They extended the investigation to mixed Mg-Fe sulfides, the formation of which should be more likely than that of pure MgS. Because of their chemical instability under oxidizing conditions, MgS grains have no chance to survive the passage through the interstellar space. Due to their greater stability, iron-rich sulfides can be expected to occur both as interstellar dust component and as a possible presolar ingredient in primitive solar system solids. The sulfide-bearing GEMS (Glasses with Embedded Metal and Sulfide; Bradley 1996, 1999) in interplanetary dust particles, supposed to be of presolar origin, support this interpretation. Keller et al. (2000) proposed that Fe-Ni-

Sulfides of GEMS and other interplanetary dust particles that show a strong band at 23.5 μm could be a candidate for the identification of a band at the same wavelength which has been observed in some ISO spectra of young stellar objects and comet Hale-Bopp.

In several carbon-rich protoplanetary nebulae, Kwok et al. (1989) detected a strong emission band at 21 μm, which, up to now, has not convincingly been identified. ISO-SWS spectra (Volk et al. 1999) show great uniformity of the profiles of this band, the peak of which is positioned within the range 20.0–20.2 μm. Those of these sources that have been studied at sufficiently long wavelengths also show the 30 μm band. However, a correlation between the strengths of the two features was not found. The identification proposals of the 21 μm feature are based on carbonaceous solids or molecular clusters (HAC, PAH, hydrogenated fullerenes; Buss et al. 1990, Webster 1995), nanodiamonds (Koike et al. 1995, Hill et al. 1998) and silicon disulfide (Goebel 1993, Begemann et al. 1996).

Chemical equilibrium calculations for S-type stars (C/O ≈ 1) and LBV (Luminous Blue Variables) by Ferrarotti et al. 2000) resulted in the conclusion that iron silicide (FeSi) could be a typical solid in these stars. They proposed that FeSi grains could be responsible for an unidentified band at 47.5 μm in AFGL4106 (Molster et al. 1999a). The explanation of the ERE phenomenon by silicon nanograins has set a new accent on the importance of silicon chemistry (cf. Gordon et al. 2000).

VI. DUST IN YOUNG CIRCUMSTELLAR DISKS AND PLANETARY SYSTEMS

VI.A. Observational Evidence for Young Circumstellar Disks

1. Evidence for Disks around Low-Mass YSOs

Circumstellar dust emission is not only restricted to evolutionarily advanced stars. Dust concentrations can be found in all types of young stellar objects (YSOs, Henning 1996): deeply embedded massive YSOs, e.g. Becklin-Neugebauer object, Herbig Ae/Be stars (intermediate-mass YSOs), and T Tauri stars (low-mass YSOs). In the following, we concentrate on dust around the less massive YSOs and around main-sequence stars (Vega phenomenon). Both types of classes of objects show evidence for disk-shaped arrangements of the circumstellar dust, often denoted as "preplanetary disks".

Direct and indirect arguments proving the presence of dusty disks around low-mass YSOs have been discussed by Beckwith (1994), André (1994) and Hanner (1995). Evolutionary aspects of disks around low-mass YSOs have been studied by André (1994), who also discussed the decrease of the circumstellar dust masses with the ages of the stars (Fig. 12).

A prominent object which was the first to clearly show a disk configuration in molecular maps has been HL Tau. The molecular results have been supple-

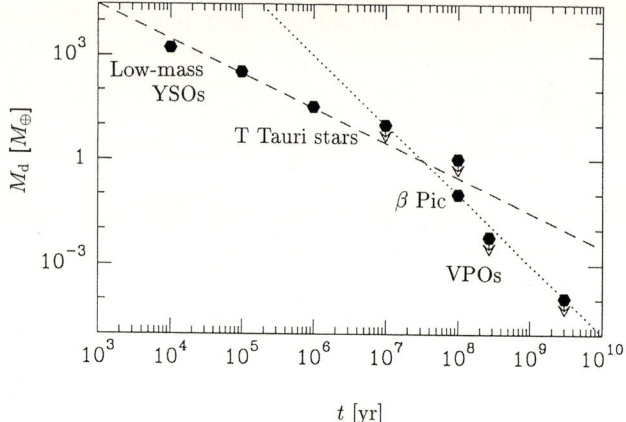

Figure 12. Circumstellar dust mass M_d as a function of the stellar age t from low-mass YSOs to young (G-type) main-sequence stars. Arrows are estimated upper limits. The dashed and dotted lines are the curves $M_d \propto t^{-1}$ and $M_d \propto t^{-2}$, respectively. After André (1994).

mented by spectrophotometry, imaging, interferometry, and polarimetry, and today HL Tau belongs to the best-studied intermediate-mass YSO disks. An overview on the manifold observational work can be found in Men'shchikov et al. (1999) who derived a detailed model of the disk.

A general break-through in detecting direct optical evidence of disks has been reached by the Hubble Space Telescope (HST). High-resolution HST images of the Orion nebula show that about 50 % of the young stars are embedded in circumstellar gas-dust disks which are mainly seen in their gas line emission excited by the UV radiation of the hottest Trapezium star θ^1 Ori (O'Dell and Wen 1994) and in some cases as dark silhouettes against the bright background of the Orion nebula (Fig. 13). Meanwhile, many investigations of single objects have been carried out which are mainly based on HST data. Here we point to the high-resolution (0.1", corresponding 15 AU) studies of optically thick disks around low-mass YSOs in the Taurus star formation region, which are seen nearly edge-on as dust silhouettes of 500–900 AU extent against scattering dust nebulosity (Padgett et al. 1999).

2. Modification of Dust Properties in YSO Disks

Disk dust around YSOs is subjected to various effects modifying the grain material: annealing and sublimation due to stellar radiation, alteration by UV irradiation, sputtering and ion implantation by the stellar wind, thermochemical effects due to the passage of shock fronts during accretion, and electrical discharges (Prinn 1993; Lenzuni et al. 1995). Moreover, the evolution of a preplanetary disk is characterized by opacity changes due to agglomeration, collisional destruction, sublimation, and recondensation of grains. Such processes have mostly been studied for the solar nebula (see, e.g. Sterzik and

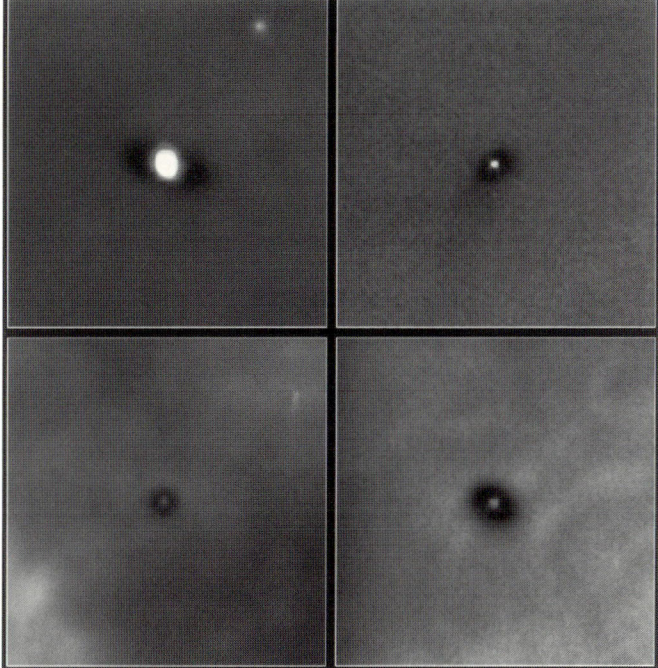

Figure 13. Four circumstellar disks around low-mass YSOs in the Orion nebula, seen under different inclination angles. These dark silhouetted disks are a minority among the great number of bright disks discovered by the HST Wide Field and Planetary Camera (O'Dell and Wen 1994). They are sufficiently far away from the Trapezium stars, the UV radiation of which ionizes the outskirts of the nearby disks. (Credit: M. McCaughrean, C. R. O'Dell, and NASA).

Morfill 1994), but they must occur in all preplanetary disks. However, for planet formation in such disks only indirect indications have been found.

The dust in disks around medium- and low-mass YSOs seems to consist of grains which, on the average, are larger than the normal ISM dust grains. This was concluded from the flatness of the dust particle emissivity curves derived in disks around T Tauri stars (Beckwith and Sargent 1991), from extinction observations of Herbig Ae/Be stars (Gorti and Bhatt 1993, Sitko et al. 1994), from very broad silicate bands (Hanner et al. 1994b) and from polarization data of YSOs (Fischer et al. 1995).

The most prominent diagnostic IR spectral features in intermediate-mass and low-mass YSO spectra are the silicate bands. Already before the "ISO revolution" that revealed the occurrence of crystalline dust phases in Herbig Ae/Be stars, evidence of fine structure in the 10 μm band was found, which was discussed as sign that a small proportion of crystalline silicate contributes to the band emission. In analogy to cometary silicate spectra (Hanner et al. 1994a, 1999) a secondary peak at 11.2 μm was interpreted as the fingerprint of olivine grains. Similarly modified 10 μm profiles have also been found in Vega excess stars (see Sec. VI.B.2). However, a secondary peak at 11.2 μm in the silicate bands of young stars does not unambiguously prove the occurrence of a crystalline olivine phase in the dust because at the same wavelength a peak at 11.2 μm could also be due to the superposition of the UIR band positioned at the same wavelength. UIR bands have been detected in intermediate- and low-mass YSOs (Hanner et al. 1994b, 1998). There is even clear evidence of the coexistence of crystalline silicate and carbonaceous dust (or molecules) responsible for the UIR bands in YSOs as well as Vega excess stars (Sylvester et al. 1996, Malfait et al. 1998b, 1999).

Modelling the spectral energy distributions of 45 Herbig Ae/Be stars Malfait et al. (1998a) found that the majority of them contained a distinct warm and cool dust component. As mentioned above Herbig Ae/Be stars for which ISO spectra became available (Malfait et al. 1998b, 1999; Bouwman et al. 2000) revealed ample evidence of crystalline dust components, and also of the coexistence of silicates and the UIR band carriers. Amorphous silicates representing the warmer dust made out the bulk of the dust mass. Among the crystalline dust Mg-rich olivine (forsterite) dominates over pyroxene. Evidence was also found of iron oxide, hydrous silicates and water ice. The ISO observations of comet Hale-Bopp (Crovisier et al. 1997) revealed striking resemblance of its crystalline silicate bands with the corresponding features of the isolated Herbig Ae/Be star HD 100546 (Malfait et al. 1998b). This similarity and also the coexistence of silicates and carbonaceous components has been interpreted as a sign that massive cometary activity could exist in YSOs and Vega excess stars (see Sec. VI.B.3.).

In contrast to intermediate- and low-mass YSOs ISO spectra of high-mass YSOs did not show any evidence of crystalline silicates. The wide and structureless bands in these objects prove the amorphous structure of the silicates. Since most of these objects are deeply embedded sources the dust absorption is the same as in the molecular cloud surrounding and/or in star-forming regions (cf. Sec. II.C.). A first discussion of ISO spectra to improve the dust models of such environments was given by Demyk et al. (1999)

VI.B. Vega-Phenomenon Dust

1. Discovery of the Vega-phenomenon

Since a disk is the natural result of the collapse of matter carrying angular momentum, it is not surprising to find disk-shaped prestellar configurations and disks around pre-main-sequence stars. However, it was a true scientific sensation when the satellite IRAS found striking evidence for the existence of optically thin dust shells around main-sequence stars, which revealed their existence by radiation excesses in the IRAS FIR passbands. This thermal emission of dust grains with a temperature of about 100 K dominates the stellar radiation flux for wavelengths larger than 25 μm. Since α Lyr (Vega) was the first object showing this excess radiation (Aumann et al. 1984), the term "Vega phenomenon" has been introduced, and the stars are called Vega-excess stars (VES).

A compendium of published lists of VESs can be found in the detailed review article by Backman and Paresce (1993). Clues to the formation of planetary systems derived from the Vega phenomenon have been summarized by Hanner (1995).

Most of the information about these objects comes from the prototypes α Lyrae, α Piscis Austrini, β Pictoris, and ε Eridani which are spatially resolved and show disk shape (Backman and Paresce 1993; Table 4). It should be noted that it was the optical detection of the β Pic disk structure that gave the decisive impact to scrutinize the IRAS data in order to search for disk indications (Smith 1994) and to recognize the significance of the VES for a better understanding of the processes during and after the formation of a planetary system (Hanner 1995, Backman et al. 1997).

A characteristic property of the VES is their heavily depleted gas content. In contrast to interstellar environments, the gas-to-dust ratio is much smaller than unity. The detection of CO lines in the prototypes listed in Table 4 failed. Gas masses via CO observations were determined for the very dusty VESs SAO 112 630 and SAO 206 462 (Coulson et al. 1998), the corresponding gas-to-dust ratios amounted to $1.2 \cdot 10^{-2}$ and $8 \cdot 10^{-3}$, respectively. In eight further dust-rich objects only upper limits in the range $1.9 \cdot 10^{-3}$ to $7.5 \cdot 10^{-1}$ could be derived. Gas-free, optically thin dust disks with grains very large compared with interstellar dust have some parallels to the present-day solar system. Backman et al. (1997) use the term "planetary debris disks".

2. Grain Properties of Vega-phenomenon Dust

The analysis of VES data shows that the grains responsible for the FIR excesses form optically and geometrically thin disks, the innermost regions of which are practically free of dust up to a distance of about 20 AU from the star (Lagage and Pantin 1994).

Clear evidence that the dust in these disks contains silicates was first found by Telesco and Knacke (1991) in the spectrum of β Pic. Knacke et al. (1993) measured the 10-μm emission profile of this object with better spectral resolution and found a striking resemblance with the structured 10-μm silicate profile of comets. In analogy to the interpretation of cometary spectra (see Hanner et al. 1994a and literature therein) the second peak at about 11.2 μm in the β Pic spectrum has been assigned to crystalline olivine. Further VESs with structured 10 μm silicate bands have been found (Fajardo-Acosta and Knacke 1995, Sylvester et al. 1996). The similar silicate band profiles suggest some relationship between Vega-phenomenon dust and the dust released from comets. In addition to silicates, UIR bands have also been found in VESs (Sylvester et al. 1996, Skinner et al. 1995, Coulson and Walther 1995).

3. The Special Case of β Pictoris

Among the VES β Pic concentrated much attention by three distinct properties that were unique in the first decade of VESs research: It showed the largest FIR excess of all VES, it was the only object showing a significant excess in the

Table 4

Parameters of the dust disks around the VES prototypes α Lyr, α PsA, β Pic, and ε Eri. PRD means Poynting-Robertson drag. Compiled with data from Backman and Paresce (1993), Yamashita et al. (1993), Weintraub and Stern (1994). The data labelled by an asterisk are taken from Lanz et al. (1995), who reevaluated the basic stellar data of β Pic.

Parameter (unit)	α Lyr	α PsA	β Pic	ε Eri
Stellar data:				
Spectral type	A0 V	A3 V	A5 IV-V*	K2 V
Luminosity ($L_\odot$)	60	13	11.3*	0.3
Mass ($M_\odot$)	2.5	2.0	1.8*	0.75
Distance (pc)	8.1	7.0	16.4	3.3
Lifetime (yr)	$4 \cdot 10^8$	$7 \cdot 10^8$	$2 \cdot 10^9$	
Dust disk data:				
BB temperature (K)	89	72	108	70
Fractional luminosity of the excess	$2 \cdot 10^{-5}$	$8 \cdot 10^{-5}$	$3 \cdot 10^{-3}$	
60-μm diameter (AU)	235	252	426	
1.3-mm diameter (AU)	≤ 1000	≤ 600		≤ 130
Minimum mass ($M_\oplus$)	$2 \cdot 10^{-3}$	$2 \cdot 10^{-3}$	$7 \cdot 10^{-3}$	$6 \cdot 10^{-4}$
Effective grain radius (μm)	80	27	1	
Central hole radius(AU)	≤ 26	30–67	20–38	≈ 4
Disk inclination (°)	0–45	≤ 70	≤ 80	
PRD removal time scale (yr)	$2 \cdot 10^5$	$6 \cdot 10^6$	$2 \cdot 10^5$	
Collisional time scale (yr)	$5 \cdot 10^5$	$6 \cdot 10^5$	$8 \cdot 10^3$	

12 μm IRAS passband, and it was the only object with clear indications for the presence of gas in the disk. Of great advantage for the study was the edge-on orientation of the disk extending outward from the star for at least 1000 AU. The outstanding significance of β Pic was stressed by the growing evidence that it is a planetary system in an advanced stage of its formation, offering the chance for a detailed investigation of this stage and challenging comparison with solar system phenomena.

The great perseverance of the spectroscopists who have devoted many observational campaigns to β Pic since 1985 was rewarded with the discovery that comets are a universal phenomenon. An extended series of basic spectroscopic studies (for references see paper XXV by H. Beust et al. 1998), laid

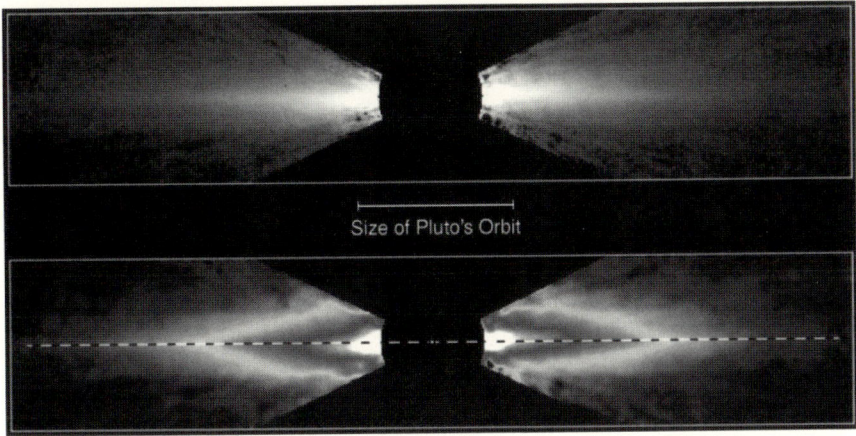

Figure 14. HST pictures in visible light of the inner part of the β Pic disk (the central star is blocked out). The false-color picture accentuates details in the warped disk. The pink-white inner edge of the disk is slightly tilted from the plane of the red-yellow-green outer region characterized by the dashed straight line (Credit: C. Burrows, J. Krist, the WFPC2 IDT team, and NASA).

the foundation for the falling evaporating body (FEB) scenario: huge swarms of star-grazing comet-like bodies feed the disk of β Pic with gas and solid debris and can account for the highly redshifted gas lines and the observed brightness variations. Today, the FEB mechanism is generally accepted and considered as a characteristic defining the subclass of the β Pictoris objects. Cometary activity has also been applied to the interpretation of the spectra of Herbig Ae/Be stars and other YSOs (Grady et al. 1997, Malfait et al. 1998b, 1999).

The second important research line of the β Pic system was modelling the dust disk in order to reproduce the IR observations, deriving dust properties in different regions of the disk and estimating time scales for dust sublimation and erosion and the necessary replenishment of the dust supply (Backman and Paresce 1993 and references therein, Artymowicz 1994, cf. Table 4). Together with the FEB mechanism, these studies stressed the general picture that VES are main-sequence stars with "planetary debris disks" (Backman et al. 1997).

Up to now, planets within the disk of β Pic could not definitely be detected. However, the existence of a cleared inner zone in the dust distribution at distances comparable to those of the planets from the sun and the warped disk of β Pic (Fig. 14) have been considered strong indirect signs for the presence of planets (Roques et al. 1994). The detection avalanche of extrasolar giant planets gives further support to this expectation (Beckwith and Sargent 1996).

REFERENCES

Allamandola, L. J., Tielens, A. G. G. M., and Barker, J. R. 1987. Infrared absorption and emission characteristics of interstellar PAHs. In *Interstellar Processes*, eds. D. J. Hollenbach, H. A. Thronson (Dordrecht: Reidel), pp. 471–489.

Allamandola, L. J., Sandford, S. A., Tielens, A. G. G. M., and Herbst, T. M. 1993. Diamonds in dense molecular clouds: a challenge to the standard interstellar medium paradigm. *Science*, **260**, pp. 64–66.

Anders, E., and Zinner, E. 1993. Interstellar grains in primitive meteorites: Diamond, silicon carbide, and graphite. *Meteoritics*, **28**, pp. 490–514.

Andersen, A. C., Jäger, C., Mutschke, H., Braatz, A., Clément, D., Henning, Th., Jørgensen, U. G., and Ott, U. 1999. Infrared spectra of meteoritic SiC grains. *Astron. Astrophys.*, **343**, pp. 933–938.

Anderson, C. M., Weitenbeck, A. J., Code, A. D., Nordsieck, K. H., Meade, M. R., Babler, B., Zellner, N. E. B., Bjorkman, K. S., Fox, K. G., and Johnson, J. J. 1996. Ultraviolet interstellar polarization of galactic starlight. I. Observations by the Wisconsin Ultraviolet Photo Polarimeter Experiment. *Astron. J.*, **112**, pp. 2726–2743.

André, Ph. 1994. Disk-like structures around young stars. In *Circumstellar Dust Disks and Planet Formation*, eds. R. Ferlet and A. Vidal-Madjar (Gif sur Yvette: Editions Frontières), pp. 115–129.

Artymowicz, P. 1994. Modeling and understanding the dust around beta Pictoris. In *Circumstellar Dust Disks and Planet Formation*, eds. R. Ferlet and A. Vidal-Madjar (Gif sur Yvette: Editions Frontières), pp. 47–65.

Aumann, H. H., Gillett, F. C., Beichmann, C. A., de Jong, T., Houck, J. R., Low, F. J., Neugebauer, G., Walker, R. G., and Wesselius, P. R. 1984. Discovery of a shell around Alpha Lyrae. *Astrophys. J.*, **278**, pp. L23-L27.

Baas, F., Grim, R. J. A., Geballe, T. R., Schutte, W., and Greenberg, J. M. 1988. The detection of solid methanol in W 33A. *Dust in the Universe*, eds. M. E. Bailey and D. A. Williams (Cambridge: Cambridge Univ. Press), pp. 55–60.

Backman, D. E. and Paresce, F. 1993. Main-sequence stars with circumstellar solid material: the Vega phenomenon. In *Protostars and Planets III*, eds. E. H. Levy, J. I. Lunine, and M. S. Matthews (Tucson: Univ. Arizona Press), pp. 1253–1304.

Backman, D. E., Werner, M. W., Rieke, G. H., and Van Cleve, J. E. 1997. Exploring planetary debris disks around solar-type stars. In *From Stardust to Planetesimals*, eds. Y. J. Pendleton and A. G. G. M. Tielens (San Francisco: Astron. Soc. Pacific), A. S. P. Conf. Ser. Vol. 122, pp. 49–66.

Baggaley, W. J. 2000. Advanced Meteor Orbit Radar observations of interstellar meteoroids. *J. Geophys. Res.*, **105 No.A5**, pp. 10353–10362.

Banhart, F. and Ajayan, P. M. 1996. Carbon onions as nanoscopic pressure cells for diamond formation. *Nature*, **382**, pp. 433–435.

Beckwith, S. V. W. 1994. Protoplanetary disks. In *NATO Advanced Research Workshop on Theory of Accretion Disks – 2.*, eds. W. J. Duschl, J. Frank, F. Meyer, E. Meyer-Hofmeister, and W. Tscharnuter (Dordrecht: Kluwer Academ. Publ.), pp. 1–18.

Beckwith, S. V. W., Sargent, A. I. 1991. Particle emissivity in circumstellar disks. *Astrophys. J.*, **381**, pp. 250–258.

Beckwith, S. V. W., Sargent, A. I. 1996. Circumstellar disks and the search for neighboring planetary systems. *Nature*, **383**, pp. 139–144.

Begemann, B., Dorschner, J., Henning, T., Mutschke H., and Thamm, E. 1994. A laboratory approach to the interstellar sulfide dust problem. *Astrophys. J.*, **423**, pp. L71–L74.

Begemann, B., Dorschner, J., Henning, Th., and Mutschke, H. 1996. Optical properties of glassy SiS_2 and the 21 micron feature. *Astrophys. J.*, **464**, pp. L195–L198.

Begemann, B., Dorschner, J., Henning, Th., Mutschke, H., Gürtler, Kömpe, C., Nass, R. 1997. Aluminum oxide and the opacity of oxygen-rich circumstellar dust in the 12-17-μm range. *Astrophys. J.*, **476**, pp. 199–208.

Bernatowicz, T. J., Cowsik, R., Gibbons, P. C., Lodders, K., Fegley, Jr., B., Amari, S., and Lewis, R. S. 1996. Constraints on stellar grain formation from presolar graphite in the Murchison meteorite. *Astrophys. J.*, **472**, pp. 760–782.

Beust, H., Lagrange, A.-M., Crawford, I. A., Goudard, C., Spyromilio, J., Vidal-Madjar, A.

1998. The β Pictoris circumstellar disk. XXV. The CaII absorption lines and the Falling Evaporating Bodies model revisited using UHRF observations. *Astron. Astrophys.*, **338**, pp 1015–1030.

Biermann, P., and Harwitt, M. 1980. On the origin of the grain-size size spectrum of interstellar dust. *Astrophys. J. Lett.*, **241**, pp. L105–L107.

Bless, R. C., and Savage, B. D. 1972. Ultraviolet photometry from the Orbiting Astronomical Observatory. II. Interstellar extinction. *Astrophys. J.*, **171**, pp. 293–308.

Blitz, L. 1993. Giant molecular clouds. In *Protostars & Planets III*, eds. E. H. Levy and J. I. Lunine (Tucson: Univ. Arizona Press), pp. 125–161.

Blum, J., Wurm, G. 2000. Experiments on sticking, restructuring, and fragmentation of preplanetary dust aggregates. *Icarus*, **143**, pp. 138–146.

Blum, J., Wurm, G., Kempf, S. + 24 authors 2000. Growth and form of planetary seedlings: results from a microgravity aggregation experiment. *Phys. Rev. Lett.*, **85**, pp. 2426–2429.

Bohren, C. F., and Huffman, D. R. 1983. *Absorption and Scattering of Light by Small Particles* (New York: Wiley).

Boogert, A. C. A., Schutte, W. A., Tielens, A. G. G. M., et al. 1996. Solid methane toward deeply embedded protostars. *Astron. Astrophys.*, **315**, pp. L377–L380.

Boulanger, F. 1994. Dust and gas in the infrared cirrus. In *The First Symposium on the Infrared Cirrus and Diffuse Interstellar Clouds*, eds. R. Cutri and W. B. Latter (San Francisco: Astron. Soc. Pacific), A. S. P. Conf. Ser., **Vol. 58**, pp. 101–114.

Boulanger, F., Baud, B., and van Albada, G. D. 1985. Warm dust in the neutral interstellar medium. *Astron. Astrophys.*, **144**, pp. L9–L12.

Boulanger, F., Prévot, M. L., and Gry, C. 1994. The contribution of small particles to the extinction curve. *Astron. Astrophys.*, **284**, pp. 956–970.

Bouwman, J., de Koter, A., van den Ancker, M. E., and Waters, L. B. F. M. 2000. The composition of the circumstellar dust around the Herbig Ae stars AB Aur and HD 163296. *Astron. Astrophys.*, **360**, pp. 213–226.

Braatz, A., Dorschner, J., Henning, Th., Jäger, C., and Ott, U. 1998. Infrared spectroscopy of presolar diamonds: The influence of chemical preparation. *Meteoritics Planet. Sci.*, **33**, p. A21.

Braatz, A., Ott, U., Henning, Th., Jäger, C., and Jeschke, G. 2000. Infrared, ultraviolet, and electron paramagnetic resonance measurements on presolar diamonds: Implications for optical features and origin. *Meteoritics Planet. Sci.*, **35**, pp. 75–84.

Bradley, J. P. 1996. The search for interstellar components in interplanetary dust. In *Physics, Chemistry, and Dynamics of Interplanetary Dust: Proc. IAU Coll. 150*, ed. B. Å. S. Gustafson (San Francisco: Astron. Soc. Pacific), A. S. P. Conf. Ser. **Vol. 104**, pp. 275–282.

Bradley, J. P. 1999. Interstellar dust: evidence from interplanetary dust particles. In *Formation and Evolution of Solids in Space*, eds. J. M. Greenberg, A. Li (Dordrecht: Kluwer Academ. Publ.), NATO ASI Ser. C, **Vol. 523**, pp. 485–504.

Bürki, P. R. 1996. Low-pressure formation routes for interstellar microdiamonds: chemical vapor deposition vs. homogeneous nucleation. *Meteoritics*, **31**, pp. A24-A25.

Buss Jr., R. H., Cohen, M., Tielens, A. G. G. M., Werner, M. W., Bregman, J. D., Witteborn, F. C., Rank, D., and Sandford, S.A. 1990. Hydrocarbon emission features in the infrared spectrum of warm supergiants. *Astrophys. J.*, **365**, pp. L23–L26.

Butchart, I., McFadzean, A. D., Whittet, D. C. B., Geballe, T. R., Greenberg, J. M. 1986. Three micron spectroscopy of the galactic center source IRS7. *Astron. Astrophys. Lett.*, **154**, pp. L5–L7.

Cardelli, J. A. 1994. Variability of interstellar extinction and its relationship to environment. In *The First Symposium on the Infrared Cirrus and Diffuse Interstellar Clouds*, eds. R. Cutri and W. B. Latter (San Francisco: Astron. Soc. Pacific), A. S. P. Conf. Ser. **Vol. 58**, pp. 24–33.

Cardelli, J. A., Clayton, G. C., Mathis, J. S. 1988. The determination of ultraviolet extinction from the optical and near-infrared. *Astrophys. J. Lett.*, **329**, pp. L33–L37.

Cesarsky, D., Jones, A. P., Lequeux, X., and Verstraete, L. 2000. Silicate emission in Orion. *Astron. Astrophys.*, **358**, pp. 708-716.

Cherchneff, I., and Barker, J. R. 1992. Polycyclic aromatic hydrocarbons and molecular equilibria in carbon-rich stars. *Astrophys. J.*, **394**, pp. 703–716.

Chiar, J. E., Tielens, A. G. G. M., Whittet, D. C. B., Schutte, W. A., Boogert, A. C. A., Lutz, D., van Dishoeck, E. F., and Bernstein, M. P. 2000. The composition and distribution of dust along the line of sight toward the Galactic center. *Astrophys. J.*, **537**, pp. 749–762.

Chlewicki, G., and Laureijs, R. J. 1988. Model of grain properties based on IRAS observations. I. Evidence for new particle populations. *Astron. Astrophys.*, **207**, pp. L11–L14.

Choi, B.-G., Huss, G. R., Wasserburg, G. J., Gallino, R. 1998. Presolar corundum and spinel in ordinary chondrites: origins from AGB stars and a supernova. *Science*, **282**, pp. 1284–1289.

Cohen, M., Tielens, A. G. G. M., and Bregman, J. D. 1989. Mid-infrared spectra of WC9 stars: The composition of circumstellar and interstellar dust. *Astrophys. J. Lett.*, **344**, pp. L13–L16.

Colangeli, L., Mennella, V., Palumbo, P., and Rotundi, A. 1999. Cosmic dust and laboratory simulation: wishes, results and open problems. In *Formation and Evolution of Solids in Space*, eds. J. M. Greenberg, A. Li (Dordrecht: Kluwer Academ. Publ.), NATO ASI Ser. C, **Vol. 523**, pp. 203–228.

Coulson, J. M., and Walther, D. M. 1995. SAO 206462 - a solar-type star with a dusty, organically rich environment. *Mon. Not. Roy. Astron. Soc.*, **274**, pp. 977–986.

Coulson, I. M., Walther, D. M., and Dent, W. R. F. 1998. Infrared and submillimetre studies of Vega-excess stars. *Monthly Not. R. Astron. Soc.*, **296**, pp. 934–942.

Cox, P. 1989. The line of sight towards AFGL 961: Detection of the librational band of water ice at 13.6 micron. *Astron. Astrophys. Lett.*, **225**, pp. L1–L4.

Cox, P. 1993. Far-infrared spectroscopy of solid state features. In *Astronomical Infrared Spectroscopy: Future Observational Directions*, ed. S. Kwok (San Francisco, Astron. Soc. Pacific), A. S. P. Conf. Ser., **Vol. 41**, pp. 163–170.

Cox, P., and Mezger, P. G. 1989. The galactic infrared/submillimeter dust radiation. *Astron. Astrophys. Rev.*, **1**, pp. 49–83.

Crovisier, J., Leech, K., Bockelee-Morvan, D., Brooke, T. Y., Hanner, M. S., Altieri, B., Keller, H. U., and Lellouch, E. 1997. The spectrum of Comet Hale-Bopp (C/1995 O1) observed with the Infrared Space Observatory at 2.9 astronomical units from the Sun. *Science*, **275**, pp. 1904–1907.

Dartois, E., Schutte, W., Geballe, T. R. Demyk, K., Ehrenfreund, P. d'Hendecourt, L. 1999. Methanol: the second most abundant ice species towards the high-mass protostars RAFGL7009S and W33A. *Astron. Astrophys.*, **342**, pp. L32–L35.

Davis, L., and Greenstein, J. L. 1951. The polarization of starlight by aligned dust grains. *Astrophys. J.*, **114**, pp. 206–240.

de Graauw, Th., Whittet, D. C. B., Gerakines, P. A. + 29 authors 1996. SWS observations of solid CO_2 in molecular clouds. *Astron. Astrophys.*, **315**, pp. L345–L348.

de Heer, W. A., and Ugarte, D. 1993. Carbon onions produced by heat treatment of carbon soot and their relation to the 217.5 nm interstellar absorption feature. *Chem. Phys. Lett.*, **207**, p. 480.

Demyk, K., Jones, A. P., Dartois, E., Cox, P., and d'Hendecourt, L. 1999. The chemical composition of the silicate dust around RAFGL7009S and IRAS 19110+1045. *Astron. Astrophys.*, **349**, pp. 267–275.

Désert, F.-X., Boulanger, F., and Puget, J. L. 1990. Interstellar dust models for extinction and emission. *Astron. Astrophys.*, **237**, pp. 215–236.

d'Hendecourt, L. B., and Jourdain de Muizon, M. 1989. The discovery of interstellar carbon dioxide. *Astron. Astrophys. Lett.*, **223**, pp. L5–L8.

d'Hendecourt, L. B. 1997. The PAH hypothesis: Infrared spectroscopic properties of PAHs. In *From Stardust to Planetesimals*, eds. Y. J. Pendleton and A. G. G. M. Tielens (San Francisco: Astron. Soc. Pacific), A. S. P. Conf. Ser., **Vol. 122**, pp. 129–145.

Donn, B. 1968. Polycyclic hydrocarbons, Platt particles, and interstellar extinction. *Astrophys. J. Lett.*, **152**, pp. L129-L133.

Dorschner, J. 1967. Theoretische Untersuchungen über den interstellaren Staub. I. Vorschlag eines Staubmodells aus meteoritischen Silikaten. *Astron. Nachr.*, **290**, pp. 171–181.

Dorschner, J. 1982. Interstellar grain size spectrum and circumstellar grain-grain collisions. *Astrophys. Space Sci.*, **81**, pp. 323–328.

Dorschner, J. 1992. Interstellar dust, subject and agent of galactic evolution. *Rev. Mod. Astron.*, **6**, pp. 117–147.

Dorschner, J. 1999. Stardust mineralogy: the laboratory approach. In *Formation and Evolu-*

tion of Solids in Space, ed. J. M. Greenberg, A. Li (Dordrecht: Kluwer Academ. Publ.), NATO ASI Ser. C, **Vol. 523**, pp. 229–264.

Dorschner, J., and Henning, Th. 1995. Dust metamorphosis in the Galaxy. *Astron. Astrophys. Rev.*, **6**, pp. 271–333.

Dorschner, J., Friedemann, C., Gürtler, J., and Schielicke, R. 1984. A catalogue of equivalent widths of the interstellar 2000 Å band. *Bull. Inf. Center Données Stellaires*, **27**, pp. 137–139.

Dorschner, J., Begemann, B., Henning, Th., Jäger, C., and Mutschke, H. 1995. Steps toward interstellar silicate mineralogy. II. Study of Mg-Fe-silicate glasses of variable composition. *Astron. Astrophys.*, **300**, pp. 503–520.

Draine, B. T. 1994. Dust in diffuse interstellar clouds. In *The First Symposium on the Infrared Cirrus and Diffuse Interstellar Clouds*, eds. R. Cutri and W. B. Latter (San Francisco: Astron. Soc. Pacific), A. S. P. **Vol. 58**, pp. 227–242.

Draine, B. T., Lee, H. M. 1984. Optical properties of interstellar graphite and silicate grains. *Astrophys. J.*, **285**, pp. 89–108.

Duley, W. W. 1988. Sharp emission lines from diamond dust in the Red Rectangle? *Astrophys. Space Sci.*, **150**, pp. 387–390.

Duley, W. W. 1993. Carbonaceous grains. In *Dust and Chemistry in Astronomy*, eds. T. J. Millar and D. A. Williams (Bristol: Inst. Physics Publ.), pp. 71–101.

Erickson, E. F., Knacke, R. F., Tokunaga, A. T., and Haas, M. R. 1981. The 45 micron H_2O ice band in the Kleinmann-Low Nebula. *Astrophys. J.*, **245**, pp. 148–153.

Fabian, D., Jäger, C., Henning, Th., Dorschner, J., and Mutschke, H. 2000. Steps toward interstellar silicate mineralogy. V. Thermal evolution of amorphous magnesium silicates and silica. *Astron.& Astrophys.*, **364**, pp. 282–292.

Fabian, D., Posch, Th., Mutschke, H., Kerschbaum, F. and Dorschner, J. 2001. Infrared optical properties of spinels. A study of the carrier of the 13, 17 and 32 μm emission features observed in ISO-SWS spectra of oxygen-rich AGB stars. *Astron. Astrophys.*, in press.

Fajardo-Acosta, S. B., and Knacke, R. F. 1995. IRAS low resolution spectra with β Pictoris-type silicate emission. *Astron. Astrophys.*, **295**, pp. 767–774.

Ferrarotti, A., Gail, H.-P., Degiorgi, L., and Ott, H. R. 2000. FeSi as a possible new circumstellar dust component. *Astron. Astrophys.*, **357**, pp. L13–L16.

Fischer, O., Henning, Th., and Yorke, H. W. 1995. Simulation of polarization maps. II. The circumstellar environments of pre-main sequence objects. *Astron. Astrophys.*, **308**, pp. 863–885.

Fitzpatrick, E. L., and Massa, D. 1986. An analysis of the shapes of ultraviolet extinction curves. I. The 2175 Å bump. *Astrophys. J.*, **307**, pp. 286–294.

Fitzpatrick, E. L., and Massa, D. 1990. An analysis of the shapes of ultraviolet extinction curves. III. An atlas of ultraviolet extinction curves. *Astrophys. J. Suppl. Ser.*, **72**, pp. 163–189.

Freivogel, P., Fulara, J., Maier, J. P. 1994. Highly unsaturated hydrocarbons as potential carriers of some diffuse interstellar bands. *Astrophys. J. Lett.*, **431**, pp. L151–L154.

Friedemann, Chr. 1969. Evolution of silicon carbide particles in the atmospheres of carbon stars. *Physica*, **41**, pp. 139–143.

Frisch, P. C., Dorschner, J. M., Geiss, J., Greenberg, J. M., Grün, E., Landgraf, M., Hoppe, P., Jones, A. P., Krätschmer, W., Linde, T. J., Morfill, G. E., Reach, W., Slavin, J. D., Svestka, J., Witt, A. N., and Zank, G. P. 1999. Dust in the local interstellar wind. *Astrophys. J.*, **525**, pp. 492–516.

Gail, H.-P., Sedlymayr, E. 1998. Inorganic dust formation in astrophysical environments. In *Chemistry and Physics of Molecules and Grains in Space*, ed. P. Sarre (London: The Faraday Division of the Royal Society of Chemistry), Faraday Discussion, **No. 109**, p. 303.

Geballe, T. R., Baas, F., Greenberg, J. M., Schutte, W. 1985. New infrared absorption features due to solid phase molecules containing sulfur. *Astron. Astrophys. Lett.*, **146**, pp. L6–L8.

Geballe, T. R., Joblin, C., d'Hendecourt, L. B., Jourdain de Muizon, M., Tielens, A. G. G. M., and Léger, A. 1994. Detection of the overtone of the 3.3 micron emission feature in IRAS 21282 +5050. *Astrophys. J. Lett.*, **434**, pp. L15–L18.

Geballe, T. R. 1997. Spectroscopy of the unidentified infrared emission bands. In *From*

Stardust to Planetesimals, eds. Y. J. Pendleton and A. G. G. M. Tielens (San Francisco: Astron. Soc. Pacific), A. S. P. Conf. Ser., **Vol. 122**, pp. 119–128.

Giard, M., Lamarre, J. M., Pajot, F., and Serra, G. 1994. The large scale distribution of PAHs in the Galaxy. *Astron. Astrophys.*, **286**, pp. 203–210.

Gibb, E. L., Whittet, D. C. B., Schutte, W. A. + 8 authors 2000. An inventory of interstellar ices toward the embedded protostar W33A. *Astrophys. J. Lett.*, **536**, pp. 347–356.

Gillett, F. C., and Forrest, W. J. 1973. Spectra of the Becklin-Neugebauer point source and the Kleinmann-Low nebula from 2.8 to 13.5 microns. *Astrophys. J.*, **179**, pp. 483–491.

Glasse, A. C. H., Towlson, W. A., Aitken, D. K., and Roche, P.F. 1986. High-resolution infrared spectroscopy: A search for the 11.52-μm graphite feature *Mon. Not. Roy. Astron. Soc.*, **220**, pp. 185–188.

Goebel, J. H. 1993. SiS_2 in circumstellar shells. *Astron. Astrophys.*, **278**, pp. 226–230.

Goebel, J. H., and Moseley, S.H. 1985. MgS grain component in circumstellar shells. *Astrophys. J. Lett.*, **290**, pp. L35-L39.

Gordon, K. D., Witt, A. N., Rudy, R. J., Puetter, R. C., Lynch, D. K., Mazuk, S., Misselt, K. A., Clayton, G. C. and Smith, T. L. 2000. Dust emission features in NGC 7023 between 0.35 and 2.5 μm: Extended Red Emission (0.7 μm) and two new emission features (1.15 and 1.5 μm). *Astrophys. J.*, in press.

Gordon, K. D., Witt, A. N., and Friedmann, B. C. 1998. Detection of extended red emission in the diffuse interstellar medium. *Astrophys. J.*, **498**, pp. 522–540.

Gorti, U., Bhatt, H. C. 1993. Anomalous dust in the environment of Herbig Ae/Be star. *Astron. Astrophys.*, **270**, pp. 426–431.

Grady, C. A., Sitko, M. L., Bjorkman, K. S., Pérez, M. R., Lynch, D. K., Russell, R. W., Hanner, M. S. 1997. The star-grazing extrasolar comets in the HD 100546 system. *Astrophys. J.*, **483**, pp. 449–456.

Greenberg, J. M. 1973. Some scattering problems of interstellar grains. In *Interstellar Dust and Related Topics: Proc. IAU Symp.*, **52**, eds. J. M. Greenberg and H.C. van de Hulst (Dordrecht: Reidel), pp. 3–9.

Greenberg, J. M. 1984. Evolution of interstellar grains. *Occasional Rep. Royal Obs. Edinburgh*, **12**, pp. 1–25.

Greenberg, J. M. 1989. The core-mantle model of interstellar grains and the cosmic dust connection In *Interstellar Dust: Proc. IAU Symp. 135*, eds. L. J. Allamandola and A. G. G. M. Tielens (Dordrecht: Kluwer Academ. Publ.), pp. 345–355.

Greenberg, J. M. and Li, A. 1995. What are the true astronomical silicates? *Astron. Astrophys.*, **309**, pp. 258–266.

Grim, R. J. A. and Greenberg, J. M. 1987. Ions in grain mantles: The 4.62 micron absorption by OCN^{--} in W33A. *Astrophys. J. Lett.*, **321**, pp. L91–L96.

Grim, R. J. A., Baas, F., Geballe, T. R., Greenberg, J. M., and Schutte, W. 1991. Detection of solid methanol in W33A. *Astron. Astrophys.*, **243**, pp. 473–477.

Grün, E., Gustafson, B., Mann, I., Baguhl, M., Morfill, G. E., Staubach, P., Taylor, A., and Zook, H. A. 1994. Interstellar dust in the heliosphere. *Astron. Astrophys.*, **286**, pp. 915–924.

Grün, E., Landgraf, M., Horányi, M., Kissel, J., Krüger, H., Srama, R., Svedhem, H., and Withnell, P. 2000. Techniques for galactic dust measurement in the heliosphere. *J. Geophys. Res.*, **105 No. A5**, pp. 10403–10410.

Gürtler, J., Henning, Th., Dorschner, J. 1989. Properties of circumstellar silicate dust (Review). *Astron. Nachr.*, **310**, pp. 319–327.

Gürtler, J., Henning, Th., Kömpe, C., Pfau, W., Krätschmer, W., and Lemke, D. 1996. Detection of an absorption feature at the position of the 4.27-μm band of solid CO_2. *Astron. Astrophys.*, **315**, pp. L189–L192.

Gürtler, J., Schreyer, K., Henning, Th., Lemke, D., and Pfau, W. 1999. Infrared spectra of young stars in Chamaeleon. *Astron. Astrophys.*, **346**, pp. 205–210.

Guillois, O., Ledoux, G., and Reynaud, C. 1999. Diamond infrared emission bands in circumstellar media. *Astrophys. J. Lett.*, **521**, pp. L133–L136.

Hall, J. S. 1949. Observations of the polarized light from stars. *Science*, **109**, p. 166.

Hallenbeck, S. L., Nuth, J. A., and Daukantas, P. L. 1998. Mid-infrared spectral evolution of amorphous magnesium silicate smokes annealed in vacuum: comparison to cometary spectra. *Icarus*, **131**, pp. 198–209.

Hanner, M. S. 1995. Dust around young stars: How related to solar system dust? *Highlights of Astronomy*, **10**, pp. 351–392.

Hanner, M. S. 1999. The silicate material in comets. *Space Sci. Rev.*, **90**, pp. 99–108.

Hanner, M. S., Lynch, D. K., and Russell, R. W. 1994a. The 8–13 micron spectra of comets and the composition of silicate grains. *Astrophys. J.*, **425**, pp. 274–285.

Hanner, M. S., Brooke, T. Y., and Tokunaga, A. T. 1994b. Silicates and aromatic hydrocarbons in the 10 micron spectrum of the Taurus dark cloud Elias 1. *Astron. J. Lett.*, **433**, pp. L97–100.

Hanner, M. S., Brooke, T. Y., and Tokunaga, A. T. 1998. 8–13 micron spectroscopy of young stars. *Astrophys. J.*, **502**, pp. 871-882.

Hecht, J. H. 1991. The nature of the dust around R Coronae Borealis stars: Isolated amorphous carbon or graphite fractals? *Astrophys. J*, **367**, pp. 635–640.

Henning, Th. 1996. Circumstellar dust around young stars. In *The Cosmic Dust Connection*, ed. J. M. Greenberg (Dordrecht: Kluwer), NATO ASI Ser. C, Vol. 487, pp. 399–412.

Henning, Th., and Salama, F. 1998. Carbon in the universe. *Science*, **282**, pp. 2204–2210.

Henning, Th., and Schnaiter, M. 1999. Carbon – From space to laboratory. In *Laboratory Astrophysics and Space Research*, eds. P. Ehrenfreund , C. Kraft, H. Kochan, and V. Pirronello (Dordrecht: Kluwer), pp. 249–277.

Henning, Th., Dorschner, J., and Gürtler, J. 1989. Size distribution of dust grains – a problem of self-similarity? In *Interstellar Dust: Contributed Papers*, NASA CP 3036, p. 395.

Henning, Th., Begemann, B., Mutschke, H., and Dorschner, J. 1995. Optical properties of oxide dust grains. *Astron. Astrophys. Suppl. Ser.*, **112**, pp. 143–149.

Herbig, G. H. 1995. The diffuse interstellar bands. *Annu. Rev. Astrophys.*, **33**, pp. 19–73.

Hill, H. G. M., Jones, A. P., d'Hendecourt, L. B. 1998. Diamonds in carbon-rich protoplanetary nebulae. *Astron. Astrophys.*, **336**, pp. L41–L44.

Hiltner, W. A. 1949. On the presence of polarization in the continuum radiation of stars. *Astrophys. J.*, **109**, pp. 471–478.

Hoffmeister, C. 1929. On the heliocentric velocity of meteors. *Astrophys. J.*, **69**, pp. 159–167.

Hoffmeister, C. 1931. Zur physikalischen Theorie der Sternschnuppen. *Astron. Nachr.*, **241**, pp. 1–8.

Hong, S. S., and Greenberg, J. M. 1978. On the size distribution of interstellar grains. *Astron. Astrophys.*, **70**, pp. 695-699.

Hoppe, P. and Ott, U. 1997. Mainstream silicon carbide grains from Meteorites. In *Astrophysical Implications of the Laboratory Study of Presolar Material*, eds. T. J. Bernatowicz and E. Zinner, AIP Conf. Ser. Proceed., **402**, pp. 27–58.

Hoppe, P. and Zinner, E. 2000. Presolar dust grains from meteorites and their stellar sources. *J. Geophys. Res.*, **105 No. A5**, pp. 10371–10385.

Hoyle, F., and Wickramasinghe, N. C. 1962. On graphite particles as interstellar grains. *Mon. Not. Roy. Astron. Soc.*, **124**, pp. 417–433.

Hron, J., Aringer, B., and Kerschbaum 1997. Semiregular variables of types SRa and SRb. Silicate dust emission features. *Astron. Astrophys.*, **322**, pp. 280–290.

Jäger, C., Mutschke, H., Begemann, B., Dorschner, J., and Henning, Th. 1994. Steps toward interstellar silicate mineralogy. I. Laboratory results of a silicate glass of mean cosmic composition. *Astron. Astrophys.*, **292**, pp. 641–655.

Jäger, C., Mutschke, H., and Henning, Th. 1998a. Optical properties of carbonaceous dust analogs. *Astron. Astrophys.*, **332**, pp. 291–299.

Jäger, C., Molster, F., Dorschner, J., Henning, Th., Mutschke, H., and Waters, L. B. F. M. 1998b. Steps toward interstellar silicate mineralogy. IV. The crystalline revolution. *Astron. Astrophys.*, **339**, pp. 904–916.

Jenniskens, P., and Désert, F.-X. 1993. A survey of diffuse interstellar bands (3800–8680 Å) *Astron. Astrophys. Suppl. Ser.*, **106**, pp. 39–78.

Jenniskens, P., and Greenberg, J. M. 1993. Environment dependence of interstellar extinction curves. *Astron. Astrophys.*, **274**, pp. 439–450.

Jones, A. P. 1997. The lifecycle of interstellar dust. In *From Stardust to Planetesimals*, eds. Y. J. Pendleton and A. G. G. M. Tielens (San Francisco: Astron. Soc. Pacific), A. S. P. Conf. Ser., **Vol. 122**, pp. 97–106.

Jones, A. P., and Williams, D. A. 1987. Interplanetary material as a guide to the composition of interstellar grains. *Mon. Not. R. Astron. Soc.*, **224**, pp. 473–479.

Jones, A. P., and Tielens, A. G. G. M. 1994. Interstellar dust – physical processes. In *The*

Cold Universe, eds. Th. Montmerle, Ch. J. Lada, I. F. Mirabel, and J. Trân Thanh Vân (Gif sur Yvette: Editions Frontières), pp. 35–44.

Jones, A. P., Tielens, A. G. G. M., and Hollenbach, D. J. 1996 Grain shattering in shocks: the interstellar grain size distribution. *Astrophys. J.*, **469**, pp. 740–764.

Kamijo, F. 1963. A theoretical study of the long period variable stars. III. Formation of solid or liquid particles in the circumstellar envelope. *Publ. Astron. Soc. Japan*, **15**, pp. 440-448.

Keller, L. P., Bradley, J. P., Bouwman, J., Molster, F. J., Waters, L. B. F. M., Flynn, G. J., Henning, T., and Mutschke, H. 2000. Sulfides in interplanetary dust particles: a possible match to the 23 μm feature detected by the Infrared Space Observatory. *Abstracts 31st Annual LPS conference*, Abstract No. 1860.

Kim, S.-H., and Martin, P. G. 1995. The size distribution of interstellar dust particles as determined from polarization: spheroids. *Astrophys. J.*, **444**, pp. 293–305.

Kim, S.-H., Martin, P. G., and Hendry, P. D. 1994. The size distribution of interstellar dust particles as determined from extinction. *Astrophys. J.*, **422**, pp. 164–175.

Knacke, R. F., Fajardo-Acosta, S. B., Telesco, C. M., Hackwell, J. A., Lynch, D. K., and Russell, R. W. 1993. The silicates in the disk of β Pictoris. *Astrophys. J.*, **418**, pp. 440-450.

Koike, C., Kimura, S., Kaito, C., Suto, H., Shibai, H., Nagata, T., Tanabe, T., and Saito, Y. 1995. Correlation between the spectral index and the degree of crystallization of carbon and graphite grains. *Astrophys. J.*, **446**, pp. 902–906.

Koike, C., Wickramasinghe, N. C., Kano, N., Yamakoshi, K., Yamamoto, T., Kaito, C., Kimura, S., and Okuda, H. 1995. The infrared spectra of diamond-like residues from the Allende meteorite. *Monthly Not. R. Astron. Soc.*, **277**, pp. 986–994.

Kwok, S., Volk, K. M., Hrivnak, B. J. 1989. A 21 micron emission feature in four protoplanetary nebulae. *Astrophys. J.*, **345**, pp. L51-L54.

Lada, E. A., Strom, K. M., and Myers, P. C. 1993. Environments of star formation: relationship between molecular clouds, dense cores, and young stars. In *Protostars & Planets III*, eds. E. H. Levy and J. I. Lunine (Tucson: Univ. Arizona Press), pp. 245–277.

Lacy, J. H., Baas, F., Allamandola, L. J., Perssons, S. E. McGregor, P. J., Lonsdale, C. J., Geballe, T. R., van de Bult, C. E. P. 1984. 4.6 micron absorption features due to solid phase CO and cyano group molecules toward compact infrared sources. *Astrophys. J.*, **276**, pp. 533–543.

Lacy, J. H., Faraji, H., Sandford, S. A., and Allamandola, L. J. 1998. Unraveling the 10 micron "silicate" feature of protostars: the detection of frozen interstellar ammonia. *Astrophys. J.*, **501**, pp. L105–L109.

Lagage, P. O., and Pantin, E. 1994. Probing dust around main-sequence stars with TIMMI. *The Messenger*, **75**, pp. 24–26.

Landgraf, M., Baggaley, W. J., Grün, E., Krüger, H., and Linkert, G 2000. Aspects of the mass distribution of interstellar grains in the solar system from in-situ measurements. *J. Geophys. Res.*, **105 No. A5**, pp. 10343–10352.

Lanz, T., Heap, S. R., and Hubeny, I. 1995. HST/GHRS observations of the β Pictoris system: basic parameters and the age of the system. *Astrophys. J.*, **447**, pp. L41–L44.

Larson, H. P., Davis, D. S., Black, J. H., and Fink, U. 1985. Interstellar absorption features toward the compact infrared source W33A. *Astrophys. J.*, **299**, pp. 873–880.

Ledoux, G., Ehbrecht, M., Guillois, O., Huisken, F., Kohn, B., Laguna, M. A., Nenner, I., Paillard, V., Papoular, R., Porterat D., and Reynaud, C. 1998. Silicon as a candidate carrier for ERE. *Astron. Astrophys.*, **333**, pp. L39–L42.

Lenzuni, P., Gail, H.-P., and Henning, Th. 1995. Dust evaporation in protostellar cores. *Astrophys. J.*, **447**, pp. 848-862.

Li, A., and Greenberg, J. M. 1997. A unified model of interstellar dust. *Astron. Astrophys.*, **323**, pp. 566–584.

Lillie, C. F., and Witt, A. N. 1976. Ultraviolet photometry from the Orbiting Astronomical Observatory. XXV. Diffuse galactic light in the 1500–4200 Å region and the scattering properties of interstellar dust grains. *Astrophys. J.*, **208**, pp. 64–74.

Linde, T.J., and Gombosi, T.I. 2000. Interstellar dust filtration at the heliospheric interfase. *J. Geophys. Res.*, **105**, pp. 10411-10417.

Loys de Chéseaux, J. P. 1744. *Traite des Cometes*.

Lutz, D., Feuchtgruber, H., Genzel, R., + 11 authors 1996. SWS observations of the Galactic

center. *Astron. Astrophys.*, **315**, pp. L269–L272.
Lynds, B. T. 1968. Dark nebulae. In *Stars and Stellar Systems.*, **Vol. VII**, eds. B. M. Middlehurst and L. H. Aller (Chicago: Univ. Chicago Press), pp. 119–139.
Malfait, K., Bogaert, E., and Waelkens, C. 1998a. An ultraviolet, optical and infrared study of Herbig Ae/Be stars. *Astron. Astrophys.*, **331**, pp. 211–223.
Malfait, K., Waelkens, C., Waters, L. B. F. M, Vandenbussche, B., Huygen, E., and De Graauw, M. S. 1998b. The spectrum of the young star HD100546 observed with the Infrared Space Observatory. *Astron. Astrophys.*, **332**, pp. L25–L28.
Malfait, K., Waelkens, C., Bouwman, J., De Koter, A., and Waters, L. B. F. M. 1999. The ISO spectrum of the young star HD142527. *Astron. Astrophys.*, **345**, pp. 181–186.
Martin, P. G. 1989. Linear and circular polarization in the diffuse interstellar medium. In *Interstellar Dust: Proc. IAU Symp.*, **135**, eds. L. J. Allamandola and A. G. G. M. Tielens (Dordrecht: Kluwer Academ. Publ.), pp. 55–65.
Martin, P. G., and Whittet, D. C. B. 1990. Interstellar extinction and polarization in the infrared. *Astrophys. J.*, **357**, pp. 113–124.
Mathis, J. S. 1993. Observations and theories of interstellar dust. *Rep. Prog. Phys.*, **56**, pp. 605–652.
Mathis, J. 1996. Dust models with tight abundance constraints. *Astrophys. J.*, **472**, pp. 643–655.
Mathis, J. 1998. The near-infrared interstellar silicate bands and grain theories. *Astrophys. J.*, **497**, pp. 824–832.
Mathis, J. S., and Whiffen, G. 1989. Composite interstellar grains. *Astrophys. J.*, **341**, pp. 808–822.
Mathis, J. S., Rumpl, W., Nordsieck, K. H. 1977. The size distribution of interstellar grains. *Astrophys. J.*, **217**, pp. 425–433.
Mattila, K., Lemke, D., Haikala, L. K., Laureijs, R.J., Léger, A., Lehtinen, K., Leinert, Ch., and Mezger, P. G. 1996. Spectrophotometry of UIR bands in the diffuse emission of the galactic disk. *Astron. Astrophys.*, **315**, pp. L353–L356.
McDonnell, J. A. M. 1988. Solar system dust as a guide to interstellar matter. In *Dust in the Universe*, eds. M.E. Bailey and D.A. Williams (Cambridge: Cambridge University Press), pp. 169–181.
McKee, C. F., and Ostriker, J. P. 1977. A theory of the interstellar medium: Three components regulated by supernova explosions in a inhomogeneous substrate. *Astrophys. J.*, **218**, pp. 148–169.
Mennella, V., Colangeli, L., Blanco, A., Bussoletti, E., Fonti, S., Palumbo, P., and Mertins, H. C. 1995a. A dehydrogenation study of cosmic carbon analogue grains *Astrophys. J.*, **444**, pp. 288–292.
Mennella, V., Colangeli, L., Bussoletti, E., Monaco G., Palumbo, P., and Rotundi, A. 1995b. On the electronic structure of small carbon grains of astrophysical interest. *Astrophys. J. Suppl. Ser.*, **100**, pp. 149–157.
Men'shchikov, A. B., Henning, Th., and Fischer, O. 1999. Self-consistent model of the dusty torus around HL Tauri. *Astrophys. J.*, **519**, pp. 257–278.
Molster, F. J. 2000. Crystalline silicates in circumstellar dust shells. Ph. D. Thesis, University of Amsterdam.
Molster, F. J., Waters, L. B. F. M., Trams, N., + 8 authors 1999a. The composition and nature of the dust shell surrounding the binary AFGL4106. *Astron. Astrophys.*, **350**, pp. 163–180.
Molster, F. J., Yamamura, I., Waters, L. B. F. M., + 9 authors 1999b. Low-temperature crystallization of silicate dust in circumstellar disks. *Nature*, **401**, p. 563.
Mutschke, H., Dorschner, J., Henning, Th., Jäger, C., and Ott, U. 1995. Facts and artifacts in interstellar diamond spectra. *Astrophys. J. Lett.*, **454**, pp. L157-L160.
Mutschke, H., Begemann, B., Dorschner, J., and Henning, Th., 1998. Steps toward interstellar silicate mineralogy. III. The role of aluminium in stardust silicates. *Astron. Astrophys.*, **333**, pp. 188–198.
Mutschke, H. , Andersen, A. C., Clément, D., Henning, Th., and Peiter, G. 1999. Infrared properties of SiC particles. *Astron. Astrophys.*, **345**, pp. 187–202.
Ney, E. P. 1977. Star dust. *Science*, **195**, pp. 541–546.
Nittler, L. R. 1997. Presolar oxide grains in meteorites. In *Astrophysical Implications of the Laboratory Study of Presolar Material*, eds. T. J. Bernatowicz and E. Zinner, AIP

Conf. Ser. Proceed., **402**, pp. 59–82.

Nittler, L. R., Hoppe, P., Alexander, C. M. O'D., + 7 authors 1995 Silicon nitride from supernovae. *Astrophys. J. Lett.*, **453**, pp. L25-L28.

Nuth, J. A. 1996. Grain formation and metamorphism. In *The Cosmic Dust Connection*, ed. J.M. Greenberg (Dordrecht: Kluwer Academ. Publ.), pp. 205–221.

Nuth, J. A., Moseley, S. H., Silverberg, R. F., Goebel, J. H., and Moore, W. H. 1985. Laboratory infrared spectra of predicted condensates in carbon-rich stars. *Astrophys. J.*, **290**, pp. L41–L43.

Nuth III, J. A., Hallenbeck, S. L., and Rietmeijer, F. J. M. 1999. Interstellar and interplanetary grains. Recent developments and new opportunities for experimental chemistry. In *Laboratory Astrophysics and Space Research*, eds. P. Ehrenfreund, C. Kraft, H. Kochan, and V. Pirronello (Dordrecht: Kluwer), pp. 143–182.

O'Dell, C. R., and Wen, Z. 1994. Post refurbishment mission Hubble Space Telescope images of the core of the Orion nebula: Proplyds, Herbig-Haro objects, and measurements of a circumstellar disk. *Astrophys. J.*, **436**, pp. 194–202.

Olbers, W. 1823. Über die Durchsichtigkeit des Weltraumes. In *Wilhelm Olbers. Sein Leben und seine Werke*, **Bd. 1**, ed. C. Schilling (Berlin: Julius Springer 1894), pp. 133–141.

Olnon, F. M., Raimond, E. (eds.) 1986. IRAS catalogues and atlases. Atlas of low-resolution spectra. *Astron. Astrophys. Suppl. Ser.*, **65**, pp. 607–1065.

Omont, A., Moseley, S. H., Forveille, T., Glaccum, W. J., Harvey, P. M., Likkel, L., Loewenstein, R. F., and Lisse, C. M. 1990. Observations of 40–70 micron bands of ice in IRAS 09371 +1212 and other stars. *Astrophys. J. Lett.*, **355**, pp. L27–L30.

Omont, A., Moseley, S. H., Cox, P., + 8 authors 1995. The 30 micron emission band in carbon-rich pre-planetary nebulae. *Astrophys. J.*, **454**, pp. 819–825.

Onaka, T., de Jong, T., and Willems, F. J. 1989. A study of M Mira variables based on IRAS LRS observations. I. Dust formation in the circumstellar shell. *Astron. Astrophys.*, **218**, pp. 169–179.

Oort, J. H., van de Hulst, H. C. 1946. Gas and smoke in interstellar space. *Bull. Astron. Inst. Netherlands*, **10**, pp. 187–204.

Öpik, E. 1929. Zur Theorie der Variation der Sternschnuppenhäufigkeit. *Astron. Nachr.*, **235**, pp. 265–268.

Öpik, E. 1931. On the physical interpretation of color-excess in early type stars. *Harvard Circ.*, **No. 359**.

Ossenkopf, V. 1993. Dust coagulation in dense molecular clouds: the formation of fluffy aggregates. *Astron. Astrophys.*, **280**, pp. 617–646.

Padgett, D. L., Brandner, W., Stapelfeldt, K. R., Strom, S. E., Terebey, S., and Koerner, D. 1999. Hubble Space Telescope/ NICMOS imaging of disk and envelopes around very young stars. *Astron. J.*, **117**, pp. 1490–1504.

Palumbo, M. E., Tielens, A. G. G. M., Tokunaga, A. T. 1995. Solid carbonyl sulphide (OCS) in W33A. *Astrophys. J.*, **449**, pp. 674–680.

Pendleton, Y. J., and Chiar, J. E. 1997. The nature and evolution of interstellar organics. In *From Stardust to Planetesimals*, eds. Y. J. Pendleton and A. G. G. M. Tielens (San Francisco: Astron. Soc. Pacific), A. S. P. Conf. Ser., **Vol. 122**, pp. 179–200.

Posch, Th., Kerschbaum, F., Mutschke, H., Fabian, D., Dorschner, J., and Hron, J. 1999. On the origin of the 13 μm feature. A study of ISO-SWS spectra of oxygen-rich AGB stars. *Astron. Astrophys.*, **352**, pp. 609–618.

Prinn, R. G. 1993. Chemistry and evolution of gaseous circumstellar disks. In *Protostars & Planets III*, eds. E. H. Levy and J. I. Lunine (Tucson: Univ. Arizona Press), pp. 1005–1028.

Puget, L. J., and Léger, A. 1989. A new component of the interstellar matter: small grains and large aromatic molecules. *Ann. Rev. Astron. Astrophys.*, **27**, pp. 161–198.

Roques, F., Scholl, H., Sicardy, B., and Smith, B. A. 1994. Is there a planet around β Pictoris? Perturbations of a planet on a circumstellar dust disk. 1. The numerical model. *Icarus*, **108**, pp. 37–58.

Rowan-Robinson, M. 1992. Interstellar dust in galaxies. *Mon. Not. Roy. Astron. Soc.*, **258**, pp. 787–799.

Sandford, S. A., Allamandola, L. J., Tielens, A. G. G. M., and Valero, G. J. 1988. Laboratory studies of the infrared spectral properties of of CO in astrophysical ices. *Astrophys. J.*, **329**, pp. 498–510.

Sandford, S. A., Pendleton, Y. J., and Allamandola, L. J. 1995. The galactic distribution of aliphatic hydrocarbons in the diffuse interstellar medium. *Astrophys. J.*, **440**, pp. 697-705.

Schalén, C. 1929. Zur Frage der allgemeinen Absorption des Lichtes im Weltraum. *Astron. Nachr.*, **236**, pp. 249–258.

Schalén, C. 1934. Untersuchungen über Dunkelnebel. *Medd. Astron. Obs. Upsala*, **No. 58**.

Schnaiter, M., Mutschke, H., Henning, Th., Lindackers, D., Strecker, M., and Roth, P. 1996. Ultraviolet spectroscopy of matrix-isolated amorphous carbon particles. *Astrophys. J.*, **464**, pp. L187–L190.

Schnaiter, M., Mutschke, H., Dorschner, J., Henning, Th., and Salama, F. 1998. Matrix-isolated nano-sized carbon grains as an analog for the 217.5 nanometer feature carrier. *Astrophys. J.*, **498**, pp. 486–496.

Schoenberg, E., and Jung, B. 1934. Über die Lichtstreuung im interstellaren Raum durch Wolken metallischere Partikel. *Astron. Nachr..*, **253**, pp. 261–272.

Schutte, W. A. 1996. Formation and evolution of interstellar icy grain mantles. In *The Cosmic Dust Connection*, ed. J. M. Greenberg (Dordrecht: Kluwer Academ. Publ.), NATO ASI Ser. C, **Vol. 487**, pp. 1–42.

Schutte, W. A. 1999. Ices in the interstellar medium. In *Laboratory Astrophysics and Space Research*, eds. P. Ehrenfreund, C. Kraft, H. Kochan, and V. Pirronello (Dordrecht: Kluwer), pp. 69–103.

Schutte, W. A., Tielens, A. G. G. M., Whittet, D. C. D., + 6 authors 1996. The 6.0 and 6.8 μm absorption features in the spectrum of NGC 7538: IRS9. *Astron. Astrophys.*, **315**, pp. L333-L336.

Schutte, W. A., van der Hucht, K. A., Whittet, D. C. B., + 8 authors 1998. ISO-SWS observations of infrared absorption bands of the diffuse interstellar medium: The 6.2 μm feature of aromatic compounds. *Astron. Astrophys.*, **337**, pp. 261–274.

Seahra, S. S., and Duley, W. W. 1999. Extended Red Emission from carbon clusters in interstellar clouds. *Astrophys. J.*, **520**, pp. 719–723.

Sedlmayr, E. 1994. From molecules to grains. In *Molecules in the Stellar Environment*, ed. U. G. Jørgensen (Berlin: Springer-Verlag), pp. 163–185.

Sellgren, K. 1989. Infrared emission from reflection nebulae. In *Interstellar Dust: Proc. IAU Symp.*, bf 135, eds. L. J. Allamandola and A. G. G. M. Tielens (Dordrecht: Kluwer Academ. Publ.), pp. 103–108.

Sellgren, K. 1994. Tiny grains, large molecules, and the infrared cirrus. In *The First Symposium on the Infrared Cirrus and Diffuse Interstellar Clouds*, eds. R. Cutri and W. B. Latter (San Francisco: Astron. Soc. Pacific), A. S. P. Conf. Ser., **Vol. 58**, pp. 243–254.

Sellgren, K., Smith, R. G., Brooke, T. Y. 1994. The 3.2–3.6 micron spectra of Monoceros R2/IRS-3 and Elias 16. *Astrophys. J.*, **433**, pp. 179–186.

Sellgren, K., Brooke, T. Y., Smith, R. G., Geballe, T. R. 1995. A new 3.25 micron absorption feature toward Monoceros R2/IRS3. *Astrophys. J.*, **449**, pp. L69–L72.

Serkowski, K. 1973. Interstellar polarization. In *Interstellar Dust and Related Topics: Proc. IAU Symp.* **52**, eds. J. M. Greenberg and H. C. van de Hulst (Dordrecht: Reidel), pp. 145–152.

Siebenmorgen, R., and Krügel, E. 1992. Dust model containing polycyclic aromatic hydrocarbons in various environments. *Astron. Astrophys.*, **259**, pp. 614–626.

Sitko, M. L., Halbedel, E. M., Lawrence, G. F., Smith, J. A., and Yanow, K. 1994. Variable extinction in the HD 45677 and the evolution of dust grains in pre-main-sequence disks. *Astrophys. J.*, **432**, pp. 753–762.

Skinner, C. J., and Whitmore, B. 1988. Circumstellar environments – IV. Mass-loss rates for carbon stars. *Mon. Not. Roy. Astron. Soc.*, **234**, pp. 79p–84p.

Sloan, G. C., and Price, S. D. 1995. Silicate emission at 10 microns in variables on the asymptotic giant branch. *Astrophys. J.*, **451**, pp. 758–767.

Sloan, G. C., and Price, S. D. 1998. The infrared spectral classification of oxygen-rich dust shells. *Astrophys. J. Suppl. Ser.*, **119**, pp. 141–158.

Smith, B. 1994. 10 years of Beta Pictoris – a personal reminiscence. In *Circumstellar Dust Disks and Planet Formation*, eds. R. Ferlet and A. Vidal-Madjar (Gif sur Yvette: Editions Frontières), pp. 1–3.

Smith, R. G., Sellgren, K., and Brooke, T. Y. 1993. Grain mantles in the Taurus dark cloud.

Mon. Not. Roy. Astron. Soc., **263**, pp. 749–766.

Sodroski, T. J., Dwek, E., Hauser, M. G., and Kerr, F. J. 1989. Dust energetics in the gas phases of the interstellar medium: The origin of the galactic large-scale far-infrared emission observed by IRAS. *Astrophys. J.*, **336**, pp. 762–779.

Sodroski, T. J., Bennett, C., Boggess, N., + 8 authors 1994. Large-scale characteristics of interstellar dust from COBE DIRBE observations. *Astrophys. J.*, **428**, pp. 638–646.

Sofia, U. J., Cardelli, J. A., and Savage, B. D. 1994. The abundant elements in the interstellar dust. *Astrophys. J.*, **430**, pp. 650–666.

Sorrell, W. F. 1990. The λ 2175-Å feature from irradiated graphite particles. *Mon. Not. Roy. Astron. Soc.*, **243**, pp. 570–587.

Spaans, M., and Ehrenfreund, P. 1999. The interstellar medium: a general introduction. In *Laboratory Astrophysics and Space Research*, eds. P. Ehrenfreund, C. Kraft, H. Kochan, and V. Pirronello (Dordrecht: Kluwer), pp. 1–36.

Speck, A. K., Barlow, M. J., and Skinner, C. J. 1997. The nature of silicon carbide in star outflows. *Monthly Not. R. Astron. Soc.*, **288**, pp. 431–456.

Speck, A. K., Barlow, M. J., Sylvester, R. J. and Hofmeister, A. M. 2000. Dust features in the infrared spectra of oxygen-rich evolved stars. *Astron. Astrophys. Suppl. Ser.*, **146**, pp. 437–467.

Stebbins, J., and Whitford, A. E. 1943. Six-color photometry of stars. I. The law of space reddening from the colors of O and B stars. *Astrophys. J.*, **98**, pp. 20–32.

Stecher, T. P. 1965. Interstellar extinction in the ultraviolet. *Astrophys. J.*, **142**, pp. 1683–1684.

Stein, W. A., and Gillett, F. C. 1969. Spectral distribution of infrared radiation from the Trapezium region of the Orion Nebula. *Astrophys. J.*, **155**, pp. L197–L199.

Sterzik, M. F., and Morfill, G. E. 1994. Evolution of protoplanetary disks with condensation and coagulation. *Icarus*, **111**, pp. 536–546.

Sylvester, R. J., Barlow, M. J., Skinner, C. J., and Mannings, V. 1996. Optical, infrared, and millimetre-wave properties of Vega-like systems *Mon. Not. R. Astron. Soc.*, **279**, pp. 925–939.

Tanaka, M., Sato, S., Nagata, T., and Yamamoto, T. 1990. Three micron ice-band features in the ρ Ophiuchi sources *Astrophys. J.*, **352**, pp. 724–730.

Taylor, A., Baggaley, W. J., and Steel, D. J. 1996. Discovery of interstellar dust entering the earth's atmosphere. *Nature*, **380**, pp. 323–325.

Telesco, C. M., and Knacke, R. F. 1991. Detection of silicates in the beta Pictoris disk. *Astrophys. J. Lett.*, **372**, pp. L29-L31.

Teixeira, T. C., Devlin, J. P., Buch, V., and Emerson, J. P. 1999. Discovery of solid HDO in grain mantles. *Astron. Astrophys.*, **347**, pp. L19–L22.

Tielens, A. G. G. M. 1990. Carbon stardust: From soot to diamonds. In *Carbon in the Galaxy: Studies from Earth and Space.*, eds. J. C. Tarter, S. Chang, and D. J. DeFrees (Washington: NASA), CP-3061, pp. 59–111.

Tielens, A. G. G. M. 1995. The interstellar medium. In *Airborne Astronomy Symposium on the Galactic Ecosystem: from Gas to Stars to Dust*, ed. M. R. Haas, J. A. Davidson, E. F. Erickson (San Francisco: Astron. Soc. Pacific), A. S. P. Conf. Ser., **Vol. 73**, pp. 3–21.

Tielens, A. G. G. M., Tokunaga, A. T., Geballe, T. R., and Baas, F. 1991. Interstellar solid CO: polar and nonpolar interstellar ices. *Astrophys. J.*, **381**, pp. 181–199.

Tielens, A. G. G. M., Wooden, D. H., Allamandola, L. J., Bregman, J., and Witteborn, F.C. 1996. The infrared spectrum of the Galactic Center and the composition of interstellar dust. *Astrophys. J.*, **461**, pp. 210–222.

Treffers, R., and Cohen, M. 1974. High-resolution spectra of cool stars in the 10- and 20-micron region. *Astrophys. J.*, **188**, pp. 545–552.

Trümpler, R. 1930. Preliminary results on the distances, dimensions, and space distribution of open star clusters. *Lick Obs. Bull.*, **No. 420**.

Van de Hulst, H. C. 1949. The solid particles in interstellar space. *Rech. Astron. Obs. Utrecht*, **11**, pt. 2.

Van Dishoeck, E. 1999. Models and observations of gas-grain interactions chemical evolution in star-forming regions. In *Formation and Evolution of Solids in Space*, ed. J. M. Greenberg, A. Li (Dordrecht: Kluwer Academ. Publ.), NATO ASI Ser. C, **Vol. 523**, pp. 229–264.

Verschuur, G. L. 1989. *Interstellar Matters. Essays on Curiosity and Astronomical Discov-*

ery (New York: Springer-Verlag).

Vrba, F. J., Coyne, G. V., and Tapia, S. 1981. Observations of grain and magnetic field properties of the R Coronae Australis dark cloud. *Astrophys. J.*, **243**, pp. 489–511.

Vrba, F. J., Coyne, G. V., and Tapia, S. 1993. An investigation of grain properties in the ρ Ophiuchi dark cloud. *Astron. J.*, **105**, pp. 1010–1026.

Volk, K., Kwok, S., and Hrivnak, B. J. 1999. High-rsolution *Infrared Space Observatory* Spectroscopy of the unidentified 21 micron feature. *Astrophys. J.*, **516**, pp. L99–L102.

Walmsley, C. M., and Schilke, P. 1993. Observations of hot molecular cores. In *Dust and Chemistry in Astronomy*, eds. T. J. Millar, and D. A. Williams (Bristol: Inst. Physics Publ.), pp. 37–52.

Waters, L. B. F. M., Molster, F. J., de Jong, T., + 34 authors 1996. Mineralogy of oxygen-rich dust shells. *Astron. Astrophys.*, **315**, pp. L361–L364.

Waters, L. B. F. M., Cami, J., De Jong, T., Molster, F. J., Van Loon, J. Th., Bouwman, J., De Koter, A., Waelkens, C., Van Winckel, H., and Morris, P. W. 1998a. An oxygen-rich dust disk surrounding an evolved star in the Red Rectangle. *Nature*, **391**, pp. 868–... .

Waters, L. B. F. M., Beintema, D. A., Zijlstra, A. A., De Koter, A., Molster, F. J., Bouwman, J., De Jong, T., Pottasch, S. R., and De Graauw, Th. 1998b. Crystalline silicates in planetary nebulae with [WC] central stars. *Astron. Astrophys.*, **331**, pp. L61–L64.

Webster, A. 1995. The lowest of the strongly infrared active vibrations of the fullerenes and an astronomical emission band at a wavelength of 21 μm. *Monthly Not. R. Astron. Soc.*, **277**, pp. 1555–1566.

Weintraub, D. A., and Stern, S. A. 1994. A reinterpretation of millimeter observations of nearby IRAS excess stars. *Astron. J.*, **108**, pp. 701–710.

Whittet, D. C. B. 1984. Interstellar grain composition: a model based on elemental depletions. *Mon. Not. Roy. Astron. Soc.*, **210**, pp. 479–487.

Whittet, D. C. B. 1992. *Dust in the Galactic Environment*, (Bristol: Inst. Physics Publ.).

Whittet, D. C. B. 1993. Observations of molecular ices. In *Dust and Chemistry in Astronomy*, eds. T. J. Millar and D. A. Williams (Bristol: Inst. Physics Publ.), pp. 9–35.

Whittet, D. C. B. 1996. Polarization of starlight by interstellar dust. In *The Cosmic Dust Connection*, ed. J. M. Greenberg (Dordrecht: Kluwer Academ. Publ.), pp. 205–221.

Whittet, D. C. B., Bode, M. F., Longmore, A. J., Adamson, A. J., McFadzean, A. D., Aitken, D. K., and Roche, P. F. 1988. Infrared spectroscopy of dust in the Taurus dark clouds: ice and silicate. *Mon. Not. Roy. Astron. Soc.*, **233**, pp. 321–326.

Whittet, D. C. B., Martin, P. G., Hough, J. H., Rouse, M. F., Bailey, J. A., and Axon, D. J. 1992. Systematic variations in the wavelength dependence of interstellar linear polarization. *Astrophys. J.*, **386**, pp. 562–577.

Whittet, D. C. B., Schutte, W. A., Tielens, A. G. G. M., Boogert, A. C. A., de Graauw, Th., Ehrenfreund, P., Gerakines, P. A., Helmich, F. P., Prusti, T., and van Dishoeck, E. F. 1996. An ISO SWS view of interstellar ices: first results. *Astron. Astrophys.*, **315**, pp. L357–360.

Whittet, D. C. B., Boogert, A. C. A., Gerakines, P. A. + 7 authors 1997. Infrared spectroscopy of dust in the diffuse interstellar medium toward Cygnus OB2 No.12. *Astrophys. J.*, **490**, pp. 729–734.

Wildt, R. 1933. Kondensation in Sternatmosphären *Zeitschr. f. Astrophys.*, **6**, pp. 345–354.

Williams, D. A. 1989. Grains in diffuse clouds: carbon-coated silicate cores. In *Interstellar Dust: Proc. IAU Symp.*, **135**, eds. L. J. Allamandola and A. G. G. M. Tielens (Dordrecht: Kluwer Academ. Publ.), pp. 367–373.

Witt, A. N. 1989. Visible/UV scattering by interstellar dust. In *Interstellar Dust: Proc. IAU Symp.*, **135**, eds. L. J. Allamandola and A. G. G. M. Tielens (Dordrecht: Kluwer Academ. Publ.), pp. 87–100.

Witt, A. N., Petersohn, J. K., Bohlin, R. C., O'Connell, R. W., Roberts, M. S., Smith, A. M., and Stecher, T. P. 1992. Ultraviolet Imaging Telescope images of the reflection nebula NGC 7023: Derivation of ultraviolet scattering properties of dust grains. *Astrophys. J. Lett.*, **395**, pp. L5–L8.

Witt, A. N., Gordon, K. D., Furton, D. G. 1998. Silicon nanoparticles: source of extended red emission? *Astrophys. J.*, **501**, pp. L111–L114.

Wolff, M. J., Clayton, G. C., Kim, S.-H., Martin, P. G. 1997. Ultraviolet interstellar linear polarization. III. Features. *Astrophys. J.*, **478**, pp. 395–402.

Wright, E. L., Mather, J. C., Bennett, C. L., + 19 authors 1991. Preliminary spectral observations of the Galaxy with a 7° beam by the Cosmic Background Explorer (COBE). *Astrophys. J.*, **381**, pp. 200–209.

Wurm, G., Blum, J. 1998. Experiments on preplanetary dust aggregation. *Icarus*, **132**, pp. 125–136.

Yamashita, T., Handa, T., Omodaka, T., Kitamura, Y., Kawazoe, E., Hayashi, S. S., and Kaifu, N. 1993. Upper limits to the CO $J = 1 - -0$ emission around Vega-like stars: gas depletion of the circumstellar ring around ε Eridani. *Astrophys. J. Lett.*, **402**, pp. L65–L67.

Zinner, E. 1997. Presolar material in meteorites: an overview. In *Astrophysical Implications of the Laboratory Study of Presolar Material*, eds. T. J. Bernatowicz and E. Zinner, AIP Conf. Ser. Proceed., **402**, pp. 3–26.

Zinner, E., Amari, S., Wopenka, B., and Lewis, R. S. 1995. Interstellar graphite in meteorites: Isotopic compositions and structural properties of single graphite grains from Murchison. *Meteoritics*, **30**, pp. 209-226.

Zubko, V. G., Smith, T. L., and Witt, A. N. 1999. Silicon nanoparticles and interstellar extinction. *Astrophys. J.*, **511**, pp. L57–L60.

Glossary

Achondrite A differentiated stony meteorite.

Aphelion The point of a heliocentric orbit farthest from the Sun.

Asteroid A rocky, carbonaceous or metallic body, smaller than a planet. Most asteroids have orbits in the Mars to Jupiter region.

Astronomical Unit (AU) 1 AU = $1.496 \cdot 10^{11}$ m, approximately the mean distance from Earth to the Sun.

β Pictoris objects Stars close to, or on the main-sequence, with gaseous dust disks indicating the occurrence of swarms of star-grazing comets and/or planet formation.

β-meteoroid A small meteoroid for which radiation pressure force is comparable to solar gravitational attraction.

Albedo, single scattering The fraction of energy in the incident beam that is transferred to the scattered field or probability that an incident photon will survive interaction with a body (and not be absorbed). The single scattering albedo equals the ratio of the scattering to extinction cross sections.

Charged particle absorption signature A reduction in high-energy radiation flux noticed in the neighborhood of the giant planets, and assumed to be due to the absorption of this radiation by satellites or rings.

CHON particle Particle who's mass spectrum is dominated by ions of the light elements H, C, N, O.

Chondrite An undifferentiated stony meteorite that contains chondrules.

Chondrules The millimeter-sized, spheroidal particles in meteorites that were once molten droplets.

Circumstellar Referring or applying to phenomena in a star's immediate surrounding.

Color temperatur The temperature of a black-body whose intensity in a given spectral range shows the same wavelength dependence as that of the observed object. The color temperature of an object that is not a good black- (or gray-) body radiator differs from its black-body temperature.

Coma The roughly spherical outflow of gas and dust from a comet nucleus.

GLOSSARY

Comet nucleus The solid part of the comet containing ices and dust that, when the comet is close to the Sun (typically within 2AU), is the source of the coma and the dust and plasma tails.

Dark cloud, dark nebula An interstellar cloud in which the dust extinction weakens or totally obliterates the light of stars and nebulae of the background.

Diffuse interstellar cloud Interstellar gas cloud in which hydrogen is completely dissociated and which is less dense and dusty than dark and molecular clouds.

Disk dust A dust population typical of protostellar (and preplanetary) disks.

Dust tail Broad, curved, diffuse comet tail composed of fine dust grains blown away from the cometary coma by sunlight radiation pressure. In the visible the spectrum of the dust tail resembles the solar spectrum.

Dust trail Narrow long trails along part or all of the orbit of some comets, most easily observed at infrared wavelengths. They consist of millimeter to centimeter sized grains that were released from the nucleus at low $(m\,s^{-1})$ speeds.

Eccentricity The measure of the departure of a body's orbit from a perfect circle. A circular orbit has eccentricity $e = 0$; an elliptic orbit has $0 < e < 1$; a parabolic orbit has $e = 1$; and a hyperbolic orbit has $e > 1$.

Ecliptic plane The plane of the Earth's orbit around the Sun.

Edgeworth-Kuiper belt (or Kuiper belt) The collection of asteroid-sized bodies in low-inclination, low-eccentricity orbits beyond Neptune's orbit.

Extinction curve The interstellar extinction (in magnitudes) plotted vs. wavenumber (in μm^{-1}).

Fractal Any of the extremely irregular curves, shapes, or patterns, for which a part is similar to a given larger or smaller part when magnified or reduced to the same size.

Fractal dimension Dimension or exponent in a power law that, unlike the dimension of a surface or the volume of a solid, is characterized by a non-integer exponent of the particle's linear dimensions.

Gegenschein An enhancement in the zodiacal light opposite to the Sun, about $20°$ across on the celestial sphere, caused by backscatter of sunlight by interplanetary dust. This German word is from *gegen* against or counter and *Schein* shine.

GLOSSARY

H I, H II region A region of interstellar gas consisting of neutral (H I) and ionized (H II) atomic hydrogen, respectively; the brilliant galactic nebulae are H II regions.

Heliosphere The cavity in the interstellar medium dominated by the solar wind and surrounding the solar system.

IDP The abbreviation for **I**nterplanetary **D**ust **P**article now commonly applied to only those particles which are captured in the stratosphere and readily available for laboratory analysis. Their sizes range from $1\mu m$ to $50\ \mu m$.

Inclination The angle between the orbit plane of a celestial body and the ecliptic plane.

Interplanetary dust particles All bodies that are larger than a molecule and smaller than an asteroid (except moons, planets and comets), which originated within the solar system.

Interstellar depletion The reduction of the abundance of an element in the interstellar gas relative to a standard abundance (usually solar system abundances).

Interstellar dust particles Dust particles larger than a molecule and smaller than an asteroid that originate outside the solar system but enter the solar system at high speed from a particular direction.

Interstellar extinction Weakening and reddening of starlight through absorption and scattering by interstellar dust.

Isotopic anomaly A deviation from the "normal" (terrestrial) isotopic elemental composition that cannot be directly related to mass fractionation effects, radioactive decay or to interaction of high energy particles (e.g. cosmic rays) with their carriers.

Laplace plane The plane about which orbits precess when they orbit a central body and are subject to gravitational perturbations caused by the central body's oblateness and the gravitational pull from other orbiting bodies. Such a plane is defined for each orbital radius and the sum of all these planes forms a warped surface.

Lorentz force Force exerted by a magnetic field on a charged particle that moves with respect to the field. Since this force generally oscillates, it can resonate with the orbital period in certain locations about planets.

Magnetosphere The volume of space around a planet which is dominated by the planet's magnetic field and associated charged particles.

Meteor The light phenomenon that results from the entry of a meteoroid into the Earth's atmosphere.

Meteorite A natural solid object of extraterrestrial origin that survived passage through Earth's atmosphere.

Meteoroid A solid object moving in interplanetary space of a size smaller than an asteroid and larger than a molecule.

Micrometeoroid Same as meteoroid except smaller than about 0.1 mm in size, also referred to as an interplanetary dust particle.

Mie theory A rigorous method of calculating the interaction of an electromagnetic wave with a homogeneous sphere. This theory finds solutions to Maxwell's equations which satisfy certain boundary conditions.

MM Micrometeorites which are extracted from Antarctica or the arctic region by melting the ice. Their sizes range from about 30 μm to mm.

Molecular cloud A dense, cool cloud of the ISM, in which the gas consists primarily of molecules (H_2, CO) and the dust grains contain ice components (condensed molecules).

Occultation The elimination of radiation from a celestial source (or spacecraft) as it passes behind another body.

Oort cloud The reservoir of long period comets which occupies a spherical region of about 50,000 AU radius around the Sun.

Optical depth The exponent characterizing the exponential decrease of radiation intensity as it passes through an absorbing or scattering medium. For very low optical depth, it is related to the fraction of surface area covered by particles.

Orbital elements Six quantities that fully describe an orbit. A typical set is 1) semi-major axis a, the mean distance from the gravitating central body; 2) eccentricity e, the fractional departure from circularity; 3) inclination i, the angle between the orbital plane and some reference plane, such as the ecliptic or the planet's equatorial plane; 4) longitude of ascending node Ω, the angle between some arbitrary line in the reference plane and the point where the body crosses the reference plane moving from south to north; 5) longitude of pericenter ω, the broken-angle from the just-mentioned reference line to the position where the body is closest to the central object; and 6) epoch T, the time of pericenter passage. These elements are constant for any orbit about a point mass.

Perhelion The point of a heliocentric orbit closest to the Sun.

Perturbations The small forces in addition to a central body's point-mass attraction. Such forces include any remaining parts to the central body's gravity, or atmospheric drag, electromagnetic forces, radiation pressure, Poynting-Robertson drag, plasma drag, and resonant charge.

Phase angle Complement to the scattering angle, the angle between the direction of propagation of scattered light and the direction of incidence.

Photometry The branch of physics and astronomy dealing with the measurement of radiation as a function of wavelength, scattering angle, polarization degree etc.; the strength and angular behavior as a function of wavelength can be interpreted in terms of surface properties and particle size.

Plasma tail Narrow straight comet tail in roughly the antisolar direction composed of ionized molecules driven out from the cometary coma by interaction with the solar wind and the interplanetary magnetic field.

Polarization Unless otherwise stated, polarization refers to the linear degree of polarization: the difference between the intensities of the linearly polarized components perpendicular and parallel to a reference plane (usually the scattering plane containing the light source, the observer, and direction of observation) divided by the total intensity.

Power law The mathematical relationship between two quantities in which one quantity depends on a power or exponent of the other.

Poynting-Robertson drag Drag on a moving particle that is due to the asymmetrical momentum distribution of scattered and absorbed light; the drag primarily affects small grains, causing them to spiral toward the central object that they orbit, whether the Sun or a planet.

Prograde The orbital or rotational motion of an object in the same sense as the orbital motion of the Earth around the Sun.

Radiation pressure force The velocity-independent component of force exerted on a body, which is illuminated by an electromagnetic radiation field.

Reflection nebula Bright nebula due to light scattering by dust grains in the neighborhood of a star.

Resonance The enhanced response of any oscillating system to an external stimulus that is characterized by a frequency that divides evenly into the natural frequency of the system or is evenly divided by that frequency. In celestial mechanics, resonances may be produced by gravitating bodies (planets or satellites), lumpy gravity fields or electromagnetic forces (Lorentz resonances).

Retrograde The opposite of prograde real motion.

Ring-moons (or Shepherd satellites) The small satellites that orbit near planetary rings, within a few planetary radii of the central planet. All of the giant planets have ring-moons that intermingle with their ring systems. These objects may supply rings (through impacts) or shepherd the rings.

Scattering angle Angle between the direction of propagation of scattered light and the direction of incidence. The scattering angle is the compliment to the phase angle.

Semimajor axis Half of the long axis of an elliptical orbit.

Solar wind The expansion of the solar corona to form supersonic plasma streaming away from the Sun.

Space debris Man-made particulates littered in space.

Sputtering Atoms and molecules ejected from a grain due to impinging energetic ions, electrons and photons.

Stardust Dust grains condensing in the outstreaming stellar wind of evolved stars. Certain types of individual particles (mostly SiC, graphite and diamonds) can be extracted from meteorites by chemical treatment. They are identified by isotopic anomalies.

Unidentified infrared bands (UIB) Vibrational bands of large interstellar molecules or dust particles; most of which are tentatively assigned to polycyclic aromatic hydrocarbon molecules.

Vega phenomenon A far-infrared excess of main-sequence stars indicating cool circumstellar particulates; first detected in the spectrum of the star α Lyrae (Vega) by the IRAS satellite.

Young stellar object (YSO) The collective name for stars in the pre-main-sequence stage, surrounded by circumstellar envelopes and disks.

Zodiacal light Sunlight scattered by interplanetary dust.

Index

β, 573
β Pic disk, 771
β Pictoris objects, 787
β meteoroid, 11, 15, 37, 316, 335, 574, 680, 787

ablation, 181
accelerator
    electrostatic dust, 390, 408, 431, 432
    plasma drag, 433, 434
    Van de Graaff, 21
accretion, 752
accretion rate of IDP, 628
achondrite, 787
activity profile, 175
Aerobee, 26
aerocapture, 185
aerofragmentation, 185
aerogel, 410
AGB stars, 765
agglomeration, 752, 759
AIBs, 730
airglow, 10, 247
albedo, 13, 29, 78, 350, 740
    single scattering, 787
aluminum oxide, 763
ammonia ice, 753
amorphization, 484
amorphous carbon, 764
amorphous SiC, 766
amorphous silicate, 761
angle of incidence, 396
angular sensitivity, 360
anhydrous mineral, 260
anomalous tail, 112

antitail, 112
aphelion, 787
Apollo
    17, 36, 415
    program, 1
aqueous alteration, 260
area time product, 193
Ariel 2, 211
aromatic interstellar band, 730
asteroid, 787
asteroid belt, 572
asteroidal band, 16, 34, 39
asteroidal dust particle, 571
asteroidal meteoroid, 355
asteroidal population, 368
astronomical silicate, 518
astronomical unit (AU), 787
asymmetry factor g, 740
atmospheric drag, 684
atmospheric entry, 254
Atmospheric Explorer C, 38

Babinet's principle, 646
ballistic limit, 203
    equation, 205
band
    aromatic interstellar, 730
    asteroidal dust, 16, 34, 39
    diagnostic IR, 729
    diffuse interstellar, 737
    silicate, 740
    solar system dust, 592
    spectral 217.5 nm, 736
    spectral 30 μm, 767
    UIR, 730
    unidentified infrared, 730, 792

beer can experiment, 296, 402, 413
Bergman's theorem, 520
boundary condition in light scattering, 528
Br overabundance, 278
Bremsat, 423
bulk density, 350
bumper shield, 401, 402

capacitor
    detector, 2
    discharge, 403
carbon
    dust, 749
    onion, 751
carbonaceous dust, 749
carbonyl sulfide, 756
Cassini mission, 46, 376, 391, 408, 411, 414, 425, 426
Cassini, Giovanni Domenico, 2
catastrophic collisions lifetime, 498
catastrophic disruption, 464
CCD arrays, 101
CDA, 408, 414, 425
CDCF, 391, 408, 409
ceramic microphone, 17
Chapman airglow mechanism, 246
charged particle absorption, 651, 663
    signature, 787
Childrey, 6
CHON particle, 787
chondrite, 787
    group, 275
    porous, 260
    smooth, 260
chondrules, 787
CIDA, 421
circular polarization, 737
circumsolar dust, 26
circumstellar, 787
circumstellar disk, 613
circumstellar system, 254
cirrus cloud, 27
classical dust model, 728
classical interstellar grain, 512

climate cycle, 628
cluster
    -cluster aggregation, 453
    analysis, 274
    model, 239
CO-ice, 754
coagulation, 752
COBE, 2, 16, 29, 38, 388
    observations, 606
CODAG, 759
coherent scattering, 529
cold cores, 732
collected dust, 22
collision, 338, 571, 686, 688, 691
    cascade, 571
    grain, 760
    lifetime, 498, 575
    molecule-molecule, 98
    particle-molecule, 96
color temperature, 542, 787
coma, 787
    morphology, 101
comet
    coma dust, 481
    dust, 46
    dust density, 513
    dust trail, 242, 788
    Jupiter family, 589
    nucleus, 788
    star-grazing, 773
    sungrazing, 27, 101
cometary dust
    composition, 45
    particle, 589
    trail, 16, 39, 634, 788
cometary meteoroid, 354
complete synchrone, 111
composition, 411, 417, 649
conjugate gradient method, 534
constituent units in an aggregate, 513
contamination process, 279
Contour, 340
convex hyperbola, 99
Copernicus, 27

core population, 368
core/mantle-grain, 744
corona, 10, 11, 26
    F, 10, 67
    F, thermal emission, 74
    K, 10
    polarization, 14
coronal brightness, 12
coronal streamer, 43
COSIMA, 427, 428
Cosmic Dust Collection Facility, CDCF, 408
cosmic ray, 760
cosmochemistry, 516
CRAF, 427
crater, 30, 460
    age, 32
    depth equation, 206
crystalline silicate, 762
cumulative distribution, 167
cumulative flux, 356
cumulative meteoroid flux, 296, 312

D2B, 29, 38
dark cloud, 727, 788
dark nebula, 788
dark nebulae, 727
Darwin, 47
DDS, 414, 421, 422, 426
deceleration, 410
density
    bulk, 350
    comet dust, 513
    dust, 41, 411
    phase space, 357
    plasma, 13, 43
    space, 79
    spatial, 14, 43, 352, 354, 357
depletion, 743
depolarization, 405, 414, 421
desorption, 760
    ionization model, 397, 399
    rate, 469
destruction, 687
detection threshold, 171

detector
    capacitor, 2
    dust, 652
    impact ionization, 2, 365
    in-situ, 297
    microphone, 1, 416
    MOS, 404
    multi-coincidence, 301
    penetration, 364, 413
    pressurized-cell, 2
diagnostic IR band, 729
diamond, 751
DIBs, 737
DIDSY, 406
dielectric constant, 515
differential ablation, 240
differential flux, 312
differential mass distribution, 167
diffuse galactic light, 740
diffuse interstellar band, 737
diffuse interstellar cloud, 788
diffuse ISM, 731
DIRBE, 29
directional flux, 380
discrete source, 101
distribution
    cumulative, 167
    differential mass, 167
    flux, 166
    lunar crater, 356
    magnitude index, 178
    mass index, 167
    MRN, 745
    of orbits, 190
    orbit, 190
    out-of-ecliptic, 39
    parameter, 358
    radius function, 734
    size, 12, 166, 571, 646, 664, 666, 676, 686
    size of fragments, 462
    spatial, 10, 11
    textbook, 358
    velocity, 168
diurnal variation, 105

Divine's interplanetary dust model, 202
Dohnanyi power law, 572
Doppler shift, 15
Drag, 683, 701
DUCMA, 405, 414, 421
Duilliers, Niccolo Fatio de, 4
dust
    -ball model, 243
    accelerator, 390, 408, 432
    analog, 744
    asteroidal particle, 571
    belt, 1, 19
    carbon, 749
    carbonaceous, 749
    charge, 99, 337, 389, 391, 408, 409, 417, 433, 435, 436, 673, 680
    circumsolar, 26
    collected, 22
    collection, 1
    color, 551
    cometary, 46
    composition, 45
    density, 41, 411
    detector, 652
    diameter, 411
    disk, 788
    dynamics, 702
    fragmentation, 114
    halo, 101
    in comet coma, 481
    in magnetosphere, 481
    in the interstellar medium, 480
    interplanetary, 254, 789
    interstellar, 331, 362, 372, 578, 789
    interstellar population, 372
    lifecycle, 733
    lifetimes, 297
    multi-component, 730
    population, 42, 730, 733
    presolar grain, 731
    refractory, 729
    silhouette, 768
    stream, 304
    tail, 108, 119, 788
    trail, 118, 788
    volatile, 731
    dust model, 744
        Bird's-Nest, 511
        Fluffy, 510

E ring, 667, 711
Earth
    gravitational focusing, 624
    resonant ring, 619
    shielding, 170
eccentric population, 368
eccentricity, 788
Echantillons, 213
ecliptic, 788
Edgeworth-Kuiper belt, 788
effective refractive index, 519-521
ejecta, 691
ejection velocity, 97
elastic scattering, 522
electromagnetic effect, 336
electronic sputtering, 486
electrostatic accelerator, 431
electrostatic dust accelerator, 390, 408, 432
electrostatic fragmentation, 99
electrostatic stress, 389, 437
element
    forced, 580
    orbital, 307, 352, 357, 693, 790
    osculating, 580
    proper, 580
ELF/VLF meteor signal, 247
equation
    ballistic limit, 205
    crater depth, 206
    hole growth, 207
    impact, 202
    Maxwell's, 528
    of motion, 699
    volume integral, 532
ERIS, 405
ESABASE, 200

EuReCa Timeband Capture Cell Experiment, TICCE, 216
evolution
....orbit, 578, 702
Explorer
....-16, 18, 211, 222, 309, 353, 402, 412, 413
....-23, 222, 309, 353
....-46, 403
exposure age, 310
Express-2, 340, 421
extended red emission, 741
extinction
....curve, 729, 788
....parameter R, 736

F-corona, 10, 67
....thermal emission, 74
fan, 101
Fe ablation trails, 244
feature
....21 μm, 767
....3.4 micron, 518
....silicate, 517
FEB mechanism, 773
field emission, 489
field-of-view, 305
Finson-Probstein model, 110
flux, 215
....cumulative, 296, 312, 356
....curve, 23, 30
....differential, 312
....directional, 380
....distributions, 166
....Grün, 193
....hazard, 36
....impact, 216, 222
....interplanetary model, 369
....isotropic, 312
....measurement, 317
....particle, 19, 359
force, 682, 701
forced element, 580
formic acid, 756
Fourier analysis, 595

fractal, 788
fractal dimension, 453, 788
fractionation, 759
fragment velocity, 464
fragmentation, 437, 460
....dust, 114
....electrostatic, 99
....secondary, 115
fullerene, 767
function
....phase, 79, 645, 646
....probability, 179
....radius distribution, 734
....scattering, 12, 740
....spectral, 520
....volume scattering, 77

G Ring, 665
galactic center, 730
Galileo mission, 2, 46, 300, 322, 361, 366, 372, 391, 414, 421
gas drag, 97
gas-to-solid condensation, 269
Gegenschein, 6, 8, 15, 788
GEMS, 265, 483, 766
generalized multiparticle Mie-solution, 530
generation of grains, 687
geometric optics, 526
Giotto, 45, 392, 406, 411, 412, 414, 420
GORID, 213, 421
GORIZONT, 213
gossamer ring, 656
Grün flux, 193
grain
....alignment, 729
....classical interstellar, 512
....collision, 760
....core/mantle, 744
....generation, 687
....mass spectrum, 45
....oxide, 763
....presolar diamond, 764
....presolar dust, 731

presolar graphite, 764
presolar SiC, 766
temperature, 474
gravitational enhancement, 167
gravitational focusing, 37, 353, 378
gravitational force, 333
gravitational perturbation, 578
gravitational scattering, 585
gravity, 677

H II, 732
Halley, 2, 46
halo population, 368
HEL, 459
Helios mission, 2, 9, 29, 39, 41, 42, 300, 307, 317, 387, 391, 411, 412, 414, 417–419
heliosphere, 789
HEOS, 2, 30, 36, 37, 211, 217, 300, 310, 313, 356, 391, 414, 416, 417
Herbig Ae/Be stars, 767
high altitude sounding rocket, 1
Hirayama family, 572
Hiten, 213, 300, 314, 391, 414, 423, 424
hole growth equation, 207
hot core, 732
Hubble Space Telescope, HST, 213
Hugoniot elastic limit, 459
Huygens probe, 425
Huygens probe, Christian, 4
hydrated phase, 260
hydrated silicate, 260
Hydrocodes, 209
hydrogenated amorphous carbon, 764
hypervelocity impact, 21
HI-, HII region, 789

icy-conglomerate model, 96
IDE, 217
IDP, 789
    accretion rate, 628
    classification, 266
    isotopic measurements, 282
    layer silicate, 269, 270
    olivine, 269
    pyroxene, 269
image, 648
    simulation, 105
imaging telescope, 388
impact
    direction, 306
    equation, 202
    flux, 216, 222
    hazard, 241
    ionization, 393, 404, 405, 414
    ionization detector, 2, 365
    restructuring, 457
    velocity, 199
in-situ detector, 297
inclination, 789
inclination dependence, 84
inclined population, 368
infrared
    astronomical satellite, IRAS, 388
    cirrus, 741
    measurement, 2
    space observatory, ISO, 739
    zodiacal light, 39
instrumentation, 385
intact capture, 410
interaction
    electromagnetic, 336
    gas-dust, 97
intermediate-mass YSO, 767
interplanetary
    dust, 254
    dust particles, 789
    electrons, 13
    flux model, 369
    magnetic field, 337
interstellar
    depletion, 789
    dust, 331, 362, 372, 578
    dust particles, 789
    dust population, 372
    extinction, 727, 789
    polarization, 728
inversion technique, 76

INDEX

ion
    mass spectrometer, 420
    time-of-flight, 411
IRAS, 2, 16, 29, 34, 38, 388
    LRS catalog, 761
    observations, 594
IRTS, 29
ISM phases, 731
ISO, 29, 739
isotopic anomaly, 282, 789
isotopic composition, 282
isotopic fractionation, 283
isotopic measurements of IDP, 282
isotropic flux, 312

Jacobian, 357
jet, 101
jet collimation, 105
Jupiter
    dust streams, 327
    family comet, 589
    ring, 653, 677, 710

K corona, 10
K factor, 171
KAO, 739
Keplerian dynamics, 357
kinematics of interplanetary dust, 15
KOSI experiment, 513
Kuiper Airborne Observatory, KAO, 739
Kuiper belt, 578
Kuiper belt), 788

laboratory astrophysics, 730
laboratory crater, 30
Laplace plane, 789
layer silicate IDP, 269, 270
LDEF, 193, 212, 380, 403
    east, 214
    observations, 577
LEAM, 415
Leonid
    filament, 238
    multi-instrument aircraft campaign, 238
libration point, 6, 15
lifecycle of dust, 733
lifetime
    catastrophic collisions, 498
    collisional, 575
    dust, 297
    Poynting-Robertson, 498
light
    curtain, 389, 409
    flash, 392
light-gas
    gun, 430, 431
Long Duration Exposure Facility, LDEF, 38
Lorentz
    force, 100, 337, 349, 578, 789
    resonance, 656, 695
Lorentz force, 680
low-mass YSO, 767
low-Ni group, 277
low-Zn group, 275
lunar crater distribution, 356
lunar ejecta, 36
lunar microcrater, 311, 364

Müller matrix, 523
M&D-SIG, 212
Mach number, 97
magnesium sulfide, 766
magnetosphere, 789
magnitude, 178
    distribution index, 178
Mairan, 6
MAP, 212
map of zodiacal light, 26
Mariner 2, 13
mass, 411
    distribution index, 167
    loss, 16
    spectrometer, 411
    spectrum of a single grain, 45
    threshold, 360
massive YSO, 767

MASTER, 201
matrix effect, 754
Maxwell Garnett mixing rule, 476
Maxwell's equations, 528
MDC, 414, 423, 424
mean motion resonance, 578
measurement
    flux, 317
    infrared, 2
    isotopy of IDP, 282
melting, 184
metal carbide, 765
Metal-Oxide-Silicon, 403
meteor, 369, 790
    jet, 239
    observation, 353
    shower, 174
    streams, 15
meteorite, 9, 790
    composition, 46
meteoroid, 165, 790
    β, 11, 15, 37, 316, 335, 574, 680, 787
    asteroidal, 355
    cometary, 354
    fragment, 240
    velocity, 354
methane, 756
method
    conjugate gradient, 534
    particle in a circle, 583
    scattering order iterative, 533
MFE, 212
micrometeorite, 254
micrometeoroid, 790
microphone, 406
    detector, 1, 416
microphonics, 296
mid- to far-infrared observations of the zodiacal light, 29
Mie theory, 14, 728, 790
milky way, 4
MM, 790
model
    Bird's-Nest, 511
    classical, 728
    cluster, 239
    desorption ionization, 397, 399
    Divine's interplanetary, 202
    dust, 744
    dust-ball, 243
    Finson-Probstein, 110
    fluffy, 510
    icy-conglomerate, 96
    interplanetary flux, 369
    NASA orbital debris dnvironment, 200
    plasma ionization, 397
    rarefied flow for meteoroids, 243
    rubble pile, 599
    standard ISM, 731
    volume ionization, 398
molecular cloud, 732, 790
momentum, 401, 406
    sensor, 406
Moon impact, 242
MOS detector, 403, 404, 412
MRN-distribution, 745
multi-coincidence dust detector, 301
multi-component dust, 730
Munich Dust Counter, 213

NASA Orbital Debris Environment Model, 200
near-infrared colors of zodiacal light, 29
near-surface breeze, 102
neckline, 110
Neptune's rings, 670, 713
nitride, 766
noctilucent cloud, 21
noise source, 301
novae, 766
Nozomi, 340
nuclear train, 116
nucleus
    precession, 101
    rotation, 101

OAO, 38

occultation, 650, 671, 790
OGO, 18, 38
Olbers' paradox, 727
olivine, 762
    IDP, 269
Oort cloud, 790
optical constant, 514
optical depth, 644, 790
optical equivalence, 523
optical properties, 686
    solar distance dependence, 81
orbit, 409
    distribution , 190
    synchronous, 674, 698
orbital elements, 307, 352, 357, 693, 790
orbital evolution, 578, 702
orbital perturbation, 701
origin of interplanetary dust, 33
osculating elements, 580
OSO, 38
out-of-ecliptic distribution of interplanetary dust, 39
outburst, 107
oxide grain, 763

packing, 478
parameter distribution, 358
particle
    -cluster aggregation, 453
    albedo, 33
    flux, 19, 359
particle in a circle method, 583
PDR, 731
Pegasus mission, 18, 211, 222, 309, 353, 412, 413
penetration, 179, 402, 403
    detector, 364, 413
    limit, 296
    rate, 296
percolation pathway, 485
perforation, 219, 402
perihelion, 791
persistent train, 245
perturbations, 791

phase
    angle, 791
    function, 79, 645, 646
    space density, 357
photo dissociation region, 731
photo emission, 492
photometer, 387
photometry, 664, 791
photomultiplier, 18
phyllosilicate MM, 257
PIA, 392, 414, 420
Pioneer 10 and 11, 2, 8, 39, 41, 44, 300, 302, 324, 388, 402, 412, 413
Pioneer 8 and 9, 2, 23, 36, 37, 217, 300, 310, 320, 356, 407, 413, 415
Planet-B, 423
planetary nebulae, 766
planetary shielding, 378
plasma, 676
    density, 13, 43
    drag, 684
    drag accelerator, 433, 434
    ionization model, 397
    tail, 791
Poisson statistics, 303
polarimetric color, 551
polarization, 522, 791
    curve, 729
    of the F corona, 14
    of zodiacal light, 8
polyvinylidene fluoride, 405
population, 376
    asteroidal, 368
    core, 368
    dust, 42, 730, 733, 788
    eccentric, 368
    halo, 368
    inclined, 368
    interstellar dust, 372
post-AGB star, 766
power law, 791
Poynting Robertson
    drag, 261

Poynting-Robertson
  decay timescale, 575
  drag, 13, 27, 37, 164, 335, 574,
    683, 791
  lifetime, 498
preplanetary disk, 769
presolar
  corundum, 763
  diamond and graphite grain, 764
  dust grain, 731
  SiC grain, 766
  silicon nitride, 766
pressurized cell, 402
  detector, 2
probability function, 179
prograde, 791
proper element, 580
PROSPERO, 37
protoplanetary nebulae, 766
protostellar object, 760
PUMA, 392, 398, 399, 414, 420
PVDF, 405, 421, 422, 425, 427
pyroxene, 762
  IDP, 269

quadrupole, 435, 436

R CrB star, 764
radar meteor, 364, 378
radial profile, 351
radiation pressure, 11, 37, 307, 334,
  349, 357, 360, 362, 372, 552,
  573, 680
  force, 791
radius distribution function, 734
random noise, 105
rarefied flow model for meteoroids,
  243
Rayleigh scattering, 526
Rayleigh-like polarization, 544
reddening of the zodiacal light, 29
reflection nebula, 740, 791
refractive index, 475, 514
refractory dust, 729
relative speed, 309

repacking, 512
repulsive force from the sun, 96
residue, 220
resonance, 694, 792
  Lorentz, 656, 695
  mean motion, 578
  scattering domain, 526
  secular, 579
resonant charge variation, 685
resonant perturbation, 578
resonant trapping, 614, 697
retrograde, 792
ring-moons, 792
Roche zone, 692
Rosetta mission, 46, 340, 389, 409,
  412, 427
rotation axis, 101

scattering, 14
  angle, 792
  cross section, 523
  function, 12, 740
  order iterative method, 533
Scoriaceous MM, 257
secondary electron emission, 490-491
secondary fragmentation, 115
secular perturbation, 579
secular precession, 581
secular resonance, 579
segregation, 485
semimajor axis, 792
shape of impact fragments, 464
shepherd satellites, 792
shielding, 353
Si_3N_4, 766
SiC, 730
silicate
  band, 740
  feature, 517
silicon
  carbide, 730
  disulfide, 767
SIMS, 273
SIMUL, 588
SIRTF, 622

size distribution, 12, 166, 571, 646, 664, 666, 676, 686
    of fragments, 462
Skylab, 38
small number statistics, 303
SOHO sungrazing comet, 101
solar corona, 10, 26
solar gravity, 349, 362
solar radiation pressure, 96
solar system dust band, 592
solar wind, 337, 417, 419, 479, 792
    drag, 575
solid methanol, 756
solid-state astrophysics, 730
sonic boom, 247
space
    debris, 189, 792
    density, 79
    weathering, 253
spatial density, 14, 352, 354, 357
    of interplanetary dust, 43
spatial distribution, 10, 11
spatial structure, 38
spectra, 649
spectral band
    217.5 nm, 736
    30 μm, 767
spectral feature
    21 μm, 767
    3.4 micron, 518
spectral function, 520
spectral information on the zodiacal light, 9
speed, 395, 407, 416
    loss, 402
sphere of influence, 333
spin, 466
    rate, 101
    vector, 105
spiral, 101
spokes, 656, 659
sporadic background, 174
sputtering, 478, 486, 688, 792
standard ISM model, 731
star-forming region, 732

star-grazing comet, 773
stardust, 37, 730, 792
    mineralogy, 762
Stardust mission, 46, 340, 421
Stokes vector, 522
streamer, 111
striae, 113
sublimation rate, 472
sulfide, 766
Sun-grazing comet, 27
supernovae, 766
surface potential, 390, 391
SXRF, 273
symmetry with respect to the ecliptic plane, 351
synchrone, 96
synchronous orbit, 698
syndyne, 96

T-matrix solution, 531
TD-1, 29, 38
terminal synchrone, 113
terminal velocity, 96
textbook distribution, 358
thermal annealing, 763
thermal emission, 72, 364, 369
thermal spike, 484
thermoionic effect, 493
thin-foil penetration, 400
TICCE, 213
timescale, 683, 685
TOF-SIMS, 280
track, 484
trail
    dust, 16, 39, 118, 242, 634, 788
    Fe ablation, 244
trailet, 235
train tubular structure, 246
trajectory, 407–409
transient recorder, 423, 424
transition, 729
trough opacity problem, 763

U-2, 22
UIR bands, 730

Ulysses mission, 2, 44, 46, 300, 323, 361, 365, 372, 391, 421
unidentified infrared bands, UIB, 730
Uranian rings, 668
Uranian system, 713

V-2, 17
Van de Graaff accelerator, 21
vapor pressure, 468
variations, 38
Vega
    excess stars, 730, 741
    phenomenon, 733, 792
VeGa mission, 392, 405, 411, 412, 414, 420
velocity, 407–409
velocity distribution, 168
Venus Flytrap, 19
volatile dust, 731
volume
    integral equation, 532
    ionization model, 398
    scattering function, 77

water ice, 470, 752
WC stars, 764

X–CN, 756
Xe-HL anomaly, 751

Yarkovsky effect, 164, 478, 578
young stellar object, 730, 733, 741, 792
YSO, 730, 733, 741, 792
    intermediate-mass, 767
    low-mass, 767
    massive, 767

zenithal hourly rate, 175
zodiacal background cloud, 606
zodiacal infrared emission, 602
zodiacal light, 1, 8–10, 15, 26, 38, 253, 364, 369, 387, 792
    annual variations, 62
    brightness, 64
    color, 27
    infrared, 39
    map, 26
    mid- to far-infrared observations, 29
    near-infrared colors, 29
    polarimetry, 9
    polarization, 8, 26, 66
    reddening, 29
    spectral information, 9
    temporal variations, 15

Printing and Binding: Stürtz AG, Würzburg